Encyclopaedia of

ENVIRONMENTALLY SUSTAINABLE GREEN SCIENCE AND TECHNOLOGY

Contents

VOL. 4

VOL. 5

Preface

This publication entitled "Encyclopaedia of Environmentally Sustainable Green Science & Technology" attempts towards creating a holistic understanding of environment and sustainable development in its totality. It defines environment and sustainable development in all its dimensions. The concepts and methodologies of ecology, environment and sustainable development have been discussed in detail. The ecosystem perspective and environmental concept of sustainable development have been highlighted, so has been the environmental methodology. An effort has been made towards defining green science and its manifestations for understanding pollution. The relevance and importance of environment impact assessment (EIA) for environmental sustainability has been analyzed. Also, the elements and dimensions of green technology have been covered in detail. The relationship among ecomovements, ecodynamics and sustainable development has been discussed in detail. Similarly, the environmental and social aspects of sustainable development, in particular, aspects related to training of people to meet the global environmental needs have been covered. The issues related to green energy, its sources, tools and technologies have been reflected upon briefly. The environmentally sustainable science and technology has been presented in brief with special reflection on climate change and global warming. Current issues and options regarding sustainability aspects of environmental management, corporate management and environmental responsibility have been covered. The aim is to develop a policy perspective. The contemporary global industrial technology has been discussed with focus on evaluating environmental performance and compliance to green laws. The applied aspects of environmentally sustainable science and technology have been outlined. Also, the green conventions enforced towards regulating global science and technology towards environmentally sustainable development has been presented, highlighting the theories of international environmental law and theory of environmental degradation.

The applications of green science and technology have been discussed along with a case of biotechnology. The green technology and environmental monitoring issues have been given considerable coverage. Select case studies have been made in this regard. Also, case studies of environmental engineering and environmental biotechnology have been incorporated with focus on applications and utilities. Future prospect of environmentally sustainable or green technology in today's world has

been analyzed briefly. Similarly, the principles, tools and techniques of environmental management, with particular reference to environmental disasters, have been covered in detail. Glossary of terms, bibliography and links has been provided to make this publication a user friendly one and to aid readers in further research and analysis.

—Editors

1

Towards Understanding Environmental and Sustainable Development in Its Totality

Introduction

Environment may refer to:

- Environment (biophysical), the physical and biological factors along with their chemical interactions that affect an organism
- Natural environment, all living and non-living things that occur naturally on Earth
- Built environment, constructed surroundings that provide the setting for human activity, ranging from the large-scale civic surroundings to the personal places
- Social environment, the culture that an individual lives in, and the people and institutions with whom they interact
- Environmental science, the study of the interactions among the physical, chemical and biological components of the environment
- Environmental psychology
- Environmental determinism
- Environmental policy
- Environmental quality
- Environmental art

In computing:

- Environment variable, the set of environments defined in a process

- Runtime environment, a virtual machine state which provides software services for processes or programmes while a computer is running
- Integrated development environment, a type of computer software that assists computer programmers in developing software
- Desktop environment, in computing, is graphical user interface to the computer.

Environment and Sustainable Development

We have learnt from this narrative how, over the last few decades, human activities have led to resource degradation, loss of biodiversity, land degradation, global climate changes etc. and release of a range of chemical, radioactive and other toxic wastes. While these activities have enabled a significant portion of the world to improve the condition of life, they are now showing their adverse effects on the quality of air, water and land, moving the ecological balance of the planet of air, water and land, moving the ecological balance of the planet to a new equilibrium. Whether this new equilibrium is conducive to human activities or whether it will prove adverse to the coming generations is a very complex question. But present indications are that there is urgency for taking corrective measures at the global as well as regional levels to prevent environmental degradation. A careful scrutiny of the economic signals currently in use shows that the way the economy is managed is very frequently the root cause of environmental degradation.

The discipline of *economics* is concerned with the production, consumption and distribution of goods and services. Adam Smith (1723-1790), David Ricardo (1772-1823), Alfred Marshal (1842-1924) were the pioneers in the development of the discipline of classical economics. But the most influential economist of all times was John Maynard Keynes (1883-1946) whose neoclassical economic theory and policy form the basis for the economic theories presently adopted by most countries, though with modifications. Keynes analysed the economic processes that led to the great depression of the 30's and described policies to avoid it. He argued convincingly that government should take measures to counter depression. His ideas helped shift emphasis away from *laissez faire,* the classical economic theory which maintains that government should not interfere in economic affairs. The basis of Keynesian economics is that the level of economic activity depends upon the total spending of consumers, business and government. If business expectations are poor, investment will be cut, causing reduction in total spending. If this should happen, the economy could move to a *depression.* To avoid a depression, Keynes urged increased government spending and easy money through increased credit and lower interest rates. These actions would encourage investment, increase employment and enable consumers to spend more. He showed that higher levels of demand were essential for both full employment and economic growth.

In the framework of the Keynesian economic model, the primary means of economic progress is measured in terms of the rate of growth of *Gross Domestic Product* (GDP), which is defined as the value of all goods and services produced within the country during a given period regardless of who owns the production facilities. Other measures of growth include:

Standard of Living which usually refers to the economic level at which an individual or a nation lives. This is sometimes measured by determining the value of goods and services produced by the individual, family or nation during a given period.

Gross National Product (GDP) which is the value of all goods and services produced by a country during a given period even if the production units are in another country.

Productive Resources are the main elements that make possible for nations to produce goods and services. These are natural resources, capital, labour and technology.

If the GDP in an industrial country (like US) for a quarter is marginally lower than the GDP in the preceding quarter, a *recession* is declared and the competence of various agencies like the government becomes the subject of a public debate. Currently the Keynesian model in different forms is the dominant feature of the national accounts systems in all non-Communist countries. With the collapse of the Communist system in USSR and other east European Countries this system will find approval in these countries too. In this model, production, consumption, savings and investment are defined and measured. The most significant omission, however, was the parameter of natural resources. This was due to the fact that the economist of the time were more preoccupied with the great depression. Further, scarcity of natural resources was of little concern during the period as the industrial Europe was flooded with foodgrains and natural resources from America, Australia and the colonies at low costs through rail and marine transport. Then came the World War-II which was followed by a *boom* (also called a period of prosperity) representing a period of rapid economic growth during which production (through advances in science and technology), consumption and employment became all-time high in the industrial countries. Development came to be measured entirely in terms of savings and investments in capital and human resources, ignoring natural resources. The Third World countries which became independent after the war adopted this model, leaving out human resources too which is considered as surplus in these countries.

But it is no longer true that natural resources (including energy resources) are abundant and that they have only a marginal value. Whether they enter the market place directly or not, they make an important contribution to long term productivity. To assume that these natural resources are *free gifts of nature* is no longer valid.

Another point to be recognised is that the economy depends not only on the inputs of energy and natural resource production and consumption but also on services generated by life-supporting systems such as, biomass, biodiversity, drinking water supply, recycling of nutrients and generation of soils. In addition, being parts of ecosystems, species sustain these crucial services. For this reason, economists have begun to realise that due consideration should be given to these factors in the overall economic system. Let us consider the simple case of an agriculturist who cuts the mango orchard which gives him a recurring income, and sells the timber to raise money to build a small house. Both the income from the timber and the investment in the house are shown in the national accounts as per the prevailing practice. Further, if he spends part of the income to go on pilgrimages, he would be poorer by this sum, which however becomes part of the national income. But the ultimate result is that he is poorer by the permanent loss of a source of recurring income though he has become the owner of a house. This is the scenario in the Third World countries which are destroying their natural resources like forests, minerals for immediate gains but becoming poorer in the long run. The true definition of income should therefore encompass the notion of sustainability at the national and global levels in terms of natural resources and life-supporting systems. It is somewhat analogous our managing the household needs with the returns on the capital without eating into it.

Environmental Economics

A study based on such approach is called *environmental economics*. The archaic definition of economics as the science or art of managing a house of household once again gains its appropriateness in this context. We are now faced with the task of managing the enormous and complex household Earth on a sustainable basis with the help of studies which involve the following:

(a) Awareness of the necessity for allocating resources wisely and economically,

(b) Knowledge of resource use and particular resource degradation,

(c) Measuring the impact, and

(d) Designing policies to combat degradation.

Environmental economics thus aims to integrate economics, ecology and ethics in an attempt to define conditions for environmental house-keeping. At the global level this becomes possible only through cooperation between the affluent developed countries which are increasingly becoming aware of their obligations to future generations and the poor Third World countries for whom their very survival is a day-to-day problem. The study of environmental economics should aim at defining policies that will lead to solutions of global environmental management problems through cooperative efforts of the entire human society.

National Account System

The United Nations has prepared a *System of National Accounts* in its model balance sheet (which only a few member nations have adopted so far) in which land, minerals, timber and other environmental resources are incorporated as economic assets to be included in the national capital stock. If the environmental damages are also accounted for in the National Account, the *Net National Product* (NNP) will measure sustainability in the service that gives the maximum *constant consumption path* that would not impair the well-being of future generations. We will consider two cases to illustrate this approach. Costa Rica has suffered devastating deterioration of its natural resources during the past twenty years. About 30 per cent of its forests were burned mostly to clean land for unproductive pastures and hill farming. The soil was eroded in torrents, an estimated 2.2 billion tons of soil having been washed away between 1970 and 1989. Water pollution and overexploitation devastated coral reefs and coastal fisheries. Estimate of changes in forest cover, mangrove area, land use, biological diversity for this period through depreciation in the value of its forests, soils and fisheries exceeded $ 4.1 billion in terms of 1984 prices. This represents more than the average value of Costa Rica's one year GDP. In the case of Indonesia, inclusion of the depreciation of three stocks of natural resources (representing the country's natural capital)—oil, timber and topsoil—resulted in the official annual growth rate of about of 7 per cent between 1971 and 1984 being reduced to 4 per cent. Analysts have now begun compiling natural resource accounts along these lines for Mexico, Bolivia, Brazil, Chile, the Philippines, China, Malaysia, India and other developing countries which are exploiting their natural resources to the detriment of future generations. The results of such studies will be of great significance to these countries in taking appropriate corrective measures to conserve their rich natural resource heritage.

Global Efforts for Sustainable Development

The foundations for a greater insight into global environment were laid at the 1972 Conference on Human Environment held at Stockholm. Late Indira Gandhi, who was one of the very few heads of governments that participated in the deliberations, emphasised that poverty is the biggest pollutor. One important outcome of the Conference was a detailed Action Plan which led to the creation of the *United Nations Environmental Programme* (UNEP). In 1983, the Brundtland Commission on Environment and Development was formed to investigate means whereby the basic needs of the planet's growing population could continue to be met. The Commission's report entitled *Our Common Future* (1987) promoted the concept of sustainable development and linked economic growth with sustainable management of natural resources. *Sustainable development* was defined as that which meets the needs of the present generation without sacrificing the welfare of the future.

In 1989, The UN General Assembly decided a convene an international conference, *United Nations Conference on Environment and Development* (UNCED) to be held in Rio de Janeiro, Brazil, in June 1992 with the prime purpose of moving the environment into the centre of development agenda and of economic and sectorial policy making. The agenda of the Conference was set out as

1. Principal environmental and developmental challenges and opportunities upto the end of the century and beyond,
2. Poverty in developing countries,
3. Unsustainable consumption patterns, and
4. Demographic pressures in a period of rapidly rising world population.

UNCED Conference (Earth Summit), 1992

Never before have so many people who are concerned about the environment gathered in one place. Official delegations from 172 countries, 1300 representatives of 44 non-governmental organisations and 7000 to 8000 members of the media attended the 12-day conference. Heads of governments of a large number of counties—India, US, UK, France, Germany, Norway, Sweden, Denmark, Canada, Pakistan and Bangladesh, to name a few—participated in the deliberations of the meet which addressed itself to a wide range of subjects of environmental importance. The conclusions of the conference were incorporated in

1. An Earth Charter,
2. Agenda 21 which is an Action Plan for the 21st century,
3. Convention on climate change and biodiversity, and
4. Agreements on forests, institutional financing and technology.

The conventions on climate change and biodiversity were signed by 154 nations and 153 nations respectively. The US was conspicuous by its refusal to sign both the conventions. The US refused to commit to targets and timetables for stabilising carbon dioxide emissions under the convention on climate change. According to Us, it has already committed itself to reduce greenhouse gas emissions to 7 to 10 per cent from what they otherwise would be in the year 2000, while other nations do not have credible plans for stabilising emissions. The convention on biodiversity says that the first beneficiaries of conservation and sustainable use of wild plant and animal species should be the rural communities (most of whom live in the developing and biodiversity rich South) and indigenous people who have conserved them for centuries. The US refused to sign this convention since according to it, "there is a flavour in the language that could allow intellectual property rights to be compromised as a result of exchange of genetic sources" and also a fear that the convention would lead to international regulation of biotechnology in which the US has invested huge funds and gained expertise.

The Conference reached agreement on Agenda 21, an Action Plan for developing the planet sustainably through the 21st century and on a broad statement of principles on how to protect forests. The statement on forests was however watered down when the Third World countries led by India and Malaysia accused the industrial North of their interest in keeping the forests in the South as a sink for carbon dioxide so that he North could continue to emit large amounts of greenhouse gases. This accusation was substantiated by the US document which said "curbing forest loss would be an extremely effective way to reduce carbon dioxide emissions" and "it is much more cost-effective to protect forests (which are mainly in the South) than to clean up emissions that cause global warming." This North-South conflict was resolved by adopting compromise language saying that the parties would consider the possibility of calling for a treaty on forest areas.

Agenda 21 has several major goals. One is that the development should proceed in a sustainable manner. To do this, it is felt that the system of incentives and penalties which motivate economic behaviour must be reoriented to become a strong force for sustainability. To this end, it recommended inclusion of environmental impacts of resource use in economic accounting. Another goal is coordinated actions to ultimately eliminate poverty throughout the world. Closely related is the aim of better managing human settlements by improving the quality of shelter, water supply, energy and transportation services, the sewage and solid-waste management. Waste minimisation for both industrial and municipal wastes is to be aimed at in this effort.

Other goals identified include, efficient use of resources, sustainable use of global and regional resources such as the atmosphere, oceans and seas and marine organisms. Another important goal concerns the managements of chemicals and wastes. These problems can be eased by reducing waste generation, recycling wastes into production activities and finding safe ways of disposing of wastes.

Agenda 21 recognises that government participation alone will not be sufficient to reach these goals. Women, youth, indigenous groups, non-governmental organisations, farmers, local authorities, trade unions, business and industrial organisations and engineers and scientists must be involved in this effort. It calls for greater international scientific cooperation and efforts to transfer environmentally sound technologies from nation to nation. Significantly, the issue of population and its relation to poverty and development did not receive much attention of the delegates. Most participants recognised the major role for technological efforts in such areas as renewable energy, materials recycling and satellite imagery for land managements. Every one agreed that environmental problems will not be solved without the application of technology. Agenda 21 points out that, to realise these goals, the industrialised world will need to provide new and additional resources to the Third World. Twenty years ago at Stockholm, all the industrialised countries except US agreed in principle that they should give 0.7 per cent of their GNP as official development assistance. But so far only Norway, Sweden, Denmark and the Netherlands reached that goal. It is estimated that if official assistance were increased from its current $ 55 billion to an overall average of 0.7 per cent of each country's GNP or a total of $ 125 billion, enough money would be generated. Unfortunately no firm commitments were entered into for a time-table for reaching the targets though it was agreed in principle to provide the funds required. The UN Earth Summit thus concluded with an agenda for action but inadequate funds. The delegates however hoped that the Conference will mark the turning point for accepting the principles for sustainable development and better protection of environment. Time only can tell whether the Conference has succeeded in creating the will to bring about changes in global development behaviours. According to Maurice Strong, Secretary General of the Conference, "Law is a method of enforcement at a national level. At the international level, political commitment is the important thing because legal instruments flow from political commitments."

Natural Environment

The natural environment, commonly referred to simply as the environment, is a terminology that comprises all living and non-living things that occur naturally on Earth or some region thereof. This term includes a few key components:

1. Complete ecological units that function as natural systems without massive human intervention, including all vegetation, animals, microorganisms, rocks, atmosphere and natural phenomena that occur within their boundaries.
2. Universal natural resources and physical phenomena that lack clear-cut boundaries, such as air, water, and climate, as well as energy, radiation, electric charge, and magnetism, not originating from human activity.

The natural environment is contrasted with the built environment, which comprises the areas and components that are strongly influenced by man. A geographical area is regarded as a natural environment (with an indefinite article), if the human impact on it is kept under a certain limited level (similar to section 1 above). This level depends on the specific context, and changes in different areas and contexts. The term wilderness, on the other hand, refers to areas without human intervention.

Challenges

It is the common understanding of *natural environment* that underlies environmentalism — a broad political, social, and philosophical movement that advocates various actions and policies in the interest of protecting what nature remains in the natural environment, or restoring or expanding the role of nature in this environment. While true wilderness is increasingly rare, *wild* nature (e.g., unmanaged forests, uncultivated grasslands, wildlife, wildflowers) can be found in many locations previously inhabited by humans.

Goals commonly expressed by environmental scientists include:

- reduction and clean up of pollution, with future goals of zero pollution;
- cleanly converting nonrecyclable materials into energy through direct combustion or after conversion into secondary fuels;
- reducing societal consumption of non-renewable fuels;
- development of alternative, green, low-carbon or renewable energy sources;
- conservation and sustainable use of scarce resources such as water, land, and air;
- protection of representative or unique or pristine ecosystems;
- preservation of threatened and endangered species extinction;
- the establishment of nature and biosphere reserves under various types of protection; and, most generally, the protection of biodiversity and ecosystems upon which all human and other life on earth depends.

Very large development projects - also called megaprojects - pose special challenges and risks to the natural environment. Major dams and power plants are cases in point. The challenge to the environment from such projects is growing because more and bigger megaprojects are being built, in developed and developing nations alike.

Recently, there has been a strong concern about climate change such as global warming caused by anthropogenic releases of greenhouse gases, most notably carbon dioxide, and their interactions with humans and the natural environment. Efforts here have focused on the mitigation of greenhouse gases that are causing climatic changes (e.g. through the Climate Change Convention and the Kyoto Protocol), and on developing adaptative strategies to assist species, ecosystems, humans, regions and nations in adjusting to the Effects of global warming.

A more profound challenge, however, is to identify the natural environmental dynamics in contrast to environmental changes not within natural variances. A common solution is to adapt a static view neglecting natural variances to exist. Methodologically this view could be defended when looking at processes which change slowly and short time series, while the problem arrives when fast processes turns essential in the object of the study.

Spaceship Earth

In the preparation for a long, deep space voyage, the brains of a group of astronauts were inadvertently altered. These alterations had a very unfortunate effect on the behaviour of the crew, and every time a member carried out a spaceship maintenance operation, an important wire leading to the life support system was disconnected. The crew did not realise the effects of these operations and did not realise the dangers of their, behaviour. The spaceship had been designed with a great degree of durability and the damages went undetected for some time. After a time, the damages became apparent; the development of problems with on-board oxygen levels, contamination of drinking water and odours from the faulty waste system alerted the crew to the problems on board. Procedures were initiated to identify the problems and to find their causes. The problems were discovered, but not their causes and the crew's damaging behaviour along with the destruction of the spaceship's life support system continued. In an attempt to control the damage, temporary patch-up repairs were initiated. However, the crew found themselves unable to keep up with the damage, which seemed to be growing worse with each day, along with their standard of living

Eventually, the captain brought his crew together for an emergency meeting to assess the damage and the future of the ship. There were many conflicting interests about what should be done. Although it was clear that the emergency patch-up operations had not worked, many believed that the emergency repair work could sustain the ship for a very long time. The outcome of the meeting was that the patch-up repair work would continue, and a detailed document called the "Agenda for Survival" was prepared, describing the responsibilities of the crew in tending to those sub-systems of the ship requiring attention. This approach did not work for long and it was decided that the operations performed by the crew should be checked and counter checked.

After a time, the true cause of the problems was traced to the minds of the crew on board and eventually to the surgical operations performed before the mission. With the equipment available the medical team was able to rectify the minds of the majority of the crew and the rest of the crew were kept under close observation. The story of the spaceship provides a metaphor for mankind's environmental impact on the Earth. As with the spaceship, the Earth is being polluted by its. , human residents, its resources are being used up, and its climate is being rapidly changed. In 1992, governments from around the world met at the Rio Earth Summit in Brazil to negotiate the "Agenda for Survival", Agenda 21. Although providing a blueprint for sustainable development and the environmental protection of the Earth, Agenda 21 has yet to really bear the fruits of success. Populations continue to grow, the Earth continues to warm, pollution is increasing and resources are consumed at an ever-increasing rate. Unlike Spaceship Earth however, such problems with mankind's development cannot be solved with a simple operation. Changes in behaviour and the way we treat the environment must come through changes in the way...we..value nature and the resources it provides us.

Ecosystems

The term 'biosphere' is used to describe the whole sum of living things found on Earth. Usually we think of the biosphere as existing within upper layers of the non-biotic components of the planet, namely the atmosphere, hydrosphere (oceans, rivers and lakes), and soil component of the lithosphere (the Earth's solid material). In fact, research suggests that more biological mass resides within the Earth's crust than in all the rest of the biosphere as described, largely consisting of bacteria and other single-celled organisms living within tiny spaces between rock minerals. Furthermore, some scientists and environmentalists speak of Earth or Gaia as a living organism in itself, engaged in a self-regulation to support its own survival. Although many have cast doubt on the merit of such an idea beyond a metaphorical application given the lack of substantial scientific evidence to support it, such a concept does reinforce the view inherent throughout sustainable development that all parts of the biosphere are interdependent, belonging as they do to a web of life.

The biosphere is composed of many ecosystems. An ecosystem consists of a community of living organisms and their local physical environment. The living and non-living elements of an ecosystem are connected through flows of energy and the cycling of chemical elements. In this context, no part of the biosphere that is smaller than an ecosystem can sustain life.

Imagine placing a single green plant in a glass jar with air, water, soil, and light from the sun. The plant could make its own food through photosynthesis (from water, carbon dioxide, and light). But eventually the plant would die as it used up all of the nutrients in the soil. It would die because it couldn't recycle the materials it needs to live; no green plant can decompose its own products into the inorganic compounds necessary for plant life. In fact, no single organism, population, or species is able to produce all of its own food and recycle all of its metabolic products. Life requires the interaction of several species in an environment that includes air and/or water to transport materials and energy. This is one of the fundamental principles of ecology; sustained life is a property of ecosystems, not organisms or species.

A forest is a common ecosystem. Forests consist of air, soil, water, nutrients, and particular species of animals, birds, insects, micro-organisms, trees, and other plant life. If too many of the trees are cut down (for example by mankind for wood or agricultural land), each of the other elements will be affected. Animals and birds may lose their habitats, soil may erode, nutrients may be displaced, and the flow of waterways may change.

There are no absolute boundaries between ecosystems - really, the biosphere is one big ecosystem in which everything is connected to everything else in some way or another. Some ecosystem borders are, however, quite well defined. For example, there is a relatively clear transition from a rocky ocean coast to the forest along its edge, or from a pond to the wood that surrounds it. Other borders are much more vague, as is the case with the gradual transition from deciduous woodland to coniferous forests in some parts of the world.

Ecosystems change continuously over time according to certain recognisable, repeatable patterns, and in response to environmental changed such as shifts in climate. This is referred to as 'succession'.

Ecosystem

An ecosystem is a natural unit consisting of all plants, animals and micro-organisms (biotic factors) in an area functioning together with all of the non-living physical (abiotic) factors of the environment. The term ecosystem was coined in 1930 by Roy Clapham, to denote the physical and biological components of an environment considered in relation to each other as a unit. British ecologist Arthur Tansley later refined the term, describing it as the interactive system established between biocoenosis (a group of living creatures) and their biotope (the environment in which they live).

Central to the ecosystem concept is the idea that living organisms are continually engaged in a set of relationships with every other element constituting the environment in which they exist. Eugene Odum, one of the founders of the science of ecology, stated: "Any unit that includes all of the organisms (ie: the "community") in a given area interacting with the physical environment so that a flow of energy leads to clearly defined trophic structure, biotic diversity, and material cycles (ie: exchange of materials between living and nonliving parts) within the system is an ecosystem." The human ecosystem concept is then grounded in the deconstruction of the human/ nature dichotomy, and the emergent premise that all species are ecologically integrated with each other, as well as with the abiotic constituents of their biotope.

Ecosystems can be bounded and discussed with tremendous variety of scope, and describe any situation where there is relationship between organisms and their environment. A system as small as a household or university, or as large as a nation state, may then be suitably discussed as a human ecosystem. While they may be bounded and individually discussed, (human) ecosystems do not exist independently, but interact in a complex web of human and ecological relationships connecting all (human) ecosystems to make up the biosphere. As virtually no surface of the earth today is free of human contact, all ecosystems can be more accurately considered as human ecosystems.

Some Examples

Examples of ecosystem include:

- Aquatic ecosystem
- Chaparral
- Coral reef
- Desert
- Greater Yellowstone Ecosystem
- Human ecosystem
- Large marine ecosystem
- Littoral zone
- Marine ecosystem
- Rainforest
- Savanna
- Subsurface Lithoautotrophic Microbial Ecosystem
- Taiga
- Tundra
- Urban ecosystem

Main Topics of Ecosystem

Classification

Ecosystems have become particularly important politically, since the Convention on Biological Diversity (CBD) - ratified by more than 175 countries - defines "the protection of ecosystems, natural habitats and the maintenance of viable populations of species in natural surroundings" as one of the binding commitments of the ratifying countries. This has created the political necessity to spatially identify ecosystems and somehow distinguish among them. The CBD defines an "ecosystem" as a "dynamic complex of plant, animal and micro-organism communities and their non-living environment interacting as a functional unit".

With the need of protecting ecosystems, the political need arose to describe and identify them within a reasonable time and cost-effectively. Vreugdenhil et al. argued that this could be achieved most effectively by using a physiognomic-ecological classification system, as ecosystems are easily recognizable in the field as well as on satellite images. They argued that the structure and seasonality of the associated vegetation, complemented with ecological data (such as elevation, humidity, drainage, salinity of water and characteristics of water bodies), are each determining modifiers that separate partially distinct sets of species. This is true not only for plant species, but also for species of animals, fungi and bacteria. The degree of ecosystem distinction is subject to the physiognomic modifiers that can be identified on an image and/or in the field. Where necessary, specific fauna elements can be added, such as periodic concentrations of animals and the distribution of coral reefs. Several physiognomic-ecological classification systems are available: Physiognomic-Ecological Classification of Plant Formations of the Earth (a system based on the 1974 work of Mueller-Dombois and Heinz Ellenberg, and developed by UNESCO), and the Land Cover Classification System (LCCS), developed by the Food and Agriculture Organisation (FAO). Several aquatic classification systems are available, and an effort is being made by the United States Geological Survey (USGS) and the Inter-American Biodiversity Information Network (IABIN) to design a complete ecosystem classification system that will cover both terrestrial and aquatic ecosystems.

Ecosystem Services

Ecosystem services are "fundamental life-support services upon which human civilisation depends," and can be direct or indirect. Example of direct ecosystem services are: pollination, wood, erosion prevention etc. Indirect services could be considered climate moderation, nutrient cycles, detoxifying natural substances and many more.

Ecosystem Legal Rights

The borough of Tamaqua, Pennsylvania passed a law giving ecosystems legal rights. The ordinance establishes that the municipal government or any Tamaqua resident can file a lawsuit on behalf of the local ecosystem. Other townships, such as Rush, followed suit and passed their own laws. This is part of a growing body of legal opinion proposing 'wild law'. Wild law, a term coined by Cormac Cullinan (a lawyer based in South Africa), would cover birds and animals, rivers and deserts.

Function and Biodiversity

From an anthropological point of view, many people see ecosystems as production units that of goods and services. Among some of the most common goods produced by ecosystems, is wood by forest ecosystems and grass for cattle by natural grasslands. Meat from wild animals, often referred to as bush meat in Africa, has proven to be extremely successful under well-controlled management schemes in South Africa and Kenya. Much less successful has been the discovery and commercialisation of substances of wild organism for pharmaceutical purposes. Services derived from ecosystems are referred to as ecosystem services. They may include (1) facilitating the enjoyment of nature, which may generate many forms of income and employment in the tourism sector, often referred to as eco-tourisms, (2) water retention, thus facilitating a more evenly distributed release of water, (3) soil protection, open-air laboratory for scientific research, etc.

A greater degree of species diversity or biological diversity - popularly referred to as Biodiversity - of an ecosystem may contribute to greater resilience of an ecosystem, because there are more species present at a location to respond to a factor of change and thus "absorb" or reduce its effects, thus reducing the effect before its structure is fundamentally changed to a different state. This is not universally the case and there is no proven relationship between the species diversity of an ecosystem and its ability to provide goods and services on a sustainable level: Humid tropical forest produce very little goods and direct services and are extremely vulnerable to change, while many temperate forests readily grow back to their previous state of development within a lifetime after felling or a forest fire. Some grasslands have been exploited sustainably for thousands of years (Mongolia, Africa, European peat and mooreland communities).

The Study of Ecosystems

Ecosystem Dynamics

Introduction of new elements, whether biotic or abiotic, into an ecosystem tend to have a disruptive effect. In some cases, this can lead to ecological collapse or "trophic cascading" and the death of many species belonging to the ecosystem in question. Under this deterministic vision, the abstract notion of ecological health attempts to measure the robustness and recovery capacity for an ecosystem; i.e. how far the ecosystem is away from its steady state.

Often, however, ecosystems have the ability to rebound from a disruptive agent. The difference between collapse or a gentle rebound is determined by two factors — the toxicity of the introduced element and the resiliency of the original ecosystem.

Ecosystems are primarily governed by stochastic (chance) events, the reactions they provoke on non-living materials and the responses by organisms to the conditions surrounding them. Thus, an ecosystem results from the sum of myriad individual responses of organisms to stimuli from non-living and living elements in the environment. The presence or absence of populations merely depends on reproductive and dispersal success, and population levels fluctuate in response to stochastic events. As the number of species in an ecosystem is higher, the number of stimuli is also higher. Since the beginning of life, in this vision, organisms have survived continuous change through natural selection of successful feeding, reproductive and dispersal behavior. Through natural selection the planet's species have continuously adapted to change through variation in their biological composition and distribution. Mathematically it can be demonstrated that greater numbers of different interacting factors tend to dampen fluctuations in each of the individual factors. Given the great diversity among organisms on earth, most of the time, ecosystems only changed very gradually, as some species would disappear while others would move in. Locally, sub-populations continuously go extinct, to be replaced later through dispersal of other sub-populations. Stochastists do recognize that certain intrinsic regulating mechanisms occur in nature. Feedback and response mechanisms at the species level regulate population levels, most notably through territorial behaviour. Andrewatha and Birch (1954) suggest that territorial behaviour tends to keep populations at levels where food supply is not a limiting factor. Hence, stochastists see territorial behaviour as a regulatory mechanism at the species level but not at the ecosystem level. Thus, in their vision, ecosystems are not regulated by feedback and response mechanisms from the (eco)system itself and there is no such thing as a balance of nature.

If ecosystems are indeed governed primarily by stochastic processes, they may be somewhat more resilient to sudden change, as each species would respond individually. In the absence of a balance of nature, the species composition of ecosystems would undergo shifts that would depend on the nature of the change, but entire ecological collapse would probably be less frequently occurring events.

The theoretical ecologist Robert Ulanowicz has used information theory tools to describe the structure of ecosystems, emphasising mutual information (correlations) in studied systems. Drawing on this methodology, and prior observations of complex ecosystems, Ulanowicz depicts approaches to determining the stress levels on ecosystems, and predicting system reactions to defined types of alteration in their settings (such as increased or reduced energy flow, and eutrophication. See also Relational order theories, as to fundamentals of life organisation.

Ecosystem Ecology

Ecosystem ecology is the integrated study of biotic and abiotic components of ecosystems and their interactions within an ecosystem framework. This science examines how ecosystems work and relates this to their components such as chemicals, bedrock, soil, plants, and animals. Ecosystem ecology examines physical and biological structure and examines how these ecosystem characteristics interact]] The relationship between systems ecology and ecosystem ecology is complex. Much of systems ecology can be considered a subset of ecosystem ecology. Ecosystem ecology also utilizes methods that have little to do with the holistic approach of systems ecology. However, systems ecology more actively considers external influences such as economics that usually fall outside the bounds of ecosystem ecology. Whereas ecosystem ecology can be defined as the scientific study of ecosystems, systems ecology is more of a particular approach to the study of ecological systems and phenomena that interact with these systems.

Systems Ecology

Systems ecology is an interdisciplinary field of ecology, taking a holistic approach to the study of ecological systems, especially ecosystems. Systems ecology can be seen as an application of general systems theory to ecology. Central to the systems ecology approach is the idea that an ecosystem is a complex system exhibiting emergent properties. Systems ecology focuses on interactions and transactions within and between biological and ecological systems, and is especially concerned with the way the functioning of ecosystems can be influenced by human interventions. It uses and extends concepts from thermodynamics and develops other macroscopic descriptions of complex systems.

The Millennium Ecosystem Assessment

In 2005, the largest ever assessment of the earth's ecosystems was conducted by a research team of over 1,000 scientists. The findings of the assessment were published in the multi volume Millennium Ecosystem Assessment, which concluded that in the past 50 years humans have altered the earth's ecosystems more than any other time in our history.

Earth Charter

The Earth Charter is a declaration of fundamental values and principles for building a just, sustainable, and peaceful global society in the 21st century. Created by the largest global consultation process ever associated with an international declaration, endorsed by thousands of organisations representing millions of individuals, the Earth Charter seeks to inspire in all peoples a sense of global interdependence and shared responsibility for the well-being of the human family and the larger living world. It is an expression of hope and a call to help create a global partnership at a critical juncture in history. The Earth Charter's inclusive ethical vision proposes that environmental protection, human rights, equitable human development, and peace are interdependent and indivisible. It provides a new framework for thinking about and addressing these issues.

History

The idea of a Charter originated in 1987, when the United Nations World Commission on Environment and Development called for a new charter to guide the transition to sustainable development. In 1992, the need for a charter was urged by then-Secretary General Boutros Boutros-Ghali at the Rio de Janeiro Earth Summit, but the time for such a declaration was not right. The Rio Declaration became the statement of the achievable consensus at that time. In 1994, Maurice Strong (Chairman of the Earth Summit) and Mikhail Gorbachev, working through organisations they each founded (Earth Council and Green Cross International respectively), restarted the Earth Charter as a civil society initiative, with the help of the Government of the Netherlands. The initial drafting and consultation process drew on hundreds of international documents.

Drafting of the Charter

The Earth Charter was created through the most open and participatory worldwide consultation process ever associated with the drafting of an international declaration. Many thousands of people and hundreds of organisations contributed to the drafting process. The drafting of the text was overseen by the independent Earth Charter Commission, which was convened by Maurice Strong and Mikhail Gorbachev with the purpose of developing a global consensus on values and principles for a sustainable future. The Commission continues to serve as the steward of the Earth Charter text.

The Earth Charter was completed in March 2000 and launched in a special ceremony at The Peace Palace in The Hague, Netherlands, on 29 June 2000. Queen Beatrix of the Netherlands attended the ceremony. The Charter has since then been formally endorsed by thousands of organisations representing millions of people, including the UNESCO Conference of Member States, the World Conservation Union of IUCN, national government ministries, national and international associations of universities, and hundreds of cities and towns in dozens of countries. It has also been endorsed by tens of thousands of individuals, and publicly supported by numerous heads of state.

Preamble to the Earth Charter

"We stand at a critical moment in Earth's history, a time when humanity must choose its future. As the world becomes increasingly interdependent and fragile, the future at once holds great peril and great promise. To move forward we must recognize that in the midst of a magnificent diversity of cultures and life forms we are one human family and one Earth community with a common destiny. We must join together to bring forth a sustainable global society founded on respect for nature, universal human rights, economic justice, and a culture of peace. Towards this end, it is imperative that we, the peoples of Earth, declare our responsibility to one another, to the greater community of life, and to future generations. "

General Principles

The Earth Charter urges environmental responsibility, peaceful coexistence, respect for life, democracy, and justice. It is organised into 16 general headings, each covering a general principle, as follows:

1. Respect Earth and life in all its diversity.
2. Care for the community of life with understanding, compassion and love.
3. Build democratic societies that are just, participatory, sustainable and peaceful.
4. Secure Earth's bounty and beauty for present and future generations.
5. Protect and restore the integrity of Earth's ecological systems, with special concern for biological diversity and the natural processes that sustain life.
6. Prevent harm as the best method of environmental protection and, when knowledge is limited, apply a precautionary approach.
7. Adopt patterns of production, consumption and reproduction that safeguard Earth's regenerative capacities, human rights and community well-being.
8. Advance the study of ecological sustainability and promote the open exchange and wide application of the knowledge acquired.
9. Eradicate poverty as an ethical, social and environmental imperative.
10. Ensure that economic activities and institutions at all levels promote human development in an equitable and sustainable manner.
11. Affirm gender equality and equity as prerequisites to sustainable development and ensure universal access to education, health care and economic opportunity.
12. Uphold the right of all, without discrimination, to a natural and social environment supportive of human dignity, bodily health and spiritual well-being, with special attention to the rights of indigenous peoples and minorities.
13. Strengthen democratic institutions at all levels, and provide transparency and accountability in governance, inclusive participation in decision-making, and access to justice.
14. Integrate into formal education and lifelong learning the knowledge, values and skills needed for a sustainable way of life.
15. Treat all living beings with respect and consideration.
16. Promote a culture of tolerance, nonviolence and peace.

Reaction

The Earth Charter has been publicly endorsed, recognized, or supported by people and organisations across a wide range of the political spectrum, from conservative to liberal, as well as from all major religious traditions. It has received support from business corporations, grassroots activists, universities, governments, and global non-governmental organisations. Overall, reaction to the document can be characterised as overwhelmingly positive.

However, the Charter has also received opposition from many groups and governments. For example, in the United States and other countries, members of religious groups, such as the Religious Right have objected to the document on the grounds that it is secular, and espouses socialism. Some groups go so far as to complain that it contains no reference to the doctrines of Judeo-Christianity. In addition, some conservatives cite an informal comment by Mikhail Gorbachev that the document is "a kind of Ten Commandments", and point to the fact that at the 2002 World Summit on Sustainable Development in Johannesburg, South Africa, a copy of the document was placed symbolically in an "Ark of Hope" — an independent project by the American artist Sally Linder. Some members of the American Religious Right infer from these incidents that the Charter is a proposed replacement for the Ten Commandments, and part of a conspiracy to establish a New World Government that replaces National Sovereignty.

Earth Charter International, the organisation responsible for promoting the Charter, argues in its literature that the Earth Charter is respectful and inclusive of all religious traditions. They state that the Charter itself makes no statements to support these claims of intent to supplant any of the world's religions or to create a world government. In their opinion, the Charter is simply a statement of common ethical values that recognises humanity's shared responsibility to the Earth and to each other. It is similar, in this respect, to the work of Jacques Maritain, who discovered that all human beings could agree on basic values, regardless of the many disparate beliefs that backed them. Some Libertarians also express numerous critiques of the Charter, including a concern that the Charter's language calling for 'economic justice' is equivalent to espousing socialism. But the Charter's leadership has stated that it does not adhere to any specific political ideology, and support for the Earth Charter has come from both traditionally "left" and "right"-leaning political leaders, in many countries.

Earth Charter Youth Initiative

The Earth Charter Youth Initiative (ECYI) (http://www.earthcharterinaction.org/youth/) is a network of young activists, youth NGOs, and partners who share a common interest in sustainable development and the Earth Charter.

Ecosystem Ecology

Ecosystematic ecology is the integrated study of biotic and abiotic components of ecosystems and their interactions within an ecosystem framework. This science examines how ecosystems work and relates this to their components such as chemicals, bedrock, soil, plants, and animals. Ecosystem ecology examines physical and also biological structures and examines how these ecosystem characteristics interact with each other. Ultimately, this helps us understand how to maintain high quality water and economically viable commodity production in this and many other ecosystems. A major focus of ecosystem ecology is on functional processes, ecological mechanisms that maintain the structure and services produced by ecosystems. These include primary productivity (production of biomass), decomposition, and trophic interactions. Studies of ecosystem function have greatly improved human understanding of sustainable production of forage, fiber, fuel, and provision of water. Functional processes are mediated by regional-to-local level climate, disturbance, and management thus ecosystem ecology provides a powerful framework for identifying ecological mechanisms that interact with global environmental problems, especially global warming and degradation of surface water. This article will describe the context of ecosystem ecology and provide an overview of the mechanisms that maintain ecosystem structure and function.

Ecosystems and Scale

Ecosystems are difficult entities to define theoretically or to delineate in space. When standing on the stream bank, one can easily see two ecosystems, an aquatic one where fish, insects, and algae interact, and the other a terrestrial one with trees, another community of insects, and perhaps herbivores and predators such as deer and coyote.

Although these communities appear distinct they interact intimately. Insects may be aquatic for certain parts of their life-cycle and emerge to become herbivores of the vegetation and prey for many predators. Riparian trees utilize stream water for growth and their leaf litter is an important flux of energy and nutrients to a rich community of benthic invertebrates. The distinction becomes even less clear when streams flood and deposit nutrient rich sediment on flood planes and scour other areas clean of biota and soil. This example demonstrates several important aspects of ecosystems:

1. Ecosystem boundaries are often nebulous and may fluctuate in time
2. Organism within ecosystems are dependent on ecosystem level biological and physical processes and
3. Adjacent ecosystems closely interact and often are interdependent for

maintenance of community structure and functional processes that maintain productivity and biodiversity.

These characteristics also introduce practical problems into natural resource management. Who will manage which ecosystem? Will timber cutting in the forest degrade recreational fishing in the stream? These questions are difficult for land managers to address while the boundary between ecosystems remains unclear even though decisions in one ecosystem will affect the other. We need better understanding of the interactions and interdependencies of these ecosystems and the processes that maintain them before we can begin to address these questions.

Ecosystem ecology is an inherently interdisciplinary field of study. An individual ecosystem is composed of populations of organisms, interacting within communities, and contributing to the cycling of nutrients and the flow of energy. The ecosystem is the principle unit of study in ecosystem ecology.

Population, community, and physiological ecology provide many of the underlying biological mechanisms influencing ecosystems and the processes they maintain. Cycling of energy and matter at the ecosystem level are often examined in ecosystem ecology but, as a whole this science is defined more by subject matter than by scale. Ecosystem ecology approaches organisms and abiotic pools of energy and nutrients as an integrated system which distinguishes it from associated sciences such as biogeochemistry.

Biogeochemistry and hydrology focus on several fundamental ecosystem processes such as biologically mediated chemical cycling of nutrients and physical-biological cycling of water. Ecosystem ecology forms the mechanistic basis for regional or global processes encompassed by landscape-to-regional hydrology, global biogeochemistry, and earth system science.

Ecosystem ecology is philosophically and historically rooted in terrestrial ecology. The ecosystem concept has evolved rapidly during the last 100 years with important ideas developed by Fredrick Clements, a botanist who argued for specific definitions of ecosystems and that physiological processes were responsible for their development and persistence. Although most of Clements ecosystem definitions have been greatly revised by contemporary ecologists, the idea that physiological processes are fundamental to ecosystem structure and function remains central to ecology. Later work by Eugene Odum and Howard T. Odum quantified flows of energy and matter at the ecosystem level, thus documenting the general ideas proposed by Clements and his contemporary Charles Elton, the intellectual father of the "food web" concept. In this model, energy flows through the whole system were dependent on biotic and abiotic interactions of each individual component (species, inorganic pools of nutrients, etc). Later work demonstrated that these interactions and flows applied to nutrient cycles, changed over the course of succession, and held powerful controls over ecosystem productivity. Transfers of energy and nutrients are innate to ecological systems regardless of whether they are aquatic or terrestrial. Thus, ecosystem ecology has emerged from important biological studies of plants, animals, terrestrial, aquatic, and marine ecosystems.

Ecosystem Services

Ecosystem services are ecologically mediated functional processes essential to sustaining healthy human societies. Water provision and filtration, production of biomass in forestry, agriculture, and fisheries, and removal of greenhouse gases such as carbon dioxide (CO_2) from the atmosphere are examples of ecosystem services essential to public health and economic opportunity. Nutrient cycling is a process fundamental to agricultural and forest production. However, like most ecosystem processes, nutrient cycling is not an ecosystem characteristic which can be "dialed" to the most desirable level. Maximising production in degraded systems is an overly simplistic solution to the complex problems of hunger and economic security. For instance, intensive fertilizer use in the midwestern United States has resulted in degraded fisheries in the Gulf of Mexico. Regrettably, a "green revolution" of intensive chemical fertilisation has been recommended for agriculture in developed and developing countries. These short-sighted strategies risk alteration of ecosystem processes that may be difficult to restore, especially when applied at broad scales without adequate assessment of impacts. Ecosystem processes may take many years to recover from significant disturbance.

For instance, large-scale forest clearance in the northeastern United States during the 18th and 19th centuries has altered soil texture, dominant vegetation, and nutrient cycling in ways that impact forest productivity in the present day. An appreciation of the importance of ecosystem function in maintenance of productivity, whether in agriculture or forestry, is needed in conjunction with plans for restoration of essential processes. Improved knowledge of ecosystem function will help to achieve long-term sustainability and stability in the poorest parts of the world.

How do Ecosystems Work?

Biomass productivity is one of the most apparent and economically important ecosystem functions. Biomass accumulation begins at the cellular level via photosynthesis. Photosynthesis requires water and consequently, global patters of annual biomass production are correlated with annual precipitation. Amounts of productivity are also dependent on the overall capacity of plants to capture sunlight which is directly correlated with plant leaf area and leaf N content. Net primary productivity (NPP) is the primary measure of biomass accumulation within an ecosystem. Net primary productivity can be calculated by a simple formula where total amount of productivity is adjusted for total productivity losses through maintenance of biological processes: Where GPP is gross primary productivity and Rplant is photosynthate (Carbon) lost via cellular respiration. NPP is difficult to measure but a new technique known as eddy co-variance has shed light on how natural ecosystems influence the atmosphere. CO_2 concentration steadily increased but within-year variation has been greater than the annual increase since measurements began in 1957. These variations were thought to be due to seasonal uptake of CO_2 during summer months. A newly developed technique for assessing ecosystem NPP has confirmed seasonal variation are driven by seasonal changes in CO_2 uptake by vegetation. This has led many scientists and policy makers to speculate that ecosystems can be managed to ameliorate problems with global warming. This type of management may include reforesting or altering forest harvest schedules many parts of the world.

Decomposition and Nutrient Cycling

Decomposition and nutrient cycling are fundamental to ecosystem biomass production. Most natural ecosystems are nitrogen (N) limited and biomass production is closely correlated with N turnover. Typically external input of nutrients is very low and efficient recycling of nutrients maintains productivity. Decomposition of plant litter accounts for the majority of nutrients recycled through ecosystems. Rates of plant litter decomposition are highly dependent on litter quality; high concentration of phenolic compounds, especially lignin, in plant litter has a retarding effect on litter decomposition. More complex C compounds are decomposed more slowly and may take many years to completely breakdown. Decomposition is typically described with exponential Globally, rates of decomposition are mediated by litter quality and climate. Ecosystems dominated by plants with low-lignin concentration often have rapid rates of decomposition and nutrient cycling (Chapin et al. 1982). However, these models do not reflect simultaneous linear and non-linear decay processes which likely occur during decomposition. For instance, proteins, sugars and lipids decompose exponentially, but lignin decays at a more linear rate Thus, litter decay is probably inaccurately predicted by the most simplistic models. Better understanding of decomposition models is an important research area of ecosystem ecology because this process is closely tied to nutrient supply and the overall capacity of ecosystems to sequester CO_2 from the atmosphere.

Trophic Dynamics

However, predators can also influence the structure of lower trophic levels from the top-down. So called top-down effects can dramatically shift dominant species in terrestrial and marine systems The interplay and relative strength of top-down vs. bottom-up controls on ecosystem structure and function is an important area of research in the greater field of ecology. Trophic dynamics can strongly influence rates of decomposition and nutrient cycling in time and in space. For example, herbivory can increase litter decomposition and nutrient cycling via direct changes in litter quality and altered dominant vegetation. Insect herbivory has been shown to increase rates of decomposition and nutrient turnover due to changes in litter quality and increased frass inputs However, insect outbreak does not always increase nutrient cycling. Stadler showed that C rich honeydew produced during aphid outbreak can result in increased N immobilisation by soil microbes thus slowing down nutrient cycling and potentially limiting biomass production. North atlantic marine ecosystems have been greatly altered by overfishing of cod. Cod stocks crashed in the 1990's which resulted in increases in their prey such as shrimp and snow crab Human intervention in ecosystems has resulted in dramatic changes to ecosystem structure and function. These changes are occurring rapidly and have unknown consequences for economic security and human well-being.

Applications: Why Does this Science Matter?

Lessons from two Central American Cities

The biosphere has been greatly altered by the demands of human societies. Ecosystem ecology plays an important role in understanding and adapting to the most pressing current environmental problems. Restoration ecology and ecosystem management are closely associated with ecosystem ecology. Restoring highly degraded resources depends on integration of functional mechanisms of ecosystems.

Without these functions intact, economic value of ecosystems is greatly reduced and potentially dangerous conditions may develop in the field. For example, areas within the mountainous western highlands of Guatemala are more susceptible to catastrophic landslides and crippling seasonal water shortages due to loss of forest resources. In contrast, cities such as Totonicapán that have preserved forests through strong social institutions have greater local economic stability and overall greater human well-being. This situation is striking considering that these areas are close to each other, the majority of inhabitants are of Mayan descent, and the topography and overall resources are similar. This is a case of two groups of people managing resources in fundamentally different ways. Ecosystem ecology provides the basic science needed to avoid degradation and to restore ecosystem processes that provide for basic human needs.

Ecosystem Valuation

Valuation can be a useful tool that aids in evaluating different options that a natural resource manager might face. Because ecological resources and services are so varied in their composition, it is often difficult to examine them on the same level. However, after they are assigned a value, an environmental resource or service can then be compared to any other item with a respective value. Ecosystem valuation is the process by which policymakers assign a value – monetary or otherwise – to environmental resources or to the outputs and/or services provided by those resources. For example, a mountain forest may provide environmental services by preventing downstream flooding.

Environmental resources and/or services are particularly hard to quantify due to their intangible benefits and multiple value options. It is almost impossible to attach a specific value to some of the experiences we have in nature, such as viewing a beautiful sunset. Problems also exist when a resource can be used for multiple purposes, such as a tree – the wood is valued differently if it is used for flood control versus if it is used for building a house. The quantity of a resource must also be taken into consideration because value can change depending how much of a resource is available. An example of this might be in preventing the first "unit" of pollution if we have a pristine air environment. Preventing the first unit of pollution is not valued very highly because the environment can easily recover. However, if the pollution continues until the air is becoming toxic to its surroundings; the value of preserving clean air by preventing additional pollution is going to be increasingly valued. Within economics, value is generally defined as the amount of alternate goods a person is willing to give up in order to get one "additional unit" of the good in question. An individual's preference for certain goods may either be stated or revealed. In the case of stated preferences, the amount of money a person is willing to pay for a good determines the value because that money could otherwise be used to purchase other goods. However, value may also be determined by simply ranking the alternatives according to the amount of benefit each will produce. Revealed preferences can be measured by examining a person's behavior when it is not possible to use market pricing.

There are typically two ways to assign value to environmental resources and services – use and non-use – and there are approaches to measuring environmental benefits based on these defined values. When environmental resources or services are being used, it is easier to observe the price consumers are willing to pay for the conservation or preservation of those resources. Market or opportunity cost pricing can be used when there are tangible products to measure, such as the amount of fish caught in a lake. Replacement cost can also be used, calculated based on any expenses incurred to reverse environmental damage. Hedonic pricing will measure the effect that negative environmental qualities have on the price of related market goods. When evaluating non-use value, contingent valuation is employed through the use of surveys that attempt to assess an individual's willingness to pay for a resource that they do not consume. The process of environmental resource or service valuation provides a way to compare alternative proposals, but it is not without problems. All valuation techniques encompass a great deal of uncertainty: flaws can exist in the methods of assigning value accurately due to a wide number of variables and it is difficult to compartmentalize and measure environmental and natural resources and/or services withing an ecosystem that functions as an interconnected web. Yet challenging, it allows policymakers to make decisions based on specific comparisons (typically monetary) rather than some other arbitrary basis. The government has placed increasing emphasis on cost-effective laws and projects; therefore, establishing a common measure by which to evaluate alternatives is essential.

Ecosystem Services

Ecosystem services are processes by which the natural environment produces resources useful to people, as to economic services. They include:

- Provision of clean water and air
- Flood control
- Pollination of crops
- Mitigation of environmental hazards
- Pest and disease control
- Carbon sequestration
- Aesthetic, cultural and ethical values associated with biodiversity.

To look at the way in which ecosystems provide goods is popular now at Stanford University. The concept of ecosystem services is similar to that of natural capital.

The Millennium Ecosystem Assessment released in 2005 showed that 60% of ecosystem services are used in a way that destroys them.

Sustainable Development

Sustainable development is a pattern of resource use that aims to meet human needs while preserving the environment so that these needs can be met not only in the present, but in the indefinite future. The term was used by the Brundtland Commission which coined what has become the most often-quoted definition of sustainable development as development that "meets the needs of the present without compromising the ability of future generations to meet their own needs." The field of sustainable development can be conceptually broken into three constituent parts: environmental sustainability, economic sustainability and sociopolitical sustainability.

Scope and Definitions

Sustainable development does not focus solely on environmental issues. The United Nations 2005 World Summit Outcome Document, refers to the "interdependent and mutually reinforcing pillars" of sustainable development as economic development, social development, and environmental protection. Indigenous peoples have argued, through various international forums such as the United Nations Permanent Forum on Indigenous Issues and the Convention on Biological Diversity, that there are *four* pillars of sustainable development, the fourth being cultural.

The Universal Declaration on Cultural Diversity (UNESCO, 2001) further elaborates the concept by stating that "...cultural diversity is as necessary for humankind as biodiversity is for nature"; it becomes "one of the roots of development understood not simply in terms of economic growth, but also as a means to achieve a more satisfactory intellectual, emotional, moral and spiritual existence". In this vision, cultural diversity is the fourth policy area of sustainable development.

Economic Sustainability: Agenda 21 clearly identified information, integration, and participation as key building blocks to help countries achieve development that recognises these interdependent pillars. It emphasises that in sustainable development everyone is a user and provider of information. It stresses the need to change from old sector-centred ways of doing business to new approaches that involve cross-sectoral co-ordination and the integration of environmental and social concerns into all development processes. Furthermore, Agenda 21 emphasises that broad public participation in decision making is a fundamental prerequisite for achieving sustainable development. According to Hasna, sustainability is a process which tells of a development of all aspects of human life affecting sustenance. It means resolving the conflict between the various competing goals, and involves the simultaneous pursuit of economic prosperity, environmental quality and social equity famously known as three dimensions (triple bottom line) with is the resultant vector being technology, hence it is a continually evolving process; the 'journey' (the process of achieving sustainability) is of course vitally important, but only as a means of getting to the destination (the desired future state). However, the 'destination' of sustainability is not a fixed place in the normal sense that we understand destination. Instead it is a set of wishful characteristics of a future system.

Green development is generally differentiated from sustainable development in that Green development prioritizes what its proponents consider to be environmental sustainability over economic and cultural considerations. Proponents of Sustainable Development argue that it provides a context in which to improve overall sustainability where cutting edge Green development is unattainable. For example, a cutting edge treatment plant with extremely high maintenance costs may not be sustainable in regions of the world with less financial resources. An environmentally ideal plant that is shut down due to bankruptcy is obviously less sustainable than one that is maintainable by the indigenous community, even if it is somewhat less effective from an environmental standpoint. Some research activities start from this definition to argue that the environment is a combination of nature and culture. The Network of Excellence "Sustainable Development in a Diverse World", sponsored by the European Union, integrates multidisciplinary capacities and interprets cultural diversity as a key element of a new strategy for sustainable development. Still other researchers view environmental and social challenges as opportunities for development action. This is particularly true in the concept of sustainable enterprise that frames these global needs as opportunities for private enterprise to provide innovative and entrepreneurial solutions. This view is now being taught at many business schools including the Center for Sustainable Global Enterprise at Cornell University and the Erb Institute for Global Sustainable Enterprise at the University of Michigan. The United Nations Division for Sustainable Development lists the following areas as coming within the scope of sustainable development:

- Agriculture
- Atmosphere
- Biodiversity
- Biotechnology
- Capacity-building
- Climate Change
- Consumption and Production Patterns
- Demographics
- Desertification and Drought
- Disaster Reduction and Management
- Education and Awareness
- Energy
- Finance
- Forests
- Fresh Water
- Health
- Human Settlements
- Indicators
- Industry
- Information for Decision Making and Participation
- Integrated Decision Making
- International Law

- International Cooperation for Enabling Environment
- Institutional Arrangements
- Land management
- Major Groups
- Mountains
- National Sustainable Development Strategies
- Oceans and Seas
- Poverty
- Sanitation
- Science
- SIDS
- Sustainable tourism
- Technology
- Toxic Chemicals
- Trade and Environment
- Transport
- Waste (Hazardous)
- Waste (Radioactive)
- Waste (Solid)
- Water

Sustainable development is an eclectic concept, as a wide array of views fall under its umbrella. The concept has included notions of weak sustainability, strong sustainability and deep ecology. Different conceptions also reveal a strong tension between ecocentrism and anthropocentrism. Thus, the concept remains weakly defined and contains a large amount of debate as to its precise definition. During the last ten years, different organisations have tried to measure and monitor the proximity to what they consider sustainability by implementing what has been called sustainability metric and indices. Sustainable development is said to set limits on the developing world. While current first world countries polluted significantly during their development, the same countries encourage third world countries to reduce pollution, which sometimes impedes growth. Some consider that the implementation of sustainable development would mean a reversion to pre-modern lifestyles.

Others have criticised the overuse of the term:

"[The] word sustainable has been used in too many situations today, and ecological sustainability is one of those terms that confuse a lot of people. You hear about sustainable development, sustainable growth, sustainable economies, sustainable societies, sustainable agriculture. Everything is sustainable (Temple, 1992)."

Environmental Sustainability

Environmental sustainability is the process of making sure current processes of interaction with the environment are pursued with the idea of keeping the environment as pristine as naturally possible based on ideal-seeking behavior. An "unsustainable situation" occurs when natural capital (the sum total of nature's resources) is used up faster than it can be replenished. Sustainability requires that human activity only uses nature's resources at a rate at which they can be replenished naturally. Inherently the concept of sustainable development is intertwined with the concept of carrying capacity. Theoretically, the long-term result of environmental degradation is the inability to sustain human life. Such degradation on a global scale could imply extinction for humanity.

Decade of Education for Sustainable Development

The United Nations has declared a Decade of Education for Sustainable Development starting in January of 2005. A non-partisan multi-sector response to the decade has formed within the U.S. via the U.S. Partnership for the Decade of Education for Sustainable Development. Active sectors teams have formed for youth, higher education, business, religion, the arts, and more. Organisations and individuals can join in sharing resources and success stories, and creating a sustainable future. Sustainable development is not just about business perspective but should be understood in such way to benefit the whole as a world.

Commission on Sustainable Development

The United Nations Commission on Sustainable Development - (CSD) - was established in December 1992 by General Assembly Resolution A/RES/47/191 as a functional commission of the UN Economic and Social Council, implementing a recommendation in Chapter 38 of Agenda 21, the landmark global agreement reached at the June 1992 United Nations Conference on Environment & Development / Earth Summit held in Rio de Janeiro.

CSD 1

CSD 1, the Organisational Session of the CSD was held in June 1993. The Organisational Session focused on a broad range of organisational and administrative issues, reflected in topics of the Commission's documents:

- Budget implications of draft decisions
- Establishing a provisional agenda & a multi-year programme of work
- National reporting on implementation of Agenda 21
- Information exchange: UN System & Donors
- UNCED follow up: international organisations & UN Coordination
- Coordination of development data
- Progress in environmentally sound technology transfer
- Initial financial commitments & flows
- Government information on financial commitments
- Urgent & major emergent issues
- UNCTAD & Agenda 21 implementation
- UNEP & Agenda 21 implementation
- Issues relating to future work of CSD
- Guidelines for national reporting
- Multi-year programme of work
- Financial commitments & financial flows
- Integrating sustainable development in the UN System

CSD 5

In its Fifth Session, the principal focus of the CSD was to prepare for the Five-Year Review of the 1992 Earth Summit, which took the form of the 19th Special Session of the General Assembly, held at UN Headquarters in New York.

CSD 10

For its Tenth Session the CSD served as the Preparatory Committee for the World Summit on Sustainable Development, held in Johannesburg in December 2002.

CSD 12

The Twelfth Session of the CSD is the first substantive session since the Johannesburg Summit - CSD-11 was an organisational session that focused on establishing priorities and an agenda for the second ten-year cycle of the Commission. Text below in "quotes" is from the introductory note from the Chair - H.E. Børge Brende, Norwegian Minister of the Environment to a description of the organisation of work during CSD-12.

"The first three days of CSD-12 will serve as the preparatory meeting for the ten-year review of the Barbados Programme of Action (BPoA) for the Sustainable Development of Small Island Developing States.

The bulk of CSD-12 will focus on water, sanitation and human settlements. "The decision of CSD-11 to focus its first Implementation Cycle on Water, Sanitation and Human Settlements bears testimony to the sense of urgency the international community ascribes to these issues."

"Today, water-borne diseases kill a person every ten seconds, the overwhelming majority children, and roughly one in two Sub-Saharan Africans and three in five South Asians lack adequate sanitation. The total number of people living in slums is close to 930 million and is growing at an accelerated rate."

"The CSD-12 Review Session will feature a number of "firsts". Thematic reviews will feature interactive discussions and will be introduced through keynote presentations, panel discussions and moderated debates. Major Group representatives will join Ministers in interactive dialogues during the High-Level Segment."

CSD 16

The Chair is H.E. Mr. Francis Nhema, Minister of Environment and Tourism (Zimbabwe).

Agenda 21

Agenda 21 is a programme run by the United Nations (UN) related to sustainable development. It is a comprehensive blueprint of action to be taken globally, nationally and locally by organisations of the UN, governments, and major groups in every area in which humans impact on the environment. The number *21* refers to the 21st century.

Development of Agenda 21

The full text of Agenda 21 was revealed at the United Nations Conference on Environment and Development (Earth Summit), held in Rio de Janeiro on June 14, 1992, where 179 governments voted to adopt the programme. The final text was the result of drafting, consultation and negotiation, beginning in 1989 and culminating at the two-week conference.

Rio+5

In 1997, the General Assembly of the UN held a special session to appraise five years of progress on the implementation of Agenda 21 (Rio +5). The Assembly recognised progress as 'uneven' and identified key trends including increasing globalisation, widening inequalities in income and a continued deterioration of the global environment. A new General Assembly Resolution (S-19/2) promised further action.

The Johannesburg Summit

The Johannesburg Plan of Implementation, agreed at the World Summit on Sustainable Development (Earth Summit 2002) affirmed UN commitment to 'full implementation' of Agenda 21, alongside achievement of the Millennium Development Goals and other international agreements.

Implementation

The Commission on Sustainable Development acts as a high level forum on sustainable development and has acted as preparatory committee for summits and sessions on the implementation of Agenda 21. The United Nations Division for Sustainable Development acts as the secretariat to the Commission and works 'within the context of Agenda 21. Implementation by member states remains essentially voluntary.

Local Agenda 21

The implementation of Agenda 21 was intended to involve action at international, national, regional and local levels. Some national and state governments have legislated or advised that local authorities take steps to implement the plan locally, as recommended in Chapter 28 of the document. Such programmes are often known as 'Local Agenda 21' or 'LA21'.

A concrete example towards this end is the initiative of the Municipality of Rosario, Batangas, Philippines which formulated what they called *Aksyon ng Bayan Rosario 2001 And Beyond*. This programme implements the framework of Agenda 21 at the municipal and barangay levels with community organising used as a strategy.

Hannover Principles

The Hannover Principles are a set of statements about designing buildings and objects with forethought about their environmental impact, their effect on the sustainability of growth, and their overall impact on society. They were first formulated by William McDonough and Michael Braungart for planning Expo 2000 in Hanover and are presented in a copyrighted 1992 document.

The principles may be summarised as:

1. Insist on human rights and sustainability.

2. Recognize the interaction of design with the environment.
3. Consider the social and spiritual aspects of buildings and designed objects.
4. Be responsible for the effect of design decisions.
5. Ensure that objects have long-term value.
6. Eliminate waste and consider the entire life-cycle of designed objects.
7. Make use of "natural energy flows" such as solar power and its derivatives.
8. Be humble, and use nature as a model for design.
9. Share knowledge, strive for continuous improvement, and encourage open communication among stakeholders.

The original document states: *"The Hannover Principles should be seen as a living document committed to the transformation and growth in the understanding of our interdependence with nature, so that they may adapt as our knowledge of the world evolves."*

Earth Summit 2002

The World Summit on Sustainable Development, WSSD or Earth Summit 2002 took place in Johannesburg, South Africa, from 26 August to 4 September 2002. It was convened to discuss sustainable development by the United Nations. WSSD gathered a number of leaders from business and non-governmental organisations, 10 years after the first Earth Summit in Rio de Janeiro. (It was therefore also informally nicknamed "Rio+10".)

Declarations

The Johannesburg Declaration was the main outcome of the Summit; however, there were several other international agreements. It laid out the Johannesburg Plan of Implementation as an action plan.

Agreements

Johannesburg, 27 August: by 2015 agreement was made to restore the world's depleted fisheries. It was agreed to by negotiators at the World Summit. But as usual the process proceeded slowly and full of ironies. Protesters outside were having fun, even dancing, whilst demanding food, work and a better environment. Annette Cotter said "If the people who were marching today were inside the meeting room representing our planet, there'd be laughter, fun, dancing, and real action.

Partnership Initiatives

Instead of new agreements between governments, the Earth Summet was organised mostly around almost 300 "partnership initiatives". These were to be the key means to achieve the Millenni n Development Goals. These are kept in a database of Partnerships for Sustainable Development.

Criticism

The absence of the United States, rendered the summit partially impotent. George W. Bush boycotted the summit and did not attend. Except for a brief appearance by Colin Powell, who hurriedly addressed the closing stages of the conference whilst his airplane taxied on the run-way of Johannesburg International, ready to take-off, the US government did not send a delegation, earning Bush praise in a letter from conservative, corporate-sponsored organisations such as Americans for Tax Reform, American Enterprise Institute, and Competitive Enterprise Institute

Earth Summit 2002 attendees have been criticised as being hypocritical for providing expensive catered food and drink for dignitaries while a few miles away there were starving South Africans. The summit organisers have also been criticised by others for excluding a variety of organisations and individuals, particularly those early founders, instrumental in the nations conservation and green history. The summit also excluded critics in the movement who believe the "sustainable development" mantra is being misused in order to greenwash economic development at the expense of long term environmental goals.

The United Nations Conference on the Human Environment, was first held in Stockholm, Sweden, in June 1972, and marked the emergence of international environmental law. The Declaration on the Human Environment also known as the Stockholm Declaration set out the principles for various international environmental issues, including human rights, natural resource management, pollution prevention and the relationship between the environment and development. The conference also led to the creation of the United Nations Environment Programme. The Brundtland Commission set up by Gro Harlem Brundtland, the pioneer of sustainable development, provided the momentum for first Earth Summit 1992 – the United Nations Conference on Environmental Development (UNCED), that was also headed by Maurice Strong, who had been a prominent member of the Brundtland Commission – and also for Agenda 21. South Africa's first National Conference on Environment and Development entitled, "Ecologise Politics, Politicise Ecology" was held at the University of the Western Cape in conjunction with the Cape Town Ecology Group and the Western Cape Branch of the World Conference on Religion and Peace in 1991. Prominent persons involved in this conference were Ebrahim Rasool, Cheryl Carolus, Faried Esack, and Julia Martin.

Strategic Sustainable Development

Strategic Sustainable Development (SSD) is a strategic approach to sustainable development and is a field of study that exists within the public domain. A number of authors, researchers and practitioners identify with this approach. Both the Blekinge Institute of Technology and The Natural Step, along with many leading experts in the sustainability arena, are contributors to it.

Advantages

SSD has the advantages of being:

- Scientific - basic concepts from physics, natural and social systems provide a foundation
- Based on systems thinking – to support decision making in situations of high complexity recognising the interdependence of the natural world and society
- Structured - there is a clear distinction between: (1) the principle functioning of the system, (2) a principled view of sustainability, (3) strategic decision making, (4) actions and (5) various support methods & tools (e.g. Life Cycle Assessment, Environmental Management Systems)
- Based on sustainability principles derived from scientific consensus
- Strategic - a clearly principled view of a future sustainable society creates a perspective on which we can base our decisions (i.e. backcasting)
- Tested by application - theories and concepts are continually applied and tested in concert with business & society
- Encouraging Creativity - a vast range of possibilities within broad but concrete constraints

The SSD approach combines a number of concepts that are published independently (e.g. backcasting as a general approach, strategic planning, systems thinking, etc.) and it is the combination of these elements that give it strength.

Sustainable Living

Sustainable living refers to an individual or society's lifestyle that can be sustained with limited exhaustion of natural resources. Its adherents most often hold true sustainability as a goal or guide, and make lifestyle tradeoffs, such as transport, housing, energy, and diet favor sustainability. Most often these tradeoffs involve making more environmentally-friendly lifestyle choices. Lester R. Brown concisely summarizes the situation as "sustaining progress depends on shifting from a fossil fuel-based, automobile-centered, economy to a renewable energy-based, diversified transport, reuse/recycle economy".

Overview

Sustainable living is a sub-division of sustainability where the prerequisites of a modern, industrialised society are left unexercised by choice for a variety of reasons. The practices and motives overlap somewhat between the movements. Sustainable living in urban areas requires a sustainable urban infrastructure.

Self-sufficiency is the principle of consuming only those things produced by oneself or one's family. It is generally a stricter lifestyle than a sustainable lifestyle in that an effort is made to limit trade with others regardless of the sustainability of such trade.

Permaculture is a design philosophy that emphasises sustainability in land use and landscaping, as well as fields such as architecture and economics (for example, encouraging the spread of Local Exchange Trading Systems (LETS)). In terms of agriculture, food production and building materials, permaculture emphasises use of well-adapted plant materials that require few inputs, especially trees, hemp, and other edible and useful perennials.

Some people are opposed to furthering mechanisation and technology in society for any reason. Adherents of sustainable living, in contrast, are willing to accept appropriate technology.

Carbon Footprint

An individual's carbon footprint represents the amount of carbon required to maintain that person's lifestyle, and what impact that lifestyle has on the earth. A higher footprint represents a lifestyle that is less sustainable, and a lower footprint represents a more sustainable lifestyle. Most carbon footprint websites have suggestions for decreasing a person's carbon footprint, and therefore increasing their sustainability.

History

- 1854 Henry David Thoreau's work *Walden* represents the earliest literature that specifically addresses the sustainable lifestyle in simple living.
- 1954 The publication of *Living the Good Life* by Helen and Scott Nearing is the modern-day beginning of the sustainability movement. The book fostered the back-to-the-land movement in the late 1960s and early 1970s. As the back-to-the-landers realised the difficulty of copying the Nearings' lifestyle, they returned to more conventional lifestyles yet incorporated self-sufficiency where they could.
- 1962 The publication of *Silent Spring* by Rachel Carson was another major milestone in the sustainability movement.

- 1972 Donella Meadows wrote the international bestselling book *The Limits to Growth*, which reported on a study of long-term global trends in population, economics, and the environment. It sold millions of copies and was translated into 28 languages.
- 1973E. F. Schumacher published a collection of essays on shifting towards sustainable living through the appropriate use of technology in his book *Small Is Beautiful*.
- The United Nations had an a series of conferences called Earth Summit in 1992, 1997, and 2002. Each of these conferences focused on increasing sustainability within societies, with the goal of conserving the world's natural resources.
- 2007 the U.N. published *Sustainable Consumption and Production, Promoting Climate-Friendly Household Consumption Patterns*, a paper that promotes sustainable lifestyles in communities and homes.

REFERENCES

Aber, J.D., and J.M., Melillo (1982). "Nitrogen immobilisation in decaying hardwood leaf litter as a function of initial nitrogen and lignin content". In: *Cannadian Journal of Botany* 60:2263-2269.

Andrewartha, H.G., and L.C. Birch. 1954. The distribution and abundance of animals. Univ. of Chicago Press, Chicago, IL.

Anil Agarwal (ed.), *State of India's Environment—A Citizen's Report,* 1982 and 1984-85, Centre for Science and Environment, New Delhi.

Barford, C. C., ea. (2001). "Factors controlling long and short term sequestration of atmospheric CO_2 in a mid-latitude forest". In: *Science* 294: 1688-1691

Belovsky, G.E. and J.B. Slade. (2000). "Insect herbivory accelerates nutrient cycling and increases plant production". In: *Proceedings of the national academy of sciences* (USA). 97:14412-14417.

Boer, P.J. den, and J. Reddingius. 1996. Regulation and stabilisation paradigms in population ecology. Population and Community Biology Series 16. Chapman and Hall, New York, p. 397.

Carpenter, S.A. (1981). "Decay of heterogeneous detritus: a general model". In: *Journal of theoretical biology* 89:539-547.

Chapin F.S. III, Matson, P.A., and Mooney, H.A. (2003). *Principles of terrestrial ecosystem ecology*. Springer-Verlag, New York, N.Y.

Chapin, F.S. III, B.H., Walker, R.J., Hobbs, D.U.,Hooper, J.H., Lawton, O.E., Sala, and D., Tilman. (1997). "Biotic control over the functioning of ecosystems". in: *Science* 277:500-504.

Chapman, S.K., Hart, S.C., Cobb, N.S., Whitham, T.G., and Koch, G.W. (2003). "Insect herbivory increases litter quality and decomposition: an extension of the acceleration hypothesis". in: *Ecology* 84:2867-2876.

Chrispeels, M.J. and Sadava, D. (1977). *Plants, food, and people*. W. H. Freeman and Company, San Francisco.

Conz, B.W. 2004. Continuity and Contestation: Conservation Landscapes in Totonicapán , Guatemala. University of Massachussetss Masters of Science thesis.

Defries, R.S., J.A. Foley, and G.P. Asner. (2004). "Land-use choices: balancing human needs and ecosystem function". in: *Frontiers in ecology and environmental science*. 2:249-257.

Ecological Society of America, Ecosytem Services, Ecological Society of America. 25 May 2007.

Ehrenfeld, J.G. and Toth, L.A. (1997). "Restoration ecology and the ecosystem perspective". in: *Restoration Ecology* 5:307-317.

Ehrlich, Paul; Walker, Brian "Rivets and Redundancy".BioScience.vol.48.no.5. May 1998. pp. 387. American Institute of Biological Sciences.

F. Powledge, *Water, The Nature, Uses and Future of our most precious and Abused Source*, Farra, Straus, Giroux, 1982.

Foster, D. R. (1992). "Land-use history (1730-1990) and vegetation dynamics in central New England, USA". In: *Journal of Ecology* 80: 753-772.

Goulden, M. L., J. W. Munger, S.-M. Fan, B. C. Daube, and S. C. Wofsy, (1996). "Effects of interannual climate variability on the carbon dioxide exchange of a temperate deciduous forest". In: *Science* 271:1576-1578

Grime, J.P. "Biodiversity and Ecosystem Function: The Debate Deepens." Science Vol. 277. no. 533029 Aug 1997 pp. 1260 - 1261. 25 May 2007

Groom , Martha J., and Gary K. Meffe. Principles of Conservation Biology. 3. Sunderland, MA: Sinauer Associates, Inc, 2006.

H.C. Perkins, *Air Pollution*, McGraw-Hill, New York, 1974.

Hagen, J.B. (1992). *An Entangled Bank: The origins of ecosystem ecology*. Rutgers University Press, New Brunswick, N.J.

Hasna, A. M. (2007). "Dimensions of sustainability". Journal of Engineering for Sustainable Development: Energy, Environment, and Health 2 (1): 47–57.

Hättenschwiler S. and P.M. Vitousek (2000). "The role of polyphenols in terrestrial ecosystem nutrient cycling". In: *Trends in Ecology and Evolution* 15: 238-243

Hunter, M.D. (2001). "Insect population dynamics meets ecosystem ecology: effects of herbivory on soil nutrient dynamics". In: *Agricultural and Forest Entomology* 3:77-84.

Huxman TE, ea.(2004). "Convergence across biomes to a common rain-use efficiency". *Nature*. 429: 651-654

J.M. Moran and others, *Introduction to Environmental Science* (2nd ed.), Freeman, New York, 1976.

Keeling, C.D. and T.P. Whorf. (2005). "Atmospheric CO_2 records from sites in the SIO air sampling network". In: *Trends: A Compendium of Data on Global Change*. Carbon

Dioxide Information Analysis Center, Oak Ridge National Laboratory, U.S. Department of Energy, Oak Ridge, Tenn., U.S.A.

L. Hodges, *Environmental Pollution* (2nd ed.), Holt, Reinhardt and Winston, New York, 1977.

Lawton, John H., *What Do Species Do in Ecosystems?*], Oikos, December, 1994. vol.71,no.3.

Lester R. Brown and others, *State of the World* (continuing series), World Watch Report (1984 to 1991) W.W. Norton, New York.

Likens, G. E., F. H. Bormann, N. M. Johnson, D. W. Fisher and R. S. Pierce. (1970). "Effects of forest cutting and herbicide treatment on nutrient budgets in the Hubbard Brook watershed-ecosystem". in: *Ecological Monographs* 40:23-47.

Lindeman, R.L. 1942. The trophic-dynamic aspect of ecology. *Ecology* 23: 399-418.

M.N. Sastri, *History of Planet Earth,* Himalaya Publishing House, Bombay, 1993.

Meentemeyer, V. 1978 "Macroclimate and lignin control of litter decomposition rates". in: *Ecology* 59:465-472.

Melillo, J.M., Aber, J.D., and Muratore, J.F. (1982). "Nitrogen and lignin control of hardwood leaf litter decomposition dynamics". In: *Ecology* 63:621-626.

Motzkin, G., D. R. Foster, A. Allen, J. Harrod, and R. D. Boone. (1996). "Controlling site to evaluate history: vegetation patterns of a New England sand plain". In: *Ecological Monographs* 66: 345-365.

Odum, E.P 1969. "The strategy of ecosystem development". in: *Science* 164:262-270.

Odum, H.T. (1971). *Environment, Power, and Society*. Wiley-Interscience New York, N.Y.

Olson, J.S. (1963). "Energy storage and the balance of producers and decomposers in ecological systems". In: *Ecology* 44:322-331.

Patten, B.C. 1959. An Introduction to the Cybernetics of the Ecosystem: The Trophic-Dynamic Aspect. *Ecology* 40, no. 2.: 221-231.

Quinones, M.A., N.E. Borlaug, C.R. Dowswell. (1997). "A fertilizer-based green revolution for Africa". In: *Replenishing soil fertility in Africa*. Soil Science Society of America special publication number 51. Soil Science Society of America, Madison, WI.

R. Carson, *The Silent Spring,* Houghton Mifflin, Boston, 1962.

R.E. Long, *The Problem of Waste Disposal,* Wilson, 1989.

Ranganathan, J & Irwin, F. (2007, May 7). Restoring Nature's Capital: An Action Agenda to Sustain Ecosystem Services

Reich, P.B., Grigal, D.F., Aber, J.D., Gower, S.T. (1997). "Nitrogen mineralisation and productivity in 50 hardwood and conifer stands on diverse soils". In: *Ecology* 78:335-347.

Robert, Karl-Henrik et al. (2002). Strategic Sustainable Development: Selection, Design and Synergies of applied tools. Journal of Cleaner Production 10(3):197-214

Robèrt, Karl-Henrik. (2002). The Natural Step Story: Seeding a Quiet Revolution. Gabriola Island, BC: New Society Publishers.

Stadler, B., Solinger, St., and Michalzik, B. (2001). "Insect herbivores and the nutrient flow from the canopy to the soil in coniferous and deciduous forests". In: *Oecologia* 126:104-113

Swank, W.T., Waide, J.B., Crossley, D.A., and Todd R.L. (1981). "Insect defoliation enhances nitrate export from forest ecosystems". In: *Oecologia* 51:297-299.

Tansley, A.G. 1935. The use and abuse of vegetational concepts and terms. *Ecology* 16: 284-307.

Tansley, A.G. 1939. The British Islands and their Vegetation. Volume 1 of 2. University Press, Cambridge, Cambridge, United Kingdom. 484 pg.

The Hindu, Survey of the Environment, 1991 and 1992.

The New Economy of Nature: The Quest to Make Conservation Profitable Gretchen C. Daily and Katherine Ellison.

United Nations Division for sustainable Development. Documents: Sustainable Development Issues Retrieved: 2007-05-12

United Nations. 1987."Report of the World Commission on Environment and Development." General Assembly Resolution 42/187, 11 December 1987. Retrieved: 2007-04-12

US Partnership for the Decade of Education for Sustainable Development. Retrieved on: 2007-07-16

Vannote, R.I., Minshall, G.W., Cummings, K.W., Sedell, J.R., and Cushing, C.E. (1980). "The river continuim concept". in: *Cannadian Journal of Fisheries and Aquatic Science* 37:120-137.

Vitousek, P.M. and Howarth, R.W. (1991). "Nitrogen limitation on land and in the sea: how can it occur?" In: *Biogeochemistry* 13:87-115.

Vreugdenhil, D., Terborgh, J., Cleef, A.M., Sinitsyn, M., Boere, G.C., Archaga, V.L., Prins, H.H.T., 2003, Comprehensive Protected Areas System Composition and Monitoring, IUCN, Gland, Switzerland. 106 pg.

Will Allen. 2007."Learning for Sustainability: Sustainable Development."

World Commission on *Our Common Future,* Oxford University Press, New York, 1987.

2

Ecology, Environment and Sustainable Developmental: Concepts and Methodologies

ECOLOGY, ECOSYSTEM AND ENVIRONMENT

The term 'ecology' was first coined in the later half of the nineteenth century by Henry David Thoreall. Though for further more explanation credit is given to Ernst Hacckel, a German Biologist; Hanns Reiter has been the first for its connotation. He combined the Greek words 'oikos'(=house) and 'Logos' (=study of) to form the term ecology. There is a consensus, however, that Hacckel first gave a meaning to the term in 1866 and put it forward in 1870. By ecology we mean the body of knowledge concerning with the economy of nature—the investigation of the total relations of the animal both to its inorganic and to its organic environment including above all, its friendly and inimical relation with those animals, and plants with which it comes directly or indirectly into contact. In a word, ecology is the study of all the complex interrelations, referred to by Darwin as the conditions of the struggle for existence.

The science of ecology, often inaccurately referred to as "biology" in a narrow sense, has thus far formed the principal component of what is commonly referred to as "Natural History". As is well known the numerous popular natural histories of both early and modern times, this subject has developed in the most close relations with systematic Zoology. The ecology of animals has been dealt with quite uncritically in natural history; but natural history has in any case had the merit of keeping alive a widespread interest in Zoology. In 1863, Isodore Geoffroy St. Hilaire a French Zoologist had proposed the term 'ethology' for "the study of the relations of the organisms within the family and society in its aggregate and in the community".

Later on, ethology became a synonym for the study of animal behaviour. In 1984, St. George Jackson Mivart an English naturalist coined the term 'hexicology', he defined it to be devoted to the study of the relations which exist between the organisms and their environment as regards the nature of the locality they frequent, the temperatures and the amounts of light which suit them, and their relations to other organisms as enemies, rivals, or accidental and involuntary benefactors. "Later on, Charles Elton, a British ecologist, defined ecology as "scientific natural history" concerned with the "sociology and economics of animals". Frederick Clement, an American plant ecologist defined ecology as "the science of the community". Karl Fraedericks, a German ecologist took it as "the science of the environment" ("unweltlehre"). In the same period Rugene Odum an American ecologist analysed it as "the study of the structure and function of nature." However, ecology is multidisciplinary and almost boundless in its concern as stated by A. MacFadyen a British ecologist:

"Ecology concern itself with the interrelationships of living orgnisms, plants or animals and their environments, these are studies with a view to discovering the principles which govern the relationships. That such principles exist is a basic assumption-and an act of faith-of the ecologist. His field of enquiry is no less wide than the totality of the living conditions of the plants and animals under observations, their systematic positions, their reactions to the environment and to each other, and the physical and chemical nature of their inanimate surroundings.... It must be admitted that the ecologist is something of a chartered libertine. He roams at will over the legitimate preserves of the plant and animal biologist, the taxonomist, the physiologist, the behaviourist, the meteorologist, the geologist the physicist, the chemist and even the sociologist; he poaches from all these and from other established and respected disciplines. It is needed a major problem for the ecologist, in his own interest, to set bounds to his divagations."

Moore defined ecology in 1863, as "is the science which treats of organism in relation to their environments; it is frequently subdivided into human ecology, animal ecology, plant ecology and bioecology, the last-named dealing with the inter relationships between animal life and plant life. Ecology lies on the frontiers of so many other subjects, including various branches of geography,..." Ecology is the branch of biology dealing with the relations between organisms their environment. As defined above, it is also known as 'ecology'. Oiko (s) is a Greek word hereby means house and 'logy' word has been derived from the word 'Logos' hereby means discussion to study. Thus ecology is the science of the study of house. There are ecological relationships between organisms of the same or different kinds and also between organisms and their environments. These relations are manifested in physico-chemical setting of non-living or 'abiotic' environmental elements. In these groups basic inorganic elements and compounds such as calcium and oxygen, water and carbon dioxide carbonates and phosphates, and an array of organic compounds, the by, products of organism activity. They also include physical factors as moisture winds currents, tides and solar radiation with its light and heat. The biotic physico-chemical environment and its particular biotic combination of plants animals and microbes got to constitute and ecological system or Ecosystem.'

Ecosystem

Ecosystem means system formed by the interaction of a community of organisms with their environment (Ecology+system). The eco system is a set of interacting, interdependent living and non-living components. Ecosystems are the subject matter of ecology, and an understanding of their structure and function is the concern of the ecologist.

Structure of Ecosystem

At the initial stage ecosystem was considered as a group of organism and the place or habitat (i.e home) they occupy and the way these are linked to form a working of functioning unit. The biosphere which is also known as 'ecosphere' is an ecosystem at one end and a single 'cell' can also be said as ecosystem. The former is too large and the latter is too small; these are rather two extremes, not very popularly used as ecosystems. Therefore, the ecosystem is a set or group of organisms and their habitat which can be visualised or can be understood as such.

There is an interplay both between various organisms themselves and with the physical and chemical components of the environment. The functional system which results from this interplay is known as an ecosystem. The ecosystem is a living complex of interlocking processes characterised by many causes effect combinations.

There are two basic elements in a given ecosystem, the living or biotic and non-living or abiotic. Our whole planet is a single ecosystem even. The portion of the earth in which ecosystem can operate, that is the soil, water and air, inhabited by organism is known as biosphere.

Ecosystem may even be called a system of the house, it is more than a human house. If human house is destroyed it can be rebuilt. In case the ecosystem is deteriorated it can hardly be brought to its previous level.

Ecosystem has a self-regulated system and it is not disturbed, it runs automatically. It has capacity to modify the condition of their international energy and materials. Under the self-regularity system, the ecosystem functions properly. The grass is eaten by the deer and deers are eaten by the lion but the lion is never eaten by any one. Grass is too much in amount, deers, are lesser and the number of the lions is the least. When lions dies the scavengers consume its body. Ecosystem and specially the biosphere can be maintained only when the biotic and abiotic components work properly.

Biomass is most important aspect of the ecosystem. It is that part of a given habitat consisting of living matter, as per unit area or per unit volume of habitat. Biomass structure moans the spatial organisation of the organisms within an ecosystem. It is of two types- vegetation biomass and animal biomass. The biomass is the standing crop of vegetation and animals. The density and distribution of both the types of biomass is affected by the (i) latitudinal location, (ii) nature of the surrounding environment (Climate and soil in particular), (iii) the type of organism (iv) the season of the year, (v) the stage of development of the ecosystem and (vi) various human activities.

The vegetation and animal biomass are interrelated and affect each other. The area under vegetation biomass depends upon the animal biomass specially from grazing point of view. The amount, type and distribution of animal biomass is affected by the plants, the more complex and species rich the plant biomass, the greater is the number of different animal species it can accommodate. However, the distribution of animal biomass is nut always directly related to that of plant biomass. The plant biomass beneath the ground provides food to the organisms. Biomass affects the soil of the region also. Nutrient-rich organic matter supports a larger and more varied soil fauna than nutrient-deficient material. A very high proportion of the above and below ground living plant material is not eaten by grazing animals. A large portion of vegetation either falls or dies to the ground surface as was to or the roots and other underground organs exists in the soil. Mixing with dead bodies of animals, these form 'dead organic matter'. (DOM).'

Scale of Ecosystem

Ecosystem can be studied at any scale as mentioned earlier from biosphere to a cell, there may be innumerable ecosystems. An ephemeral drop of water on a pavement or even a crack in a wall, to the equatorial rain forest form perfect ecosystem. The delimitation of the boundaries is usually difficult, as there is a gradual change in the physical environment instead of abrupt fall. Groups of organisms or communities merge with the other zones aof transition, forming 'ecotones' of various extensions. However, waterbodies specially water ponds form the well defined ecosystems having clear boundaries. Ecosystems are real-like a pond, a field, a forest, an ocean, or even an aquarium. There is a great diversity in the types of actual ecosystems—from small to large, terrestrial to fresh-water and to marine and field to laboratory. There are also typical combinations of particular abiotic and biotic components in any particular ecosystem.

Function of Ecosystem

In spite of differences among the ecosystems there are certain common characters regarding structure, functional attributes that are recognizable, analyzable and predictable. The earth ecosystem (or ecosphere) is to a great extent a closed system with regard to matter and unlikc energy, there is not seemingly an infinite external supply. However, there although is a continuous circulation of materials within individual ecosystems, they rarely act as closed systems. Thus, they are all open systems in that they take in energy one form or another in order to convert it into organic material and give off waste products. As a result, there are two basic types of links or flows between the component parts of all ecosystems, viz. (i) energy flow and (ii) biological cycling.

(i) Energy Flow

The sun rays in the form of radiant energy provides an eternal character to the ecosystems. The plants by consuming carbon dioxide prepare their food with the help of only this energy through the photosynthesis process. Through this energy carbon dioxide is assimilated into energy-rich carbon compounds. The organisms participating in this function are known 'producers'. The organisms, i.e. producers, are the chlorophyll-bearing plants, the algae of a pond, the grass, the trees of the forest. Besides, there are certain bacteria which play their role in each ecosystem. The purple bacteria assimilate carbon dioxide from inorganic compounds with the help of solar energy. These are known as photosynthetic bacteria. Another group of bacteria is of chemosynthetic producers group. These obtain energy by oxidizing simple inorganic compounds. The former bacteria are account upto 25 per cent of the total photosynthesis in certain lakes, rich in hydrogen sulphide whereas latter bacteria are rather insignificant but play a significant role in the movement of mineral nutrients in ecosystems. As a matter of fact producers are the converters or transformers, converting radiant energy into a chemical form, i.e. the carbohydrates.

(ii) Biological Cycling

This process depends on the flow of energy through the ecosystem. The organic (or biotic) and the inorganic (or abiotic) components of the ecosystem interact. These are linked by the regular processes of growth and decay, of synthesis and breakdown, of biological productivity and degradation. The inorganic elements and compounds of carbon, oxygen, hydrogen, nitrogen, .phosphorus and potassium together with those of essential minerals, are converted into complex substances. These comprise plant and animal tissues which eventually die and undergo bacterial decay and decomposition. In this process of biological degradation, the inorganic elements are eventually released and can be refused by the living plants. Due to this there is a continuous circulation of elements from an inorganic to an organic forms and back again. This cycle is known as bio-geo-chemical cycling.

Environment

Environment means the aggregate of surrounding things or conditions or influences. It is also a state of being environed. It carries different meanings as habitat, locale, milieu etc. Ecosystem is the matter of process and system of getting the different things. Ecosystem is mainly concerned with the process of happenings whose, base is the radiant energy. It has a cause and effect procedure of different conditions of influences. Environment is a mater of resources and incidental factors. Environment is the total or the whole around us in which the process of ecosystem works to run the former. With this analogy environment can be named as house and house does not mean a four walled tenement. It is the area in which everything lives, i.e. the house of human being, the house of fauna and in the same way the house of flora. Individually all the houses are the integrated parts of the 'one palacial house' i.e. biosphere. This one house has its close contact with atmosphere, hydrosphere, and lithosphere itself on which it is resting. The house and is other sub-house are being affected by all these spheres rather the very existence depends upon these. Thus the house and spheres go to make the environment. the environment is an envelope in which every house is safe and intact, every house regulates its activity individually and collectively; affecting each other being the integral part of the environment.

Under the influence of the stimuli provided by phsical environment, or as a result of the efforts man makes for satisfying his wants, the material or cultural environment is created. thus, human being, finds himself between the two environments, the physical and the cultural. In these environments man occupies the central position. The first is the created by nature, while the second is the creation of his activities. Physical environment offers him stimuli of various kinds and of different degrees. The climatic conditions or other elements of nature, compels him to wear the clothes and even to seek any shelter. Similarly, to satisfy his basic necessities, he must acquire food and or drink. this whole process leads to development. Land is cultivated, irrigated is provided, industries develop, rivers are bridged, roads and railways are built, towns grow with population and civic amenities—with this whole development, the exploitation of the resources simultaneously increased. Due to over and unmindful exploitation, ecological imbalance is created, resulting in the deterioration of the environment.

Human society adjusts in an environment which is created by biological activities. the oxygen breathe is released by green plants by consuming carbon dioxide. Therefore, environment is a self-cleaning system. Likewise, there are systems in the soil which enable it to absorb human waste and convert it into manure. Manure is then able to support the growth of plants which in turn feed people. This is termed as a ecological cycle. In industrialised nation the cycle is broken by dumping the human waste into water, where it integrates with the water cycle. Consequently, it is not returned to the soil cycle. It breaks the ecological set-up of both the soil cycle and the water cycle. This is an example in which we are beginning to break down the self-contained, self-regulated properties of the environment.

Ecologically the biosphere is unitary in which the natural life-support systems of the species are highly interlinked. The preservation of the delicately balanced ecological system is in the interest of the whole mankind. As a matter of fact, the human group must be considered in a stable equilibrium not merely with reference to temperature, humidity, sunshine, altitude etc. but also to their indirect effects; the inter-woven chain of biotic communities to which it is inextricably linked, the plants it cultivates, the animals it breeds and even the insects which are indigenous to the region.

Environmental scientists, the world over, are profoundly concerned over man's abuse of nature in the name of progress. There is standing warning that if the present rate of air, water, and soil pollution persisted, the biological rhythm on the earth would be irreversibly upset and man himself might face a long-drawn-out process of extinction. The industrial and agricultural progress which leads to a higher standard of living has undesirable effects on the physical environment. Industry is belching more than 100 million tons of pollutants, like soot, noxious fumes, gases and chemical substances, per year into the air, and dumping out millions of tonnes of chemical effluents into rivers, lakes, the seas, and oceans. Over the last 100 years, it is estimated that 1.25 million tonnes of flint; 1.4 tones of arsenic, 6 million tonnes of antimony, one million tonne of cobalt and about 0.8 million tonne of nickel-all extremely poisonous substances-have entered the atmosphere.

Through the burning of fossil fuel the last one hundred years, more than 240 billion tonnes of oxygen was exhausted and about 360 billion tonnes of carbon dioxide was discharged into the atmosphere. This has led to a higher carbon dioxide content and oxygen deficiency in the atmosphere. The disturbance in the ozone sphere is the serious threat to the mankind.

The basic reason for the ecological crisis is man's greed, lead into different attitude towards nature and his adaptability to the deteriorating environment. At present, the exploding population and the technological pressure to sustain them, spells new danger to the biosphere. The projection of over 10 billion people on earth by the year 2000 A.D. (twice as many as now) may bring a global catastrophe. It took the world a million year to reach the first billion mark of population. The second billion was added in another 100 years and the third in just 60 years whereas 4th billion must have hardly taken 25 years. Till 1990 the 5th billion will be touched. The population upsurge of such magnitude could bring strain on the earth's dwindling resources and endanger the stability of ecosystem that provide food, oxygen and water, besides creating uncontrollable socio-economic problems. The degradation of the environment all over the world has become a matter of great concern to everyone. Although much damage has already been done and the process of degeneration is continuing unabated to some extent; fortunately an awareness has also been grown everywhere about the imminent threat that looms large because of reckless exploitation of the resources. A positive approach towards understanding the environment and initiating action towards its improvement by undesirable human interference and its preservation by planned development are being adopted by most of the countries.

In India the establishment of the Department of Environment in November, 1980 by the central government has brought the activeness in this sphere. This department is a controlling agency for all the environmental issues in the country. It has already sponsored innumerable projects on different aspects of environment, studying various problems and suggesting appropriate measures to rectify the deficiencies. Full attention has to be paid for maintaining the environmental qualities. The impact of the industrial activity and reaction of other economic activities including social activities towards industrial development have to be studied interdependently to avoid the environmental catastrophe.

Human being is most active and an important part of the natural environment. Besides, having adequate food to fulfil all daily necessities of life, he needs space to live and to organise mental and psychological activities. In this connection, economic and social dimensions must be considered not to neglect the third dimension of development, i.e. the environment.

The economic activity located in the region has been viewed only in economic and social context so far as economic development is concerned but the improvement of environmental qualities are considered secondary. It is prevailing nusnomer that the quality of life does not enter into GNP, rather it involves additional cost and does not provide material gain. This idea is false as proper functioning of economic activities depends upon heal thy environment and more so, investment on environment enriches the economic activities and enhances in achieving social goals.

It is not enough to understand the existing complex relationship among various entities of environment but concerted and planned efforts have to be made to manage the resource and to create a better relationship with nature by locating economic activities in a planned way. This will not have an irretrievably adverse impact on the ecosystem. The country is passing through a phase, haphazard unplanned and undesirable industrial development, which is leading to deterioration of ecosystems.

ENVIRONMENTAL CONCEPT OF SUSTAINABLE DEVELOPMENT

Introduction

For the first time the issues related to the environment were globalised. The heads of the Government had assembled to discuss the human assault on the biosphere. Not that the citizens of the earth were not aware of the benefits of good environment. They were, in fact, quite concerned about polluted rivers and denuded mountains. They knew the importance of green vegetation and clean seashores. However, for the first time it was realised at Stockholm that greening one's own backyard will not solve the problem. The issue needed to cross the fence, the borders of the nation. Borders of a nation probably aid in preserving man's historical heritage owned by that particular nation, but they do not contribute to the preservation of the environmental heritage that belongs to all living beings-present and future. It was crucial Principle number 21 of the Declaration of Stockholm Conference that rang the bell. Though it was basically related to marine environment, it summed up the trend, i.e., States have, in accordance with the charter of the United Nations and the principles of international law, the sovereign right to exploit their own resources pursuant to their own environmental policies and the responsibility to ensure that activities within their jurisdiction or control do not cause damage to the environment of other States or of areas beyond the limits of natural jurisdiction. Thousands of years ago great Indian sages in *Prithvi Sukta* (Hymn to the Land) stated the importance of keeping the earth free from pollution and any disturbance caused to its equilibrium. The message of environment protection is thus not new. Man in his pursuit of 'development at any cost' forgot the significance of this message. Now he is saking up to the realities and devising ways to sustain development and life on earth.

Sustainable Development

The concept of sustainable development could be termed development without destruction. It integrates environment and development and can be defined as— eeting the needs of the present without compromising the ability of future generations to meet their own needs.

A key question for the longer term is how to reconcile sound environmental activities with economic growth. Economic growth has traditionally been linked with depletion of the earth's natural resources and is now placing growing pressure on the earth's assimilative capacity.

Debate between Developed and Developing Countries

Since the summer of 1972, the attitudes and approaches of developed and developing countries have undergone a seachange. Developing countries considered environmental protection a luxury that only developed countries could afford. Alleviating poverty, satisfying the basic needs and improving the standard of living of people were their priority areas. Environment protection appeared at the end of their agenda. Developed countries, on the other hand, wanted 'global action' towards environmental protection, notwithstanding the fact that they were the main contributors to environmental damage and wished to continue their development and thereby assault on the earth. Such attitudes and general differences, in approach characterised the debate at the 1972 Stockholm Conference.

Churning of various debates led to the establishment of the UNEP (United Nations Environment Programme). Today, the developing countries squarely recognise the relevance of environmental issues, having faced the conspicuous evidence of soil erosion and deforestation. They are also concerned about the rapid population growth and its effect on the environment. Environmental protection is now recognised as essential for sustainable economic development and not as an alternative to it. Countries now realise that environmental interdependence is as vital a parameter for the development process as economic and financial interdependence.

Developing countries, having understood the necessity of moving environment up on the agenda, also realised the historical developments, i.e., exploitation by the mighty few, flow of resources from South to North and the indiscriminate and blatant use of resources by developed countries. Pinning the responsibility for damage was followed by awareness of the right to improve the standard of living of tilling millions and the right to the financial and technical resources of the earth's foremost exploiters.

Main Issues

What are the environmental issues today? They are wide ranging and all-encompassing: deforestation, biodiversity, soil erosion, climate change, pesticide build-up, industrial and municipal pollution. All have been caused by anthropogenic interventions in the natural and self-sustaining cycles. The human ambition for a higher living standard has hammered 'consumerism', ignoring the lasting and adverse impact of this hammer on earth's limited resources.

Land

The most urgent environmental problems facing the developing countries are those relating to the use of land. Of particular concern are deforestation, devegetation and desertification. All these are outcomes of many interrelated activities. Overgrazing, land clearing and commercial logging rob the soil of its cover and fertility and reduce agricultural yields. They also increase flooding, sedimentation and the silting up of dams and reservoirs. The problems are worse where the inadequacy of fuelwood compels rural families to burn agricultural residues and dung, thus the source of soil fertility and accelerating erosion. Such deforestation deprives people of other essential products for their daily survival and threatens wildlife reserves and fragile ecosystems. Resources that in principle are renewable, are rapidly disappearing as a result of overexploitation. And the extinction of plant and animal species, proceeding all too rapidly, may have serious and as yet unknown consequences for the future of mankind.

The problem has also been complicated by the fact that deforestation takes place to earn much needed foreign exchange by way of exporting wood. Thus the natural resources travel from South to North. This temporarily helps the South to reduce its balance of payments. It also helps the North by meeting its wood requirements. But today South and North have come to realise that this mutually beneficial trade in wood has had a devastating impact on the global environment.

Pollution

Environmental pollution, although typically associated with industrialisation, is a great and growing concern in developing countries. Air and water pollution in such cities as Calcutta, Bombay and Delhi are as in most developed countries critical as and, as demonstrated by the Bhopal Disaster, Industrial catastrophes are not confined to the industrial world. But one aspect of the environmental problem-unsanitary living conditions due to inadequate water supply and improper waste disposal-is observed almost exclusively in the poorer countries. Throughout the developing world, water polluted by pesticides and by municipal and industrial wastes causes disease, malnutrition and death.

Not only industry, but agriculture likewise contributes to pollution. The use of chemical fertilizers for example, pollutes the environment in various ways:

(a) Fertilizers not used by the soil are washed away by water and contaminate the rivers.

(b) Manufacture and distribution of fertilizers are energy-intensive processes and pollute the environment.

(c) Many agricultural activities lead to generation of greenhouse gases.

Gas	*Source*
CO_2	• Burning of fossil fuel
	• Deforestation
Water vapour	• After-effect of warming from oceans

Methane	• Paddy fields
	• Ruminating animals
	• Decay and burning of biomass
CFC	• Refrigerator equipment
	• Aerosols
	• Comfort and insulation foams
Nitrous oxide	• Synthetic fertilisers
	• Burning of fuel and biomass
Tropospheric ozone	• Atmospheric reactions with certain gases

The problem acquires another dimension when one considers the fact that most of the pollution arising out of the industrial processes is the result of technology that has been transferred from North to South.

Global Concerns

Acid Rain

Over the past few years air pollution has been seen increasingly as a regional or global problem, not a local one. Acid rain may fall to earth thousands of miles away from the places of emission of sulphur dioxide and nitrogen oxide. Thus the cloud generated in the developed world may rain in the territory of the developing world. The destruction of forests in northern Europe is perhaps the clearest consequence.

Ozone Depletion and Global Warming

Other global phenomena of unprecedented dimension are damage to the ozone layer by Chlorofluorocarhons and the 'greenhouse' effect-the warming of the planet. The consequences of ozone depletion are highly uncertain but potentially catastrophic. The responsibility for polluting the stratosphere due to emission of CFCs largely rests with the developed world. The smoking gun is clearly visible in the hands of the North. The contributions by the developing countries are miniscule. Global warming is caused by the emission of greenhouse gases (carbon dioxide, methane, nitrous oxides etc.) which trap more energy emitted by the earth than required for the purpose of a 'sustained and habitable' environment. Such gases are emitted from natural as well as anthropogenic sources. Anthropogenic sources are both agricultural and industrial. The question of sharing the responsibility in emission of greenhouse gases is more tricky than complex. The owning of 'global sinks' which help in absorbing the greenhouse gases is at the root of this issue.

Global Impacts of Environmental Degradation

The causes of environmental degradation are as varied as its manifestation. But the root cause is materialistic exploitation. Exploiters ravaged the earth without understanding the likely damages.

Land

What are the effects? The effects of environmental degradation are difficult to establish with precision, but physical indicators alone give grounds for concern. For example, desertification worldwide is proceeding at 6 million hectares-more than the size of Punjab—a year, and tropical deforestation at 10 million hectares—almost the size of West Bengal-per vear. Every year 20 million hectares of agricultural land-almost the size of Gujarat-are lost to soil erosion. And over the next twenty years a fifth of the world's plant and animal species may become extinct.

Ozone Depletion

The predicted impact due to depletion of the ozone layer is being professed with near certainty, though experimental evidence is not fully supportive. It is stated by the Assessment Panel reports of UNEP (1991) that:

- By the year 2000, ozone depletion between 5 and 10% is likely for mid-latitudes in summer. According to present data, a sustained 10% loss of ozone would lead to an increase in the incidence of skin cancers by 26%.
- A 1% decrease of ozone, other things being equal, would lead to between 100,000 and 150,000 additional cases of cataract-induced blindness worldwide.
- UV-B radiation has a profound influence on the immune system in humans as well as in animals. There is concern that changes in the immune system might lead to an increase in infectious diseases.
- UV-B radiation, even at present levels, has a detrimental effect on many plants and aquatic organisms. This raises the concern that enhanced UV-B radiation may have a negative effect on world food production.

Global Warming

The impact of greenhouse warming has been sketched by the Intergovernment Panel of Climate Change (1990) amidst varied and interesting uncertainties. The uncertainties are:

- Though global warming is correlated to rise in CO_2 concentration, the relatively cool period during 1945-1975 has yet to be explained.

- The effect of cloud formation due to global warming may give rise to global cooling.
- There is a possibility of 'warm oceans' acting as 'more efficient sinks' of carbon dioxide.

Amidst these uncertainties the predicted impacts of global warming are catastrophic:

Rise in Sea Level

Due to expansion of the water mass in the oceans, there will be a rise in sea-level of one metre due to a 3° Celsius rise in global temperature. This is expected by the turn of the century. The rise in sea-level will inundate the low-lying coastal areas of Bangladesh and the Philippines and sink islands such as the Maldives. The 'eco-refugees' will migrate towards the highlands.

Reduction in Agricultural Productivity

The reduction in agricultural productivity is another familiar impact. Changes in regionwise climate have been predicted . A major change appears to be expansion of grasslands and desert. In the tropics, the climate will be wetter and the cloud cover will be prolonged. The present 'wheat mines, (i.e., USA and Mediterranean), due to warmer and drier weather, will shift polewards. This means Russia will become the major producer and exporter of wheat and corn. Tundra will disappear. Thus there will be a 'shift' in the agricultural regions and the agricultural pattern. India will have a wetter climate with more cloud coverage. All this does not necessarily mean reduced grain production. UNEP/GEMS documents says: 'CO_2 is essentially a fertilizer. Its presence encourages plants to increase their carbohydrate levels'. It is therefore predicted that plants will grow bigger. A doubling of CO_2 is likely to increase growth and yield of some plants by 5 to 50 per cent.

Impact in Developing Countries

Environmental issues arise in both developed and developing countries but their significance varies depending on the country's stage of development and mix of economy. Industrial pollution is a menace in the developed countries and an emerging threat in many cities in the developing world as well. Soil erosion and deforestation are worse in developing counties due to their tropical environment.

Countries and continents do not share the burden of environmental degradation evenly. Two-fifths of Africa's non-deserts land risks being turned into desert, as does a third of the non-desert in Asia and a fifth in Latin America. Tropical deforestation is massive in south-east Asia, Africa and the Amazon basin. In Nepal, deforestation and soil erosion have taken most of the productive land. Not all soil erosion involves a loss, but in turkey more than three-quarters of the land has been adversely affected and in India one-quarter.

Degradation of land and water dearly threatens economic development in many countries. The evidence for this is fragmentary but telling. Because of the scarcity of fuelwood, an estimated 400 million tonnes of dung are burned each year. That robs the soil of its fertility and depresses each year's grain harvest by 20 million tonnes, enough to feed 100 million people for a year. The alternatives to burning crop residues and dung are to travel farther and farther afield to obtain fuelwood or to pay more for it. But in Gambia and Tanzania households already spend 250 to 300 days of labour a year gathering wood. In Addis Ababa fuelwood costs up to a fifth of a household's income. Between now and the end of the century the number of developing countries that are net exporters of forest products will fall from over thirty to fewer than ten, and developing countries' exports of industrial forest products will drop from $7 billion to less than $2 billion.

Soil erosion in India removes nutrients worth $6 billion (Rs. 18,000 Cr.) a year in chemical fertilizer replacements. In parts of Mexico erosion reduces maize yields from 3.8 t/ha to 0.6 tonnes and in parts of Nigeria from 6.5 t/ha to 1 tonne. Other sectors pay for soil erosion as well. The flood-prone area in India doubled from 20 million to 40 million hectares between 1970 and 1980, and in the Ganges Valley alone the annual cost of flood damage to crops and other property averages about $1 billion (Rs. 3,000 Cr.). Reduction of sedimentation in hydroelectric facilities will cost the developing countries about $3 billion over the next fifteen years. And the disruption of major water flows because of deforestation threatens water supply systems throughout the world a grim prospect since acute water shortages combined with pollution now cause 25,000 deaths a day.

Necessity of Short-term and Long-term Actions

Perspective of Developing Countries

Sustaining growth, alleviating poverty and protecting the environment are mutually supportive objectives in the long run. Rather than address environmental issues in isolation, decision-makers in governments and international institutions should consider the preservation of the environment along with other issues central to the formulation of development policy. In the short run, however, the objectives are not always compatible and decision-makers often confront difficult choices in pursuing them simultaneously. Those affected most adversely by environmental degradation are the disadvantaged—the poorer, less educated, and politically less powerful dements of society. The urban poor live in unsanitary conditions without safe drinking water. They are hurt most by noise, flooding and water pollution and cannot afford to pay for better services or to defend themselves against damage.

Priorities in Developing Countries

Because the problem of many developing countries are so extreme and urgent, real trade-offs often arise between protecting the environment and satisfying basic short-term needs. For example, given the high rates of population growth in many countries, it is hard in the short run to find alternative fuels to stem the rapid depletion of fuelwood stocks or the burning of dung. Similarly, levels of air and water pollution that would be intolerable in industrial countries persist in many cities in the developing world, in part because the immediate costs of abatement are deemed to outweigh immediate improvements in health and the quality of life and in part because long-term benefits are heavily discounted.

The rural poor farm flood-prone valleys and swiftly eroding hillsides. They cannot afford to invest in conservation measures or to grow crops which take a long time to mature; they cannot move to more productive land; and they must immediately spend any income on consumption goods.

Poverty-of people and of countries-is thus a major cause of environmental degradation. That makes it essential, if environmental degradation is not to become completely unmanageable, to devise policies oriented toward economic growth, with special emphasis on improving the incomes of the poor.

Without growth, all other things suffer in developing countries. Nevertheless economic growth may also destroy the environment and further jeopardise the already tenuous lives of the poor. For example, agricultural development that further concentrates landholdings may drive the poor onto increasingly marginal land, unchecked urban commercial growth may create shanty towns, and roads may pave over the small properties of the poor. Thus, although growth is imperative for alleviating poverty, it may also adversely affect the poor and the environment if inadequate attention is paid to the poor and their needs.

CFCs and other ozone-depleting substances must be phased out as early as possible if we are to hold the ozone level in the stratosphere of a safe level. Developing countries use the CFCs mainly for refrigeration and air-conditioning, whereas in developed countries it is mainly used for comfort-related applications.

Our core developmental programmes, such as food preservation, vaccination, rural health, manufacture of life-saving drugs, depend on the use of CFCs. Our consumption compared to any of the developed countries is miniscule. Though we need to participate in the global action by joining the Montreal Protocol, we need to consider the socio-economic impact of phase-out of CFCs in the right perspective.

Our Finance Minister, Dr. Manmohan Singh, addressing the International Conference on Environment and Development in New Delhi stated:

> Tropical countries like ours, where there is much greater need for refrigeration in food preservation than in cold countries, need very special consideration. Developed countries' efforts to control production of CFCs in areas and in countries where relative consumption should be allowed to increase, therefore represent a rather inefficient and one-sided way of dealing with the problem.

We need to make amends to the Protocol, after becoming a party to it, to remove certain inequities. Dr. Manmohan Singh further stated in his address:

> Equity in the use of CFCs requires that per capita consumption be equated between the developing countries and the developed countries and the average worldwide level of consumption in developed countries be reduced. This means that a per capita consumption in developed countries must go down, while that in the developing countries must go up but certainly not to the levels currently prevailing in the developed countries.

Suggested Approaches

Without doubt there are uncertainties in understanding environmental science, cause-effect relationship and prediction of impacts. Uncertainties, however, should not preclude preventive measures and should not be an excuse for inaction.

Need to Understand Uncertainties

While it is impossible to carry out independent research, India should actively participate in modelling and worldwide research and in monitoring conducted by WMO (World Meteorological Organisation), GEMS (Global Environmental Monitoring System) etc. While it may be tempting to our scientists, semi-scientists and bureaucrats to 'fashion' the popular beliefs and prophesy disasters regarding greenhouse warming, it is absolutely essential to understand the uncertainties in totality. Otherwise, we may overreact and land up taking disproportionately severe actions. True, one does not wait for disease to emerge, but one does not start with a blood transfusion or open-heart surgery either!

Energy Conservation

This is the single most important preventive measure that India can take with immediate effect. By improved energy efficiency and judicious use we can contribute greatly to reducing CO_2 emission in the atmosphere. This measure has many beneficial 'spin-offs': reducing balance of payment gap, keeping plan expenditure for energy generation low, alleviating atmospheric pollution etc.

Some major areas in which a massive drive to conserve energy should start immediately are:

- Minimising transmission and distribution losses of electricity.
- Encouraging co-generation.
- Increasing pump-set efficiencies in the agricultural sector.
- Improving energy efficiencies of refrigerators.
- Changing to sodium lamps for lighting.
- Phasing out all vehicles which give lower fuel efficiencies.
- Converting caustic soda plants to membrane cells.

Subsidies on use of energy may discourage conservation. A need to link subsidies to efficient use of energy is therefore obvious.

Alternate Sources of Energy

Nuclear, solar and wind energies need to be reviewed to give environmental orientation to our planning process. The hazards of Nuclear plants and that of greenhouse warming need to be weighted to understand which one is more manageable. Greater use of natural gas would produce lesser CO_2.

Modified Agricultural Techniques

Increased use of synthetic fertilizers contribute to greenhouse, warming in more than one way:

- Utilised nitrogen in the soil is converted to nitrous oxide, a greenhouse gas.
- Organis manure from livestock and plants remains unutilised. The CO_2 and CH_4 produced from it enter the atmosphere and contribute to global warming.
- Increased production of synthetic fertilizers requires additional energy to manufacture and distribute them. This adds to CO_2 emission, a greenhouse gas.

Thus 'organic farming' should be encouraged along with optimum use of chemical fertilizers, with the use of natural fertilisers, better crop protection and minimum tillage. Flooded paddy fields produce more methane. It is proposed by agricultural scientists that flooding the fields may not be necessary and that yield of rice could be adequately increased just by wetting the soil.

Afforestation

Aggressive policies to conserve existing forests and to create new forests would contribute to absorption of CO_2. Afforestation would yield other environmental benefits, including control of soil erosion and preservation of plant species and wildlife.

Evaluation of Environment Friendly Technologies

Developing countries are negotiating for financial and technical assistance from developed countries. They also require environment friendly substitutes and the technologies to produce them. However, before the transfer of such environment friendly technologies, it is necessary that we evaluate their 'environment friendly' characteristics in an integrated and unified way.

The critical need is to select truly environment friendly technologies. This work needs to be done by the developing countries depending on the socioeconomic needs and conditions existing in those countries. If selection is not proper, we may again get carried over by 'market' forces and resort to technologies which may alleviate one environmental hazard but give rise to another.

Some of the 'ozone friendly' substitutes have proven to be not necessarily 'environment friendly' as they consume more energy, which results in emitting more CO_2 to the atmosphere and contributing to global warming. Some other substitutes, though '*ozone* benign', have been found to be more toxic.

Ecodiplomacy: Need of the Time

All the above measures have multiple benefits and it would be short of prudence not to undertake them immediately. Bargaining is on the side of developing countries, as they have caused little of the global environmental damage. Major polluters are developed countries. Negotiations on the Montreal Protocol-in the case of ozone-depleting substances-have shown us the path, i.e., principle of equity as regards control of emissions. We should use the same bargaining position in the case of greenhouse gas emissions and to improve conservation techniques in the energy sector. What is required is the political will to make it happen.

'Polluters to pay' is the compensating principle now accepted at the international level. Developing countries will need environment friendly technologies and the incremental financial assistance. These based on acceptable principles, have to come from developed countries. Putting forward the case of developing countries would need an altogether different approach- now termed ecodiplomacy-devoid of any intransigent attitudes leading to isolation.

Conclusion

As more people make greater demands on the natural resource base that ultimately sustains much of the economic activities, the issue of sustainable development will become increasingly complex. Every problem, however, gives rise to an opportunity. In response, we need to adopt a more integrated and holistic approach to the issue of resources. Much can be accomplished by taking advantage of this opportunity to establish policies that promote better resource management and conservation and improve economic productivity. There is no conflict whatsoever between good resource management and sound development policy.

Role of Decision-makers

Economic analysis can and should help in making the decisions related to the environment. Although monetary values alone are typically inadequate as criteria for decisions, the benefit-cost approach, which systematically sets out the good and bad consequences of alternative courses of action, is generally applicable. Where the objectives of environmental protection and sustainable economic growth coincide, governments face politically difficult choices. Public interventions to address environmental issues are likely to have costs for some parties and benefits for others. For example, as is implied by the tendency of the poor to suffer most from environmental degradation, policy reform might have distributional consequences that powerful interests might resist. The urgency of current needs, the uncertainty about the future, the legitimate differences in value judgements, and the short-term perspective of many decision-makers, make it clear that a major political-commitment is required to improve substantially the management of natural resources. But such a commitment would be only the first step in a long process.

ENVIRONMENTAL METHODOLOGY

To be an effective tool in the planning process the environmental assessment segment of the EIS must be a comprehensive and systematic analysis that encompasses an interdisciplinary approach to the environment and its many related aspects. A plant (although it could be any living organism) is shown as it is acted upon by the many environmental factors. Each factor is shown to have a direct impact on the plant and also an impact on the other, factors which, in turn, results in an indirect impact on the plant. Thus, the environment is a complex interaction of many factors and a change in one aspect of the environment will effect changes of the entire system. Therefore, a comprehensive analysis of the environmental impact of a proposed project is necessary to understand the physical and biological effects and their influences on the social, cultural and aesthetic concerns of man.

The analysis must also be systematic and interdisciplinary because an environmental assessment encompasses a variety of physical biological and social sciences and is examined and evaluated by numerous special interest groups. It is subjected to a public hearing and expert testimony and must withstand careful scientific scrutinisation approach to the EIS. Each project will affect each of these environmental segments to a greater or lesser degree. More detail can then be given to the segment that requires it, and the project planner or decision-maker, without too much difficulty, can consider the various aspects and impacts of a project.

A description of each subdivision and the parameters to consider in the environmental assessment follow. Further, the types of analysis that could be made are included.

Physical/Chemical Category

Water

Water is one of the necessities to life as we know it. It is used by all flora and fauna life cycles and as a sink for wastes. These wastes have now begun to inhibit the use of water for its intended purpose. Table 1 lists the major types of water pollution and the water uses most affected by them. Thus, water of a region must be assessed as to its quality, quantity and availability. Typical considerations might include:

- What are the nonpoint pollution sources, such as pesticides in agricultural runoff?
- Is a stream-flow reduction affected by diversions set up by the proposed project?
- Would BOD, COD and suspended solids be reduced to meet EPA standards with water treatment? How?
- What are the ground water effects due to land application waste disposal?

Land

Land has always been a cherished commodity for man. Its uses range from agricultural production to mineral ore supplier, from housing foundations to industrial sites, from open space beauty to waste disposal sites. Proper land use management is imperative to reaching sound enviromental goals. Typical considerations might include:

- What types of land weathering take place and what are the causes?
- Is the project suitable for local land use ordinances and regional planning criteria?
- What are the solid and liquid waste disposal practices and are they environmentally safe?

TABLE 1
Major Types of Water Pollution and the Water Uses Most Affected by them

Types of Pollution	*Water Quality Uses Most Affected*			
	Recreation	*Aquatic Life*	*Municipal Water*	*Industrial Use Supply*
Bacteria	X		X	
Turbidity (muddy water)	X	X	X	X
Lack of Dissolved Oxygen		X		
Inorganic materials				
(iron and manganese)		X	X	
(copper and lead)	X	X	X	
Phenols (smelly, bad taste substances)	X	X	X	
Organic materials (oil, pesticides, exotics)	X	X		
Radioactivity	X	X	X	X
Heat	X	X		X
Nutrients (phosphorus and nitrogen)	X	X	X	X
Dissolved Solid			X	X

Air

The pollutants of our atmosphere are known to have deterimental effects on humans and vegetation; we now know that they contribute to the deterioration of the stratospheric ozone layer resulting in the admission of the cancer-causing ultraviolet rays of the sun. Anything that is emitted to the atmosphere that ordinarily would not be there is a contaminant. Typical considerations might include:

- What is the ambient air quality and how will it be affected by the proposed project?
- As a result of project-induced population shifts, are sulphur dioxide emissions increased due to increased power requirements?
- Will solid waste disposal sites produce objectionable odours?

Noise

Noise is usually considered to be unwanted sounds. It may consist of truck engines or screeching subways or many other undesirable sounds which may have a detrimental effect on man. Typical considerations might be:

- What will be the increase in noise intensity produced by construction blasting?
- Will increased truck and rail traffic to the proposed project increase the noise frequency to surrounding areas?
- Will the natural sounds of the environment (e.g., birds, flowing streams) be drowned out by increased noise levels of the project?
- Are sonic booms going to disrupt everyday life styles in adjacent areas?

Ecological Category

Species and Populations

The local environment's ability to support the life of an organism makes it possible for that organism to live. Species and populations are the basis for the ecological world, and their existence can be greatly influenced by changes in their environments. Care should be taken to describe those species and populations that can be identified. However, many species cannot be identified and how they will be affected by environmental changes is unknown. This should be discussed.

Further, certain species are used a indicators of healthy and unhealthy human environments as well as the status of the general environments. These should be discussed in detail. Typical considerations might be:

- With the removal of natural feeding sites, is the animal life of the area adversely affected?
- Does the effluent of the proposed project suggest impairment to aquatic life?
- Are the flora and fauna of the region significantly affected?

Communities and Habitats

A group of species is a community. A habitat is the environment of a community. The existence of a variety of species and communities living together is the result of numerous and complex interactions. Major habitats include streams, lakes, estuaries, swamps, deserts, marshlands, forests, all components of some river basin. Illustrates the variation in tree species with mountain altitude. Different temperatures and overall climates affect different habitats for the various trees. Typical considerations might be:

- Is the proposed action going to result in the destruction of rare and endangered species habitat (such as filling in a swamp)?
- Are project-induced population shifts eliminating important biological communities of the local ecosystem?
- Are soil communities destroyed due to leachate at waste disposal sites?

Ecosystems

An ecosystem is the entire natural life system of a local environment. River basins, forests and plains are typical natural systems in which ecosystems function. The fundamentals of most ecosystems are known, but data are lacking on the numerous processes and interactions that occur. Thus, in the ecosystem analysis, assumptions are made and the justifications for such assumptions are described.

- Does the addition of nutrients from the project's wastewater result in lake eutrofication?
- Does the urbanisation of a region due to the proposed project result in loss of prime agricultural land and other land productivity?
- Does the elimination of vegetation result in a disruption of the energy flow through an ecosystem?

Aesthetics

Land

The scenic beauty produced by a region's land forms is not necessarily dependent on mountain ranges or flowing brooks, but can also include the vast metropolis or rustic rural community which make their own contributions to life quality. Typical considerations might be:

- Does project-induced urbanisation result in the alteration of the region's land forms?
- Is the clearing and excavation of trees necessary for project-related power lines going to add unsightly scars to the countryside?

Air

Air has aesthetic qualities that at times can go unnoticed or taken for granted. Temperature, humidity, wind speed, and pollutants all add or detract from the air aesthetics. Odours, smog, dirty streets, and buildings are some of the air effects of pollutant emissions. Noise, as an air medium, can affect the tranquility of an area. Typical considerations might be:

- What is the smog potential in reducing the visibility of a skyline due to stack emissions of the proposed project?
- Are any project chemical processes likely to release odour emissions?
- Will project-related sounds drown out bird calls?

Water

The beauty of water is unceasing. In its many forms—rain, lakes, streams, geysers, waterfalls, oceans, ponds, estuaries—it creates an almost incomparable beauty. Typical considerations might be:

- Does the diversion of a river by a dam eliminate any natural sites?
- Will the project's effluent introduce any pollutants to a stream, thereby changing is colour?
- Will project-related sludge dumped offshore result in unsightly and unusable beaches?

Biota

The flora and fauna of a region contribute greatly to its aesthetic qualities. The elimination of vegetation not only removes its beauty, but also any animal life it may support. Typical considerations might be:

- Does the clearing of vegetation for a highway eliminate a natural animal habitat?
- Will vegetation types differ as a result of project-induced land development?

Man-Made Objects

Man-made objects such as art figures, skyscrapers, bridges, monuments and other buildings and structures create their own aesthetics through combination of beauty, uniqueness, age and history. The visual effect of the Manhattan skyline is at times awe-inspiring as would be a lonely bridge connecting opposite ridges above a steep ravine. Typical consideration would be:

- Does the design of the proposed project conform to the visual landscaping as it now exists?
- Would further development induced by the project conform to the present environment?

Social

Individual Environmental Interests

Individual aspirations must be dealt with here. They can add new dimensions and qualities to individual life styles. These factors may include historical settings, recreational features, or university facilities. Typical considerations might include:

- Does the proposed project infringe upon a historical setting disrupting its continuity?
- Are recreation facilities endangered by accidents such as oil spills from a tanker?

Individual Well-Being

People are greatly affected both physically and psychologically by their environment. Like other living things, changes, in their environment can cause adverse effects on many individuals. Typical considerations might include:

- What are the psychological effects of frequent sonic booms, e.g., the SST?
- What health problems will result from short and long term exposure to project effected emissions?

Social Interactions

The organisation of individuals into various groups and communities is a social function. Of primary concern is the environment of the society and changes that may occur in the environment. This environment includes not only physical parameters, but job opportunities, adequate housing, social services, and recreational and educational opportunities. Typical] considerations might be:

- Will the proposed project result in large population shifts?
- What are the economic and social opportunities created by this project?
- What will be the changes in the existing population's life styles?

Community Well-Being

The individual and social interactions that make up a community are in part governed by its environment. Any changes in the environment will produce changes in the community. Typical considerations might be:

- In what manner is the community disrupted by the proposed project?
- It the overall environment of the community, as a result of the project, improved, deteriorated or unchanged?

The environmental aspects to be considered in an environmental assessment have been summarised. Each of these aspects should be considered fully with regard to:

1. the existing environment;
2. the impacts of the proposed project on the environment;
3. each altenative to the proposed project; and,
4. the irreversible and irrevocable uses of the environment.

Interactions of the Environmental Components

We have shown the interactions of the natural environment to be highly complex. The natural environment interacts with man and his social settings, these intricate interrelationships. In most cases, a proposed action's impact is in the physical/chemical sector of the environment— such as some type of air, water or noise pollution. These, in turn, have their impacts in the various ecological, social and aesthetic sectors of the environment. Interactions occur among sectors and between sectors. Thus, in the making of an environmental assessment, the writer must be cognizant of these interrelationships and be able to analyze each component as it exists separately and as it fits into the whole environmental picture.

Environmental Factors

Once the environment has been segmented into the previously identified components, each component can be subdivided into actual parameters to be measured and assessed. These parameters are listed in Table 3 and, depending on the proposed action, the list under each environmental component may expand or decrease.

The environment is systematically evaluated through the segmenting of the whole into major categories. A further subdivision into components establishes the ecological impacts conceptually. Then the paramaters of each environmental component are identified. This identification process for each parameter would be followed again for:

1. the existing environment;
2. the impacts of the proposed project on the environment;
3. each alternative to the proposed project; and
4. the irreversible and irrevocable impacts.

This parameter assessment would include:

- Data gathering or other measurements to determine the quality of each parameter and the existing environment.
- Extrapolation of each parameter and/or group of parameters to estimate the future environmental status with and without the project.
- Differentiate between each parameter and/or group of parameters and there extrapolation to determine the beneficial and adverse environmental impacts.

TABLE 2

Environmental Parameter Assessment List

Physical / Chemical	
Water	Land
BOD	Soil Erosion
Ground Water Flow	Flood Plain Usage
Dissolved Oxygen	Buffer Zones

Fecal Coliforms
Inorganic Carbon
Inorganic Nitrogen
Inorganic Phosphate
Heavy Metals
Pesticides
Petrochemicals
pH
Stream Flow
Temperature
Total Dissolved Solids
Toxic Substances
Turbidity

Noise
Intensity
Duration
Frequency

Soil Suitability for Use
Compatibility of Land Uses
Solid Waste Disposal

Air
Carbon Monoxide
Hydrocarbons
Nitrogen Oxides
Particulate Matter
Photochemical Oxidants
Sulphur Oxides
Methane
Hydrogen and Organic Sulphides
Other

ECOLOGICAL

Species and Populations
Game and Nangame Animals
Natural Vegetation
Managed Vegetation
Resident and Migratory Birds
Sport and Commercial
Fisheries
Pest Species
Energe Flow

Habitats and Communities
Species Diversity
Rate and Endangered Species
Food Chain Index
Ecosystems
Productivity
Biogeochemical Cycling

AESTHETIC

Land
Geological Surface Material
Relief and Topography

Air
Odour
Visunal
Sounds

Water

Biota
Animals—Wild and Domestic
Vegetation Type
Vegetation Diversity

Man-Made Objects
Man-Made Objects
Consonance with Environment

Composition

Flow	Composite Effect
Clarity	Unique Composition
Interface Land and Water	Mood Atmosphere
Floating Materials	

SOCIAL

Individual Environmental	Individual Well-Being
Interests	
Educational/Scientific	Physiological Health
Cultural	Physiological Health
Historical	Safety
Leisure/Recreation	Hygiene
Social Interactions	Community Well-Being
Political	Community Well-Being
Socialisation	
Religious	
Family	
Economic	

Thus, each parameter is analyzed. Each is compared with respect to changes in itself should the proposed project or any of its alternatives be implemented. If these changes are negligible than the environmental impact due to this parameter is nonexistent. If these changes are evident, then their size and significance must be determined, since an impact would occur.

The significance of an impact due to each parameter is assessed as to its contribution to the entire environmental impact of the proposed project. Alternative parameters are compared to the original project parameters, and an evaluation is made as to the desirability of the alternative. Alternatives can then be ranked with respect to the magnitude and significance of their environmental impact.

REFERENCES

Allee, W. C., Emerson, A. E., Park, O., Park, T., and Schmidt, K. P., 1949, Principles of Animal Ecology: Philadelphia, Saunders, 837 p.

Auffenburg, W., 1981, The Behavorial Ecology of the Komodo Monitor: Gainesville, Florida, University of Florida Presses.

Bakker, R. T., 1971, Ecology of the brontosaurs: Nature, v. 229, p. 172-174.

Bartlett, M. S., 1960, Stochastic Population Models in Ecology and Epidemiology: London, Methuen, 90 p.

Berry, R. J., and Jakobson, M. E., 1975, Ecological genetics of an island population of the house mouse *(Mus musculus)*: Journal of Zoology, v. 175, p. 532-540.

Birch, L. C., 1948, The intrinsic rate of natural increase of an insect population: Journal of Animal Ecology, v. 16, p. 15-26.

Black, C. A., 1968, Soil-Plant Relationships [2nd ed.]: New York, Wiley, 792 p.

Brown, J. L., 1970, Seeds of change. The green revolution and development in the 1970's: New York, Praeger.

Cailliet, G., Setzer, P., and Love, M., 1971, Everyman's Guide to Ecological Living: New York, 119 p.

Campbell, R. R., and Wade, J. L., 1972, Society and Environment: The Coming Collision: Boston, Allyn and Bacon, 375 p.

Carson, R., 1962, Silent Spring: Boston, Houghton-Mifflin.

Caswell, H., Koenig, H. E., Resh, J. A., and Ross, Q. E., 1972, An Introduction to Systems Science for Ecologists, *in* Patten, B., ed., Systems Analysis and Simulation in Ecology: New York, Academic Press, v. 2, p. 4-78; 592 pp.

Caswell, H., Reed, F., Stephenson, S. N., and Werner, P. A., 1973, Photosynthetic pathways and selective herbivory: a hypothesis: American Naturalist, v. 107, p. 465-480.

Caughley, G., 1966, Mortality patterns in mammals: Ecology, v. 47, p. 906-918.

Clark, L. R., Geier, P. W., Hughes, R. D., and Morris, R. F., 1967, The Ecology of Insect Populations in Theory and Practice: London, Methuen, 232 p.

Clarke, G., 1954, Elements of Ecology: New York, Wiley, 560 p.

Clements, F. E., 1949, Dynamics of Vegetation: New York, Hafner, 296 p.

Cody, M. L., 1966, A general theory of clutch size: Evolution, v. 20, p. 174- 184.

Colinvaux, P. A., 1973, Introduction to Ecology: New York, Wiley, 621 p.

Collier, B., Cox, G. W., Johnson, A. W., and Miller, P. C., 1973, Dynamic Ecology: Englewood Cliffs, New Jersey, Prentice-Hall, 563 p.

Colwell, R. K., and Futuyma, D. J., 1971, On the measurement of niche breadth and overlap: Ecology, v. 52, p. 567-576.

Cowles, R. P., and Brambel, C. E., 1936, A study of the environmental conditions in a bog pond with special reference to the diurnal vertical distribution of *Gonyostomum semen*: Biological Bulletin, Marine Biological Laboratory, Woods Hole, Mass., v. 71, p. 286-298.

Dale, M. B., 1970, Systems analysis and ecology: Ecology, v. 51, p. 2-16.

Dammerman, K. W., 1948, The fauna of Krakatau: 1883-1933: Verhandel. Kon-Inkl. Ned. Akad. Wetenschap. Afdel. Natuurk., v. 44, p. 1-594.

Ellenberger, P., 1974, Le Stormberg superior-I le biome de la zone B/1: Palaeovert., Mem. extraord. Montpellier, v. 5, p. 1-141.

Elton, C. S., 1927, Animal ecology: London, Sidgwick and Jackson, 209 p.

—, 1958, The ecology of invasions by animals and plants: London, England, Methuen, 181 p.

Houstan, D. C., 1979, The Adaptations of Scavengers, *in* Sinclair, A. R. E., and Norton-Griffiths, M., eds., Serengeti: Dynamics of an Ecosystem: Chicago, University of Chicago Press, p. 263-286.

Jablonski, D., 1986, Larval ecology and macroevolution in marine invertebrates: Bulletin of Marine Science, v. 39, p. 565-587.

Ladd, H. S., 1959, Ecology, paleontology and stratigraphy: Science, v. 129, p. 69-78.

Leather, D., 1986, The tale of the Tasaday: The Geographical Magazine, v. 58, no. 10, p. 490-491.

Morman, R. H., 1979, Distribution and ecology of lampreys in the lower peninsula of Michigan, 1957-1975. Great Lakes Fishery Commission, Technical Report No. 33, 59 pp.

Peters, R. H., 1976, Tautology in Evolution and Ecology: American Naturalist, v. 110, p. 1-12.

Stanley, S. M., 1973, An ecological theory for the sudden origin of multicellular life in the Late Precambrian: Proceedings of the National Academy of Sciences, v. 70, p. 1486-1489.

Tappen, H., 1971, Microplankton, ecological succession and evolution: North American Paleontological Convention, Proceedings, p. 1058-1103; Part H.

3

Towards Defining Green Science and its Manifestations for Understanding Pollution

Introduction

Green science requires an increasing number of odd redefinitions. "Organic" no longer refers to carbon compounds. Tomatoes do not have "genes" unless biotechnologists put them there. A "chemical" is something synthesized by humans and doesn't exist in nature — ditto for toxins. "Organic" water is bottled and sold (there is now a debate as to whether it should be decertified as organic). And now DNA is a "complex protein." A proposed ordinance to ban the deadly chemical dihydrogen monoxide was on the agenda for a California city council meeting until someone realized that it was H2O, a.k.a. water. The odd items in this "new science" seemingly go on without end. To the plant physiologist, microbiologist, or biotechnologist, the supposedly "scientific" claims of the anti-GMO [genetically modified organism] movement are as absurd as any of the foregoing.

Natural resources are naturally occurring substances that are considered valuable in their relatively unmodified (natural) form. A natural resource's value rests in the amount of the material available and the demand for it. Natural resource are natural capital converted to commodity inputs to infrastructural capital processes. They include soil, timber, oil, minerals, and other goods taken more or less from the Earth. Both extraction of the basic resource and refining it into a purer, directly usable form, (e.g., metals, refined oils) are generally considered natural-resource activities, even though the latter may not necessarily occur near the former. A nation's natural resources often determine its wealth in the world economic system, by determining its political influence. Developed nations are those which are less dependent on natural resources for wealth, due to their greater reliance on infrastructural capital for production. However, some see a resource curse whereby easily obtainable natural resources could actually hurt the prospects of a national economy by fostering political corruption. Political corruption can negatively impact the national economy because time is spent giving bribes or other economically unproductive acts instead of the generation of generative economic activity. There also tends to be concentrations of ownership over specific plots of land that have proven to yield natural resources.

In recent years, the depletion of natural capital and attempts to move to sustainable development have been a major focus of development agencies. This is of particular concern in rainforest regions, which hold most of the Earth's natural biodiversity - irreplaceable genetic natural capital. Conservation of natural resources is the major focus of natural capitalism, environmentalism, the ecology movement, and Green Parties. Some view this depletion as a major source of social unrest and conflicts in developing nations.

What is the Pollution of Envirement ?

Pollution is the release of chemical, physical, biological or radioactive contaminants to the environment. Principal forms of pollution include: air , water....

Air pollution is a broad term applied to any chemical, physical (e.g. particulate matter), or biological agent that modifies the natural characteristics of the atmosphere. The atmosphere is a complex, dynamic natural system that is essential to support life on planet earth. Stratospheric ozone depletion due to air pollution has long been recognised as a threat to human health as well as to the earth's ecosystems.

Worldwide air pollution is responsible for large numbers of deaths and cases of respiratory disease. Setting of air quality standards, such as the U.S. Clean Air Act, has begun to improve air quality for some pollutants. While major stationary sources are often identified with air pollution, the greatest source of emissions are actually mobile sources, principally the automobile. There are many available air pollution control technologies and urban planning strategies available to reduce air pollution; however, worldwide costs of addressing this issue are high. The most immediately available method of improving air quality is adoption of greater use of ethanol fuel (and other bio fuels) and increased reliance upon hybrid vehicle technology.

Evironmental Pollution

Pollution - Environmental pollution is any discharge of material or energy into water, land, or air that causes or may cause acute (short-term) or chronic (long-term) detriment to the Earth's ecological balance or that lowers the quality of life. Pollutants may cause primary damage, with direct identifiable impact on the environment, or secondary damage in the form of minor perturbations in the delicate balance of the biological food web that are detectable only over long time periods.

Until relatively recently in humanity's history, where pollution has existed, it has been primarily a local problem. The industrialisation of society, the introduction of motorised vehicles, and the explosion of the human population, however, have caused an exponential growth in the production of goods and services. Coupled with this growth has been a tremendous increase in waste by-products. The indiscriminate discharge of untreated industrial and domestic wastes into waterways, the spewing of thousands of tons of particulates and airborne gases into the atmosphere, the "throwaway" attitude toward solid wastes, and the use of newly developed chemicals without considering potential consequences have resulted in major environmental disasters, including the formation of smog in the Los Angeles area since the late 1940s and the pollution of large areas of the Mediterranean Sea. Technology has begun to solve some pollution problems (see pollution control), and public awareness of the extent of pollution will eventually force governments to undertake more effective environmental planning and adopt more effective antipollution measures.

Different Types of Pollution

Water Pollution

Water pollution is the introduction into fresh or ocean waters of chemical, physical, or biological material that degrades the quality of the water and affects the organisms living in it. This process ranges from simple addition of dissolved or suspended solids to discharge of the most insidious and persistent toxic pollutants (such as pesticides, heavy metals, and nondegradable, bioaccumulative, chemical compounds).

Conventional

Conventional or classical pollutants are generally associated with the direct input of (mainly human) waste products. Rapid urbanisation and rapid population increase have produced sewage problems because treatment facilities have not kept pace with need. Untreated and partially treated sewage from municipal wastewater systems and septic tanks in unsewered areas contribute significant quantities of nutrients, suspended solids, dissolved solids, oil, metals (arsenic, mercury, chromium, lead, iron, and manganese), and biodegradable organic carbon to the water environment.

Conventional pollutants may cause a myriad of water pollution problems. Excess suspended solids block out energy from the Sun and thus affect the carbon dioxide-oxygen conversion process, which is vital to the maintenance of the biological food chain. Also, high concentrations of suspended solids silt up rivers and navigational channels, necessitating frequent dredging. Excess dissolved solids make the water undesirable for drinking and for crop irrigation.

Although essential to the aquatic habitat, nutrients such as nitrogen and phosphorus may also cause overfertilisation and accelerate the natural aging process (eutrophication) of lakes. This acceleration in turn produces an overgrowth of aquatic vegetation, massive algal blooms, and an overall shift in the biologic community—from low productivity with many diverse species to high productivity with large numbers of a few species of a less desirable nature. Bacterial action oxidizes biodegradable organic carbon and consumes dissolved oxygen in the water. In extreme cases where the organic-carbon loading is high, oxygen consumption may lead to an oxygen depression: (less than 2 mg/l compared with 5 to 7 mg/l for a healthy stream) is sufficient to cause a fish kill and seriously to disrupt the growth of associated organisms that require oxygen to survive.

Nonconventional

The nonconventional pollutants include dissolved and particulate forms of metals, both toxic and nontoxic, and degradable and persistent organic carbon compounds discharged into water as a by-product of industry or as an integral part of marketable products. More than 13,000 oil spills of varying magnitude occur in the United States each year. Thousands of environmentally untested chemicals are routinely discharged into waterways; an estimated 400 to 500 new compounds are marketed each year. In addition, coal strip mining releases acid wastes that despoil the surrounding waterways. Nonconventional pollutants vary from biologically inert materials such as clay and iron residues to the most toxic and insidious materials such as halogenated hydrocarbons (DDT, kepone, mirex, and polychlorinated biphenyls—PCB). The latter group may produce damage ranging from acute biological effects (complete sterilisation of stretches of waterways) to chronic sublethal effects that may go undetected for years. The chronic low-level pollutants are proving to be the most difficult to correct and abate because of their ubiquitous nature and chemical stability.

Thermal Pollution

Thermal pollution is the discharge of waste heat via energy dissipation into cooling water and subsequently into nearby waterways. The major sources of thermal pollution are fossil-fuel and nuclear electric-power generating facilities and, to a lesser degree, cooling operations associated with industrial manufacturing, such as steel foundries, other primary-metal manufacturers, and chemical and petrochemical producers.

The discharge temperatures from electric-power plants generally range from 5 to

11 C degrees (9 to 20 F degrees) above ambient water temperatures. An estimated 90% of all water consumption, excluding agricultural uses, is for cooling or energy dissipation.

The discharge of heated water into a waterway often causes ecologic imbalance, sometimes resulting in major fish kills near the discharge source. The increased temperature accelerates chemical-biological processes and decreases the ability of the water to hold dissolved oxygen. Thermal changes affect the aquatic system by limiting or changing the type of fish and aquatic biota able to grow or reproduce in the waters. Thus rapid and dramatic changes in biologic communities often occur in the vicinity of heated discharges.

Land Pollution

Land pollution is the degradation of the Earth's land surface through misuse of the soil by poor agricultural practices, mineral exploitation, industrial waste dumping, and indiscriminate disposal of urban wastes.

Soil Misuse

Soil erosion—a result of poor agricultural practices—removes rich humus topsoil developed over many years through vegetative decay and microbial degradation and thus strips the land of valuable nutrients for crop growth. Strip mining for minerals and coal lays waste thousands of acres of land each year, denuding the Earth and subjecting the mined area to widespread erosion problems. The increases in urbanisation due to population pressure presents additional soil-erosion problems; sediment loads in nearby streams may increase as much as 500 to 1,000 times over that recorded in nearby undeveloped stretches of stream. Soil erosion not only despoils the Earth for farming and other uses, but also increases the suspended-solids load of the waterway. This increase interferes with the ecological habitat and poses silting problems in navigation channels, inhibiting the commercial use of these waters.

Solid Waste

In the United States in 1988 municipal wastes alone—that is, the solid wastes sent by households, business, and municipalities to local landfills and other waste-disposal facilities—equaled 163 million metric tons (1980 million U.S. tons), or 18 k (40lb) per person, according to figures released by the Environmental Protection Agency. Additional solid wastes accumulate from mining, industrial production, and agriculture. Although municipal wastes are the most obvious, the accumulations of other types of wastes are the most obvious, the accumulations of other types of waste are far greater, in many instances are more difficult to dispose of, and present greater environmental hazards.

The most common and convenient method of disposing of municipal solid wastes is in the sanitary landfill. The open dump, once a common eyesore in towns across the United States, attracted populations of rodents and other pests and often emitted hideous odors; it is now illegal. Sanitary landfills provide better aesthetic control and should be odor-free. Often, however, industrial wastes of unknown content are commingled with domestic wastes. Groundwater infiltration and contamination of water supplies with toxic chemicals have recently led to more active control of landfills and industrial waste disposal. Careful management of sanitary landfills, such as providing for leachate and runoff treatment as well as daily coverage with topsoil, has alleviated most of the problems of open dumping. In many areas, however, space for landfills is running out and alternatives must be found.

Recycling of materials is practical to some extent for much municipal and some industrial wastes, and a small but growing proportion of solid wastes is being recycled. When wastes are commingled, however, recovery becomes difficult and expensive. New processes of sorting ferrous and nonferrous metals, paper, glass, and plastics have been developed, and many communities with recycling programmes now require refuse separation. Crucial issues in recycling are devising better processing methods, inventing new products for the recycled materials, and finding new markets for them.

Incineration is another method for disposing of solid wastes. Advanced incinerators use solid wastes as fuel, burning quantities of refuse and utilising the resultant heat to make steam for electricity generation. Wastes must be burned at very high temperatures, and incinerator exhausts must be equipped with sophisticated scrubbers and other devices for removing dioxins and other toxic pollutants. Problems remain, however: incinerator ash contains high ratios of heavy metals, becoming a hazardous waste in itself, and high-efficiency incinerators may discourage the use of recycling and other waste-reduction methods.

Composting is increasingly used to treat some agricultural wastes, as well as such municipal wastes as leaves and brush. Composting systems can produce usable soil conditioners, or humus, within a few months (see compost).

Pesticide Pollution

Pesticides are organic and inorganic chemicals originally invented and first used effectively to better the human environment by controlling undesirable life forms such as bacteria, pests, and foraging insects. Their effectiveness, however, has caused considerable pollution. The persistent, or hard, pesticides, which are relatively inert and nondegradable by chemical or biologic activity, are also bioaccumulative; that is, they are retained within the body of the consuming organism and are concentrated with each ensuing level of the biologic food chain. For example, DDT provides an excellent example of cumulative pesticide effects. (Although DDT use has been banned in the United States since 1972, it is still a popular pesticide in much of the rest of the world.) DDT may be applied to an area so that the levels in the surrounding environment are less than one part per billion. As bacteria or other microscopic organisms ingest and retain the pesticide, the concentration may increase several hundred- to a thousandfold. Concentration continues as these organisms are ingested by higher forms of life—algae, fish, shellfish, birds, or humans. The resultant concentration in the higher life forms may reach levels of thousands to millions of parts per billion.

Many pesticides are nondiscriminatory; that is, they are not specific for a particular plant or organism. A dramatic example of this effect is DDE (a product of the breakdown of DDT), which effectively inhibits the ability of birds to provide sufficient calcium deposits for their eggs, producing fragile shells and a high percentage of nested eggs that break prematurely. Another reported side effect of pesticides is their effect on the nervous system of animals and fish; they can cause instability, disorientation, and, in some cases, death. These examples are generally a result of relatively high body residuals producing acute (short-term) readily recordable results.

The long-term (chronic) effects of persistent pesticides are virtually unknown, but many scientists believe they are as much an environmental hazard as are the acute effects. Nonpersistent (readily degradable) pesticides or substitutes, insect sterilisation techniques, hormone homologues that check or interfere with maturation stages, and introduction of animals that prey on the pests present a potentially brighter picture for pest control with significantly reduced environmental consequences.

Radiation Pollution

Radiation pollution is any form of ionizing or nonionising radiation that results from human activities. The most well-known radiation results from the detonation of nuclear devices and the controlled release of energy by nuclear-power generating plants (see nuclear energy). Other sources of radiation include spent-fuel reprocessing plants, by-products of mining operations, and experimental research laboratories. Increased exposure to medical X rays and to radiation emissions from microwave ovens and other household appliances, although of considerably less magnitude, all constitute sources of environmental radiation.

Public concern over the release of radiation into the environment greatly increased following the disclosure of possible harmful effects to the public from nuclear weapons testing, the accident (1979) at the Three Mile Island nuclear-power generating plant near Harrisburg, Pa., and the catastrophic 1986 explosion at Chernobyl, a

Soviet nuclear power plant. In the late 1980s, revelations of major pollution problems at U.S. nuclear weapons reactors raised apprehensions even higher.

The environmental effects of exposure to high-level ionizing radiation have been extensively documented through postwar studies on individuals who were exposed to nuclear radiation in Japan. Some forms of cancer show up immediately, but latent maladies of radiation poisoning have been recorded from 10 to 30 years after exposure. The effects of exposure to low-level radiation are not yet known. A major concern about this type of exposure is the potential for genetic damage.

Radioactive nuclear wastes cannot be treated by conventional chemical methods and must be stored in heavily shielded containers in areas remote from biological habitats. The safest of storage sites currently used are impervious deep caves or abandoned salt mines. Most radioactive wastes, however, have half-lives of hundreds to thousands of years, and to date no storage method has been found that is absolutely infallible.

Noise Pollution

Noise pollution has a relatively recent origin. It is a composite of sounds generated by human activities ranging from blasting stereo systems to the roar of supersonic transport jets. Although the frequency (pitch) of noise may be of major importance, most noise sources are measured in terms of intensity, or strength of the sound field. The standard unit, one decibel (dB), is the amount of sound that is just audible to the average human. The decibel scale is somewhat misleading because it is logarithmic rather than linear; for example, a noise source measuring 70 dB is 10 times as loud as a source measuring 60 dB and 100 times as loud as a source reading 50 dB. Noise may be generally associated with industrial society, where heavy machinery, motor vehicles, and aircraft have become everyday items. Noise pollution is more intense in the work environment than in the general environment, although ambient noise increased an average of one dB per year during the 1980s. The average background noise in a typical home today is between 40 and 50 decibels. Some examples of high-level sources in the environment are heavy trucks (90 dB at 15 m/50 ft), freight trains (75 dB at 15 m/50 ft), and air conditioning (60 dB at 6 m/20 ft).

The most readily measurable physiological effect of noise pollution is damage to hearing, which may be either temporary or permanent and may cause disruption of normal activities or just general annoyance. The effect is variable, depending upon individual susceptibility, duration of exposure, nature of noise (loudness), and time distribution of exposure (such as steady or intermittent). On the average an individual will experience a threshold shift (a shift in an individual's upper limit of sound detectability) when exposed to noise levels of 75 to 80 dB for several hours. This shift will last only several hours once the source of noise pollution is removed. A second physiologically important level is the threshold of pain, at which even short-term exposure will cause physical pain (130 to 140 dB). Any noise sustained at this level will cause a permanent threshold shift or permanent partial hearing loss. At the uppermost level of noise (greater than 150 dB), even a single short-term blast may cause traumatic hearing loss and physical damage inside the ear.

Although little hard information is available on the psychological side effects of increased noise levels, many researchers attribute increased irritability, lower productivity, decreased tolerance levels, increased incidence of ulcers, migraine headaches, fatigue, and allergic responses to continued exposures to high-level noises in the workplace and the general environment.

Air Pollution

Air pollution is the accumulation in the atmosphere of substances that, in sufficient concentrations, endanger human health or produce other measured effects on living matter and other materials. Among the major sources of pollution are power and heat generation, the burning of solid wastes, industrial processes, and, especially, transportation. The six major types of pollutants are carbon monoxide, hydrocarbons, nitrogen oxides, particulates, sulfur dioxide, and photochemical oxidants.

Local and Regional

Smog has seriously affected more persons than any other type of air pollution. It can be loosely defined as a multisource, widespread air pollution that occurs in the air of cities. Smog, a contraction of the words smoke and fog, has been caused throughout recorded history by water condensing on smoke particles, usually from burning coal. The infamous London fogs—about 4,000 deaths were attributed to the severe fog of 1952—were smog of this type. Another type, ice fog, occurs only at high latitudes and extremely low temperatures and is a combination of smoke particles and ice crystals.

As a coal economy has gradually been replaced by a petroleum economy, photochemical smog has become predominant in many cities. Its unpleasant properties result from the irradiation by sunlight of hydrocarbons (primarily unburned gasoline emitted by automobiles and other combustion sources) and other pollutants in the air. Irradiation produces a long series of photochemical reactions (see photochemistry). The products of the reactions include organic particles, ozone, aldehydes, ketones, peroxyacetyl nitrate, and organic acids and other oxidants. Sulfur dioxide, which is always present to some extent, oxidizes and hydrates to form sulfuric acid and becomes part of the particulate matter. Furthermore, automobiles are polluters even in the absence of photochemical reactions. They are responsible for much of the particulate material in the air; they also emit carbon monoxide, one of the most toxic constituents of smog.

All types of smog decrease visibility and, with the possible exception of ice fog, are irritating to the respiratory system. Statistical studies indicate that smog is a contributor to malignancies of many types. Photochemical smog produces eye irritation and lacrimation and causes severe damage to many types of vegetation, including important crops. Acute effects include an increased mortality rate, especially among persons suffering from respiratory and coronary ailments. Air pollution also has a deleterious effect on works of art (see art conservation and restoration).

Air pollution on a regional scale is in part the result of local air pollution—including that produced by individual sources, such as automobiles—that has spread out to encompass areas of many thousands of square kilometers. Meteorological conditions and landforms can greatly influence air-pollution concentrations at any given place, especially locally and regionally. For example, cities located in bowls or valleys over which atmospheric inversions form and act as imperfect lids are especially likely to suffer from incidences of severe smog. Oxides of sulfur and nitrogen, carried long distances by the atmosphere and then precipitated in solution as acid rain, can cause serious damage to vegetation, waterways, and buildings.

Global

Humans also pollute the atmosphere on a global scale, although until the early 1970s little attention was paid to the possible deleterious effects of such pollution. Measurements in Hawaii suggest that the concentration of carbon dioxide in the atmosphere is increasing at a rate of about 0.2% every year. The effect of this increase may be to alter the Earth's climate by increasing the average global temperature. Certain pollutants decrease the concentration of ozone occurring naturally in the stratosphere, which in turn increases the amount of ultraviolet radiation reaching the Earth's surface. Such radiation may damage vegetation and increase the incidence of skin cancer. Examples of stratospheric contaminants include nitrogen oxides emitted by supersonic aircraft and chlorofluorocarbons used as refrigerants and aerosol-can propellants. The chlorofluorocarbons reach the stratosphere by upward mixing from the lower parts of the atmosphere (see ozone layer). It is believed that these chemicals are responsible for the noticeable loss of ozone over the polar regions that has occurred in the 1980s.

Environmental Disaster

An environmental disaster is a disaster that is due to human activity and should not be confused with natural disasters. In this case, the impact of humans' alteration of the ecosystem has led to widespread and/or long-lasting consequences. It can include the deaths of animals (including humans) and plant systems, or severe disruption of human life, possibly requiring migration.

Difficulty in Categorisation

Some might view, for example, the Three Gorges Dam as an environmental disaster, requiring the migration of 1 million people. Others might see it as beneficial to stop flooding. Some might see the destruction of most of the North American forests as beneficial, as it cleared land for farming and other uses. In Ireland, the clearing of forest led to the formation of bogs, which some people like for their beauty, as well as their products such as peat moss. Others might see this deforestation as negative. More cynical examples would be, for example, that Saddam Hussein felt it was beneficial to get rid of the Madan people by draining the Al-Hawizeh marsh, because they had joined the United States in the first Gulf War. Another example is the depopulation of the American Bison. It was thought by General William Sherman and others to be a good way to get rid of the American Indians living in the Great Plains, and would make way for the exploding population of the United States of America to take over the area. The pulverisation of the Twin Towers with the Collapse of the World Trade Center and the release of Ground Zero dust propelled fine toxic dust into the vicinity; some have challenged whether this was an *industrial* disaster since this was a commercial site.

Water Resources

Definition and Terminology

Water resources are sources of water that are useful or potentially useful to humans. Uses of water includes agricultural, industrial, household, recreational and environmental activities. Virtually all of these human uses require fresh water. 97.5% of water on the Earth is salt water, leaving only 2.5% as fresh water of which over two thirds is frozen in glaciers and polar ice caps. The remaining unfrozen freshwater is mainly found as groundwater, with only a small fraction present above ground or in the air. Fresh water is a renewable resource, yet the world's supply of clean, fresh water is steadily decreasing. Water demand already exceeds supply in many parts of the world, and as world population continues to rise at an unprecedented rate, many more areas are expected to experience this imbalance in the near future. The framework for allocating water resources to water users (where such a framework exists) is known as water rights.

One of the thesauri to find terminology about water resources is Agrovoc).

Water and Conflict

The only known example of an actual inter-state conflict over water took place between 2500 and 2350 BC between the Sumerian states of Lagash and Umma. Yet, despite the lack of evidence of international wars being fought over water alone, water has been the source of various conflicts throughout history. When water scarcity causes political tensions to arise, this is referred to as water stress. Water stress has led most often to conflicts at local and regional levels. Using a purely quantitative methodology, Thomas Homer-Dixon successfully correlated water scarcity and scarcity of available arable lands to an increased chance of violent conflict.

Water stress can also exacerbate conflicts and political tensions which are not directly caused by water. Gradual reductions over time in the quality and/or quantity of fresh water can add to the instability of a region by depleting the health of a population, obstructing economic development, and exacerbating larger conflicts.

Conflicts and tensions over water are most likely to arise within national borders, in the downstream areas of distressed river basins. Areas such as the lower regions of China's Yellow River or the Chao Phraya River in Thailand, for example, have already been experiencing water stress for several years. Additionally, certain arid countries which rely heavily on water for irrigation, such as China, India, Iran, and Pakistan, are particularly at risk of water-related conflicts.. Political tensions, civil protest, and violence may also occur in reaction to water privatisation. The Bolivian Water Wars of 2000 are a case in point.

Sources of Fresh Water

Surface water

Surface water is water in a river, lake or fresh water wetland. Surface water is naturally replenished by precipitation and naturally lost through discharge to the oceans, evaporation, and sub-surface seepage.

Although the only natural input to any surface water system is precipitation within its watershed, the total quantity of water in that system at any given time is also dependent on many other factors. These factors include storage capacity in lakes, wetlands and artificial reservoirs, the permeability of the soil beneath these storage bodies, the runoff characteristics of the land in the watershed, the timing of the precipitation and local evaporation rates. All of these factors also affect the proportions of water lost.

Human activities can have a large impact on these factors. Humans often increase storage capacity by constructing reservoirs and decrease it by draining wetlands. Humans often increase runoff quantities and velocities by paving areas and channelising stream flow.

The total quantity of water available at any given time is an important consideration.

Some human water users have an intermittent need for water. For example, many farms require large quantities of water in the spring, and no water at all in the winter. To supply such a farm with water, a surface water system may require a large storage capacity to collect water throughout the year and release it in a short period of time. Other users have a continuous need for water, such as a power plant that requires water for cooling. To supply such a power plant with water, a surface water system only needs enough storage capacity to fill in when average stream flow is below the power plant's need. Nevertheless, over the long term the average rate of precipitation within a watershed is the upper bound for average consumption of natural surface water from that watershed. Natural surface water can be augmented by importing surface water from another watershed through a canal or pipeline. It can also be artificially augmented from any of the other sources listed here, however in practice the quantities are negligible. Humans can also cause surface water to be "lost" (i.e. become unusable) through pollution. Canada is the country estimated to have the largest supply of fresh water in the world, followed by Brazil and Russia.

Sub-Surface Water

Sub-Surface water, or groundwater, is fresh water located in the pore space of soil and rocks. It is also water that is flowing within aquifers below the water table. Sometimes it is useful to make a distinction between sub-surface water that is closely associated with surface water and deep sub-surface water in an aquifer (sometimes called "fossil water"). Sub-surface water can be thought of in the same terms as surface water: inputs, outputs and storage. The critical difference is that due to its slow rate of turnover, sub-surface water storage is generally much larger compared to inputs than it is for surface water. This difference makes it easy for humans to use sub-surface water unsustainably for a long time without severe consequences. Nevertheless, over the long term the average rate of seepage above a sub-surface water source is the upper bound for average consumption of water from that source.

The natural input to sub-surface water is seepage from surface water. The natural outputs from sub-surface water are springs and seepage to the oceans. If the surface water source is also subject to substantial evaporation, a sub-surface water source may become saline. This situation can occur naturally under endorheic bodies of water, or artificially under irrigated farmland. In coastal areas, human use of a sub-surface water source may cause the direction of seepage to ocean to reverse which can also cause soil salinisation. Humans can also cause sub-surface water to be "lost" (i.e. become unusable) through pollution. Humans can increase the input to a sub-surface water source by building reservoirs or detention ponds.

Water in the ground are in sections called aquifers. Rain rolls down and comes into these. Normally an aquifer is near to the equilibrium in its water content. The water content of an aquifier normally depends on the grain sizes. This means that the rate of extraction may be limited by poor permiability.

Desalination

Desalination is an artificial process by which saline water (generally sea water) is converted to fresh water. The most common desalination processes are distillation and reverse osmosis. Desalination is currently expensive compared to most alternative sources of water, and only a very small fraction of total human use is satisfied by desalination. It is only economically practical for high-valued uses (such as household and industrial uses) in arid areas. The most extensive use is in the Persian Gulf.

Frozen Water

Several schemes have been proposed to make use of icebergs as a water source, however to date this has only been done for novelty purposes. Glacier runoff is considered to be surface water.

Water Stress

The concept of water stress is relatively simple: According to the World Business Council for Sustainable Development, it applies to situations where there is not enough water for all uses, whether agricultural, industrial or domestic. Defining thresholds for stress in terms of available water per capita is more complex, however, entailing assumptions about water use and its efficiency. Nevertheless, it has been proposed that when annual per capita renewable freshwater availability is less than 1,700 cubic meters, countries begin to experience periodic or regular water stress. Below 1,000 cubic meters, water scarcity begins to hamper economic development and human health and well-being.

Population Growth

In 2000, the world population was 6.2 billion. The UN estimates that by 2050 there will be an additional 3 billion people with most of the growth in developing countries that already suffer water stress. Thus, water demand will increase unless there are corresponding increases in water conservation and recycling of this vital resource.

Increased Affluence

The rate of poverty alleviation is increasing especially within the two population giants of China and India. However, increasing affluence inevitably means more water consumption: from needing clean fresh water 24 hours a day, 7 days a week and basic sanitation service, to demanding water for gardens and car washing, to wanting jacuzzis or private swimming pools.

Expansion of Business Activity

Business activity ranging from industrialisation to services such as tourism and entertainment continues to expand rapidly. This expansion requires increased water services including both supply and sanitation, which can lead to more pressure on water resources and natural ecosystems.

Rapid Urbanisation

The trend towards urbanisation is accelerating. Small private wells and septic tanks that work well in low-density communities are not feasible within high-density urban areas. Urbanisation requires significant investment in water infrastructure in order to deliver water to individuals and to process the concentrations of wastewater – both from individuals and from business. These polluted and contaminated waters must be treated or they pose unacceptable public health risks. In 60% of European cities with more than 100,000 people, groundwater is being used at a faster rate than it can be replenished. Even if some water remains available, it costs more and more to capture it.

Climate Change

Climate change could have significant impacts on water resources around the world because of the close connections between the climate and hydrologic cycle. Rising temperatures will increase evaporation and lead to increases in precipitation, though there will be regional variations in rainfall. Overall, the global supply of freshwater will increase. Both droughts and floods may become more frequent in different regions at different times, and dramatic changes in snowfall and snowmelt are expected in mountainous areas. Higher temperatures will also affect water quality in ways that are not well understood. Possible impacts include increased eutrophication. Climate change could also mean an increase in demand for farm irrigation, garden sprinklers, and perhaps even swimming pools.

Depletion of Aquifers

Due to the expanding human population, competition for water is growing such that many of the worlds major aquifers are becoming depleted. This is due both for direct human consumption as well as agricultural irrigation by groundwater. Millions of small pumps of all sizes are currently extracting groundwater throughout the world. Irrigation in dry areas such as northern China and India is supplied by groundwater, and is being extracted at an unsustainable rate. Cities that have experienced aquifer drops between 10 to 50 meters include Mexico City, Bangkok, Manila, Beijing, Madras and Shanghai.

Pollution and Water Protection

Water pollution is one of the main concerns of the world today. The governments of many countries have striven to find solutions to reduce this problem. Many pollutants threaten water supplies, but the most widespread, especially in underdeveloped countries, is the discharge of raw sewage into natural waters; this method of sewage disposal is the most common method in underdeveloped countries, but also is prevalent in quasi-developed countries such as China, India and Iran.

Sewage, sludge, garbage, and even toxic pollutants are all dumped into the water. Even if sewage is treated, problems still arise. Treated sewage forms sludge, which is sent out into the sea and dumped. In addition to sewage, chemicals dumped by industries and governments are another major source of water pollution.

Uses of Fresh Water

Uses of fresh water can be categorised as consumptive and non-consumptive (sometimes called "renewable"). A use of water is consumptive if that water is not immediately available for another use. Losses to sub-surface seepage and evaporation are considered consumptive, as is water incorporated into a product (such as farm produce). Water that can be treated and returned as surface water, such as sewage, is generally considered non-consumptive if that water can be put to additional use.

Agricultural

It is estimated that 69% of world-wide water use is for irrigation, with 15-35% of irrigation withdrawals being unsustainable. In some areas of the world irrigation is necessary to grow any crop at all, in other areas it permits more profitable crops to be grown or enhances crop yield. Various irrigation methods involve different trade-offs between crop yield, water consumption and capital cost of equipment and structures. Irrigation methods such as most furrow and overhead sprinkler irrigation are usually less expensive but also less efficient, because much of the water evaporates or runs off. More efficient irrigation methods include drip or trickle irrigation, surge irrigation, and some types of sprinkler systems where the sprinklers are operated near ground level. These types of systems, while more expensive, can minimize runoff and evaporation. Any system that is improperly managed can be wasteful. Another trade-off that is often insufficiently considered is salinisation of sub-surface water. Aquaculture is a small but growing agricultural use of water. Freshwater commercial fisheries may also be considered as agricultural uses of water, but have generally been assigned a lower priority than irrigation (see Aral Sea and Pyramid Lake). As global populations grow, and as demand for food increases in a world with a fixed water supply, there are efforts underway to learn how to produce more food with less water, through improvements in irrigation methods and technologies, agricultural water management, crop types, and water monitoring.

Industrial

It is estimated that 15% of world-wide water use is industrial. Major industrial users include power plants, which use water for cooling or as a power source (i.e. hydroelectric plants), ore and oil refineries, which use water in chemical processes, and manufacturing plants, which use water as a solvent. The portion of industrial water usage that is consumptive varies widely, but as a whole is lower than agricultural use.

Household

It is estimated that 15% of world-wide water use is for household purposes. These include drinking water, bathing, cooking, sanitation, and gardening. Basic household water requirements have been estimated by Peter Gleick at around 50 liters per person per day, excluding water for gardens.

Recreation

Recreational water use is usually a very small but growing percentage of total water use. Recreational water use is mostly tied to reservoirs. If a reservoir is kept fuller than it would otherwise be for recreation, then the water retained could be categorised as recreational usage. Release of water from a few reservoirs is also timed to enhance whitewater boating, which also could be considered a recreational usage. Other examples are anglers, water skiers, nature enthusiasts and swimmers. Recreational usage is usually non-consumptive. Golf courses are often targeted as using excessive amounts of water, especially in drier regions. It is, however, unclear whether recreational irrigation (which would include private gardens) has a noticeable effect on water resources. This is largely due to the unavailability of reliable data. Some governments, including the Californian Government, have labelled golf course usage as agricultural in order to deflect environmentalists' charges of wasting water. However, using the above figures as a basis, the actual statistical effect of this reassignment is close to zero. Additionally, recreational usage may reduce the availability of water for other users at specific times and places. For example, water retained in a reservoir to allow boating in the late summer is not available to farmers during the spring planting season. Water released for whitewater rafting may not be available for hydroelectric generation during the time of peak electrical demand.

Environmental

Explicit environmental water use is also a very small but growing percentage of total water use. Environmental water usage includes artificial wetlands, artificial lakes intended to create wildlife habitat, fish ladders around dams, and water releases from reservoirs timed to help fish spawn. Like recreational usage, environmental usage is non-consumptive but may reduce the availability of water for other users at specific times and places. For example, water release from a reservoir to help fish spawn may not be available to farms upstream.

World Water Supply and Distribution

Food and water are two basic human needs. However, global coverage figures from 2002 indicate that, of every 10 people:

- roughly 5 have a connection to a piped water supply at home (in their dwelling, plot or yard);
- 3 make use of some other sort of improved water supply, such as a protected well or public standpipe;
- 2 are unserved;
- In addition, 4 out of every 10 people live without improved sanitation.

At Earth Summit 2002 governments approved a Plan of Action to:

- Halve by 2015 the proportion of people unable to reach or afford safe drinking water. The Global Water Supply and Sanitation Assessment 2000 Report (GWSSAR) defines "Reasonable access" to water as at least 20 liters per person per day from a source within one kilometer of the user's home.
- Halve the proportion of people without access to basic sanitation. The GWSSR defines "Basic sanitation" as private or shared but not public disposal systems that separate waste from human contact.

As the picture shows, in 2025, water shortages will be more prevalent among poorer countries where resources are limited and population growth is rapid, such as the Middle East, Africa, and parts of Asia. By 2025, large urban and peri-urban areas will require new infrastructure to provide safe water and adequate sanitation. This suggests growing conflicts with agricultural water users, who currently consume the majority of the water used by humans.

Generally speaking the more developed countries of North America, Europe and Russia will not see a serious threat to water supply by the year 2025, not only because of their relative wealth, but more importantly their populations will be better aligned with available water resources. North Africa, the Middle East, South Africa and northern China will face very severe water shortages due to physical scarcity and a condition of overpopulation relative to their carrying capacity with respect to water supply. Most of South America, Sub-Saharan Africa, Southern China and India will face water supply shortages by 2025; for these latter regions the causes of scarcity will be economic constraints to developing safe drinking water, as well as excessive population growth.

Economic Considerations

Water supply and sanitation require a huge amount of capital investment in infrastructure such as pipe networks, pumping stations and water treatment works. It is estimated that OECD nations need to invest at least USD 200 billion per year to replace aging water infrastructure to guarantee supply, reduce leakage rates and protect water quality. International attention has focused upon the needs of the developing countries. To meet the Millennium Development Goals targets of halving the proportion of the population lacking access to safe drinking water and basic sanitation by 2015, current annual investment on the order of USD 10 to USD 15 billion would need to be roughly doubled. This does not include investments required for the maintenance of existing infrastructure. Once infrastructure is in place, operating water supply and sanitation systems entails significant ongoing costs to cover personnel, energy, chemicals, maintenance and other expenses. The sources of money to meet these capital and operational costs are essentially either user fees, public funds or some combination of the two. But this is where the economics of water management start to become extremely complex as they intersect with social and broader economic policy. Such policy questions are beyond the scope of this article, which has concentrated on basic information about water availability and water use. They are, nevertheless, highly relevant to understanding how critical water issues will affect business and industry in terms of both risks and opportunities.

Business Response

The World Business Council for Sustainable Development in its H2OScenarios engaged in a scenario building process to:

- Clarify and enhance understanding by business of the key issues and drivers of change related to water.
- Promote mutual understanding between the business community and non-business stakeholders on water management issues.
- Support effective business action as part of the solution to sustainable water management.

It concluded that

- Business cannot survive in a society that thirsts
- You don't have to be in the water business to have a water crisis.
- Business is part of the solution, and its potential is driven by its engagement.
- Growing water issues and complexity will drive up costs.

Freshwater

Lakes are one of the planet's most important freshwater resources providing water for domestic, agricultural and industrial uses for much of the world's population. We are all familiar with the image of Earth as the "blue planet" when seen from space. It gives the impression that water is plentiful and indeed it is. However, pictures can be deceiving. Freshwater only accounts for 2% of all the Earth's water. But even that percentage is deceiving because 99% of all surface freshwater is locked away in continental ice. Freshwater resources are vital for meeting basic human needs and inadequate protection of the quality and the supply of freshwater can set important limits to sustainable development.

Freshwater pollution can be divided into two main categories: non-persistent and persistent. Non-persistent pollutants are degradable; they can be broken down by chemical reactions or by natural bacteria into simple, non-polluting substances such as carbon dioxide and nitrogen. Organic waste is an example of a non-persistent pollutant. The breakdown of organic waste can lead to low oxygen levels and 'eutrophication' (a process of nutrient enrichment), but the damage is reversible. Organic waste may also contain microorganisms which are the waterborne agents of diseases such as cholera, typhoid, and dysentery. Sources of non-persistent pollution include domestic sewage, fertilisers, some household cleaners, and some industrial wastes.

Freshwater acidification is another environmental problem. Natural acidification of freshwater environments has been taking place since the last Ice Age. However, the recent rapid acidification of many of lakes throughout the world can not be attributed to natural causes, but instead to the effects of acidic pollution from the burning of fossil fuels by mankind. Lakes and streams that are generally regarded as acidified contain very nutrient-poor water. Groundwater is water found in the tiny spaces between soil particles or in cracks in bedrock, much like the water in a sponge. The underground areas of soil or rock where substantial quantities of water are found are called 'aquifers'. These aquifers are the source of wells and springs. It is the top of the water in these aquifers that form the 'water table'. Groundwater can also be also polluted by the inflow of agricultural pesticides. Sewage from poorly maintained septic tanks and leachate from older landfill sites are also causes for concern. Because groundwater flows so slowly, contaminants are not carried away and diluted as rapidly as they are in rivers or oceans. It also purifies itself very slowly since the microbes that normally break down organic pollutants require oxygen, and groundwater is cut off from the atmosphere. For these reasons, polluted groundwater may remain contaminated for centuries.

Oceans

Covering about 70% of the earth's surface, the oceans are a highly productive system which continuously recycles chemicals, nutrients and water through the "hydrological cycle", which powers climate and weather, and which regulates global temperature by acting as a giant heat reservoir from the sun. About two-thirds of the world's population live within 60 kilometres of the coast, and almost half of the world's cities with more than one billion people are sited in and around the tide-washed river mouths known as estuaries. From a human point of view, oceans are also a major source of food and employment, provide natural routes for communication, transportation and trade. The year 1998 was declared International Year of the Ocean by the United Nations General Assembly in order to draw attention to the essential, but limited, resources of the ocean. The benefits from oceans and seas granted to the humanity have been underestimated, and are vulnerable because of the degree of pollution, degradation and overexploitation, which can ultimately threat the coastal human community, their economy and society in general. Foremost, care, management and use of oceans and seas and their natural resources is a common responsibility, demanding the maximum effort from all countries in order for it to be achieved sustainably.

Water Pollution

Water pollution is the contamination of water bodies such as lakes, rivers, oceans, and groundwater caused by human activities, which can be harmful to organisms and plants which live in these water bodies. Although natural phenomena such as volcanoes, algae blooms, storms, and earthquakes also cause major changes in water quality and the ecological status of water, water is typically referred to as polluted when it impaired by anthropogenic contaminants and either does not support a human use (like serving as drinking water) or undergoes a marked shift in its ability to support its constituent biotic communities. Water pollution has many causes and characteristics. The primary sources of water pollution are generally grouped into two categories based on their point of origin. Point-source pollution refers to contaminants that enter a waterway through a discrete "point source". Examples of this category include discharges from a wastewater treatment plant, outfalls from a factory, leaking underground tanks, etc. The second primary category, non-point source pollution, refers to contamination that, as its name suggests, does not originate from a single discrete source. Non-point source pollution is often a cumulative effect of small amounts of contaminants gathered from a large area. Nutrient runoff in stormwater from sheet flow over an agricultural field, or metals and hydrocarbons from an area with high impervious surfaces and vehicular traffic are examples of non-point source pollution. The primary focus of legislation and efforts to curb water pollution for the past several decades was first aimed at point sources. As point sources have been effectively regulated, greater attention has come to be placed on non-point source contributions, especially in rapidly urbanising/suburbanising or developing areas.

The specific contaminants leading to pollution in water include a wide spectrum of chemicals, pathogens, and physical or sensory changes. While many of the chemicals and substances that are regulated may be naturally occurring (iron, manganese, etc) the concentration is often the key in determining what is a natural component of water, and what is a contaminant. Many of the chemical substances are toxic. Pathogens can produce waterborne diseases in either human or animal hosts. Alteration of water's physical chemistry include acidity, electrical conductivity, temperature, and eutrophication. Eutrophication is the fertilisation of surface water by nutrients that were previously scarce. Water pollution is a major problem in the global context. It has been suggested that it is the leading worldwide cause of deaths and diseases, and that it accounts for the deaths of more than 14,000 people daily.

Contaminants

Contaminants may include organic and inorganic substances.

Some organic water pollutants are:

- Insecticides and herbicides, a huge range of organohalide and other chemicals
- Bacteria, often is from sewage or livestock operations
- Food processing waste, including pathogens
- Tree and brush debris from logging operations
- VOCs (volatile organic compounds), such as industrial solvents, from improper storage
- DNAPLs (dense non-aqueous phase liquids), such as chlorinated solvents, which may fall at the bottom of reservoirs, since they don't mix well with water and are more dense
- Petroleum Hydrocarbons including fuels (gasoline, diesel, jet fuels, and fuel oils) and lubricants (motor oil) from oil field operations, refineries, pipelines, retail service station's underground storage tanks, and transfer operations. Note: VOCs include gasoline-range hydrocarbons.
- Detergents
- Various chemical compounds found in personal hygiene and cosmetic products
- Disinfection by-products (DBPs) found in chemically disinfected drinking water

Some inorganic water pollutants include:

- Heavy metals including acid mine drainage

- Acidity caused by industrial discharges (especially sulfur dioxide from power plants)
- Pre-production industrial raw resin pellets, an industrial pollutant
- Chemical waste as industrial by products
- Fertilizers, in runoff from agriculture including nitrates and phosphates
- Silt in surface runoff from construction sites, logging, slash and burn practices or land clearing sites

Transport and Chemical Reactions of Water Pollutants

Most water pollutants are eventually carried by the rivers into the oceans. In some areas of the world the influence can be traced hundred miles from the mouth by studies using hydrology transport models. Advanced computer models such as SWMM or the DSSAM Model have been used in many locations worldwide to examine the fate of pollutants in aquatic systems. Indicator filter feeding species such as copepods have also been used to study pollutant fates in the New York Bight, for example. The highest toxin loads are not directly at the mouth of the Hudson River, but 100 kilometers south, since several days are required for incorporation into planktonic tissue. The Hudson discharge flows south along the coast due to coriolis force. Further south then are areas of oxygen depletion, caused by chemicals using up oxygen and by algae blooms, caused by excess nutrients from algal cell death and decomposition. Fish and shellfish kills have been reported, because toxins climb the foodchain after small fish consume copepods, then large fish eat smaller fish, etc. Each successive step up the food chain causes a stepwise concentration of pollutants such as heavy metals (e.g. mercury) and persistent organic pollutants such as DDT. This is known as biomagnification which is occasionally used interchangeably with bioaccumulation.

The big gyres in the oceans trap floating plastic debris. The North Pacific Gyre for example has collected the so-called "Great Pacific Garbage Patch" that is now estimated at two times the size of Texas. Many of these long-lasting pieces wind up in the stomachs of marine birds and animals. This results in obstruction of digestive pathways which leads to reduced appetite or even starvation.

Many chemicals undergo reactive decay or chemically change especially over long periods of time in groundwater reservoirs. A noteworthy class of such chemicals are the chlorinated hydrocarbons such as trichloroethylene (used in industrial metal degreasing and electronics manufacturing) and tetrachloroethylene used in the dry cleaning industry (note latest advances in liquid carbon dioxide in dry cleaning that avoids all use of chemicals). Both of these chemicals, which are carcinogens themselves, undergo partial decomposition reactions, leading to new hazardous chemicals (including dichloroethylene and vinyl chloride). Groundwater pollution is much more difficult to abate than surface pollution because groundwater can move great distances through unseen aquifers. Non-porous aquifers such as clays partially purify water of bacteria by simple filtration (adsorption and absorption), dilution, and, in some cases, chemical reactions and biological activity: however, in some cases, the pollutants merely transform to soil contaminants. Groundwater that moves through cracks and caverns is not filtered and can be transported as easily as surface water. In fact, this can be aggravated by the human tendency to use natural sinkholes as dumps in areas of Karst topography. There are a variety of secondary effects stemming not from the original pollutant, but a derivative condition. Some of these secondary impacts are:

- Silt bearing surface runoff from can inhibit the penetration of sunlight through the water column, hampering photosynthesis in aquatic plants.
- Thermal pollution can induce fish kills and invasion by new thermophilic species. This can cause further problems to existing wildlife.

Sampling & Monitoring

Sampling water can take several forms depending on the accuracy needed and the characteristics of the contaminant. Many contamination events are temporal and most commonly in association with rain events. For this reason 'grab' samples can be used as indicators, but are often inadequate for fully accessing contaminant concerns in a water body. Scientists gathering this type of data often employ auto-sampler devices that pump increments of water at either time or discharge intervals.

Regulatory Framework

In the UK there are common law rights (civil rights) to protect the passage of water across land unfettered in either quality of quantity. Criminal laws dating back to the 16th century exercised some control over water pollution but it was not until the *River (Prevention of pollution)Acts 1951 - 1961* were enacted that any systematic control over water pollution was established. These laws were strengthened and extended in the *Control of Pollution Act 1984* which has since been updated and modified by a series of further acts. It is a criminal offense to either pollute a lake, river, groundwater or the sea or to discharge any liquid into such water bodies without proper authority. In England and Wales such permission can only be issued by the Environment Agency and in Scotland by SEPA.

In the USA, concern over water pollution resulted in the enactment of state anti-pollution laws in the latter half of the 19th century, and federal legislation enacted in 1899. The Refuse Act of the federal Rivers and Harbors Act of 1899 prohibits the disposal of any refuse matter from into either the nation's navigable rivers, lakes, streams, and other navigable bodies of water, or any tributary to such waters, unless one has first obtained a permit. The Water Pollution Control Act, passed in 1948, gave authority to the Surgeon General to reduce water pollution.

Growing public awareness and concern for controlling water pollution led to enactment of the Federal Water Pollution Control Act Amendments of 1972. As amended in 1977, this law became commonly known as the Clean Water Act. The Act established the basic mechanisms for regulating contaminant discharge. It established the authority for the United States Environmental Protection Agency to implement wastewater standards for industry. The Clean Water Act also continued requirements to set water quality standards for all contaminants in surface waters. Further amplification of the Act continued including the enactment of the Great Lakes Legacy Act of 2002.

In 2004, the United States Environmental Protection Agency tested drinking water quality on commercial airline's aircraft and found that 15 percent of tested aircraft tested positive for total coliform bacteria, according to a press release issued on Friday March 28, 2008.

Water Pollution and Socity

Introduction

Comprising over 70% of the Earth's surface, water is undoubtedly the most precious natural resource that exists on our planet. Without the seemingly invaluable compound comprised of hydrogen and oxygen, life on Earth would be non-existent: it is essential for everything on our planet to grow and prosper. Although we as humans recognize this fact, we disregard it by polluting our rivers, lakes, and oceans. Subsequently, we are slowly but surely harming our planet to the point where organisms are dying at a very alarming rate. In addition to innocent organisms dying off, our drinking water has become greatly affected as is our ability to use water for recreational purposes. In order to combat water pollution, we must understand the problems and become part of the solution.

Point and non-point Sources

According to the American College Dictionary, pollution is defined as: `to make foul or unclean; dirty.' Water pollution occurs when a body of water is adversely affected due to the addition of large amounts of materials to the water. When it is unfit for its intended use, water is considered polluted. Two types of water pollutants exist; point source and nonpoint source. Point sources of pollution occur when harmful substances are emitted directly into a body of water. The Exxon Valdez oil spill best illustrates a point source water pollution. A nonpoint source delivers pollutants indirectly through environmental changes. An example of this type of water pollution is when fertilizer from a field is carried into a stream by rain, in the form of run-off which in turn effects aquatic life. The technology exists for point sources of pollution to be monitored and regulated, although political factors may complicate matters. Nonpoint sources are much more difficult to control. Pollution arising from nonpoint sources accounts for a majority of the contaminants in streams and lakes.

Causes of Pollution

Many causes of pollution including sewage and fertilizers contain nutrients such as nitrates and phosphates. In excess levels, nutrients over stimulate the growth of aquatic plants and algae. Excessive growth of these types of organisms consequently clogs our waterways, use up dissolved oxygen as they decompose, and block light to deeper waters.

This, in turn, proves very harmful to aquatic organisms as it affects the respiration ability or fish and other invertebrates that reside in water.

Pollution is also caused when silt and other suspended solids, such as soil, washoff plowed fields, construction and logging sites, urban areas, and eroded river banks when it rains. Under natural conditions, lakes, rivers, and other water bodies undergo Eutrophication, an aging process that slowly fills in the water body with sediment and organic matter. When these sediments enter various bodies of water, fish respirationbecomes impaired, plant productivity and water depth become reduced, and aquatic organisms and their environments become suffocated. Pollution in the form of organic material enters waterways in many different forms as sewage, as leaves and grass clippings, or as runoff from livestock feedlots and pastures. When natural bacteria and protozoan in the water break down this organic material, they begin to use up the oxygen dissolved in the water. Many types of fish and bottom-dwelling animals cannot survive when levels of dissolved oxygen drop below two to five parts per million. When this occurs, it kills aquatic organisms in large numbers which leads to disruptions in the food chain.

Polluted River in the United Kingdom

The pollution of rivers and streams with chemical contaminants has become one of the most crutial environmental problems within the 20th century. Waterborne chemical pollution entering rivers and streams cause tramendous amounts of destruction.

Pathogens are another type of pollution that prove very harmful. They can cause many illnesses that range from typhoid and dysentery to minor respiratory and skin diseases. Pathogens include such organisms as bacteria, viruses, and protozoan. These pollutants enter waterways through untreated sewage, storm drains, septic tanks, runoff from farms, and particularly boats that dump sewage. Though microscopic, these pollutants have a tremendous effect evidenced by their ability to cause sickness.

Additional Forms of Water Pollution

Three last forms of water pollution exist in the forms of petroleum, radioactive substances, and heat. Petroleum often pollutes waterbodies in the form of oil, resulting from oil spills. The previously mentioned Exxon Valdez is an example of this type of water pollution. These large-scale accidental discharges of petroleum are an important cause of pollution along shore lines. Besides the supertankers, off-shore drilling operations contribute a large share of pollution. One estimate is that one ton of oil is spilled for every million tons of oil transported. This is equal to about 0.0001 percent. Radioactive substances are produced in the form of waste from nuclear power plants, and from the industrial, medical, and scientific use of radioactive materials. Specific forms of waste are uranium and thorium mining and refining. The last form of water pollution is heat. Heat is a pollutant because increased temperatures result in the deaths of many aquatic organisms. These decreases in temperatures are caused when a discharge of cooling water by factories and power plants occurs.

Demonstrators Protest Drilling

Oil pollution is a growing problem, particularly devestating to coastal wildlife. Small quantities of oil spread rapidly across long distances to form deadly oil slicks. In this picture, demonstrators with "oil-covered" plastic animals protest a potential drilling project in Key Largo, Florida. Whether or not accidental spills occur during the project, its impact on the delicate marine ecosystem of the coral reefs could be devastating.

Oil Spill Clean-up

Workers use special nets to clean up a California beach after an oil tanker spill. Tanker spills are an increasing environmental problem because once oil has spilled, it is virtually impossible to completely remove or contain it. Even small amounts spread rapidly across large areas of water. Because oil and water do not mix, the oil floats on the water and then washes up on broad expanses of shoreline. Attempts to chemically treat or sink the oil may further disrupt marine and beach ecosystems.

Classifying Water Pollution

The major sources of water pollution can be classified as municipal, industrial, and agricultural. Municipal water pollution consists of waste water from homes and commercial establishments. For many years, the main goal of treating municipal wastewater was simply to reduce its content of suspended solids, oxygen-demanding materials, dissolved inorganic compounds, and harmful bacteria. In recent years, however, more stress has been placed on improving means of disposal of the solid residues from the municipal treatment processes. The basic methods of treating municipal wastewater fall into three stages: primary treatment, including grit removal, screening, grinding, and sedimentation; secondary treatment, which entails oxidation of dissolved organic matter by means of using biologically active sludge, which is then filtered off; and tertiary treatment, in which advanced biological methods of nitrogen removal and chemical and physical methods such as granular filtration and activated carbon absorption are employed. The handling and disposal of solid residues can account for 25 to 50 percent of the capital and operational costs of a treatment plant. The characteristics of industrial waste waters can differ considerably both within and among industries. The impact of industrial discharges depends not only on their collective characteristics, such as biochemical oxygen demand and the amount of suspended solids, but also on their content of specific inorganic and organic substances. Three options are available in controlling industrial wastewater. Control can take place at the point of generation in the plant; wastewater can be pretreated for discharge to municipal treatment sources; or wastewater can be treated completely at the plant and either reused or discharged directly into receiving waters.

Wastewater Treatment

Raw sewage includes waste from sinks, toilets, and industrial processes. Treatment of the sewage is required before it can be safely buried, used, or released back into local water systems. In a treatment plant, the waste is passed through a series of screens, chambers, and chemical processes to reduce its bulk and toxicity. The three general phases of treatment are primary, secondary, and tertiary. During primary treatment, a large percentage of the suspended solids and inorganic material is removed from the sewage. The focus of secondary treatment is reducing organic material by accelerating natural biological processes. Tertiary treatment is necessary when the water will be reused; 99 percent of solids are removed and various chemical processes are used to ensure the water is as free from impurity as possible.

Agriculture, including commercial livestock and poultry farming, is the source of many organic and inorganic pollutants in surface waters and groundwater. These contaminants include both sediment from erosion cropland and compounds of phosphorus and nitrogen that partly originate in animal wastes and commercial fertilizers. Animal wastes are high in oxygen demanding material, nitrogen and phosphorus, and they often harbor pathogenic organisms. Wastes from commercial feeders are contained and disposed of on land; their main threat to natural waters, therefore, is from runoff and leaching. Control may involve settling basins for liquids, limited biological treatment in aerobic or anaerobic lagoons, and a variety of other methods.

Ground Water

Ninety-five percent of all fresh water on earth is ground water. Ground water is found in natural rock formations. These formations, called aquifers, are a vital natural resource with many uses. Nationally, 53% of the population relies on ground water as a source of drinking water. In rural areas this figure is even higher. Eighty one percent of community water is dependent on ground water. Although the 1992 Section 305(b) State Water Quality Reports indicate that, overall, the Nation's ground water quality is good to excellent, many local areas have experienced significant ground water contamination. Some examples are leaking underground storage tanks and municipal landfills.

Legislation

Several forms of legislation have been passed in recent decades to try to control water pollution. In 1970, the Clean Water Act provided 50 billion dollars to cities and states to build wastewater facilities. This has helped control surface water pollution from industrial and municipal sources throughout the United States. When congress passed the Clean Water Act in 1972, states were given primary authority to set their own standards for their water. In addition to these standards, the act required that all state beneficial uses and their criteria must comply with the `fishable and swimmable' goals of the act. This essentially means that state beneficial uses must be able to support aquatic life and recreational use. Because it is impossible to test water for every type of disease-causing organism, states usually look to identify indicator bacteria. One for a example is a bacteria known as fecal coliforms. These indicator bacteria suggest that a certain selection of water may be contaminated with untreated sewage and that other, more dangerous, organisms are present. These legislations are an important part in the fight against water pollution. They are useful in preventing Environmental catastrophes. The graph shows reported pollution incidents since 1989-1994. If stronger legislations existed, perhaps these events would never have occurred.

Global Water Pollution

Estimates suggest that nearly 1.5 billion people lack safe drinking water and that at least 5 million deaths per year can be attributed to waterborne diseases. With over 70 percent of the planet covered by oceans, people have long acted as if these very bodies of water could serve as a limitless dumping ground for wastes. Raw sewage, garbage, and oil spills have begun to overwhelm the diluting capabilities of the oceans, and most coastal waters are now polluted. Beaches around the world are closed regularly, often because of high amounts of bacteria from sewage disposal, and marine wildlife is beginning to suffer.

Perhaps the biggest reason for developing a worldwide effort to monitor and restrict global pollution is the fact that most forms of pollution do not respect national boundaries. The first major international conference on environmental issues was held in Stockholm, Sweden, in 1972 and was sponsored by the United Nations (UN). This meeting, at which the United States took a leading role, was controversial because many developing countries werc fearful that a focus on environmental protection was a means for the developed world to keep the undeveloped world in an economically subservient position. The most important outcome of the conference was the creation of the United Nations Environmental Programme (UNEP).

UNEP was designed to be `the environmental conscience of the United Nations,' and, in an attempt to allay fears of the developing world, it became the first UN agency to be headquartered in a developing country, with offices in Nairobi, Kenya. In addition to attempting to achieve scientific consensus about major environmental issues, a major focus for UNEP has been the study of ways to encourage sustainable development increasing standards of living without destroying the environment. At the time of UNEP's creation in 1972, only 11 countries had environmental agencies. Ten years later that number had grown to 106, of which 70 were in developing countries.

Water Quality

Water quality is closely linked to water use and to the state of economic development. In industrialised countries, bacterial contamination of surface water caused serious health problems in major cities throughout the mid 1800's. By the turn of the century, cities in Europe and North America began building sewer networks to route domestic wastes downstream of water intakes. Development of these sewage networks and waste treatment facilities in urban areas has expanded tremendously in the past two decades. However, the rapid growth of the urban population (especially in Latin America and Asia) has outpaced the ability of governments to expand sewage and water infrastructure. While waterborne diseases have been eliminated in the developed world, outbreaks of cholera and other similar diseases still occur with alarming frequency in the developing countries. Since World War II and the birth of the `chemical age`, water quality has been heavily impacted worldwide by industrial and agricultural chemicals. Eutrophication of surface waters from human and agricultural wastes and nitrification of groundwater from agricultural practices has greatly affected large parts of the world. Acidification of surface waters by air pollution is a recent phenomenon and threatens aquatic life in many area of the world. In developed countries, these general types of pollution have occurred sequentially with the result that most developed countries have successfully dealt with major surface water pollution. In contrast, however, newly industrialised countries such as China, India, Thailand, Brazil, and Mexico are now facing all these issues simultaneously.

Conclusion

Clearly, the problems associated with water pollution have the capabilities to disrupt life on our planet to a great extent. Congress has passed laws to try to combat water pollution thus acknowledging the fact that water pollution is, indeed, a seriousissue. But the government alone cannot solve the entire problem. It is ultimately up to us, to be informed, responsible and involved when it comes to the problems we face with our water. We must become familiar with our local water resources and learn about ways for disposing harmful household wastes so they don`t end up in sewage treatment plants that can`t handle them or landfills not designed to receive hazardous materials. In our yards, we must determine whether additional nutrients are needed before fertilizers are applied, and look for alternatives where fertilizers might run off into surface waters. We have to preserve existing trees and plant new trees and shrubs to help prevent soil erosion and promote infiltration of water into the soil. Around our houses, we must keep litter, pet waste, leaves, and grass clippings out of gutters and storm drains. These are just a few of the many ways in which we, as humans, have the ability to combat water pollution. As we head into the 21st century, awareness and education will most assuredly continue to be the two most important ways to prevent water pollution. If these measures are not taken and water pollution continues, life on earth will suffer severely.

Global environmental collapse is not inevitable. But the developed world must work with the developing world to ensure that new industrialised economies do not add to the world's environmental problems. Politicians must think of sustainable development rather than economic expansion. Conservation strategies have to become more widely accepted, and people must learn that energy use can be dramatically diminished without sacrificing comfort. In short, with the technology that currently exists, the years of global environmental mistreatment can begin to be reversed.

Protect Water Quality

Water quality depends upon well-defined water quality standards and criteria, consistent and well-constructed monitoring and evaluation programmes, and a legal framework that produces effective incentives to control public and private activities that may result in water pollution. ELI's Water Quality Protection activities focus on the laws and implementation strategies that influence:

- The effectiveness of the federal Clean Water Act in improving water quality. ELI's work includes projects to preserve broad national authority, improve water quality standards, identify impaired waters and address Total Maximum Daily Loads, strengthen permitting programmes, control nonpoint sources and stormwater, and resolve wet weather issues.

- Innovation in state and local programmes that can improve water quality. ELI examines state laws and local ordinances that address sectors such as agriculture, forestry, and land development; the connection between groundwater and surface water and its uses; and activities in coastal and riparian areas. We identify opportunities for state laws dealing with water quality, taxation, land use, funding, and other subjects to support water quality objectives.

Water Pollution Prevention

Measures to prevent water pollution essentially strive to conserve and protect water quality - in terms of its use reduction and disposal, waste water treatment, procedural changes and recycling. Water conservation programmes have also included the substitution/reduction of hazardous materials and the generation of hazardous wastes. Employee awareness, education and training in pollution identification and reduction is critical in achieving successful results.

Water Pollution Prevention and Control

China's Policies and Successful Experiences

The Government of China pays high attention to the prevention and control of water pollution. Ever since 1996, China has designated water pollution prevention as the essential work of environmental protection of the entire country. Achievements have been made in the water pollution prevention in the recent decade, with some good experiences and performances as follows:

1. The laws and regulations on water pollution prevention have been continuously improved; and the enforcement of the laws is fairly strict. Since the promulgation of the Water Pollution Prevention Law by National People's Congress, China has formulated or amended seventeen laws, administrative laws and regulations, sector norms and technical policies, such as Water Pollution Prevention, Measures of the Implementation of Water Pollution Prevention, Temporary Rules of Water Pollution Prevention in Huai River Basin, and Technical Policy of Urban Sewage Treatment and Pollution Prevention. State Environmental Protection Administration has issued 21 standards on water environmental protection such as Standards of Ambient Surface Water Quality, Standards of Ground Water Quality, and Comprehensive Standards of Sewage Discharge. The local governments have also, in line with the local situation, formulated a series of local laws, rules regulations and standards for water pollution prevention. Up till now, a legal framework for ambient water protection and water pollution prevention based on the Law of Water Pollution Prevention has been set up. Large-scale inspection on environmental enforcement is carried out in the entire country every year, including inspection on the implementation

of laws and regulations on water pollution prevention. In addition, special inspection on certain areas and sectors has been conducted.

2. The programme of water pollution prevention has been enhanced and the prioritised action plans have been identified. There is a large population and enormous rapid economic growth in the areas of Three Rivers (Huai River, Liao River and Hai River) and Three Lakes (Tai Lake, Dianchi Lake and Cao Lake). In the meantime, the resources consumption is highly concentrated and the environmental pollution is comparative seriously in this area, which has become a factor restricting the further economic development in this area, with serious damages on the public health, and strong complaints from the public. The central government has identified the water pollution prevention in the Three Rivers and Three Lakes as the important work of the environmental protection of the country. The 9th and 10th Plan of Water Pollution Prevention as well as the pollution treatment programmes for Three Gorges Dam area and the east rout project of Water Diversion project have been formulated and implemented. Concrete stipulations on the targets, tasks, measures and deadlines of pollution treatment have been identified. Water pollution prevention has been carried out on large scale. The water pollution treatment in major river basins has promoted on full scale the water pollution prevention in the entire country. Local governments have also, in line with local situation, identified local prioritised action plan.

3. Transit of policies and measures for water pollution prevention: transit from point-source pollution to integrated treatment of river basins, from end-of-pipe treatment to head-source and entire process control, from concentration control to the combination control of total amount and concentration, and from pollution prevention as main measure to paying attention on both pollution prevention and ecological conservation. The main measures include:

a) To control the total amount of pollutant discharge. Local governments and various sector control the total industrial and urban domestic sewage discharge in an active way in connection with the industrial restructuring;

b) To implement strictly the Environmental Impact Assessment and "Three-Simultaneities" regimes, and veto the new construction projects that do not fit in with the environmental requirements and industrial policies;

c) To phase out and close down a number of production capacity, processes, products and enterprises with serious environmental pollution and waste of resources;

d) To promote cleaner production in enterprises in connection with technology innovation. The discharge of the main pollution in industrial pollution sources have attained the standards;

e) To devote great efforts to the development of ecological agriculture and organic agriculture, and the integrated environmental treatment in small river basins;

f) To turn the sewage into resource. The sewage with treatment to certain standards is reused;

g) To speed up the infrastructure construction of urban sewage treatment plants, and promote water pollution prevention with economic instruments. In order to solve the problem of urban sewage, State Environmental Protection Administration, jointly with relevant departments, has issued the Notice on Enhancing the Levy of Sewage Treatment Fee and Establishing Benign Operation Mechanism of Urban Sewage Discharge and Centralised Treatment of Urban Sewage, as well as the Notice of Levy Mechanism of Urban Sewage and Industrialisation of Garbage Treatment. Relevant systems have been set up in local areas to guide the social funds to invest the construction and operation of environmental infra-structure in various forms, which has strengthened the urban environmental treatment and infra-structure construction;

h) To increase investment in pollution prevention. Since 1998, environmental protection has been regarded as an important point of the active financial policies. Over 69 billion Yuan national bonds have been used in the key engineering projects of pollution prevention, most of which is directly or indirectly used in the treatment of pollution prevention, with effective impact on the financial investment in environmental protection by local governments and society.

In the context of per capita GDP less than 800 US dollars, China started pollution treatment in river basins on large scale, which rarely happens in the world and is the first case in the developing countries. With efforts for many years, achievements have been made in the water pollution treatment in major river basins. The industrial pollution sources have basically attained the discharge standards. A large number of enterprises with serious pollution and low profits have been phased out. A great deal of urban sewage treatment plants have been completed and put to operation. In the context of rapid economic growth and population increase in the river basin, the trend of rapid water quality degradation in the river basin is under basic control. In 2003, percentage of the sections under control in Huai River, Hai river and Liao River that have attained the water quality target of the 10th Five Year Plan is 56%, 43% and 46% respectively, among which the water quality in Huai River Basin has been improved significantly, and the section attaining the standards has increase by 30% over 2000. The organic pollution and eutrophication in "Three Lakes" area has been improved to certain extent, and the water quality in the main lake areas of Tai Lake has been improved significantly.

Pollution Prevention

Water pollution is a complex issue - from the source of poluution to its impacts, both short-term and long-term on human and other species that depend on water. Point and non-point sources of pollution have a profound impact on the degree of pollution, and many times the actual causes are hidden behind more 'visible' causes. Understanding these causes-behind-causes is critical in developing appropriate responses to reduce pollution. Broad-based awareness of the sources and impacts of pollution - involving a number of stakeholders on the water continuum - is also important to effect lasting solutions.

Resources specific to river pollution are aslo included - River pollution is a result of a complex combination of processes that reduce overall river water quality. Acid rain, industrial pollution, agricultural pollution contribute to river pollution, but so do everyday activities that drain untreated pollutants and leachate into rivers and streams. Transportation has a role to play too - where carbon and one-drop-at-a-time 'oil spills' can also cause pollution through storm run-off. A holistic and integrative understanding of the cause-effect cycles of river pollution is an effective starting point to improve river water quality.

Soil

Soil is the naturally occurring, unconsolidated or loose covering of broken rock particles and decaying organic matter (humus) on the surface of the Earth, capable of supporting life. In simple terms, soil has three components: solid, liquid, and gas. The solid phase is a mixture of mineral and organic matter. Soil particles pack loosely, forming a soil structure filled with voids. The solid phase occupies about half of the soil volume. The remaining void space contains water (liquid) and air (gas). Soil is also known as earth: it is the substance from which our planet takes its name.

Characteristics

Soil color is the first impression one has when viewing soil. Striking colors and contrasting patterns are especially memorable. The Red River in Louisiana carries sediment eroded from extensive reddish soils like Port Silt Loam in Oklahoma. Soil color results from chemical and biological weathering. As the primary minerals in parent material weather, the elements combine into new and colorful compounds. Iron forms secondary minerals with a yellow or red color; organic matter decomposes into brown compounds; and manganese, sulfur and nitrogen can form black mineral deposits. Soil structure is the arrangement of soil particles into aggregates. These may have various shapes, sizes and degrees of development or expression. Soil texture refers to sand, silt and clay composition. Sand and silt are the product of physical weathering while clay is the product of chemical weathering. Clay content is particularly influential on soil behavior due to a high retention capacity for nutrients and water.

Formation

Soil formation, or pedogenesis, is the combined effect of physical, chemical, biological, and anthropogenic processes on soil parent material resulting in the formation of soil horizons. Soil is always changing. The long periods over which change occurs and the multiple influences of change mean that simple soils are rare. While soil can achieve relative stability in properties for extended periods of time, the soil life cycle ultimately ends in soil conditions that leave it vulnerable to erosion. Little of the soil continuum of the earth is older than Tertiary and most no older than Pleistocene. Despite the inevitability of soils retrogression and degradation, most soil cycles are long and productive. How the soil "life" cycle proceeds is influenced by at least five classic soil forming factors: regional climate, biotic potential, topography, parent material, and the passage of time.

An example of soil development from bare rock occurs on recent lava flows in warm regions under heavy and very frequent rainfall. In such climates plants become established very quickly on basaltic lava, even though there is very little organic material. The plants are supported by the porous rock becoming filled with nutrient bearing water, for example carrying dissolved bird droppings or guano. The developing plant roots themselves gradually breaks up the porous lava and organic matter soon accumulates but, even before it does, the predominantly porous broken lava in which the plant roots grow can be considered a soil.

In Nature

Biogeography is the study of spatial variations in biological communities. Soils are a restricting factor as to what plants can grow in which environments. Soil scientists survey soils in the hope of understanding controls as to what vegetation can and will grow in a particular location

Geologists also have a particular interest in the patterns of soil on the surface of the earth. Soil texture, color and chemistry often reflect the underlying geologic parent material and soil types often change at geologic unit boundaries. Buried paleosols mark previous land surfaces and record climatic conditions from previous eras. Geologists use this paleopedological record to understand the ecological relationships in past ecosystems. According to the theory of biorhexistasy, prolonged conditions conducive to forming deep, weathered soils result in increasing ocean salinity and the formation of limestone.

Geologists use soil profile features to establish the duration of surface stability in the context of geologic faults or slope stability. An offset subsoil horizon indicates rupture during soil formation and the degree of subsequent subsoil formation is relied upon to establish time since rupture.

Soil examined in shovel test pits is used by archaeologists for relative dating based on stratigraphy (as opposed to absolute dating). What is considered most typical is to use soil profile features to determine the maximum reasonable pit depth than needs to be examined for archaeological evidence in the interest of cultural resources management.

Soils altered or formed by man (anthropic and anthropogenic soils) are also of interest to archaeologists. An example is Terra preta do Indio.

Uses

Soil material is a critical component in the mining and construction industries. Soil serves as a foundation for most construction projects. Massive volumes of soil can be involved in surface mining, road building, and dam construction. Earth sheltering is the architectural practice of using soil for external thermal mass against building walls.

Soil resources are critical to the environment, as well as to food and fiber production. Soil provides minerals and water to plants. Soil absorbs rainwater and releases it later thus preventing floods and drought. Soil cleans the water as it percolates. Soil is the habitat for many organisms.

Waste management often has a soil component. Septic drain fields treat septic tank effluent uses aerobic soil processes. Landfills use soil for daily cover. Organic soils, especially peat, serve as a significant fuel resource. Both humans in many cultures and animals occasionally eat soil.

Degradation

Land degradation is a human induced or natural process which impairs the capacity of land to function. Soils are the critical component in land degradation when it involves acidification, contamination, desertification, erosion, or salination.

While soil acidification of alkaline soils is beneficial, it degrades land when soil acidity lowers crop productivity and increases soil vulnerability to contamination and erosion. Soils are often initially acid because their parent materials were acid and initially low in the basic cations (calcium, magnesium, potassium, and sodium). Acidification occurs when these elements are removed from the soil profile by normal rainfall or the harvesting of crops. Soil acidification is accelerated by the use of acid-forming nitrogenous fertilizers and by the effects of acid precipitation.

Soil contamination at low levels are often within soil capacity to treat and assimilate. Many waste treatment processes rely on this treatment capacity. Exceeding treatment capacity can damage soil biota and limit soil function. Derelict soils occur where industrial contamination or other development activity damages the soil to such a degree that the land cannot be used safely or productively. Remediation of derelict soil uses principles of geology, physics, chemistry, and biology to degrade, attenuate, isolate, or remove soil contaminants and to restore soil functions and values. Techniques include leaching, air sparging, chemical amendments, phytoremediation, bioremediation, and natural attenuation.

Desertification is an environmental process of ecosystem degradation in arid and semi-arid regions, or as a result of human activity. It is a common misconception that droughts cause desertification. Droughts are common in arid and semiarid lands. Well-managed lands can recover from drought when the rains return. Soil management tools include maintaining soil nutrient and organic matter levels, reduced tillage and increased cover. These help to control erosion and maintain productivity during periods when moisture is available. Continued land abuse during droughts, however, increases land degradation. Increased population and livestock pressure on marginal lands accelerates desertification.

Soil erosional loss is caused by wind, water, ice, movement in response to gravity. Although the processes may be simultaneous, erosion is distinguished from weathering. Erosion is an intrinsic natural process, but in many places it is increased by human land use. Poor land use practices include deforestation, overgrazing, and improper construction activity. Improved management can limit erosion using techniques like limiting disturbance during construction, avoiding construction during erosion prone periods, intercepting runoff, terrace-building, use of erosion suppressing cover materials and planting trees or other soil binding plants.

A serious and long-running water erosion problem is in China, on the middle reaches of the Yellow River and the upper reaches of the Yangtze River. From the Yellow River, over 1.6 billion tons of sediment flow each year into the ocean. The sediment originates primarily from water erosion in the Loess Plateau region of northwest China.

Soil piping is a particular form of soil erosion that occurs below the soil surface. It is associated with levee and dam failure as well as sink hole formation. Turbulent flow removes soil starting from the mouth of the seep flow and subsoil erosion advances upgradient. The term sand boil is used to describe the appearance of the discharging end of an active soil pipe.

Soil salination is the accumulation of free salts to such an extent that it leads to degradation of soils and vegetation. Consequences include corrosion damage, reduced plant growth, erosion due to loss of plant cover and soil structure, and water quality problems due to sedimentation. Salination occurs due to a combination of natural and human caused processes. Aridic conditions favor salt accumulation. This is especially apparent when soil parent material is saline. Irrigation of arid lands is especially problematic. All irrigation water has some level of salinity. Irrigation, especially when it involves leakage from canals, often raise the underlying water table. Rapid salination occurs when the land surface is within the capillary fringe of saline groundwater. Salinity control involves flushing with higher levels of applied water in combination with tile drainage..

Soil

Soil is an integral part of the environment. It is essential for the production of food and other crops, for maintaining biodiversity, for the landscape. It contains much archaeological evidence of our history.

Soil quality is not a major problem in the UK, although there are localised problems of erosion, acidification and other contamination. But there are long term pressures on soils. Minimising the loss of soils to new development and land use presents a particular challenge. In the past, soil protection has received less attention than the protection of air and water. The Government will ensure that soil protection receives equal priority in future. A draft soil strategy for England and Wales will be available for public consultation shortly. Taking forward strategies for Scotland and Northern Ireland will be the responsibility of the devolved administrations.

The strategy will identify pressures on soils and set out objectives and measures for soil protection and include indicators of progress. It will also set out action to improve understanding of soil processes and to increase public awareness of the importance of soils.

Soil Contamination

Soil contamination is caused by the presence of man-made chemicals or other alteration in the natural soil environment. This type of contamination typically arises from the rupture of underground storage tanks, application of pesticides, percolation of contaminated surface water to subsurface strata, oil and fuel dumping, leaching of wastes from landfills or direct discharge of industrial wastes to the soil. The most common chemicals involved are petroleum hydrocarbons, solvents, pesticides, lead and other heavy metals. This occurrence of this phenomenon is correlated with the degree of industrialisation and intensity of chemical usage.

The concern over soil contamination stems primarily from health risks, both of direct contact and from secondary contamination of water supplies. Mapping of contaminated soil sites and the resulting cleanup are time consuming and expensive tasks, requiring extensive amounts of geology, hydrology, chemistry and computer modeling skills.

It is in North America and Western Europe that the extent of contaminated land is most well known, with many of countries in these areas having a legal framework to identify and deal with this environmental problem; this however may well be just the tip of the iceberg with developing countries very likely to be the next generation of new soil contamination cases.

The immense and sustained growth of the People's Republic of China since the 1970s has exacted a price from the land in increased soil pollution. The State Environmental Protection Administration believes it to be a threat to the environment, to food safety and to sustainable agriculture. According to a scientific sampling, 150 million mi (100,000 square kilometres) of China's cultivated land have been polluted, with contaminated water being used to irrigate a further 32.5 million mi (21,670 square kilometres) and another 2 million mi (1,300 square kilometres) covered or destroyed by solid waste. In total, the area accounts for one-tenth of China's cultivatable land, and is mostly in economically developed areas. An estimated 12 million tonnes of grain are contaminated by heavy metals every year, causing direct losses of 20 billion yuan (US$2.57 billion).

The United States, while having some of the most widespread soil contamination, has actually been a leader in defining and implementing standards for cleanup. Other industrialised countries have a large number of contaminated sites, but lag the U.S. in executing remediation. Developing countries may be leading in the next generation of new soil contamination cases.

Each year in the U.S., thousands of sites complete soil contamination cleanup, some by using microbes that "eat up" toxic chemicals in soil, many others by simple excavation and others by more expensive high-tech soil vapor extraction or air stripping. At the same time, efforts proceed worldwide in creating and identifying new sites of soil contamination, particularly in industrial countries other than the U.S., and in developing countries which lack the money and the technology to adequately protect soil resources.

Microanalysis of Soil Contamination

To understand the fundamental nature of soil contamination, it is necessary to envision the variety of mechanisms for pollutants to become entrained in soil. Soil particulates may be composed of a gamut of organic and inorganic chemicals with variations in cation exchange capacity, buffering capacity, and redox poise. For example, at the extremes, one has a sand component, a coarse grained, inert, and totally inorganic substance; whereas peat soils are dominated by a fine organic material, made of decomposing organic material and highly active. Most soils are mixtures of soil subtypes and thus have quite complex characteristics. There is also a great diversity of soil porosity, ranging from gravels to sands to silt to clay (in increasing order of porosity), pore size, and pore tortuosity (both in decreasing order). Finally there is a wide spectrum of chemical bonding or adhesion characteristics: each contaminant has a different interaction or bonding mechanism with a given soil type.

On balance, some contaminants may literally drain through soils such as sand and gravel and move to other soils or deeper aquifers, while polar or organic chemicals discharged into a clay soil will have a very high adsorption. Thus most soil contamination is the result of pollutants adhering to the soil particle surface, or lodging in interstices of a soil matrix. Clearly the equilibrium reached is a dynamic one, where new pollutants may lodge on new soil particles and the action of groundwater movement may over time transport some of the soil contaminants to other locations or depths.

Soil contamination results when hazardous substances are either spilled or buried directly in the soil or migrate to the soil from a spill that has occurred elsewhere. For example, soil can become contaminated when small particles containing hazardous substances are released from a smokestack and are deposited on the surrounding soil as they fall out of the air. Another source of soil contamination could be water that washes contamination from an area containing hazardous substances and deposits the contamination in the soil as it flows over or through it.

Health Effects

The major concern is that there are many sensitive land uses where people are in direct contact with soils such as residences, parks, schools and playgrounds. Other contact mechanisms include contamination of drinking water or inhalation of soil contaminants which have vaporized. There is a very large set of health consequences from exposure to soil contamination depending on pollutant type, pathway of attack and vulnerability of the exposed population. Chromium and obsolete pesticide formulations are carcinogenic to populations. Lead is especially hazardous to young children, in which group there is a high risk of developmental damage to the brain,while to all populations kidney damage is a risk.

Chronic exposure to at sufficient concentrations is known to be associated with higher incidence of leukemia. Obsolete pesticides such as mercury and cyclodienes are known to induce higher incidences of kidney damage, some irreversible; cyclodienes are linked to liver toxicity. Organophosphates and carbamates can induce a chain of responses leading to neuromuscular blockage. Many chlorinated solvents induce liver changes, kidney changes and depression of the central nervous system. There is an entire spectrum of further health effects such as headache, nausea, fatigue (physical), eye irritation and skin rash for the above cited and other chemicals.

Ecosystem Effects

Not unexpectedly, soil contaminants can have significant deleterious consequences for ecosystems. There are radical soil chemistry changes which can arise from the presence of many hazardous chemicals even at low concentration of the contaminant species. These changes can manifest in the alteration of metabolism of endemic microorganisms and arthropods resident in a given soil environment. The result can be virtual eradication of some of the primary food chain, which in turn have major consequences for predator or consumer species. Even if the chemical effect on lower life forms is small, the lower pyramid levels of the food chain may ingest alien chemicals, which normally become more concentrated for each consuming rung of the food chain. Many of these effects are now well known, such as the concentration of persistent DDT materials for avian consumers, leading to weakening of egg shells, increased chick mortality and potentially species extinction.

Effects occur to agricultural lands which have certain types of soil contamination. Contaminants typically alter plant metabolism, most commonly to reduce crop yields. This has a secondary effect upon soil conservation, since the languishing crops cannot shield the earth's soil mantle from erosion phenomena. Some of these chemical contaminants have long half-lives and in other cases derivative chemicals are formed from decay of primary soil contaminants.

Regulatory Framework

United States of America

Until about 1970 there was little widespread awareness of the worldwide scope of soil contamination or its health risks. In fact, areas of concern were often viewed as unusual or isolated incidents. Since then, the U.S. has established guidelines for handling hazardous waste and the cleanup of soil pollution. In 1980 the U.S.Superfund/ CERCLA established strict rules on legal liability for soil contamination. Not only did CERCLA stimulate identification and cleanup of thousands of sites, but it raised awareness of property buyers and sellers to make soil pollution a focal issue of land use and management practices.

While estimates of remaining soil cleanup in the U.S. may exceed 200,000 sites, in other industrialised countries there is a lag of identification and cleanup functions. Even though their use of chemicals is lower than industrialised countries, often their controls and regulatory framework is quite weak. For example, some persistent pesticides that have been banned in the U.S. are in widespread uncontrolled use in developing countries. It is worth noting that the cost of cleaning up a soil contaminated site can range from as little as about $10,000 for a small spill, which can be simply excavated, to millions of dollars for a widespread event, especially for a chemical that is very mobile such as perchloroethylene.

China

China, an economy that regularly records double digit annual economic growth, has little or no legislation to protect the environment. Currently, given the amount of land in question (up to one-tenth of China's cultivatable land may be polluted), the degree of the pollution in specific locations is unclear, making both prevention and remedy difficult. There are no laws or environmental standards regarding soil. Funding is limited, too, so there is little advanced scientific study of China's soil taking place. The severity of the pollution is not known by the public or business population, and the situation is most likely worsening as a result.

United Kingdom

Generic guidance commonly used in the UK are the Soil Guideline Values published by DEFRA and the Environment Agency. These are screening values that demonstrate the minimal acceptable level of a substance. Above this there can be no assurances in terms of significant risk of harm to human health. These have been derived using the Contaminated Land Exposure Asseeement Model (CLEA UK). Certain input parameters such as Health Criteria Values, age and land use are fed into CLEA UK to obtain a probablistic output.

Guidance by the Inter Departmental Committee for the Redevelopment of Contaminated Land (ICRCL) has been formally withdrawn by the Department for Environment, Food and Rural Affairs (DEFRA), for use as a prescriptive document to determine the potential need for remediation or further assessment. Therefore, no further reference is made to these former guideline values. Other generic guidance that exists (to put the concentration of a particular contaminant in context), includes the United States EPA Region 9 Preliminary Remediation Goals (US PRGs), the US EPA Region 3 Risk Based Concentrations (US EPA RBCs) and National Environment Protection Council of Australia Guideline on Investigation Levels in Soil and Groundwater.

However international guidance should only be used in the UK with clear justification. This is because foreign standards are usually particular to that country due to drivers such as political policy, geology, flood regime and epidemiology. It is generally accepted by UK regulators that only robust scientific methods that relate to the UK should be used. The CLEA model published by DEFRA and the Environment Agency (EA) in March 2002 sets a framework for the appropriate assessment of risks to human health from contaminated land, as required by Part IIA of the

Environmental Protection Act 1990. As part of this framework, generic Soil Guideline Values (SGVs) have currently been derived for ten contaminants to be used as "intervention values". These values should not be considered as remedial targets but values above which further detailed assessment should be considered. Three sets of CLEA SGVs have been produced for three different land uses, namely

- residential (with and without plant uptake)
- allotments
- commercial/industrial

It is intended that the SGVs replace the former ICRCL values. It should be noted that the CLEA SGVs relate to assessing chronic (long term) risks to human health and do not apply to the protection of ground workers during construction, or other potential receptors such as groundwater, buildings, plants or other ecosystems. The CLEA SGVs are not directly applicable to a site completely covered in hardstanding, as there is no direct exposure route to contaminated soils.

To date, the first ten of fifty-five contaminant SGVs have been published, for the following: arsenic, cadmium, chromium, lead, inorganic mercury, nickel, selenium ethyl benzene, phenol and toluene. Draft SGVs for benzene, naphthalene and xylene have been produced but their publication is on hold. Toxicological data (Tox) has been published for each of these contaminants as well as for benzo[a]pyrene, benzene, dioxins, furans and dioxin-like PCBs, naphthalene, vinyl chloride, 1,1,2,2 tetrachloroethane and 1,1,1,2 tetrachloroethane, 1,1,1 trichloroethane, tetrachloroethene, carbon tetrachloride, 1,2-dichloroethane, trichloroethene and xylene. The SGVs for ethyl benzene, phenol and toluene are dependent on the soil organic matter (SOM) content (which can be calculated from the total organic carbon (TOC) content). As an initial screen the SGVs for 1% SOM are considered to be appropriate.

Cleanup Options

Cleanup or remediation is analysed by environmental scientists who utilize field measurement of soil chemicals and also apply computer models for analysing transport and fate of soil chemicals. Thousands of soil contamination cases are currently in active cleanup across the U.S. as of 2006. There are several principal strategies for remediation:

- Excavate soil and remove it to a disposal site away from ready pathways for human or sensitive ecosystem contact. This technique also applies to dredging of bay muds containing toxins.
- Aeration of soils at the contaminated site (with attendant risk of creating air pollution)

- Bioremediation, involving microbial digestion of certain organic chemicals. Techniques used in bioremediation include landfarming, biostimulation and bioaugmentation soil biota with commercially available microflora.
- Extraction of groundwater or soil vapor with an active electromechanical system, with subsequent stripping of the contaminants from the extract.
- Containment of the soil contaminants (such as by capping or paving over in place).

Forest Protection

Forest protection is a general term describing methods purported to preserve or improve a forest threatened or affected by abuse. There is considerable debate over the effectiveness of forest protection methods.

One simple type of forest protection is the purchasing of land in order to secure it, or in order to plant trees (afforestation). It can also mean forest management or the designation of areas such as natural reservoirs which are intended to be left to themselves.

However, merely purchasing a piece of land does not prevent it from being used by others for poaching and illegal logging. A better way to protect a forest, particularly old growth forests in remote areas, is to obtain a part of it and to live on and monitor the purchased land. Even in the USA, these measures sometimes don't suffice because arson can burn a forest to the ground, leaving burnt areas free for different use. Enforcement of laws regarding purchased forest land is weak or non-existent in most parts of the world. In the increasingly dangerous South America, home of major rainforests, officials of the Brazilian National Agency for the Environment (IBAMA) have recently been shot during their routine duties. Another issue about living on purchased forest-land is that there may not be a suitable site for a standard home without clearing land, which defies the purpose of protection. Alternatives include building a treehouse or an earthhouse. This is being done currently by indigenous people in South America to protect large reservoirs. In former times, North American Native Americans used to live in tipies or mandan earthhouses, which also require less land. An undertaking to develop modern treehouses is being taken by a company from Germany called "True School treehouses" A compromise is to conduct agriculture and stock farming, or sustainable wood management. This ascribes different values to forest land and farmland, for which many areas are clear felled. A number of less successful methods of forest protection have been tried, such as the trade in certified wood. The types of abuse that forest protection seeks to prevent include:

- Aggressive or unsustainable farming and logging

- Expanding city development caused by population explosion and the resulting urban sprawl

Protecting a small section of land in a larger forest may also have limited value. For example, tropical rainforests can die if they decrease in size, since they are dependent on the moist microclimate which they create. A recent discovery in Europe relating to forest protection is that urban areas have forests of their own. Many cities have tens of thousands of trees which constitute forests. In addition the air in the cities is lately becoming better, providing conditions favourable for small associated species such as mosses and lichens.

Wildlife

Wildlife and wildlife habitats are increasingly under threat from the environmental consequences of mnakind's unsustainable activities. Global warming, acid rain and air pollution all pose serious threats to many faunal and floral species, in additions to the problems they face as a result of deforestation and other land-use changes for agriculture and urbanisation. It is believed by many scientists that global biodiversity is diminishing as a result of mankind's impacts.

Freshwater acidification can be directly harmful to many species of fish, amphibians and water-dwelling invertebrates, with indirect consequences for many mammals and birds. Air pollutants can also be harmful to wildlife directly regardless of the effects of acid rain. Species particularly at risk include soft-bodied invertebrates like worms, moist skinned amphibians and various types of vegetation. Perhaps the most threatening of environmental impacts to wildlife is climate change. Global average surface temperatures is projected to rise by 2 to 3°C by the end of the 21st century, or 0.2 to 0.3°C per decade. It is currently believed that most ecosystems can withstand at most a 0.1°C global temperature change per decade, before experiencing severe ecological stresses, leading in some cases to species extinction.

The Convention on Biological Diversity, negotiated at the Rio Earth Summit in 1992, provides the framework for nation states to implement strategies to protect wildlife and their habitats. Following the implementation of the Convention in 1994, the UK's national Biodiversity Action Plan has co-ordinated activity to conserve and enhance biodiversity in this country. Wildlife protection has often focused on special reserves which contain habitats and species which must be maintained. But there is a risk that populations become smaller, fragmented and vulnerable to extinction. Their long-term survival, and overall enhancement of wildlife, depends on action in cities, towns and the countryside as a whole.

Earth's Atmosphere

The Earth's atmosphere is a layer of gases surrounding the planet Earth and retained by the Earth's gravity. It contains roughly (by molar content/volume) 78.08% nitrogen, 20.95% oxygen, 0.93% argon, 0.038% carbon dioxide, trace amounts of other gases, and a variable amount (average around 1%) of water vapor. This mixture of gases is commonly known as air. The atmosphere protects life on Earth by absorbing ultraviolet solar radiation and reducing temperature extremes between day and night.

There is no definite boundary between the atmosphere and outer space. It slowly becomes thinner and fades into space. Three quarters of the atmosphere's mass is within 11 km of the planetary surface. In the United States, people who travel above an altitude of 80.5 km (50 statute miles) are designated astronauts. An altitude of 120 km (~75 miles or 400,000 ft) marks the boundary where atmospheric effects become noticeable during re-entry. The Kármán line, at 100 km (62 miles or 328,000 ft), is also frequently regarded as the boundary between atmosphere and outer space.

Temperature and Layers

The temperature of the Earth's atmosphere varies with altitude; the mathematical relationship between temperature and altitude varies among five different atmospheric layers (ordered highest to lowest, the ionosphere is part of the thermosphere):

- Exosphere: from 500 – 1000 km (300 – 600 mi) up to 10,000 km (6,000 mi), free-moving particles that may migrate into and out of the magnetosphere or the solar wind.

Exobase Boundary

- Ionosphere: is the part of the atmosphere that is ionised by solar radiation. It plays an important part in atmospheric electricity and forms the inner edge of the magnetosphere. It has practical importance because, among other functions, it influences radio propagation to distant places on the Earth. It is located in the thermosphere and is responsible for auroras.

Thermopause Boundary

- Thermosphere: from 80 – 85 km (265,000 – 285,000 ft) to 640+ km (400+ mi), temperature increasing with height.

Mesopause Boundary

- Mesosphere: The mesosphere extends from about 50 km (160,000 ft) to the range of 80 to 85 km (265,000 – 285,000 ft), temperature decreasing with height. This is also where most meteors burn up when entering the atmosphere.

Stratopause Boundary

- Stratosphere: From the Latin word "stratus" meaning a spreading out. The stratosphere extends from the troposphere's 7 to 17 km (23,000 – 60,000 ft) range to about 50 km (160,000 ft). Temperature increases with height. The stratosphere contains the ozone layer, the part of the Earth's atmosphere which contains relatively high concentrations of ozone. "Relatively high" means a few parts per million—much higher than the concentrations in the lower atmosphere but still small compared to the main components of the atmosphere. It is mainly located in the lower portion of the stratosphere from approximately 15 to 35 km (50,000 – 115,000 ft) above Earth's surface, though the thickness varies seasonally and geographically.

Tropopause Boundary

- Troposphere: From the Greek word "ôñÝðù" meaning to turn or change. The troposphere is the lowest layer of the atmosphere; it begins at the surface and extends to between 7 km (23,000 ft) at the poles and 17 km (60,000 ft) at the equator, with some variation due to weather factors. The troposphere has a great deal of vertical mixing due to solar heating at the surface. This heating warms air masses, which makes them less dense so they rise. When an air mass rises the pressure upon it decreases so it expands, doing work against the opposing pressure of the surrounding air. To do work is to expend energy, so the temperature of the air mass decreases. As the temperature decreases, water vapor in the air mass may condense or solidify, releasing latent heat that further uplifts the air mass. This process determines the maximum rate of decline of temperature with height, called the adiabatic lapse rate. It contains roughly 80% of the total mass of the atmosphere. 50% of the total mass of the atmosphere is located in the lower 5 km of the troposphere.

The average temperature of the atmosphere at the surface of Earth is 15 °C (59 °F).

Pressure and Thickness

The average atmospheric pressure, at sea level, is about 101.3 kilopascals (about 14.7 psi); total atmospheric mass is 5.1480×10^{18} kg. Atmospheric pressure is a direct result of the total weight of the air above the point at which the pressure is measured. This means that air pressure varies with location and time, because the amount (and weight) of air above the earth varies with location and time. However the *average* mass of the air above a square meter of the earth's surface is known to the same high accuracy as the total air mass of 5148.0 teratonnes and area of the earth of 51007.2 megahectares, namely 5148.0/510.072 = 10.093 metric tonnes or 14.356 lbs (mass) per square inch. This is about 2.5% below the officially standardised unit atmosphere (1 atm) of 101.325 kPa or 14.696 psi, and corresponds to the mean pressure not at sea level but at the mean base of the atmosphere as contoured by the earth's terrain.

Atmospheric pressure decreases with height, dropping by 50% at an altitude of about 5.6 km (18,000 ft). For comparison: the highest mountain, Mount Everest, is higher, at 8.8 km, which is why it is so difficult to climb without supplemental oxygen. Equivalently, about 50% of the total atmospheric mass is within the lowest 5.6 km. This pressure drop is approximately exponential, so that pressure decreases by approximately half every 5.6 km. However, because of changes in temperature throughout the atmospheric column, as well as the fact that the force of gravity begins to decrease at great altitudes, a single equation does not model atmospheric pressure through all altitudes (it is modeled in 7 exponentially decreasing layers, in the equations given above). Even in the exosphere, the atmosphere is still present (as can be seen for example by the effects of atmospheric drag on satellites). The equations of pressure by altitude in the above references can be used directly to estimate atmospheric thickness. However, the following published data are given for reference:-

- 50% of the atmosphere by mass is below an altitude of 5.6 km.
- 90% of the atmosphere by mass is below an altitude of 16 km. The common altitude of commercial airliners is about 10 km.
- 99.99997% of the atmosphere by mass is below 100 km. The highest X-15 plane flight in 1963 reached an altitude of 354,300 ft (108,000 m).

Therefore, most of the atmosphere (99.9997%) is below 100 km, although in the rarefied region above this there are auroras and other atmospheric effects. The mean molar mass of air is 28.97 g/mol. Note that the composition figures above are by volume-fraction (V%), which for ideal gases is equal to mole-fraction (that is, fraction of total molecules). By contrast, *mass-fraction* abundances of gases, particularly for gases with significantly different molecular (molar) mass from that of air will differ from those by volume. For example, in air, helium is 5.2 ppm by *volume-fraction* and *mole-fraction*, but only about (4/29) × 5.2 ppm = 0.72 ppm by *mass-fraction*.

Heterosphere

Below the turbopause at an altitude of about 100 km (not far from the mesopause), the Earth's atmosphere has a more-or-less uniform composition (apart from water vapor) as described above; this constitutes the homosphere. However, above about

100 km, the Earth's atmosphere begins to have a composition which varies with altitude. This is essentially because, in the absence of mixing, the density of a gas falls off exponentially with increasing altitude, but at a rate which depends on the molar mass. Thus higher mass constituents, such as oxygen and nitrogen, fall off more quickly than lighter constituents such as helium, molecular hydrogen, and atomic hydrogen. Thus there is a layer, called the heterosphere, in which the earth's atmosphere has varying composition. As the altitude increases, the atmosphere is dominated successively by helium, molecular hydrogen, and atomic hydrogen. The precise altitude of the heterosphere and the layers it contains varies significantly with temperature. After loss of the hydrogen, helium and other hydrogen-containing gases from early Earth due to the Sun's radiation, primitive Earth was devoid of an atmosphere. The first atmosphere was formed by outgassing of gases trapped in the interior of the early Earth, which still goes on today in volcanoes.

Density and Mass

The density of air at sea level is about 1.2 kg/m^3(1.2 g/L). Natural variations of the barometric pressure occur at any one altitude as a consequence of weather. This variation is relatively small for inhabited altitudes but much more pronounced in the outer atmosphere and space due to variable solar radiation. The atmospheric density decreases as the altitude increases. This variation can be approximately modeled using the barometric formula. More sophisticated models are used by meteorologists and space agencies to predict weather and orbital decay of satellites. The average mass of the atmosphere is about 5 quadrillion metric tons or 1/ 1,200,000 the mass of Earth. According to the National Center for Atmospheric Research, "The total mean mass of the atmosphere is 5.1480×10^{18} kg with an annual range due to water vapor of 1.2 or 1.5×10^{15} kg depending on whether surface pressure or water vapor data are used; somewhat smaller than the previous estimate. The mean mass of water vapor is estimated as 1.27×10^{16} kg and the dry air mass as $5.1352 \pm 0.0003 \times 10^{18}$ kg."

Evolution on Earth

The history of the Earth's atmosphere prior to one billion years ago is poorly understood and an active area of scientific research. The following discussion presents a plausible scenario. The modern atmosphere is sometimes referred to as Earth's "third atmosphere", in order to distinguish the current chemical composition from two notably different previous compositions. The original atmosphere was primarily helium and hydrogen. Heat from the still-molten crust, and the sun, plus a probably enhanced solar wind, dissipated this atmosphere. About 4.4 billion years ago, the surface had cooled enough to form a crust, still heavily populated with volcanoes which released steam, carbon dioxide, and ammonia. This led to the early "second atmosphere", which was primarily carbon dioxide and water vapor, with some nitrogen but virtually no oxygen. This second atmosphere had approximately 100 times as much gas as the current atmosphere, but as it cooled much of the carbon dioxide was dissolved in the seas and precipitated out as carbonates. The later "second atmosphere" contained largely nitrogen and carbon dioxide. However, simulations run at the University of Waterloo and University of Colorado in 2005 suggest that it may have had up to 40% hydrogen. It is generally believed that the greenhouse effect, caused by high levels of carbon dioxide and methane, kept the Earth from freezing.

One of the earliest types of bacteria was the cyanobacteria. Fossil evidence indicates that bacteria shaped like these existed approximately 3.3 billion years ago and were the first oxygen-producing evolving phototropic organisms. They were responsible for the initial conversion of the earth's atmosphere from an anoxic state to an oxic state (that is, from a state without oxygen to a state with oxygen) during the period 2.7 to 2.2 billion years ago. Being the first to carry out oxygenic photosynthesis, they were able to produce oxygen while sequestering carbon dioxide in organic molecules, playing a major role in oxygenating the atmosphere.

Photosynthesising plants would later evolve and continue releasing oxygen and sequestering carbon dioxide. Over time, excess carbon became locked in fossil fuels, sedimentary rocks (notably limestone), and animal shells. As oxygen was released, it reacted with ammonia to release nitrogen; in addition, bacteria would also convert ammonia into nitrogen. But most of the nitrogen currently present in the atmosphere results from sunlight-powered photolysis of ammonia released steadily over the aeons from volcanoes. As more plants appeared, the levels of oxygen increased significantly, while carbon dioxide levels dropped. At first the oxygen combined with various elements (such as iron), but eventually oxygen accumulated in the atmosphere, resulting in mass extinctions and further evolution. With the appearance of an ozone layer (ozone is an allotrope of oxygen) lifeforms were better protected from ultraviolet radiation. This oxygen-nitrogen atmosphere is the "third atmosphere". 200 – 250 million years ago, up to 35% of the atmosphere was oxygen (as found in bubbles of ancient atmosphere were found in an amber). This modern atmosphere has a composition which is enforced by oceanic blue-green algae as well as geological processes. O_2 does not remain naturally free in an atmosphere, but tends to be consumed (by inorganic chemical reactions, and by animals, bacteria, and even land plants at night), and CO_2 tends to be produced by respiration and decomposition and oxidation of organic matter. Oxygen would vanish within a few million years due to chemical reactions and CO_2 dissolves easily in water and would be gone in millennia if not replaced. Both are maintained by biological productivity and geological forces seemingly working hand-in-hand to maintain reasonably steady levels over millions of years.

Air Pollution

Air pollution is a chemical, physical (e.g. particulate matter) or biological agent that modifies the natural characteristics of the atmosphere in an unwanted way. Stratospheric ozone depletion due to air pollution (chiefly from chlorofluorocarbons) has long been recognised as a threat to human health as well as to the earth's ecosystems.

Worldwide air pollution is responsible for large numbers of deaths and cases of respiratory disease. Enforced air quality standards, like the Clean Air Act in the United States, have reduced the presence of some pollutants. While major stationary sources are often identified with air pollution, the greatest source of emissions is actually mobile sources, principally the automobile. Gases such as carbon dioxide, methane, and fluorocarbons contribute to global warming, and these gases, or excess amounts of some emitted from fossil fuel burning, have recently been identified by the United States and many other countries (see Kyoto accord), as pollutants.

Air

Air means Earth's atmosphere. It is the clear gas we live in and breathe in. It has no color or smell. It has weight. Air creates atmosphere pressure. There is no air in the vacuum and cosmos. Air is a mixture of 78.03% Nitrogen, 20.99% Oxygen, 0.94% Argon, 0.03% Carbon Dioxide, 0.01% Hydrogen, 0.00123% Neon, 0.0004% Helium, 0.00005% Krypton, 0.000006% Xenon. Humans need the oxygen in the air to live. In our bodies, the lungs give oxygen to the blood, and give back carbon dioxide to the air. Plants and animals also need air to live. Air is very important to everyone. The wind is moving air. Air can be polluted from gas, smoke, and ash. This pollution may be one of the causes of global warming.Air is also something that planes fly through. There are three things in air, Nitrogen (79%), oxygen (20%), and other gasses (1%) warm air goes up while cold air goes down.

Particulate

Particulates, alternatively referred to as particulate matter (PM) or fine particles, are tiny particles of solid or liquid suspended in a gas. In contrast, aerosol refers to particles and the gas together. Sources of particulate matter can be man made or natural. Some particulates occur naturally, originating from volcanoes, dust storms, forest and grassland fires, living vegetation, and sea spray. Human activities, such as the burning of fossil fuels in vehicles, power plants and various industrial processes also generate significant amounts of aerosols. Averaged over the globe, *anthropogenic* aerosols—those made by human activities—currently account for about 10 percent of the total amount of aerosols in our atmosphere. Increased levels of fine particles in the air are linked to health hazards such as heart disease, altered lung function and lung cancer.

Scale Classification

Among the most common categorisations imposed on particulates are those with respect to size, referred to as fractions. As particles are often non-spherical (for example, Asbestos fibers), there are many definitions of particle size. The most widely used definition is the *aerodynamic diameter*. A particle with an aerodynamic diameter of 10 micrometers moves in a gas like a sphere of unit density (1 gram per cubic centimeter) with a diameter of 10 micrometers. PM diameters range from less than 10 nanometers to more than 100 micrometers. These dimensions represent the continuum from a few molecules up to the size where particles can no longer be carried by a gas. The notation PM_{10} is used to describe particles of 10 micrometers or less and $PM_{2.5}$ represents particles less than 2.5 micrometers in aerodynamic diameter. But because no sampler is perfect in the sense that no particle larger than its cutoff diameter passes the inlet, all reference methods allow a high margin of error. These are also sometimes referred to with other equivalent numeric values. These are the formal definitions. Depending on the context, alternative definitions may be applied. In some specialised settings, each fraction may exclude the fractions of lesser scale, so that PM_{10} excludes particles in a smaller size range, e.g. $PM_{2.5}$, usually reported separately in the same work. Such a case is sometimes emphasised with the difference notation, e.g. PM_{10}-$PM_{2.5}$. Other exceptions may be similarly specified. This is useful when not only the upper bound of a fraction is relevant to a discussion. The facts that some particle size ranges require greater filter strength and the smallest ones can outstrip the body's ability to keep them out of cells both serve to guide understanding of related public policy, environment, and health topics.

Sources

There are both natural and human sources of atmospheric particulates. The biggest natural sources are dust, volcanoes, and forest fires. Sea spray is also a large source of particles though most of these fall back to the ocean close to where they were emitted. The biggest human sources of particles are combustion sources, mainly the burning of fuels in internal combustion engines in automobiles and power plants, and wind blown dust from construction sites and other land areas where the water or vegetation has been removed. Some of these particles are emitted directly to the atmosphere (primary emissions) *and some are emitted as gases and form particles in the atmosphere* (secondary emissions).

Composition

The composition of aerosol particles depends on their source. Wind-blown mineral dust tends to be made of mineral oxides and other material blown from the Earth's crust; this aerosol is light-absorbing. Sea salt is considered the second-largest contributor in the global aerosol budget, and consists mainly of sodium chloride originated from sea spray; other constituents of atmospheric sea salt reflect the composition of sea water, and thus include magnesium, sulfate, calcium, potassium, etc. In addition, sea spray aerosols may contain organic compounds, which influence their chemistry. Sea salt does not absorb.

Secondary particles derive from the oxidation of primary gases such as sulfur and nitrogen oxides into sulfuric acid (liquid) and nitric acid (gaseous). The precursors for these aerosols—i.e. the gases from which they originate—may have an anthropogenic origin (from fossil fuel combustion) and a natural biogenic origin. In the presence of ammonia, secondary aerosols often take the form of ammonium salts; i.e. ammonium sulfate and ammonium nitrate (both can be dry or in aqueous solution); in the absence of ammonia, secondary compounds take an acidic form as sulfuric acid (liquid aerosol droplets) and nitric acid (atmospheric gas). Secondary sulfate and nitrate aerosols are strong light-scatterers. This is mainly because the presence of sulfate and nitrate causes the aerosols to increase to a size that scatters light effectively.

Organic matter (OM) can be either primary or secondary, the latter part deriving from the oxidation of VOCs; organic material in the atmosphere may either be biogenic or anthropogenic. Organic matter influences the atmospheric radiation field by both scattering and absorption. Another important aerosol type is constitude of elemental carbon (EC, also known as black carbon, *BC): this aerosol type includes strongly light-absorbing material and is thought to yield large positive radiative forcing. Organic matter and elemental carbon together constitute the carbonaceous fraction of aerosols.ii*

The chemical composition of the aerosol directly affects how it interacts with solar radiation. The chemical constituents within the aerosol change the overall refractive index. The refractive index will determine how much light is scattered and absorbed.

Removal Processes

In general, the smaller and lighter a particle is, the longer it will stay in the air. Larger particles (greater than 10 micrometers in diameter) tend to settle to the ground by gravity in a matter of hours whereas the smallest particles (less than 1 micrometer) can stay in the atmosphere for weeks and are mostly removed by precipitation.

Radiative Forcing From Aerosols

Aerosols, natural and anthropogenic, can affect the climate by changing the way radiation is transmitted through the atmosphere. Direct observations of the effects of aerosols are quite limited so any attempt to estimate their global effect necessarily involves the use of computer models. The Intergovernmental Panel on Climate Change, IPCC, says: While the radiative forcing due to greenhouse gases may be determined to a reasonably high degree of accuracy... the uncertainties relating to aerosol radiative forcings remain large, and rely to a large extent on the estimates from global modelling studies that are difficult to verify at the present time.

A graphic showing the contributions (at 2000, relative to pre-industrial) and uncertainties of various forcings is available here.

Sulfate Aerosol

Sulfate aerosol has two main effects, direct and indirect. The direct effect, via albedo, is to cool the planet: the IPCC's best estimate of the radiative forcing is -0.4 watts per square meter with a range of -0.2 to -0.8 W/m^2 but there are substantial uncertainties. The effect varies strongly geographically, with most cooling believed to be at and downwind of major industrial centres. Modern climate models attempting to deal with the attribution of recent climate change need to include sulfate forcing, which appears to account (at least partly) for the slight drop in global temperature in the middle of the 20th century. The indirect effect (via the aerosol acting as cloud condensation nuclei, CCN, and thereby modifying the cloud properties -albedo and lifetime-) is more uncertain but is believed to be a cooling.

Black Carbon

Black carbon (BC), or carbon black, or elemental carbon (EC), often called soot, is composed of pure carbon clusters, skeleton balls and buckyballs, and is one of the most important absorbing aerosol species in the atmosphere. It should be distinguished from organic carbon (OC): clustered or aggregated organic molecules on their own or permeating an EC buckyball. BC from fossil fuels is estimated by the IPCC in the Fourth Assessment Report of the IPCC, TAR, to contribute a global mean radiative forcing of +0.2 W/m^2 (was +0.1 W/m^2 in the Second Assessment Report of the IPCC, SAR), with a range +0.1 to +0.4 W/m^2.

All aerosols both absorb and scatter solar and terrestrial radiation. If a substance absorbs a significant amount of radiation, as well as scattering, it is called absorbing. This is quantified in the *Single Scattering Albedo* (SSA), the ratio of scattering alone to scattering plus absorption (*extinction*) of radiation by a particle. The SSA tends to unity if scattering dominates, with relatively little absorption, and decreases as absorption increases, becoming zero for infinite absorption. For example, sea-salt aerosol has an SSA of 1, as a sea-salt particle only scatters, whereas soot has an SSA of 0.23, showing that it is a major atmospheric aerosol absorber.

Health Effects

The effects of inhaling particulate matter has been widely studied in humans and animals and include asthma, lung cancer, cardiovascular issues, and premature death. The size of the particle is a main determinant of where in the respiratory tract the particle will come to rest when inhaled. Larger particles are generally filtered in the nose and throat and do not cause problems, but particulate matter smaller than about 10 micrometres, referred to as PM_{10}, *can settle in the bronchi and lungs and cause health problems. The 10 micrometer size does not represent a strict boundary between respirable and non-respirable particles, but has been agreed upon for monitoring of airborne particulate matter by most regulatory agencies. Similarly, particles smaller than 2.5 micrometres,* $PM_{2.5}$, *tend to penetrate into the gas-exchange regions of the lung, and very small particles (< 100 nanometers) may pass through the lungs to affect other organs. In particular, a study published in the* Journal of the American Medical Association *indicates that* $PM_{2.5}$ *leads to high plaque deposits in arteries, causing vascular inflammation and atherosclerosis — a hardening of the arteries that reduces elasticity, which can lead to heart attacks and other cardiovascular problems. Researchers suggest that even short-term exposure at elevated concentrations could significantly contribute to heart disease.*

Researchers at the Johns Hopkins Bloomberg School of Public Health have conducted the largest nationwide study on the acute health effects of coarse particle pollution. Coarse particles are airborne pollutants that fall between 2.5 and 10 microns in diameter. The study, published in the May 14, 2008, edition of JAMA, found evidence of an association with hospital admissions for cardiovascular diseases but no evidence of an association with the number of hospital admissions for respiratory diseases. After taking into account fine particle levels, the association with coarse particles remained but was no longer statistically significant.

*The smallest particles, less than 100 nanometers (nanoparticles), may be even more damaging to the cardiovascular system. There is evidence that particles smaller than 100 nanometres can pass through cell membranes and migrate into other organs, including the brain. It has been suggested that particulate matter can cause similar brain damage as that found in Alzheimer patients. Particles emitted from modern diesel engines (commonly referred to as Diesel Particulate Matter, or DPM) are typically in the size range of 100 nanometres (*0.1 micrometres*). In addition, these soot particles also carry carcinogenic components like benzopyrenes adsorbed on their surface. It is becoming increasingly clear that the legislative limits for engines, which are in terms of emitted mass, are not a proper measure of the health hazard. One particle of 10 μm diameter has approximately the same mass as 1 million particles of 100 nm diameter, but it is clearly much less hazardous, as it probably never enters the human body - and if it does, it is quickly removed. Proposals for new regulations exist in some countries, with suggestions to limit the particle surface area or the particle number.*

The large number of deaths and other health problems associated with particulate pollution was first demonstrated in the early 1970s and has been reproduced many times since. PM pollution is estimated to cause 22,000-52,000 deaths per year in the United States (from 2000) and 200,000 deaths per year in Europe.

Regulation

Due to the health effects of particulate matter, maximum standards have been set by various governments. Many urban areas in the U.S. and Europe still frequently violate the particulate standards, though urban air on these continents has become cleaner, on average, with respect to particulates over the last quarter of the 20th century.

United States

The United States Environmental Protection Agency (EPA) sets standards for PM_{10} and $PM_{2.5}$ concentrations in urban air. (See National Ambient Air Quality Standards.) EPA regulates primary particulate emissions and precursors to secondary emissions (NOx, sulfur, and ammonia).

Affected Areas

The most concentrated particulate matter pollution tends to be in densely populated metropolitan areas in developing countries. The primary cause is the burning of fossil fuels by transportation and industrial sources.

Aerosol Science

The field of aerosol science and technology has grown in response to the need to understand and control natural and manmade aerosols.

Sulfur Dioxide

Sulfur dioxide is the chemical compound with the formula SO_2. SO_2 is produced by volcanoes and in various industrial processes. Since coal and petroleum often contain sulfur compounds, their combustion generates sulfur dioxide. Further oxidation of SO_2, usually in the presence of a catalyst such as NO_2, forms H_2SO_4, and thus acid rain. This is one of the causes for concern over the environmental impact of the use of these fuels as power sources.

Preparation

Sulfur dioxide can be prepared by burning sulfur:

$$S_8 + 8\ O_2\ \text{'!}\ 8\ SO_2$$

The combustion of hydrogen sulfide and organosulfur compounds proceeds similarly.

$2\ H_2S(g) + 3\ O_2(g)$ '! $2\ H_2O(g) + 2\ SO_2(g)$

The roasting of sulfide ores such as iron pyrites, sphalerite (zinc blende) and cinnabar (mercury sulfide) also releases SO_2:

$4\ FeS_2(s) + 11\ O_2(g)$ '! $2\ Fe_2O_3(s) + 8\ SO_2(g)$

$2\ ZnS(s) + 3\ O_2(g)$ '! $2\ ZnO(s) + 2\ SO_2(g)$

$HgS(s) + O_2(g)$ '! $Hg(g) + SO_2(g)$

Sulfur dioxide is a by-product in the manufacture of cement: $CaSiO_3$ and $CaSO_4$ is heated with coke and sand in this process:

$2\ CaSO_4(s) + 2SiO_2(s) + C(s)$ '! $2\ CaSiO_3(s) + 2\ SO_2(g) + CO_2(g)$

Action of hot sulfuric acid on copper turnings produces sulfur dioxide.

$Cu(s) + 2H_2SO_4(aq)$ '! $CuSO_4(aq) + SO_2(g) + 2H_2O(l)$

Structure and Bonding

SO_2 is a bent molecule with C_{2v} symmetry point group. In terms of electron-counting formalisms, the sulfur atom has an oxidation state of +4, a formal charge of 0, and is surrounded by 5 electron pairs and can be described as a hypervalent molecule. From the perspective of molecular orbital theory, most of these valence electrons are engaged in S-O bonding.

The S-O bonds are shorter in SO_2 (143.1 pm) than in sulfur monoxide, SO (148.1 pm), whereas the O-O bonds are longer in O_3 (127.8 pm) than in dioxygen, O_2 (120.7 pm). The mean bond energy is greater in SO_2 (548 kJ mol^{+1}) than in SO (524 kJ mol^{+1}), whereas its is less in O_3 (297 kJ mol^{+1}) than in O_2 (490 kJ mol^{+1}). These pieces of evidence lead chemists to conclude that the S-O bonds in sulfur dioxide have a bond order of at least 2, unlike the O-O bonds in ozone, which have a bond order of 1.5.

Reactions

Treatment of basic solutions with sulfur dioxide affords sulfite salts:

$SO_2 + 2\ NaOH$ '! $Na_2SO_3 + H_2O$

Featuring sulfur in the +4 oxidation state, sulfur dioxide is a reducing agent. It is oxidised by halogens such as chlorine to give the sulfuryl halides:

$SO_2 + Cl_2$ '! SO_2Cl_2

However, on rare occasions, it can also act as an oxidising agent: in the Claus process, sulfur dioxide is reduced by hydrogen sulfide to give elemental sulfur:

$$SO_2 + 2\,H_2S\ \text{'!}\ 3\,S + 2\,H_2O$$

Sulfur dioxide can act as a metal binding ligand, typically where the transition metal is in oxidation state 0 or +1. Up to 9 different bonding modes have been determined which include:

- S donation -planar and pyramidal ç[1]
- O donation ç[1]
- ð donation- side on ç[2]
- S bridging across two metal centres or two ends of a metal-metal bon
- O-S bridging to two metal centres
- bridging- one metal centre ð donation- side on, the other metal center O donation
- bridging over three metal centres

Uses

As a Preservative

Sulfur dioxide is sometimes used as a preservative for dried apricots and other dried fruits due to its antimicrobial properties, it is sometimes called E220 when used in this way. The preservative is used to maintain the appearance of the fruit and prevent rotting. Its presence can give fruit a distinctive chemical taste.

In Winemaking

Sulfur dioxide is a very important compound in winemaking, and is designated as parts per million in wine, E number: E220. It is present even in so-called unsulphurated wine at concentrations of up to 10 milligrams per litre. It serves as an antibiotic and antioxidant, protecting wine from spoilage by bacteria and oxidation. It also helps to keep volatile acidity at desirable levels. Sulfur dioxide is responsible for the words "contains sulfites" found on wine labels. Wines with SO_2 concentrations below 10ppm do not require "contains sulfites" on the label by US and EU laws. The upper limit of SO_2 allowed in wine is 350ppm in US, in the EU is 160 ppm for red wines and 210 ppm for white and rosé wines. In low concentrations SO_2 is mostly undetectable in wine, but at over 50ppm, SO_2 becomes evident in the nose and taste of wine.

SO_2 is also a very important element in winery sanitation. Wineries and equipment must be kept very clean, and because bleach cannot be used in a winery, a mixture of SO_2, water, and citric acid is commonly used to clean hoses, tanks, and other equipment to keep it clean and free of bacteria.

As a Reducing Bleach

Sulfur dioxide is also a good reductant. In the presence of water, sulfur dioxide is able to decolorize substances. Specifically it is a useful reducing bleach for papers and delicate materials such as clothes. This bleaching effect normally does not last very long. Oxygen in the atmosphere reoxidizes the reduced dyes, restoring the color.

Precursor to Sulfuric Acid

Sulfur dioxide is also used to make sulfuric acid, being converted to sulfur trioxide, and then to oleum, which is made into sulfuric acid. Sulfur dioxide for this purpose is made when sulfur combines with oxygen. The method of converting sulfur dioxide to sulfuric acid is called the contact process.

Biochemical and Biomedical Roles

Sulfur dioxide is toxic in large amounts. It or its conjugate base bisulfite is produced biologically as an intermediate in both sulfate-reducing organisms and in sulfur oxidizing bacteria as well. Sulfur dioxide has no role in mammalian biology. Sulfur dioxide blocks nerve signals from the pulmonary stretch receptors (PSR's) and abolishes the Hering-Breuer inflation reflex.

As a Refrigerant

Being easily condensed and with a high heat of evaporation, sulfur dioxide is a candidate material for refrigerants. Prior to the development of freons, sulfur dioxide was used as a refrigerant in home refrigerators.

As a Reagent and Solvent

Sulfur dioxide is a versatile inert solvent that has been widely used for dissolving highly oxidizing salts. It is also used occasionally as a source of the sulfonyl group in organic synthesis. Treatment of aryldiazonium salts with sulfur dioxide affords the corresponding aryl sulfonyl chloride.

Dechlorination

In municipal wastewater treatment sulfur dioxide is used to treat chlorinated wastewater prior to release. Sulfur dioxide reacts with free and combined chlorine to form negatively charged chloride ions.

Emissions

According to the U.S. EPA (as presented by the *2002 World Almanac* or in chart form), the following amount of sulfur dioxide was released in the U.S. per year, measured in thousands of short tons:

1999	18,867
1998	19,491
1997	19,363
1996	18,859
1990	23,678
1980	25,905
1970	31,161

Due largely to the US EPA's Acid Rain Programme, the U.S. has witnessed a 33 percent decrease in emissions between 1983 and 2002. This improvement resulted from flue gas desulfurisation, a technology that enables SO_2 to be chemically bound in power plants burning sulfur-containing coal or oil. In particular, calcium oxide (lime) reacts with sulfur dioxide to form calcium sulfite:

$$CaO + SO_2 \text{ '! } CaSO_3$$

Aerobic oxidation converts this $CaSO_3$ into $CaSO_4$, gypsum. Most gypsum sold in Europe comes from flue gas desulfurisation. New fuel additive catalysts, such as ferox, are being used in gasoline and diesel engines in order to lower the emission of sulfur oxide gases into the atmosphere. This is also done by forcing the sulfur into stable mineral salts and mixed mineral sulfates as opposed to sulfuric acid and sulfur oxides. As of 2006, China is the world's largest sulfur dioxide polluter, with 2005 emissions estimated to be 25.49 million tons. This amount represents a 27% increase since 2000, and is roughly comparable with U.S. emissions in 1980. Al-Mishraq, an Iraqi sulfur plant, was the site of a 2003 disaster resulting in the release of massive amounts of sulfur dioxide into the atmosphere.

Temperature Dependence of Aqueous Solubility

22 g/100ml (0 °C)	15 g/100ml (10 °C)
11 g/100ml (20 °C)	9.4 g/100 ml (25 °C)
8 g/100ml (30 °C)	6.5 g/100ml (40 °C)
5 g/100ml (50 °C)	4 g/100ml (60 °C)
3.5 g/100ml (70 °C)	3.4 g/100ml (80 °C)
3.5 g/100ml (90 °C)	3.7 g/100ml (100 °C)

- The values are tabulated for 101.3 kPa partial pressure of SO_2. Solubility of gas in a liquid depends on the gas partial pressure according to Henry's law.
- The solublity is given for "pure water", i.e., water that contains only SO_2 in the amount at equilibrium with the gas phase. This "pure water" is going to be acidic. The solublity of SO_2 in neutral (or alkaline) water is generally going to be higher because of the pH-dependent speciation of SO_2 in the solution with the production of bisulfite and some sulfite ions.

Threats to Health

Sulfur dioxide acts as an acid. Inhalation results in labored breathing, coughing, and/or a sore throat and may cause permanent pulmonary damage. When mixed with water and contacted by skin, frostbite may occur. When it makes contact with eyes, redness and pain will occur.

Nitrogen Oxide

The term nitrogen oxide typically refers to any binary compound of oxygen and nitrogen, or to a mixture of such compounds:

- Nitric oxide (NO), nitrogen(II) oxide
- Nitrogen dioxide (NO_2), nitrogen(IV) oxide
- Nitrous oxide (N_2O), nitrogen (I) oxide
- Dinitrogen trioxide (N_2O_3), nitrogen(II, IV) oxide
- Dinitrogen tetroxide (N_2O_4), nitrogen(IV) oxide
- Dinitrogen pentoxide (N_2O_5), nitrogen(V) oxide

(Note that the last three are unstable.)

Chemical reactions that produce nitrogen oxides often produce several, the proportions depending on the specific reaction and conditions. This is one reason why secondary production of N_2O is undesirable; the other two stable oxides — which are extremely toxic — are liable to be produced.

NO_x

This section refers to the chemical term for nitrogen oxides produced during combustion. For other definitions see Nox

NO_x is a generic term for mono-nitrogen oxides (NO and NO_2). These oxides are produced during combustion, especially combustion at high temperatures. At ambient temperatures, the oxygen and nitrogen gases in air will not react with each other. In an internal combustion engine, combustion of a mixture of air and fuel produces combustion temperatures high enough to drive endothermic reactions between atmospheric nitrogen and oxygen in the flame, yielding various oxides of nitrogen. In areas of high motor vehicle traffic, such as in large cities, the amount of nitrogen oxides emitted into the atmosphere can be quite significant.

In the presence of excess oxygen (O_2), nitric oxide (NO) will be converted to nitrogen dioxide (NO_2), with the time required dependent on the concentration in air as shown below:

NO concentration in air (ppm)	Time required for half NO to be oxidised to NO_2 (min)
20,000	0.175
10,000	0.35
1,000	3.5
100	35
10	350
1	3500

When NO_x and volatile organic compounds (VOCs) react in the presence of sunlight, they form photochemical smog, a significant form of air pollution, especially in the summer. Children, people with lung diseases such as asthma, and people who work or exercise outside are susceptible to adverse effects of smog such as damage to lung tissue and reduction in lung function.

Mono-nitrogen oxides eventually form nitric acid when dissolved in atmospheric moisture, forming a component of acid rain. The following chemical reaction occurs when nitrogen dioxide reacts with water:

$2NO_2 + H_2O$ '! $HNO_2 + HNO_3$
(nitrogen dioxide + water '! nitrous acid + nitric acid).

Nitrous acid then decomposes as follows:

$3HNO_2$ '! $HNO_3 + 2NO + H_2O$
(nitrous acid '! nitric acid + nitric oxide + water),

where nitric oxide will oxidize to form nitrogen dioxide that again reacts with water, ultimately forming nitric acid:

$4NO + 3O_2 + 2H_2O$ '! $4HNO_3$
(nitric oxide + oxygen + water '! nitric acid).

Mono-nitrogen oxides are also involved in tropospheric production of ozone.

NO_x should not be confused with NOS, a term used to refer to nitrous oxide (N_2O) in the context of its use as a power booster for internal combustion engines.

Definition of NO_x and NO_y in Atmospheric Chemistry

In atmospheric chemistry the term NO_x is used to mean the total concentration of NO plus NO_2. During daylight NO and NO_2 are in equilibrium with the ratio NO/NO_2 determined by the intensity of sunshine (which converts NO_2 to NO) and the concentration of ozone (which reacts with NO to give back NO_2). NO and NO_2 are also central to the formation of tropospheric ozone. This definition excludes other oxides of nitrogen such as nitrous oxide (N_2O). NO_y (reactive odd nitrogen) is defined as the sum of NO_x plus the compounds produced from the oxidation of NO_x which include nitric acid and peroxyacetyl nitrate. In this context nitrous oxide and ammonia are not considered as reactive nitrogen compounds.

Industrial Sources of NO_x

The three primary sources of NO_x in combustion processes:

- thermal NO_x
- fuel NO_x
- prompt NO_x

Thermal NO_x formation, which is highly temperature dependent, is recognised as the most relevant source when combusting natural gas. Fuel NO_x tends to dominate during the combustion of fuels, such as coal, which have a significant nitrogen content, particularly when burned in combustors designed to minimise thermal NO_x. The contribution of prompt NO_x is normally considered negligible. A fourth source, called *feed NO_x* is associated with the combustion of nitrogen present in the feed material of cement rotary kilns, at between 300° and 800°C, where it is also a minor contributor.

Thermal NO_x

Thermal NO_x refers to NO_x formed through high temperature oxidation of the diatomic nitrogen found in combustion air. The formation rate is primarily a function of temperature and the residence time of nitrogen at that temperature. At high temperatures, usually above 1600°C (2900°F), molecular nitrogen (N_2) and oxygen (O_2) in the combustion air disassociate into their atomic states and participate in a series of reactions.

The three principal reactions producing thermal NO_x are:

(Extended Zeldovich Mechanism)

- N_2 + O '! NO + N
- N + O_2 '! NO + O
- N + OH '! NO + H

all 3 reactions are reversible. Zeldovich was the first to suggest the importance of the first two reactions. The last reaction of atomic Nitrogen with Hydroxyl radical, OH, was added by Lavoie, Heywood and Keck to the mechanism and makes a significiant contribution to the formation of thermal NO_x.

Fuel NO_x

The major source of NO_x production from nitrogen-bearing fuels such as certain coals and oil, is the conversion of fuel bound nitrogen to NO_x during combustion. During combustion, the nitrogen bound in the fuel is released as a free radical and ultimately forms free N_2, or NO. Fuel NO_x can contribute as much as 50% of total emissions when combusting oil and as much as 80% when combusting coal.

Although the complete mechanism is not fully understood, there are two primary paths of formation. The first involves the oxidation of volatile nitrogen species during the initial stages of combustion. During the release and prior to the oxidation of the volatiles, nitrogen reacts to form several intermediaries which are then oxidised into NO. If the volatiles evolve into a reducing atmosphere, the nitrogen evolved can readily be made to form nitrogen gas, rather than NO_x. The second path involves the combustion of nitrogen contained in the char matrix during the combustion of the char portion of the fuels. This reaction occurs much more slowly than the volatile phase. Only around 20% of the char nitrogen is ultimately emitted as NO_x, since much of the NO_x that forms during this process is reduced to nitrogen by the char, which is nearly pure carbon.

It is possible revert NO_x emitted from a diesel engine back into Nitrogen and Water by using an SCR Selective Catalytic Reduction Unit. Ths requires the addition of a urea based liquid re-agent namely AdBlue to initialise the chemical reaction.

Prompt NO_x

This third source is attributed to the reaction of atmospheric nitrogen, N_2, with radicals such as C, CH, and CH_2 fragments derived from fuel, where this cannot be explained by either the aforementioned thermal or fuel processes. Occurring in the earliest stage of combustion, this results in the formation of fixed species of nitrogen such as NH (nitrogen monohydride), HCN (hydrogen cyanide), H_2CN (dihydrogen cyanide) and CN- (cyano radical) which can oxidize to NO. In fuels that contain nitrogen, the incidence of prompt NO_x is especially minimal and it is generally only of interest for the most exacting emission targets.

Regulation and Emission Control Technologies

The Kyoto Protocol, ratified by 54 nations in 1997, classifies N_2O as a greenhouse gas, and calls for substantial worldwide reductions in its emission.

As discussed above, atmospheric NO_x eventually forms nitric acid, which contributes to acid rain. NO_x emissions are regulated in the United States by the Environmental Protection Agency, and in the UK by the Department for Environment, Food and Rural Affairs.

Technologies such as flameless oxidation (FLOX) and staged combustion significantly reduce thermal NO_x in industrial processes. Bowin low NO_x technology is a hybrid of staged-premixed-radiant combustion technology with a major surface combustion preceded by a minor radiant combustion. In the Bowin burner, air and fuel gas are premixed at a ratio greater than or equal to the stoichiometric combustion requirement. Water Injection technology, wherby water is introduced into the combustion chamber, is also becoming an important means of NO_x reduction through increased efficiency in the overall combustion process. Alternatively, the water (e.g. 10 to 50%) is emulsified into the fuel oil prior to the injection and combustion. This emulsification can either be made in-line (unstabilized) just before the injection or as a drop-in fuel with chemical additives for long term emulsion stability (stabilized). Other technologies, such as selective catalytic reduction (SCR) and selective non-catalytic reduction (SNCR) reduce post combustion NO_x.

The use of Exhaust gas recirculation and catalytic converters in motor vehicle engines have significantly reduced emissions.

Volatile Organic Compound

Volatile organic compounds (VOCs) are organic chemical compounds that have high enough vapor pressures under normal conditions to significantly vaporize and enter the atmosphere. A wide range of carbon-based molecules, such as aldehydes, ketones, and other light hydrocarbons are VOCs. The term often is used in a legal or regulatory context and in such cases the precise definition is a matter of law. These definitions can be contradictory and may contain "loopholes"; e.g. exceptions, exemptions, and exclusions. The United States Environmental Protection Agency defines a VOC as any organic compound that participates in a photoreaction; others believe this definition is very broad and vague as organics that are not volatile in the sense that they vaporize under normal conditions can be considered volatile by this EPA definition. The term may refer both to well characterised organic compounds and to mixtures of variable composition.

Sources of VOCs

The most common VOC is methane, a greenhouse gas sometimes excluded from analysis of other VOCs using the term non-methane VOCs, or NMVOCs. Other terms used in pollution regulations include NMOG - "non-methane organic gas", and NMHC - "non-methane hydrocarbons".

Major worldwide sources of atmospheric methane include wetlands, ruminants such as cows, energy use, rice agriculture, landfills, and burning biomass such as wood. Methane is the primary component of natural gas. Common artificial VOCs include paint thinners, dry cleaning solvents, and some constituents of petroleum fuels (eg. gasoline and natural gas). Trees are also an important biological source of VOC; it is known that they emit large amounts of VOCs, especially isoprene and terpenes. Another significant source of VOC emission is crude oil tanking. Both during offloading and loading of crude oil tankers VOC are released to the atmosphere. Lately, there has been an environmental focus on this issue resulting in improved VOC handling on newer tankers, and crude oil loading terminals. Considered a factor in indoor air quality issues such as sick building syndrome, VOCs "are generated by photocopiers, carpets, and furnishings as they are used or when components oxidize.... One irritant, formaldehyde, present in hundreds of office components, including wood and laminated furniture, shelving, and wall covers. It also evaporates from paints, varnishes, and chemicals used for sealing and finishing walls." Tobacco smoke can contribute high levels of VOCs. VOCs including halogenide and sulfide are emitted through human respiration, and formaldehyde is emitted at a lower rate from the surface of the human body.

Environmental Effects

VOCs are sometimes accidentally released into the environment, where they can damage soil and groundwater. Vapours of VOCs escaping into the air contribute to air pollution.

VOCs are an important outdoor air pollutant. In this field they are often divided into the separate categories of methane ((CH_4) and non-methane (NMVOCs). Methane is an extremely efficient greenhouse gas which contributes to enhanced global warming. Other hydrocarbon VOCs are also significant greenhouse gases via their role in creating ozone and in prolonging the life of methane in the atmosphere, although the effect varies depending on local air quality. Within the NMVOCs, the aromatic compounds benzene, toluene and xylene are suspected carcinogens and may lead to leukemia through prolonged exposure. 1,3-butadiene is another dangerous compound which is often associated with industrial uses.

Some VOCs also react with nitrogen oxides in the air in the presence of sunlight to form ozone. Although ozone is beneficial in the upper atmosphere because it absorbs UV thus protecting humans, plants, and animals from exposure to dangerous solar radiation, it poses a health threat in the lower atmosphere by causing respiratory problems. In addition high concentrations of low level ozone can damage crops and buildings.

Contribution to Indoor Air Pollution

Many VOCs found around the house, such as paint strippers and wood preservatives, contribute to sick building syndrome because of their high vapour pressure. VOC's are often used in paint, carpet backing, plastics, and cosmetics. The United States Environmental Protection Agency (EPA) has found concentrations of VOCs in indoor air to be 2 to 5 times greater than in outdoor air. During certain activities indoor levels of VOCs may reach 1,000 times that of the outside air. Not all organic compounds are volatile; many plastics (polymers) and other large molecules may not have significant vapor pressure at normal temperatures. Air quality with reference to Volatile Organic Compound Emission.

Terminology and Legal Definitions

There are a number of different ways to collectively refer to those chemical compounds that participate in photochemical reactions. That is, those that react with other pollutants, in the presence of sunlight, to form tropospheric ozone.

Some of the more common terms are:

- NMHC — Non-Methane Hydrocarbons
- NMOG — Non-Methane Organic Gases
- NMVOC — Non-Methane Volatile Organic Compounds
- ROG — Reactive Organic Gases
- SVOC — Semi-Volatile Organic Compounds
- TOG — Total Organic Gases
- TVOC — Total Volatile Organic Compounds
- VOC — Volatile Organic Compounds

While all these terms are used, it is not always clear which pollutants are included in each term. The term "VOC" has the advantage of having precise definitions codified by regulators such as the European Parliament and the US EPA.

Worldwide, legal definitions of the term "VOC" are in many respects, more a matter of policy than a matter of science. For example, because the US EPA Code of Federal Regulations (CFR) has characterised a compound as having "negligible photochemical reactivity" it does not necessarily imply that it is, at any particular time, less reactive than those compounds which are not on the list. Since first establishing the list of exempt compounds in 1977, the EPA has added several to the list, and frequently has several petitions undergoing review.

The traditional US standard to determine if a compound is a non-VOC is to compare its reactivity to that of ethane, which was the least reactive compound on the original list. Unfortunately, this is a very difficult comparison to make as it is frequently impossible to duplicate the real-world conditions in a laboratory. To complicate the issue, typical real-world conditions are different from day to day and from place to place. However, there is ongoing study on the use of a compound's reactivity as a better tool for pollution control regulation than the "is or isn't" approach currently in use.

United States Definition

The U.S. Environmental Protection Agency definition of VOCs is published in the Code of Federal Regulations. It defines VOCs as "any compound of carbon, excluding carbon monoxide, carbon dioxide, carbonic acid, metallic carbides or carbonates, and ammonium carbonate, which participates in atmospheric photochemical reactions," but also includes a list of dozens of exceptions for compounds "determined to have negligible photochemical reactivity."

European Union Definition

Under European law, the definition of "VOC" is based on evaporation into the atmosphere, rather than reactivity. For example European Union Directive 2004/42/CE which covers VOC emissions from paints and varnishes defines a VOC as any organic compound having an initial boiling point less than or equal to 250°C measured at a standard atmospheric pressure of 101.3 kPa. Directive 94/63/EC which regulates VOC emissions from storage and distribution of petrol simply defines *vapours* as any gaseous compound which evaporates from petrol.

UK Coatings Classification

The British coatings industry has adopted a VOC labelling scheme for all decorative coatings to inform customers about the levels of organic solvents and other volatile materials present. Coatings manufacturers use standard terminology, text and categories for all products. Information is provided according to five 'bands,' and manufacturers may label products with either a British Coatings Federation text box on the back panel, or a graphical globe symbol, the latter subject to licensing from B&Q plc. Both styles of labels contain the same text, and warn that VOCs contribute to atmospheric pollution.

The five bands are:

1. Minimal - VOC content 0% to 0.29%
2. Low - VOC content 0.3% to 7.99%
3. Medium - VOC content 8% to 24.99%
4. High - VOC content 25% to 50%
5. Very High - VOC content more than 50%

An example of text box labelling for the Minimal band is shown below, while examples of the graphical globe symbols may be seen on websites of some British coatings companies.

Air Quality Index

The Air Quality Index (AQI) is a standardised indicator of the air quality in a given location. It measures mainly ground-level ozone and particulates (except the pollen count), but may also include sulfur dioxide, and nitrogen dioxide. Various agencies around the world measure such indices, though definitions may change between places.

The United States Environmental Protection Agency (EPA) and the Meteorological Service of Canada (MSC) differ on what AQI structure and health classification is used:

Health classifications used by the EPA:

- 0-50 Good is usually green
- 51-100 Moderate is usually yellow
- 101-150 Unhealthy for sensitive groups is usually orange
- 151-200 Unhealthy is usually red
- 201-300 Very unhealthy is usually purple
- 301-500 Hazardous is usually maroon

The EPA's AQI 100 corresponds to 0.08 ppm ozone, and to other levels for other pollutants. The AQI standards in Canada are relatively more stringent. The current health classifications used by the Meteorological Service of Canada (MSC) are as follows:

- 0-25: Good (green)
- 26-50: Fair (yellow)
- 51-100: Poor (orange)
- 101+: Very poor (red)

In Ontario, 31 is the upper limit for good and 32 the lower limit for moderate. Zero to 15 is classified as very good, and is given the color blue. In June 2007, the EPA proposed a slight possible tightening of the pollution standards associated with smog after an independent EPA scientific board said that the standard "needs to be substantially reduced" and that there is "no scientific justification" for retaining the current, weaker standard. In light of the new scientific findings, one should expect adjustments in the AQI such that pollution currently denoted as "moderate" will in the future be recognised as "unhealthy." The AQI can worsen (go up) due to lack of dilution with fresh air. Stagnant air, often caused by an anticyclone or temperature inversion, or other lack of winds lets air pollution remain in a local area. On these days, the news media may ask the public to carpool or use public transport, or take other air pollution prevention measures such as teleworking.

Other Indices

Hong Kong

The Air Pollution Index (API) levels for Hong Kong are related to the measured concentrations of ambient respirable suspended particulate (RSP), sulfur dioxide (SO_2), carbon monoxide (CO), ozone (O_3) and nitrogen dioxide (NO_2) over a 24-hour period based on the potential health effects of air pollutants.

An API level at or below 100 means that the pollutant levels are in the satisfactory range over 24 hour period and pose no acute or immediate health effects. However, air pollution consistently at "High" levels (API of 51 to 100) in a year may mean that the annual Hong Kong "Air Quality Objectives" for protecting long-term health effects could be violated. Therefore, chronic health effects may be observed if one is persistently exposed to an API of 51 to 100 for a long time.

"Very High" levels (API in excess of 100) means that levels of one or more pollutant(s) is/are in the unhealthy range. The Hong Kong Environmental Protection Department provides advice to the public regarding precautionary actions to take for such levels.

API	*Air Pollution Level*	*Health Implications*
0 - 25	Low	Not expected.
26 - 50	Medium	Not expected for the general population.
51 - 100	High	Acute health effects are not expected but chronic effects may be observed if one is persistently exposed to such levels.

100 - 200	Very High	People with existing heart or respiratory illnesses may notice mild aggravation of their health conditions. Generally healthy individuals may also notice some discomfort.
201 - 500	Severe	People with existing heart or respiratory illnesses may experience significant aggravation of their symptoms. There may also be widespread symptoms in the healthy population (e.g. eye irritation, wheezing, coughing, phlegm and sore throats).

Malaysia

The air quality in Malaysia is reported as the API or Air Pollution Index. Four of the index's pollutant components (i.e., carbon monoxide, ozone, nitrogen dioxide and sulfur dioxide) are reported in ppmv but PM_{10} particulate matter is reported in ìg/ m^3. Unlike the American AQI, the index number can exceed 500. Above 500, a state of emergency is declared in the reporting area. Usually, this means that non-essential government services are suspended, and all ports in the affected area closed. There may also be a prohibition on private sector commercial and industrial activities in the reporting area excluding the food sector.

Singapore

Singapore uses the Pollutant Standards Index to report on its air quality.

United Kingdom

The Met Office of the United Kingdom (UK) issues air quality forecasts wherein the level of pollution is described either as an index (ranging from 1 to 10) or as a banding (low, moderate, high or very high). These levels are based on the health effects of each pollutant.

Index	*Banding*	*Health Effect*
1 - 3	Low	Effects are unlikely to be noticed even by individuals who know they are sensitive to air pollutants.
4-6	Moderate	Mild effects, unlikely to require action, may be noticed amongst sensitive individuals.

7-9	High	Significant effects may be noticed by sensitive individuals and action to avoid or reduce these effects may be needed (e.g. reducing exposure by spending less time in polluted areas outdoors). Asthmatics will find that their 'reliever' inhaler is likely to reverse the effects on the lung.
10	Very High	The effects on sensitive individuals described for 'High' levels of pollution may worsen.

United States

The United States Environment Protection Agency (USEPA) developed the Pollutant Standards Index (PSI) to provide accurate, timely and easily understandable information about daily levels of air pollution. It is no longer in use, having been replaced by the AQI, which is more sensitive. For example, particulate matter with an aerodynamic diameter less than 2.5 micrometres (PM2.5) is a sub index, replacing the less sensitive PM10 component of the PSI.

National Ambient Air Quality Standards

The National Ambient Air Quality Standards (NAAQS) are standards established by the United States Environmental Protection Agency that apply for outdoor air throughout the country. Primary standards are designed to protect human health, with an adequate margin of safety, including sensitive populations such as children, the elderly, and individuals suffering from respiratory disease. Secondary standards are designed to protect public welfare from any known or anticipated adverse effects of a pollutant (e.g. building facades, visibility, crops, and domestic animals).

NAAQS requires the EPA to set standards on six criteria air contaminants:

1. Ozone (O_3)
2. Particulate Matter
   - o PM_{10}, coarse particles: 2.5 micrometers (ìm) to 10 ìm in size (although current implementation includes all particles 10 ìg or less in the standard)
   - o $PM_{2.5}$, fine particles: 2.5 ìm in size or less
3. Carbon monoxide (CO)
4. Sulfur dioxide (SO_2)
5. Nitrogen oxides (NO_x)
6. Lead (Pb)

Standards

The standards are listed in Title 40 of the Code of Federal Regulations Part 50.

Pollutant	*Type*	*Standard*	*Averaging Time*[a]	*Regulatory Citation*
SO_2	Primary	0.14 ppm (365 ìg/m^3)	24-hour	40 CFR 50.4(b)
SO_2	Primary	0.030 ppm (80 ìg/m^3)	annual	40 CFR 50.4(a)
SO_2	Secondary	0.5 ppm (1,300 ìg/m^3)	3-hour	40 CFR 50.5(a)
PM_{10}	Primary and Secondary	150 ìg/m^3	24-hour	40 CFR 50.6(a)
$PM_{2.5}$	Primary and Secondary	35 ìg/m^3	24-hour	40 CFR 50.7(a)
$PM_{2.5}$	Primary and Secondary	15 ìg/m^3	annual	40 CFR 50.7(a)
CO	Primary	35 ppm (40 mg/m^3)	1-hour	40 CFR 50.8(a)(2)
CO	Primary	9 ppm (10 mg/m^3)	8-hour	40 CFR 50.8(a)(1)
O_3	Primary and Secondary	0.12 ppm (235 ìg/m^3)	1-hour[b]	40 CFR 50.9(a)
O_3	Primary and Secondary	0.075 ppm (235 ìg/m^3)	8-hour	40 CFR 50.10(a)
NO_2	Primary and Secondary	0.053 ppm (100 ìg/m^3)	annual	40 CFR 50.11(a) and (b)
Pb	Primary and Secondary	1.5 ìg/m^3	quarterly	40 CFR 50.12

Note a: Each standard has its own criteria for how many times it may be exceeded, in some cases using a three year average.

Note b: As of June 15, 2005, the 1-hour ozone standard no longer applies to areas designated with respect to the 8-hour ozone standard (which includes most of the United States, except for portions of 10 states).

Acid Gas

Acid gas is natural gas or any other gas mixture which contains significant amounts of hydrogen sulfide (H_2S), carbon dioxide (CO_2), or similar contaminants. The terms *acid gas* and *sour gas* are often incorrectly treated as synonyms. Strictly speaking, a sour gas is any gas that contains hydrogen sulfide in significant amounts; an acid gas is any gas that contains significant amounts of acidic gases such as carbon dioxide (CO_2) or hydrogen sulfide. Thus, carbon dioxide by itself is an acid gas but not a sour gas. Before a raw natural gas containing hydrogen sulfide and/or carbon dioxide can be used, the raw gas must be treated to reduce impurities to acceptable levels and this is commonly done with an amine gas treating process. The removed H_2S is most often subsequently converted to by-product elemental sulfur in a Claus process. Processes within oil refineries or natural gas processing plants that remove mercaptans and/or hydrogen sulfide are commonly referred to as *sweetening* processes because they result in products which no longer have the sour, foul odors of mercaptans and hydrogen sulfide. Hydrogen sulfide is a toxic gas. It also restricts the materials that can be used for piping and other equipment for handling sour gas, as many metals are sensitive to sulfide stress cracking.

Air Pollution

One of Earth's most important natural resources is its atmosphere. The atmosphere contains air without which plants and animals could not survive. It contains greenhouse gases which keep the planet naturally warmer than it would be otherwise, maintaining an average global temperature above freezing that allows water to exist in its liquid state, a necessary condition for most life. If mankind is to protect and preserve this unique natural resource for future generations as well as other ecosystems, it must continue to address the problem of air pollution which affects the atmosphere from the local to the global scale.

Air Pollution and concern about air quality are not new. Complaints were recorded in the 13th century when coal was first used in London. Since the middle of the 19th century, the atmosphere of the major British cities was regularly polluted by coal smoke in winter, giving rise to an infamous mixture of fog and smoke known as smog.

In the 1970s the transboundary effects of industrial air pollution become known as acid rain. Acid rain is a widespread term used to describe all forms of acid precipitation such as rain and snow. Atmospheric pollutants, particularly oxides of sulphur and nitrogen react with water in the atmosphere to form sulphuric and nitric acid, causing precipitation to become more acidic when converted to sulphuric and nitric acids, hence the term acid rain. Acid deposition, acid rain and acid precipitation all relate to the chemistry of air pollution and moisture in the atmosphere. Scientists generally use the term acid deposition but all three terms relate to the same issue.

Acid rain is particularly harmful to vegetation as the acid in the rain changes the pH of the soil and leeches away important minerals. The amount of acid rain that an ecosystem can tolerate is known as its critical load. After the sulphuric and nitric acids have formed in the atmosphere, they can travel long distances with the wind before being deposited. In fact, much of the acid rain found in Scandinavia originated from the UK. In response to these problems, a number of agreements, including the Convention on Long-Range Transboundary Air Pollution (Geneva, 1979), have set out to reduce acid rain. Although acid rain remains an international concern, poor air quality in cities is now considered to be a more pressing environmental problem. Today poor air quality can be mainly attributed to the production and consumption of energy, industrial processes and road transport. The main pollutants associated with poor air quality are sulphur dioxide (SO_2), nitrogen oxides (NOx), particulates

(PM_{10}), and volatile organic compound (VOCs). These pollutants can lead to respiratory problems, particularly amongst the young and elderly.

Transport in particular is becoming an increasing source of air pollution. The transport related problems experienced now are likely to worsen according to the projected increase in traffic. Individually, a vehicle engine is not a particularly important source of pollution. Collectively however, vehicles represent a major source of air pollution in the UK and throughout the world. The best way to reduce the threat of air pollution is to use cleaner fuel and less of it, and to adopt more sustainable modes of transport like public transport, walking and cycling. The UK Government has recognised the problems that air pollution causes for the environment and for society and in 1997 published the first National Air Quality Strategy, setting air quality standards for today and making commitments to achieve new air quality objectives by 2005.

Scrubber

Scrubber systems are a diverse group of air pollution control devices that can be used to remove particulates and/or gases from industrial exhaust streams. Traditionally, the term "scrubber" has referred to pollution control devices that used liquid to "scrub" unwanted pollutants from a gas stream. Recently, the term is also used to describe systems that inject a dry reagent or slurry into a dirty exhaust stream to "scrub out" acid gases. Scrubbers are one of the primary devices that control gaseous emissions, especially acid gases.

Removal and Neutralisation

The exhaust gases of combustion may at times contain substances considered harmful to the environment, and it is the job of the scrubber to either remove those substances from the exhaust gas stream, or to neutralize those substances so that they cannot do any harm once emitted into the environment as part of the exhaust gas stream...

Wet Scrubbing

A wet scrubber is used to clean air or other gases of various pollutants and dust particles. Wet scrubbing works via the contact of target compounds or particulate matter with the scrubbing solution. Solutions may simply be water (for dust) or complex solutions of reagents that specifically target certain compounds. Removal efficiency of pollutants is improved by increasing residence time in the scrubber or by the increase of surface area of the scrubber solution by the use of a spray nozzle, packed towers or an aspirator. Wet scrubbers will often significantly increase the proportion of water in waste gases of industrial processes which can be seen in a stack plume.

Typical Wet Scrubber

Compliance agencies typically place minimum DP thresholds on wet scrubber.

Dry Scrubbing

A dry or semi-dry scrubbing system, unlike the wet scrubber, does not saturate the flue gas stream that is being treated with moisture. In some cases no moisture is added; while in other designs only the amount of moisture that can be evaporated in the flue gas without condensing is added. Therefore, dry scrubbers do not have a stack steam plume or wastewater handling/disposal requirements. Dry scrubbing systems are used to remove acid gases (such as SO_2 and HCl) primarily from combustion sources.

There are a number of dry type scrubbing system designs. However, all consist of two main sections or devices: a device to introduce the acid gas sorbent material into the gas stream and a particulate matter control device to remove reaction products, excess sorbent material as well as any particulate matter already in the flue gas. Dry scrubbing systems can be categorised as dry sorbent injectors (DSIs) or as spray dryer absorbers (SDAs). Spray dryer absorbers are also called semi-dry scrubbers or spray dryers. Dry sorbent injection involves the addition of an alkaline material (usually hydrated lime or soda ash) into the gas stream to react with the acid gases. The sorbent can be injected directly into several different locations: the combustion process, the flue gas duct (ahead of the particulate control device), or an open reaction chamber (if one exists). The acid gases react with the alkaline sorbents to form solid salts which are removed in the particulate control device. These simple systems can achieve only limited acid gas (SO_2 and HCl) removal efficiencies. Higher collection efficiencies can be achieved by increasing the flue gas humidity (i.e., cooling using water spray). These devices have been used on medical waste incinerators and a few municipal waste combustors.

In spray dryer absorbers, the flue gases are introduced into an absorbing tower (dryer) where the gases are contacted with a finely atomised alkaline slurry. Acid gases are absorbed by the slurry mixture and react to form solid salts which are removed by the particulate control device. The heat of the flue gas is used to evaporate all the water droplets, leaving a non-saturated flue gas to exit the absorber tower. Spray dryers are capable of achieving high (80+%) acid gas removal efficiencies. These devices have been used on industrial and utility boilers and municipal waste combustors.

Mercury Removal

Mercury has no known beneficial uses in nature, but it is a common substance found in coal that must also be removed. Wet scrubbers are only effective for mercury removal under certain conditions. Mercury vapor in its elemental form, Hg^0, is insoluble in the scrubber slurry and not removed. Oxidised mercury, Hg^{2+}, compounds are more soluble in the scrubber slurry and can be captured. The type of coal burned as well as the presence of a selective catalytic reduction unit both affect the ratio of elemental to oxidised mercury in the flue gas and thus the degree to which the mercury is removed.

Scrubber Waste Products

One side effect of scrubbing is that the process only moves the unwanted substance from the exhaust gases into a solid paste or powder form. If there is no useful purpose for this solid waste, it must be either contained or buried to prevent environmental contamination. Limestone-based scrubbers can produce a synthetic gypsum of sufficient quality that can be used to manufacture drywall and other industrial products.

Mercury removal results in a waste product that either needs further processing to extract the raw mercury, or must be buried in a special hazardous wastes landfill that prevents the mercury from seeping out into the environment.

Bacteria Spread

Until recently, scrubbers have not been associated with health risks involving bacteria spread as a result of inadequate cleaning, unlike other devices such as cooling towers. However, a 2005 outbreak of Legionnaires' disease in Norway was proven to emanate from a scrubber, causing ten deaths and more than fifty cases of infection as it spread the bacteria through the air during a period of only two weeks. This particular system had undergone regular cleaning routines every three weeks. This case is believed to be the first documented case of a scrubber being the source of such bacteria spread.

Ozone Depletion

Ozone is a form of oxygen. In the stratosphere, around 25 km above the Earth's surface, there is a layer of ozone that absorbs ultra-violet light from the Sun. Ultra-violet light is known to cause skin cancer in humans, and to damage plants. The ozone layer forms a protective shield around the Earth, and without it most life on Earth would not be likely to survive.

In the 1970s scientists realised that a hole was appearing in the ozone layer over Antarctica during the spring. They realised that this was due to the build-up of chlorofluorocarbons (CFCs) - chemicals used in aerosol sprays, packaging and air conditioning systems. One molecule of CFC can remove up to 100,000 ozone molecules, and the widespread use of CFCs before the damage was noticed has produced a serious problem. Life on the planet will not be sustainable if the ozone layer is destroyed, because damage caused to plants will severely reduce the global food supply.

International action has been taken to relieve the problem. In 1987 the Montreal Protocol was agreed which requires countries to take steps to eliminate CFCs and other substances which cause ozone depletion. If CFC levels are reduced considerably, particularly in the developing world, then the ozone layer will probably fully repair itself by around 2050. Agenda 21 suggests that the ozone layer can be viewed as a vital resource for life, and should be protected for sustainable development to be achieved.

Ozone Depletion

Ozone depletion describes two distinct, but related observations: a slow, steady decline of about 4 percent per decade in the total amount of ozone in Earth's stratosphere since the late 1970s; and a much larger, but seasonal, decrease in stratospheric ozone over Earth's polar regions during the same period. The latter phenomenon is commonly referred to as the ozone hole. In addition to this well-known stratospheric ozone depletion, there are also tropospheric ozone depletion events, which occur near the surface in polar regions during spring.

The detailed mechanism by which the polar ozone holes form is different from that for the mid-latitude thinning, but the most important process in both trends is catalytic destruction of ozone by atomic chlorine and bromine. The main source of these halogen atoms in the stratosphere is photodissociation of chlorofluorocarbon (CFC) compounds, commonly called freons, and of bromofluorocarbon compounds known as halons. These compounds are transported into the stratosphere after being emitted at the surface. Both ozone depletion mechanisms strengthened as emissions of CFCs and halons increased.

CFCs and other contributory substances are commonly referred to as ozone-depleting substances (ODS). Since the ozone layer prevents most harmful UVB wavelengths (270–315 nm) of ultraviolet light (UV light) from passing through the Earth's atmosphere, observed and projected decreases in ozone have generated worldwide concern leading to adoption of the Montreal Protocol banning the production of CFCs and halons as well as related ozone depleting chemicals such as carbon tetrachloride and trichloroethane. It is suspected that a variety of biological consequences such as increases in skin cancer, damage to plants, and reduction of plankton populations in the ocean's photic zone may result from the increased UV exposure due to ozone depletion.

Ozone Cycle Overview

Three forms (or allotropes) of oxygen are involved in the ozone-oxygen cycle: Oxygen atoms (O or atomic oxygen), oxygen gas (O_2 or diatomic oxygen), and ozone gas (O_3 or triatomic oxygen). Ozone is formed in the stratosphere when oxygen molecules photodissociate after absorbing an ultraviolet photon whose wavelength is shorter than 240 nm. This produces two oxygen atoms. The atomic oxygen then combines with O_2 to create O_3. Ozone molecules absorb UV light between 310 and 200 nm, following which ozone splits into a molecule of O_2 and an oxygen atom. The oxygen atom then joins up with an oxygen molecule to regenerate ozone. This is a continuing process which terminates when an oxygen atom "recombines" with an ozone molecule to make two O_2 molecules: O + O_3 '! 2 O_2

The overall amount of ozone in the stratosphere is determined by a balance between photochemical production and recombination.

Ozone can be destroyed by a number of free radical catalysts, the most important of which are the hydroxyl radical (OH·), the nitric oxide radical (NO·) and atomic chlorine (Cl·) and bromine (Br·). All of these have both natural and anthropogenic (manmade) sources; at the present time, most of the OH· and NO· in the stratosphere is of natural origin, but human activity has dramatically increased the high in oxygen chlorine and bromine. These elements are found in certain stable organic compounds, especially chlorofluorocarbons (CFCs), which may find their way to the stratosphere without being destroyed in the troposphere due to their low reactivity. Once in the stratosphere, the Cl and Br atoms are liberated from the parent compounds by the action of ultraviolet light, e.g. ('h' is Planck's constant, 'í' is frequency of electromagnetic radiation)

$CFCl_3$ + hí '! $CFCl_2$ + Cl

The Cl and Br atoms can then destroy ozone molecules through a variety of catalytic cycles. In the simplest example of such a cycle, a chlorine atom reacts with an ozone molecule, taking an oxygen atom with it (forming ClO) and leaving a normal oxygen molecule. A free oxygen atom then takes away the oxygen from the ClO, and the final result is an oxygen molecule and a chlorine atom, which then reinitiates the cycle. The chemical shorthand for these gas-phase reactions is:

Cl + O_3 '! ClO + O_2

ClO + O '! Cl + O_2

The net reaction is: O_3 + O '! 2 O_2, the "recombination" reaction given above.

The overall effect is to increase the rate of recombination, leading to an overall decrease in the amount of ozone. For this particular mechanism to operate there must be a source of O atoms, which is primarily the photo dissociation of O_3; thus this mechanism is only important in the upper stratosphere where such atoms are abundant. More complicated mechanisms have been discovered that lead to ozone destruction in the lower stratosphere as well.

A single chlorine atom would keep on destroying ozone for up to two years (the time scale for transport back down to the troposphere) were it not for reactions that remove them from this cycle by forming reservoir species such as hydrogen chloride (HCl) and chlorine nitrate ($ClONO_2$). On a per atom basis, bromine is even more efficient than chlorine at destroying ozone, but there is much less bromine in the atmosphere at present. As a result, both chlorine and bromine contribute significantly to the overall ozone depletion. Laboratory studies have shown that fluorine and iodine atoms participate in analogous catalytic cycles. However, in the Earth's stratosphere, fluorine atoms react rapidly with water and methane to form strongly-bound HF, while organic molecules which contain iodine react so rapidly in the lower atmosphere that they do not reach the stratosphere in significant quantities. Furthermore, a single chlorine atom is able to react with 100,000 ozone molecules. This fact plus the amount of chlorine released into the atmosphere by chlorofluorocarbons(CFCs) yearly demonstrates how dangerous CFCs are to the environment.

Quantitative Understanding of the Chemical Ozone Loss Process

New research on the breakdown of a key molecule in these ozone-depleting chemicals, dichlorine peroxide (Cl2O2), calls into question the completeness of present atmospheric models of polar ozone depletion. Specifically, chemists at NASA's Jet Propulsion Laboratory in Pasadena, California, found in 2007 that the temperatures, and the spectrum and intensity of radiation present in the stratosphere created conditions insufficient to allow the rate of chemical-breakdown required to release chlorine radicals in the volume necessary to explain observed rates of ozone depletion. Instead, laboratory tests, designed to be the most accurate reflection of stratospheric conditions to date, showed the decay of the crucial molecule almost a magnitude lower than previously thought.

Observations on Ozone Layer Depletion

The most pronounced decrease in ozone has been in the lower stratosphere. However, the ozone hole is most usually measured not in terms of ozone concentrations at these levels (which are typically of a few parts per million) but by reduction in the total *column ozone*, above a point on the Earth's surface, which is normally expressed in Dobson units, abbreviated as "DU". Marked decreases in column ozone in the Antarctic spring and early summer compared to the early 1970s and before have been observed using instruments such as the Total Ozone Mapping Spectrometer (TOMS).

Reductions of up to 70% in the ozone column observed in the austral (southern hemispheric) spring over Antarctica and first reported in 1985 (Farman et al 1985) are continuing. Through the 1990s, total column ozone in September and October have continued to be 40–50% lower than pre-ozone-hole values. In the Arctic the amount lost is more variable year-to-year than in the Antarctic. The greatest declines, up to 30%, are in the winter and spring, when the stratosphere is colder.

Reactions that take place on polar stratospheric clouds (PSCs) play an important role in enhancing ozone depletion. PSCs form more readily in the extreme cold of Antarctic stratosphere. This is why ozone holes first formed, and are deeper, over Antarctica. Early models failed to take PSCs into account and predicted a gradual global depletion, which is why the sudden Antarctic ozone hole was such a surprise to many scientists. In middle latitudes it is preferable to speak of ozone depletion rather than holes. Declines are about 3% below pre-1980 values for 35–60°N and about 6% for 35–60°S. In the tropics, there are no significant trends.

Ozone depletion also explains much of the observed reduction in stratospheric and upper tropospheric temperatures. The source of the warmth of the stratosphere is the absorption of UV radiation by ozone, hence reduced ozone leads to cooling. Some stratospheric cooling is also predicted from increases in greenhouse gases such as CO_2; however the ozone-induced cooling appears to be dominant. Predictions of ozone levels remain difficult. The World Meteorological Organisation Global Ozone Research and Monitoring Project - Report No. 44 comes out strongly in favor for the Montreal Protocol, but notes that a UNEP 1994 Assessment overestimated ozone loss for the 1994–1997 period.

Chemicals in the Atmosphere

CFCs in the Atmosphere

Chlorofluorocarbons (CFCs) were invented by Thomas Midgley in the 1920s. They were used in air conditioning/cooling units, as aerosol spray propellants prior to the 1980s, and in the cleaning processes of delicate electronic equipment. They also occur as by-products of some chemical processes. No significant natural sources have ever been identified for these compounds — their presence in the atmosphere is due almost entirely to human manufacture. As mentioned in the *ozone cycle overview* above, when such ozone-depleting chemicals reach the stratosphere, they are dissociated by ultraviolet light to release chlorine atoms. The chlorine atoms act as a catalyst, and each can break down tens of thousands of ozone molecules before being removed from the stratosphere. Given the longevity of CFC molecules, recovery times are measured in decades. It is calculated that a CFC molecule takes an average of 15 years to go from the ground level up to the upper atmosphere, and it can stay there for about a century, destroying up to one hundred thousand ozone molecules during that time.

Verification of Observations

Scientists have been increasingly able to attribute the observed ozone depletion to the increase of anthropogenic halogen compounds from CFCs by the use of complex chemistry transport models and their validation against observational data (e.g. SLIMCAT, CLaMS). These models work by combining satellite measurements of chemical concentrations and meteorological fields with chemical reaction rate constants obtained in lab experiments. They are able to identify not only the key chemical reactions but also the transport processes which bring CFC photolysis products into contact with ozone.

The Ozone Hole and its Causes

The Antarctic ozone hole is an area of the Antarctic stratosphere in which the recent ozone levels have dropped to as low as 33% of their pre-1975 values. The ozone hole occurs during the Antarctic spring, from September to early December, as strong westerly winds start to circulate around the continent and create an atmospheric container. Within this polar vortex, over 50% of the lower stratospheric ozone is destroyed during the Antarctic spring.

As explained above, the overall cause of ozone depletion is the presence of chlorine-containing source gases (primarily CFCs and related halocarbons). In the presence of UV light, these gases dissociate, releasing chlorine atoms, which then go on to catalyze ozone destruction. The Cl-catalysed ozone depletion can take place in the gas phase, but it is dramatically enhanced in the presence of polar stratospheric clouds (PSCs).

These polar stratospheric clouds form during winter, in the extreme cold. Polar winters are dark, consisting of 3 months without solar radiation (sunlight). Not only lack of sunlight contributes to a decrease in temperature but also the polar vortex traps and chills air. Temperatures hover around or below -80 °C. These low temperatures form cloud particles and are composed of either nitric acid (Type I PSC) or ice (Type II PSC). Both types provide surfaces for chemical reactions that lead to ozone destruction. The photochemical processes involved are complex but well understood. The key observation is that, ordinarily, most of the chlorine in the stratosphere resides in stable "reservoir" compounds, primarily hydrogen chloride (HCl) and chlorine nitrate ($ClONO_2$). During the Antarctic winter and spring, however, reactions on the surface of the polar stratospheric cloud particles convert these "reservoir" compounds into reactive free radicals (Cl and ClO). The clouds can also remove NO_2 from the atmosphere by converting it to nitric acid, which prevents the newly formed ClO from being converted back into $ClONO_2$. The role of sunlight in ozone depletion is the reason why the Antarctic ozone depletion is greatest during spring. During winter, even though PSCs are at their most abundant, there is no light over the pole to drive the chemical reactions. During the spring, however, the sun comes out, providing energy to drive photochemical reactions, and melt the polar stratospheric clouds, releasing the trapped compounds. Most of the ozone that is destroyed is in the lower stratosphere, in contrast to the much smaller ozone depletion through homogeneous gas phase reactions, which occurs primarily in the upper stratosphere.

Warming temperatures near the end of spring break up the vortex around mid-December. As warm, ozone-rich air flows in from lower latitudes, the PSCs are destroyed, the ozone depletion process shuts down, and the ozone hole heals.

Interest in Ozone Layer Depletion

While the effect of the Antarctic ozone hole in decreasing the global ozone is relatively small, estimated at about 4% per decade, the hole has generated a great deal of interest because:

- The decrease in the ozone layer was predicted in the early 1980s to be roughly 7% over a sixty-year period.
- The sudden recognition in 1985 that there was a substantial "hole" was widely reported in the press. The especially rapid ozone depletion in Antarctica had previously been dismissed as a measurement error.
- Many were worried that ozone holes might start to appear over other areas of the globe but to date the only other large-scale depletion is a smaller ozone "dimple" observed during the Arctic spring over the North Pole. Ozone at middle latitudes has declined, but by a much smaller extent (about 4–5% decrease).
- If the conditions became more severe (cooler stratospheric temperatures, more stratospheric clouds, more active chlorine), then global ozone may decrease at a much greater pace. Standard global warming theory predicts that the stratosphere will cool.
- When the Antarctic ozone hole breaks up, the ozone-depleted air drifts out into nearby areas. Decreases in the ozone level of up to 10% have been reported in New Zealand in the month following the break-up of the Antarctic ozone hole.

Consequences of Ozone Layer Depletion

Since the ozone layer absorbs UVB ultraviolet light from the Sun, ozone layer depletion is expected to increase surface UVB levels, which could lead to damage, including increases in skin cancer. This was the reason for the Montreal Protocol. Although decreases in stratospheric ozone are well-tied to CFCs and there are good theoretical reasons to believe that decreases in ozone will lead to increases in surface UVB, there is no direct observational evidence linking ozone depletion to higher incidence of skin cancer in human beings. This is partly due to the fact that UVA, which has also been implicated in some forms of skin cancer, is not absorbed by ozone, and it is nearly impossible to control statistics for lifestyle changes in the populace.

Increased UV

Ozone, while a minority constituent in the earth's atmosphere, is responsible for most of the absorption of UVB radiation. The amount of UVB radiation that penetrates through the ozone layer decreases exponentially with the slant-path thickness/density of the layer. Correspondingly, a decrease in atmospheric ozone is expected to give rise to significantly increased levels of UVB near the surface.

Increases in surface UVB due to the ozone hole can be partially inferred by radiative transfer model calculations, but cannot be calculated from direct measurements because of the lack of reliable historical (pre-ozone-hole) surface UV data, although more recent surface UV observation measurement programmes exist (e.g. at Lauder, New Zealand).

Because it is this same UV radiation that creates ozone in the ozone layer from O_2 (regular oxygen) in the first place, a reduction in stratospheric ozone would actually tend to increase photochemical production of ozone at lower levels (in the troposphere), although the overall observed trends in total column ozone still show a decrease, largely because ozone produced lower down has a naturally shorter photochemical lifetime, so it is destroyed before the concentrations could reach a level which would compensate for the ozone reduction higher up.

Biological Effects of Increased UV and Microwave Radiation from a Depleted Ozone Layer

The main public concern regarding the ozone hole has been the effects of surface UV on human health. So far, ozone depletion in most locations has been typically a few percent and, as noted above, no direct evidence of health damage is available in most latitudes. Were the high levels of depletion seen in the ozone hole ever to be common across the globe, the effects could be substantially more dramatic. As the ozone hole over Antarctica has in some instances grown so large as to reach southern parts of Australia and New Zealand, environmentalists have been concerned that the increase in surface UV could be significant.

Effects of Ozone Layer Depletion on Humans

UVB (the higher energy UV radiation absorbed by ozone) is generally accepted to be a contributory factor to skin cancer. In addition, increased surface UV leads to increased tropospheric ozone, which is a health risk to humans. The increased surface UV also represents an increase in the vitamin D synthetic capacity of the sunlight.

The cancer preventive effects of vitamin D represent a possible beneficial effect of ozone depletion. In terms of health costs, the possible benefits of increased UV irradiance may outweigh the burden.

1. Basal and Squamous Cell Carcinomas — The most common forms of skin cancer in humans, basal and squamous cell carcinomas, have been strongly linked to UVB exposure. The mechanism by which UVB induces these cancers is well understood — absorption of UVB radiation causes the pyrimidine bases in the DNA molecule to form dimers, resulting in transcription errors when the DNA replicates. These cancers are relatively mild and rarely fatal, although the treatment of squamous cell carcinoma sometimes requires extensive reconstructive surgery. By combining epidemiological data with results of animal studies, scientists have estimated that a one percent decrease in stratospheric ozone would increase the incidence of these cancers by 2%.

2. Malignant Melanoma — Another form of skin cancer, malignant melanoma, is much less common but far more dangerous, being lethal in about 15% - 20% of the cases diagnosed. The relationship between malignant melanoma and ultraviolet exposure is not yet well understood, but it appears that both UVB and UVA are involved. Experiments on fish suggest that 90 to 95% of malignant melanomas may be due to UVA and visible radiation whereas experiments on opossums suggest a larger role for UVB. Because of this uncertainty, it is difficult to estimate the impact of ozone depletion on melanoma incidence. One study showed that a 10% increase in UVB radiation was associated with a 19% increase in melanomas for men and 16% for women. A study of people in Punta Arenas, at the southern tip of Chile, showed a 56% increase in melanoma and a 46% increase in nonmelanoma skin cancer over a period of seven years, along with decreased ozone and increased UVB levels.

3. Cortical Cataracts — Studies are suggestive of an association between ocular cortical cataracts and UV-B exposure, using crude approximations of exposure and various cataract assessment techniques. A detailed assessment of ocular exposure to UV-B was carried out in a study on Chesapeake Bay Watermen, where increases

in average annual ocular exposure were associated with increasing risk of cortical opacity. In this highly exposed group of predominantly white males, the evidence linking cortical opacities to sunlight exposure was the strongest to date. However, subsequent data from a population-based study in Beaver Dam, WI suggested the risk may be confined to men. In the Beaver Dam study, the exposures among women were lower than exposures among men, and no association was seen. Moreover, there were no data linking sunlight exposure to risk of cataract in African Americans, although other eye diseases have different prevalences among the different racial groups, and cortical opacity appears to be higher in African Americans compared with whites.

4. Increased Tropospheric Ozone — Increased surface UV leads to increased tropospheric ozone. Ground-level ozone is generally recognised to be a health risk, as ozone is toxic due to its strong oxidant properties. At this time, ozone at ground level is produced mainly by the action of UV radiation on combustion gases from vehicle exhausts.

Effects on Crops

An increase of UV radiation would be expected to affect crops. A number of economically important species of plants, such as rice, depend on cyanobacteria residing on their roots for the retention of nitrogen. Cyanobacteria are sensitive to UV light and they would be affected by its increase.

Effects on Plankton

Research has shown a widespread extinction of plankton 2 million years ago that coincided with a nearby supernova. There is a difference in the orientation and motility of planktons when excess of UV rays reach earth. Researchers speculate that the extinction was caused by a significant weakening of the ozone layer at that time when the radiation from the supernova produced nitrogen oxides that catalysed the destruction of ozone (plankton are particularly susceptible to effects of UV light, and are vitally important to marine food webs).

Public Policy in Response to the Ozone Hole

The full extent of the damage that CFCs have caused to the ozone layer is not known and will not be known for decades; however, marked decreases in column ozone have already been observed (as explained above).

After a 1976 report by the U.S. National Academy of Sciences concluded that credible scientific evidence supported the ozone depletion hypothesis, a few countries, including the United States, Canada, Sweden, and Norway, moved to eliminate the use of CFCs in aerosol spray cans. At the time this was widely regarded as a first step towards a more comprehensive regulation policy, but progress in this direction slowed in subsequent years, due to a combination of political factors (continued resistance from the halocarbon industry and a general change in attitude towards environmental regulation during the first two years of the Reagan administration) and scientific developments (subsequent National Academy assessments which indicated that the first estimates of the magnitude of ozone depletion had been overly large). The European Community rejected proposals to ban CFCs in aerosol sprays while even in the U.S., CFCs continued to be used as refrigerants and for cleaning circuit boards. Worldwide CFC production fell sharply after the U.S. aerosol ban, but by 1986 had returned nearly to its 1976 level. In 1980, DuPont closed down its research programme into halocarbon alternatives.

The US Government's attitude began to change again in 1983, when William Ruckelshaus replaced Anne M. Burford as Administrator of the United States Environmental Protection Agency. Under Ruckelshaus and his successor, Lee Thomas, the EPA pushed for an international approach to halocarbon regulations. In 1985 20 nations, including most of the major CFC producers, signed the Vienna Convention which established a framework for negotiating international regulations on ozone-depleting substances. That same year, the discovery of the Antarctic ozone hole was announced, causing a revival in public attention to the issue. In 1987, representatives from 43 nations signed the Montreal Protocol. Meanwhile, the halocarbon industry shifted its position and started supporting a protocol to limit CFC production. The reasons for this were in part explained by "Dr. Mostafa Tolba, former head of the UN Environment Programme, who was quoted in the June 30, 1990 edition of The New Scientist, '...the chemical industry supported the Montreal Protocol in 1987 because it set up a worldwide schedule for phasing out CFCs, which [were] no longer protected by patents. This provided companies with an equal opportunity to market new, more profitable compounds.'"

At Montreal, the participants agreed to freeze production of CFCs at 1986 levels and to reduce production by 50% by 1999. After a series of scientific expeditions to the Antarctic produced convincing evidence that the ozone hole was indeed caused by chlorine and bromine from manmade organohalogens, the Montreal Protocol was strengthened at a 1990 meeting in London. The participants agreed to phase out CFCs and halons entirely (aside from a very small amount marked for certain "essential" uses, such as asthma inhalers) by 2000. At a 1992 meeting in Copenhagen, the phase out date was moved up to 1996.

To some extent, CFCs have been replaced by the less damaging hydro-chloro-fluoro-carbons (HCFCs), although concerns remain regarding HCFCs also. In some applications, hydro-fluoro-carbons (HFCs) have been used to replace CFCs. HFCs, which contain no chlorine or bromine, do not contribute at all to ozone depletion although they are potent greenhouse gases. The best known of these compounds is probably HFC-134a (R-134a), which in the United States has largely replaced CFC-12 (R-12) in automobile air conditioners. In laboratory analytics (a former "essential" use) the ozone depleting substances can be replaced with various other solvents.

Ozone Diplomacy, by Richard Benedick (Harvard University Press, 1991) gives a detailed account of the negotiation process that led to the Montreal Protocol. Pielke and Betsill provide an extensive review of early US government responses to the emerging science of ozone depletion by CFCs.

Current Events and Future Prospects of Ozone Depletion

Since the adoption and strengthening of the Montreal Protocol has led to reductions in the emissions of CFCs, atmospheric concentrations of the most significant compounds have been declining. These substances are being gradually removed from the atmosphere. By 2015, the Antarctic ozone hole would have reduced by only 1 million km^2 out of 25 (Newman *et al.*, 2004); complete recovery of the Antarctic ozone layer will not occur until the year 2050 or later. Work has suggested that a detectable (and statistically significant) recovery will not occur until around 2024, with ozone levels recovering to 1980 levels by around 2068.

There is a slight caveat to this, however. Global warming from CO_2 is expected to cool the stratosphere. This, in turn, would lead to a relative increase in ozone depletion and the frequency of ozone holes. The effect may not be linear; ozone holes form because of polar stratospheric clouds; the formation of polar stratospheric clouds has a temperature threshold above which they will not form; cooling of the Arctic stratosphere might lead to Antarctic-ozone-hole-like conditions. But at the moment this is not clear.

Even though the stratosphere as a whole is cooling, high-latitude areas may become increasingly predisposed to springtime stratospheric warming events as weather patterns change in response to higher greenhouse gas loading. This would cause PSCs to disappear earlier in the season, and may explain why Antarctic ozone hole seasons have tended to end somewhat earlier since 2000 as compared with the most prolonged ozone holes of the 1990s.

The decrease in ozone-depleting chemicals has also been significantly affected by a decrease in bromine-containing chemicals. The data suggest that substantial natural sources exist for atmospheric methyl bromide (CH_3Br).

The 2004 ozone hole ended in November 2004, daily minimum stratospheric temperatures in the Antarctic lower stratosphere increased to levels that are too warm for the formation of polar stratospheric clouds (PSCs) about 2 to 3 weeks earlier than in most recent years.

The Arctic winter of 2005 was extremely cold in the stratosphere; PSCs were abundant over many high-latitude areas until dissipated by a big warming event, which started in the upper stratosphere during February and spread throughout the Arctic stratosphere in March. The size of the Arctic area of anomalously low total ozone in 2004-2005 was larger than in any year since 1997. The predominance of anomalously low total ozone values in the Arctic region in the winter of 2004-2005 is attributed to the very low stratospheric temperatures and meteorological conditions favorable for ozone destruction along with the continued presence of ozone destroying chemicals in the stratosphere.

A 2005 IPCC summary of ozone issues observed that observations and model calculations suggest that the global average amount of ozone depletion has now approximately stabilized. Although considerable variability in ozone is expected from year to year, including in polar regions where depletion is largest, the ozone layer is expected to begin to recover in coming decades due to declining ozone-depleting substance concentrations, assuming full compliance with the Montreal Protocol.

Temperatures during the Arctic winter of 2006 stayed fairly close to the long-term average until late January, with minimum readings frequently cold enough to produce PSCs. During the last week of January, however, a major warming event sent temperatures well above normal — much too warm to support PSCs. By the time temperatures dropped back to near normal in March, the seasonal norm was well above the PSC threshold. Preliminary satellite instrument-generated ozone maps show seasonal ozone buildup slightly below the long-term means for the Northern Hemisphere as a whole, although some high ozone events have occurred. During March 2006, the Arctic stratosphere poleward of 60 degrees North Latitude was free of anomalously low ozone areas except during the three-day period from March 17 to 19 when the total ozone cover fell below 300 DU over part of the North Atlantic region from Greenland to Scandinavia.

The area where total column ozone is less than 220 DU (the accepted definition of the boundary of the ozone hole) was relatively small until around 20 August 2006. Since then the ozone hole area increased rapidly, peaking at 29 million km^2 September 24. In October 2006, NASA reported that the year's ozone hole set a new area record with a daily average of 26 million km^2 between 7 September and 13 October 2006; total ozone thicknesses fell as low as 85 DU on October 8. The two factors combined, 2006 sees the worst level of depletion in recorded ozone history. The depletion is attributed to the temperatures above the Antarctic reaching the lowest recording since comprehensive records began in 1979.

The Antarctic ozone hole is expected to continue for decades. Ozone concentrations in the lower stratosphere over Antarctica will increase by 5%–10% by 2020 and return to pre-1980 levels by about 2060–2075, 10–25 years later than predicted in earlier assessments. This is because of revised estimates of atmospheric concentrations of Ozone Depleting Substances — and a larger predicted future usage in developing countries. Another factor which may aggravate ozone depletion is the draw-down of nitrogen oxides from above the stratosphere due to changing wind patterns.

History of the Research

The basic physical and chemical processes that lead to the formation of an ozone layer in the earth's stratosphere were discovered by Sydney Chapman in 1930. These are discussed in the article Ozone-oxygen cycle — briefly, short-wavelength UV radiation splits an oxygen (O_2) molecule into two oxygen (O) atoms, which then combine with other oxygen molecules to form ozone. Ozone is removed when an oxygen atom and an ozone molecule "recombine" to form two oxygen molecules, i.e. $O + O_3$ '! $2O_2$. In the 1950s, David Bates and Marcel Nicolet presented evidence that various free radicals, in particular hydroxyl (OH) and nitric oxide (NO), could catalyze this recombination reaction, reducing the overall amount of ozone. These free radicals were known to be present in the stratosphere, and so were regarded as part of the natural balance – it was estimated that in their absence, the ozone layer would be about twice as thick as it currently is.

In 1970 Prof. Paul Crutzen pointed out that emissions of *nitrous* oxide (N_2O), a stable, long-lived gas produced by soil bacteria, from the earth's surface could affect the amount of *nitric* oxide (NO) in the stratosphere. Crutzen showed that nitrous oxide lives long enough to reach the stratosphere, where it is converted into NO. Crutzen then noted that increasing use of fertilizers might have led to an increase in nitrous oxide emissions over the natural background, which would in turn result in an increase in the amount of NO in the stratosphere. Thus human activity could have an impact on the stratospheric ozone layer. In the following year, Crutzen and (independently) Harold Johnston suggested that NO emissions from supersonic aircraft, which fly in the lower stratosphere, could also deplete the ozone layer.

The Rowland-Molina Hypothesis

In 1974 Frank Sherwood Rowland, Chemistry Professor at the University of California at Irvine, and his postdoctoral associate Mario J. Molina suggested that long-lived organic halogen compounds, such as CFCs, might behave in a similar fashion as Crutzen had proposed for nitrous oxide. James Lovelock (most popularly known as the creator of the Gaia hypothesis) had recently discovered, during a cruise in the South Atlantic in 1971, that almost all of the CFC compounds manufactured since their invention in 1930 were still present in the atmosphere.

Molina and Rowland concluded that, like N_2O, the CFCs would reach the stratosphere where they would be dissociated by UV light, releasing Cl atoms. (A year earlier, Richard Stolarski and Ralph Cicerone at the University of Michigan had shown that Cl is even more efficient than NO at catalysing the destruction of ozone. Similar conclusions were reached by Michael McElroy and Steven Wofsy at Harvard University. Neither group, however, had realised that CFC's were a potentially large source of stratospheric chlorine — instead, they had been investigating the possible effects of HCl emissions from the Space Shuttle, which are very much smaller.)

The Rowland-Molina hypothesis was strongly disputed by representatives of the aerosol and halocarbon industries. The Chair of the Board of DuPont was quoted as saying that ozone depletion theory is "a science fiction tale...a load of rubbish...utter nonsense". Robert Abplanalp, the President of Precision Valve Corporation (and inventor of the first practical aerosol spray can valve), wrote to the Chancellor of UC Irvine to complain about Rowland's public statements (Roan, p 56.) Nevertheless, within three years most of the basic assumptions made by Rowland and Molina were confirmed by laboratory measurements and by direct observation in the stratosphere. The concentrations of the source gases (CFC's and related compounds) and the chlorine reservoir species (HCl and $ClONO_2$) were measured throughout the stratosphere, and demonstrated that CFCs were indeed the major source of stratospheric chlorine, and that nearly all of the CFCs emitted would eventually reach the stratosphere. Even more convincing was the measurement, by James G. Anderson and collaborators, of chlorine monoxide (ClO) in the stratosphere. ClO is produced by the reaction of Cl with ozone — its observation thus demonstrated that Cl radicals not only were present in the stratosphere but also were actually involved in destroying ozone. McElroy and Wofsy extended the work of Rowland and Molina by showing that Bromine atoms were even more effective catalysts for ozone loss than chlorine atoms and argued that the brominated organic compounds known as halons, widely used in fire extinguishers, were a potentially large source of stratospheric bromine. In 1976 the U.S. National Academy of Sciences released a report which concluded that the ozone depletion hypothesis was strongly supported by the scientific evidence. Scientists calculated that if CFC production continued to increase at the going rate of 10% per year until 1990 and then remain steady, CFCs would cause a global ozone loss of 5 to 7% by 1995, and a 30 to 50% loss by 2050. In response the United States, Canada, Sweden and Norway banned the use of CFCs in aerosol spray cans in 1978. However, subsequent research, summarised by the National Academy in reports issued between 1979 and 1984, appeared to show that the earlier estimates of global ozone loss had been too large.

Crutzen, Molina, and Rowland were awarded the 1995 Nobel Prize in Chemistry for their work on stratospheric ozone.

The Ozone Hole

The discovery of the Antarctic "ozone hole" by British Antarctic Survey scientists Farman, Gardiner and Shanklin (announced in a paper in *Nature* in May 1985) came as a shock to the scientific community, because the observed decline in polar ozone was far larger than anyone had anticipated. Satellite measurements showing massive depletion of ozone around the south pole were becoming available at the same time. However, these were initially rejected as unreasonable by data quality control algorithms (they were filtered out as errors since the values were unexpectedly low); the ozone hole was detected only in satellite data when the raw data was reprocessed following evidence of ozone depletion in *in situ* observations. When the software was rerun without the flags, the ozone hole was seen as far back as 1976.

Susan Solomon, an atmospheric chemist at the National Oceanic and Atmospheric Administration (NOAA), proposed that chemical reactions on polar stratospheric clouds (PSCs) in the cold Antarctic stratosphere caused a massive, though localised and seasonal, increase in the amount of chlorine present in active, ozone-destroying forms. The polar stratospheric clouds in Antarctica are only formed when there are very low temperatures, as low as -80 degrees C, and early spring conditions. In such conditions the ice crystals of the cloud provide a suitable surface for conversion of unreactive chlorine compounds into reactive chlorine compounds which can deplete ozone easily.

Moreover the polar vortex formed over Antarctica is very tight and the reaction which occurs on the surface of the cloud crystals is far different from when it occurs in atmosphere. These conditions have led to ozone hole formation in Antarctica. This hypothesis was decisively confirmed, first by laboratory measurements and subsequently by direct measurements, from the ground and from high-altitude airplanes, of very high concentrations of chlorine monoxide (ClO) in the Antarctic stratosphere.

Alternative hypotheses, which had attributed the ozone hole to variations in solar UV radiation or to changes in atmospheric circulation patterns, were also tested and shown to be untenable.

Meanwhile, analysis of ozone measurements from the worldwide network of ground-based Dobson spectrophotometers led an international panel to conclude that the ozone layer was in fact being depleted, at all latitudes outside of the tropics. These trends were confirmed by satellite measurements. As a consequence, the major halocarbon producing nations agreed to phase out production of CFCs, halons, and related compounds, a process that was completed in 1996.

Since 1981 the United Nations Environment Programme has sponsored a series of reports on scientific assessment of ozone depletion. The most recent is from 2007 where satellite measurements have shown the hole in the ozone layer is recovering and is now the smallest it has been for about a decade.

Controversy Regarding Ozone Science and Policy

That ozone depletion takes place is not seriously disputed in the scientific community. There is a consensus among atmospheric physicists and chemists that the scientific understanding has now reached a level where countermeasures to control CFC emissions are justified, although the decision is ultimately one for policy-makers.

Despite this consensus, the science behind ozone depletion remains complex, and some who oppose the enforcement of countermeasures point to some of the uncertainties. For example, although increased UVB has been shown to constitute a melanoma risk, it has been difficult for statistical studies to establish a direct link between ozone depletion and increased rates of melanoma. Although melanomas did increase significantly during the period 1970–1990, it is difficult to separate reliably the effect of ozone depletion from the effect of changes in lifestyle factors (e.g., increasing rates of air travel).

Ozone Depletion and Global Warming

Although they are often interlinked in the mass media, the connection between global warming and ozone depletion is not strong. There are four areas of linkage:

- The same CO_2 radiative forcing that produces near-surface global warming is expected to cool the stratosphere. This cooling, in turn, is expected to produce a relative *increase* in polar ozone (O_3) depletion and the frequency of ozone holes.
- Conversely, ozone depletion represents a radiative forcing of the climate system. There are two opposing effects: Reduced ozone causes the stratosphere to absorb less solar radiation, thus cooling the stratosphere while warming the troposphere; the resulting colder stratosphere emits less long-wave radiation downward, thus cooling the troposphere. Overall, the cooling dominates; the IPCC concludes that "*observed stratospheric O_3 losses over the past two decades have caused a negative forcing of the surface-troposphere system*" of about "0.15 ± 0.10 watts per square meter (W/m^2).
- One of the strongest predictions of the greenhouse effect is that the stratosphere will cool. Although this cooling has been observed, it is not trivial to separate the effects of changes in the concentration of greenhouse gases and ozone depletion since both will lead to cooling. However, this

can be done by numerical stratospheric modeling. Results from the National Oceanic and Atmospheric Administration's Geophysical Fluid Dynamics Laboratory show that above 20 km (12.4 miles), the greenhouse gases dominate the cooling.

- Ozone depleting chemicals are also greenhouse gases. The increases in concentrations of these chemicals have produced 0.34 ± 0.03 W/m^2 of radiative forcing, corresponding to about 14% of the total radiative forcing from increases in the concentrations of well-mixed greenhouse gases.
- The long term modeling of the process, its measurement, study, design of theories and testing take decades to both document, gain wide acceptance, and ultimately become the dominant paradigm. Several theories about the destruction of ozone, were hyphtosised in the 1980s, published in the late 1990s, and are currently being proven. Dr Drew Schindell, and Dr Paul Newman, NASA Goddard, proposed a theory in the late 1990s, using a SGI Origin 2000 supercomputer, that modeled ozone destruction, accounted for 78% of the ozone destroyed. Further refinement of that model, accounted for 89% of the ozone destroyed, but pushed back the estimated recovery of the ozone hole from 75 years to 150 years. (An important part of that model is the lack of staratospheric flight due to depletion of fossil fuels.)

Misconceptions About Ozone Depletion

A few of the more common misunderstandings about ozone depletion are addressed briefly here; more detailed discussions can be found in the ozone-depletion FAQ.

CFCs are "too Heavy" to Reach the Stratosphere

It is sometimes stated that since CFC molecules are much heavier than nitrogen or oxygen, they cannot reach the stratosphere in significant quantities. But atmospheric gases are not sorted by weight; the forces of wind (turbulence) are strong enough to fully intermix gases in the atmosphere. CFCs are heavier than air, but just like argon, krypton and other heavy gases with a long lifetime, they are uniformly distributed throughout the turbosphere and reach the upper atmosphere.

Man-Made Chlorine is Insignificant Compared to Natural Sources

Another objection occasionally voiced is that *It is generally agreed that natural sources of tropospheric chlorine (volcanoes, ocean spray, etc.) are four to five orders of magnitude larger than man-made sources.* While strictly true, *tropospheric* chlorine is irrelevant; it is *stratospheric* chlorine that matters to ozone depletion. Chlorine from ocean spray is soluble and thus is washed out by rainfall before it reaches the stratosphere. CFCs, in contrast, are insoluble and long-lived, which allows them to reach the stratosphere. Even in the lower atmosphere there is more chlorine present in the form of CFCs and related haloalkanes than there is in HCl from salt spray, and in the stratosphere the halocarbons dominate overwhelmingly. Only one of these halocarbons, methyl chloride, has a predominantly natural source, and it is responsible for about 20 percent of the chlorine in the stratosphere; the remaining 80% comes from manmade compounds.

Very large volcanic eruptions can inject HCl directly into the stratosphere, but direct measurements have shown that their contribution is small compared to that of chlorine from CFCs. A similar erroneous assertion is that soluble halogen compounds from the volcanic plume of Mount Erebus on Ross Island, Antarctica are a major contributor to the Antarctic ozone hole.

An Ozone Hole was First Observed in 1956

G.M.B. Dobson (Exploring the Atmosphere, 2nd Edition, Oxford, 1968) mentioned that when springtime ozone levels over Halley Bay were first measured, he was surprised to find that they were ~320 DU, about 150 DU below spring levels, ~450 DU, in the Arctic. These, however, were the pre-ozone hole normal climatological values. What Dobson describes is essentially the *baseline* from which the ozone hole is measured: actual ozone hole values are in the 150–100 DU range.

The discrepancy between the Arctic and Antarctic noted by Dobson was primarily a matter of timing: during the Arctic spring ozone levels rose smoothly, peaking in April, whereas in the Antarctic they stayed approximately constant during early spring, rising abruptly in November when the polar vortex broke down.

The behavior seen in the Antarctic ozone hole is completely different. Instead of staying constant, early springtime ozone levels suddenly drop from their already low winter values, by as much as 50%, and normal values are not reached again until December.

If the theory were correct, the ozone hole should be above the sources of CFCs

CFCs are well mixed in the troposphere and the stratosphere. The reason the ozone hole occurs above Antarctica is not because there are more CFCs there but because the low temperatures allow polar stratospheric clouds to form. There have been anomalous discoveries of significant, serious, localised "holes" above other parts of the globe.

The "Ozone Hole" is a Hole in the Ozone Layer

When the "ozone hole" forms, essentially all of the ozone in the lower stratosphere is destroyed. The upper stratosphere is much less affected, however, so that the overall amount of ozone over the continent declines by 50 percent or more. The ozone hole does not go all the way through the layer; on the other hand, it is not a uniform 'thinning' of the layer either. It's a "hole" in the sense of "a hole in the ground", a depression, not in the sense of "a hole in the windshield."

World Ozone Day

In 1994, the United Nations General Assembly voted to designate September 16 as "World Ozone Day", to commemorate the signing of the Montreal Protocol on that date in 1987.

Dangerous Goods

A dangerous good is any solid, liquid, or gas that can harm people, other living organisms, property, or the environment. An equivalent term, used almost exclusively in the United States, is *hazardous material* (hazmat). Dangerous goods may be radioactive, flammable, explosive, toxic, corrosive, biohazardous, an oxidizer, an asphyxiant, a pathogen, an allergen, or may have other characteristics that render it hazardous in specific circumstances.

Mitigating the risks associated with hazardous materials may require the application of safety precautions during their transport, use, storage and disposal. Most countries regulate hazardous materials by law, and they are subject to several international treaties as well.

Persons who handle dangerous goods will often wear protective equipment, and metropolitan fire departments often have a response team specifically trained to deal with accidents and spills. These teams train with different organisations at a variety of specialised locations. Some of the most well-known in the U.S. and Canada include the California Specialised Training Institute, the Texas A&M TEEX Academy, Signet North America, the Justice Institute of British Columbia, and the U.S. National Fire Academy. Persons who handle or potentially come into contact with dangerous goods as part of their work are also often subject to monitoring or health surveillance to ensure that their exposure does not exceed occupational exposure limits.

Laws and regulations on the use and handling of hazardous materials may differ depending on the activity and status of the material. For example one set of requirements may apply to their use in the workplace while a different requirements may apply to spill response, sale for consumer use, or transportation. Most countries regulate some aspect of hazardous materials.

The most widely applied regulatory scheme is that for the transportation of dangerous goods. The Committee of Experts on the Transport of Dangerous Goods of the United Nations Economic and Social Council issues Model Regulations on the Transportation of Dangerous Goods. Most regional and national regulatory schemes for hazardous materials are harmonised to a greater or lesser degree with the UN Model Regulation. For instance, the International Civil Aviation Organisation has developed regulations for air transport of hazardous materials that are based upon the UN Model but modified to accommodate unique aspects of air transport. Individual airline and governmental requirements are incorporated with this by the International Air Transport Association to produce the widely used IATA Dangerous Goods Regulations. Similarly, the International Maritime Organisation has developed the IMO Dangerous Goods Regulations for transportation on the high seas. Many individual nations have also structured their dangerous goods transportation regulations to harmonize with the UN Model in organisation as well as in specific requirements.

Dangerous goods are divided into classes on the basis of the specific chemical characteristics producing the risk. The graphics and text in this article representing the dangerous goods safety marks are derived from the United Nations-based system of identifying dangerous goods. Not all countries use precisely the same graphics (label, placard and/or text information) in their national regulations. Some use graphic symbols, but without English wording or with similar wording in their national language. Refer to the Dangerous Goods Transportation Regulations of the country of interest. For example, see the Dangerous Goods Safety Marks in the Canadian Transportation of Dangerous Goods Regulations. The statement above applies equally to all the Dangerous Goods classes discussed in this article.

Classification and Labeling Summary Tables

Class 1: Explosives

Information on this graphic changes depending on which, "Division" of explosive is shipped. Explosive Dangerous Goods have compatibility group letters assigned to facilitate segregation during transport. The letters used range from A to S excluding the letters I, M, O, P, Q and R. The example above shows an explosive with a compatibility group "A" (shown as 1.1A). The actual letter shown would depend on the specific properties of the substance being transported.

For example, the Canadian Transportation of Dangerous Goods Regulations provides a description of compatibility groups.

- 1.1 Explosives with a mass explosion hazard
  - o Ex: TNT, dynamite, nitroglycerine.
- 1.2 Explosives with a severe projection hazard.

- 1.3 Explosives with a fire, blast or projection hazard but not a mass explosion hazard.
- 1.4 Minor fire or projection hazard (includes ammunition and most consumer fireworks).
- 1.5 An insensitive substance with a mass explosion hazard (explosion similar to 1.1)
- 1.6 Extremely insensitive articles.

The United States Department of Transportation (DOT) regulates hazmat transportation within the territory of the US.

1.1 — Explosives with a mass explosion hazard. (nitroglycerin/dynamite)

1.2 — Explosives with a blast/projection hazard.

1.3 — Explosives with a minor blast hazard. (rocket propellant, display fireworks)

1.4 — Explosives with a major fire hazard. (consumer fireworks, ammunition)

1.5 — Blasting agents.

1.6 — Extremely insensitive explosives.

Class 2: Gases

Gases which are compressed, liquefied or dissolved under pressure as detailed below. Some gases have subsidiary risk classes; poisonous or corrosive.

- 2.1 Flammable Gas: Gases which ignite on contact with an ignition source, such as acetylene and hydrogen.
- 2.2 Non-Flammable Gases: Gases which are neither flammable nor poisonous. Includes the cryogenic gases/liquids (temperatures of below -100°C) used for cryopreservation and rocket fuels, such as nitrogen and neon.
- 2.3 Poisonous Gases: Gases liable to cause death or serious injury to human health if inhaled; examples are fluorine, chlorine, and hydrogen cyanide.

Class 3 Flammable Liquids

Flammable liquids included in Class 3 are included in one of the following packing groups:

- Packing Group I, if they have an initial boiling point of 35°C or less at an absolute pressure of 101.3 kPa and any flash point, such as diethyl ether or carbon disulfide;

- Packing Group II, if they have an initial boiling point greater than 35°C at an absolute pressure of 101.3 kPa and a flash point less than 23°C, such as gasoline (petrol) and acetone; or
- Packing Group III, if the criteria for inclusion in Packing Group I or II are not met, such as kerosene and diesel.

Other Hazardous Materials Labels (CHIP)

Australia

Australia uses the standard international UN numbers with a few slightly different signs on the back, front and sides of vehicles carrying hazardous substances. The country uses the same "HAZCHEM" as the UK HAZCHEM Code to provide advisory information to emergency services personnel in the event of an emergency situation.

Canada

Transportation of dangerous goods (hazardous materials) in Canada by road is normally a provincial jurisdiction. The federal government has jurisdiction over air, most marine, and most rail transport. The federal government acting centrally created the federal transportation of dangerous goods act and regulations, which provinces adopted in whole or in part via provincial transportation of dangerous goods legislation. The result is that all provinces use the federal regulations as their standard within their province; some small variances can exist because of provincial legislation. Creation of the federal regulations was coordinated by Transport Canada. Hazard classifications are based upon the UN Model.

Europe

The European Union has passed numerous directives and regulations to avoid the dissemination and restrict the usage of hazardous substances, the most famous being the Restriction of Hazardous Substances Directive and the REACH regulation. There are also long standing European treaties such as ADR and RID that regulate the transportation of hazardous materials by road, rail, river and inland waterways, following the guide of the UN Model Regulation.

United States

Due to the increased threat of terrorism in the early 21st century, funding for greater HAZMAT-handling capabilities was increased throughout the United States, in recognition of the fact that flammable, poisonous, explosive, or radioactive substances in particular could make attractive weapons for terrorist attacks.

The United States Department of Transportation (DOT) regulates hazmat transportation within the territory of the US. The regulations are found in 49 CFR (Title 49 of the Code of Federal Regulations).

The U.S. Occupational Safety and Health Administration (OSHA) regulates the handling of hazardous materials in the workplace as well as response to hazardous materials-related incidents, most notably through HAZWOPER (HAZ-ardous W-aste OP-erations and E-mergency R-esponse) regulations found at 29 CFR 1910.120.

The Environmental Protection Agency regulates hazardous materials as they may impact the community and environment, including specific regulations for environmental cleanup and for handling and disposal of waste hazardous materials.

The Consumer Product Safety Commission regulates hazardous materials that may be used in products sold for household and other consumer uses.

Hazard Classes for Materials in Transport

Following the UN Model, the DOT divides regulated hazardous materials into nine classes, some of which are further divided into divisions. Hazardous materials in transportation must be placarded and have specified packaging and labelling. Some materials must always be placarded, others may only require placarding in certain circumstances.

Trailers of goods in transport are usually marked with a four digit UN (United Nations) number. This number can be referenced by first responders (Firefighters, Police Officers, and ambulance personnel) who can find information about the material in the *Emergency Response Guidebook*.

Fixed Facilities

Different standards usually apply for handling and marking HAZMATs at fixed facilities, including NFPA 704 diamond markings (a consensus standard often adopted by local governmental jurisdictions), OSHA regulations requiring chemical safety information for employees, and CPSC requirements requiring informative labeling for the public, as well as wearing Hazmat suits when handling hazardous materials.

Hazardous Materials

Chemicals are found everywhere. They purify drinking water, increase crop production, and simplify household chores. But chemicals also can be hazardous to humans or the environment if used or released improperly. Hazards can occur during production, storage, transportation, use, or disposal. You and your community are at risk if a chemical is used unsafely or released in harmful amounts into the environment where you live, work, or play.

Hazardous materials in various forms can cause death, serious injury, long-lasting health effects, and damage to buildings, homes, and other property. Many products containing hazardous chemicals are used and stored in homes routinely. These products are also shipped daily on the nation's highways, railroads, waterways, and pipelines.

Chemical manufacturers are one source of hazardous materials, but there are many others, including service stations, hospitals, and hazardous materials waste sites.

Varying quantities of hazardous materials are manufactured, used, or stored at an estimated 4.5 million facilities in the United States—from major industrial plants to local dry cleaning establishments or gardening supply stores.

Hazardous materials come in the form of explosives, flammable and combustible substances, poisons, and radioactive materials. These substances are most often released as a result of transportation accidents or because of chemical accidents in plants.

Hazardous Waste

The term hazardous waste comprises all toxic chemicals, radioactive materials, and biologic or infectious waste. These materials threaten workers through occupational exposure and the general public in their homes, communities, and general environment. Exposure to these materials can occur near the site of generation, along the path of its transportation, and near their ultimate disposal sites. Most hazardous waste results from industrial processes that yield unwanted byproducts, defective products, and spilled materials. The generation and disposal of hazardous wastes is controlled through a variety of international and national regulations.

Hazardous waste was formerly known as 'special' waste.

Hazardous Waste

In addition to releasing gases and particles into the atmosphere, humans produce waste that is dumped on the environment. Often, this waste is hazardous and dangerous to both nature and human life. The levels of dangerous wastes continue to grow. Industries and individuals continue to be largely unaware of this major environmental problem. As a result, many people and industries are failing to prevent the creation of hazardous waste or to limit the negative effects it produces. Individuals often throw out goods without realising that they are headed for a landfill and could be dangerous for the environment. No matter where people put these hazardous waste materials, there is always a chance that they could find their way into the ground, and eventually into our bodies. Corporations usually want to avoid the costs associated with having to limit creation of hazardous waste. Consequently, they build landfills on site and fill them with waste, or sometimes pay to have their waste removed. Often, hazardous materials are transported to areas that accept money to take the waste. It may prove very difficult to reduce hazardous waste in the future. Unlike many other environmental problems, waste creation is something people do not often think about. In the future, people may have to reduce not only their generation of hazardous waste, but also their consumption of many products that end up in landfills.

Environmental Effects of Pesticides

Over 98% of sprayed insecticides and 95% of herbicides reach a destination other than their target species, including nontarget species, air, water, bottom sediments, and food. Pesticide contaminates land and water when it escapes from production sites and storage tanks, when it runs off from fields, when it is discarded, when it is sprayed aerially, and when it is sprayed into water to kill algae. The amount of pesticide that migrates from the intended application area is influenced by the particular chemical's properties: its propensity for binding to soil, its vapor pressure, its water solubility, and its resistance to being broken down over time. Factors in the soil, such as its texture, its ability to retain water, and the amount of organic matter contained in it, also affect the amount of pesticide that will leave the area. Some pesticides contribute to global warming and the depletion of the ozone layer.

Air

Pesticides can contribute to air pollution . Pesticide drift occurs when pesticides suspended in the air as particles are carried by wind to other areas, potentially contaminating them. Pesticides that are applied to crops can volatilize and may be blown by winds into nearby areas, potentially posing a threat to wildlife. Also, droplets of sprayed pesticides or particles from pesticides applied as dusts may travel on the wind to other areas, or pesticides may adhere to particles that blow in the wind, such as dust particles. Ground spraying produces less pesticide drift than aerial spraying does.Farmers can employ a buffer zone around their crop, consisting of empty land or non-crop plants such as evergreen trees to serve as windbreaks and absorb the pesticides, preventing drift into other areas. Such windbreaks are legally required in the Netherlands.

Pesticides that are sprayed onto fields and used to fumigate soil can give off chemicals called volatile organic compounds, which can react with other chemicals and form a pollutant called ozone, accounting for an estimated 6% of the total ozone production.

Water

In the United States, pesticides were found to pollute every stream and over 90% of wells sampled in a study by the US Geological Survey. Pesticide residues have also been found in rain and groundwater. Studies by the UK government showed that pesticide concentrations exceeded those allowable for drinking water in some samples of river water and groundwater. Pesticide impacts on aquatic systems are often studied using a hydrology transport model to study movement and fate of chemicals in rivers and streams. As early as the 1970s quantitative analysis of pesticide runoff was conducted in order to predict amounts of pesticide that would reach surface waters. There are four major routes through which pesticides reach the water: it may drift outside of the intended area when it is sprayed, it may percolate, or leach, through the soil, it may be carried to the water as runoff, or it may be spilled, for example accidentally or through neglect. They may also be carried to water by eroding soil. Factors that affect a pesticide's ability to contaminate water include its water solubility, the distance from an application site to a body of water, weather, soil type, presence of a growing crop, and the method used to apply the chemical. Maximum limits of allowable concentrations for individual pesticides in public bodies of water are set by the Environmental Protection Agency in the US. Similarly, the government of the United Kingdom sets Environmental Quality Standards (EQS), or maximum allowable concentrations of some pesticides in bodies of water above which toxicity may occur. The European Union also regulates maximum concentrations of pesticides in water.

Soil

Many of the chemicals used in pesticides are persistent soil contaminants, whose impact may endure for decades and adversely affect soil conservation.

The use of pesticides decreases the general biodiversity in the soil. Not using the chemicals results in higher soil quality, with the additional effect that more organic matter in the soil allows for higher water retention. This helps increase yields for farms in drought years, when organic farms have had yields 20-40% higher than their conventional counterparts. A smaller content of organic matter in the soil increases the amount of pesticide that will leave the area of application, because organic matter binds to and helps break down pesticides.

Plants

Nitrogen fixation, which is required for the growth of higher plants, is hindered by pesticides in soil. The insecticides DDT, methyl parathion, and especially pentachlorophenol have been shown to interfere with legume-rhizobium chemical signaling. Reduction of this symbiotic chemical signaling results in reduced nitrogen fixation and thus reduced crop yields. Root nodule formation in these plants saves the world economy $10 billion in synthetic nitrogen fertilizer every year. Pesticides can kill bees and are strongly implicated in pollinator decline, the loss of species that pollinate plants, including through the mechanism of Colony Collapse Disorder, in which worker bees from a beehive or Western honey bee colony abruptly disappear. Application of pesticides to crops that are in bloom can kill honeybees, which act as pollinators. The USDA and USFWS estimate that US farmers lose at least $200 million a year from reduced crop pollination because pesticides applied to fields eliminate about a fifth of honeybee colonies in the US and harm an additional 15%.

Persistent Organic Pollutants

Persistent organic pollutants (POPs) are compounds that resist degradation and thus remain in the environment for years Some pesticides, including aldrin, chlordane, DDT, dieldrin, endrin, heptachlor, hexachlorobenzene, mirex, and toxaphene, are considered POPs. POPs have the ability to volatilize and travel great distances through the atmosphere to become deposited in remote regions. The chemicals also have the ability to bioaccumulate and biomagnify, and can bioconcentrate (i.e. become more concentrated) up to 70,000 times their original concentrations. POPs may continue to poison non-target organisms in the environment and increase risk to humans by disruption in the endocrine, reproductive, and immune systems; cancer; neurobehavioral disorders, infertility and mutagenic effects, although very little is currently known about these chronic effects. Some POPs have been banned, while others continue to be used.

Animals

Pesticides inflict extremely widespread damage to biota, and many countries have acted to discourage pesticide usage through their Biodiversity Action Plans. Animals may be poisoned by pesticide residues that remain on food after spraying, for example when wild animals enter sprayed fields or nearby areas shortly after spraying.

Widespread application of pesticides can eliminate food sources that certain types of animals need, causing the animals to relocate, change their diet, or starve. Poisoning from pesticides can travel up the food chain; for example, birds can be harmed when they eat insects and worms that have consumed pesticides. Some pesticides can bioaccumulate, or build up to toxic levels in the bodies of organisms that consume them over time, a phenomenon that impacts species high on the food chain especially hard. The USDA and USFWS estimate that about 20% of the endangered and threatened species in the US are jeopardised by use of pesticides.

Birds

Birds are common examples of nontarget organisms that are impacted by pesticide use. Rachel Carson's landmark book *Silent Spring* dealt with the topic of loss of bird species due to bioaccumulation of pesticides in their tissues. There is evidence that birds are continuing to be harmed by pesticide use. In the farmland of Britain, populations of ten different species of birds have declined by 10 million breeding individuals between 1979 and 1999, a phenomenon thought to have resulted from loss of plant and invertebrate species on which the birds feed. Throughout Europe, 116 species of birds are now threatened. Reductions in bird populations have been found to be associated with times and areas in which pesticides are used. In another example, some types of fungicides used in peanut farming are only slightly toxic to birds and mammals, but may kill off earthworms, which can in turn reduce populations of the birds and mammals that feed on them. Some pesticides come in granular form, and birds and other wildlife may eat the granules, mistaking them for grains of food. A few granules of a pesticide is enough to kill a small bird. The herbicide paraquat, when sprayed onto bird eggs, causes growth abnormalities in embryos and reduces the number of chicks that hatch successfully, but most herbicides do not directly cause much harm to birds. Herbicides may endanger bird populations by reducing their habitat. The USDA and USFWS estimate that over 67 million birds are killed by pesticides each year in the US.

Aquatic Life

Fish and other aquatic biota may be harmed by pesticide-contaminated water. Pesticide surface runoff into rivers and streams can be highly lethal to aquatic life, sometimes killing all the fish in a particular stream. For example, in Montague P.E.I., nine "fish kills" happened in one year: every fish, snake, and snail was killed in a river called Sutherland's Hole near potato farms from which herbicides, insecticides, and fungicides ran off after heavy rains. Pesticide-related fish kills are frequently unreported and likely underestimated.

Application of herbicides to bodies of water can cause fish kills when the dead plants rot and use up the water's oxygen, suffocating the fish. Some herbicides, such as copper sulfite, that are applied to water to kill plants are toxic to fish and other water animals at concentrations similar to those used to kill the plants. Repeated exposure to sublethal doses of some pesticides can cause physiological and behavioral changes in fish that reduce populations, such as abandonment of nests and broods, decreased immunity to disease, and increased failure to avoid predators.

Application of herbicides to bodies of water can kill off plants on which fish depend for their habitat. Pesticides can accumulate in bodies of water to levels that kill off zooplankton, the main source of food for young fish. Pesticides can kill off the insects on which some fish feed, causing the fish to travel farther in search of food and exposing them to greater risk from predators. The USDA and USFWS estimate that between 6 and 14 million fish are killed by pesticides each year in the US. The faster a given pesticide breaks down in the environment, the less threat it poses to aquatic life. Insecticides are more toxic to aquatic life than herbicides and fungicides.

Amphibians

Some scientists believe that certain common pesticides already exist at levels capable of killing amphibians in California. They warn that the breakdown products of these pesticides can be 10 to 100 times more toxic to amphibians than the original pesticides. Direct contact of sprays of some pesticides (either by drift from nearby applications or accidental or deliberate sprays) can be highly lethal to amphibians. A Canadian study showed that exposing tadpoles to endosulfan, an organochloride pesticide at levels that are likely to be found in habitats near fields sprayed with the chemical kills the tadpoles and causes behavioral and growth abnormalities.

US scientists have found that some pesticides used in farming disrupt the nervous systems of frogs, and that use of these pesticides is correlated with a decline in the population of frogs in the Sierra Nevada. In the past several decades, decline in amphibian populations has been occurring all over the world, for unexplained reasons which are thought to be varied but of which pesticides may be a part. Being downwind from agricultural land on which pesticides are used has been linked to the decline in population of threatened frog species in California.

Mixtures of multiple pesticides appear to have a cumulative toxic effect on frogs Tadpoles from ponds with multiple pesticides present in the water take longer to metamorphose into frogs and are smaller when they do, decreasing their ability to catch prey and avoid predators.

In Minnesota, pesticide use has been causally linked to congenital deformities in frogs such as eye, mouth, and limb malformations Researchers in California found that similar deformities in frogs in the US and Canada may have been caused by breakdown products from pesticides which themselves did not pose a threat.

The herbicide atrazine has been shown to turn male frogs into hermaphrodites, decreasing their ability to reproduce.

Pest Resistance

Pests may evolve to become resistant to pesticides. Many pests will initially be very susceptible to pesticides, but some with slight variations in their genetic makeup are resistant and therefore survive to reproduce. Through natural selection, the pests may eventually become very resistant to the pesticide. Pest resistance to a pesticide is commonly managed through pesticide rotation, which involves alternating among pesticide classes with different modes of action to delay the onset of or mitigate existing pest resistance. Tank mixing pesticides is the combination of two or more pesticides with different modes of action in order to improve individual pesticide application results and delay the onset of or mitigate existing pest resistance.

Pest Rebound and Secondary Pest Outbreaks

Non-target organisms, organisms that the pesticides are not intended to kill, can be severely impacted by use of the chemicals. In some cases, where a pest insect has some controls from a beneficial predator or parasite, an insecticide application can kill both pest and beneficial populations. A study comparing biological pest control and use of pyrethroid insecticide for diamondback moths, a major cabbage family insect pest, showed that the insecticide application created a rebounded pest population due to loss of insect predators, whereas the biocontrol did not show the same effect. Likewise, pesticides sprayed in an effort to control adult mosquitoes, may temporarily depress mosquito populations, however they may result in a larger population in the long run by damaging the natural controlling factors. This phenomenon, wherein the population of a pest species rebounds to equal or greater numbers than it had before pesticide use, is called pest resurgence and can be linked to elimination of predators and other natural enemies of the pest.

Loss of predator species can also lead to a related phenomenon called secondary pest outbreaks, an increase in problems from species which were not originally very damaging pests due to loss of their predators or parasites. An estimated third of the 300 most damaging insects in the US were originally secondary pests and only became a major problem after the use of pesticides. In both pest resurgence and secondary pest outbreaks, the natural enemies have been found to be more susceptible to the pesticides than the pests themselves, in some cases causing the pest population to be higher than it was before the use of pesticide.

Radiation

The Industrial Pollution and Radiochemical Inspectorate (IPRI) is responsible for controlling the keeping and use of radioactive material and the disposal of radioactive waste under the Radioactive Substances Act 1993. Certificates of Registration are issued to permit the keeping and use of radioactive material and Certificates of Authorisation are used to permit the accumulation and disposal of radioactive waste. The Inspectorate also enforces the statutory control of the transport of radioactive material. As well as regulatory responsibilities, IPRI also carries out environmental monitoring of the impact of nuclear discharges into the Irish Sea on the Northern Ireland coastline. The Inspectorate has an important role to play in the Northern Ireland response should there be an overseas peacetime nuclear incident. During radioactive emergencies IPRI would work closely with other Northern Ireland Departments likely to be involved. We are also responsible for mapping levels of radon, a naturally occurring radioactive gas, in dwellings throughout Northern Ireland. We are also responsible for mapping levels of Radon, a naturally occurring Radioactive Gas in dwellings throughout Northern Ireland.

Radon

Radon is a naturally occurring radioactive gas formed as a result of the radioactive decay of uranium, which is present to some extent in all rocks and soils, but amounts vary from place to place. Radon rises from the soil into the air, outdoors, radon is diluted into the air and the risk it poses is negligible. Where it enters buildings the levels can vary from property to property and even between neighboring dwellings. Radon can build up in homes and, at high concentrations, leads to an increased risk of lung cancer. Smokers who are in a radon affected house are at an increased risk. Radon dominates the radioactive dose to the population contributing about 50% of dose at the average level in houses. Radioactivity is measured in Becquerels. The Government has established an Action Level for radon in homes of 200 Becquerels per cubic metre (Bq/m3), based on advice from the Health Protection Agency (HPA).

The Radon in Dwellings Report produced in 1999 by the Environment and Heritage Service includes a radon map of Northern Ireland and the risk and information on radon tests carried out in the Province. To date there have been 15700 houses tested in Northern Ireland. The average level is low at 19 Bq/m3, similar to the rest of the United Kingdom. The radon map shows that there are areas in the West of the Province and the Southeast where radon risk is elevated and it is estimated that 4000 homes in Northern Ireland are likely to exceed the action level. Free booklets can be obtained from the Industrial Pollution and Radiochemical Inspectorate (IPRI) in the Klondyke Building, Cromac Avenue, Gasworks Business Park, Lower Ormeau Road Belfast. Telephone: (028) 9056 9299. The booklets are

1. Radon - A Householder's Guide
2. Radon - A guide to reducing levels in your home
3. Radon - A guide to homebuyers and sellers
4. Radon Factsheet

If you wish to discuss any technical matters about radon please contact (028) 9056 9305. If anyone feels that they would like a test done on their house they could contact HPA (Health Protection Agency), Chilton, Didcot, Oxfordshire, Telephone freephone 0800 614529 and they will issue the test kits which cost in the region of £38.00, HPA will also confirm the cost. The control of exposure to radon in commercial and public buildings, including schools, is the responsibility of the Health and Safety Executive for Northern Ireland (Ladas Drive, Belfast) of the Department of

Enterprise, Trade and Investment or the appropriate District Council. Anyone wishing to view maps showing levels of Radon in dwellings in the Republic of Ireland should visit the website of the Radiological Protection Institute of Ireland.

Radioactivity

The Industrial Pollution and Radiochemical Inspectorate (IPRI) is responsible for controlling the keeping and use of radioactive material and the disposal of radioactive waste under the Radioactive Substances Act 1993. Certificates of Registration are issued to permit the keeping and use of radioactive material and Certificates of Authorisation are used to permit the accumulation and disposal of radioactive waste. The Inspectorate also enforces the statutory control of the transport of radioactive material.

As well as regulatory responsibilities, IPRI also carries out environmental monitoring of the impact of nuclear discharges into the Irish Sea on the Northern Ireland coastline.

The Inspectorate has an important role to play in the Northern Ireland response should there be an overseas peacetime nuclear incident. During radioactive emergencies IPRI would work closely with other Northern Ireland Departments likely to be involved. We are also responsible for mapping levels of radon, a naturally occurring radioactive gas, in dwellings throughout Northern Ireland.

We are also responsible for mapping levels of Radon, a naturally occurring Radioactive Gas in dwellings throughout Northern Ireland.

Radioactive Pollution

Surface waters are a powerful factor that causes migration of radionuclides across the territory of Belarus. For this reason it is essential to take into due account the transit role of rivers in the transportation of radionuclides, including transboundary transfer. In watercourses and flowing water bodies concentration of radionuclides are reducing every year, but they tend to accumulate in sttatic water bodies (lakes, ponds, reservoirs, especially in bottom sediments).

Due to the accident at the Chernobyl nuclear power station (CNPS) one quarter of the country's territory was contaminated by caesium-137 (23%), strontium-90 (10%) and plutonium (about 2 %). Concentrations of caesium-137 by the year 2001.

According to the monitoring data the radiation situation in the Dnieper-Sozh and Pripyat basins is stable. The annual average concentration of caesium-137 has decreased significantly in large and small rivers for the observed 1987-2000 period. Exceeding of the National permissible levels (NPL-96, 99) for caesium-137 and strontium-90 were not observed. However, caesium-137 concentration in surface waters are closely connected with the annual volume of river flow, as can be seen from the increase in concentrations of caesium-137 in some rivers, where water supply was lower than the perennial average values.

The data analysis of concentration of caesium-137 during the spring flood in the Pripyat basin in 1999 shows that concentration of caesium-137 in a dissolved form in the Pripyat basin (the city of Mozyr) remains at the level of the average indices for the prior period (1996-1998). Yet concentration of this radionuclide has considerably increased in dredges. This means that caesium-137 is washed out and transported by flood flow with sediments.

According to the measurements of the total radioactivity of caesium-137, transferred by rivers during 12 years (1987-1999), 85% were transferred by the Ipyt during 2 years, 81% by the Sozh during 3 years, the Besed during 4 years, the Pripyat during 6 years, and 71% by the Dnieper during 9 years. The comparison between the indices of caesium-137 transfer for the period of 1987-1993 at the end-point monitoring checkpoints on the Dnieper and the Pripyat in the Ukraine and at the monitoring checkpoints in Belarus shows that only 26% of the total transfer of radioactive caesium-137 were generated within the 30-kilometer Chernobyl zone. Thus, in the river runoff formation large watersheds had greater impact on secondary river pollution within the first few years (3-4 years) after the Chernobyl accident than surface river runoff from highly contaminated territories with relatively small watersheds.

Since the radiation situation has stabilized, transboundary transport of radioactive elements through river flow has significantly decreased. Mainly it is the Pripyat that transports radionuclides, in particular strontium-90, as they are washed out from the 30-kilometer Chernobyl zone.

Due attention is devoted to studies of the radiation state of small rivers, which are tributaries of the Pripyat (the Braginka, Nesvich and Slovechna Rivers) and the Sozh (the Lipa and the Senna) in the most contaminated areas of the Gomel and Mogilev regions. Over the years radioactivity in water tends to decrease. The exception is the dissolved strontium -90, which is a specific feature for the CNPS zone.

Annual data on the content of caesium-137 and strontium-90 (in soluble and suspended forms) in the bottom sediments and water biota suggest that bottom sediments and water biota are significant contributors to the total radioactivity of surface water systems. A tendency towards a reduction in radioactivity of bottom sediments and water biota is minor.

During spring high water and summer-autumn flood, migration of radionuclides into open water systems occurs both in soluble and in absorbed forms on organic and mineral carrieres.

The ratio between concentrations of caesium-137 and strontium-90 at the end checkpoints of the Braginka and Senna Rivers suggests that starting from1992-1993, the concentration of strontium-90 has begun to exceed that of caesium-137. That phenomenon is characteristic for surface watercourses close to the CNPS zone and is explained by an increase in migration ability of strontium-90 due to its release from active particles.

Radiation Monitoring

Background levels of radiation are monitored at various locations in the City by Pollution Team officers. Additionally, a radiation monitor continuously registers background levels at Walbrook Wharf. The data from the Walbrook Wharf monitor can be seen on the internet at www.weatherprobe.com. Officers are immediately alerted by the equipment if there are any significant changes in the background data.

Environment Monitoring

The Environment Agency (EA) is responsible under the Radioactive Substances Act 1993 for authorising discharges from Sellafield. The Centre for Environment, Fisheries and Aquaculture Science (CEFAS), an Executive Agency within the Ministry of Agriculture, Fisheries and Food (MAFF), has been sampling seawater, fish, shellfish and sediments in the Irish Sea since the early 1950's. In Northern Ireland the Environment and Heritage Service's Industrial Pollution and Radiochemical Inspectorate monitors the impact of Sellafield discharges on the Northern Ireland coastal environment.

The Industrial Pollution & Radiochemical Inspectorate arranges for samples of seaweed, sediment, fish, nephros, and winkles to be collected quarterly and forwarded to the CEFAS Research Laboratory at Lowerstoft. The seaweeds are collected in the Ards Peninsula and at Portrush; the marine life samples are obtained as far as possible from commercial landings at Kilkeel and Portavogie. Sediment samples are collected from the five marine loughs.

The Northern Ireland results are published annually in a report entitled "Radioactivity in Food and the Environment" (pdf - 3.32 MB) issued by the Environment Agency, Environment and Heritage Service, Food Standards Agency and the Scottish Environment Protection Agency. The results are also published in the Northern Ireland Abstract of Statistics. The levels of radioactivity measured indicate that they are of negligible radiological significance. In addition to this programme, the IPRI monitors the gamma dose rate in air over inter-tidal sediments in each district council area which has a coastline. The results to date indicate minimal radioactivity deposition and are consistent with normal background levels.

Radioactive Emergency Response

Following the Chernobyl accident in April 1996 the Northern Ireland Emergency Committee set up a specialised group, the Northern Ireland Technical Advice Group (NITAG). Civil Contingencies Policy Branch (CCPB) ensure that the most efficient and effective response can be made to assist the public during, and in the aftermath of a civil emergency.

NITAG operates under the Chairmanship of the Chief Radiochemical Inspector with professional membership of the group drawn from the Inspectorate and those Northern Departments likely to be involved in an overseas peacetime nuclear emergency. Under the National Response Plan, the Radiation Incident Monitoring Network (RIMNET) is in place throughout the UK. There are 92 monitoring stations continuously measuring gamma radiation dose rates throughout the United Kingdom with 5 stations in Northern Ireland. The 5 stations are at Ballypatrick, Aldergrove, Glenanne, St Angelo and Carnmoney.

The Inspectorate play a major role in NITAG activities locally, participates in the Radiation Incident Monitoring Co-ordinating Committee (RIMCC) and also take part in various exercises using the RIMNET system to test its reaction capability and effectiveness.

Near-Surface Disposal Facilities for Radioactive Waste.

In 1997, we published guidance on regulating the disposal of low and intermediate level radioactive waste to specialised facilities on land. We have now decided to update this guidance because of developments in UK law and policy, revised international advice and recommendations and also recognition that specific guidance for near-surface facilities was required. We are jointly publishing this guidance with the Environment Agency and the Scottish Environment Protection Agency. The guidance sets out the principles and requirements that we would expect a developer or operator of a radioactive waste disposal facility to meet to ensure protection of people and the environment, both now and in the future. We are consulting on the updated guidance between 15 May and1 September 2008. The consultation version of the guidance can be downloaded using the following link http://www.environment-agency.gov.uk/yourenv/consultations/2040401/ We have identified a number of questions in the consultation document, to which we invite you to respond. We would also be interested in any other matters or concerns relating to the consultation document that you might wish to draw to our attention.

Deep Geological Facilities for Radioactive Wastes

In 1997, we published guidance on regulating the disposal of low and intermediate level radioactive waste to specialised facilities on land. We have now decided to update this guidance because of developments in UK law and policy, revised international advice and recommendations and the need to address disposal of higher activity wastes, which were not covered by the 1997 guidance. We are jointly publishing this guidance with the Environment Agency. The guidance sets out the principles and requirements that we would expect a developer or operator of a radioactive waste disposal facility to meet to ensure protection of people and the environment, both now and in the future.

Radioactive Waste

Radioactive wastes are waste types containing radioactive chemical elements that do not have a practical purpose. They are sometimes the products of nuclear processes, such as nuclear fission. However, industries not directly connected to the nuclear industry can produce large quantities of radioactive waste. It has been estimated, for instance, that the past 20 years the oil-producing endeavors of the United States have accumulated eight million tons of radioactive wastes. The majority of radioactive waste is "low-level waste", meaning it contains low levels of radioactivity per mass or volume. This type of waste often consists of used protective clothing, which is only slightly contaminated but still dangerous in case of radioactive contamination of a human body through ingestion, inhalation, absorption, or injection. In the United States alone, the Department of Energy states that there are "millions of gallons of radioactive waste" as well as "thousands of tons of spent nuclear fuel and material" and also "huge quantities of contaminated soil and water". Despite these copious quantities of waste, the DOE has a goal of cleaning all presently contaminated sites successfully by 2025. The Fernald, Ohio site for example had "31 million pounds of uranium product", "2.5 billion pounds of waste", "2.75 million cubic yards of contaminated soil and debris", and a "223 acre portion of the underlying Great Miami Aquifer had uranium levels above drinking standards". The United States currently has at least 108 sites it currently designates as areas that are contaminated and unusable, sometimes many thousands of acres. The DOE wishes to try and clean or mitigate many or all by 2025, however the task can be difficult and it acknowledges that some will never be completely remediated, and just in one of these 108 larger designations, Oak Ridge National Laboratory, there were for example at least "167 known contaminant release sites" in one of the three subdivisions of the 37,000-acre (150 km^2) site.Some of the U.S. sites were smaller in nature, however, and cleanup issues were simpler to address, and the DOE has successfully completed cleanup, or at least closure, of several sites.

The issue of disposal methods for nuclear waste was one of the most pressing current problems the international nuclear industry faced when trying to establish a long term energy production plan, yet there was hope it could be safely solved. A recent research report on the Nuclear Industry perspective of the current state of scientific knowledge in predicting the extent that waste would find its way from the deep burial facility - back to soil and drinking water (such that it presents a direct threat to the health of human beings - as well as to other forms of life) is presented in a document from the IAEA (The International Atomic Energy Agency) - which was published in October 2007 This document states "The capacity to model all the effects involved in the dissolution of the waste form, in conditions similar to the disposal site, is the final goal of all the research undertaken by many research groups over many years. As we will see in this report, this kind of investigation is far from being finished". In the United States, the DOE acknowledges much progress in addressing the waste problems of the industry, and successful remediation of some contaminated sites, yet also major uncertainties and sometimes complications and setbacks in handling the issue properly, cost effectively, and in the projected time frame. In other countries with lower ability or will to maintain environmental integrity the issue would be even more problematic.

The Nature and Significance of Radioactive Waste

Radioactive waste typically comprises a number of radioisotopes: unstable configurations of elements that decay, emitting ionizing radiation which can be harmful to human health and to the environment. Those isotopes emit different types and levels of radiation, which last for different periods of time.

Physics

The radioactivity of all nuclear waste diminishes with time. All radioisotopes contained in the waste have a half-life - the time it takes for any radionuclide to lose half of its radioactivity and eventually all radioactive waste decays into non-radioactive elements. Certain radioactive elements (such as plutonium-239) in "spent" fuel will remain hazardous to humans and other living beings for hundreds of thousands of years. Other radioisotopes will remain hazardous for millions of years. Thus, these wastes must be shielded for centuries and isolated from the living environment for hundreds of millennia. Some elements, such as Iodine-131, have a short half-life (around 8 days in this case) and thus they will cease to be a problem much more quickly than other, longer-lived, decay products but their activity is much greater initially. The two tables show some of the major radioisotopes, their half-lives, and their radiation yield as a proportion of the yield of fission of Uranium-235.

The faster a radioisotope decays, the more radioactive it will be. The energy and the type of the ionizing radiation emitted by a pure radioactive substance are important factors in deciding how dangerous it will be. The chemical properties of the radioactive element will determine how mobile the substance is and how likely it is to spread into the environment and contaminate human bodies. This is further complicated by the fact that many radioisotopes do not decay immediately to a stable state but rather to a radioactive decay product leading to decay chains.

Chemistry

The chemical properties of the radioactive substance and the other substances found within (and near) the waste store has a great effect upon the ability of the waste to cause harm to humans or other organisms. For instance TcO_4^- tends to adsorb on the surfaces of steel objects which reduces its ability to move out of the waste store in water.

Pharmacokinetics

Exposure to high levels of radioactive waste may cause serious harm or death. Treatment of an adult animal with radiation or some other mutation-causing effect, such as a cytotoxic anti-cancer drug, may cause cancer in the animal. In humans it has been calculated that a 1 sievert dose has a 5% chance of causing cancer and a 1% chance of causing a mutation in a gamete which can be passed to the next generation. If a developing organism such as an unborn child is irradiated, then it is possible to induce a birth defect but it is unlikely that this defect will be in a gamete or a gamete forming cell.

Depending on the decay mode and the pharmacokinetics of an element (how the body processes it and how quickly), the threat due to exposure to a given activity of a radioisotope will differ. For instance Iodine-131 is a short-lived beta and gamma emitter but because it concentrates in the thyroid gland, it is more able to cause injury than cesium-137 which, being water soluble, is rapidly excreted in urine. In a similar way, the alpha emitting actinides and radium are considered very harmful as they tend to have long biological half-lives and their radiation has a high linear energy transfer value. Because of such differences, the rules determining biological injury differ widely according to the radioisotope, and sometimes also the nature of the chemical compound which contains the radioisotope.

Philosophy

The main objective in managing and disposing of radioactive (or other) waste is to protect people and the environment. This means isolating or diluting the waste so that the rate or concentration of any radionuclide returned to the biosphere is harmless. To achieve this the preferred technology to date has been deep and secure burial for the more dangerous wastes; transmutation, long-term retrievable storage, and removal to space have also been suggested. Management options for waste are discussed in this study. Radioactivity by definition reduces over time, so in principle the waste needs to be isolated for a particular period of time until its components have decayed such that it no longer poses a threat. In practice this can mean periods of hundreds of thousands of years, depending on the nature of the waste involved. Though an affirmative answer is often taken for granted, the question as to whether or not we should endeavor to avoid causing harm to remote future generations, perhaps thousands upon thousands of years hence, is essentially one which must be dealt with by philosophy.

Sources of Waste

Radioactive waste comes from a number of sources. The majority originates from the nuclear fuel cycle and nuclear weapon reprocessing. However, other sources include medical and industrial wastes, as well as naturally occurring radioactive materials (NORM) that can be concentrated as a result of the processing or consumption of coal, oil and gas, and some minerals.

Nuclear Fuel Cycle

Front End

Waste from the front end of the nuclear fuel cycle is usually alpha emitting waste from the extraction of uranium. It often contains radium and its decay products.

Uranium dioxide (UO_2) concentrate from mining is not very radioactive - only a thousand or so times as radioactive as the granite used in buildings. It is refined from yellowcake (U_3O_8), then converted to uranium hexafluoride gas (UF_6). As a gas, it undergoes enrichment to increase the U-235 content from 0.7% to about 4.4% (LEU). It is then turned into a hard ceramic oxide (UO_2) for assembly as reactor fuel elements.

The main by-product of enrichment is depleted uranium (DU), principally the U-238 isotope, with a U-235 content of ~0.3%. It is stored, either as UF_6 or as U_3O_8. Some is used in applications where its extremely high density makes it valuable, such as the keels of yachts, and anti-tank shells. It is also used (with recycled plutonium) for making mixed oxide fuel (MOX) and to dilute highly enriched uranium from weapons stockpiles which is now being redirected to become reactor fuel. This dilution, also called downblending, means that any nation or group that acquired the finished fuel would have to repeat the (very expensive and complex) enrichment process before assembling a weapon.

Back End

The back end of the nuclear fuel cycle, mostly spent fuel rods, contains fission products that emit beta and gamma radiation, and actinides that emit alpha particles, such as uranium-234, neptunium-237, plutonium-238 and americium-241, and even sometimes some neutron emitters such as californium (Cf). These isotopes are formed in nuclear reactors.

It is important to distinguish the processing of uranium to make fuel from the reprocessing of used fuel. Used fuel contains the highly radioactive products of fission (see high level waste below). Many of these are neutron absorbers, called neutron poisons in this context. These eventually build up to a level where they absorb so many neutrons that the chain reaction stops, even with the control rods completely removed. At that point the fuel has to be replaced in the reactor with fresh fuel, even though there is still a substantial quantity of uranium-235 and plutonium present. In the United States, this used fuel is stored, while in countries such as the United Kingdom, France, and Japan, the fuel is reprocessed to remove the fission products, and the fuel can then be re-used. This reprocessing involves handling highly radioactive materials, and the fission products removed from the fuel are a concentrated form of high-level waste as are the chemicals used in the process.

Proliferation Concerns

When dealing with uranium and plutonium, the possibility that they may be used to build nuclear weapons is often a concern. Active nuclear reactors and nuclear weapons stockpiles are very carefully safeguarded and controlled. However, high-level waste from nuclear reactors may contain plutonium. Ordinarily, this plutonium is reactor-grade plutonium, containing a mixture of plutonium-239 (highly suitable for building nuclear weapons), plutonium-240 (an undesirable contaminant and highly radioactive), plutonium-241, and plutonium-238; these isotopes are difficult to separate. Moreover, high-level waste is full of highly radioactive fission products. However, most fission products are relatively short-lived. This is a concern since if the waste is stored, perhaps in deep geological storage, over many years the fission products decay, decreasing the radioactivity of the waste and making the plutonium easier to access. Moreover, the undesirable contaminant Pu-240 decays faster than the Pu-239, and thus the quality of the bomb material increases with time (although its quantity decreases during that time as well). Thus, some have argued, as time passes, these deep storage areas have the potential to become "plutonium mines", from which material for nuclear weapons can be acquired with relatively little difficulty. Critics of the latter idea point out that the half-life of Pu-240 is 6,560 years and Pu-239 is 24,110 years, and thus the relative enrichment of one isotope to the other with time occurs with a half-life of 9,000 years (that is, it takes 9000 years for the *fraction* of Pu-240 in a sample of mixed plutonium isotopes, to spontaneously decrease by half— a typical enrichment needed to turn reactor-grade into weapons-grade Pu). Thus "weapons grade plutonium mines" would be a problem for the very far future (>9,000 years from now), so that there remains a great deal of time for technology to advance to solve this problem, before it becomes acute.

Pu-239 decays to U-235 which is suitable for weapons and which has a very long half life (roughly 10^9 years). Thus plutonium may decay and leave uranium-235. However, modern reactors are only moderately enriched with U-235 relative to U-238, so the U-238 continues to serve as denaturation agent for any U-235 produced by plutonium decay.

One solution to this problem is to recycle the plutonium and use it as a fuel e.g. in fast reactors. But in the minds of some, the very existence of the nuclear fuel reprocessing plant needed to separate the plutonium from the other elements represents a proliferation concern. In pyrometallurgical fast reactors, the waste generated is an actinide compound that cannot be used for nuclear weapons.

Nuclear Weapons Reprocessing

Waste from nuclear weapons reprocessing (as opposed to production, which requires primary processing from reactor fuel) is unlikely to contain much beta or gamma activity other than tritium and americium. It is more likely to contain alpha emitting actinides such as Pu-239 which is a fissile material used in bombs, plus some material with much higher specific activities, such as Pu-238 or Po.

In the past the neutron trigger for a bomb tended to be beryllium and a high activity alpha emitter such as polonium; an alternative to polonium is Pu-238. For reasons of national security, details of the design of modern bombs are normally not released to the open literature. It is likely however that a D-T fusion reaction in either an electrically driven device or a D-T fusion reaction driven by the chemical explosives would be used to start up a modern device.

Some designs might well contain a radioisotope thermoelectric generator using Pu-238 to provide a longlasting source of electrical power for the electronics in the device.

It is likely that the fissile material of an old bomb which is due for refitting will contain decay products of the plutonium isotopes used in it, these are likely to include alpha-emitting Np-236 from Pu-240 impurities, plus some U-235 from decay of the Pu-239; however, due to the relatively long half-life of these Pu isotopes, these wastes from radioactive decay of bomb core material would be very small, and in any case, far less dangerous (even in terms of simple radioactivity) than the Pu-239 itself.

The beta decay of Pu-241 forms Am-241; the in-growth of americium is likely to be a greater problem than the decay of Pu-239 and Pu-240 as the americium is a gamma emitter (increasing external-exposure to workers) and is an alpha emitter which can cause the generation of heat. The plutonium could be separated from the americium by several different processes; these would include pyrochemical processes and aqueous/organic solvent extraction. A truncated PUREX type extraction process would be one possible method of making the separation.

Medical

Radioactive medical waste tends to contain beta particle and gamma ray emitters. It can be divided into two main classes. In diagnostic nuclear medicine a number of short-lived gamma emitters such as technetium-99m are used. Many of these can be disposed of by leaving it to decay for a short time before disposal as normal trash. Other isotopes used in medicine, with half-lives in parentheses:

- Y-90, used for treating lymphoma (2.7 days)
- I-131, used for thyroid function tests and for treating thyroid cancer (8.0 days)
- Sr-89, used for treating bone cancer, intravenous injection (52 days)
- Ir-192, used for brachytherapy (74 days)
- Co-60, used for brachytherapy and external radiotherapy (5.3 years)
- Cs-137, used for brachytherapy, external radiotherapy (30 years)

Industrial

Industrial source waste can contain alpha, beta, neutron or gamma emitters. Gamma emitters are used in radiography while neutron emitting sources are used in a range of applications, such as oil well logging.

Naturally Occurring Radioactive Material (NORM)

Processing of substances containing natural radioactivity; this is often known as NORM. A lot of this waste is alpha particle-emitting matter from the decay chains of uranium and thorium. The main source of radiation in the human body is potassium-40 (^{40}K).

Coal

Coal contains a small amount of radioactive uranium, barium, thorium and potassium, but, in the case of pure coal, this is significantly less than the average concentration of those elements in the Earth's crust. However, the surrounding strata, if shale or mudstone, often contains slightly more than average and this may also be reflected in the ash content of 'dirty' coals. The more active ash minerals become concentrated in the fly ash precisely because they do not burn well. However, the radioactivity of fly ash is still very low. It is about the same as black shale and is less than phosphate rocks, but is more of a concern because a small amount of the fly ash ends up in the atmosphere where it can be inhaled.

Oil and gas

Residues from the oil and gas industry often contain radium and its daughters. The sulphate scale from an oil well can be very radium rich, while the water, oil and gas from a well often contains radon. The radon decays to form solid radioisotopes which form coatings on the inside of pipework. In an oil processing plant the area of the plant where propane is processed is often one of the more contaminated areas of the plant as radon has a similar boiling point as propane.

Types of Radioactive Waste

Although not significantly radioactive, uranium mill tailings are waste. They are byproduct material from the rough processing of uranium-bearing ore. They are sometimes referred to as 11(e)2 wastes, from the section of the U.S. Atomic Energy Act that defines them. Uranium mill tailings typically also contain chemically-hazardous heavy metals such as lead and arsenic. Vast mounds of uranium mill tailings are left at many old mining sites, especially in Colorado, New Mexico, and Utah.

Low level waste (LLW) is generated from hospitals and industry, as well as the nuclear fuel cycle. It comprises paper, rags, tools, clothing, filters, etc., which contain small amounts of mostly short-lived radioactivity. Commonly, LLW is designated as such as a precautionary measure if it originated from any region of an 'Active Area', which frequently includes offices with only a remote possibility of being contaminated with radioactive materials. Such LLW typically exhibits no higher radioactivity than one would expect from the same material disposed of in a non-active area, such as a normal office block. Some high activity LLW requires shielding during handling and transport but most LLW is suitable for shallow land burial. To reduce its volume, it is often compacted or incinerated before disposal. Low level waste is divided into four classes, class A, B, C and GTCC, which means "Greater Than Class C".

Intermediate level waste (ILW) contains higher amounts of radioactivity and in some cases requires shielding. ILW includes resins, chemical sludge and metal reactor fuel cladding, as well as contaminated materials from reactor decommissioning. It may be solidified in concrete or bitumen for disposal. As a general rule, short-lived waste (mainly non-fuel materials from reactors) is buried in shallow repositories, while long-lived waste (from fuel and fuel-reprocessing) is deposited in deep underground facilities. U.S. regulations do not define this category of waste; the term is used in Europe and elsewhere.

High level waste (HLW) is produced by nuclear reactors. It contains fission products and transuranic elements generated in the reactor core. It is highly radioactive and often thermally hot. LLW and ILW accounts for over 95% of the total radioactivity produced in the process of nuclear electricity generation. The amount of HLW worldwide is currently increasing by about 12,000 metric tons every year, which is the equival to about 100 double-decker busses or a two-story structure built on top of a basketball court.

Transuranic waste (TRUW) as defined by U.S. regulations is, without regard to form or origin, waste that is contaminated with alpha-emitting transuranic radionuclides with half-lives greater than 20 years, and concentrations greater than 100 nCi/g (3.7 MBq/kg), excluding High Level Waste. Elements that have an atomic number greater than uranium are called transuranic ("beyond uranium"). Because of their long half-lives, TRUW is disposed more cautiously than either low level or intermediate level waste. In the U.S. it arises mainly from weapons production, and consists of clothing, tools, rags, residues, debris and other items contaminated with small amounts of radioactive elements (mainly plutonium).

Under U.S. law, TRUW is further categorised into "contact-handled" (CH) and "remote-handled" (RH) on the basis of radiation dose measured at the surface of the waste container. CH TRUW has a surface dose rate not greater than 200 mrem per hour (2 mSv/h), whereas RH TRUW has a surface dose rate of 200 mrem per hour (2 mSv/h) or greater. CH TRUW does not have the very high radioactivity of high level waste, nor its high heat generation, but RH TRUW can be highly radioactive, with surface dose rates up to 1000000 mrem per hour (10000 mSv/h). The United States currently permanently disposes of TRUW generated from nuclear power plants and military facilities at the Waste Isolation Pilot Plant.

Management of Waste

Nuclear waste requires sophisticated treatment and management in order to successfully isolate it from interacting with the biosphere. This usually necessitates treatment, followed by a long-term management strategy involving storage, disposal or transformation of the waste into a non-toxic form.

Initial Treatment of waste

Vitrification

Long-term storage of radioactive waste requires the stabilisation of the waste into a form which will not react, nor degrade, for extended periods of time. One way to do this is through vitrification. Currently at Sellafield the high-level waste (PUREX first cycle raffinate) is mixed with sugar and then calcined. Calcination involves passing the waste through a heated, rotating tube. The purposes of calcination are to evaporate the water from the waste, and de-nitrate the fission products to assist the stability of the glass produced.

The 'calcine' generated is fed continuously into an induction heated furnace with fragmented glass. The resulting glass is a new substance in which the waste products are bonded into the glass matrix when it solidifies. This product, as a molten fluid, is poured into stainless steel cylindrical containers ("cylinders") in a batch process. When cooled, the fluid solidifies ("vitrifies") into the glass. Such glass, after being formed, is very highly resistant to water. According to the ITU, it will require about 1 million years for 10% of such glass to dissolve in water.

After filling a cylinder, a seal is welded onto the cylinder. The cylinder is then washed. After being inspected for external contamination, the steel cylinder is stored, usually in an underground repository. In this form, the waste products are expected to be immobilised for a very long period of time (many thousands of years).

The glass inside a cylinder is usually a black glossy substance. All this work (in the United Kingdom) is done using hot cell systems. The sugar is added to control the ruthenium chemistry and to stop the formation of the volatile RuO_4 containing radio ruthenium. In the west, the glass is normally a borosilicate glass (similar to Pyrex), while in the former Soviet bloc it is normal to use a phosphate glass. The amount of fission products in the glass must be limited because some (palladium, the other Pt group metals, and tellurium) tend to form metallic phases which separate from the glass. In Germany a vitrification plant is in use; this is treating the waste from a small demonstration reprocessing plant which has since been closed down.

Ion Exchange

It is common for medium active wastes in the nuclear industry to be treated with ion exchange or other means to concentrate the radioactivity into a small volume. The much less radioactive bulk (after treatment) is often then discharged. For instance, it is possible to use a ferric hydroxide floc to remove radioactive metals from aqueous mixtures. After the radioisotopes are absorbed onto the ferric hydroxide, the resulting sludge can be placed in a metal drum before being mixed with cement to form a solid waste form. In order to get better long-term performance (mechanical stability) from such forms, they may be made from a mixture of fly ash, or blast furnace slag, and portland cement, instead of normal concrete (made with portland cement, gravel and sand).

Synroc

The Australian Synroc (synthetic rock) is a more sophisticated way to immobilize such waste, and this process may eventually come into commercial use for civil wastes (it is currently being developed for U.S. military wastes). Synroc was invented by the late Prof Ted Ringwood (a geochemist) at the Australian National University. The Synroc contains pyrochlore and cryptomelane type minerals. The original form of Synroc (Synroc C) was designed for the liquid high level waste (PUREX raffinate) from a light water reactor. The main minerals in this Synroc are hollandite ($BaAl_2Ti_6O_{16}$), zirconolite ($CaZrTi_2O_7$) and perovskite ($CaTiO_3$). The zirconolite and perovskite are hosts for the actinides. The strontium and barium will be fixed in the perovskite. The caesium will be fixed in the hollandite.

Long Term Management of Waste

We are talking here of durations ranging from 10,000 to 1,000,000 years, according to studies based on the effect of estimated radiation doses. It is worthwhile noting that state of the art only allows geological considerations for such long periods. Researchers suggest that forecasts of health detriment for such periods *should be examined critically*. Practical studies only consider up to 100 years as far as effective planning and cost evaluations are concerned.

Storage

High-level radioactive waste is stored temporarily in spent fuel pools and in dry cask storage facilities. This allows the shorter-lived isotopes to decay before further handling.

In 1997, in the 20 countries which account for most of the world's nuclear power generation, spent fuel storage capacity at the reactors was 148,000 tonnes, with 59% of this utilized. However, a number of nuclear power plants in countries that do not reprocess had nearly filled their spent fuel pools, and resorted to Away-from-reactor storage (AFRS). AFRS capacity in 1997 was 78,000 tonnes, with 44% utilized, and annual additions of about 12,000 tonnes. AFRS cannot be expanded forever, and the lead times for final disposal sites have proven to be unpredictable (see below).

In 1989 and 1992, France commissioned commercial plants to vitrify HLW left over from reprocessing oxide fuel, although there are adequate facilities elsewhere, notably in the United Kingdom and Belgium. The capacity of these western European plants is 2,500 canisters (1000 t) a year, and some have been operating for 18 years.

Geological Disposal

The process of selecting appropriate deep final repositories for high level waste and spent fuel is now under way in several countries with the first expected to be commissioned some time after 2010. However, many people remain uncomfortable with the immediate stewardship cessation of this management system. In Switzerland, the Grimsel Test Site is an international research facility investigating the open questions in radioactive waste disposal). Sweden is well advanced with plans for direct disposal of spent fuel, since its Parliament decided that this is acceptably safe, using the KBS-3 technology. In Germany, there is a political discussion about the search for an *Endlager* (final repository) for radioactive waste, accompanied by loud protests especially in the Gorleben village in the Wendland area, which was seen ideal for the final repository until 1990 because of its location next to the border to the former German Democratic Republic. Gorleben is presently being used to store radioactive waste non-permanently, with a decision on final disposal to be made at some future time. The U.S. has opted for a final repository at Yucca Mountain in Nevada, but this project is widely opposed and is a hotly debated topic, with some of the main concerns being the long distance transportation of the waste from across the United States to this area, and the possibility of accidents over time that could occur. The Waste Isolation Pilot Plant in the United States is the world's first underground repository for transuranic waste. There is also a proposal for an international HLW repository in optimum geology, with Australia or Russia as possible locations, although the proposal for a global repository for Australia has raised fierce domestic political objections.

The Canadian government, for example, is seriously considering this method of disposal, known as the *Deep Geological Disposal* concept. Under the current plan, a vault is to be dug 500 to 1000 meters below ground, under the Canadian Shield, one of the most stable landforms on the planet. The vaults are to be dug inside geological formations known as *batholiths*, formed about a billion years ago. The used fuel bundles will be encased in a corrosion-resistant container, and further surrounded by a layer of *buffer material*, possibly of a special kind of clay (bentonite clay). The case itself is designed to last for thousands of years, while the clay would further slow the corrosion rates of the container. The batholiths themselves are chosen for their low ground-water movement rates, geological stability, and low economic value.

The Finnish government has already started building a vault to store nuclear waste 500 to 1000 meters below ground, not far from the Olkiluoto Nuclear Power Plant. In the EU, Covra is negotiating about a European-wide waste disposal system with single disposal sites that can be used by several EU-countries. This EU-wide storage possibility is being researched under the SAPIERR-2 programme. Storing high level nuclear waste above ground for a century or so is considered appropriate by many scientists. This allows for the material to be more easily observed and any problems detected and managed, while the decay over this time period significantly reduces the level of radioactivity and the associated harmful effects to the container material. It is also considered likely that over the next century newer materials will be developed which will not break down as quickly when exposed to a high neutron flux thus increasing the longevity of the container once it is permanently buried. Sea-based options for disposal of radioactive waste include burial beneath a stable abyssal plain, burial in a subduction zone that would slowly carry the waste downward into the Earth's mantle, and burial beneath a remote natural or human-made island. While these approaches all have merit and would facilitate an international solution to the vexing problem of disposal of radioactive waste, they are currently not being seriously considered because of the legal barrier of the Law of the Sea and because in North America and Europe sea-based burial has become taboo from fear that such a repository could leak and cause widespread damage. Dumping of radioactive waste from ships has reinforced this concern, as has contamination of islands in the Pacific. However, sea-based approaches might come under consideration in the future by individual countries or groups of countries that cannot find other acceptable solutions.

Article 1 (Definitions), 7., of the 1996 Protocol to the Convention on the Prevention of Marine Pollution by Dumping of Wastes and Other Matter, (the London Dumping Convention) states:

"Sea" means all marine waters other than the internal waters of States, as well as the seabed and the subsoil thereof; it does no include sub-seabed repositories accessed only from land." A subduction zone accessed from land as is the embodiment of the subductive waste disposal method http://www3.telus.net/subductionservices/ is not nor ever has been prohibited by international agreement. This method has been described as the most viable means of disposing of radioactive and the state-of-the-art in nuclear waste disposal technology. Another approach termed Remix & Return would blend high-level waste with uranium mine and mill tailings down to the level of the original radioactivity of the uranium ore, then replace it in empty uranium mines. This approach has the merits of totally eliminating the problem of high-level waste, of providing jobs for miners who would double as disposal staff, and of facilitating a cradle-to-grave cycle for all radioactive materials.

Transmutation

There have been proposals for reactors that consume nuclear waste and transmute it to other, less-harmful nuclear waste. In particular, the Integral Fast Reactor was a proposed nuclear reactor with a nuclear fuel cycle that produced no transuranic waste and in fact, could consume transuranic waste. It proceeded as far as large-scale tests but was then canceled by the U.S. Government. Another approach, considered safer but requiring more development, is to dedicate subcritical reactors to the transmutation of the left-over transuranic elements.

Transmutation was banned in the US on April 1977 by President Carter due to the danger of plutonium proliferation however, President Reagan rescinded the ban in 1981. Due to the economic losses and risks, construction of reprocessing plants during this time did not resume. Due to high energy demand, work on the method has continued in the EU. This has resulted in a practical nuclear research reactor called Myrrha in which transmutation is possible. Additionally, a new research programme called ACTINET has been started in the EU to make transmutation possible on a large, industrial scale. According to President Bush's Global Nuclear Energy Partnership (GNEP) of 2007, the US is now actively promoting research on transmutation technologies needed to markedly reduce the problem of nuclear waste treatment.

There have also been theoretical studies involving the use of fusion reactors as so called "actinide burners" where a fusion reactor plasma such as in a tokamak, could be "doped" with a small amount of the "minor" transuranic atoms which would be transmuted (meaning fissioned in the actinide case) to lighter elements upon their successive bombardment by the very high energy neutrons produced by the fusion of deuterium and tritium in the reactor. It was recently found by a study done at MIT, that only 2 or 3 fusion reactors with parameters similar to that of the International Thermonuclear Experimental Reactor (ITER) could transmute the entire annual minor actinide production from all of the light water reactors presently operating in the United States fleet while simultaneously generating approximately 1 gigawatt of power from each reactor.

Reuse of Waste

Another option is to find applications of the isotopes in nuclear waste so as to reuse them. . Already, caesium-137, strontium-90 and a few other isotopes are extracted for certain industrial applications such as food irradiation and radioisotope thermoelectric generators. While re-use does not eliminate the need to manage radioisotopes, it may reduce the quantity of waste produced.

Space Disposal

Space disposal is an attractive notion because it permanently removes nuclear waste from the environment. However, it has significant disadvantages, not least of which is the potential for catastrophic failure of a launch vehicle. Furthermore, the high number of launches that would be required — due to the fact that no individual rocket would be able to carry very much of the material relative to the material needed to be disposed of—makes the proposal impractical (for both economic and risk-based reasons). To further complicate matters, international agreements on the regulation of such a programme would need to be established.

It has been suggested that through the use of a stationary launch system many of the risks of catastrophic launch failure could be avoided. A promising concept is the use of high power lasers to launch "indestructible" containers from the ground into space. Such a system would require no rocket propellant, with the launch vehicle's payload making up a near entirety of the vehicle's mass. Without the use of rocket fuel on board there would be little chance of the vehicle exploding.

Another form of safe removal would possibly be the space elevator. Encasing the waste in glassified form inside a steel shell 9 inches (230 mm) thick, which in turn is tiled with shuttle tile to its exterior. If the launch vehicle fails just before reaching orbit, the waste ball will safely re-enter the earth's atmosphere. The steel shell would deform on impact, but would not rupture due to the density of the shell. Also, this would potentially allow the waste to be shot into the Sun.

Accidents Involving Radioactive Waste

A number of incidents have occurred when radioactive material was disposed of improperly, shielding during transport was defective, or when it was simply abandoned or even stolen from a waste store. In the former Soviet Union, waste stored in Lake Karachay was blown over the area during a dust storm after the lake had partly dried out. At Maxey Flat, a low-level radioactive waste facility located in Kentucky, containment trenches covered with dirt, instead of steel or cement, collapsed under heavy rainfall into the trenches and filled with water. The water that invaded the trenches became radioactive and had to be disposed of at the Maxey Flat facility itself. In other cases of radioactive waste accidents, lakes or ponds with radioactive waste accidentally overflowed into the rivers during exceptional storms.

Scavenging of abandoned radioactive material has been the cause of several other cases of radiation exposure, mostly in developing nations, which may have less regulation of dangerous substances (and sometimes less general education about radioactivity and its hazards) and a market for scavenged goods and scrap metal. The scavengers and those who buy the material are almost always unaware that the material is radioactive and it is selected for its aesthetics or scrap value. Irresponsibility on the part of the radioactive material's owners, usually a hospital, university or military, and the absence of regulation concerning radioactive waste, or a lack of enforcement of such regulations, have been significant factors in radiation exposures. For an example of an accident involving radioactive scrap originating from a hospital.

Transportation accidents involving spent nuclear fuel from power plants are unlikely to have serious consequences due to the strength of the spent nuclear fuel shipping casks.

Radioactive Waste in Fiction and Popular Culture

In fiction, radioactive waste is often cited as the reason for gaining super-human powers and abilities. An example of this fictional scenario is the 1981 movie "Modern Problems" in which actor Chevy Chase portrays a jealous, harried air traffic controller

Max Fiedler; Max Fiedler, recently dumped by his girlfriend, comes into contact with nuclear waste and is granted the power of telekinesis, which he uses to not only win her back, but to gain a little revenge.

In reality, of course, exposure to radioactive waste instead would lead to illness and/or death. In the science fiction television series, "Space: 1999," a massive nuclear waste dump on the Moon explodes, hurtling the Moon, and the inhabitants of "Moonbase Alpha" out of the Solar System at interstellar speeds. In the television comedy series Family Guy, the Griffin family all get super-human powers from toxic waste. When the local mayor Adam West tries to do the same thing, he gets lymphoma. In The Simpsons, many mutant three-eyed fish live near the Springfield Nuclear Power Plant.

Radioactive Contamination

Radioactive contamination is the uncontrolled distribution of radioactive material in a given environment. The amount of radioactive material released in an accident is called the source term.

Sources of Contamination

Radioactive contamination is typically the result of a spill or accident during the production or use of radionuclides (radioisotopes), an unstable nucleus which has excessive energy. Contamination may occur from radioactive gases, liquids or particles. For example, if a radionuclide used in nuclear medicine is accidentally spilled, the material could be spread by people as they walk around. Radioactive contamination may also be an inevitable result of certain processes, such as the release of radioactive xenon in nuclear fuel reprocessing. In cases that radioactive material cannot be contained, it may be diluted to safe concentrations. Nuclear fallout is the distribution of radioactive contamination by a nuclear explosion. For a discussion of environmental contamination by alpha emitters please see actinides in the environment. Containment is what differentiates radioactive material from radioactive contamination. Therefore, radioactive material in sealed and designated containers is not properly referred to as contamination, although the units of measurement might be the same.

Measurement

Radioactive contamination may exist on surfaces or in volumes of material or air. In a nuclear power plant, detection and measurement of radioactivity and contamination is often the job of a Certified Health Physicist.

Surface Contamination

Surface contamination is usually expressed in units of radioactivity per unit of area. For SI, this is becquerels per square meter (or Bq/m^2). Other units such as picoCuries per 100 cm² or disintegrations per minute per square centimeter (1 dpm/cm² = 166 2/3 Bq/m²) may be used. Surface contamination may either be fixed or removable. In the case of fixed contamination, the radioactive material cannot by definition be spread, but it is still measurable.

Hazards

In practice there is no such thing as zero radioactivity. Not only is the entire world constantly bombarded by cosmic rays, but every living creature on earth contains significant quantities of carbon-14 and most (including humans) contain significant quantities of potassium-40. These tiny levels of radiation are not any more harmful than sunlight, but just as excessive quantities of sunlight can be dangerous, so too can excessive levels of radiation.

Low Level Contamination

The hazards to people and the environment from radioactive contamination depend on the nature of the radioactive contaminant, the level of contamination, and the extent of the spread of contamination. Low levels of radioactive contamination pose little risk, but can still be detected by radiation instrumentation. In the case of low-level contamination by isotopes with a short half-life, the best course of action may be to simply allow the material to naturally decay. Longer-lived isotopes should be cleaned up and properly disposed off, because even a very low level of radiation can be life-threatening when in long exposure to it. Therefore, whenever there's any radiation in an area, many people take extreme caution when approaching.

High Level Contamination

High levels of contamination may pose major risks to people and the environment. People can be exposed to potentially lethal radiation levels, both externally and internally, from the spread of contamination following an accident (or a deliberate initiation) involving large quantities of radioactive material. The biological effects of external exposure to radioactive contamination are generally the same as those from an external radiation source not involving radioactive materials, such as x-ray machines, and are dependent on the absorbed dose.

Biological Effects

The biological effects of internally deposited radionuclides depend greatly on the activity and the biodistribution and removal rates of the radionuclide, which in turn depends on its chemical form. The biological effects may also depend on the chemical toxicity of the deposited material, independent of its radioactivity. Some radionuclides may be generally distributed throughout the body and rapidly removed, as is the case with tritiated water. Some radionuclides may target specific organs and have much lower removal rates. For instance, the thyroid gland takes up a large percentage of any iodine that enters the body. If large quantities of radioactive iodine are inhaled or ingested, the thyroid may be impaired or destroyed, while other tissues are affected to a lesser extent. Radioactive iodine is a common fission product; it was a major component of the radiation released from the Chernobyl disaster, leading to many cases of pediatric thyroid cancer and hypothyroidism. On the other hand, radioactive iodine is used in the diagnosis and treatment of many diseases of the thyroid precisely because of the thyroid's selective uptake of iodine.

Means of Contamination

Radioactive contamination can enter the body through ingestion, inhalation, absorption, or injection. For this reason, it is important to use personal protective equipment when working with radioactive materials. Radioactive contamination may also be ingested as the result of eating contaminated plants and animals or drinking contaminated water or milk from exposed animals. Following a major contamination incident, all potential pathways of internal exposure should be considered.

Phosphate Fertilizers & Water Pollution

Phosphates are a major source of pollution in lakes and streams, and high phosphate levels support over-production of algae and water weeds. However, many of us have misconceptions regarding the source of polluting phosphates, and many homeowners unknowingly contribute to the problem.

Lawn and Garden fertilizers often are implicated as the major source of phosphate pollution. However, research clearly demonstrates that with proper application, fertilizer does not pollute. When phosphates are applied to soils, they quickly bind to soil particles, much like a magnet picks up paper clips. Soil-bound phosphates contribute to pollution only when soil erosion occurs. Research studies found little or no difference in the phosphate content of storm runoff from lawn fertilizers with and without phosphate. Problems arise, however, when fertilizer is over-spread or spilled onto hard surfaces like streets, driveways and sidewalks. Here, the phosphate washes with rain through the storm drains into lakes and rivers. Likewise, grass clippings and leaves that fall on hard surfaces release their phosphorous into water sources. Research studies indicate that 80 percent of the phosphorous from urban settings comes from lawn clippings and leaves that end up in street gutters. While a few grass clippings mowed into the street look rather innocent, collectively they have a major impact on our water quality. Using a mulching lawn mower to keep lawn clippings on your lawn is especially useful. Another important source of phosphate pollution comes from soil erosion caused by water or wind. When soil moves, it takes the soil-bound phosphates with it. Construction sites and sparsely vegetated ground also cause erosion. You can take an active part in protecting Colorado's water quality by keeping fertilizers, leaves and grass clippings off streets and driveways. Prevent soil erosion by planting slopes with grass or other plants. We all have a part in protecting our water, one of Colorado's most valuable resources!

For more information, see the following Colorado State Extension fact sheet(s).

- Soil, Water and Plant Testing
- Lawn Care
- Composting Yard Waste
- Organic Fertilizers
- Xeriscaping: Ground Cover Plants
- Ornamental Grasses
- Xeriscaping: Retrofit Your Yard
- Vegetable Garden: Soil Management and Fertilisation

REFERENCES

Bingham, S (2007), Pesticides in rivers and groundwater. Environment Agency, UK. Retrieved on 2007-10-12.

Blankenship, Karl (1997-10). NOx in the Air: Multiple Effects. Chesapeake Bay Journal. Retrieved on 2008-06-04.

Bob Joynt & Stephen Wu, *Nitrogen oxides emissions standards for domestic gas appliances background study* Combustion Engineering Consultant; February 2000

Brassard, PG Wetlands and Water Pollution BOSTON COLL. ENVIRON. AFF. LAW REV., vol. 23, no. 4, pp. 885-919, 1996

Buol, S. W.; Hole, F. D. and McCracken, R. J. (1973). Soil Genesis and Classification, First, Ames, IA: Iowa State University Press.

Centers for Disease Control and Prevention. Pesticides. cdc.gov. Retrieved on September 15, 2007.

China has its worst spell of acid rain, United Press International.

Coal Combustion - ORNL Review Vol. 26, No. 3&4, 1993

Cone M (December 6, 2000), A wind-borne threat to Sierra frogs: A study finds that pesticides used on farms in the San Joaquin Valley damage the nervous systems of amphibians in Yosemite and elsewhere. *L.A. Times* Retrieved on September 17, 2007.

Cornell University. Pesticides in the environment. Pesticide fact sheets and tutorial, module 6. Pesticide Safety Education Programme. Retrieved on 2007-10-11.

Current EU approved additives and their E Numbers, The Food Standards Agency website.

D. Fowler, et al. (1998). "The atmospheric budget of oxidised nitrogen and its role in ozone formation and deposition". New Phytologist 139: 11-23. doi:10.1046/j.1469-8137.1998.00167.x.

D.A. Crossley, *Roles of Microflora and fauna in soil systems*, International Symposium on Pesticides in Soils, Feb. 25, 1970, University of Michigan

Daly H, Doyen JT, and Purcell AH III (1998), *Introduction to insect biology and diversity*, 2nd edition. Oxford University Press. New York, New York. Chapter 14, Pages 279-300.

Decommissioning costs of WWER-440 nuclear power plants. International Atomic Energy Agency (November 2002).

Dooley, Alan (June, 2006). Sandboils 101: Corps has experience dealing with common flood danger. US Army Corps of Engineers. Retrieved on 2008-05-14.

E.A. Davidson & W. Kingerlee (1997). "A global inventory of nitric oxide emissions from soils". Nutrient Cycling in Agroecosystems 48: 37-50. doi:10.1023/A:1009738715891.

Environment Canada (September-October 2001), Agricultural pesticides and the atmosphere. Retrieved on 2007-10-12.

EU-wide centralised geological waste disposal sites

Fox, JE, Gulledge, J, Engelhaupt, E, Burrow, ME, and McLachlan, JA (2007). "Pesticides reduce symbiotic efficiency of nitrogen-fixing rhizobia and host plants". Proceedings of the National Academy of Sciences of the USA 104: 10282-10287.

Gannon, RW; Osmond, DL; Humenik, FJ; Gale, JA; Spooner, J Agricultural Water Quality WATER RESOUR BULL, vol. 32, no. 3, pp. 437-450, 1996

Gerrard, Michael B. (2007-09-25). Global Climate Change and U.S. Law. American Bar Association, 38. ISBN 1590318161.

Gilliom, RJ, Barbash, JE, Crawford, GG, Hamilton, PA, Martin, JD, Nakagaki, N, Nowell, LH, Scott, JC, Stackelberg, PE, Thelin, GP, and Wolock, DM (February 15, 2007), The Quality of our nation's waters: Pesticides in the nation's streams and ground water, 1992–2001. Chapter 1, Page 4. US Geological Survey. Retrieved on September 13, 2007.

GlobalSecurity.org, Chelyabinsk-65/Ozersk, retrieved September 2007

Graeme Murphy (December 1 2005), Resistance Management - Pesticide Rotation. Ontario Ministry of Agriculture, Food and Rural Affairs. Retrieved on September 15, 2007.

Greenwood, N. N.; Earnshaw, A. (1997). Chemistry of the Elements, 2nd Edition, Oxford:Butterworth-Heinemann. p. 700

Greenwood, N. N.; Earnshaw, A. (1997). Chemistry of the Elements, 2nd Edition, Oxford:Butterworth-Heinemann.

Groundwater – the processes and global significance of aquifer degradation

Hackenberg D (2007-03-14). Letter from David Hackenberg to American growers from March 14, 2007 (English). Plattform Imkerinnen — Austria. Retrieved on 2007-03-27.

Haefeker, Walter (2000-08-12). Betrayed and sold out – German bee monitoring (English). Retrieved on 2007-10-10.

Health and Environmental Impacts of NOx. United States Environmental Protection Agency. Retrieved on 2007-12-26.

Helfrich, LA, Weigmann, DL, Hipkins, P, and Stinson, ER (June 1996), Pesticides and aquatic animals: A guide to reducing impacts on aquatic systems. Virginia Cooperative Extension. Retrieved on 2007-10-14.

Hogan,, CM, Patmore L, Latshaw, G, Seidman, H, et al. (1973), *Computer modeling of pesticide transport in soil for five instrumented watersheds*, U.S. Environmental Protection Agency Southeast Water laboratory, Athens, Ga. by ESL Inc., Sunnyvale, California.

Holleman, A. F.; Wiberg, E. "Inorganic Chemistry" Academic Press: San Diego, 2001.

Homer-Dixon, Thomas. "Environment, Scarcity, and Violence." Princeton University Press. (1999).

How is high-level nuclear waste managed in Canada?. *The Canadian Nuclear FAQ*. Retrieved on June 28, 2006.

http://reports.eea.europa.eu/92-826-5409-5/en Europe's Environment: The Dobris Assessment]

http://www.worldwater.org/data.html The World's Water 2006-2007 Tables, Pacific Institute.

IAEA Waste Management Database: Report 3 - L/ILW-LL. International Atomic Energy Agency (2000-03-28).

International Atomic Energy Agency, *The radiological accident in Goiânia*, 1988, retrieved September 2007

International Atomic Energy Agency, *The radiological accident in Goiânia*, 1988, retrieved September 2007

Issues relating to safety standards on the geological disposal of radioactive waste. International Atomic Energy Agency (2001-06-22). Retrieved on 2008-06-06..

J.N. Galloway, et al. (Sep 2004). "Nitrogen cycles: past, present, and future". Biogeochemistry 70 (2): 153-226. doi:10.1007/s10533-004-0370-0.

James A. Danoff-Burg, Columbia University The Terrestrial Influence: Geology and Soils

Johnston, AE (1986). "Soil organic-matter, effects on soils and crops". Soil Use Management 2: 97-105.

Jones, J. A. A. (1976). "Soil piping and stream channel initiation". Water Resources Research 7 (3): 602 - 610.

Kellogg RL, Nehring R, Grube A, Goss DW, and Plotkin S (February 2000), Environmental indicators of pesticide leaching and runoff from farm fields. United States Department of Agriculture Natural Resources Conservation Service. Retrieved on 2007-10-03.

Kerbs JR, Wilson JD, Bradbury RB, and Siriwardena GM (August 12, 1999), The second silent spring. Commentary in *Nature*, Volume 400, Pages 611-612.

Krivtsov, A.I., 2006, Geoenvironmental Problems of Mineral Resources Development, in *Geology and Ecosystems*, Zekster (Ru), Marker (UK), Ridgeway (UK), Rogachevskayarochmaninoff (Ru), & Vartanyan (Ru), 2006 Springer Inc.,

Lave, Lester B.; Eugene P. Seskin (1973). "An Analysis of the Association Between U.S. Mortality and Air Pollution". J. Amer. Statistical Association 68: 342.

Lindsey, T; Neese, S; Thomas, D Pollutin Prevention WATER QUAL. INT., pp. 32-36, 1996

Lotter DW, Seidel R, and Liebhardt W (2003). "The performance of organic and conventional

cropping systems in an extreme climate year". American Journal of Alternative Agriculture 18: 146-154.

Lutgens, Frederick K. and Edward J. Tarbuck (1995) *The Atmosphere*, Prentice Hall, 6th ed., pp14-17, ISBN 0-13-350612-6

M. I. Ojovan, W.E. Lee. *An Introduction to Nuclear Waste Immobilisation*, Elsevier Science Publishers B.V., Amsterdam, 315pp. (2005)

MacDonnell, LJ Water Quality LAND WATER LAW REV., vol. 31, no. 2, pp. 329-348, 1996

MacKenzie, SH The Ecosystem ISLAND PRESS, CA (USA), 1996, 240 pp

Marathon Resources Ltd :: Our Business :: Uranium Industry :: Nuclear Waste

Meersman T (October 25, 1999), Studies link frog deformities to pesticides. *Star Tribune* Retrieved on September 18, 2007.

Michael Hogan, Leda Patmore, Gary Latshaw and Harry Seidman *Computer modelng of pesticide transport in soil for five instrumented watersheds*, prepared for the U.S. Environmental Protection Agency Southeast Water laboratory, Athens, Ga. by ESL Inc., Sunnyvale, California (1973)

Miller GT (2004), *Sustaining the Earth*, 6th edition. Thompson Learning, Inc. Pacific Grove, California. Chapter 9, Pages 211-216.

Mokdad, Ali H.; et al. (2004). "Actual Causes of Death in the United States, 2000". J. Amer. Med. Assoc. 291 (10): 1238.

Muckenfuss AE, Shepard BM, Ferrer ER, Natural mortality of diamondback moth in coastal South Carolina Clemson University, Coastal Research and Education Center.

National Park Service. US Department of the Interior. (August 1, 2006), Sequoia & Kings Canyon National Park: Air quality — Airborne synthetic chemicals. Nps.gov. Retrieved on September 19, 2007.

National Policy Analysis #396: The Separations Technology and Transmutation Systems (STATS) Report: Implications for Nuclear Power Growth and Energy Sufficiency - February 2002

National Research Council *(1995). Technical Bases for Yucca Mountain Standards. Washington, D.C.: National Academy Press.* cited in in The Status of Nuclear Waste Disposal. The American Physical Society (January 2006).

National Trends in Sulfur Dioxide Levels, United States Environmental Protection Agency.

Negraa, Christine; Donald S. Rossa and Antonio Lanzirottib (2005). "Oxidizing Behavior of Soil Manganese: Interactions among Abundance, Oxidation State, and pH". *Soil Science Society of America Journal 1 (69): 97–95.*

Newswise: National Study Examines Health Risks of Coarse Particle Pollution

NOx Removal. Branch Environmental Corp. Retrieved on 2007-12-26.

Nuclear Information and Resource Service,Radioactive Waste Project, retrieved September 2007

Onsdorff, KA Pollution; Water ENVIRON. LAW PRACT., vol. 4, no. 1, pp. 14-18, 1996

Palmer, WE, Bromley, PT, and Brandenburg, RL. Wildlife & pesticides - Peanuts. North Carolina Cooperative Extension Service. Retrieved on 2007-10-11.

Papendick RI, Elliott LF, and Dahlgren RB (1986), Environmental consequences of modern production agriculture: How can alternative agriculture address these issues and concerns? *American Journal of Alternative Agriculture*, Volume 1, Issue 1, Pages 3-10. Retrieved on 2007-10-10.

Pedersen, TL (June 1997), Pesticide residues in drinking water. extoxnet.orst.edu. Retrieved on September 15, 2007.

Pesticide Action Network North America (June 4, 1999), Pesticides threaten birds and fish in California. PANUPS. Retrieved on 2007-09-17.

Pope, C Arden*; et al. (2002).* "Cancer, cardiopulmonary mortality, and long-term exposure to fine particulate air pollution". *J. Amer. Med. Assoc. 287: 1132-1141.* doi:10.1001/jama.287.9.1132.

Poppe, Wayne; Hurst, Renee Water Pollution WATER QUAL INT, pp. 39-43, 1997

Postel, S. L. and A. T. Wolf. "Dehydrating Conflict." Foreign Policy. 126. (2001): 60-67.

Public Health and Environmental Radiation Protection Standards for Yucca Mountain, Nevada; Proposed Rule. Environmental Protection Agency (2005-08-22). Retrieved on 2008-06-06..

R. B. Brown (September 2003). Soil Texture. *Fact Sheet SL-29*. University of Florida, Institute of Food and Agricultural Sciences. Retrieved on 2007-12-02.

R. V. Hoffman "m-Trifluoromethylbenzenesulfonyl chloride" Organic Syntheses, Collected Volume 7, p.508 (1990). http://www.orgsyn.org/orgsyn/pdfs/CV7P0508.pdf.

Rainer Stegmann, *Treatment of Contaminated Soil: Fundamentals, Analysis, Applications*, Springer Verlag, Berlin 2001

Raloff, J (September 5, 1998) Common pesticide clobbers amphibians. Science News, Volume 154, Number 10, Page 150. Retrieved on 2007-10-15.

Rasler, Karen A. and W. R. Thompson. "Contested Territory, Strategic Rivalries, and Conflict Escalation." International Studies Quarterly. 50. 1. (2006): 145-168.

Removal of Silicon from High Level Waste Streams via Ferric Flocculation

Reynolds, JD (1997), International pesticide trade: Is there any hope for the effective regulation of controlled substances? *Florida State University Journal of Land Use & Environmental Law*, Volume 131. Retrieved on 2007-10-16.

Richman, M IND. Water Pollution WASTEWATER, vol. 5, no. 2, pp. 24-29, 1997

Risk Assessment Guidance for Superfund, Human Health Evaluation Manual, Office of Emergency and Remedial Response, U.S. Environmental Protection Agency, Washington D.C. 20450

Rockets, Rusty (June 8, 2007), Down On The Farm? Yields, Nutrients And Soil Quality. Scienceagogo.com. Retrieved on September 15, 2007.

S.K. Gupta, C.T. Kincaid, P.R. Mayer, C.A. Newbill and C.R. Cole, "A multidimensional finite element code for the analysis of coupled fluid, energy and solute transport", Battelle Pacific Northwest Laboratory PNL-2939, EPA contract 68-03-3116 (1982)

Science Daily (February 3, 2006), Pesticide combinations imperil frogs, probably contribute to amphibian decline. Sciencedaily.com. Retrieved on 2007-10-16.

Science Daily (May 4, 1998), Pesticides linked to widespread cases of deformed frogs. Sciencedaily.com. Retrieved on 2007-10-12.

Science Daily (November 19, 1999), Evergreens help block spread of pesticide from crop fields. Sciencedaily.com. Retrieved on September 19, 2007.

ScienceDaily (June 25, 2007), Breakdown products of widely used pesticides are acutely lethal to amphibians, study finds. Sciencedaily.com. Retrieved on September 17, 2007.

ScienceDaily (November 28, 2002), More evidence to link pesticide use with amphibian decline. Sciencedaily.com. Retrieved on September 17, 2007.

Scientific Facts on Water: State of the Resource. GreenFacts Website. Retrieved on 2008-01-31.

Soil Survey Division Staff (1993). Soil Structure. *Handbook 18. Soil survey manual*. Retrieved on 2006-04-11.

Source for figures: Carbon dioxide, NASA Earth Fact Sheet, (updated 2007.01). Methane, IPCC TAR table 6.1, (updated to 1998). The NASA total was 17 ppmv over 100%, and CO2 was increased here by 15 ppmv. To normalize, N2 should be reduced by about 25 ppmv and O2 by about 7 ppmv.

Space Disposal of Nuclear Wastes Eric E. Rice Battelle Memorial Institute

States of Jersey (2007), Environmental protection and pesticide use. Retrieved on 2007-10-10.

Survey & Identification of NORM Contaminated Equipment

Tashkent (1998), Part 1. Conditions and provisions for developing a national strategy for biodiversity conservation. Biodiversity Conservation National Strategy and Action Plan of Republic of Uzbekistan. Prepared by the National Biodiversity Strategy Project Steering Committee with the Financial Assistance of The Global Environmental Facility (GEF) and Technical Assistance of United Nations Development Programme (UNDP). Retrieved on September 17, 2007.

Taylor, S. A., and G. L. Ashcroft. 1972. Physical Edaphology

Tchobanoglous, George. Wastewater Engineering. 3rd ed. New York: Mc Graw Hill, 1979.

Terry, LA Water Pollution ENVIRON. LAW PRACT., vol. 4, no. 1, pp. 19-29, 1996

The Color of Soil. United States Department of Agriculture - Natural Resources Conservation Service. Retrieved on 2007-11-25.

The cost of meeting the Johannesburg targets for drinking water

Tibbetts, J Anon. Water ENVIRON. SOLUTIONS, vol. 9, no. 8, p. 26, 1996

Toughill K (1999), The summer the rivers died: Toxic runoff from potato farms is poisoning P.E.I. Originally published in *Toronto Star Atlantic Canada Bureau*. Retrieved on September 17, 2007.

U.S. Department of Energy Environmental Management - "Department of Energy Five Year Plan FY 2007-FY 2011 Volume II." Retrieved on 8 April 2007.

U.S. ENVIRONMENTAL PROTECTION AGENCY, WASHINGTON, DC (USA), 1996, 253 pp

U.S. Geological Survey, Radioactive Elements in Coal and Fly Ash: Abundance, Forms, and Environmental Significance, *Fact Sheet* FS-163-1997, October 1997, retrieved September 2007

UC IPM Online. (August 11, 2006), What's up, Doc? Maybe less air pollution. Statewide IPM Programme, Agriculture and Natural Resources, University of California. Ipm.ucdavis.edu. Retrieved on 2007-10-15.

Ungate, CD Clean Water Initiative J. ENVIRON. PLANN. MANAGE., vol. 39, no. 1, pp. 113-122, 1996

Uranium Information Centre, Synroc, *Nuclear Issues Briefing Paper* 21, retrieved September 2007

US Environmental Protection Agency (September 11th, 2007), Pesticide registration (PR) notice 2001-X Draft: Spray and dust drift label statements for pesticide products. Epa.gov. Retrieved on September 19, 2007.

US EPA, WASHINGTON, D.C. 20460 (USA), 1996, v.p.

USEPA (2007), Sources of common contaminants and their health effects. Epa.gov. Retrieved on 2007-10-10.

Vercheval, J. The thermosphere: a part of the heterosphere. (offline, see Internet Archive copy)

Voroney, R. P., 2006. The Soil Habitat *in* Soil Microbiology, Ecology and Biochemistry, Eldor A. Paul ed.

Wells, M. "Vanishing bees threaten US crops", *www.bbc.co.uk*, *BBC News*, March 11, 2007. Retrieved on 2007-09-19. (English)

Wolf, Aaron T. "Water and Human Security." Journal of Contemporary Water Research and Education. 118. (2001): 29

World population to reach 9.1 billion in 2050, UN projects

Zeissloff, Eric (2001). Schadet imidacloprid den bienen (German). Retrieved on 2007-10-10.

4

EIA: Relevance and Importance for Environmental Sustainability

ENVIRONMENTAL IMPACT

Introduction

Real and potential environmental quality problems have accompanied the changes in agricultural productivity in recent decades. Difficulties in obtaining precise relationships between agricultural practices and environmentai problems occur because of the diffuse source of residuals from agriculture and the many factors that are involved, some of which are not controllable by man. The factors include topography, precipitation, cover crop, timing and location of chemical or fertiliser application, and cultivation practices. The variability of waste discharges complicates assessment of the environmental impact of agricultural production. Agricultural wastes frequently are not discharged on a regular basis; food processing wastes tend to be seasonal with production a function of crop maturity, feed-lot runoff is a function of rainfall frequency and intensity, fertiliser applications are timed for ease of distribution and to maximise crop production, and land disposal of wastes are related to need for disposal and ability to travel on the land.

Available information suggests that potential environmental quality problems due to agricultural operations may be more dependent upon the production practices and waste management techniques utilised by farmers and processors than the size of the operation, the number of animals fed, or the amount of waste involved.

All of the environmental impacts described in this chapter that result from agriculture can be minimised by the use of suitable production and waste management methods. Common sense and the elevation of the concept of environmental impact to a higher level of consciousness in agricultural operations are important first steps in the control of real or potential environmental quality problems caused by agriculture.

This chapter will attempt to summarise the environmental problems that can or have been associated with agricultural production, to put these problems in perspective with those caused by other segments of society, and to indicate procedures to minimise these problems. Detailed waste management methods are described in subsequent chapters.

Water Quality

Nutrients

One of the most challenging problems is that of excessive nutrients and the conditions they can cause. Of particular concern are nitrogen and phosphorus compounds. Although these elements are needed in small amounts for all living matter, excessive amounts in surface waters can result in overfertilisation and can accelerate the process of eutrophication. Other concerns include excessive amounts of nitrates in groundwaters and surface waters, ammonia toxicity to 6sh, altered effectiveness of chlorination by ammonia, and the nitrogenous oxygen demand of reduced nitrogen compounds in surface waters.

Until recently the major reason for investigating the nutrient content of the soil water below the crop root depth, was to evaluate the loss of fertilizer nutrients and to estimate the loss of production or the need for more fertilizer to adjust for the losses. The increased concern on environmental quality has changed this emphasis to the study of nutrients and other contaminants in soil water because of potential water, pollution. The quantity of rainfall or drainage water ani the permeability of the soil are key factors in the leaching of contaminants from the soil.

The processes governing the transformation and movement of nitrogen and phosphorus must be understood as waste management alternatives are considered. A schematic summary of nitrogen in the soil-plant system illustrates some of the important factors and transformations.

The nitrogen in sludges, food processing wastes, manure and other biological residues is primarily in the organic and ammonium form. Fertilizer nitrogen may be in the ammonia ammonium, nitrate, or urea form. The largest quantity of nitrogen in soil occurs as organic nitrogen. Microbial decomposition of the organic matter results, in the production of ammonia nitrogen which with good aeration and suitable temperatures can be oxidised to nitrites and then to nitrates. Unless inhibition of the nitrate-forming bacteria occurs, practically no nitrite accumulation will occur. Nitrates can be reduced to gaseous nitrogen compounds and lost to the atmosphere if the nitrates are held under conditions of poor aeration and if suitable carbon sources are available for denitrification. Ammonium ions are held to the cation exchange sites in soils and the ammonium ion concentration in soil solution and leaching to the groundwater is, low. Nitrate ions are soluble in the soil water. The form of nitrogen of greatest environmental quality concern in the soil is nitrate nitrogen since it is subject to leach-ing and will move with the soil water. Losses of organic nitrogen are related to surface runoff.

A generalised summary of phosphorus interactions in soils and plants indicates the important inputs, transformations and outputs. Fertilizers and organic wastes add to the organic and inorganic forms of phosphorus in the soil. The inorganic forms mainly are iron and aluminum phosphates in acid soils and calcium phosphates in alkaline soils. Any phosphorus added as fertilizer or released in the decomposition of organic matter rapidly is converted to one of these compounds. All inorganic forms of phosphate in soils are extremely insoluble and concentrations of phosphorus in soil water solutions are low, generally less than 0.2 mg/liter. The phosphorus of environmental quality concern is associated with the soil itself and with the inter-change of phosphorus, from bottom deposit in bodies of water with the, upper waters.

Soil management practices can reduce the losses of phosphorus caused by erosion. Losses of soil nitrogen are more difficult to control since soluble nitrate nitrogen can be transported through the soil. Soil conservation can reduce the quantity of organic nitrogen reaching surface waters by erosion.

Nutrient budget estimates of agricultural operations have indicated that in the humid East and parts of the West nutrients added to cropland are generally taken off in crops. Nitrogen applied in excess of the crop requirements can end up in surface waters or groundwaters or lost to the atmosphere by denitrification. In the prairie and plains area, residual fertility remains high and more nutrients are removed by the crops than are added as fertilizer. These high fertility soils can contribute nitrates to surface waters and groundwaters even in the absence of added fertilizers. In western states where irrigation is practiced, salts including nitrates tend to build up in the ground-water. Finally in the semiarid and western states where intensive agriculture is not practiced, nutrients added to land may be about equal to the amount in the crop or brush (2).

Fish Kills

Actual pollution caused by agriculture is difficult to document. The effect of agricultural contaminants can be observed in the number of fish kills that have occurred in the United States (Table 1). Comparable data from other countries is less available. For the period of 1962-1967, a total of 1218 fish kills were reported in fishing waters of Switzerland (3). Manure liquids caused the greatest number of fish kills (30.3%). The readily identified agricultural sources, manure liquids, silage juice, and insecticides and herbicides, caused 39.4% of the fish kills. A similar assessment in Switzerland covering 1952-1967 noted that manure liquids caused 32.7% and silage juice 4.8% of the fish kills in this period. An element of uncertainty in citing data on fish kills is the importance of each kill in terms of the size and value of the fish and the effect on the total ecology of the streams. The role of agriculture in contributing to the fish killed by insecticides and herbicides also is unclear.

TABLE 1

Source of Fish Kills in the United States[a]

Source	*1964*	*1965*	*1968*	*1969*	*1970*	*1971*	*1972*	*1973*
Total fish kill reports, all sources	590	625	542	594	635	860	760	749
Reports attributable to Farming								
Insecticides	95	74	51	80	63	63	57	91
Fertilizers	5	4	5	5	6	8	10	18
Manure, silage drainage	29	29	21	29	38	41	46	52
Food products	41	60	35	37	32	25	31	19
Fish killed by farming sources (%)	8	12	2.5	15	20	1.5	12	6
Primary source of fish killed								
Industrial	1	2	2	1	1	4	2	4
Municipal	2	1	1	4	2	1	1	2
Agricultural (farming[b])	3	3	4	2	4	5	3	3

a. From Ref. (4),

b. Wastes from food processing operations are classified as industrial wastes.

The data in Table 1 represent an assessment of the source of pollution by the investigating organisations and as such are based on oral investigations and studies conducted after fish kills were noted. In the official reports, the term agriculture was used to describe fish kills that resulted from farming operations, i.e., use of insecticides and herbicides, fertilizers, and manure-silage drainage. Problems due to food production were included in the industrial waste category. In keeping with the terminology used in this book agricultural sources noted in Table 1 are more broadly defined to include both farming and food processing sources.

In 1969, pollution from farming practices was the second largest source of pollution as recorded by the number of fish killed. Fertilizers were a minor cause of problems. Insecticides, food products, manure, and silage drainage caused greater problems than did fertilizers with insecticides and food products causing the greater problems in most years.

In the United States, municipal and industrial pollution sources clearly are a more important source of water pollution, as measured by fish kills, than are agricultural sources.

Groundwater

Agricultural practices have the capacity to produce wastes that can enter the groundwater and alter its quality. The magnitude of agricultural contamination of groundwater is difficult to ascertain. There is limited information on the amount of agricultural contaminants that enter subsurface waters under given cropping and disposal practices and on the changes in groundwater quality caused by various agricultural contaminants. In addition, there is a scarcity of adequate groundwater monitoring programmes to investigate these factors.

Of the rural water supply samples examined in Missouri, 50-75% contained sufficient nitrogen to be of concern in livestock production (5). The main contaminating source both in distribution and concentration was indicated to be waste matter at sites of animal habitation. Soil containing 200 to 4000 Ib of nitrate nitrogen per acre was found below certain feedlots. Standard surface soils contained-about 50 to 150 Ib of nitrate nitrogen per acre. The increased nitrates remained after an area was abandoned from animal use.

TABLE 2

Soil Nitrogen Concentration in Agricultural Land

Land use	*Average nitrate nitrogen (Ib / acre)*[a]
Irrigated land, alfalfa	80
Native grassland	90
Nonirrigated cropland	260
Irrigated cropland not alfalfa	506
Feedlots	1436

Land use	*Nitrate nitrogen (Ib / acre)* [b/c]	*Ammonium nitrogen (I b / acre)* [b/c]
Virgin soil-Jack Pine	56	21
Cultivated	142	31
Barnyard	407	*2200*

a. From Stewart *et al.* (6).

b. 20-ft cores.

c. From Beatty *et al.* (7).

Comparative studies in Colorado and Wisconsin have indicated that land use affected the nitrogen content of soils (Table 2). The highest soil nitrate nitrogen occurred under feedlots. The nitrogen content in virgin soil and under native grassland represents natural decomposition of organic matter as well as natural deposits in the soil rather than man's activities. The lower quantities below alfalfa are because of the ability of the crop to remove nitrates deep in the soil and the fact that this crop rarely is fertilised with nitrogen. Data of this type need not reflect groundwater nitrate concentrations since that concentration is a function of the water moving through the soil and the hydraulic characteristics of the aquifer. The nitrate nitrogen concentrations in the groundwaters under the lands studied in Colorado (Table 2) (6,7) ranged from zero to more than 10 mg/liter.

Water from shallow wells is more likely to contain contaminants than water from deeper wells, particularly shallow wells near barnyard, feeding lots, and manure piles. Occasionally, deep drilled wells contain considerable nitrate, either entering the well by surface leakage through poor well seals or from nitrogen-rich deposits within lower soil zones.

At present, groundwater contamination from agricultural sources is, for the most part, below levels that have been demonstrated to cause disease, removal, or cause aesthetic nuisance. In establishing enlarged agricultural operations and in choosing sites for the land disposal of wastes, the possibility of groundwater contamination should receive thorough investigation.

Sediment

About four billion tons of sediment wash into the streams and rivers of the United States each year. Approximately 75% of this material originates from forested and agricultural lands. This erosion loss is not distributed evenly throughout the United States. Losses are relatively small in areas of high density natural vegetation and larger in areas where agriculture is more intense.

The rates of erosion and resultant sediment have been accelerated by human use and management of lands, vegetation, and streams. Iowa data indicate that the soil loss in one county averaged 10 ton/acre/year since the land has been used agriculturally 125 years ago, whereas, before agricultural use the soil loss was about 1 ton/acre/year (8). In regions of erosive soils and well-defined drainage systems, 10-30 tons/acre/year are eroded if vegetative cover is poor. Generally, cultivated soils in warmer and more humid regions are more susceptible to erosion than soils in cooler, less humid areas. Erosion rates can increase due to agricultural activities because of removal of the natural vegetative cover of the land, plowing and cultivation which decrease the erosion resistance of the soil, and the trend to large planting and harvesting equipment which may not be compatible with soil conservation practices on small acreage and sloping land.

Sediment can reduce the recreational value of water, is a carrier of plant nutrients, crop chemicals, and plant and animal bacteria, and increases water treatment costs. Proper erosion control is a positive solution to sediment problems.

Soil erosion and sediment yield is a function of rainfall, soil properties, land slope length, steepness, and cropping practices. While little can be done to change the amount, distribution, and intensity of rainfall, measures can be taken to reduce the ability of rainfall to cause erosion. Examples include decreasing the amount and velocity of overland How and decreasing, the impact of the rain on the soil. Both flow and impact can be reduced by maintaining vegetative cover on the land through proper crops or mulch. The incorporation of crop residue and animal wastes in the soil increases soil porosity and aggregation and decreases runoff. Slowing runoff by vegetation cover and land modification (terraces, ponds waterways) offers opportunities to provide greater infiltration time, conserve moisture, and decrease erosion. Wise land use planning and careful use and management of land can reduce erosion and control certain water quality problems arising from agricultural activities. Practices and structures to conserve soil and water are identical to those producing a reduction of pollutants found in agricultural runoff.

Land Use

Each watershed will have a unique combination of land uses, which will contribute to the flow and constituents in a streams. Precipitation also will contribute to the constituents as well as being an important factor in the quantity of runoff that occurs. The contribution of each land use will vary.

In the following sections, the characteristics of several point and nonpoint sources are reported in concentration units (mg/ liter) and potential area yield rates (kilograms/ hectare/year). The characteristics represent the reported data of many investigators and should not be utilised or extended beyond the conditions of the original study. Each situation will have its own runoff or leachate characteristics as well as potential control approaches.

The area yield rate information should be interpreted as a potential yield rate. It would be incorrect to multiply the yield rates by the area of a particular land use category and infer that the result is the actual amount of material entering a specific surface water body. The actual amount that enters a specific water body is a function of interrelated hydrologic, geochemi-cal, agricultural, and human activities on the land

The characteristics are presented only to indicate relative concentrations and rates and to compare the contributions from various sources: Application of the reported average characteristics not likely to indicate the effect of an actual situation.

Forest Runoff. Forested areas represent areas that have not been grossly contaminated by man's activities. The runoff from these areas can indicate characteristics that result from natural conditions. However, the increased world demand for wood fiber has accelerated cultural production practices in managed forested areas. Practices such as forest fertilisation, block cutting of mature trees, and other forest management practices will alter the characteristics of runoff from forested areas

The effect of forest fertilisation on surface water quality has been reviewed (9). The leaching of nitrates into ground water following fertilisation was insignificant. Major loss of fertilizers to surface waters occurred by inadvertent application to water courses and compacted areas such as roads and trails. Nutrient problems caused by forest fertilisation do not appear to be a problem under current conditions and practices.

Rang Land: Where rainfall is sparse or the land not very fertile, intensive crop production is not practiced and few animals can be supported per acre of land. Some domestic animals are permitted to range over the land in search of available food. Range land, with its low level of animals and fertilised acreage, represents a natural situation. Runoff from these lands can represent background or "natural" contamination.

Runoff from range land that had low intensify agriculture and no evidence of chemical fertilizers contained 0.65 kg NO_3-N, 0.76 kg total P, and 0.024 kg soluble P per hectare per year (10) The characteristics of the runoff were due to the erosion and leaching of the soil in the area.

Rural and Crop Land: Rural land refers to land that is or could be under agricultural production. Examples include grass-land and idle farm land. Crop land refers to land under cultivation.

Constituents of the runoff from rural and crop lands originate in rainfall, wastes from wildlife, leaf and plant residue decay, applied nutrients, nutrients and organic matter initially in the soil, and wastes from animals on pasture. It is complex to separate the natural from the other contaminant sources. To do so requires that water volumes be accurately measured, storm characteristics described, and representative sampling initiated. Constituents are released at varying rates from all soils and exposed geological formations. The natural weathering of rocks and minerals and the oxidation and leaching of organic matter will contribute to runoff and leachate characteristics even in the absence of man's activities.

Some nutrients leave crop lands. It does not necessarily follow that the same amount reaches surface waters and lakes. Any intermediate vegetation may remove most of the material in suspension and part of the material in solution.

The time distribution of nitrates in a number of rivers suggests that the nitrate concentration and load is the greatest when the flow is high, i.e., when rainfall has the greatest opportunity to leach the soluble nitrates from the soil. A seasonal variation in nitrate concentrations was found with the highest levels in surface waters found in spring and late summer. At many sampling stations, the nitrate concentrations closely paralleled stream flow (11). The relatively direct correlation between the two parameters indicated that the nitrate reached the stream seasonally by way of surface runoff, drainage, or interflow and originated from diffuse sources.

Similar correlations have been observed in an Iowa stream in which during low flow periods domestic and industrial waste waters were the principal sources of nitrogen to the stream (12). During high runoff periods large quantities of nitrogen entered the streams over extended periods. During the wet periods, phosphorus levels rose to as high as 19 times the level expected from municipal wastewater alone. At low river flows, the phosphorus levels in the river were below that in the discharges of domestic and industrial wastewaters. The phosphorus was bound to soil sediment and removed.

Data from a small agricultural watershed in Kansas indicated a relationship between total phosphate and nitrate concentrations and stream flow (13). Total phosphate concentrations were from 0.2 to 0.4 mg/liter during low flows but increased to maximums of 2.4 to 4.0 mg/liter during storm runoff. The phosphate load followed the turbidity pattern. The nitrate concentrations ranged from 0.2 to 4.2 mg/liter as nitrate in low flows but increased to maximums of 40 to 45 mg/liter during storm runoff.

Many factors affect the release of nitrate from the soil organic matter and the existing nitrogen pool in the soil. Conditions that favour decomposition of organic matter and oxidation of nitrogen are neutral to slightly basic conditions, warm temperatures, and aeration. When drainage losses increase so does the rate of loss of soluble constituents, such as nitrate, from the soil. A study of nitrate concentrations in English rivers (14) was unable to correlate nitrogen fertilizer usage in adjacent areas with the nitrate concentrations. A high proportion of the total nitrogen was carried by the rivers in a few periods of high flow.

In many cases, the nutrient concentration in a stream will vary with the flow In a predictable manner. A detailed study in an agricultural watershed noted that an equation of the type $L=AQ^{(B+1)}$ could be used to relate the load *(L)* carried by a stream in terms of kilograms/day to the stream flow *(Q*, ft^3/sec) (15). *A* and *B* were regression coefficients for the data. This equation was useful to describe the variation of magnesium, chlorides, total and soluble phosphorus, ammonia nitrogen, and soluble and particulate organic nitrogen. The variation of nitrate and nitrite nitrogen in the stream flow was related by an equation of the form $L=(A+BQ)$ (Q). The watershed contained agricultural land including dairy farms (50 per cent of the area), forest (30%), and the remaining 15 per cent was water, residential, commercial, and recreational lands.

Fertilised Lands: One of the requirements for satisfactory crop production is the availability of nutrients when the crop is growing. Intensive farming operations are designed to satisfy this requirement either by the addition of fertilizers or by the incorporation of readily decomposable organic material such as manures in the soil.

Crop fertilizers have been blamed as major contributors to nutrients in surface waters. Generally only 10-30 per cent of the fertilizer phosphorus added to a soil is taken up by the following crop. The remaining applied phosphorus is converted to insoluble forms and may become a source of available phosphorus for crops in subsequent years. Rarely does phosphorus from commercial fertilizers or spread manures occur in groundwater except with very porous soils in areas with heavy rainfall or irrigation. Phosphorus added to soils as a fertilizer or released in the decomposition of organic matter will be rapidly converted to iron and aluminum phosphates in acid soils and to calcium phosphates in alkaline soils. With proper erosion control measures, little of the applied phosphorus should reach surface waters.

Efforts have been made to attempt a direct relationship between fertilizer nitrogen applied and the concentration of nitrogen in stream and drainage waters. A direct relationship is not likely because of a large preexisting reservoir of nitrogen in the soil and the rate of biological transformation the reservoir can undergo. In the upper Rio Grande River, the nitrate concentration in the river had not increased although fertilizer use increased considerably over a 30-year period (16). Crop usage and denitrification were probable explanations for the disappearance of the applied nitrogen. In other areas where organic and inorganic nitrogen has accumulated for centuries, the nitrate concentration of the drainge waters are more a result of natural conditions than of fertilizer applications

A study was made of the extent to which the increase in fertilizer usage in the Netherlands has affected the quality of groundwater over the years. It was not possible to determine whether observed increases in chlorides and nitrates were caused by the increased use of fertilizers or by the increase in well water demand. In the period from 1920 to 1966 the average chloride and nitrate concentration of unpurified groundwater in fertilised sandy soils increased by about 3 and 2.5 mg/

liter, respectively. Over two-thirds of wells had no nitrates in their water during this period (17). During the same period nitrogen fertilizer usage increased from 28 to 178 kg/ha/year. Denitrification was suggested as a principal reason for the low nitrate content of the groundwater. The change in nitrate and chloride content of the deeper groundwater in the Netherlands during a period of increased fertilizer usage was not large enough to reduce the quality of potable water prepared from the groundwater.

Conditions required to leach nitrates from soil depend upon the water percolating through the soil and upon the presence of excessive nitrate in the soil when percolation occurs. Situations conducive to nitrate leaching include conditions where infiltration exceeds water lost by evaporation and transpiration, where soils have low water-holding capacities, where irrigation is practiced, and where soils have high infiltration rates.

The best way to add fertilizer is as it is needed for crop growth. Attention has been called to the fact that fall application of fertilizers may not be used as effectively as that applied at crop planting time. Greater amounts of nitrogen may be needed in the fall than are required if the nitrogen is applied at planting and sidedressed (18). The nitrogen not incorporated into the crop will be lost in some manner and it is reasonable to expect that a fraction of the nitrogen will end up in the surface water or groundwater. The greater the excess of nitrogen above plant and soil bacteria needs, the greater the amount that may be in percolate water. A study on sandy soils indicated that about 83 per cent of the fall added fertilizer nitrogen was lost to the succeeding spring crop (19).

It is evident that factors such as natural nitrogen content of the soil, soil type and topography, crop type and rate of fertilizer application, and climatic conditions complicate the relationship between fertilizer use and water quality changes. It is inevitable that there is a risk of nitrate loss during runoff and leaching. It is difficult to conceive of any approach that will reduce that risk to zero. It is possible to decrease the risk by proper application of inorganic fertilizers in periods when the crops are growing coupled with the use of slow release organic fertilizers.

Cropland Subsurface Drainage: The soils of many farms have poor drainge characteristics. Subsurface tile drains can be installed to permit better water movement and crop yields. The drainage from the lands will contain soluble constituents from the soil and materials added to the soild (20). The drainage can enter surface steams at many places and essentially is a non point source of contaminants for the streams.

Pasture and Manured Lands: These lands contain either animals for production or are used for disposal of animal wastes. In either case, the runoff of these lands can contain contaminants from animal wastes. Where animals have direct access to streams, animal urine and feces can bo discharged directly to these waters.

Studies on the relative pollutional effects of manure disposal and pastured animals (21) indicated that the extent of water pollution from farm animal production units is more dependent on waste management methods than on the volume of the waste involved. To minimise water pollution from pastured animals, points of animal concentration should be located away from streams and away from hillsides leading directly to streams and vegetation should be provided between areas of animal concentration and drainage paths or surface waters to intercept contaminants.

The access of animals and animal wastes to surface waters represents a possible controllable nonpoint pollution source. Animal manures should not be disposed of where rainfall or snow melt will result in their direct discharge to watercourses. The manure should be integrated with crop or disposal land shortly after disposal.

Manure Storage: Where manure cannot be disputed on land routinely, such as in the winter, it may be stacked or stored until conditions permit land disposal and integration with crop production. During storage, seepage can result which can contaminate surface waters. Table 3 (22) indicates the magnitude of contaminants from stacked dairy manure. Although the volume of seepage is small, the quantity of contaminants is not insignificant.

The release of stored manure seepage to surface waters should not be permitted. The seepage can be controlled by retention ponds and distribution on crop land in a nonpolluting manner. Even though manure seepage can occur in a large number of locations throughout the country, and as such approximates a nonpoint source of pollution, it can be considered as a controllable source of pollution.

TABLE 3

Characteristics of Seepage from Stacked Dairy Cattle Manure and Bedding[1]

Parameter	*Winter*		*Summer*	
	Average	*Range*	*Average*	*Range*
Total solids (%)	2.8	1.8-4.3	2.3	1.7-2.9
Volatile solids (% TS)	55	52-59	53	50-58
Suspended solids (%)	0.35	0.2-0.8	0.24	0.2-0.3
BOD (mg/liter)	13,800	4,200-31,000	10,300	4,400-21,700
COD (mg/liter)	31,500	21,000-41,000	25,900	16,400-33,300
Total N (mg/liter as N)	2,350	1,500-2,900	1,800	1,200-2.770
NH_4-N (mg/liter)	1,600	980-1,580	1,330	780-2,200

Total P (mg/liter as P)	280	64-560	190	90-340
Potassium (mg/liter as K)	4,700	3,000-7,200	3,900	3,000.4,900
Total precipitation (in.)	15.0		9.4	
Seepage volume (gal/cow/day)	3.0		1.2	

1. From Cramer *et. al.*, (22).

Animal Feedlots: In many areas of the United States, animals remain an integral part of cropping systems and their wastes are returned to the fields that produce feed for the animals. In the Midwest and West, large feedlots now produce animals for slaughter and the manure tends to accumulate in the feeding areas. Until these wastes enter groundwaters or surface waters, they do not represent a serious water pollution problem. The characteristics of the livestock wastes are such that they will stay within a confinement area until the area is cleaned or until runoff washes them away. The water pollution problem associated with livestock wastes is a drainage problem. Runoff from uncovered confinement areas and from land used for disposal of waste occurs during and following rainfall.

Surface drainage from cattle feedlots and other areas where confined animals are housed in the open has caused considerable concern. In the early development of open confined animal operations, feedlots were situated in locations where the natural drainage facilitated the removal of the wastes. The water pollution potential was given little if any consideration.

The quantity and quality of feedlot runoff will depend upon previous weather conditions, the number of cattle per feedlot area, the method of feedlot operation, soil characteristics, topography of the area, and intensity of rainfall. Studies that have included natural and simulated rainfall events have provided data on the characteristics of feedlot runoff under a variety of environmental conditions.

An early investigation illustrated the water quality changes that can occur when feedlot runoff reaches surface waters (23). Considerable length of streams were devoid of oxygen due to the runoff. BOD and ammonia concentrations were as high as 90 and 12 mg/liter, respectively (Table 4). Ammonia concentrations from the runoff were detectable before other parameters. Such runoff provides little warning to downstream users and can trap game fish in the polluted waters.

Detailed studies on the environmental factors affecting the quantity and quality of feedlot runoff have been conducted. The greatest pollutant concentrations occurred during warm weather, during periods of low rainfall intensity, and when the manure had been wet.

Part of the organic matter in animal manure is stabilised by bacterial action after it is deposited in the feedlot. Some nitrogen contained in the manure and urine can be lost to the atmosphere. About 50 per cent of the organic matter can be decompored into carbon dioxide and water. The amount of decomposition that takes place is dependent upon temperature and moisture conditions. The greater the bacterial action and the moisture content of the wastes, the greater the degree of solubilisation and the greater the amount of soluble constituents in subsequent runoff. Because of a decrease in bacterial action in the winter, there will be a greater accumulation of wastes in an open feedlot during this period than during the rest of the year.

The loss of carbon and nitrogen from the waste by bacterial action increases the mineral content, reduces the organic content, and results in an accumulation of undegradable material. This residual material is not readily amenable to treatment by biological treatment methods and will result in a high COD/BOD ratio. While the manure is decomposing on the lot. it is constantly being mixed with fresh wastes and with soil if the feedlot is unsurfaced. The characteristics of the actual waste in a feedlot will be a mixture of these physical and biochemical changes.

TABLE 4

Water Quality Changes Caused by Feed lot Runoff, Fox Creek, Kansas (mg/liter)[1]

Time	*DO*[2]	*BOD_5*	*COD*	*Cl*	*NH_3*
Dry weather average	8.4	2	29	11	0.C6
After rainfall (hr)					
13	7.2	8	37	19	12.0
20	0.8	90	283	50	5.3
26	5.9	22	63	35	—
46	6.8	5	40	31	0.44
69	4.2	7	43	26	0.02
117	6.2	3	22	25	0.08

1. From Smith and Miner (23).
2. DO, dissolved oxygen.

The highest concentration of pollutants in the feedlot runoff can occur in the initial runoff and decreases to "equilibrium" conditions as the runoff continues. Illustrates the pattern for a number of rainfall rates and antecedent feedlot conditions. The "equilibrium" conditions were covered by the runoff continually dissolving the surface layer of manure on the lot. Once the ground surface was covered with manure, the depth of manure was not an important parameter of runoff water quality.

The characteristics of runoff from other beef cattle feedlots even larger contaminant concentrations. Data from additional studies corroborate the strength of feedlot runoff. Reported characteristics are presented in Tables 5 (26-31) and 6. An acre foot of feedtot runoff may contain 5-60 Ib of NH_4-N, 2-23 Ib of NO_3-N and about 100 lb of P (32).

In addition to the beef cattle feedlots, there are a large number of cattle herds of less than 100 head of beef cattle. These herds are confined in barn lots in the winter and pastured in the summer. The barn lots have intermixed mud and manure as a surface in the wet, cool seasons and a hard packed surface in the warm, dry periods. The runoff from barn lots also can be a source of surface water contamination. The runoff from a small barnlot housing 60 beef steers had the following range of characteristics: total solids, 0.1-1.2 per cent on a wet basis; volatile solids, 17-38 per cent of total solids; COD, 350-5650 mg/liter; and BOD, 9-560 mg/liter (33).

TABLE 5

Beef Cattle Feedlot Runoff Characteristics

Location and condition	*Total solids*	*Volatile solids*	*COD*	*BOD*
1	2	3	4	5
Concentration (mg / liter)				
Nebraska				
Snowmelt runoff	•	—	41.000[1] (1 [1],100-77,000)[2]	—
Rainfall runoff	•	—	3.100 (1,300-8 200)	—
Texas				
Dirt lots	•	—	9,500 (2,900-28,000)	1,460 (1,010-2,200) 8,000
Concrete lots	•	—	21,500 (8,400-32,800)	8,000 (3,300-12,700)
Colorado	•	100-7,000	•	300-6,000
Kansas	10,000-25,000	•	4,000-40,000	1,000-11,000
Texas	3,100-28,800	•	1,440-16,320	1,075-3,450
Area yield rate (kg / ha / year)				
Nebraska				
Snowmelt runoff				
100 ft²/head				
1969	60,000	31,000	•	—
1970	1,100	770	•	—
200 ft²/head				
1969	14,000	7,200	•	—
1970	640	450	•	—

Rainfall runoff				
100 ft^2/head				
1969	7,500	3,400	•	—
1970	27,000	16,000	•	—
200 ft^2/head				
1969	7,300	3,400	•	—
1970	18,000	10,000	•	—
South Dakota[3]	10,500	4,900	7,200	1,560
	(1,600-23,200)	(800-10,000)	(720-16,000)	(135-3,800)

1. Average.
2. Range of values.
3. Runoff form beeffeeding, dairy confinement, and lamb finishing operations.

	TKN	*NH_4-N*	*NO_3-N*	*Total*	*Total P*	*Alkalinity*	*Reference*
1	6	7	8	9	10	11	12
Concentration (mg/liter)							
Nebraska							
Snowmelt runoff	•	780	17	2,100	290	•	(26)
		(6-2,020)	(0-280)	(190-6,530)	(5-920)		
Rainfall runoff	•	140	10	920	360	•	(26)
		(2-1,240)	(0 220)	(11-8,590)	(4-5,200)		
Texas							(27)
Dirt lots	138	56	23	•	—	730	(Aug.-
	(9-280)	(2-85)	(0-103)			(100-1,400)	Nov. 1969)
Concrete lots	550	350	220	•	—	1.250	(27)
	(70-1,070)	(33-775)	(0-880)			(85-2,400)	
Colorado	•	—	•	—	•	—	(28)
Kansas	200-450	•	—	•	—	•	(29)
Texas	•	4-173	0-2.3	•	21-223	•	(30)
Area yield rate (kg/ha/year)							
Nebraska							
Snowmelt runoff							
100ft^2/head							
1969	•	—	•	1,600	620	•	(26)
1970	•	—	•	100	10	—	
200ft^2/head							
1969	•	—	•	450	200	•	(26)
1970	•	—	•	360	60	—	
Rainfall runoff							
100ft^2/head							
1969	•	—	•	330	22	•	(26)
1970	•	—	•	900	130	—	
200ft^1/head							
1969	•	—	•	200	33	•	(26)
1970	•	—	•	350	70	•	
South Dakota	510	•	—	•	130	•	(31)
	(46-3,110)				(9.470)		

Runoff from feedlots can have a detrimental effect on surface waters. Runoff collection ponds can reduce the sediment load, in such runoff and can produce an effluent with less variable characteristics. The quality of the effluent from runoff retention ponds is inadequate for release to streams or reservoirs.

Although climatic conditions and size of beef feedlots are different in Canada, studies on the quality of foodlot runoff in Ontario have produced data similar to that noted in the United States. Runoff from the Canadian feedlots contained from 800 to 7500 mg/liter BOD_5, 265 to 3400 mg/liter total Kjeldahl nitrogen, and 93 to 2180 mg/liter phosphorus as P_3O_5. The events causing the runoff were 0.3-to 1-in. rains with the higher concentrations occurring shortly after the larger rains (34).

The concentrations of pollutants in feedlot runoff are relatively independent of the type of ration fed the animals. In the semiarid regions of the southwestern United States, cattle wastes are dehydrated in a short time and remain that way until wetted by precipitation. The wetting may reconstitute the wastes to almost original composition. In more humid climates, the wastes may remain moist for longer periods of time before natural drying occurs. The longer the wastes remain moist, the greater the opportunity for bacterial action and solubilisation of the solid matter.

TABLE 6

Major Ions in Cattle Feedlot Runoff

Location	*Ion*	*Mean (mg/liter)*	*Range (mg/liter)*
Nebraska[a]	Na	840	40-2750
	K	2520	50-8250
	Ca	790	75-3460
	Mg	490	30-2350
	Zn	110	1-415
	Cu	7.6	0.6-28
	Fe	765	24-4170
	Mn	27	0.5-146
Texas[b]	Na	408	130-655
	K	760	226-1350
	Ca	700	194-1620
	Mg	69	28-89

(a) From McCalla *et al.* (26).

(b) From Kreis *et el.* (30).

Pollution from an uncovered livestock area, such as a feed-lot, is related to the fraction of precipitation that becomes runoff and reaches surface streams. Only after a portion of the rainfall soaks into the manur does runoff occur. This fraction will depend upon previous weather and percipitation conditions. The relationship between precipitation and runoff at feedlots can be used to estimate the quantity of feedlot runoff that will occur for a given storm and therefore to design control systems for the runoff.

The general relationship relating precipitation (inches) and runoff (inches) from cattle feedlot surfaces is,

$$\text{Runoff} = \text{A X rainfall} — \text{B}$$

Results obtained from small surfaced and unsurfaced cattle feedlots are shown in Table 6. The data represent both natural and simulated rainfall events occurring over separate two- to three-hour periods from June to November in Kansas.

Other precipitation—runoff data at cattle feedlots have been analysed in separate studies. In general, runoff was shown to occur only after precipitation exceeded 0.2 to 0.5 inch. The proportion of the precipitation that ends up as runoff will vary at different locations but generally is greater than 30 per cent and has ranged to 90 per cent.

The rainfall—runoff relationships at beef feedlots indicate that, after a minimum amount of rainfall, much of the rainfall ends up running off the feedlot. The surface of the feedlot acts milch the same as a paved surface in this regard. A dry hard crust frequently is observed below the manure surface. The silt, clay, and manure mixtures of feedlots result in the clogging of the porous soil voids and in the development of a less permeable layer.

These conditions suggest that infiltration to groundwaters can be neglected in determining the total runoff from operating feedlots. When feedlots are abandoned, the accumulated wastes may remain, the hard crust break up, and greater downward motion of pollutants may occur.

Runoff and percolation are not the only ways that feedlot contaminants can affect water quality. Beef cattle feedlots can contribute ammonia, amines, and odorous sulphur compounds to the atmosphere (35, 36). Such compounds are potential pollutants to waters in the area and can have an effect on nearby plants and animals. Aliphatic amines were a significant fraction of the volatilised nitrogen. Similar results can be expected from other accumulations of organic wastes exposed to the atmosphere.

The water pollution potential of livestock feedlots is related to the waste production per animal, the number of animals in the confinement unit, days confined, frequency of cleaning, climate, waste characteristics, and waste degradation in the lots. The contribution of feedlot runoff to surface water pollution will be a function of the temperature, magnitude of rainfall, slope of the confinement area, surface area of the feedlot, type of lot surface, and management practices. Range-fattened cattle represent a smaller runoff pollution problem than feedlot cattle since they are more widely distributed on the land. As the density of animals per acre decreases, the wastes are less concentrated and the nature can absorb more of the wastes. The tend, however, is toward confinement livestock feeding.

Irrigation Return Flows: Irrigation return flow is any water that finds its way back to a water supply source after it has been diverted for irrigation purposes. The term includes:

(a) tailwater runoff, the portion of the applied water that runs off the land surface;

(b) seepage, water seeping from the distribution system;

(c) deep percolation, applied water that contributes to groundwater recharge or to a subsurface drainage system; and

(d) bypass water, water that is diverted but which is returned directly to the source of supply without being used for irrigation.

The soluble salt concentration in the irrigation return flows is the major contaminant of concern. The environmental irrigation return flow problems primarily are located in the arid areas. However, both the salt and nutrient content of irrigation return flows can contribute to the contaminant load of water sources wherever irrigation is practiced.

An increase in soluble salts is an unavoidable result of the use of water for irrigation. As water is transpired by the crop and evaporated from the soil, the salts in the applied water are concentrated. Most of the salts in the applied irrigation water remain soluble in the soil solution. The amount of mineral uptake by crops rarely is significant in the overall salt balance of irrigated agriculture. Almost the entire amount of salts present in the applied irrigation water is contained in the soil solution. A favourable salt balance is obtained when the output of salt in the drainage water equals or exceeds the input to the soil. To maintain a favourable salt balance in the root zone of the irrigated crops, i.e., to prevent an accumulation of soluble salts in the soil, a more saline water must leave the land by percolating below the root zone to the groundwater or to a natural or artificial drain age system. The quantity of water applied for irrigation includes the amount required by the crop and a sufficient amount to leach the soluble salts from the soil.

The characteristics of surface and subsurface irrigation return flows will be different. Surface runoff from irrigated land generally will have the following characteristics: total dissolved salts only slightly greater than in the applied water; presence of precipitated or adsorbed contaminants will be a function of the amount of erosion; colloidal and sediment load will be greater than in subsurface flow; and characteristics will be variable for any given area.

The general characteristics of subsurface return flows will include: a greater concentration of dissolved solids than in the applied water; little colloidal or particulate matter; a different distribution of cations and anions than those in the applied water; a nitrate content that will depend upon the concentration in the applied water, the nitrogen content of the soil, and the nitrogen uptake by the crop; and low phosphorus concentrations.

Fruit and Vegetable Processing: The food processing industry creates considerable quantities of liquid and solid wastes for disposal. Food processing wastes not disposed of as solid waste are disposed of as part of the plant waste-water with or without treatment. On a national basis, about 10 per cent of the solid waste from food processing operations is disposed of in plant wastewater.

These liquid wastes are discharged in the following manner; about 35-38 per cent to public treatment systems, about 11-17 per cent to land or ground disposal systems, and about 42-51 per cent to surface waters (37). Discharges to the land are primarily by spray irrigation. Wastewater discharged to surface waters usually received some degree of treatment before discharge. Plants in urban areas send almost all of their wastes to city systems, whereas plants in nonurban areas primarily use spray irrigation or treatment units such as oxidation ponds or aerated lagoons.

Food processing wastes are organic and can be treated by processes generally used at municipal treatment plants. If, however, these plants are not designed to treat the variable and high strength food processing wastes, the municipal plants may be overloaded and unable to attain the efficiency for which they were designed. Excessive food processing wastes in municipal plants can tax the solids handling and disposal facilities of a municipal waste treatment plant as well as the oxygen capacity of aerobic units.

The seasonal nature of the food processing industry compounds the problem of designing treatment facilities for the wastes or of joint treatment of the wastes in municipal treatment plants. The preponderance of processing activity takes place during the summer and fall months.

Dairy Product Processing: The dairy food industry represents an important part of the food industry and contributes significant liquid wastes for treatment and disposal. There has been a decrease in the number of plants processing dairy food with increased waste loads per plant. About 11 per cent of the total dairy plants in the country process about 65 per cent of the total milk supply. Waste loads per plant range from 2000 to 10,000 lb BOD/day (39).

Wastes from dairy plants consist primarily of varying quantities of milk solids, sanitisers, detergents, and human wastes. The wastes produced in urban plants are sent to municipal waste treatment facilities with a minimum of prior treatment. Dairy plant wastes produced in nonurban areas are treated in aerated treatment facilities, by irrigation on the land, or by uncontrolled disposal to land and surface waters.

Approximately 32 billion gal of wash water, 3.5 billion gal of cheese whey, and 65 billion gal of condenser water are discharged from the nation's dairy plants each year. Whey wastes contain about 4 per cent solids and have a high oxygen demand (33,000—45,000 mg/liter BOD) which makes disposal to streams undesirable. Only about one-third of the whey produced each year is used to any extent. The remainder is discharged from the plants. The nutritive content of why offers interesting possibilities for by-product recovery and utilisation if suitable equipment and markets are available.

Nonagricultural Sources

Agriculture is not the only source of contaminants in a watershed. Effluent from sewage treatment systems, septic tank drainage, urban land runoff, precipitation, and the effluent from industries also contribute to the contaminants in the watershed. This section will attempt to present the characteristics, of other nonpoint sources, so that agricultural nonpoint sources can be placed in perspective.

Precipitation: The constituents of precipitation are influenced by both man-made and natural events. Fuel burning, automobiles, manufacturing operations, forest and other fires, volcanic eruptions, and wind erosion are examples of activities that contribute to the constituents in dry, dust fallout and in rain or snow that are deposited on the sample collecting surfaces. The distribution of contaminants between dry fallout and rainfall is rarely indicated by various investigations. Data presented in Table 7 (20, 40-45) summarise, the area input rates for nutrients. Variable characteristics are reported since they are a function of precipitation, which varies with both time and location, of relative proximity to sources of gases and particulates, and of relationship to air movement patterns.

Sulphate concentrations in precipitation have increased probably because of pollution of the air, masses passing over an area (43) and can result from sulphur dioxide and other sulphur products added by combustion. The sulphate and other ions placed in the atmosphere by man's activities at one location will return at other locations. Air movement patterns can assist in evaluating where these contaminants will be deposited by precipitation.

The sulphur in precipitation collected close to industrial locations was 10 to 15 times greater than that collected from urban locations. Sulphur in precipitation followed seasonal variations being lower in the summer and increasing in the winter (46). The amount of ammonia and organic nitrogen in precipitation was highest during the spring and lowest during the winter months and was higher in areas adjacent to barnyards than in areas removed from the barnyards. The distribution of nitrogen and sulphur in rural and urban situations are shown in Table 8.

Precipitation is a variable and intermittent source of constituents, in surface waters. Once contaminants are in precipitation, they are uncontrollable. Man, however, can exert some control over the contaminants that are released to the atmosphere through management practices that minimise particulates and gases from combustion processes, particulates from disturbing the land, and volatile gases from industrial operations.

TABLE 7

Precipitation Characteristics—Average Nitrogen and Phosphorus Area Yield Rates[a]

Location	Total N	NOa + NH4-N	P	Reference
World mean	6.2			
	(0.8-7.0)[c]	•	—	(40) (before 1954)
World mean	8.7			
	(1.8-22.2)	•	—	(40) (1905)
Temperate zone	•	6.8	•	(40) (1938)
Humid temperate zone	5.6	•	—	(40) (I960)
Europe and United States	•	0.8-2.1	•	(40) (before 1952)
United Kingdom				
Upland	8.2	•	0.27	(41)
Northern	8.7-19	•	0.2-1.0	(41)
Netherlands				
Rural	8.5	•	—	(20)
Industrial	16-100	•	—	(20)
Canada				
Hamilton, Ontario	•	6.0	•	(42) (1948)
Ottawa	7.7	•	—	
	(4.8-12.9)			(42) (1924-1925)

Ceylon	•	12.9	•	(42) (1941)
New York	10.0	•	—	(40) (before 1948)
Ithaca, New York	•	7.4	0.05	(43)
Aurora, New York	•	7.6	0.06	(43)
Geneva, New York	•	8.3	0.05	(43)
Hubbard Book, New Hampshire	•	5.8	0.10	(44)
Cincinnati, Ohio[b]	9.6	5.2	0.6	(45)

(a) In kilograms/hectare/year; 1.12 lb/acre=1.0 kg/ha.

(b) Average United States rainfall of 30 in/year assumed.

(c) Data in parentheses indicate range of data.

Urban Land Drainage: Street litter, gas combustion products, ice control chemicals, rubber and metals lost from vehicles, decaying vegetation, domestic pet wastes, fallout of industrial and residential combustion products, and chemicals applied to lawns and parks can be sources of contaminants in urban runoff. A portion of the urban runoff can drain to sewerage systems while the remainder may reach surface waters by natural drainage channels without receiving treatment. The contaminant load to streams or sewage treatment plants results in a slug load effect that the receptors may be unable to handle. Besides the conventional water pollution parameters, a number of other contaminants such as fecal coliform bacteria, chlorinated hydrocarbon and organic phosphate compounds, a number of heavy metals, and polychlorinated biphenyls have been found in urban runoff (47).

TABLE 8

Average Sulphur and Nitrogen Loadings Due to Precipitation[1,2]

	Rural		*Urban*	*Industrial*
	Adjacent to barnyard	*Removed from barnyard*		
Sulphur, as S	16		42	108
Nitrogen				
Organic N	14.4	7.5	6.2	—
NH_4— N	12.2	2.9	3.6	—
$N0_3$—N	3.5	2.7	3.7	—
Total N	30.2	13.1	13.5	—

1. From Hoeft *et al.* (46).
2. Measured in kilograms/hectare/year.

Comparison of Sources

It is difficult to compare data from various studies because of the variations in sampling methods, analytical methods, and the fact that all studies did not measure comparable parameters. The characteristics of non-point sources are the result of complex interactions in and on the soil making definitive comparisons of these non-point sources difficult. For a comparison of these sources, the order of magnitude of the characteristics and the differences between sources are more significant than the values themselves. The range of values reported for many of the sources has been summarised in Table 9 for different parameters and total nitrogen and total phosphorus. All of the comparisons as presented in terms of concentration or yield rate potentially contributed to a surface water and are not in terms of the constituents in a surface water.

Comparing the data in terms of concentration units, precipitation, forest land runoff, and surface irrigation return flow have comparable values. The wide range in crop land phosphorus is due to the effect of erosion. With erosion control, the total phosphorus of cropland runoff should be near the lower end of the range. Subsurface irrigation return flow nitrogen concentrations are greater than those from surface irrigation return flows, undoubtedly because of the leaching of nitrogen compounds from the soil. Cropland tile drainage has nitrogen and phosphorus concentrations comparable to subsurface irrigation return flows.

The Total phosphorus concentration of cropland runoff, irrigation return flows, cropland tile drainage, and urban land drainage span comparable ranges. Because of the difficulties of comparing different studies, it is impossible to state that the concentrations of total phosphorus in these five non-point surfaces are likely to be grossly different. There is, however, a wide range of values for all non-point sources.

Animal feedlot runoff and manure see page have characteristics that are many orders of magnitude greater than those of the other non-point sources.

TABLE 9

Summary of Non-point Source Characteristics[1]

Source	*Concentration (mg/liter)*				
	COD	*BOD*	*NO_3-N*	*Total N*	*Total P*
1	2	3	4	5	6
Precipitation	9-16	12-13	0.14-1.1	1.2-1.3	0.02-0.04
Forested land	•	—	0.1-1.3	.0.3-1.8	0.01.0,11
Range land	•	—	•	—	
Agricultural crop land	80	7	0.4	9	0-02-1.7

Land receiving manure	•	—	•	—	—
Irrigation tile drainage. western United States					
Surface flow	•	—	0.4-1.5	0.6-2.2	0.2-0.4
Subsurface drainage	•	—	1.8-1.9	2.1-1.9	0.1-0.3
Crop land tile drainage	•	—	•	10-25	0.02-0.7
Urban land drainage	85-110	12-160	•	3	0.2-1.1
See page from stacked manure	25,900-31,500	10,300-13,800	•	1,800-2,350	190-280
Feedlot runoff	3,100-41,000	1,000-11,000	20-23	920-2,100	290-360

1. Data do not reflect the extreme ranges caused by improper waste management or extreme storm conditions; the data represent the range of average values reported.

Source	*Area yield rate (Kg/ ha/year)*					
	COD	*BOD*	*NO_3-N*	*Total N*	*Total P*	*Surface area of interest*
1	7	8	9	10	11	12
Pricipitation	124	•	1.5-4.1	5.6-10	0.05-0.06	Total land area
Forested land	•	—	0.7-8.8	3-13	0.03-0.9	Forest area
Range land	•	—	0.7	•	0.08	Range land
Agricultural crop land	•	—	•	0.1-13	0.06-29	Active crop land
Land receiving manure	•	—	•	4-13	0.8-2.9	Crop or unused land used for disposal
Irrigation tile drainage. western United States	•	—	•	3-27	1.0-4.4	Irrigated western soils
Surface flow						
Subsurface drainage	•	—	83	42-186	3-10	Irrigated western soils
Crop land tile drainage	•	—	•	0.3-13	0.01-0.3	Active crop land requiring drainage
Urban land drainage	220-310	30-50	•	7-9	1.1-5.6	Urban land areas
See page from stacked manure	•	—	•	—	•	Manure holding area
Feedlot runoff	7,200	1,560	•	100-1,600	10-620	Confined, unenclosed animal holding areas

The water pollution effect on a surface water can be evaluated better when the data are in terms of a yield per unit area since although constituent concentrations may be large, the volume of flow may be low. The data in again illustrates the range of reported data but permits a reasonable comparison of the reltive contribution from the various non-point sources. The total nitrogen yield from precipitation, forest land, crop land, land receiving manure, surface irrigation return flows, and urban land drainage span a comparable range. The total phosphorus yield of precipitation and range land are comparable and the total phosphorus yield of land receiving manure, surface irrigation return flows, and urban land drainage are comparable. The total phosphorus yield from forest land and cropland span a wide but comparable range, again undoubtedly due to the effect of erosion.

The nitrogen and phosphorus yield of animal feedlot runoff also is many orders of magnitude greater than that of the other non-point sources.

Although these comparisons cover wide ranges and may be considered gross, they do permit an assessment of the non-point sources that may require control. Actual decisions on the control of non-point sources should be made on the basis of the relative importance of the respective sources in specific locations and on what is technically and logically controllable.

Other comparisons illustrate the contribution of phosphorus from various sources (Tables 10 and 11) (48, 49). The data from these sources fit within the range of data noted in Table 9 reasonably well. The data in Table 11 resulted from twenty months of measurements in the watersheds plus an analysis of the data reported in the literature.

TABLE 10

Phosphorus from Sources in South Central Wisconsin[a]

Source	*Amount[b]*
Domestic wastewater	0.9 kg/capita/yr
Urban storm runoff	0.7 kg/ha'yr
Crop land and pasture runoff	0.035 kg/ha/yr
Woodlands—	0.0027 kg/ha/yr
Manured lands (100 cows/squre mile)	0.9 kg/ha/yr
Drained marshes	18 kg/acre

(a) From Lee (48).

(b) 1.12 Ib/acre equals 1.0 kg/ha.

TABLE 11

Effect of Geology and Land Use on Total Phosphorus Export from Watersheds[a, b]

Watershed type	*Average*	*Range*
Granitic, igneous geology[c]		
Forested	0.05	0.007-0.09

Forested and pasture	0.1	0.06 -0.16
Sedimentary geology[c]		
Forested	0.12	0.07 -0.18
Forested and pasture	0.23	0.11 -0.37
Intensive agriculture	0.46	—
Urban	•	1.1 -16.6

(a) From Dillin and Kirchner (49).

(b) Measure in kilograms/hectare/year, 1.12 lb/acre=1.0 kg/ha.

(c) Results from 34 watersheds in Southern Ontario, Canada.

Summary

The review of available data illustrates the difficulty of quantifying the contribution of agricultural operations to characteristics of surface waters. The differences between the types of agricultural situations are apparent. Forest land runoff and range land runoff represent baseline, natural conditions that do not need to be controlled. The concentrations and yield of constituents from these sources are comparable to that of precipitation. Precipitation, forest runoff, and range land runoff can be considered as background non-point contaminant sources.

Non-point agricultural sources that may require control include cropland runoff, land receiving manure, cropland tile drainage, and irrigation return flow. Routine control of cropland runoff and runoff from land receiving manure technically is difficult and may not he needed in all cases. Each situation requires individual evaluation before deciding upon the need for control and the type of control if it is needed. In most cases agricultural nonpoint sources can be controlled through proper management of the land rather than through the application of treatment techniques. When erosion of crop land is controlled, the contaminants in crop land runoff should be comparable to that of background nonpoint sources. If manure is disposed of on land under good crop and land management conditions, contaminants from these lands should be comparable to that of cropland.

Cropland tile drainage and irrigation return flows are difficult to intercept and control. The nitrogen content of this drainage is the major item of concern.

Nonpoint agricultural sources that should be controlled include manure seepage, feedlot runoff, and the direct discharge of animal wastes to surface waters. Nonsophisticated control procedures are available for the control of these sources and should be included in water pollution control policies.

Urban land runoff and the effluent from municipal and industrial waste treatment plants can be major sources of contaminants in urban and suburban areas. In rural areas, septic tank leakage, manure runoff, and crop land runoff can be important contributors to contaminants in surface waters. The sources in each watershed need to be evaluated, relative contributions identified, and relative costs estimated before pollution control policies are implemented. Uniform policies in each watershed may not be cost-effective non produce the desired pollution abatement.

Bacteria

The list of infectious disease organisms common to man and livestock is lengthy, including a number that can be water borne. When drainage or runoff from animal production units reach a watercourse, a potential chain for the spread of disease has been initiated. Salmonella organisms have been isolated from animal fecal specimens, runoff from animal confinement operations, car-casses of dead animals, and from waterholes from which the animals drank (50-52). Two organisms, Salmonella dublin and S. typhimurium were the salmonella organisms most commonly found in the cattle and contaminated water investigated S. dublin is essentially a pathogen of cattle but can cause meningitis and septicemia in humans. Children are more susceptible than adults. S. typhimurium can infect practically all species of birds, animals, and man. Animal wastes contain other organisms, such as Staphylococcus aureus, which also can be human pathogens. Evidence indicates, however, that the animal waste, environment of the soil environ-ment is not very hospitable to the survival of these pathogens.

The disease potential inherent in the disposal of animal wastes on land is unknown but is considered not to be critical. Organism, survival in soil is determined by the viability of organisms in the soil environment, their interaction with the soil, and the nature of their transport. Bacteria and virtuses in soil have moved from inches to several hundred feet from the point of disposal on or in the land. Salmonella species have survived over 40 days in soils with high organic content. Survival was less in soils of lower organic content (53).

A study of the effectiveness of soil to remove bacteria in animal wastes indicated that coliform and enterococcus bacteria were removed by adsorption during soil percolation and by die-off because of the inability to compete against the established soil microflora (54). Greater than 98 per cent removal occurred in 14 in, of soil. It was concluded that there was little concern that the bacteria would move any great distance from the point of application of the manures or manure water on undisturbed soil.

The ratio of fecal coliform to fecal streptococci has been used to delineate between human and animal pollution. The ratio in human and domestic wastes was shown to be over 4 while for fecal material from animals, the ratio was less than 0.6. A ratio of 0.2 has been noted in cattle feedlot runoff (24). The bacterial types found in a lagoon treating bovine wastes have led one group of investigators to conclude that the pollu-tional potential for true E. coli and Streptococcus faecalis in lagoon waters is minor (55).

Agricultural operations such as livestock and crop production are natural contributors to bacteria in surface waters. Runoff from cropland can increase the bacteria in surface waters which must be removed in water treatment plants to produce potable water. The possible presence of pathogens in runoff and effluent from livestock waste treatment systems suggests that caution should be exercised in reusing such water without bacterial control.

Air Quality

In the broad sense, air pollution can be defined as the presence of air contaminants under conditions that are injurious to human, plant, or animal life or property or which unreasonably interfere with the enjoyment of life and property. Agriculture can be a recipient of the air contaminants of municipalities and industries and sustain the associated injuries. It is rare that agriculture is the producer of air contaminants that are injurious to plants, animals, humans, or property. Agriculture can, however, produce odours and dusts that interfere with the enjoyment of life and property and thus, be a source of localised air pollution. An agricultural industry that offends its neighbours with the production of air pollutants can be forced to cease if the nuisances are not corrected.

Odours

The subjective nature of odours is a difficulty in evaluating their offensiveness and effects. Individuals have varying personal response to odors based upon age, sex, smoking habits, occupation, mental attitudes, and respiratory differences. Each can have varying thresholds of detection. Regular and prolonged exposure to an odour can raise the detection threshold. However for any group of people, as the concentration of odour increases, the odour eventually will be detected and become obnoxious or discomforting.

In spite of the sophisticated technical equipment that has been developed, the human odour sensing system remains the most suitable detector available. Most odours are a complex mixture of chemical compounds. In some odours a specific compound such as ammonia or hydrogen sulphide can be predominant and detected instrumentally to obtain a measure of its strength. Because most odours are caused by small amounts of a number of compounds, a human odour panel remains one of the best odour detection and evaluation systems. Methods to develop odour evaluation panels have been standardised and such panels are helpful for comparable studies.

Odours from animal wastes are a persistent problem arising from the confinement of large numbers of animals. The problem is especially prevalent near feedlots, in and around confined and enclosed animal production operations, and where field spreading of unstabilised waste is practiced.

The conditions under which the odorous compounds are produced offer suggestions as to possible control measures. With current practice, most agricultural wastes are held in piles or in storage tanks or pits prior to disposal. The storage period varies from weeks to years depending on the flexibility desired by the operator. The environmental conditions desires such storage are controlled and anaerobic products such as ammonia, sulphides, mercaptans, amines, organic acids, and methane will result. Many of the unoxidised compounds have their own unique and, even in low concentrations, objectionable odours.

The fact that the odour-causing materials are reduced organic compounds suggests the possibility of an oxidative microbially process for odour control. If the stored wastes are aerated adequate oxygen can be added so that anaerobic end products are not produced. Under these conditions the primary oderous compound will be ammonia since during the aeration of annual wastes, ammonia can be released from the mixture. If sufficient oxygen is added to the aerated mixture the ammonia can be microbially oxidised to nitrites and nitrates eliminating ammonia as an air pollution problem in these circumstances. The key in reducing anaerobically caused odours is to be sure that the oxygen supplied is equal to or greater than the oxygen demand. If the oxygen supply is less than that of the demand objectionable odours may still result.

The greater part of the odour associated with confined animal housing originates from the manure. Liquid manure handling and disposal have generated many of the odor nuisance conditions that have resulted in complaints. Anaerobic liquid systems, in which the waste is diluted with water for ease of handling, provide a better environment for the production of the odorous compounds than does a "dry" or an adequately aerated system. Odours from diluted, anaerobic "liquid" mannure are more offensive than odour from the undiluted, fresh mannure.

Compounds identified as part of the odours from anaerotic poultry manure included ammonia, hydrogen sulphate, twice to five-carbon organic acids, indole, skatole, diketones, and sulphides. An odour panel indicated that the organic acids, mercaptans, and sulphides were the major malodourous components of the odour. Ammonia, methylamine, ethylamine, and triethylamine, were components of hog house odours (56). Odours causing compounds exist not only as gases but also as volatile material associated with particulate matter.

Gases

The gases generated by microbial degradation of the wastes are of concern to the health of animals and humans in confined livestock environments. The gases of greatest concern are carbon dioxide, ammonia, hydrogen sulphide, and methane. Under conditions with adequate ventilation, there has been no evidence that these gases reached harmful concentrations for animals or humans. Under abnormal conditions, gas concentrations have exceeded threshold values and have bad adverse effects on animals.

The conditions leading to accidents have been animals in close proximity to anaerobic wastes, inadequate ventilation, and the release of anaerobic gases as the anaerobic wastes were agitated for movement or removal. Toxic gas concentrations can exist in a waste storage unit and the unit should be adequately ventilated prior to human entry after mixing anaerobic wastes.

ENVIRONMENT IMPACT ASSESSMENT

Meaning and Scope

From ancient time human societies have established certain norms of social behaviour for the purpose of ensuring their survival and protection against environment. The threat and emergencies for which these norms were designed originally were purely natural such as floods and dangerous animals etc. However, they are not adequate in the present day as environmental risks caused today are partly or wholly by human action or reaction with the nature such as water pollution, air pollution or dangerous man-made products. Hence it requires additional protective norms. To cope with the emerging problem of the environment, a wide range of administrative and legislative measures have been taken by several nations. The common characteristics of these provisions included their defensive and protective functions to preserve specific quality of standards of human life against specific environmental risks and which was later termed as "Environment Impact Assessment."

The process of Environmental Impact Assessment (hereafter referred to as EIA), thus means an examination, analysis and assessment of planned activities with a view to ensuring environmentally sound and sustainable development. It is essentially a process or procedure whereby information on likely environmental impact, and possible alternatives and mitigating measures, becomes a pre-requisite to developmental decision making. In this way EIA as a planning tool seeks to upgrade the place of environmental factors in the decision making process so that unnecessary damage to the environment can be avoided. This does not imply that environmental concerns should occupy a paramount position *vis-a-vis* the economic, technological or financial, but that they should rather receive due and balanced consideration before final decision is taken about any plan. Thus implementation of EIA depends upon the relationship of environment and development. This developed a new term 'Sustainable development'.

Objectives of E.I.A.

EIA an important aspect of environment protection has some significant objectives:

(i) To ensure that before decisions are taken by the competent authorities to undertake or authorise activities that are likely significantly effect the environment, the environmental effects of these activities should be taken fully into account;

(ii) To provide the implementation of appropriate procedures in all countries consistent with national laws and decision making process, through which the foregoing may be realised; and

(iii) To encourage the development of reciprocal procedures for information exchange, notification and consultation between States when proposed activities are likely to have significant transboundary effects on the environment of these states.

Problems in Implementation of E.I.A.

States are facing many problems in the proper implementation of E.I.A., to prevent pollution at national, and international levels both. These problems are though seems to be very general but are full of complications. Some of such problems are:

(a) The question, how to determine that pollution has occurred or that it has reached a degree unacceptable to society, raises more puzzling problems than the identification of pollutant. Scientific experts today are quite often unable to agree on the elementary scientific facts. In the past, it was common to approach the question of pollution through the legal concept of nuisance, which was often determined by reference to the value criteria of human health and welfare, property rights and commercial interests. Whether the conduct complained of as the nuisance really amounted to an actionable nuisance, based upon whether the conduct complained of was appropriate in the neighbourhood where it was carried on. In such cases it was right to say that right and wrongs in law (as in morals) were relative to the circumstances in which the question arise. But in the light of new trends in political philosophy, we are most likely to consider these criteria in pollution cases also as obligatory, falling within the powers of governments and public agencies as a matter of social responsibility. The question of this social responsibility of States is of great significance in International Law. What should be the responsibility of States is disputed;

(b) The question as to who will assess the pollution is a matter of global concern. Of course, they are not lawyers, judges or jurists who can on their own be expected to provide the criteria for pollution assessment.

Basically, they are scientists, who can be approached on inter-disciplinary basis, for this act. However; lawyers, scientists and other specialists if would do jointly in their respective sphere, they can find out the criteria for the prevention and control of pollution. How to bring them together for their working in proper prospective is a matter of concern; and

(c) The establishment of criteria of pollution assessment, as distinguished from criteria of pollution control is also a crucial issue. Is it sufficient to adopt a simple test that enable the potential polluter to determine the critical point that whether his activity involves pollution ? Whether pollution control treaties in the regard can be enforced only through national penal legislation or can sometimes be strengthened by the inclusion of definitions containing criteria that consist of simple empirical tests?

All these problems alongwith many other issues required uniform principles of international law for developed and developing countries both. The UNEP has made a lot of efforts to evolve some guidelines and principles of international law in this regard which includes removal of some of genuine complication faced by States in implementation of EIA. Some of them are discussed as follows.

E.I.A. AND INTERNATIONAL LAW

The formulation of international legal rules for environmental protection must respond to the needs and requirements of the different concerned countries. While framing such legal rules, States should take into their account, actual state of legal institutional development. Subsequently, giving effect to such rules (international agreements, guidelines, or principles) inevitably requires further action in terms of implementing legislation and administrative measures that are adopted to local circumstances, and to developing countries in particular.

The need for national measures in this field had already been forecast by "The Founex Report" on 'Development' and 'Environment'. The Report recommends that a range of specific policies be made available to developing countries for choosing them in accordance with their requirement and preference. The Stockholm Conference 1972, while agreeing with the importance of EIA proclaimed that a point has been reached in history when we must shape our actions through out the World with a more prudent care for our environmental consequences. The conference has also worked that industrialised countries should think above the development of the third World and reduce the gap between themselves and developing countries. The U.N. General Assembly under its resolution 3436 (XXX), of 9, December, 1975, called on the executive Director of UNEP to provide technical assistance to the developing countries at their request for the purpose of strengthening their environmental legislation and institutions. This mandate was reaffirmed by the

Governing Council of UNEP in 1976 and by the 1981 Montevideo Meeting of Senior Government Officials Experts in Environmental Law. The UNEP under this authority has adopted many guide lines and principles from time to time. Some of which have been discussed below:

(a) Environmental Law, Guidelines and Principles Shared Natural Resources (1978)

The draft principles of conduct, in this note referred to as the principles have been drawn up for the guidance of States in the field of the environment with respect to the conservation and harmonious utilisation of natural resources shared by the two or more States. The principles refer to such conduct of individual States as is considered Conducive to the attainment of the said objective in a manner which does not adversely affect the environment. Moreover, the principles aim to encourage States sharing a natural resource, to co-operate in the field of the environment.

(b) Environmental Guidelines and Principles on Weather Modification, 1980

Provisions for cooperation between States in weather modifications have been adopted by UNEP Governing Council while recalling, Stockholm Declaration, 1972, relevant decisions of the Seventh and Eighth World Meteorological Conference held in 1975 and 1979 and UN General Assembly Resolution 31/12 of 10 December, 1976, On the Convention on the Prohibition of Military or any other Hostile use of Environmental Modification techniques, while recommending many obligation upon it said that States should ensure that an assessment is made of the environmental consequences of prospective weather modification activities under their jurisdiction or Control which are likely to have an affect on areas outside their national jurisdiction, and either directly or through World Meteorological organisations (hereafter referred to as WMO) made the result of such assessments available to all concerned States.

(c) Environmental Law, Guidelines and Principled on Offshore Mining and Drilling, 1982

The Governing Council of UNEP on the study of legal aspects concerning the environment related to offshore mining and drilling within the limits of national jurisdiction have laid down some general provisions.These are: (1) States should, either individually or jointly, by all appropriate means take preventive measures against, limit and in so far as possible reduce pollution and other adverse effect on the environment resulting from offshore exploration for and exploitation of hydrocarbons and other minerals, and related activities, within the limits of national jurisdiction (hereinafter referred to as "operations"). To this end, States should, in particular adopt legislative and regulatory measures which are not less effective than international rules, and recommended practice and to provide appropriate machinery. In addition to these, the Charter also laid down that States should enact harmonious laws and regulations and act directly or through the competent international organisations.

To implement above provisions it has adopted various provisions viz. authorisation system, environmental assessment, procedures for information and consultation on tranfrontier environmental impact, safety measure, contingency planning and implementation measures including measures for liability and compensation.

(d) World Charter for Nature, 1982

The UN General Assembly while recognising the degradation of natural systems owing to excessive consumption of natural resources resolved this charter, in order to maintain essential ecological processes and Support System. This has given significance to EIA process in case of exploiting the nature or other activities in this regard.

(e) Environmental Law, Guidelines and Principles on Banned and Severely Restricted Chemicals, 1984

The Governing Council of UNEP in this scheme framed provisional notification for banned and severely restricted chemicals, for exchange of information in international trade. The purpose of this notification scheme is to provide experience which would assist the Adhoc working Group of Experts for the exchange of information in international trade in its future work.

(f) Montreal Guideline for the Protection of the Marine Environment Against Pollution from Land Based Sources

This set of guidelines is addressed to Governments of States with a view to assisting them in the process of developing appropriate bilateral, regional and multilateral agreements and national legislation for the protection of the marine environment against Pollution from land-based sources. They have been prepared on the basis of common elements and principles drawn from relevant existing agreements, drawing upon experience already gained through their preparation and implementation.

These guidelines are suggested as a broad framework for the development of similar agreements in those regions where such agreements are called for, for the guidance of Governments in areas which are not at present covered by any regional agreements; and for the preparation in the long term, Should the need arise of a global Convention on Pollution from land-based sources designed to strengthen international institutional agreements to ensure the harmonisation and application of global and regional rules, criteria, standards and recommended practices and procedures and to review the effectiveness of measures taken.

The guidelines are of a recommendatory nature. They are presented as a check-list of basic provision rather than a model agreement, which governments may select from, adopt or elaborate upon, as appropriate, to meet the needs of specific regions. They are without prejudice to the elaboration of cross-sectoral guidelines/ principles within the framework of the Montevideo Programme for the Development and Periodic Review of Environmental Law, as recommended by the UNEP Ad hoc Meeting of Senior Government Officials Expert in Environmental Law.

(g) Cairo Guidelines and Principles for the Environmentally Sound Management of Hazardous Wastes, 1987

These guidelines cover the management of hazardous wastes from their generation to their final disposal, and in particular, the problem of transfrontier movement of such wastes, which calls for international co-operation between exporting and importing countries in the light of their joint responsibility for the protection of the global environment.

This set of guidelines and principles is addressed to Governments with a view to assisting them in the process of developing policies for the environmentally sound management of hazardous wastes. They have been prepared on the basis of common elements and principles derived from relevant existing bilateral, regional and global agreements and national regulations, drawing upon, experiences already gained through their preparation and implementation. Social importance is attached to respect for the balance achieved in Principle 21 of the 1971 Stockholm Declaration on the Human Environment, between the rights and duties of Slates concerning their natural resources and the environment.

Although these guidelines have not been prepared specifically to address the situation of developing countries, they nevertheless provide a framework for effective and environmentally sound hazardous waste management policies in those countries. Implementation of the guidelines Should thus help them to avoid serious and costly environmental problems due to mismanagement of hazardous wastes. By implementing the guidelines, countries could incorporate a sound waste management policies. On the basis of this set of guidelines Convention on the Control of Transboundary Movement of Hazardous Wastes, 1989, was signed.

(h) Environmental Law, Guidelines and Principles on Environmental Impact Assessment (EIA) 1987

EIA frames general goals and principles to ensure environmentally sound and sustainable development. The document is consists of three goals and 13 principles. The first goal seeks to establish that before decisions are taken by the Competent authority or authorise the activities that are likely to significantly effect the development, the environmental effects should be fully taken into account; the second goal promotes implementation of appropriate national procedure in order to implement EIA; and the third goal encourages countries to develop reciprocal procedures for information exchange, notification and consultation on activities that are likely to have significant transboundary environmental effects of these States. The thirteen specified principles further elaborate the goals. The UNEP Governing Council while adopting these goals and principles also requested the Executive Director to bring this document to the attention of all States and relevant international organisations for considering it as a basis for preparing national measures in the field as well as for international cooperation including preparation of international agreements.

(i) London Guidelines for the Exchange of Information on Chemicals in International Trade Amended, 1989

This set of guidelines is addressed to Governments with a view to assisting them in the process of increasing chemical safety in all countries through the exchange of information on chemicals in international trade. They have been developed on the basis of common elements and principles derived from relevant existing bilateral, regional and global instruments and national regulations drawing upon experience already gained through their preparation and implementation.

These guidelines are general in nature and are aimed at enhancing the sound management of chemicals through the exchange of scientific, technical, economic and legal information. Special provisions have been included regarding the exchange of information on banned or severely restricted chemicals in international trade, which call for cooperation between exporting and importing countries in the light of their joint responsibility for the protection of human health and the environment at the global level. To this end, all references in these guidelines to a Government or Governments shall be deemed to apply equally to regional economic integration organisations for matters falling within their areas of competence. These guidelines were complementary to existing instruments developed by the United Nations, the World Health Organisation (WHO) and to the international code of conduct on the distribution and use of pesticides of the Food and Agriculture Organisation (FAO). These were primary guidance for the management of pesticides internationally. These guidelines should implement in a non-duplicative manner for different classes of chemicals covering by existing instruments. Although the guidelines have not been prepared specifically to address the situation of developing countries, they nevertheless provide a framework for the establishment of procedures of the effective use of information on chemicals in these countries. Implementation of these guidelines should thus help them to avoid serious and costly health and environmental problem due to ignorance about the risks associated with the case of chemicals, particularly those that have been banned or severally restricted in other State.

(j) United Nations Conference on Protection of Environment and Development (The Earth Summit), 1992

The Earth Summit held in 1992, discussed many environmental issues including the need to integrate policies for the environment and the development. The Rio Declaration of the Summit is quite significant towards the process of EIA as it sets out many principles about it. However, Rio Declaration's main features are given below:

Rio Declaration

Rio Declaration on Principles of General Rights and Obligations on Environment Protection, has laid down many provisions on EIA in its different principles as follows:

(i) To achieve sustainable development and higher quality of life for all people, States should reduce and eliminate unsustainable patterns of production and consumption and promote appropriate demographic policies (Principle 9).

(ii) States should cooperate to strengthen indigenous capacity building for sustainable development by improving scientific understanding through exchanges of scientific and technological knowledge, and by enhancing the development adaptation, diffusion and transfer of technologies including new incentive technologies, (Principle 9).

(iii) States shall enact effective environmental legislation, environmental standards, management objectives and priorities should reflect the environmental and developmental context to which they apply. Standards applied by some countries, may be inappropriate and of unwarranted economic and social costs to other countries, in particular developing countries (Principle 11).

(iv) States should cooperate to promote supportive and open international economic system that would lead to economic growth and sustainable development in all countries to better address the problems of environmental degradation. Trade policy measures for environmental purposes should not constitute a means of arbitrary or unjustifiable discrimination or a disguised restriction on international trade. Unilateral actions to deal with environmental challenges outside the jurisdiction of the importing country should be avoided. Environmental measures addressing transboundary or global environmental problem should, as far as possible be based on an international Consensus. (Principle 12).

(v) Environmental Impact Assessment (EIA), as a national instrument, shall be undertaken for proposed activities that are likely to have a significant adverse impact on the environment and are subject to a discussion of a competent national authority.

Forgoing Principles of Rio Declaration suggest that basing developmental and environmental policies on a comparison of benefits and costs and on careful macroeconomic analysis will strengthen environmental protection and lead to rising and sustainable levels of welfare. Foregoing discussions Clarify that all nations are in efforts to develop an uniform principle of international law about EIA. However, if these Principles which got recognition in many international conventions and agreements etc., are applied by all nations, the problem of global pollution would be minimised to a larger extent.

States Implementing EIA

Almost all States of the world endeavoured to implement EIA process before finalising any major developmental projects. Such procedures have been integrated into the land use Planning procedure or licensing procedure. Principles of legislation governing EIA, as described above indicated as clearly as possible that various projects of States are subject to EIA procedure. In this way above principles or guidelines laid down by UNEP establish effective review and dispute settlement procedure to avoid unnecessary delay in decision making. These guidelines also tried to provide adequate enforcement powers, including the power of the review agency to halt proposed action until the EIA cleared it off.

However a review of the environmental legislation existing in developed and developing countries presents a very diverse picture. Generally speaking, the developed countries have for the most part well-established and highly developed legal systems, which when necessary, can assimilate new legal mechanism and legislation for environmental management without undue turbulence. In contrast, there are many developed countries, who have not made environmental protection, a part of the fundamental law that exists at present. A perusal of the legal systems adopted by developed and the developing Countries, as to EIA procedure is discussed as follows:

(A) EIA and Developed Countries; and

(B) EIA and Developing Countries.

(A) EIA and Developed Countries

EIA process adopted by some of the developed countries are as follows:

(a) United States of America

The enactment of the National Environmental Policy Act (hereafter referred to as NEPA) 1970, is an important U.S. policy adopted for the prevention of environmental pollution. Draft of NEPA made a deliberate attempt to establish a new environmental perspective in Federal decision making, rather than formulate the details of a procedure to achieve this aim. The U.S. Council on Environmental Quality (hereafter referred to as CEQ) was made responsible to clarify procedural details about implementation of NEPA. The CEQ issued several guidelines in 1971 and 1973, to resolve the problem of converting the general provisions of NEPA into a working system for reviewing the potential environment effects of development proposals. The basic presume of NEPA was that environmental factors were not the major concern of mission oriented' Federal Agencies and it (NEPA) would ensure that environmental matters be given sufficient consideration in decision making. The main problem which later on arose was that who shall review the adequacy of Agency compliance with the Act? Naturally this responsibility fell upon the Congress. But the Congress failed on reviewing individual proposals as appeared from the cases decided by the U.S. Courts to interpret NEPA.

Consequently, this function was thrust upon the courts. The Courts traditionally played this role with respect to non-environmental matters. Judicial review resulted in delay. The rulings that have been important in establishing the courts power of review were also discussed and grouped under several categories for threshold determinations that whether an impact statement required the adequacy of statements and substantive review of agency decisions. The courts have also ruled on the regulatory activities of individual agencies under other States. But the review of such impact of statements through such statutes were not very clear for want of specific regulations for the same.

(b) Canada

The evolution of environmental awareness in Canada Closely paralleled events in the United States as close interaction existed between Canadian non-governmental bodies and their U.S counterparts. Consequently during the early 1970s, a period of open debate on the potential role of impact analysis in environmental planning' took place in Canada. Many Canadians looked to the U.S. experience to evaluate the relevance of U.S. procedures in formulating a system for Canada. But because of certain significant differences between two governments (Canada and U.S.), the direct transfer of U.S procedures to Canada was not acceptable. So it was suggested that measures would be introduced to strengthen and improve the existing functions of Government. In this back ground an Environmental Assessment Panel (hereafter referred to as EAP) was constituted to study environmental issues. The reports of the recommendations of an EAP were public documents. By March, 1978, 24 EAP's were appointed and each one considered single major development proposal. Each project is described in the Feberal Environmental Assessment Review Office Publication, which also included details of the composition of each Panel Guidance on implementation of aspects of the review process. In addition Canada Government also issued several guidelines for environmental assessments for a different types of projects.. These included power station, water related development, linear transmission system, oil and gas exploration and production, airports, ports, mining and industrial development. These guidelines were designed to aid appraisal by describing the types of impact that should be considered in an environment impact system.

(c) United Kingdom

The increase in environmental awareness which occurred in the United States also took place in most industrialised countries, and the United Kingdom was no exception. In the last 10 to 15 years increasing concern has been expressed about the effect of new development on the quality of environment and the ability of the U.K. Planning system to take these effects into account. New towns, motorways, and industrial developments have encroached on the countryside and resulting pollution and loss of land had been an important factor in increasing public concern.

Two events, the inquiry into the third London airport and the discovery of North Sea oil and gas, have been significant in the history of environmental impact analysis in the United Kingdom. The demand for impact analysis in the United Kingdom is felt due to more events in the country rather than the influence of U.S. experience. U.K. legislation covering environmental quality, planning, and, the activities of statutory undertaking were required to be studied in relation to impact analysis. The inquiry into alternative sites for the proposed third London airport held in 1970, demonstrated the inadequacies of traditional cost benefit analysis in providing a balanced assessment of economic, social and environmental impacts. The experience of assessing the impact of the proposed airport led many planners to consider alternative means of analysing the impacts of major development proposals.

Further, two main factors relate to the introduction of environmental impact analysis in the United Kingdom. The concerns the desirability of introducing mandatory system on the patterns of U.S practice and, the second-consideration should be given to the type of system to be used. It was argued that there should be qualified acceptance of environmental impact analysis. In spite of some danger in its adoption as experiences show that changes in policy were difficult to effect in advance. Consequently, the results of impact appraisal were likely to be overturned on political considerations. The public must be educated to evaluate environmental impact statements rationally, otherwise the effectiveness of the approach will be undermined.

The United Kingdom is a member of the European Economic Community (hereafter referred to as EEC) and is subject to its directions. The EEC Commission has sponsored a number of studies into aspects of environmental impact analysis. The most important reports on environmental impact analysis within the EEC have been produced by Lee and Wood.They examined current practice in Member States and the possible scope for introducing impact statements in the decision and planning processes within the community and in Member States. In the light of these reports, the EEC issued a directive to U.K. on the introduction of a system for environmental impact analysis, which might be modified to accommodate to the community requirement.

The assessment of policies as seen are required to be modified in the United Kingdom. The lengthy 'and expensive public inquiry, to examine a proposal to reprocess nuclear fuel at wind scale has shown the desirability of assessing the implications of national policies prior to examine its local environmental, social, and economic implications, of projects whereby desired policies are implemented. Similar issue's, have been raised in relation to the proposal by the National Coal Board to extract coal from a rural agricultural area of the East Midlands.

(d) Australia

Australia, like the United States and Canada, is a Federation. The Federal government has introduced legislation to cover proposals over which it was jurisdiction either as developer or as an authorising body. The Environment Protection (Impact of Proposals) Act was introduced in December, 1974 and, subsequently, amended in 1975. Regulations and guidelines for implementation of the legislation were formulated in May and June, 1978, respectively. Prior to the introduction of the legislation, the Department of the Environment and Conservation had acquired expertise as a result of involvement in ad hoc studies of the impact of developments proposals. The legislation was formulated to take account of this experience.

There was some dissatisfactions over the procedures of the Environment Protection (Impact and Proposal) Act. One major cause of such dissatisfactions was the discretionary nature of the power to 'subject a particular proposal to an impact assessment. This power rested with' the Federal administration which was not an independent authority. Formly, argued that, this factor was one of the major reasons why the Act had been implemented slowly. In addition, it has resulted 'in a relatively low level of expertise in the preparation of impact statements. The lack of cleanly defined procedures, which were set in motion automatically when a particular type of proposal was formulated means that Impact analysis utilised in frequently. Meseley stated that the environment impact assessment procedures had not been used for certain projects, regarded by many as controversial. The discretionary powers were used deliberately to prevent public examination of these proposals. The one other factor which influenced implementation of the federal legislation was the lack of the process of judicial review of the Act. Finally there was no legal obligation to implement measures to mitigate adverse effects even though they might be considered an Impact.

In this way, it may be observed that all developed countries have laid down policies regarding the implementation of EIA process. Undoubtedly few of such policies create confusion regarding their adequate implementation. But they may be modified to the extent of their applicability. Since developed countries have evolved new technologies, they don't have any such problem which would restrict them to implement EIA provision.

B. EIA and Developing Countries

The general nature of environmental problems in the developing countries differ from the developed countries. In the developing countries despite the rhetoric of Nairobi Conference, 1987, where they came forward to fill up some of the gaps left by the United States, their leadership strongly raised their views that environmental protection controls were part of western capitalists, neocolonialist plot to prevent the development of developing countries. Similar reactions came from the, developing world when the 1972 Stockholm Conference, was proposed by Sweden in 1968. There was general feeling that environmental concerns, were less of a priority than the real problems facing them, i.e. to further economic development for alleviation of poverty. They also fear that these concerns of the developed countries are designed to detract them from their own path toward economic development thus perpetuating the existing levels of economic inequality between the north and the south.

The developing countries insisted, on the other hand, that a major portion of their environmental problems was a result of under development. Founex Report also submitted that conditions of poverty in both rural and urban areas, not only affected the "quality of life", but "life" itself was endangered by poor water, housing, sanitation and nutrition, by sickness and diseases, and by natural disaster.

Environmental concerns have therefore been reviewed by the developing countries as a shift away from their economic developed goals. Their preferences has been to place economic development and economic growth before the preservation of the environment. But at the same time, they recognised the reality that the character and extent of future economic development was subject to ecologied principles and constraints. This led to the idea of sustainable development. In fact, the sustainable development reduced sharp dichotomy that had existed between environment and development before the 1972 Stockholm Conference (UNCHE), and acknowledged the truism that the problem of undevelopment plaguing many countries such as poverty, malnutrition, ill health, natural calamities, and others are as much environmental as developmental. Thus for environmental legislation can be seen as a tool for planning and management. Referred to here, as an example, the legal requirement on the environment impact assessment (EIA) techniques which can improve the quality of life and at the same time avoid the long term negative environmental effects of a project or an undertaking. To illustrate further, this can be accomplished through the maintenance of specific environmental standards or through the establishment of executive or co-ordinating bodies in order to integrate programmes of environmental protection with those of economic and social development. Realising the feasibility of the concept of sustainable development, many developing countries have enacted legislations while incorporating the process of EIA in accordance with the principles of international law to protect the environment. However, EIA process adopted by some of the developing countries are discussed below.

(a) India

In India several measures were adopted to protect environment from pollution after Independence. In this regard, legislations including constitutional arrangements and judicial pronouncements had done remarkable work. As far as legislations were concerned some of them adopted EIA process specifically 'while others had their indirect bearings. Important legislative efforts may be alluded to (1) The Wild Life Protection, Act, 1972, which provided that the State Governments were empowered to declare any area to be sanctuary or as a national park, if they considered that such area would be of adequate ecological, faunal, floral, geomorphological, natural or zoological significance for the purpose of protecting, propagating or developing wild life or its environments; (2) The Water (Prevention and Control of Pollution) Act, 1974, made provisions for the establishment of a Central Board as also of State Boards to take all necessary measures.

However to emphasise more; about the implementation relating to environmental protection and for the prevention and control of water pollution, the State Bards may make survey of any area and the flow or volume of any river, stream or well. It may also take samples of water from any river, stream or well or sewage or trade efficient for the purpose of making the necessary analysis. A prohibition has been imposed on every person to cause or permit any poisonous, noxious or polluting matter to enter into any river, stream or well. Previous consent of the State Board is required to bring into use any new or altered outlet for the discharge of sewage or trade effluent. The State Board can also take emergency measures where immediate action is required pertaining to any pollution matter. Provisions have been made for the establishment of Central Water Laboratory and State Laboratories for testing and analysing the polluted materials. The Water (Prevention and Control of Pollution) Cess Act, 1977 has also been passed for the levy and collection of a cess or water consumed by industries and local authorities, especially as a measures to augment the financial resources of the Central Board and the State Boards. The Forest (Conservation) Act, 1980, also prohibits forest land to be used for non-forest purposes, except with the prior approval of the Central Government. The Act has been passed to prevent deforestation which results in ecological imbalance and environmental deterioration. The Air (Prevention and Control of Pollution) Act, 1981, has been passed to control and abatement of air pollution. The Act adopts the Central Water Board and State Water Boards as the Boards for its own purposes. The Central Board had been entrusted with the functions of improving the quality of air and preventing, controlling, and abating air pollution in the country. The State Boards have also been entrusted with various functions, inter alia, to inspect industrial plants and the manufacturing processes and also air pollution control areas. They can also prescribe standards of emission of air pollutants into the atmosphere from industrial plants and automobiles. The previous consent of the State Board has been made necessary for a person to operate any industrial plant in air pollution control area. Provisions were also made for entry and inspection to ensure that the provisions of the Act were being observed.

The Environment (Protection) Act, 1986 which refers to Stockholm Conference expressed concern about the decline in the environmental quality, increasing pollution, loss of vegetal quality cover and biological diversity, excessive concentrations of harmful chemicals in the ambient atmosphere, growing' risks of environmental accidents and threats of life support systems. While the existing laws relating to environment touched specific aspects, this Act, 1986 was general in character. This Act gave all omnibious powers to the Central Government to take the necessary measures for protecting and improving the quality of the environment, preventing, controlling and abating environmental pollution. The Government were given the power to lay down standards for emission or discharge of environmental pollutant from any source and providing the necessary safeguards for the prevention of accidents which might cause environmental pollution. The Government was empowered to make inspection of any premises, plant, equipment, machinery or manufacturing process for the purposes of the Act. For determining the nature and extent of environmental pollutants, environmental laboratories could also be established. In case of any environmental pollution, the persons concerned were bound to take the necessary steps to prevent and to mitigate the same and to intimate the said fact to the prescribed authorities.

Further Indian awareness, towards environment protection could also be observed through their passing of the Constitution, (42nd Amendment) Act, 1976, which was inserted as one of the Directive Principles of State Policy to 'empower the States to take all necessary measures to protect and save the environments. Similarly a new part IV-A by way of fundamental duties were also added by the same amendment and sub-clause (g) of the Article 51 casts a duty on every citizen of India to protect and improve the natural environment including forests, lakes, rivers and wild life and to have compassion for, living creatures.

In addition to above Indian judiciary also played an important role in this regard. In Municipal Council Ratlam *v.* Vardhichand and others, the Supreme Court compelled the local authorities, through the country to discharge their obligations in respect of public health, public safety and public convenience. This decision implies that local authorities should observe EIA process while discharging their duties. In Rural Litigation and Entitlement Kendra *v.* State of Uttar Pradesh, the court had directed the closure of mining operations through blasting in the Doon Valley. It held that although closure would cause hardships to the persons affected, but it was a price that had to be paid for protecting and safeguarding the rights of the people to live in the healthy environment with minimal disturbance of ecological balance and affection of air, water, and environment. Subsequently while delivering detailed reasons in this case, the court observed the importance of Himalaya, as being the source of important rivers, and regulating the monsoons and rainfall in the Indo-gangetic belt, recorded that on account of erratic, irrational and uncontrolled mining operations, the landscape had subsequently lost its green cover. Mining observations had to be stopped because they led to cutting down of the forest, shaking the natural soil and loosening the rocky structures. The court expressly referred to the Stockholm Conference and held that with the growth of population and scientific developments, man had started encroaching upon the nature and its assets, the consequences of which were serious. The courts cautioned that the Himalayas and the Forest growth on the mountain range should be left uninterfered with, so that there might be sufficient quantity of rain. In another important pronouncement, M.C. Mehta, V. Union of India, the Court after express reference to, the Water (Prevention and Control of Pollution) Act, 1974, and The Environment Protection Act, 1986, as also to Stockholm Conference, 1972, directed for the stopping of the working in the tanneries which were discharging effluents in Ganga and which did not set up primary treatment plants for the proper treatment of the waste. While passing further orders in this case, the Court observed that Municipal authorities had the statutory duty to prevent public nuisance caused by the pollution of the river Ganga. The Court issued various directions, *inter-alia,* including non-issue of licences for establishing it of new industries unless adequate provisions were made for the treatment of trade effluents. In this connection M.C. Mehta *v.* Union of India, may be referred to in this case, the Court held that victims who suffered on account of leakage of oleum gas from one of the units of the large corporation had right to compensation as every industry engaged in an inherently hazardous activity, had strict and absolute liability. The Court while passing orders, in this case, directed that for restarting a chlorine plant, a high Powered Authority for purposes of overseeing the functioning of various other hazardous industries be set up. In this way it may be observed that in the aforesaid cases, the Supreme Court of India not applied only international principles of EIA process but also extended it to protect the environment from pollution. Of EIA process in India, it is pertinent to point out that major hydro-electric and irrigation projects, thermal power projects, including refineries and petrochemicals are subject to EIA guidelines Environmental Appraisal Committees have been constituted in the department of environment with the specific purpose of analysing the environmental impact of projects. The Department is now contemplating, establishing, monitoring committees to oversee the implementation of safeguards or mitigating measures and extend EIA to other development project.

(b) Nepal

The basic development plan of 1980-85 of Nepal, clearly recognised the relationship between environmental and natural resources issues and development. It states, problems like population pressure, limited cultivable land and destruction of natural resources will adversely tell the whole development process itself.

(c) Philippines

The environmental policy of the Philippines is outlined in Presidential Decree (hereafter referred to as PD). Its PD 1151 and PD 1152 refer specifically to the institutionalisation of EIA system. Subsequent Presidential issuances of PD's further strengthened the system by simplifying the procedure, clarifying issues and making the process fully operational. Now EIA is required for public and private projects of three types, i.e., heavy industry, resource extraction and infrastructure and for twelve environmentally critical areas, e.g. mangroves, coral reefs.

(d) Africa

In Africa situation appears to be considerably different. An attempt was made in 1979 and 1980, by UNEP and ECA (Economic Commission for Africa) to ascertain and analyse the environmental legislation of 15 African countries, which were selected on the basis that their environmental problems could be considered as representatives of the problems of the other African countries. Generally, environmental problems were dealt with sectorally by various ministries such as Health, Agriculture, Commerce and Industry, Forestry, or Lands and Natural Resources, without the presence of any central coordinating machinery. There may, however, be an ad hoc inter departmental coordinating committee, which invariably has only advisory functions.

(e) Columbia

It is important to mention about the modification of the environment and natural resources laws of Columbia (1974) which have been, implemented by specific regulations on forests, national parks, environmental education, Oceanic resources, wild life, inland waters, fisheries and marine resources. It does not, however, deal with subjects such as environmental sanitation or chemical and waste management which are covered by the subsequent Public Health Code (1979) and its subsidiary regulations; Senegal's Environmental Code (1983) on the other hand, is confined to pollution control, classifying establishments into two depending on the seriousness of the adverse effects which their operations may entail.

(f) Iran

EIA requirements in Iran is based on its Environmental Enhancement Act (1974). It authorises the Department of Environment to determine whether the implementation of any development is consistent with the laws and regulations covering protection of the environment. If not, the department may notify the ministry or an agency concerned so that the difficulty can be corrected. Accordingly, in response to this authority, the Department of Environment has established an Environmental Impact Studies Bureau within the human Environment Division of the Department, while the Bureau deals with large-scale development projects, small undertakings are normally referred to the clean Air and clean water bureau for review. The bureau has identified six major development categories requiring EIA.

(g) Other Developing Countries

In addition to above, other developing countries also operate under rational developmental plans prepared by their national economic and development authorities. In the preparation of these plans, countries have incorporated environmental policies and resource management principles, brought about by the growing awareness of the interconnections among sustainable development, a healthy environment and natural resources conservation. In some regional development plans which also identify policy priorities, resource needs and constraints assign responsibility and identify the time frame within which policy goals are anticipated to be achieved, have likewise been developed. Additionally, functional plans are prepared for tourism, agriculture, human settlements, land use, energy, etc.

Master plans for environmental management are sometimes designed and formulated as well. Malayasia, for instance, has master plans for sewerage, and water resources development and urban development. Though developed by the Federal Government, master plans are supposed to be implemented by local authorities. For a federal system like Malayasia, policy decisions by state agencies are required to be consistent with policy statements in the national plan. Municipal or local government plans are required to be consistent with national regional plans.

In the Island States of the South Pacific such as environmental policy statements in plan documents is one of the more important forms in which such commitment to environmental protection is enshrined.

National Conservation plans and strategies have likewise been prepared (e.g., Botswana, Togo, Senegal) or are being prepared (e.g. Rwand, Kenya) to ensure that the objectives of conservation are achieved as expeditiously as possible and to speed up integration of conservation with development. The purpose of these strategies is to focus attention on the relevant priority requirements for conservation, stimulate appropriate action, raise public consciousness, and overcome any apathy there might be to taking the action needed.

In giving effect to an EIA requirement, the developing countries developed their approaches, appropriate to their constitutional and legal systems and socio-economic requirements. There are atleast 44 developing countries, whose official portfolio titles expressly refers to the environment as a rough indicator of political priority currently accorded to this subject. In addition 72 countries have impaired responsibilities of environmental offices either in their President or Prime Minister, Central Administrative Agencies or Governmental Institutions or in other Sectoral Ministries.

While EIA is slowly being accepted as a planning tool for environment management, very few developing countries have specific legislation on the same. As such enactment of laws to effect environmental policy by developing countries is utmost necessity of the day.

Enactment of Environmental Legislation

(a) The Formulation of Environmental Legislation

Lately, some countries have made efforts to enact environmental legislations through the environmental agencies who drafted it. Further, the environmental law drafting, however, is not confined to governmental agencies. In Costa Rica, preparation of an environmental bill is being undertaken by the school of Law of the State University, while in Equador, the Nature Foundation an NGO, is promoting the preparation of a bill on the subject. Parliamentary commissions in charge of deliberation on bills on many countries of Latin America usually have formal consultations with scientists and technicians, experts in such matters and with representatives of society most affected by an environmental legislation under consideration. Non-governmental organisations (NGOs), academics and professional people are also allowed to participate in environmental policy making bodies in Nepal, Senegal and Tunisia.

The Implementation of Environmental Legislation—Some Constraints

On account of extreme diversity of developing countries, in terms of size, level of development, resources endowment, legal system and cultures and the like, it is very difficult to generalise the constraints to environmental law implementation.

However, it is possible to identify a number of factors relating to governmental policy, institutions and planning that lie at the root of the problem. These factors do not operate in isolation and should be interpreted as relating to situations where they function as parts of a system of interacting elements.

The following factors have been identified as constraints in the implementation of environmental laws.

(a) Lack of Political will to Apply/Implement Environmental Law

This is the most commonly identified cause of lack of effectiveness of environmental legislation in developing countries. In such situation, the problems are plentiful, the social needs are many but the material and human resources are so limited that the need for environmental improvements must be perceived by the leadership as of highest priority before there can be a will to meet the need. It is as if the leadership is of the view that any kind of economic development, no matter how environmentally destructive, is welcome and fails to appreciate the case for preventive action and the requirements of conservation that flow directly from economic development.

As a result, there is no mechanism to enforce the environmental legislation.

(b) Lack of Public Concern, Awareness and Participation

Again, on account of the diversity of the social and political systems in developing countries, some were able to evolve a tradition of public involvement while others have very little understanding about the importance of participation in governmental affairs particularly at the village level. In particular, the complexity of modem technology like nuclear power plants and chemical manufacturing complexes work against meaningful public understanding of the technical aspects and related policy and legal issues involved in such technologies. Most often, the populace do not understand the importance of observing environmental regulations because there is not even a campaign to increase awareness by dissemination of legislative information. Little is also known of public participation in decisions being taken by governments on setting of power and industrial plants and the development of industrial zones and complexes. Nor are there people especially trained in environmental law.

(c) Insufficient Funds in Both the Short and Long Terms

Insufficiency of fund is often mentioned as an important reason for poor progress in environmental law implementation. Yet, in many cases, the principal problem is not so much the lack of funds as their effective utilisation. Whatever, it is the task of those concerned with environmental programme to ensure that information on environmental needs is available in the proper form and at the proper time in the government's overall programme and budgetary process. This would ensure appropriate support to its programme of equipment acquisition, training of law enforcers, etc. To illustrate penal sanctions in environmental laws are only effective if sufficient effort is directed to detection and prosecution offenders. Environmental law requires the development of special resources, of trained, personnel and equipment, for this purpose. Without the necessary funds, there is no way public authorities could secure compliance of permit holders and of the general Public to regulations.

(d) Inadequately Trained Personnel

In countries where considerable industrialisation and increase in urban population are evident, the need is primarily for personnel trained in the highly specialised environment problems associated with industrialisation such as hazards from toxic chemicals, radiation, pesticides and other air-borne and water-borne pollution. Enforcement of anti-pollution laws and regulations needs a strong back-up by trained technical personnel.

(e) Insufficiency of Environmental Legislation

Legislation may be inadequate because it does not regulate the situation in question or its does not sufficiently regulate the situation. Or, the regulation is inadequate because of the lack of essential scientific elements to make it effective which hinders its enforcement. Often times, there are no detailed regulations, guidelines, criteria and standards to implement the environmental framework law which is the foundation for a regulatory programme and permit system for the country (e.g. Botswana's Atmospheric Pollution (Prevention), Act.)., This entails a large amount of specialised, detailed work.

Inadequacies of legislation, however, should not be considered a major causal factor in the slow progress toward environmental improvement in developing countries. This is so, because in most cases, there is an adequate basis in existing legislation to get started. Legislations in this regard are of two types:

(i) laws which are obviously environment—related but which do not contain the environmental aspects, and

(ii) laws which are environment-related with no explicit reference to environmental consideration in the implementation but are broadly worded to admit such considerations in the implementation of the purposes of the Act.

The problem is that the environmental implications of these categories of legislation are not exploited to the fullest possible extent. The reason is perhaps the lack of will on the part of policy makers and implementors. Nevertheless, there are also instances when an environmental legislation based on foreign models cannot be implemented because the country does not have the administrative, judicial and professional resources to implement and enforce the laws which have been developed in the places where the necessary infrastructure.

(f) Ineffectiveness of Natural Systems and Environmental Organisations

Some developing countries are experiencing difficulties in achieving urgently needed economic development while conserving and sustaining the productivity of natural resources and environmental quality on account of fragmented bureaucracies. Specifically, the management problems being experienced by these organisations are fragmentation of responsibility, inadequate communication among agencies, poor co-ordination, and inter-agency conflicts over policies, jurisdictional responsibility and specific resources uses. A special deficiency is the lack of adequate co-ordination between agencies in charge of economic and social development and those concerned with the environment. It aggravates the lack of human, material and financial resources for enforcing and supervising the enforcement of environmental legislation.

(g) Inadequacies in the Planning

To the extent that environmental consideration are not reflected in the overall national development plan and the specific economic development programme, they are not likely to be reflected in the annual national budget nor in the budgets and financial plans for specific projects. Without, funds, environmental programmes and environmental components of developments projects cannot proceed and results in failure.

The situation exists because there is no well-developed administrative basis for preparing an environmental quality sector plan and where such plan exists, there is often no well-developed means integrating the environmental quality planning with the overall economic and social development planning activity of the government. In short, environmental analysis of long-range economic development programmes and specific development projects is not built into the planning process which produces the programmes and projects.

As discussed above, sustainable development and conservation are mutually dependent and the latter is a prerequisite to the former. In turn, the aims of conservation can only be achieved in a lasting way if they are supported by the necessary legislation. It follows that environmental and conservation law must be viewed as a necessary integral part of development, and it is essential that developing countries take the steps needed to strengthen their legislation in this field.

Environmental law now covers both preventive and remedial measures for all aspects of resources use. These are:

(1) regulation of predominantly quantitative aspects of natural resources use (e.g., water, wildlife forestry, fishery),

(2) regulation of predominantly qualitative aspects of such use (e.g., anti-pollution legislation), and

(3) regulation of the general framework for such use through allocation of development, land use planning and other environmental planning tools. These aspects are very much inter—linked to one another.

Thus, to achieve optimal environmental legislation means not only having the necessary and appropriate sectoral legislation but also providing for the adequate inter-sectoral mechanisms. Moreover, environmental legislation should explicitly take into account the uncertainties associated with development factors changing technology, resource availability, population changes, probability of damage from natural phenomena. Developed countries have no problem to achieve all these things but for many developing countries to achieve these things have become a milestone unless they are furnished them by developed countries.

REFERENCES

Allen, P. (Ed) 1993. *Food for the Future: Conditions and Contradictions of Sustainability*. ISBN 0-471-58082-1 Paperback. 344 pages.

AtKisson, A. 1999. *Believing Cassandra, An Optimist looks at a Pessimist's World*, Chelsea Green Publishing., White River Junction, VT

Bartlett, A. 1998. "Reflections on Sustainability, Population Growth, and the Environment—Revisited" revised version (January 1998) of paper first published in Population & Environment, Vol. 16, No. 1, September 1994, pp. 5-35.

Benyus, J. 1997. *Biomimicry: Innovations Inspired by Nature*, William Morrow, New York

Bookchin, M. 2005. The Ecology of Freedom: the emergence and dissolutioni of hierarchy. Oakland, CA: AK Press.

Brown, M.T. and Ulgiati, S 1999. *Emergy Evaluation of Natural Capital and Biosphere Services*, AMBIO, Vol.28, No.6, Sept. 1999.

Brundtland, G.H. (ed.), (1987), *Our common future: The World Commission on Environment and Development*, Oxford, Oxford University Press.

Costanza, Robert, Lisa J. Graumlich, and Will Steffen (eds.), (2007), *Sustainability or Collapse? An Integrated History and Future of People on Earth*, The MIT Press. ISBN 978-0-262-03366-4.

Dalal-Clayton, B. (1993) *Modified Eia And Indicators Of Sustainability: First Steps Towards Sustainability Analysis*, Environmental Planning Issues No.1, International Institute For Environment And Development, Environmental Planning Group.

Daly H. 1996. *Beyond Growth: The Economics of Sustainable Development*. Boston: Beacon Press. ISBN 0-8070-4709-0

Daly H. and J. Cobb. 1989. *For the Common Good: Redirecting the Economy Toward Community, the Environment, and a Sustainable Future.* Boston: Beacon Press. ISBN 0-8070-4705-8 Review

Dean, J. W. (2006). *Conservatives Without Conscience.* New York: Viking Penguin.

Ekins, P. (Ed.). (1986). *The Living Economy.* London: Routledge and Kegan Paul.

Hargroves, K. and M. Smith (Eds.) 2005. *The Natural Advantage of Nations: Business Opportunities, Innovation and Governance in the 21st Century.* ISBN 1-84407-121-9, 525 pages. Earthscan/James&James. (See the books online companion at www.thenaturaladvantage.info)

Hawken, Paul, Lovins, Amory and Lovins, L. H. 1999. *Natural Capitalism: Creating the Next Industrial Revolution*, Earthscan, London (Downloadable from www.natcap.org)

International Institute for Sustainable Development (1996) *Global Tomorrow Coalition Sustainable Development Tool Kit: A Sample Policy Framework*, Chapter 4.

Jarzombek, Mark. "Sustainability—Architecture: between Fuzzy Systems and Wicked Problems," *Blueprints* 21/1 (Winter 2003), pp. 6-9; and LOG 8 (Summer 206) 7-13.

Kriegman, O. 2006. *Dawn of the Cosmopolitan: The Hope of a Global Citizens Movement.* Boston: Tellus Institute.

Kull, Kalevi; Kukk, Toomas; Lotman, Aleksei 2003. When culture supports biodiversity: The case of the wooded meadow. In: Roepstrorff, Andreas; Bubandt, Nils; Kull, Kalevi (eds.), *Imagining Nature: Practices of Cosmology and Identity.* Aarhus: Aarhus University Press, 76-96.

Lane, R. E. (1991). The Market Experience. New York: Cambridge University Press.

Marks, N., Simms, A., Thompson, S., and Abdallah, S. (2006).The (Un)happy Planet Index. London: New Economics Foundation. Downloadable from www.neweconomics.org

McDonough, W. & Braungart, M. (2002). Cradle to Cradle. North Point Press

Nelson, E. H. (1986). *New Values and Attitudes Throughout Europe.* Epsom, England: Taylor-Nelson.

Rajan, C. 2006. *Global Politics and Institutions.* Boston: Tellus Institute.

Raskin, P., Banuri, T., Gallopin, G., Gutman, P., Hammond, A., Kates, R., and Swart, R. 2002. *Great Transition: The Promise and Lure of the Times Ahead.* Boston: Tellus Institute.

Raven, J. (1995). The New Wealth of Nations: A New Enquiry into the Nature and Origins of the Wealth of Nations and the Societal Learning Arrangements Needed for a Sustainable Society. Unionville, New York: Royal Fireworks Press; Sudbury, Suffolk: Bloomfield Books.

Richardson, B.J. and Wood, S. (eds) (2006). *Environmental Law for Sustainability: A Reader*, Hart Publishing, Oxford.

Robèrt, Karl-Henrik. (2002). *The Natural Step Story: Seeding a Quiet Revolution.* Gabriola Island, BC: New Society Publishers.

Rolando, Leveious (2008). *Children and Youth in Sustainable Development.* New York: HUGS Movement.

Shah, H., & Marks, N. (2004). *A Well-being Manifesto for a Flourishing Society*. London: New Economics Foundation.

Sinclair, Fiona, D. (2007). "What is Sustainability"

Steffen, Alex (2006). Worldchanging: A User's Guide to the 21st Century. Abrams, New York.

Trainer, F. E. (Ed.). (1990). A rejection of the Brundtland Report. *International Foundation for the Development of Alternatives Dossier, 77*, May-June, 71-85.

Unruh, G. (2000). *Understanding Carbon Lock-in*, Energy Policy, Volume 28, Issue 12, October, 817–830.

Unruh, G. (2002). *Escaping Carbon Lock-in*, Energy Policy, Volume 30, Issue 4, March, 317-325.

Yankelovitch, D., Zetterberg, H., Strumpel, B., Shanks, M., et al. (1983). Work and Human Values. New York: Aspen Institute for Humanistic Studies.

Young, L. & J. Hamshire 2000. *Promoting Practical Sustainability*. Australian Agency for International Development (AusAID), Canberra Australia, ISBN 0-642-45058-7. Free copies available at AusAID Public Affairs, GPO Box 887, Canberra, ACT 2601, Australia.

ANNEXURE I

"UNITED NATIONS CONFERENCE ON THE HUMAN ENVIRONMENT, 1972"

The United Nations Conference on the Human Environment Having met at Stockholm from 5 to 16 June, 1972.

Having considered the need for a common outlook and for common principles to inspire and guide the peoples of the world in the preservation and enhancement of the human environment.

Proclaims that:

1. Man is both creature and moulder of his environment, which gives him physical sustenance and affords him the opportunity for intellectual, moral, social and spiritual growth. In the long and tortuous evolution of the human race on this planet a stage has been reached when, through the rapid acceleration of science and technology, man has acquired the power to transform his environment in countless ways and on an unprecedented scale. Both aspects of man's environment, the natural and the man made, are essential to his well-being and to the enjoyment of basic human rights even the right to life itself.
2. The protection and improvement of the human environment is a major issue which affects the well-being of peoples and economic development throughout the world; it is the urgent desire of the peoples of the whole world and the duty of all Governments.
3. Man has constantly to sum up experience and go on discovering, inventing, creating and advancing. In our time man's capability to transform his surroundings, if used wisely, can bring to all peoples the benefits of development and the opportunity to enhance the quality of life. Wrongly or heedlessly applied, the same power can do incalculable harm to human beings and the human environment. We see around us growing evidence of man-made harm in many regions of the earth; dangerous levels of pollution in water, air, earth and living beings; major and undesirable disturbances to the ecological balance of the biosphere; destruction and depletion of irreplaceable resources; and gross deficiencies, harmful to the physical, mental and social health of man, in the man made environment, particularly in the living and working environment.
4. In the developing countries most of the environmental problems are caused by underdevelopment. Millions continue to live far below the minimum levels required for a decent human existence deprived of adequate

food and clothing, shelter and education, health and sanitation. Therefore, the developing countries must direct their events to development, bearing in mind their priorities and the need to safeguard and improve the environment. For the same purpose, the industrialised countries should make efforts to reduce the gap between themselves and the developing countries. In the industrialised countries, environmental problems are generally related to industrialisation and technological development.

5. The natural growth of population continuously presents problems for the preservation of the environment, and adequate polices and measures should be adopted, as appropriate, to face these problems. Of all things in the world, people are the most precious. It is the people that propel social progress, create social wealth, develop science and technology and, through their hard work, continuously transform the human environment. Along with social progress and the advance of production, science and technology, the capability of man to improve the environment increases with each passing day.

6. A point has been reached in history when we must shape our actions throughout the world with a more prudent care for their environmental consequences. Through ignorance or indifference we can do massive and irreversible harm to the earthly environment on which our life and well-being depend. Conversely, through fuller knowledge and wiser action, we can achieve for ourselves and our posterity a better life in an environment more in keeping with human needs and hopes. There are broad vistas for the enhancement of environmental quality and the creation of good life. What is needed is an enthusiastic but calm state of mind and intense but orderly work. For the purpose of attaining freedom in the world of nature, man must use knowledge to build, in collaboration with nature, a better environment. To defend and improve the human environment for present and future generations has become imperative goal for mankind—a goal to be pursued together with, and in harmony with, the established and fundamental goals of peace and of world-wide economic and social development.

7. To achieve this environmental goal will demand the acceptance of responsibility by citizens and communities and by enterprises and institutions at every level, all sharing equitably in common efforts. Individuals in all walks of life as well as organisations in many fields, by their values and the sum of their actions, will shape the world environment of the future. Local and national governments will bear the greatest burden for large-scale environmental policy and action within their jurisdictions. International co-operation is also needed in order to raise resources to support the developing countries

in carrying out their responsibilities in this field. A growing class of environmental problems, because they are regional or global in extent or because they effect the common international realm, will require extensive co-peration among nations and action by international organisations in the common interest. The Conference calls upon Governments and Peoples to exert common efforts for the preservation and improvement of the human environment, for the benefit of all the people and for their posterity.

II

PRINCIPLES

States the common conviction that:

Principle 1

Man has the fundamental right to freedom, equality and adequate conditions of life, in an environment of a quality that permits a life of dignity and well-being, and he bears a solemn responsibility to protect and improve the environment for present and future generations. In this respect, policies promoting or perpetuating apartheid, racial segregation, discrimination, colonial and other forms of oppression and foreign domination and stand condemned and must be eliminated.

Principle 2

The natural resources of the earth, including the air, water, land, flora and fauna and especially representative samples of natural ecosystems, must be safeguarded for the benefit of present and future generations through careful planning or management, as appropriate.

Principle 3

The capacity of the earth to produce vital renewable resources must be maintained and, wherever practicable, restored or improved.

Principle 4

Man has a special responsibility to safeguard and wisely manage the heritage of wildlife and its habitat, which are now gravely imperilled by a combination of adverse factors. Nature conservation, including wildlife, must therefore receive importance in planning for economic development.

Principle 5

The non-renewable resources of the earth must be employed in such a way as to guard against the danger of their future exhaustion and to ensure that benefits from such employment are shared by all mankind.

Principle 6

The discharge of toxic substances or of other substance and the release of heat, in such quantities or concentrations as to exceed the capacity of the environment to render them harmless, must be halted in order to ensure that serious or irreversible damage is not inflicted upon ecosystems. The just serious or irreversible damage is not inflicted upon ecosystems. The just struggle of the peoples of all countries against pollution should be supported.

Principle 7

States shall take all possible steps to prevent pollution of the seas by substances that are liable to create hazards to human health, to harm living resources and marine life, to damage amenities or to interfere with other legitimate uses of the sea.

Principle 8

Economic and social development is essential for ensuring a favourable living and working environment for man and for creating conditions on earth that are necessary for the improvement of the quality of life.

Principle 9

Environmental deficiencies generated by the conditions of under development and natural disasters pose grave problems and can best be remedied by accelerated development through the transfer of substantial quantities of financial and technological assistance as a supplement to the domestic effort of the developing countries and such timely assistance as may be required.

Principle 10

For the developing countries, stability of prices and adequate earnings for primary commodities and raw materials are essential to environmental management since economic factors as well as ecological processes must be taken into account.

Principle 11

The environmental policies of all States should enhance and not adversely effect the present or future development potential of developing countries, nor should they hamper the attainment of better living conditions for all, and appropriate steps should be taken by States and international organisations with a view to reaching agreement on meeting the possible application of environmental measures.

Principle 12

Resources should be made available to preserve and improve the environment, taking into account the circumstance and particular requirements of developing countries and any costs which may emanate from their incorporating environmental safeguard into their development planning and the need for making available to them, upon their request, additional international technical and financial assistance for this purpose.

Principle 13

In order to achieve a more rational management of resources and thus to improve the environment, States should adopt an integrated and co-ordinated approach to their development is compatible with the need to protect and improve environment for the benefit of their population.

Principle 14

Rational Planning constitutes an essential tool for reconciling any conflict between the needs of development and the need to protect and improve the environment.

Principle 15

Planning must be applied to human settlements and urbanisation with a view to avoiding adverse affects on the environment and obtaining maximum social, economic and environmental benefits for all. In this respect, projects which are designed for colonialist and racist domination must be abandoned.

Principle 16

Demographic policies which are without prejudice to basic human rights and which are deemed appropriate by Governments concerned should be applied in those regions where the rate of population growth or excessive population concentrations are likely to have adverse effects on the environment of the human environment and impede development.

Principle 17

Appropriate national institutions must be entrusted with the task of planning, managing or controlling the environmental resources of States with a view to enhancing environmental quality.

Principle 18

Science and technology, as part of their contribution to economic and social development, must be applied to the identification, avoidance and control of environmental risks and the solution of environmental problems and for the common good of mankind.

Principle 19

Education in environmental matters, for the younger generation as well as adults, giving due consideration to the underprivileged, is essential in order to broaden the basis for an enlightened opinion and responsible conduct by individuals, enterprises and communities in protecting and improving the environment in its full human dimension. It is also essential that mass media of communications avoid contributing to the deterioration of the environment, but on the contrary, disseminate information of an educational nature on the need to protect and improve the environment in order to enable man to develop in every respect.

Principle 20

Scientific research and development in the context of environmental problems, both national and multi-national, must be promoted in all countries, especially the developing countries. In this connexion, the free flow of up-to-date scientific information and transfer of experience must be supported and assisted, to facilitate the solution of environmental problems; environmental technologies should be made available to developing countries on terms which would encourage their wide dissemination without constituting an economic burden on the developing countries.

Principle 21

States have, in accordance with the charter of the United Nations and the principles of international law, the sovereign right to exploit their own resources pursuant to their own environmental policies, and the responsibility to ensure that activities within their jurisdiction or control do not cause damage to the environment of other States or of areas beyond the limits of national jurisdiction.

Principle 22

States shall co-operate to develop further the international law regarding liability and compensation for the victims of pollution and other environmental damage caused by activities within the jurisdiction or control of such States to areas beyond their jurisdiction.

Principle 23

Without prejudice to such criteria as may be agreed upon by the international community, or to standards which will have to be determined nationally, it will be essential in all cases to consider the systems of values prevailing in each country, and the extent of the applicability of standards which are valid for the most advanced countries but which may be inappropriate and of unwarranted social cost for the developing countries.

Principle 24

International matters concerning the protection and improvement of the environment should be handled in a co-operative spirit by all countries, big and small, on an equal footing. Co-operation through multilateral or bilateral arrangements or other appropriate means is essential to effectively control, prevent, reduce and eliminate adverse environmental effects resulting from activities conducted in all spheres, in such a way that due account is taken of the sovereignty and interests of all States.

Principle 25

States shall ensure that international organisations play a co-ordinated, efficient and dynamic role for the protection and improvement of the environment.

Principle 26

Man and his environment must be spared the effects of nuclear weapons and all other means of mass destruction. States must strive to reach agreement, in the relevant international organs, on the elimination and complete destruction of such weapons.

ANNEXURE II

VIENNA CONVENTION FOR THE PROTECTION OF OZONE LAYER, 1985

I

RESEARCH AND SYSTEMATIC OBSERVATIONS

1. The Parties to the Convention recognise that the major Scientific issues are:

(a) Modification of the Ozone layer which would result in a change in the amount of solar ultra-violet radiation having biological effects (UV-B) that reaches the Earth's surface and the potential consequence for human health, for organisms, ecosystems and materials useful to mankind;

(b) Modification of the vertical distribution of Ozone, which could change the temperature structure of the atmosphere and the potential consequences for weather and climate.

2. The Parties to the Convention, in accordance with article 3, shall co-operate in conducting research and systematic observations and in formulating recommendations for future research and observation in such areas as:

(a) Research Into the Physics and Chemistry of the Atmosphere

(i) Comprehensive theoretical models; further development of models which consider the interaction between radiative, dynamic and chemical processes; studies of the simultaneous effects of various man-made and naturally occurring species upon atmospheric ozone; evaluation of satellite and non-satellite measurement data sets; evaluation of trends in atmospheric and geophysical parameters, and the development of methods for attributing changes in these parameters to specific causes;

(ii) Laboratory studies of: rate coefficients, absorption Cross-sections and mechanisms of tropospheric and stratospheric chemical and photochemical processes; spectroscopic data to support field measurement in all relevant spectral regions;

(iii) Field measurements; the concentration and fluxes of key sources gases of both natural and anthropogenic origin; atmospheric dynamics studies; simultaneous measurements of photochemically-related species down to the planetary boundary layer, using in *situ* and remote sensing instruments; intercomparison of different sensors, including co-ordinated correlative measurements for satellite instrumentation; three-dimensional fields of key atmospheric trace constituents, solar spectral flus and meteorological parameters;

(iv) Instrument development, including satellite and non-satellite sensors for atmospheric trace constituents, solar flux and meteorological parameters;

(b) Research into Health, Biological and Photodegradation Effects

(i) The relationship between human exposure to visible and ultra-violet solar radiation, and

(a) the development of both non-melanoma and melanoma skin cancer, and

(b) the effects on the immunological system;

(ii) Effects of UV-B radiation, including the wavelength dependence, upon:

(a) agricultural crops, forests and other terrestial ecosystems, and

(b) the aquatic food web and fisheries, as well as possible inhibition of oxygen production by marine phytoplankton;

(iii) The mechanisms by which UV-B radiation acts on biological materials, species and ecosystems, including; the relationship between does, dose rate, and response; photorepair, adaptation, and protection;

(iv) Studies of biological action spectra and the spectral response using polychromatic radiations in order to include possible interactions of the various wavelength regions;

(v) The influence of UV-B radiation on: the sensitivities and activities of biological species important to the biospheric balance; primary processes such as photosynthesis and biosynthesis;

(vi) The influence of UV-B radiation on the photodegradation of pollutants, agricultural chemicals and other materials;

(c) Research on Effects on Climate

(i) Theoretical and observational studies of the radiative effects of ozone and other trace species and the impact on climate parameters, such as land and ocean surface temperature, precipitation, patterns, the exchange between the troposphere and stratosphere;

(ii) The investigation of the effect of such climate impacts on various aspects of human activity;

(d) Systematic Observations on

(i) The status of the Ozone layer (i.e., the spatial and temporal variability of the total column content and vertical distribution) by making the Global Ozone Observing

System, based on the integration of satellite and ground-based Systems. Fully operational;

(ii) The tropospheric and stratospheric concentrations of source gases for the HO_X, NO_X, CLO_X and Carbon families;

(iii) The temperature from the ground to the mesosphere, utilising both ground-based and satellite systems;

(iv) Wavelength-resolved solar flux reaching, and thermal radiation leaving, the Earth's atmosphere, utilising satellite measurements;

(v) Wavelength-resolved solar flux reaching the Earth's surface in the ultra-violet range having biological effects (UV-B);

(vi) Aerosol properties and distribution from the ground to the mesosphere, utilising ground-based, airborne and satellite systems;

(vii) Climatically important variables by the maintenance of programme of high-quality meterological surface measurements;

(viii) Trace species, temperatures, solar flux and aerosole utilising improve methods for analysing globalal data.

3. The Parties to the Convention shall Co-operate, taking into account the particular needs of the developing countries, in promoting the appropriate scientific and technical training required to participate in the research and systematic observations outlined in this annex. Particular emphasis should. be given to the inter-calibration of observational instrumentation and methods with a view to generating comparable or standardised scientific data sets.

4. The following chemical substances of natural and anthropogenic origin, not listed in order of priority, are thought to have the potential to modify the chemical and physical properties of the ozone layer.

(a) *Carbon substance*

(i) Carbon monoxide has significant natural and anthropogenic sources, and is thought to play a major direct role in tropospheric photochemistry, and an indirect role in stratospheric photochemistry.

(ii) *Carbon dioxide* (CO_2)

Carbon dioxide has significant natural and anthropogenic sources, and effects stratospheric ozone by influencing the thermal structure of the atmosphere.

(iii) *Methane* (CH_4)

Methane has both natural and anthropogenic sources, and affects both tropospheric and stratospheric ozone.

(iv) *Non-Methane hydrocarbon species*

Non-methane hydrocarbon species; which consist of a large number of chemical substances, have both natural and anthropogenic sources, and play a direct role in tropospheric photochemistry and an indirect role in stratospheric photochemistry.

(b) *Nitrogen substances*

(i) Nitrous Oxide (N_2O)

The dominant sources of N_2O are natural, but anthropogenic contributions are becoming increasingly important. Nitrous oxide is the primary source of stratospheric NO_X, which play a vital role in controlling the abundance of stratospheric ozone.

(ii) *Nitrogen Oxides* (NO_X)

Ground-level sources of NO_X play a major direct role only in tropospheric photochemical processes and an indirect role in stratosphere photochemistry, whereas injection of NO_X close to the tropopause way lead directly to a change in upper tropospheric and stratospheric ozone.

(c) *Chlorine substances*

(i) Fully halogenated alkanes, e.g. CCl_4, $CFCl_3$, (CFC-11) CF_2Cl_2 (CFC-12) $C_2F_3Cl_3$ (CFC-113), $C_2F_4Cl_2$ (CFC-114)

Fully halogenated alkanes are anthropogenic and act as a source of $C1O_X$, which plays a vital role in ozone photochemistry, especially in the 30-50 km altitude region.

(ii) Partially halogenated alkanes, e.g. CH_3Cl, CHF_2Cl

(CFC-22), CH_3CCl_3, $CHFCl_2$ (CFC-21)

The sources of CH_3 Cl are natural whereas the other partially halogenated alkanes mentioned above are anthropogenic in origin. These gases also act as a source of stratospheric $C10_X$.

(d) *Bromine substances*

Fully halogenated alkanes, e.g. CF_3Br

These gases are anthropogenic and act as a source of BrO_X, which behave in, a manner similar to $C10_X$.

(e) *Hydrogen substances*

(i) Hydrogen (H_2)

Hydrogen, the source of which is natural and anthropogenic, plays a minor role in stratospheric photochemistry.

(ii) *Water* (H_2O)

Water, the source of which is natural, a plays a vital role in both trospheric an stratospheric photochemistry. Local sources of water vapour in the stratosphere include the oxidation of methane and, to a lesser extent, of hydrogen.

II

INFORMATION EXCHANGE

1. The Parties to the Convention recognise that the collection and sharing of information is an important means of implementing the objectives of this Convention and of assuring that any actions that may be taken are appropriate and equitable. Therefore, Parties shall exchange scientific, technical, socio-economic, business, commercial and legal information.

2. The parties to the Convention, in deciding what information is to be collected and exchanged, should take into account the usefulness of the information and the cost of obtaining it. The Parties further recognise that co-operation under this annex has to be consistent with national laws, regulations and practices regarding patents, trade secrets, and protection of confidential and proprietary information.

3. Scientific informations

This includes information on:

(a) Planned and ongoing research, both governmental and private, to facilitate the co-ordination of research programmes so as to make the most effective use of available national and international resources;

(b) The emission data needed for research;

(c) Scientific results published in poor reviewed literature on the understanding of the physics and chemistry of the Earth's atmosphere and of its susceptibility to change, in particular on the state of the ozone layer and effects on human health, environment and climate which would result from changes on all time-scales in either the total column content or the vertical distribution of ozones;

(d) The assessment of research results and the recommendations for future research.

4. Technical information

This includes information on:

(a) The availability and cost of chemical substitutes and of alternative technologies to reduce the emissions of ozone-modifying substances and related planned and ongoing research;

(b) The limitations and any risks involved in using chemicals or other substitutes and alternative technologies.

5. Socio-economic and Commercial information on the substances referred to in Ann. I

This includes information on:

(a) Production and production capacity;

(b) Use and use patterns;

(c) Impts/exports;

(d) The costs, risks and benefits of human activities which may indirectly modify the ozone layer and of the impact of regulatory actions taken or being considered to control these activities.

6. Legal information

This includes information on:

(a) National laws, administrative measures and legal research relevant to the protection of the ozone layer;

(b) International agreements, including bilateral agreements, relevant to the protection of the ozone layer;

(c) Methods and terms of licencing and availability of patents relevant to the protection of the ozone layer.

ANNEXURE III

"MONTREAL PROTOCOL ON SUBSTANCES THAT DEPLETE THE OZONE LAYER, AS ADJUSTED AND AMENDED, 1990"

Annex. A (Group I and II)

CONTROLLED SUBSTANCES

Group	*Substance*	*Ozone Depleting Potential**
Group I		
$CFCl_3$ (CFC-11)	1.0	
CF_2Cl_2 (CFC-12)	1.0	
$C_2F_3Cl_3$	(CFC-113)	0.8
$C_2F_4Cl_2$	(CFC-114)	1.0
C_2F_5Cl (CFC-115)	0.6	
Group II		
CF_2BrCl	(halon-1211)	3.0
CF_3Br (halon-1301)	10.0	
$C_2F_4Br_2$	(halon-2402)	6.0

* These ozone depleting potentials are estimates based on existing knowledge and will be reviewed and revised periodically.

ANNEXURE IV

(Group I, II and III)

"MONTREAL PROTECTION SUBSTANCES THAT DEPLETE THE OZONE LAYER, AS ADJUSTED AND AMENDED, 1990

Controlled Substances

Group	*Substance*	*Ozone-depleting potential*
Group I		
CF_3Cl (CFC-13)	1.0	
C_2FCl_5 (CFC-111)	1.0	
$C_2F_2Cl_4$	(CFC-112)	1.0
C_3FCl_7 (CFC-211)	1.0	
$C_3F_2Cl_6$	(CFC-212)	1.0
$C_3F_2Cl_6$	(CFC-213)	1.0
$C_3F_3Cl_5$	(CFC-213)	1.0
$C_3F_4Cl_4$	(CFC-214)	1.0
$C_3F_5Cl_3$	(CFC-215)	1.0
$C_3F_6Cl_2$	(CFC-216)	1.0
C_3F_7Cl (CFC-217)	1.0	
Group II		
CCl_4 Carbon tetrachloride	1.1	
Group III		
$C_2H_3Cl_3$*	1,1,1-trichloroethane	
(methyl chloroform)	Methyl chloroform	0.1

* This formula does not refer to 1,1,2-trichloroethane.

5

Elements and Dimensions of Green Technology

Some *environmental technologies* that retain *sustainable development* are; recycling, water purification, sewage treatment, remediation, flue gas treatment, solid waste management, and renewable energy. Some technologies assist directly with energy conservation, while other technologies are emerging that help the environment by reducing the amount of waste produced by human activities. Energy sources such as solar power create less problems for the environment than traditional sources of energy like coal and petroleum. Scientists continue to search for clean energy alternatives to our current power production methods. Some technologies such as anaerobic digestion produce renewable energy from waste materials. The global reduction of greenhouse gases is dependent on the adoption of energy conservation technologies at industrial level as well as this clean energy generation. That includes using unleaded gasoline, solar energy and alternative fuel vehicles, including plug-in hybrid and hybrid electric vehicles. Since electric motors consume 60% of all electricity generated[*citation needed*], advanced energy efficient electric motor (and electric generator) technology that are cost effective to encourage their application, such as the brushless wound-rotor doubly-fed electric machine and energy saving module, can reduce the amount of carbon dioxide (CO_2) and sulfur dioxide (SO_2) that would otherwise be introduced to the atmosphere, if electricity is generated using fossil fuels. Greasestock is an event held yearly in Yorktown Heights, New York which is one of the largest showcases of environmental technology in the United States.

Some groups, including green anarchists, have criticised the concept of environmental technology. From their viewpoint, technology is seen as a system rather than a specific physical tool. Technology, it is argued, *requires* the exploitation of the environment through the creation and extraction of resources, and the exploitation of people through labour, specialisation and the division of labor. There is no "neutral" form of technology, as things are always created in a certain context with certain aims and functions. Thus, green technology is rejected as an attempt to reform this exploitative system, merely changing it on the surface to make it seem environmentally friendly, despite continued unsustainable levels of human and natural exploitation.

The term "technology" refers to the application of knowledge for practical purposes.

The field of "green technology" encompasses a continuously evolving group of methods and materials, from techniques for generating energy to non-toxic cleaning products.

The present expectation is that this field will bring innovation and changes in daily life of similar magnitude to the "information technology" explosion over the last two decades. In these early stages, it is impossible to predict what "green technology" may eventually encompass.

The goals that inform developments in this rapidly growing field include:

Sustainability - meeting the needs of society in ways that can continue indefinitely into the future without damaging or depleting natural resources. In short, meeting present needs without compromising the ability of future generations to meet their own needs.

"Cradle to cradle" design - ending the "cradle to grave" cycle of manufactured products, by creating products that can be fully reclaimed or re-used.

Source reduction - reducing waste and pollution by changing patterns of production and consumption.

Innovation - developing alternatives to technologies - whether fossil fuel or chemical intensive agriculture - that have been demonstrated to damage health and the environment.

Viability - creating a center of economic activity around technologies and products that benefit the environment, speeding their implementation and creating new careers that truly protect the planet.

Examples of Green Technology Subject Areas

Energy: Perhaps the most urgent issue for green technology, this includes the development of alternative fuels, new means of generating energy and energy efficiency.

Green building: Green building encompasses everything from the choice of building materials to where a building is located.

Environmentally preferred purchasing: This government innovation involves the search for products whose contents and methods of production have the smallest possible impact on the environment, and mandates that these be the preferred products for government purchasing.

Green chemistry: The invention, design and application of chemical products and processes to reduce or to eliminate the use and generation of hazardous substances.

Green nanotechnology: Nanotechnology involves the manipulation of materials at the scale of the nanometer, one billionth of a meter. Some scientists believe that mastery of this subject is forthcoming that will transform the way that everything in the world is manufactured. "Green nanotechnology" is the application of green chemistry and green engineering principles to this field.

Carbon Disclosure Project

The Carbon Disclosure Project is an organisation which works with shareholders and corporations to disclose the Greenhouse Gas Emissions of major corporations. In 2007, it published the emissions data for 2400 of the world's largest corporations, accounting for 26% of global anthropogenic emissions. The CDP represents institutional investors, with a combined $57 trillion under management.

Background

International agreements such as the Kyoto Protocol have proved problematic, and individual governments have been reluctant to develop stringent national limits on emissions for fear of big companies relocating their factories and jobs to nations with laxer regulatory regimes. The Carbon Disclosure Project (CDP) attempts to side step these national interests by focusing on individual companies rather than on nations. CDP brings together institutional investors to focus attention on carbon emissions, energy usage and reduction – where ever companies and assets may be located. Some corporations have higher Greenhouse gas emissions than individual nation states. Some leading companies have moved to become carbon neutral, but for others there is scope to reduce energy usage and greenhouse gas emissions by adoption of energy efficiency methods and business planning.

Mechanism

CDP works with 3,000 of the largest corporations in the world to help them ensure that an effective carbon emissions / reductions strategy is made integral to their business. This effort is taken seriously because of the size of the shareholdings backing CDP - 315 institutional investors with $57 trillion under management. CDP operates from New York and London and has partners in 18 of the world's major economies which help deliver the programme globally. It has:

- Established the world's largest repository of GHG emissions and energy use data accounting for some 26% of global anthropogenic CO_2
- Started to establish a globally used standard for emissions and energy reporting
- Examined the 265 major electric utilities globally (high GHG emitters)
- Obtained backing from blue chip investors including HSBC, JPMorgan Chase, Bank of America, Merrill Lynch, Goldman Sachs, American International Group, and State Street Corp.
- Active staff or partner organisations in the United States, Japan, Germany, China, United Kingdom, France, Canada, Netherlands, Brazil, Australia, New Zealand, among others
- Works with corporations including WalMart, Tesco, Cadbury Schweppes, Procter and Gamble, and many others to measure emissions through the supply chain.

Much of the data elicited has never been collected before. This information is helpful to investors, corporations and to regulators in making informed decisions which take into account corporate risk from future government legislation, possible future lawsuits and shifts in consumer's perceptions towards heavy emitters. An estimated $27 billion will be spent over the next 30 years on energy related capital developments (new power stations, fuel distillation plants, etc). It is vital that the right technologies are adopted. In particular:

- Giving higher priority to energy efficient design in new capital projects
- Declaring firm targets for capping and reducing corporate emissions
- Identifying new low carbon business opportunities
- Pricing in how, under different scenarios, the price of carbon emissions will shift the economics of alternative sources of energy.

The process of companies having to respond to CDP delivers real changes in business practice resulting in lower energy use. In many cases this leads to a higher proportion of energy being sourced from renewables.

CDP has 501(c)3 charitable status through Rockefeller Philanthropy Advisors in New York and is in the process of registering as a charitable organisation in the United Kingdom.

Participants

- Dell
- Hewlett Packard
- L'Oreal
- PepsiCo
- Cadbury Schweppes
- Nestlé
- Procter & Gamble
- Tesco
- Unilever

Clean Technology

Clean technology includes the renewable energy (wind power, solar power, biomass, hydropower, biofuels), information technology, green transportation, electric motors, lighting, and many other appliances that are now more energy efficient. It is a means to create electricity and fuels with a smaller environmental footprint. And it is the need to make green buildings both more energy efficient and environmentally benign. Environmental finance is a method by which new clean technology projects that have proven that they are "additional" or "beyond business as usual" can obtain financing through the generation of carbon credits. A project that is developed with concern for climate change mitigation (such as a Kyoto Clean Development Mechanism project) is also known as a carbon project.

While there is no standard definition of "clean technology," it has been described by Clean Edge, a clean-tech research firm, as "a diverse range of products, services, and processes that harness renewable materials and energy sources, dramatically reduce the use of natural resources, and cut or eliminate emissions and wastes." It notes that "Clean technologies are competitive with, if not superior to, their conventional counterparts. Many also offer significant additional benefits, notably their ability to improve the lives of those in both developed and developing countries"

Investments in clean technology have grown considerably since coming into the spotlight around 2000. According to the United Nations Environment Program, wind, solar and biofuel companies received a record $148 billion in new funding in 2007 as rising oil prices and climate change policies encouraged investment in renewable energy. $50 billion of that funding went to wind power. Overall, investment in clean-energy and energy-efficiency industries rose 60 percent from 2006 to 2007. In the United States, the clean tech industry is largely based in Silicon Valley

Permaculture

Permaculture design principles extend from the position that "The only ethical decision is to take responsibility for our own existence and that of our children" (Mollison, 1990). The intent was that, by rapidly training individuals in a core set of design principles, those individuals could become designers of their own environments and able to build increasingly self-sufficient human settlements — ones that reduce society's reliance on industrial systems of production and distribution that Mollison identified as fundamentally and systematically destroying the earth's ecosystems. While originating as an agro-ecological design theory, permaculture has developed a large international following of individuals who have received training through intensive two week long 'permaculture design courses'. This 'permaculture community' continues to expand on the original teachings of Mollison and his associates, integrating a range of alternative cultural ideas, through a network of training, publications, permaculture gardens, and internet forums. In this way permaculture has become both a design system as well as a loosely defined philosophy or lifestyle ethic.

Origins

In the mid 1970s, two Australians, Dr. Bill Mollison and David Holmgren, started to develop ideas that they hoped could be used to create stable agricultural systems. This was a result of their perception of a rapidly growing use of destructive industrial-agricultural methods. They saw that these methods were poisoning the land and water, reducing biodiversity, and removing billions of tons of soil from previously fertile landscapes. A design approach called "permaculture" was their response and was first made public with the publication of *Permaculture One* in 1978. The term *permaculture* initially meant "permanent agriculture" but this was quickly expanded to also stand for "permanent culture" as it was seen that social aspects were an integral part of a truly sustainable system. Mollison and Holmgren are widely considered to be the co-originators of the modern permaculture concept.

After the publication of *Permaculture One*, Mollison and Holmgren further refined and developed their ideas by designing hundreds of permaculture sites and organizing this information into more detailed books. Mollison lectured in over eighty countries and his two-week Design Course was taught to many hundreds of students. By the early 1980s, the concept had moved on from being predominantly about the design of agricultural systems towards being a more fully holistic design process for creating sustainable human habitats.

By the mid 1980s, many of the students had become successful practitioners and had themselves begun teaching the techniques they had learned. In a short period of time permaculture groups, projects, associations, and institutes were established in over one hundred countries. In 1991 a four-part Television documentary by ABC productions called 'The Global Gardener' showed permaculture applied to a range of world-wide situations, bringing the concept to a much broader public. Excerpts are available online through YouTube. Permaculture has developed from its origins in Australia into an international 'movement'. English permaculture teacher Patrick Whitefield, author of *The Earth Care Manual* and *Permaculture in a Nutshell*, suggests that there are now two strands of permaculture: a) Original and b) Design permaculture. Original permaculture attempts to closely replicate nature by developing edible ecosystems which closely resemble their wild counterparts. Design permaculture takes the working connections at use in an ecosystem and uses them as its basis. The end result may not look as "natural" as a forest garden, but still has an underlying design based on ecological principles. Through close observation of natural energies and flow patterns efficient design systems can be developed. This has become known as Natural Systems Design. (Dr. M Millington and A Sampson-Kelly)

Elements of Permaculture Design

Permaculture principles draw heavily on the practical application of ecological theory to analyze the characteristics and potential relationships between design elements. Each element of a design is carefully analyzed in terms of its needs, outputs, and properties. For example a chicken needs water, moderated microclimate, food and other chickens, and produces meat, eggs, feathers and manure while doing a lot of scratching. Design elements are then assembled in relation to one another so that the products of one element feed the needs of adjacent elements. Synergy between design elements is achieved while minimizing waste and the demand for human labour or energy. Exemplary permaculture designs evolve over time, and can become extremely complex mosaics of conventional and inventive cultural systems that produce a high density of food and materials with minimal input. While techniques and cultural systems are freely borrowed from organic agriculture, sustainable forestry, horticulture, agroforestry, and the land management systems of indigenous peoples, permaculture's fundamental contribution to the field of ecological design is the development of a concise set of broadly applicable organizing principles that can be transferred through a brief intensive training.

Permaculture of Modern Time

Modern permaculture is a system design tool. It is a way of

- 1. looking at a whole system or problem
- 2. seeing connections between key elements (parts)
- 3. observing how the parts relate,
- 4. planning to mend sick systems by applying ideas learnt from long-term sustainable working systems.

In permaculture, we are learning from the working systems of nature to plan to fix the sick landscapes of human agricultural and city systems. We can apply systems thinking to the design of a kitchen tool as easily to the re-design of a farm. In permaculture we apply it to everything we need in order to build a sustainable future. Commonly, "Initiatives that are taken tend to evolve from strategies that focus on efficiency (for example, more accurate and controlled uses of inputs and minimisation of waste) to substitution (for example, from more to less disruptive interventions, such as from biocides to more specific biological controls and other more benign alternatives) to redesign — fundamental changes in the design and management of the operation (Hill & MacRae 1995, Hill et al 1999)." "Permaculture is about helping people make redesign choices: setting new goals and a shift in thinking that affects not only their home but their actions in the workplace, borrowings and investments" (A Sampson-Kelly and Michel Fanton 1991). Examples include the design and employment of complex transport solutions, optimum use of natural resources such as sunlight, "radical design of information-rich, multi-storey polyculture systems" (Mollison & Slay 1991). "This progression generally involves a shift in the nature of one's dependence — from relying primarily on universal, purchased, imported, technology-based interventions to more specific locally available knowledge and skill-based ones. This usually eventually also involves fundamental shifts in world-views, senses of meaning, and associated lifestyles (Hill 1991). My experience is that although efficiency and substitution initiatives can make significant contributions to sustainability over the short term, much greater longer-term improvements can only be achieved by redesign strategies; and, furthermore, that steps need to be taken at the outset to ensure that efficiency and substitution strategies can serve as stepping stones and not barriers to redesign..." (Hill 2000)

Influences

The term *permanent agriculture* was first coined by Franklin Hiram King in his classic book from 1911, *Farmers of Forty Centuries: Or Permanent Agriculture in China, Korea and Japan*. In this context, permanent agriculture is understood as agriculture that can be sustained indefinitely.

This definition was supported by Australian P. A. Yeomans (*Water for Every Farm*, 1973) who introduced an observation-based approach to land use in Australia in the 1940s, based partially on his understanding of geology. Yeomans introduced Keyline Design as a way of managing the supply and distribution of water of a site. Holmgren based his EcoVillage design on the keyline principle, (see WikiMapia view)

The work of Howard T. Odum was also an early influence, especially for Holmgren . Odum's work focused on system ecology, in particular the Maximum power principle, which examines the energy of a system and how natural systems tend to maximise the energy embodied in a system. For example, the total calorific value of woodland is very high with its multitude of plants and animals. It is an efficient converter of sunlight into biomass. A wheat field, on the other hand, has much less total energy and often requires a large energy input in terms of fertiliser. Another early influence was the work of Esther Deans where she pioneered No-Dig Gardening methods. Other recent influences include the VAC system in Vietnam which is a government supported system to build Vegetable Aquaculture and Animal enClosures that cycle resources.

Core Values

Permaculture is a broad-based and holistic approach that has many applications to all aspects of life. At the heart of permaculture design and practice is a fundamental set of 'core values' or ethics which remain constant whatever a person's situation, whether they are creating systems for town planning or trade; whether the land they care for is only a windowbox or an entire forest. These 'ethics' are often summarised as;

- Earthcare – recognising that the Earth is the source of all life (and is possibly itself a living entity- see Gaia theory) and that we recognise and respect that the Earth is our valuable home and we are a part of the Earth, not apart from it.
- Peoplecare – supporting and helping each other to change to ways of living that are not harming ourselves or the planet, and to develop healthy societies.
- Fairshare (or placing limits on consumption) - ensuring that the Earth's limited resources are utilised in ways that are equitable and wise.

Everyone needs to eat and drink, and it is the issue of food production where permaculture had its origins. It started with the belief that for people to feed themselves sustainably they need to move away from reliance on industrialised agriculture. Where industrial farms use fossil fuel (gasoline, diesel, natural gas..) driven technology specialising in each farm producing high yields of a single crop, permaculture stresses the value of low inputs into the land and diversity in terms of what is grown. The model for this was an abundance of small scale market and home gardens for food production with food miles being a primary issue.

Permaculture Design Innovation

The core of permaculture has always been in supplying a design toolkit for human habitation. This toolkit helps the designer to model a final design based on an observation of how ecosystems themselves interact. A simple example of this is how the Sun interacts with a plant by providing it with energy to grow. This plant may then be pollinated by bees or eaten by deer. These may disperse seed to allow other plants to grow into tall trees and provide shelter to these creatures from the wind. The bees may provide food for birds and the trees provide roosting for them. The tree's leaves will fall and rot, providing food for small insects and fungus. There will be a web of intricate connections that allow a diverse population of plant life and animals to survive by giving them food and shelter. One of the innovations of permaculture design was to appreciate the efficiency and productivity of natural ecosystems and seek to apply this to the way human needs for food and shelter are met. One of the most notable proponents of this design system has been David Holmgren, who based much of his permaculture innovation on zone analysis.

O'BREDIM Design Methodology

O'BREDIM is a mnemonic based on Observation, Boundaries, Resources, Evaluation, Design, Implementation and Maintenance.

- Observation allows you first to see how the site functions within itself, to gain an understanding of its initial relationships. Some people recommend a year-long observation of a site before anything is planted. During this period all factors, such as lay of the land, natural flora and so forth, can be brought into the design. A year allows the site to be observed through all seasons, although it must be realised that, particularly in temperate climates, there can be substantial variations between years.
- Boundaries refer to physical ones as well as to those your neighbours might place on you, for example.
- Resources would include the people involved, funding, as well as what you can grow or produce in the future.
- Evaluation of the first three will then allow you to prepare for the next three. This is a careful phase of taking stock of what you have at hand to work with.
- Design is always a creative and intensive process, and you must stretch your ability to see possible future synergetic relationships.
- Implementation is literally the ground-breaking part of the process when you carefully dig and shape the site.
- Maintenance is then required to keep your site at a healthy optimum, making minor adjustments as necessary. Good design will preclude the need for any major adjustment.

Patterns

The use of patterns both in nature and reusable patterns from other sites is often key to permaculture design. This echoes the Pattern language of Christopher Alexander used in architecture which has been an inspiration for many permaculture designers. All things, even the wind, the waves and the earth on its axis, moving around the Sun, form patterns. In pattern application, permaculture designers are encouraged to develop: 1. Awareness of the patterns that exist in nature (and how these function) 2. Application of pattern on sites in order to satisfy specific design needs. "The application of pattern on a design site involves the designer recognising the shape and potential to fit these patterns or combinations of patterns comfortably onto the landscape" Sampson-Kelly. We can use branching for the direction of our paths, rather than straight paths with square angles. Or we may use lobe-like paths of the main path (these are known as keyhole paths) that minimise waste and compaction of the soil.

Permaculture Zones

Permaculture zones are a way of organizing design elements in a human environment based on the frequency of human use. Frequently manipulated or harvested elements are located close to the house in zones one and two, while less frequently manipulated elements of the design are farther away from the house.

Links and connections

Also key to the permacultural design model is that useful connections are made between components in the final design. The formal analogy for this is a natural mature ecosystem. So, in much the same way as there are useful connections between Sun, plants, insects and soil there will be useful connections between different plants and their relationship to the landscape and humans. Another innovation of the permaculture design is to design a landuse or other system that has multiple outputs. In terms of Holmgren's application of H.T. Odum's work, a useful connection is viewed as one that maximises power: that is, maximizes the rate of useful energy transformation. A comparison which illustrates this is between a wheat field and a forest. "It is not the number of diverse things in a design that leads to stability, it is the number of beneficial connections between these components" Mollison 1988.

Layers/'Stacking'

In permaculture and forest gardening, seven layers are identified:

1. The canopy
2. Low tree layer (dwarf fruit trees)
3. Shrubs

4. Herbaceous
5. Rhizosphere (root crops)
6. Soil Surface (cover crops)
7. Vertical Layer (climbers, vines)

The 8th layer, or Mycosphere (fungi), is often included in layering.

A mature ecosystem such as ancient woodland has a huge number of relationships between its component parts: trees, understory, ground cover, soil, fungi, insects and other animals. Plants grow at different heights. This allows a diverse community of life to grow in a relatively small space. Plants come into leaf and fruit at different times of year. For example, in the UK, wild garlic comes into leaf on the woodland floor in the time before the top canopy re-appears with the spring. A wood suffers very little soil erosion as there are always roots in the soil. It offers a habitat to a wide variety of animal life which the plants rely on for pollination and seed distribution. The productivity of such a forest in terms of how much new growth it produces exceeds the most productive wheat field. It is in this observation of how more productive a wood may be on far less input of fertilizers that the potential productivity of a permaculture design is modelled. The many connections in a wood contribute together to a proliferation of opportunities for amplifier feedbacks to evolve that in turn maximise energy flow through the system.

Here is a photo of a layered warm temperate garden in NSW, Australia (courtesy of PermacualtureVisions). There are several layers: the canopy layer is Inga Edulis (Ice Cream Bean, the middle stratum contains Plum and Peach, Mango, Mulberry and nurse plants such as native wattle. There are shrubs such as sage and woody herbs, ground covers such as sweet potato and vines such as passionfruit and kiwi fruit. The tubers consist of onions and taro.

Polyculture

Polyculture is agriculture using multiple crops in the same space, in imitation of the diversity of natural ecosystems, and avoiding large stands of single crops, or monoculture. It includes crop rotation, multi-cropping, and inter-cropping. Alley cropping is a simplification of the layered system which typically uses just two layers, with alternate rows of trees and smaller plants.

Perennial Plants

Perennial plants are often used in permaculture design. As they do not need to be planted every year they require less maintenance and fertilisers. They are especially important in the outer zones and in layered systems. Ken Fern of Plants For A Future has spent many years investigating suitable perennial plants.

Guilds

Permaculture Guilds are groups of plants, animals and microbacteria which work particularly well together. These can be those observed in nature such as the White Oak guild which centers on the White Oak tree and includes 10 other plants. Native communities can be adapted by substitution of plants more suitable for human use. The Three Sisters of maize, squash and beans is a well known guild. The British National Vegetation Classification provides a comprehensive list of plant communities in the UK. Guilds can be thought of as an extension of companion planting.

Increase Edge

Permaculturists maintain that where vastly differing systems meet, there is an intense area of productivity and useful connections. The greatest example of this is the coast. Where the land and the sea meet there is a particularly rich area that meets a disproportionate percentage of human and animal needs. This is evidenced by the fact that the overwhelming majority of humankind lives within 100 km of the sea. So this idea is played out in permacultural designs by using spirals in the herb garden or creating ponds that have wavy undulating shorelines rather than a simple circle or oval. Edges between woodland and open areas have been claimed to be the most productive.

Animals

Many permaculture designs involve animals. Chickens can be used as a method of weed control and also as a producer of eggs, meat and fertiliser. Some types of agroforestry systems combine trees with grazing animals. Some projects are critical of the use of animals (see vegan organic gardening). However not all permaculture sites farm the animals. The animals are pets and can be treated as co-habitators and co-workers of the site, eating foods normally unpalatable to people such as slugs, termites, being an integral part of the pest management by eating some pests, supplying fertiliser through their droppings and controlling some weed species.

Annual monoculture (anti-pattern)

Annual monoculture such as a wheatfield can be considered a pattern to be avoided in terms of space (height is uniform) and time (crops grow at the same rate until harvesting). During growth and especially after harvesting the system is prone to soil erosion from rain. The field requires a hefty input of fertilizers for growth and machinery for harvesting. The work is more likely to be repetitive, mechanised and rely on fossil fuels. No pattern should be hard and fast and depending on the design considerations they can be broken. An example of this is broadscale permaculture practiced at Ragmans Lane Farm, which has a component of annual farming. Here the amount of human involvement is a key factor influencing the design.

Energy

Applying these values means using fewer non-renewable sources of energy, particularly petroleum based forms of energy. Burning fossil fuels contributes to greenhouse gases and global warming; however, using less energy is more than just combatting global warming. Food production should be a fully renewable system; but using current agricultural systems this is not the case. Industrial agriculture requires large amounts of petroleum, both to run the equipment, and to supply pesticides and fertilizers. Permaculture is in part an attempt to create a renewable system of food production that relies upon minimal amounts of energy. For example permaculture focuses on maximizing the use of trees (agroforestry) and perennial food crops because they make a more efficient and long term use of energy than traditional seasonal crops. A farmer does not have to exert energy every year replanting them, and this frees up that energy to be used somewhere else. Traditional pre-industrial agriculture was labor intensive, industrial agriculture is fossil fuel intensive and permaculture is design and information intensive and petrofree. Partially permaculture is an attempt to work smarter, not harder; and when possible the energy used should come from renewable sources such as wind power, passive solar designs or biofuels.

A good example of this kind of efficient design is the chicken greenhouse. By attaching the chicken coop to a greenhouse you can reduce the need to heat the greenhouse by fossil fuels, as the chicken's bodies heat the area. The chickens scratching and pecking can be put to good use to clear new land for crops. Their manure can be used to fertilise the soil. Feathers could be used in compost or as a mulch. In a conventional factory situation all these chicken outputs are seen as a waste problem. So in factories cooled by huge air conditioners, the chicken waste is extracted. All the energy is focused on egg production. Thus it is a further principle of permaculture that "pollution is energy in the wrong place".

Holmgren's 12 Design Principles

These restatements of the principles of permaculture appear in David Holmgren's *Permaculture:*

1. Observe and interact - By taking the time to engage with nature we can design solutions that suit our particular situation.
2. Catch and store energy - By developing systems that collect resources when they are abundant, we can use them in times of need.

3. Obtain a yield - Ensure that you are getting truly useful rewards as part of the work that you are doing.
4. Apply self-regulation and accept feedback - We need to discourage inappropriate activity to ensure that systems can continue to function well.
5. Use and value renewable resources and services - Make the best use of natures abundance to reduce our consumptive behaviour and dependence on non-renewable resources.
6. Produce no waste - By valuing and making use of all the resources that are available to us, nothing goes to waste.
7. Design from patterns to details - By stepping back, we can observe patterns in nature and society. These can form the backbone of our designs, with the details filled in as we go.
8. Integrate rather than segregate - By putting the right things in the right place, relationships develop between those things and they work together to support each other.
9. Use small and slow solutions - Small and slow systems are easier to maintain than big ones, making better use of local resources and produce more sustainable outcomes.
10. Use and value diversity - Diversity reduces vulnerability to a variety of threats and takes advantage of the unique nature of the environment in which it resides.
11. Use edges and value the marginal - The interface between things is where the most interesting events take place. These are often the most valuable, diverse and productive elements in the system.
12. Creatively use and respond to change - We can have a positive impact on inevitable change by carefully observing, and then intervening at the right time.

Permaculture Design for Ecologinomic Ethics

A basic principle is thus to "add value" to existing crops. A permaculture design therefore seeks to provide a wide range of solutions by including its main ethics (see above) as an integral part of the final value-added design. Crucially, it seeks to address problems that include the economic question of how to either make money from growing crops or exchange crops for labour such as in the LETS scheme. Each final design therefore should include economic considerations as well as give equal weight to maintaining ecological balance, making sure that the needs of people working on the project are met and that no one is exploited.

Community economics require a balance between the three aspects that comprise a community, justice, environment and economics aka the triple bottom line. This approach is also referred to as the Triple E (EEE)which stands for ecological-economics-ethics. A cooperative farmer's market could be an example of this structure. The farmers are the workers and owners. Additionally, all economics are limited by their ecology. No economic system stands apart independently from its eco-system; therefore, all external costs must be considered when discussing economics.

Examples

One way of doing this is through designing a system that has "multiple outputs" For example, a wheat field interspersed with walnuts will reduce soil erosion, act as a windbreak and provide a walnut crop as well as a wheat crop. As there are two crops to manage the work will be more interesting. Here the system comes into conflict with conventional agriculture and economics. By interplanting trees in wheat fields there is a reduction in the wheat yield. The field is also harder to harvest using machinery, as the operator has to drive around the trees. Most farms specialise in a few crops at a time and seek to maximise surplus in order to increase profit. This surplus can only be maintained with a massive injection of fossil fuels. So, as things stand it is quite hard for a permaculture farm to compete with a "conventional " farm in order to grow basic fruit and vegetables.

Critiques

John Robin has been one the strongest critics of permaculture, criticising it for its potential to spread environmental weeds, reflecting a divide between native plant advocates and permaculture. Permaculture in the tropics, as expressed in 'Permaculture: A Designer's Manual', did not produce significant amounts of food or fruit when applied in Northern New South Wales and Queensland, largely since the closed canopy is not conducive to fruit production. The system, whilst very healthy in and of itself, yielded very little produce.

Bill Mollison himself has also been critical of *itinerant teachers* of permaculture who would go on to teach after only a short course. At one point Mollison unsuccessfully tried to trademark the term permaculture to prevent this practice. Perhaps the strongest criticism of permaculture is to be found in the Review of Toby Hemenway's book *Gaia's Garden*, which was published in the Winter 2001 edition of the *Whole Earth Review*. In it, Greg Williams critiques the view that woods were more highly productive than farmland based on the theory of ecological succession which says *that net productivity declines as ecosystems mature*. He also criticised the lack of scientifically respectable data and questions whether permaculture is applicable to more than a small number of dedicated people. Hemenway's response in the same magazine disputes Williams's claim on productivity as focusing on climax rather than on maturing forests, citing data from ecologist Robert Whittaker's book *Communities and Ecosystems*. Hemenway is also critical of Williams's characterisation of permaculture as simply forest gardening.

Contemporary examples

In the years since its conception, permaculture has become a successful approach to designing sustainable systems. Its adaptability and emphasis on meeting human needs means that it can be utilized in every climatic and cultural zone. However, at the moment the large proportion of practitioners are only likely to be inspired individuals and there is a distinct lack of broadscale permaculture projects. Nevertheless, permaculture has also been used successfully as a development tool to help meet the needs of indigenous communities at risk of vastly improved standards of living from exposure to free-market economics.

There are some examples of permaculture sites (see also a wiki-map of permaculture sites) :

Europe

Cyprus

Two acres of land at Ayia Skepi Therapeutic Centre in Filani village, a drug rehabilitation centre about 25 km from Nicosia, are being developed by Emily Markides, Julia Yelton, Charles Yelton and the residents of the detoxification centre.

France

There are several examples of Permaculture. The site "La Permaculture" has maps and links.

United Kingdom

There are a number of example permaculture projects in the UK, including:

- Ragmans Lane, a 60 acre farm in the Forest of Dean in Gloucestershire .
- Tir Penrhos Isaf, near Dolgellau, developed by Chris and Lyn Dixon since 1986.
- Plants for a Future is a vegan-organic project based at Lostwithiel in Cornwall that is researching and trialing edible and otherwise useful plant crops for sustainable cultivation. Their online database currently features over 7,000 such species that can be grown within the UK. A collaborative version of the database is in development by the

permaculture.info project. The project has a second, larger property in North Devon, for which it is seeking a new group to take over.

- *Prickly Nut Woods* is a ten acre woodland near Haslemere, Surrey that is managed by Ben Law. He is using a 'whole system' permacultural approach, utilising a wide variety of woodland products and documenting a complex web of relationships. He has built a house almost entirely using products from the woodland, which was featured in Channel 4's Grand Designs TV series .
- Agroforestry Research Trust, a not for profit organisation based in Dartington, Devon that runs a 2 acre forest garden and publishes the journal Agroforestry News
- Middlewood Trust, a permculture based farm in North Lancashire running courses in permaculture, crafts, forestry and sustainability
- The RISC Roof Garden, on top of a development education centre in Reading city centre and inspired by Robert Hart's permaculture forest garden in Shropshire, is an excellent example of urban permaculture design. . It is used by schools, educators and designers as an educational resource for sustainable development and is a member of the National Gardens Scheme. The garden is comprised of dense plantings of over 180 species of edible and medicinal plants and is fed by rainwater and composted waste from the centre.
- Chickenshack Housing co-op, is a fully mutual housing co-op established in 1995 using permaculture design principles. Based in rural North Wales the community has 4 dwellings and 6 residents on a 5 acre site. Features include a biomass and solar district heating scheme, a half acre forest garden and various wildlife conservation and habitat creation strategies. The community are very active in regional sustainability projects such as the Machynlleth Transition Towns initiative. They run occasional courses in permaculture design and regularly receive visits from interested parties.

Other projects tend to be more community oriented, particularly in urban areas. These include Naturewise, a north London based group who tend a number of forest gardens and allotments as well as running regular permaculture introductory and design courses ; and Organiclea, a workers cooperative who are involved in developing local food growing and distribution initiatives around the Walthamstow area of east London . The Transition Towns movement initiated in Totnes and Kinsale by Rob Hopkins is underpinned by permaculture design principles in its attempts to visualise sustainable communities beyond peak oil

The UK Permaculture Association publishes an extensive directory of other projects and example sites throughout the country .

Oceania

Australia

The development of permaculture co-founder David Holmgren's home plot at Melliodora, Central Victoria, has been well documented at his website and published in e-book format . Designed from permaculture principles, Crystal Waters is a socially and environmentally responsible, economically viable rural subdivision north of Brisbane, Australia. Crystal Waters was designed by Max Lindegger, Robert Tap, Barry Goodman and Geoff Young, and established in 1987. It received the 1996 World Habitat Award (assessed by Dr Wally N'Dow) for its "pioneering work in demonstrating new ways of low impact, sustainable living". Eighty-three freehold residential and two commercial lots occupy 20% of the 259ha (640 acre) property. The remaining 80% is the best land, and is owned in common. It can be licensed for sustainable agriculture, forestry, recreation and habitat projects.

Africa

Zimbabwe has sixty schools designed using permaculture, with a national team working within the schools' curriculum development unit. The UN High Commissioner for Refugees (UNHCR) has produced a report on using permaculture in refugee situations after successful use in camps in Southern Africa and Macedonia. The Biofarming approach applied in Ethiopia has very similar features and can be considered as permaculture. It is mainly promoted by BEA (NGO) based in Addis Ababa.

South America

Brazil

IPEC - Ecocentro at the Instituto de Permacultura e Ecovilas do Cerrado - the Institute of Permaculture and Ecovillage of the Cerrado

IPCP - Instituto de Permacultura Cerrado-Pantanal (Permaculture Institute of Cerrado-Pantanal), Campo Grande, MS. Specializing in interactive teaching of Permaculture and direct work with Indigenous communities within the Cerrado biome.

North America

United States of America

- The Northeastern Permaculture Network brings together enthusiasts in the northeastern United States and eastern Canada.

- The Seattle Permaculture Guild is active in that city.
- Promoting urban permaculture in Los Angeles is Path to Freedom .
- The Edible Plant Project implements & promotes elements of Permaculture through a nonprofit nursery & workshops in Gainesville, FL (Home of U of FL).
- In Santa Fe, New Mexico, the Permaculture Institute utilizes a hands on approach to education on topics such as landscape and building design as well as water systems.
- The Regenerative Design Institute (RDI) is a non-profit educational organization in Bolinas, California .
- The Portland Permaculture Guild (PPG) is very active in Portland, Oregon. There are many PC gardens in and around the Portland area. Additionally, there are many fine teachers in Portland and in Oregon, including teachers OF PC teachers. Also of note is that Toby Hemenway, noted Permaculture author and teacher, lives in the Portland area.

Cuba

Cuba has in the past 18 years transformed their food production using biodynamic agriculture and permaculture. Havana produces up to 50% of its food requirements from within the city limits, all of it is organic and produced by people in their homes, gardens and in municipal spaces.

Asia

Indonesia

The Indonesian Development of Education and Permaculture assisted in disaster relief in Aceh, Indonesia after the 2004 Tsunami . They have also developed Wastewater Gardens, a small-scale sewage treatment systems similar to Reedbeds.

Thailand

The Panya Project , located in Mae Taeng, Chiang Mai, Thailand, is a sustainable living project implementing permaculture principals and hosting workshops in English and Thai. In fall 2006, the project hosted a PDC taught by Geoff Lawton of the Permaculture Research Institute of Australia, and subsequently installed over 500 meters of swales and a 2 million liter dam. The Panya Project used permaculture to help regenerate what used to be a monocrop mango plantation, transforming it into what is called a "biodiverse food forest, organic farm and education center". The Panya Project also incorporates what they call "natural building" into their design, e.g., wattle, cob and adobe brick.

Green Anarchy

Green Anarchy is a magazine published by a collective located in Eugene, Oregon. The magazine's focus is on primitivism, post-left anarchy, radical environmentalism, anarchist resistance, indigenous resistance, earth and animal liberation, anti-capitalism and supporting political prisoners. It has a circulation of 8,000.

The subtitle of the magazine is "The Anti-Civilization Journal of Theory and Action". According to their web site, *Green Anarchy* was established in the summer of 2000 by one of the founders of *Green Anarchist*, "a similar, yet more rudimentary, publication from the United Kingdom" and "began as a basic introduction to eco-anarchist and radical environmental ideas, but after a year of limited success, members of the current collective took on the project with an enthusiastic trajectory."

Encyclopaedia of

ENVIRONMENTALLY SUSTAINABLE GREEN SCIENCE AND TECHNOLOGY

Encyclopaedia of

ENVIRONMENTALLY SUSTAINABLE GREEN SCIENCE AND TECHNOLOGY

Volume 2

R.B. Patnaik

ANMOL PUBLICATIONS PVT. LTD.

NEW DELHI - 110 002 (INDIA)

ANMOL PUBLICATIONS PVT. LTD.
H.O.: 4374/4B, Ansari Road, Daryaganj,
New Delhi-110 002 (India)
Ph.: 23278000, 23261597
B.O.: No. 1015, Ist Main Road, BSK IIIrd Stage
IIIrd Phase, IIIrd Block,
Bangalore - 560 085 (India)
Visit us at: www.anmolpublications.com

Encyclopaedia of Environmentally Sustainable Green Science and Technology

First Edition, 2009
ISBN 978-81-261-3842-5 (Set)

PRINTED IN INDIA

Printed at Mehra Offset Press, Delhi.

Contents

VOL. 4

VOL. 5

Preface

This publication entitled "Encyclopaedia of Environmentally Sustainable Green Science & Technology" attempts towards creating a holistic understanding of environment and sustainable development in its totality. It defines environment and sustainable development in all its dimensions. The concepts and methodologies of ecology, environment and sustainable development have been discussed in detail. The ecosystem perspective and environmental concept of sustainable development have been highlighted, so has been the environmental methodology. An effort has been made towards defining green science and its manifestations for understanding pollution. The relevance and importance of environment impact assessment (EIA) for environmental sustainability has been analyzed. Also, the elements and dimensions of green technology have been covered in detail. The relationship among ecomovements, ecodynamics and sustainable development has been discussed in detail. Similarly, the environmental and social aspects of sustainable development, in particular, aspects related to training of people to meet the global environmental needs have been covered. The issues related to green energy, its sources, tools and technologies have been reflected upon briefly. The environmentally sustainable science and technology has been presented in brief with special reflection on climate change and global warming. Current issues and options regarding sustainability aspects of environmental management, corporate management and environmental responsibility have been covered. The aim is to develop a policy perspective. The contemporary global industrial technology has been discussed with focus on evaluating environmental performance and compliance to green laws. The applied aspects of environmentally sustainable science and technology have been outlined. Also, the green conventions enforced towards regulating global science and technology towards environmentally sustainable development has been presented, highlighting the theories of international environmental law and theory of environmental degradation.

The applications of green science and technology have been discussed along with a case of biotechnology. The green technology and environmental monitoring issues have been given considerable coverage. Select case studies have been made in this regard. Also, case studies of environmental engineering and environmental biotechnology have been incorporated with focus on applications and utilities. Future prospect of environmentally sustainable or green technology in today's world has

been analyzed briefly. Similarly, the principles, tools and techniques of environmental management, with particular reference to environmental disasters, have been covered in detail. Glossary of terms, bibliography and links has been provided to make this publication a user friendly one and to aid readers in further research and analysis.

—Editors

6

Ecomovements, Ecodynamics and Sustainable Development

ENVIRONMENTAL MOVEMENTS

Environmental movement is an organised social activity consciously directed towards sustainable use of natural resources. It aims at halting environmental degradation or bringing about environmental restoration or regeneration. India is witness to quite a wide range of environmental movements involving such categories as *omnivores, ecosystem people* and *ecological refugees*. In its biomass society, the bulk of the poor, and not the well-off, scratch the earth and depend on rains for growing their own food, gather wood or dung for cooking purpose. They build their own huts with bamboo or sticks and draw their sustenance from nature: such people depend on natural environment of their own habitat to meet their material needs. Raymond Desmann calls them *ecosystem people.* About four-fifth of India's rural people, that is, over half of the total population are ecosystem people under the pressures of modernisation, technological use and material growth the natural world is receding. This shrinks the capacities of local systems to sustain and support these people. For example, industries, dams, and mining have physically displaced millions of forest dwellers, tribals and peasants in India during the last forty-five years. So also because of scarcity of water and depletion of natural resources and recurrent droughts people have to leave their habitat like forests. These people are called *ecological refugees* or *ecorefugees.* They live on margin of islands of prosperity such as sugarcane harvesters in South Gujarat and western Maharashtra. As many as one-third of the Indian population is considered as ecorefugees. They live today such a life as displaced, unable to freely pickup from nature.

The remaining one-sixth of the population of India are real beneficiaries of economic development which takes place at the cost of nature. They are big landowners with access to irrigation, modern entrepreneurs in the industrial belts, the blue-collar workers and white-collar workers in the organised sector (which is about 9 per cent on India's workforce) and urban professionals (teachers, lawyers, medical community and bankers etc.), contractors and the employees in government and semi-government organisations. Unlike the first two, they have the cash and the purchasing power to buy consumer goods. They can manipulate the state for subsidies, dearness allowances and raw materials. Dashmann calls them *biosphere people* as they enjoy produce of the entire biosphere. In contrast, the ecosystem people have a very limited catchment of natural resources. Madhav Gadgil and Ramchandra Guha call them *omnivores* as they devour every thing produced all over the earth. In India, unlike the poor and the deprived people comprising of eco-system people and ecorefugees, the omnivores, who are quite small in numbers, exercise tremendous power to influence state policy. Again, unlike the first two categories, the omnivores exercise high degree of control over their own lives. Gadgil and Guha call them ecological and sociological categories. It is the ecosystem people and ecorefugees who are in the forefront of the environmental movement, and they enter a sort of social conflict directly or indirectly, against the omnivores or the government and the industries who are the protagonists of the dominant model of development.

Silent Valley

Silent Valley in Kerala has a rich 89 sq km biological treasure drove in the vast expanse of tropical virgin forests on the green rolling hills. Here, in 1980s, a 200 MW hydroelectric dam on the crystal clear river Kunthipuzha under the Kundremukh project was to come up. The Kerala government was very keen to see it through. But a combination of scientists, environmental activists and the local people made out a well reasoned case against it as it was a risk for certain endangered species and rare flower plants. It would also seriously disturb the age-old ecosystem preserved since centuries near the southern ranges of the Western Ghat. The state government, supported by an equally persuasive group of other scientists, engineers and some citizen groups put up a strong front against the anti-project movement. Finally, Prime Minister Indira Gandhi, informed by the spirit of the Stockholm Conference on Human Environment, intervened in this controversy and stopped this ambitious project to the chagrin of the Communist government in the state. The decade-long nationwide movement against the hydel dam became the bedrock of India's green movement.

TABLE 1

Ideological Preference of the Various Strands of the Indian Environmental Movement

	Crusading Gandhians	*Ecological Marxists*	*Appropriate Technologists*	*Scientific Conservation*	*Wilderness Enthusiasts*
Polity	Highly decentralised democracy, 'village republics'	Dictatorship of the proletariat	Decentralised democracy, with women, low-caste participation	No firm view	No firm view
Decision making	Highly dispersed power of decision making	Centralised planning	Decentralised planning	Centralised planning	Strongly centralised administration
Society	No firm view	Economically equitable, but centralised political power	Economic and political equity	No firm view	No firm view
Economy	Mixed economy	State occupying 'commanding heights'	Mixed economy	Mixed economy	No firm view
Scale of economic enterprises	Predominantly smali, village level	Predominantly large	Focus on small, complemented by large	No firm view	No firm view
Appetite for consumption	Limited through moral choice	Limited only on grounds of equity	Limited on grounds of both equity and ecology	Unlimited	Unlimited
Linkages to global economy	Weak	Weak	Weak	Weak	No firm view
Rate of technological change	Exceedingly low	High	Moderate	No firm view	No firm view
Commitment to military expenditure	Very weak	strong	Weak	No firm view view	No firm

Source: Madhav Gadgil and Ramachandra Guha, *Ecology and Equity* (Penguin, 1995).

Sumi Krishna states, "environment-development trade-offs are rarely so clear cut as the case of Silent Valley. It cannot always be gauged by impact assessments and cost-benefit analyses. Even more difficult is to judge whether some loss of natural diversity is the price We have to pay for a more modern and equitable society society. And if so, how do we decide how much diversity, at what price and who pays".

Appiko Movement

The Chipko movement has produced ripple effect in other parts of India. In Western Ghats, in the Uttara Kannada region of Karnataka, the Chipko has inspired villagers to start Appiko Chalewali movement. Here, the destruction of the forest was caused by commercial interests for obtaining timber. Natural forests were clear-felled by the contractors, which led to severe soil erosion and drying up of perennial water resources. In the Salkani village in Sirsi, the people were deprived of the only patch of forest left near this and surrounding villages to obtain fuelwood, fodder and honey, etc. In September 1983, this led youth and women to launch a Chipko type of movement in South India. Youth and women from Salkani and surrounding villages walked five miles to a nearby forest and hugged trees there. They stopped the axe-men who were felling trees pursuing the orders of the forest department of the state.

The people demanded ban on felling of green trees. They were ready to sacrifice their lives for this cause. The protest within the forest continued for 38 days which forced the government to finally withdrew the felling orders. In the beginning the government had completely ignored people's sentiments, but after the visit of the forest officers and experts of the area of the conflict, people's involvement in the decisions regarding the felling of trees was promised.

However, felling of trees in this forest started again. This again invited the local people's reaction who resumed the movement by hugging the trees. There was full support of local people. So popular was this protest movement that the labourers of forest contractors also left the felling sites.

In October, the movement entered into its second place. Here, there was a problem of forest with mix tropical semi-evergreen trees on hilly slopes. Here most of the people were tribals who were dependent on the forest for their survival and subsistence. Disappearance of bamboo deprived them of the source to make such things as baskets, mats, etc., the main source of income. But the contractors continued to serve their commercial interests. This led the people to start the movement in Bengaon forest.

Significantly, it was launched by the people spontaneously. This time the tribals hugged the trees in order to save them. Soon this movement gathered momentum.

The forest department as a response had to give in. Another movement started in the deciduous forest of Husri. Here also the greed of the contractors was revetted on timber. Again Appiko inspired the local people to launch the movement. Here about a hundred men and women joined their hands and stopped the trees felling through. While the contractors aimed at commercial species the people wanted the trees for fuelwood needs. Disappearance of trees also made the labourer finding it difficult to get wood for agricultural implements. Again the monoculture of teak and eucalyptus affected their agricultural yield. Only mix varieties of trees could yield them biomass for fodder, fertiliser and fibre, etc. The ranger officer of the forest department visited the forest and asked the people to give up the movement and allow felling of trees, but undaunted at the sight of once an awe-inspiring symbol of power, the people protected trees by hugging them in front of him. Appiko thus became a replica of the Chipko in the South.

In November, the movement spread to Nidgod (Siddapur taluka) where 300 people collected. It was a deciduous forest with laterite soil with steep slopes. The forest was clearfelled of its 10 acres by the government and instead, monoculture trees were to be planted. But it was the only patch of missed forest left near the Nidgod village. And the people wanted the forest for their need of fuelwood, while the government wanted to take the woods of the fallen trees for the use of the nearby Siddapur town. The local people succeeded in stopping the felling of the trees. Its message reverberated in the adjacent taluka. The Organisation of Appiko spread the movement there. In the same month, another forest movement was launched in Kelagin Jaddi of Siddapur taluka where the same type of forest existed. A plywood company wanted to get plywood raw material from this forest. In pursuit of commercial purpose, the company damaged 542 trees in the process of felling 51 trees marked by the forest department. The villagers saw that because of such an action they will be deprived of the traditional source which gave them fuelwood, green manure and fodder, etc. This movement also was launched spontaneously by the local people. In the same way, within a fortnight, people started the movement in Parsi forests. Industrial timber was to be obtained from this deciduous forests. But the people then would be left with no fuelwood and such item of subsistence, as there will be meagre forest cover left. About 300 people succeeded in the stopping the felling of the trees.

In December, the movement was launched in Bilgal forests where clear felling of the trees was attempted. The timber wood will be used for commercial purpose and fuelwood will be provided to nearby town Sirsi. This would again deprive the local people their only source of sustenance and this was the only mixed forest in India. About 200 people including 100 women succeeded in stopping the clear felling.

The Appiko movement became the trend-setter of the like movement in the

South India, which exposed the greed of the commercial interests like contractors who would fell more trees than actually permitted and marked by the forest department. But the forest officers turned a blind eye to such an action. The collusions of the contractors, the forest department and the politicians who gave them a cover were also exposed. The Appiko movement succeeded in its three-fold objectives: (1) protecting the existing forest cover, (2) regeneration of trees in denuded land, and (3) utilising forest wealth with proper consideration to conservation of natural resources. The locally established popular organisation known as Parisar Samrakshana Kendras (Environmental Conservation Centres) worked as the vehicle to carry out these objectives. The Appiko movement saved the basic life sources for the people— trees like bamboo useful for making handcrafted items which they could sell for earning a few rupees. It also saved medicinal trees for their use by the local people. The movement created awareness among the villagers throughout the Western Ghats about the ecological danger posed by the commercial and industrial interests to their forest which was the main source of sustenance. It spread the Chipko philosophy that the basic product of the forests of the Western Ghats as in the Himalayas, such as soil, water and pure air must remain the basic in their hand. Like the Chipko and the Silent Valley movements, the Appiko continued our search for a strategy for conservating natural resources—in fact, the entire life-support system and withit the option for human survival. Steeped in the peaceful method of powerful protect, 'Chip Appiko' revived the Gandhian way of mobilisation of the people for a sustainable society in which neither man nor nature exploit/destroy each other. As Vandana Shiva observes, "it marked the beginning of this civilisational response to the threat of human survival".

Chipko: Mother of Movements

It was in the Garhwal region of the Haryana range in Uttar Pradesh that Miraben, a close disciple of Gandhi, traced the links between deforestation and water crisis. Learning from the old people in the region, she related the environmental degeneration there with the disappearance of abundant forest of *banj* trees. The replacement of the *banj* and mixed forests by commercially attractive pine was a major reason for the increasing deprivation of Garwahli women, as the pine could not perform the ecological and economic functions of the indigenous trees like *banj*. The major flood in the Alaknanda (1970), a major tributary of Ganga, and a disaster in its catchment area(1987) when its second tributary was blocked by a landslide flooding its basin all the way to Calcutta prompted mobilisation of once docile local people against logging. The latter was the watershed in the ecological history of the Himalayas. Even the official thinking was cruelly shaken with these experiences which conceded "to the Chipko view that the main function of the Himalayan forest was to manage water, not to yield timber". The government accepted that the largo scale erosion and floods were the unintended result of what was called 'development'—deforestation, road building, dams and mining in the mountain catchments. In the early 1980 the Chipko *hawa* also spread to other fields. For example, the movement against mining activities in Doon Valley: The local people started the Chipko blockade at a mine at Nati khala, which was not closed in sprite of a Supreme Court order for doing so. It was the fight for the life of nature, and therefore, of human beings. The song graphically brings out the people's perception of close links between catchments, forests and mountains. This movement as a continuing revolution for two decades contributed to empowerment of illiterate women. The world's most celebrated, the Chipko became an ecological epic. It inspired the formidable popular upsurge against the big hydroelectric project at Vishnu-prayag near the Valley of Flowers and the Tehri dam (the latter also is opposed on the grounds of its earthquake prone possibilities).

Private Sector-Government Nexus

Paper and pulp factories as well as rayon factories are one of the most notorious industrial polluters. The Birlas had built three units of Gwalior Rayons on the river Chaliyar in Kerala. The effluents of the factory adversely affected the health and economic welfare of the villagers staying in the downstream area of this river. The Kerala Sastra Sahitya Parishad, an NGO with an excellent record of creating science awareness and people-oriented development, studied this case of pollution and brought out the extent of environmental hazards caused to the downstream villages. This created a spirited movement against this factory and forced the factory owners to close down their unit. The same parent company was running Hariar Polyfibres in Karnataka. This company was discharging untreated effluents into the river Tungabhadra. The people complained of being hit by new diseases, depleting fish yield and reduced availability of irrigation water. This created a well-organised protest movement against the company. Thus, these cases illustrate the conflict between the rival factors of profit and the public good. Another variant of such an issue is provided by the Maharashtra State Industrial Development Corporation. About forty industrial units working on the industrial estate owned by the corporation used to be discharged untreated effluents in the Raigad district. The nearby Revadanda creek area was polluted because of the effluents. The water could not be used by the farmers for their crops and fishstock was depleted and the fishes which they caught and sold were not found hygenically eatable. This provoked the farmers and fisherfolk from the creek area. They accused the state corporation of neglecting the tasks of treating effluents before discharging them into a nearby river. In October 1987, the enraged farmers inserted a wooded log into the discharge pipeline. In the same way, when South Asia's largest distillery located on the Western Maharashtra Industrial Corporation destroyed the land and polluted water by its discharge, the villagers got united and opposed the Corporation and the Maharashtra State Pollution Control Board. They filed a suit in the Bombay High

Court and claimed rupees 10 million in damages from the offending distillery, the State Corporation and the State Pollution Control Board (August 1988).This is a case of the farmers and fishermen versus a combination of private sector and the governmental authority.

Bhopal Gas Tragedy

The Bhopal gas tragedy has become known the world over for the severe environmental disaster it caused and the role of a multinational corporation in this case. The Union Carbide Corporation, a US-based multinational chemical corporation, was operating a giant pesticide plant on the outskirt of Bhopal city. On 3 December, 1984, an accident occurred, probably caused by the introduction of water into the toxic methyl isocynate storage tank. This resulted in a powerful and uncontrollable reaction generated by the liberation of heat. As a result the deadly methyle isocynate escaped in the form of a gas. The gas killed thousands of human beings, blinded many more and also harmed the people living nearby in different ways. Living around the Union Carbide plant was something like a slum area where a large mass of 'ecological refugees' were living. As the gas began to escape, within minutes about 3000 people died on the spot. About 50,000 sustained chronic lung damage. The human rights activists, the environmentalists and humanitarian groups spontaneously arose against the callousness of the MNC authorities who tried to play down the extent of the tragedy. But its negligence was exposed by the NGOs. Supported by the national and international media, they expressed their anger against the administration and the medical authorities which were suspected to have resorted to the stratagem of underestimating the magnitude of the disaster. Regional, national and international NGOs, jurists and environmental/human right activists networked in a concerted way and brought pressure to bear upon the initial hesitant state and union governments. The documents, videos and other evidences bared the blatant underplaying of the reality and the ghastly sufferings of mostly the poor people and workers. They suffered because of this accident "caused by a multinational corporation who could not operate in this field in its own country". Is the life of the people of a Third World country cheaper than that of the citizen of rich country of the North ? Is the life of the common man expendable for profiteering foreign owners ?

The Bhopal tragedy is considered as, perhaps, the tragic episode of the poisoning of the environment in human history with international dimension. It was an ecocide caused in India by a unrepentant multinational company. The Bhopal tragedy has raised the issues of the environmental prudence, human right issues and MNC-Indian State relationship, the powerful international industrial lobby backed by its government and the need for networking the human rights/humanitarian NGOs across the nation.

located near the Lushpaddish of Nandikar in the shadow of the rainforests of the Western Ghats. It is one of the seven fast-track foreign power projects, for which, like the Enron (Dhabol Power Corporation) project, the Central Government has assured counterguarantee. As Karnataka like the rest of India, needs massive energy supply to meet its growing need of electricity in the wake of economic liberalisation, the need for such a power project is appreciated by all.

But the way the procedure of facilitating the meeting the demands of MPC was expedited by the Government of Karnataka and the Central Government has raised political problems and environmental concerns in the state and outside. So also the content of the power purchase agreement (PPA) between the MPC and Karnataka State Electricity Board (KSEB) relating to the payment of income tax by the company and the withholding tax of 5 per cent that the company must pay when it transfers its earnings to USA has raised critical questions about PPA.

The Congress government of Karnataka and Congentrix signed a memorandum of understanding in July 1992.Thereafter, the PPA was signed in September 1994. In July 1995, the State Pollution Control Board gave 'no-objection certificate' to the company but not before laying down 26 stringent conditions. One of the safeguards described by the MEF in its environmental clearance letter is that the company should ensure all discharges from the plant like gaseous emissions, flyash or liquid emissions are fully treated before released. Also prescribed is a limit of 0.82 per cent of sulphur content in imported coal, so that sulphur dioxide emission which can cause acid rain are minimised. The MEF has also laid down that flyash and coal storage area must be suitably lined so that leaking into the ground water can be avoided. In the clearance letter, it is prescribed that liquid effluents be recycled and air quality be constantly monitored. Again a green belt of "adequate width and density with suitable selected species" is to be developed around the plant. Significantly, the nearby Mulki river is guarded from the environmental hazards of this project. The project is not allowed to draw in the Mulki river for its soft water needs. The clearance also stipulates that the company must set aside funds for environmental protection measures. On 18 January, 1996, the PPA was revised and initialed (not signed) by MPC and KSEB after receiving clarifications from Central Electricity Authority (CEA). On 22 April 1996, the CEA gave conditional techne-clearance to this project and on 11 June, 1996 the MEF gave its final clearance. This was a year after the company submitted the mandatory rapid EIA report.

The one year period, however, was marked by a growing number of environmentally conscious people and activists who articulated their fears about the danger that this project would cause damage to the fragile ecosystem of coastal Karnataka.

The environmental clearance is the core issue in the Cogentrix. The energy plant is located right next to the Western Ghats which is one of the world's major 18 biodiversity regions. If the plant fails to strictly and honestly conform to the strict environmental conditions as above, the green and rich forests and fauna of the Western Ghats would suffer unmitigated and irreversible damage. Fears of the environmentalists is particularly focused on the acid rain and the treatment of effluents and its disposal. In 1995, the former Karnataka's Environment Secretary Yellappa Reddy felt strongly about the project and even went in public to record his protest against this aspect of the project. As Chairman of the Environment Clearance Committee (ECC) of the State Government, Reddy warned in his report (July 1995) that "effluents of the Cogentrix plant will certainly lead to concentrated acid rain in the Dakshina Kannada area". He stated, "the whole site is an ecologically sensitive area. It has been designated as the twelfth World Heritage Biodiversity Site. There are three highly sensitive ecosystems that are intimately integrated—the marine, the inland aquatic and the terrestrial. The region has a number of estuaries; these can be called cradles of the oceans as 60 per cent of marine animal species came to estuaries for breeding. We have spent between 1991 and 1996 Rs. 15-20 crores to restore the endangered mangrove ecosystems in this area. The entire zone has a very productive biomass with more than five lakh people depending upon it."

An energy expert in Karnataka stated: "There is no point in placing a limit on the sulphur content of imported coal. Instead, there should be a limit placed on the sulphur content of the gas that escapes into the atmosphere. There is no thermal plant in the US which does not have this fairly simple technology." He wondered, why the company is not using such technologies when all costs are going to be passed on to the purchasing agency by way of tariff. "We are willing to pay for the preservation of biodiversity." The Prime Minister Deve Gowda favouring it is determined to push it on.

Fisherfolk's Agitation

The Dakshina Kannada district in Karnataka is considered as one of 34 ecologically sensitive areas in the world. But the Karnataka government has, all along, transformed it into an 'industrial zone More than 30 big industries like Mangalore Refinery and Petrochemicals Limited (MRPL), Cogentrix. Mangalore Chemical Industries and other industries coming up—some of which helped by Deve Gowda as the Chief Minister of Karnataka and as the Prime Minister. The sea close by is the major locational reason for such industries. The Mangalore Chemical Industries and the Kudremukh are alleged to have together discharged untreated effluents of seven lakh litres into the sea. Moreover, sewage from the Mangalore city and untreated effluents released from small industries have exerted together huge ecological strains. All this have brought down the fish catch by the local, small scale fisherfolk. Income of such large number of fishing families has gone down. These along with the low-income people whose staple diet is fish have suffered because of the tapering supply of fish.

The State Pollution Control Board (SPCB) had cleared the MRPL project on the points raised regarding water and air pollution. But the National Institute for Oceanography (NIO) has dismissed such a claim. The NIO has asserted that contrary to the claims of the SPCB and the MRPL, 0.01 mg mercury and 2.0 mg cadmium are found in the discharged effluents. It has also directed the MRPL to discharge them only after they are 'duly treated'.

Such a situation has reduced the supply and catch offish as they are killed outright. Moreover, the fish become toxic thus harming the humans who eat them. Mercury, cadmium, copper, cyanide, sulphide and the like cause quite a broad spectrum of diseases and illness from vomiting and diarrhoea to damaging the central nervous system.

All this has hit hard the fisherfolk. about 4,00,000 fisherfolk have been in dire economic condition because of such ecological degeneration caused by ecounfriendly authorities of the state

There is another dimension of this problem. This is caused by the Central Government's new policy of economic liberalisation and globalisation. Among the 15 major marine fishing regions on this earth, as the World Watch Report states, the productivity in all except the two regions has gone down. Only in the Indian Ocean, the output of fisheries is still on increase. But it also is seen to be reaching its limits.

Such global changes in the world of fisheries have prompted the fisheries, particularly of the developed world to turn to the Indian Ocean for ravishing fish harvests. Multinational companies of these countries which have almost exhausted the once rich Pacific coast catch have now turned to the Indian Ocean. The Government of India, with its new policy of economic reforms, has made easy the commercial entry in India's coastal and oceanic region. They have the technical means and resources as MNCs can engage bigger boats with sophisticated means to fish in the mid-sea. With the help of highly sophisticated trawlers and miniaturised electronic equipments they can catch up to 1,200 to 2,000 tonnes offish. They have become a great threat to the fisherfolk of the Mangalore coast (as also to those of the coast of Kerala). They leave lesser catch for thousands of local fisherman. With modern, large equipment of storage and packing, the large fishing corporations export their huge catch to international markets. The Indian government has given licenses to 175 such companies for deep-sea fishing inside the Indian waters.

As Sudha Sitaraman observes on the basis of Latin American experience, "export-oriented fishing has severe implications for low-income people and subsistence culture who rely on fish as their staple diet. When governments subvert traditional and local control of fishing grounds, severe ecological as well as social consequences can ensue". Such developments have prompted thousands of fisherfolk to protest against the MRPL. In September 1995, when the MRPL started laying pipeline, they forced the construction work to halt. The company agreed to reuse the effluents for agricultural and horticultural purposes. But the government was found inclined towards the MRPL. As the agreement between the two remained unsatisfactory, in December 1995 the reactivated movement reached its climax. Earlier in August 1995, about one lakh people participated in a huge rally against the company. Various NGOs have also started action against the potential environmental disaster.

All this forced the state government to invite the representatives of the NGOs and fisherfolk at a meeting to discuss then demands with the company. The meeting was attended by the MPs and MLAs. As a result, the government issued an order directing the company to install suitable equipment of treatment of effluent, recycling of waste and disposal of sludge.

In a way this was a notable victory of the ecostruggle of the small fishermen of Dakshina Kannada district. However, the MRPL has dithered under the excuse of financial resources and time needed to install such a plant. With an obliging government, the company continues to pump effluents into the sea detrimental to the interest of the fisherfolk. The collusion of the industrial interests and government as evident in this case is stalling the move of the NGOs and organised community of the fisherfolk. It is the economic justice and ecodevelopment which are being jeopardised. The struggle is still on. The time is on the side of the people.

People Versus State

The Harihar Polyfibres was started by Birla in 1972 on the bank of river Tungabhadra in Dharwar district, Karnataka. Since its inception, the company has been polluting the river which has "adversely affected the live of one lakh villagers in untold ways", mainly fishermen, shepherds, farmers and labourers. There has large scale fish kills and people drinking the water of this once pristine river suffered from strange diseases. The industrialists found it cheaper to silence those making noise against it than abiding by the conditionality of effluent treatment. Local people called the foul area as 'Birla Perfume', and the polluted water as 'Birla Teertha' (Birla water). Only after continued strong protects that the Karnataka State Pollution Control Board took Harihar Polyfibres to court. This is for the first time anywhere in the country that a PCB has taken a Birla owned industry to court.

In 1984, in partnership with the Birlas, the Karnataka government launched a joint sector company called Karnataka Pulpwood Ltd. (KPL) for growing eucalyptus on forest and village common lands for the captive consumption of Harihar Polyfibres.

Under this dispensation, 30,0000 hectares of forest land and grazing lands were made over to KPL on lease for 40 years at Re. 1 per hectare per year. This was totally unjustified. It was a blow for poor people as these lands were used for generations by the people now converted into plantation areas. In the process, all the laws of the land, particularly the Forest Conservation Act, have been given a go by. Traditional and natural rights of the local communities were compromised on the altar of interests of big business and sources of subsistence were sought to be used for commercial purpose. The local people began a series of protests and *satyagrahas*. Newspapers picked up these events and gave them wide publicity. Voluntary bodies like 'Samaj Parivartan Samudaya', Dharwad led by S.R. Hiremath and others filed a public interest case in the Supreme Court.

On one plane, the people organised a non-violent direct action culminating in *Kittiko-Hachchiko* (pluck and plant) *satyagraha*. And, on the other, it was being fought in the Supreme Court where it was pending for final verdict for years. The Supreme Court issued a stay order in 1987. The Karnataka government and the KPL filed affidavits to the effect that, as on that date, the KPL was in possession of only 3,590 hectares. A lot more lands (over 14,000 hectares) were handed over on the sly to the KPL which carried on plantation in violation of the stay order. For which a contempt petition has also been filed.

The documents produced in evidence are a veritable scoop. And bring into focus the machinations of the highest officers in collusion with their political bosses. Questions have been raised in the Karnataka Legislative Assembly and the Forest Minister has admitted violation of the ICA and stay order on the floor of the house. More than seventy legislators have appealed to the Chief Minister and the Forest Minister to cancel the KPL agreement. "Why no action is taken, in spite of positive proof is an open secret?"

In disregard of the stay order the Karnataka government released to KPL a few more thousands of hectares of land, for which a contempt of court petition was filed in the Supreme Court. The partnership deed made available to the court revealed that though the Karnataka government owned 51 per cent of the shares, it was to get only 12.5 per cent of the profits. Whereas the Birlas having 49 per cent of the shares enjoyed 87.5 per cent of the profits! It does not stop here. The Karnataka government was 100 per cent guarantor to KPL's heavy borrowing, leaving the Birlas free from any liability.

The spirited *satyagraha* by the people and the consequent pressure of public opinion and the writ in the Supreme Court was too much to swallow for the Karnataka Government. There was no other go but to close down KPL. Even after its closure the Government tried to pay to the Birlas a compensation of one crore of rupees.

Besides depriving common people of their natural rights to minor forest produce, the much favoured Harihar Polyfibres has caused irreparable damage to the environment by polluting with impunity the waters of Tungabhadra as also the air around. The Karnataka State Pollution Control Board, instead of taking firm action, had been supporting this polluting industry by going out of the way and suggesting ways and means of circumvention. "What are the governments and boards for, if they do not enforce their own laws?"

For the first time an industrial giant is put on the mat, related misdeeds are exposed by and, above all people are alerted about their rights. The media, regional and national, gave the strength and a great boost. Deccan Herald, Business India, Times of India, BBC, etc., supported it and exposed the claims of the industry and the government. It is an eye-opener not merely to Karnataka but to the whole of India and is a pointer to what people can achieve if they mean to. A major fallout of the movement has been that another wood-based industry that was to come up in Karnataka backed out looking to the rough weather the KPL has run into. So far so good for the trees. In fact, this is an achievement that the NGOs and all organisations and individuals supporting this cause can legitimately be proud of.

There was little hope of success against the combined might of an industrial giant and a state government: "What with industry capable of computing anything and everything in this ancient land of ours where seers prayed for the happiness of mankind." But the struggle has finally achieved a rare success so as to hearten all environmental activists.

The final order of the Supreme Court in this case was issued on 26 March, 1992 based on the Government of Karnataka's order of 24 October, 1991 confirming the winding up of Karnataka Pulpwood Ltd. (KPL) and a detailed affidavit of 12 March, 1992 recognising the people's rights over the common lands. This order of the highest court of the land sets an important precedent that can be effectively used anywhere in India for protection of common lands against encroachment by the vested interests and for meeting people's needs.

The struggle of village for village commons was a successful quest for justice. It is a classic case of 'struggle and reconstruction', like the Anil Patel's ARCH-Vahini which fought against the displaced tribal's rights(1980-1987) and then worked for their just rehabilitation in the case of Sardar Sarovar project(1987-1995) and then again alienating from the project and returning to struggle on the issue of the tribal rights to forest produce in Dediapara Sanctuary and unsatisfactory progress of resettlement. SPS is also a good case of fine blending of people's mobilisation and legal action. It also brings out the collusion of a big industrial house and the state government and the village community's struggle to change the favourable power equation with the help of the media and judication.

Bailiadila Controversy

The National Mineral Development Corporation (NDMC) started major project at Bailiadila in Bastar tribal region (Madhya Pradesh) in 1968. The Bailiadila range of iron ore deposits in Bastar was a rich forest area. It was claimed that the project would creats employment and help develop this backward area. In the wake of such a clear forests were cleared and lands of the tribals acquired.

In the initial stage local tribals got jobs for the labour in the mining which required manual work, but their number began to shrink in the latter stage of the project. Recession in the steel industry, the main importer of these iron ore resulted in the slowing down of manual mining at Bailiadila (which means not locks hump). these caused laying off more than seven thousand labourers. The workers protested resulting into police firing and killing of some tribals.

The project also caused environmental problems. the eve was washed in Sakhine—a tributary of the river Indravati. It water was polluted for over 40 km. Its polluted deep rust-red water was used by villagers for irrigation. The district authority tried to persuade the NDMC to reduce the pollution. But the villagers began to doubt whether the new dam would be worth submersion of their land. Sumi Krishna quotes a wisened villager: "They took our lands and our women, and brought us liquor and red river, why should I even go there? They have no work for me."

"With the progress of the project well placed officers from outside who came their created social problems as well project officials lured our young tribal girls." this created hostile social environment.

All such developments have created resistance in the tribal region during the regimes of the Congress as well as BJP in the state. The matter was also raised in parliament with shrill voice and the governments under Narsimha Rao, and Deve Gowda were hard put to satisfy the irate MPs. The Central Government is forced to state that it will abandon this project under the combined pressure of hostile tribals, environmentalists and some political leaders who are sensitive to the environmental concerns.

Maradia Industry versus People/Government

In Surendranagar district of Gujarat an unprecedented anti-pollution movement was recorded in July 1996. Maradia Industrial Complex producing chemicals in Sayla taluka is spread over 1,300 acres of land. It started production in 1991.As pollution control arrangements were not provided for and the authorities remained oblivious of this stinking effluence which started polluting the wells of nearby villages, the people began to murmur against it. When the waters of wells and other reservoirs nearby began to be coloured the people got consternated particularly during the rainy season." The managers of the complex saw to it that along with the rain water untreated liquid effluents to be flowed into the seep nearby small ponds. About 8 litres of the polluted liquid deposited in the ponds will then evaporate during the sunshine. But many cattle who drank such water had died, and the people were denied of clean drinking water.

There was another trouble chasing the people of the surrounding villages. Everyday, during the night from 12 to 4 a.m. the managers used to dispose off the gas produced during the production process. The pungent smell of the gas used to make them tipsy and uneasy and cause suffocation. Even the leaves of such big trees like 'babal' were burnt by such gas. In the beginning, the politicians and the officers of the state and district administration remained soft to the Maradias and continued to disregard complaints of the villagers. The Gujarat Pollution Control Board (GPCB) asked the industrialists to establish the effluent treatment plants. Its owner Rasikbhai Maradia claimed that it had spent Rs.11 crores for the plant. He claimed, "we are not releasing a single drop of the polluted liquid outside. Rather, after the treatment, we give this treated water to raise one lakh trees which we have planted in this complex."

The share of highly polluting industries, especially those that generate hazardous and toxic wastes, has increased substantially. These include electricity, chemicals and petroleum refining, basic metals, non-metallic minerals, paper, and leather industries, whose share in industrial value added rose from 45 to 52 per cent between 1975 and 1989. The heavy rain caused the water of this pond to overflow which then reached to Saburi dam, about 6 km away. The dam water became red and unworthy for drinking purpose. If the rain had continued the water from the Saburi dam would have flowed to a nearby Nayika dam and Dholidhaja dam. Two lakh people of towns Surendranagar and Wadhvan and other two lakh people living in nearby 35 villages get their drinking water from these two dams. About 40 per cent of the district population depend upon these two dams for their drinking water. Had the heavy rain continued waters of these two dams also would have been polluted. This.made people lose their patience. Their protest blurred the party line. This stringency of the protest prompted the District Collector. Under Article 143 of the Indian Penal Code, the Collector gave a provisional order for closing down 17 plants of the complex. The water samples of the Saburi dam, and the surrounding wells and dunkies were tested in the laboratories. Then five out of six samples of water were found to be completely undrinkable; even the waters of the dunkies were found red, that means waters at the deep down the ground were also polluted. The water was found to have very high level of Chemical Oxygen Demand (COD). A case was filed against the Maradia Complex as public interest litigation because the vital interest of many people was involved. In 1995, the Gujarat Ecology Commission (CEC) had appointed a committee of two experts to study this problem in the context of growing complaints of the people. Its report clearly brought out that in spite of the GPCB, right from the very first day, this industry had flouted the environmental laws on a large scale.

TABLE 2

Status of Pollution Control in Identified Polluting Industries

Industry and Particulars of Compliance	*Number of Plants*	*Per cent of Total*
OIL INDUSTRY	12	
Plants complying with diluent and emission standards	6	50
FERTILISER INDUSTRY	110	
Plants complying with emission standards only	49	44.5
THERMAL POWER PLANTS	70	
Plants complying with emission standards	25	35.7
Plants complying with the standards for ashpond effluents	33	47
INTEGRATED IRON AND STEEL PLANS	7	
Plants partially complying with effluent and emission standards	6	85.7
Plants not complying with emission standards	1	14.3
CEMENT INDUSTRY	94	
Plants complying with emission standards	62	66
PULP AND PAPER INDUSTRY	336	
Plants having fully operational ETPs	118	35.1
Plants having partially operational ETPs	49	14.6
SUGAR INDUSTRY	365	
Plants having fully operational ETPs	180	49.3
Plants having partially operational ETPs	31	8.5
DISTILLERY INDUSTRY	176	
Plants having fully operational ETPs	74	42
Plants having partially operational ETPs	32	18.2

Note: ETP means effluent treatment plant.
Source: Bowonder and Rao (1994).

The former member of the Central Pollution Control Board (CPCB) P.R. Gharekhan and the former chairman of the GPCB, R.C. Trivedi threw a bombshell. They stated that the Maradia Complex had started production without getting the permission of GPCB. Shockingly, the company obtained the permission of GPCB only in 1995. And even when permission for producing naptheline-based chemical, which was hazardous to environment, was not given to it, the company "merrily went on to manufacture the chemical".

In 1995, the committee has also cautioned that the Maradia complex is situated near Nayika and Dholidhaja dams and the topography of the region is like an inverted saucer. The hazardous waste and effluents, when overflowed, will proceed downwards to these dams and pollute their waters meant for drinking. The committee was of view that the GPCB was responsible for giving approval for such a location to the complex. It also felt that GPCB had not taken enough caution or prior investigation about the environmental impact of such chemical complex. The effluent is also not treated according to the standards of the GPCB. The polluted water is directly released on the land which then sips underneath and mixes with the subterrain water. It also pollutes the air, but the company remains indifferent to the environmental hardships of the people. The committee has recommended to this industry the adoption of modern technology for treating the polluted water. In committee's view, "even though such a technology is costly (for the industry), nothing is costlier than the life of the people". It was no wonder then that the Collector had to give the interim order to close down the plants. Significantly, even thereafter, the people observed bandh.

The closure of the complex is not taken by the Maradia Group in its stride. The Chairman and Managing Director, Rasik Maradia, stoutly defended his industry. He protested that if the complex will be permanently locked out, then 1,380 permanent employees and 1,200 daily workers will lose their jobs. The company incurs a loss of Rs. 40 lakhs everyday because of the lockout. His cosy relations with atop leader was one of the factors which enabled to manipulate the regulatory power structures of the state like the GPCB. But at long last such a brazen display of insensitivity causing serious hazards to the health of the people reached its denouement. Shailaben Bhatt, the President of Surendranagar Municipality and the BJP leader demanded that the best solution will be the permanent closure of the complex. Arvindbhai Acharya, the former MLA of the Congress and the former President of Surendranagar Zilla Panchayat also came down heavily on the complex. Sanat Mehta, the new Member of the Parliament from the area (1996), also reflected the people's mood and forcefully demanded to take action against this defiant industry. Politically cushioned Rasik Maradia blamed the heavy rain for all this. He challenged the people to prove if he had committed the breach of any law related to environment. He is even ready to face the CBI inquiry. He threatened that the government has given the order for closure without giving him any chance to state his case. He claims to have been heavily shocked because of such an order given in such a way (but one cannot forget the soft treatment so far enjoyed by him from official quarters). Maradia even threatened that if this continues then no one will come to invest in Gujarat.

Actually, the GPCB continued to turn a deaf ear to the recurrent complaints and accusation that the Maradia industry was causing serious environmental pollution.

Only when the dam water was polluted, in order to repair the people's wrath and avoiding the onus of fault, the GPCB awoke from its benign slumber. The GPCB then asked the industry to close down four plants which were manufacturing environmentally hazardous, naptheline-based chemicals like H-acid, and J-acid. In the last year, the GPCB had ordered the company not to do the process of treating some chemicals collectively. Maradia did not conform to this order. Even then, the GPCB did not take action against it. Only the uprising of the people during June-July could make breakthrough. This time leaders of both the main political parties and people of all the stratas have joined together to constitute 'Maradia Hatao Ladat Samiti' (the committee to fight for removal of Maradia). The villager Khima Magan of Sorimbhada, Sarapanch Pratapsinh Jhala, Shailaben Bhatt, Sanat Mehta as well as Navin Jobanputra of Wadhvan Industries Association are united. When Maradia employees packed in two buses coming towards Ahmedabad to attend the court case scheduled for hearing, they were stopped by a group of villagers on their way.

Sayla villagers erected road blocks on the highway towards Ahmedabad and an altercation took place. However, the employees outnumbered by the villagers had to retreat with their buses to the factory premises situated between Sitagarth and Aya village. When they reached the factory they were gheraoed by another strong mob of 500 villagers of Sayla. The unruly mob vacated the buses and set them on fire, besides a private vehicle parked outside the factory in Maradianagar. The irate mob also cut telephone lines of the factory.

Meanwhile, about 2,000 Maradia workers thronged the Gujarat High Court premises as the company filed an application, requesting the court to reopen its closed units in Limbdi taluka of Surendranagar.

Following this, Maradia moved the High Court. Clashing with this, another petition was filed by the farmers of the areas adjoining the chemical units pleading for continuance of the closure orders.

The counsel for Maradia pleaded before the court that it be permitted to run its closed units as it had invested crores of rupees and the closure will affect the company's financial condition. He further contended, it will also affect the workers and assured the court of immediate steps to control pollution.

The counsel for GPCB pointed out that the workers were brought to pressurise the court. Taking note of this, the court directed the company not to adopt such tactics.

The GPCB also brought it to the court's notice that the industry was a highly polluting one and so it had recommended the closure of its units to the government under Section 5 of the Environment Protection Act.

The farmers' application contended that the effluents from the chemical units had polluted the water and soil. They claimed the pollution had affected their agriculture and cattle and after the rains, the effluents had entered the Saburi, Naika and Dholidhaja dams.

The government ultimately has started taking back the land from Maradia unit signalling the end of this anti-people and ecohostile enterprise. A case of people-politicians-government combination against a stubborn industry.

An inquiry team of the Central Pollution Control Board from New Delhi has found that pollution control measures were lacking at the plants of Maradia Chemicals Limited in Surendranagar district. According to an official statement here, the inquiry team suggested that the closed chemical factory could not be allowed to resume production until pollution control measures were taken up.

The Collector said that the effluent treatment plant in the factory was not effective. Thus, the claims of the company that a treatment plant was installed had no relevance. He said that till the date (October) the company had failed to install a treatment plant for solid waste. The inquiry team had also taken samples of water from the nearby villages and found them contaminated.

The Collector said that the inquiry had also made certain recommendations such as proper drainage system having cement plastered lakes to prevent polluted water from percolating in the soil, installing facilities to treat solid waste as well as water. The team suggested the factory could not start production unless it followed the recommendations.

The Maradia case is an example where all have been up against an industry with powerful political connection. The government will have to now turn its face off the economic interest and would have to respond to the pressures from the ground. The insensitive and anti-people industry hostile to environment and carrying out its activities in the name of development is challenged by the cause of environment and the health of the people. The ugly face of dominant pattern of development has been fully exposed.

The Maradia case has indirectly brought out the point that the real development has to be people-oriented and ecofriendly in order to be humane and legitimate.

Women and Ecomovement

The Doon Valley in the Himalayan foothills was once like a massive storage tank providing water throughout the year. However, in the name of development, Doon mountains were mined for chemical grade limestones. The forest range around was cutdown, debris was thrown down the slopes, which caused the destruction of the age-old natural system of water storage. All this disturbed the water cycle and had the streams and rivers dried up. In the wake of the mining of the rocks of limestones, actually nature's reservoir of water itself was mined.

In the Doon Valley, the rural women were put to much hardship. They had to go a long way everyday to fetch water. But they were living in the ethos of the Chipko movement, which had been going on in Garhwal region. The Doon Valley women replicated their own Chipko movement, by laying blockade in mining operations in their area on the banks of the stream Sinsyaru Khala, which was the lifeline of the village, but whose source had been mined for twenty years. Mining had killed the rich forest of *Banj, Chir, Ringal, Tun,* and *Sinsyaru.*

Women like Chamundevi and Iswairi Devi had seen the gradual devastation of the natural resources. Iswairi Devi guided the local Chipko movement. Such women who cohabited with the elements like the forest, the river, could not helplessly witness the nature's destruction in their everyday lives along with the steady erosion of sustenance which they drew from nature. Participation in nature gave them different kind of knowledge and power as opposed to the knowledge and power that use of modern technology for mining and dams, that go in the name of 'destructive development'. This is succinctly captured by Vandan Shiva who quotes Iswairidevi: "*Shakti* (strength) comes to us from these forests and green lands, we watch them grow, year in and year out through their internal *shakti* and we derive our strength from her... That is why 'primitive', 'backward' women who do not buy their needs from the market but produce them themselves are leading Chipko. Our power is nature's power, our *shakti* comes from *prakriti.* Our power against the contractor comes from these inner sources and is strengthened by his trying to oppress and bully us with his false power of money and muscle. We have offered ourselves, even at the cost of our lives for a peaceful protest to close the mine, to challenge and oppose the power that the government represents..."

In another incident, the quarry workers of the contractor had removed the Chipko blocked in the region and the trucks were climbing up the mountain to the limestone quarry, in November 1986. Once, Chamundevi, while collecting fodder in the forest, heard the climbing of the trucks, she raced down the slope and stood in the front of trucks. The truck driver dragged her along but had to turn back.

The lease of Nahilkala quarry mine had expired in 1982, but the government had not implemented its own laws for five years since then. The local people were protesting against the contractor who was interested in the mine who assaulted the peaceful protest but women along with men and children continued their protest. They initiated the protest and continued it with their own strength without the lead given by male leaders from outside for ten months. Iswairidevi and Chamundevi, the ordinary women, became the extraordinary leaders in the process. The women exhibited impressive staying power of Chipko—the movement which continued more than two decades of evolution, and assumed various forms of hugging tress, embracing living mountains, saving limestone mines and the like, and replicated in many forms!

On the World Environment Day, 1979 hundreds of Chipko women, mostly illiterates, met in Tehri with empty water pots to protest against the nagging hardship caused by water scarcity. They made the District Collector to hear that the failure of water supply schemes was a failure of engineering as symbolised by metal pipes and concrete tanks and the male engineers, who fitted pipes and designed schemes. In the simple but telling words, they stated: "Nature is the primary source of water, and we are the providers for our families. Unless the mountains are clothed with forests, the springs will not come alive, and unless the springs are alive, the taps will remain dry. It is the living springs, not the dry taps which feel our pots. If you want to solve out water problems, please plan for water not for pipes. Women thus proved to be better experts of water science. If water supply schemes are to succeed than nature cycle should not be disrupted by arrogant dependence on mere science. Engineering works cannot replace the work of the women and nature. The exclusion of people from water resource management leads to mal-development of water resources. The self-interest of India's ecosystem people is congruent for the most part with the good husbanding of natural resources, at least in their own localities. The real solution for the long-term health of the environment thus lies in passing effective political power to the people."

Concerns for Wildlife

A number of environmental and human rights activists have made efforts to bring together local communities and local government agencies in India.

In January-February 1995, a "Jungle Jivan Bachao Yatra" (Save Forest Life Journey) was taken up by 35 villagers, activists and researchers. Travelling through 18 national parks and sanctuaries in western and central India, they initiated a dialogue among local communities living in and around national parks and sanctuaries, and more importantly, between these communities and the forest department. They stressed the fact of rapidly depleting wildlife habitats, notwithstanding the existing wildlife protection legislation. They also explored various forms of conservation in which forest-dwellers have participated while maintaining security of their livelihood. They noted that on one side the local communities had strong feeling of alienation owing to curtailment of their traditional access to the forests, while at the same time, they were quite ready to participate in conservation and management of local natural resources.

They found out, however, the hypocritical nature of government policies with regard to wildlife habitats. Local forest-dwelling communities are denied traditional rights and access to forest resources in the name of wildlife conservation, while the same areas were being opened up to commercial uses and tourist taste of the elites. Observes Ashish Kothari: "New and vocal alliances which cut across the divide between conventional conservation and human rights consisting of villagers, conservations, concerned academics, social activists and forest officers are raising these concerns, throwing up answers and learning from each other. Their activism is helping to shift India's conservation strategy in a direction which gives hope for its wildlife and diverse cultural communities."

Extreme Leftist Conflicts

Naxalism, a new socio-political force, has intermittently been in violent action in the forests of Alidabad (north Andhra Pradesh) and adjoining districts of Maharashtra and Madhya Pradesh. However, the new Joint Forest Management Programme of MEF which attracted cooperation between the hostile forces of the tribals and forest officials in the task of regenerating the forests in Alidabad achieved some successes in the area.

The crux of the tribal conflict abetted by the Naxalites is the environmental and adverse economic consequences felt in the wake of tribal law and forests. The tribals were therefore struggling since more than half a decade to liberate them from poverty and from socio-economic oppression and political repression.

So also the Jharkhand (meaning the forest land) movement in southern Bihar has been closely linked to the loss of the local tribals' land and forest. The development policy of the central and state governments resulted in exploitation of its forest and mineral wealth but without any benefit to the Jharkandi tribals, e.g., the government policy of encouraging commercial *sag*-teak plantations (useful as building and furniture material) by replacing the indigenous *sal* by the Bihar State Forest Corporation resulted in the violent tribal struggle against such official policy. Their protest was graphically ringing in the popular slogan, *"Sal* for Jharkhand, *Saguwan* (teakwood) for Bihar".

The slogan succinctly expressed the tribals' understanding of the linkages between the economy and ecology of the region, a problem felt by the tribals and forest-dwellers elsewhere in India from Bihar to the Dediapada (in Gujarat where the tribals helped by the pro-tribal NGO ARCH-Vahini are carrying on a struggle against the state government and commercial pulp industry), from Bastar and Bailiadila to Baliapal. Their anger at the neglect of *sal,* their own best resources remind the Gujarat tribals' fight for right to have access to indigenous *Mahuva* tree (an exotic source of drink and medicine). "Yet, the movement was much more than a question of trees, and its economic and political focus through many years."

Environmentalists generally, however, have not been very comfortable with the 'tree war' carried on by Gonds and Jharkhandi tribals in Bihar. Unlike the Chipko, it has not received the kind of wide reportage on the Gandhian tactics related to the Chipko movement. Nor like the sustained multifaceted struggle against the Narmada project. The environmentalists, almost by reflex action, have supported movements against big dams like Narmada, Tehri Ghatprabha and Subarnarekha. But those supporting struggles against big development projects have been found to be cool in identifying with local struggles for political and structural change. This is because, most environmentalists are allergic to getting involved in party politics. Again, middle class urban environmentalists pay lip service to the tribal cause, but are uncomfortable with the armed tribal militancy guided by the extremist, Left (Marxist-Leninist) groups like the Naxals, People's War Group and Indian People's Front. Moreover, as Sumi Krishna comments, "environmentalists are reluctant to give up their wider out reach for causes which seem to be more narrowly political". Such a perception also explains why such environmentalists (of India and abroad) have rather evinced and supported the anti-Narmada movement almost instinctly, and consistently without having cared to see the data of ever improving resettlement process and environmental promotion actions which enable the Sardar Sarovar project, a development project with a clean break with the past experience. One thing is, however, clear. Most environmentalists, like many genuine development activists, do recognise that decisions related to development and environment, in process and essence, are *political* decisions. And, therefore, while devising their strategy from fast and *dharna* to mass mobilisation and public interest litigation, they relate their action in a way that should influence the political system, the policy (formulation-modification and implementation) process. Swami Agnivesh, Medha Patkar, Chunibhai Vaidya, Anil Patel, Ela Bhatt, B.D. Sharma and the like are the most audio-visual cases in the point; the most illustrative role model, perhaps, can be referred to Medha Patkar.

Chilika Movement

Chilika is a brackish water lagoon, the largest in Asia situated in Orissa, with length of 72 km and width of 25 km, covering approximately 1,000 sq km.

Chilika is the home of migrating birds of 158 kinds of species and commercially important species of the tiger prawn. Chilika is also a home as well as an earning source of 192 villages, dependent on fishing and particularly prawn fishing as their major source of livelihood. More than 50,000 fishermen and over two lakhs total population depend upon the Chilika for their living and livelihood.

Fishing has been the traditional occupation in the Chilika region for centuries. Fishermen have their right to fishing in the Chilika since Afghan rule. The right was protected even during the British who aided the fishermen by setting fishermen's cooperatives. Thus, Chilika has a rare and inspiring history of fish producers, cooperatives and village level democracy since early times. In post-independence India the situation continued.

Ecologically, Chilika is a stupendous collector of water and sunlight. The solar energy is converted by the process of photosynthesis by grass and phytoplankton into food for brackish water fish and grass for cattle, in turn providing food and livelihood to the thousands of traditional dependents of Chilika. This is a well-balanced sustainable, self-regenerating system needing no outside high cost and non-renewable inputs—no' imported' dollars or fossil fuel or US-European technology. It is a system developed by years of experience of living in the region by the poor people of Chilika.

The year 1977-78 saw the development of prawns as an important export commodity. The name Chilika became synonymous all over with prawns—and money. The whole area became a virtual gold mine. And, with this transformation, the commercial invaders started appearing. First, the traders and middleman, then the politicians with their musclemen, some big business families of Orissa through local middlemen and mafia, and finally, the big industrial houses with the blessings of the state government, which was now very keen to usher in 'development' to the region.

The Chilika struggle is a struggle of the poor against development'—a form of development which threatens to leave them destitute and exiles from the land which has supported them and coexisted with them for centuries.

In 1986 the then J.B. Pattnaik government decided to lease out 1,400 hectares of prime prawn area from the Chilika to a company to be formed jointly by the Tatas and the Orissa government. Then the resistance of fishermen was joined by the opposition political party—the Janata Dal. The resistance gave the Janata Dal leader Biju Pattnaik and the Janata Dal all the five assembly seats in the region. However, in 1989, with the Janata Dal coming to power the situation once again took a change. The Tatas once again made their appearance. This illustrates the power of the economic elites manipulating any political party and the duplicity of the political parties. In the beginning, Banke Bihari Das, a Congress youth leader gave a boost to this struggle. But afterwards, in order to gam prominence in his party, he tried to cause a cleavage in the rank of the protesters. It was alleged that elements close to the ruling Congress rased a small village in fire. Das lost the confidence of other leaders.

In 1991, the Janata Dal invited the Tatas to form a joint sector company for prawn development in the Chilika. For them it was correct to develop all export-oriented initiatives to augment foreign exchange reserves of the nation. It did not matter if it gave a death blow to the livelihood of 50,000 fishermen and dependency of two lakhs people, and to their traditional right. And also, it did not matter, if the beautiful, low cost, high efficiency, balanced ecosystem of the Chilika was being destroyed in the process for ever.

But the year 1991 also saw the birth of an intense struggle. The fisher people from 192 villages of Chilika united under the banner of "Matsyasibi Mahasangh" (Fishermen Association) to fight for their rights. They were joined in this struggle by a meet the students—a group of students from Utkal University. And it was this united struggle which on 15 January, 1992 at Gopinathpur village, saw the formation of a people's movement. The "Chilika Bachao Andolan" against the pattern of development which threatens peace, development, ecology and livelihood of the poor of Chilika, to protect"it from commercial exploitation by big business operating under political umbrella, and to restore to the people their right to manage Chilika.

Protesting against the livelihood versus dollars problem, the inhabitants of the surrounding 192 villages of Chilika area came to meet the Chief Minister Biju Pattnaik sometime in 1992. He assured the huge mass not to establish any prawn manufacturing company there against their livelihood. But politics entered into the matter and things changed. They waited few days for the actions from the government, but nothing was undertaken. So all the mass of the Chilika area with the active support of the "Krantidarshi Yuva Sangam" (about 35 students of Utkal University) started to break the dam which was built by the Tatas inside the Chilika. The leader of the movement Chittarangan Sarangi was arrested by the police. After agitation by the students, he was freed. Keeping this in view, the Government of Orissa thought over the issue and withdrew the lease and finally it was cancelled in December 1992. This was comparatively one of the shortest-time ecostruggles. But it will be noted for its success as a movement on the line of Gandhian principle of *satyagraha*—a militant non-violent way.

Narmada Project: Democratic Development

Narmada river project covering four major states of western India provides a most important case to study in terms of maturation of environmental movement and dynamics related to politics of development. No major project in India has brought the ecodevelopment problems to such a high level of informed debate, emotive rhetorics, political mobilisation, and grass roots activism as,this project has. The controversy around this project has engulfed horizontal and vertical levels of government and authority and helped forge linkages between NGOs operating at grass roots to national and international levels. It has also contributed to the political discourse on development alternative—in fact, becoming the ideo-political nursery for greening the politics of India.

Sardar Sarovar Project (SSP)—The inter-state multipurpose project with a terminal major dam in Gujarat, is being built on river Narmada—the fifth largest river in India. About 96 per cent of the unharnessed water of this 1,312 km long river flows down into the sea every day. Narmada Valley Project (NVP), with its two mega-projects, viz., SSP and NSP (Narmada Sagar Project in Madhya Pradesh), is the largest single river valley project, with the world's largest man-made lake, ever to be planned in India. In terms of canal network, it is the longest in the world.

The reservoir will submerge about 37,000 hectares of lands, of which about 11,000 hectares are classified as forests. It will displace about one lakh persons of 248 villages (19 of Gujarat, 36 of Maharashtra and 193 of Madhya Pradesh). The anti-project activists give more than double this figure as the project-affected persons (PAPs).

The Narmada project has struck an instant emotional chord. In Gujarat, being admittedly as one of the worst water-starved regions in India with a growing need of water for its domestic use, commercial, agriculture and burgeoning industry, its three-year drought of 1985-88 has reinforced this need of the districts of northern and Saurashtra parts of the state.

On the other hand, the Narmada project has been seen as "the world's worst man-made ecological disaster" by it critics. On financial, sociological and ecological grounds, SSP and other Narmada Projects have been strongly criticised as 'unviable'. The SSP (Narmada) project has a long and chequered career of more than forty years. Originally conceived only as an irrigation project of a 161-feet high dam, the technological potential of harnessing its waters made it a multipurpose dam with its level to be raised to 455 feet. And for financial and other help, the authorities began to look beyond the centre to the World Bank.

During 1980-87, the leader of the pro-tribal NGO—ARCH-Vahini—Anil Patel successfully fought against the Gujarat government in Gujarat High Court and the Supreme Court on the issue of right of the tribals/PAPs to just rehabilitation. Responding to the severe indictment of the Supreme Court and its directive to the Gujarat government (applicable to other states as well), the Chief Minister of Gujarat Amarsingh Choudhary came out with a liberal policy of Resettlement and Rehabilitation (R and R) (December 1987).

Earlier, the state governments of Maharashtra and Madhya Pradesh had evinced lack of political will to enforce the relevant legislation (respectively of 1976 and 1985) in this regard. Such socio-political constraints working in the state political system had to be removed by cooperation with the governments where they demonstrated response and competence with the passage of time as in case of Gujarat government/SS P authorities, or by persistent pressure when necessary. On the other hand, the pro-tribal Narmada Bachao Andolan (NBA) led by Medha Patkar has effectively exposed the lacuna and weaknesses in the design of SSP and implementing its R and R policy.

One important by-product from the viewpoint of democratic development has been NGOs' effective fight against the application of the Official Secrets Act in the vicinity of Kevadia colony and the dam site as enforced by Gujarat government (1988-89). This issue contributed to three major developments. It brought civil rights activists of the nation to have an intimate understanding of the problem of PAPs and the status of R and R. A new stream of NGOs joined the local activists working for and with PAPs. It exposed the Gujarat government's arbitrary attitude in this matter. And it underlined the importance of people's right to information which the authorities had to concede under media and popular pressure.

It must be noted that anti-dam lobby itself has also tried to black out from the tribals of Madhya Pradesh the information about the liberal Rand R policy of Gujarat and its implementation. A few pro-dam activists and government functionaries were threatened by the anti-dam group at Badwani that they will be drowned in the river if they again visit the villages to be submerged. In Badwani NBA's stall exhibiting the prospects of the Narmada project held out for the Madhya Pradesh and Gujarat was damaged by the miscreant anti-dammers in January 1991.

When around mid-1980s, the Indian State and the state governments were getting a bit relieved on the R and R front, a new challenge developed against the Narmada project. Environmentalists in India were becoming aware of India's fast declining environment. The Government of India passed the Forest Conservation Act in 1980 which stipulated that the state governments in the country were required to get the permission of the Central Government while taking up any project or action that involved the use of reserved forest land for purposes other than afforestation.

It is on all such grounds that the government planners of the Narmada project(s) and its supporters were branded as those whose thinking is a product of 'resource illiteracy' or 'ecological illiteracy'. In process the very approach of our dominant model is questioned.

The Environment Defence Fund and two other NGOs of USA represented against the Narmada project before the Senate subcommittee of the US Congress. The environmental groups there also tried to pressurise World Bank to stop financing this and such projects. It was followed by climactic exercise in political mobilisation back home at Harsud, Madhya Pradesh.

The Harsud Convention (September 1989) was attended by about 45,000 people including PAPs and activists belonging to about 200 NGOs. Thus, the Narmada issue brought many leading organisations on one national forum. Again, it was the first ever meet of public activists working in such varied fields as environment, oustees, tribals, social justice, civil rights movement, consumer movement, women's movement, anti-nuclear peace action workers, and anti-hi-tech Gandhians. The Harsud rally was actively and widely reported in national and foreign media. It was by far the biggest ever environmental protest of the PAPs as well as public-spirited interest groups led by leading activists in the country. It was not only damning of a dam, but a protest burst against what the Harsud Convention called 'destructive development'. Never was an environmental event in India received such a world-wide attention.

The Harsud Convention tried to bring in a macro perspective to their individual, single issue/local micro movements. Harsud, in that context, was a high watermark with international ramifications.

The whole phenomenon could be likened to the heat of the NGOs (pro-and anti-dam) generated from outside a pressure cooker of democratic system which in the process created the steam through the World Bank from outside the political system (the pressure cooker) to produce a wholesome policy package by the state government and project authorities. Constitution of the Independent Review Team under Bradford Morse by the World Bank and politics of pressure involving the international environmentalists and the World Bank are the recent developments illustrating this metaphor.

The pro-project argument, however, that cannot be convincingly countered, on point of facts or by offering a viable alternative by anti-dam lobby, is the stark reality that Gujarat is the water-starved area. In terms of all-India indicators of drought affected area, districts and population, Gujarat is far more unfortunate on this count. The Planning Commission study carried out by Y.K. Alagh classified the region in 13 agroclimate zones and persuasively argued that 75 per cent of the command area (SSP) in Gujarat is drought-prone; the entire command in Rajasthan is also drought-prone. Similarly, 54 per cent of the agriculture community (projected beneficiaries of irrigation) in the command area belongs to marginal and small farmers, most of them belonging , to OBCs. Minor irrigation works suggested as an alternative to the SSP cannot survive without recharge from adjacent major irrigation works. Even submergence by major irrigation works is hardly 5 per cent of the total forest disappearance in the country.

Again the medium and minor irrigation and traditional water conservation sources will not get filled in as the large part of the command where they are proposed are arid, drought-prone or noted for decreasing pattern of rainfall. Contrasted to systematically resettled PAPs with secure today and a better tomorrow, 20 million 'ecorerugees' of Gujarat's arid regions are troubled with a nagging migration syndrome. Their right to environment and water-induced development can be assured only by the SSP.

From time to time Medha Patkar and her NGOs have shifted their position, or changed their emphasis on issues related to the project. Earlier proper rehabilitation was the goal for which they were rightly agitating. After December 1987, "rehabilitation is *not* the goal but a means with which to beat the project itself (Anil Patel). They have also made environment as the major issue for sometime, while at another point of time they used it to beat the governments and the Sardar Sarovar Narmada Nigam (SSNN). They maintained that the whole task of R and R was a 'mission impossible'. They are not ready to candidly take note of the substantial improvement in R and R, in terms of physical amenities and participation of the PAPs, thanks to the groundwork and monitoring of the pro-dam activists and cooperation which has grown between them and the authorities.

All along the tactics of anti-dam NBA has remained negative and obstructive. Earlier, some of such NGOs argued that there never was enough land to spare, acquire or buy from private landowners. When such land did come by, this lobby began to propagate that the Gujarat government was "giving quite high price to capitalist landowners" or giving uncultivable land.

Thereafter, the anti-dammers changed their tone. Even if the SSNN will be able to rehabilitate the 4,500 PAPs from Gujarat, they would not be able to do so in case of PAPs from the neighbouring states. But, out of total 7,500 PAPs resettled by 1995, more than 2,375 PAPs from Madhya Pradesh and 656 from Maharashtra have resettled in Gujarat. Most of the PAPs are living in the Narmada basin in Madhya Pradesh region, where Baba Amte and Patkar have concentrated and given rigorous agitational programmes, and where "Hamara Gaon, Hamara Shasan" (our village, our rule) is vigorously forced making it quite difficult for government functionaries and pro-dam NGOs to enter. Again, the SSNN has welltaken the points of anti-dam lobby regarding the utility of the minor irrigation and the traditional way of water storage and integrated such methods in its overall plan of the project's water resources management. *But such schemes can supplement the major scheme; they cannot supplant the mega-dam altogether.*

The linkages developed between the grass roots organisations, state (Gujarat) government and (till 1993) the World Bank are quite significant. It is for the first time in India that such local-regional-international patterns of pressures as well as cooperation on both sides of the divide has been forged, and implicitly formalised for a development project. The battle line was thus clearly drawn and the wildfire of political communication blazed the controversy even outside India. For the first time a Chief Minister of Gujarat asserted about the development project of the state before the Prime Minister - and in public in - New Delhi (1990). May 1990 marked the high point of Gujarat's solidarity abroad, and in India's capital and Bombay where 1,00,000 strong Gujaratis rallied almost spontaneously in an unprecedented demonstration of support to the project. It was Gujarat's answer—mobilisation for mobilisation, activists like Chunibhai Vaidya with moral image against Baba Amte; Anil Patel versus Medha Patkar (1987-1995).

In December 1990, in their carefully prepared "Sangharsh Yatra", and by inviting foreign environmentalists, the Narmada Bachao Andolan proposed to make it "perhaps the most explosive event in the history of the movement". The pro-dam NGOs of Gujarat countered the movement by "Shanti Yatra". Farmers, tribals and a cross-section of informed citizens countered the NBA at Ferkuwa on the MP-Gujarat border. The one-month long 'face-off between both the sides was marked by massive rallies on Gujarat side. The anti-dam marchers had to withdraw from the Ferkuwa front, without gaining in substance. Before starting the Sangharsh Yatra, Baba and Patkar had claimed that the police and the government of Gujarat "will resort to force, repression and violences". During the whole one-month period of this face-off nothing of this sort had happened.

Certain parochially marked comments then on Baba Amte and Medha Patkar appeared in a section of Gujarati press and in some public statements made by pro-dam leaders. This showed lack of modesty and propriety. Tactics and strategies of Medha may be questionable, but not the high salience of commitment and intellect of this ideologue. The English press by and large played up, giving excessive weightage, to the anti-dam views and programmes. It provides a productive case study in misinformation and disinformation. Anyway, in terms of opinion building in India and abroad, Patkar and her allies have clearly put the SSP in bad light and its supporters on the defensive.

After setback at Ferkuwa the NBA cultivated Manibeli as the major front of PAPs resistance. Assiduously cultivated by Medha Patkar and her NBA, some villagers resisted their shifting by the authorities during the monsoon of 1991. "We will drown, but will not move" was their slogan. Meanwhile, the NBA got a tremendous boost and propaganda mileage out of the Independent Review Team (Morse) report which has come down heavily on the Narmada project and recommended World Bank to 'step back' from the SSP (June 1992). Subsequent developments led India to let go the World Bank loan(March 1993).

But in view of the visible progress of the SSP, money spent (Rs. 5,000 crore plus) and dire need of large masses of water-starved arid region of Gujarat, the people and the government of Gujarat did not and will not agree to abandon this project altogether. There are limits to environmental agitation in certain cases as in Gujarat case, and any talk of getting oneself drowned in the waters if the dam ever saw the light of the day will be a questionable action according to strict Gandhian norms of *satyagraha*.

For example, Patkar gave ultimatum to the government that if it did not concede to 'review' the project, NBA activists will resort to *jalsamarpan* on 6 August, 1993. On 6 August, the way Medha Patkar folded up her programme sounded like a denouement of a high profile drama on a development theme. In the process Patkar emerged more as a conventional strategist than an activist of international awards stature; or a genuine *Satyagrahi,* ready to die for a cause. More significant were the critical comments she has received from the Indian press, which had usually unilaterally given anti-project coverage and has played a crucial role in the build-up of her charisma in India and abroad.

Ironically, Patkar's moves resulted in forcing even the antagonistic forces of Gujarat to combine—which produced the unanimous resolution in the State Assembly to that effect. This is unprecedented in the legislative history of all the states of India. And not once, but thrice.

At the theoretical level also, this has shown that in a liberal democracy such as India, as there are limits to the state's power, there are also limits to dissent and protest by the citizens. A liberal democracy must be guarded by its vigilant citizens against the two extremes—authoritarianism of the state and anarchism of an individual or group even formed around a cause, or an important issue of public concern. The state should be sensitive to the citizen's right to dissent and protest and human rights in general. On the other side, the group with the air of self-righteousness should not wear the moral trappings of self-sacrifice like *jalsamarpan* when there is a strong counterpoint/reasoned popular position on the other side as well. Patkar's recent moves "crossed the limits of legitimate and radical protest, took a form of extremist posture and blurred the areas of political pressure and threshold of militant activism" as a critic then observed.

Both the sides must remain open for re-evaluation of the progress related to the project as well as one's own position, even ideological. Development with human face and equity will have to be accepted in this context. The Narmada project (particularly its Gujarat part) is a classic case of environmentalists versus the public, 'a doomsday scenario' versus the present-day compulsive needs.

Viewed as a whole, the SSP may turn out to be the best of its kind in India. *The democratic process of the political system has enabled the grass roots activists of both sides to insert the maximum corrective inputs in the 'revised' project plan through commendable intellectual home work and persuasive preservance and national-international networking.* They will do well to have institutionalised arrangement to monitor its implementation at every stage, in every area from R and R to environmental package. For the governments have the tendency to be arbitrary. The Indian State has a long history of having been manipulated by economic elite (like contractors and well-off cash crop farmers) and the bureaucratic/political corruption and lethargy in the proper implementation of a liberal policy are too evident to need any emphasis. The rabid opponents of SSP, like its supporters, will have to work for monitoring the process of the project. Again, even when the environmentalists and other anti-dam activists seem to be losing their ground in case of the SSP, in a way, they have gained. They have in the process generated public consciousness and official recognition that will prevent the governments' proverbial "going back on their promises" kind of the behaviour as in the past (Hirakund, Bhakra, Nagarjuna Sagar).

In the process, NBA has effectively questioned the prevailing dominant development paradigm. The conflict approach or sheer "no dam at any cost" of NBA carried on even with sophisticatedly couched statements and arrived at with laboured processes of argumentation is rather easier than adopting *conflict with cooperation* approach adopted by ARCH-Vahini, Anand Niketan and Shramik Vikas Sangathan. During 1995-96, Anil Patel, due to the apathetic state governments under Chief Ministers Chhabildas Mehta, Keshubhai Patel and Suresh Mehta, has withdrawn from his path of cooperation. Will the Chief Minister Shankersinh Vaghela (as he explicitly expressed before a group of activists including this author) reactivate the link between the government and NGOs?

Since mid-1994, Medha Patkar concentrated to cultivate her front in Madhya Pradesh. In November Patkar opened two levels— political and legal. Under the pressure generated by Patkar's three-weeks long fast in Bhopal, Digvijay Singh, Chief Minister of Madhya Pradesh conceded to her demands and stopped whatever action needed to expedite the resettlement process in the Narmada Valley. This also suited his policy of keeping the well-off farmers of the fertile Nimad region satisfied who did not want to shift if submergence of their land would take place. In order to save their land, he declared that the height of the Sardar Sarovar dam should be brought down from 455 feet (as stipulated in the Narmada Tribunal Award of 1979) to 436 feet.

With convergence of his interest with NBA's strategy, Singh took a position in the Supreme Court that helped NB As obstructionist tactics regarding stopping the construction of dam. In response to NBA's writ, the Supreme Court gave a stay against further construction of the dam in May 1995, which remain stagnant at 80.3 metre since then.

Meanwhile, due to political uncertainties in the post-Chimanbhai Patel regimes (1994-1996), Gujarat was in a disarray and stymied to push the progress of the project at political and legal levels in New Delhi.

However, the context changed with Deve Gowda becoming the Prime Minister of the United Front government. He called the meeting of the Chief Ministers of the four concerned states on 16 July, 1996. An agreement was reached in the meeting for implementing the R and R speedily and properly.

As a farmer, Deve Gowda is inclined to appreciate the need for big dams. Again, the new Chief Minister of Gujarat, Vaghela has shown agility and political skill to pick up the thread left since Chimanbhai Patel's death. Soon recharging a resonant wave length with the Prime Minister (November 1996), he has tried to reactivate the centre's pressure on Singh who has made Madhya Pradesh, the Achilles Heel for Gujarat. Gujarat anxiously waits to raise the height to 110 metres when the project can start delivering some of its benefits of drinking water, irrigation and power. Much depends on the role of Supreme Court as well.

Whether Gujarat can cross this hump, and make a clear break through for SSP by persuading Madhya Pradesh government for resolving the most difficult issue of R and R of its the remaining PAPs is a difficult question which can substantively contribute to the discourse of just and democratic ecodevelopment.

With whatever political uncertainties, faults or lacunae in the SSP, the dynamics, on the whole, have led to development. From the process of identification of their own land by PAPs themselves, the Land Purchase Committee represented by NGOs on their behalf, attempts to form villagers' cooperative to look after their affairs in new villages and the proposal to form the water users'/farmers' cooperative in the command area by Gujarat's authorities and NGOs are indicators of people's participation (some of them novel) which is a vital component of development. So also distribution of good land with assured irrigation and the whole package of inputs and thinking at the level of the Nigam's planners to devise mechanism for equitable distribution of water and its fair management is another component of development. One subtle aspect of this developmental dynamics, as noticed by perceptive activists, is the growth of awareness among the tribals about their rights, increase in their capacity to struggle against the state and develop confidence to meet the problems which the PAP-turned-resettlers have to face on certain points. The PAPs in Gadhed, Vadgam or Manibeli, in Ferakadi, Golagamdi, Ambawadi or Simalia of 1984 have 'politically' metamorphosed during 1990-95. With the support of pro-and anti-project NGOs, the resettled PAPs have demonstrated confidence and skill to fight the R and R authorities and resolve their problems. This is the immeasurable aspect of development. Again, growth of awareness and readiness of some NGOs and youth to take to afforestation and thus greening of the region as well as NGOs monitoring the SSP—all this constitute what can be called *development.* The SSP thus provides a ground of convergence between those advocating seemingly different development paradigms.

Now almost every aspect of this project has been fully debated and politicised at every level of political process. The SSP seeking to dam the river called *Abhayanvayi* has empowered the poor to be come *abhaya* (fearless). It should now become *Ubhayanvayi.* The politics of contention be replaced by the politics of dialogue that of confrontation by that of consensus focusing on developing a new centre in a place of misplaced polarities. The project has carved out its place in the history of development asking the people and the state in India to set an agenda, from grass roots perspective for broadening the concern and (dialogue, *praxis* and action for a 'new model of ecodevelopment.

TNC: Second Salt Satyagraha

Cargill is a giant transnational corporation, the world's largest grain trader and the largest private company in the United States. It is also the world's largest producer of malting barley, the largest oilseed processor, and the second largest producer of phosphate fertiliser.

Its influence on the US agricultural policy has been highly detrimental to farmers in both North America and the Third World and effects of its policy on the environment have also been severe

Its success as a dominant global company (like entering new product markets in many different localities) is because of its capacity for identifying key political actors and politically relevant business groups. It speaks of 'beachheads' for entering and penetration in areas of its business. Hybrid seed has proved particularly attractive as a 'beachhead product' because it requires virtually no capital investment.

It became visible in India in early 1990s. But it has been its spadework quite earlier since 1983, when it decided to enter seed business in India by setting up a joint venture company—Cargill Seeds India—with Tedco, a Tala subsidiary. In July 1988, the Government of India with a "New Policy on Seed Development" reduced the duty on imported seeds from 95 per cent to 15 per cent. Thereafter, Cargill became more active. It established its office in Bangalore and in early 1993, it began building a seed processing factory at Bellary in north Karnataka with a claim to develop modern agriculture, and with a schedule to start production in October 1993. To some agricultural scientists-academicians it was known how this giant company, with a combination of its resources, agro-technology, shrewd trade practices and 'current' political lobbying in the native countries, had captured the market and driven out farmers and small seed traders out of work.

The presence of Cargill in India in the context of the Uruguay round of the GATT negotiations (criticised by the forces of the Right and the Left in Indian politics and the Gandhians) prompted a significant section of the people to start a popular campaign against this TNC. In July 1993, local farmers led by the farmers' union gathered at the Cargill site and demolished the facility with their bare hands. They started the Seed Satyagraha, which Vandana Shiva called the "Second Salt Satyagraha" in India. Lakhs of people had gathered in one rally against the Cargill, which demonstrated the depth and expanse of the popular anger and the deep sense of distrust and insecurity among the farmers which the giant image and stories about its neo-colonial shadow over the Third World and Japan began to be known through well-organised campaign of the Raitu Sangham of the state led by a dynamic professor Neduchad Swamy.

The Cargill had also managed to purchase a large tract of land in Kutch to have its own jetties, and other infrastructure to produce and market industrial salt for which Kutch region was quite productive. However, resistance slowly built up against this project. Because mangroves would be affected, acqua life will be disturbed and small salt manufacturers and traders feared that they will be pushed out of the market once this giant TNC pitches in. Such a foreign presence in the strategic Indo-Pak border region was not advisable from the viewpoint India's security either. So *satyagrahas* were held in the Kundla region, one led by George Fernandes who as the minister in Janata Party government in 1977 had played then a crucial role in ousting Coca Cola from India. Gandhians like Chunibhai Vaidya held protest meetings in Gujarat, questioning the wisdom of granting land and other facilities by the Chief Minister Chimanbhai Patel and the local salt traders association of Kundla also approached the Gujarat High Court against the agreement between the State and the Cargill. The TNC tried to manipulate the Central Government (relevant ministry) and persuade the local traders for some sort of understanding so that the latter may relent. The TNC also let people know how it was set to win in this litigation. For a while, the local traders felt shaky in carrying their fight as a David against the Goliath. But the public opinion was becoming too hot for the Cargill bosses. Finally, they relented and packed off from Kandla.

It is significant that when Indian farmers protested in 1993 against Cargill's proposals to open this export-oriented salt facility in Kandla, Gujarat contributed to the denouement of the TNC and forced it to drop the project. It was Gujarat where Gandhi staged his Salt Satyagraha in the early 1930 against the British—a symbol that tantalised the whole nation against the mighty empire of a 'nation of traders'.

The National Alliance of People's Movement had brought out the significance of this struggle through its posture in which Gandhi was shown picking up salt (on the shore of Dandi village). Printed on the posture was: "Save our Sovereignty, Save our Seeds, Save our Salts".

And the call given was, "Boycott Cargill and other TNCs".

Brewster Kneen aptly observes: "Powerful though Cargill appears from its balancesheet and its political contacts, there are clearly many things that it cannot do. Cargill and other transnational corporations have the wealth, skill and political cover to outflank or overpower virtually any organisation that attacks its head-on in a game which is rigged in their favour. They cannot, however, force people-either farmers or the general public—to play their game".

The Indian farmers, in their numbers, have demonstrated a hard line of resistance to the Cargill. The consumers, who are conscious, have refused to eat highly processed food (like the Kentucky fried chicken) that has travelled from a centralised production facility and an increasing number of farmers have rejected the growing industrial monocultures.

"Around these old affirmations and new beginnings social movements and their allies are making common cause worldwide to lay the grounds for socially just and environmentally sound alternatives to the global production systems which Cargill exemplifies. New forms of social organisations are emerging which thrive on and generate diversity and inclusivity. It is hard to imagine a place for Cargill in such communities."

However, it is hard to nurse such a hope. In the present context of liberalisation and globalisation, the Indian experience with the Enron, Cogentrix, KFC, etc., illustrates that the MNCs/TNCs manage to step and stay put in India with the support of the governments in the states and at the Centre.

Positive Approach

Environmental movements are not confined only to conflict against the state, governmental departments and the commercial interests. They have developed positive aspects such as conservation and preservation of natural resources, e.g., afforestation and water resource management.

In Bhitardhari and Marangmali village in Bihar, the villagers had seen calamities like excessive exploitation of local forests, drought, or clashes with villages in neighbourhood. Deprived of the once green forest cover, the villagers had to plan to leave their villages.

However, they tried to save the situation by regeneration and protection of local forests. With the help of the forest department they undertook to plan saplings and raised them into trees. They provided water, throughout the year by small check dams, which were connected as a network of small canals. They also raised fish and ducks and ran a poultry farm and a bee breeding programme. Some villagers also raised a few cows which will provide milk.

The Oraon tribals of a small village near Ranchi lost their forest when it was declared as 'reserved' by the government in 1952. Having lost their traditional rights to their forest, the villagers began plundering the forest. But later on they decided to work through their panchayats, which could decide the needs of every family and their duties to guard the habitat. They chased away the contractors with bows and arrows and protected the forest, the major source of their sustenance, from commercial exploitation. The panchayat made rules for cutting a tree to meet the real need for extra wood. Fruit trees are protected and green grass can be harvested for personal use or grazing allowed. But the grass cannot be cut for the purpose of sale. They live a life of simplicity and community discipline which demands the minimum from nature. Thus, people's management of forest has brought them livelihood and peace.

In some villages in the valley of south Bihar, the villagers have responded to protect their forest in response to severe drought caused by depletion of forests. They have formed *Van Suraksha Samiti* (forest protection committee) to manage their forests. A sense of collective ownership had emerged with the involvement of the entire village community in this task. It is significant that many such villages can be identified in backward states of Bihar and Orissa, and they belong to Kurmi peasants, as well as tribals and other backward communities like Mahatos.

They have more commonsense and traditional knowledge to realise that disappearance or degradation of the green cover means natural calamity like scanty rain and drought. And afforestation is an effective answer to this challenge. Their forest management techniques based on their experience and intimate knowledge of the forest need to be integrated in the forest policy of the state. The new Union Minister of State for Environment—Nisadh—a tribal leader from Bihar—should know better.

Eco-Task Force

Like tribals on the forest and the marginal farmers in the interior of south Bihar, the ex-servicemen from the army also can constitute 'ecoforce' to fight the environmental degradation problem. The disciplined ex-serviceman were reemployed to constitute 'Eco-Task Force' which is involved in tree planting, conservation of soil through the construction of plug dams in the Himalayas. A task force also works for the stabilisation of sand dunes in the desert of Rajasthan. The costs of their maintenance are refunded to defense budget by Ministry of Environment & Forests.

A number of success stories like water conservation by building checkdams, watershed development, deepening of ponds and recharging of aquifers need to be recorded. For example, Raol Gaon of Nana Saheb Hazare, Salunke's Pani Panchayats in Maharashtra, Vivekanand Foundation in Kutch, Pani Chetna in Bihar, Lok Manch in Rajkot and the impressive local NGO initiative of latent underwater checkdams—a series of water conservation low-tech dams built by the Gujarat Lok Samiti led by Gandhian activist Chunibhai Vaidya along the river Banas in the arid Gujarat have proved to be path finders to guide the people to grapple with increasingly formidable problem of acute water shortage felt in different parts of India. Hazare, Salunke, Vaidya have shown how once arid or devegetated rural areas can be greened again, with increase in rural people's income through irrigated farming and nursing the impoverished villages back to health. The local people, after seeing the rewards of such a regeneration, have started giving support to such activists, for they have understood clearly that such initiatives means development of the local community itself.

Environmental 'Vigilantism'

A counterpoint to such 'mainstream' view is examined here. The advancement of the human lot has never come on the wings of inevitability, says Charles Birch, a Professor of Environmental Biology in Sydney. It has always depended on human choice and public policy.

Since all the world's nations have economic growth as their goal they come up against what Birch calls the 'Impossibility Theorem'. It states that if all the world's people used resources and polluted the environment at the rate of the developed world we would run out of resources and destroy the life support systems of this planet within the foreseeable future. This view is countered by an alternative ecological model that has, as its goal, "healthy people with healthy relations to a healthy environment. While it is not possible to retreat from an unknown future based on science and technology, it may be possible to choose between alternative technologies":

1. Developed countries have woken up to the fact that environmental care is necessary only after their own economic prosperity has been assured at the expense of that very same environment.
2. If reaching the goal and staying at it can be achieved by alternative technologies, why were they not adopted earlier? Why should less developed countries with less resources adopt this more expensive and untried route to prosperity ?

3. And this is something which few people are willing to talk about despite the fact that there is increasing awareness about it. Are environmentalist activist groups like Greenpeace developing a vested interest in the environmentalist cause?

There has been no question about the fact that pollution is caused by the inefficient use of energy and raw material and that control of pollution counts on steps in two directions: one, in increasing the efficiency of the use of raw material including sources of energy; and, two, active pollution prevention measures. In the long run it also makes sound economic sense.

Much of the pollution is caused by the burgeoning population and the increasing demand for goods and services as a result of rising standards of living.

The environmentalist movement has spawned special interest groups of lawyers, waste management specialists, management consultants and various non-governmental bodies who do well out of these new pressures. Much more than in the West such special interest groups like to use any muddle-headed do-gooding environmental activist who offers to lend his name to anything remotely connected with the environmentalist cause, his sole qualification being his crusading zeal.

Industrial pollution on a large scale, if and when it occurs, whether it is the Bhopal gas disaster or the effect of the Mathura Refinery on the Taj, has invariably taken place because of the use of obsolescent rather primitive plant and machinery. The government is aware of this problem and has, therefore, introduced stiff penalties for infringement of the new laws which it has passed to prevent pollution. It has also insisted that large industrial units should take active preventive measures to minimise, if not eliminate altogether, the effects of the pollutants which come off as an inevitable result of the production process. In the setting up of new industrial units it is now mandatory to make pollution prevention and treatment a part of the basic infrastructure of the plant and machinery. The thinking that has gone into this stance on the part of the government owes its inspiration to the efforts of environment protection agencies in various countries.

With liberalisation a large number of big industrial houses in India and multinationals investing in this country are importing the latest plant and machinery to set up units all over the country. This new equipment already meets the specifications of pollution control insisted upon in the countries from which they originate, which is much more stringent than what the Government of India mandates. In fact, in most cases, equipment that does not meet such specifications is no longer manufactured in the countries of origin. Moreover, the new units have to strictly follow the norms of environmental protection laid down, in the immediate vicinity of the plant.

While, therefore, "vigilance to prevent industrial pollution is absolutely necessary, mindless vigilantism by environmentalist crusaders and special interest groups is both infructuous and unnecessary, particularly when the economic health of entire state and huge districts is involved." Such a counterpoint, however, reflects a corporate culture. Our profile of the role of industries, large and small, clearly belie the claims and expectations of this alternative ecological model.

Uninterrupted Revolution

The Chipko movement in Garhwal district of the Himalayan region has become a classic example for its multifaceted, long term implications. Started in 1973, it continued for more than two decades assuming different forms and adding new content in the hill people's struggle against the establishment, challenge to the dominant paradigm of development and in the process empowering the once government-fearing *girijans* and women.

Initially started as a struggle to save the Garhwal forests from commercial exploitation by contractors in league with the forest functionaries, the Chipko movement developed the demands of the management of the forests and other local resources by the local community. From Sunderlal Bahuguna's emphasis on the conservation of nature to Chandi Prasad Bhatt's emphasis on management of local resources by the *girijans* or indigenous people themselves; and then gradually it evolved as the demand for ownership of natural resources by the local people and their use for the local people. Unlike the West, where the middle class dominated the environmental movement and defined ecopolitics, in India the Chipko initiated the women, *girijans* and the poor in it. It was these marginalised communities who participated in the movement, developed it as a major ecological movement and women particularly emerged as the key actors in the evolving development process. As a trendsetter and role model of such movements as Appiko, Bedthi, Kharikot, Vishnuprayag, Tehri and Narmada, Chipko became the catalyst for the development discourse, and an alternative paradigm of participative (democratic) development.

Gandhian Roots and Route

Significantly, such a major movement had its roots in the ideas and work of Gandhian centres in the Himalayan area where such leaders as Sarladevi, Bahuguna and Radhadevi had prepared workers for respecting nature and cultivate a harmonious relationship between man and nature. Sarvodaya leaders like Bahuguna and Bhatt profiled the process of exploitation of nature for commercial and consumerist use of the urban well-offs, and questioned the dominant development paradigm which met a modern, urban industrial civilisation.

The Chipko inspired environmental movements in other areas and itself became an epic in the history of grass roots ecological struggles in India. As Harsh Sethi has observed: "Spread over two decades, thousands of hill people kept up their non-violent resistance movement, almost without a unified leadership and in the absence of a centralised organisation." During it long course, it formed a new language of public communication and new patterns of ecopolitical mobilisation in which women emerged as the key actors, with a sense of confidence to manage the movement and even their households in the absence of their men, who, deprived of a sustainable ecology nearer home, had gone away to far away cities to earn their living. Also with Chipko, other movements like Appiko, Silent Valley, Subarnarekha, Doon Valley, Tehri and Narmada changed the very terms of political discourse in India.

Struggles over land use (say, against limestone quarrying in Kharikot and near Dehradun), in the Western Ghats, Central India and Gandhmardhan in Orissa, Singrauli in Madhya Pradesh, Appiko, Bedthi and Sea Bird Project in Karnataka, five-star entertainment centres in Goa or Konkan Railway and the like have articulated the problems of the marginal agrarian and rural people, faulty irrigation and tourism and fisheries projects, schemes causing waterlogging industries and mining and generally callous approach to the problems of the displaced. Studies of socio-ecological costs of Green Revolution have helped in the formulation of the ecological critique of agricultural modernisation. As most of such projects affected the common mass including women, tribals, forest—hill dwellers or those who draw their sustenance from natural resources like the forests, lakes and public land or common resources property, it were these vast mass of the people who got agitated and mobilised. Such struggles and ecopolitics that was shaped around them, therefore, acquired a large mass base, which the political elite in a representative-democratic system could not ignore.

The struggles against big dams, river projects, forest management and so-called land development have created major conflicts and generated ecological debate in India. Bedthi, Ichampalli and Narmada projects in the west; Tehri and Vishnuprayag (near the Valley of Flowers) in the north; Thungbhadra and Ghataprabha in the south; and Gandak, and Koel Karo schemes in the east are major cases in point. Thus, from the Silent Valley in the south to the Valley of Flowers in the north; from anti-Kakrapar atomic plant in the west to the Chilika lake in the east—the anti-big dam or other environmental movements have generated all the issues related to justice to the project-affected people, the design of the dam and all environmental problems like the destruction of the forest in the catchment area, sanctuaries, damage to wildlife, rare species, and biodiversity, etc. The policy of linkage with foreign agencies and networking with international NGOs working for environment and human rights and humanitarian concerns like OXFAM, Friends of the Earth, Environment Defense Fund and the like have brought pressures on the Indian State and states in India, and providing inputs to the ongoing discourse on democratic development.

In the larger perspective, political philosophy has developed from grass roots questions and in the course of responding to the problems of the people residing in the valleys of the Narmada, Tehri and Subarnarekha and forests of Kerala, Madhya Pradesh, Bihar and the Himalayas. The issue-based struggles have, during the last few years, broadened to the questioning of the model of development. Distributive aspects of development in the contemporary dominant model are also getting into focus. In the process ecopolitics has shaped itself, asking for redefining or expanding the conceptions of rights of different strata in the domain of political economy (like who gets, what and how in the share, control and use of natural resources ?) for environmental correctives for better and informed resource use.

Recently, cases, not of people's conflicts but of their own initiative to become self-reliant, have made out the case for "participant, decentralised development". Case studies of local initiatives for development in the Asia Pacific region (India, China, Sri Lanka, South-East Asia, etc.) underline this point. In the whole gamut of development history, the reassuring trend is that organisational and processual components of people's participation have been crucial factor in their success in fighting the authority at various levels of the political system as illustrated in the two volumes on *Building from Below.* In most cases of empirical studies in these countries of Asia, the poor and the disadvantaged have taken significant initiative either on their own or when motivated by a catalytic agent from outside in organising themselves; identifying their specific development needs, formulating plans and liasing with relevant authority structures. As the volumes, *Building from Below* demonstrate, such an approach to development is quite contrary to the conventional view of 'decentralisation', which was perceived in terms of top-down transfer of authority within a governmental framework. As Anil Bhatt shows, decentralisation occurs more effectively when people at grass roots become conscious enough to organise themselves, to fulfil their development needs, and succeed in mobilising their own resources and effectively extracting supplementary resources from outside. In such a turning around of the traditional theory of "democratic, decentralised development" is implied the process of *empowerment* and development of self-reliant communities, which ultimately makes the whole process an ecofriendly and people-centred development. Development, as it emerges currently, is a process of people's participation, building up of a self-reliant, self-confidant community, a people-centred process and all this through "learning from the people" and guided by environmental prudence.

The revival of Panchayati Raj in India also envisions increased local initiative and participation including conservation of local environmental resources. But, in practice, it is difficult to come by.

Environment movements in India have developed mainly with the issue of equity in the focus. The movements have largely taken the route of conflicts between the omnivores who have gained disproportionality from the development policy and projects and the large ecosystem people whose livelihood have been seriously compromised because of the use of national resources, biased against the latter affected by environmental degradation.

Gadgil and Guha give four broad strands within the environmental movements:

The one, which emphasises the moral imperative of restraining overuse, and doing justice to the poor. Mainly Gandhians belong to this strand.

The second, which emphasises the need to dismantle the unjust social order, through struggle. Marxists follow this strand.

The third and fourth strands emphasise the reconstruction, employing technologies appropriate to the context and the times. These strands arise out of the concerns of scientists or spontaneous village level efforts of the community to protect and sustainably use local wood lots or ponds or the right to pursue environment-friendly agricultural practices.

Unlike in the West, the environmental movements in India have mainly involved the women, the poor and the disadvantaged masses, who have been the victims of environment degradation. Thus, these movements have been political expression of the struggle of the victims of environmental degradation or abuse of resources through the organisation of social action group.

ECODYNAMICS OF SYSTEM OPERATION

A political system is a persistent pattern of interaction between its interrelated structures and processes through which values are allocated authoritatively. The Indian political system has provided a framework of interaction between its formal institutions like parliament, governments and courts at different levels of the federation and also between such structures like interest groups like industry and environmental NGOs and local communities. They have interacted through such processes like pressures, persuasions, mobilisation and public interest litigation. In the process operational lvalues like environment and development have been decided upon and sought to be binding on its components like industries, governmental agencies, citizens, etc. Such an interaction between its structures has taken patterns of cooperation as between the industry/forest department and governments, or conflict between forest department/industry and the local/tribal communities. In both cases, it is adjudicated by the judiciary and sometimes, the environmental conflict is mediated by the government (its agencies).

The Indian political system has also to manage the larger economic environmental constraints set by external forces like multinational corporations (MNCs) or foreign corporate power or within by forces of economy like consideration of material production and employment generation offered by new industries (even though polluting) and development projects like the big dams or huge construction programmes. The media also plays a supportive role in favour of the local forest communities in rural India or against pollutant industries in urban India.

During fifty years of India's independence, the value of environment has been largely overshadowed by the value of dominant model of development. However, during the last decade, the NGOs, local communities and ecoadvocacy groups, through conflict against the state and through intervention of the courts, have sought to change this equation in favour of environment. External factors like international environmental awareness have also created a new psycho-environment to put environmental prudence in the priority list of the concerns of India's political system. Some of the vignettes of such dynamics of interaction are narrated and discussed in this chapter.

Colonial Period

Apart from forest laws, legislation under colonial India in the nineteenth century also partially regulated two other aspects of India's environment—water pollution and wildlife. These laws, however, had a narrow purpose with a limited territorial reach.

The Indian Penal Code, (1860) imposed a fine on a person who voluntarily "fouls water of any public spring or reservoir". The Indian Easements Act (1882) protected riparian owners against 'unreasonable' pollution by upstream users. The Indian Fisheries Act (1897) penalised the killing offish by poisoning water as a result of using explosives. Legislative provisions regulating the discharge of oil into port waters (Indian Port Act of 1908) and prohibiting the poisoning of water in the forest (Indian Forest Act of 1927) were also enacted in the pre-independence regime.

The Forest Act (1948) required all factories to make 'effective arrangements' for disposal of wastes and empowered state governments to frame rules for implementing this directive. In pre-independence period, river boards established under the River Boards Act for regulation and development of inter-state rivers and river valleys were empowered to prevent water pollution. For both these pre-independence laws, prevention of water pollution was only incidental to the principal objective of the enactment.

The earliest laws aimed at controlling air pollution were the Bengal Smoke Nuisance Act (1905) and the Bombay Smoke Nuisance Act (1912). In the field of wildlife protection, early legislation was limited to specific areas and particular species, like preservation of elephants. The same aimed at regulating of hunting and did not regulate trade in wildlife and wildlife products. This resulted in the decline of Indian wildlife. In 1936, the Hailey National Park Act was enacted under which (Hailey Chow Jim Corbett) National Park was established in Uttar Pradesh.

Thus, "the early legislative efforts were piecemeal and inadequate. Not until the 1970s did the Central Government begin enacting comprehensive environmental laws."

It is since early 1970s that we see "the transformation of government policy, from environmental *indifference* to environmental *concern,* which has guided India into an era of comprehensive environmental legislation".

Controversy Around Enviro-Legislation

The growth of environmental awareness caused by deforestation, its adverse effect on the plight of the forest dwellers and escalating social conflict over forest resources have prompted the Forest Departments, both at the central and state levels, to arrest deterioration on this front. Conflicts between villagers and industry (Bedthi in Karnataka and Dediapada in Gujarat) and between the tribals and the Forest Department have come to occupy more space in the press and exposure of the nexus between the forest officers, industry and contractors has also created public opinion against such a development. They have forced the Forest Departments from both within the parliament and the government and outside from the environmental NGOs and the press to monitor the legislation. The government had to listen to a barrage of criticism for its narrow, commercially oriented policy. It was hard put to answer for the neglect of environmental safeguards and social needs which were compromised by the government policy allowing maximisation of the output of the commercially profitable species of trees like timber.

It was in this context that the Union Government responded positively to several points of criticism. It introduced a new policy in December 1988, which has "an explicitly ecological thrust." It also rejected two central planks of earlier policies, viz., the emphasis on revenue generation and industrial wood production from natural forests respectively. But even such a policy is far from satisfactory. It has, in the process, been subject to widespread criticism and protest from forest dwellers and environmentalists.

Under the new phase of privatisation, the former Environment & Forest Minister, Kamal Nath had (August 1994) proposed to lease out two million hectares of deforested land to the corporate sector. Nath had claimed that this would attract funds to expand forest cover and also meet the increasing threat of encroachment on forest land.

But continuous pressure of tribals and supportive NGOs on Ministry of Environment & Forests (MEF) in the political system has finally worked. This is borne out in the statement of the Minister for Environment & Forest (himself a tribal), Jai Narain Prasad Nishad in the United Front government in September 1996. He made it clear that the Union Government had no intention of reviving his predecessor's plan to lease degraded forest lands to industry. He said: "The local people who depend on forest for their livelihood should be the ones to benefit from this activity, not industry. This can be possible only if the land is given to these people themselves." In a bid to give the ongoing afforestation schemes a more 'human face' the Union Environment & Forest Ministry has decided to form 'forest cooperative societies' all over the country.

Thus, the MEF will make 'an all out effort' to form co-operatives in all such places. Degraded forest lands will be leased out to these societies. Nishad claimed, "efforts will be made to involve each and every individual of the local communities in the scheme."

Nishad also assured that the government would welcome funds for afforestation from public sector banks, the National Bank for Agriculture and Rural Development (NABARD) and corporate sector.

The MEF officials, however, are undecided about the execution of the plan. They have sought the help of the Chairman of the National Dairy Development Board, Dr. Verghese Kurien, the world renowned cooperative wizard in this matter. But in the face of strong opposition from the wildlife enthusiasts and environmentalists the proposal was given a temporary burial which now seems to have been made permanent.

Plans and Enviro-Development

The year 1972, the year of UN Conference on Human Environment in which Prime Minister Indira Gandhi participated, also marks a watershed in the history of environmental policy and management in India. Prior to 1972 environmental concerns such as sewage disposal, sanitation and public health were dealt with by different ministries of the Government of India. Each department/ministry pursued these objectives without any adequate coordination system established at the federal or inter-governmental level.

It was in response to the UN Conference on Human Environment that India set up a National Committee on Environmental Planning and Coordination (NCEPC) in the Department of Science and Technology. The first committee under Pitamber Pant enjoyed some political clout, but it waned in the subsequent committees consequently, when the willing cooperation of other relevant departments in Government of India began to wear. This was because the different departments started viewing this committee as a competitor rather than a planner and coordinator.

It was in the Fourth Plan (1969-74) that for the first time concern for preserving the quality of life and promoting the environment, while promoting development, was emphasised. The plan recognised the necessity to introduce the environmental aspect into planning and development. The Fifth Plan (1974-79) and the Sixth Plan (1980-85) turned such a concern into concrete actions by launching several programmes which were to enhance the quality of life, at the same time incorporating environmental norms while assessing economic and technical feasibility of a development project. For example, during 1979-80, thirty hydroelectric and irrigation projects and fifteen thermal power generation schemes were appraised from the environmental point and, as a result, appropriate safeguards were recommended.

Soon after the come back of Indira Gandhi to power in 1980, the Congress (I) government appointed a committee, with N.D. Tiwari as chairperson, to recommend legislative measures and administrative machinery for ensuring environmental protection. The Tiwari Committee completed its task fairly well in time (in September 1980) and many made far-reaching recommendations like creation of a Department of Environment at the centre that could explicitly recognise the pivotal role of environmental conservation in sustainable national development. The same was upgraded as a Ministry of Environment & Forests (MEF) under Rajiv Gandhi's regime.

The working of MEF has witnessed to some interesting political decision patterns. It has been continuously visited from outside by the business-industrial interests on one side and environmentalists/human rights activists on the other. Within the governmental system also, the nature of MEF's powers making it deal with other concerned ministries/departments of the Union Government like power, industry, agriculture, commerce made them uncomfortable and take umbrage at the MEF "imposing environmental conditions" on the projects or cases which come under their jurisdiction. Cases of friction between such departments and MEF have been reported, some of which have also required the Prime Minister's mediation.

In early 1980s, Digvijay Singh was the Deputy Minister in the then Department of Environment under the Prime Minister Indira Gandhi. In the case of the Silent Valley controversy, when the matter was referred to Indira Gandhi, she gave her decision in favour of anti-project environmentalists to the chagrin of the left democratic government of Kerala. When the UP government planned the huge Vishnuprayag

Dam Project, adjoining the Valley of Flowers in Garhwal region, aimed at meeting all the needs of the northern states of for power the environmentalists organised a well-coordinated protest against the project in early 1980s. The pro-project lobby argued that Rs. 10 crores had already been spent after the infrastructure needed for the execution of this project. The environmentalists lobbied with the Prime Minister on suggestion of Digvijay Singh. Indira Gandhi again conceded to the environmentalists' demand. Even as the UP government of was displeased, this project was stalled.

The subject of environment has given enough scope to the centre to rein in the states which require the sanction of the centre, particularly of the MEF, for their projects. The states have to get prior environmental clearance before allowing a new industry to come up in their region. Certain types of industries have to get 'no objection certificate' from the state government concerned and environmental clearance from the MEF. A 1994 notification made environmental impact assessment (EIA) statutory for 29 sectors such as industries, irrigation, power, transport, tourism, etc. Again, each of these ministers are further linked with ministries of commerce, energy finance and others. Therefore, it is the Prime Minister who, in such a situation, finally decides on all contentious issues like Enron, Cogentrix, or river projects like Cauveri, Upper Krishna and Tehri. The Prime Minister's intervention in Narmada project proved crucial at some points of its continuing controversy. Thus, almost all matters that the MEF deals with are the inter-ministerial. The status of the occupant of MEF is only that of a Minister of State. All this results in curtailing the powers of the MEF meant to protect environmental interests of the society. When Maneka Gandhi was the Minister of State for Environment & Forest under Nilamani Routray, the Cabinet Minister for Environment & Forest in V.P. Singh's government at the centre, she consciously used this power. She blocked the proposals for starting industries which were likely to cause considerable pollution. Herself an informed environmentalist, she subjected many an ecohostile industries of powerful corporate leaders or prestigious projects of some Chief Ministers to rigorous check up by expert committee in her ministry. The powerful industrial lobby sometimes supported by the governments of the concerned states lobbied with Routray and sometimes with the Prime Minister Singh himself.

The powerful MEF Secretary Prasad also sometimes handled the files of such controversial projects or proposals in such a way that Maneka Gandhi was embarrassed. Prasad could not have behaved in this way unless he was supported or asked to play such a role by Nilamony Routray. The cases of certain industries, and review of projects like Tehri and Narmada became subject of friction between Maneka Gandhi and Prasad, or Maneka Gandhi and Routray. In fact, Maneka Gandhi's attempts to become an ecoactivist minister drew flak from other ministries, leading to induction of Routray at cabinet level in MEF to keep a check on her. Routray succinctly expressed this when he said: "Maneka can look after the Delhi zoo, and I will look after the environment." Sometimes such a matter was publicised making Maneka Gandhi feel that the matter was tactfully leaked out to the press with a view to embarrassing her. Her senior, Routray, as the cabinet minister of MEF, sometimes tried to skip her and directly handle such files when she was found to sit on them. When she approached the Prime Minister Singh, he was generally found cool to her or asked her to meet Nilamony Routray and sort out the differences. Narmada and Tehri projects and some industrial projects continued in such a context, even when Maneka Gandhi wanted them to be delayed till conditionalities were met by them. (Maneka Gandhi had remained present at the anti-Narmada project rally at Harsud in September 1989. She was, however, politely asked by the organisers of the rally not to be on the dias, as she was affiliated to a political party, Janata Dal). Maneka Gandhi had promoted the idea of having environmental assessment committee comprising independent experts, etc. She had also fervently tried to establish environmental courts for monitoring and speedy disposal of cases against polluting industries and controversial cases in this field. However, the Prime Minister's role was not helpful for this crusading minister. While striving for setting strong environmental norms and policy, she found herself isolated—and sometimes ineffective—before the combination of her senior minister and the strong bureaucrats supported by Routray. For example, in May 1990, when she tried to support Baba Amte and Medha Patkar seeking to stop the Sardar Sarovar project and put the matter through with V.P. Singh, she could not do much as the Prime Minister condescendingly inclined later on towards the pro-project group from Gujarat (NGOs and the Chief Minister Chimanbhai Patel) asking for clearing the project.

On continuous pressure of the powerful anti-Narmada dam lobby, the ministers like Maneka Gandhi and thereafter Kamal Nath, as the head of MEF and V.C. Shukla as the Cabinet Minister of Water Resources, insisted on full compliance of the terms of environmental clearance and favourably responded to Patkar's demands. The pro-dam forces were equally determined to pressurise Prime Ministers V.P. Singh, Chandrashekhar and Narsimha Rao not to give in to anti-dam pressures. The same scenario was repeated in case of Deve Gowda (July-August 1996). Thus, the MEF and the Prime Minister became the venue of conflict of interests and an arena of cross-pressures. The attitude of the political decision makers in this case depended upon the prevailing equation between the Prime Minister and the Chief Minister of the state; and the changing equation between the project-and anti-project lobbies and the ministers concerned. The policy, therefore, kept on shifting, sometimes in a subtle way and such very policy shifts (if not change) and changing equations encouraged lobbies to make the MEF and Prime Minister's office the arena of politics of environment and development. In fact, the commitment of the JD government under V.P. Singh to environment got diluted because of intra-governmental wranglings. The case of another JD Prime Minister is rather disappointing. When Deve Gowda was the Chief Minister of Karnataka, bitter controversy was raging regarding Co-gentrix power project. When he was interviewed by Supriya Akarkar (Centre for Science and Environment), Gowda stated: "Often, people are being misled by environmentalists who tend to be anti-development" He then advised her that a young lady like her should "not be led by such people". In September 1996, the Deve Gowda announced that all fast track projects of less than Rs. 1,000 crores will henceforth be cleared (not by the Centre) but by state governments.

Deve Gowda's government has also belied the UF government's claim of keeping environmental sensitivity. An illustration may be given on this point. Most of the mining leases were given before Valmiki Wildlife Sanctuary in west Champaran in Bihar was notified in 1990. Earlier, in 1989, a mining lease on over 20 hectares, of forest land was given to the rich and influential businessman, Vinay Verma. The lease was valid till June 1994.

In April 1991, the Regional Chief Conservator of Forests, Bhubaneshwar, who was monitoring the leases reported to the ministry that in violation of the agreement, Verma was digging a metre deep into the riverbed and collecting stones and pebbles. The state government canceled the lease at the direction of the centre. Then, Vinay Verma appealed to the High Court and was allowed to continue extraction till the end of original lease period in 1994.

R.P. Verma and R.B.M. Sharma also had mining leases in Valmiki. In July 1996 R.P. Verma's 10-year lease of 1994 (which officials of the ministry now admit should not have been given for it) for a notified areas was canceled for various violations.

The I.G. Forests has also sought the cancellation of the 10-year lease given to R.B.M. Sharma in 1994 for collection of pebbles and stones from the riverbed inside the sanctuary. There were several reports with the ministry on Sharma's violation of the lease terms.

On 25 June, soon after taking over as Environment Minister, Jai Narain Prasad Nishad himself had endorsed the cancellation of lease to R.P. Verma's 42 hectares of forest land in Pandal riverbed for collection of floats, pebbles, etc. While endorsing cancellation, he said information should be speedily collected in respect of all other leases also and similar action should be taken if there has been any violation. "We should not wait for third parties to file complaints but should have a strong system of monitoring."

However, after a letter to him from Bhola Ram Toofani, Minister for Animal Husbandry and Fisheries in Bihar, Nishad has had a 'change of heart'. On the other hand, both the I.G., Forests and the Additional I.G., Wildlife have stood by their stand that leases should not be renewed.

Angry about wildlife and forest officials in his ministry expressing their inability to renew mining leases right inside the Valmiki Wildlife Sanctuary and Tiger Reserve in his home state, Nishad first turned down the trip of the Additional I.G., Wildlife to Geneva for an IUCN meeting, then shuffled the inconvenient pack of forest officials. Choubey, who had made a strong defence against renewing the mining leases to R.P. Verma, Vinay Verma and other influential people of the state, has gone on protest leave.

With several cases against the MEF in the courts, officials have now become cautious about mining leases in protected areas. S.C. Dey, the Additional I.G. for Wildlife, had in fact, filed an affidavit before the courts "strongly endorsing the suggestion to treat national parks and sanctuaries as 'no development zones'."

Environmental Management in the States

Following the recommendations of the NCEPC, most of the states and the union territories have constituted environmental boards on the pattern of the national committee. With the Chief Minister as the chairman, these boards have attained some stature and authority. The state level Department of Environment is represented on the board. The board is to perform the following functions:

1. To be the focal point for initiating, reviewing, and monitoring all environmental matters in the respective states.
2. To undertake environmental education and create awareness.
3. To appraise projects at an early stage.
4. To site industries and incorporate environmental safeguards of development projects.
5. To serve as a liaison with the Department of Science & Technology, and NCEPC at the centre and other government departments, and with research and educational institutions, and industries in respective states.

The lack of technical expertise at the state level and the inability of state governments to impart an integrated treatment to environmental problems have made many of the state boards ineffective or non-starters.

The state governments have been found wanting in providing sufficient staff and infrastructure to these boards. As a result, the states have been found wanting in monitoring environmental problems and their timely solution. Again, most of the Chief Ministers have not found environmental problems that pressing as other matters. "At present, only the State Boards of Karnataka, Tamilnadu, Kerala,

Gujarat, Maharashtra, Andhra Pradesh and Uttar Pradesh are involved in some important activities."

MNC and Indian State

It is therefore left to the informed citizen groups and environmental NGOs to prompt these boards and expose or fight against the erring industries and activities causing environmental degradation. More than such boards, it is the citizen initiative which has mostly moved the court to seek its intervention in such cases when the projects are sanctioned by the government without taking care of their environmental impact. Again, even in ghastly cases of environmental hazards, the governments are found insensitive and willing for just actions. One case may be briefly mentioned here. The Bhopal gas tragedy which had caused instant death of hundreds of people and agonizing diseases to thousands owing to the negligence of the Union Carbide, an MNC, which occurred in December 1984. However, more than a decade has lapsed and the problem is not resolved in a just and fair manner to the satisfaction of the victims. Thus, in spite of protests from human rights and environmental NGOs at local, national and international levels, judicial intervention and media reports critical of the insensitive MNC and a fumbling government criss-crossed in the cross pressures of the victims and their supportive NGOs on one side and the pressures of the MNC supported by the governments of USA. Even the Supreme Court though giving order for compensation to the victims of this unprecedented tragedy, did not give rigorous stricture against the MNC or allowing prosecution against and extradition of the three top executives of the Union Carbide India Limited. An NGO, Sangathan, is thinking of filing a review petition with the Supreme Court on the decision quashing the charges framed by the Additional Sessions Judge. But most callous is the role of the State Government of Madhya Pradesh where this tragedy, with its untold miseries of the poor, is still lingering for the last 13 years. It is in this context that 13 years after the tragedy, the Bhopal Gas Peedit Udyog Sangathan has accused the Madhya Pradesh government of ignoring the cause of victims of Union Carbide gas disaster. Its convener Abdul Jabbar has alleged that the government under the Digvijay Singh "has left the previous governments far behind when it comes to ignoring the cause of the gas victims". Jabbar has said that Digvijay Singh, earlier as President of the Madhya Pradesh Congress Committee, when the Bhartiya Janata Party was in power, had always come forward with assurances on the demands of the gas victims. But, in the past three years, since he became the Chief Minister, none of these assurances has been fulfilled. Jabbar cited an example of the advisory committee set up by the state government for gas relief and rehabilitation, which last met on 16 November 1993. After that it met again in September 1996, "that too, after pressure had been mounted for holding this meeting." He charged that "the state government has failed its responsibility to improve the environment and provide healthy living conditions in the gas-affected areas".

However, on the whole, the judiciary has played an assertive role in safeguarding the concept of social justice (as enshrined in the Indian Constitution) because, in the final analysis, it is the poor sections, the tribals and the women who suffer most from the poor environmental quality or scarcity of natural resources like wood, water and the like.

Role: Centre-State-Local Levels

As seen earlier, the Prime Minister Indira Gandhi's participation in the Stockholm Conference on Human Environment heralded the onset of environment policy, legislation and structures in India. Her celebrated statement, "poverty is the biggest polluter" implying that the poor are the main adversary (and victims) of environment influenced Congress's policies on environment, at least, in rhetorics. Ms. Gandhi drew her eco-ideas from middle class environmental groups and she extended her support to their projects like WWF-1 and Project Tiger. "In return, she cultivated a small constituency that saw her as being above petty politics." Her eco-role "helped legitimise the centralisation and personalisation of decision making power" in this area.

During her regime, the Project Tiger was launched aiming at protecting the tigers and their habitat which led to eviction of people from reserve areas. In 1976, the subject of 'forest' was brought into the Concurrent List of the Constitution. The Central Government passed Forest Conservation Act, 1980 which prohibited the diversion of any forest land for 'non-forest purposes'. This gave the centre power over the states which required to approach MEF for clearance of their projects requiring forest-related projects. As noted, it was at her initiative that a separate Department of Environment was created in November 1980. Later on it was upgraded to the status of Ministry of Environment and Forests in 1986, ironically with Bhajan Lal as its first minister in Rajiv Gandhi's government.

Under Rajiv Gandhi environment was given an independent reference in 1984 Congress manifesto which stated that the party "will optimise use of natural resources so that while meeting the current needs of growth, the resource base is oriented towards sustainable development". But it ignored the rights of the local communities to forest resources, failing thus in forging the organic link between environment and local people's lives. Rajiv Gandhi, pursued by Sam Pitroda, initiated a National Drinking Water Mission to provide water to all villages by digging more wells and through water resource management schemes.

It was during Rajiv Gandhi's regime that some important actions were taken.

For example, he sponsored the Ganga Action Plan in 1986 for cleaning up this once pristine pure river turned much polluted river bearing the sears of urban industrial civilisation and the rituals of an orthodox religious mentality. Introduced with much fanfare and expectations, the plan however got stranded and slowed down just after its take-off under bureaucratic procedures. It was again this young Prime Minister who created a National Wasteland Development Board (NWDB) in 1985 under Kamala Choudhari for reclaiming and greening India's vast tracts of wasteland. Its object was to bring five million hectares of land every year under fuelwood and fodder plantations.

However, this well meant measure also succeeded only partially. Rajiv Gandhi also evinced his interest in making his administration eco-prudent, when he invited renowned eco-activist Anil Agarwal to orient his cabinet colleagues and members of subcommittees on environment with his lectures. "However, the exercise demonstrated that not many of the ministers were really interested in these issues." Again, his period was marked out by cynicism. The Rashtriya Chemical Fertilisers Ltd., which was manufacturing urea in Chembur-Bombay, notorious for its polluting effects, was awarded the Priya Darshini Award for Environment Protection by his government!

In 1987, the government under Rajiv Gandhi formulated a National Water Policy to handle water use and management and prioritised importance of drinking water needs.

In mid-1980s, under Gujarat Chief Minister, Amarsinh Chowdhry's pressure, the centre gave lease approval to the brother of B.K. Gadhvi, the then Union Minister of State for Finance, enabling his company to mine 144 hectares of forest lands where precious marble rocks were available. This was cited by non-Congress Chief Ministers, whose projects for their states were either delayed or turned down when they sought clearance from MEF. Thus, even while environmentalists had welcomed this act, it also created centre-state conflicts on the point of use of non-forest land for 'development'.

In 1991 Kamal Nath became the Minister for Environment and Forests in Congress government under Narasimha Rao. The high profiled Kamal Nath floated a draft Forest Bill (1992) proposing more powers to forest bureaucracy. In 1995 Nath also proposed to entrust degraded forest lands to the paper and pulp industry which meant the industry would be enabled to grow its captive raw material. In 1994, a notification on environment impact assessment (EIA) was issued which required appraisal of new projects of over Rs. 50 crore in 29 sectors (as listed) by committees of experts constituted by the MEF. However, this move of Nath ignored to include public hearing clause, which would have given scope for people's participation in assessing the ecological damages that an industry can cause. To the further consternation of the environmentalists, in May 1994, an amendment equipped the MEF with discretionary powers to disregard experts, environment groups and demands of the affected people. It even allowed some specific industries to do away with EIA procedures altogether.

Meanwhile, the parliament had to respond to widespread demand for environmental courts, earlier initiated by Maneka Gandhi. In May 1995 Rao government sponsored the bill for establishing environment tribunals. However, it was criticised as a 'piecemeal measure' as it did not ensure local communities' right to information regarding industrial hazards, and as it exempted the government and its agencies from the purview of the tribunal.

Kamal Nath evinced his dynamism by his jet-setting tours when he participated in a number of international meets such as World Forestry Congress (1991), Kuala Lumpur (1992), UNCED-Earth Summit at Rio (1992), UN General Assembly meet to consider the UNCED report (1992), Montreal Protocol meeting at Copenhagen (1992), UNEP-Nairobi (1993), Globe 1994 at Vancouver (1994), UNESCO meet on drought and desertification at Geneva (1994), Convention on Biodiversity at Nassau-Bahama (1994), Climate Change Convention at Berlin (1995) and many others related to environment held New Delhi. He gave a powerful voice to the Third World concerns and networking with NGOs of India on national and global issues and setting strategies on which the future of humankind rests. He also sought "to strike a balance between the needs of national development and needs of planetary conservation".

As a response to widespread and sustained agitation against the proposed draft Forest Bill carried on by pro-tribal NGOs, the MEF conceded to recognise rights of the forest dwellers to the forest produce. Kamal Nath himself remained present at the ceremony in the eastern tribal belt of Gujarat to give away land entitlement deeds to thousands of tribals. On the other hand, Nath's proposal in 1995 to give degraded forest lands to paper and pulp industry for growing captive raw material was branded as submission to commercial interests at the cost of environment and tribals' rights.

Village Panchayats

Rajiv Gandhi had initiated a process to devolve powers to local levels to panchayati raj institutions (PRIs). Social programmes were implemented with involvement of NGOs and PRIs. When Rajesh Pilot replaced Kamal Nath as the Minister of Environment & Forest in Rao government, he mooted a proposal to appoint *Van Mukhias* by Forest Departments in each panchayat, who would act as middlemen between the departments and Gram Suraksha Samitis to protect the local resources. Environmentalists reacted to it by describing it as "a retrograde step which would take whatever voice local communities have in the management of local forest". It was also a step away from the spirit of Rajiv Gandhi's Jawahar Rozgar Yojana (JRY) which sought to make panchayats the channel of development works. Such a decentralised arrangement had, at least, the potential to bring together the employment generation and ecological healing. It is another matter that most panchayats could not choose their project in such a way that environmental priorities could be clearly made out. In such a situation, JRY could not move PRIs far enough in linking food security with ecological degradation at the local level. Initiated by Rajiv Gandhi ("Power to the People" approach) and completed by Rao, PRIs were given more powers through the 73rd amendment to the Indian Constitution. Schedule XI of the Constitution lists certain functions related to managing land, water resources and local forestry to PRIs. It is a challenge as well as an opportunity to the decentralised democratic system to guard whatever environmental resources are still available and even regenerate the degraded resources like the land, village forest and water resources in and surrounding villages.

It is, however, the Bhuria Committee whose recommendations like handing over natural resources management in tribal areas to local communities need to be adopted and implemented with honesty by the Union Government. In spite of the promise of the Congress government of Rao and thereafter, of the United Front government of Deve Gowda, and the pressure by the NGOs working in the field of tribals and environment, this report, vital to tribal interests and environmental protection, is being ignored.

The Congress has consistently emphasised on going in for large irrigation dams like Bhakra, Nagarjun Sagar, Hirakud, Subarnarekha, Tehri and Narmada but neglecting small water development and water harvesting systems for conserving water locally.

During 1989-1994 the Congress governments allowed development projects that caused havoc to the fragile ecosystem of the Western Ghats. Even the opinion of the Yellappa Reddy Commission not to build mega projects on the western coast was ignored. The Cogentrix and Nagarjuna Steels in Karnataka have created acute controversy vehemently criticised by the environmentalists nationwide. Environmentalists have also trained their guns on the Kundremukh Iron and Ore Company Project, Indal Copper industry, Harihar Polyfibres and a chain of pollutant distilleries. Though Karnataka's Chief Minister Moiley claimed to have adopted clean technologies so that "there are no major polluting industries in the state today", in March 1994, large scale death offish was noted in the Tungabhadra river, apparently due to effluents released by Harihar Polyfibres Company. Again, large scale granite mining has caused severe problem of ground water depletion being faced by the State. The Reddy Commission also indicted the Moiley government by illustrating the criminal-politician nexus in illegal tree felling. Activists of the Appiko movement have criticised the Congress policy leading to destruction of the biodiversity of the once rich Western Ghats. They allege that the local communities, depending on the forests for their subsistence, are deprived access to their natural, age-old habitat.

It is ironical that almost all the political parties have neglected the issue of environment. Only since mid-1980s, they have begun to take explicit cognition of environmental problems by including them in their manifestoes. However, most sensitive to environmental problems was Janata DaI(JD) whose manifesto in 1989 parliamentary elections stated: "With the majority of country people still being poor, the degradation of environment is leading to the degradation of resource base with which alone the basic needs of people can be met." It promised pollution control and equitable use of natural resources with emphasis on the interests and role of village communities. JD's electoral promises was an environmentalist's dream come true. In 1990, Singh government took the first step by promoting 'forest management' which recognised limited rights of local communities over forest produce in return for protecting them. However, it is the same JD government during whose regime, in July 1990 alone, as much as 1,11,116 hectares, of forests were felled, that is, nearly the equivalent of forests cut in the entire past decade!

Centre-State Politics

Energy is one of the best environmental issues that graphically illustrates the dynamics of centre-state politics disoriented to environmental sensitivity.

Farmers, being influential pressure group and politically mobile community, are cajoled and appeased by most political parties. Congress, for long in power at the centre as well as in states, has also devised its energy policies charaterised by heavy subsidies. Liberalisation and globalisation trends have further boosted building up of power projects in the states, e.g., the Cogentrix power project being set up in Dakshina Kannada. States which are high on speedy process of industrialisation argue for installation of more, huge power projects. They cite that delay, if caused because of their processing on grounds of social and ecological considerations, will result in loss of income or of that project itself. Even though Janata Dal claims to be committed to environmental sensitivity and some of its leaders like V.P. Singh and Maneka Gandhi (member of JD till July 1996) were quite responsive to some well-known intellectual/activists in the field of environmentalism, it is the JD-ruled states which have been hyperactive in getting the power projects through. JD leadership in Karnataka remained cool to the presentations of the forces who have opposed the Kaiga nuclear power project in Uttar Kannada. Even the Chief Minister remained insensitive to complaints made during the faulty or hazardous technical aspects of this project. All the Chief Ministers of the state from Hegde, to Deve Gowda, irrespective of their party affiliation, have been soft to power projects. In this context, it is significant that Deve Gowda as the Prime Minister has helped the process of transfer of vital decision regarding clearance of all power projects from the centre to the states in response to the vocal controversy raised on this point by the Chief Ministers in the recent past.

An inter-ministerial expert committee had been set up in late August 1996 at the prodding of the power ministry to do away with central environmental clearance for all power projects. The committee, comprising representatives from the power, coal, mining, environment and industry ministries, in itsreport submitted to the Prime Minister's office, cautioned that given the level of expertise and the manpower available in the states this was clearly not possible.

At a meeting of state ministers of state industry and environment departments with Central Pollution Control Board as well as union environment ministry (September 1996), it was repeatedly stressed that the state environmental authorities are under-staffed and underfunded.

At the meeting of state power ministers along with Chief Ministers in mid-October 1996, the Prime Minister clearly disregarded the advice of the expert committee and announced that green clearance would now come from the states. He said both the Environment Protection Act as well as the Forest Conservation Act would be amended to facilitate the process. The Union Minister of State for Power Venugopalachari even added that if necessary, the centre would table a bill in parliament to ensure that states wen granted the power to give the green signal. In addition, the centn too would have an 'empowered committee' which would iron out all procedural bottlenecks between the departments of power, expenditure, planning and economic affairs.

This move might lead to, what environmental experts call; "race to the bottom", that is, a lowering of standards (by acts of omission or commission) between states to attract the projects!

Experience in mid-1990s has shown that both the recent mega power projects—one promoted by Dabhol Power Company (Enron) in Maharashtra and the other by Cogentrix it Karnataka—were mired by environment controversies which erupted after the state environment departments accorded then clearances. According to the sources in the power ministry, "competition among states to attract projects is rising rapidly and there is a genuine fear that certain norms may be overlooked".

That the states are only too willing to give environmental clearance to projects without much security in evident from the Cogentrix case. After the Karnataka authorities gave it the green signal, the MEF in Rao government had kept the project pending till the promoters promised to set up a desulphurisation plant. But, predictably, the Deve Gowda's government cleared it.

Case of Coastal Protection

In the last decade of the twentieth century, the centre has formulated quite a few laws and rules but the states have been found wanting in their implementation. In April 1996, the Supreme Court has asked to stop any type of construction within 500 metres of the line of tide along the coastal line. This was a historical step for the protection of the country's long coastal area. Along with, the Supreme Court also gave a directive that India's coastal states should constitute 'Coastal Area Management Authority' to look after environment. Moreover, it asked the states of Gujarat, Andhra Pradesh, Karnataka and Kerala to frame such scheme for coastal management and send it to the Union Government by June 1996. In fact, Gujarat had not sent such a scheme to the Union Government in spite of the reminder by the latter to this effect. This case of the state's lack of action is not an exception. But, in some cases on such points, the centre is also found wanting. It is here that advocacy groups have to be vigilant and should detect such cases at various levels, bring them to light through media, activate the judiciary and try to put it across the executive in the Vidhan Sabhas or parliament through their elected representatives. But all this requires an adequate understanding of relevant laws, policies and operative authority structures in the rather new, emerging field of environment.

States as Ecoculprits

More or less, all political parties are factually insensitive to eco-issues. The government of Karnataka, disregarding popular protest, considered reviving of the controversial 210 MW Bedthi hydel power project. The project required submergence of 6,065 hectares of land including 4,216 hectares forests.

In Bangalore, the government has conceded to builders' lobby to converting more than 100 tanks into posh residential complexes. In Bihar, the JD Chief Minister Laloo Prasad Yadav has been personally taking interest in giving fillip to the controversial big projects like Koel Karo. In Orissa Chief Ministers (J.B. Patnaik of Congress and Patnaik of JD) campaigned for Tata's controversial Rs.20 crore Project—Chilika Aquatic Farms—in the Chilika lake against the well-articulated protest of environmentalists and small local fisherfolk. In 1989, in Karnataka, the Chief Minister R.K. Hegde invited Birla-owned Harihar Polyfibres which then became notorious for polluting the river Tungbhadra with its effluents. Earlier in 1984, during his regime, Karnataka Pulpwood Limited came up with the lease of 75,000 hectares of forests for growing eucalyptus in captive plantations for the Harihar Polyfibres. All talks of following such principle as consultations with local people and independent experts as well as 'compulsory public hearings' before sanctioning a 'development' project have remained on paper.

In Rajasthan, the BJP Chief Minister, Shekhawat is making overturns to MNCs for power plants for building the second phase of the Suratgarh thermal power plant. In spite of the uproar that the state's Dholpur thermal power plant will emit SO_2 emissions which would cause acid rains and affect Taj Mahal in Agra, and despite the MEF's opposition to its hot water emissions project in Dholpur, the Rajasthan Chief Minister's constituency, his personal interest in the project persuaded the MEF to finally clear it in 1995.

The BJP, in its manifestoes, has considered protection of environment as a 'sacred responsibility'. It has stated "our land, air and water, our hills, forests, flora and fauna—all as sacred". But ironically, the party which thus takes pride in Indian culture, has prescribed solutions of environmental problems as mainly technological (anti-pollution devices, improve auto-exhausts and "developing indigenous pollution-free technology"). Its concerns are hardly premised on people's traditions that preserve ecological well-being. In practice, the BJP governments in some states have not acquitted themselves creditably. For example, Chief Minister Shekhawat had actively supported the illegal mining of sandstone and dolomite inside the state's well-known Sariska Tiger Reserve itself, ignoring the local protests led by the NGO, Tarun Bharat Sangh. It is reported that the Shekhawat government has clandestinely given about 364 bighas of the reserve land on lease for commercial purpose to the Heritage Resorts for building a resort in contravention to the Forest Conservation Act, 1980.

Shanta Kumar, the former BJP Chief Minister of Himachal Pradesh, was also criticised for launching an official forest land grab movement (Van Lagao, Rozi Kamao Yojana) with his afforestation through people's participation scheme in the early 1991. When the Congress replaced the BJP government in this state, it shelved this politicised scheme.

Shanta Kumar had announced the setting up of four large and 12 small cement plants within a narrow area. It also got Gujarat Ambuja Cement Plant near a wildlife sanctuary, part of which was denotified to provide land for this project. All this invited local people's wrath as cement plants cause serious damage to crops and grass under corroding coating of limestone and dust and cause fodder crisis. Interestingly, the Chief Minister. Virbhadrasingh's Congress, which had criticised these plants, continued most of them.

Shekhawat had also opposed the MEF's modifications for environment impact assessment in 1994, which aimed at regulating development activities in all states in the interest of environment. He has also asked for more powers for states in this area. His government has not shifted the toxic-waste releasing factory Pesticides India near Udaipur, which is owned by a nephew of Ashok Singhal—the Vishwa Hindu Parishad leader.

Environment did not figure as an explicit concern in the priorities of CPI (M) till 1989. Only, in 1991, its election manifesto promised to give an environment policy to be in tune "with the needs of rapid and sustainable development". But, in earlier elections, it was the economic issues on which its manifestoes are based. Like the Congress and the BJP, it is also not against major development like Narmada and Tehri projects per se.

The CPI (M)-led government in Kerala has planned 700 MW power project on Pooyamkutti river. Environmentalists have opposed it on the ground that it will destroy more than 3,000 hectares of natural forests of the Western Ghats, and deprive thousands of their forest-based livelihood. They also put the conservative estimate of environmental loss at Rs. 3,334.65 crores.

The Marxist trade union CITU has not really used its influence to force the management of polluting industries into taking sound pollution control measures. To the CITU the Supreme Court judgments on the pollution caused to the Ganga by tanneries and other industrial units in Calcutta are impractical and unrealistic. After 1990 most of about four lakh hectares and their closure of paddy land were brought under Shrimp cultivation in the 24 Paraganas endangering the fragile ecosystem of the Sunderbans—India's best mangroves. The Marxist government of Kerala's persistence to carry out its prestigious Silent Valley project (till it was turned down by Indira Gandhi) is too well-known. All this have exposed the CPI(M)'s apathy towards ecology and preference of growth over environment.

No Further Ecodecentralisation

On one side, as a senior Congressman has reported, political leaders do not see the relevance of panchayats to the environment issue. "Villagers have been instrumental in destroying the forests. Giving the governance of forests to the local people or panchayats will lead to their further degradation." He demanded stricter policing and governance of panchayats in their area. "Decentralisation of powers in these matters will lead to disastrous results. The local people are not committed to environmental issues." Chief Ministers who were opposed to the Forest Conservation Bill claiming that the Congress government at the centre was usurping the powers of the states, have not recognised, in the same vein, the rights of local communities to say whether they want development project to the detriment of their forest or environment or not.

The local people, however, view this problem rather differently. In many places, the villagers have seen through the problems arising out of degradation of environment—indiscrete tree felling, overuse of local water resources and the like.

On the other side, numerous examples of NGOs can be cited which show their effective programme for effective environment awareness campaigns, greening of degenerated or wasteland and management of local water resources. For example, from Balakedarara Vedika (Dakshina Kannada), Pani Panchayat (Salunke-Maharashtra), hidden check dams *(gupta bandh)* by Lok Samiti and Swadhyaya (Gujarat), Ravalgaon (Hazare-Maharashtra), Pani Chetna (Bihar), etc. In Palamau, Bihar has been the witness to the tragedy of a vanishing forest cover and destruction of water harvesting structures. But Pani Chetna Manch, a local NGO and Chhatra Yuva Sangharsh Vahini influenced by Jayaprakash Narayan have developed an alternative water management system. However, such efforts in many states (except in Gujarat and Maharashtra) ate quite scanty and scattered.

Political leaders and the people both evince a very low level of environmental awareness. Such issues have not become politically hot or electorally crucial in India. Again, while state governments are vocal in arguing for dilution of the centre's powers related to environment in favour of the states (Rajasthan's Shekhawat, Andhra Pradesh's Nayadu, Karnataka's Hegde and Maharashtra's Joshi are quite vocal), the states themselves are cool in extending similar powers to local communities through the panchayati raj institutions.

State governments have been found to be uncritically responsive to the pressures of industrial elites in giving away land and other facilities to even overtly polluting industries. Ambition of the state rulers to push their state to the top rank in the country on the map of industrialisation, covert deal or dishonest benefits that accrue to them out of approval of new projects and manipulative power of the industry (Indian and also foreign since the adoption of liberalisation) are the main reasons.

On the other hand, we have success stories of time-tested indigenous water harvesting system waiting to be replicated. For example, in Rajasthan, the traditional water harvesting systems of *johads* and *khadinas* (check dams for collecting water from small catchment areas) have been revived. Such systems have served as water storage tanks, drinking water sources and minor irrigation. In Alwar, 565 *Johads* have raised water levels in local wells, and help increase the yields of maize, wheat, mustard and fodder to at least 10-fold. Important role in this process of rejuvenation is played by Gram Sabhas which are formed to manage *johads* in villages. They now work as local people's councils for protection of 3 "Js", viz., *jal, jungle* and *jameen* (water, forests and lands).

Under the new strengthened panchayat raj system, Gram Panchayats have been given powers to manage their environment including land maintenance, water resources, local forest and health and sanitation. New industries require approval of local panchayat in its jurisdiction. Actually, a village panchayat in Goa in mid-1996 has stalled the commencement of an industry by MNC even when this powerful entity managed to get the approval of the union government and the state government! Such cases are rare. The chirping of one sparrow does not herald a spring. Again, how far the state government and local elected body become partners in environmental protection or degradation remains to be seen. In absence of sustained grass roots activism, vibrant NGOs and enviro-conscious panchayati bodies, decentralised environmental management or ecodevelopment will remain more a populist rhetoric and a distant dream.

Ecodevelopment

The case of industrial pollution in India's national capital succinctly illustrates the conflict between ecology and economy in one classical environment-development framework. It also disturbs the political system of Delhi. The heavy pollution in Delhi has, since long, been a cause of concern and consternation for its elites and middle class population. It has been widely reported and commented upon in the media. The Dengue syndrome (August-October 1996) has set off the popular detonation. It is in this context that the non-industry people have welcomed the Supreme Court ultimatum of 24 October, 1996.

The pollution created by Delhi's 39,000 industries was extensively reported and investigated time and again by the press in the 1990s. The case was with the Supreme Court since long, which gave an interim order alerting the industries to either control their polluting process by proper effluent treatment or relocate themselves outside the city as proposed by the government.

On 24 October, 1996 the Supreme Court gave a five-day ultimatum to the Delhi government to find a suitable site and create a new industrial belt to relocate 39,000 industries from the residential arete of the capital. The court had directed all these industries to close down by the year end. Justice Kuldip Singh and Justice Saghir Ahmed of the divisional bench warned that "if the state government cannot carry out the court's direction, we may have to direct the government to pay up the compensation to the workers of these units which are required to be closed down by the earlier order". Though, the state had earlier stated that it would clear residential areas of unauthorised industries at the earliest, but now in response to the Court's order of 24 October, it has reversed its earlier stand. It now says that it cannot take up the work at once.

Meanwhile, the industries became tense with the deadline as the last after earlier warnings giving them a breathing time to correct themselves. The court now set the deadline for shifting their units, fast approaching by the end of 1996. Again, the apex court had left no scope for a judicial review. Even then the industry as a whole was not as yet prepared to leave the site where they have settled so far. The industrialists who are actually planning to relocate their units argue that it would take at least a year to set up their new operations a new site after clearing problems related to shifting.

Apart from bringing pressures on political leaders and government influentials, they also have been holding desperate public meetings to evoke sympathy. They also pleaded for a phased relocation that can be matched with linked incremental generation needed for such relocation. The industrial interests also blame the government stating that industry has become "the softest target", for it is easy to tell them to leave within such a short time even though it were the political leaders who have helped them to establish. They argue, above 70 per cent of the air pollution in this metropolits is caused by vehicular pollution. The cars of elites, buses and the three-wheelers are the main culprits. Many of them do not confirm to rules to use deleaded petrol. "But these polluters are spared because they are influential."

On the other side are workers who are worst off in such a shifting, or else closing down of the industries. They either land on the streets with a small amount of compensation or bonus. The well-off middle class and the elite are happy that such an ultimatum of the apex court will cleanse the capital of the urban tumor of a polluting industrial settlement and help it get a green lung to breath better to help them become rather healthier.

But the problem of ecology is joined by economy. As the PHD Chamber of Commerce has noted on the whole affair with a warning: "Unemployment of lakhs of workers would lead to law and order problems coupled with a rise in the crime rate. Apart from the displacement of workers and their families, this would result in great hardship including loss of academic year for their children."

However, the industry itself is not united in its reaction to the implications of the court's ultimatum. About half of the industry, the big industrial houses like the textile mill such as Birla Textile Mill, DCM, Silk Mills, Swatantra Bharat and others expect to use this ultimatum to get rid of their large number of workers quite cheaply. For they have to pay only two years wages as compensation as compared to the amount which they would have to give if the issue had been open for negotiation between the management and workers. This will be a small amount for them to pay to the workers for closing down their mills. But the workers in such big units demand a 10 years consolidated wage as compensation and also a share from the sale of 32 per cent of the vacated land which these big units are entitled to dispose of at the current appreciated price. On the other hand, the small units which are making profits do not want to close down. But they have the dilemma. This unit either have to buckle down with heavy loss of shifting to the place of relocation but a disgruntled workforce wary of many problems of living their present location, do not want to leave the city. Ultimately the small profit making units and the workers feel that they will be war affected in this "make-green-soon' policy. On the other hand the state will have to immediately help the shifting of the industrial limits such a way that proper relocation takes place in shares possible time and cause minimum hardships and uncertainty to fine workers. The state in a democratic political system needs to fire imaginative policy to balance environmental concerns with development imperatives.

Media as Ecocatalyst

The Indian media, particularly the press, has player analytic role in creating awareness by reporting the cases of polling industries and tree felling, struggles of action groups, grass root movement and voluntary agencies. Also, in informing about the trends and perspectives at the national and global levels and reprogramming out court decisions and in opinion building functions.

One of the first issues projected in the media way rack in early 1970s was threat to that rare piece of architecture and historical monument—the Taj Mahal—as a result of location of mill refinery at Mathura at a distance of barely 40 km from Agra. Immediately, after the foundation stone of the refinery was laid by the Prime Minister, Indira Gandhi, in October 1973, muted protest by several public-spirited persons were reported. Protest letter were written to the newspapers, questions were raised in parliament and public opinion was sought to be created for the preservation of the aesthetic features of the Taj. The awareness created due in sustained efforts through the press and series of articles and editorials that appeared in leading newspapers then made the government to sit up and appoint an expert committee to examine the dangers to the Taj. Such sustained efforts, mainly through the press and by co-activist lawyer M.C. Metha in the court, forced the government to agree to closing down the two thermal power stations locate in Agra and shifting the existing small scale industrial units and foundries to an area beyond the Taj, and also for establishing a green belt between Agra and Mathura.

The media, particularly the Hindi press in Uttar Pradesh and English press in New Delhi, also played a supportive role in transmitting the message inherent in the 'Chipko' movement since late 1970s. The movement received sustained media attention in the country as probably the world's most well-known grass roots ecodevelopment movement. Through a very friendly press and radio, Chipko activists like Sunderlal Bahuguna and C.P. Bhatt have attracted the attention of environmentalists, Gandhians, and women activists interested in the process of social change and the media through out the world.

Its non-violent action has greatly helped unite the tribals, women and focus attention on the mismanagement of forest resources and alternative approach of management of local resources. The media also emphasised that no conservation is possible without the cooperation of the people concerned in ecodevelopment strategies. "They will inspire the poor when they are based not just on ecology but also in the needs in of the people."

The media has also drawn attention to danger to life from foul environmental conditions as in the Chembur belt of Bombay, chemical industries in Ahmedabad, Vadodara and New Delhi, or the degradation of living conditions because of foul water and air as a result of industrial effluents dumped irresponsibly everywhere, and magazines have regular columns on subjects like environment, ecology, wildlife, conservation and pollution, and development which have been recurrent features of dailies and periodicals.

An interesting feature of the environmental journalism is to publish a sort of investigative reports on industries releasing effluents without treatment and belching hazardous gases in violation of anti-pollution laws and adequate reporting on public interest litigation launched by advocacy groups to protect green belts in towns from encroachment, at the hands of builders and municipal authorities, clearing lands for founding polluting units like oil, cement and chemicals. For example, the insensitive attitude of Maradia Chemicals Industries and petitions against Sanghi Cement Project near Narayan Sarovar in Kutch (1996), and Chinkara Sanctuary (1994) have been widely reported in *The Times of India, The Indian Express* and *India Today* (English and Gujarati editions).

The press in India has succeeded in highlighting the ill-effects of big dams which resulted in the decision of the government to abandon the Bodhghat project on the Indravati river in Madhya Pradesh pending a fresh reappraisal of its impact on environment. The three month old campaign, "Save the Western Ghats March" during 1987-88, a unique 3,000km *padyatra,* covering Gujarat, Maharashtra, Goa, Karnataka, Kerala and Tamil Nadu and involving more than 200 NGOs, was well received in the press.

The *Jan Satta,* (Hindi national daily) *The Indian Express,* and *The Pioneer* have a weekly column devoted exclusively to development and environment. Some other dailies have published several articles on air pollution in New Delhi and pollution caused by stone crushing industries and thermal plants in areas of their immediate vicinity.

The long-drawn struggle by the common people of Kusnur village of Dharwar, Karnataka for their age-old rights to local produce against Karnataka government which favoured the Birla industry trampling over these rights received very strong support in a variety of newspapers like *Deccan Herald, The Times of India,* and *Indian Express* in the form of articles apart from frequent reporting. People "fighting for common lands which are vital for their own survival and protection of the local ecology" was commented upon in *Earthscan Feature* (U.K.) which lauded their important victory to main common lands. Charlie Pye-Smith, a regular presenter of the BBC World Service's programme *Global Concerns,* commented in his programme on the Karnataka government's acceptance to wind up the Karnataka pulpwood project: "Nowhere in India has the power of the meek to change the course of history been more brilliantly demonstrated as in Kusnur village."

No movement has received such a boost and a continuous supportive coverage in reports, photographs and articles as the anti-Narmada project movement. It has helped build up a sort of charisma around Medha Patkar, giving disproportionately loaded support and put the Gujarat project authorities and dam supportive NGOs on defensive. But it did put the governments on the defensive and contributed towards corrections of many of its weaknesses or deficiencies leading it to ever higher level of performance.

The Times of India (Ahmedabad) published several informed pieces on pollution generating industrial estates in Gujarat.

Earlier, the Kerala newspaper *Mathru Bhoomi* has supported the campaign to preserve Silent Valley as a 'gene pool' to study and a preserve of rare species of flora.

The programmes telecast on Doordarshan are informed and educational in nature. The two well received documentaries, viz., David Attenborough's 'Our Living Planet' and 'Survival' provided rich information on nature, wild and the ecosystem as a whole. It has stimulated environmental awareness among the viewers. One of the serials in Hindi, *Rakshak,* sponsored by a leading Indian corporate house pointed out graphically and in detail the threat posed to the scenic Dal Lake in Srinagar.

The purpose of promoting environmental awareness is not widely served if such programmes are continued to be telecast only in English. Again, programmes on deforestation failed to point out the responsibility of the state governments in the area of afforestation. The state governments are known to show more of lip sympathy than real concrete action on the ground in protecting whatever forests have remained there.

But unless environment has political overtones, it still does not make news and headlines. For sustained coverage, environmental story has to have political ramifications, such as exposure of politician-contractor or industry nexus or a movement.

Besides, being an environmental journalist, one has to be an 'activist' who is prepared to go beyond the confines of his-her reportage desk. He/she has to have interaction with grass roots activists, environmental actions groups and checking on sites the two sides of data and arguments, besides meeting with members of ministry or agencies concerned. Usha Rai, Bharat Dogra, Kalpana Sharma, Anju Sharma, Rajat Banerji, Max Martin, Madhu Mita Dutta and Himanshu Thakkar are cases in the point.

The environment is no longer a concern of merely environmentalists, women, tribals, citizens suffering from acute pollution and the like. But also a growing band of eco-active journalists, most of them young, who expose the nexus of industrialists, politicians, bureaucrats and builders and thereby bringing pressures on the government and its agencies to correct their policy or action.

B. Verghese, Vandana Shiva, B.B. Vohra, Ashish Kothari, Darryl D'Monte and Bittu Sehgal have in their own way contributed in significantly interpreting the environmental issues and exposing the hazards, as well as criticising the environmental policies of the state. Thereby they have stimulated environmental awareness of the people. Besides having published well-researched works and contributing to national and international fora, Verghese, Shiva, Agarwal and Kothari, through their well-argued pieces on environmental policies and controversies in influential dailies and popular magazines, have created public opinion to bear pressure on the policy making and implementing power structures. Shiva has even worked up such new and difficult issues like biodiversity and intellectual property rights intelligible to informed readers.

As an issue, environment has always remained a feature in media like TV and regular weekly dailies like the *Indian Express, The Pioneer, The Hindu* and particularly frequently featured in the Sunday review editions of papers like *The Hindustan Times* and *The Times of India, Sunday Observer* and prestigious magazines like *Indian Today* and *Frontline, Illustrated Weekly*. Every year *The Hindu* brings out an annual supplement on environment and development with a focus on the problem of environment. Its each supplement is noted for its rich data of survey of the status of the problem and analytical pieces on this issue by informed environmental scholars and activists.

The rich contribution in the.field of environment and develop-ment made by Anil Agarwal and Sunita Narain and their able team of correspondents in *Down to Earth*

(New Delhi) deserves emphatic mention. Alongwith their earlier discussive pieces in "Green Politics", column of *The Economic Times,* and excellent features, elegantly prepared and displayed, Agarwal and Narain have generated public consciousness on various environmental problems from local to international, with their irrefutable data which discuss in politically relevant way. With his closeness to decision makers and access to MEF heads like Kamal Nath, Agarwal has also, with his own but persuasive style and critical analysis, influenced the environmental policy process in New Delhi.

All this provide arguments to environmental struggles and public interest litigations. Sometimes such press reports have prompted the courts to take up cases of breach of environmental degradation as *suo moto* or the like cases.

As more people will find themselves affected by environmental degradation and exploitation of natural resources and will understand to relate their plight and ill-health with such destructive aspects of 'development'(projects, etc.), more interest they will take in reading about or responding to reports about environment in the press. And this will spur the media to add inputs of reports and articles in their publications. The media, particularly the press-has played the role of stimulating environmental awareness, sometimes even as catalyst activating people as in case of the Narmada project, Tehri, Silent Valley, and public concern against Mathura Refinery, and air and industrial pollution in New Delhi and Gujarat.

In the political system, the formidable force of industrial interests, forest contractors in collusion with the bureaucrats of the departments concerned and political class are countered by advocacy groups like environmentalists, advocates and grass roots movements of the marginalised groups. Governments are rather willing accomplice to the so-called developmental ists to the detriment of environment. It is the media who helps the people's struggles against such development policies and projects and the entrenched interests—economic and political. And the judiciary has tilted the balance in favour of the former in the unequal fight between the affected people and the industry and political interests. Finally, it depends on the extent to which the general public become aware and active which could use a catalytic media and an adjudicating judiciary to generate political dynamics on the system in favour of ecodevelopment.

Judiciary: Active Player

If media has played a role of a catalyst or conveyor-belt in some of the major cases of environmental degradation, the judiciary has worked as a critic and monitor of defaulters like the government and economic interests to press for environmental conservation. In fact, it has become an influential player in the political system. Some examples cited here. The Gujarat High Court has takén *suo moto* notice of media reports on hazard industrial pollution and summoned the defaulters to be present before it.

The number of vehicles in Delhi were 28 lakhsin 1996, which was more than the number of vehicles registered in other three met-ros in India. According to the World Bank study reported in October 1996, Delhi was afflicted with 7,500 deaths annually due to air pollution alone. The other more striking feature is that an additional four million admissions in hospitals are registered due to contaminants in the air.

Taking a *suo moto* notice of media reports on health hazards posed by vehicular pollution, the Supreme Court bench comprising Justice Kuldip Singh and Justice B.L. Hansaria has suggested sweeping restrictions ranging from suspension of vehicle licenses to keeping automobiles off the road. The apex court has called for scrapping of licenses of vehicles which are more than seven years old and the non-renewal of licenses for three-wheelers after their permits expire. It has also asked the authorities to suspend registration of new vehicles for a few years.

With such a position taken, the Supreme Court has gone a step further of its positive response in many PILs related to complaints against unchecked pollution. It has become an eco-activist, for the time being, when the executive has failed to check the air pollution caused by a large bulk of insensitive citizenry. Reacting to the court directive, the Transport Minister Rajendra Gupta had to admit:

> "Air pollution is a serious threat and is going to be the most important issue in the next elections. Whether we like it or not, there are no soft option in controlling air pollution." Thus, the Supreme Court's proposals "may turn Delhi green, and motorists red!" Any way, the apex court's turning red eyes to degraded urban environment in India's major cities, and directives for making it bearable—this task is put on the top of the agenda of political system by a sensitive and active judiciary. Whether the Court succeeds to set constraints on the government and the effluents to significant extent remains to be seen. Much depends on the role of eco-sensitive citizen's groups who should follow up the policy line set forth by the judiciary.

Check on Manipulation

The judiciary has played an exemplary role mainly in effectively intervening against the erring polluting industries. Alongwith environmental groups and public litigation activists, the judiciary has prevented India's political system from being unilaterally manipulated by industrial interests, quarry contractors and developers alongwith the state as a condescending witness or a latecomer enforcer of environmental laws. On this point, some cases may be further cited here.

The Gujarat Pollution Control Board (GPCB) had listed some units as pollutants, say 23 units in Vapi and 756 units in Ahmedabad. There has also been reference to the 'politically motivated' list which included only 'the small fries' (units owned and run by small entrepreneurs) but leaving out big companies like United Phosphorous Ltd. As claimed by the complainant advocate Ajit Padiwal, the latter is said "to belong to a brother of Dawood Ibrahim".

The Government of Gujarat ordered the closure of 756 units in Ahmedabad in 1995. But more than 70 per cent of these have managed to procure a stay order from the Gujarat High Court in May 1995, on the plea that they will have their own treatment plant to be set up within 60 days. The High Court ordered the government to deal with the remaining 30 per cent of the units as it deemed fit. But, by the end of July 1995, only 34 units closed on their own. Moreover, only 78 of the original 756 units (listed as defaulters) had been shut down by the state government.

Therefore, in the monsoon of 1995, in spite of cut on power and water supplies applied by the authorities, many of the units were found operating with generators and bore wells, although both these improvisations were further damaging the environment! Padiwal had to be busy with collecting evidence of the contempt of the court (against such units side lining or sabotaging the court order of closing down). That meant another round of court hearing, causing delay; while the defaulting units will "merrily go on with their dirty operations", and make money.

In such a context, as the High Court observed, the GPCB was obviously not doing its job. Its Member-Secretary, F.D. Salunia, however, claimed that the GPCB had filed 2,500 cases in the courts. *"That is about 50 per cent of pollution cases in the whole country.* If nothing is happening, it is the judiciary to be blamed." And nothing was happening, and in its hearing on 31 August, the High Court asked the state authorities to closedown all the 756 defaulting units working in Ahmedabad by 1 January, 1996, unless they rectify their operations. In the process, on 1 September, 26 units in Nandesari were also ordered to clean up or close down.

Doubts were however expressed, not without reason, whether the state would abide by the Court's order. For example, Maradia Chemicals of Surendranagar was still going on producing hazardous chemicals. The Gujarat Ecological Commission (GEC) took up the case of this unit with GPCB. But the latter certified that "the effluents are within limits".

How did GPCB, the state agency interpret this position? Salunia stated: "Even if all the units around Ahmedabad were to treat their effluents up to the GPCB norms, still it would not help, as long as the Gujarat Industrial Development Corporation (GIDC) does not construct drainage channels to take them away."

And the GIDC had its own reasoning, “The Ahmedabad Municipal Corporation had not provided land and such infrastructure to lay down the channels.” Was it not the famous case of passing the buck or shifting the onus from one power structure to the other within the political system? or a sort of refined eco-game!

West Bengal also presented the venue of callous neglect of environmental law and lack of concern against industrial pollution.

In east Calcutta, some 70 odd tanneries, power producing units, the Howrah station and about hundreds of industries had been found causing pollution. They were ordered to shift from the Tanga and Topsia areas east of this metropolis. But only a handful of industries were closed down on the charge of being polluters. Since 1993, the Supreme Court has been ordering closures. But till September 1995, only two industries, Remington Rand and Crooksons, abided by the court order. In fact, Remington Rand rather welcomed it as the company had been sick and it suited them to close down.

In such a situation of non-conformity with the court order by a large number of industrial units, the state government and the CITU (Marxist trade union) proved to be biggest obstacles.

The Public

Finally, the public should be convinced that only by protecting the environment and controlling pollution, its health will be safeguarded and life will not be endangered. Otherwise, all the well-intentioned policies, legal measures and administrative apparatus will remain on paper. Environmental awareness and public participation, the debate and action against the cases related to environmentally threatening development activities are the key elements of environmental policy planning and management. These call for the mass media, educational institutions and judiciary to provide leadership. Environmental awareness programme should commence from the junior school level. Mass media will have to keep up its initiative to report and expose the events of environmental degradation like illegal or indiscreet felling of trees, callousness of the authorities in cleansing the city or indolence to despoliation of urban environment and industrial pollution and the like. And the judiciary will have to act upon the environmental provisions forcing the erring interest groups, the builders and the authorities at local and state levels to scrupulously follow up its rulings in the environmental cases. At the same time, it should be borne in mind that even the judiciary has certain limitations. For example, it cannot take executive action to follow up its green verdicts.

Much depends on continuous informed action of citizen's groups. Deve Gowda, the nation's chief executive, has stated that “environmentalists mislead people”. It reveals the government's attitude which is a matter of concern to environmentalists. As Anil Agarwal sees, "Deve Gowda's stance shows that the country is going back on its environmental commitments. And the environmental movement has failed to acquire enough power to force the political system to accept the value of its arguments.

Only well-formulated and concerted environment that appeals to people and a green strategy for checkmating policies and projects hostile to environment can promote the value of genuine eco-development in the Indian political system.

REFERENCES

Anil Bhatt et. al. (eds.), *Building from Below: Local Initiatives for Decentralised Development in Asia and Pacific* (Kuala Lumpur, 1983).

Ashish Kothari et. al., "People and Protected Areas: Rethinking Conservation in India", *The Ecologist* (UK), September-October 1995, p. 193.

Brewster Kneen, "Invisible Giant: Cargill and its Transnational Strategies", The *Geologist* (UK), September-October, 1995, p.199.

Madhav Gadgil and Ramachandra Guha, *Ecology and Equity* (Penguin, 1995), "Introduction."

Madhav Gadgil and Ramachandra Guha, *Ecology and Equity: Tke Use and Abuse of Nature in Contemporary India* (Penguin, 1995), p.190.

Ponnu Wignaraja (Ed.), *New Social Movements in the South: Empowering the People* (Vistaar Publications, New Delhi, 1993).

Pravin Sheth, *Narmada Project Politics of Ecodevelopment* (New Delhi, 1994).

Sadanand Kanwalli, *Quest for Justice.* (Samaj Parivartana Samudaya, Dharwar, 1993).

Sharad Bailur, "Costing the Earth," *The Economic Times,* 28 September, 1995.

Shiva, *Staying Alive* (Zed Books, London, 1989).

Sumi Krishna: *Environmental Politics: People's Lives and Development Choices* (Sage, 1996), p. 146.

7

Environmental and Social Aspects of Sustainable Development

GLOBALISATION, SUSTAINABLE DEVELOPMENT AND ENVIRONMENT

The current debate on globalisation and the environment has revolved around tour major themes, all informed by the multitude and complexity of a set of environmental concerns dubbed, even by the most conservative commentators, as symptoms of a global environmental crisis. From climate change to biodiversity loss and from air pollution to fresh water contamination, the symptoms of a global environmental crisis have often been treated from an essentialist position. This theme acknowledges the significance of the natural interdependence of the environmental life support system (air, water, and soil) on a global scale. To that extent, the most common theme in the debate on globalisation and the environment is that, if it is environmental then it must imply global consequences on the human and non-human world. This assumption implies that the impact of any large-scale development interventions (industrialisation, raw material extraction and processing, mega-irrigation projects, timber logging, mining; fossil energy exploration, production and processing, and consumption, etc.) in the environment would influence the quality of the environmental life support system. Such large-scale environmental interventions are part of an ecological modernisation process that has contributed to an unprecedented modification of the global environment. In essence ecological modernisation carries the insignia of globalisation and its capacity to produce momentous transformation in the way human lives are ordered and reordered, which in turn results in further modification of the global environment. Ecological modernisation can therefore be defined as a manifestation of a complex and far-reaching development or process or modernisation in one of its most common interactions with the natural environment.

If the use and abuse of technology is an inseparable part of development and environmental modification then it is only logical that technology has also been conceived as the saviour from global environmental destruction. In the process the quest for ecological modernisation becomes a global imperative resulting in ecological damage as well as ecological repair. Second, ecological modenisation is associated with high per capita energy consumption and, also as a restut of industrialisation, the penetration of technology into almost every aspect of life, from' household appliances to transport and the manufacturing of goods and services or development redefined as ecological modernisation.

Paradoxically, under the ecological modernisation project, a major reduction in energy consumption (the main contributor to air pollution, climate change, ozone layer depletion, and greenhouse effect, etc.) may not be possible without major alterations to today's lifestyles. The view that technological advancement and raw material fabrication are our potential saviours from an impending global environmental crisis is not supported by the current expansion of new environmentally harmful products. The implications of such developments for ecological modernisation are immense, particularly in the event of further compression of the environmental space and the increase of ecological debt incurred by the industrially advanced world.

The second theme in the debate is more concerned with the environmental consequences of the production of the goods and services required for the satisfaction of an increasing world population. The views expressed by social scientists attached to the Club of Rome are commonly known. For instance, Meadows et al. (1972) argue that if the current trends of industrialisation, population growth, and depletion of non-renewable natural resources continue unchecked, the earth would reach *The Limits to Growth* within a hundred years. What is needed to reverse these trends; not only to curb population growth, but also to work towards a sustainability revolution (Meadows et al. 1995)? Global economic growth would then set the limits to the capacity of the environment to provide the raw materials and energy resources needed to maintain economic growth. Generally, this theme is common in multilateral environmental initiatives and the manner in which they have responded to what has been dubbed the global environmental crisis. These initiatives are exemplified by a long sequence of global environmental conferences considered by the development lobby as an anti-climax to civilisational progression. Within this theme, the Group of Lisbon (1995: xvii), takes the view that the key to responding to global environmental change is not one of merely getting 'carrying capacity' right, but that "the response to present and immediate future needs and opportunities demands a system of cooperative governance. Only by linking the multitude of socio-economic networks at various territorial levels around visible targets and common objectives can one realistically hope to achieve social justice, economic efficiency, environmental sustainability, and political democracy, as well as avoid the many possible sources (economic, political, ethnic) of global implosion."

The most doomsday-like scenario treats exponential development as a threat to long-term human survival in a healthy environment and transforms the daily struggles for survival to struggles over progress. This possible scenario would result in a global implosion summed up by Beck's (1995) risk society, characterised by uncertainty and lack of control over its future destiny. In the risk society: 1) Environmental hazards are global and cannot be delimited spatially, temporally, or socially; they encompass nation-states, military alliances, and all social classes, and, by their very nature, present wholly new kinds of challenge to the institutions designed for their control and the legitimatisation of their anonymity. 2) The established rules of attribution and liability - casualty and guilt -break down. 3) The hazards can only be minimised by technological means, never entirely eliminated. The technocracy of hazard squirms in the thumbscrews of the safety guarantees which it is forced to impose on itself. 4) The lack of provision for catastrophe prevention or prediction plainly exposes the paradigm error; the bewitchment of reason caused by the false belief that the 20th century was only a continuation of the 19th. The security system, which anticipates social provisions for the worst conceivable case, brokedown with the advent of large-scale (nuclear, chemical, ecological, genetic) hazards. Although Beck has moderated his position vis-a-vis the risk society thesis, it still poses some serious questions concerning what type of development developing countries should pursue.

The third theme is advocated by a large number of multilateral institutions and social scientists, mostly pro-South in their views on the relationship between the developing countries and the dominant global economic forces.Their basic argument is that the development of globalisation and the liberalisation of trade and finance are not environmentally neutral, and often do not operate in a world free of national and regional interests. The implication here is the rich North is responsible for the bulk of global pollution. If that is the case what is needed then is a differentiated rather than uniform approach to whether curbing economic growth should be adopted equally by poor and richer countries. Pro-South social scientists also place a larger proportion of the blame for the global environmental crisis on the countries of the North. For example, the South Commission (1990:218) argues that,

The countries of the South are today victims of the delirious environmental effects of policies and patterns of development in the North. These include such global phenomena as the thinning of the ozone layer, nuclear radiation, and the green house effect, as well as such direct acts as the dumping of hazardous wastes and the location of polluting industries in the South. Attempts by the developing countries to bring the global commons - in particular the oceans and outer-space - under effective international jurisdiction have been defeated in practice by the lack of co-operation of the developed countries.

On the other hand, social scientists of various radical environmental persuasions argue that, the global sustainable development movement usurps a neo-colonial ideology aim to frustrate the development of the South. This view, however, is not new; it originated in historical materialism and Engels' (1940) thesis on the *Dialectics of Nature* and Marx's contribution to the relationship between the expansion of capitalism and the destruction of the natural environment.

The fourth theme found expression in Agenda 21, as well as a variety of United Nations conventions, treaties, and declarations espouse, and a number of 'establishment' scholars who work within this dominant paradigm. The common view here is that global environmental protection could be realised by adopting sustainable development as a global ideal. As a global imperative, sustainable development is built around six axes: (i) reviving growth; (ii) changing the quality of growth; (iii) meeting essential needs for jobs, energy, water, and sanitation; (iv) conserving and enhancing the resource base; (v) reorienting technology and managing risk and (vi) merging environment and economics in decision-making.

The main thrust of sustainable development, therefore, is that global environmental protection and economic growth should proceed hand in hand. More importantly, sustainable development hinges on the premise that global environmental institutions and governance should manage global environmental problems. The creation of the Sustainable Development Commission and Global Environmental Facility (GEF), the Global Governance initiatives after the Earth Summit of 1992, are but a few pointers in the direction of creating a multi-layered integration of management of global environmental change and the economic, political and technological developments responsible of this change.

Although some social scientists and activists were quick to dub sustainable development's emphasis on reviving growth as a contradiction in terms, others began the search for methods and theories that could reconcile economic growth and global environmental protection. Some of the most stimulating research in this area is what Eckersley (1995:15) refers to as free market environmentalism, which conceives of sustainable development within the dominant free market paradigm. In this paradigm, nature itself is, to some extent, treated as a commodity subject to market principles linked to its ability to supply environmental goods and services.

However, for other social scientists such as Wilson and Bryant (1997:107-8), (i) the market attaches a commonly recognised economic value to environmental resources. This value is dependent on the importance of the resource in question to environmental managers, (ii) The market provides the means by which environmental resources can be exchanged, (iii) As a direct response to valuation and exchange, the market has a direct bearing on the operation of multi-layered environmental management, (iv) The market serves as an important means by which resource scarcity is regulated, (v) Externalities highlight what can be termed market failure; that is, that the market does not capture the full environmental implications of human-environmental interactions. Free market environmentalism has to deal with this not at a resource, sector or national level, but all of these levels as well as integrating them at global level.

Ironically, while the debate on the global nature of the environmental crisis continues, the constellation of the world political and economic order began to shift away from global hegemony of two superpowers to unipolarity. In Axford's (1995) words, "global hegemony has slipped beyond the capacity of any single state, thus ushering in the emergence of a truly global managerial entrepreneurial class whose transnational economic transactions are efficiently enhanced by global private banks, financial markets and corporations". The advocacy of a rather muted free market principles and trade in production, services, and finance, has also worked its way to replace the primes of exchange relations informed by the sovereignty of national economies controlled by the state. The global scale operations that have doubled the World trade in new production materials in less than two decades could not have been possible without the formidable technological and organisational development.

New trends in globalisation, including robotisation, the dematerialisation of production (media and communication), and post-Fordism (franchising the great corporate conglomerations) are not environmentally neutral. The political economy of these new trends is exemplified by the structural power of mobile capital and private investment, which has contributed to the decline of the state and its ability to control economic activities and the factors of production, including environment and natural resources. Two observations merit some attention here: first, while the protection of the global environment requires a globally concerted action, the prevailing economic globalisation and its manifestation in trade, the information revolution and the accompanying technological advancements are growth driven. Second, globalisation is born in the womb of a neo-liberal ideology hostile to attempts aiming at market regulation or control over the new entrepreneurial class's capacity to expand production, extract materials and exploit the riches of nature wherever they prove profitable. The question that begs a serious answer is whether it is possible to reconcile the ethos of neo-liberal globalisation and the imperative of global environmental protection.

While the introductory section of this chapter has been devoted to the environment-development nexus seen as a process of ecological modernisation, the following section takes issue with the debate on globalisation and the environment, with specific reference to the debate on free market environmentalism and its potential contribution to global environmental protection. The section argues that development research has, to some extent, not been able to cope with the enormity of the configuration of the intricate relationship between globalisation, sustainable development, and the quest for economic growth.

Is there a Way Out?

The picture that I have painted so far is a gloomy one in which global environmental actors as well as the development lobby seem to have been locked into a losing battle against the powerful tides of globalisation and an impending environmental crisis. The question that follows is whether the social sciences have managed, somehow, to leave their imprints on the debate by explaining the 'social' and its relationship to 'nature' and how they influence each other on a global-scale and what policy issues have likewise preoccupied them. Social sciences have contributed to the debate in the field of environmental policies and respective institutional mechanisms responsible for the mitigation of global environmental problems. What can environmental policies and institutions deliver in such a situation that requires more than a balancing act, and what are their limitations? These are difficult questions to answer and are certainly beyond the scope of this chapter. However, I will attempt to delineate some of the main factors that influence global environmental policies and institutions and their response to the processes that simultaneously influence globalisation and sustainable development. Environmental policies are part of international policies. This was first defined by Vig (1999:17) as joint responses to common problems that two or more national governments work out with one another, often with the active participation of inter-governmental organisations and non-governmental organisations. In the field of international environmental policy, Vig defines them as ranging from broad policy declarations and informal agreements to highly articulated conventions and treaties. According to Vig (1999: Ibid),

The range of international environmental policies currently in force is vast, covering protection of endangered plants and animals; protection against transboundary pollution of air, water, and soil; protection of the atmosphere against acidification, ozone depletion, and climate change; protection of the oceans against oils spills and the dumping of radioactive and other hazardous chemicals, pesticides, and hazardous wastes, measures to combat desertification; and protection of Antarctica.

There is no gainsaying that international environmental policies represent a direct response to the consequences of ecological modernisation. On a global scale this can be seen in the expansion of Western consumption patterns and life-styles that have subsequently strained the functioning of the fragile environmental life support system on which the production of globally consumed goods and services depend. However, international environmental policies require global environmental institutions with transnational mandates to safeguard the protection of the global commons.

The major institutions entrusted with the implementation of international environmental polices are the United Nations General Assembly (UNGA), the United Nations Environmental Programme (UNEP), the Commission on Sustainable Development (CSD), the Global Environmental Facility (GEF). The three Bretton Wood Institutions (the International Bank for Reconstruction and Development, the IMF, and the World Trade Organisation) have adopted reforms in the face of strong criticism for pursuing economic missions in ways that have further aggravated environmental problems (Soroos 1999:47). Soroos (1999:48) laments by conceding that:

> It is difficult to conceive of circumstances in which states would be willing to relinquish or pool their sovereignty to substantially strengthen global institutions charged with mounting a more effective response to the deepening environmental crisis confronting humanity. If anything, there is growing public disillusionment with global institutions and policies in both industrial and developing countries because of their perceived failure to be responsive to local needs and preferences.

The international environmental policies and institutions together constitute a diverse body of global environmental governance. The strong and active presence of Bretton Wood institutions and their quest for the liberalisation of the world economy elucidates that global environmental governance is part of the globalisation process. If traced from an evolutionary perspective of sustainable development, the Earth Summit and Agenda 21 have laid the foundations for the advocacy of global environmental governance within the wider concept of global governance. According to these foundations, there is no single model, form, or institution of global governance, nor is there a single structure or set of structures. However, recognising the systemic nature of the issues with which it deals, it promotes systemic approaches to dealing with major global concerns. The aims of global governance include reforming and strengthening the existing system of intergovernmental institutions, and improving its means of collaboration with private and independent groups. Among other things, global governance will strive to subject the rule of arbitrary power — economic, political, or military — to the rule of law within global society.

I argue elsewhere (Mohamed Salih 1997a) that global environmental governance operates within the confines of, and consistent with, the principles laid down in global governance legal instruments, conventions and international environmental policy proclamations. Our *Global Neighbourhood* (1995:216) indicates that global environmental governance emphasises the need: 1) to put into action global legal, intellectual, and institutional groundwork for a concerted effort to achieve sustainable development; 2) to establish a Trusteeship of the Global Commons (TGC) and to be made responsible for acting on behalf of all nations, including the administration of environmental treaties; to enact an Earth Charter for world endorsement, pending the outcome of the global efforts exerted by the Earth Council and Green Cross to achieve this endeavour.

Considering the ultimate association between environment and development, it has become obvious that one of the most conspicuous outcomes of the Earth Summit is that the stage has been set for global governance aided by diverse trends of globalisation to become the dominant paradigm and the most influential actor on the global scale. Global environmental policies, according to Rich (1994:273), have evolved around the premise that,

Global environmental management can work. The project as it emerged out of Rio has two principal approaches - a growing body of international environmental treaties among nation-states and increased foreign aid for environmental protection and management as well as poverty alleviation, channelled and managed mainly through the Global Environmental Facility (GEF) and other financing arms of the World Bank.

Notions such as global environmental policies imply not only that the natural economy has already been rendered subject to global capital operations, but also that sustainable development has been defined according to the logic of global capital operations and that of the dominant global governance institutions that support it. Seven years after the Rio Conference proclaimed Agenda 21 (1992) as the guiding principle for global-sustainable development, the question remains as to why contradictory Northern political discourses have converged to agree on the assumption that what is ecological is by necessity global. Erroneously, however, that which is ecological is global, but that which is global concerns a socially, economically, and politically differentiated humanity. Global environmental governance has therefore been an integral part of the economic, social, and political processes responsive to Agenda 21 and other international environmental policy proclamations, conventions, and treaties.

The question that continues to beg for an answer is whether such grand global projects as environment and sustainable development, free trade, and even globalisation are reconcilable within the new global context of development. The global environmental movement is explicit about the need to establish a global resource management regime responsible for protecting the global commons from abuse by irresponsible nation-states or greedy corporate interests. Managing the global commons therefore requires global environmental policies, respected and implemented by all nation-states. Neo-liberal globalisation, on the other hand, is premised on the need for trade liberalisation and opening up new markets and reaching out for new consumers.

The global environmental conventions, treaties, and laws, which resulted from the deliberations of multilateral environmental forums and negotiations, have contributed to the consummation of sustainable development as a global ideal. As I have mentioned earlier, sustainable development advocates the need to integrate environment and development in decision-making. In my view, sustainable development meets free market environmentalism half way in defending decentralised private choice of the market as a superior economic management system to the command and control approach. According to Eckersley (1995:15-16), free market environmentalism asserts that bureaucratic regulations are inefficient and open to corruption in many environmental regulatory regimes concerned with the management of forests, water rights, and public lands. It also argues for the removal of government subsidies and intervention in the form of externality taxes to correct market failures.

Because free market environmentalism has much in common with the current drive towards economic policy reforms (deregulation, free trade, and the market liberalisation and open door economic policy), not surprisingly it argues for rolling back the state and assigning it a minimal role in environmental protection. Hence, as part of the core values of the global environmental agendas agreed upon by states (Agenda 21, 1992) and other international conventions, it advocates growth as an engine of sustainable development and hence environmental protection.

More than the good-will of global finance institutions that try to integrate the environment into their investment operations, the protection of the globalenvironment will depend on the capacity of strong states to implement economic policy instruments capable of harnessing/mitigating or abating the consequences of economic growth on the environment. Legal instruments or command and control policies, although important, will not be in a position to deliver results without persuasive socio-economic incentives.

The current environmental policy principles developed for OECD countries and applied in the developing countries have been summarised by Opschoor and Turner (1994:4) as follows:

(i) The 'polluter pays' principle: i.e. the polluters pay the cost of meeting socially acceptable environmental quality standards.

(ii) The prevention or precautionary principle: this explicitly recognises the existence of uncertainty (environmental and social) and seeks to avoid irreversible damages in relation to the imposition of a safety margin into policy; it also seeks to prevent waste generation at source, as well as retaining some end-of-pipe measures.

(iii) The economic efficiency/cost effectiveness principle: this applies both to the setting of standards and the design of the policy instruments for attaining them.

(iv) The decentralisation principle: to assign environmental decisions and enforcement to the lowest level of government capable of handling it, without significant residual externalities.

(v) The legal efficiency principle: this seeks to preclude the passage of regulations that cannot be realistically enforced.

Reflections on these environmental policy principles and their practice in various parts of the world have produced mixed results, even in their proponents' view. Business and international trade operators often misuse economic policy instruments, including market-based instruments such as (a) taxes on emissions and tradable emissions permits and (b) resource use-permits. However, I equally doubt whether the economic policy instruments without legal instruments to enforce them could provide sufficient safeguards for effective environmental quality control. My reasons are as follows:

(i) Environmental taxes keep down the cost of compliance since the market determines that those who can most afford to act do so, and they deal with the total amount and distribution of the problem.

(ii) Environmental taxes act as a continuous irritant to the polluters, who therefore have a continuing incentive to avoid the financial cost and taxes by introducing cleaner technology.

(iii) The price of polluting products will tend to be higher than the price of clean products, and thereby the consumer will encourage producers to choose the latter.

(iv) Environmental taxes can be used in a fiscally neutral manner to reduce other distorting taxes in the economy.

(v) Reinforcement of environmental standards and the 'polluter pays' principle. If the influential regional trading blocs (EU, ASEAN, and NAFTA) adopt reasonable environmental standards and adhere to the 'polluter pays' principle, they may be able to ensure that pollution control and environmental costs are internalised into enterprise costs and product prices.

However, it is doubtful given the current variations in levels of economic development within and between states that the affluent ones would agree to sacrifice affluence for the sake of global environmental gains. In this respect, developing and small industrial countries seem fearful of the implications of implementing global environmental protection policies. On the other hand, enforcing compliance with global environmental legal instruments such as trade restriction may encourage non-co-opcration in global environmental protection. However, without global environmental co-operation, international environmental laws are ineffective regulatory instruments.

Obviously, this discussion illustrates that social scientists are at the forefront of attempts to understand the forces behind globalisation and environmental destruction, as well as attempts to contribute to the policy debate. In the process, social scientists have become more aware of the intricacies that may hinder the advancement of firm conclusions, such as those of the natural sciences. Essentially, the globalisation-environment debate has questioned the ability of disciplinary boundaries to withstand the complex nature of environment and development and their multi-layered consequences on the totality of the human condition. The richness of the debate and its capacity to traverse several social and even natural science disciplines is in my view, to a large extent forced into the social sciences by the paradigmatic shift from development to sustainable development. This shift however speaks more for the ability of the social sciences to respond to new challenges than it does for whether the social scientists agree or disagree with the ethos of sustainable development as a global ideal.

The debate also integrates theoretical-philosophical policy and activists' research. As lamented by Redclift and Benton (1994:7-8), the relationship between structure and agency became more focused and there is hardly any contribution that can be described as purely theoretical or philosophical in nature. This is particularly true even for contributions from environmental ethics and environmental philosophy, which are defined as practical philosophies that engage the environmental consciousness of individuals and groups to restructure their relationship with the environment.

The social sciences have more than ever before been confronted with a myriad of data collected on a global level and that has little to say about the socio-economic and environmental conditions of the societies from which it has been collected. While some social sciences (e.g. economics) can generate data on a global scale, others are more concerned with the local consequences of globalisation, better known as the global-local nexus.

Although globalisation, environment, and development are not entirely new subjects, the intensity and complexity of their interaction pose new and complex questions that have structurally forced disciplinary social scientists to abandon their earlier hard-nosed attitudes towards interdisciplinary research.

The questions posed by globalisation in its various manifestations, including sustainable development, are difficult to answer within the confines of any single social science discipline. Disciplinary approaches unconscious of their own limitations (and those of the social sciences as a whole to integrate development and the environment outside the policy realms) have not succeeded in hindering our understanding of the complex relationships linking powerful and overloaded concepts, such as globalisation and the environment. As the current debate illustrates, the uncertainty that has evolved around the future of development has also manifested itself in the uncertainty about our environmental future. Unfortunately, post development is also not an answer as globalisation eats its way, sometimes violently, through the globe creating new dynamisms that traverse environment in its physical sense and development in its most celebrated economic sense. A new creative and all-encompassing understanding of the complex relationship between the global and the environmental of necessity requires a better understanding of the dynamic forces that steer the future of human destiny. The social sciences will betray their historical responsibility, if (and here I am ameliorating Beck 1995) they continue to treat the social and environmental problems of the 21st century as if they are a continuation of the 20th century. Global environmental consciousness has become a dominant factor influencing the definitions not only of what is local and what is global, but also the linkages between the local, the global mediated by the local-global nature of the environment. It has in a sense become increasingly difficult to manage the balancing act that satisfies the requirements of a liveable world within the core values linking globalisation perceived by some as an economic ideal and a sustainable development ethos perceived by others as a global ideal.

Globalisation and the Environment

One of the questions we face is whether global resource competition for high and sustained growth would foster or hinder the realisation of global projects such as free trade and global environmental protection. Is it practically possible to reconcile globalisation, free trade, and global environmental protection? Even if it is possible to reconcile free trade and the environment protection on a global scale, what institutional mechanisms are needed to address these issues? In this respect, globalisation and the environment can be explored by examining one of the most influential trends, which (if a perfect market ever existed) should ideally transform global trade and conversely the relationship between trade and environment in a globally open trade regime.

In essence, I argue that free market environmentalism is an extension of the ethos of neo-liberal globalisation and therefore both treat global environmental protection as subservient to global capital flows and the expansion of trade and economic growth. However, paradoxically, the globalisation of the environmental issues and the quest for free trade constitutes the foundation stone of open globalisation, which, if properly implemented, would allow for greater solidarity of a disenfranchised and diverse environmental movement (Martinez and Guha 1997). This explains why all regional trading blocs have signed protocols and conventions, which have committed them to regional and global environmental protection laws.

However, there is no agreement as to the exact impact of the liberalisation of trade and free market globalisation on the environment. The anti free market lobby, including environmentalists, argue that trade liberalisation is harmful to the environment because it puts more emphasis on growth and development at the expense of global environmental protection. They also argue for greater decentralisation of economic and political' activities to empower people and enable them to care for their local environments. Trade liberalisation without proper environmental protection policies, legal instruments and economic incentives (or liabilities) often results in land degradation and environmental disasters in the marginal and fragile global commons. This will hold globalisation responsible for the dangers that threaten our global future.

In a globally free trade regime, one of the institutions that will have a direct effect on the environment is the World Trade Organisation (WTO), which represents the triumph of the trade liberalisation lobby in conceiving the General Agreement on Tariff and Trade (GATT). The WTO is hailed as the epitome of how trade liberalisation should lead to growth and therefore generates financial resources, which could be utilised to protect the environment. The WTO aims to liberalise four areas of economic activity previously considered to be the exclusive domain of the national legislation. These are: (a) trade in services, (b) intel-lectual property rights, (c) international investment flows, and d) agriculture.

The WTO presumes that, by concentrating resources within each country into the most productive and competitive sectors, (i) free trade will enable countries to maximise their competitive advantage; (ii) a globally increased trade is essentially good for development, and the more people trade, the better off they will be. Brack (1998:1) supports this view by arguing that,

In principle, the pursuit of both (free trade and global environmental protection), are compatible. According to the theory of comparative advantage, trade allows countries to maximise output from a given input of resources, which is a movement in the direction of environmental sustainability. Furthermore, trade liberalisation can help to remove distortionary subsidies and pricing policies, improving efficiency of resource allocation, and can encourage the spread of environmentally friendly technology.

Other aspects of global trade and the environment are articulated by Junior and Welford (1997:138), who argue that trade and economic globalisation have greatly expanded the opportunities for the rich to pass on their environmental burdens to the poor by exporting both waste and polluting factories. Trade has a secondary impact on issues of equity as it results in impacts (positive or negative) on poverty, health, employment, human rights, democracy, labour laws, and self-determination. I add to these two points, one aspect of globalisation that may have some positive result on the environment.

The pro-free trade lobby's assumptions have been contested by environmentalists who argue defiantly that the nature of comparative advantage in a globalised world has more to do with absolute price advantage than differences in comparative advantage. For instance, although many developing countries have comparative advantage in the production of tropical crops such as bananas, tea, and coffee, they continue to decrease in the international market due to their weak bargaining position in the world economy. In Eckersley's (1995) words,

One obvious objection to free market environmentalism is that full-scale privatisation of the commons is not possible. After all, many environmental assets (e.g. clean air) are public goods, which, by definition are not amenable to being captured, commodified and bought and sold (i.e. they are non-excludable and able to withstand non-rival consumption).

The difficulty of integrating all environmental goods and services into the market economy raises issues of internalising externalities that are not amenable to economic valuation, as the case of clean air cited earlier.

The meagre attention given to the environment in the GATT process would suggest that no or little value would be attached to the potential environmental destruction caused by excessive use of natural resources in response to such policies. This is particularly true for developing countries, which depend on primary commodity production which, in 1992, insisted on the inclusion of the environmental issues in the GATT process and negotiations. From an environmental point of view free trade and the race for comparative advantage will lower environmental standards for the sake of free trade rather than raise them. Article I of GATT rules out discrimination between trading partners. Articles III and XVI, by forbidding the use of trade subsidies or restrictions for environmental purposes, mean that GATT already has difficulty in acknowledging the legitimacy of trade restrictions (for instance, on polluting substances) in the pursuit of global environmental protection.

However, it is difficult to analyze the development of world trade and its consequences on the environment in isolation from its differential effects on local communities, states, and regional entities. Wilson and Bryan (1996) explored the interdependence of local, national, regional, and global trade transactions and illustrated that multi-layered trade operations would result in the shrinking of environmental space. Earlier Opschoor et al. (1994) concluded that the expansion of global trade operations has resulted in an increased ecological debt and the shrinking of the environmental space available for the maintenance of the global environmental life support system. This process contributes, according to Opschoor et al. (1994), to the shrinking of the environmental space and the increase of the ecological debt, i.e. an increased 'environmental reach' through far-reaching 'environmental foot steps'. A global condition in which the environmental space is shrinking at an accelerating rate is detrimental to the well-being of the industrial world as well as the developing world, where polluting industries have often been relocated in places of lax legal systems. The interdependence of the environmental life support system and its response to large-scale environmental damage or degradation, emanating from an increasing trade in primary products and goods and services, cannot be more revealing.

TRAINING PEOPLE TO MEET THE GLOBAL NEED THE GLOBAL NEED

Education or Training

The word *education* comes from the Latin and means "to draw out." As commonly used, education means drawing out the potential ability of students so that they are prepared for life—for the whole of their future. On the other hand *training* generally means preparing people for a particular task or function or job. All training should also to some extent draw out the ability of the trainees; it should be part of their total education. But training is more specific than education. As a Tanzanian wrote recently, regardless of the organisation the training programme must be based on a known requirement [Swere, 1981]."

The known requirement of the training discussed in this chapter is improvement of trainees' ability to satisfy the global need for water and sanitation. Emphasis is on the Water Decade, the International Drinking Water Supply and Sanitation Decade, 1981-1990.

The Decade

Decade Objectives

The objective of the Decade as set out after the Water Conference held at Mar del Plata in 1977 was that all the world's population should be provided with reasonable access to safe drinking water by 1990. To this was added the aim of giving everyone adequate sanitation. Taking the people who were then without water and sanitation, and allowing for the expected increase in population, gave an estimated global total of 1833 million people to be provided with water and 2400 million to be provided with sanitation by the end of the Decade. However, although the Decade is international, each country is formulating its own programme and many countries have set realistic targets which assume much less than 100% coverage.

New Suppliers and Other Needs

The most attractive and politically rewarding aspect of the Decade is to give a new well or pipeline to people who previously obtained their water from distant or scarce or polluted sources. The technology is understood and the result can be seen. Of course, the technology should be appropriate; all *good* engineering is appropriate (Pickford, 1977). In developing countries appropriate water supply and sanitation is usually low cost and simple. It should also be long lasting, easily operated, and easily maintained. The local people should want to use it and should think of it as their own, not as something imposed by outsiders.

Of course new schemes are politically attractive, but if the emphasis is only on providing new water supplies and sanitation for as many people as possible there is a danger of installing systems which only last a short time. In terms of the Decade, all water and sanitation provided during the 1980s should still be in good order in 1990. In terms of "health for all" there should be no failure by the year 2000. This requires good operation and maintenance. Money spent on training for operation and maintenance is a better investment than money spent on new systems which soon break down.

Renovation is a way of providing water and sanitation which is often ignored. There are thousands of schemes throughout the Third World which are out of use because they were not properly operated or maintained in the past. In some places renovation would be easy. Nearly always it would be cheaper than providing, a new system. However, renovation usually requires thought and skill and care. It therefore depends on having well-trained people to plan it and to carry it out.

Beyond the Decade

The Decade is intended to concentrate global efforts to provide good water and sanitation for those without them. These efforts must not end in 1990. Not all targets will be reached and many targets are short of 100% coverage. Populations will still grow. Operation and maintenance of old and new installation must go on. There will be a need for incremental up-grading. So there will continue to be a need for trained people, and the need for training is likely to grow rather than decrease.

Affluence Versus Poverty

The Decade is largely concerned with developing countries, but their present level of prosperity ranges from $110 per person in Bangladesh through $1070 in the Ivory coast and $1450 in Malaysia to the heights of Kuwait at $20,250 per person per year. Between 1970 and 1979 Botswana's GNP per person rose by 15% whereas nearby Angola's fell by nearly 10% (World Bank, 1982).

Geographical

Developing countries tend to be hot at some time of the year but temperature varies with latitude, season, and elevation. More pertinent, rainfall may be excessive or scarce, seasonal or perennial. Density of population; the slope of the ground; the underlying geology; evaluation, quality, and quantity of groundwater and surface water; ease of communication; local sources of pipes and construction materials; local industries and their wastes—all these influence water and sanitation and hence training needs.

Specific Needs

There is a global need for water and sanitation, and training is needed everywhere. But within the overall requirements for water and sanitation there are great variations which are reflected in different training needs.

Political

The variety of governmental organisation—or disorganisation—is a continual source of amazement. Water and sanitation may be dealt with by one, two, or several ministries, departments, boards, corporations, or councils. Titles of these include water, water resources, health, works, public works, public utilities, environment, agriculture, development, local government, housing, and so on. Control may be by direct government administration, through state or regional agencies, or through parastatal or locally elected bodies, or by a combination of these. Water may be administratively combined with or separated from sanitation; urban and rural sectors may be together or apart.

Sectoral

Water and sanitation, rural and urban are divisions that give four broad sectors—rural water, urban water, rural sanitation, and urban sanitation. In general, each sector uses different technical methods and has different problems. Sectoral boundaries are not easily defined. Should the urban- rural limit be 10,000, 20,000, or 100,000 people? Villages can be densely populated, towns open and widespread. Squatters may bring rural practices to periurban areas. Water supply influences sanitation and sanitation affects water. Nevertheless, the four-sector idea is useful and will be referred to as we consider training.

Who and Why?

Training for Everyone?

The Water Decade calls for some kind of training for nearly everyone. Even the general public in the "North" (developed countries), already well served with water and sanitation, need to be made aware of the situation in the "South" (developing countries). Much has already been done in this direction by UNICEF and other UN agencies, by non-government organisations such as Oxfam and Christian Aid, by church and women's groups, and by organisations such as Rotary International. As a result the rich are "trained" to give to the poor.

At the other end of the "have-have not" scale are those who are now without access to safe water or adequate sanitation and for whom nothing is planned in the Decade. Some form of training can help these most underprivileged of people. If there is fuel they can boil drinking water and the most insanitary practices can be discouraged.

Obviously training is most suitable for the people in developing countries whose jobs are connected with the Decade—those who provide new facilities or keep existing systems in working order. These people may be divided into four categories—policymakers, professional implementers, subprofessional implementers, and users.

Policymakers

It has been said that a first requirement for Decade success is "political will." This applies particularly at the national level because national politicians control the purse strings in nearly all countries. Further down the political scale are states within nations (as in India and Nigeria), regions, and local and municipal councils. At all levels politicians who determine policy and financial priorities should be trained to make decisions that give the greatest benefit.

Although politicians make decisions, or at least think they make decisions, their outlook is affected by others. Senior administrative civil servants often have the greatest influence. Technical chief officers not directly connected with water or sanitation schemes also wield power which can be crucial. The author experienced this in an African country where the top people in the Ministry of Health were experienced medical doctors. They argued that the capital city should be sewered, which was economically impossible. They were totally opposed to pit latrines, which were locally successful. They should have received training to appreciate the economic and other merits of on-site sanitation and to know about new improved latrines.

Public opinion influences political decisions in various ways and to varying extents. Therefore training of the people as a whole is valuable. It is often said that there are votes in water. A common expression of this is politicians promising water supplies to voters in their constituencies. Voters and party members can also exert a more subtle influence. For example, pressure from the public might lead to priority for congested slums or rural water supplies. The well-off might be "trained" to pay more for their own water to subsidize poor people

Professional Implementers

The Engineer in Charge

The people most important to the success of the Decade programme are good engineers in charge. These men and women convert objectives into achievements. They take others' ideas and general policies, initiate detailed plans, and ensure that their own plans are implemented. They are "doers" (Pickford 1982). Their greatest satisfaction is in providing water and sanitation for people.

Good Decade engineers know about the necessary technology, which ranges from the hydrogeology of groundwater to clearing blocked drains. They make the technology appropriate for the people they serve, which often entails having an understanding of their traditions, economic status, ambitions, and way of life. They know about diseases whose incidence an be reduced by engineering measures, and they can make best use organisations vested with responsibility for providing water and sanitation. They control expenditure and get the utmost value from whatever funds are available and make the best use of resources like building materials, human and mechanical input, and local skills. They are ultimately responsible for organising and training other members of the team—professional colleagues, subprofessional subordinates, political masters, and users. They are motivated to work for the continued functioning of any facility as much as for its installation.

Technical Specialists

In many situations the engineer in charge has no professional colleagues and must cope with the whole technological field alone. On the other hand, in a large organisation or center the engineer in charge may have many specialist supporters. They reflect the range of technology involved in water and waste engineering. Depending on the sector (water-sanitation, rural-urban) the full professional team may include hydrogeologists for finding groundwater; hydrologists; soil mechanics people for building earth dams; surveyors; distribution engineers; structural engineers; chemists and micro- biologists for developing treatment; mechanical and electrical engineers for designing plant. All need some training to relate their specialisations to the jobs in hand.

Non-technical Specialists

The large organisation also has professionals in the nontechnical fields. There are often accountants to look after the money, administrators and lawyers to look after the paperwork, social scientists to deal with community relations.

Subprofessional Implementers

Technicians, Artisans, and Supervisors

A shortage of trained technicians in developing countries is often mentioned as a major hindrance to progress, including progress in water supply and sanitation. It is seldom explained what is meant by *technician.* In some places a technician does work similar to that of a professional engineer, but at a slightly lower level with more routine work and less demand for initiative. Surveyors, draughtsmen, and laboratory analysts are often classified as technicians. Elsewhere a "technician" may be an artisan, like a pipe fitter, mechanic, or carpenter. Titles vary a lot. Without a detailed job description it is impossible to know what training is necessary for a technical assistant, a field officer, or a water inspector. What about health inspectors, some of whom claim to be professionals?

The divisions between professionals, and technicians, and artisans cannot *easily* be defined in terms of length of training or level of qualification or even type of work. Graduates with degrees of some universities are only equivalent to entrants to other universities. The training of some craftsmen is longer than that of some professionals. Fortunately in many countries there are professional institutions or societies with rules for membership based on education and experience. So a professional can be defined for convenience as someone with membership in a national professional body accepted in other countries.

Categorising nonprofessionals is not as easy. One possible division is between those whose main value is their *knowledge,* those with valuable *skills,* and those whose work is primarily *supervision* of other people. Divisions like these are far from watertight. For example a senior draughtsman has knowledge and skill and supervises other draughtsmen. A good carpenter knows about joints and what size timber to use, and is also skilled in making joints with saws and chisels. A concreter may have to supervise the unskilled laborers who dig foundations and mix sand and cement. Although the boundaries are not clear-cut, this three-part division of sub-professional work is useful when considering training.

Attendants

In water supply and sanitation there are a large number of workers who cannot be reasonably grouped with the technicians and artisans. Their activities are mainly concerned with operation and maintenance. Some training can make their work much more effective, but the duration of training may be measured in days rather than months or years. These people do jobs that require little skill or knowledge, but are often absolutely essential. Reliability is probably the main qualification for success. An example is a borehole pump attendant at an isolated site. If the pump has an electric motor with automatic switching the operator may have nothing to do unless something goes wrong. Then a message must be sent for help immediately. Without automatic controls the attendant may have to turn the pump on when the water level in a storage tank is low and turn it off when the tank is full. With a diesel or gasoline motor the fuel tank must be topped off. For a hand pump a caretaker's main responsibilities may be to ensure that the pump is not abused and that spilled water is drained away.

In almost all programmes for water supply and communal sanitation these "attendants" are important members of the team, 50 must be trained.

The Users

Users and Beneficiaries

Often users do not fully benefit from water and sanitation programmes for one or more of the following reasons.

1. *Misunderstanding of the reasons* for improvement particularly applies when the primary purpose is to reduce disease. For example, people may not connect water-borne diseases with a polluted source replaced by a new scheme. The old source may be preferred because of taste, convenience, or habit.
2. *Misunderstanding of how to use* a facility occurred in a water-short squatter area where pour-flush latrines were provided. Householders would not use the latrines because of the effort of getting water until they were shown how to flush with dirty water from washing clothes and bathing.
3. *Misuse* of a communal system can harm other people. Waste of water by allowing taps to flow continuously is common even when supplies are very limited. Putting hard material such as newspaper down water closets is a frequent cause of blocked drains and sewers.

Participation in Implementation

In most community participation or self-help programmes the participants need training. In fact there are countless examples of schemes that have failed completely, or have been less effective than they should have been, because the community did not really know what to do.

How Many Need Training?

Global Totals

Since nearly everyone, even the rich in industrial countries, should be given some kind of training, the total global need for training may be taken as the global population, some 4 billion people. The number of professionals and subprofessionals required to provide water and sanitation has been estimated by UN agencies. Generally global figures are far from accurate and are therefore of little help. For example, in the 1970 WHO Survey the figures for people said to be involved in water supply and sanitation showed the South East Asia Region (dominated by India) to have 40 times as many engineers as the rest of the Third World put together. For specialised artisans South East Asia claimed 230 times as many as all other developing countries (Pineo and Subrahmanyam, 1975).

The Implementers

Professionals

One method of estimating the required number of engineers is to study expenditure and devise an engineer-expenditure ratio. Another rule of thumb is to relate the number of engineers to the population. For example, it has been suggested that in Africa one sanitary engineer is needed for every 250,000 people (WHO AFRO, 1973). If a "sanitary engineer" specializes exclusively in waste-water treatment, as implied by some U.S. authors, this ratio may be reasonable. But obviously for engineers involved in water supply as well as high-cost and low-cost sanitation, and in operation, maintenance, and renovation as well as design and construction, the ratio needs to be very much greater—possibly one engineer for between 10,000 and 50,000 people.

Subprofessionals

The number of subprofessionals is sometimes related to the number of engineers by a simple formula, although this is no more satisfactory than rules of thumb for deciding professional needs. In two parallel documents we read that "a hundred technicians are needed for every engineer (IDRC, 1981b)" and that a suitable engineer-technician ratio is 1:4 (IDRC 1981a). For engineers, technicians, and craftsmen there is apparently a "known standard ratio of 1:5:25 [Mlingwa, 1982]."

Human Resources Surveys

Since training should be related to real needs, an obvious and logical first step in any training programme is to find out as exactly as possible the shortage of trained people. A pitfall to be avoided is drawing a hierarchical responsibility diagram working from the top down: a "boss" needs so many deputy bosses, each of whom needs assistants, and so on down to the lowest grades. Ideally the survey should start with the actual jobs to be done, such as opening valves or joining pipes. These jobs should be defined as accurately as possible in terms of skill and knowledge, and related to the men and women doing the tasks. The supervision needed by these people should then be decided so that the skills and knowledge required by the supervisors can be specified. The process is repeated up the scale to the boss.

Having decided the total needs of man-and-woman-power in each relevant category, the number of people already effectively employed should be deducted. The deficit can be made up either by obtaining already trained people "in the market" or by training untrained employees. In this way training is specifically related to the gap between the existing ability in the available pool of untrained people and the unfilled jobs for which people are required after training.

What kind of Training?

Programmes

Unplanned training can be wasteful of effort. So when precise training needs have been decided a programme should be devised to provide for these needs in the most efficient way. An organisation's training programme obviously varies according to the type of organisation, the sectors involved (water-sanitation, rural-urban), the level of general education, and so on. In training there are spirals of "have-have not" which afflict so much Third World development. A country or organisation that is already well staffed with reasonably trained and competent people has the resources for providing in-house training and organising field training. On the other hand an organisation whose staff strength is already low either in numbers or in terms of previous training may only be able to cope with a dismally inadequate training programme.

Fortunately many international agencies and bilateral donors are keen to help with training. During times of financial stringency it is seen as a way of using a small input effectively. Of course, having a motley collection of trained people derived from many countries creates difficulties. Just as a country that accepts a few pumps from each of several donors faces serious maintenance and spare-parts problems, so training provided by several foreign countries may be so diverse that trainees cannot work together. At the artisan or crafts level such diversity is less troublesome, and there is something to be said for accepting all relevant foreign training in the purely technical field and insisting on a single training pattern for management and supervisory skills.

Whatever the level or duration or form or location of training for the Decade there are many common issues—enough to fill several books if fully examined. Some will be touched on next.

Some Aspects of Implementers' Training

Rational and Emotional Thinking

A Kenyan has suggested that the objects of training are "to achieve excellence, commitment, dedication, devotion and loyalty in 'service to the nation' [Shikwe, 1981]." Apart from excellence, which describes *how* people work, these aims refer to qualities which determine *why* people work. Skill and knowledge can be enhanced by emotional involvements. In a paper on the achievement of engineering excellence Hanny (1981) points out the importance of the interface between two parts of the brain, the interbrain and the cerebral cortex. The nerves of the interbrain control feelings and emotions; the cerebral cortex is a storage unit for information and is the center of rational thought and speech. The proper combination between rational and emotional thinking is like a sailing boat supported by the water (rational) and propelled by the wind (emotional).

Past and Future

A good teaching method is to move from the known to the unknown. So training at all levels should build on the previous knowledge and experience of the trainees. Most (but not all) training should develop the trainees' present effectiveness and future usefulness. In other words, on completion of training trainees should immediately be able to do the work for which they have been trained, and their potential to be useful throughout the rest of their working lives should be increased.

Technology and Health

Elimination of disease is such an important feature of Decade activity that almost all technological training should include some health education. In particular the possible health hazards resulting from poor work should be emphasized. As an example at the lowest level, hand-pump caretakers after training should know something about the diseases likely to be transmitted if they allow puddles of spilled water to form around their pumps.

Status and Job Satisfaction

People often submit to training for personal or selfish reasons. They may simply want to get away from home and work for a time, or may expect greater status, promotion, or higher pay. All good training motivates trainees to work well because of the satisfaction that good work gives. Increased status resulting from the training should be seen as a way of improving effectiveness. Taking the example of hand-pump caretakers again, the higher status accruing to the trained caretaker (maybe including an el borate certificate) should give him or her authority to prevent abuse of the pump by other users.

Training as Trainers

As far as possible everyone completing a period of training should be able to train other people.

Elements of Implementers' Training

Courses

These may be defined as periods during which trainees are free from normal work for intensive training. A variety of methods may be used, including lectures, question-and-answer sessions, discussion, and practical work of all kinds. In general the more the trainees have to do (rather than being passive listeners or watchers) the more they learn.

On the Job

This is essential. On-the-job training is ideal if there is plenty of time, if experienced supervisors are available (preferably with a one-to-one ratio), and if the trainees can be given a variety of experiences. At its worst trainees learn to appreciate the difficulties of the job.

Continuous Training

No one should be considered too old to learn. An individual's training should continue throughout the working life, revising what is half-forgot- ten, increasing ability, and adapting to new conditions.

Evaluation

At the individual level, the effect of training on each trainee should be measured by some kind of test or examination, or by observation of performance on the job. At the programme level, training staff must continually review the content and methodology in order to effect improvements.

Training of Polycymakers

Bearing in mind that policymakers (in the sense used here) are those not directly involved in implementation, the training they receive should concentrate on motivation and providing sufficient information to enable them to make sound decisions. Consequently it may be to some extent a "public relations" effort. As with commercial advertising, success may be achieved by constant repetition of similar themes. Fortunately there is a plethora of printed material available from international bodies and in magazines and books (including this one). If possible this should be supplemented by local information and personal contact. In many places seminars have been successfully used to put over the entire Decade message or some particular aspect of it, such as the advantages of low-cost sanitation. In the Philippines the national water authority A WAS organises courses for politicians and members of local committees.

Training of Professionals

Current Malpractice

The most common entry to professional water and waste engineering is through a degree in civil engineering. A fascinating indictment of engineering degree courses in Africa (Gecaga, 1981) suggests that "the implicit objective of university courses seems to be the production of researchers. The appointment and promotion of staff is based upon research criteria and it is almost impossible to appoint experienced engineers because they seldom have research qualifications [p.149]." Unfortunately similar criticism can also be leveled at most universities outside Africa—to make matters worse civil engineering is dominated by theory of structures and many students have no practical water engineering instruction included in the syllabus. In those few universities where "sanitary engineering," "public health engineering," or "environmental engineering" have a place, the lecturer's background is likely to be in research in a Sophisticated technique such as the kinetics of activated sludge, which is irrelevant to Third World needs. It says a lot for the enthusiasm of graduates with this kind of education that many of them quickly pick up an understanding of practical engineering from their subsequent on-the job experience.

Postgraduate Courses

Around the world many institutions provide postgraduate courses in water supply and sanitation. Some suit Third World needs well. For example, hundreds of engineers have graduated from the 2 year master's programme at the All India Institute of Hygiene and Public Health (AIIHPH) in Calcutta. The training is appropriate for Indian conditions with emphasis being placed on rural water supply and sanitation. Staff and students are also concerned with urban problems such as Calcutta's slum improvement schemes and the collection of refuse. The international courses at Delft in the Netherlands have also been operating for many years, attracting students from all parts of the world. There is less emphasis on low-cost lechnology as students come from the "North" as well as the "South," but all benefit from exchange of ideas and experience.

The majority of master's courses concentrate on research-orientated high level technology. In India alone there are many courses each with a handful of students learning from professors who gained their Ph.D.'s doing sophisticated experimental research in Europe and North America. The Asian Institute of technology near Bangkok has a reputation for good research and has modern buildings with excellent laboratories. In England the public health engineering courses at Imperial College,

London, and the University of Newcastle upon Tyne have been running since the 1950s and usually have some Third World students of high caliber. Their former students now hold leading Positions throughout the world. Many universities in the United States and Canada draw large numbers of students from developing countries. They are even better known for their Ph.D. programmes, which are particularly attractive to the staff of Third World universities and polytechnic Several Australian universities accept large numbers of students from Southeast Asia and the southern Pacific. In Africa master's courses have recently started at Nairobi University and Ahmadu Bello University in Nigeria, but from the start both had serious staffing problems. A former member of Ahmadu Bello's staff has proposed regional Environmental Technology and Research Institutes (Iwugo, 1981).

There is a different emphasis in the master's course in water and waste engineering for developing countries (WEDC) at Loughborough, England.

All candidates for registration must have had at least 2 years' practical experience in a developing country. Teaching staff (the WEDC Group) are selected for their experience rather than their research records; most have worked as engineers or scientists in Africa and Asia for several years. The aim of the course is quite clearly to train engineers for work in their own countries on water supply and sanitation projects and programmes. This gives a bonus to Third World employers as WEDC graduates have not been trained for work in the "North." They do not contribute to the "brain drain" to Europe and North America.

In the WEDC course, practical design, construction, operation, and maintenance are given more attention than theory and analysis. Successful completion of the course should ensure that graduates have a thorough mastery of essential technology. In addition the syllabus includes health and disease, integrated urban and rural development, socioeconomic factors and community participation, finance, and management. Because of the variety of conditions in the Third World students are able to specialize for part of the course in the technology and related topics that are most appropriate for their future employment—whether it is for conventional urban water supply or low-cost sanitation, for populations that are below the poverty line or those that are comparatively affluent.

The WEDC Group also takes well-qualified technicians (such as those who gained a distinction in a relevant higher diploma examination) and converts them into professional water and waste engineers by an extended course. It has been suggested that for developing countries the more practical training of a technician may be better preparation for professional work than the "academic" education provided by many universities.

Although the majority of engineers are male, water and waste engineering is quite suitable for women. More women graduate as chemists and biologists than as civil, mechanical, or electrical engineers. Women scientist who "convert" to water and waste engineering often do well. In developing countries women are usually responsible for domestic water supply and sanitation, and for hygiene and health education. It is therefore appropriate that they should be involved in these things at the professional level. Training of women professionals should have an especially high priority.

Training of Specialists

In the section entitled, "Who and Why?" the crucial role played by engineers in charge was emphasized, and so it has been reasonable to devote a great deal of attention to their training. Large organisations or centers may also have specialists. Their work may be restricted to their own specialisation and the extra training they should receive is directed to ensuring that they appreciate the work of others and understand how their part fits into the whole. For example, a mechanical engineer responsible only for the maintenance of a water board's vehicles has very limited training needs regarding water supply and sanitation.

On the other hand if the board needs hydrologists or hydrogeologists to investigate catchment or aquifer yield it may have to sponsor the training of people with suitable potential. The most appropriate training for someone who will be mainly involved in research-type investigations may be research in a university leading to a M.Sc., M.Phil. or Ph.D. Often the training of specialists can be readily provided by home or overseas universities because most university teachers are themselves specialists.

Training of Sub-proFessionals

Variation of Practice

Division of Functions

As discussed in the section entitled "Who and Why?" the titles given to subprofessionals vary considerably. The word *technician* has a whole range of connotations. One rough and not entirely satisfactory division of subprofessionals and their training needs is into those who use knowledge, those who use skills, and those who supervise.

Health Personnel

Another complication arises from variation in departmental responsibility. In many countries all pipeless systems for both water supply and sanitation come under a ministry of health with medical doctors rather than engineers in control at the professional level. Titles of subprofessional then include *health aides, health assistants,* and others incorporating the fashionable word *environmental.*

Training for Knowledge

The Quest for Qualifications

In a number of countries the courses offered to technicians seem to be determined by the institutions' pursuit of status. A technical school aspires to be a technical college, a technical college desires to be a polytechnic, and a polytechnic wants to be a university. More and "better" qualifications are offered with little regard to the work students will do when qualified. In some of these places the staff have the failings of universities: appointment and promotion depend on research records and the ambition is to make courses "academically respectable." The contents of some courses are watered-down versions of university courses, even when they are two or three steps below degree standard. The syllabus may become absurd when the objective of a course is to obtain a foreign qualification, which sometimes applies even to training of plumbers and fitters. An extreme example in another field was an examination paper taken by students in tropical Africa: seven questions related to heating of buildings and only one to air-conditioning. In examinations set in temperate climates about waste-water treatment few questions relate to waste stabilisation ponds, which should be the first option considered in the tropics. To many commentators the obsession with examinations is the bane of education at all levels from primary school upward. In the 1950s and 1960s much was spoken and written about the harm caused, especially in the Indian subcontinent, by the colonial examination system. Despite this, since national independence the influence of examinations on education has actually increased as educational opportunities increase.

Sandwich Courses

For water and waste technicians knowledge imparted in the classroom is usually supplemented by that learned on the job. For example in Kenya a training school maintained by the Ministry of Water Development takes "0" level candidates and gives them a preselection course which covers such diverse subjects as water law, accounting, hydrology, mechanical work, and drawing. After selection *water supply inspectors* are given 2 years' field training followed by a refresher course and examination. Those selected as the best students spend 4 years training to be *engineering assistants* during which 8 terms are spent at the polytechnic and the rest of the time is spent on the job (Ichung'wa, 1981).

A Courses for Health Assistants

In another African country, after completion of primary school, students can be accepted to a 1-year course for health assistants. On completion they know something about construction of wells and pit latrines, operation and maintenance of rural water supplies, pest and malaria control, and a number of other subjects related to personal and public health. It is likely that with such a wide coverage the knowledge gained is superficial. Learning in this way is often confined to the application of simple rules. Many sanitation programmes have failed because rules were too rigidly applied, without regard for variations in soil, groundwater, user-preference, and other factors that determine the most appropriate latrines for particular situations.

Training in Skills

Informal and Formal Training

In developing countries many skills are learned entirely on the job. The diligence and ability of artisans working on their own or in small groups belies the absurd notion that Africans or Asians "do not like to get their hands dirty." In the slums in and around the towns and cities a great variety of articles are produced from waste material, and wayside motor mechanics and panel-beaters can "fix" a car or lorry in a quarter of the time taken by a reputable garage. True craftsmanship is often found, and is passed on to others, in the informal sector (King, 1977). On the other hand attendance at trade schools and the like often produces results that are quite contrary to that intended by their founders. By continually trying to achieve higher standards these schools train students who feel above their crafts and who are only willing to take jobs in government departments or in large multinational corporations.

Barefoot Technicians

The concept of the "barefoot doctor" is familiar because of its often- reported success in mainland China. The "barefoot technician" idea is similar and has three components:

1. *Training:* The barefoot technician has limited training obtained on the job (working with another barefoot technician) or during a short course. This training is sufficient for the work to be done but does not go beyond it. Consequently training resources are used to the full by dealing quickly with a large number of trainees rather than by giving extended training to a few.
2. *Communal:* Barefoot technicians work among people they know with whom they live. In rural areas they probably have their own farms raising subsistence crops. Their craftsmanship, whether in the construction of latrines or cleaning of filters, becomes part of the local tradition, fitting in in the same way as the carpenter who repairs bullock carts or the man who makes sandals from worn-out tires.

3. *Referral:* They know the limit of their own capability, know when outside help is needed, and know how to get outside help.

Conventional Skills

Motor mechanics and electricians for pumps, masons, concreters, carpenters, drain layers, plumbers, and pipe fitters have their place in the larger water and waste organisations, especially those dealing with urban water supply and water-carried sanitation. All these tradesmen are also likely to be employed by other government departments and by private industry such as building and civil engineering contracting. A water and waste authority is then in the same position as other employers with the same obligation to [train new artisans and the same opportunities to take on artisans trained by others. However, at least three factors increase the training efforts which an authority should make.

1. There may be little requirement outside the authority for specialist trades such as pipe fitting, so nobody else provides training.
2. Rates of pay that the authority can offer may be lower than in industry; or overtime, bonus payments, or conditions of work are inferior to those in the private sector. The authority is then unable to attract trained artisans from outside and has to train *all* its employees, with the probability that a large proportion will move into private industry soon after their training is complete. This problem of skill drain from authorities is increased in some countries by emigration of artisans to more affluent nations (e.g., to the United States or Canada from the Caribbean, to the Middle East from Asia and East Africa).
3. Conditions of service, particularly regarding continuity of employment, may be better with the authority than with other employers. Then the authority can reasonably expect a higher standard of workmanship from its artisans, and may introduce special training to achieve this.

Training for Supervision

The roles of foremen, chargehands, and other supervisors is often crucial to the success of projects and programmes. Their functions vary tremendously, from a man or woman leading a gang of unskilled laborers while working with them to the works superintendent responsible to the engineer for all the subprofessional employees on a large scheme. Three points about their training are worth noting.

1. The supervisors' emotional thinking is often more important than their rational thinking. Training should concentrate on their motivation, their attitude to their work, and their attitude and understanding of the people they supervise.

2. Supervisors may be chosen either by their knowledge or by their skill and experience, and there is something to be said for both methods. An example of the former is when secondary school leavers are specifically trained in colleges or in special courses for supervisory functions. An example of the latter is when a fitter is selected because of his or her qualities of leadership or skill to be head fitter.
3. An important function of supervision is to act as an intermediary between those in higher authority and the work force. Supervisors must therefore be able to understand instructions, and interpret them and pass them on. Communication of this kind may be by word of mouth, or by writing or drawing; good communication can be encouraged by training.

Attendants

In the operation and maintenance of many water and waste facilities, essential tasks can be carried out by unskilled people. The duties of a latrine attendant or the "caretaker" of a village hand pump can be learned in a few minutes, and the routine operation of small treatment plants may take little longer. Nevertheless training can perform two useful functions, and it is worthwhile to take trainee attendants away from their homes for 2 or 3 days (and then awarding them certificates at the end). The first function refers to the attendants themselves. A training course can be used to impress them with the importance of the job and the ill effects of failure to do simple tasks regularly and reliably. Consequently basic health education forms part of the course content. The second function is to give the attendant status in the eyes of his or her neighbors so that they will accept instructions from the attendant.

Training of the Community

Community Participation

When the community is involved in self-help schemes some training is necessary. At least part of this may be in the form of "public relations." For the rural water supply projects in Malawi, one of the most successful examples of participation, the community was involved at all levels and through the whole cycle of planning implementation. (Robertson, 1981).

Training of Users

The success of completed schemes depends on proper use. For example, the benefits to health of latrine construction programmes are lost if people continue to defecate in the open, and piped water supplies can be positive health dangers if nothing is done to get rid of surplus water. Training of the community may be the responsibility of community health workers or development officers, possibly in collaboration with committees of local people or traditional meetings. Schools can be effective media for general communal training, the children passing on what they have learned to other members of the family. Engineers and others involved in project implementation can contribute usefully to training of the whole community.

Points to Ponder

- Training for the Decade must be for operation and maintenance as much as for new water supply and sanitation schemes.
- Beyond the Decade there will still be much to be done, and training in the 1980s will prepare people for the 1990s.
- Training must be appropriate for Decade activities, which must themselves be appropriate for all the variations of conditions and varieties of people in the Third World.
- Training for emotional thinking (motivation) may be more important than for rational thinking (technology).
- Training of graduates should be related to their future work rather than to the research interests of University staff members.
- Artisans' quests for qualifications sometimes result in their losing interest in what they are able to do best-using their skills.
- The most useful function of Decade training may be to instill qualities of reliability in every job that has to be done.

REFERENCES

Adams, W. (1990) *Greening Development: Sustainable Development in the South,* London and New York, Routledge.

Agarwal, A. and Narian, S. (1991) *Global Warning in an Unequal World: A Case Study of Environmental Colonialism,* Delhi, Centre for Science and Environment.

Axford, B. (1995) *The Global System: Economics, Politics and Culture,* Cambridge, Polity Press.

Beck (1993) in Risk *Society* and Giddens (1996) in "Urbanism, Globalisation and Environmental Politics".

Beck, U. (1995) *Ecological Politics in an Age of Risk,* Cambridge, Polity Press.

Bierstreker, T.J. (1996) "The Triumph of Liberal Economic Ideas in the Developing World". In Stallings, B. (ed.) *Global Change and Regional Response: The New International Context of Development,* Cambridge, Cambridge University Press.

Blaikie, P. M. and Brookfield (1989) *Land Degradation and Society,* London, Methuen.

Brack, Duncan (ed.) (1998) *Trade and Environment: Conflict or Compatibility,* London, The Royal Institute of International Affairs.

Brenton, T. (1994) *The Greening of Machiavelli, The Evolution of International Environmental Policies,* London, The Royal Institute of International Affairs.

Brundtland Report (1987) *Our Common Future,* Oxford, Oxford University Press. World Commission on Environment and Development.

Charonvitz, S. (1994) "Free Trade, Fair Trade, Green Trade: Defogging", *Cornell International Law,* 27:3, pp. 498-505.

Clark, L Gordon (1995) "Global Competition and Environmental Regulation: Is the Race to the Bottom Inevitable." In Eckersley, R. (ed.) *Markets, the State and the Environment,* Basingstoke, Macmillan.

Daly, H. (1996) *Free Trade, Capital Mobility and Growth versus Environment and Community,* The Hague, Institute of Social Studies.

Dieren, Wouter v. (ed.) (1995) *Taking Nature into Account: A Report to the Club of Rome,* New York, Copernicus.

Eckersley, R. (1995) "Markets, the State and the Environment". In Eckersley, R. (ed.) *Markets, the State and the Environment,* Basingstoke, Macmillan.

Ehrlich, P. and Ehrlich A. (1996) *Betrayal of Science and Reason: How Anti-Environmental Rhetoric Threatens Our Future,* Washington DC, Island Press.

Ekins, Paul (1995) *Harnessing Trade to Sustainable Development,* Oxford, Green College.

Engels, Frederick (1940) *The Dialectics of Nature,* New York, International Publishers.

Gecaga, J. (1981). "Training of civil engineers in Kenya." In *Sanitation in Developing Countries,* p. 148-151. IDRC, Ottawa.

Giddens, A. (1996) "Urbanism, Globalisation and Environmental Politics". In Goldblatt, D. (ed.) *Social Theory and the Environment,* Cambridge, Polity Press.

Guha, R. and Martinez-Alier, J. (1997) *Varieties of Environmentalism:* Essays *North and South,* London, Earthscan.

Hannigan, J. A. (1995) *Environmental Sociology: A Social Constructionist Perspective,* London and New York, Routledge.

Hanny, J. (1981), Interfaces: The achievement of engineering excellence. *Sulzer Technical Review,* 77-81.

Hirst, P. and Thompson, G. (1996) *Globalisation in Question: The International Economy and the Possibility of Governance,* Cambridge, Polity Press.

Ichung'wa, S.K. (1981), Role of operation and maintenance in raining. In "Rural Water Supply in Developing Countries," pp. 66-88. IDRC, Ottawa.

IDRC (1981a). *Rural Water Supply in Developing Countries.* IDRC, Ottawa.

Iwugo, K.O. (1981), Public health engineering needs in Africa during the Water and Sanitation Decade. *Public Health Engineer* 10, 123-126.

Kinerade, P. (1995) "Towards Ecologically Sustainable Development: The Role and Shortcomings of the Market". In Eckersley, R. (ed.) *Markets, the State and the Environment,* Basingstoke, Macmillan.

King, K. (1977), *The African artisan: Education and the informal sector in Kenya*. Heineman, London.

Meadows, D. (1972) *The Limits to Growth: A Global Challenge; a Report to the Club of Rome,* New York, University Books.

Meadows, D.H., Meadows, D.L., and Randers, J. (1992) *Beyond the Limits: Global Collapse or a Sustainable Future,* London, Earthscan.

Merchant, C. (1992) *Radical Ecology: The Search for a Liveable World,* London and New York, Routledge.

Milingwa, G. (1982), The National Construction Council of Tanzania, *Construction Management and Technology News* (ILO, Geneva) 4, 8.

Mohamed Salih, M. A. (1997a) "Global Ecologism and Its Critics". In Thomas, C., and Wilkin, Peter (eds.) *Globalisation and the South,* Basingstoke, Macmillan.

Mohamed Salih, M. A. (1997b) "Natural Capital". In O'Hara, P. (ed.) *Encyclopaedia of Political Economy,* London and New York, Routledge.

Mohamed Salih, M. A. (1999) "Introduction", Mohamed Salih, M.A., and Tedla, S. (eds.) *Environmental Planning, Polices and Politics in Eastern and Southern Africa,* Basingstoke, Macmillan.

North, R. D. (1995) *Life on a Modern Planet: A Manifesto for Progress,* Manchester and New York, Manchester University Press.

Opschoor, J. B. (1999) "Foreword", Mohamed Salih, M.A., and Tedla, S. (eds..) *Environmental Planning, Polices and Politics in Eastern and Southern Africa,* Basingstoke, Macmillan.

Opschoor, J. B. and Turner, R. K. (1994) "Environmental Economics and Environmental Policy Incentives: Introduction and Overview". In Opschoor, J.B., and Turner, R.K. (eds.), *Economic Incentives and Environmental Policies: Principles and Practice,* Kluwer Academic Publishers.

Pickford, J. (1977). Appropriate technology for Third World water supply and sanitation. *Environmental Pollution Management* 7, 50-51.

Pickford, J. (1982), Aspects of human resource management, *African Water Technology Conference* (Nairobi, Kenya), Paper No. 13.

Pineo, C.S., and Subrahmanyam, D.V. (1975). *Community Water Supply and Excreta Disposal Situation in the Developing Countries: A Commentary*. World Health Organisation, Geneva.

Redclift, M. (1984) *Development and the Environment; Red or Green Alternatives,* London, Methuen.

Redclift, M. (1987) *Sustainable Development: Exploring the Contradictions,* London, Methuen.

Redclift, M. and T. Benton (eds.) (1994) *Social Theory and the Global Environment,* London and New York, Routiedge.

Rich, Bruce (1994) *Mortgaging the Earth: The World Hank, Environmental Impoverishment and the Crisis of Development,* London, Earthscan.

Robertson, L.H. (1981), The development of self-help gravity piped water projects in Malawi, In *Appropriate Technology in Civil Engineering*, pp 208-210. Thomas Telford, London.

Sachs, W. (1994) "The Blue Planet: An Ambiguous Modern Icon", *The Ecologist,* Vol. 24, No. 5. September.

Sachs, W. (ed.) (1993) "Introduction", *Global Ecology: A New Arena of Political Conflict,* London and New Jersey, Zed Books.

Shikwe, R.C. (1981), The planning and organisation of training for water development in Kenya, In *Rural Water Supply in Developing Countries,* pp. 110-116, IDRC, Ottawa.

Shiva, V. (1993) "The Greening of the Global Reach; Global Environment or Green Imperialism." In Sachs, W., (ed.) *Global Ecology: A New Arena of Political Conflict,* London and New Jersey, Zed Books.

Soroos, M. S. (1999) "Global Institutions and the Environment". In Vig, Norman J., and Axelrod, R.S. (eds.) *The Global Environment: Institutions Law and Policy,* London, Earthscan.

South Commission (1990) *Challenge to the South,* The Report of the South Commission, Oxford, Oxford University Press.

Stallings, B. (ed.) (1996) *Global Change and Regional Response: The New International Context of Development,* Cambridge, Cambridge University Press.

Swere, R.M.A./ (1981). Planning and organising training in Tanzania, *In* "Rural Water supply in Developing Countries," pp. 207-109. IDRC, Ottawa.

Taylor, Annie and Thomas, C. (1999) *Global Trade and Social Issues,* London and New York Routledge.

Thomas, C. (1992) *Environment in International Relations,* London, Royal Institute of International Relations.

Tolba M. and EI Kholy (1992) The- *World Environment 1992* Two Decades for Challenge, provide a comprehensive review of the environmental problems considered global by the United Nations Environmental Programme (UNEP). These include climate change, biodiversity loss, population growth, fresh water pollution/contamination, air pollution, greenhouse effect, deforestation, land degradation/desertification.

Tolba, M. K. and El Kholy, O. (1992) *The World Environment 1972 - 7992: Two Decades for Challenge,* London, Chapman and Hall (United Nations Environmental Programme).

Tylor, G. R (1970) *The Doomsday Book: Can the World Survive,* London, Thames and Hudson.

Vig, Norman J. and Axelrod, R.S. (1999) *The Global Environment: Institutions, Law and Policy,* London, Earthscan.

Weizsacker and Jestinghaus (1992) *Ecological Tax Reform: A Policy Proposal for Sustainable Development.*

Weizsacker, E. U. von and Jestinghaus, J. (1992) *Ecological Tax Reform: A Policy Proposal for Sustainable Development,* London and New Jersey, Zed Books.

Welford, R. (1997) *Hijacking Enviromentalism: Corporate Responses to Sustainable Development,* London, Earthscan.

WHO AFRO (1973). "Health Progress in Africa 1968-1973." Technical Paper No. 6 AFRO, Brazzaville.

Wilson, G. A. and Bryant, R. (1997) *Environmental Management: New Directions for the Twenty-First Century,* London, University College London.

World Bank (1982), *1981 World Bank Atlas*. World Bank, Washington, D.C.

World Bank (1993) *Environmental Briefs,* Washington DC, World Bank.

8

Green Energy, Sources, Tools and Technology

Soft Energy Path

In 1976 Amory Lovins coined the term "soft path" to describe an alternative future where efficiency and appropriate renewable energy sources steadily replace a centralised energy system based on fossil and nuclear fuels.

Soft vs Hard

As physicist/consultant/lobbyist Amory Lovins describes it, the "hard energy path" (with which the soft path contrasts) is based on the assumption that the more energy we use the better off we are. It involves inefficient liquid-fuel automotive transport, as well as giant, centralised electricity-generating facilities, often burning fossil fuels (e.g., coal or petroleum) or harnessing a nuclear fission reaction (see nuclear power). The hard path is not simply a matter of energy sources, though, because it is greatly augmented and complicated by wastage and loss of electricity and other common, directly usable forms of energy. The "soft energy path" assumes that energy is but a means to social ends, and is not an end in itself. Soft energy paths involve efficient use of energy, diversity of energy production methods (matched in scale and quality to end uses), and special reliance on co-generation and "soft technologies" (i.e., alternative technology) such as solar energy, wind energy, biofuels, geothermal energy, etc.

Soft Energy Technologies

Soft energy technologies (appropriate renewables) have five defining characteristics (Lovins, 1977). They (1) rely on renewable energy resources, (2) are diverse and designed for maximum effectiveness in particular circumstances, (3) are flexible and relatively simple to understand, (4) are matched to end-use needs in terms of scale, and (5) are matched to end-use needs in terms of quality.

Residential solar energy technologies are prime examples of soft energy technologies and rapid deployment of simple, energy conserving, residential solar energy technologies is fundamental to a soft energy strategy. *Active* residential solar technologies use special devices to collect and convert the sun's rays to useful energy and are located near the users they supply. *Passive* residential solar technologies involve the natural transfer (by radiation, convection and conduction) of solar energy without the use of mechanical devices. Lovins argued that besides environmental benefits, global political stresses might be reduced by Western nations committing to the soft energy path. In general, soft path impacts are seen to be more "gentle, pleasant and manageable" than hard path impacts. These impacts range from the individual and household level to those affecting the very fabric of society at the national and international level (see Lovins, 1977).

Implementation

Facts have validated some of Lovins' assertions. While U.S. federal commitments in the energy field have varied and no ongoing, official resolve to adhere to the soft energy path has been made at the national level, in the Winter 1998 edition of the *Whole Earth Catalog*, Amory Lovins pointed out that choices made by industry and citizens had resulted in a national energy use curve *just slightly lower* than the levels he had projected for the U.S. 'soft energy path' in 1976. In this sense, the success of the soft energy path for the country exceeded expectation, and there has been a renewed interest in the soft energy path due to the recent popularisation of "peak oil" theories. At the same time, proponents of renewable energy sources complain that the fraction of energy derived from renewable sources has declined slightly in a number of nations (though grown slightly in others) if, say, the years 1985 and 1994 are compared.

Still, some proponents put much hope in the commercial emergence of hydrogen fuel cells, which appear to have possibilities as an efficient energy-storage method. The utilisation of both wind and solar power is generally predicted to grow significantly, and is already doing so in many regions. Further, the European Commission of 1997 predicted a significant growth in all of the main sources of renewable electricity generation except hydro-electric by 2010 (see renewable energy development). One criticism leveled at decentralised energy technologies is that, generally, the large, centralised methods produce energy much more efficiently than small, distributed plants. The energy decentralists counter that this is a generalisation, that new developments are in the works, and even today there are sometimes exceptions (see the discussion in the renewable energy article). In Lovins' analysis, large-scale electricity production facilities have an important place, but it is a place that they were already filling by the mid-1970s. At that time, Lovins felt that more centralized, large-scale "conventional" energy production facilities would not generally be needed.

Rocky Mountain Institute

As an example of the thinking at Rocky Mountain Institute (Amory Lovins and colleagues) in light of the end of the "cheap oil" era, the following has been proposed as a transitional strategy:

- Double the efficiency of oil utilisation. This, they say, could be accomplished mainly through improved vehicle design (the development of improved hybrid cars, ultralight designs, streamlined large transport vehicles, etc.).
- Apply creative business models. They suggest a "new industrial cluster" focusing on certain advanced technologies and on lightweight materials.
- Substitute 25% of U.S. oil needs via a major domestic biofuels industry. They see a huge economic boost for the rural regions that would supply the plant material for carbohydrate derivation.
- Make natural gas again abundant and affordable through wider utilisation of well-established efficiency techniques.

Composting

Composting is the aerobic decomposition of biodegradable organic matter, producing compost. (Or in a simpler form: Composting is the decaying of food, mostly vegetables or manure.) The decomposition is performed primarily by facultative and obligate aerobic bacteria, yeasts and fungi, helped in the cooler initial and ending phases by a number of larger organisms, such as springtails, ants, nematodes and oligochaete worms. Composting can be divided into home composting and industrial composting. Essentially the same biological processes are involved in both scales of composting, however techniques and different factors must be taken into account.

Importance

Composting recycles or "downcycles" organic household and yard waste and manures into an extremely useful humus-like, soil end-product called compost. Examples are fruits, vegetables and yard clippings. Ultimately this permits the return of needed organic matter and nutrients into the foodchain and reduces the amount of "green" waste going into landfills. Composting is widely believed to speed up the natural process of decomposition appreciably as a result of the raised temperatures that often accompany it. The elevated heat results from exothermic processes, and the heat in turn reduces the generational time of microorganisms and thereby speeds the energy and nutrient exchanges taking place. It is a popular misconception that composting is a "controlled" process; if the right environmental circumstances are present the process virtually runs itself. Hence a popular expression, "compost happens". It is nonetheless very necessary to provide as optimal circumstances as possible for large amounts of organic waste to break down properly. This is especially so when it is accompanied by heating, since at elevated temperatures oxygen within the piles is consumed more rapidly, and if not controlled, will lead to malodor.

Decomposition similar to composting occurs throughout nature as garbage dissolves in the absence of all the conditions that modern composters talk about; however, the process can be slow. For example, in the forest bark, wood and leaves break down into humus over 3-7 years. In restricted environments, for example, vegetables in a plastic trash container, decomposition with a lack of air encourages growth of anaerobic microbes, which produce disagreeable odors. Another form of degradation practiced deliberately in absence of oxygen is called anaerobic digestion- an increasingly popular companion to composting as it enables capture of residual energy in the form of biogas, whereas composting releases the majority of bound carbon-energy as excess heat (which helps sanitize the material) as well as copious amounts of biogenic CO_2 to the atmosphere.

It is important to distinguish between terms such as "biodegradable", "compostable", and "compost-compatible".

- A biodegradable material is capable of being broken down completely under the action of microorganisms into carbon dioxide, water and biomass. It may take a very long time for a material to biodegrade depending on its environment (e.g. hardwood in an arid area), but it ultimately breaks down completely.
- A compostable material biodegrades substantially under composting conditions, into carbon dioxide, methane, water and compost biomass. Compost biomass refers to the portion of the material that is metabolised by the microorganisms and which is incorporated into the cellular structure of the organisms or converted into humic acids etc. Compost biomass residues from a compostable material are fully biodegradable. "Compostable" is thus a subset of "biodegradable". The size of the material is a factor in determining compostability because it affects the rate of degradation. Large pieces of hardwood may not be compostable under a specific set of composting conditions, whereas sawdust of the same type of wood may be.
- A compost-compatible material does not have to be compostable or even biodegradable. It may biodegrade too slowly to be compostable itself, or it may not biodegrade at all. However, it is not readily distinguishable from

the compost on a macroscopic scale and does not have a deleterious effect on the compost (e.g. it is not a biocide). Compost-compatible materials are generally inert and are present in compost at relatively low levels. Examples of compost-compatible materials include sand particles and inert particles of plastic.

Reducing Waste

Although composting has historically focussed on creating garden-ready soil, it is becoming more important as a tool for reducing solid waste. More than 60 percent of household waste in the United States is recyclable or compostable. But Americans only compost 8 percent of their waste. Surveys have shown that the #1 reason Americans don't compost their waste is because they feel the process is complicated, time-consuming or requires special equipment. However, especially in rural areas, much of the solid waste could be removed from the waste stream by promoting "extremely passive composting" where consumers simply discard their yard waste and kitchen scraps on their own land, regardless of whether the material is ever re-used as "compost". (http://www.nrdc.org/cities/recycling/fover.asp)

Materials

Many different materials are suitable for composting organisms. Composters often refer to "C:N" requirements; some materials contain high amounts of carbon in the form of cellulose which the bacteria need for their energy. Other materials contain nitrogen in the form of protein, which provide nutrients for the energy exchanges. It would however be an over-simplification to describe composting as about carbon and nitrogen, as is often portrayed in popular literature. Elemental carbon - such as charcoal - is not compostable nor is a pure form of nitrogen, even in combination with carbon. Not only this, but a great variety of man-made, carbon-containing products, including many textiles and polyethylene, are not compostable - hence the push for biodegradable plastics.

Suitable ingredients with relatively high carbon content include:

- Dry, straw-type material, such as cereal straws
- Autumn leaves
- Sawdust and wood chips
- paper and cardboard (such as corrugated cardboard or newsprint with soy-based inks)

Ingredients with relatively high nitrogen content include:

- Green plant material (fresh or wilted) such as crop residues, hay, grass clippings, weeds
- Manure of poultry and herbivorous animals such as horses, cows and llamas
- Fruit and vegetable trimmings

The most efficient composting occurs by seeking to obtain an initial C:N mix of 25-30 to 1 by dry chemical weight. Grass clippings have an average ratio of 10-19 to 1 and dry autumn leaves from 55-100 to 1. Mixing equal parts by volume approximates the ideal range.

Poultry manure provides much nitrogen but with a ratio to carbon that is imbalanced. If composted alone, this results in excessive N-loss in the form of ammonia - and some odor. Horse manure provides a good mix of both, although in modern stables, so much bedding may be used as to make the mix too carbonaceous.

For home-scale composting, mixing the materials as they are added increases the rate of decomposition, but it can be easier to place the materials in alternating layers, approximately 15 cm (6 in) thick, to help estimate the quantities. Keeping carbon and nitrogen sources separated in the pile can slow down the process, but decomposition will still occur.

Some people put special materials and activators into their compost. A light dusting of agricultural lime (not on animal manure layers) can curb excessive acidity, especially with food waste. Seaweed meal provides a ready source of trace elements. Finely pulverised rock (rock flour or rock dust) can also provide minerals, while clay and leached rock dust are poor in trace minerals.

Composting in the form of bioremediation can break down petroleum hydrocarbons, TNT and a variety of toxic compounds. It is the bacterial and in some cases fungal content of the compost that possess the enzymatic properties to de-polymerize the complex man-made molecules. In other words, there is nothing about the composting process *per se* that adds or detracts from this, unless as noted above, by warming, to increase the metabolic rate of the constituent organisms.

Some materials are best left to a high-rate thermophilic composting system, as they decompose slower, attract vermin and require higher temperatures to kill pathogens than backyard composting provides. These materials include meat, dairy products, eggs, restaurant grease, cooking oil, manure and bedding of non-herbivores, and residuals from the treatment of wastewater and drinking water. Meat and dairy products can be recycled using bokashi, a fermentation method.

Human waste can be composted by industrial, high-heat methods and also composting toilets, even though most composting toilets do not allow for the thermophilic decomposition believed to be necessary for a rapid kill of pathogens, such as Salmonella. This is not a problem, however, since composting toilets also incorporate the essential element of time required to reduce the available substrate on which pathogens can feed, while increasing the growth of competing microbes. If high temperatures are reached, the resulting compost can be safely used as a fertilizer for food crops and even directly edible crops provided it is not illegal in the regions where the humanure is applied. Careful filtration of the compost also prevents contamination. Humanure fertilizer is, however, used throughout the less developed world and is becoming more accepted as a garden amendment in the developed world (see humanure).

Approaches

There are two major approaches to composting: active and passive. These terms are somewhat of a misnomer since both active and passive composts can attain high heating, which increases the rate of biochemical processes. But the terms active and passive are appropriate descriptions for the nature of human intervention used.

Active

Active (hot) composting is composting at close to ideal conditions, allowing aerobic bacteria to thrive. Aerobic bacteria break down material faster and produce less odor and fewer pathogens and destructive greenhouse gases than anaerobic bacteria. Commercial-grade composting operations actively control the composting conditions such as the carbon-to-nitrogen ratio. For backyard composters, the charts of carbon and nitrogen ratios in various ingredients and the calculations required to get the ideal mixture can be intimidating, so many rules of thumb exist for approximating it.

Pasteurisation is a misnomer in composting, as no compost will become truly sterilised by high temperatures alone. Rather, in a very hot compost where the temperature exceeds 55 °C (130 °F) for several days, the ability of organisms to survive is greatly compromised. Nevertheless, there are many organisms in nature that can survive extreme temperatures, including the group of pathogenic Clostridium, and so no compost is completely safe. To achieve the elevated temperatures, the compost bin must be kept warm, insulated and damp.

Aerated Composting is an efficient form of composting from the chemical point of view as it produces ultimately only energy in the form of waste heat and CO_2 and H_2O . With aerated composting, fresh air (i.e. oxygen) is introduced throughout the mix of materials using any appropriate mechanism. The air stimulates the microorganisms that are already in the mix, and their by-product is heat. In a properly operated compost system, pile temperatures are sufficient to stabilize the raw material, and the oxygen-rich conditions within the core of the pile eliminate offensive odors. High temperatures also destroy fly larvae and weed seeds, yielding a safe, high-quality finished product.

Finally, aeration expedites the composting process through the mechanism of heating insofar as the elevated heat will drive biochemical processes faster, so that a finished product can be rendered in 60 to 120 days. Aerated compost is an excellent source of macro- and micro-nutrients as well as stable organic matter, all of which support healthy plant growth. In addition, the micro-organisms in compost aid in the suppression of plant pathogens. Finally, compost retains water extremely well resulting in improved drought resistance, a longer growing season, and reduced soil erosion.

In Thailand this system has been used by farmer groups for more than 445 sites (May, 2008). The process needs only 30 days to finish without turning need and 10 metric tons of compost is obtained each time. A blower (15 inch squirrel-caged blower with 3 hp motor) is needed to force the air through 10 static piles of compost twice a day and 15 minutes each. The raw materials consist of agricultural wastes and animal manure in the ratio of 3 : 1 by volume. For more information please visit www.compost.mju.ac.th/eng

Passive

Passive composting is composting in which the level of physical intervention is kept to a minimum, and often as a result the temperatures never reach much above 30°C (86 °F). It is slower but is the more common type of composting in most domestic garden compost bins. Such composting systems may be either enclosed (home container composting, industrial in-vessel composting) or in exposed piles (industrial windrow composting). Kitchen scraps are put in the garden compost bin and left untended. This scrap bin can have a very high water content which reduces aeration, and so becomes odorous. To improve drainage and airflow, a gardener can mix in wood chips, small pieces of bark, leaves or twigs, or make physical holes through the pile.

Natural

An unusual form of natural composting in nature is seen in the case of the mound-builders (megapodes) of eastern Indonesia, New Guinea, and Australia as well as in the case of bowerbirds of New Guinea and Australia. These Megapodes are fowl-sized birds famous for building nests in the form of huge compost heaps containing leaf litter, in which they incubate their eggs. The birds work constantly to maintain the correct, almost exact, incubation temperatures, by adding and removing leaves from the compost pile. In effect, this teaches us that thermophilic high-temperature composting is not man-made.

Home Composting

Home composters use a range of techniques, varying from extremely passive (throw everything in a pile and leave it for a year or two) to extremely active (monitor the temperature, turn the pile regularly, and adjust the ingredients over time). Some composters use mineral powders to absorb smells, although a well-maintained pile seldom has bad odors. It is usually located in the back garden.

Moisture and Heat

An effective compost pile is about as damp as a well wrung-out sponge. This provides the moisture that all life requires. Microorganisms vary by their ideal temperature and the heat they generate as they digest. Mesophilic bacteria survive best at temperatures of 20 to 44 °C (70 to 120 °F). thermophilic (heat-surviving) bacteria grow optimally at around 55°C (130 °F), and can attain the fastest decomposition, since metabolic processes proceed more rapidly under higher temperatures. Elevated temperature is also preferred since it causes the most rapid pathogen reduction, and is more destructive of weed seeds. To minimally achieve it, the heap should be about 1 m (3 ft) wide, 1 m (3 ft) tall, and as long as is practicable. This provides enough insulating mass to build up heat but also allows aeration. The center of the pile heats up the most.

If the pile does not heat up, common reasons include that:

- The heap is too wet, limiting the oxygen which bacteria require
- The heap is too dry for the bacteria to survive and reproduce
- There is insufficient protein (nitrogen-rich material)

The necessary material should be added, or the pile should be turned to aerate it and bring the outer layers inside and vice versa. You should add water at this time to help keep the pile damp. One guideline is to turn the pile when the high temperature has begun to drop, indicating that the food source for the fastest-acting bacteria (in the center of the pile) has been largely consumed. When turning the pile does not cause a temperature rise, it brings no further advantage. When all the material has turned into dark brown crumbly matter, it is ready to use.

Worm Composting

Worm composting or vermicomposting is a method of composting using Red Wiggler worms in a container. Food waste and moistened bedding are added, and the worms and micro-organisms eventually convert them to rich compost. The worms excrete a soil-nutrient material called worm castings. Worm composting can be done indoors, allowing year-round composting, and providing apartment dwellers with a means of composting. Worms are low in the food chain, and so are critical to healthy soil. This is why farmers have historically wanted healthy worm populations to live in their fields. The nutrients and micro-organisms can be concentrated in liquid form called worm tea, made by running distilled water through worm castings. When it is poured into the soil, the microorganisms multiply, creating a healthy growing environment for plants.

Industrial Composting

Industrial composting systems are increasingly being installed as a waste management alternative to landfills, along with other advanced waste processing systems. Industrial composting or anaerobic digestion combined with mechanical sorting of mixed waste streams is called mechanical biological treatment increasingly used in Europe due to stringent new regulations controlling the amount of organic matter allowed in landfills. Treating biodegradable waste before it enters a landfill reduces global warming from fugitive methane; untreated waste breaks down anaerobically in a landfill, producing landfill gas that contains methane, a greenhouse gas even more potent than carbon dioxide.

Most commercial and industrial composting operations use active composting techniques. These ensure that the process does not get out of control especially with the high through-put demand imposed by contracted, incoming waste. This means that as short as possible a processing time must be maintained to keep the facility properly functioning (see compost windrow turner). Partly for this reason composters have declined to support compost maturity standards if it would increase the required holding time. The greatest amount of technological control of composting is seen in systems using an enclosed vessel and controlling its temperature, air flow, moisture and other parameters.Large-scale composting systems are used by many urban centers around the world. Co-composting is a technique which combines solid waste with de-watered biosolids, which originated in the 1960s and has fallen somewhat out of favor due to difficulties controlling inert and plastic contamination from MSW. In Europe, mixed waste composting is virtually illegal. The world's largest MSW co-composter is the Edmonton Composting Facility in Edmonton in Alberta, Canada, which turns 220,000 tonnes of residential solid waste and 22,500 dry tonnes of biosolids per year into 80,000 tonnes of compost. The facility is 38,690 square metres (416,500 ft^2) large (equivalent to 4½ Canadian football fields), and the aeration building alone is the largest stainless steel building in North America, the size of 14 NHL rinks.

Management of Composting

Proper and regular additions of on-farm organic wastes such as animal manure and crop residues are of utmost importance in maintaining the tilth, fertility, and productivity of agricultural soils; in protecting them from wind and water erosion; and in preventing nutrient losses through runoff and leaching. The restoration and rehabilitation of degraded soils to an acceptable level of productivity can be enhanced by using various off-farm sources of organic wastes such as sewage sludge, municipal solid wastes, crop wastes, and agricultural and industrial processing wastes. However, the quality and acceptability of these materials as soil amendments can be greatly improved through composting, or even by co-composting them with available on-farm wastes.

Sustainable agriculture is increasingly viewed as a long-term goal that seeks to overcome problems and constraints on the economic viability, environmental soundness, and social acceptance of agricultural production systems worldwide. Soil quality has focused on soil productivity, food safety and quality, human and animal health, and environmental quality. It may also play a major role in plant health and in the nutritional quality of the food that is produced.

Composting is a viable means of transforming various organic wastes into products that can be used safely and beneficially as biofertilizers and soil conditioners. It can resolve a number of problems associated with the use of raw and unstable organic wastes as soil amendments such as malodors, human pathogens, and undesirable chemical and physical properties. During the composting process, organic wastes are decomposed, plant nutrients are mineralised into plant-available forms, pathogens are destroyed, and malodors are abated (Parr and Hornick 1992; Yang 1997b). Composting is a microbiological process that depends on the growth and activity of mixed populations of bacteria, actinomycetes, and fungi that are indigenous to the wastes being composted. It can be done by either aerobic or anaerobic methods. However, the aerobic method is generally preferred, since it proceeds more rapidly and provides greater pathogen reduction because higher temperatures are attained.

Management of Composting Materials

Many biomass materials show active decomposition accompanied by elevated temperature. However, some of these materials can be used for high-value utilisation and are considered unsuitable for composting. For example, grass, straw, and foliage are generally more valuable as feeding materials for livestock. Therefore, other by-products and disposable products are generally considered for composting. These materials range from animal wastes, with a high fertilizer value, to straw and husk, with a minimal fertilizer content (Yang 1994, 1997a). These materials are sometimes processed separately, but more often they are processed as mixed materials to adjust the C:N ratio, bulk density, and moisture content.

Selection of composting materials is important since it directly influences composting quality. There is less possibility of increasing harmful materials in composting of rice straw, but increasing use of livestock manure and industrial and municipal wastes creates concerns for composting. Therefore, selection of good raw materials for composting is crucial in quality control of composting.

Hog manure has low C:N ratio, high bulk density, and bad aeration. Adding straw and corncob to hog manure will improve the C:N ratio and bulk density. Adding nitrogen and phosphate will also improve the composting. The C:N ratios of the composts of chicken manure, hog wastes, soybean oil extract residues with mushroom growth medium wastes decreased from 30 to 15 during composting. Rice hull contained SiO2 (11.20-14.10%), nitrogen (0.48-0.50%), and potassium (0.31-0.68%); the contents of other elements were low. Corn and sorghum stalks contained potassium (1.34-1.61%) and nitrogen (0.73-0.91%). Soybean and peanut stems had high contents of nitrogen, potassium, calcium, and magnesium. Although bark had high contents of nitrogen, calcium, and potassium, it also had high content of phenolic compound. Therefore, composting of bark is necessary.

Tobacco leaves and tobacco factory wastes had high contents of nitrogen, potassium, calcium, magnesium, and nicotine (Hsieh and Hsu 1993. Vegetable wastes had high crude protein and crude fat contents, while bamboo shoot wastes had high cellulose and hemicellulose contents. Food wastes contained high total carbon and total nitrogen contents, while C:N ratio was low. Cow feces contained high crude fiber and ash contents, while total nitrogen, crude protein, phosphate, and potassium contents were low. Hog feces had average total nitrogen, crude protein, phosphate, and potassium contents; while chicken feces had high ash, total nitrogen, crude protein, phosphate, and potassium contents (Yang *et al.* 1991; Tsai and Yang 2004.

Biosolids, municipal solid wastes, yard wastes, and food wastes contain pathogens. In yard wastes, the major source of pathogens is domestic animal feces; in food wastes, eggs, chicken parts, and other contaminated sources can result in significant levels of pathogens. Composting, if carried out properly, is very effective in destroying pathogens. This is primarily the result of temperature-time relationships. However, other factors contribute to the demise of pathogens such as antagonistic organisms and ammonia. Regrowth of salmonellae is not a serious problem as these organisms die very quickly during curing and storage. Knoll (1961) described several experiments on different *Salmonella* strains subjected to composting temperature at the composting plant. After 14 days of reactor time with temperatures of 55o-60oC and a moisture content of 40-60%, the product did not contain pathogens.

Pig manure has a considerable copper content coming from feed materials. Sewage sludge has a relatively high nutrient content and can be used as an organic fertilizer. However, it has a high content of heavy metals and, possibly, harmful organic materials. Human manure waste sludge had high phosphate and lead contents (Um and Lee 2001). Food wastes mainly come from agricultural products and can be used as raw materials for composting. The problem with food wastes was high salt (4.10%) and fat (3.53-8.02%) contents (Tsai and Yang 2004). According to company type, harmful materials in sludge are different. Sludge from food processing companies has a lead content and that from dairy industries has a low metal content. Sludge from oil industries has chromium and lead contents, and some harmful effects on plant growth were reported. Sludge from industrial wastewater treatment process used as raw materials for composting needs safety examination.

Some say that mixing harmful material with sound material can dilute adverse effects. But it is necessary to exclude such contaminant in the selection of raw materials. There are specific inspection standards for the preinspection of required usable materials. These standards include effective constituents of classified harmful materials. In Korea, the standards of composting materials are as follows: organic material, >60%; As, <50 mg kg^{-1}; Hg, <2 mg kg^{-1}; Pb, <150 mg

kg^{-1}; Cd, <5 mg kg^{-1}; Cu, <500 mg kg^{-1}; Cr, <300 mg

kg^{-1}; Zn, <900 mg kg^{-1}; and Ni, < 300 mg kg^{-1} (Um and Lee 2001).

Co-composting is a waste treatment method where different types of wastes are treated together. As an attractive method of resource recovery and waste disposal, it offers many advantages. Co-composting is expected to cost less than separate treatment systems, mainly due to the low cost per volume treated at large treatment plants. Co-composting of potato starch sludge and piggery manure as a substitute for sawdust was established. Composting by the aerated static pile method for 10-14 days and curing process with stacking the composts in 10-12 bags high can obtain a good product (Yang *et al.* 2001).

Municipal solid wastes contain 74-84% organic matter, 1.4% nitrogen, 0.1% phosphorus, and 0.1% potassium in Jakarta, Indonesia. The addition of nitrogenous compounds to adjust the C:N ratio can often accelerate the process by alleviating the microbial demand for nitrogen. Urea addition accelerated the composting process (47 days for urea addition and 54 days for control) (Prihatini and Kurnia 2001).

Problems in the Use of Composts

High Price

Average price of commercial compost is about $2 for a 20-kg package in Korea and about $3 in Taiwan. When compared with nitrogen supply only, the price of compost is 20 times higher than that of chemical fertilizer. Although most farmers agree on the positive effects of composts, some low-grade composts are very popular for their low price and some crop damages are attributed to industrial wastes used as raw materials.

Unbalanced Nutrient Content

Livestock manure composts are the major commercial composts in Korea, while both livestock manure composts and vegetable-market waste composts are the major composts in Taiwan. However, some companies provide sewage sludge as raw materials for composting. These materials are mixed with livestock manure for composting and have a high phosphate content. Therefore, compost is applied with a mixture of chemical fertilizers to get appropriate ratios in N, P, and K.

Immature Compost

Composting is minimising the damage from intermediate compounds in the degradation of organic materials. It takes about 6 months in a natural environment and about 80 days in an industrial facility. Some immature composts are being sold in the market. Biological indicators or regulations should be established for maturity determination.

Salt Accumulation and Environmental Pollution

When livestock manure compost is applied as N requirement to crops, phosphate input becomes an excess of 200-954% over standard input. Food waste compost has high concentrations of salt and fat. Excess application of such compost damages crops by salt accumulation in the soil. Generally, such soil shows unbalanced excess accumulation of phosphate, potassium, and nitrate (Um and Lee 2001; Tsai and Yang 2004).

Greenhouse Gases Emission

Unsuitable handling or treatment of livestock wastes has caused the emission of harmful gases such as ammonia from livestock production systems. In addition, livestock wastes contribute significantly to the emissions of CO_2, CH_4, and N_2O, which are greenhouse gases. Some efforts have already been made to quantify the emissions from livestock waste stores and treatment systems. One cubic meter of swine dung heap compost emitted 400-970 g of ammonia, 0.6-386 g methane, and 1.9-71.9 g nitrous oxide. During composting for eight days, each cubic meter of compost produced 2 g of CO_2, 239 mg CH_4, and 660 mg NH_3. Each ton of hog wastes, cow wastes, and chicken wastes produced 22.1 g, 16.3 g, and 12.9 g of methane, respectively, during a year of composting. Methane emission rates from the compost of the mixture of hog wastes, chicken wastes, and sawdust ranged from 0.12 mg^{-2} h^{-1} to 707 mg^{-2} h^{-1} depending on the stage of maturity. Nitrous oxide emission rates were between 0.15 mg m^{-2} h^{-1} and 25.0 mg m^{-2} h^{-1}, and those of carbon dioxide ranged from 1,036 mg m^{-2} h^{-1} to 19,558 mg m^{-2} h^{-1} (Osada *et al.* 2000; Tang and Wang 2000; Osada 2001; Chen *et al.* 2003; Lai *et al.* 2003).

Quality Control

Compost manufacturing standards are based on the supply of organic materials to the soil. There are no regulations on nutrient content. The commercial organic fertilizer and commercial compost companies are small in scale compared with other manufacturing industries. They thus have difficulties in keeping their high-quality manpower and in maintaining their composting facilities. Special fresh industrial wastes that are supplied by the waste disposal companies have many serious problems. Most commercial organic products and composts give the composition on the package label. But there is no such information for organic products or composts in bulk, so it is difficult to get practical information for using them.

Heavy Metal Content

There are some regulations on heavy metals in composts in order to protect soil environment. Allowable heavy metal concentrations on a dry weight basis are used in Taiwan, Europe, the USA, and Japan, and on a fresh weight basis in Korea.Although sewage sludge contains considerable amounts of plant nutrients such as organic matter, N, P, and K, it is not used in agricultural practices because of its high concentration of heavy metals. The Cu contents are generally high in poultry and hog composts _ from 9 mg kg^{-1} to 394 mg kg^{-1} and from 6 mg kg^{-1} to 301 mg kg^{-1}, respectively. The Zn contents are between 56 mg kg^{-1} and 1,147 mg kg^{-1} and between 35 mg kg^{-1} and 623 mg kg^{-1}, respectively. The Cr contents are often high in organic manure of animal origin, in hog compost, and in compound organic manure. The contents are 2 mg kg^{-1}-54,324 mg kg^{-1}, 3 mg kg^{-1}-8,486 mg kg^{-1}, and 2 mg kg^{-1}-9,196 mg kg^{-1}, respectively. The product of hog compost contains nitrogen, 2.8%; P_2O_5, 1.7%; K_2O, 0.23%; organic matter, 86%; Zn, 349 mg kg^{-1}; and Cu, 140 mg kg^{-1}. Rice straw contains potassium (1.41-1.44%) and nitrogen (0.48-0.67%); the contents of other components are low. The high content of Cr in the manure is conjectured to be related to the admixing of animal skin powder which is the by-product of the leather industry (Lian and Lee 1994).

The sludge of final clarifier has the highest Cu content, the primary clarifier of aerobic treatment comes next, and the anaerobic treatment has the lowest. However, the sludge of anaerobic treatment has the highest Fe and Mn contents (Fu and Chen 1990). The heavy metal contents of sludge depend on the feed supplement.

Organic Compounds

The presence of toxic organics in compost depends on the type of feedstock involved. Biosolids can contain organic compounds as a result of the disposal of industrial, commercial, and household wastes. Pesticides can be found in yard wastes and food wastes. Pthalates are found in plastics along with other organic dynes and compounds. Household wastes discharged into the municipal solid waste stream contain oils, solvents, pesticides, and many other toxic organic compounds. Paper products may contain toxic organics as a result of printing inks and ash discharged from incinerators or boilers may contain dioxins.

Compost has been shown to be an effective method of microbial degradation of many toxic organics. Thus, many organisms are capable of decomposing toxic organics and use the C as an energy source. Many of the studies were conducted under laboratory conditions. Therefore, research is needed to develop techniques for cost-effective composting of recalcitrant organic compounds. These may include inoculating specific organisms and providing the environment that sustains them.

Distribution and Use of Composts

The major reasons for non-adoption of rice straw composting include time-consuming interference with various farming activities, difficulty in applying, and burning as a convenient method of disposal. In addition, there are some problems in both quality and handling of standard animal waste composts, namely: content of fertilizer nutrients and fertilizer efficiency are unclear and their seasonal measurement fluctuates notably; maturity of compost is indefinite; special spreader machine and large storage space are neces-sary, as well as high forwarding charges, because the specific gravity of compost is small and the moisture content is usually high. To solve these problems, the following criteria must be set (Hara 2001):

- Production of high-quality animal waste composts that fit the needs of the cultivating farms by guaranteeing the quality and fertilizer efficiency, and adjustment of the content of plant nutrients of the compost produced for the cultivating plant.
- Production of animal waste composts which can withstand distribution over distances by decreasing the volume and moisture content of produced compost.
- Production of animal waste composts which can be spread easily.
- Creation of manuals of animal waste compost fertilisation.
- The handling of animal waste compost, which is produced in granular or pellet form, is better than that of conventional compost.

Shelf Life

The major consideration under this item is how to prolong the shelf life of the compost for agricultural applications by the inoculation of biological microbes.

Inoculation of Microbes

Microbes can dissolve insoluble phosphate compounds in soils, fix nitrogen to biomass from the air, mineralize organic compounds to release nutrients, and secrete plant growth-promoting substances. Inoculation of appropriate microbes into composts could improve the quality of composts and shorten the period of maturity. Inoculation of phosphate-solubilising *Klebsiella pneumoniae* subsp. and nitrogen-fixing *Azospirillum brasilense* into the compost of livestock wastes and mushroom waste media increased the soil phosphate and nitrogen concentrations, and the growth of kale, cabbage, and corn. Inoculation of *Rhizobium japonicum* and *Bradyrhizobium japonicum* for nitrogen fixation supported the growth of food crops and legumes, and increased the yields of soybean, peanut, and alfalfa by 6-21%, 25-69%, and 123-179%, respectively (Kang and Ha 1994; Chien 2001).

The application of microbial inoculants as microbiofertilizers can accelerate the decomposition of organic residues in various processes with a concomitant release of plant nutrients through mineralisation; facilitate the uptake of plant nutrients (e.g., mycorrhizal association); increase the nitrogen content of plants through symbiosis (e.g., *Rhizobium*) and other associative N_2-fixing systems (e.g., *Azospirillum*); and improve plant growth and vigor by providing plant growth-promoting substances. The well-nodulated legumes reduce soil erosion and increase soil fertility. The young rubber trees also benefit from the mineralizable nitrogen released during the decomposition of the legume residue. Nitrogen fixation by *Azorhizobium, Azospirillum* in association with several plants had been well documented: maize, kallar grass, sorghum, millet, sweet potato, and oil palm. It increased the number, yield, and nitrogen concentration of the crops. Inoculation with VAM (vesicular-arbuscular mycorrhiza) had also been found in rubber, cocoa, *Calopogonium caeruleum*, chili, and groundnut. *Bacillus pumilus* could decompose cellulose and grow at high temperature (Shamsuddin 1994).

In Taiwan, nitrogen-fixing microbes, *Rhizobium*, had been applied in lupins, alfalfa, peanuts, crotelarias, and soybeans to increase crop yields; phosphate-solubilising bacteria, *Bacillus*, had been used to improve the phosphate availability in the soil; VA-mycorrhizal fungi, *Glomus*, had been inoculated to enhance the nutrition of the host plant (Chang and Young 1992; Tseng *et al.* 1993; Young 1994; Chang *et al.* 2001). For the inoculation of microbes for biofertilizer preparation, it is necessary to investigate the following items: 1) selection of effective and competitive biofertilizers for the crops; 2) inoculant production and application to the field to ensure the benefits of plant-microorganisms symbiosis; 3) study of microbial persistence of biofertilizers in soil environments including stresses on production; 4) agronomic, soil, and economic evaluation of biofertilizers in agricultural production; and 5) study of rhizosphere environment and microbial interaction.

The quality of the municipal solid waste compost could be improved by inoculation with beneficial microorganisms such as antagonists. Inoculation of non-pathogenic Pseudomonas spp. to control *Pythium* and *Sclerotium* in tomatoes reduced the damage intensity by the pathogen from 63.0% to 38.9% and from 41.7% to 16.7%, respectively (Iswandi 1994a, b). Antagonistic fungi *Gliocladium frimbiatum* or *Trichoderma hamatum* inoculation into municipal solid waste compost inhibited the *Rhizoctonia*, *Sclerotium*, *Pythium*, and *Fusarium* diseases in peanuts (Sinaga 1992). Inoculation of non-pathogenic *Pseudomonas* spp. LIES and LD reduced the infection of peanuts by Sclerotium rolfsii (Iswandi 1994b).

Inoculation of *Trichoderma* sp. SS_{33} into the compost hastened the decomposition of rice straw-chicken manure mixture, *Azotobacter* sp. H1BFA 4b promoted the biological nitrogen fixation and enriched the nitrogen content, and *Trichoderma-Azotobacter* increased the decomposition rate and nitrogen fixation activity. The biofertilizer prepared with the inoculation increased the soil nitrogen, phosphate, potassium, organic matter, and rice yield (Espiritu and dela Torre 2001).

Values of Composts

Composts are considered substitutes for chemical fertilizers. Compost amendments increased cation exchange capacity, buffering capacity, chelating capacity, soil aggregation, aggregate stability, water-holding capacity, soil porosity, water infiltration, water percolation, nutrient availability, and earthworm population; they decreased soil crusting, bulk density, and plant pathogens. In addition, compost application alleviated acidic and alkaline conditions, and stimulated beneficial microorganisms to produce polysaccharides and/or antibiotics (Parr *et al.* 1994). Among the other values of composts are animal feed, water gain, soil conservation, and soil-carbon sequestration.

The values of four common composts (i.e., cattle manure, crop residues, sewage sludge, and municipal solid wastes) based on their average macronutrient contents are shown in. The important consideration is good quality composts and residues. Biofertilizers and soil conditioners have far greater values than just their macronutrient contents. These materials have a much greater residual effect on soil tilth and fertility than most chemical fertilizers because of the slow-release character of their nitrogen and phosphorus components. Colacicco (1982) estimated that the cumulative agronomic and economic value of some organic materials applied to agricultural soils could be more than five times greater in the post-application period than the value realised during the year of application. Application of compost improved the development of root systems, increased the diversity of root fungal flora, promoted the growth of plants, reduced the incidence of soil-borne diseases, and depressed the propagation of pathogens (Nitta 1994).

Effects of Composts on Crop Yields and Soil Qualities

Application of compost to the soil improved the soil organic matter, P, and Zn contents. The average organic matter, P, and Zn contents of compost application were 2.85%, 8.14 ppm, and 3.53 ppm, respectively; while the values of farmers' practice were 2.18%, 6.14 ppm, and 2.62 ppm, respectively. In addition, rice plant with compost application contained higher N, P, and Zn than that of farmers' practice. More recently, soil quality includes soil productivity, food safety and quality, human and animal health, and environmental quality (Parr and Hornick 1992; Evangelista 2001).

The yields of cabbage in a cattle compost plot with 1.0, 2.0, and 3.0 N application rates were 55-61%, 75-82%, and 84-91% as those with chemical fertilizer application, respectively; while the yields of rice at first crop season were 72-74%, 80-83%, and 88-91% as those with chemical fertilizer application, respectively (Huang and Lin 2001). The relative N efficiencies of compost for cabbage and rice were 0.219-0.329 and 0.254-0.292, respectively. N-value in compost of cattle manure had about 25% of chemical fertilizer. The relative N-value of chicken manure was the highest with 0.37 of chemical fertilizer in rice cultivation, cattle manure (0.34) was the second, and hog manure (0.32) was the lowest. In cabbage cultivation, hog manure (0.37) was the highest, chicken manure (0.29) was the second, and cattle manure (0.25) was the lowest. In peanut production, cattle manure (0.21) was the highest, hog manure (0.19) was the second, and chicken manure (0.11) was the lowest. Mineralisation rate and amount of N-release varied with soil texture and kind of manure. However, N-value of compost increased gradually year by year in continuous application conditions. Soil Zn contents in all compost plots were higher than those in chemical fertilizer plots. The hog manure compost plot had a higher Cu accumulation (Huang and Lin 2001).

When composts and residues are applied to soils, many of the introduced microorganisms can function as biocontrol agents by controlling or suppressing soil-borne plant pathogens through their competitive and antagonistic activities (Kloepper *et al.* 1989). The possible mechanisms are the shifts in soil microbiological equilibrium following the addition of microbial inoculants and organic amendments. These include antibiosis, competition, parasitism, detoxification, and inhibition.

Microbiology of Composting

Various microorganisms participate in composting, and the microflora population changes successively at different stages of composting. In the early stages of composting, mesophilic bacteria and fungi are dominant and consume sugar, starch, and protein. As the temperature rises above 40oC, these are replaced by thermophilic bacteria, actinomycetes, and thermophilic fungi. At this stage, lipids, hemicellulose, and cellulose are decomposed. Finally, as the temperature falls, mesophilic bacteria and fungi reappear. In other words, sugar and starch are the first to decompose during composting, followed by hemicellulose, and, finally, lignin. During the composting of sewage sludge, the number of mesophilic actinomycetes and fungi was around 10_2 cells g-1 of dry compost (Kubota and Nakasaki 1994).

Inoculation or seeding has been reported to be effective in some composting procedures. On the other hand, some negative results have also been reported about inoculation. The rate of composting of sewage sludge is mainly controlled by the degradability of the solid substrate, and not by the kinds of microorganisms inhabiting the compost. The seeding of compost products at the beginning of sewage sludge composting is unlikely to have any appreciable effect on the rate of composting or on the quality of the final product. However, the contribution of thermophilic bacteria and thermophilic actinomycetes to the CO_2 evaluation rate depends heavily on the amount of inoculum added. In the fertilizer N efficiency, *Azolla* and *Azolla* compost score 3 kg rice per kg N applied as compared with 10 for urea and supergranulated urea. However, 1/2 *Azolla* + 1/2 the recommended rate for urea increased fertilizer N efficiency to 9, which is comparable to that of urea (dela Cruz 1994).

Odor Control

Ventilation

Many of the materials used for composting are liable to become putrid and stinky. These materials are readily decomposed by microorganisms. Ventilation should provide enough air to avert anaerobic conditions during the temperature build-up stage, but should not remove so much heat that temperatures cannot rise to biologically inhibitive levels. Vacuum has been used as an odor-control measure, so that the exhaust gas can be vented through a scrubber pile. This involves placing a water condenser trap between the composting mass and the scrubber pile. In tests which applied this method, materials composted at the highest temperature gave off the most unpleasant odor, while materials composted at the lowest temperature were judged the least unpleasant. Temperatures which maximised the decomposition rate gave a final product with the least odor.

Biofiltration

Biofiltration involves passing an odorous airflow through a layer of filter material (compost, filamentous peat, etc.), followed by the biodegradation of the captured odor components. The odor components are transferred from the gas to the liquid and solid particles in the filter material. The microbial degradation of the odor components then takes place on these particles. The pressure drop across the biofilters depends not only on the applied air load, but also on the nature and the composition of the filter material.

Evaluation of Compost Maturity

Compost maturity is important in assessing both the quality of the end product and its possible uses. Some uses require very mature compost (horticulture, market gardening, plant nurseries, etc.). Others require fresh compost (hot beds in the greenhouse and for mushroom production, etc.). Fresh compost generates heat, while mature compost prevents any adverse effects on plants from the heat of decomposition, a high C:N ratio, or phytotoxic compounds.

Several parameters are available which indicate compost maturity. These include temperature; odor; texture; C:N ratio; pH value; gas production; cation exchange capacity; level of ammonium, nitrate, and immobilised nitrogen; total organic carbon; level of hydrogen sulfide; polysaccharide; adenosine triphosphate; chromatographic tests; colorimetric tests (after extraction of humic components); polymerisation of humic substances; hydrolase activity; respiratory activity test; behavior of earthworms; and phytotoxicity tests (germination and growth tests). Although many methods have been proposed, none of these have been pursued far enough to allow full appreciation of their potential value.

Temperature

A temperature rise is necessary not only to accelerate the decomposition of organic constituents, but also to inactivate harmful organisms. Swine manure reached a maximum temperature of 74ºC and had an average temperature of 50º-70ºC over more than two weeks in a container with a volume of 120 L. In a container of only 12-70 L, the maximum temperature was only 56º-58ºC, and temperatures higher than 50ºC lasted only two to six days. A temperature higher than 70ºC does not favor the growth of fungi in the compost and reduces the rate of decomposition. Therefore, the volume of the different substrates used for composting should be adjusted to give a maximum temperature of around 70ºC and temperatures higher than 50ºC for more than two weeks. With appropriate aeration, the temperature of swine manure increased on the second day and remained at 50º-70ºC for more than two weeks. After 17 days of composting, the volume of the substrate had fallen by 30%. If the compost was not aerated, the maximum temperature was lower than 60ºC, and temperatures mainly ranged from 40° to 50°C. Temperatures in the upper layer were higher than those in the middle and bottom layers. Temperature fluctuations in different parts of the substrate were not significantly different in either case.

The maximum temperature of pig farm sludge during composting exceeded 68°C. After 30 days' incubation, thc fcrmented sludge had become good-quality compost (Fu and Chen 1990). The maximum temperature of a mixture of bagasse and swine manure was 63°C, while temperatures remained between 50° and 60°C for more than one week. In the case of a mixture of bagasse and alcoholic slops, the maximum temperature was 60°C, while temperatures remained above 50°C for only two or three days (Huang *et al.* 1994). Straw undergoes rapid decomposition when temperatures reach around 70°C, although not as rapidly as animal wastes. When straw had been chopped and placed in a pile for 55 days, temperatures reached their maximum level on the third day, and remained at 60°-65°C during incubation. Temperatures rose again when the substrate was turned, but over time, the maximum temperature gradually decreased.

When pig farm sludge was composted with rice hull, the temperature increased to 62°C on the second day, eventually rising to a maximum of 68°C. When the same sludge was mixed with composted pig manure and made into compost, the maximum temperature was 74°C on the second day. In both cases, temperatures remained above 55°C for two weeks. The temperature of compost with mushroom growth medium and chicken manure, soybean residues, bone meal, swine manure, urea, or calcium superphosphate is 50°-60°C for three weeks during incubation. Turning the compost will stimulate the decomposition of organic matter and produce heat from the fermentation (Lin 1994).

Odor

Mature composts should smell like forest soil (typical soil odor is caused by actinomycetes). Soil smell is primarily the result of two gases, geosmin and 2-methylisobornol, which are by-products of fungi and actinomycetes. If these two gases are present in compost, it is possible that they could be used to determine maturity (Becker 1995).

Microcalorimeter Analysis

The promotion effect in the decomposition of organic matter was evaluated by analysing the thermogram. The decomposition of compost was promoted by the addition of microbial materials (Owa 1994).

Biomass Production

Three kinds of parameters may be used for determining the maturity of composts:

respirometry, analysis of the biodegradable constituents (total organic carbon and polysaccharides), and biochemical activity (ATP and enzyme activity).

1. Respirometric Methods

Respirometry measures the oxygen uptake or carbon dioxide emission, either in pure compost or in compost mixed with soil. It indicates a marked drop in respiratory activity as th compost matures. Compost is considered relatively mature when the respiration rate is less than 5 mg CO_2-C/g compost C. Rates over this amount reflect different stages of immaturity. Oxygen uptake and composting time have a significant negative correlation. Oxygen uptake less than 1.0 mg O_2/g VS/h is considered relatively mature (Owa 1994; E&A Environmental Consultants 1994). Methane emission from compost below 0.4 mg m^{-2} h^{-1} can be used as the index of maturity (Chen *et al.* 2003).

2. Degradable Organic Substances

A number of physico-chemical parameters are useful indicators because of their high correlation with compost degradability. The correlation between the respiration of a soil and compost mixture, and the dry matter production of ryegrass in a similar mixture, allows us to define two classes of maturity. Compost with a value of less than 2.4, corresponding to a low rate of respiration in the soil and compost mixture, may be considered mature. At the other end of the scale, a value of more than 2.7 shows that the compost is still highly degradable and will stunt vegetable growth, so, therefore, is not mature (Morel *et al.* 1979).

3. Biochemical Parameters

The concentration of ATP and the hydrolytic enzyme activity (proteolytic, amylolytic, and cellulolytic activities) can be used as indications of compost maturity.

Chemical Analysis

During composting, the organic constituents undergo transformations, which lead to biologically more stable components. Apart from the loss of total organic carbon and easily degradable components, an increase in highly polymerised organic products may be found. Total organic carbon, C:N ratio, polysaccharide content, and state of humic substances can all be considered suitable measurements of compost maturity. In addition, cation exchange capacity, water-holding capacity, and ash content may be useful parameters for determining maturity.

1. Total Organic Carbon and C:N ratio

As composting proceeds, the carbon content falls, while the nitrogen content increases, so that the C:N ratio falls. The C:N ratio changes with the ageing of compost, finally reaching values which are characteristic of a stable organic material.

The maturity of compost made from urban refuse could be defined in two classes according to the organic matter/nitrogen ratio: semi-mature urban compost (OM:N < 60) and mature urban compost (OM:N < 50). Organic matter content was ?20 g/ 100 g dry matter (Owa 1994). As Juste and Pommel (1977) indicated, it is preferable to keep a constant check on changes in the C:N ratio during fermentation rather than measure these occasionally. The C:N ratio and the final C:N/ initial C:N ratio provide good indications of the maturity of compost.

Swine manure has a low C:N ratio, a high bulk density, and poor aeration. The addition of straw and corncob to the swine manure will improve all three factors, while the addition of carbon and phosphate will improve the composting process. The C:N ratio of compost made from chicken manure, swine manure, soybean cake, and mushroom growth waste medium fell from 30 to 15 during composting (Lin *et al.* 1994). Composted swine manure contained 2.8% N, 1.7% P_2O_5, 0.23% K_2O, 86% organic matter, 349 mg kg^{-1} Zn, and 140 mg kg^{-1} Cu (Wang *et al.* 1992). Compost made from husk and bark has a C:N ratio of around 35, and may have certain quality problems. Although bark contains a high level of N, Ca, and K, it also has a high level of phenolic compounds. Tobacco leaves and tobacco factory wastes had a high content of nicotine as well as of N, K, Ca, and Mg. Composting to reduce the nicotine content was therefore necessary (Hsieh and Hsu 1993).

The bulk density and water-holding capacity, and the levels of ash, total N, and 1.0 N HCI insoluble nitrogen increased in the course of composting corncob, while the total C and C:N ratio decreased (Chung *et al.* 1993). The bulk density of soil amendments and composts was ?20 g/100 g dry matter (Owa 1994).

2. Nitrate Content

In the earlier stages of composting, ammonium is produced by the decomposition of nitrogenous compounds such as protein. As the compost matures, the ammonium is oxidised into nitrate by the action of ammonium-oxidizing bacteria and nitrate-oxidizing bacteria. Consequently, nitrate accumulates in the mature compost. The presence of nitrate can be detected with diphenylamine. A diphenylamine solution dissolved in concentrated sulfuric acid is added to water extracted from the compost. If nitrate is contained in the extract, the solution turns blue. This method can be used to test the maturity of cattle manure compost but not of swine manure compost and poultry manure compost, which produce only a very small quantity of nitrate even when they are mature.

3. Organic Matter Content

During composting, the organic matter content decreases, while the ash content increases. The organic matter content remains constant when the compost is mature and stable. Swine manure and cattle manure take three to four weeks to become stable, while stabilisation of chicken manure takes two weeks. The decomposition of organic matter during the composting process is characterised by the changes in residual rate (i.e., the percentage of organic matter which remains compared with the original amount). Poultry wastes are more easily decomposed than those of cattle and pig. The decomposition rate of cattle wastes is similar to that of pig wastes.

4. Polysaccharides

When composting begins, the simple polysaccharide content of urban refuse is high (20% of the total organic material). However, this falls to only 4-10% after 240 days. Water-soluble polysaccharides undergo a similar change. Simple polysaccharide components, therefore, are progressively decomposed by microflora during the different stages ofthermogenesis and maturation. Water-soluble sugars, consisting mainly of mono- or di-saccharides, disappear much faster than hydrolysable sugars. Genuine polysaccharides, however, take far longer to decompose because of their structure. At the end of the composting period, the quantity of polysaccharides extractable in acid is relatively high. This seems to correspond with newly synthesised stabilised microbial components rather than fractions, which are not yet decomposed.

5. Ash content

The ash content of the substrate is a constant value during composting. Organic matter decomposition leads to a relative increase in the ash content. Therefore, the ash content of the substrate in relation to the total weight of the substrate can be used as a parameter of the maturity of the compost. Compost maturity based on ash content is calculated as follows:

$$Wt = Wo\ (1 - Ao\%/At\%)$$

where Wt = weight of substrate at time t

Wo = weight of substrate at time zero
At = ash content of substrate at time t
Ao = ash content of substrate at time zero
(1 - Ao%/At%) = the degree of maturity

6. Cation Exchange Capacity (CEC)

The negatively charged CEC of organic matter increases as compost matures. Conventional CEC tests used for soil are not suitable for compost samples. Harada (1995) had developed a suitable and simple method to determine the CEC of compost. The CEC of composted cattle manure increased to 110 cmol(+)/kg over 4-5 weeks, and thereafter remained constant. A highly significant correlation was observed between the CEC and the C:N ratio (r = -0.992), total carbon (r = -0.968), total nitrogen (r = 0.995), and ash content (r = 0.992) in composted cattle wastes. Thus, since the CEC reflects the changes in the constituents during maturation, it is a useful parameter for estimating the degree of maturity of the compost. The CEC of organic amendments and composts was <50 cmol(+)/kg dry matter (Owa 1994).

7. Organic acid Generation

Acetic acid, propionic acid, and butyric acid generated with decomposition of organic matter were measured. The concentration of acetic acid in soil solution was increased by the addition of biological soil amendments and composts.

8. pH

Changes in pH have been noted to occur during the composting period and, therefore, have been considered as a possible indicator of biological activity. Generally, the pH drops during the very early stages of composting and then increases to a range of 6.5-7.5. Acid pH values indicate a lack of maturity due to short composting time or occurrence of anaerobic conditions.

9. Organic Chemical Constituents

Cellulose content yields a good index of the degree of maturity of the compost. Cellulose decreased with duration of composting. Inoko *et al.* (1979) reported that hemicellulose and cellulose (reducing sugars) decreased from about 36% of the total dry weight to about 20% after 60 days. Amino acids and low fatty acids greatly decreased during the composting of refuse and garbage. Zhang *et al.* (1992) suggested that reactive carbon decreased from 33 mg/100 mg sample to about 11 mg/100 mg for 160 days of composting.

Chromatography Test

The chromatography test uses humic substances, which are extracted from the compost and isolated using a filter paper. The slightly polymerised components move towards the periphery of the filter paper, while the highly polymerised components stay at the center. Chromatograms are read by the shape and color of three separate zones. This technique is essentially qualitative, and gives a quick indication of the state of maturity of the compost. However, the interpretation of chromatograms may sometimes be difficult. It is quite simple to classify compost from the same plant, so the chromatography test is very suitable for compost control when the raw material is known to be homogeneous. Paper chromatograms showed that the time needed for compost to reach maturity was around 10 days in the case of chicken manure, 15 days for swine manure, and 14-21 days for sludge from pig farms (Lin 1994).

Gel chromatography, monitored by UV absorbance at 280 nm of the water extracted from solid samples during composting, can also be used for the evaluation of maturity. Compounds with a relatively high molecular weight correspond to the peak with the least-elution volume. During the composting of sludge, organic compounds in the water extract changed to those with a high molecular weight. Gel chromatograms, using Sephadex G-15 gel and a compost of mixed urban waste and bark, gave different results. The UV 280 nm absorbance of the relatively high molecular weight compounds, corresponding to the peak with the least-elution volume, increased considerably in one-week-old compost, indicating that the organic compounds in the water extract had shifted to a higher molecular weight. On the other hand, the differences between compost samples after 1 week, 2 months, and 3 months of composting were small (Kubota and Nakasaki 1994).

Humification Parameters

Humification ratio was the percentage of total extractable humic-C as related to the total organic-C. Inbar (1979) reported that the total content of humic material extracted from separated cattle manure compost increased from 377 g/kg to 710 g/kg organic matter. Humic material increased rapidly during the first 60 days, and from 60 to 140 days, a very gradual change was noted.

Colorimetric Method

Mature composts should be dark-brown to black regardless of the feedstock. Alkaline extraction of the organic material from compost has a brown hue in a range of shades, depending on the age of the compost. The hue is light at the beginning of composting, and tends to darken as the compost matures. These color variations provide the basis for a test to determine maturity, using periodic measurements of the optical density of alkaline extracts from the product. This technique does not require sophisticated equipment, and can easily be used by composting plants. The compost has reached maturity when the optical density of the alkaline extract remains a stable tint.

The 2N KCl extract of composts at different stages was eluted with TSK HW-55(F). The absorbance at wavelengths of 235 nm, 280 nm, and 490 nm is shown in Fig. 2. The absorbance of raw substrate was very low in all tests, while the absorbance of compost after 17 days' incubation at wavelengths of 235 nm and 280 nm was high. The molecular weight of composted pig manure was around 100,000, while it was only around 10,000 for the control. Molecular weight with the absorption at wavelengths of 235 nm and 280 nm increased as composting continued. There was no significant difference between a supplement of 5% sludge in swine manure, and one of 10%. The odor was the same in both treatments. The color changed from yellowish to darkish brown after 40 days' incubation. The absorbance at wavelength 490 nm represents the humic substances. The absorption of the extract of compost at 490 nm was very weak before 17 days' incubation, while the absorption peak was high after one-month incubation. Therefore, absorption at a wavelength of 490 nm can be used as a parameter for the presence of humic substances and the degree of stability of compost (Sheen and Wang 1994).

Near Infrared Spectroscopy Analysis

A rapid estimate of the quality of cattle waste compost can be obtained by using near infrared spectroscopy analysis. The levels of total C, total N, ash, cellulose, hemicellulose, and lignin of the compost, and its CEC, can be measured by this method. Its biological activity, in terms of inhibition of seed germination, was assayed using Brassica rapa seeds, and an attempt was made to estimate this by near infrared spectroscopy. Although the accuracy of the estimation was not high, at least severely inhibitory activity could be detected by near infrared spectroscopy (Harada 1995).

Volatile Gases

During the composting of swine manure, the concentration of volatile gases is highest in the upper part of the pile and lowest at the bottom of the pile. After 40 days' incubation, the gas concentration remained constant, showing that the compost had stabilised (Sheen and Wang 1994).

Phytotoxicity Test

This test shows the presence of phytotoxic properties in composted urban wastes. Two kinds of test plants, maize and bean, are grown in the compost, and their growth compared with that of plants growing in non-enriched peat with a pH raised by the addition of calcium carbonate. E&A Environmental Consultants, Inc. (1990) conducted a study using compost produced from municipal biosolids with eucalyptus or sawdust. Rooted plant cuttings of *Pittosporium* tovira were transplanted into pots containing compost soil mixtures.

Germination Test

The analysis of metabolic toxicity may supply a wide range of information on the stabilisation process, thus providing a useful instrument for industrial, agricultural, and environmental analyses. Germination tests are a simple and rapid type of bioassay, which can be quickly carried out with simple instrumentation. Cress (*Lepidium sativum* L.) is often selected as the test seed because of its rapid response (24 h). According to one study, maximum sensitivity is at 27oC and after 24 h incubation. This index has proved to be a most sensitive parameter, able to account for both low toxicity (10-7M, or 25 ppb, abscisic acid) which affects root growth, and toxicity which affects germination (Zucconi *et al.* 1985).

Germination of cabbage seed was 50-80% in a composted mushroom growth medium, rising to 80-90% when the medium was composted with a mixture of chicken manure, swine manure, and soybean residues (Lin *et al.* 1994). The germination rate of cabbage grown in composted swine wastes was 88% (Fu and Chen 1990). Germination of radish seed on chicken manure before composting was 4.2%, rising to 91.6% after 10 days of composting, and 95.6% after 32 days of composting. This shows that composted animal wastes or crop residues can be used directly in the field without any damage to growth.

Conductivity

The level of electrical conductivity shows the overall salinity of the substrate. Early in the composting process, the salinity of compost made from wood shavings, and composted urban wastes and grape marc (skins and seeds left after pressing grapes for wine) compost was very high. Levels diminished progressively as composting proceeded. The level of electrical conductivity remained stable at a very low level in sewage sludge and bark compost. Changes in the conductivity do not always follow the same pattern in different piles, but in cases of excessive salinity at the beginning of composting, it defines the point after which the compost may be used. This parameter, which is easy to calculate, seems therefore worth taking into account during the preparation of compost. The electrical conductivity and cation exchange capacity of sewage sludge compost increased and the pH gradually decreased during incubation (Chung *et al.* 1992).

It can be seen that there are many methods available for assessing compost maturity. Some of these, such as the measurements of biomass activity, require complex laboratory equipment and are not suitable for small compost plants or farms. Others, such as the seed germination tests, are easily carried out with simple equipment and a minimum of scientific training. A major problem in the quality control of composts is the range of variability in the raw materials. Many of the testing methods available are suited to some materials, but not to others. Very few, if any, of the testing techniques described can be applied with equal success to the whole range of materials used in modern commercial compost production.

Sustainable Agricultural Technologies, Inc.

When Vermicompost is used for the starter to Compost Tea the result is a soil amendment that is teaming with beneficial bacteria and fungi. The Compost Tea can then be applied giving all the benefits of compost in a liquid form. When Vermicompost in the liquid form of Compost Tea is applied as a soil drench the microbial population is quickly available to the soil as well as the plant root. Compost Tea also allows for foliar application as well, providing protection for the plant surface from pest and disease.

Traditional methods of vermicomposting have been based on beds or windrows on the ground containing waste up to 18" deep, but such methods have numerous drawbacks. They require large areas of land for large-scale production and are relatively labor intensive, even when machinery is used for adding waste to beds. More importantly, such systems process wastes relatively slowly, taking anywhere from 6 to 18 months for processing to be complete. There is good evidence that a large proportion of the essential plant nutrients, that are in a relatively soluble form, are washed out and also a significant proportion can volatilize during this long processing period. Such nutrient losses are undesirable, particularly in relation to ground water pollution and result in a poor finished product. Our flow through bins require no compost turning and require no separating of worms and partially decomposed waste from the finished compost.

Water Consumption

Our daily consumption of water is 180-200 litres which translates to approximately 70,000 litres of drinking water/person/year. Ordinary flush toilets are responsible for 15-20,000 litres of this amount. Composting toilets do away with this water wastage and at the same time keep feces, paper and even urine out of the sewage system.

The Composting Toilet

When one mentions the word "composting toilet" or "mulch latrine", people immediately protest and come up with such comments as, "you don't mean you expect us to go back to the use of stinky latrines ?!" and "doesn't it smell like slurry ?!" But on this matter I must be firm - today we have an odourless and comfortable solution very suitable for installation inside the home. In contrast to many so-called "ingenious" solutions such as incineration toilets, freezer-bag toilets, dry toilets and the like, which often require the use of large amounts of energy, a genuine composting toilet consists of a large container in which a composting process takes place before the contents are removed, and a toilet stool + seat very similar in appearance to that of an ordinary flush toilet. The end product of the composting toilet is an odourless humus material totally unlike manure or the contents of a latrine. (there is no danger in handling it and it can be used as fertilizer for shrubs and decorative plants as long as it has had soil contact for 19 months)

Denmark's Neighboring Countries

During the past few decades, individuals in Norway and Sweden have been continuously working on the development of composting toilets, so that today we have a new generation of composting toilets. In 1982, according to the Nordic Ministercounsil, there were 300,000 composting toilets installed throughout

Scandinavia. I suppose that Denmark accounts for only about 100 of these. Development has continued at a fast rate in the neighbouring countries and the latest encouraging news is that Tannum "municipality", near Goteborg, Sweden, has recently forbidden flush toilets from the year 2000 and currently builds 70% of their new housing without flush toilets.

Development in Denmark

Because of the fact that 7% of Danish housing (in rural areas) aren't connected to central sewage systems, some research on composting toilets has been undertaken at The Danish Technical University and the most current news in this field is that a project called "Kompost-toiletter 1994" by Simon Wris-Berg and Peter Friis Moller has been awarded a prize by Denmark's agricultural school. Some Danish firms have experienced an increase in inquiries about composting toilets during the past few years due both to increasing environmental awareness as well as ever increasing water costs.

The Design of a Composting Toilet

Composting toilets vary in appearance depending upon the manufacturer, but common for them all, is that they have a collection container for feces, paper, in some cases kitchen garbage, while some separate urine into a second container. The contents of the feces container are composted by aerobic process, which takes about one year, into a humus like material. In the company "Renewable Energy", we have developed a year-round model called the "Eco-composter" for family homes. The collection container of our toilets is made of fibre reinforced concrete which makes for a reduction in use of fibreglass and epoxy. The toilet stool, which is made of either fibre-concrete or porcelain, is designed so that urine and feces are kept separate. The feces, paper and kitchen garbage drop vertically down into one tank, while the urine is collected at the front - a design appropriate for both female and male use. In this way the urine can, without coming into contact with the feces, be collected in a tank and because there is no danger of disease, be used as liquid fertilizer when mixed with 4-5 times its amount in water. The urine can also, of course, be mixed with household "grey waste water". But this must be regarded as the second solution. Our porcelain stool is designed in such a way that the urine compartment can be flushed with 0.1 litres of water. The tank has ventilation channels to enable the oxidizing process to take place, and all air is drawn up through a vent which reaches up through and over the roof of the house. This means that the composting toilet itself functions as a ventilation system for the entire toiletroom, keeping out unwanted smell. This makes for quite an advantage - there is never any unpleasant smell connected with a "visit" to the toiletroom. The components of our composting toilet take up the same amount of space as a traditional flush toilet in the toiletroom itself, while the collection-container and ante-chamber are located beneath the floor of the toiletroom or from a connected "lightbox" located outdoors.

Infection and the Authorities

The Danish authorities require that the same minimum criteria are fullfilled as with sludge delivery, due to a risk of disease-carrying bacteria and parasites, although the compost can always be spread on the ground because the amount of heavy metals in human feces + urine is extremely low when not mixed with other wastewater (see table 1) The compost-humus could of course be pasteurised at 70 degrees for a minimum of 1 hour, but this holds no interest for us, because we want to avoid the use of large amounts of energy and retain the good balance with nutrients which are present in the compost-humus. We don't want to use energy for no reasons. We require that the compost is brought in contact with soil bacteria and kept behind a little fence for yet another year before being spread out on gardens or fields. This requirement is easy to live with in cases where here is a complete plan for the composting toilet and the "grey wastewater", for instance first via a microbiological cleaning in a sandfilter and finally by means of gravitation in an "energy-forest draining system" so that we really can put the compost in the "energy-forest". Today we are no longer confronted with the same resistance to this installation design as 10 years ago, due to greater environmental awareness and stricter wastewater treatment requirements.

The Future

We will experience fast and hopefully creative development in the area of composting toilets combined with more research which initially can find support in the results of our neighbouring countries. Because the lack of fresh water is a global problem, these solutions are quite suitable for export to other countries.

Windrow Composting

In agriculture, windrow composting is the production of compost by piling organic matter or biodegradable waste, like animal manure and crop residues, in long rows (*windrows*). This method is suited to producing large volumes of compost. These piles are generally turned to improve porosity and oxygen content, mix in or remove moisture, and redistribute cooler and hotter portions of the pile. Windrow composting is the most commonly used of farm scale composting methods. Process control parameters include the initial ratios of carbon and nitrogen rich materials, the amount of bulking agents added to assure air porosity, the pile size, moisture content, and turning frequency.

Compost Windrow Turners

Compost windrow turners were developed to produce compost on a large scale.

They are a traditionally a large machine that straddles a windrow of approximately eight feet high by 13 feet across. Although smaller machines exist for smaller windrows, most operations need the larger machines to be efficient. Turners drive through the windrow at a slow rate of forward movement. They have a steel drum with paddles that are rapidly turning. As the turner moves through the windrow, fresh air containing oxygen is injected into the compost by the drum/paddle assembly and waste gases produced by harmful bacteria are eliminated. The oxygen feeds the beneficial composting bacteria and thus speeds the eventual composting process. This process is then extended by windrow dynamics.

Utilisation

To properly utilize a compost windrow turner, it is ideal to compost on a concrete pad; haul in the feed stocks, and place them into windrows. Heavy-duty compost windrow turners can be built that allow the user to obtain optimum results. By using 4-wheel-drive or tracks the windrow turner is capable of turning compost in windrows located in remote locations. With a self-trailering option this allows the compost windrow turner to convert itself into a trailer to be pulled by a semi-truck tractor. These two options combined allow the compost windrow turner to be easily hauled anywhere and to turn the compost in muddy and wet locations.

Vermicompost

Vermicompost (also called worm compost, vermicast, worm castings, worm humus or worm manure) is the end-product of the breakdown of organic matter by some species of earthworm. Vermicompost is a nutrient-rich, natural fertilizer and soil conditioner. The process of producing vermicompost is called *vermicomposting* . The earthworm species (or composting worms) most often used are Red Wigglers (*Eisenia foetida*) or Red Earthworms (*Lumbricus rubellus*). These species are commonly found in organic rich soils throughout Europe and north America and especially prefer the special conditions in rotting vegetation, compost and manure piles. Composting worms are available from nursery mail-order suppliers or angling (fishing) shops where they are sold as bait. Small-scale vermicomposting is well-suited to turn kitchen waste into high-quality soil, where space is limited. Together with bacteria, earthworms are the major catalyst for decomposition in a healthy vermicomposting system, although other soil species also play a contributing role: these include insects, other worms and molds.

Bins

Vermicomposting bins (also known as worm bins) vary drastically depending on the desired size of the system. Garden-scale containers can be made out of bricks arranged in the shape of a box. Bins for an apartment or similar dwelling can be anything from reused plastic buckets to purpose-built commercial containers.

Small Scale

Small-scale systems may use a wide variety of bins. Often, small-scale composters build their own bins. Companies also sell such bins. Commonly, bins are made of old plastic containers, wood, Styrofoam containers, or metal containers. Some materials are less desirable than others in bin construction. Styrofoam is believed to release toxins into the earthworms' environment. Metal containers often conduct heat too readily, are prone to rusting, and may release heavy metals into compost. Bins should have holes in the sides to allow air to flow, and a spout that can be opened or closed or holes in the bottom to drain into a collection tray. Plastic bins require more drainage than wooden ones because they are non-absorbent. The design of a small bin usually depends on where an individual wishes to store the bin and how they wish to feed the worms. Most small bins can be grouped into three categories:

- Non-continuous – an undivided container. A layer of bedding materials is placed in the bin, lining the bottom. Worms are added and organic matter for composting is added in a layer above the bedding. Another layer is added on top of the organic matter and the worms will start to compost the organic matter and bedding. This type of bin is often used because it is small and easy to build. But it is relatively difficult to harvest because all the materials and worms must be emptied out when harvesting.
- Continuous vertical flow – a series of trays stacked vertically. The bottom-most tray is filled first, in a similar fashion to any other bin, but is not harvested when it is full. Instead, a thick layer of bedding is added on top and the tray above is used for adding organic material. Worms finish composting the materials in the bottom tray and then migrate to the one above. When a sufficient number of worms have migrated, the vermicompost in the bottom tray can be collected and should be relatively free of worms. These bins provide an easier method of harvesting.
- Continuous horizontal flow – a series of trays lined horizontally. This method too relies on the earthworms migrating towards a food source in order to ease the process of harvesting. The bin is usually constructed to be similar to a non-continuous bin but longer horizontally. It is divided in half, usually by a large gauge screen of chicken wire. One half is used until it becomes full, then the other half is filled with bedding and organic matter. In time, the worms migrate to the side with the food and the compost can be collected. These bins are larger than a non-continuous system but still small enough to be convenient.

Large Scale

There are two main methods of large-scale vermiculture. Some systems use a windrow, which consists of bedding materials for the earthworms to live in (see bedding below) and acts as a large bin; organic material is added to it. Although the windrow has no physical barriers to prevent worms from escaping, in theory they should not due to an abundance of organic matter for them to feed on. Often windrows are used on a concrete surface to prevent predators from killing the worm population. When large scale windrows are fed on one side consistently, a wave motion is generated over time.

The second type of large-scale vermicomposting system is the raised bed or flow-through system. Here the worms are fed an inch of "worm chow" across the top of the bed, and an inch of castings are harvested from below by pulling a breaker bar across the large mesh screen which forms the base of the bed. Because red worms are surface dwellers and are constantly moving towards the new food source, the flow-through system eliminates the need to separate worms from the castings before packaging. Flow-through systems are well suited to indoor facilities, making them the preferred choice for operations in colder climates. One major Australian company wormswork has successfully commissioned this continous flow system at their manufacturing plant in Millicent South Australia. with the capacity of producing 12,000 tonnes of finished solid organic fertilizer and 6 million litres of organic liquid fertilizer annually. This can be seen at www.wormswork.com.au

Starting Off

When beginning a vermicomposting bin, moist bedding is put into the bin and the worms are added. In hot climates, the bin is placed away from direct sunlight. Appropriate waste can be added daily or weekly. At first, the worms are fed at most half their body weight per day. After they have established themselves, they can be fed up to their entire body weight. It is best not to add new food on top of old food until the old food has been processed by the worms. However, new food can be added in a different location in the bin.

Bedding

Bedding is the living medium and also a food source for the worms. It is material high in carbon and made to mimic decaying dried leaves on the forest floor, the worms' natural habitat. The bedding should be moist (often similar to the consistency of a wrung-out sponge) and loose to enable the worms to breathe and to facilitate aerobic decomposition of the food that is buried in it. A wide variety of bedding materials can be used, including shredded newspaper, sawdust, hay, cardboard, coir, burlap coffee sacks, peat moss, pre-composted (aged) manure, and dried leaves. Cat litter, and pet and human waste should not be used. Most vermicomposters avoid using glossy paper from newspapers and magazines, junk mail, and shredded paper from offices, because they may contain toxins which may disrupt the system.

Also, coated cardboard that contains wax or plastic, such as milk boxes, cannot be used. Newspaper and phone books printed on regular, non-glossy paper with non-toxic soy ink are safe for use, and decompose relatively quickly. Some bedding is easier to use and add food scraps to than others.

Temperature

Worms used in composting systems prefer temperatures of 55 to 70 degrees Fahrenheit (12-21 degrees Celsius). The temperature of the bedding should not drop below freezing or above 89.6 °F (32 °C). This temperature range means that indoor vermicomposting is suitable for homes in all but tropical climates.

Food

Worms and other composting organisms have a preferred ratio of carbon to nitrogen (C:N), approximately 30:1. As some waste is richer in carbon and others in nitrogen, waste must be mixed to approximate the ideal ratio. "Brown matter", or wood products such as shredded papers, is rich in carbon. "Green matter", such as food scraps, has more nitrogen, which is related to the amount of protein in the waste.If the waste is mostly vegetable and fruit scraps, and does not regularly include animal products or high-protein vegetable foods like beans, the resulting vermicompost and waste liquid will be low in nitrogen.

Suitable

Kitchen waste suitable for worms includes coffee grounds and filters, tea bags and plate scrapings, as well as rotting fruit (including citrus fruit but NOT citrus peel), vegetable peels, leftovers, moldy bread, etc. These materials can be raw or cooked. They do not have to be ground up, as the micro-organisms in the bin will gradually soften them. However, if a large quantity of dry food (e.g., moldy bread) is added and covered with bedding, pour a little purified water over the bedding to moisten the mixture. If too much kitchen waste is added, the bin mixture putrifies before the worms can process it and becomes harmful to the worms. High-protein foods like beans are particularly susceptible. Check the bin at least once a week, give the materials a stir to oxygenate, and add bedding if the bin appears too moist. Soft garden wastes such as carrot tops and tomato leaves are suitable foods. An occasional sprinkling of garden soil in the bin gives the worms the grit they need to digest food. It's not harmful to throw in an entire plant, but the worms will not process the woody parts or large roots and these will have to be hand-removed later from the finished vermicompost. Compostable plates, cups, etc. are also suitable, but in small bins they should be torn first into smaller pieces so as not to block oxygen flow.

Unsuitable

High-water-content materials like watermelon rinds provide very little food for the worms while disrupting the moisture level of the system. They should be added sparingly. Grass clippings and other products sprayed with pesticides should be avoided. Some banana peels are heavily sprayed, and can kill everything if added to a small bin. Although worms can digest proteins and fats in meat scraps, these materials can attract scavengers. Too much oil or fat can hinder the breathing of the worms, as they breathe through their skin. Worms cannot break down bone and are said to dislike highly spiced foods such as onions, garlic, and salt. If possible, sticky food labels, rubber bands, tea bag staples, and other inedibles should be removed before placing the food in the worm bin, as these items will not decompose. Fruit pits need not be removed from decaying fruit before adding, as the worms will eat all the soft matter. Compostable cutlery takes too long to degrade in vermicompost.

Bin Maintenance

Worms and other composting microorganisms require oxygen, so the bin must "breathe". This can be accomplished by regularly removing the composted material, adding holes to the bin, or using a continuous-flow bin. If insufficient oxygen is available, the decay becomes anaerobic, like that in swamps and bogs, producing a strong odor and creating a toxic environment for the worms.

The moisture level and oxygen flow in a home worm bin should be checked at least once a week. Over the long term, care should be taken to maintain optimum moisture levels. In a non-continuous-flow vermicomposting bin, excess liquid can be drained via a tap and used as plant food. A continuous flow bin does not retain excess liquid and, depending on the foods used, may require sprinklings of water to keep the bedding moist. The pH should be slightly alkaline. Alkalinity can be increased by occasionally adding a handful of calcium carbonate, sold as "garden lime." Do not confuse calcium carbonate with regular lime (Calcium oxide), which is far too alkaline and will kill worms. Adding many citrus peels can hinder the worms, but probably due not to acidity but to d-limonene, a fragrant chemical present in the rind of citrus fruits. Coffee grounds have sometimes been blamed for acidity, but analysis shows they are only mildly acidic, with a pH of 6.2.

Feeding

There are two methods of adding matter to the bin.

- Top feeding — organic matter is placed directly on top of the existing layer of bedding in a bin and then covered with another layer of bedding. This is repeated every time the bin is fed.
- Pocket feeding — a top layer of bedding is maintained and food is buried

beneath. The location of the food is changed each time, rotating around the bin to give the worms time to decompose the food in the previously fed pockets. The top layer of bedding is replenished as necessary.

Vermicomposters often use a combination of both methods. Sometimes unburied food can attract fruit flies, so food should be buried under at least one inch of bedding material.

Harvesting

Vermicompost is ready for harvest when it contains few to no scraps of uneaten food or bedding. Even a properly composted mixture will contain large items that should be discarded, such as peach or date pits, glassine-like sheets from melon skins, and twigs. Small seeds from composted food such as tomatoes and apples cannot be removed from the vermicompost and may sprout later in the seed-starting pots or garden.

There are several methods of harvesting, depending on the purpose for which the vermicompost will be used, and whether or not the composter wishes to salvage as many worms and worm eggs as possible from the vermicompost.

Vermicompost Properties

Vermicompost, also known as worm castings and vermicast, is richer in many nutrients than compost produced by other composting methods. It is also rich in microbial life which helps break down nutrients already present in the soil into plant-available forms. Unlike other compost, worm castings also contain worm mucus which keeps nutrients from washing away with the first watering and holds moisture better than plain soil. For this reason, some fruit and seed pits are reported to germinate in vermicompost easily. Vermicompost made from ordinary kitchen scraps will contain small seeds, especially those of tomatoes, peppers, and eggplants, that may sprout weeks later.

Vermicompost benefits soil by

- improving its physical structure;
- enriching soil in micro-organisms, adding plant hormones such as auxins and gibberellic acid, and adding enzymes such as phosphatase and cellulase;
- attracting deep-burrowing earthworms already present in the soil;
- improving water holding capacity;
- enhancing germination, plant growth, and crop yield; and
- improving root growth and structure.

Vermicompost can be used to make compost tea (worm tea), by mixing some vermicompost in water and steeping for a number of hours or days. The resulting liquid is used as a fertilizer.

The dark brown waste liquid that drains into the bottom of some vermicomposting systems, as water-rich foods break down, is also excellent as fertilizer. However, the pH and nutrient contents of these liquids (as well as vermicompost) varies, depending on the food fed to the worms and whether or not lime has been added to the system. pH and nitrogen, phosphorus, and potassium (NPK) measurements should be taken periodically to determine the fertilizer composition before use. Home kits for testing are sold in hardware stores and nurseries.

Problems

Odor, usually due to overabundance of "greens" (wet waste) in the bin, results from too much nitrogen combining with hydrogen to form ammonia. To neutralize the odors, add a fair amount of shredded newspaper or other "browns" to the mix to absorb excess moisture, remove the smelly waste, and stop adding food to the bin until a substantial portion of the uneaten food has been turned into compost. The carbon will balance the nitrogen and form a compound that is not smelly. The higher level of carbon means that decomposition will be slower. Also, always add new material deep in the bin to disallow access to would-be pests. Consistently doing so will greatly reduce any nuisance from odor and undesirable organisms.

Pests such as rodents and flies may be attracted by certain materials and odors, especially lots of kitchen waste and especially meat. This problem is largely avoided if a sealed bin is used where the pests cannot access the material, although many proponents recommend having ample access to air. This promotes natural decomposition, as worms and beneficial bacteria require oxygen. Fruit flies are common in warm weather if the food is not thoroughly covered with bedding.

Commercially Available Vermicomposting Systems

Home vermicompost systems or worm bins are available from many retailers. Clive Roberts BSc invented the very first 'Wormery' in the late 1980s and in 1991 patented his design for The Original Wormery, in the UK. In the United States, state Cooperative Extension Services and local governments may provide assistance and training in home vermicomposting.

OSU Research Update

Today most floriculture crops are produced in soilless media that often contains an organic component. Certainly peat is one of the most widely employed organic materials contained in commercial soilless media. However, as peat supplies diminish because of over- harvesting, prices will inevitably rise. Therefore, identification of comparably priced substitutes for peat is imperative. One approach is to amend peat-based soilless media with other organic materials such as carbonised bark and compost. While these substances are not intended to substitute for peat, they do reduce the amount of peat used by 10 to 20%, while at the same time alter the physical and chemical properties of the media to provide optimal growing conditions.

At the same time there is growing environmental concern about the disposal of organic wastes such as sewage sludge, yard and garden waste, and animal manure. Therefore, bio-solid wastes must either be contained or converted to a non-hazardous form that will not pollute the environment. This, of course, has led to the skyrocketing of the cost of disposal. Composting such wastes is one environmentally sound way to convert biological waste into a form that can be easily disposed and provide a mechanism to recoup some of the costs incurred.

Incorporation of compost into potting and container media is one potential use for this material. Indeed, the addition of compost has been shown to contribute positively to the chemical and physical properties of potting media. But there are downsides as well; foremost of which is the difficulty in producing a reproducible and stabilised product. The primary reason for this is because the composting process is dependent on the aerobic action of soil microorganisms that break down the complex organic molecules to simple substances that can be taken up and utilised by plants. This bio-oxidative process is highly dependent on variable environmental conditions such as temperature and the availability of oxygen that are hard to control.

Another "composting" process that has not received as much attention is vermicomposting that relies on the action of earthworms to break down complex organic compounds. The worm castings or fecal material produced by the worms is a rich source of a variety of essential plant nutrients. In contrast to conventional composts, vermicomposts are less variable and much more stable. In addition, the time to complete the composting process is much shorter for vermicomposts: one pound of worms can convert one pound of pig manure to compost in 48 hours!

Could vermicompost be an economical and useful amendment to soilless potting media? To answer this question, Gary Bachman, a graduate student of mine, and I have been collaborating with Drs. Clive Edwards and Scott Subler from the Entomology Department at The Ohio State University. We have been examining the effects of vermicompost, derived from pig manure, on the growth of bedding plants and assessing whether or not the use of vermicompost in commercial soilless media is economically viable.

In initial studies we found that incorporation of vermicompost in a commercial potting mix resulted in faster growth, increased height, and higher fresh and dry weights for a variety of bedding plant species. The optimal amount of vermicompost need to produce positive results was relatively low: between 10 and 20% by volume. We found that the incorporation of vermicompost in the media essentially eliminated the need for additional fertilizer in the production tomato plugs. These results suggest that the routine incorporation of vermicompost in soilless media may provide the grower with significant savings in plug production by reducing the costs associated with fertilising.

We also observed that the positive effects of the vermicompost on plant quality were sustained for a substantial period of time after seedling emergence. In other words, incorporation of vermicompost in media used to grow bedding plants may help to maintain crop quality and salability even after the plants reach the garden center. Additional evidence for this was obtained through a consumer preference study conducted prior to an OSU football game involving nearly 400 adults. An overwhelming majority (around 80%) of the participants said that they would choose to buy tomato plants grown in media amended with 10% vermicompost over those grown using standard fertilisation methods. Although much of the effect of vermicompost on plant growth can be attributed to the addition of important nutrients, there are other unknown factors involved. Plants grown in vermicompost-amended media still grow better than those grown in unamended media with similar nutritional levels. Furthermore, the promotive effect of vermicompost on growth is lost if it is sterilised and can not be restored by adding additional nutrients. We are presently trying to determine the nature of the factor(s) contained in vermicompost that promote plant growth. Regardless, the use of vermicompost in commercial soilless media holds great promise if the economics hold out. The widespread use of vermicompost as an amendment will ultimately depend on its cost of production. However, with the increased pressure to find alternate ways to safely dispose of animal manure, the techniques for vermicomposting on a large scale are now being developed that can economically process enormous amounts of animal waste.

Butanol Fuel

Butanol may be used as a fuel in an internal combustion engine. Because its longer hydrocarbon chain causes it to be fairly non-polar, it is more similar to gasoline than is ethanol. Butanol has been demonstrated to work in some vehicles designed for use with gasoline without any modification. It can be produced from biomass (as "biobutanol") as well as fossil fuels (as "petrobutanol"); both biobutanol and petrobutanol have the same chemical properties

Production of Butanol from Biomass

Butanol from biomass is called biobutanol. It can be produced by fermentation of biomass by the A.B.E. process. The process uses the bacterium *Clostridium acetobutylicum,* also known as the *Weizmann organism.* It was Chaim Weizmann who first used this bacteria for the production of acetone from starch (with the main use of acetone being the making of Cordite) in 1916. The butanol was a by-product of this fermentation (twice as much butanol was produced). The process also creates a recoverable amount of H_2 and a number of other by-products: acetic, lactic and propionic acids, acetone, isopropanol and ethanol.

The difference from ethanol production is primarily in the fermentation of the feedstock and minor changes in distillation. The feedstocks are the same as for ethanol: energy crops such as sugar beets, sugar cane, corn grain, wheat and cassava as well as agricultural byproducts such as straw and corn stalks (reference needed). According to DuPont, existing bioethanol plants can cost-effectively be retrofitted to biobutanol production.

Algae Butanol

Biobutanol can be made entirely with solar energy, from algae (called Solalgal Fuel) or diatoms.

Centia Process

Centia is based on a three-step thermal, catalytic, and reforming process that has the potential to turn virtually any lipidic compound—e.g., vegetable oils, oils from animal fat and oils from algae—into 1-for-1 replacements for petroleum jet fuel, diesel, and gasoline. The three steps are:

- Hydrolytic conversion.
- Decarboxylation.
- Reforming long-chain alkanes.

Distribution

Butanol better tolerates water contamination and is less corrosive than ethanol and more suitable for distribution through existing pipelines for gasoline. In blends with diesel or gasoline, butanol is less likely to separate from this fuel than ethanol if the fuel is contaminated with water. There is also a vapor pressure co-blend synergy with butanol and gasoline containing ethanol, which facilitates ethanol blending. This facilitates storage and distribution of blended fuels.

Energy Content and Effects on Fuel Economy

Switching a gasoline engine over to butanol would in theory result in a fuel consumption penalty of about 10% but butanol's effect on mileage is yet to be determined by a scientific study. While the energy density for any mixture of gasoline and butanol can be calculated, tests with other alcohol fuels have demonstrated that the effect on fuel economy is not proportional to the change in energy density.

Octane Rating

The octane rating of n-butanol is similar to that of gasoline but lower than that of ethanol and methanol. n-Butanol has a RON (Research Octane number) of 96 and a MON (Motor octane number) of 78 while t-butanol has octane ratings of 105 RON and 89 MON. t-Butanol is used as an additive in gasoline but cannot be used as a fuel in its pure form because its relatively high melting point of 25.5 °C causes it to gel and freeze near room temperature.

A fuel with a higher octane rating is less prone to knocking (extremely rapid and spontaneous combustion by compression) and the control system of any modern car engine can take advantage of this by adjusting the ignition timing. This will improve energy efficiency, leading to a better fuel economy than the comparisons of energy content different fuels indicate. By increasing the compression ratio, further gains in fuel economy, power and torque can be achieved. Conversely, a fuel with lower octane rating is more prone to knocking and will lower efficiency. Knocking can also cause engine damage.

Air-Fuel Ratio

Alcohol fuels, including butanol and ethanol, are partially oxidised and therefore need to run at richer mixtures than gasoline. Standard gasoline engines in cars can adjust the air-fuel ratio to accommodate variations in the fuel, but only within certain limits depending on model. If the limit is exceeded by running the engine on pure butanol or a gasoline blend with a high percentage of butanol, the engine will run lean, something which can damage it. Compared to ethanol, butanol can be mixed in higher ratios with gasoline for use in existing cars without the need for retrofit as the air-fuel ratio and energy content are closer to that of gasoline.

Specific Energy

Alcohol fuels have less energy per unit weight and unit volume than gasoline. To make it possible to compare the net energy released per cycle a measure called the fuels specific energy is sometimes used. It is defined as the energy released per air fuel ratio. The net energy released per cycle is higher for butanol than ethanol or methanol and about 10% higher than for gasoline.

Viscosity

The viscosity of alcohols increase with longer carbon chains. For this reason, butanol is used as an alternative to shorter alcohols when a more viscous solvent is desired. The kinematic viscosity of butanol is several times higher than that of gasoline and about as viscous as high quality diesel fuel.

Heat of Vaporisation

The fuel in an engine has to be vaporised before it will burn. Insufficient vaporisation is a known problem with alcohol fuels during cold starts in cold weather. As the latent heat of vaporisation of butanol is less than half of that of ethanol, an engine running on butanol should be easier to start in cold weather than one running on ethanol or methanol.

Potential Problems with the Use of ButanolFuel

The potential problems with the use of butanol are similar to those of ethanol:

- To match the combustion characteristics of gasoline, the utilisation of butanol fuel as a substitute for gasoline requires fuel-flow increases (though butanol has only slightly less energy than gasoline, so the fuel-flow increase required is only minimal, maybe 10%, compared to 40% for ethanol.)
- Alcohol-based fuels are not compatible with some fuel system components.
- Alcohol fuels may cause erroneous gas gauge readings in vehicles with capacitance fuel level gauging.
- While ethanol and methanol have lower energy densities than butanol, their higher octane number allows for greater compression ratio and efficiency. Higher combustion engine efficiency allows for lesser greenhouse gas emissions per unit motive energy extracted.

As an advantage, butanol production from biomass could be more efficient (i.e. unit engine motive power delivered per unit solar energy consumed) than ethanol or methanol routes. Also, some bacteria that produce butanol are able to digest cellulose, not just starch and sugars.

Possible Butanol Fuel Mixtures

Standards for the blending of ethanol and methanol in gasoline exist in many countries, including the EU, the US and Brazil. Approximate equivalent butanol blends can be calculated from the relations between the stochiometric fuel-air ratio of butanol, ethanol and gasoline. Common ethanol fuel mixtures for fuel sold as gasoline currently range from 5% to 10%. The share of butanol can be 60% greater than the equivalent ethanol share, which gives a range from 8% to 32%. "Equivalent" in this case refers only to the vehicle's ability to adjust to the fuel. Other properties such as energy density, viscosity and heat of vaporisation will vary and may further limit the percentage of butanol that can be blended with gasoline.

Current use of Butanol in Vehicles

Currently no production vehicle is known to be approved by the manufacturer for use with 100% butanol, though any model that is able to run 10% ethanol blends should be able to use butanol without any problems. David Ramey drove from Blacklick, Ohio to San Diego, California using butanol in an unmodified 1992 Buick Park Avenue. Although further long term testing must be done, it is highly likely that most late model cars can run on 100% butanol safely with no modifications. Justification for this conclusion is based on data for RON in comparison of n-Butanol with Gasoline. Also, modern ECU-injected motorcar piston engines are designed to be flexible enough to deliver good performance with 91-RON fuels, which n-Butanol exceeds in RON rating.

Research Challenges

The key research challenge that must be resolved is that butanol production inhibits microbial growth even at low concentrations. The result is that the product of the fermentation is less than 2% butanol. The overwhelming majority of the fermentation broth is water, so an energy-intensive distillation step is required for purification. This may be acceptable if the goal is to produce butanol for use as a solvent, but if butanol is to gain traction as a motor fuel, energy inputs into the process need to be minimized.

The Swiss company Butalco GmbH uses a special technology to modify yeasts in order to produce butanol instead of ethanol. Yeasts as production organisms for butanol have decisive advantages compared to bacteria.

Biogas

Biogas typically refers to a gas produced by the biological breakdown of organic matter in the absence of oxygen. Biogas originates from biogenic material and is a type of biofuel. One type of biogas is produced by anaerobic digestion or fermentation of biodegradable materials such as biomass, manure or sewage, municipal waste, and energy crops. This type of biogas comprises primarily methane and carbon dioxide. The other principal type of biogas is wood gas which is created by gasification of wood or other biomass. This type of biogas is comprised primarily of nitrogen, hydrogen, and carbon monoxide, with trace amounts of methane.

The gases methane, hydrogen and carbon monoxide can be combusted or oxidised with oxygen. Air contains 21% oxygen. This energy release allows biogas to be used as a fuel. Biogas can be used as a low-cost fuel in any country for any heating purpose, such as cooking. It can also be utilised in modern waste management facilities where it can be used to run any type of heat engine, to generate either mechanical or electrical power. Biogas is a renewable fuel and electricity produced from it can be used to attract renewable energy subsidies in some parts of the world.

Production

Depending on where it is produced, biogas can also be called swamp, marsh, landfill or digester gas. A biogas plant is the name often given to an anaerobic digester that treats farm wastes or energy crops.

Biogas can be produced utilising anaerobic digesters. These plants can be fed with energy crops such as maize silage or biodegradable wastes including sewage sludge and food waste. Landfill gas is produced by organic waste decomposing under anaerobic conditions in a landfill. The waste is covered and compressed mechanically and by the weight of the material that is deposited from above. This material prevents oxygen from accessing the waste and anaerobic microbes thrive. This gas builds up and is slowly released into the atmosphere if the landfill site has not been engineered to capture the gas. Landfill gas is hazardous for three key reasons. Landfill gas becomes explosive when it escapes from the landfill and mixes with oxygen within lower and higher explosive limits. The methane in biogas forms explosive mixtures in air. The lower explosive limit is 5% methane and the upper explosive limit is 15% methane. The methane contained within biogas is 20 times more potent as a greenhouse gas than carbon dioxide. Therefore uncontained landfill gas which escapes into the atmosphere may significantly contribute to the effects of global warming. In addition to this volatile organic compounds (VOCs) contained within landfill gas contribute to the formation of photochemical smog. Sweden produces biogas from confiscated alcoholic beverages.

Composition

The composition of biogas varies depending upon the origin of the anaerobic digestion process. Landfill gas typically has methane concentrations around 50%. Advanced waste treatment technologies can produce biogas with 55-75%CH_4

In some cases biogas contains siloxanes. These siloxanes are formed from the anaerobic decomposition of materials commonly found in soaps and detergents. During combustion of biogas containing siloxanes, silicon is released and can combine with free oxygen or various other elements in the combustion gas. Deposits are formed containing mostly silica (SiO_2) or silicates (Si_xO_y) and can also contain calcium, sulfur, zinc, phosphorus. These white mineral deposits build to a surface thickness of several millimetres and must be removed by chemical or mechanical means.

Applications

Biogas can be utilised for electricity production, cooking, space heating, water heating and process heating. If compressed, it can replace compressed natural gas for use in vehicles, where it can fuel an internal combustion engine or fuel cells.

Methane within biogas can be concentrated to the same standards as natural gas, when it is, it is called biomethane. If the local gas network permits it the producer of the biogas may be able to utilize the local gas distribution networks. Gas must be very clean to reach pipeline quality, and must be of the correct composition for the local distribution network to accept. Carbon dioxide, Water, hydrogen sulfide and particulates must be removed if present. If concentrated and compressed it can also be used in vehicle transportation. Compressed biogas is becoming widely used in Sweden, Switzerland and Germany. A biogas-powered train has been in service in Sweden since 2005.

Bates' and his biogas car were the subject of a short documentary film called 'Sweet as a Nut' in 1974, at which point he had run his car for 17 years on gas he had produced by processing pig manure. Bates, an inventor, lived in Devon, UK and in the film talks through the simple process and benefits of running a car on biogas. The conversion was simply made with an adapter attached to any combustion engine.

Biogas in Developing Nations

In India biogas produced from the anaerobic digestion of manure in small-scale digestion facilities is called Gober gas. In India biogas is generated at an estimated 2 million+ household facilities. The digester is an airtight circular pit made of concrete with a pipe connection. The manure is directed to the pit, usually directly from the cattle shed. The pit is then filled with a required quantity of wastewater. The gas pipe is connected to the kitchen fire place through control valves. The combustion of this biogas has very little odour or smoke. Owing to simplicity in implementation and use of cheap raw materials in villages, it is one of the most environmentally sound energy sources for rural needs. Biogas is used extensively throughout rural China and where wastewater treatment and industry coincide. The Biogas Support Programme in Nepal has installed over 100,000 biogas plants in rural areas. Vietnam's Biogas Programme for Animal Husbandry Sector has led to the installation of over 20,000 plants throughout that country. Biogas is also in use in rural Costa Rica. In Colombia experiments with diesel engines-generator sets partially fuelled by biogas demonstrated that biogas could be used for power generation, reducing elecricity costs by 40% compared with purchase from the regional utility.

Legislation

The European Union presently has some of the strictest legislation regarding waste management and landfill sites called the Landfill Directive. The United States legislates against landfill gas as it contains these VOCs. The United States Clean Air Act and Title 40 of the Code of Federal Regulations (CFR) requires landfill owners to estimate the quantity of non-methane organic compounds (NMOCs) emitted. If the estimated NMOC emissions exceeds 50 tonnes per year the landfill owner is required to collect the landfill gas and treat it to remove the entrained NMOCs. Treatment of the landfill gas is usually by combustion. Because of the remoteness of landfill sites it is sometimes not economically feasible to produce electricity from the gas.

Gober Gas

Gober gas is a biogas generated out of cow dung. In India, gober gas is generated at the countless number of micro plants (an estimated more than 2 million) attached to households. The gober gas plant is basically an airtight circular pit made of concrete with a pipe connection. The manure is directed to the pit (usually directed from the cattle shed). The pit is then filled with a required quantity of water (usually waste water). The gas pipe is connected to the kitchen fire place through control valves. The flammable methane gas generated out of this is practically odorless and smokeless. The residue left after the extraction of the gas is used as biofertiliser. Owing to its simplicity in implementation and use of cheap raw materials in the villages, it is often quoted as one of the most environmentally sound energy source for the rural needs.

Compost

Compost is the aerobically decomposed remnants of organic matter. It is used in landscaping, horticulture and agriculture as a soil conditioner and fertiliser. It is also useful for erosion control, land and stream reclamation, wetland construction, and as landfill cover. Compost serves as a *growing medium*,or a porous, absorbent material that holds moisture and soluble minerals, providing the support and nutrients in which most plants will flourish. To maximize plant growth, it is sometimes necessary to dilute compost with soil or peat to reduce salinity or to add neutralisers to bring the pH closer to 7, or additional nutrients like fertilisers or manure, wetting agents, and materials to improve drainage and aeration, such as sand, grit, bark chips, vermiculite, perlite, or clay granules.

Composting as an Alternative to Landfill

As concern about landfill space increases, worldwide interest in recycling by means of composting is growing, since composting is a widely accepted process for converting decomposable wastes of natural origin into stable, sanitised products useful for horticulture. Modern composting originated in European organic farming in the early 20th century. However, the more recent application of composting for large-scale waste reduction has very little in common with organic farming. The 1999 European Landfill Directive put pressure on European states to meet specified targets for landfill reduction, principally by establishing alternate disposal and treatment of organic materials. While certain countries such as Belgium, Holland, Germany and Austria readily achieved the mandated targets, other countries such as the UK, Spain and Italy have not. Indeed, it is commonly accepted that the UK, despite its early important contributions to organic farming and John Innes Compost, started taking composting seriously only after Brussels threatened a penalty for states not attaining the required reduction targets. A recent National Audit Office report for England warned that councils were in danger of missing EU targets to cut the amount of waste at landfills. The NAO's report stated that to meet European targets for 2010, a reduction in the UK of at least 3.5m tonnes of biodegradable waste sent to landfill was needed. A reduction of a further 3.7m tonnes was needed by 2013. Less than one-half this has been currently met.

Modern large-scale composting should therefore not be confused with an idealistic, organic-oriented goal to recycle and improve soils; since, for most western countries now, it is virtually the law. These factors could lead to a conflict between required production of composts and the quality of the product.

Compost Ingredients

Given enough time, all biodegradable material will compost, and the primary objective in the modern push to compost is to capture readily degradable materials so they do not enter landfills. However, most small-scale domestic systems will not reach sufficiently high temperatures to kill pathogens and weed seeds or deter vermin, so pet droppings, scraps of meat, and dairy products are often best left to operators of high-rate, thermophilic composting systems. Hobby animal manure (horses, goats), vegetable kitchen and garden waste are nevertheless all excellent raw material for home composting. Early roots of composting as a treatment for municipal solid waste were spurred by awareness of the trash crisis as early as the 1950s, and the rise worldwide of large MSW composting plants in the 1960s into the 1970s was virtually unregulated. Public outcry in Europe against contamination of soils on farms and vineyards from MSW compost containing residues of plastic, metals and glass triggered a shakeup of the industry, and in the 1980s a phasing out of MSW composting,

European Composting Standards

An overview of European efforts to attain compost standardisation can be seen on the European Compost Network (ECN). The British Composting Association has established very recently a set of guidelines for compost, called the BSI PAS 100 listed by the British Standards Institute (PAS stands for "Publicly Available Specification" and is not necessarily an adopted or certified standard). There are a variety of such voluntary industry standards in Europe and worldwide, such as the German Bundegütegemeinschaft Kompost e.V. (BGK) German Compost Association RAL-standard for compost developed 10 years prior to the British standard, and updated recently to include separate standards for fermented by-products(from biogas reactors) and sludge. In America, Procter & Gamble Company sponsored the USCC in the early 1990s to develop compost process and product standards called "TMECC", still in a draft state. These standarisation programmes (guidelines would be a better word than standards to describe the objective) are intended to provide structure in the composting community for handling the entire composting process from raw materials and production methods, through quality control and lab testing. Swiss compost guidelines recognize distinct end-uses of composts, as determined by specific laboratory assays (see VKS-ASIC-ASAP-ASCP Swiss Compost Association).

Compost Types and Ingredients

There are different ways to compost, starting with layers of 'brown' and 'green' biodegradable waste mixed with garden soil. 'Brown' waste refers to old straw, tough vegetable stems and hedge clippings. 'Green' waste refers to biodegradable waste that breaks down faster, such as fruit, coffee grounds, cut flowers, and grass clippings.

There is also Vermicomposting, which uses worms to help break down the organic waste.

Compostable Materials

- Biodegradable waste
- Coffee grounds
- Cocoa shell
- Green waste
- Humanure
- Leaf mold
- Manure
- Spent mushroom compost
- Tree bark
- Egg shells

- Brown waste
- Fruit
- Vegetables

Inorganic Additives

- Loam
- Grit
- Urea
- Vermiculite
- Perlite

Compost end Uses

Compost is almost universally recommended as a soil amendment. It is principally intended as a blend with soil or other matrices such a coir and peat (although it may also be used to make compost tea). High rates of mixture (e.g. 80–100%) of compost have been occasionally noted in growing media, but generally direct seeding into a compost is not recommended. It is very common to see blends of 20–30% compost used for transplanting seedlings at cotyledon stage or later. The primary factors controlling how well a compost blend performs are salinity and maturity, which singly and together can trigger phytotoxicity symptoms. It is well known that high salt content in growing media will affect water relations of plants, especially in early stages of growth. The effects or symptoms of damage can be yield reduction, leaf deformation and tip-burning or even plant epinasty.

These effects can also be attributed to a variety of other factors that may be present in active or finished composts, depending on ingredients. Such elements include pesticides, presence volatile fatty acids which are by-products of anaerobic conditions or residues of anaerobic digestion, ammonia associated with high manure content, heavy metals such as copper from farm ingredients and sludge, and ethylene oxide from plant debris, any of which can trigger some form of stunting and other phytotoxicity traits. In container-mix studies, it has been demonstrated that immature compost deprives the soil of oxygen content for a significant period of time, resulting in stunting of roots.

As a result of these numerous challenges, the introduction of compost products into professional horticulture as a competition to peat and soil-based products has been significantly less successful than originally hoped for. A Jan 2008 consumer report in the UK severely criticised compost quality, showing that only one out of 24 composts tested against 4 cultivars in actual growing media trials could be recommended as viable "peat-free" product. Nevertheless, the broad popularity of composts and their long term beneficial effects for soils and crops mean that demand will continue to grow worldwide.

What is a Composting Toilet System and How Does it Compost?

Composting toilet systems (sometimes called biological toilets, dry toilets and waterless toilets) contain and control the composting of excrement, toilet paper, carbon additive, and, optionally, food wastes. Unlike a septic system a composting toilet system relies on unsaturated conditions (material cannot be fully immersed in water), where aerobic bacteria and fungi break down wastes, just as they do in a yard waste composter. Sized and operated properly, a composting toilet breaks down waste to 10 to 30 percent of its original volume. The resulting end-product is a stable soil-like material called "humus," which legally must be either buried or removed by a licensed seepage hauler in accordance with state and local regulations in the United States. In other countries, humus is used as a soil conditioner on edible crops.

The primary objective of the composting toilet system is to contain, immobilize or destroy organisms that cause human disease (pathogens), thereby reducing the risk of human infection to acceptable levels without contaminating the immediate or distant environment and harming its inhabitants.

This should be accomplished in a manner that

- is consistent with good sanitation (minimising both human contact with unprocessed excrement and exposure to disease vectors, such as flies).
- produces an inoffensive and reasonably dry end-product that can be handled with minimum risk.
- minimizes odor.

A secondary objective is to transform the nutrients in human excrement into fully oxidized, stable plant-available forms that can be used as a soil conditioner for plants and trees.

The main components of a composting toilet are:

- a composting reactor connected to one or more dry or micro-flush toilets;
- a screened exhaust system (often fan-forced) to remove odors, carbon dioxide, water vapor, and the by-products of aerobic decomposition;
- a means of ventilation to provide oxygen (aeration) for the aerobic organisms in the composter;
- a means of draining and managing excess liquid and leachate;

- process controls, such as mixers, to optimize and manage the process; and
- an access door for removal of the end-product.

The composting reactor should be constructed to separate the solids from the liquids and produce a stable, humus material with less than 200 most probable number (MPN) per gram of fecal coliform.

General Types of Composting Toilet Systems

Composting toilet systems can be classified in several ways:

Self-Contained versus Centralized

Composting toilet systems are either *self-contained,* whereby the toilet seat and a small composting reactor are one unit (typically small cottage models), or *centralised* or *remote,* where the toilet connects to a composting reactor that is somewhere else.

Manufactured versus Site-Built

One can either purchase a *manufactured* composting toilet system or have a *site-built* composting toilet system constructed (however, the latter can be difficult to get permitted by local health agents).

Batch (Multiple-Chamber) versus Continuous (Single-Chamber)

Most composting toilet systems use one of two approaches to manage the composting process: either *single-chamber continuous* composting or *multi-chamber batch* composting processes

A continuous composter (including Clivus Minimuses and such brands as CTS, Clivus Multrum, Phoenix, BioLet, Sun-Mar) features a single chamber into which excrement is added to the top, and the end-product is removed from the bottom.

A batch composter (such as the EcoTech Carousel, all Vera systems, BioLet NE, and many site-built composters) utilizes two or more interchangeable composting reactors. One is filled at a time, then allowed to cure while another reactor fills, just as with two- and three-bin yard composters.

Proponents of continuous composting maintain that it is simple (takes place in one fixed reactor), allows urine to constantly moisten the process, and allows the center of the mass to heat up through uninterrupted microbial activity.

Advocates of the batch-composting approach say that by not continually adding fresh excrement and urine to older, more advance material, the material composts more thoroughly, uninterrupted by the added nutrients, pathogens, salts and ammonia in fresh excrement. Also, by dividing the material, it can have more surface area, and thus better aeration. Batch composting also offers an opportunity for unlimited capacity, as one simply adds compost reactors to the process to add capacity.

Batch systems require monitoring the level of the composter to determine when a chamber has filled and a new one must be moved into place. However, because there is more surface area and the material is divided, there is often less or no mixing and raking of the material.

What is true is that complete composting needs time to ecologically cascade all the by-products of the myriad organisms in the composting food web until all of the organic matter is finally transformed into safe, stable humus.

In continuous composters, urine or flush water can leach fresh excrement into the finished compost removal area. Segregating the material into batches reduces the risk of having living disease organisms in the finished products. (One of the authors [Del Porto] has designed, sold and serviced both types of composters. He prefers batch composting-based systems, because they are more forgiving and users seem to be happier with them. However, he would never say that single-chamber continuous systems do not work. They do! It is difficult to generalize about which process affords the greatest opportunity for complete processing and minimizes the potential for pathogen survival. In a batch system, a finite supply of nutrients is cycled and recycled through microbe populations until the nutrients, both the free ones and those bound in microbial protoplasm and cell walls, are ultimately converted to stable, fully oxidised forms, and the fungi have performed their work on the remaining lignin and cellulose compounds, releasing antibiotics in the process. Definitive research is needed in this area.

Active versus Passive

As with solar systems, composting systems are usually either *passive* or *active.* Passive systems are usually simple moldering reactors in which ETPA (excrement, toilet paper and additive) is collected and allowed to decompose in cool environments without active process control (heat, mixing, aeration).

Active systems may feature automatic mixers, pile-leveling devices, tumbling drums, thermostat-controlled heaters, fans, and so forth. The trend in the composting of municipal solid waste (garbage and trash), sewage sludge and yard and agricultural residues is toward active systems. By making the process active, the size of the composter can be reduced, because composting efficiency is speeded up (and the volume of the material reduced faster).

Passive systems are designed to optimize the process by design, not mechanical action, allowing only time, gravity, ambient temperature and the shape of the container to control the process. Passive composters are often referred to as moldering toilets, as the process at work is natural uncontrolled decay at cool in-ground temperatures at or below 68° F. In this cool environment, molds (fungi and actinomycetes) are the primary biological decomposers, because it is a bit too cool for the faster-acting mesophilic and thermophilic bacteria.

Air Resources Laboratory

The Air Resources Laboratory (ARL) is an air quality and climate laboratory in the Office of Oceanic and Atmospheric Research (OAR) which is an operating unit within the National Oceanic and Atmospheric Administration (NOAA). In October 2005, the Surface Radiation Research Branch of the ARL was merged with five other NOAA labs to form the Earth System Research Laboratory.

The Air Resources Laboratory (ARL) studies processes and develops models relating to climate and air quality, including the transport, dispersion, transformation and removal of pollutants from the ambient atmosphere. The emphasis of the ARL's work is on data interpretation, technology development and transfer. The specific goal of ARL research is to improve and eventually to institutionalize prediction of trends, dispersion of air pollutant plumes, air quality, atmospheric deposition, and related variables.

ARL provides scientific and technical advice to elements of NOAA and other Government agencies on atmospheric science, environmental problems, emergency assistance (Homeland Security), and climate change.

ARL operates with research divisions in Idaho Falls, Idaho; Research Triangle Park, North Carolina; Las Vegas, Nevada; Oak Ridge, Tennessee; and Silver Spring, Maryland.

The Atmospheric Sciences Modeling Division

The Atmospheric Sciences Modeling Division (ASMD) of the ARL was established in 1955 to collaborate with the U.S. Environmental Protection Agency (EPA) and its predecessor agencies in developing advanced air quality models that can simulate the formation, transport, dispersion and fate of pollutants in the atmosphere. The ASMD develops advanced modeling and decision support systems for effective forecasting and management of the air quality in the United States. The Division is located in Research Triangle Park, North Carolina.

The ASMD develops and evaluates predictive atmospheric models on all spatial and temporal scales for forecasting air quality, and for assessing changes in air quality and air pollutant exposures, as affected by changes in ecosystem management and regulatory decisions. The ASMD is responsible for providing a sound scientific and technical basis for regulatory policies to improve ambient air quality. The air quality and pollution dispersion models developed by the ASMD are being used by the EPA, NOAA, and the air pollution community in understanding and forecasting not only the magnitude of the air pollution problem, but also in developing emission control policies and regulations.

Air pollution Dispersion Terminology

Air pollution dispersion terminology describes the words and technical terms that have a special meaning to those who work in the field of air pollution dispersion modeling. Governmental environmental protection agencies (local, state, province and national) of many countries have also adopted and used much of the terminology in their laws and regulations regarding air pollution control.

It should be noted that some of the words and technical terms in Air pollution dispersion terminology quite often have other special meanings when used in fields of activity other than air pollution dispersion modeling.

Air Pollution Emission Plumes

- Buoyant plumes — Plumes which are lighter than air because they are at a higher temperature and lower density than the ambient air which surrounds them, or because they are at about the same temperature as the ambient air but have a lower molecular weight and hence lower density than the ambient air. For example, the emissions from the flue gas stacks of industrial furnaces are buoyant because they are considerably warmer and less dense than the ambient air. As another example, an emission plume of methane gas at ambient air temperatures is buoyant because methane has a lower molecular weight than the ambient air.
- Dense gas plumes — Plumes which are heavier than air because they have a higher density than the surrounding ambient air. A plume may have a higher density than air because it has a higher molecular weight than air (for example, a plume of carbon dioxide). A plume may also have a higher density than air if the plume is at a much lower temperature than the air. For example, a plume of evaporated gaseous methane from an accidental release of liquefied natural gas (LNG) may be as cold as -161 °C.
- Passive or neutral plumes — Plumes which are neither lighter or heavier than air.

Air Pollution Dispersion Models

There are five types of air pollution dispersion models, as well as some hybrids of the five types:

- Box model — The box model is the simplest of the model types. It assumes the airshed (i.e., a given volume of atmospheric air in a geographical region) is in the shape of a box. It also assumes that the air pollutants inside the box are homogeneously distributed and uses that assumption to estimate the average pollutant concentrations anywhere within the airshed. Although useful, this model is very limited in its ability to accurately predict dispersion of air pollutants over an airshed because the assumption of homogeneous pollutant distribution is much too simple.
- Gaussian model — The Gaussian model is perhaps the oldest (circa 1936) and perhaps the most commonly used model type. It assumes that the air pollutant dispersion has a Gaussian distribution, meaning that the pollutant distribution has a normal probability distribution. Gaussian models are most often used for predicting the dispersion of continuous, buoyant air pollution plumes originating from ground-level or elevated sources. Gaussian models may also be used for predicting the dispersion of non-continuous air pollution plumes (called *puff models*). The primary algorithm used in Gaussian modeling is the *Generalised Dispersion Equation For A Continuous Point-Source Plume*.
- Lagrangian model — a Lagrangian dispersion model mathematically follows pollution plume parcels (also called particles) as the parcels move in the atmosphere and they model the motion of the parcels as a random walk process. The Lagrangian model then calculates the air pollution dispersion by computing the statistics of the trajectories of a large number of the pollution plume parcels. A Lagrangian model uses a moving frame of reference as the parcels move from their initial location. It is said that an observer of a Lagrangian model follows along with the plume.
- Eulerian model — an Eulerian dispersions model is similar to a Lagrangian model in that it also tracks the movement of a large number of pollution plume parcels as they move from their initial location. The most important difference between the two models is that the Eulerian model uses a fixed three-dimensional Cartesian grid as a frame of reference rather than a moving frame of reference. It is said that an observer of an Eulerian model watches the plume go by.
- Dense gas model — Dense gas models are models that simulate the dispersion of dense gas pollution plumes (i.e., pollution plumes that are heavier than air). The three most commonly used dense gas models are:
  - The DEGADIS model developed by Dr. Jerry Havens and Dr. Tom Spicer at the University of Arkansas under commission by the US Coast Guard and US EPA.

- o The SLAB model developed by the Lawrence Livermore National Laboratory funded by the US Department of Energy, the US Air Force and the American Petroleum Institute.
- o The HEGADAS model developed by Shell Oil's research division.

Air Pollutant Emission Sources

The types of air pollutant emission sources are commonly characterised as either point, line, area or volume sources:

- Point source — A point source is a single, identifiable source of air pollutant emissions (for example, the emissions from a combustion furnace flue gas stack). Point sources are also characterised as being either elevated or at ground-level. A point source has no geometric dimensions.
- Line sources — A line source is one-dimensional source of air pollutant emissions (for example, the emissions from the vehicular traffic on a roadway).
- Area source — An area source is a two-dimensional source of diffuse air pollutant emissions (for example, the emissions from a forest fire, a landfill or the evaporated vapors from a large spill of volatile liquid).
- Volume source — A volume source is a three-dimensional source of diffuse air pollutant emissions. Essentially, it is an area source with a third (height) dimension (for example, the fugitive gaseous emissions from piping flanges, valves and other equipment at various heights within industrial facilities such as oil refineries and petrochemical plants). Another example would be the emissions from an automobile paint shop with multiple roof vents or multiple open windows.

Other air pollutant emission source characterisations are:

- Sources may be characterised as either stationary or mobile. Flue gas stacks are examples of stationary sources and busses are examples of mobile sources.
- Sources may be characterised as either urban or rural because urban areas constitute a so-called *heat island* and the heat rising from an urban area causes the atmosphere above an urban area to be more turbulent than the atmosphere above a rural area.
- Sources may be characterised by their elevation relative to the ground as either surface or ground-level, near surface or elevated sources.
- Sources may also be characterised by their time duration:

- o puff or intermittent: short term sources (for example, many accidental emission releases are short term puffs)
- o continuous: a long term source (for example, most flue gas stack emissions are continuous)

Characterisation of Atmospheric Turbulence

The amount of turbulence in the ambient atmosphere has a major effect on the dispersion of air pollution plumes because turbulence increases the entrainment and mixing of unpolluted air into the plume and thereby acts to reduce the concentration of pollutants in the plume (i.e, enhances the plume dispersion). It is therefore important to categorize the amount of atmospheric turbulence present at any given time.

The Pasquill Atmospheric Stability Classes

The oldest and, for a great many years, the most commonly used method of categorising the amount of atmospheric turbulence present was the method developed by Pasquill in 1961. He categorised the atmospheric turbulence into six stability classes named A, B, C, D, E and F with class A being the most unstable or most turbulent class, and class F the most stable or least turbulent class. Table 1 lists the six classes and Table 2 provides the meteorological conditions that define each class.

Historical stability class data, known as the Stability Array (STAR) data, for sites within the USA can be purchased from the National Climatic Data Center (NCDC).

Advanced Methods of Categorising Atmospheric Turbulence

Many of the more advanced air pollution dispersion models do not categorize atmospheric turbulence by using the simple meteorological parameters commonly used in defining the six Pasquill classes as shown in Table 2. The more advanced models use some form of Monin-Obukhov similarity theory.

For example, the US EPA's most advanced model, AERMOD, no longer uses the Pasquill stability classes to categorize atmospheric turbulence. Instead, it uses the surface roughness length and the Monin-Obukhov length.

As another example, the United Kingdom's most advanced model, ADMS 3, uses the Monin-Obukhov length, the boundary layer height and the windspeed to categorize the atmospheric turbulence.

The detailed explanation of the mathematical formulation for the turbulence categorisation methods used in AERMOD, ADMS 3 and other advanced air pollution dispersion models is very complex and beyond the scope of this article. More detailed explanations are available on the Internet.

Miscellaneous Other Terminology

(Work on this section is continuously in progress)

- Building effects or downwash: When an air pollution plume flows over nearby buildings or other structures, turbulent eddies are formed in the downwind side of the building. Those eddies cause a plume from a stack source located within about five times the height of a nearby building or structure to be forced down to the ground much sooner than it would if a building or structure were not present. The effect can greatly increase the resulting near-by ground-level pollutant concentrations downstream of the building or structure. If the pollutants in the plume are subject to depletion by contact with the ground (particulates, for example), the concentration increase just downstream of the building or structure will decrease the concentrations further downstream.
- Deposition of the pollution plume components to the underlying surface can be defined as either dry or wet deposition:
  - o Dry deposition is the removal of gaseous or particulate material from the pollution plume by contact with the ground surface or vegetation (or even water surfaces) through transfer processes such as absorption and gravitational sedimentation. This may be calculated by means of a *deposition velocity*, which is related to the resistance of the underlying surface to the transfer.
  - o Wet deposition is the removal of pollution plume components by the action of rain. The wet deposition of radionuclides in a pollution plume by a burst of rain often forms so called *hot spots* of radioactivity on the underlying surface.
- Inversion layers: Normally, the air near the Earth's surface is warmer than the air above it because the atmosphere is heated from below as solar radiation warms the earth's surface, which in turn then warms the layer of the atmosphere directly above it. Thus, the atmospheric temperature normally decreases with increasing altitude. However, under certain meteorological conditions, atmospheric layers may form in which the temperature increases with increasing altitude. Such layers are called inversion layers. When such a layer forms at the earth's surface, it is called a surface inversion. When an inversion layer forms at some distance

above the earth, it is called an inversion aloft (sometimes referred to as a *capping inversion*). The air within an inversion aloft is very stable with very little vertical motion. Any rising parcel of air within the inversion soon expands, thereby adiabatically cooling to a lower temperature than the surrounding air and the parcel stops rising. Any sinking parcel soon compresses adiabatically to a higher temperature than the surrounding air and the parcel stops sinking. Thus, any air pollution plume that enters an inversion aloft will undergo very little vertical mixing unless it has sufficient momentum to completely pass through the inversion aloft. That is one reason why an inversion aloft is sometimes called a capping inversion.

- Mixing height: When an inversion aloft is formed, the atmospheric layer between the earth's surface and the bottom of the inversion aloft is known as the mixing layer and the distance between the earth's surface and the bottom of inversion aloft is known as the mixing height. Any air pollution plume dispersing beneath an inversion aloft will be limited in vertical mixing to that which occurs beneath the bottom of the inversion aloft (sometimes called the *lid*). Even if the pollution plume penetrates the inversion, it will not undergo any further significant vertical mixing. As for a pollution plume passing completely through an inversion layer aloft, that rarely occurs unless the pollution plume's source stack is very tall and the inversion lid is fairly low.

Atmospheric Chemistry

Atmospheric chemistry is a branch of atmospheric science in which the chemistry of the Earth's atmosphere and that of other planets is studied. It is a multidisciplinary field of research and draws on environmental chemistry, physics, meteorology, computer modeling, oceanography, geology and volcanology and other disciplines. Research is increasingly connected with other areas of study such as climatology.

The composition and chemistry of the atmosphere is of importance for several reasons, but primarily because of the interactions between the atmosphere and living organisms. The composition of the Earth's atmosphere has been changed by human activity and some of these changes are harmful to human health, crops and ecosystems. Examples of problems which have been addressed by atmospheric chemistry include acid rain, photochemical smog and global warming. Atmospheric chemistry seeks to understand the causes of these problems, and by obtaining a theoretical understanding of them, allow possible solutions to be tested and the effects of changes in government policy evaluated.

Atmospheric Composition

Average composition of dry atmosphere, by volume

Gas	*per NASA*
Nitrogen, N_2	78.084%
Oxygen, O_2	20.946%
Argon, Ar	0.934%
Carbon Dioxide, CO_2	383
Neon, Ne	18.18
Helium, He	5.24
Methane, CH_4	1.7
Krypton, Kr	1.14
Hydrogen, H_2	0.55

Notes: the concentration of CO_2 and CH_4 vary by season and location. ppmv represents parts per million by volume.

The mean molecular mass of air is 28.97 g/mol.

History

The ancient Greeks regarded air as one of the four elements, but the first scientific studies of atmospheric composition began in the 18th century. Chemists such as Joseph Priestley, Antoine Lavoisier and Henry Cavendish made the first measurements of the composition of the atmosphere.

In the late 19th and early 20th centuries interest shifted towards trace constituents with very small concentrations. One particularly important discovery for atmospheric chemistry was the discovery of ozone by Christian Friedrich Schoenbein in 1840.

In the 20th century atmospheric science moved on from studying the composition of air to a consideration of how the concentrations of trace gases in the atmosphere have changed over time and the chemical processes which create and destroy compounds in the air. Two particularly important examples of this were the explanation of how the ozone layer is created and maintained by Sydney Chapman and Gordon Dobson, and the explanation of Photochemical smog by Haagen-Smit.

In the 21st century the focus is now shifting again. Atmospheric Chemistry is increasingly studied as one part of the Earth system. Instead of concentrating on atmospheric chemistry in isolation the focus is now on seeing it as one part of a single system with the rest of the atmosphere, biosphere and geosphere. An especially important driver for this is the links between chemistry and climate such as the effects of changing climate on the recovery of the ozone hole and vice versa but also interaction of the composition of the atmosphere with the oceans and terrestrial ecosystems.

Methodology

Observations, lab measurements and modeling are the three central elements in atmospheric chemistry. Progress in atmospheric chemistry is often driven by the interactions between these components and they form an integrated whole. For example observations may tell us that more of a chemical compound exists than previously thought possible. This will stimulate new modelling and laboratory studies which will increase our scientific understanding to a point where the observations can be explained.

Observation

Observations of atmospheric chemistry are essential to our understanding. Routine observations of chemical composition tell us about changes in atmospheric composition over time. One important example of this is the Keeling Curve - a series of measurements from 1958 to today which show a steady rise in of the concentration of carbon dioxide. Observations of atmospheric chemistry are made in observatories such as that on Mauna Loa and on mobile platforms such as aircraft (e.g. the UK's Facility for Airborne Atmospheric Measurements), ships and balloons. Observations of atmospheric composition are increasingly made by satellites with important instruments such as GOME and MOPITT giving a global picture of air pollution and chemistry. Surface observations have the advantage that they provide long term records at high time resolution but are limited in the vertical and horizontal space they provide observations from. Some surface based instruments e.g. LIDAR can provide concentration profiles of chemical compounds and aerosol but are still restricted in the horizontal region they can cover. Many observations are available on line in Atmospheric Chemistry Observational Databases.

Lab Measurements

Measurements made in the laboratory are essential to our understanding of the sources and sinks of pollutants and naturally occurring compounds. Lab studies tell us which gases react with each other and how fast they react. Measurements of interest include reactions in the gas phase, on surfaces and in water. Also of high importance is photochemistry which quantifies how quickly molecules are split apart by sunlight and what the products are plus thermodynamic data such as Henry's law coefficients.

Modeling

In order to synthesise and test theoretical understanding of atmospheric chemistry, computer models (such as chemical transport models) are used. Numerical models solve the differential equations governing the concentrations of chemicals in the atmosphere. They can be very simple or very complicated. One common trade off in numerical models is between the number of chemical compounds and chemical reactions modelled versus the representation of transport and mixing in the atmosphere. For example, a box model might include hundreds or even thousands of chemical reactions but will only have a very crude representation of mixing in the atmosphere. In contrast, 3D models represent many of the physical processes of the atmosphere but due to constraints on computer resources will have far fewer chemical reactions and compounds. Models can be used to interpret observations, test understanding of chemical reactions and predict future concentrations of chemical compounds in the atmosphere. One important current trend is for atmospheric chemistry modules to become one part of earth system models in which the links between climate, atmospheric composition and the biosphere can be studied.

Some models are constructed by automatic code generators. In this approach a set of constituents are chosen and the automatic code generator will then select the reactions involving those constituents from a set of reaction databases. Once the reactions have been chosen the ordinary differential equations (ODE) that describe their time evolution can be automatically constructed.

Atmospheric Studies Group

The Atmospheric Studies Group (ASG) is a part of TRC Solutions, a subsidiary of TRC Companies, Inc. headquartered in Lowell, Massachusetts. The ASG provides research and consulting services in the environmental and physical sciences with major emphasis on:

- Atmospheric pollution dispersion model development
- Atmospheric boundary layer research
- Air quality permitting, licensing, and regulatory consulting.

The ASG also provides intensive training programmes in the use of the CALPUFF dispersion model.

The ASG Staff

ASG's principals and supporting staff are very experienced scientists and engineers who have developed and evaluated numerous meteorological and dispersion models for both public agencies and private industries. The most notable dispersion model developed by the ASG staff is the CALPUFF model.

The ASG staff is committed to developing and using the latest scientific advances relative to air quality and air pollution dispersion modeling.

Most of the ASG principals were originally with the Sigma Research Company where they developed the CALPUFF model. Sigma Research subsequently became the dispersion modeling staff of the Earth Tech Company. In April of 2006, the dispersion modeling staff of Earth Tech became a part of TRC Solutions.

Atmospheric Dispersion Modeling

Atmospheric dispersion modeling is the mathematical simulation of how air pollutants disperse in the ambient atmosphere. It is performed with computer programmes that solve the mathematical equations and algorithms which simulate the pollutant dispersion. The dispersion models are used to estimate or to predict the downwind concentration of air pollutants emitted from sources such as industrial plants and vehicular traffic. Such models are important to governmental agencies tasked with protecting and managing the ambient air quality. The models are typically employed to determine whether existing or proposed new industrial facilities are or will be in compliance with the National Ambient Air Quality Standards (NAAQS) in the United States and other nations. The models also serve to assist in the design of effective control strategies to reduce emissions of harmful air pollutants.

The dispersion models require the input of data which includes:

- Meteorological conditions such as wind speed and direction, the amount of atmospheric turbulence (as characterised by what is called the "stability class"), the ambient air temperature and the height to the bottom of any inversion aloft that may be present.
- Emissions parameters such as source location and height, source vent stack diameter and exit velocity, exit temperature and mass flow rate.
- Terrain elevations at the source location and at the receptor location.
- The location, height and width of any obstructions (such as buildings or other structures) in the path of the emitted gaseous plume.

Many of the modern, advanced dispersion modeling programmes include a pre-processor module for the input of meteorological and other data, and many also include a post-processor module for graphing the output data and/or plotting the area impacted by the air pollutants on maps. The atmospheric dispersion models are also known as atmospheric diffusion models, air dispersion models, air quality models, and air pollution dispersion models.

Gaussian Air Pollutant Dispersion Equation

The technical literature on air pollution dispersion is quite extensive and dates back to the 1930's and earlier. One of the early air pollutant plume dispersion equations was derived by Bosanquet and Pearson. Their equation did not assume Gaussian distribution nor did it include the effect of ground reflection of the pollutant plume. Sir Graham Sutton derived an air pollutant plume dispersion equation in 1947 which did include the assumption of Gaussian distribution for the vertical and crosswind dispersion of the plume and also included the effect of ground reflection of the plume. Under the stimulus provided by the advent of stringent environmental control regulations, there was an immense growth in the use of air pollutant plume dispersion calculations between the late 1960s and today. A great many computer programmes for calculating the dispersion of air pollutant emissions were developed during that period of time and they were called "air dispersion models". The basis for most of those models was the Complete Equation For Gaussian Dispersion Modeling Of Continuous, Buoyant Air Pollution Plumes. The above equation not only includes upward reflection from the ground, it also includes downward reflection from the bottom of any inversion lid present in the atmosphere.

The sum of the four exponential terms in g_3 converges to a final value quite rapidly. For most cases, the summation of the series with $m = 1$, $m = 2$ and $m = 3$ will provide an adequate solution. It should be noted that $ó_z$ and $ó_y$ are functions of the atmospheric stability class (i.e., a measure of the turbulence in the ambient atmosphere) and of the downwind distance to the receptor. The two most important variables affecting the degree of pollutant emission dispersion obtained are the height of the emission source point and the degree of atmospheric turbulence. The more turbulence, the better the degree of dispersion. The resulting calculations for air pollutant concentrations are often expressed as an air pollutant concentration contour map in order to show the spatial variation in contaminant levels over a wide area under study. In this way the contour lines can overlay sensitive receptor locations and reveal the spatial relationship of air pollutants to areas of interest.

The Briggs Plume Rise Equations

The Gaussian air pollutant dispersion equation (discussed above) requires the input of H which is the pollutant plume's centerline height above ground level—and H is the sum of H_s (the actual physical height of the pollutant plume's emission source point) plus ÅH (the plume rise due the plume's buoyancy).

To determine ÄH, many if not most of the air dispersion models developed between the late 1960s and the early 2000s used what are known as "the Briggs equations." G.A. Briggs first published his plume rise observations and comparisons in 1965. In 1968, at a symposium sponsored by CONCAWE (a Dutch organisation), he compared many of the plume rise models then available in the literature. In that same year, Briggs also wrote the section of the publication edited by Slade dealing with the comparative analyses of plume rise models. That was followed in 1969 by his classical critical review of the entire plume rise literature, in which he proposed a set of plume rise equations which have become widely known as "the Briggs equations". Subsequently, Briggs modified his 1969 plume rise equations in 1971 and in 1972.

Briggs divided air pollution plumes into these four general categories:

- Cold jet plumes in calm ambient air conditions
- Cold jet plumes in windy ambient air conditions
- Hot, buoyant plumes in calm ambient air conditions
- Hot, buoyant plumes in windy ambient air conditions

Briggs considered the trajectory of cold jet plumes to be dominated by their initial velocity momentum, and the trajectory of hot, buoyant plumes to be dominated by their buoyant momentum to the extent that their initial velocity momentum was relatively unimportant. Although Briggs proposed plume rise equations for each of the above plume categories, *it is important to emphasize that "the Briggs equations" which become widely used are those that he proposed for bent-over, hot buoyant plumes*.

In general, Briggs's equations for bent-over, hot buoyant plumes are based on observations and data involving plumes from typical combustion sources such as the flue gas stacks from steam-generating boilers burning fossil fuels in large power plants. Therefore the stack exit velocities were probably in the range of 20 to 100 ft/s (6 to 30 m/s) with exit temperatures ranging from 250 to 500 °F (120 to 260 °C).

Clean Air Act

A Clean Air Act describes one of a number of pieces of legislation relating to the reduction of smog and air pollution in general. The use of governments to enforce clean air standards has contributed to an improvement in human health and longer life spans. Critics argue it has also sapped corporate profits and contributed to outsourcing, while defenders counter that improved environmental air quality has generated more jobs than it has terminated.

Clean Air Acts

Canada

Former Canadian Environment Minister Rona Ambrose introduced in mid-October 2006, a Clean Air Act with mostly measures to fight smog pollution and greenhouse emissions.. On October 19, 2006, Ambrose revealed details of the plan which would include reducing the greenhouse emissions levels of 2003 by about 45 to 65% for the year 2050. There are plans for regulations on vehicle fuel consumption for 2011 as well as targets for ozone and smog levels for 2025. The effectiveness of this act has been challenged by the opposition parties, with Jack Layton of the New Democratic Party stating that the act does little to prevent climate change and that more must be done. After threatening to make this into an election issue the Conservative Party agreed to rework the act with the opposition parties.

United Kingdom

In response to the Great Smog of 1952, the British Parliament introduced the Clean Air Act 1956. This act legislated for zones where smokeless fuels had to be burnt and relocated power stations to rural areas. The Clean Air Act 1968 introduced the use of tall chimneys to disperse air pollution for industries burning coal, liquid or gaseous fuels.

United States

United States, the Congress passed the Clean Air Act in 1963, the Air Quality Act in 1967, the Clean Air Act Extension of 1970, and Clean Air Act Amendments in 1977 and 1990. Numerous state and local governments have enacted similar legislation, either implementing federal programmes or filling in locally important gaps in federal programmes.

The Clean Air Act Amendments of 1990 proposed emissions trading, added provisions for addressing acid rain, ozone depletion and toxic air pollution, and established a national permits programme. The amendments once approved also established new auto gasoline reformulation requirements, set Reid Vapor Pressure (RVP) standards to control evaporative emissions from gasoline and mandated that the new gasoline formulations be sold from May-September in many states.

In May 2007, President Bush issued an executive order to cut greenhouse gas emissions from motor vehicles, spurred by a Supreme Court ruling that the EPA must take action under the Clean Air Act to regulate GHG emissions from motor vehicles. The President proposed the 20-in-10 bill, a goal to reduce gasoline consumption by 20 percent over the next ten years.

Bush sent the Congress a proposal that would meet it in two steps:

- Firstly to set a mandatory fuel standard, requiring 35 billion gallons of renewable and other alternative fuels by 2017. This amounts to nearly five times the target previously set for 2012, and was intended to displace 15 percent of projected annual gasoline use in the United States.
- Secondly, to continue the efforts to increase fuel efficiency by reforming and modernising Corporate Average Fuel Economy (CAFE) Standards for cars, and extending the current Light Truck Rule, with the intention of

reducing projected annual gasoline use in the United States by up to 8.5 billion gallons by 2017. This would bring a further 5 percent reduction that, in combination with increasing the supply of renewable and other alternative fuels, would bring the total reduction in projected annual gasoline use to 20 percent.

Features and Capabilities

Some of the basic features and capabilities of DISPLAY21 are:

- Source types: Multiple point, area , and volume sources as well as street canyons.
- Source releases: Surface, near surface and elevated sources.
- Source locations: Urban or rural locations.
- Plume types: Continuous or intermittent buoyant plumes
- Plume dispersion treatment: Gaussian model treatment using Green's functions and includes multiple reflections.
- Terrain types: Simple terrain with no more than 10 degree slopes.
- Building effects: Building downwash algorithms are included.
- Meteorological data: The model includes a preprocessor to produce the meteorological parameters needed to characterize the atmospheric turbulence as well as to produce wind speed and direction profiles.
- The street canyon module includes some atmospheric chemistry for photochemical reactions.

Low-Carbon Emission

The main components of automobile exhaust are carbon dioxide (CO_2) and water vapor (H_2O). Carbon dioxide is the most important anthropogenic greenhouse gas (GHG) and the most significant Greenhouse Gas emitted in the U.S. (with 82-84% of all U.S. emissions). Increases in carbon dioxide concentration are due primarily to fossil fuel use and land-use change.

The CO_2 emission standards can be referred to the fuel or to the vehicle :

- The European Union is moving towards fuel mandatory CO_2 standards and USA has reflected it in the Greenhouse Gas Score.
- There are information, but no mandatory, standards for CO_2 limits in vehicle (engine) emissions, excepting some state regulations (i.e. California). See hybrid vehicles.

CO_2 mass emission in vehicles is measured in g/km (g/mi * 0.621371192 = g/km).

Low Carbon Fuel Standard

Transportation accounts for forty percent of California's annual greenhouse gas emissions, and California relies on petroleum-based fuels for 96 percent of its transportation needs. The Californian Governor Arnold Schwarzenegger announced an Executive Order that California adopt so-called Low Carbon Fuel Standard (LCFS). The LCFS requires fuel providers to ensure that the mix of fuel they sell in the California Market meets, on average, a declining target for greenhouse gas emissions measured in CO2-equivalent grams per unit of fuel energy sold. By 2020, the LCFS will produce a 10 % reduction in GHG emissions from production and use of fuel in passenger vehicles in California.

Puff Model

The Puff model is a volcanic ash tracking model developed at the University of Alaska Fairbanks. It requires windfield data on a geographic grid covering the area over which ash may be dispersed. Representative ash particles are initiated at the volcano's location and then allowed to advect, diffuse, and settle within the atmosphere. The location of the particles at any time after the eruption can be viewed using the post-processing software included with the model. Output data is in netCDF format and can also be viewed with a variety of software.

History

Puff was initially conceived and developed by Prof. H. Tanaka as a novel method for simulating ash cloud trajectories during the eruption of Mt. Redoubt, 1989. Dr. Craig Searcy rewrote and modified the Puff code in C++, and created the initial GUI so the programme could be used operationally for volcano monitoring in the early and mid 1990's. His version of the programme is running at the National Weather Service (NWS), Anchorage, Alaska, although updated versions of Puff are also available at the NWS.

The Alaska Volcano Observatory (AVO) provided support for Puff through a post doctorate position (Drs. Mark Servilla and Jon Dehn) during the late 1990's to support analysis of volcanic clouds during eruptions.

In a joint programme called University Partnering for Operational Support (UPOS) between the University of Alaska Fairbanks and the John Hopkins Applied Physics Laboratory (early 2000's), Puff was integrated into the U.S. Air Force Weather Agency (AFWA) volcano monitoring system by Rorik Peterson and David Tillman. UPOS support resulted in the testing of the sensitivity of Puff and the development of WebPuff, and new modules including the capability to model stratospheric eruptions, non-point source events (e.g. fires) and tracking of volcanic clouds from multiple eruptions simultaneously by Dr. Rorik Peterson. The utility of the multiple eruption capability became evident during the 13 January 2006 eruption of Augustine Volcano where the movement of six volcanic clouds across the Gulf of Alaska were tracked simultaneously.

Starting in 2006, the Arctic Region Supercomputing Center (ARSC) provided support for Puff through a Post Doctorate position occupied by Dr. Peter Webley. Puff is now in use at AVO, Anchorage Volcanic Ash Advisory Center (VAAC), AFWA, and other national agencies worldwide as well as at other universities. Professor Ken Dean has been the principal scientist leading the development of Puff since Professor Tanaka returned to Japan in the early 1990's. .

RIMPUFF

RIMPUFF is a local-scale puff diffusion model developed by Risø National Laboratory, Denmark. It is an emergency response model to help emergency management organisations deal with chemical, nuclear, biological and radiological releases to the atmosphere.

RIMPUFF is in operational use in several European national emergency centres for preparedness and prediction of nuclear accidental releases (RODOS, EURANOS), chemical gas releases (ARGOS), and for airborne Foot-and Mouth Disease virus spread

Description

RIMPUFF builds from parameterised formulas for puff diffusion, wet and dry deposition, and gamma dose radiation. Its range of application covers distances up to ~1000 km from the point of release. RIMPUFF calculates the instantaneous atmospheric dispersion taking into account the local wind variability and the local turbulence levels. The puff sizes represent instantaneous relative diffusion (no averaging) and is calculated from similarity scaling theory. Puff diffusion is patameterised for travel times in the range from a few seconds and up to ~1 day. Wet and dry deposition is also calculated as a function of local rain intensity and turbulence levels.

Line Source

A line source is a source of air, noise, water contamination or electromagnetic radiation that emanates from a linear (one-dimensional) geometry. The most prominent linear sources are roadway air pollution, aircraft air emissions, roadway noise, certain types of water pollution sources that emanate over a range of river extent rather than from a discrete point, elongated light tubes, and electromagnetic antennas. While point sources of pollution were studied since the late nineteenth century, linear sources did not receive much attention from scientists until the late 1960s, when environmental regulations for highways and airports began to emerge. At the same time, computers with the processing power to accommodate the data processing needs of the computer models required to tackle these one-dimensional sources became more available.

In addition, this era of the 1960s saw the first emergence of environmental scientists who spanned the disciplines required to accomplish these studies. For example, meteorologists, chemists, and computer scientists in the air pollution field were required to build complex models to address roadway air dispersion modeling. Prior to the 1960's, these specialities tended to work within their own disciplines, but with the advent of NEPA, the Clean Air Act, the Noise Control Act in the USA, and other seminal legislation, the era of multidisciplinary environmental science had begun.

For electromagnetic linear sources, the principal early advances in computer modeling arose in the Soviet Union and USA when the end of World War II and the Cold War were fought partially by progress in electronic warfare, including the technologies of active antenna arrays.

Linear Air Pollution Source

Air pollution levels near major highways and urban arterials are in violation of U.S. National Ambient Air Quality Standards where millions of Americans live or work. Even the interior of a building does not really protect inhabitants from adverse exterior air quality, since the exterior air is the intake supply, and it is well known that indoor air quality is typically worse than exterior air.

A roadway travelled by motor vehicles can be idealised by a line source emitting air pollutants. This mathematical problem was first solved in 1970 by a collaboration of physics, mathematics and computer science. The original theory assumed steady-state traffic conditions and meteorology on a perfectly straight roadway. Currently the models have evolved to treat variable meteorology, time-variant traffic operations and complex roadbed geometries. Current technology allows highway designers and city planners to analyze alternative roadway development plans and assess air quality impacts. The same basic model theory can be applied to airport operations, since the linear source is merely an inclined line. In the early 1970s these ESL models were refined into area source models to account for the finite width of the roadway.

Linear Noise Source

Roadway noise is the most important example of a linear noise source, since it comprises about 80 percent of the environmental noise exposure for humans worldwide. In the 1960s, when computer modeling of this phenomenon was perfected, the first applications of linear source noise modeling became systematic. After passage of the National Environmental Policy Act and Noise Control Act, the demand for detailed analysis soared, and decision makers began to look to acoustical scientists for answers regarding the planning of new roadways and the design of noise mitigation. The intensity of roadway noise is governed by the following variables: traffic operations (speed, truck mix, age of vehicle fleet), roadway surface type, tire types, roadway geometrics, terrain, micrometeorology and the geometry of area structures.

Due to the complexity of the variables, a line source acoustic model must be a computer model that can analyze sound levels in the vicinity of roadways. The first meaningful models arose in the late 1960s and early 1970s. Two of the leading research teams were BBN in Boston and ESL Inc. of Sunnyvale, California. Both of these groups developed complex mathematical models to allow the study of alternate roadway designs, traffic operations and noise mitigation strategies in an arbitrary setting. Later model alterations have come into widespread use among state Departments of Transportation and city planners, but the accuracy of early models has had little change in 40 years.

Generally line source acoustic models trace sound ray bundles and calculate spreading loss along with ray bundle divergence (or convergence} from refractive phenomena. Diffraction is usually addressed by establishing secondary emitters at any points of topographic or anthropomorphic "sharpness" (such as noise barriers or building surfaces. Meteorology can be addressed in a statistical manner allowing for actual wind rose and wind speed statistics (along with thermocline data).

Water Pollution Line Source

Less common are line source applications in the field of water pollutant dispersal. This phenomenon generally arises when surface runoff scours soil contaminants from upper soil layers and transports these pollutants to a linear receiving water, such as a river. The underlying land management practices which lead to such sources of water pollution are logging, pesticide application, construction grading, slash-and-burn activity and urban stormwater runoff.

Again computer models are needed to address the complexity of such an extended linear discharge into a dynamic medium such as flowing water. The resulting surface runoff water carrying pollutants may be considered a line source discharging into a river or stream. The chemical composition of this surface runoff may be characterised by a surface runoff model such as the USGS runoff precipitation algorithm, while the instream transport may be analysed by a dynamic river pollutant model such as DSSAM.

Light Emission Line Source

In the study of illumination, a variety of sources are linear in nature, most commonly the fluorescent tube, During the process of interior lighting design it is important to calculate the light intensity at work stations or other user areas, not only to insure sufficient light is present, but more importantly to avoid over-illumination and its attendant energy wastage as well as adverse health effects. Thus the scientists involved in light transmission calculations employ computer models that recognize linear sources when fluorescent fixtures are used. In a typical setting there may be hundreds of finite length light sources that comprise the light output in an office environment. On a larger scale, an illuminated roadway may act as a line source of light pollution.

MEMO Model

The MEMO Model (version 6.2) is a Eulerian non-hydrostatic prognostic mesoscale model for wind flow simulation. It was developed by the Aristotle University of Thessaloniki in collaboration with the Universität Karlsruhe. The MEMO Model together with the photochemical dispersion model MARS are the two core models of the European Zooming Model (EZM). This model belongs to the family of models designed for describing atmospheric transport phenomena in the local-to-regional scale, frequently referred to as mesoscale air pollution models.

History

Initially, EZM was developed for modelling the transport and chemical transformation of pollutants in selected European regions in the frame of the EUROTRAC sub-project EUMAC and therefore it was formerly called the EUMAC Zooming Model (EUROTRAC, 1992). EZM has evolved to be one of the most frequently applied mesoscale air pollution model systems in Europe. It has been already successfully applied for various European airsheds including the Upper Rhine valley and the areas of Basel, Graz, Barcelona, Lisbon, Madrid, Milano, London, Cologne, Lyon, The Hague, Athens (Moussiopoulos, 1994; Moussiopoulos, 1995) and Thessaloniki. More details are to be found elsewhere (Moussiopoulos 1989), (Flassak 1990), (Moussiopoulos et al. 1993).

Model Equations

The prognostic mesoscale model MEMO describes the dynamics of the atmospheric boundary layer. In the present model version, air is assumed to be unsaturated. The model solves the continuity equation, the momentum equations and several transport equations for scalars (including the thermal energy equation and, as options, transport equations for water vapour, the turbulent kinetic energy and pollutant concentrations).

Transformation to Terrain-Following Coordinates

The lower boundary of the model domain coincides with the ground. Because of the inhomogeneity of the terrain, it is not possible to impose boundary conditions at that boundary with respect to Cartesian coordinates. Therefore, a transformation of the vertical coordinate to a terrain-following one is performed. Hence, the originally irregularly bounded physical domain is mapped onto one consisting of unit cubes.

Numerical Solution of the Equation System

The discretised equations are solved numerically on a staggered grid, i.e. the scalar quantities ñ, p and è are defined at the cell centre while the velocity components u, v and w are defined at the centre of the appropriate interfaces.

Temporal discretisation of the prognostic equations is based on the explicit second order Adams-Bashforth scheme. There are two deviations from the Adams-Bashforth scheme: The first refers to the implicit treatment of the nonhydrostatic part of the mesoscale pressure perturbation p_{nh}. To ensure non-divergence of the flow field, an elliptic equation is solved. The elliptic equation is derived from the continuity equation wherein velocity components are expressed in terms of p_{nh}. It should be noted that, since the elliptic equation is derived from the discrete form of the continuity equation and the discrete form of the pressure gradient, conservativity is guaranteed (Flassak and Moussiopoulos, 1988). The discrete pressure equation is solved numerically with a fast elliptic solver in conjunction with a generalised conjugate gradient method. The fast elliptic solver is based on fast Fourier analysis in both horizontal directions and Gaussian elimination in the vertical direction (Moussiopoulos and Flassak, 1989).

The second deviation from the explicit treatment is related to the turbulent diffusion in vertical direction. In case of an explicit treatment of this term, the stability requirement may necessitate an unacceptable abridgement of the time increment. To avoid this, vertical turbulent diffusion is treated using the second order Crank-Nicolson method.

On principle, advective terms can be computed using any suitable advection scheme. In the present version of MEMO, a 3-D second-order total-variation-diminishing (TVD) scheme is implemented which is based on the 1-D scheme proposed by Harten (1986). It achieves a fair (but not any) reduction of numerical diffusion, the solution being independent of the magnitude of the scalar (i.e., preserving transportivity).

Parameterisations

Turbulence and radiative transfer are the most important physical processes that have to be parameterised in a prognostic mesoscale model. In the MEMO model, radiative transfer is calculated with an efficient scheme based on the emissivity method for longwave radiation and an implicit multilayer method for shortwave radiation (Moussiopoulos 1987).

The diffusion terms may be represented as the divergence of the corresponding fluxes. For turbulence parameterisations, K-theory is applied. In case of MEMO turbulence can be treated either with a zero-, one- or two-equation turbulence model. For most applications a one-equation model is used, where a conservation equation for the turbulent kinetic energy is solved.

Initial and Boundary Conditions

In MEMO, initialisation is performed with suitable diagnostic methods: a mass-consistent initial wind field is formulated using an objective analysis model and scalar fields are initialised using appropriate interpolating techniques (Kunz, R., 1991). Data needed to apply the diagnostic methods may be derived either from observations or from larger scale simulations.

Suitable boundary conditions have to be imposed for the wind velocity components u, v and w, the potential temperature è and pressure p at all boundaries. At open boundaries, wave reflection and deformation may be minimised by the use of so-called *radiation conditions* (Orlanski 1976). According to the experience gained so far with the model MEMO, neglecting large scale environmental information might result in instabilities in case of simulations over longer time periods. For the nonhydrostatic part of the mesoscale pressure perturbation, homogeneous Neumann boundary conditions are used at lateral boundaries. With these conditions, the wind velocity component perpendicular to the boundary remains unaffected by the pressure change. At the upper boundary, Neumann boundary conditions are imposed for the horizontal velocity components and the potential temperature. To ensure non-reflectivity, a radiative condition is used for the hydrostatic part of the mesoscale pressure perturbation p_h at that boundary. Hence, vertically propagating internal gravity waves are allowed to leave the computational domain (Klemp and Durran 1983). For the nonhydrostatic part of the mesoscale pressure perturbation, homogeneous staggered Dirichlet conditions are imposed. Being justified by the fact that nonhydrostatic effects are negligible at large heights, this condition is necessary, if singularity of the elliptic pressure equation is to be avoided in view of the Neumann boundary conditions at all other boundaries. The lower boundary coincides with the ground (or, more precisely, a height above ground corresponding to its aerodynamic roughness). For the nonhydrostatic part of the mesoscale pressure perturbation, inhomogeneous Neumann conditions are imposed at that boundary. All other conditions at the lower boundary follow from the assumption that the Monin-Obukhov similarity theory is valid. The one way interactive nesting facility is possible within MEMO.

Thus, successive simulations on grids of increasing resolution are possible. During these simulations, the results of the application to a coarse grid are used as boundary conditions for the application to the finer grid (Kunz and Moussiopoulos, 1995).

Grid Definition

The governing equations are solved numerically on a staggered grid. Scalar quantities as the temperature, pressure, density and also the cell volume are defined at the centre of a grid cell and the velocity components *u*, *v* and *w* at the centre of the appropriate interface. Turbulent fluxes are defined at different locations: Shear fluxes are defined at the centre of the appropriate edges of a grid cell and normal stress fluxes at scalar points. With this definition, the outgoing fluxes of momentum, mass, heat and also turbulent fluxes of a grid cell are identical to incoming flux of the adjacent grid cell. So the numerical method is conservative.

Topography and Surface Type

For calculations with MEMO, a file must be provided which contains orography height and surface type for each grid location The following surface types are distinguished and must be stored as percentage:

- water (type: 1)
- arid land (type: 2)
- few vegetation (type: 3)
- farmland (type: 4)
- forest (type: 5)
- suburban area (type: 6)
- urban area (type: 7)

Only surface types 1 to 6 have to be stored. Type 7 is the difference between 100% and the sum of types 1 to 6. If the percentage of a surface type is 100%, then write the number 10 and for all other surface types the number 99.

The orography height is the mean height for each grid location above sea level in meter.

Meteorological Data

The prognostic model MEMO is a set of partial differential equations in three spatial directions and in time. To solve these equations, information about the initial state in the whole domain and about the development of all relevant quantities at the lateral boundaries is required.

Initial State

To generate an initial state for the prognostic model, a diagnostic model (Kunz, R., 1991) is applied using measured temperature and wind data. Both data can be provided as:

- surface measurements i.e. single measurements directly above the surface (not necessary)
- upper air soundings (i.e., soundings that consist of two or more measurements at different heights) at a constant geographical location is required (with at least one sounding for temperature and wind velocity).

Time-Dependent Boundary Conditions

Information about quantities at the lateral boundaries can be taken into account as surface measurements and upper air soundings. Therefore, a key word and the time when boundary data is given must occur in front of a set of boundary information.

Nesting Facility

In MEMO, a one-way interactive nesting scheme is implemented. With this nesting scheme a coarse grid and a fine grid simulation can be nested. During the coarse grid simulation, data is interpolated and written to a file. A consecutive fine grid simulation uses this data as lateral boundary values.

Met Office

The Met Office (originally an abbreviation for Meteorological Office, but now the official name in itself), which has its headquarters at Exeter in Devon, is the United Kingdom's national weather service and a subsidiary of the Ministry of Defence. Part of the Met Office complex in Exeter is the Met Office College, which handles the training for internal personnel and many forecasters from around the world. The current chief executive is John Hirst who replaced Mark Hutchinson on 17 September 2007.

History

Established in 1854 as a small department within the Board of Trade under Robert FitzRoy as a service to mariners. The loss of the passenger vessel the Royal Charter and 459 lives off the coast of Anglesey in a violent storm in October 1859 led to the first gale warning service. In 1861 FitzRoy had established a network of 15 coastal stations from which visual gale warnings could be provided for ships at sea.

The development of the electric telegraph in the 1870's led to the more rapid dissemination of warnings and also led to the development of an observational network which could then be used to provide synoptic analyses.

In 1879 the Met Office started providing forecast to Newspapers.

Following the First World War the Met Office later became part of the Air Ministry in 1920. In 1936 the Met Office split with services to the Royal Navy being provided by their own forecasting services.

It currently holds a quasi-governmental role, being required to act commercially but also remaining an executive agency of the Ministry of Defence. A little known branch of the Met Office known as the Mobile Met Unit (MMU) accompany forward units in times of conflict advising the armed forces of the prevailing conditions for battle, particularly the RAF. The Hadley Centre for Climate Prediction and Research is also part of the Met Office.

The Shipping Forecast

One of the British stalwarts — the Shipping Forecast — is produced by the Met Office and broadcast on BBC Radio 4. The Shipping Forecast has long been of real interest to, and vital to the safety of, Mariners traversing the Sea Areas around the British Isles and its broadcast on radio is still avidly listened to. Less vitally, the Shipping Forecast has been the subject of both books and song lyrics.

Weather Forecasts and Warnings

At the Met Office, they have the responsibility to issue severe weather warnings for the United Kingdom through the National severe weather warning service (NSWWS). These are weather events that may affect transport infrastructure and endanger people's lives. In March 2008, the system was improved and a new stage of warning was introduced, the 'Advisory'.

Their main role is to produce forecast models by gathering all the information from satellites in space and observations on earth, then processing it using supercomputers. If necessary, forecasters may then make adjustments to it. This main bulk of data is then passed on to companies who acquire it. In particular, two of the main media companies, the BBC and ITV produce forecasts using the Met Office's data. At the BBC Weather Centre, they are continuously updated on the latest information arriving by computer, or by fax and e-mail. The BBC's new graphics are used on all of their television weather broadcasts, but ITV use animated weather symbols. This is mainly how the public are informed of weather events which may affect day-to-day life.

Met Office Locations

In 2003 the Met Office moved its headquarters to Exeter, in Devon from its previous location of Bracknell in Berkshire and it has a worldwide presence —

including a forecasting centre in Aberdeen and offices in Gibraltar and on the Falklands. Other outposts lodge in establishments such as the Joint Centre for Mesoscale Meteorology (JCMM) at University of Reading in Berkshire, the Joint Centre for Hydro-Meteorological Research (JCHMR) site at Wallingford in Oxfordshire and there is also a Met Office presence at many Army and Air Force bases within the UK and abroad. Royal Navy weather forecasts are generally provided by naval officers, not Met Office personnel.

World Area Forecast Centre

The Met Office is also one of only two World Area Forecast Centres or WAFCs, and is referred to as WAFC London. The other WAFC is located in Missouri, USA but known as WAFC Washington. WAFC data are used daily to safely and economically route aircraft, particularly on long-haul journeys. The data provide details of wind speed and direction, air temperature, cloud type and tops, and other features of interest to the aviation community, such as volcanic ash eruptions.

Air Quality Forecasts

The Met Office issues air quality forecasts made using NAME, the Met Office's medium-to-long-range atmospheric dispersion model. It was originally developed as a nuclear accident model following the Chernobyl accident in 1986, but has since evolved into an all-purpose dispersion model capable of predicting the transport, transformation and deposition of a wide class of airborne materials. NAME is used operationally by the Met Office as an emergency response model as well as for routine air quality forecasting.

In the air quality forecasts, the level of pollution is described either as an index (ranging from 1 to 10) or as a banding (low, moderate, high or very high). These levels are based on the health effects of each pollutant as shown just below.

Index	*Banding*	*Health Effect*
1–3	Low	Effects are unlikely to be noticed even by individuals who know they are sensitive to air pollutants.
4–6	Moderate	Mild effects, unlikely to require action, may be noticed amongst sensitive individuals.
7-9	High	Significant effects may be noticed by sensitive individuals and action to avoid or reduce these effects may

		be needed (e.g. reducing exposure by spending less time in polluted areas outdoors). Asthmatics will find that their 'reliever' inhaler is likely to reverse the effects on the lung.
10	Very High	The effects on sensitive individuals described for 'High' levels of pollution may worsen.

The forecast is produced for a number of different pollutants and their typical health effects are shown in the following table.

Pollutant	*Health Effects at High Level*
Nitrogen dioxideOzoneSulphur dioxide	These gases irritate the airways of the lungs, increasing the symptomsof those suffering from lung diseases.
Particulates	Fine particles can be carried deep into the lungs where they can causeinflammation and a worsening of heart and lung diseases

High Performance Computing

Due to the large amount of computation needed for Numerical Weather Prediction the Met Office has had some of the most powerful supercomputers in the world. In November 1997 the Met Office supercomputer was ranked third in the world.

Weather Stations

Reports (observations) from weather stations vary considerably. They can be automatic (totally machine produced), semi-automatic (part-machine and part manual), or manual. Some stations produce manual observations during business hours and revert to automatic observations outside these times. Many stations now also feature recent innovations such as "present weather" sensors, CCTV, etc.

Some stations have limited reporting times, while other report continuously, mainly RAF and Army Air Corps stations where a manned met office is provided for military operations. The "standard" is a once-hourly reporting schedule, but automatic stations can often be "polled" as required, while stations at airfields regularly report twice-hourly, with additional (often frequent in times of bad weather) special reports as necessary to inform airfield authorities of changes to the weather that may affect aviation operations.

Some stations report only CLIMAT data (e.g maximum and minimum temperatures, rainfall totals over a period, etc.) and these are usually recorded at 0900 and 2100 hours daily. Weather reports are often performed by Observers not specifically employed by the Met Office, e.g. Air traffic control staff, Coastguards, University staff, etc.

This list is incomplete; you can help by expanding it.

- Penkridge weather station
- Wye weather station
- RAF Lossiemouth weather station
- RAF Kinloss weather station
- RAF Leuchars weather station
- RAF Leeming weather station
- RAF Little Rissington weather station (supported by RAF Brize Norton)
- RAF Marham weather station
- RAF Cranwell weather station
- RAF Waddington weather station
- RAF Odiham weather station
- RAF Brize Norton weather station
- AAC Middle Wallop weather station

Compressed-Air Vehicle

A compressed air vehicle is powered by an air engine, using compressed air, which is stored in a tank. Instead of mixing fuel with air and burning it in the engine to drive pistons with hot expanding gases, *compressed air vehicles* (CAV) use the expansion of compressed air to drive their pistons. One manufacturer claims to have designed an engine that is 90 percent efficient.

Compressed air propulsion may also be incorporated in hybrid systems, e.g., battery electric propulsion and fuel tanks to recharge the batteries. This kind of system is called a hybrid-pneumatic electric propulsion. Additionally, regenerative braking can also be used in conjunction with this system.

Technology

Engine

One can buy the vehicle with the engine or buy an engine to be installed in the vehicle. Typical air engines use one or more expander pistons. In some applications it is advantageous to heat the air, or the engine, to increase the range or power.

Tanks

The tanks must be designed to safety standards appropriate for a pressure vessel, such as ISO 11439.

The storage tank may be made of:

- steel,
- aluminium,
- carbon fiber,
- Kevlar,
- other materials, or combinations of the above.

The fiber materials are considerably lighter than metals but generally more expensive. Metal tanks can withstand a large number of pressure cycles, but must be checked for corrosion periodically.

One company stores air in tanks at 4,500 pounds per square inch (about 30 MPa) and hold nearly 3,200 cubic feet(around 90 cubic metres) of air.

The tanks may be refilled at a service station equipped with heat exchangers, or in a few hours at home or in parking lots, plugging the car into the electric grid via an on-board compressor. The cost of driving such a car is typically projected to be around •0.75 per 100 km, with a complete refill at the "tank-station" at about US$3.

Compressed Air

Compressed air has a low energy density. In 300 bar containers, about 0.1 MJ/L and 0.1 MJ/kg is achievable, comparable to the values of electrochemical lead-acid batteries. While batteries can somewhat maintain their voltage throughout their discharge and chemical fuel tanks provide the same power densities from the first to the last litre, the pressure of compressed air tanks falls as air is drawn off. A consumer-automobile of conventional size and shape typically consumes 0.3-0.5 kWh at the drive shaft per mile of use, though unconventional sizes and may perform with significantly less.

Emission Output

Like other non-combustion energy storage technologies, an air vehicle displaces the emission source from the vehicle's tail pipe to the central electrical generating plant. Where emissions-free sources are available, net production of pollutants can be reduced. Emission control measures at a central generating plant may be more effective and less costly than treating the emissions of widely-dispersed vehicles.

Since the compressed air is filtered to protect the compressor machinery, the air discharged has less suspended dust in it, though there may be carry-over of lubricants used in the engine.

History

Compressed air has been used since the 19th century to power mine locomotives, and was previously the basis of naval torpedo propulsion.

In 1863, Jules Verne wrote a novel called *Paris in the 20th Century* about a world of glass skyscrapers, high-speed trains, and air-powered automobiles.

In 1903, the Liquid Air Company located in London England manufactured a number of compressed air and liquified air cars. The major problem with these cars and all compressed air cars is the lack of torque produced by the "engines" and the cost of compressing the air. Reference: http://www.didik.com/ev_hist.htm

Recently several companies have started to develop compressed air cars, although none have been released to the public, or have been tested by third parties.

Disadvantages

Disadvantages of compressed-air vehicles are less well known, since the vehicles are currently at the pre-production stage and have not been extensively tested by independent observers.

- When the air is expanded in the engine, it will cool down via adiabatic cooling and lose pressure thus its ability to do work at colder temperatures. It is difficult to maintain or restore the air temperature by simply using a heat exchanger with ambient heat at the high flow rates used in a vehicle, thus the ideal isothermic energy capacity of the tank will not be realised. Cold temperatures will also encourage the engine to ice up.
- Compressed air vehicles likely will be less robust than typical vehicles of today. Which poses a danger to users of compressed air vehicles sharing the road with larger, heavier and more rigid vehicles.
- Very little is known about air powered vehicles thus far.

Advantages

The advantages are well publicised since the developers need to make their machines attractive to investors. Compressed-air vehicles are comparable in many ways to electric vehicles, but use compressed air to store the energy instead of batteries. Their potential advantages over other vehicles include:

- Much like electrical vehicles, air powered vehicles would ultimately be powered through the electrical grid. Which makes it easier to focus on reducing pollution from one source, as opposed to the millions of vehicles on the road.
- Transportation of the fuel would not be required due to drawing power off the electrical grid. This presents significant cost benefits. Pollution created during fuel transportation would be eliminated.
- Compressed air technology reduces the cost of vehicle production by about 20%, because there is no need to build a cooling system, fuel tank, spark plugs or silencers.
- Air, on its own, is non-flammable.
- High torque for minimum volume.
- The mechanical design of the engine is simple and robust.
- Low manufacture and maintenance costs as well as easy maintenance.
- Compressed-air tanks can be disposed of or recycled with less pollution than batteries.
- Compressed-air vehicles are unconstrained by the degradation problems associated with current battery systems.
- The tank may be able to be refilled more often and in less time than batteries can be recharged, with re-fueling rates comparable to liquid fuels.
- Lighter vehicles would mean less abuse on roads. Resulting in longer lasting roads.
- The price of fueling air powered vehicles will be significantly cheaper than current fuels.

Other Effects

- Many entities would suffer from the adoption of air powered vehicles. Federal, state and local governments possibly would lose significant income which was gained from taxes on high priced fuel. Fueling stations would have to invest in new dispensers. The oil industry would suffer the effects of less demand.
- Adoption of a new fuel would have a turbulent effect. Jobs would be lost and gained.

Possible Improvements

- Other common users of compressed gas fuels, such as a propane barbecue or a liquid oxygen rocket engine, use some of the heat of combustion to reheat the gas prior to its use, compressed air offers no reheating. Some auxiliary reheating energy input might be needed to increase efficiency. Some designs obtain this heat from the ambient air and from the passenger compartment for air conditioning.
- When the air is being compressed at reasonable speeds, it heats up. This heat needs to be retained in the air; otherwise the overall efficiency of the cycle is reduced. The heat given off during compression could be reclaimed for space heating or water heating, or used in a stirling engine.

Vehicles

Mopeds

Jem Stansfield, an English inventor has been able to convert a regular scooter to a compressed air moped. This has been done by equipping the scooter with a compressed air engine and air tank.

Cars

Several companies are investigating and producing prototypes, and others plan to offer cars in the beginning of 2008.

Buses

MDI makes MultiCATs vehicle that can be used as buses or trucks. RATP has also already expressed an interest in the compressed-air pollution-free bus.

Locomotives

Compressed air locomotives have been historically used as mining locomotives.

Trams

Various compressed-air-powered trams were trialled, starting in 1876.

Watercraft and Aircraft

Currently, no water or air vehicles exist that make use of the air engine. Historically certain torpedoes were propelled by compressed air engines.

Compressed-Air Car

A Compressed air car is an alternative fuel car that uses a motor powered by compressed air. This can also be combined (as in a hybrid electric vehicle) with gasoline/diesel/ethanol or electric plant and regenerative braking.

Several companies are investigating and producing prototypes and other ones, including Tata Motors. "The first cars are promoted as being available on the market in summer 2008 (promotions which were preceded by similar promotions in earlier years)"

Overview

The compressed air car has been mentioned in *Popular Mechanics* and the *Yahoo Green* blog as being the true car of tomorrow, with a range comparable to an electric vehicle or fuel-cell car. It claims to offer zero emissions at the tailpipe without batteries or hydrogen fuel.

Technology

Engines

You will be able to buy the vehicle with the engine, or buy an engine to be installed in a vehicle.

Compressed air cars are powered by engines, fueled by compressed air, which is stored in a tank under high pressure such as 30 MPa (4500 psi or 300 bar), following ISO 11439. The storage tank is likely to be made of carbon-fiber in order to reduce its weight while achieving the necessary strength. Instead of mixing fuel with air and burning it to drive pistons with hot expanding gases; *compressed air cars* use the expansion of compressed air to drive their pistons.

The idea is not new. It has been used since 1930s in cars and in Formula 1 as a start-up engine, as well as for the propulsion of torpedoes. While gasoline or diesel fuel tanks have the same amount of energy per litre of fuel from the first to the last litre, compressed air motors rely on the pressure within the tank, which decreases as air is drawn off. Because the air flows through a pressure regulator before going to the engine, no reduction in power would be noticed until the tank was nearly exhausted. This is the case for electric vehicles, too. This process of pressure regulation will reduce the energy available due to the Second Law of Thermodynamics.

Storage Tanks

Compressed air is a heavy way of storing fuel, 300l air at 30 MPa (300 Bar) about 16 kWh (the equivalent of 1.7 liters (0.44 gallons) of gasoline, assuming a 100% efficiency of the engine). During rupture testing, the tank cracks, but does not break up, producing no splinters or fragments.

All four major manufacturers that are developing air cars have designed safety features into their containers, in contrast to hydrogen's issues of damage and danger involved in high-impact crashes. Air, on its own, is non-flammable. Though no company has yet demonstrated the effectiveness of an imploding engine (ZAP) vs. a quick release (MDI) standard, and other safety designs, it is expected that large-scale production may lead specific governments to set their own standards. It was reported on Discovery's *Beyond Tomorrow* that on its own, carbon-fiber is brittle and splits, but creates no shrapnel.

Emission Output

Compressed air cars are emission-free at the 'tailpipe', but the energy, usually electric, required to recharge the compressed air tanks may come from the combustion of fossil fuels. Unless all of the energy comes from non-fossil sources, like nuclear power, biomass, wind power, solar power, or hydroelectricity, compressed air cars will still cause net emissions of greenhouse gas, proportional to the ratio of energy produced via fossil fuels. However, emissions would be relocated from city streets to remote smokestacks. CO_2 can potentially be sequestered from large plants but not from individual vehicles. Of course a solar and/or wind recharging system can be installed in one's home, in the car or power plants, thereby eliminating the dependence on the consumption of nuclear-fossil electricity. When energy from biomass is used as the energy source with CO_2 removal, neutral emissions of CO_2 would — in theory — be the result. The same is true of traditional IC vehicles run on biofuels.

An air motor also releases air cleaner than its intake, due to the presence of an air filter to keep contaminants out of the mechanism. Consequently, this can be viewed as air purification instead of air pollution.

Advantages

The principal advantages of an air powered vehicle are:

- Much like electrical vehicles, air powered vehicles would ultimately be powered through the electrical grid. That makes it easier to focus on reducing pollution from one source, as opposed to the millions of vehicles on the road.
- Compressed air engine reduces the cost of vehicle production by about 20%, because there is no need to build a cooling system, fuel tank, spark plugs or silencers.
- Most compressed air engines do not need a transmission, just a flow control.
- Transportation of the fuel would not be required due to drawing power off the electrical grid. This presents significant cost benefits. Pollution created during fuel transportation would be eliminated.
- Very low self-discharge rate (most batteries will deplete their charge

without external load at a rate determined by the chemistry, design, and size, while compressed gas storage will have an extremely low leakage rate)

- Long storage lifetime device (electric vehicle batteries have a limited useful number of cycles, and sometimes a limited calendar lifetime, irrespective of use). This means that batteries in operation are much more expensive than compressed air storage, and are more polluting because a lot more pollutant material needs to be used (typical car batteries are made from sulfuric acids and lead).
- Lower initial cost than battery electric vehicles when mass produced. One estimate is •3,000 less.
- Compressed air not subject to fuel tax at present (one taxing method would apply also to electric cars by taxing the electricity used for compression or charging).
- Expansion of the compressed air reduces its temperature and heat from the passenger compartment may be cooled using a heat exchanger, providing both relief from hot weather by air conditioning and increased efficiency.
- Zero pollutant emissions from the vehicle itself.
- Compressed air is not a fire hazard, only a rapid depressurisation hazard (this can be mitigated by using carbon fiber tanks).
- Air turbine technology, closely related to steam turbine technology, is a practice over 50 years old. It is simple to achieve with low tech materials. This would mean that developing countries, and rapidly growing countries like China and India, could easily implement a less polluting means of personal transportation than an internal combustion engine automobile.
- Possibility to refill air tank at home (using domestic power socket).
- Lighter vehicles would mean less abuse on roads. Resulting in longer lasting roads.
- The price of fueling air powered vehicles will be significantly cheaper than current fuels.

Disadvantages

Having solved most of the high pressure storage and handling problems, the main remaining disadvantages are related to the thermodynamics.

- When the air is expanded in the engine, it will cool down via adiabatic cooling and lose pressure thus its ability to do work at colder temperatures. It is difficult to maintain or restore the air temperature by simply using a heat exchanger with ambient heat at the high flow rates used in a vehicle, thus the ideal isothermic energy capacity of the tank will not be realised. Cold temperatures will also encourage the engine to ice up.

- Compressed air vehicles likely will be less robust than typical vehicles of today. Which poses a danger to users of compressed air vehicles sharing the road with larger, heavier and more rigid vehicles.
- Using energy to compress air is less efficient than charging a battery with that same energy - compressed-air car's overall efficiency is approximately one third of a comparable electric car.
- The major problem of all compressed air or liquified air cars is the lack of torque provided by the compressed gas and the high cost to compress the gas. This was shown to be the case starting in 1903 with the Liquified Air Company car of London England
- Long refill time when refilled using a home or low-end system (circa 4 hours). However, at a commercial station refilling would take only 3 minutes. But for that, fueling stations would have to invest in new dispensers.
- Governments would either lose tax income or have to adapt their taxation strategies.

Models

Various companies are investing in the research, development and deployment of *Compressed air cars*. Overoptimistic reports of impending production date back to at least May 1999. For instance, the MDI Air Car made its public debut in South Africa in 2002, and was predicted to be in production "within six months" in January 2004. Most of the cars under development also rely on using similar technology to Low-energy vehicles in order to increase the range and performance of their cars.

The AirCar

Air Car Factories SA is proposing to develop and build a compressed air engine.

Energine

The Energine Corporation was a South Korean company that claimed to deliver fully-assembled cars running on a hybrid compressed air and electric engine. These cars are more precisely named pneumatic-hybrid electric vehicles.. Engineers from this company made, starting from a Daewoo Matiz, a prototype of a hybrid electric/ compressed-air engine (Pne-PHEV, pneumatic plug-in hybrid electric vehicle). The compressed-air engine is used to activate an alternator, which extends the autonomous operating capacity of the car.

The CEO is the first compressed air car promoter to be arrested for fraud.

A similar (but only for braking energy recovery) concept using a pneumatic accumulator in a largely hydraulic system has been developed by U.S. government research laboratories and industry, and is now being introduced for certain heavy vehicle applications such as refuse trucks.

EngineAir

EngineAir, an Australian company, is developing a rotary engine powered by compressed air.

K'Airmobiles

K'Airmobiles has presented two running prototypes of VPA (Vehicles with Pneumatic Assistance). Their leaders now seek to gain the means of developing several projects of urban or leisure VPP (Vehicles with Pneumatic Propulsion). K'Airmobiles propose a different technology with their *VPP* , which may allow a reasonable range, generally with compressed air tanks of about 50L-100L/3000 psi capacity only.

These ecological vehicles use the technology of the compressed-air engine K'Air, developed in France by a small group of researchers, which thus proposes a range of projects around an idea: that of the urban or leisure compressed-air vehicles.

K'Airmobiles is the name given to a set of projects relating to "VPA" (Vehicles with Pneumatic Assistance) and "VPP" (Vehicles with Pneumatic Propulsion), aiming to escape the constraints from thermodynamics. To do so, these models are conceived like ultra light vehicles (limited to 250 kg max.), and their consumption of compressed air was calculated to remain lower than 120 L/min., although developing a dynamic push able to reach 4kN.

Two VPA prototypes are operational today, the "K'AirBike" and the K'AirKart. Two new VPP prototypes, the one-seater "K'AirTrike" and the three-seater "K'AirMobile Max" are intended for public presentation in October and November 2007 respectively.

The technical concept of the K'Air pneumatic engines returns to direct conversion of what makes the fundamental characteristic of compressed air, namely:

- the pushing force of compressed air is exclusively exploited for conversion into kinetic energy of translation,
- itself is simultaneously converted into induced power of rotation of the axis and
- thus gives to the engine a particularly imposing torque while needing only a very low "fuel" consumption.

To simplify, one can compare the principle to that of the rotary jacks:

- the energy of the fluid (compressed air) is directly transformed into rotational movement;
- the double-acting jacks involve a pinion-toothed rack system;
- the cyclic angle of rotation can vary between 90 and 360°;
- it supports hydraulic supercharging systems.

MDI and Tata Motors

MDI proposes a range of vehicles developed on an identical concept, made up of MiniCATs and CityCATs.

The MDI MiniCat has a range of up to 1000 miles when fitted with an internal combustion engine, but that drops to 30 miles when used in ZEV mode at low speeds in cities. OneCAT, priced in a range ($5,100 to $7,800) within reach of consumers in a developing economy, such as India. The ultralight bodies of the vehicles would be made of glued-together fiberglass and injected foam, and the aluminum chassis would also be glued, not welded, to simplify manufacturing. The engine is available in two versions. The Mono Energy air engine is a true air engine. The Dual Energy engines are Internal combustion engines, which use hydrocarbon fuels. They have licensed 12 factories : 5 in Mexico, 3 in Australia and New Zealand, 1 in South Africa and 3 in France. MDI Andina S.A is going to sell the car in Colombia, Peru, Ecuador and Panama. MDI has entered into an agreement with Tata Motors, to produce air cars in India. Zero Pollution Motors will be the first company to manufacture and sell a car in the U.S. called the "Air Car" using MDI technology. ZPM will begin taking reservations in early 2008 for US deliveries in early 2010.

Psycho-Active

Team Psycho-Active (TPA) has competed in the Automotive X PRIZE. TPA is working to develop the EPOCH: a highly customizable and serviceable pneumatic / multi-fuel hybrid. At its core is the DBRE engine. The DBRE's unique characteristics allow the EPOCH to use most commercially available fuels and/or compressed air.

Quasiturbine

The Quasiturbine is a prototype for a hybrid engine which, according to its originators, could function with an explosive fuel and with compressed air alike.

Hinged Rotor Internal Combustion Engine

Despite its name, this patent (US 6,718,938) describes an engine very similar to the Quasiturbine, and equally applicable to compressed air technology. The main difference is the much simpler method used to transfer rotor power to the central driveshaft.

Tata

As of May 2007 Tata Motors of India planned to launch an MDI air car in 2008.

Electro-Tech Enterprises

Electro-Tech Enterprises is a small company that specializes in electric air vehicles using a new technology discovered by themselves that is currently in the proto-type phase.

Fireless Locomotive

A fireless locomotive was a type of locomotive designed for use under conditions restricted by either the presence of flammable material (such as in mines) or the need for cleanliness (such as at a food factory). Thus a traditional steam locomotive is ruled out because of its open fire and the possibility of hot embers ejected from its chimney.

There are two types of fireless locomotive – fireless steam locomotives and compressed air locomotives.

Types

Fireless Steam Locomotives

A fireless steam locomotive is similar to a conventional steam locomotive, but has a reservoir, known as a steam accumulator, instead of a boiler. This reservoir is partly filled with water and charged with steam from a stationary boiler. The locomotive can then work on the stored steam until the pressure has dropped to a minimum level, after which it must be recharged.

European fireless steam locomotives usually have the cylinders at the back, while American ones often have the cylinders at the front, as in a conventional locomotive. Major builders of fireless steam locomotives in the UK included Andrew Barclay and W.G. Bagnall.

The last commercially-operated steam locomotive in UK industry was a fireless steam locomotive. It operated at the Glaxo factory in Ulverston, Cumbria. A few fireless steam locomotives are still at work in Germany.

Compressed Air Locomotives

Compressed air locomotives are used mainly in mines, but have also been used on tramways.

Hybrid Locomotives

Several hybrid locomotives have been built that have either used a fire for part of the time, e.g., Fowler's Ghost of the Metropolitan Railway, or have used a fire to superheat stored steam, such as the Receiver Locomotives built by Sentinel Waggon Works. None has been a success.

Wheel Arrangements

Most fireless locomotives have been of 0-4-0 or 0-6-0 wheel arrangement but there have been some 0-8-0 and even a few 0-10-0. Some 600 mm gauge 0-10-0 fireless locomotives from the German company Henschel were used in the construction of the Baghdad Railway, probably to avoid the risk of carbon monoxide poisoning during the boring of tunnels.

Another German company, Hohenzollern, built some articulated fireless steam locomotives with a cab at each end. Only one of the bogies was powered, making the wheel arrangement B-2.

Preservation

Although a large proportion of fireless locomotives have been preserved, very few have been operated. This is due to the low power of the locomotives, the long time needed to charge a locomotive from cold and the low steam pressures available for charging. Perhaps the only exception was "Lord Ashfield" (Andrew Barclay works no. 1989 of 1930) at the Museum of Science and Industry in Manchester that ran for a while in the 1990s sharing a steam supply with the stationary exhibits in their exhibition hall.

The Future

Fireless steam locomotives might, possibly, make a comeback because of their ability to use steam generated by rubbish incinerators or solar collectors. See "Solar Steam Train" link below. They could even use steam generated by nuclear reactors.

Alkaline Fuel Cell

The alkaline fuel cell (AFC), also known as the Bacon fuel cell after its British inventor, is one of the most developed fuel cell technologies and is the cell that flew Man to the Moon. NASA has used alkaline fuel cells since the mid-1960s, in Apollo-series missions and on the Space Shuttle. AFCs consume hydrogen and pure oxygen producing potable water, heat, and electricity. They are among the most efficient fuel cells, having the potential to reach 70%.

Chemistry

The fuel cell produces power through a redox reaction between hydrogen and oxygen. At the anode, hydrogen is oxidised according to the reaction: producing water and releasing two electrons. The electrons flow through an external circuit and return to the cathode, reducing oxygen in the reaction: producing hydroxide ions. The net reaction consumes one oxygen molecule and two hydrogen molecules in the production of two water molecules. Electricity and heat are formed as by-products of this reaction.

Electrolyte

The two electrodes are separated by a porous matrix saturated with an aqueous alkaline solution, such as potassium hydroxide (KOH). Aqueous alkaline solutions do not reject carbon dioxide (CO_2) so the fuel cell can become "poisoned" through the conversion of KOH to potassium carbonate (K_2CO_3). Because of this, alkaline fuel cells typically operate on pure oxygen, or at least purified [[air] and would incorporate a 'scruber' into the design to clean out as much of the carbon dioxide as is possibe. Because the generation and storage requirements of oxygen make pure-oxygen AFCs relatively expensive, there are few companies engaged in active development of the technology. There is, however, some debate in the research community over whether the poisoning is permanent or reversible. The main mechanisms of poisoning are blocking of the pores in the cathode with K_2CO_3, which is not reversible, and reduction in the ionic conductivity of the electrolyte, which may be reversible by returning the KOH to its original concentration. An alternate method involves simply replacing the KOH which is practically cheaper than water which returns the cell back to its original output.

Basic Designs

Because of this poisoning effect, two main variants of AFCs exist: static electrolyte and flowing electrolyte. Static, or immobilized, electrolyte cells of the type used in the Apollo space craft and the Space Shuttle typically use an asbestos separator saturated in potassium hydroxide. Water production is managed by evaporation out the anode, as pictured above, which produces pure water that may be reclaimed for other uses. These fuel cells typically use platinum catalysts to achieve maximum volumetric and specific efficiencies.

Flowing electrolyte designs use a more open matrix that allows the electrolyte to flow either between the electrodes (parallel to the electrodes) or through the electrodes in a transverse direction (the ASK-type or EloFlux fuel cell). In parallel-flow electrolyte designs, the water produced is retained in the electrolyte, and old electrolyte may be exchanged for fresh, in a manner analogous to an oil change in a car . In the case of "parallel flow" designs, greater space is required between electrodes to enable this flow, and this translates into an increase in cell resistance, decreasing power output compared to immobilised electrolyte designs. A further challenge for the technology is that it is not clear how severe is the problem of permanent blocking of the cathode by K_2CO_3, however, some published reports indicate thousands of hours of operation on air. These designs have used both platinum and non-noble metal catalysts, trading off volumetric and specific efficiencies with cost. The EloFlux design, with its transverse flow of electrolyte, has the advantage of low-cost construction and replaceable electrolyte, but so far has only been demonstrated using oxygen. Further variations on the alkaline fuel cell include the metal hydride fuel cell and the direct borohydride fuel cell.

Commercial Prospects

AFCs are, however, the cheapest of fuel cells to manufacture. The catalyst required for the electrodes can be any of a number of different chemicals that are relatively inexpensive compared to those required for other types of fuel cells. The commercial prospects for AFCs lie largely with the recently developed bi-polar plate version of this technology, considerably superior in performance to earlier mono-plate versions. The world's first Fuel Cell Ship (HYDRA) was realised by Christian Machens and used an AFC system with 6.5 kW net output. Another very interesting recent development (though not necessartily for high power applications) is the solid-state alkaline fuel cell, utilising anion-exchange membranes rather than a liquid. This work is pioneered at the University of Surrey in the United Kingdom. See details at http://mypages.surrey.ac.uk/chs1jv/.

Bio-Nano Generator

A bio-nano generator is a nanoscale electrochemical device, like a fuel cell or galvanic cell, but drawing power from blood glucose in a living body, much the same as how the body generates energy from food. To achieve the effect, an enzyme is used that is capable of stripping glucose of its electrons, freeing them for use in electrical devices. The average person's body could, theoretically, generate 100 watts of electricity using a bio-nano generator. However, this estimate is only true if all food was converted to electricity, and the human body needs some energy consistently, so possible power generated is likely much lower. The electricity generated by such a device could power devices embedded in the body (such as pacemakers), or sugar-fed nanorobots. A similar technology was presented in the Matrix series of science fiction major motion pictures, with robots shown enslaving mankind for its bio-energy. At the time, bio-nano generators were unheard of in popular culture, and the plot device of "a form of fusion" was used to make the harnessing of bio-electricity more plausible to the audience. Much of the research done on bio-nano generators is still experimental, with Panasonic's Nanotechnology Research Laboratory among those at the forefront.

Direct Borohydride Fuel Cell

Direct borohydride fuel cells (DBFCs) are a subcategory of alkaline fuel cells that use a solution of sodium borohydride for fuel. The advantage of sodium borohydride over conventional hydrogen in an alkaline fuel cell is that the highly alkaline fuel and waste borax prevents poisoning of the fuel cell from carbon dioxide (CO_2) in the air.

Sodium borohydride could potentially be used in more conventional hydrogen fuel cell systems as a means of storing hydrogen. The hydrogen can be regenerated for a fuel cell by catalytic decomposition of the borohydride:

$NaBH_4 + 2H_2O$ ”! $NaBO_2 + 4H_2$

Direct borohydride fuel cells decompose and oxidize the borohydride directly, side-stepping hydrogen production and even producing slightly higher energy yields:

Cathode: $2O_2 + 4H_2O + 8e$” ”! $8OH$” ($E^0 = +0.4V$)

Anode: $NaBH_4 + 8OH$” ”! $NaBO_2 + 6H_2O + 8e$” ($E^0 = 1.24$ V)

Total $E^0 = +1.64V$

DBFCs could be produced more cheaply than a traditional fuel cell because they do not need expensive platinum catalysts. In addition, they have a higher power density. Unfortunately, DBFCs do produce some hydrogen from a side step reaction of $NaBH_4$ with water heated by the fuel cell. This hydrogen can either be piped out to the exhaust or piped to a conventional hydrogen fuel cell. Either fuel cell will produce water, and the water can be recycled to allow for higher concentrations of $NaBH_4$.

After releasing its hydrogen and being oxidized, $NaBO_2$ or borax is produced. Borax is a common detergent and soap additive and is relatively non-toxic. Borax can be hydrogenated back into borohydride fuel by several different techniques, some of which require nothing more than water and electricity or heat. These techniques are still in active development.

Sodium borohydride costs US$50 per kg, but with borax recycling and mass production projected prices for the fuel are as low as US$1/kg.

Direct Methanol Fuel Cell

Direct-methanol fuel cells or DMFCs are a subcategory of proton-exchange fuel cells where, the fuel, methanol (CH_3OH), is not reformed, but fed directly to the fuel cell. Because methanol is fed directly into the fuel cell, complicated catalytic reforming is unneeded. Storage of methanol is much easier than that of hydrogen because it does not need to be done at high pressures or low temperatures, as methanol is a liquid from -97.0 °C to 64.7 °C (-142.6 °F to 148.5 °F). The energy density of methanol, the amount of energy contained in a given volume of methanol, is an order of magnitude greater than even highly compressed hydrogen.

The efficiency of current direct-methanol fuel cells is low due to the high permeation of methanol through the membrane materials used, which is known as methanol crossover. A new kind of membrane (polymer electrolyte thin films, assembled "layer by layer") was shown to reduce this problem dramatically.

Other problems include the management of carbon dioxide created at the anode and the sluggish dynamic behaviour. Current DMFCs are limited in the power they can produce, but can still store a high energy content in a small space. This means they can produce a small amount of power over a long period of time. This makes them presently ill-suited for powering vehicles (at least directly), but ideal for consumer goods such as mobile phones, digital cameras or laptops.

Methanol is toxic and flammable. However, the International Civil Aviation Organisation's (ICAO) Dangerous Goods Panel (DGP) voted in November 2005 to allow passengers to carry and use micro fuel cells and methanol fuel cartridges when aboard airplanes to power laptop computers and other consumer electronic devices. On September 24th, 2007, the US Department of Transportation issued a proposed rulemaking to allow airline passengers to carry fuel cell cartridges on board. The Department of Transportation issued a final ruling on April 30, 2008, permitting passengers and crew to carry an approved fuel cell with an installed methanol cartridge and up to two additional spare cartridges. It is worth noting that 200 ml maxium methanol cartridge volume allowed in the final ruling is double the 100 ml limit on liquids allowed by the Transportation and Security Administration in carry-on bags.

Reaction

The DMFC relies upon the oxidation of methanol on a catalyst layer to form carbon dioxide. Water is consumed at the anode and is produced at the cathode. Positive ions (H^+) are transported across the proton exchange membrane (often Nafion) to the cathode where they react with oxygen to produce water. Electrons are transported through an external circuit from anode to cathode providing power to external devices.

The half-reactions are:

Anode: $CH_3OH + H_2O$ '! $CO_2 + 6H^+ + 6e^-$

Cathode: $(3/2)O_2 + 6H^+ + 6e^-$ '! $3H_2O$

Overall reaction: $CH_3OH + (3/2)O_2$ '! $CO_2 + 2H_2O$

The methanol is adsorbed on a catalyst, usually made of platinum particles, and deprotonised until carbon dioxide is formed. Usually, the catalyst consists of another metallic component, usually ruthenium, which is used to catalyze methanol oxidation. Because water is consumed at the anode in the reaction, pure methanol cannot be used without provision of water via either passive transport such as back diffusion (osmosis), or active transport such as pumping. The need for water limits the energy density of the fuel. Currently, platinum is used as a catalyst for both half-reactions. This contributes to the loss of cell voltage potential, as any methanol that is present in the cathode chamber will oxidize. If another catalyst could be found for the reduction of oxygen, the problem of methanol crossover would likely be significantly lessened. Furthermore, platinum is very expensive and contributes to the high cost per kilowatt of the fuel cell. In one of the steps of the methanol oxidation reaction, a CO species is produced, which adsorbs strongly on the platinum catalyst, reducing the surface area for the catalyst reaction.

Blue Energy

Blue energy, osmotic power or salinity gradient power is the energy retrieved from the difference in the salt concentration between seawater and river water. Two practical methods for this are Reverse Electrodialysis (RED), or Pressure Retarded Osmosis (PRO). Both processes rely on osmosis with ion specific membranes. The key waste product is brackish water. This byproduct is the result of natural forces that are being harnessed: the flow of fresh water into seas that are made up of salt water. The technologies have been confirmed in laboratory conditions. They are being developed into commercial use in the Netherlands (RED) and Norway (PRO). The cost of the membrane has been an obstacle. A new, cheap membrane, based on an electrically modified polyethylene plastic, made it fit for potential commercial use

Availabilty and Status

It is estimated that each year 1600 TWh could be generated world wide, and 12 TWh in Norway, sufficient to meet 10% of Norway's total demand for electricity. A prototype 2-4 kW plant is being constructed in Buskerud, Norway in 2008.

PRO (Pressure Retarded Osmosis)

In PRO, the water potential between fresh water and sea water corresponds to a pressure of 26 bars. This pressure is equivalent to a column of water (hydraulic head) 270 meters high. However, the optimal working pressure is only half of this, 11 to 15 bar. In the Netherlands, for example, more than 3,300 m^3 fresh water runs into the sea per second on average. The membrane halves the pressure differences which results in a water column of approx. 135 meters.

RED (Reverse ElectroDialysis)

In reverse electrodialysis (RED) a salt solution and fresh water are let though a stack of alternating cathode and anode exchange membranes. The chemical potential difference between salt and fresh water generates a voltage over each membrane and the total potential of the system is the sum of the potential differences over all membranes. It is important to remember that the process works though difference in ion concentration instead of an electric field, this has implications for the needed properties for a suitable membrane. In RED, as in a fuel cell, the cells are stacked. A module with a capacity of 250 kW has the size of a shipping container.

Testing

With RED, in 2006 a 50 kW plant in planned located at a coastal test site in Harlingen, the Netherlands, the focus is on prevention of biofouling on the anode, cathode and membranes and increasing the membrane performance. In 2007 the Directorate for Public Works and Water Management, Redstack and ENECO signed a declaration of intent for development of a clean plant in Afsluitdijk dam in Netherlands. To start with, the conditions for the construction of an installation in or on the Afsluitdijk dam will be investigated. This study should be concluded by the middle of 2008. Subsequently a small 10-50 kilowatt installation will be built to test the RED technology under actual working conditions. These tests will run from mid 2008 until 2010. After that, the capacity will be expanded to 1000 kW and the system will be optimized. It is to be expected that after this phase the installation will be further expanded to a final capacity of 200 MW. In the case of PRO, Statkraft in Norway has announced that it will build an osmotic power plant prototype in Hurum in Buskerud. The prototype is planned to produce 2-4 kW at the start in 2008.

Solar Energy

Solar energy is energy from the Sun in the form of radiated heat and light. It drives the climate and weather and supports life on Earth. Solar energy technologies make controlled use of this energy resource. Solar power is a synonym of solar energy or refers specifically to the conversion of sunlight into electricity by photovoltaics, concentrating solar thermal devices or various experimental technologies. In building design, daylighting techniques use sunlight to provide interior lighting and to warm the air instead of using electric lighting and powered heating. Solar water heaters heat swimming pools and provide domestic hot water. In agriculture, greenhouses grow specialty crops and photovoltaic-powered pumps bring water to grazing animals. Evaporation ponds find applications in the commercial and industrial sectors where they are used to harvest salt and clean waste streams of contaminants.

Solar distillation and disinfection techniques produce potable water for millions of people worldwide. Family-scale solar cookers and larger solar kitchens concentrate sunlight for cooking, drying and pasteurisation. More sophisticated concentrating technologies magnify the rays of the Sun for high temperature material testing, metal smelting and industrial chemical production. A range of prototype solar vehicles provide ground, air and sea transportation.

Energy From the Sun

Earth continuously receives 174 PW of incoming solar radiation (insolation) at the upper atmosphere. Approximately 30% is reflected back to space while the rest is absorbed by the atmosphere, oceans and land masses. After passing through the atmosphere, the insolation spectrum is mostly split between the visible and infrared ranges with a small part in the ultraviolet. The absorption of solar energy by atmospheric convection (sensible heat transport) and evaporation and condensation of water vapor (latent heat transport) powers the water cycle and drives the winds. Sunlight absorbed by the oceans and land masses keeps the surface at an average temperature of 14 °C. The conversion of solar energy into chemical energy via photosynthesis produces food, wood and the biomass from which fossil fuels are derived. Solar radiation along with secondary solar resources such as wind and wave power, hydroelectricity and biomass account for over 99.9% of the available flow of renewable energy on Earth. The flows and stores of solar energy in the environment are vast in comparison to current human energy needs.

- The total solar energy absorbed by Earth's atmosphere, oceans and land masses is approximately 3,850 zettajoules (ZJ) per year.
- Global wind energy at 80 m is estimated at 2.25 ZJ per year.
- Photosynthesis captures approximately 3 ZJ per year in biomass.
- Worldwide electricity consumption was approximately 0.0567 ZJ in 2005.
- Worldwide primary energy consumption was 0.487 ZJ in 2005.

Applications of Solar Energy Technology

Solar energy technologies use solar radiation for practical ends. Technologies that use secondary solar resources such as biomass, wind, waves and ocean thermal gradients can be included in a broader description of solar energy but only primary resource applications are discussed here. Because the performance of solar technologies varies widely between regions, solar technologies should be deployed in a way that carefully considers these variations. Solar technologies such as photovoltaics and water heaters increase the supply of energy and may be characterised as supply side technologies. Technologies such as passive design and shading devices reduce the need for alternate resources and may be characterised as demand side. Optimising the performance of solar technologies is often a matter of controlling the resource rather than simply maximising its collection.

Architecture and Urban Planning

Sunlight has influenced building design since the beginning of architectural history. Fully developed solar architecture and urban planning methods were first employed by the Greeks and Chinese who oriented their buildings toward the south to provide light and warmth. The elemental features of passive solar architecture are Sun orientation, compact proportion (a low surface area to volume ratio), selective shading (overhangs) and thermal mass. When these features are tailored to the local climate and environment they can produce well-lit spaces that stay in a comfortable temperature range. Socrates' Megaron House is a classic example of passive solar design. The most recent approaches to solar design use computer modeling to tie together solar lighting, heating and ventilation systems in an integrated solar design package. Active solar equipment such as pumps, fans and switchable windows can also complement passive design and improve system performance. Urban heat islands (UHI) are metropolitan areas with higher temperatures than the surrounding environment. These higher temperatures are the result of urban materials such as asphalt and concrete that have lower albedos and higher heat capacities than the natural environment. A straightforward method of counteracting the UHI effect is to paint buildings and roads white and plant trees. Using these methods, a hypothetical "cool communities" programme in Los Angeles has projected that urban temperatures could be reduced by approximately 3 °C at an estimated cost of US$1 billion, giving estimated total annual benefits of US$530 million from reduced air-conditioning costs and healthcare savings.

Agriculture and Horticulture

Agriculture inherently seeks to optimize the capture of solar energy, and thereby plant productivity. Techniques such as timed planting cycles, tailored row orientation, staggered heights between rows and the mixing of plant varieties can improve crop yields. While sunlight is generally considered a plentiful resource, there are exceptions which highlight the importance of solar energy to agriculture. During the short growing seasons of the Little Ice Age, French and English farmers employed fruit walls to maximize the collection of solar energy. These walls acted as thermal masses and accelerated ripening by keeping plants warm. Early fruit walls were built perpendicular to the ground with a south facing orientation but over time sloping walls were developed to make better use of sunlight. In 1699, Nicolas Fatio de Duillier even suggested using a tracking mechanism which could pivot to follow the Sun. Solar energy is also used in many areas of agriculture aside from growing crops. Applications include pumping water, drying crops, brooding chicks and drying chicken manure. Greenhouses control the use of solar heat and light to grow plants in controlled environments, enabling year-round production and the growth of specialty crops and other plants not naturally suited to the local climate. Primitive greenhouses were first used during Roman times to grow cucumbers year-round for the Roman emperor Tiberius. The first modern greenhouses were built in Europe in the 16th century to conserve exotic plants brought back from explorations abroad. Greenhouses remain an important part of horticulture today, while plastic transparent materials have also been used to similar effect in polytunnels and row covers.

Solar Lighting

The history of lighting is dominated by the use of natural light. The Romans recognised the Right to Light as early as the 6th century and English law echoed these judgments with the Prescription Act of 1832. In the 20th century artificial lighting became the main source of interior illumination. Daylighting systems collect and distribute sunlight to provide interior illumination. These systems directly offset energy use by replacing artificial lighting, and indirectly offset non-solar energy use by reducing the need for air-conditioning. Although difficult to quantify, the use of natural lighting also offers physiological and psychological benefits compared to artificial lighting. Daylighting design carefully selects window type, size and orientation and may also consider exterior shading devices. Individual features include sawtooth roofs, clerestory windows, light shelves, skylights and light tubes. These features may be incorporated into existing structures but are most effective when integrated in a solar design package that accounts for factors such as glare, heat flux and time-of-use. When daylighting features are properly implemented they can reduce commercial lighting-related energy requirements by 25%. Hybrid solar lighting (HSL) is an active solar method of using sunlight to provide illumination. HSL systems collect sunlight using focusing mirrors that track the Sun and use optical fibers to transmit the light into a building's interior to supplement conventional lighting. In single-story applications, these systems are able to transmit 50% of the direct sunlight received. Although daylight saving time is promoted as a way to use sunlight to save energy, recent research is limited and reports contradictory results: several studies report savings, but just as many suggest no effect or even a net loss, particularly when gasoline consumption is taken into account. Electricity use is greatly affected by geography, climate and economics, making it hard to generalize from single studies.

Solar Thermal

Solar thermal technologies can be used for water heating, space heating, space cooling and process heat generation.

Water Heating

Solar hot water systems use sunlight to heat water. When sited in low latitudes (below 40 degrees), solar heating system can provide around 60 to 70% of domestic hot water use with temperatures up to 60 °C. The most common types of solar water heaters are evacuated tube collectors (44%) and glased flat plate collectors (34%) generally used for domestic hot water; and unglazed plastic collectors (21%) used mainly to heat swimming pools. As of 2007, the total installed capacity of solar hot water systems is approximately 154 GW. China is the world leader in the deployment of solar hot water with 70 GW installed as of 2006 and a long term goal of 210 GW by 2020. Israel is the per capita leader in the use of solar hot water with 90% of homes using this technology. In the United States, Canada and Australia, heating swimming pools is the dominant application of solar hot water, with an installed capacity of 18 GW as of 2005.

Heating, Cooling and Ventilation

In the United States, heating, ventilation and air conditioning (HVAC) systems account for over 25% (4.75 EJ) of the energy used in commercial buildings and nearly 50% (10.1 EJ) of the energy used in residential buildings. Solar heating, cooling and ventilation technologies can be used to offset a portion of this energy. Thermal mass, in the most general sense, is any material that has the capacity to store heat. In the context of solar energy, thermal mass materials are used to store heat from the Sun. Common thermal mass materials include stone, cement and water. These materials have historically been used in arid climates or warm temperate regions to keep buildings cool by absorbing solar energy during the day and radiating stored heat to the cooler atmosphere at night, but they can also be used in cold temperate areas to maintain warmth. The size and placement of thermal mass should consider several factors such as climate, daylighting and shading conditions. When properly incorporated, thermal mass can passively maintain comfortable temperatures without consuming energy. A solar chimney (or thermal chimney) is a passive solar ventilation system composed of a vertical shaft connecting the interior and exterior of a building. As the chimney warms, the air inside is heated causing an updraft that pulls air through the building. Performance can be improved by using glazing and thermal mass materials in a way that mimics greenhouses. These systems have been in use since Roman times and remain common in the Middle East. Deciduous trees and plants can be used to provide heating and cooling. When planted on the southern elevation of the building, the leaves can provide shade during the summer while the bare limbs allow light and warmth to pass during the winter.

Process Heat

Concentrating solar technologies such as parabolic dish, trough and Scheffler reflectors can provide process heat for commercial and industrial applications. The first commercial process heating project was the Solar Total Energy Project (STEP) in Shenandoah, Georgia where a field of 120 parabolic dishes provided 50% of the process heating, air conditioning and electrical requirements for a clothing factory. This system generated 400 kW of electricity and 3 MW of thermal energy in the form of steam, and had a thermal storage system which allowed for peak-load shaving. A prototype Scheffler reflector is currently being constructed in India for use in a solar crematorium. Evaporation ponds are shallow pools that concentrate dissolved solids through evaporation. The use of evaporation ponds to obtain salt from sea water is one of the oldest applications of solar energy. Modern uses include concentrating brine solutions used in leach mining and removing dissolved solids from waste streams. Clothes lines, clotheshorses, and clothes racks dry clothes through evaporation. These devices use wind and sunlight instead of electricity or natural gas. Florida legislation specifically protects the 'right to dry' and similar solar rights legislation has been passed in Utah and Hawaii. Unglazed transpired collectors (UTC) are perforated sun-facing walls used for preheating ventilation air. UTCs can raise the incoming air temperature up to 22 °C and deliver outlet temperatures of 45-60 °C. The short payback period of transpired collectors (3 to 12 years) makes them a more cost-effective alternative than glazed collection systems. As of 2003, over 80 systems with a combined collector area of 35,000 m² had been installed worldwide,including an 860 m² collector in Costa Rica used for drying coffee beans and a 1,300 m² collector in Coimbatore, India used for drying marigolds.

Cooking

Solar cookers use sunlight for cooking, drying and pasteurisation. Solar cooking offsets fuel costs, reduces demand for fuel or firewood, and improves air quality by reducing the generation of smoke. The simplest type of solar cooker is the box cooker first built by Horace de Saussure in 1767. A basic box cooker consists of an insulated container with a transparent lid. These cookers can be used effectively with partially overcast skies and will typically reach temperatures of 50-100 °C. Concentrating solar cookers use reflectors to focus light on a cooking container. The most common reflector geometries are flat plate, disc and parabolic trough type. These designs reach temperatures up to 315 °C but require direct light to function properly and must be repositioned to track the Sun. The solar bowl is a unique concentrating technology employed by the Solar Kitchen in Auroville, India. The solar bowl is a stationary spherical reflector that focuses light along a line perpendicular to the sphere's surface and a computer control system moves the receiver to intersect this line. Steam is produced in the receiver at temperatures reaching 150 °C and then used for process heat in the kitchen.

A reflector developed by Wolfgang Scheffler in 1986 is used in many solar kitchens. Scheffler reflectors are flexible parabolic dishes that combine aspects of trough and power tower concentrators. Polar tracking is used to follow the Sun's daily course and the curvature of the reflector is adjusted for seasonal variations in the incident angle of sunlight. These reflectors can reach temperatures of 450-650 °C and have a fixed focal point which improves the ease of cooking. The world's largest Scheffler reflector system in Abu Road, Rajasthan, India is capable of cooking up to 35,000 meals a day. By early 2008, over 2,000 large Scheffler cookers had been built worldwide.

Desalination and Disinfection

Solar distillation is the production of potable water from saline or brackish water using solar energy. The first recorded use was by 16th century Arab alchemists. The first large-scale solar distillation project was constructed in 1872 in the Chilean mining town of Las Salinas. This 4,700 m² still could produce up to 22,700 L per day and operated for 40 years. Individual still designs include single-slope, double-slope (or greenhouse type), vertical, conical, inverted absorber, multi-wick and multiple effect. These stills can operate in passive, active or hybrid modes. Double slope stills are the most economical for decentralised domestic purposes while active multiple effect units are more suitable to large-scale applications. Solar water disinfection (SODIS) is a method of disinfecting water by exposing water-filled plastic PET bottles to several hours of sunlight. Exposure times vary according weather and climate from a minimum of six hours to two days during fully overcast conditions. SODIS is recommended by the World Health Organisation as a viable method for household water treatment and safe storage. Over two million people in developing countries use SODIS for their daily drinking water needs.

Solar Electricity

Solar power technologies convert sunlight into electricity using photovoltaics, concentrating solar thermal devices, or various experimental technologies. PV has mainly been used to generate power for small and medium-sized applications, from the calculator powered by a single solar cell to off-grid homes powered by a photovoltaic array. For large-scale generation, concentrating solar thermal power plants like SEGS have been the norm but multi-megawatt PV plants are becoming more common. Completed in 2007, the 14 MW power station in Clark County, Nevada and the 20 MW site in Beneixama, Spain are characteristic of the trend toward larger photovoltaic power stations in the US and Europe.

Photovoltaics

A solar cell (or photovoltaic cell) is a device that converts light into direct current using the photoelectric effect. The first solar cell was constructed by Charles Fritts in 1883. Although the prototype selenium cells converted less than 1% of incident light into electricity, both Ernst Werner von Siemens and James Clerk Maxwell recognised the importance of this discovery. Following the fundamental work of Russell Ohl in the 1940s, researchers Gerald Pearson, Calvin Fuller and Daryl Chapin created the silicon solar cell in 1954. These early solar cells cost 286 USD/watt and reached efficiencies of 4.5-6%. The earliest significant application of solar cells was as a back-up power source to the Vanguard I satellite, which allowed the satellite to continue transmitting for over a year after its chemical battery was exhausted. The successful operation of solar cells on this mission was duplicated in many other Soviet and American satellites, and by the late 1960s PV had become the established source of power for satellites. Photovoltaics went on to play an essential part in the success of early commercial satellites such as Telstar and continue to remain vital to the telecommunications infrastructure today. The high cost of solar cells limited terrestrial uses throughout the 1960s. This changed in the early 1970s when prices reached levels that made PV generation competitive in remote areas without grid access. Early terrestrial uses included powering telecommunication stations, off-shore oil rigs, navigational buoys and railroad crossings. These and other off-grid applications have proven very successful and accounted for over half of worldwide installed capacity until 2004. The 1973 oil crisis stimulated a rapid rise in the production of PV during the 1970s and early 1980s. Economies of scale which resulted from increasing production along with improvements in system performance brought the price of PV down from 100 USD/watt in 1971 to 7 USD/watt in 1985. Steadily falling oil prices during the early 1980s led to a reduction in funding for photovoltaic R&D and a discontinuation of the tax credits associated with the Energy Tax Act of 1978. These factors moderated growth to approximately 15% per year from 1984 through 1996. Since the mid-1990s, leadership in the PV sector has shifted from the US to Japan and Germany. Between 1992 and 1994 Japan increased R&D funding, established net metering guidelines, and introduced a subsidy programme to encourage the installation of residential PV systems. As a result, PV installations in the country climbed from 31.2 MW in 1994 to 318 MW in 1999, and worldwide production growth increased to 30% in the late 1990s.

Germany has become the leading PV market worldwide since revising its Feed-in tariff system as part of the Renewable Energy Sources Act. Installed PV capacity has risen from 100 MW in 2000 to approximately 4,150 MW at the end of 2007. Spain has become the third largest PV market after adopting a similar feed-in tariff structure in 2004, while France, Italy, South Korea and the US have also seen rapid growth recently due to various incentive programmes and local market conditions. By 2006 more polysilicon was used in solar panels than in computer chips, and the limited supply caused prices to flatten. In 2008 new factories were built to meet the demand, and thin film technologies were developed which use a printing process to make solar panels, and do not use polysilicon.

Concentrating Solar Power

Concentrated sunlight has been used to perform useful tasks since the time of ancient China. A legend claims Archimedes used polished shields to concentrate sunlight on the invading Roman fleet and repel them from Syracuse. In 1866, Auguste Mouchout used a parabolic trough to produce steam for the first solar steam engine, and subsequent developments led to the use of concentrating solar-powered devices for irrigation, refrigeration and locomotion. Concentrating Solar Power (CSP) systems use lenses or mirrors and tracking systems to focus a large area of sunlight into a small beam. The concentrated light is then used as a heat source for a conventional power plant. A wide range of concentrating technologies exist; the most developed are the solar trough, parabolic dish and solar power tower. These methods vary in the way they track the Sun and focus light. In all these systems a working fluid is heated by the concentrated sunlight, and is then used for power generation or energy storage. A solar trough consists of a linear parabolic reflector that concentrates light onto a receiver positioned along the reflector's focal line. The reflector is made to follow the Sun during the daylight hours by tracking along a single axis. Trough systems are the most developed CSP technology. The SEGS plants in California and Acciona's Nevada Solar One near Boulder City, Nevada are representatives of this technology. A parabolic dish system consists of a stand-alone parabolic reflector that concentrates light onto a receiver positioned at the reflector's focal point. The reflector tracks the Sun along two axes. Parabolic dish systems give the highest efficiency among CSP technologies. The Big Dish in Canberra, Australia is an example of this technology. A solar power tower uses an array of tracking reflectors (heliostats) to concentrate light on a central receiver atop a tower. Power towers are less advanced than trough systems but offer higher efficiency and better energy storage capability. The Solar Two in Daggett, California and the Planta Solar 10 in Sanlucar la Mayor, Spain are representatives of this technology.

Experimental Solar Power

A solar updraft tower (also known as a solar chimney or solar tower) consists of a large greenhouse that funnels into a central tower. As sunlight shines on the greenhouse, the air inside is heated and expands. The expanding air flows toward the central tower where a turbine converts the air flow into electricity. A 50 kW prototype was constructed in Ciudad Real, Spain and operated for eight years before decommissioning in 1989.

A solar pond is a pool of salt water (usually 1-2 m deep) that collects and stores solar energy. Solar ponds were first proposed by Dr. Rudolph Bloch in 1948 after he came across reports of a lake in Hungary in which the temperature increased with depth. This effect was due to salts in the lake's water, which created a "density gradient" that prevented convection currents. A prototype was constructed in 1958 on the shores of the Dead Sea near Jerusalem. The pond consisted of layers of water that successively increased from a weak salt solution at the top to a high salt solution at the bottom. This solar pond was capable of producing temperatures of 90 °C in its bottom layer and had an estimated solar-to-electric efficiency of two percent.

Thermoelectric devices convert a temperature difference between dissimilar materials into an electric current. First proposed as a method to store solar energy by solar pioneer Mouchout in the 1800s, thermoelectrics reemerged in the Soviet Union during the 1930s. Under the direction of Soviet scientist Abram Ioffe a concentrating system was used to thermoelectrically generate power for a 1 hp engine. Thermogenerators were later used in the US space programme as an energy conversion technology for powering deep space missions such as Cassini, Galileo and Viking. Research in this area is focused on raising the efficiency of these devices from 7–8% to 15–20%.

Solar Chemical

Solar chemical processes use solar energy to drive chemical changes. These processes offset energy that would otherwise be required from an alternate source and can convert solar energy into a storable and transportable fuel. Solar chemical reactions are diverse but can generically be described as either thermochemical or photochemical.

Hydrogen production technologies have been a significant area of solar chemical research since the 1970s. Aside from electrolysis driven by photovoltaic or photochemical cells, several thermochemical processes have also been explored. The seemingly most direct of these routes uses concentrators to split water at high temperatures (2300-2600 °C), but this process has been limited by complexity and low solar-to-hydrogen efficiency (1-2%). A more conventional approach uses process heat from solar concentrators to drive the steam reformation of natural gas thereby increasing the overall hydrogen yield. Thermochemical cycles characterised by the decomposition and regeneration of reactants present another avenue of hydrogen production. The Solzinc process under development at the Weitzman Institute is one such method. This process uses a 1 MW solar furnace to decompose zinc oxide (ZnO) at temperatures above 1200 °C. This initial reaction produces pure zinc which can subsequently be reacted with water to produce hydrogen.

Sandia's Sunshine to Petrol (S2P) technology uses the high temperatures generated by concentrating sunlight along with a zirconia/ferrite catalyst to break down atmospheric carbon dioxide into oxygen and carbon monoxide (CO). The CO may then be used to synthesize fuels such as methanol, gasoline and jet fuel.

Photoelectrochemical cells or PECs consist of a semiconductor, typically titanium dioxide or related titanates, immersed in an electrolyte. When the semiconductor is illuminated an electrical potential develops. There are two types of photoelectrochemical cells: photoelectric cells that convert light into electricity and photochemical cells that use light to drive chemical reactions such as electrolysis.

A photogalvanic device is a type of battery in which the cell solution (or equivalent) forms energy-rich chemical intermediates when illuminated. These chemical intermediates then react at the electrodes to produce an electric potential. The ferric-thionine chemical cell is an example of this technology.

Solar Mechanical

Passive solar tracking devices use imbalances caused by the movement of a low-boiling-point fluid to respond to the movement of the Sun. Tracking PV systems can generally produce 25% more electricity than fixed tilt PV systems. Shading systems that respond to the movement of the Sun can also be used in buildings to maximize natural lighting during winter, lessen summer glare, and reduce cooling loads associated with unwanted solar gain.

A light mill or Crookes radiometer consists of a glass bulb containing a set of vanes mounted on a spindle. Each vane has a dark side and a white side. When illuminated, the dark side becomes warm due to absorbing light but the white side reflects light and stays cool. The vanes rotate due to the motion of gases from the hot to the cool side of each vane.

Solar Vehicles

Development of a solar powered car has been an engineering goal since the 1980s. The center of this development is the World Solar Challenge, a biannual solar-powered car race in which teams from universities and enterprises compete over 3,021 kilometres (1,877 mi) across central Australia from Darwin to Adelaide. In 1987, when it was founded, the winner's average speed was 67 kilometres per hour (42 mph). The 2007 race included a new challenge class using cars which could be a practical proposition for sustainable transport with little modification. The winning car averaged 90.87 kilometres per hour (56.46 mph). The North American Solar Challenge and the planned South African Solar Challenge are comparable competitions that reflect an international interest in the engineering and development of solar powered vehicles.

In 1975, the first practical solar boat was constructed in England. By 1995, passenger boats incorporating PV panels began appearing and are now used extensively. In 1996, Kenichi Horie made the first solar powered crossing of the Pacific Ocean, and the *sun21* catamaran made the first solar powered crossing of the Atlantic

Ocean in the winter of 2006–2007. Plans to circumnavigate the globe in 2009 are indicative of the progress solar boats have made.

In 1974, the unmanned *Sunrise II* inaugurated the era of solar flight. In 1980, the *Gossamer Penguin* made the first piloted flights powered solely by photovoltaics. This was quickly followed by the *Solar Challenger* which demonstrated a more airworthy design with its crossing of the English Channel in July 1981. Developments then turned back to unmanned aerial vehicles with the *Pathfinder* (1997) and subsequent designs, culminating in the *Helios* which set the altitude record for a non-rocket-propelled aircraft at 29,524 metres (96,860 ft) in 2001. The *Zephyr*, developed by BAE Systems, is the latest in a line of record-breaking solar aircraft, making a 54-hour flight in 2007, and month-long flights are envisioned by 2010.

A solar balloon is a black balloon that is filled with ordinary air. As sunlight shines on the balloon, the air inside is heated and expands, causing an upward buoyancy force, much like an artificially-heated hot air balloon. Some solar balloons are large enough for human flight, but usage is limited to the toy market as the surface-area to payload-weight ratio is relatively high.

Solar sails are a proposed form of spacecraft propulsion using large membrane mirrors to exploit radiation pressure from the sun. Unlike rockets, solar sails require no fuel. Although the thrust is small compared to rockets, it continues as long as the Sun shines onto the deployed sail and in the frictionless vacuum of space significant speeds can eventually be achieved.

Thermal and Electrical Storage

Storage is an important issue in the development of solar energy because modern energy systems usually assume continuous availability of energy. Solar energy is not available at night, and the performance of solar power systems is affected by unpredictable weather patterns; therefore, storage media or back-up power systems must be used.

Thermal mass systems can store solar energy in the form of heat at domestically useful temperatures for daily or seasonal durations. Thermal storage systems generally use readily available materials with high specific heat capacities such as water, earth and stone. Well-designed systems can lower peak demand, shift time-of-use to off-peak hours and reduce overall heating and cooling requirements.

Phase change materials such as paraffin wax and Glauber's salt are another thermal storage media. These materials are inexpensive, readily available, and can deliver domestically useful temperatures (approximately 64 °C). The "Dover House" (in Dover, Massachusetts) was the first to use a Glauber's salt heating system, in

1948. Solar energy can be stored at high temperatures using molten salts. Salts are an effective storage medium because they are low-cost, have a high specific heat capacity and can deliver heat at temperatures compatible with conventional power systems. The Solar Two used this method of energy storage, allowing it to store 1.44 TJ in its 68 m^3 storage tank with an annual storage efficiency of about 99%. Off-grid PV systems have traditionally used rechargeable batteries to store excess electricity. With grid-tied systems, excess electricity can be sent to the transmission grid. Net metering programmes give these systems a credit for the electricity they deliver to the grid. This credit offsets electricity provided from the grid when the system cannot meet demand, effectively using the grid as a storage mechanism.

Development, Deployment and Economics

Beginning with the surge in coal use which accompanied the Industrial Revolution, energy consumption has steadily transitioned from wood and biomass to fossil fuels. The early development of solar technologies starting in the 1860s was driven by an expectation that coal would soon become scarce, but solar development stagnated in the early 20th century in the face of the increasing availability, economy, and utility of fossil fuels such as coal and petroleum.

The 1973 oil embargo and 1979 energy crisis caused a reorganisation of energy policies around the world and brought renewed attention to developing solar technologies. Deployment strategies focused on incentive programmes such as the Federal Photovoltaic Utilisation Programme in the US and the Sunshine Programme in Japan. Other efforts included the formation of research facilities in the US (SERI, now NREL), Japan (NEDO), and Germany (Fraunhofer Institute for Solar Energy Systems ISE).

Between 1970 and 1983, photovoltaic installations grew rapidly, but dropping oil prices in the early 1980s moderated the growth of PV from 1984 to 1996. Since 1997, PV development has accelerated due to supply issues with oil and natural gas, global warming concerns (see Kyoto Protocol), and the improving economic position of PV relative to other energy technologies. Photovoltaic production growth has averaged 40% per year since 2000 and installed capacity reached 10.6 GW at the end of 2007. Since 2006 it has been economical for investors to install photovoltaics for free in return for a long term power purchase agreement. 50% of commercial systems were installed in this manner in 2007 and it is expected that 90% will by 2009. Nellis Air Force Base is receiving photoelectric power for about 2.2 ¢/kWh and grid power for 9 ¢/kWh. Commercial solar water heaters began appearing in the United States in the 1890s. These systems saw increasing use until the 1920s but were gradually replaced by cheaper and more reliable heating fuels. As with photovoltaics, solar water heating attracted renewed attention as a result of the oil crises in the 1970s but interest subsided in the 1980s due to falling petroleum prices. Development in the solar water heating sector progressed steadily throughout the 1990s and growth rates have averaged 20% per year since 1999. Although generally underestimated, solar water heating is by far the most widely deployed solar technology with an estimated capacity of 154 GW as of 2007.

Commercial concentrating solar power (CSP) plants were first developed in the 1980s. CSP plants such as SEGS project in the United States have a LEC of 12-14 cents/kWh. The 11 MW PS10 power tower in Spain, completed in late 2005, is Europe's first commercial CSP system and a total capacity of 300 MW is expected to be installed in the same area by 2013.

REFERENCES

Amory B. Lovins, *Soft Energy Paths: Toward a Durable Peace*, Penguin Books, 1977.

Amory Lovins *Soft Energy Paths: Towards a Durable Peace* (1977).

Amory Lovins, E. Kyle Datt, *et al. Winning the Oil Endgame* (2005).

Australian Drought and Climate Change, retrieved on June 7th 2007.

Becker, H. 1995. Good earth. Agriculture Research 42(6): 16.

Beychok, Milton R. (2005). *Fundamentals Of Stack Gas Dispersion*, 4th Edition, author-published.

Bosanquet, C.H. and Pearson, J.L., "The spread of smoke and gases from chimneys", Trans. Faraday Soc., 32:1249, 1936

Briggs, G.A., "A plume rise model compared with observations", JAPCA, 15:433-438, 1965

Briggs, G.A., "CONCAWE meeting: discussion of the comparative consequences of different plume rise formulas", Atmos. Envir., 2:228-232, 1968

Briggs, G.A., "Discussion: chimney plumes in neutral and stable surroundings", Atmos. Envir., 6:507-510, 1972

Briggs, G.A., "Plume Rise", USAEC Critical Review Series, 1969

Briggs, G.A., "Some recent analyses of plume rise observation", Proc. Second Internat'l. Clean Air Congress, Academic Press, New York, 1971

Chang, C. H., C.Y. Hsieh, and S.S. Yang. 2001. Effect of cultural media on the phosphate-solubilising activity of thermo-tolerant bacteria. Journal of the Biomass Energy Society of China 20: 79-90.

Chang, F. P. and C.C. Young. 1992. Effects of VA mycorrhizal fungi and phosphprus-solubilising bacteria inoculated on growth of tea cuttings in plastic bag. Taiwan Tea Research Bulletin 11: 79-89.

Chen, I. C., C.H. Chang and S.S. Yang. 2003. Greenhouse gases emissions and microbial populations during the composting. pp. 157-178. In: Yang, S. S. (ed.). Flux and Mitigation of Greenhouse Gases Emissions (IV). Global Change Research Center and Department of Agricultural Chemistry of National Taiwan University, and Institute of Biotechnology of National Pingtung University of Science and Technology, Taiwan.

Chien, S. Y.. 2001. Using microorganisms to promote compost quality. In: Proceedings of the International Workshop on Recent Technologies of Composting and Their Application. Food and Fertilizer Technology Center for the Asian and Pacific Region and Department of Soil and Environmental Sciences of National Chung-Hsing University, Taiwan. pp. 11-1_11-22.

Chung, R. S., M.Y. Yap, H.C. Lin and T.C. Chang. 1993. Effect of composting of corncob on the availability of rock phosphate and the physico-chemical properties of the composts. Journal of the Chinese Agricultural Chemical Society 31: 220-228.

Chung, R. S., S.Y. Yuan, Y.S. Wang and H.C. Lin. 1992. Fertilizer characteristics of cabbage and city refuse compost. Journal of the Biomass Energy of China 11: 21-30.

Climate Change 2001: Working Group II: Impacts, Adaptation and Vulnerability IPCC

Climate change, agricultural policy and poverty reduction – how much do we know?. Overseas Development Institute (2007).

Colacicco, D.. 1982. Economic aspects of composting. BioCycle 23(5): 26-30.

Compost Marketing in Switzerland Schliess, K. 2002. Kompostvermarktung (in German) Report to the Swiss Agency for Environment, Bern.

Daudin, D. and Michelot, P. 1985. Experimental use of organic by-products as culture substrates. In: Gasser, J.K.R. (ed.). Composting of Agricultural and Other Wastes. Elsevier Applied Science Publishers, London, pp. 216-226.

dela Cruz, R. E.. 1994. Status of microbial fertilizers in the Philippines. In: International Seminar on the Use of Micro Organic Fertilizers for Agricultural Production. Rural Development Administration of Korea and Food and Fertilizer Technology Center, Suwon, Korea. pp. 279-306.

E & A Environmental Consultants, Inc. 1990. Results of a growth trial examining the pytotoxicity of eucalyptus-swedge sludge compost and sawdust-sewage sludge. Report to Las Virgenes Municipal WaterDistrict. Las Virgenes, CA.

E & A Environmental Consultants, Inc. 1994. In-house data.

Espiritu, B. M. and dela Torre, J. C. 2001. Combination of microbial inoculants and compost in rice production. pp. 12-1_12-16. In: Proceedings of the International Workshop on Recent Technologies of Composting and Their Application. Food and Fertilizer Technology Center for the Asian and Pacific Region and Department of Soil and Environmental Sciences of National Chung-Hsing University, Taiwan.

European Compost Network European Compost Network

Evangelista, P. 2001. Organic-based fertilisation for rice production in the Philippines. pp. 9-1_9-21. In: Proceedings of the International Workshop on Recent Technologies of Composting and Their Application. Food and Fertilizer Technology Center for the Asian and Pacific Region and Department of Soil and Environmental Sciences of National Chung-Hsing University, Taiwan.

Federal Service for Hydrometeorology and Environmental Monitoring 5Roshydromet), Strategic

Forecast of Climate Change in the Russian Federation 2010–2015 and Its Impact on Sectors of the Russian Economy (Moscow 2005)

Fischer G., Shah M. and van Velthuizen H. (2002) "Climate Change and Agricultural Vulnerability". International Institute for Applied Systems Analysis. Report prepared under UN Institutional Contract Agreement 1113 for World Summit on Sustainable Development. Laxenburg, Austria

Fu, C. M. and Chen, S. Y. 1990. Characteristic and utilisation of excess sludge compost on the pig farm. Journal of the Biomass Energy Society of China 9: 137-143.

German Research Indicates Warming in Siberia, Global Warming Today, Global Warming Today

Global warming 'will hurt Russia', 14:23 03 October 2003, NewScientist.com news service

Hara, M. 2001. Improvement of physical and chemical properties of composts and its application. pp. 15-1_9-13. In: Proceedings of the International Workshop on Recent Technologies of Composting and Their Application. Food and Fertilizer Technology Center for the Asian and Pacific Region and Department of Soil and Environmental Sciences of National Chung-Hsing University, Taiwan.

Harada, Y. 1995. The Composting of Animal Wastes. Extension Bulletin No. 408, Food and Fertilizer Technology Center for the Asian and Pacific Region, Taipei, Taiwan.

Hsieh, C. F. and K.N. Hsu. 1993. A survey on the mineral nutrient content of the different kinds of organic material in Taiwan. In: Research Report No. 302. Taichung District Agricultural Improvement Station, Changhua, Taiwan.

Huang, C. M., Lee, S. W. and Chen, W. C. 1991. New method for hog waste composting. pp. 77-90. In: Proceedings of Symposium on Pig Waste Treatment and Composting. II. The Biomass Energy Society of China. Taipei, Taiwan.

Huang, C. M., W.C. Chen, H.K. Sheen, S.J. Chang, S.W. Lee and L.H. Wang. 1994. Mass production of bagasse compost in an economic way. pp. 122-130. In: Proceedings of Symposium on Technology and Utilisation of Composting. The Biomass Energy Society of China. Taipei, Taiwan.

Huang, S. N. and Lin, C. C. 2001. Current status of organic materials recycling in southern Taiwan. pp. 14-1_9-25. In: Proceedings of the International Workshop on Recent Technologies of Composting and Their Application. Food and Fertilizer Technology Center for the Asian and Pacific Region and Department of Soil and Environmental Sciences of National Chung-Hsing University, Taiwan.

Husted, S. 1994. Seasonal variation in methane emission from stored slurry and solid wastes. Journal of Environmental Quality 23: 585-592.

Inbar, Y. 1979. Formation of humic substances during the composting of agricultural wastes and characterisation of their physicochemical properties. Ph. D. Dissertation. Hebrew University of Jerusalem, Isreal.

Inoko, A., K. Miyamatsu, K. Sugahara and Y. Harada. 1979. On some organic constituents of city refuse composts produced in Japan. Soil Science and Plant Nutrition 25: 225-234.

Intergovermental Panel on Climate Change Special Report on Emissions Scenarios retrieved 26 Jun 2007

IPCC Technical Summary retrieved 25 June 2007

Iswandi, A.. 1994a. Biofertilizers and their application in Indonesia. pp. 85-116. In: International Seminar on the Use of Micro Organic Fertilizers for Agricultural Production. Rural Development Administration of Korea and Food and Fertilizer Technology Center, Suwon, Korea.

Iswandi, A.. 1994b. Potential of municipal solid waste compost for agriculture in Indonesia. pp. 117-126. In: International Seminar on the Use of Micro Organic Fertilizers for Agricultural Production. Rural Development Administration of Korea and Food and Fertilizer Technology Center, Suwon, Korea.

Itävaara et al. Compost maturity - problems associated with testing. in Proceedings of Composting. Innsbruck Austria 18-21.10.2000

J. Heckman. The Roots of Modern Organic Farming *in* Renew. Agric. and Food Systems 21, 143 (2006).

Juste, C. and B. Pommel. 1977. La Valorisation Agricole de Dechets: 1. Le Compost Urbain. Misistère de la Culture et de l'Environnement, Ministère de l'Agricole, France. pp. 75.

Kang, U. G. and Ha, H. S. 1994. Selection and use of effective strains of Rhizobium in Korea. pp. 127-145. In: International Seminar on the Use of Micro Organic Fertilizers for Agricultural Production. Rural Development Administration of Korea and Food and Fertilizer Technology Center, Suwon, Korea.

Kloepper, J. W., Lifshitz, R. and Zablotowicz, R. M. 1989. Free-living bacterial inocula for enhancing crop productivity. Trends in Biotechnology 7: 39-44.

Knoll, K. H. (1961). Public health and refuse disposal. Compost Science 2: 35-40.

Kubota, H. and Nakasaki, K. 1994. Our composting research since 1975: Composting reaction and product evalution. pp. 1-34. In: Proceedings of Symposium on Technology and Utilisation of Composting. The Biomass Energy Society of China. Taipei, Taiwan.

Lai, C. M., Feng, C. F. and Yang, S. S. 2003. Flux and mitigation of nitrous oxide from composting of livestock wastes. pp. 179-198. In: Yang, S. S. (ed.) Flux and Mitigation of Greenhouse Gases Emissions (IV). Global Change Research Center and Department of Agricultural Chemistry of National Taiwan University, and Institute of Biotechnology of National Pingtung University of Science and Technology, Taiwan.

Let's Recycle.com review of Which? Consumer Safety Group -Gardening Div. report

Lian, S. and Lee, Y. C. 1994. Metal contents of organic manures in Taiwan and the current criteria of regulation. pp. 158-173. In: Proceedings of the Symposium on Soil and Fertilizer Pollution. Chinese Soil and Fertilizer Society, Taichung, Taiwan.

Lin, J. K. 1994. Effect of hog waste compost on the soil microflora and activity in different environmental conditions. In: Final Report of Soil and Fertilizer Research 1993. Department of Agriculture and Forest, Taiwan Provincial Government, Nan-tou, Taiwan.

Lobell DB, Burke MB, Tebaldi C, Mastrandrea MD, Falcon WP, Naylor RL (2008). "Prioritising

climate change adaptation needs for food security in 2030". Science 319 (5863): 607–10. doi:10.1126/science.1152339. PMID 18239122.

Morel, J. L., Jacquin, F., Guckert, A. and Barthel, C. 1979. Contribution a la determination de tests de la maturite des compost urbains. Compte-rendu de fin de contrat no 75 124. Ministere de l'Environnement et du Cadre de Vie. ENSAIA. Nancy, France. pp. 32.

Morel, P. and Guillemain, G. 2004. Assessment of the possible phytotoxicity of a substrate using an easy and representative biotest. Acta Horticulture 644:417–423

Morrison, D.E., and Lodwick D.G. (1981). "The social impacts of soft and hard energy systems", *Annual Review of Energy,* 6, 357-378.

Nitta, T. 1994. Roles of organic fertilizers in promoting soil microbiological fertility. pp. 69-84. In: International Seminar on the Use of Micro Organic Fertilizers for Agricultural Production. Rural Development Administration of Korea and Food and Fertilizer Technology Center, Suwon, Korea.

NOAA (2006-01-30). "NOAA reports 2005 global temperature similar to 1998 record warm year". Press release.

Osada, T. 2001. Gas emission from composting animal waste. pp. 4-1_4-15. In: Proceedings of the International Workshop on Recent Technologies of Composting and Their Application. Food and Fertilizer Technology Center for the Asian and Pacific Region and Department of Soil and Environmental Sciences of National Chung-Hsing University, Taiwan.

Osada, T., Kuroda, K. and Yonada, M. 2000. Nitrous oxide, methane and ammonia emissions from composting process of swine waste. The Japanese Society of Wate Management Experts 2: 51-56.

Owa, N. 1994. Microbial material and soil amendment in Japan. pp. 167-174. In: International Seminar on the Use of Micro Organic Fertilizers for Agricultural Production. Rural Development Administration of Korea and Food and Fertilizer Technology Center, Suwon, Korea.

Ozone layer most fragile on record *Brown, Paul* The Guardian April 2005

Parr, J. F. and Hornick, S. B. 1992. Utilisation of municipal wastes. In: Metting, F. B. (ed.) Soil Microbial Ecology: Applications in Agricultural and Environmental Management. Marcel Dekker Inc., New York. pp. 545-559.

Parr, J. F., Hornick, S. B. and Kaufman, D. D. 1994. Use of microbial inoculants and organic fertilizers in agricultural production. pp. 1-40. In: International Seminar on the Use of Micro Organic Fertilizers for Agricultural Production. Rural Development Administration of Korea and Food and Fertilizer Technology Center, Suwon, Korea.

Prihatini, T. and Kurnia, U. 2001. Organic waste management by means of environment friendly technology. pp. 6-1_6-13. In: Proceedings of the International Workshop on Recent Technologies of Composting and Their Application. Food and Fertilizer Technology Center for the Asian and Pacific Region and Department of Soil and Environmental Sciences of National Chung-Hsing University, Taiwan.

Royal Society (2005) Impact of climate change on crops worse than previously thought http://royalsociety.org/news.asp?id=3084

Shamsuddin, Z. H. 1994. Application of microbial inoculants for crop production. pp. 219-231. In: International Seminar on the Use of Micro Organic Fertilizers for Agricultural Production. Rural Development Administration of Korea and Food and Fertilizer Technology Center, Suwon, Korea.

Sheen, S. Y. and Wang, H. H. 1994. Study on stabilisation parameters of hog waste compost. pp. 1-34. In: Proceedings of Symposium on Technology and Utilisation of Composting. TheBiomass Energy Society of China. Taipei, Taiwan.

Sinaga, M. S. 1992. Evaluation of compost to control crop diseases. Agricultural Potential of Citywaste Compost. Research Collaboration between Bogor Agricultural University with Center for Policy and Implementation Studies, Jakarta.

Slade, D.H. (editor), "Meteorology and atomic energy 1968", Air Resources Laboratory, U.S. Dept. of Commerce, 1968

Sommer, S. G., Sibbesen, E., Nielsen, T., Schjoerring, J. K. and Olsen, J. E. 1996. A passive flux samplers for ammonia volatilisation from manure storage facilities. Journal of Environmental Quality 25: 241-247.

Sutton, O.G., "The problem of diffusion in the lower atmosphere", QJRMS, 73:257, 1947 and "The theoretical distribution of airborne pollution from factory chimneys", QJRMS, 73:426, 1947

Swiss Compost Guidelines Swiss Compost Association

Tang, C. C. and Wang, Y. P. 2000. Methane emission and mitigation from the composting and application of livestock wastes. pp. 95-109. In: Yang, S. S. (ed.) Flux and Mitigation of Greenhouse Gases Emissions (II). Global Change Research Center, Department of Agricultural Chemistry and Agricultural Exhibition Hall of National Taiwan University, Taipei, Taiwan.

Tsai, H. S. and S.S. Yang. 2004. Microbial conversion of food waste for biofertilizer preparation with thermophilic and lipolytic microbes. Renewable Energy (in press).

Tseng, C. N., J.J. Hua and C.C. Young. 1993. Studies on gentic relationships among isolates of Rhizopus from Taiwan by Multilocus enzyme electrophoresis. Journal of Chinese Agricultural Chemical Society 31: 120-129.

Turner, D.B. (1994). *Workbook of atmospheric dispersion estimates: an introduction to dispersion modeling*, 2nd Edition, CRC Press.

Um, M. H. and Lee, Y. 2001. Evaluation of organic wastes for composting and quality control of commercial composts in Korea. pp. 8-1_8-18. In: Proceeding of the International Workshop on Recent Technologies of Composting and Their Application. Food and Fertilizer Technology Center for the Asian and Pacific Region and Department of Soil and Environmental Sciences of National Chung-Hsing University, Taiwan.

Wang, L. F., S.W. Lee, C.M. Huang, Y.D. Chung, S.J. Chang and W.C. Chen. 1992. Secondary treatment of hog waste and manufacture of rapid stabilisation compost. pp. 119-127. In: Experiment Report of Taiwan Sugar Company 1991-1992. Research Institute of Taiwan Sugar Company, Tainan, Taiwan.

Yang, S. S. 1994. Composts and agricultural production in Taiwan. Soils and Fertilizers in Taiwan, 1994: 29-62.

Yang, S. S. 1997a. Preparation and characterisation of compost. Journal of the Biomass Energy of China, 16: 47-62.

Yang, S. S. 1997b. Preparation of compost and evaluating its maturity. Food and Fertilizer Technology Center Extension Bulletin 445: 1-23.

Yang, S. S., C.B. Wei, K. Koo and S.S. Tsai. 1991. Food and agricultural wastes produced in Taiwan area. Journal of the Biomass Energy Society of China 10: 70-87.

Yang, S. S., H.T. Yang, W.R. Chen and C.B. Wei. 2001. Properties of livestock and poultry wastes during composting in the farms of northern Taiwan. Journal of the Chinese Environmental Society 24: 8-25.

Young, C. C. 1994. Selection and application of biofertilizers in Taiwan Agriculture. pp. 147-165. In: International Seminar on the Use of Micro Organic Fertilizers for Agricultural Production. Rural Development Administration of Korea and Food and Fertilizer Technology Center, Suwon, Korea.

Zhang, L., F. Chung and M.A. Cole. 1992. A simple chemical assay for estimating compost maturity. ASA Annual Meeting, Minneapolis, MN.

Zucconi, F., A. Monaco and M. Forte. 1985. Phytotoxins during the stabilisation of organic matter. pp. 73-86. In: Gasser, J. K. R. (ed.) Composting of Agricultural and Other Wastes. Elsevier Applied Science Publishers, London.

9

Environmentally Sustainable Science and Technology: Special Reflection w.r.t. Climate Change and Global Warming

Industrial Revolution

The Industrial Revolution began in the late 18th and early 19th centuries in Britain, before spreading around the world. Coal, oil and gas (collectively termed fossil fuels) offered levels of energy production previously undreamed of, leading to shifts towards factory-based systems and the mass production of goods such as cotton. Fossil fuels, principally coal at the beginning of the Industrial Revolution, were primarily used to generate steam power and electricity, but their applications were vast, with many industries becoming automated, hence increasing their output. In the search for a better standard of living, many people moved from the countryside to the cities to find work in the new factories. The burning of fossil fuels led to a massive increase in urban air pollution, although most people felt that such a disadvantage was not significant in the context of their new found prosperity.

In addition to urban air pollution however, other impacts of industrialisation were felt. There were drastic changes to land use with the construction of new buildings, including factories and houses for employees, and transport facilities, including new roads and rail tracks. Areas of countryside were destroyed and replaced by industrial developments. In order to make best use of the remaining land, agricultural machinery was modernised to make the production of food more efficient.

Today, industrialisation continues in the less well developed areas of the world like Africa and parts of Asia. We have gradually become aware that there are many environmental impacts as a consequence of industrialisation, and that we have the ability to take the appropriate action. The main impacts of concern are pollution, resource consumption (including energy resources) and population growth.

Climate Change

Climate change is any long-term significant change in the "average weather" that a given region experiences. Average weather may include average temperature, precipitation and wind patterns. It involves changes in the variability or average state of the atmosphere over durations ranging from decades to millions of years. These changes can be caused by dynamic process on Earth, external forces including variations in sunlight intensity, and more recently by human activities. In recent usage, especially in the context of environmental policy, the term "climate change" often refers to changes in modern climate (see global warming). For information on temperature measurements over various periods, and the data sources available, see temperature record. For attribution of climate change over the past century, see attribution of recent climate change.

Climate Change Factors

Climate changes reflect variations within the Earth's atmosphere, processes in other parts of the Earth such as oceans and ice caps, and the effects of human activity. The external factors that can shape climate are often called climate forcings and include such processes as variations in solar radiation, the Earth's orbit, and greenhouse gas concentrations.

Variations within the Earth's climate

Weather is the day-to-day state of the atmosphere, and is a chaotic non-linear dynamical system. On the other hand, *climate* — the average state of weather — is fairly stable and predictable. Climate includes the average temperature, amount of precipitation, days of sunlight, and other variables that might be measured at any given site. However, there are also changes within the Earth's environment that can affect the climate.

Glaciation

Glaciers are recognised as being among the most sensitive indicators of climate change, advancing substantially during climate cooling (e.g., the Little Ice Age) and retreating during climate warming on moderate time scales. Glaciers grow and collapse, both contributing to natural variability and greatly amplifying externally forced changes. For the last century, however, glaciers have been unable to regenerate enough ice during the winters to make up for the ice lost during the summer months (see glacier retreat).

The most significant climate processes of the last several million years are the glacial and interglacial cycles of the present ice age. Though shaped by orbital variations, the internal responses involving continental ice sheets and 130 m sea-level change certainly played a key role in deciding what climate response would be observed in most regions. Other changes, including Heinrich events, Dansgaard–Oeschger events and the Younger Dryas show the potential for glacial variations to influence climate even in the absence of specific orbital changes.

Ocean Variability

On the scale of decades, climate changes can also result from interaction of the atmosphere and oceans. Many climate fluctuations — including not only the El Niño Southern oscillation (the best known) but also the Pacific decadal oscillation, the North Atlantic oscillation, and the Arctic oscillation — owe their existence at least in part to different ways that heat can be stored in the oceans and move between different reservoirs. On longer time scales ocean processes such as thermohaline circulation play a key role in redistributing heat, and can dramatically affect climate.

The Memory of Climate

More generally, most forms of internal variability in the climate system can be recognised as a form of hysteresis, meaning that the current state of climate reflects not only the inputs, but also the history of how it got there. For example, a decade of dry conditions may cause lakes to shrink, plains to dry up and deserts to expand. In turn, these conditions may lead to less rainfall in the following years. In short, climate change can be a self-perpetuating process because different aspects of the environment respond at different rates and in different ways to the fluctuations that inevitably occur.

Non-Climate Factors Driving Climate Change

Effects of CO2 on Climate Change

Current studies indicate that radiative forcing by greenhouse gases is the primary cause of global warming. Greenhouse gases are also important in understanding Earth's climate history. According to these studies, the greenhouse effect, which is the warming produced as greenhouse gases trap heat, plays a key role in regulating Earth's temperature.

Over the last 600 million years, carbon dioxide concentrations have varied from perhaps >5000 ppm to less than 200 ppm, due primarily to the effect of geological processes and biological innovations. Royer et al. have used the CO_2-climate correlation to derive a value for the climate sensitivity. There are several examples of rapid changes in the concentrations of greenhouse gases in the Earth's atmosphere that do appear to correlate to strong warming, including the Paleocene–Eocene thermal maximum, the Permian–Triassic extinction event, and the end of the Varangian snowball earth event.

During the modern era, the naturally rising carbon dioxide levels are implicated as the primary cause of global warming since 1950. According to the Intergovernmental Panel on Climate Change (IPCC), 2007, the atmospheric concentration of CO_2 in 2005 was 379 ppm^3 compared to the pre-industrial levels of 280 ppm^3. Thermodynamics and Le Chatelier's principle explain the characteristics of the dynamic equilibrium of a gas in solution such as the vast amount of CO_2 held in solution in the world's oceans moving into and returning from the atmosphere. These principles can be observed as bubbles which rise in a pot of water heated on a stove, or in a glass of cold beer allowed to sit at room temperature; gases dissolved in liquids are released under certain circumstances.

Plate Tectonics

On the longest time scales, plate tectonics will reposition continents, shape oceans, build and tear down mountains and generally serve to define the stage upon which climate exists. During the Carboniferous, tectonics may have triggered the large-scale storage of Carbon and increased glaciation. More recently, plate motions have been implicated in the intensification of the present ice age when, approximately 3 million years ago, the North and South American plates collided to form the Isthmus of Panama and shut off direct mixing between the Atlantic and Pacific Oceans.

Solar Variation

The sun is the ultimate source of essentially all heat in the climate system. The energy output of the sun, which is converted to heat at the Earth's surface, is an integral part of shaping the Earth's climate. On the longest time scales, the sun itself is getting brighter with higher energy output; as it continues its main sequence, this slow change or evolution affects the Earth's atmosphere. It is thought that, early in Earth's history, the sun was too cold to support liquid water at the Earth's surface, leading to what is known as the Faint young sun paradox..

On more modern time scales, there are also a variety of forms of solar variation, including the 11-year solar cycle and longer-term modulations. However, the 11-year sunspot cycle does not manifest itself clearly in the climatological data. Solar intensity variations are considered to have been influential in triggering the Little Ice Age, and for some of the warming observed from 1900 to 1950. The cyclical nature of the sun's energy output is not yet fully understood; it differs from the very slow change that is happening within the sun as it ages and evolves..

Orbital Variations

In their effect on climate, orbital variations are in some sense an extension of solar variability, because slight variations in the Earth's orbit lead to changes in the distribution and abundance of sunlight reaching the Earth's surface. Such orbital variations, known as Milankovitch cycles, are a highly predictable consequence of basic physics due to the mutual interactions of the Earth, its moon, and the other planets. These variations are considered the driving factors underlying the glacial and interglacial cycles of the present ice age. Subtler variations are also present, such as the repeated advance and retreat of the Sahara desert in response to orbital precession.

Volcanism

A single eruption of the kind that occurs several times per century can affect climate, causing cooling for a period of a few years. For example, the eruption of Mount Pinatubo in 1991 affected climate substantially. Huge eruptions, known as large igneous provinces, occur only a few times every hundred million years, but can reshape climate for millions of years and cause mass extinctions. Initially, scientists thought that the dust emitted into the atmosphere from large volcanic eruptions was responsible for the cooling by partially blocking the transmission of solar radiation to the Earth's surface. However, measurements indicate that most of the dust thrown in the atmosphere returns to the Earth's surface within six months.

Volcanoes are also part of the extended carbon cycle. Over very long (geological) time periods, they release carbon dioxide from the earth's interior, counteracting the uptake by sedimentary rocks and other geological carbon dioxide sinks. However, this contribution is insignificant compared to the current anthropogenic emissions. The US Geological Survey estimates that human activities generate more than 130 times the amount of carbon dioxide emitted by volcanoes.

Human Influences on Climate Change

Anthropogenic factors are human activities that change the environment and influence climate. In some cases the chain of causality is direct and unambiguous (e.g., by the effects of irrigation on temperature and humidity), while in others it is less clear. Various hypotheses for human-induced climate change have been debated for many years, though it is important to note that the scientific debate has moved on from scepticism, as there is consensus that human activity is beyond reasonable doubt as the main explanation for the current rapid climate changes. Consequently in politics, the debate has largely shifted onto ways to reduce human impact and adapt to change that is already 'in the system.'

The biggest factor of present concern is the increase in CO_2 levels due to emissions from fossil fuel combustion, followed by aerosols (particulate matter in the atmosphere), which exert a cooling effect, and cement manufacture. Other factors, including land use, ozone depletion, animal agriculture and deforestation, also affect climate.

Fossil Fuels

Beginning with the industrial revolution in the 1850s and accelerating ever since, the human consumption of fossil fuels has elevated CO_2 levels from a concentration of ~280 ppm to more than 380 ppm today. These increases are projected to reach more than 560 ppm before the end of the 21st century. It is known that carbon dioxide levels are substantially higher now than at any time in the last 750,000 years. Along with rising methane levels, these changes are anticipated to cause an increase of 1.4–5.6 °C between 1990 and 2100 (see global warming).

Aerosols

Anthropogenic aerosols, particularly sulphate aerosols from fossil fuel combustion, exert a cooling influence. This, together with natural variability, is believed to account for the relative "plateau" in the graph of 20th-century temperatures in the middle of the century.

Cement Manufacture

Cement manufacturing is the third largest cause of man-made carbon dioxide emissions. Carbon dioxide is produced when calcium carbonate ($CaCO_3$) is heated to produce the cement ingredient calcium oxide (CaO, also called *quicklime*). While fossil fuel combustion and deforestation each produce significantly more carbon dioxide (CO_2), cement-making is responsible for approximately 2.5% of total worldwide emissions from industrial sources (energy plus manufacturing sectors).

Land Use

Prior to widespread fossil fuel use, humanity's largest effect on local climate is likely to have resulted from land use. Irrigation, deforestation, and agriculture fundamentally change the environment. For example, they change the amount of water going into and out of a given location. They also may change the local albedo by influencing the ground cover and altering the amount of sunlight that is absorbed. For example, there is evidence to suggest that the climate of Greece and other Mediterranean countries was permanently changed by widespread deforestation between 700 BC and 1 AD (the wood being used for shipbuilding, construction and fuel), with the result that the modern climate in the region is significantly hotter and drier, and the species of trees that were used for shipbuilding in the ancient world can no longer be found in the area.

A controversial hypothesis by William Ruddiman called the early anthropocene hypothesis suggests that the rise of agriculture and the accompanying deforestation led to the increases in carbon dioxide and methane during the period 5000–8000 years ago. These increases, which reversed previous declines, may have been responsible for delaying the onset of the next glacial period, according to Ruddimann's overdue-glaciation hypothesis.

In modern times, a 2007 Jet Propulsion Laboratory study found that the average temperature of California has risen about 2 degrees over the past 50 years, with a much higher increase in urban areas. The change was attributed mostly to extensive human development of the landscape.

Livestock

According to a 2006 United Nations report, Livestock's Long Shadow, livestock is responsible for 18% of the world's greenhouse gas emissions as measured in CO_2 equivalents. This however includes land usage change, meaning deforestation in order to create grazing land. In the Amazon Rainforest, 70% of deforestation is to make way for grazing land, so this is the major factor in the 2006 UN FAO report, which was the first agricultural report to include land usage change into the radiative forcing of livestock. In addition to CO_2 emissions, livestock produces 65% of human-induced nitrous oxide (which has 296 times the global warming potential of CO_2) and 37% of human-induced methane (which has 23 times the global warming potential of CO_2).

Interplay of Factors

If a certain forcing (for example, solar variation) acts to change the climate, then there may be mechanisms that act to amplify or reduce the effects. These are called positive and negative feedbacks. As far as is known, the climate system is generally stable with respect to these feedbacks: positive feedbacks do not "run away". Part of the reason for this is the existence of a powerful negative feedback between temperature and emitted radiation: radiation increases as the fourth power of absolute temperature.

However, a number of important positive feedbacks do exist. The glacial and interglacial cycles of the present ice age provide an important example. It is believed that orbital variations provide the timing for the growth and retreat of ice sheets. However, the ice sheets themselves reflect sunlight back into space and hence promote cooling and their own growth, known as the ice-albedo feedback. Further, falling sea levels and expanding ice decrease plant growth and indirectly lead to declines in carbon dioxide and methane. This leads to further cooling. Conversely, rising temperatures caused, for example, by anthropogenic emissions of greenhouse gases could lead to decreased snow and ice cover, revealing darker ground underneath, and consequently result in more absorption of sunlight.

Water vapor, methane, and carbon dioxide can also act as significant positive feedbacks, their levels rising in response to a warming trend, thereby accelerating that trend. Water vapor acts strictly as a feedback (excepting small amounts in the stratosphere), unlike the other major greenhouse gases, which can also act as forcings.

More complex feedbacks include heat movement from the equatorial regions to the northern latitudes and involve the possibility of altered water currents with in the oceans or air currents with in the atmosphere. A significant concern is that melting glacial ice from Greenland may interfere and change the thermohaline circulation of water in the North Atlantic, affecting the Gulf Stream which brings warmer water to replace sinking colder water; which would change the distribution of heat to Europe and the east coast of the United States.

Other potential feedbacks are not well understood and may either inhibit or promote warming. For example, it is unclear whether rising temperatures promote or inhibit vegetative growth, which could in turn draw down either more or less carbon dioxide. Similarly, increasing temperatures may lead to either more or less cloud cover. Since on balance cloud cover has a strong cooling effect, any change to the abundance of clouds also affects climate.

Monitoring the Current Status of Climate

Testing for spatial dependence between independently measured values in an ordered set is based on applying Fisher's F-test to the variance of a set and the first variance term of the ordered set. Charting statistically significant variance terms gives a sampling variogram that shows where spatial dependence in our sample space of time dissipates into randomness. The lag of a sampling variogram is a statistically robust measure for a change in a climate statistic.

Scientists use "Indicator time series" that represent the many aspects of climate and ecosystem status. The time history provides a historical context. Current status of the climate is also monitored with climate indices.

Evidence for Climatic Change

Evidence for climatic change is taken from a variety of sources that can be used to reconstruct past climates. Most of the evidence is indirect—climatic changes are inferred from changes in indicators that reflect climate, such as vegetation, dendrochronology, ice cores, sea level change, and glacial retreat.

Pollen Analysis

Palynology is the science that studies contemporary and fossil palynomorphs, including pollen. Palynology is used to infer the geographical distribution of plant species, which vary under different climate conditions. Different groups of plants have pollen with distinctive shapes and surface textures, and since the outer surface of pollen is composed of a very resilient material, they resist decay. Changes in the type of pollen found in different sedimentation levels in lakes, bogs or river deltas indicate changes in plant communities; which are dependent on climate conditions.

Beetles

Remains of beetles are common in freshwater and land sediments. Different species of beetles tend to be found under different climatic conditions. Knowledge of the present climatic range of the different species, and of the age of the sediments in which remains are found, allows past climatic conditions to be inferred.

Glacial Geology

Advancing glaciers leave behind moraines and other features that often have datable material in them, recording the time when a glacier advanced and deposited a feature. Similarly, by tephrochronological techniques, the lack of glacier cover can be identified by the presence of datable soil or volcanic tephra horizons. Glaciers are considered one of the most sensitive climate indicators by the IPCC, and their recent observed variations provide a global signal of climate change. See Retreat of glaciers since 1850.

Examples of Climate Change

Climate change has continued throughout the entire history of Earth. The field of paleoclimatology has provided information of climate change in the ancient past, supplementing modern observations of climate.

1. Climate of the deep past
   - o Faint young sun paradox
   - o Snowball earth
   - o Oxygen Catastrophe
2. Climate of the last 500 million years
   - o Phanerozoic overview
   - o Paleocene–Eocene Thermal Maximum
   - o Cretaceous Thermal Maximum
   - o Permo–Carboniferous Glaciation
   - o Ice ages

3. Climate of recent glaciations
   - o Dansgaard–Oeschger event
   - o Younger Dryas
   - o Ice age temperatures
4. Recent climate
   - o Holocene Climatic Optimum
   - o Medieval Warm Period
   - o Little Ice Age
   - o Year Without a Summer
   - o Temperature record of the past 1000 years
   - o Global warming
   - o Hardiness Zone Migration

Climate Change and Biodiversity

The life cycles of many wild plants and animals are closely linked to the passing of the seasons; climatic changes can lead to interdependent pairs of species (e.g. a wild flower and its pollinating insect) losing synchronisation, if, for example, one has a cycle dependent on day length and the other on temperature or precipitation. In principle, at least, this could lead to extinctions or changes in the distribution and abundance of species. One phenomenon is the movement of species northwards in Europe. A recent study by Butterfly Conservation in the UK, has shown that relatively common species with a southerly distribution have moved north, whilst scarce upland species have become rarer and lost territory towards the south. This picture has been mirrored across several invertebrate groups. Drier summers could lead to more periods of drought, potentially affecting many species of animal and plant. For example, in the UK during the drought year of 2006 significant numbers of trees died or showed dieback on light sandy soils. In Australia, since the early 90s, tens of thousands of flying foxes (Pteropus) have died as a direct result of extreme heat. Wetter, milder winters might affect temperate mammals or insects by preventing them hibernating or entering torpor during periods when food is scarce. One predicted change is the ascendancy of 'weedy' or opportunistic species at the expense of scarcer species with narrower or more specialised ecological requirements. One example could be the expanses of bluebell seen in many woodlands in the UK. These have an early growing and flowering season before competing weeds can develop and the tree canopy closes. Milder winters can allow weeds to overwinter as adult plants or germinate sooner, whilst trees leaf earlier, reducing the length of the window for bluebells to complete their life cycle. Organisations such as Wildlife Trust, World Wide Fund for Nature, Birdlife International and the Audubon Society are actively monitoring and research the effects of climate change on biodiversity and advance policies in areas such as landscape scale conservation to promote adaptation to climate change.

Climate Change and Agriculture

Climate change and agriculture are interrelated processes, both of which take place on a global scale. Global warming is projected to have significant impacts on conditions affecting agriculture, including temperature, precipitation and glacial run-off. These conditions determine the carrying capacity of the biosphere to produce enough food for the human population and domesticated animals. Rising carbon dioxide levels would also have effects, both detrimental and beneficial, on crop yields. The overall effect of climate change on agriculture will depend on the balance of these effects. Assessment of the effects of global climate changes on agriculture might help to properly anticipate and adapt farming to maximize agricultural production.

At the same time, agriculture has been shown to produce significant effects on climate change, primarily through the production and release of greenhouse gases such as carbon dioxide, methane, and nitrous oxide, but also by altering the earth's land cover, which can change its ability to absorb or reflect heat and light, thus contributing to radiative forcing. Land use change such as deforestation and desertification, together with use of fossil fuels, are the major anthropogenic sources of carbon dioxide; agriculture itself is the major contributor to increasing methane and nitrous oxide concentrations in earth's atmosphere.

Impact of Climate Change on Agriculture

Despite technological advances, such as improved varieties, genetically modified organisms, and irrigation systems, weather is still a key factor in agricultural productivity, as well as soil properties and natural communities. The effect of climate on agriculture is related to variabilities in local climates rather than in global climate patterns. Consequently, agronomists consider any assessment has to be individually consider each local area.

On the other hand, agricultural trade has grown in recent years, and now provides significant amounts of food, on a national level to major importing countries, as well as comfortable income to exporting ones. The international aspect of trade and security in terms of food implies the need to also consider the effects of climate change on a global scale.

A study published in *Science* suggest that, due to climate change, "southern

Africa could lose more than 30% of its main crop, maize, by 2030. In South Asia losses of many regional staples, such as rice, millet and maize could top 10%".

The 2001 IPCC Third Assessment Report concluded that the poorest countries would be hardest hit, with reductions in crop yields in most tropical and sub-tropical regions due to decreased water availability, and new or changed insect pest incidence. In Africa and Latin America many rainfed crops are near their maximum temperature tolerance, so that yields are likely to fall sharply for even small climate changes; falls in agricultural productivity of up to 30% over the 21st century are projected. Marine life and the fishing industry will also be severely affected in some places. Climate change induced by increasing greenhouse gases is likely to affect crops differently from region to region. For example, average crop yield is expected to drop down to 50% in Pakistan according to the UKMO scenario whereas corn production in Europe is expected to grow up to 25% in optimum hydrologic conditions. More favourable effects on yield tend to depend to a large extent on realisation of the potentially beneficial effects of carbon dioxide on crop growth and increase of efficiency in water use. Decrease in potential yields is likely to be caused by shortening of the growing period, decrease in water availability and poor vernalisation.

In the long run, the climatic change could affect agriculture in several ways :

- *productivity*, in terms of quantity and quality of crops
- *agricultural practices*, through changes of water use (irrigation) and agricultural inputs such as herbicides, insecticides and fertilizers
- *environmental effects*, in particular in relation of frequency and intensity of soil drainage (leading to nitrogen leaching), soil erosion, reduction of crop diversity
- *rural space*, through the loss and gain of cultivated lands, land speculation, land renunciation, and hydraulic amenities.
- *adaptation*, organisms may become more or less competitive, as well as humans may develop urgency to develop more competitive organisms, such as flood resistant or salt resistant varieties of rice.

They are large uncertainties to uncover, particularly because there is lack of information on many specific local regions, and include the uncertainties on magnitude of climate change, the effects of technological changes on productivity, global food demands, and the numerous possibilities of adaptation. Most agronomists believe that agricultural production will be mostly affected by the severity and pace of climate change, not so much by gradual trends in climate. If change is gradual, there may be enough time for biota adjustment. Rapid climate change, however, could harm agriculture in many countries, especially those that are already suffering from rather poor soil and climate conditions, because there is less time for optimum natural selection and adaption.

Shortage in Grain Production

Between 1996 and 2003, grain production has stabilised slightly over 1800 millions of tons. In 2000, 2001, 2002 and 2003, grain stocks have been dropping, resulting in a global grain harvest that was short of consumption by 93 millions of tons in 2003. The earth's average temperature has been rising since the late 1970s, with nine of the 10 warmest years on record occurring since 1995. In 2002, India and the United States suffered sharp harvest reductions because of record temperatures and drought. In 2003 Europe suffered very low rainfall throughout spring and summer, and a record level of heat damaged most crops from the United Kingdom and France in the Western Europe through Ukraine in the East. Bread prices have been rising in several countries in the region. (see w:fr:canicule 2003).

Poverty Impacts

Researchers at the Overseas Development Institute (ODI) have investigated the potential impacts climate change could have on agriculture, and how this would affect attempts at alleviating poverty in the developing world. They argued that the effects from moderate climate change are likely to be mixed for developing countries. However, the vulnerability of the poor in developing countries to short term impacts from climate change, notably the increased frequency and severity of adverse weather events is likely to have a negative impact. This, they say, should be taken into account when defining agricultural policy.

Crop Development Models

Models for climate behavior are frequently inconclusive. In order to further study effects of global warming on agriculture, other types of models, such as *crop development models*, *yield prediction*, quantities of *water or fertilizer consumed*, can be used. Such models condense the knowledge accumulated of the climate, soil, and effects observed of the results of various agricultural practices. They thus could make it possible to test strategies of adaptation to modifications of the environment. Because these models are necessarily simplifying natural conditions (often based on the assumption that weeds, disease and insect pests are controlled), it is not clear whether the results they give will have an *in-field* reality. However, some results are partly validated with an increasing number of experimental results. Other models, such as *insect and disease development* models based on climate projections are also used (for example simulation of aphid reproduction or septoria (cereal fungal disease) development).

Scenarios are used in order to estimate climate changes effects on crop development and yield. Each scenario is defined as a set of meteorological variables, based on generally accepted projections. For example, many models are running simulations based on doubled carbon dioxide projections, temperatures raise ranging from 1°C up to 5°C, and with rainfall levels an increase or decrease of 20%. Other parameters may include humidity, wind, and solar activity. Scenarios of crop models are testing farm-level adaptation, such as sowing date shift, climate adapted species (vernalisation need, heat and cold resistance), irrigation and fertilizer adaptation, resistance to disease. Most developed models are about wheat, maize, rice and soybean.

Temperature Potential Effect on Growing Period

Duration of crop growth cycles are above all, related to temperature. An increase in temperature will speed up development. In the case of an annual crop, the duration between sowing and harvesting will shorten (for example, the duration in order to harvest corn could shorten between one and four weeks). The shortening of such a cycle could have an adverse effect on productivity because senescence would occur sooner.

Potential Eeffect of Atmospheric Carbon Dioxide on Yield

Carbon dioxide is essential to plant growth. Rising CO_2 concentration in the atmosphere can have both positive and negative consequences.

Increased CO_2 is expected to have positive physiological effects by increasing the rate of photosynthesis. Currently, the amount of carbon dioxide in the atmosphere is 380 parts per million. In comparison, the amount of oxygen is 210,000 ppm. This means that often plants may be starved of carbon dioxide, being outnumbered by the photosynthetic pollutant oxygen. The effects of an increase in carbon dioxide would be higher on C3 crops (such as wheat) than on C4 crops (such as maize), because the former is more susceptible to carbon dioxide shortage. Under optimum conditions of temperature and humidity, the yield increase could reach 36%, if the levels of carbon dioxide are doubled. However, other studies also show a change in harvest quality. The growth improvement in C3 plants could favor vegetative biomass on grain biomass; thus leading to a decrease in grain production yield. Further, few studies have looked at the impact of elevated carbon dioxide concentrations on whole farming systems. Most models study the relationship between CO_2 and productivity in isolation from other factors associated with climate change, such as an increased frequency of extreme weather events, seasonal shifts, and so on. Moreover, it is reasonable to expect that weed productivity will increase in parallel with that of crop and pasture plants; potentially raising the cost of defensive expenditures like herbicides.

In 2005, the Royal Society in London concluded that the purported benefits of

CO_2 fertilisation are "likely to be far lower than previously estimated" when factors such as increasing ground-level ozone are taken into account."

Effect on Quality

According to the IPCC's TAR, "The importance of climate change impacts on grain and forage quality emerges from new research. For rice, the amylose content of the grain—a major determinant of cooking quality—is increased under elevated CO_2" (Conroy et al., 1994). Cooked rice grain from plants grown in high-CO2 environments would be firmer than that from today's plants. However, concentrations of iron and zinc, which are important for human nutrition, would be lower (Seneweera and Conroy, 1997). Moreover, the protein content of the grain decreases under combined increases of temperature and CO_2 (Ziska et al., 1997)." Studies have shown that higher CO_2 levels lead to reduced plant uptake of nitrogen (and a smaller number showing the same for trace elements such as zinc) resulting in crops with lower nutritional value. This would primarily impact on populations in poorer countries less able to compensate by eating more food, more varied diets, or possibly taking supplements. Reduced nitrogen content in grazing plants has also been shown to reduce animal productivity in sheep, which depend on microbes in their gut to digest plants, which in turn depend on nitrogen intake.

Agricultural Surfaces and Climate Changes

Climate change is likely to increase the amount of arable land in high-lattitude region by reduction of the amount of frozen lands. A 2005 study reports that temperature in siberia has increased three degree celcius in average since 1960 (much more than the rest of the world). However, reports about the impact of global warming on russian agriculture indicate conflicting probable effects : while they expect an northward extension of farmable lands, they also warn of possible productivity losses and increased risk of drought. Sea levels are expected to get up to one meter higher by 2100, though this projection is disputed. A rise in the sea level would result in an agricultural land loss, in particular in areas such as South East Asia. Erosion, submergence of shorelines, salinity of the water table due to the increased sea levels, could mainly affect agriculture through inundation of low-lying lands.

Erosion and Fertility

With global warming, soil degradation is more likely to occur, and soil fertility would probably be affected by global warming. However, because the ratio of carbon to nitrogen is a constant, a doubling of carbon is likely to imply a higher storage of nitrogen in soils as nitrates, thus providing higher fertilising elements for plants, providing better yields. The average needs for nitrogen could decrease, and give the opportunity of changing often costly fertilisation strategies.

Due to the extremes of climate that would result, the increase in precipitations would probably result in greater risks of erosion, whilst at the same time providing soil with better hydration, according to the intensity of the rain. The possible evolution of the organic matter in the soil is a highly contested issue: while the increase in the temperature would induce a greater rate in the production of minerals, lessening the soil organic matter content, the atmospheric CO_2 concentration would tend to increase it.

Potential Effects of Global Climate Change on Pests, Diseases and Weeds

A very important point to consider is that weeds would undergo the same acceleration of cycle as cultivated crops, and would also benefit from carbonaceous fertilisation. Since most weeds are C3 plants, they are likely to compete even more than now against C4 crops such as corn. However, on the other hand, some results make it possible to think that weedkillers could gain in effectiveness with the temperature increase.

Global warming would cause an increase in rainfall in some areas, which would lead to an increase of atmospheric humidity and the duration of the wet seasons. Combined with higher temperatures, these could favor the development of fungal diseases. Similarly, because of higher temperatures and humidity, there could be an increased pressure from insects and disease vectors.

Glacier Retreat and Disappearance

The continued retreat of glaciers will have a number of different quantitative impacts. In areas that are heavily dependent on water runoff from glaciers that melt during the warmer summer months, a continuation of the current retreat will eventually deplete the glacial ice and substantially reduce or eliminate runoff. A reduction in runoff will affect the ability to irrigate crops and will reduce summer stream flows necessary to keep dams and reservoirs replenished.

According to a UN climate report, the Himalayan glaciers that are the principal dry-season water sources of Asia's biggest rivers - Ganges, Indus, Brahmaputra, Yangtze, Mekong, Salween and Yellow - could disappear by 2035 as temperatures rise. Approximately 2.4 billion people live in the drainage basin of the Himalayan rivers. India, China, Pakistan, Afghanistan, Bangladesh, Nepal and Myanmar could experience floods followed by severe droughts in coming decades. In India alone, the Ganges provides water for drinking and farming for more than 500 million people.

Ozone and UV-B

Some scientists think agriculture could be affected by any decrease in stratospheric ozone, which could increase biologically dangerous ultraviolet radiation B. Excess ultraviolet radiation B can directly effect plant physiology and cause massive amounts of mutations, and indirectly through changed pollinator behavior, though such changes are difficult to quantify. However, it has not yet been ascertained whether an increase in greenhouse gases would decrease stratospheric ozone levels.

In addition, a possible effect of rising temperatures is significantly higher levels of ground-level ozone, which would substantially lower yields.

Impact of Agriculture on Climate Change

The agricultural sector is a driving force in the gas emissions and land use effects thought to cause climate change. In addition to being a significant user of land and consumer of fossil fuel, agriculture contributes directly to greenhouse gas emissions through practices such as rice production and the raising of livestock; according to the Intergovernmental Panel on Climate Change, the three main causes of the increase in greenhouse gases observed over the past 250 years have been fossil fuels, land use, and agriculture.

Land Use

Agriculture contributes to greenhouse gas increases through land use in four main ways:

- CO_2 releases linked to deforestation
- Methane releases from rice cultivation
- Methane releases from enteric fermentation in cattle
- Nitrous oxide releases from fertilizer application

Together, these agricultural processes comprise 54% of methane emissions, roughly 80% of nitrous oxide emissions, and virtually all carbon dioxide emissions tied to land use.

The planet's major changes to land cover since 1750 have resulted from deforestation in temperate regions: when forests and woodlands are cleared to make room for fields and pastures, the albedo of the affected area increases, which can result in either warming or cooling effects, depending on local conditions. Deforestation also affects regional carbon reuptake, which can result in increased concentrations of CO_2, the dominant greenhouse gas. Land-clearing methods such as slash and burn compound these effects by burning biomatter, which directly releases greenhouse gases and particulate matter such as soot into the air.

Livestock

Livestock and livestock-related activities such as deforestation and increasingly fuel-intensive farming practices are responsible for over 18% of human-made greenhouse gas emissions, including:

- 9% of global carbon dioxide emissions
- 35-40% of global methane emissions (chiefly due to enteric fermentation and manure)
- 64% of global nitrous oxide emissions (chiefly due to fertilizer use.)

Livestock activities also contribute disproportionately to land-use effects, since crops such as corn and alfalfa are cultivated in order to feed the animals. Worldwide, livestock production occupies 70% of all land used for agriculture, or 30% of the land surface of the Earth.

Kyoto Protocol

The Kyoto Protocol is a protocol to the international Framework Convention on Climate Change with the objective of reducing greenhouse gases that cause climate change.

It was adopted on 11 December 1997 by the 3rd Conference of the Parties, which was meeting in Kyoto, and it entered into force on 16 February 2005. As of June 2008, 182 parties have ratified the protocol. Of these, 36 developed cg countries (plus the EU as a party in its own right) are required to reduce greenhouse gas emissions to the levels specified for each of them in the treaty (representing over 61.6% of emissions from Annex I countries), with three more countries intending to participate. One hundred thirty-seven (137) developing countries have ratified the protocol, including Brazil, China and India, but have no obligation beyond monitoring and reporting emissions. The United States has not ratified the treaty. Among various experts, scientists, and critics, there is debate about the usefulness of the protocol, and there have been cost-benefit studies performed on its usefulness.

Description

The Kyoto Protocol is an agreement made under the United Nations Framework Convention on Climate Change (UNFCCC). Countries that ratify this protocol commit to reducing their emissions of carbon dioxide and five other greenhouse gases (GHG), or engaging in emissions trading if they maintain or increase emissions of these green house gases.

The Kyoto Protocol now covers more than 170 countries globally but only 60% of countries in terms of global greenhouse gas emissions. As of December 2007, the US and Kazakhstan are the only signatory nations not to have ratified the act. The first commitment period of the Kyoto Protocol ends in 2012, and international talks began in May 2007 on a subsequent commitment period.

At its heart, the Kyoto Protocol establishes the following principles:

- Kyoto is underwritten by governments and is governed by global legislation enacted under the UN's aegis.
- Governments are separated into two general categories: developed countries, referred to as Annex I countries (who have accepted greenhouse gas emission reduction obligations and must submit an annual greenhouse gas inventory), and developing countries, referred to as Non-Annex I countries (who have no greenhouse gas emission reduction obligations but may participate in the Clean Development Mechanism).
- Any Annex I country that fails to meet its Kyoto obligation will be penalised by having to submit 1.3 emission allowances in a second commitment period for every ton of greenhouse gas emissions they exceed their cap in the first commitment period (i.e., 2008-2012).
- As of January 2008, and running through 2012, Annex I countries have to reduce their greenhouse gas emissions by a collective average of 5% below their 1990 levels (for many countries, such as the EU member states, this corresponds to some 15% below their expected greenhouse gas emissions in 2008). While the average emissions reduction is 5%, national limitations range from an 8% average reduction across the European Union to a 10% emissions increase for Iceland; but, since the EU's member states each have individual obligations, much larger increases (up to 27%) are allowed for some of the less developed EU countries (see below #Increase in greenhouse gas emission since 1990). Reduction limitations expire in 2013.
- Kyoto includes "flexible mechanisms" which allow Annex I economies to meet their greenhouse gas emission limitation by purchasing GHG emission reductions from elsewhere. These can be bought either from financial exchanges, from projects which reduce emissions in non-Annex I economies under the Clean Development Mechanism (CDM), from other Annex 1 countries under the JI, or from Annex I countries with excess allowances. Only CDM Executive Board-accredited Certified Emission Reductions (CER) can be bought and sold in this manner. Under the aegis of the UN, Kyoto established this Bonn-based Clean Development Mechanism Executive Board to assess and approve projects ("CDM Projects") in Non-Annex I economies prior to awarding CERs. (A similar scheme called "Joint Implementation" or "JI" applies in transitional economies mainly covering the former Soviet Union and Eastern Europe).

In practice this means that Non-Annex I economies have no GHG emission restrictions, but when a greenhouse gas emission reduction project (a "Greenhouse Gas Project") is implemented in these countries the project will receive Carbon Credits, which can then be sold to Annex I buyers. These Kyoto mechanisms are in place for two main reasons:

- there were fears that the cost of complying with Kyoto would be expensive for many Annex I countries, especially those countries already home to efficient, low greenhouse gas emitting industries, and high prevailing environmental standards. Kyoto therefore allows these countries to purchase (cheaper) carbon credits on the world market instead of reducing greenhouse gas emissions domestically, and
- this is seen as a means of encouraging Non-Annex I developing economies to reduce greenhouse gas emissions through sustainable development, since doing so is now economically viable because of the investment flows from the sale of Carbon Credits.

All the Annex I economies have established Designated National Authorities to manage their greenhouse gas portfolios under Kyoto. Countries including Japan, Canada, Italy, the Netherlands, Germany, France, Spain and many more are actively promoting government carbon funds and supporting multilateral carbon funds intent on purchasing Carbon Credits from Non-Annex I countries. These government organisations are working closely with their major utility, energy, oil & gas and chemicals conglomerates to try to acquire as many Greenhouse Gas Certificates as cheaply as possible. Virtually all of the Non-Annex I countries have also set up their own Designated National Authorities to manage the Kyoto process (and specifically the "CDM process" whereby these host government entities decide which Greenhouse Gas Projects they do or do not wish to support for accreditation by the CDM Executive Board). The objectives of these opposing groups are quite different. Annex I entities want Carbon Credits as cheaply as possible, whilst Non-Annex I entities want to maximize the value of Carbon Credits generated from their domestic Greenhouse Gas Projects.

Objectives

The objective is to achieve "stabilisation of greenhouse gas concentrations in the atmosphere at a level that would prevent dangerous anthropogenic interference with the climate system."

The Intergovernmental Panel on Climate Change (IPCC) has predicted an average global rise in temperature of 1.4°C (2.5°F) to 5.8°C (10.4°F) between 1990 and 2100.

Proponents also note that Kyoto is a first step as requirements to meet the

UNFCCC will be modified until the objective is met, as required by UNFCCC Article 4.2(d).

Status of the Agreement

The treaty was negotiated in Kyoto, Japan in December 1997, opened for signature on March 16, 1998, and closed on March 15, 1999. The agreement came into force on February 16, 2005 following ratification by Russia on November 18, 2004. As of April 2008, a total of 178 countries and other governmental entities have ratified the agreement (representing over 61.6% of emissions from Annex I countries).

According to article 25 of the protocol, it enters into force "on the ninetieth day after the date on which not less than 55 Parties to the Convention, incorporating Parties included in Annex I which accounted in total for at least 55% of the total carbon dioxide emissions for 1990 of the Parties included in Annex I, have deposited their instruments of ratification, acceptance, approval or accession." Of the two conditions, the "55 parties" clause was reached on May 23, 2002 when Iceland ratified. The ratification by Russia on 18 November 2004 satisfied the "55%" clause and brought the treaty into force, effective February 16, 2005.

Details of the Agreement

According to a press release from the United Nations Environment Programme:

> "The Kyoto Protocol is an agreement under which industrialised countries will reduce their collective emissions of greenhouse gases by 5.2% compared to the year 1990 (but note that, compared to the emissions levels that would be expected by 2010 without the Protocol, this limitation represents a 29% cut). The goal is to lower overall emissions of six greenhouse gases - carbon dioxide, methane, nitrous oxide, sulfur hexafluoride, hydrofluorocarbons, and perfluorocarbons - averaged over the period of 2008-2012. National limitations range from 8% reductions for the European Union and some others to 7% for the US, 6% for Japan, 0% for Russia, and permitted increases of 8% for Australia and 10% for Iceland.

It is an agreement negotiated as an amendment to the United Nations Framework Convention on Climate Change (UNFCCC, which was adopted at the Earth Summit in Rio de Janeiro in 1992). All parties to the UNFCCC can sign or ratify the Kyoto Protocol, while non-parties to the UNFCCC cannot. The Kyoto Protocol was adopted at the third session of the Conference of Parties to the UNFCCC (COP3) in 1997 in Kyoto, Japan. Most provisions of the Kyoto Protocol apply to developed countries, listed in Annex I to the UNFCCC. Emission figures exclude international aviation and shipping.

Common but Differentiated Responsibility

The United Nations Framework Convention on Climate Change agreed to a set of a "common but differentiated responsibilities." The parties agreed that:

1. the largest share of historical and current global emissions of greenhouse gases has originated in developed countries;
2. per capita emissions in developing countries are still relatively low, and
3. the share of global emissions originating in developing countries will grow to meet their social and development needs.

In other words, China, India, and other developing countries were not included in any numerical limitation of the Kyoto Protocol because they were not the main contributors to the greenhouse gas emissions during the pre-treaty industrialisation period. However, even without the commitment to reduce according to the Kyoto target, developing countries do share the common responsibility that all countries have in reducing emissions.

Financial Commitments

The Protocol also reaffirms the principle that developed countries have to pay billions of dollars, and supply technology to other countries for climate-related studies and projects. This was originally agreed in the UNFCCC.

Emissions Trading

Kyoto is a 'cap and trade' system that imposes national caps on the emissions of Annex I countries. On average, this cap requires countries to reduce their emissions 5.2% below their 1990 baseline over the 2008 to 2012 period. Although these caps are national-level commitments, in practice most countries will devolve their emissions targets to individual industrial entities, such as a power plant or paper factory. One example of a 'cap and trade' system is the 'EU ETS'. Other schemes may follow suit in time. This means that the ultimate buyers of credits are often individual companies that expect their emissions to exceed their quota (their Assigned Allocation Units, AAUs or 'allowances' for short). Typically, they will purchase credits directly from another party with excess allowances, from a broker, from a JI/CDM developer, or on an exchange.

National governments, some of whom may not have devolved responsibility for meeting Kyoto obligations to industry, and that have a net deficit of allowances, will buy credits for their own account, mainly from JI/CDM developers. These deals are occasionally done directly through a national fund or agency, as in the case of the Dutch government's ERUPT programme, or via collective funds such as the World Bank's Prototype Carbon Fund (PCF). The PCF, for example, represents a consortium of six governments and 17 major utility and energy companies on whose behalf it purchases Credits.

Since allowances and carbon credits are tradeable instruments with a transparent price, financial investors can buy them on the spot market for speculation purposes, or link them to futures contracts. A high volume of trading in this secondary market helps price discovery and liquidity, and in this way helps to keep down costs and set a clear price signal in CO2 which helps businesses to plan investments. This market has grown substantially, with banks, brokers, funds, arbitrageurs and private traders now participating in a market valued at about $60 billion in 2007. Emissions Trading PLC, for example, was floated on the London Stock Exchange's AIM market in 2005 with the specific remit of investing in emissions instruments.

Although Kyoto created a framework and a set of rules for a global carbon market, there are in practice several distinct schemes or markets in operation today, with varying degrees of linkages among them. Kyoto enables a group of several Annex I countries to join together to create a market-within-a-market. The EU elected to be treated as such a group, and created the EU Emissions Trading Scheme (ETS). The EU ETS uses EAUs (EU Allowance Units), each equivalent to a Kyoto AAU. The scheme went into operation on 1 January 2005, although a forward market has existed since 2003.

The UK established its own learning-by-doing voluntary scheme, the UK ETS, which ran from 2002 through 2006. This market existed alongside the EU's scheme, and participants in the UK scheme have the option of applying to opt out of the first phase of the EU ETS, which lasts through 2007. The sources of Kyoto credits are the Clean Development Mechanism (CDM) and Joint Implementation (JI) projects. The CDM allows the creation of new carbon credits by developing emission reduction projects in Non-Annex I countries, while JI allows project-specific credits to be converted from existing credits within Annex I countries. CDM projects produce Certified Emission Reductions (CERs), and JI projects produce Emission Reduction Units (ERUs), each equivalent to one AAU. Kyoto CERs are also accepted for meeting EU ETS obligations, and ERUs will become similarly valid from 2008 for meeting ETS obligations (although individual countries may choose to limit the number and source of CER/JIs they will allow for compliance purposes starting from 2008). CERs/ERUs are overwhelmingly bought from project developers by funds or individual entities, rather than being exchange-traded like allowances. Since the creation of Kyoto instruments is subject to a lengthy process of registration and certification by the UNFCCC, and the projects themselves require several years to develop, this market is at this point largely a forward market where purchases are made at a discount to their equivalent currency, the EUA, and are almost always subject to certification and delivery (although up-front payments are sometimes made). According to IETA, the market value of CDM/JI credits transacted in 2004 was EUR 245 m; it is estimated that more than EUR 620 m worth of credits were transacted in 2005.

Several non-Kyoto carbon markets are in existence or being planned, and these are likely to grow in importance and numbers in the coming years. These include the New South Wales Greenhouse Gas Abatement Scheme, the Regional Greenhouse Gas Initiative and Western Climate Initiative in the United States, the Chicago Climate Exchange and the State of California's recent initiative to reduce emissions. These initiatives, taken together may create a series of partly-linked markets, rather than a single carbon market. The common theme across most of them is the adoption of market-based mechanisms centered on carbon credits that represent a reduction of CO_2 emissions. The fact that some of these initiatives have similar approaches to certifying their credits makes it conceivable that carbon credits in one market may in the long run be tradeable in other schemes. This would broaden the current carbon market far more than the current focus on the CDM/JI and EU ETS domains. An obvious precondition, however, is a realignment of penalties and fines to similar levels,since these create an effective ceiling for each market.

Revisions

The protocol left several issues open to be decided later by the sixth Conference of Parties (COP). COP6 attempted to resolve these issues at its meeting in the Hague in late 2000, but was unable to reach an agreement due to disputes between the European Union on the one hand (which favoured a tougher agreement) and the United States, Canada, Japan and Australia on the other (which wanted the agreement to be less demanding and more flexible). In 2001, a continuation of the previous meeting (COP6bis) was held in Bonn where the required decisions were adopted. After some concessions, the supporters of the protocol (led by the European Union) managed to get Japan and Russia in as well by allowing more use of carbon dioxide sinks. COP7 was held from 29 October 2001 through 9 November 2001 in Marrakech to establish the final details of the protocol.

The first Meeting of the Parties to the Kyoto Protocol (MOP1) was held in Montreal from November 28 to December 9, 2005, along with the 11th conference of the Parties to the UNFCCC (COP11). See United Nations Climate Change Conference. The 3rd of December 2007, Australia ratified the protocol during the first day of the COP13 in Bali.

Enforcement

If the Enforcement Branch determines that an Annex I country is not in compliance with its emissions limitation, then that country is required to make up the difference plus an additional 30%. In addition, that country will be suspended from making transfers under an emissions trading programme.

Current Positions of Governments

Australia

Despite Australia being one of the biggest emitters on a per capita basis, the country was granted a limitation of an 8% increase. This was because of considerations specified in Article 4, section 8(h) of the Convention. The Australian Prime Minister at the time, John Howard (Liberal Party), declined to ratify the Agreement, arguing that the protocol would cost Australians jobs, due to countries with booming economies and massive populations such as China and India not having any reduction obligations. Further, it was claimed that Australia was already doing enough to cut emissions; having pledged $300 million to reduce Greenhouse gas emissions over three years.

Australia's new government formed by the Australian Labor Party after the November 2007 election fully supports the protocol and Prime Minister Kevin Rudd signed the instrument of ratification immediately after assuming office on 3 December 2007, just before the meeting of the UN Framework Convention on Climate Change; it took effect in March, 2008. When still in Opposition, Kevin Rudd commissioned Professor Ross Garnaut to report into the economic issues of reducing greenhouse gas emissions. Garnaut's report is due to be handed to the Australian Government in September 2008, with a draft in June 2008.

Analysis has projected Australia's greenhouse gas emissions at 109% of the 1990 emissions level over the period 2008–12, calculated including the effects of Land use, land-use change and forestry (LULUCF). This is slightly above its 108% Kyoto Protocol limitation. As of 2007, the UNFCCC is reporting that Australia's 2004 greenhouse gas emissions were at 125.6% of 1990 levels, calculated without the LULUCF correction. http://unfccc.int/files/inc/graphics/image/gif/graph3_2007_ori.gif The previous Australian Government, along with the United States, agreed to sign the Asia Pacific Partnership on Clean Development and Climate at the ASEAN regional forum on 28 July 2005. Furthermore, the Australian state of New South Wales (NSW) commenced The NSW Greenhouse Gas Abatement Scheme (GGAS). This mandatory greenhouse gas emissions trading scheme commenced on 1 January 2003 and is currently being trialled by the state government in NSW alone. Uniquely this scheme allows Accredited Certificate Providers (ACP) to trade emissions from householders in the state. As of 2006 the scheme is still in place despite the outgoing Prime Minister's clear dismissal of emissions trading as a credible solution to climate change. Following the example of NSW, the National Emissions Trading Scheme (NETS) has been established as an initiative of State and Territory Governments of Australia, all of which have Labor Party governments. The focus of

NETS is to bring into existence an intra-Australian carbon trading scheme and to coordinate policy developments to this end. According to the Constitution of Australia, environmental matters are under the jurisdiction of the States, and the NETS is intended to facilitate ratification of the Kyoto Protocol by the incoming Labor Government.

Greenpeace have called Clause 3.7 of the Kyoto Protocol the *"Australia Clause"*, as Australia was the major beneficiary. The clause allows for Annex 1 countries with high rates of land clearing in 1990 to consider that year a base level. Greenpeace argues that Australia had extremely high levels of land clearing in 1990, and that this meant that Australia's "baseline" was unusually high compared to other countries.

Canada

On December 17, 2002, Canada ratified the treaty that came into force in February 2005, requiring it to reduce emissions to 6% below 1990 levels during the 2008-2012 commitment period. At that time, numerous polls showed support for the Kyoto protocol at around 70%. Despite strong public support, there was still some opposition, particularly by the Canadian Alliance, precursor to the governing Conservative Party, some business groups, and energy concerns, using arguments similar to those being used in the US. In particular, there was a fear that since US companies would not be affected by the Kyoto Protocol that Canadian companies would be at a disadvantage in terms of trade. In 2005, the result was limited to an ongoing "war of words", primarily between the government of Alberta (Canada's primary oil and gas producer) and the federal government. As of 2003, the federal government claimed to have spent or committed 3.7 billion dollars on climate change programmes. By 2004, CO_2 emissions had risen to 27% above 1990 levels (which compares unfavorably to the 16% increase in emissions by the United States during that time).

In January 2006, a Conservative minority government under Stephen Harper was elected, who previously has expressed opposition to Kyoto, and in particular to the plan to participate in international emission trading. Rona Ambrose, who replaced Stéphane Dion as the environment minister, has since endorsed some types of emission trading, and indicated interest in international trading. On April 25, 2006, Ambrose announced that Canada would have no chance of meeting its targets under Kyoto, and would look to participate in U.S. sponsored Asia-Pacific Partnership on Clean Development and Climate. "We've been looking at the Asia-Pacific Partnership for a number of months now because the key principles around [it] are very much in line with where our government wants to go," Ambrose told reporters. On May 2, 2006, it was reported that environmental funding designed to meet the Kyoto standards had been cut, while the Harper government develops a new plan to take its place. As the co-chair of UN Climate Change Conference in Nairobi in November 2006, Canada and its government received criticism from environmental groups and from other governments for its climate change positions. On January 4, 2007, Rona Ambrose moved from the Ministry of the Environment to become Minister of Intergovernmental Affairs. The Environment portfolio went to John Baird, the former President of the Treasury Board.

Canada's federal government has introduced legislation to set mandatory emissions targets for industry, but it will not take effect until an estimated 2050. The government has since begun working with opposition parties to improve the legislation.

A private member's bill, was put forth by Pablo Rodriguez, Liberal, aiming to force the government to "ensure that Canada meets its global climate change obligations under the Kyoto Protocol." With the support of the Liberals, the New Democratic Party and the Bloc Québécois, and with the current minority situation, the bill passed the House of Commons on 14 February 2007 with a vote of 161-113, and is now being considered by the Senate. If passed, the bill would give the government 60 days to form a detailed plan of action. The government has flatly refused to abide by the bill, which may spark a constitutional crisis, lawsuit, or non-confidence motion once the bill becomes law, as is expected.

In May 2007 Friends of the Earth sued the Canadian federal government for failing to meet its Kyoto Protocol obligations to cut greenhouse gas emissions linked to global warming. This was based on a clause in the Canadian Environmental Protection Act that requires Ottawa to "prevent air pollution that violates an international agreement binding on Canada". Canada's obligation to the treaty began in 2008.

Regardless of the national position, some individual provinces are pursuing policies to restrain emissions, including Quebec and British Columbia and Manitoba as part of the Western Climate Initiative.

People's Republic of China

In 2004 the total greenhouse gas emissions from the People's Republic of China were about 54% of the USA emissions. However, China is now building on average one coal-fired power plant every week, and plans to continue doing so for years. Various predictions see China overtaking the US in total greenhouse emissions between late 2007 and 2010, and according to many other estimates, this already occurred in 2006.

The Chinese government insists that the gas emissions level of any given country is a multiplication of its per capita emission and its population. Because China has put into place population control measures while maintaining low emissions per capita, it claims it should therefore in both of the above aspects be considered a contributor to the world's environment. In addition, the country's energy intensity - measured as energy consumption per unit of GDP - was lowered by 47 per cent between 1991 and 2005; from 1950 to 2002, China's carbon dioxide emissions from fossil sources accounted for only 9.33% of the global total in the same period, and in 2004, its per-capita emission of carbon dioxide from fossil sources was 3.65 tons, which is 87% of the world average and 33 per cent of that of Organisation for Economic Co-operation and Development countries.

In June of 2007, China unveiled a 62-page climate change plan and promised to put climate change at the heart of its energy policies but insisted that developed countries had an "unshirkable responsibility" to take the lead on cutting greenhouse gas emissions and that the "common but differentiated responsibility" principle, as agreed up in the UNFCCC should be applied. In response to critics of the nation's energy policy, China responded that those criticisms were unjust, while studies of carbon leakage suggest that nearly a quarter of China's emissions result from exports for consumption by developed countries.

European Union

On May 31, 2002, all fifteen then-members of the European Union deposited the relevant ratification paperwork at the UN. The EU produces around 22% of global greenhouse gas emissions, and has agreed to a cut, on average, by 8% from 1990 emission levels. On 10 January 2007, the European Commission announced plans for a European Union energy policy that included a unilateral 20% reduction in GHG emissions by 2020.

The EU has consistently been one of the major nominal supporters of the Kyoto Protocol, negotiating hard to get wavering countries on board.

In December 2002, the EU created an emissions trading system in an effort to meet these tough targets. Quotas were introduced in six key industries: energy, steel, cement, glass, brick making, and paper/cardboard. There are also fines for member nations that fail to meet their obligations, starting at •40/ton of carbon dioxide in 2005, and rising to •100/ton in 2008. Current EU projections suggest that by 2008 the EU will be at 4.7% below 1990 levels.

Transport CO_2 emissions in the EU grew by 32% between 1990 and 2004. The share of transport in CO_2 emissions was 21% in 1990, but by 2004 this had grown to 28%.

The position of the EU is not without controversy in Protocol negotiations, however. One criticism is that, rather than reducing 8%, all the EU member countries should cut 15% as the EU insisted a uniform target of 15% for other developed countries during the negotiation while allowing itself to share a big reduction in the former East Germany to meet the 15% goal for the entire EU. Also, emission levels of former Warsaw Pact countries who now are members of the EU have already been reduced as a result of their economic restructuring. This may mean that the region's 1990 baseline level is inflated compared to that of other developed countries, thus giving European economies a potential competitive advantage over the U.S.

Both the EU (as the European Community) and its member states are signatories to the Kyoto treaty.

Greece, however was excluded from the Kyoto Protocol on Earth Day (April 22, 2008) due to unfulfilled commitment of creating the adequate mechanisms of monitoring and reporting emissions, which is the minimum obligation, and delivering false reports by having no other data to report.

Germany

Germany has reduced greenhouse gas emissions by 17.2% between 1990 and 2004. On June 28, 2006, the German government announced it would exempt its coal industry from requirements under the EU internal emission trading system. Claudia Kemfert, an energy professor at the German Institute for Economic Research in Berlin said, "For all its support for a clean environment and the Kyoto Protocol, the cabinet decision is very disappointing. The energy lobbies have played a big role in this decision."

United Kingdom

The energy policy of the United Kingdom fully endorses goals for carbon dioxide emissions reduction and has committed to proportionate reduction in national emissions on a phased basis. The United Kingdom is a signatory to the Kyoto Protocol.

On March 13, 2007, a draft Climate Change Bill was published after cross-party pressure over several years, led by environmental groups. Informed by the Energy White Paper 2003, The Bill aims to put in place a framework to achieve a mandatory 60% cut in the UK's carbon emissions by 2050 (compared to 1990 levels), with an intermediate target of between 26% and 32% by 2020. If approved, the United Kingdom is likely to become the first country to set such a long-range and significant carbon reduction target into law.

The UK currently appears on course to meet its Kyoto limitation for the basket of greenhouse gases, assuming the Government is able to curb rising carbon dioxide emissions between now (2007) and the period 2008-2012. Although the UK's overall greenhouse gas emissions have fallen, annual net carbon dioxide emissions have risen by around 2% since The Labour Party came to power in 1997. As a result it now seems highly unlikely that the Government will be able to honour its manifesto pledge to cut carbon dioxide emissions by 20% from 1990 levels by the year 2010, unless immediate and drastic action is taken under after the passing of the Climate Change Bill.

France

In 2004, France shut down its last coal mine, and now gets 80% of its electricity from nuclear power and therefore has relatively low CO_2 emissions

Norway

Between 1990 and 2006, Norway's carbon emissions increased by almost 8%. As well as directly reducing their own greenhouse gas emissions, Norway's idea for carbon neutrality, is that they will finance reforestation in China, which is still legal under the Kyoto protocol.

India

India signed and ratified the Protocol in August, 2002. Since India is exempted from the framework of the treaty, it is expected to gain from the protocol in terms of transfer of technology and related foreign investments. At the G-8 meeting in June 2005, Indian Prime Minister Manmohan Singh pointed out that the per-capita emission rates of the developing countries are a tiny fraction of those in the developed world. Following the principle of *common but differentiated responsibility*, India maintains that the major responsibility of curbing emission rests with the developed countries, which have accumulated emissions over a long period of time. However, the U.S. and other Western nations assert that India, along with China, will account for most of the emissions in the coming decades, owing to their rapid industrialisation and economic growth.

Russia

Vladimir Putin approved the treaty on November 4, 2004 and Russia officially notified the United Nations of its ratification on November 18, 2004. The issue of Russian ratification was particularly closely watched in the international community, as the accord was brought into force 90 days after Russian ratification (February 16, 2005).

President Putin had earlier decided in favour of the protocol in September 2004, along with the Russian cabinet, against the opinion of the Russian Academy of Sciences, of the Ministry for Industry and Energy and of the then president's economic adviser, Andrey Illarionov, and in exchange to EU's support for Russia's admission in the WTO. As anticipated after this, ratification by the lower (22 October 2004) and upper house of parliament did not encounter any obstacles.

The Kyoto Protocol limits emissions to a percentage increase or decrease from their 1990 levels. Since 1990 the economies of most countries in the former Soviet Union have collapsed, as have their greenhouse gas emissions. Because of this, Russia should have no problem meeting its commitments under Kyoto, as its current emission levels are substantially below its limitations.

It is debatable whether Russia will benefit from selling emissions credits to other countries in the Kyoto Protocol.

United States

The United States (U.S.), although a signatory to the Kyoto Protocol, has neither ratified nor withdrawn from the Protocol. The signature alone is symbolic, as the Kyoto Protocol is non-binding on the United States unless ratified. The United States was, as of 2005, the largest single emitter of carbon dioxide from the burning of fossil fuels. The "Lieberman-Warner Climate Security Act of 2008",also more commonly referred to in the U.S. as the "Cap and Trade Bill", was proposed for greater U.S. allignment with the Kyoto standards and goals. The current bill is almost 500 pages long, and provides for establishment of a federal bureau of Carbon Trading, Regulation, and Enforcement with mandates which some authorities suggest will amount to the largest tax increase in the history of the United States. http://lieberman.senate.gov/documents/amendment.pdf

On July 25, 1997, before the Kyoto Protocol was finalised (although it had been fully negotiated, and a penultimate draft was finished), the U.S. Senate unanimously passed by a 95–0 vote the Byrd-Hagel Resolution (S. Res. 98), which stated the sense of the Senate was that the United States should not be a signatory to any protocol that did not include binding targets and timetables for developing as well as industrialised nations or "would result in serious harm to the economy of the United States". On November 12, 1998, Vice President Al Gore symbolically signed the protocol. Both Gore and Senator Joseph Lieberman indicated that the protocol would not be acted upon in the Senate until there was participation by the developing nations. The Clinton Administration never submitted the protocol to the Senate for ratification.

The Clinton Administration released an economic analysis in July 1998, prepared by the Council of Economic Advisors, which concluded that with emissions trading among the Annex B/Annex I countries, and participation of key developing countries in the "Clean Development Mechanism" — which grants the latter business-as-usual emissions rates through 2012 — the costs of implementing the Kyoto Protocol could be reduced as much as 60% from many estimates. Other economic analyses, however, prepared by the Congressional Budget Office and the Department of Energy, Energy Information Administration (EIA), demonstrated a potentially large loss to GDP from implementing the Protocol of up to 4.2% (EIA).

The current President, George W. Bush, has indicated that he does not intend to submit the treaty for ratification, not because he does not support the Kyoto principles, but because of the exemption granted to China (the world's largest emitter of carbon dioxide). Bush also opposes the treaty because of the strain he believes the treaty would put on the economy; he emphasizes the uncertainties which he believes are present in the climate change issue. Furthermore, the U.S. is concerned with broader exemptions of the treaty. For example, the U.S. does not support the split between Annex I countries and others. Bush said of the treaty:

This is a challenge that requires a 100% effort; ours, and the rest of the world's. The world's second-largest emitter of greenhouse gases is the People's Republic of China. Yet, China was entirely exempted from the requirements of the Kyoto Protocol. India and Germany are among the top emitters. Yet, India was also exempt from Kyoto ... America's unwillingness to embrace a flawed treaty should not be read by our friends and allies as any abdication of responsibility. To the contrary, my administration is committed to a leadership role on the issue of climate change ... Our approach must be consistent with the long-term goal of stabilising greenhouse gas concentrations in the atmosphere."

In June 2002, the United States Environmental Protection Agency (EPA) released the "Climate Action Report 2002". Some observers have interpreted this report as being supportive of the protocol, although the report itself does not explicitly endorse the protocol. At the G-8 meeting in June 2005 administration officials expressed a desire for "practical commitments industrialised countries can meet without damaging their economies". According to those same officials, the United States is on track to fulfill its pledge to reduce its carbon intensity 18% by 2012. The United States has signed the Asia Pacific Partnership on Clean Development and Climate, a pact that allows those countries to set their goals for reducing greenhouse gas emissions individually, but with no enforcement mechanism. Supporters of the pact see it as complementing the Kyoto Protocol while being more flexible, but critics have said the pact will be ineffective without any enforcement measures.

The Administration's position is not uniformly accepted in the U.S. For example, Paul Krugman notes that the target 18% reduction in carbon intensity is still actually an increase in overall emissions. The White House has also come under criticism for downplaying reports that link human activity and greenhouse gas emissions to climate change and that a White House official, former oil industry advocate and current Exxon Mobil officer, Philip Cooney, watered down descriptions of climate research that had already been approved by government scientists, charges the White House denies. Critics point to the Bush administration's close ties to the oil and gas industries. In June 2005, State Department papers showed the administration thanking Exxon executives for the company's "active involvement" in helping to determine climate change policy, including the U.S. stance on Kyoto. Input from the business lobby group Global Climate Coalition was also a factor.

In 2002, Congressional researchers who examined the legal status of the Protocol advised that signature of the UNFCCC imposes an obligation to refrain from undermining the Protocol's object and purpose, and that while the President probably cannot implement the Protocol alone; Congress can create compatible laws on its own initiative.

Local Governments

As of January 18, 2007, eight Northeastern US states are involved in the Regional Greenhouse Gas Initiative (RGGI), which is a state level emissions capping and trading programme. It is believed that the state-level programme will indirectly apply pressure on the federal government by demonstrating that reductions can be achieved without being a signatory of the Kyoto Protocol.

- Participating states: Maine, New Hampshire, Vermont, Connecticut, New York, New Jersey, Delaware, Massachusetts, and Maryland (these states represent over 46 million people).
- Observer states and regions: Pennsylvania, District of Columbia, Rhode Island.

On August 31, 2006, the California Legislature (representing over 33 million Californians) reached an agreement with Governor Arnold Schwarzenegger to reduce the state's greenhouse-gas emissions, which rank at 12th-largest in the world, by 25% by the year 2020. This resulted in the Global Warming Solutions Act which effectively puts California in line with the Kyoto limitations, but at a date later than the 2008-2012 Kyoto commitment period.

As of December 4, 2007, 740 US cities in 50 states, the District of Columbia and Puerto Rico, representing over 76 million Americans support Kyoto after Mayor Greg Nickels of Seattle started a nationwide effort to get cities to agree to the protocol. On October 29, 2007, it was reported that Seattle met their target reduction in 2005, reducing their greenhouse gas emissions by 8 percent since 1990.

Support

Advocates of the Kyoto Protocol state that reducing these emissions is crucially important, as carbon dioxide is causing the earth's atmosphere to heat up. This is supported by attribution analysis. No country has passed national legislation requiring compliance with their treaty obligation. The governments of all of the countries whose parliaments have ratified the Protocol are supporting it. Most prominent among advocates of Kyoto have been the European Union and many environmentalist organisations. The United Nations and some individual nations' scientific advisory bodies (including the G8 national science academies) have also issued reports favoring the Kyoto Protocol. An international day of action was planned for 3 December 2005, to coincide with the Meeting of the Parties in Montreal. The planned demonstrations were endorsed by the Assembly of Movements of the World Social Forum. A group of major Canadian corporations also called for urgent action regarding climate change, and have suggested that Kyoto is only a first step. In the United States, there is at least one student group, Kyoto Now!, which aims to use student interest to support pressure towards reducing emissions as targeted by the Kyoto Protocol compliance.

Opposition

Some public policy experts who are skeptical of global warming, although few, see Kyoto as a scheme to either slow the growth of the world's industrial democracies or to transfer wealth to the third world in what they claim is a global socialism initiative. Others argue the protocol does not go far enough to curb greenhouse emissions (Niue, The Cook Islands, and Nauru added notes to this effect when signing the protocol). Some environmental economists have been critical of the Kyoto Protocol. Many see the costs of the Kyoto Protocol as outweighing the benefits, some believing the standards which Kyoto sets to be too optimistic, others seeing a highly inequitable and inefficient agreement which would do little to curb greenhouse gas emissions. Finally, some economists as Gwyn Prins and Steve Rayner think that an entirely different approach needs to be followed than the approach suggested by the Kyoto-protocol.

Further, there is controversy surrounding the use of 1990 as a base year as well as not using per capita emissions as a basis. Countries had different achievements in energy efficiency in 1990. For example, the former Soviet Union and eastern European countries did little to tackle the problem and their energy efficiency was at its worst level in 1990; the year just before their communist regimes fell. On the other hand, Japan, as a big importer of natural resources, had to improve its efficiency after the 1973 oil crisis and its emissions level in 1990 was better than most developed countries. However, such efforts were set aside, and the inactivity of the former Soviet Union was overlooked and could even generate big income due to the emission trade. There is an argument that the use of per capita emissions as a basis in the following Kyoto-type treaties can reduce the sense of inequality among developed and developing countries alike, as it can reveal inactivities and responsibilities among countries.

Cost-Benefit Analysis

Economists have been trying to analyze the overall net benefit of Kyoto Protocol through cost-benefit analysis. There is disagreement due to large uncertainties in economic variables. Some of the estimates indicate either that observing the Kyoto Protocol is more expensive than not observing the Kyoto Protocol or that the Kyoto Protocol has a marginal net benefit which exceeds the cost of simply adjusting to global warming. However, a study by De Leo *et al.* found that "*accounting only for local external costs, together with production costs, to identify energy strategies, compliance with the Kyoto Protocol would imply lower, not higher, overall costs.*" The recent Copenhagen consensus project found that the Kyoto Protocol would slow down the process of global warming, but have a superficial overall benefit. Defenders of the Kyoto Protocol argue, however, that while the initial greenhouse gas cuts may have little effect, they set the political precedent for bigger (and more effective) cuts in the future. They also advocate commitment to the precautionary principle. Critics point out that additional higher curbs on carbon emission are likely to cause significantly higher increase in cost, making such defense moot. Moreover, the precautionary principle could apply to any political, social, economic or environmental consequence, which might have equally devastating effect in terms of poverty and environment, making the precautionary argument irrelevant. The Stern Review (a UK government sponsored report into the economic impacts of climate change) concluded that one percent of global GDP is required to be invested in order to mitigate the effects of climate change, and that failure to do so could risk a recession worth up to twenty percent of global GDP.

Discount Rates

One problem in attempting to measure the "absolute" costs and benefits of different policies to global warming is choosing a proper discount rate. Over a long time horizon such as that in which benefits accrue under Kyoto, small changes in the discount rate create very large discrepancies between net benefits in various studies. However, this difficulty is generally not applicable to "relative" comparison of alternative policies under a long time horizon. This is because changes in discount rates tend to equally adjust the net cost/benefit of different policies unless there are significant discrepancies of cost and benefit over time horizon.

While it has been difficult to arrive at a scenario under which the net benefits of Kyoto are positive using traditional discounting methods such as the Shadow Price of Capital approach, there is an argument that a much lower discount rate should be utilized; that high rates are biased toward the current generation. This may appear to be a philosophical value judgment, outside the realm of economics, but it could be equally argued that the study of the allocation of resources does include how those resources are allocated over time.

Increase in Greenhouse Gas Emission Since 1990

Below is a list of the change in greenhouse gas emissions from 1990 to 2004 for some countries that are part of the Climate Change Convention as reported by the United Nations. Below is a table of the changes in CO_2 emission of some other countries which are large contributors, but are not required to meet numerical limitations.

Germany	-17%	-18.2%	-21%	-8%
Canada	+27%	+26.6%	N/A	-6%
Australia	+25%	+5.2%	N/A	+8%
Spain	+49%	+50.4%	+15%	-8%
Norway	+10%	-18.7%	N/A	+1%
New Zealand	+21%	+17.9%	N/A	0%
France	-0.8%	-6.1%	0%	-8%
Greece	+27%	+25.3%	+25%	-8%
Ireland	+23%	+22.7%	+13%	-8%
Japan	+6.5%	+5.2%	N/A	-6%
United Kingdom	-14%	-58.8%	-12.5%	-8%
Portugal	+41%	+28.9%	+27%	-8%
EU-15	-0.8%	-2.6%	N/A	-8%

Comparing total greenhouse gas emissions in 2004 to 1990 levels, the US emissions were up by 16%, with irregular fluctuations from one year to another but a general trend to increase. At the same time, the EU group of 23 (EU-23) Nations had reduced their emissions by 5%. In addition, the EU-15 group of nations (a large subset of EU-23) reduced their emissions by 0.8% between 1990 and 2004, while emission rose 2.5% from 1999 to 2004. Part of the increases for some of the European Union countries are still in line with the treaty, being part of the cluster of countries implementation (see objectives in the list above).

As of year-end 2006, the United Kingdom and Sweden were the only EU countries on pace to meet their Kyoto emissions commitments by 2010. While UN statistics indicate that, as a group, the 36 Kyoto signatory countries can meet the 5% reduction target by 2012, most of the progress in greenhouse gas reduction has come from the stark decline in Eastern European countries' emissions after the fall of communism in the 1990s.

Successor

In the non-binding 'Washington Declaration' agreed on February 16, 2007, Heads of governments from Canada, France, Germany, Italy, Japan, Russia, United Kingdom, the United States, Brazil, China, India, Mexico and South Africa agreed in principle on the outline of a successor to the Kyoto Protocol. They envisage a global cap-and-trade system that would apply to both industrialised nations and developing countries, and hoped that this would be in place by 2009. On June 7, 2007, leaders at the 33rd G8 summit agreed that the G8 nations would 'aim to at least halve global CO_2 emissions by 2050'. The details enabling this to be achieved would be negotiated by environment ministers within the United Nations Framework Convention on Climate Change in a process that would also include the major emerging economies. A round of climate change talks under the auspices of the United Nations Framework Convention on Climate Change (UNFCCC) (Vienna Climate Change Talks 2007) concluded in 31 August 2007 with agreement on key elements for an effective international response to climate change. A key feature of the talks was a United Nations report that showed how energy efficiency could yield significant cuts in emissions at low cost. The talks are meant to set the stage for a major international meeting to be held in Nusa Dua, Bali, which started on 3 December, 2007.

Asia Pacific Partnership on Clean Development and Climate

The Asia Pacific Partnership on Clean Development and Climate is an agreement between seven Asia-Pacific nations: Australia, Canada, China, India, Japan, South Korea, and the United States. Between them, these seven countries are responsible for more than half of the world's carbon dioxide emissions.

The partnership had its official launch in January 2006 at a ceremony in Sydney, Australia. The alliance states that member nations have initiated nearly 100 projects aimed at clean energy capacity building and market formation since then. Building on these activities, long-term projects are scheduled to deploy clean energy and environment technologies and services. The pact allows those countries to set arbitrary goals for reducing greenhouse gas emissions individually, without any enforcement mechanism for these goals.

Supporters of the pact see it as "complementing the Kyoto Protocol" whilst being more flexible. Critics have said the pact will be ineffective without any enforcement measures and is a means to undermine the negotiations leading to the Protocol scheduled to replace the current Kyoto Protocol (negotiations started in Montreal in December 2005). U.S. Senator John McCain said the partnership "[amounted] to nothing more than a nice little public relations ploy.", while the Economist described the partnership as "patent fig-leaf for the refusal of America and Australia to ratify Kyoto".

PROTECTION OF THE GLOBAL CLIMATE

By the late 1980s warmings of global warning, from scientists and from green non-governmental organisations (NGOs), had become too loud and extensive to be ignored by governments and politicians around the world. The 'greenhouse' issue had such a strong appeal to media and public opinion that it led to an outpouring of rhetoric and another round in the 'green beauty contest' among major political figures. Decision-makers soon realised, however, that the calls for protection of the atmospheric were not as innocent as they sounded: The actions that might follow could be as disastrous for certain national interests as inaction for the global climate. Utmost caution was necessary, in view of the potentially extremely high stakes.

In this article we shall first briefly review how the international community of states has responded to the 'greenhouse' challenge and then analyse the perceptions and interests underlying the reactions and policy of major governmental actors. To date, very little action has been taken in this field, so we shall refer to public statements outlining the positions of various actors, but primarily rely on our own interpretation of their interests and affectedness, on the basis of available information. Our focus will be on the link between carbon dioxide (CO_2) emissions and production and consumption of energy, as this has been a central point in the international climate debate. We do not intend to analyse the current negotiations on a climate convention, but will instead emphasise the underlying interests, on the one hand, and possible outcomes in the form of a global regime, on the other. In the conclusion we will assess the consequences of the interest configuration for future solutions to the problems of climate change, and present some suggestions for overcoming the obstacles to international co-operation.

International Response

When a new issue with serious implications rises from 'political obscurity to the stage of world politics with astounding rapidity', the instinctive response of governments is to ask for more knowledge—and for time. This can be done in several ways—for instance by requesting more expert advice, or by holding diplomatic conferences to discuss the matter. Recent years have seen hectic activity in both respects: a number of intergovernmental conferences on climate change have taken place and, more importantly, hundreds of scientists—and scores of government officials have been assigned by the UN to assess the relevant issues within the framework of the Intergovernmental Panel on Climate Change.

The Intergovernmental Panel on Climate Change—IPCC

The report from the panel (Working Group 1) established that there already exists a natural greenhouse effect, making the earth warmer than it would otherwise have been. Emissions of greenhouse gases resulting from human activities "are substantially increasing the atmospheric concentration of the greenhouse gases". The scientists stated with certainty that "these increases will enhance the greenhouse effect, resulting on average in an additional warming of the Earth's surface." They calculated with confidence that "carbon dioxide has been responsible for over half the enhanced greenhouse effect, and is likely to remain so in the future", and that the long-lived gases (carbon dioxide, nitrous oxide and the CFCs) "would require immediate reductions in emissions from human activities of over 60 per cent to stabilise their concentrations at today's levels".

Under its "business-as-usual" scenario (in which no special measures are implemented to limit emissions) the IPCC predicted that, by the end of the next century, the global average temperature could increase by 3 degrees Celsius, which in turn could lead to a sea-level rise of 65 centimetres. The panel underlined the wide remaining uncertainties, especially regarding the uptake of carbon dioxide in the oceans and natural feedback mechanisms, both of which are poorly understood. They also stressed that "unequivocal detection of the enhanced greenhouse effect from observations is not likely for a decade or more." Despite aggressive criticism from certain individual scientists, the IPCC consensus still commands wide recognition and respect. It is noteworthy that the climate scientists succeeded in producing an assessment (of a highly politicised issue) that appears both independent and legitimate. Attempts to exert political influence on the scientists invariably failed. A revised assessment under publication in early 1992 reinforces the conclusions of Working Group 1. The scientists now maintain that the man-made greenhouse effect is probably larger than earlier anticipated, but the actual warming has been temporarily mitigated by ozone depletion and sulphur dioxide (SO_2) emissions.

Working Group 2 of the IPCC established that the speed of climate changes will be decisive for how serious and unpredictable the effects will be. The conclusions of the work in this group are fraught with difficulties in dealing with uncertainty. Serious effects of climate change can occur in some areas—for example, reduction in food production in countries and regions like Brazil, Peru, the Sahel belt in Africa, South-East Asia, the Central Asian part of the former Soviet Union, and China. Generally, the potentially productivity increase is expected to be limited mainly to the northern hemisphere. The communities with the least ability to adapt to new climatic conditions will be the ones most vulnerable to the negative consequences of the greenhouse effect.

The projected sea-level rise will lead to a loss of large areas of land. The Maldives, for instance, and low-lying parts of other countries may literally sink into the sea. A rise in the sea level of 30-50 centimetres (envisaged by the year 2050) will threaten low-lying island nations and coastal zones. A rise of 1 metre would render some island states uninhabitable, drive tens of millions of people away from their homes, represent a serious threat to low-lying urban districts, flood productive land, contaminate freshwater sources, and alter coastlines.

The difficulties experienced by Working Group 3 in arriving at concrete proposals for action illustrate how difficult it is to proceed from scientific consensus to political agreement. (Its deliberations were largely controlled by representatives of governments, not scientists). This goes a long way to explain why so little progress was made in identifying policy measures or in solving problems related to burden-sharing between countries. However, the group did recommend that the industrialised countries should adjust their economy to future agreements to reduce emissions and transfer financial resources, appropriate technology, and technical assistance to developing countries. Perhaps the most valuable result from the work of Group 3 was in demonstrating the conflicts of interests that arise when concrete proposals for action are approached.

From Assessment to Negotiation

The IPCC report made it clear that the prospect of climate change could no longer be brushed aside as mere speculation. The consensus among the natural scientists had to be taken seriously by governments: they had to start considering both how their own interest should best be protected, and how the world as a whole could respond. The first indications of such response emerged during the Second World Climate Conference in October-November 1990, only a few months after the first IPCC assessment was concluded. Here, representatives of 137 governments agreed that "the ultimate global objective should be to stabilise greenhouse-gas concentrations at a level that would prevent dangerous anthropogenic interference with climate". They welcomed the commitments made by several OECD countries to stabilise national emissions. As regards all other governments (which remain sceptical of the need of national action) the Ministerial Declaration is full of understanding: other developed countries are urged "to analyse the feasibility of and options for, and as, appropriate in light of these analyses, to develop programmes, strategies and/or targets" for reducing emissions. The need for "adequate and additional financial resources" to be channelled to developing countries is duly noted—as are the special difficulties of the fossil-fuel dependent countries. Finally, the declaration calls for increased assistance to developing low-lying coastal and island countries. This statement illustrates the content of intergovernmental consensus on the greenhouse issue: A lofty principle, without any tangible commitments, plus a recognition of the position expressed by the major groups of states. This was the starting-point for the negotiations that began in early 1991 within the framework of the UN-appointed International Negotiating Committee on climate change (INC). Whatever the eventual outcome of this bargaining process, it has at least given good indications of the positions of major governments—to which we shall now turn.

State Perceptions and Interests: The USA

With by the far the largest national emissions of both greenhouse gases in general and carbon dioxide in particular, the United States holds the key to any future climate policy on the global level. Initially, the Bush Administration was eager to present a positive image on environmental matters, both at home and abroad. Reflecting the president's much-quoted campaign statement—'those who think we are powerless to do something about the greenhouse effect, are forgetting about the White House effect' the administration joined in the overall concern in the late 1980s about threats to the climate. This culminated in the alarmist statement from the annual economic summit of the Group of Seven in 1989, where the major powers "strongly advocate common efforts to limit emissions of carbon dioxide and other greenhouse gases".

Only a year later this harmony was shattered: at the subsequent meeting in Houston in July 1990, the US strongly opposed European attempts to include concrete commitments on abatement measures in the final statement. Since then, US and EC representatives have often opposed each other during international climate discussions. As of the beginning of 1992 the US and Turkey are the only OECD countries which have not made a commitment to stabilise or reduce carbon dioxide emissions. Despite numerous efforts at persuasion from Europe and Japan, the Bush administration has adamantly held on to its overall policy line, which can be summarised as follows.

First, there remains considerable uncertainty about the nature and especially the consequences of future man-made climate change. In several official studies serious attempts have been made to identify and, whenever possible, quantify the impacts on US society of global warning. So far, the bottom line is inconclusive: Much of the estimated damage can either be mitigated with reasonable costs or will be compensated by gains in other areas. The scientific dispute following the publication of the IPCC report has also encouraged the US government. When pressed for a reaction to the IPCC consensus, President Bush has taken refuge by referring to other scientists, who consistently hold a different view.

Secondly, the economic impacts of carbon dioxide stabilisation, not to speak of reductions, are considered prohibitively large. Influential US economists have produced estimates of these costs ranging into hundreds of billion dollars. Again, uncertainty is acute: In a recent report by the congressional Office of Technology Assessment the annual costs of a 35 per-cent reduction of carbon dioxide emissions are estimated as somewhere between plus 20 and minus 150 billion dollars.

Given this wide range of uncertainty, the optimal strategy is seen as being to invest in research but defer major policy action until a sounder scientific basis has been established. The administration has consequently sought a major increase in funding of research on global warming and climate change—upto 1.2 billion dollars for fiscal year 1992, close to half the total research efforts in the world in the field. In an attempt to placate international and domestic criticism the US government has adopted a climate strategy aimed at stabilising emissions of all greenhouse gases taken together. As the US has already made a commitment to phase out CFCs (the second most important greenhouse gas at the moment), this policy will permit an increase of about 15 per cent in carbon dioxide emissions by the year 2000 and 23 per cent by 2005, according to the Environmental Protection Agency. The new findings from IPCC referred above, could seriously undermine this strategy.

US policy was presented as outright rejection of the conclusions drawn by climate scientists within IPCC and the arguments made by environmental NGOs (often with backing from the EC) as long as former White House chief of staff, John Sununu, was in charge of climate policy. Since his departure at the end of 1991, the position of the US government has been presented in more diplomatic terms, but the substance seems basically unchanged. An indiction of the lack of priority accorded to this issue was given in President Bush's State of the Union Address for 1992, where environmental matters were hardly mentioned. It remains to be seen, however, what impact—if any—the presidential election campaign will have on the climate policy of the US government.

Reasons behind US Climate Policy

1. The US economy is based on large, continuous, and relatively cheap supplies of fossil fuels. Energy consumption per capita is more than twice as high as the West European average—and about ten times that of developing countries. Since more than 80 per cent of US energy consumption derives from fossil fuels, per-capita emissions of carbon dioxide area also much higher than in the rest of the world—again twice the level of Western Europe (see Fig. 2). The carbon intensity of the US economy is, however, on a par with Western Europe. With its vast natural resources the US has supplied most of its fossil-fuel requirements from its own sources. However, even this endowment is not infinite: oil production in the United States has fallen consistently since the mid-1980s to a level where it can barely cover one-half of consumption. With a tenuous reserve situation the decline seems bound to continue, even if the rate is uncertain. With consumption fairly stable since the late 1980s, import dependence is set to grow. The reserve situation is not much better for natural gas, even if production is still slowly increasing. In stark contrast, US coal reserves are abundant—one-fourth of world total and large enough to last over 200 years with current production. This is why coal is considered the key to secure long-term energy supplies by the Bush administration—and these concerns prevailed over climate-change considerations in the formulation of the new national energy strategy, made public in 1991.

2. The American way of life is closely linked in with the profligate use of fossil energy. A life-style with emphasis on individual consumerism and high mobility (coupled with long distances) is, at least in the near future, inconceivable without continuous fuel supplies at present levels, or higher. Policy-makers in the government, led by Sununu, camc to regard the environmental argument as a plot designed to "cut off our use of coal, oil and natural gas". The life-style implication of rejecting that argument is equally clear: "I don't think Americans want not to be able to use their automobiles." When the issues are formulated in that manner, the answers are obvious.

3. The overriding economic philosophy of the Bush administration has set clear guide-lines for its approach to ecological problems: Environmental policy "must be consistent with economic growth and free-market principles in all countries". As mentioned above, carbon dioxide stabilisation is considered detrimental to economic development. In addition, any programme to implement a specific emission target lower than what market forces will produce would require political intervention in the form of taxation or industrial policy, and that is anathema to the prevailing ideology in Washington.

4. At the international level, the US government has always been sceptical of grand political bargains between North and South. Ever since the governments of developing countries presented their first demands for large transfers to resources and "a new international economic order", US officials have tended to brand such ideas as either unwarranted interference with world markets or bottomless welfare payments. This goes a long way towards explaining why the US government has yet to accept in practice that global environmental problems warrant hefty new transfers of money to the Third World.

The uncertainty outlined above gives decision-makers considerable room for moneuvre. The climate issues are still so diffuses that a clear-cut, uncontestable cost-benefit analysis is impossible. Official response and subsequent policy reaction will then be determined by ideological reflexes and the mobilisation of interest groups fearful of losing their position in society. Admittedly, there are also counter-trends in US politics—among environmental NGOs, in Congress, and not least at the state level—but these are of less important to the administration, in terms of ideology and affectedness: any gains from reductions of greenhouse gases are long-term, uncertain, and widely scattered, whereas the losses are visible and tangible in the near term for groups that can be clearly identified.

The Former USSR

The Soviet Union used to be the world's largest producer and second-largest consumer of energy. During the Communist era an extremely energy-intensive economy was established, as a result of the forced, centralised model of industrialisation—and the country's rich natural resource endowment. As of the end of 1990, the territories of the former Union held more than one-third of the gas reserves, almost one-fourth of the coal reserves, and 5.6 per cent of the oil reserves in the world. Today, the result of the "Soviet model" of development can be clearly seen in all its striking paradox: a miserable standard of living combined with very high emissions of pollutants. With regard to greenhouse gases, the former Union is the second largest emitter in the world, accounting for 12 per cent of the global total. If we concentrate on carbon dioxide—the major anthropogenic greenhouse gas—the former USSR alone accounts for roughly one-sixth of world emissions (see Fig. 1). The corresponding level of per-capita emissions is well above the West European average, which may make one wonder how this is possible in a country where the standard of living is so much lower. There is, for example, only one car for every twenty persons in the former Union, barely a tenth of the European level. The explanations to this seeming paradox can be found in the economic system with ensuing structural distortions of the economy:

- priority has always been given to physical output. Indeed, the system had no way of measuring efficiency adequately. For the energy industries this has meant larger and larger investments in production, which in turn deprived the energy complex itself and the rest of the economy of resources badly needed for modernisation;
- with prices at artificially low levels and a rigid energy-allocation system that gave very limited freedom of choice to energy consumers, there has been no incentive for energy efficiency in the system—and a corresponding culture of waste and carelessness;
- the military complex has consumed immense volumes of energy (and other natural resources), again without any consideration for economic efficiency.

The Soviet experience indicates that there is hardly any limit to how little welfare for the population an economic system can produce with enormous input of resources, if the system itself is skewed enough. But what happens when the system breaks down? Carbon dioxide emissions from the former Union have declined slightly since 1988, but by no means as much as the fall in industrial production. In the energy sector the legacy of the Communist era is more visible and harder to reform than in other parts of society. The structure of energy consumption is so ingrained in society that it will not be altered without profound changes in the basis of the economic system.

International Climate Policy

At the scientific level, Soviet representatives participated actively in the proceedings of the IPCC. Some of them acquired a reputation for their deviating views as to the consequences of climate change—notably the theory expressed by Professor M. Budyko which involved the notion that global warning might be beneficial to the country. However, it is unclear to what extent that perception has taken root in official circles. Apart from these scientific disputes, the Soviet Union kept a very low profile in international discussions of climate change. It is known that its representatives have been very sceptical about making any concrete commitments concerning future emissions, but gladly leave it to other governments to take the brunt of defending that position in public. The effects, and perhaps the memory, of the rhetoric on "environmental security" in earlier statements from Gorbachev and Shevardnadze, seem to have faded away.

Future Prospects

There is no reason to expect global climate change to have any noticeable impact on the energy policies of the former Soviet republics in the near future. Their governments will for a long time be preoccupied with the far more acute difficulties of economic restructuring. Concern for the global climate in the next century cannot be expected to reach the already crowded agendas of the key decision-makers. Nevertheless, the future direction of economic development in the new Common wealth in Independent States will in itself have major effects of world emissions of greenhouse gases. This follows from the above description of the sheer size of the energy complex. We underline that the economic future of the former Union will be a consequence not of global concerns, but of the outcome of the political bargaining over economic reforms.

In this battle, domestic ecological damage will be a factor to reckon with. The devastating impacts on the natural and human environment of the Soviet way of industrialisation are now well-known. For example, the concentration air pollution (mainly from energy production) had by 1989 reached "a level ten times the maximum permissible in 88 Soviet cities with a combined population of 42 million people". However, the capital and human resources available for urgent environmental tasks remain very scarce.

International environment concerns will carry much less weight than domestic ones. This does not mean that the governments of the new states will be immune to arguments from outside, but they will have to be backed up by money. That would mean one might envisage joint efforts to reduce greenhouse-gas emissions, in particular in Russia (which accounts for roughly two-thirds of the old Union in this respect), that is, financial and technical support from the West.

The European Community (EC)

EC member states account for 12.8 per cent of world energy-related carbon dioxide emissions. Although the per-capita level is less than half of the US, it is still almost twice the world average. Like the US, the European Community embarked upon climate discussions in the late 1980s with resounding declarations of concern for the global environment. But unlike their US counterparts, EC politicians and officials have maintained their basic policy line, at least at the level of rhetoric:

- the Community has made it clear that it intends to play a leading role in international climate negotiations;
- the EC has advocated a global climate regime with considerable powers, and it has accepted an obligation for the industrialised countries to transfer additional resources to developing countries. This corresponds to the active role played by EC member states in setting up and funding the Global Environment Facility. Two-thirds of the core contributions have been pledged by EC governments;
- the Community has committed itself to stabilising its carbon dioxide emissions by the year 2000, at the level of 1990. This objective applies to the member states as a collective, not to individual countries. (In addition, the EC has decided to phase out CFCs earlier than required under the revised Montreal Protocol). It has been indicated that this will be following up at a later stage with more advanced commitments to reduce carbon dioxide emissions.

This ambitious policy requires concrete, internal implementation if the Community is to maintain its credibility in international fora. In fact, key policy-makers within the EC have skillfully exploited the global agenda to bring pressure to bear on reluctant governments and parts of the Eurobureaucracy. Every new international conference or negotiating session is used to speed up the internal process. Had it not been for the global bargaining on climate issues it would have taken the Community much longer to reach common positions on its own greenhouse policy.

However, as the policy discussion approaches the phase where concrete measures must be formulated and decided upon, opposition and controversy arise: the EC Commission has concluded that the stabilisation objective requires additional taxation on energy. Simply relying on market forces, technical innovation, regulation, and standard-setting will not suffice. So the "green forces" within the Community—the governments of Denmark, Germany, the Netherlands (with tactical support from others), the Environment Directorate with increasing support from the EC Commission—have placed all their weight behind a proposed energy/carbon tax. The idea is to introduce a new levy all over the Community (but collected by national governments) that will be large enough to de-link the price of fossil fuels from world-market levels. A figure of 10 dollars per barrel of crude oil (and its equivalent in other markets), often mentioned as an appropriate level to be reached in 2000, would, for example, imply a price hike of 60 per cent for coal, 40 per cent for heavy fuel oil and 30 per cent for natural gas. The Commission estimates that this would be sufficient to reach the stabilisation objective, even if exemptions would be given to energy-intensive industries. Their future emissions will probably be regulated primarily by voluntary agreements.

The proposed energy/carbon tax is controversial in several ways: The most important conflict is among the various EC government, reflecting differences in economic development and energy-supply systems. The proposal is favoured by governments which already have a national stabilisation objective, like the ones mentioned above; by France, which counts on a boost for its exports of nuclear electricity; and by countries with large reserves of natural gas, like the UK and the Netherlands. The major opposition comes from the southern European governments, who fear that the tax will hamper economic growth.

This north-south divide is further exacerbated by the distribution of the emerging market for energy-efficient products and services. There is little doubt that this will follow the trend of the "green markets" in general, with the bulk being taken by German, French and British industries. So the northern member states will have the least to lose and the most to win from a Community-wide energy taxation that improves efficiency markedly, while the southern governments perceive their prospects in the opposite way.

The second conflict runs along transnational lines: the Brussels lobbies of the electricity industry and the large industrial users of energy, like the chemical, cement, and steel industries, have mobilised fierce resistance to the proposed taxation. They argue that new environmental taxes will put them at a competitive disadvantage and, hence, lead to a loss of jobs in Europe.

Whether this opposition is strong enough to thwart the Commission's proposal is hard to predict, but it will most likely lead to modifications and compensation to the industries and governments with the strongest grievances. The question is whether the bargaining to come will make the taxation more symbolic than real, thereby reducing the likelihood of reaching the stabilisation objective. This brings us to the interests and forces behind EC climate policy.

Explaining the EC Position

Since the mid-1980s public opinion has evinced a steadily growing concern about environmental issues in public opinion. Even if it is stronger in northern Europe, the upward trend applies across the Community, probably reflecting deeply rooted value changes. Environmental protection is now generally rated at the top of political priorities, or very close to it, by all European electorates. No politician can afford to ignore such a powerful trend. So ecology issues have been brought to the forefront of

EC policy-making in recent years—partly because political leaders want to present a "green" image, partly because the problems cannot be solved on a national basis.

As the "green mobilisation has unequal strength in northern and southern Europe, it creates an environmental divide, which makes co-ordination within the Community extremely difficult. The affluent troika in the North Denmark, Germany and the Netherlands—is not willing to reduce its ecological ambitions out of concern for the internal market process. These member states urge the Community to act on concrete environmental issues—if not, they will take unilateral action.

In this way, environmental problems have become a real threat to the internal market, making it imperative even for sceptical governments to take a stand on these issues. They—and parts of the Eurobureaucracy—may not care so much about the environment, but they are all worried about the cohesion of the Community. So part of the argument behind the proposed energy/carbon tax is the need to forestall national measures of a similar kind, which would inevitably distort the internal market.

The Community has clear ambitions of playing a more importance role in world politics. When the environmental issues rose to the top of the political agenda in the late 1980s, this presented the EC with both a challenge and an opportunity—to avoid internal disagreement and to seize on a popular cause which could demonstrate its ability to play "a leading role" in the world. The climate issue suits EC needs perfectly: it has a strong popular appeal and ample media attention. Moreover, because of US reluctance, there is a leadership vacuum to be filled. In addition, initially it looked as if the issue was primarily *symbolic* with few concrete implications—another good reason to go for the leading part.

So the activist role on the world stage suited actors with very different motives: some governments primarily interested in the climate issues as such, other governments seizing the opportunity to enhance European prestige in the world and finally the Commission dearly wanting to carve out a larger place for itself in the global arena.

In contrast to the case of the US, it is remarkable how far the EC has come in reconciling energy and climate concerns at a general policy level. The key to this unexpected, but happy marriage is *energy efficiency*. This is the crux of the EC approach to reducing carbon emissions; simultaneously, it enhances energy security by constraining future demand for imported fossil fuels, especially oil from the Middle East. (Anxiety for future supplies from Russia is likely to reinforce such concern.) If the proposed energy taxation works roughly as presumed, it will as an extra benefit significantly reduce (possibly eliminate) EC vulnerability to upheavals in international energy markets, which the Community has so far failed to achieve by other means.

Again in contrast to the US, Western Europe cannot seek energy security by expanding its own supplies of fossil fuels. They are either too scarce (oil), already largely committed to growing markets (natural gas) or too expensive and pollutive (coal). Most importantly, the lack of a large fossil-industrial complex based on indigenous resources in Europe makes it easier for policy-makers to advocate a transition to a more energy-efficient economy. The main losers—the producers of primarily fossil fuels—are largely outsiders with little political clouts, while some domestic industries stand to gain from new markets in energy-efficient technologies and services. It remains to be seen whether the new "eco-economic" philosophy proves attractive enough to carry the day when faced with the opposition from large and small consumers of energy, but the absence of a US-style energy complex does make the task much easier.

On the other hand, the EC has a serious international disadvantage: the Commission is not empowered to negotiate on behalf of the member states, and is formally merely an observer at INC meetings. Member states have traditionally been very reluctant to transfer to the Community the right to conclude environmental treaties. The Montreal Protocol is apparently an exception, as the Community has been granted exclusive competence regarding CFC production. However, as this activity is to be phased out within a few years anyway, that concession does not count for much and will hardly set a precedent for the climate negotiations.

Developing Countries

There are great differences among developing countries regarding energy-related carbon dioxide emissions—and thus their relative importance in a greenhouse context. China alone accounts for approximately 11 per cent of total energy-related carbon dioxide emissions; all other developing countries add up to about 14 per cent. Together with china, countries like Brazil, India, and Mexico attract most attention in international climate negotiations. This is first of all due to their large emissions of greenhouse gases, and the exceptions of a dramatic growth in their emissions, but their importance is also related to the fact that their governments serve as spokesmen for other Third World countries in the negotiations.

There are many reasons why developing countries should participate in a global climate convention. Though their relative importance in terms of anthropogenic emissions is low today, population growth, social, and economic development will radically change this picture. The largest increase in energy use is expected to take place in developing countries, and primarily from fossil fuels. So without the participation of developing countries, OECD policies to abate climate change will have little value in the long run.

Perceptions and Climate Policy

The IPCC report concluded that the effects of global warming are likely to vary greatly from one region to another, but that the areas and communities that are the most vulnerable today will probably be hardest hit by climate change. For example, the most negative effects on agriculture are expected to occur in regions which already have severe problems in meeting basic needs for food. Nevertheless, to judge from attitudes expressed in the climate negotiations, many developing country governments seen uncertain about, or at least not frightened by, the possible consequences of climate change. There are some important exceptions, namely the low-lying countries and island states. However, other developing countries have either not voiced much concern, or they have expressed strong resistance against any climate regime that would constrain their own plans for social and economic development.

Global justice is one major argument often cited by Third World spokesmen. They maintain that the main responsibility for meeting the challenge of global warming lies in the North, which is currently responsible for three-fourths of energy related emissions. Historically, their responsibility is even larger, as their industrialisation was built on the burning of fossil fuels. During the IPCC process and the climate negotiations, representatives of LDC (less developed countries) governments have time and again stressed the need for resource transfers from North to South to finance climate-related measures. Moreover, these transfers must be additional, that is, they must come on top of existing aid transfers. This principle of additionality must be respected in order to avoid a situation where global climate measures are implemented at the expense of social and economic development in the Third World.

The demand for additional resource transfers is closely linked to the development of North-South relations over the past decade and the deteriorating position of most developing countries in the international economy. There has been a significant decline in net transfers from North to South, and the debt situation of many developing countries has become a major obstacle to economic and social development. To many LDC governments, the greenhouse issue appears to be another negative factor in their relations with industrialised countries. A international climate convention that would commit the South to stabilise or reduce its greenhouse gas emissions is seen as just another hindrance to Third World development. Hence, the greenhouse issues have been linked to the old North-South agenda for a “new international economic order”.

Also vital is the question of “green conditionality”. Efforts by industrialised countries to promote environmental protection through development assistance has already become a delicate issue in North-South relations. Many governments in the

South regard this as interference in internal affairs and in conflict with national sovereignty over natural resources. The climate issue has provoked similar reactions.

Deforestation has proven a particularly difficult problem. The governments of countries like Brazil and Malaysia have argued very strongly that figures on carbon dioxide emissions from deforestation in their countries have been exaggerated by actors in the North. Estimating these emissions and their impact on the climate has become a very touchy political issue, and estimates used by Northern NGOs and governments are seen as deliberate attempts to put the blame for the greenhouse effect on developing countries, thereby reducing the responsibility of the West. Proposals put forward by the North regarding the management of tropical forests have been considered by many Third World governments as improper interference. The protection of national sovereignty is among the key issues reflected in the Beijing Declaration of June 1991, a joint statement by LDC governments on environment and development.

The follow-up report from IPCC seems to conclude that the importance of carbon dioxide emissions from deforestation is somewhat less than had been estimated in the first assessment. Such revision of earlier estimates of variables important to developing countries may bolster the legitimacy of the work of IPCC in the eyes of LDC governments. It was a problem from the very outset of the process that most climate research had taken place in the North, and that developing countries felt they were under-represented in the IPCC. This was the background for the establishment of a special working group to handle the issue of LDC participation in the panel.

The Alliance of Small Island States

Third World governments have succeeded in building a joint platform on the issues of transfer of additional resources. However, developing countries do not comprise a homogenous group: level of development, access to energy resources, and vulnerability to climate change vary significantly. This situation is reflected in the formation of sub-groups such as the Alliance of Small Island States (AOSIS). Together with other low-lying developing countries, these states are among the most vulnerable to sea-level rise and other efforts of climate change—floods, hurricanes, cyclones etc. Many of them are also very poor, and have hardly contributed to the man-made greenhouse effect at all. In contrast to many other developing countries, governments of these countries really feel threatened by the consequences of climate change and have consequently joined efforts to safeguard their interests. They have demanded earmarked funding from the international community to protect their countries against the impact of global warming.

China

China is an actor of utmost importance in a greenhouse context. Unless it succeeds in limiting its growth in energy-related carbon dioxide emissions, efforts made by other countries will not suffice to stabilise global emissions. China ranks as number three in the world in total energy consumption, and follows the USA and the former USSR in energy-related carbon dioxide emissions. As noted earlier, China accounts for approximately 11 per cent of total energy-related emissions of carbon dioxide. Energy consumption in 1989 was equivalent to 672 million tonnes of oil. While coal accounted for about 76 per cent of primary energy consumption in the past decade, the share of oil was 17 per cent, natural gas 2 per cent, and hydropower 5 per cent of the primary energy mix. Biomass is an important non-commercial fuel in China, primarily in the household sector.

The industrial sector consumes a remarkably high proportion of the final energy demand—more than 60 per cent. Coal dependence contributes to China's high energy intensity. In order to promote economic development, a major objective of official policy has been to reduce energy intensity, and official reports indicate that conservation measures have had some success during the 1980s. But unless profound political and economic reforms are implemented, there is every reason to expect a dramatic increase in energy consumption and energy-related carbon dioxide emissions in the decades to come. Population growth and economic development, combined with the energy reserve situation of the country all point in the same direction.

China has 15.4 per cent of the world's total coal reserves: it is hard thus to see how an economy already heavily dependent on this energy source should be able to change the energy mix to the extent necessary from an environmental perspective.

If major changes are implemented by the authorities, some hope is represented by China's great hydropower potential. Hydropower resources in China are estimated to be the largest in the world, with are exploitable potential four times the current total production of electricity. But there are major obstacles to the exploitation of this resource, one of them another environmental concern: if the hydropower potentials to be exploited to such an extent that a major part of coal consumption could be replaced by this clean energy source, enormous land areas would have to be flooded and millions of people would have to leave their homes. There are also major financial and technical obstacles to large-scale production of hydro-electric power in China.

Future Prospects

Waste of energy in the production process and in the transport sector in many developing countries represents a significant potential for improving energy efficiency. In many countries there is also a considerable potential for transition from fossil fuels to alternative energy sources. However, specific economic, technological, and political conditions will have to be met in order to make use of this potential. Transfers of financial resources and development of appropriate technology are needed, but also major economic and political reforms. Energy prices must reflect real costs, including adverse environmental effects. Consumers dependent on pollutive energy sources need to be given a cleaner alternative without additional expenses. Such reforms have proved to be politically very difficult and indeed threatening to governments which already suffer from lack of legitimacy and popular support. Vested interests in supply-oriented energy systems and current development models are major barriers. If reforms in the energy and industrial sector seem difficult now, there is no reason to believe that profound changes will be easier later. The choice of development paths and energy strategies made today will have a bearing on these societies for decades to come. It may prove more costly, perhaps impossible, in twenty years to reverse the present pattern of energy consumption.

OPEC

Reductions in carbon dioxide emissions on the scale recommended by most climate scientists would have disastrous effects for fossil-fuel demand. The countries that stand to lose the most from such a development are those whose economies depend on exports of coal, oil or natural gas. The higher and the more enduring the dependence, the larger the risk of a market backlash induced by climate policies (whether national, regional or global). The member states of the Organisation of Petroleum Exporting Countries (OPEC) are the prime examples of this category: most of them receive the bulk of their export earnings and government revenues from oil exports. Most critical is the situation for the producers in the Gulf-Iran, Iraq, Kuwait, Saudi Arabia, and the United Arab Emirates. These countries are completely petroleum-dependent and still hold very large reserves, in contrast to other major oil producers in the world. These governments have looked forward to a long period when they will be able to take advantage of their anticipated control over world oil supplies. Their reaction to the message from the climate scientists has, not surprisingly, been one of total rejection. Headed by Saudi Arabia, they have started a long fight in UN and other fora to delete or constrain any reference to carbon dioxide as an environmental problem. OPEC as an organisation has also warned against unilateral measures, such as the proposed EC carbon/energy tax, and even issued thinly disguised threats of retaliation from the oil producers. In their view, an international climate convention should be "realistic, comprehensive, balanced, and without infringement on the sovereignty of any nation".

Conclusions

In concluding this article we will discuss the implications of the pattern of interests we have drawn up above, we will use this as a basis for assessing the prospects of an effective climate regime and the idea of a "clearing-house". Finally, we shall summarise what we consider the main obstacles to international problem-solving in this area—and consider how they might be overcome. We need to establish the main features of an "effective" global climate regime: one that would stand a chance of handling the greenhouse problems as defined by IPCC. In practice, this will mean developing and implementing a long-term plan for large and systematic reductions in world carbon dioxide emissions, barring a complete reversal of the scientific evidence on global warning. For such a purpose the following elements are of critical importance.

First, national commitments or targets. Each government must undertake to contribute to the common goal, either in the form of a quantified target or in another meaningful and measurable way. These commitments must make sense in relation to the overall global objectives, and they must be verifiable at specified future dates. The latter is the key to the distinction between rhetoric and expression of good intentions on the one hand, and commitments on the other.

Secondly, international funding. Some climate-policy measures will need extra infusion of capital in the short term to become realistic under present circumstances. There is a bill to be shared, primarily among the industrialised countries. Without serious commitments on this score, the chances of attracting co-operation from LDC governments are nil.

Thirdly, institutional follow-up. Some independent body must be able to ascertain whether governments comply with their commitments or not. Let us presume, along the lines above, that these are formulated in a way that makes a verdict on compliance possible (in principle akin to national legislation). The next step is agreed procedures for national reporting and review of those reports. Crucial here is independent, scientific input, as well as transparency. The public must have a right to access to the information submitted by governments and to the subsequent reviews. (Without such provisions, governmental representatives would have a propensity to shield non-compliance from the outside world). The main outcome of such a review procedure would be regular reports revealing which governments follow up on their commitments and which do not. If the process succeeds in acquiring an independent, legitimate status, the published reports are likely to have an impact on non-complying governments. This "power to embarrass" may be the best means of influence a global regime can aspire to.

Is an Effective Regime Possible?

From this configuration of perceptions and interests it follows that he prospects of winning widespread support for an effective global climate regime—as defined by us are bleak. First, there is the *asymmetry of perceived costs and benefits:* Those who stand to lose from climate change, have the least ability to do anything about it—whereas some of those governments whose contribution is imperative, either see no benefit in doing so or are preoccupied with more pressing problems. Secondly, the complexity of the greenhouse problem, especially the uncertainty concerning the scale and timing of adverse effects, makes governments reluctant to undertake commitments that may prove very costly. Thirdly, climate negotiations are closely linked to a number of unsettled North-South conflicts. This has created distrust and suspicion among the parties from the outset.

Clearing-House

As an attempt to improve the prospects of consensus in climate negotiations, the idea has been launched that the UN could establish a clearing-house to facilitate an exchange of investments from the North against emission permits from the South and the East. This would channel capital to projects with the largest pay-off in terms of carbon dioxide reductions per dollar, which are often found in the Third World. The industrialised countries would, in return, be allowed to increase their emissions by a volume corresponding to the reductions they have financed abroad.

It is not necessary to be an economist to see that such a system is on paper far more cost-effective than national targets or pro-rata reductions—but will it work in practice? In real life, the clearing-house idea encounters the very problems we have outlined above. Most importantly, without verifiable national commitments and independent international follow-up, emission trading (or its equivalent under other labels) through a clearing-housing is likely to produce a chaotic situation, where it becomes impossible to identify the net environmental effects. If a government has not drawn up a national climate strategy—and won international acceptance for it—how can anybody tell the impact of its buying or selling permits to reduce or increase its emissions? In other words: If a clearing-house is not to give the green light for large-scale cheating, it must be clearly confined to governments with internationally sanctioned climate policies—and this brings us back to the basis of an effective regime, as outlined above. However, once an effective institutional follow-up is in place, emission-trading can play a very constructive role, that is, through a clearing-house.

Can the Obstacles be Overcome?

As most governments are unwilling to take on concrete commitments, like carbon dioxide stabilisation or reduction, a more fruitful approach might be first to search for normative standards of behaviour that can be widely supported. The case of energy efficiency immediately springs to mind: nobody is against it, and it can be defined in a way that provides guidelines for action. Any meaningful definition would, for example, imply reducing or eliminating public subsidies for fossil fuel production and use. Standards could also be set for technical efficiency in different applications, taking into account a country's level of development. Our point here is not to develop this concept *as such,* but to illustrate an approach to the problem of international co-operation a controversial area. This process could start by 1. *defining a principle or a norm* that all, or most, governments can adhere to, and 2. inferring from it specified *standards of behaviour,* against which future action can be measured or at least evaluated.

The model we have in mind is the UN regime for protection of civil and political rights—a process which started in the 1940s with principles, and ended up with specifics a generation later. The pace of this development should serve as a warning to climate negotiators, but such an approach to global collaboration in highly sensitive matters is promising. This example, furthermore, underlines that the critical link (often the 'missing link") between agreed principles and standards on the one hand and governmental action on the other is *institutional follow-up.* This is a major challenge to governments and other actors interested in an effective regime: How to overcome the ingrained resistance against transferring real powers to a UN body? It should be kept in mind that the very independence that in our view is crucial for any regime, will often be perceived by governments as a potential threat. If those who hold power in a particular country are used to ruling in an autocratic manner, this is not without reason: Independent international institutions will by their very nature strengthen the role of science and non-governmental segments of society, that is, unruly actors beyond the control of governments. This line of thinking cannot, however, be exposed to the world—which is why it is normally camouflaged under the banner of "national sovereignty". This kind of resistance can probably be overcome if the financial rewards are large enough. Therefore, Western governments should begin to commit contributions to a global climate fund, while making it clear that its resources will be restricted to states that accept the underlying principles, standards and independent follow-up. To avoid allegations of undue conditionality, these governments should themselves submit to the same regime—without any side-payment in their case. Indeed, as most countries in the West have large inefficiencies, subsidies, and supply biases in their own energy systems, such adherence would be no mere symbolism. If the international follow-up works along the lines sketched here, it will expose Western governments to a public scrutiny which most of them would have been more comfortable without.

Overcoming National Obstacles?

Thus it is possible to envisage how governments could agree on the basis of what could eventually become an effective regime. That, however, does not suffice to overcome the remaining obstacles on the national level. If the strongest sanction that can come out of UN bodies is bad publicity, governments—possibly with popular support in democratic societies—may simply choose to ignore it. Even if a country is told in clear language that its behaviour is harmful to the atmosphere, that may not be enough to carry the day. As climate change will remain a complex issue with diffuse effects, the norm of protecting the atmosphere badly needs more tangible allies if it is to become a political force to be reckoned with. We are convinced that the climate problem, as currently defined, will never be solved without such "enabling combinations", for which there are, fortunately, numerous candidates:

- in the East, restoring the natural environment will require a total economic and industrial restructuring. Even if the priority of the people, and their governments, is a clean local environment and protection of natural resources for productive use, the by-product could be a radical reduction of pollution, energy use and—with it—carbon dioxide emissions;
- in the West, local air pollution, noise, and congestion are serious environmental and economic problems, which large parts of the population want to solve for purely selfish reasons. It is also well-known that energy efficiency makes good economic sense for the countries as a whole, albeit not for narrowly focused interest groups. Again, solutions to these problems will inevitably contribute to mitigating global warming;
- in the South, local air pollution is in many cases far worse than in the West, and environmental awareness is on the rise. Even more important, the current energy path of most Third World countries is no longer sustainable, for purely economic reasons. Financial necessity will force a reorientation on governments, investors and multilateral agencies alike. As pointed out by Amulya Reddy in his contribution to this volume and elsewhere, an energy strategy designed to meet basic needs does not have not have to run counter to global environmental concerns.

In sum, protecting the global climate for its own sake may well look like an environmentalist Utopia, as far as we can see today (but that may not be very far!). There seems to be a long way to go, but that is no excuse for not taking the first steps. The longer we wait, the steeper the road ahead may turn out to be—and the more Utopian the goal will appear. At this point the sceptical reader will probably ask: if it is all so easy, why isn't it happening? However, we are not saying that it is easy for everybody—for powerful interests and individuals across the world, a transition to a sustainable energy development is very risky indeed. Hence the resistance. We also admit that the merger of climate concerns and local interests may not be sufficient to stabilise the global warming trend. However, in our view it is the only viable beginning of a reorientation away from the present path.

REFERENCES

Ackerman, David M. (2002-10-01). Global Climate Change - Selected Legal Questions About the Kyoto Protocol (pdf). Congressional Research Service. Retrieved on 2006-11-07.

Amanda Griscom Little, "Pact or Fiction? New Asia-Pacific climate pact is long on PR, short on substance", Grist.org article, August 4, 2005]

Canadian Parliament Bill C-288 "An Act to ensure Canada meets its global climate change obligations under the Kyoto Protocol". House of Commons of Canada.

Catherine Brahic. "China's emissions may surpass the US in 2007", New Scientist, April 25, 2007. Retrieved on 2007-05-20.

China overtakes US as world's biggest CO2 emitter | EnergyBulletin.net | Peak Oil News Clearinghouse

Cyprus and Malta will participate indirectly through membership of the EU ETS. Kazakhstan has commented that it intends to become an Annex I party after its ratification of the Protocol]

Eilperin, Juliet. "Climate Plan Splits U.S. and Europe", *Washington Post*, 2005-07-02. Retrieved on 2006-11-05.

Energy White Paper 2003, Department of Trade and Industry, published February 2002, accessed 19 May 2007

Executive Summary. Chapter 9: Projections of Future Climate Change. *Climate Change 2001: The Scientific Basis*. Retrieved on November 15, 2005.

Farber, Daniel A.; Paul A. Hemmersbaugh (1993). "The Shadow of the Future: Discount Rates, Later Generations, and the Environment". *Vanderbilt Law Review 46: 267-304.*

Friends Of The Earth sue Canada for breaching Kyoto Protocol. Seed Magazine. Retrieved on 2007-06-01.

GHG DATA 2006 – Highlights from Greenhouse Gas (GHG) Emissions Data for 1990-2004 for Annex I Parties. United Nations Framework Convention on Climate Change (2006-10-19). Retrieved on 2007-08-06.

Graves, Frank; Boucher, Christian. Public Attitudes Towards the Kyoto Protocol. Ekos Research Associates. Retrieved on November 15, 2005.

Greenhouse Gas Abatement Scheme. NSW Greenhouse Gas Abatement Scheme. Retrieved on 2006-10-31.

Greenhouse gas emission trends and projections in Europe 2006. European Environment Agency.

Greenhouse Gas Emissions By Country. Carbon Planet. Retrieved on 2007-05-20.

Guardian Unlimited: Global leaders reach climate change agreement.

Gwyn Prins and Steve Rayner calling for radical rethink of Kyoto-protocol approach

Hilsenrath, Jon E.. "Environmental Economists Debate Merit of U.S.'s Kyoto Withdrawal", *Wall Street Journal*, 2001-08-07. Retrieved on 2006-11-07.

Industrial world losing sight of Kyoto target. Retrieved on 2007-05-28.

International Carbon Dioxide Emissions and Carbon Intensity. The Energy Information Administration.

J.T. Houghton, G.J. Jenkins, and J.J. Ephraums (eds.), "Climate Change", in *The IPCC Scientific Assessment* (Cambridge University Press, Cambridge, 1990), p. xi.

Kare Dahl Martinsen, *Sovjetunionen og klimaendringene (The Soviet Union and climate change),* EED Report 1990/13 (Fridtjof Nansen Institute, Oslo, 1990).

Keith Bradsher. "China to Pass U.S. in 2009 In Emissions", The New York Times, November 7, 2006. Retrieved on 2007-05-20.

Krugman, Paul. "Ersatz Climate Policy (requires login)", *New York Times*. Retrieved on November 15.

Kyoto bill sparks constitutional questions. "Canadian Broadcasting Corporation. Retrieved on 2007-02-15.

Kyoto protocol status(pdf). UNFCCC. Retrieved on 2006-11-07.

Kyoto Protocol: Status of Ratification, 10 July 2006 (PDF). UNFCC. Retrieved on October 30, 2006.

Leiv Lunde, *Science or Politics in the Global Greenhouse? A Study of the Development Towards the Scientific Consensus on Climate Change,* EED Report 1991/8 (Fridtjof Nansen Institute, Oslo, 1991).

Ma Kai. "China is shouldering its climate change burden", June 3, 2007. Retrieved on 2007-06-04.

Mark Clayton. "New coal plants bury 'Kyoto'", The Christian Science Monitor, December 23, 2004. Retrieved on 2007-05-20.

Mayors Climate Protection Center, List of Cities. Retrieved on 2007-08-31.

McKitrick, Ross (2000-09-26). Submission to the Joint Standing Committee on Treaties Inquiry into the Kyoto Protocol (pdf).

Mendelsohn, Robert O. (2005-02-18). An Economist's View of the Kyoto Climate Treaty. NPR. Retrieved on 2006-11-07.

Michael Grubb, *Energy Policies and the Greenhouse Effect* (Dartmouth, Aldershot, 1991), 293.

National Emissions Trading Taskforce. Retrieved on 2006-10-31.

New Bill and strategy lay foundations for tackling climate change, Department for Environment, Food and Rural Affairs, published 13 March 2007, accessed 13 March 2007.

news.bbc.co.uk/2/hi/europe/3651881.stm. Retrieved on 2007-05-28.

Newton, Peter W. (2001). Australia State of the Environment 2001 Report, Human Settlements Theme Report. Australian Government Department of the Environment and Heritage.

Opposition MPs pass Kyoto bill despite Tory resistance. Canadian Broadcast Corporation. Retrieved on 2007-02-15.

Québec, leader en changements climatiques

R. Caron Cooper, Kare Dahl Martinsen, Arild Moe, and Lee Schipper, in *Energy Conservation and Reduction of Greenhouse Gas Emissions in the Soviet Union: A Role for Western Companies?* (Fridtjof Nansen Institute, Oslo, 1992).

R.A. Reinstein, chief US negotiator, confirmed in a statement to *International Environment Reporter,* 25 Sept. 1991, that the energy strategy of the United States relied on the use of its abundant coal reserves.

Radical rethinking of approach needed says Steve Rayner and Gwyn Prins

Regional Greenhouse Gas Initiative. Retrieved on 2006-11-07.

Renewable energy alternatives for electricity generation. ABC Radio website (2006-06-10).

Richard McGregor. "China urges rich nations to lead on climate", Financial Times, June 4, 2007. Retrieved on 2007-06-04.

Saeed Shah. "China to pass US greenhouse gas levels by 2010", The Independent, November 8, 2006. Retrieved on 2007-05-20.

Statistics Norway, *Strong increase in greenhouse gas emissions*

Stern, Nicholas (2006-10-30). Stern Review: The Economics of Climate Change, Summary of Conclusions (pdf).

Tao Wang & Jim Watson: *Who Owns China's Carbon Emissions?* Tyndall Centre Briefing Note No. 23 October 2007

Text of the Byrd-Hagel Resolution (1997-07-25). Retrieved on 2006-11-05.

The full text of the convention. *The United Nations Framework Convention on Climate Change*. Retrieved on November 5, 2006.

The Impact of the Kyoto Protocol on U.S. Economic Growth and Projected Budget Surpluses. Retrieved on 2005-11-15.

Torleif Haughland and Kjell Roland, *Energy, Environment and Development in China,* EED Report 1990/17 (Fridtjof Nansen Institute, Oslo, 1990), 29.

United Nations (2007-08-31). "Vienna UN conference shows consensus on key building blocks for effective international response to climate change" (PDF).

United Nations Environment Programme (1997-12-11). "Industrialised countries to cut greenhouse gas emissions by 5.2%". Press release. Retrieved on 2007-08-06.

United Nations Framework Convention on Climate Change: Changes in GHG emissions from 1990 to 2004 for Annex I Parties.

United States Country Analysis Brief. US Energy Information Administration (2005). Retrieved on December 5, 2006.

USEPA, *The Potential Effects of Global Climate Change on the United States.* Report to Congress, Dec. 1989; and US National Academy of Science, *Policy Implications of Greenhouse Warming* (1991).

Vidal, John. "Revealed - how oil giant influenced Bush", *Guardian*, 2005-06-08. Retrieved on 2006-11-07.

Wigley, Tom (Spring 2006). The effect of the Kyoto Protocol on global warming. Retrieved on 2006-10-30.

William U. Chandler *et al.,* "Energy for the Soviet Union, Eastern Europe and China", *Scientific American* (Sept. 1990), 122.

Wu Zonxin and Wei Zhiong, "Policies to promote energy conservation in China", *Energy Policy* (Dec. 1991), 934.

Encyclopaedia of

ENVIRONMENTALLY SUSTAINABLE GREEN SCIENCE AND TECHNOLOGY

Encyclopaedia of

ENVIRONMENTALLY SUSTAINABLE GREEN SCIENCE AND TECHNOLOGY

Volume 3

R.B. Patnaik

ANMOL PUBLICATIONS PVT. LTD.
NEW DELHI - 110 002 (INDIA)

ANMOL PUBLICATIONS PVT. LTD.

H.O.: 4374/4B, Ansari Road, Daryaganj,
New Delhi-110 002 (India)
Ph.: 23278000, 23261597

B.O.: No. 1015, Ist Main Road, BSK IIIrd Stage
IIIrd Phase, IIIrd Block,
Bangalore - 560 085 (India)
Visit us at: www.anmolpublications.com

Encyclopaedia of Environmentally Sustainable Green Science and Technology

First Edition, 2009

ISBN 978-81-261-3842-5 (Set)

PRINTED IN INDIA

Printed at Mehra Offset Press, Delhi.

Contents

VOL. 4

VOL. 5

Preface

This publication entitled "Encyclopaedia of Environmentally Sustainable Green Science & Technology" attempts towards creating a holistic understanding of environment and sustainable development in its totality. It defines environment and sustainable development in all its dimensions. The concepts and methodologies of ecology, environment and sustainable development have been discussed in detail. The ecosystem perspective and environmental concept of sustainable development have been highlighted, so has been the environmental methodology. An effort has been made towards defining green science and its manifestations for understanding pollution. The relevance and importance of environment impact assessment (EIA) for environmental sustainability has been analyzed. Also, the elements and dimensions of green technology have been covered in detail. The relationship among ecomovements, ecodynamics and sustainable development has been discussed in detail. Similarly, the environmental and social aspects of sustainable development, in particular, aspects related to training of people to meet the global environmental needs have been covered. The issues related to green energy, its sources, tools and technologies have been reflected upon briefly. The environmentally sustainable science and technology has been presented in brief with special reflection on climate change and global warming. Current issues and options regarding sustainability aspects of environmental management, corporate management and environmental responsibility have been covered. The aim is to develop a policy perspective. The contemporary global industrial technology has been discussed with focus on evaluating environmental performance and compliance to green laws. The applied aspects of environmentally sustainable science and technology have been outlined. Also, the green conventions enforced towards regulating global science and technology towards environmentally sustainable development has been presented, highlighting the theories of international environmental law and theory of environmental degradation.

The applications of green science and technology have been discussed along with a case of biotechnology. The green technology and environmental monitoring issues have been given considerable coverage. Select case studies have been made in this regard. Also, case studies of environmental engineering and environmental biotechnology have been incorporated with focus on applications and utilities. Future prospect of environmentally sustainable or green technology in today's world has

been analyzed briefly. Similarly, the principles, tools and techniques of environmental management, with particular reference to environmental disasters, have been covered in detail. Glossary of terms, bibliography and links has been provided to make this publication a user friendly one and to aid readers in further research and analysis.

—Editors

10

Currents Issues and Options Regarding Sustainability Aspects of Environment Management

GLOBAL ENVIRONMENTAL MANAGEMENT, SUSTAINABILITY AND RELATED ISSUES

Increasing Environemntal Understanding

Behind the rising prosperity in the developing world lurks the shadow of lethal air pollution from motor vehicles, smokestacks, and hearths. All these sources and others emit tiny airborne particles that lodge deep in the lungs, causing severe and sometimes fatal respiratory problems. Every year in four Chinese cities alone—Chongqing, Beijing, Shanghai, and Shenyang—10,000 people die prematurely from exposure to these particles. And throughout the developing world such pollution kills hundreds of thousands of people and seriously damages the health of millions more. Workdays lost to respiratory illness run into the hundreds of millions and the corresponding economic costs into the billions of dollars.

These losses were once viewed as the price of economic development. Fortunately, the countries that have pioneered environmental protection in the developing world have shown this view to be gravely mistaken. Operating in very different political and institutional cultures, they have used pollution charges, grassroots community pressure, and command-and-control regulation to contain or even roll back air pollution at supportable cost. Across their diverse approaches lies an important unifying factor: the global community's accumulation and dissemination of knowledge—knowledge about measuring air quality, about evaluating health risks, about identifying pollution sources, about estimating abatement costs, about setting enforcement priorities, and about designing cost-effective regulatory instruments. As knowledge has accumulated and environmental policy strengthened, air quality has stabilised and even improved in many rapidly industrialising areas.

What is true for air pollution is true for many other types of environmental damage. Each year, diarrheal diseases from contaminated water kill about 2 million children and cause 900 million episodes of illness, most of Which oral rehydration therapy and basic sanitation could have prevented. Annual losses from soil erosion range from 0.5 to 1.5 per cent of GNP in many countries, but well-informed policies can trim these costs substantially. Greenhouse gas emissions will double within a generation at current growth rates, but effective policies and dissemination of information can greatly reduce them.

There is now general agreement about the importance of limiting environmental damage. We also have a good understanding of why government action is needed to preserve the environment. For example, pollution has adverse effects on others, but the polluter typically does not have to compensate them. When such spillover effects, or externalities, occur, the cost of pollution to society is greater than the cost to the polluter. There is then too much pollution because individuals and firms do not have the right incentives to reduce it. A factory that discharges pollutants into a river has no incentive to consider the damage inflicted on those downstream. A user of freon-based air conditioning has no incentive to take account of the damage freon causes to the atmosphere's ozone layer. Drivers have no incentive to reduce exhausts and improve air quality for their neighbors.

Effective public policies provide incentives to reduce pollution and natural resource degradation by aligning social and private costs. In some cases, legal systems can produce such alignment without direct government action. For instance, laws in some countries require polluters to compensate others for certain kinds of pollution damage. The assignment of property rights can also reduce the scope for environmental degradation; for example, a lake that has one owner is not likely to be overfished. But the holders of chose rights may incur large transactions costs in enforcing them and the assignment of property rights sometimes is simply not feasible-who, for example, should own the atmosphere?

Problems such as these have forced governments to do more to protect the environment. In some cases government regulators can give polluters the right incentive to clean up by charging them for the damage their activities cause. When information on discharges or the extent of their damage is not available, systems that monitor the more easily observable actions of polluters, such as the required installation of pollution control devices, may be desirable. With the right information, however, pollution charges are superior. Unlike technology standards, they put firms under continuous pressure to reduce pollution.

Information can also encourage pollution reduction. A government's regulations (including pollution charges) apply across its whole jurisdiction and may not be appropriate for all areas—environmental, social, and economic conditions may differ. In such cases, public disclosure of a polluters emissions can complement formal regulation. Informed consumers may then buy fewer products from heavily polluting firms. Investors concerned about liability may become reluctant to finance them. And neighboring communities may insist that they improve their environmental performance.

Disseminating information on the implications of environmental degradation can also offer opportunities for improvement, but the impact of better information depends on people's ability and willingness to use it. That returns us to this Report's two main themes: narrowing knowledge gaps and addressing information problems.

This chapter discusses two main issues:

- *The importance of knowledge and information for environmental management.* Better environmental outcomes require more knowledge about environmental impacts and about technologies, as well as information about environmental performance such as the pollution generated by particular polluters.
- *The design of appropriate institutions for environmental management.* Effective management requires knowledge about the impact of alternative institutional arrangements, their information requirements, and the circumstances in which they will work well.

Global Environmental Politics

Pattern of Global Consumption

The Brundtland Report of 1987 states that, 'The Earth is one but the world is not. We all depend on one biosphere for sustaining our lives. Yet each community, each country, strives for survival and prosperity with little regard for its impact on others'. This highlights the fact that most environmental issues by their very nature are transboundary and can not forcibly be pushed into narrow state conduits. Many environmental problems such as desertification, pollution (of air, ground-water, river and high seas), deforestation, transportation of hazardous substances and atomic leakages from power plants have a spillover effect to neighbouring states. The technological development used by a country can contribute to the global climate change. Also what is primarily the interest of the state may not necessarily be the interest of the people or that of the Civil Society composed of all non-state actors, communities and NGOs. The globalisation of economy and the awakening about the environmental threats have together contributed to the global debates on environment and development.

Global environmental Politics is designed upon three fundamental debates of Political Science: first, the relationship between the State and Civil Society. Second, the relationship between the North and the South and lastly the linkages between the economic ideology and the resources that a country possesses at a given time in its history.

The difference between the State and the Civil Society is porous and overlapping in global environmentalism. Any decision in one sphere affects the other. This has resulted into innumerable non-state actors participating into global environmental debates. Although there is no definite understanding to the civil society yet Hegel has given an analysis which can be safely used for exploring issues in global environmental debates. The civil society by its basic nature has been placed by Hegel above the family and below the state and it includes economy within its domain. Later writers like Anatonio Gramsci [1971] and John Keane [1988] have systematically differentiated it from both the state and the economy and which functions independent of both. The politicisation of the civil society in the environmental sphere has led to the weakening of these distinctions and it is placed closer to the understanding of Hegel. 'Therefore global civil society, is that slice of associational life that exists above the individual and below the state, but also across national boundaries' [Wapner 1996:4]. To affect changes in the global policies environmental activism works through their national governments.

The immense gap that exists between the North and the South is another crucial point that tremendously influences the global environmental politics. The process of globalisation with improved electronic interdependence has however turned the world into what McLuhan, the Canadian academician in 1962 had written as a 'Global Village'. It is somewhat also a deep cultural interchange which is leading towards the homogenisation of the world. This is in no case a just integration since the North is a great consumer of the resources from the South and this is being consistently maintained by regulating the world trade on laws and institutions which legitimizes the regimes of the North. The global business expands into the South but in a manner that the South is made to roll up its sustainability of resource use. Global economy has been functioning in an inequitable institutional framework and this is evident in the policies which the North has pursued towards the South.

The linkages between the economic ideology and the resources of the country are manifested in the hidden externalities of the market. The resource demands of the North cannot be met with their own resource base so the markets are created which perpetuate the consumer patterns of the North while at the same time enabling the flow of raw resources from South to the North in a subsidised manner. This is maintained by promoting an economic ideology that subsidizes the biomass resources and conceals the debt burden or the amount in the GNP that is actually not the earning that the country has made but it is that amount which has to be returned to the rich debtor country of the North. Southern states continue to fuel their economies through debt and then more debt to repay the earlier debt. India, as an example of the fast liberalising market has been heading towards a debt trap or 'is heading towards debt vicious circle'. This is due to the fact that since the borrowings are on an up trend to meet the fiscal deficit of an all time high of Rs. 1,11,275 crore, the government will be repaying a higher debt to the tune of Rs. 40,478 crore in 2000-2001 as compared to Rs. 26553 crore in 1999-2000 [Statesman, 15.3.2000]. Thus as Redclift has observed, 'Development is sustained by economic ideology rather than resources' [1984:21].

The developmental paradigm for a sustainable consumption pattern are available but the government does not consider them in the policy structures. 'The limits on international society's ability to solve environmental problems are not technical but political and economic, many of them stemming from the structure of demands in the developed countries and their relations with the underdeveloped world' [Redclift 1984:20]. It is because the resources are never accounted or never seriously accounted through comprehensive ecological methods. Nations take recourse to policies which sustain a particular pattern of consumption in the society and use GNP as a tool to legitimize high growth whereas the GNP is infact representing only the measure of decay, degeneration and depletion of resources for perpetuating that system. Yet the State continues to show the inflated GNP ratings as a measure of country's progress.

The 23 per cent of the world's population living in the industrial countries is earning 85 per cent of world income and consuming most of its resources. The energy consumption per capita measured in giga joules is just 12 in Africa and in some African countries it is not more than one. It is 21 in Asia, 30 in Latin America but it shoots up to 110 in Japan, 165 in Germany and to the explosive 280 in USA. The energy consumption of one citizen of USA is equal to the energy consumption of more than 160 Tanzanians or 900 Nepalese. This affluence of the North has also resulted in the irresponsible production of the CFC gasses to the extent of more than 28 per cent in just one country USA in contrast to a meager 8 per cent produced by the whole of Third World taken together. The Human Development Report 1992 has also estimated that due to the partisan policies of the North poor nations of the South are denied $500 billion worth of market opportunities annually in the international market which is more than 10 times the annual flow of foreign assistance they receive.

Conditions of International market influences the Third World countries more than the developed countries. There is a pressure for overproduction when the international market suffers the crash of commodity prices and the poor countries step up production to sell their goods at a very high tariff rates, higher than the countries of the North. There is excessive demand upon the tropical forests of the South due to the wasteful affluence of the West. Global 2000 Report, from USA in 1982 had predicted the disappearance of all the physically accessible forests of the South by 2020 A.D. This has serious consequences for the poor countries of the South since the livelihood including the home and the hearth of the three quarters of the world's poor living in the South is totally dependent upon the forests. When there is a sudden increased demand in the North for certain commodities the market prices of the raw material increases and this stimulates the poor country to exploit it more. This happened when in the 1960s and the 1970s the increased demand for timber in Japan raised the timber prices in the international market leading Philippine and Indonesia to step up their deforestation to the extent of the virtual elimination of their major forest areas. Much of the deforestation in the South is taking place to meet the high energy demands of the North, or for meeting their rising demand of meat. The forests of Latin American countries have been brought down to make place for grazing grounds for cattle. The waste generated by the affluent living in the North is so enormous that it is cheaper to dump it to the poor countries of the South than to recycle and reuse it. So these countries have become a hazardous dump houses for the waste from the North. Despite the widening gulf of disparity between the North and the South the latter always held a common view with the North on global environmental issues.

The marginalisation of the South has been achieved systematically by the North through their exclusive control over the world markets. According to the study by Lant Pritchett of the World Bank, in the Summer 1997 issue of Journal of Economic Perspectives, in 1870 the world's richest industrial countries, Britain and United States had an income per head roughly nine times that of the poorest country. In 1990, America's income per head was more than 45 times that of Chad or Ethiopia. Mr. Pritchett calculates that in 1870 the average income per head of the world's 17 richest countries was 2.4 times that of all the other countries; In 1990, the same group was 4.5 times as rich as the rest. The gap is still widening *[The Economist* 1999:23]. The catching up with the West has been an arduous journey for the South and most countries are trying to do it by prematurely encashing their natural reserves. Since 1974 when the South enthusiastically called for a New International Economic Order (NIEO) at the UN General Assembly to their torpedoing of the Seattle Conference of the WTO in December 1999 all efforts to restructure the global economic system had been directed to make the international trade just and equitable. The maturing of the South from the tables of GATT negotiations to the signing of the WTO in Marrakech (Morocco) and finally to the turning of the tables on the North at Seattle's Ministerial Conference is an indication of the awakening which would change the undemocratic and the non-transparent nature of international trade.

Poverty and Environmental Degradation

Global population is projected to increase by nearly two-thirds over the next thirty years (World Resources Institute 1994). Expanding populations which exacerbate economic, social, and ecological impoverishment make all of the existing environmental problems more critical (Caldwell 1984). The increase has several components, including a desire for large families, unwanted fertility, and "population momentum," the tendency for population to increase for some time after fertility has stabilised (Bongaarts 1994). The environmental impacts of continued population growth will be felt through increases in energy demand, production, consumption, and waste. The implications of population growth for property rights regimes are also sobering. There is a limit to the extent that systems of property rights, however well-specified, consistent, and well-enforced, can protect environmental resources in the face of extreme population pressure. The need to understand the relationship between poverty, population, and the environment is critical.

Individual decisions about child-bearing are fundamental to the issue of population growth. In all countries, children are in some sense durable "consumer goods," which require expenditures and yield a flow of services over their lifetime. The services include those of love and affection, companionship, and extension of personal genetic material. In poor countries children provide additional services to the household: they contribute to household production and provide security for parents in old age. These two services provide poor households with a significant motivation for having children, a factor recognised in the literature (Cain 1982; 1983; Mueller 1976; Dasgupta and Maler 1991; Nerlove and Myer 1991; Dasgupta 1992; Bongaarts 1994)

The motivation for fertility has its origins in the combination of private property rights to children's productive services and a paucity of basic services in rural areas of poor countries. Poor countries are for the most part biomass-based subsistence economies. Households typically do not have the easy access to sources of household energy and water that are standard for households in industrialised economies. Children perform services of collecting water and energy sources (fuelwood, dung, fodder) for the household, services which are necessary for survival and complementary to other productive activities. Several hours per day may be required for obtaining the minimum essential levels of water and energy sources, resulting in much higher relative costs for these basic factors than are faced by households in more industrialised economies. Children generally add more to household income than they cost to rear, creating a direct incentive for households to produce children as workers (Dasgupta and Maler 1993). The number of basic activities required to survive means that each household needs many hands, and under certain circumstances the usefulness of each hand increases with declining resource availability. The further the distance to find fuel or water, the larger the number of gatherers required.

In resource-degraded areas, a high rate of fertility and population growth causes further damages to the environmental resource base, creating further need for large families, causing more damage to environmental resources, until some countervailing factor stops the spiral. The spiral may be stopped by public policy or by the feedback of lowered productivity of children. The problem is that by the time feedback is registered, many people have suffered. Such a spiral may be set off by a number of sources: government or private usurpation of community property rights to resources, labor immobility, breakdown of compliance with environmental rules, and the introduction of new technology (Dasgupta 1995; Gadgil and Rao 1995; Jodha 1995; Pradham and Parks 1995).

Environment-Poverty Nexus Revisited: Linkages and Policy Options

Introduction

The eradication of poverty is a major development challenge. Among the millennium development goals contained in the Millennium Declaration in General Assembly resolution 55/2 of 18 September 2002, adopted by 147 heads of State and Government and 189 Member States, the eradication of extreme poverty and hunger was underscored as the paramount goal. Rapid economic growth is often seen as the key foundation for achieving poverty reduction. Recent findings by the World Bank provide additional evidence which strengthens this assertion. While the linkage between economic growth and poverty reduction is generally obvious, the relationship between economic growth and improvement in the environment remains unclear. In recent times, the fast economic growth in many countries, including developing countries of the Asian and Pacific region, where two thirds of the world's poor live, has often been accompanied by a range of adverse environmental impacts. Apparently, there is a trade-off between economic growth and environmental improvement; achieving both of these goals simultaneously may not be easy, especially in the early stage of development.

For many developing countries, the enhancement of economic growth is often the major macro-objective of economic development although the associated adverse environmental impacts are recognized. It is expected that sustainable poverty reduction will take place through a trickle-down of economic growth: the benefits of economic growth will percolate down to the poorer section of the population and increase their income. Under this premise, the tempo of economic growth is maintained as much as possible in order to maximize the pace of poverty reduction. In certain cases, policies and programmes are formulated to mitigate the adverse environmental impacts of economic growth and reverse environmental degradation. Against the background of possible positive linkages between the livelihood income and health of the poor and the environment, such action would lead to a reduction in poverty. In addition, there are certain poverty reduction policies which have a beneficial impact on the environment and certain environmental interventions targeted at the poor which enhance their livelihood income and health status. The aim of this chapter is to review and analyse the poverty-environment nexus against the background of the linkages between economic growth and the environment and the associated policies. An attempt is also made to identify modalities for strengthening the impact of the policies and to draw the attention of policy makers of the Asian and Pacific countries to the fact that environmental improvement strengthens the impact of economic growth on poverty reduction and that improving the environment is an important element of a pro-poor economic strategy.

This chapter has three specific objectives. First, an attempt will be made to identify, classify and analyse the major linkages between growth and the environment and between the environment and poverty (income and health of the poor). This will serve as a framework for understanding the objectives and impact of the relevant policies and programmes. Second, policy options will be presented, elaborated and analysed, giving examples from Asian and Pacific countries. The conclusion will focus on suggestions for enhancing the effectiveness of policies and programmes in the context or achieving the millennium development goals following the Plan of Implementation of the World Summit on Sustainable Development.

Although absolute poverty in the Asian and Pacific region is predominantly a rural phenomenon, rapid urbanisation and the expansion or urban-based economic activities have recently led to very substantial rural-urban migration. The migration process has, to some extent, helped to reduce population pressure on agricultural land and contributed to increasing agricultural productivity and reducing rural poverty. However the migration has put pressure on housing and other services, leading to the development of slums, thus having an impact on the environment. The nature of poverty-environment interlinkages in urban areas is therefore somewhat different from that in a rural setting. The policies, especially those associated with the environment, also address different phenomena and issues. For example, vehicular pollution is an urban environmental problem while land degradation affects the lives of the rural poor. Rural-urban differences will be highlighted in appropriate places in this chapter and linkages and policies differentiated accordingly.

It is not possible to incorporate every aspect of the environment-poverty nexus in this limited review. For example, climate change is not dealt with, despite the realisation that it has a major adverse impact on the poor, especially in the island developing countries. The review is selective and based on observations with regard to the broad priority issues and constraints faced by the member countries in formulating and implementing relevant policies for improving the environment, which contributes to the reduction of poverty.

Coping with Climate Change

Spurred by a quadrupling of carbon emissions during the past half-century alone, Earth's atmosphere is warming at an increasing rate. The hottest 14 years since systematic measurements began in the 1860s have all occurred in the past two decades; the summer of 1998 was the hottest on record, and the winter of 1999-2000 may turn out to be the warmest. Average temperatures are projected to increase further, by 1.2° to 3.5° C (2° to 6° F) over the course of the present century—which would melt glaciers and the polar ice caps, raise sea levels and pose threats to hundreds of millions of coastal dwellers while drowning low-lying islands altogether.

Portents of this future are already visible. As the warming trend has accelerated, weather patterns have become more volatile and more extreme, while the severity of weather-related disasters has escalated. The cost of natural disasters in 1998 alone exceeded the cost of all such disasters in the entire decade of the 1980s. Tens of thousands of mostly poor people were killed that year, and an estimated 25 million "environmental refugees" were forced from their homes. The damage wrought by these disasters has been exacerbated by unsustainable environmental practices and the fact that more and more poor people have little choice but to live in harm's way—on flood plains and unstable hillsides and in unsafe buildings.

Reducing the threat of global warming requires, above all, that carbon emissions be reduced. The burning of fossil fuels, which still provide more than 75 per cent of energy worldwide, produces most of these emissions. The rapidly expanding number of automobiles around the globe threatens an even greater escalation in emissions. The need to promote energy-efficiency and greater reliance on renewable resources is obvious.

Further development of fuel cell, wind turbine, photovoltaic and cogeneration technologies will help. In the developing world, particularly in rural areas that are not connected to energy grids, the rapidly falling costs of solar cells and wind power have the potential to bring energy to the poor at reasonable costs, thereby also enhancing agricultural productivity and generating income.

Stabilising levels of carbon dioxide in the atmosphere to a range that is considered safe will require overall reductions on the order of 60 per cent or more in the emission of the "greenhouse gases" that are responsible for global warming. Thus far, the international community has not found the political will needed to make the necessary changes.

BOX-1

Using Economic Incentives to Reduce Global Warming and Promote Investment in Developing Countries

Addressing the challenge of climate change is one of the most important tasks of the twenty-first century. It will require major reductions in emissions of the so-called greenhouse gases that cause global warming. Reductions can be achieved in a number of ways. One of the most ingenious of these, the Clean Development Mechanism (CDM), provides benefits for both industrial and developing countries.

The CDM allows industrial countries to gain emissions credits for climate-friendly investments in developing countries where these would reduce pre-existing levels of greenhouse emissions. Emission credits count towards the reduction targets that the industrial states have to meet.

The prospect of gaining emission credit provides incentives for rich countries to make energy-saving investments in poor countries. The fact that the emissions savings have to be verified and certified provides incentives to create a new service industry dedicated to this task. The climate-friendly investment helps build sustainability in the developing country.

The CDM and other Kyoto mechanisms seek to use incentives to engage the private sector in the vital task of reducing global warming. They are very much in tune with the spirit of the times.

Implementing the 1997 Kyoto Protocol would mark a significant advance by binding the industrialised countries to verifiable emission limitation and reduction targets averaging 5 per cent below 1990 levels, to be achieved over the period 2008-2012. Recognising the economic roots of the climate change problem, the Protocol seeks to engage the private sector in the search for solutions. It does so by the use of market mechanisms that provide incentives for cutting emissions, and which stimulate investment and technology flows to developing countries that will help them achieve more sustainable patterns of industrialisation (see box 2).

Although the first generation of Kyoto targets represent just one step towards what is needed to reduce global warming, their achievement would result in a sharp reduction in current rates of increase of greenhouse gas emissions by the industrialised countries. Early action is essential. Without success, there will be little incentive for the further rounds of emission limitations that must follow, in which the developing countries will need to become progressively engaged.

> I call upon the Millennium Summit to promote the adoption and implementation of the Kyoto Protocol. Specifically, I urge those states whose ratifications are needed to bring it into effect to take the necessary action in time for entry into force by 2002, as a fitting celebration of our progress since Stockholm in 1972 and Rio in 1992.

In several other areas, there are severe challenges for which we still lack remotely adequate responses.

Major Conferences—Stockholm to Seattle (Environment and Development)

United Nations Conference on Human Environment Stockholm, Sweden, 1972

The rise of environmentalism in the last phase of sixties led Sweden to initiate United Nations to address the issue of global deterioration of human environment. The UNO in 1972 convened the first international Conference on Environment in Stockholm. Although the initiative was muted and narrow yet delegates from 114 states joined hands on 26 principles which launched the first 'Action Plan" for environment. This period was infested with cold war politics inside UNO therefore when USA proposed an immediate moratorium on commercial whaling USSR along with Japan, Norway, Iceland, Chile and Peru unanimously defeated the proposal. It was also important because it led General Assembly to begin a new office called United Nations Environment Programme (UNEP) immediately after this conference. Since then all future trends in population, economic growth and natural resources have got linked and this has created a niche for environment in all deliberations on development. The conference was followed by 5 major Conventions; Convention on the prevention of Marine Pollution by Dumping of Waste, (1972), Convention Concerning the Protection of World Cultural and Natural Heritage, (1972) Convention for the Conservation of Antarctic Seals, International Convention for the Prevention of Pollution from Ships (MARPOL, 1973) and The Convention on International Trade in Endangered Species (CITES, 1973). At Stockholm Conference 'Poverty' surfaced erroneously as the major cause for environmental deterioration since the rich North was sufficiently able to conceal their consumption pattern which was later found to be the major cause of environmental deterioration. However the factor of poverty led to the declaration on the much required establishment of a 'New International Economic order' in the Sixth Special Session of General Assembly. The major achievement of Stockholm lies in not what it was able to do but the direction which it was able to provide to all future action. In 1972 only 25 countries possessed national environmental agencies, with only 11 in the developing world, whereas today there are over 140 national environmental agencies with 110 of them in the developing world [Mathews 1991:25].

The decade of eighties began with the new awakening on climate since the discovery of 'Ozone Hole' had created panic about Global Warming. The first World Climate Conference in Geneva in 1979 led UNEP and International Union for Conservation of Nature (IUCN) to launch the first World Conservation Strategy in 1980. The Vienna Convention for the Protection of Ozone Layer was signed in 1985. As the experts were seriously negotiating the technological restructuring that would subsequently be undertaken due to Global Warming the Chernobyl blast in the Soviet nuclear plant led to transboundary radioactive pollution over Europe and Japan. This accelerated the global efforts for environmental governance. The Brundtland Commission of 1987 focused the environmental debate in the context of development, poverty and consumption patterns prevailing in the states of North and South.

United Nations Conference on Environment and Development, Rio de Janeiro, Brazil, 1992

The decade of eighties had consolidated belief that the existing pattern of growth would not continue forever. Sustainable development emerged as the new paradigm of development. This new paradigm brought together 110 heads of state and many thousands of NGOs under the Agenda-21 of the United Nations Conference on Environment and Development held in Rio de Janeiro (Brazil) in June 1992.

The main achievement of the Rio was preceded by the fallout of three major events, first, the economic strains that followed the Gulf war, second the uncertainties about climate change and global warming and the third, the persistent animosity between the North and the South on issues of population and inequitable trade. This made deliberations at Rio a virtual debate between the donor countries of the North and the debt ridden Third World countries.

Rio explored an expanded meaning to the term 'Sustainable Development'. It was found to be indicating four directions of growth mainly, environmental, social, cultural and economic sustainability. This was a comprehensive and holistic treatment to the issue of development which brought people out as leaders and repositories of indigenous knowledge. Agenda-21 emerged as that primary document discussed at the conference which provides a comprehensive framework for activity into the 21st century. It addresses the combined issues of environment protection and fair and equitable development for all. The agenda comprises 40 chapters each addressing specific aspects of issues and strategies for an environmentally responsible development. Most problems would be addressed at the local and the national levels but there were still a number of areas in which an international commitment to change was needed. These are set out in Agenda-21 they include:

1. Allocation of international aid to programmes with high returns for poverty alleviation and environmental health, such as providing sanitation and clean water, reducing indoor air pollution and meeting basic needs.
2. Investing in research and extension to reduce soil erosion and degradation and put agricultural practices on a sustain-able footing.
3. Allocating more resources to family planning and to primary and secondary education, especially for girls.

4. Supporting governments in their attempts to remove distortions and macroeconomic imbalances that damage the environment.
5. Providing finance to protect natural habitat and biodiversity.
6. Investing in research and development of non-carbon energy alternatives to respond to climate change.
7. Resisting protectionists pressures and ensuring that international markets for goods and services, including finance and technology remain open.

The three visible outcomes of the Summit were the Climate Convention, the Biodiversity Convention and the establishment of Commission on Sustainable Development (CSD) which was to report to UN Economic and Social Council (ECOSOC). Its other major accomplishment was the impetus that the Summit provided to the grassroot organisations all over the world. There amidst the carnival acts, prophets of Armageddon, and environmentally correct T-shirt salesmen, they engaged in what may have been the most productive business of the entire Summit connecting environmental activists and entrepreneurs with networking organisations, exchanging ideas and strategies for mobilising political action and renewing faith among members of grassroot environmental groups that environmental protection was worth dedicating one's life to' [Hempel 1996:44]. The Rio Summit faced many setbacks also. It brought out the differences within the countries of the North on issues of Forests and Biodiversity. President of USA George Bush played the role of the big brother trying to impose its consumption requirements upon the South by emphasising deforestation and population as the means to sidetrack the major issue of inequitable trade. However Netherlands and the Nordic countries were more favourable to the demands of the Southern states. The Southern states had tried to revive the UN goal of transferring the 0.7 per cent of the GDP of the rich countries as official development assistance (ODA) to the poor countries of the South yet the US intransigence and its virtual sell off to the wealthiest of the transnational companies of the North for the protection of its more than $50 billion annual germplasm trade disgusted the Third World. The Malaysian Prime Minister Mahathir Mohammed remarked that, 'At Rio, the eco-imperialism of the North ought to be put to rest once and for all' [1992:B7].

The Rio Cluster

The Rio set the pace for further action on related issues of environment like Human Rights, Population, Social Development Women, Human Settlements and Food Security. The following United Nations proceedings—that can be loosely described as the Rio Cluster are all focussed in one way or other on issues of sustainability.

- World Conference on Human Rights, Vienna, June 1993
- International Conference on Population and Development, Cairo, September 1994

- World Summit for Social Development, Copenhagen, March 1995
- Fourth World Conference on Women, Beijing, September 1995
- Habitat II, Second United Nations Conference on Human Settlements, Istanbul, June 1996
- World Food Summit, Rome, November 1996.

Some of the important landmarks in the direction of sustainable development may be discussed as follows.

Cairo Conference, September 1994

The International Conference for Population Development (ICPD) was considered by the 48th session of the UN General Assembly on 4-5 November 1993. During the discussion, delegates raised a number of key points, including the rights of the individual, the empowerment of women, means of implementation, issues of consumption and lifestyles, the perspective and needs of countries in transition, indigenous people, and the role of NGOs.

ICPD Secretary General Nafis Sadik, opened the Conference on Monday, September 5,1994. UN Secretary General Boutros Boutros-Ghali, welcomed delegates to Cairo and noted that the ICPD is a turning point for the relationship between population and development in addressing many important issues, including poverty, the role of women, environment and development. He highlighted three principles for the ICPD concerning the demands made in a world of growing population, tolerance regarding the ethical and religious issues, population measures that must be displayed on a mutual basis, and conscience that would allow us the right to carry out our lives, but with full respect for other rights, particularly those of women.

His Excellency Mohamed Hosni Mubarak, President of Egypt, as President of the Conference hoped that the ICPD would be a bridge between North, South, East and West. Prime Minister, Gro Harlem Brundtland (Norway), asked participants to turn away from the media dramatisation of the Conference and to focus on the issues of education and health, particularly for women. She highlighted the importance of the ICPD in addressing the spread of STDs and also clarified that the term "reproductive health" could not be construed as supporting abortion.

The nominations for Vice-Chairs from the Asian Group were, Bangladesh, China, Indonesia, Japan, the Marshall Islands and Pakistan. Vice-Chairs from the Eastern European Group were Hungary, Romania and the former Yugoslav Republic of Macedonia. During the five and a half days that followed, the general debate involved many countries. Representatives from 183 countries and numerous intergovernmental organisations, NGOs and the media heard over 230 speakers, including 155 UN member States and observers, 24 UN agencies, programmes and funds, 15 intergovernmental organisations and 37 non-governmental organisations.

As compared to the earlier two population meets at Bucharest and Mexico City, Cairo Conference did not have to face either the Soviet and Vatican opposition or the anti-abortion campaigners of Family Planning. The threats from the Islamic fundamentalist groups did not deter the aggressive lobbying by the Women's Caucus along with the groups which had been part of the Rio Summit. They were able to change the conventional paradigm of population control through family planning methods to 'reproductive health and women's empowerment.' It was expected that the Beijing Conference coming up an year later would consolidate the position on this front. The conference was not an easy sailing for most organisers. In fact, Maher Mahran, the Egypt's population minister and an obstetrician by training had remarked that he had never been involved in such "a difficult and protracted delivery" [Hempel 1996:46].

The World Summit for Social Development, Copenhagen, March 1995

It was part of a series of global conferences convened by the United Nations with the objective of enriching the international agenda and to raise awareness on the major issues and concerns of the time. Three core issues were addressed eradication of poverty, promotion of full employment, and fostering of social integration. One hundred and seventeen Heads of State and government supported by ministers representing another 69 countries adopted the Copenhagen Declaration on Social Development and the Programme of Action. The 10 commitments, embodied in the Declaration, signalled the emergence of a collective determination to treat social development as one of the highest priorities of national and international policies and to place the human being at the centre of development.

The General Assembly of the United Nations decided in 1995, on the recommendation of the Summit to hold a Special Session in the year 2000 for a review and appraisal of the implementation of the outcome of the Summit and to decide on further initiatives to strengthen the effectiveness of implementation. The Assembly also later accepted the invitation of the Government of Switzerland to hold the Special Session at the United Nations Office in Geneva from 26-30 June, 2000.

Social development is inseparable from the cultural, ecological, economic, political and spiritual environment in which it takes place and cannot be pursued as a sectoral initiative. Social Development requires an orientation of values, objectives and priorities towards the well-being of all and the strengthening and promotion of conducive institutions and policies. Global community is threatened by environmental degradation, severe food crises, epidemics, all forms of racial discrimination, xenophobia, various forms of intolerance, violence and criminality and the risk of losing the richness of cultural diversity. Governments increasingly recognize that their responses to changing circumstances and their desires to achieve sustainable development and social progress will require increased solidarity, expressed through appropriate multilater programmes and strengthened international cooperation. Such cooperation is particularly crucial to ensure that countries in need of assistance, such as those in Africa and the least developed countries, can benefit from the process of globalisation.

Empowerment and participation are essential for democracy, harmony and social development. All members of society should have the opportunity and be able to exercise the right and responsibility to take an active part in the affairs of the community in which they live. Gender equality, equity and the full participation of women in all economic, social and political activities are essential. It is necessary to change the prevailing social paradigm of gender to usher in a new generation of women and men working together to create a more humane world order.

A people-centred approach to sustainable development, was suggested in the summit. The main features are broad-based participation and involvement of civil society in the formulation and implementation of decisions determining the functioning and well-being of our societies. Broad-based patterns of sustained economic growth and sustainable development and the integration of population issues into economic and development strategies, equitable and non-discriminatory distribution of the benefits of growth among social groups and countries, interaction of market forces conducive to efficiency and social development, public policies that seek to overcome socially divisive disparities, pluralism and diversity, a supportive and stable political and legal framework that promotes the mutually reinforcing relationship between democracy, development and all human rights and fundamental freedoms became the direction of all future environmental action.

World Conference on Women, Beijing, September 1995

Women have an essential role to play in the development of sustainable and ecologically sound consumption and production patterns. The fundamental framework of action for natural resource management, was already recognised at the United Nations Conference on Environment and Development and the International Conference on Population and Development and reflected throughout Agenda-21.

The deterioration of natural resources displaces communities, especially women from income generating activities, while greatly adding to unremunerative work. In both urban and rural areas, environmental degradation results in negative effects on the health, well-being and quality of life of the population at large, especially girls and women of all ages. Particular attention and recognition should be given to the role and special situation of women living in rural areas and those working in the agricultural sector, where access to occupational training, natural and productive resources, credit development programmes and cooperative structures can help them increase their participation in sustainable development. Environmental risks in the home and workplace may have a disproportionate impact on women's health because of women's different susceptibilities to the toxic effects of various chemicals. These risks to women's health are particularly high in urban areas, as well as in low-income areas where there is a high concentration of polluting industries.

Governments have expressed their commitment to creating a new development paradigm that integrates environmental sustainability with gender equality and justice within and between generations as contained in chapter 2 of Agenda-21. Women remain largely absent at all levels of policy formulation and decision-making in natural resource and environmental management, conservation, protection and rehabilitation, and their experience and skills in advocacy for and monitoring of proper natural resource management too often remains marginalised in policy-making and decision-making bodies. Women are rarely trained as professional natural resource managers with policy-making capacities, such as land-use planners, agriculturalists, foresters, marine scientists and environmental lawyers. Women have often played leadership roles or taken the lead in promoting an environmental ethic, reducing resource use, and reusing and recycling resources to minimize waste and excessive consumption. Women can have a particularly powerful role in influencing sustainable consumption decisions. In addition, women's contributions to environmental management, including through grassroots and youth campaigns to protect the environment, have often taken place at the local level. Women, especially indigenous women, have particular knowledge of ecological linkages and fragile ecosystem management. Women in many communities provide the main labour force for subsistence production. In certain regions, women are generally the most stable members of the community as men often pursue work in distant locations, leaving women to safeguard the natural environment and ensure adequate and sustainable resource allocation within the household and the community.

The strategic actions needed for sound environmental management require a holistic, multi-disciplinary and intersectoral approach. Women's participation and leadership are essential to every aspect of that approach. The recent United Nations global conferences on development, as well as regional preparatory conference for the Fourth World Conference on Women, have all acknowledged that sustainable development policies that do not involve women and men alike will not succeed in the long run. They have called for the effective participation of women in the generation of knowledge and environmental education in decision-making and management at all levels. Women's experiences and contributions to an ecologically sound environment must therefore be central to the agenda for the twenty first century. Sustainable development will be an elusive goal unless women's contribution to environmental management is recognised and supported. In addressing the lack of adequate recognition and support for women's contribution to conservation and management of natural resources and safeguarding the environment, Governments and other actors should promote an active and visible policy of mainstreaming a gender perspective in all policies and programmes. The three basic objectives set on priority are as follows.

(i) Involve women actively in environmental decision-making at all levels.

(ii) Integrate gender concerns and perspectives in policies and programmes for sustainable development.

(iii) Strengthen or establish mechanisms at the national, regional and international levels to assess the impact of development and environmental policies on women.

United Nations Conference on Human Settlements (Habitat II) Istanbul, Turkey, June 1996

It endorsed the universal goals of ensuring adequate shelter for all and making human settlements safer, healthier and more equitable, sustainable and productive. The two major themes of the Conference, adequate shelter for all and sustainable human settlements development in an urbanising world, have been inspired by the Charter of the United Nations and are aimed at reaffirming existing and forging new partnerships for action at the international, national and local levels to improve living environment. The participants recalled the first United Nations Conference on Human Settlements, held at Vancouver, Canada, the celebration of the International Year of Shelter for the Homeless and the Global Strategy for Shelter to the Year 2000, all of which have contributed to increased global awareness of the problems of human settlements and called for action to achieve adequate shelter for all. Recent United Nations world conferences including, in particular, the United Nations Conference on Environment and Development, has provided a comprehensive agenda for the equitable attainment of peace, justice and democracy built on economic development, social development and environmental protection as interdependent and mutually reinforcing components of sustainable development. The outcomes of these conferences is integrated into the Habitat Agenda.

The conference highlighted that the Rural and Urban development were interdependent. In addition to improving the urban habitat, one must also work to extend adequate infrastructure, public services and employment opportunities to rural areas in order to enhance their environment. Small and Medium sised towns need special focus so that migration to the metropolitan cities could be minimized. A global partnership was suggested to conserve, protect and restore the health and integrity of the Earth's ecosystem. In view of different contributions to global environmental degradation countries have common but differentiated responsibilities. Local action should be guided and stimulated through local programmes based on Agenda-21, the Habitat Agenda, or any other equivalent programme. The enabling strategy includes a responsibility for Governments to implement special measures for members of disadvantaged and vulnerable groups when appropriate.

This Conference in Istanbul marks a new era of cooperation, an era of a culture of solidarity. Its declaration said, 'As we move into the twenty first century, we offer a positive vision of sustainable human settlements, a sense of hope for our common future and an exhortation to join a truly worthwhile and engaging challenge, that of building together a world where everyone can live in a safe home with the promise of a decent life of dignity, good health, safety, happiness and hope'.

World Trade Organisation Ministerial Conference, Seattle (USA), December 1999

World Trade Organisation (WTO) was established in 1995 at Marrakech (Morocco). It was the culmination of the process that was inherent in the demand to solve the North-South inequitable trade which was initiated in the UNCTAD in the sixties by Raul Prebisch. The Dunkel Draft prepared to establish the GATT/WTO regime had been through three earlier Conferences at Uruguay (1988), Montreal (1989) and at Brussels (1990). The three decisive persons at the Seattle ministerial meet was the Director General Mike Moore, US trade representative Charlene Barshefsky and the European Union's Trade Commissioner Pascal Lamy. However at the frenzied end of the Meet the real victory was that of the Third World which had 100 members in the 135 member WTO but had been given no representation in the so called Green Room meetings organised by the three top decision-makers. The Director General Moore insisted that the 'WTO is democratic and Globalisation is not a choice'. He also assured that the Organisation of Economic Cooperation and Development (OECD) had projected that a new round would boost output by more than US $1.2 trillion and that the developing countries would benefit the most [Oh, 2000:27].

The Seattle Conference failed miserably as a 'Theatre of the Absurd' [Raghavan 2000:16]. The following reasons contributed to the collapse of the talks prematurely even before any formal declaration could be made:

1. The Third World nations felt they were completely sidelined by the undemocratic and secretive negotiating process of the ministerial dominated by the big three Bershefsky, Moore and Lamy. They had closed door Green Room meetings with their selected NGOs and carefully picked out developing countries leaving the majority aside.

2. There were street protests by the civil society groups and the aggressive activism against the inclusion of the labour linked and environment linked trade by the North. This resulted in a joint statement by over 2000 NGOs and social movements with the central message, 'No new round only a turnaround to WTO'.
3. The USA and EU differences on the issue of agricultural subsidies led to the open tiff between USA and the EU ministers of Environment and Agriculture.
4. The statement issued by the UN Committee on Economic, Social and Cultural Rights which urged the WTO to ensure the human rights obligations while deciding. This committee has been established by the UN Economic and Social Council to oversee implementation of the International Covenant on Economic, Social and Cultural Rights (the Covenant) which is a legally binding agreement. The resolution [1999, 30th August] of the UN Sub Commission on the Promotion and Protection of Human Rights called for the integration of human rights in the WTO agenda and a proper study to precede the social impact of the liberalisation process. The Human Development Report 1999 had already warned of the Trade Related Aspects of the Intellectual Property Rights (TRIPS).
5. UNO felt that there is no need for WTO to ponder over issues of labour and environment. UN Secretary General Kofi Annan's suggestion that the ILO and the UNEP are fully capable and empowered to deal with these issues. The Singapore Meet in 1996 had already agreed to tackle the issues through the UN Agencies. He said, 'The UN and its specialised agencies are charged with advancing the cause of development, the environment, human rights and labour. We can be part of the solution' [TWN/SUNS 4564].
6. The Women's Caucus, a powerful Women's Rights NGO declared that the Seattle Meet was the undoing of all the earlier gains that women had achieved through the UNCED, Vienna Human Rights Declaration, Copenhagen Summit on Social Development and the Fourth World Conference on Women and Habitat II.
7. A Declaration of the indigenous people at Seattle strongly condemned the WTO Agreements due their total contradiction of their 'Core Values', 'Spirituality and World Views'. They demanded amendment of Article 27.3 (b) of the TRIPS for the patenting of life forms. They reiterated the assurances given to them in the Convention on Biological Diversity in Articles 8 (j) 10 (c), 17.2 and 18.4 on the protection of indigenous and traditional knowledge, innovation and practices.

Seattle turned out to be just another round of the economic domination of the South by the North which has intricate and catastrophic consequences for the environment of the South. 'The ultimate goal (of the WTO) is to permanently codify trade issues and corporate profits making them the primary standards of a new form of global governance' [Barker and Mander 2000:37].

The linkages of the environmental governance with the Global trade issues is inseparable. The future of these international negotiations largely would depend upon the way the Third World is able to hold itself together against the concerted G-7. Guyana, which at present holds the Chairmanship of G-77 has asked for a thorough reform of the WTO and has urged the developing countries to use the breathing space provided by the collapse of WTO to push the reform agenda in accordance with a more just system of international trade.

The forthcoming UNCTAD X conference in Bangkok and the South Summit in Havana would hopefully provide further support to the Third World Communities to unite and propose meaningful reforms in the WTO.

Earth Summit+5: Progress on the Road from Rio

THE 1992 "EARTH SUMMIT" in Rio de Janeiro may be remembered as a turning point in humanity's thinking about development. Officially known as the United Nations Conference on Environment and Development (or UNCED), it brought together more than 100 heads of state and government, and more than 20,000 NGO representatives. After intensive negotiations, it reached an unprecedented consensus on Agenda 21, its far-reaching programme for sustainable development into the next century.

FIVE YEARS AFTER, how far have individual nations and the world community come in implementing Agenda 21? That question was discussed at a high-level Special Session of the United Nations General Assembly from 23 to 27 June 1997.

THE FOOD AND AGRICULTURE ORGANISATION (FAO) is Task Manager for four of Agenda 21's key chapters—planning and management of land resources, combating deforestation, sustainable mountain development, and sustainable agriculture and rural development. It is actively involved in the implementation of others—such as combating desertification and drought, and conservation of biological diversity—and in broader areas covered by Agenda 21, particularly climate change and energy.

FAO PREPARED for the UN special session a series of Agenda 21 Progress Reports, published here, describing the challenges addressed by Agenda 21, progress since the Earth Summit, key issues to be resolved, future directions, FAO's role and contact points.

Agenda 21 Progress Reports (FAO, June 1997)

- Planning and management of land resources
- Combating deforestation
- Combating desertification and drought
- Sustainable mountain development
- Sustainable agriculture and rural development

MABnet

THE MAN AND THE BIOSPHERE (MAB) PROGRAMME

Draft Synthetic Report on Biosphere Reserves

Call for Contributions

A report is being prepared on the biosphere reserve concept and its implementation. Please find the draft in pdf and Word format. Readers are invited to provide corrections and suggestions for additional written and graphic materials, by 1 December.

The Seville+5 International Expert Meeting took place in Pamplona, Spain (23-27 October 2000). The recommendations from this meeting will be submitted to the MAB-ICC at its 16th session. They will be posted on the MABnet thereafter.

United Nations Development Programme

Sustainable Development Networking Programme

This is a trangram, a puzzle in which the player uses 7 shapes to create a picture like the trangram, the SDNP is based on the concept of building solutions. Countries design their own unique plans to achieve sustainable development and the SDNP provides the decision makers in government and civil society with information that is relevant for both design and implementation of the development plan.

What is SDNP?

- SDNP is a network of national development actors in over 40 developing countries,
- It's origins lie in the 1992 UNCHD Conference and it's focus is sustainable development issues.
- Every SDNP node is a country-owned, non-profit network that is open to all the stakeholders in sustainable development.
- Each SDNP crafts it's own programme in keeping with the needs of its stakeholders (the tangram)

- The multi-national SDNP family has produced multi-lingual resources on sustainable development as well as on the technical aspects of networking.

Take a tour of Some SDNP innovations .....

- Guatemala: Empowering indigenous women, preserving traditional crafts via Guatemala based e-commerce
- Honduras: Coordinating disaster relief and linking communities via public access centers and people networks
- Lebanon: An Environmental Web Director)' and lots of other information on Lebanon.
- Colombia: E-commerce for Small Enterprises
- Cameroon: Empowering Women Through IT
- Pakistan: Disaster Relief through IT Tools
- SDNP IT News Site
- Highlights.

United Nations: General Assemly Resolution A/RES/S-19/2 of the 19th Special Session on Progress Achieved Towards Meeting Objectives of the Earth Summit with Annex of a Programme for the Further Implementaiton of Agenda 21, [June 28, 1997]

I. Statement of Commitment

1. At the nineteenth-special session of the United Nations General Assembly, we-heads of State or Government and other heads of delegations, together with our partners from international institutions and non-governmental organisations—have gathered to review progress achieved over the five years that have passed since the United Nations Conference on Environment and Development and to re-energize our commitment to further action on goals and objectives set out by the Earth Summit.

2. The United Nations Conference on Environment and Development was a landmark event. At that Conference, we launched a new global partnership for sustainable development—a partnership that respects the indivisibility of environmental protection and the development process. It is founded on a global consensus and political commitment at the highest level. Agenda 21, adopted at Rio de Janeiro, addresses the pressing environment and development problems of today and also aims at preparing the world for the challenges of the next century in order to attain the long-term goals of sustainable development.

3. Our focus at this special session has been to accelerate the implementation of Agenda 21 in a comprehensive manner and not to renegotiate its provisions or to be

selective in its implementation. We reaffirm that Agenda 21 remains the fundamental programme of action for achieving sustainable development. We reaffirm all the principles contained in the Rio Declaration on Environment and Development and the Forest Principles. We are convinced that the achievement of sustainable development requires the integration of its economic, environmental and social components. We recommit to working together-in the spirit of global partnership—to reinforce our joint efforts to meet equitably the needs of present and future generations.

4. We acknowledge that a number of positive results have been achieved, but we are deeply concerned that the overall trends with respect to sustainable development are worse today than they were in 1992. We emphasize that the implementation of Agenda 21 in a comprehensive manner remains vitally important and is more urgent now than ever.

5. Time is of the essence in meeting the challenges of sustainable development as set out in the Rio Declaration and Agenda 21. To this end, we recommit ourselves to the global partnership established at the United Nations Conference on Environment and Development and to the continuous dialogue and action inspired by the need to achieve a more efficient and equitable world economy, as a means to provide a supportive international climate for achieving environment and development goals. We therefore, pledge to continue to work together, in good faith and in the spirit of partnership, to accelerate the implementation of Agenda 21. We invite everyone throughout the world to join us in our common cause.

6. We commit ourselves to ensuring that the next comprehensive review of Agenda 21 in the year 2002 demonstrates greater measurable progress in achieving sustainable development. The present Programme for the Further Implementation of Agenda 21 is our vehicle for achieving that goal. We commit ourselves to fully implementing this Programme.

II. Assessment of Progress Made Since the United Nations Conference on Environment and Development

7. The five years that have elapsed since the United Nations Conference on Environment and Development have been characterised by the accelerated globalisation of interactions among countries in the areas of world trade, foreign direct investment and capital markets. Globalisation presents new opportunities and challenges. It is important that national and international environmental and social policies be implemented and strengthened in order to ensure that globalisation trends have a positive impact on sustainable development, especially in developing countries. The impact of recent trends in globalisation on developing countries has been uneven. A limited number of developing countries have been able to take advantage of those

trends, attracting large inflows of external private capital and experiencing significant export-led growth and acceleration of growth in per capita gross domestic product. Many other countries, however, in particular African countries and the least developed countries, have shown slow or negative growth and continue to be marginalized. As a result, they generally experienced stagnating or falling per capita gross domestic product through 1995. In these and in some other developing countries, the problems of poverty, low levels of social development, inadequate infrastructure and lack of capital have prevented them from benefiting from globalisation. While continuing their efforts to achieve sustainable development and to attract new investments, these countries still require international assistance in their efforts directed towards sustainable development. In particular the least developed countries continue to be heavily dependent on a declining volume of official development assistance for the capacity-building and infrastructure development required to provide for basic needs and more effective participation in the globalising world economy. In an increasingly interdependent world economy, the responsible conduct of monetary and other macroeconomic policies requires that their potential impact on other countries be taken into account. Since the Conference, the countries with economies in transition have achieved significant progress in implementing the principles of sustainable development. However, the need for full integration of these countries into the world economy remains one of the crucial problems on their way towards sustainable development. The international community should continue to support these countries in their efforts to accelerate the transition to a market economy and to achieve sustainable development.

8. Although economic growth—reinforced by globalisation—has allowed some countries to reduce the proportion of people in poverty, for others marginalisation has increased. Too many countries have seen economic conditions worsen and public services deteriorate; the total number of people in the world living in poverty has increased. Income inequality has increased among countries and also within them, unemployment has worsened in many countries, and the gap between the least developed countries and other countries has grown rapidly in recent years. On a more positive note, population growth rates have been declining globally, largely as a result of expanded basic education and health care. That trend is projected to lead to a stable world population in the middle of the twenty-first century. There has also been progress in social services, with expanding access to education, declining infant mortality and increasing life expectancy in most countries. However, many people, particularly in the least developed countries, still do not have access to adequate food and basic social services or to clean water and sanitation. Reducing current inequities in the distribution of wealth and access to resources, both within and among countries, is one of the most serious challenges facing humankind.

9. "Five years after the United Nations Conference on Environment and Development, the state of the global environment has continued to deteriorate, as noted in the Global Environment Outlook of the United Nations Environment Programme, and significant environmental problems remain deeply embedded in the socio-economic fabric of countries in all regions. Some progress has been made in terms of institutional development, international consensus-building, public participation and private sector actions and, as a result, a number of countries have succeeded in curbing pollution and slowing the rate of resource degradation. Overall, however, trends are worsening. Many polluting emissions, notably of toxic substances, greenhouse gases and waste volumes are continuing to increase although in some industrialised countries emissions are decreasing. Marginal progress has been made in addressing unsustainable production and consumption patterns. Insufficient progress has also been identified in the field of environmentally sound management and adequate control of transboundary movements of hazardous and radioactive wastes. Many countries undergoing rapid economic growth and urbanisation are also experiencing increasing levels of air and water pollution, with accumulating impacts on human health. Acid rain and transboundary air pollution, once considered a problem only in the industrialised countries, are increasingly becoming a problem in many developing regions. In many poorer regions of the world, persistent poverty is contributing to accelerated degradation of natural resources and desertification has spread. In countries seriously affected by drought and/or desertification, especially those in Africa, their agricultural productivity, among other things, is uncertain and continues to decline, thereby hampering their efforts to achieve sustainable development. Inadequate and unsafe water supplies are affecting an increasing number of people worldwide, aggravating problems of ill health and food insecurity among the poor. Conditions in natural habitats and fragile ecosystems, including mountain ecosystems, are still deteriorating in all regions of the world, resulting in diminishing biological diversity. At the global level, renewable resources, in particular fresh water, forests, topsoil and marine fish stocks, continue to be used at rates beyond their viable rates of regeneration; without improved management, this situation is clearly unsustainable.

10. While there has been progress in material and energy efficiency, particularly with reference to non-renewable resources, overall trends remain unsustainable. As a result, increasing levels of pollution threaten to exceed the capacity of the global environment to absorb them, increasing the potential obstacles to economic and social development in developing countries.

11. Since the United Nations Conference on Environment and Development, extensive efforts have been made by Governments and international organisations to integrate environmental, economic and social objectives into decision-making by

elaborating new policies and strategies for sustainable development or by adapting existing policies and plans. As many as one hundred and fifty countries have responded to the commitments established at the Conference through national-level commissions or coordinating mechanisms designed to develop an integrated approach to sustainable development.

12. The major groups have demonstrated what can be achieved by taking committed action, sharing resources and building consensus, reflecting grass-roots concern and involvement. The efforts of local authorities are making Agenda 21 and the pursuit of sustainable development a reality at the local level through the implementation of "local Agenda 21s" and other sustainable development programmes. Non-governmental organisations, educational institutions, the scientific community and the media have increased public awareness and discussion of the relations between environment and development in all countries. The involvement, role and responsibilities of business and industry, including transnational corporations, are important. Hundreds of small and large businesses have made "green business" a new operating mode. Workers and trade unions have established partnerships with employers and communities to encourage sustainable development in the workplace. Farmer—led initiatives have resulted in improved agricultural practices contributing to sound resource management. Indigenous people have played an increasing role in addressing issues affecting their interests and particularly concerning their traditional knowledge and practices. Young people and women around the world have played a prominent role in galvanising communities into recognising their responsibilities to future generations. Nevertheless, more opportunities should be created for women to participate effectively in economic, social and political development as equal partners in all sectors of the economy.

13. Among the achievements since the United Nations Conference on Environment and Development have been the entry into force of the United Nations Framework Convention on Climate Change, the Convention on Biological Diversity and the United Nations Convention to Combat Desertification in those Countries Experiencing Serious Drought and/or Desertification, particularly in Africa; the conclusion of the Agreement on the Implementation of the Provisions of the United Nations Convention on the Law of the Sea of 10 December 1982 Relating to the Conservation and Management of Straddling Fish Stocks and Highly Migratory Fish Stocks; the adoption of the Programme of Action for the Sustainable Development of Small Island Developing States; the elaboration of the Global Programme of Action for the Protection of the Marine Environment from Land-based Activities; and the entry into force of the United Nations Convention on the Law of the Sea. Implementation of these important commitments and of others adopted before the United Nations Conference on Environment and Development by all the parties to them, remains however, to be carried out, and in many cases further strengthening of their provisions

is required as well as the mechanisms for putting them into effect. The establishment, restructuring, funding and replenishment of the Global Environment Facility were a major achievement. However, its levels of funding and replenishment have not been sufficient fully to meet its objectives.

14. Progress has been made in incorporating the principles contained in the Rio Declaration on Environment and Development—including the principle of common but differentiated responsibilities, which embodies the important concept of and basis for international partnership; the precautionary principle; the polluter pays principle; and the environmental impact assessment principle—in a variety of international and national legal instruments. While some progress has been made in implementing United Nations Conference on Environment and Development commitments through a variety of international legal instruments, much remains to be done to embody the Rio principles more firmly in law and practice.

15. A number of major United Nations conferences have advanced international commitment for the achievement of long-term goals and objectives directed towards sustainable development.

16. Organisations and programmes of the United Nations system have played an important role in the progress made in the implementation of Agenda 21. The Commission on Sustainable Development was established to review progress achieved in the implementation of Agenda 21, advance global dialogue and foster partnerships for sustainable development. The Commission has catalysed new action and commitments and has contributed to the deliberations on sustainable development among a wide variety of partners within and outside the United Nations system. Although much remains to be done, progress has also been made at the national, regional and international levels in implementing the United Nations Conference on Environment and Development Forest Principles, including through the Commission's Ad Hoc Intergovernmental Panel on Forests.

17. Provision of adequate and predictable financial resources and the transfer of environmentally sound technologies to developing countries are critical elements for the implementation of Agenda 21. However, while some progress has been made, much remains to be done to activate the means of implementation set out in Agenda 21, in particular in the areas of finance and technology transfer, technical assistance and capacity-building.

18. Most developed countries have still not reached the United Nations target, reaffirmed by most countries at the United Nations Conference on Environment and Development, of committing 0.7 per cent of their gross national product to official development assistance or the United Nations target, as agreed, of committing 0.15 per cent of gross national product as official development assistance to the least

developed countries. Regrettably, on average, official development assistance as a percentage of the gross national product of developed countries has drastically declined in the post—Conference period, from 0.34 per cent in 1992 to 0.27 per cent in 1995, but official development assistance has taken more account of the need for an integrated approach to sustainable development.

19. In other areas, results have been encouraging since the United Nations Conference on Environment and Development. There has been a sizeable expansion of private flows of financial resources from developed to a limited number of developing countries and, in a number of countries, efforts have been made in support of domestic resource mobilisation, including the increasing use of economic instruments to promote sustainable development.

20. In many developing countries, the debt situation remains a major constraint on achieving sustainable development. Although the debt situation of some middle-income countries has improved, there is a need to continue to address the debt problems of the heavily indebted poor countries, which continue to face unsustainable external debt burdens. The recent World Bank/International Monetary Fund Heavily Indebted Poor Countries Initiative could help to address that issue with the cooperation of all creditor countries. Further efforts by the international community are still required to remove debt as an impediment to sustainable development.

21. Similarly, technology transfer and technology-related investment from public and private sources, which are particularly important to developing countries, have not been realised as outlined in Agenda 21. Although increased private flows have led to investments in industry and technology in some developing countries and economies in transition, many other countries have been left behind. Conditions in some of these countries have been less attractive to private sector investment and technological change has been slower, thus limiting their ability to meet their commitments to Agenda 21 and other international agreements. The technology gap between developed countries and, in particular, the least developed countries has widened.

III. Implementation of Agenda 21 in Areas Requirement Urgent Action

22. Agenda 21 and the principles contained in the Rio Declaration on Environment and Development established a comprehensive approach to the achievement of sustainable development. While it is the primary responsibility of national Governments to achieve the economic, social and environmental objectives of Agenda 21, it is essential that international cooperation be reactivated and intensified, recognising, *inter alia*, the principle of common but differentiated responsibilities as set forth in Principle 7 of the Rio Declaration. This requires the mobilisation of stronger political will and the invigoration of a genuine new global partnership, taking into account

the special needs and priorities of developing countries. Such an approach remains as relevant and as urgently needed as ever. It is clear from the assessment above that, although progress has been made in some areas, a major new effort will be required to achieve the goals established at the United Nations Conference on Environment and Development, particularly in areas of cross-sectoral matters where implementation has yet to be achieved. The proposals set out in sections A to C below outline strategies for accelerating progress towards sustainable development. The sections are equally important and must be considered and implemented in a balanced and integrated way.

A. Integration of Economic, Social and Environmental Objectives

23. Economic development, social development and environmental protection are interdependent and mutually reinforcing components of sustainable development. Sustained economic growth is essential to the economic and social development of all countries, in particular developing countries. Through such growth, which should be broadly based so as to benefit all people, countries will be able to improve the standards of living of their people through the eradication of poverty, hunger, disease and illiteracy and the provision of adequate shelter and secure employment for all, and the preservation of the integrity of the environment. Growth can foster development only if its benefits are fully shared. It must therefore also be guided by equity, justice and social and environmental considerations. Development, in turn, must involve measures that improve the human condition and the quality of life itself. Democracy, respect for all human rights and fundamental freedoms, including the right to development, transparent and accountable governance in all sectors of society, as well as effective participation by civil society, are also an essential part of the necessary foundations for the realisation of social and people-centred sustainable development.

24. Sustainable development strategies are important mechanisms for enhancing and linking national capacity so as to bring together priorities in social, economic and environmental policies. Hence, special attention must be given to the fulfilment of commitments in the areas set out below, in the framework of an integrated approach towards development, consisting of mutually reinforcing measures to sustain economic growth, as well as to promote social development and environmental protection. Achieving sustainable development cannot be carried out without greater integration at all policy-making levels and at operational levels, including the lowest administrative levels possible. Economic sectors, such as industry, agriculture, energy, transport and tourism, must take responsibility for the impact of their activities on human well-being and the physical environment. In the context of good governance, properly constructed strategies can enhance prospects for economic growth and employment and at the same time project the environment. All sectors of society should be involved in their development and implementation, as follows:

(a) By the year 2002, the formulation and elaboration of national strategies for sustainable development that reflect the contributions and responsibilities of all interested parties should be completed in all countries, with assistance provided, as appropriate, through international cooperation, taking into account the special needs of the least developed countries. The efforts of developing countries in effectively implementing national strategies should be supported. Countries that already have national strategies should continue their efforts to enhance and effectively implement them. Assessment of progress achieved and exchange of experience among Governments should be promoted. Local Agenda 21s and other local sustainable development programmes, including youth activities, should also be actively encouraged;

(b) In integrating economic, social and environmental objectives, it is important that a broad package of policy instruments, including regulation, economic instruments, internalisation of environmental costs in market prices, environmental and social impact analysis, and information dissemination, be worked out in the light of country-specific conditions to ensure that integrated approaches are effective and cost-efficient. To this end, a transparent and participatory process should be promoted. This will require the involvement of national legislative assemblies, as well as all actors of civil society, including youth and indigenous people and their communities, to complement the efforts of Governments for sustainable development. In particular, the empowerment and the full and equal participation of women in all spheres of society, including participation in the decision-making process, are central to all efforts to achieve such development;

(c) The implementation of policies aiming at sustainable development, including the 'Combating poverty' and 'Strengthening the role of workers and their trade unions' of Agenda 21, may enhance the opportunities for job creation, thus helping to achieve the fundamental goal of eradicating poverty.

An Enabling International Economic Climate

25. A mutually supportive balance between the international and the national environment is needed in the pursuit of sustainable development. As a result of globalisation, external factors have become critical in determining the success or failure of developing countries in their national efforts. The gap between developed and developing countries points to the continued need for a dynamic and enabling international economic environment supportive of international cooperation, particularly in the fields of finance, technology transfer, debt and trade, if the momentum for global progress towards sustainable development is to be maintained and increased.

26. To foster a dynamic and enabling international economic environment favourable to all countries is in the interest of all countries. Moreover, issues, including environmental issues, that bear on the international economic environment can be approached effectively only through a constructive dialogue and genuine partnership on the basis of mutuality of interests and benefits, taking into account that, in view of the different contributions to global environmental degradation. States have common but differentiated responsibilities.

Eradicating Poverty

27. Given the severity of poverty, particularly in developing countries, the eradication of poverty is one of the fundamental goals of the international community and the entire United Nations system, as reflected in Commitment 2 of the Copenhagen Declaration on Social Development, and is essential for sustainable development. Poverty eradication is thus an overriding theme of sustainable development for the coming years. The enormity and complexity of the poverty issue could very well endanger the social fabric, undermine economic development and the environment, and threaten political stability in many countries. To achieve poverty eradication, efforts of individual Governments and international cooperation and assistance should be brought together in a complementary way. Eradication of poverty depends on the full integration of people living in poverty into economic, social and political life. The empowerment of women is a critical factor for the eradication of poverty. Policies that promote such integration to combat poverty, in particular policies for providing basic social services and broader socio-economic development, are effective as well since enhancing the productive capacity of poor people increases both their well-being and that of their communities and societies, and facilitates their participation in resource conservation and environmental protection. The provision of basic social services and food security in an equitable way is a necessary condition for such integration and empowerment. The 20/20 initiative as referred to in the Programme of Action of the World Summit for Social Development is, among other things, a useful means for such integration. However, the five years sinde the Rio Conference have witnessed an increase in the number of people living in absolute poverty, particularly in developing countries. In this context, there is an urgent need for the timely and full implementation of all the relevant commitments, agreements and targets already agreed upon since the Rio Conference by the international community, including the United Nations system and international financial institutions. Full implementation of the Programme of Action of the World Summit for Social Development is essential. Priority actions include:

(a) Improving access to sustainable livelihoods, entrepreneurial opportunities and productive resources, including land, water, credit, technical and administrative training, and appropriate technology, with particular efforts to broaden the human and social capital basis of societies so as to reach the rural poor and the urban informal sector;

(b) Providing universal access to basic social services, including basic education, health care, nutrition, clean water and sanitation;

(c) Progressively developing, in accordance with the financial and administrative capacities of each society, social protection systems to support those who cannot support themselves, either temporarily or permanently; the aim of social integration is to create a "society for all";

(d) Empowering people living in poverty and their organisations by involving them fully in the formulation, implementation and evaluation of strategies and programmes for poverty eradication and community development and by ensuring that these programmes reflect their priorities;

(e) Addressing the disproportionate impact of poverty on women, in particular by removing legislative, policy, administrative and customary barriers to women's equal access to productive resources and services, including access to and control over land and other forms of property, credit, including micro-credit, inheritance, education, information, health care and technology. In this regard, full implementation of the Beijing Platform for Action is essential;

(f) Working together of interested donors and recipients to allocate increased shares of official development assistance to poverty eradication. The 20/20 initiative is an important principle in this respect, as it is based on a mutual commitment among donors and recipients to increasing resources allocated to basic social services;

(g) Intensifying international cooperation to support measures being taken in developing countries to eradicate poverty, to provide basic social protection and services, and to approach poverty eradication efforts in an integral and multidimensional manner.

Changing Consumption and Production Patterns

28. Unsustainable patterns of production and consumption, particularly in the industrialised countries, are identified in Agenda 21 as the major cause of continued deterioration of the global environment. While unsustainable patterns in the industrialised countries continue to aggravate the threats to the environment, there remain huge difficulties for developing countries in meeting basic needs such as food, health care, shelter and education for people. All countries should strive to promote sustainable consumption patterns; developed countries should take the lead in achieving sustainable consumption patterns; developing countries should seek to achieve sustainable consumption patterns in their development process, guaranteeing the provision of basic needs for the poor, while avoiding those

unsustainable patterns, particularly in industrialised countries, generally recognised as unduly hazardous to the environment, inefficient and wasteful, in their development processes. This requires enhanced technological and other assistance from industrialised countries. In the follow-up of the implementation of Agenda 21, the review of progress made in achieving sustainable consumption patterns should be given high priority. Consistent with Agenda 21, the development and further elaboration of national policies and strategies, particularly in industrialised countries, are needed to encourage changes in unsustainable consumption and production patterns, while strengthening, as appropriate, international approaches and policies that promote sustainable consumption patterns on the basis of the principle of common but differentiated responsibilities, applying the polluter pays principle, and encouraging producer responsibility and greater consumer awareness. Eco-efficiency, cost internalisation and product policies are also important tools for making consumption and production patterns more sustainable. Actions in this area should focus on:

(a) Promoting measures to internalize environmental costs and benefits in the price of goods and services, while seeking to avoid potential negative effects for market access by developing countries, particularly with a view to encouraging the use of environmentally preferable products and commodities. Governments should consider shifting the burden of taxation onto unsustainable patterns of production and consumption; it is of vital importance to achieve such an internalisation of environmental costs. Such tax reforms should include a socially responsible process of reduction and elimination of subsidies to environmentally harmful activities;

(b) Promoting the role of business in shaping more sustainable patterns of consumption by encouraging, as appropriate, the voluntary publication of environmental and social assessments of its own activities, taking into account specific country conditions, and actions as an agent of change in the market, and actions in its role as a major consumer of goods and services;

(c) Developing core indicators to monitor critical trends in consumption and production patterns, with industrialised countries taking the lead;

(d) Identifying best practices through evaluations of policy measures with respect to their environmental effectiveness, efficiency and implications for social equity, and disseminating such evaluations;

(e) Taking into account the linkages between urbanisation and the environmental and developmental effects of consumption and production patterns in cities, thus promoting more sustainable patterns of urbanisation;

(f) Promoting international and national programmes for energy and material efficiency with timetables for their implementation, as appropriate. In this regard, attention should be given to studies that propose to improve the efficiency of resource use, including consideration of a 10-fold improvement in resource productivity in industrialised countries in the long term and a possible factor-four increase in industrialised countries in the next two or three decades. Further research is required to study the feasibility of these goals and the practical measures needed for their implementation. Industrialised countries will have a special responsibility and must take the lead in this respect. The Commission on Sustainable Development should consider this initiative in the coming years in exploring policies and measures necessary to implement eco-efficiency and, for this purpose, encourage the relevant bodies to adopt measures aimed at assisting developing countries in improving energy and material efficiency through the promotion of their endogenous capacity-building and economic development with enhanced and effective international support;

(g) Encouraging Governments to take the lead in changing consumption patterns by improving their own environmental performance with action-oriented policies and goals on procurement, the management of public facilities and the further integration of environmental concerns into national policy-making. Governments in developed countries, in particular, should take the lead in this regard;

(h) Encouraging the media, advertising and marketing sectors to help shape sustainable consumption patterns;

(i) Improving the quality of information regarding the environmental impact of products and services and, to that end, encouraging the voluntary and transparent use of eco-labelling;

(i) Promoting measures favouring eco-efficiency; however, developed countries should pay special attention to the needs of developing countries, in particular by encouraging positive impacts, and to the need to avoid negative impacts on export opportunities and on market access for developing countries and, as appropriate, for countries with economies in transition;

(j) Encouraging the development and strengthening of educational programmes to promote sustainable consumption and production patterns;

(k) Encouraging business and industry to develop and apply environmentally sound technology that should aim not only at increasing competitiveness but also at reducing negative environmental impacts;

(m) Giving balanced consideration to both the demand side and the supply side of the economy in matching environmental concerns and economic factors, which could encourage changes in the behaviour of consumers and producers. A number of policy options should be examined; they include regulatory instruments, economic and social incentives and disincentives, facilities and infrastructure, information, education, and technology development and dissemination.

Making Trade and Environment Mutually Supportive

29. In order to accelerate economic growth, poverty eradication and environmental protection, particularly in developing countries, there is a need to establish macroeconomic conditions in both developed and developing countries that favour the development of instruments and structures enabling—all countries, in particular developing countries, to benefit from globalisation. International cooperation and support for capacity-building in trade, environment and development should be strengthened through renewed system-wide efforts, and with greater responsiveness to sustainable development objectives, by the United Nations, the World Trade Organisation and the Bretton Hoods institutions, as well as by national Governments. There should be a balanced and integrated approach to trade and sustainable development, based on a combination of trade liberalisation, economic development and environmental protection. Trade obstacles should be removed with a view to contributing to the achieving of more efficient use of the earth's natural resources in both economic and environmental terms. Trade liberalisation should be accompanied by environmental and resource management policies in order to realise its full potential contribution to improved environmental protection and the promotion of sustainable development through the more efficient allocation and use of resources. The multilateral trading system should have the capacity to further integrate environmental considerations and enhance its contribution to sustainable development, without undermining its open, equitable and non-discriminatory character. The special and differential treatment for developing countries, especially the least developed countries, and the other commitments of the Uruguay Round of maltilateral trade negotiations should be fully implemented in order to enable those countries to benefit from the international trading system, while conserving the environment. There is a need for continuing the elimination of discriminatory and protectionist practices in international trade relations, which will have the effect of improving access for the exports of developing countries. This will also facilitate the full integration of economies in transition into the world economy. In order to make trade, environment and development mutually supportive, measures need to be taken to ensure transparency in the use of trade measures related to the environment, and should address the root causes of environmental degradation so as not to result in disguised barriers to trade. Account should be taken of the fact that environmental standards valid for developed countries may have unwarranted social and economic costs in other countries, in particular developing countries. International cooperation is needed and unilateralism should be avoided. The following actions are required:

(a) There should be timely and full implementation of the results of the Uruguay Round of multilateral trade negotiations and full use of the Comprehensive and Integrated World Trade Organisation Plan of Action for the Least Developed Countries;

(b) An open, non-discriminatory, rule-based, equitable, secure, transparent and predictable multilateral trading system should be promoted. In this context, effective measures are called for to achieve the complete integration of developing countries and countries with economies in transition into the world economy and the new international trading system. In this connection, there is a need to promote the universality of the World Trade Organisation and to facilitate the admission to membership in that organisation, in a mutually beneficial way, of developing countries and countries with economies in transition applying for membership. Actions should be taken to maximize the opportunities and to minimize the difficulties of developing countries, including the net food-importing ones, especially the least developed countries, and of countries with economies in transition, in adjusting to the changes introduced by the Uruguay Round. Decisions on further liberalisation of trade should take into account effects on sustainable development and should be consistent with an open, rule-based, non-discriminatory, equitable, secure and transparent multilateral trading system. The relationship between multilateral environmental agreements and the World Trade Organisation rules should be clarified;

(c) Implementation of environmental measures should not result in disguised barriers to trade;

(d) Within the framework of Agenda 21, trade rules and environmental principles should interact harmoniously;

(e) Further analysis of the environmental effects of the international transport of goods is warranted;

(f) Cooperation and coordination between the United Nations Conference on Trade and Development, the United Nations Industrial Development Organisation the World Trade Organisation, the United Nations Environment Programme and other relevant institutions should be strengthened on various issues, including (i) the role of positive measures in multilateral environmental agreements as part of a package of measures including, in certain cases, trade measures; (ii) the special conditions and needs of small and medium-sized enterprises in the trade and environment interface; (iii) trade and environment issues at the regional and subregional levels, including within the context of regional economic and trade as well as environmental agreements;

(g) Cooperation and coordination between the United Nations Conference on Trade and Development and other relevant bodies within their existing respective mandates should be enhanced, *inter alia,* on environment and sustainable development issues, without prejudice to the clear understanding in the World Trade Organisation that future negotiations, if any, regarding a multilateral agreement on investment will take place only after an explicit consensus decision, future agreements on investments should take into account the objectives of sustainable development and, when developing countries are parties to these agreements, special attention should be given to their needs for investment;

(h) National Governments should make every effort to ensure policy coordination on trade, environment and development at the national level in support of sustainable development;

(i) There is a need for the World Trade Organisation, the United Nations Environment Programme and the United Nations Conference on Trade and Development to consider ways to make trade and environment mutually supportive, including through due respect for the objectives and principles of the multilateral trading system and for the provisions of multilateral environmental agreements. Such considerations should be consistent with an open, rule-based, non-discriminatory, equitable, secure and transparent multilateral trading system.

Population

30. The impact of the relationship among economic growth, poverty, employment, environment and sustainable development has become a major concern. There is a need to recognize the critical linkages between demographic trends and factors and sustainable development. The current decline in population growth rates must be further promoted through national and international policies that promote economic development, social development, environmental protection, and poverty eradication, particularly the further expansion of basic education, with full and equal access for girls and women, and health care, including reproductive health care, including both family planning and sexual health, consistent with the report of the International Conference on Population and Development.

Health

31. The goals of sustainable development cannot be achieved when a high proportion of the population is afflicted with debilitating illnesses. An overriding goal for the future is to implement the Health for All strategy and to enable all people, particularly the world's poor, to achieve a higher level of health and well-being, and to improve their economic productivity and social potential. Protecting children from environmental

health threats and infectious disease is particularly urgent since children are more susceptible than adults to those threats. Top priority should be attached to supporting the efforts of countries, particularly developing countries, and international organisations to eradicate the major infectious diseases, especially malaria, which is on the increase, to improve and expand basic health and sanitation services, and to provide safe drinking water. It is also important to reduce indigenous cases of vaccine-preventable diseases through the promotion of widespread immunisation programmes, promote accelerated research and vaccine development and reduce the transmission of other major infectious diseases, such as dengue fever, tuberculosis and human immunodeficiency virus/acquired immunodeficiency syndrome (HIV/AIDS). Given the severe and irreversible health effects of lead poisoning, particularly on children, it is important to accelerate the process of eliminating unsafe uses of lead, including the use of—lead in gasoline worldwide, in light of country-specific conditions and with enhanced international support and assistance to developing countries, particularly through the timely provision of technical and financial assistance and the promotion of endogenous capacity-building. Strategies at the regional, national and local levels for reducing the potential risk due to ambient and indoor air pollution should be developed, bearing in mind their serious impacts on human health, including strategies to make parents, families and communities aware of the adverse environmental health impacts of tobacco. The clear linkage between health and the environment needs to be emphasised and the lack of information on the impact of environmental pollution on health should be addressed. Health issues should be fully integrated into national and subnational sustainable development plans and should be incorporated into project and programme development as a component of environmental impact assessments. Important to efforts at national levels is international cooperation in disease prevention, early warning, surveillance, reporting, training and research, and treatment.

Sustainable Human Settlements

32. Sustainable human settlements development is essential to sustainable development. The need to intensify efforts and cooperation to improve living conditions in the cities, towns, villages and rural areas throughout the world is recognized. Approximately half the world's population already lives in urban settlements, and by early in the next century the majority—more than 5 billion people—will be urban residents. Urban problems are concerns common to both developed and developing countries, although urbanisation is occurring most rapidly in developing countries. Urbanisation creates both challenges and opportunities. Global urbanisation is a cross-sectoral phenomenon that has an impact on all aspects of sustainable development. Urgent action is needed to implement fully the commitments made at the United Nations Conference on Human Settlements (Habitat II) consistent with its report, and in Agenda 21. New and additional financial resources from various

sources are necessary to achieve the goals of adequate shelter for all and sustainable human settlements development in an urbanising world. Transfer of expertise and technology, capacity-building, decentralisation of authority through, *inter alia,* strengthening of local capacity and private-public partnerships to improve the provision and environmentally sound management of infrastructure and social services should be accelerated to achieve more sustainable human settlements development. Local Agenda 21 programmes should also be actively encouraged. Global targets could be established by the Commission on Sustainable Development to promote local Agenda 21 campaigns and to deal with obstacles to local Agenda 21 Initiatives.

B. Sectors and Issues

33. The present section identifies a number of specific areas that are of widespread concern since failure to reverse current trends in these areas, notably in resource degradation, will have potentially disastrous effects on social and economic development, on human health and on environmental protection for all countries, particularly developing countries. All sectors covered by Agenda 21 are equally important and thus deserve attention by the international community on an equal footing. The need for integration is important in all sectors, including the areas of energy and transport because of the adverse effects that developments in those areas can have on human health and ecosystems; the areas of agriculture and water use, where inadequate land-use planning, poor water management and inappropriate technology can result in the degradation of natural resources and human impoverishment and where drought and desertification can result in land degradation and soil loss; and the area of management of marine resources, where competitive overexploitation can damage the resource base, food supplies and the livelihood of fishing communities, as well as the environment. The recommendations made in each of the sectors take into account the need for international cooperation in support of national efforts, within the context of the principles of the United Nations Conference on Environment and Development, including, *inter alia,* the principle of common but differentiated responsibilities. It is likewise understood that these recommendations do not in any way prejudice the work accomplished under legally binding conventions, where they exist, concerning these sectors.

Fresh Water

34. Water resources are essential for satisfying basic human needs, health and food production, and the preservation of ecosystems, as well as for economic and social development in general. It is a matter of urgent concern that more than one fifth of all people still do not have access to safe drinking water and more than one half of humanity lacks adequate sanitation. From the perspective of developing countries, fresh water is a priority and a basic need, especially taking into account

that in many developing countries fresh water is not readily available for all segments of the population, *inter alia,* owing to lack of adequate infrastructure and capacity, water scarcity, and technical and financial constraints. Moreover, fresh water is also crucial for developing countries in order to satisfy the basic needs of their population in the areas of agricultural irrigation, industrial development, hydroelectric generation, and so forth. In view of the growing demands on water, which is a finite resource, it will become a major limiting factor in socio-economic development unless early action is taken. There is growing concern regarding the increasing stress on water supplies caused by unsustainable use patterns, affecting both water quality and quantity, and the widespread lack of access to safe water supply and suitable sanitation in many developing countries. Because the commitments of the International Drinking Water Supply and Sanitation Decade of the 1980s have not been fully met, there is still a need to ensure the optimal use and protection of all fresh-water resources, so that the needs of everyone on this planet, including access to safe drinking water and sanitation, can be met. This calls for the highest priority to be given to the serious fresh-water problems facing many regions, especially in the developing world. There is an urgent need:.

(a) To assign high priority, in accordance with specific national needs and conditions, to the formulation and implementation of policies and programmes for integrated watershed management, including issues related to pollution and waste, the interrelationship between water and land, including mountains, forests, upstream and downstream users, estuarine environments, biodiversity and the preservation of aquatic ecosystems, wetlands, climate and land degradation and desertification, recognising that subnational, national and regional approaches to fresh-water protection and consumption following a watershed basin or river basin approach offer a useful model for the protection of fresh-water supplies;

(b) To strengthen regional and international cooperation for technological transfer and the financing of integrated water resources programmes and projects, in particular those designed to increase access to safe water supply and sanitation;

(c) To ensure the continued participation of local communities, and women in particular in the management of water resources development and use;

(d) To provide an enabling national and international environment that encourages investments from public and private sources to improve water supply and sanitation services, especially in fast growing urban and peri-urban areas, as well as in poor rural communities in developing countries; and for the international community to adopt and implement commitments to support the efforts to assist developing countries in achieving access to safe drinking water and sanitation for all;

(e) To recognize water as a social and economic good with a vital role in the satisfaction of basic human needs, food security, poverty alleviation and the protection of ecosystems. Economic valuation of water should be seen within the context of its social and economic implications, reflecting the importance of meeting basic needs. Consideration should be given to the gradual implementation of pricing policies that are geared towards cost recovery and the equitable and efficient allocation of water, including the promotion of water conservation, in developed countries; such policies could also be considered in developing countries when they reach an appropriate stage in their development, so as to promote the harmonious management and development of scarce water resources and generate financial resources for investment in new water supply and treatment facilities. Such strategies should also include programmes assigned to minimize wasteful consumption of water;

(f) To strengthen the capability of Governments and international institutions to collect and manage information, including scientific, social and environmental data, in order to facilitate the integrated assessment and management of water resources, and foster regional and international cooperation for information dissemination and exchange through cooperative approaches among United Nations institutions, including the United Nations Environment Programme, and centres for environmental excellence. In this regard, technical assistance to developing countries will continue to be important;

(g) For the international community to give support to the efforts of developing countries, with their limited resources, to shift to higher-value, less water-intensive modes of agricultural and industrial production and to develop the educational and informational infrastructure necessary to improve the skills of the labour force required for the economic transformation that needs to take place if use of fresh-water resources is to be sustainable. International support for the integrated development of water resources in developing countries, and appropriate innovative initiatives and approaches at the bilateral and regional levels are also required;

(h) To encourage watercourse States to develop international watercourses with a view to attaining sustainable utilisation and appropriate protection thereof and benefits therefrom, taking into account the interests of the watercourse States concerned.

35. Considering the urgent need for action in the field of fresh water, and building on existing principles and instruments, arrangements, programmes of action and customary uses of water, Governments call for a dialogue under the aegis of the Commission on Sustainable Development, beginning at its sixth session,

aimed at building a consensus on the necessary actions, and in particular, on the means of implementation and on tangible results, in order to consider initiating a strategic approach for the implementation of all aspects of the sustainable use of fresh water for social and economic purposes, including, *inter alia,* safe drinking water and sanitation, water for irrigation, recycling, and waste-water management, and the important role fresh water plays in natural ecosystems. This intergovernmental process will be fully fruitful only if there is a proved commitment by the international community to the provision of new and additional financial resources for the goals of this initiative.

Oceans and Seas

36. Progress has been achieved since the United Nations Conference on Environment and Development in the negotiation of agreements and voluntary instruments for improving the conservation and management of fishery resources and for the protection of the marine environment. Furthermore, progress has been made in the conservation and management of specific fishery stocks for the purpose of securing the sustainable utilisation of these resources. Despite this, the decline of many fish stocks, high levels of discards, and rising marine pollution continue. Governments should take full advantage of the challenge and opportunity presented by the International Year of the Ocean in 1998. There is a need to continue to improve decision-making at the national, regional and global levels. To address the need for improving global decision-making on the marine environment, there is an urgent need for Governments to implement decision 4/15 of the Commission on Sustainable Development, in which the Commission, *inter alia,* called for a periodic intergovernmental review by the Commission of all aspects of the marine environment and its related issues, as described, Agenda 21, and for which the overall legal framework was provided by the United Nations Convention on the Law of the Sea. There is a need for concerted action by all countries and for improved cooperation to assist developing countries in implementing the relevant agreements and instruments in order that they may participate effectively in the sustainable use, conservation and management of their fishery resources, as provided for in the Convention and other international legal instruments, and achieve integrated coastal zone management. In that context, there is an urgent need for:

(a) All Governments to ratify or to accede to the relevant agreements as soon as possible and to implement effectively such agreements as well as relevant voluntary instruments;

(b) All Governments to implement General Assembly resolution 51/189 of 16 December 1996, including the strengthening of institutional links to be established between the relevant intergovernmental mechanisms involved in the development and implementation of integrated coastal zone management. Following progress on the United Nations Convention on the Law of the Sea, and bearing in mind Principle 13 of the Rio Declaration on Environment and Development, there is a need to strengthen the implementation of existing international and regional agreements on marine pollution, with a view in particular to ensuring better contingency planning, response, and liability and compensation mechanisms;

(c) Better identification of priorities for action at the global level to promote the conservation and sustainable use of the marine environment, as well as better means for integrating such action;

(d) Further international cooperation to support the strengthening, where needed, of regional and subregional agreements for the protection and sustainable use of the oceans and seas;

(e) Governments to prevent or eliminate overfishing and excess fishing capacity through the adoption of management measures and mechanisms to ensure the sustainable management and utilisation of fishery resources and to undertake programmes of work to achieve the reduction and elimination of wasteful fishing practices, wherever they may occur, especially in relation to large-scale industrialised fishing. The emphasis given by the Commission on Sustainable Development at its fourth session to the importance of effective conservation and management of fish stocks, and in particular to eliminating overfishing, in order to identify specific steps at national or regional levels to prevent or eliminate excess fishing capacity, will need to be carried forward in all appropriate international forums including, in particular, the Committee on Fisheries of the Food and Agriculture Organisation of the United Nations;

(f) Governments to consider the positive and negative impact of subsidies on the conservation and management of fisheries through national, regional and appropriate international organisations and, based on these analyses, to consider appropriate action;

(g) Governments to take actions, individually and through their participation in competent global and regional forums, to improve the quality and quantity of scientific data as a basis for effective decisions related to the protection of the marine environment and the conservation and management of marine living resources; in this regard, greater international cooperation is required to assist developing countries, in particular small island developing States, to operationalize data-networks and clearing houses for information-sharing on oceans. In this context, particular emphasis must be placed on the collection of biological and other fisheries-related information and the resources for its collation, analysis and dissemination.

Forests

37. The management, conservation and sustainable development of all types of forests are a crucial factor in economic and social development, in environmental protection and in the planet's life-support system. Forests are one of the major reservoirs of biological diversity; they act as carbon sinks and reservoirs; and they are a significant source of renewable energy, particularly in the least developed countries. Forests are an integral part of sustainable development and are essential to many indigenous people and other forest-dependent people practising traditional lifestyles, forest owners and local communities, many of whom possess important traditional forest-related knowledge.

38. Since the adoption of the Forest Principles at the Rio Conference, tangible progress has been made in sustainable forest management at the national, subregional, regional and international levels and in the promotion of international cooperation on forests. The proposals for action contained in the report of the Ad Hoc Intergovernmental Panel on Forests on its fourth session, which were endorsed by the Commission on Sustainable Development at its fifth session, represent significant progress and consensus on a wide range of forest issues.

39. To maintain the momentum generated by the Intergovernmental Panel process and to facilitate and encourage the holistic, integrated and balanced intergovernmental policy dialogue on all types of forests in the future

Atmosphere

48. Ensuring that the global climate and atmosphere are not further damaged, with irreversible consequences for future generations, requires political will and concerted efforts by the international community in accordance with the principles enshrined in the United Nations Framework Convention on Climate Change. Under the Convention, some first steps have been taken to deal with the global problem of climate change. Despite the adoption of the Convention, the emission and concentration of greenhouse gases continue to rise, even as scientific evidence assembled by the Intergovernmental Panel on Climate Change and other relevant bodies continues to diminish the uncertainties and points ever more strongly to the severe risk of global climate change. So far, insufficient progress has been made by many developed countries in meeting their aim to return greenhouse gas emissions to 1990 levels by the year 2000. It is recognised as one critical element of the Berlin Mandate that the commitments set out under article 4, paragraph 2 (a.) and (b) of the Convention are inadequate and that there is therefore a need to strengthen them. It is most important that the Conference of the Parties to the Convention, at its third session, to be held at Kyoto, Japan in December 1997, adopt a protocol or other legal instrument that fully encompasses the Berlin Mandate. The Geneva Ministerial

Declaration, which was noted without formal adoption, but which received majority support among ministers and other heads of delegation attending the second session of the Conference of the Parties, also called for, *inter alia,* the acceleration of negotiations on the text of a legally binding protocol or other legal instrument.

49. At the present nineteenth special session of the General Assembly, the international community has confirmed its recognition of the problem of climate change as one of the biggest challenges facing the world in the next century. The leaders of many countries have stressed the seriousness of this problem in their statements to the Assembly, and have outlined the actions they had in hand to respond to the challenge, both in their own countries and internationally.

50. The ultimate goal that all countries share is to achieve stabilisation of greenhouse gas concentrations in the atmosphere at a level that would prevent dangerous anthropogenic interference with the climate system. This requires efficient and cost-effective policies and measures that will be sufficient to result in a significant reduction in emissions. At the present session, countries reviewed the status of the preparations for the third session of the Conference of the Parties to the United Nations Framework Convention on Climate Change. All agreed that a satisfactory result was vital.

51. The position of many countries with respect to these negotiations is still evolving, and it was agreed that it would not be appropriate to seek to predetermine the results; however, useful discussions on evolving positions took place.

52. There is already widespread but not universal agreement that it will be necessary to consider legally binding, meaningful, realistic and equitable targets for countries listed in annex I to the Convention that will result in significant reductions in greenhouse gas emissions within specified time frames, such as 2005, 2010 and 2020. In addition to establishing targets, there is also widespread agreement that it will be necessary to consider ways and means of achieving them and to take into account the economic, adverse environmental and other effects of such response measures on all countries, particularly developing countries.

53. International cooperation in the implementation (Agenda 21) in particular in the transfer of technology to and capacity-building in developing countries, is also essential to promote the effective implementation of the United Nations Framework Convention on Climate change.

54. There is also a need to strengthen systematic observational networks so as to identify the possible onset and distribution of climate change and assess potential impacts, particularly at the regional level.

55. The ozone layer continues to be severely depleted and the Montreal Protocol on Substances that Deplete the Ozone Layer needs to be strengthened. The Copenhagen Amendment to the Protocol needs to be ratified. The recent successful conclusion of the replenishment negotiations with respect to the Montreal Protocol Multilateral Fund is welcomed. This has made available funds for, among other things, earlier phase-out of ozone-depleting substances, including methyl bromide, in developing countries. Future replenishment should also be adequate to ensure timely implementation of the Montreal Protocol. An increased focus on capacity-building programmes in developing countries within multilateral funds is also needed, as well as the implementation of effective measures against illegal trade in ozone-depleting substances.

56. Rising levels of transboundary air pollution should be countered, including through appropriate regional cooperation to reduce pollution levels.

Toxic Chemicals

57. The sound management of chemicals is essential to sustainable development and is fundamental to human health and environmental protection. All those responsible for chemicals throughout their life cycle bear the responsibility for achieving this goal. Substantial progress on the sound management of chemicals has been made since the United Nations Conference on Environment and Development, in particular through the establishment of the Intergovernmental Forum on Chemical Safety and the Inter-Organisational Programme for the Sound Management of Chemicals. In addition, domestic regulations have been complemented by the Code of Ethics on the International Trade in Chemicals and by voluntary industry initiatives, such as Responsible Care. Despite substantial progress, a number of chemicals continue to pose significant threats to local, regional and global ecosystems and to human health. Since the United Nations Conference on Environment and Development, there has been an increased understanding of the serious damage that certain toxic chemicals can cause to human health and the environment. Much remains to be done and the environmentally sound management of chemicals should continue to be an important issue well beyond 2000. Particular attention should also be given to cooperation in the development and transfer of technology of safe substitutes and in the development of capacity for the production of such substitutes. The decision concerning the sound management of chemicals adopted by the Governing Council of the United Nations Environment Programme at its nineteenth session should be implemented in accordance with the agreed timetables for negotiations on the conventions relating to prior informed consent and persistent organic pollutants. It is noted that inorganic chemicals possess roles and behaviour that are distinct from organic chemicals.

Hazardous Wastes

58. Substantial progress has been made in the implementation of the Basel Convention, the Bamako Convention, the Fourth Lome Convention and other regional conventions, although more remains to be done. Important initiatives aimed at promoting the environmentally sound management of hazardous wastes under the Basel Convention, include (a) activities undertaken to prevent illegal traffic in hazardous wastes; (b) the establishment of regional centres for training and technology transfer regarding hazardous waste minimisation and management; and (c) the treatment and disposal of hazardous wastes as close as possible to their source of origin. These initiatives should be further developed. It is also important and urgent that work under the Basel convention be completed to define which hazardous wastes are controlled under the Convention and to negotiate, adopt and implement a protocol on liability and compensation for damage resulting from the transboundary movement and disposal of hazardous wastes. Land contaminated by the disposal of hazardous wastes needs to be identified and remedial actions put in hand. Integrated management solutions are also required to minimize urban and industrial waste generation and to promote recycling and reuse.

Radioactive Wastes

59. Radioactive wastes can have very serious environmental and human health impacts over long periods of time. It is therefore essential that they be managed in a safe and responsible way. The storage, transportation, transboundary movement and disposal of radioactive wastes should be guided by all the principles of the Rio Declaration on Environment and Development and by Agenda 21. States that generate radioactive wastes have a responsibility to ensure their safe storage and disposal. In general, radioactive wastes should be disposed of in the territory of the State in which they are generated as far as is compatible with the safety of the management of such material. Each country has the responsibility of ensuring that radioactive wastes that fall within its jurisdiction are managed properly in accordance with internationally accepted principles, taking fully into account any transboundary effects. The international community should make all efforts to prohibit the export of radioactive wastes to those countries that do not have appropriate waste treatment and storage facilities. The international community recognizes that regional arrangements or jointly used facilities might be appropriate for the disposal of such wastes in certain circumstances. The management of radioactive wastes should be undertaken in a manner consistent with international law, including the provisions of relevant international and regional conventions, and with internationally accepted standards. It is important to intensify safety measures with regard to radioactive wastes. States, in cooperation with relevant international organisations, where appropriate, should not promote or allow the storage or disposal of high-level,

intermediate-level or low-level radioactive wastes near the marine environment unless they determine that scientific evidence, consistent with the applicable internationally agreed principles and guidelines, shows that such storage or disposal poses no unacceptable risk to people or the marine environment and does not interfere with other legitimate uses of the sea. In the process of the consideration of that evidence, appropriate application of the precautionary approach principle should be made. Further action is needed by the international community to address the need for enhancing awareness of the importance of the safe management of radioactive wastes, and to ensure the prevention of incidents and accidents involving the uncontrolled release of such wastes.

60. One of the main recommendations of Agenda 21 and of the Commission on Sustainable Development at its second session in this area was to support the ongoing efforts of the International Atomic Energy Agency, the International Maritime Organisation and other relevant international organisations. The Joint Convention on the Safety of Spent Fuel Management and on the Safety of Radioactive Waste Management currently being negotiated under the auspices of the Agency is now close to completion. It will provide a comprehensive codification of international law and a guide to best practices in this area. It will rightly be based on all the principles of best practice for this subject that have evolved in the international community, including the principle that, in general, radioactive wastes should be disposed of in the State in which they were generated as far as is compatible with the safety of the management of such material. Governments should finalize this text and are urged to ratify and implement it as soon as possible so as to further improve practice and strengthen safety in this area. Transportation of irradiated nuclear fuel and high-level waste by sea should be guided by the INF Code, which should be considered for development into a mandatory instrument. The issue of the potential transboundary environmental effects of activities related to the management of radioactive wastes and the question of prior notification, relevant information and consultation with States that could potentially be affected by such effects, should be further addressed within the appropriate forums.

61. Increased global and regional cooperation, including exchange of information and experience and transfer of appropriate technologies, is needed to improve the management of radioactive wastes. There is a need to support the clean-up of sites contaminated as a result of all types of nuclear activity and to conduct health studies in the regions around those sites, as appropriate, with a view to identifying where health treatment may be needed and should be provided. Technical assistance should be provided to developing countries, recognising the special needs of small island developing States in particular, to enable them to develop or improve procedures for the management and safe disposal of radioactive wastes deriving from the use of radionuclides in medicine, research and industry.

Land and Sustainable Agriculture

62. Land degradation and soil loss threaten the livelihood of millions of people and future food security, with implications for water resources and the conservation of biodiversity. There is an urgent need to define ways to combat or reverse the worldwide accelerating trend of soil degradation, using an ecosystem approach, taking into account the needs of populations living in mountain ecosystems and recognising the multiple functions of agriculture. The greatest challenge for humanity is to protect and sustainably manage the natural resource base on which food and fibre production depend, while feeding and housing a population that is still growing. The international community has recognised the need for an integrated approach to the protection and sustainable management of land and soil resources, as stated in decision III/11 of the Conference of the Parties to the Convention on Biological Diversity, including identification of land degradation, which involves all interested parties at the local as well as the national level, including farmers, small-scale food producers, indigenous people(s), non-governmental organisations and, in particular, women, who have a vital role in rural communities. This should include action to ensure secure land tenure and access to land, credit and training, as well as the removal of obstacles that inhibit farmers, especially small-scale farmers and peasants, from investing in and improving their lands and farms.

63. It remains essential to continue efforts for the eradication of poverty through, *inter alia,* capacity-building to reinforce local food systems, improving food security and providing adequate nutrition for the more than 800 million undernourished people in the world, located mainly in developing countries. Governments should formulate policies that promote sustainable agriculture as well as productivity and profitability. Comprehensive rural policies are required to improve access to land, combat poverty, create employment and reduce rural emigration. In accordance with the commitments agreed to in the Rome Declaration on World Food Security and the World Food Summit Plan of Action, adopted by the World Food Summit, sustainable food security for both the urban and the rural poor should be a policy priority, and developed countries and the international community should provide assistance to developing countries to this end. To meet these objectives, Governments should attach high priority to implementing the commitments of the Rome Declaration and Plan of Action, especially the call for a minimum target of halving the number of undernourished people in the world by the year 2015. Governments and international organisations are encouraged to implement the Global Plan of Action for the Conservation and Sustainable Utilisation of Plant Genetic Resources for Food and Agriculture, adopted by the International Technical Conference on Plant Genetic Resources held at Leipzig, Germany from 17 to 23 June 1996. At the sixth session of the Commission on Sustainable Development, in 1998, the issues of sustainable agriculture and land use should be considered in relation to freshwater. The challenge

for agricultural research is to increase yields on all farmlands while protecting and conserving the natural resource base. The international community and Governments must continue or increase investments in agricultural research because it can take years or decades to develop new lines of research and put research findings into sustainable practice on the land. Developing countries, particularly those with high population densities, will need international cooperation to gain access to the results of such research and to technology aimed at improving agricultural productivity in limited spaces. More generally, international cooperation continues to be needed to assist developing countries in many other aspects of basic requirements of agriculture. There is a need to support the continuation of the reform process in conformity with the Uruguay Round agreements, particularly article 20 of the Agreement on Agriculture, and to fully implement the World Trade Organisation Decision on Measures Concerning the Possible Negative Effects of the Reform Programme on Least-Developed and Net Food-Importing Developing Countries.

Desertification and Drought

64. Governments are urged to conclude (by signing and ratifying, accepting, approving and/or acceding to) and to implement as soon as possible the United Nations Convention to Combat Desertification in those Countries Experiencing Serious Drought and/or Desertification, particularly in Africa, which entered into force on 26 December 1996, and to support and actively participate in the first session of the Conference of the Parties to the Convention, which is to be held in Rome in September 1997.

65. The international community is urged to recognize the vital importance and necessity of international cooperation and partnership in combating desertification and mitigating the effects of drought. In order to increase the effectiveness and efficiency of existing financial mechanisms, the international community, in particular developed countries, should therefore support the global mechanism that would have the capacity to promote actions leading to the mobilisation and channelling of substantial resources for advancing the implementation of the Convention and its regional annexes, and to contribute to the eradication of poverty, which is one of the principal consequences of desertification and drought in the majority of affected countries. Another view was that the international community, in particular developed countries, should provide new and additional resources towards the same ends. The transfer to developing countries of environmentally sound, economically viable and socially acceptable technologies relevant to combating desertification and/or mitigating the effects of drought, with a view to contributing to the achievement of sustainable development in affected areas, should be undertaken without delay on mutually agreed terms.

Biodiversity

66. There remains an urgent need for the conservation and sustainable use of biological diversity and the fair and equitable sharing of benefits arising from the utilisation of components of genetic resources. The threat to biodiversity stems mainly from habitat destruction, over-harvesting, pollution and the inappropriate introduction of forcign plants and animals. There is an urgent need for Governments and the international community, with the support of relevant international institutions, as appropriate:

(a) To take decisive action to conserve and maintain genes, species and ecosystems with a view to promoting the sustainable management of biological diversity;

(b) To ratify the Convention on Biological Diversity and implement it fully and effectively together with the decisions of the Conference of the Parties, including recommendations on agricultural biological diversity and the Jakarta Mandate on Marine and Coastal Biological Diversity, and pursue urgently other tasks identified by the Conference of the Parties at its third meeting under the work programme on terrestrial biological diversity, within the context of the ecosystems approach adopted in the Convention;

(c) To undertake concrete actions for the fair and equitable sharing of the benefits arising from the utilisation of genetic resources, consistent with the provisions of the Convention and the decisions of the Conference of the Parties on, *inter alia,* access to genetic resources and. the handling of biotechnology and its benefits;

(d) To pay further attention to the provision of new and additional financial resources for the implementation of the Convention;

(e) To facilitate the transfer of technologies, including biotechnology, to developing countries, consistent with the provisions of the Convention;

(f) To respect, preserve and maintain knowledge, innovations and practices of indigenous and local communities embodying traditional lifestyles, and encourage the equitable sharing of the benefits arising from traditional knowledge so that those communities are adequately protected and rewarded, consistent with the provisions of the Convention on Biological Diversity and in accordance with the decisions of the Conference of the Parties;

(g) To complete rapidly the biosafety protocol under the Convention on Biological Diversity, on the understanding that the United Nations Environment Programme International Technical Guidelines for Safety in Biotechnology may be used as an interim mechanism during its development, and to complement it after its conclusion, including the recommendations on capacity-building related to biosafety;

(h) To stress the importance of the establishment of a clearing-house mechanism by Parties to the Convention, consistent with the provisions of the Convention;

(i) To recognize the role of women in the conservation of biological diversity and the sustainable use of biological resources;

(j) To provide the necessary support to integrate the conservation of biological diversity and the sustainable use of biological resources into national development plans;

(k) To promote international cooperation to develop and strengthen national capacity-building, including human resource development and institution-building;

(l) To provide incentive measures at the national, regional and international levels to promote the conservation and sustainable use of biological diversity, and to consider means to enhance developing countries' capabilities to compete in the emerging market for biological resources, while improving the functioning of that market.

Sustainable Tourism

67. Tourism is now one of the world's largest industries and one of its fastest growing economic sectors. The expected growth in the tourism sector and the increasing reliance of many developing countries, including small island developing States, on this sector as a major employer and contributor to local, national, subregional and regional economies highlights the need to pay special attention to the relationship between environmental conservation and protection and sustainable tourism. In this regard, the efforts of developing countries to broaden the traditional concept of tourism to include cultural and eco-tourism merit special consideration as well as the assistance of the international community, including the international financial institutions.

68. There is a need to consider further the importance of tourism in the context of Agenda 21. Tourism, like other sectors, uses resources, generates wastes and creates environmental, cultural and social costs and benefits in the process. For sustainable patterns of consumption and production in the tourism sector, it is essential to strengthen national policy development and enhance capacity in the areas of physical planning, impact assessment, and the use of economic and regulatory instruments, as well as in the areas of information, education and marketing. A particular concern is the degradation of biodiversity and fragile ecosystems, such as coral reefs, mountains, coastal areas and wetlands.

69. Policy development and implementation should take place in cooperation with all interested parties, especially the private sector and local and indigenous communities. The Commission should develop an action-oriented international programme of work on sustainable tourism, to be defined in cooperation with the World Tourism Organisation, the United Nations Conference on Trade and Development, the United Nations Environment Programme, the Conference of the Parties to the Convention on Biological Diversity and other relevant bodies.

70. The sustainable development of tourism is of importance for all countries, in particular for small island developing States. International cooperation is needed to facilitate tourism development in developing countries—including the development and marketing of eco-tourism, bearing in mind the importance of the conservation policies required to secure long-term benefits from development in this sector—in particular in small island developing States, in the context of the Programme of Action for the Sustainable Development of Small Island Developing States.

Small Island Developing States

71. The international community reaffirms its commitment to the implementation of the Programme of Action for the Sustainable Development of Small Island Developing States. The Commission on Sustainable Development carried out a mid-term review of selected programme areas of the Programme of Action at its fourth session, in 1996. At its sixth session, in 1998, the Commission will undertake a review of all the outstanding chapters and issues of the Programme of Action. A full and comprehensive review of the Programme of Action, consistent with the review of other United Nations global conferences, is scheduled for 1999. The Commission, at its fifth session, adopted a resolution on modalities for the full and comprehensive review of the Programme of Action, in which it recommended that the General Assembly hold a two-day special session immediately preceding its fifty-fourth session for an in-depth assessment and appraisal of the implementation of the Programme of Action. The full implementation of the decision would represent a significant contribution to achieving the objectives of the Global Conference for the Sustainable Development of Small Island Developing States.

72. Considerable efforts are being made at the national and regional levels to implement the Programme of Action. These efforts need to be supplemented by effective financial support from the international community. External assistance for building the requisite infrastructure and for national capacity-building, including human and institutional capacity, and for facilitating access to information on sustainable development practices and the transfer of environmentally sound technologies, in accordance with paragraph 34.14 (b) of Agenda 21, is crucial for small island developing States to effectively attain the goals of the Programme of

Action. To assist national capacity-building, the small island developing States information network and small island developing States technical assistance programme should be made operational as soon as possible, with support for existing regional and subregional institutions.

Natural Disasters

73. Natural disasters have disproportionate consequences for developing countries, in particular small island developing States and countries with extremely fragile ecosystems. Programmes for sustainable development should give higher priority to the implementation of the commitments made at the World Conference on Natural Disaster Reduction held at Yokohama, Japan from 23 to 27 May 1994. There is a particular need for capacity-building for disaster planning and management and for the promotion and facilitation of the transfer of early-warning technologies to countries prone to disasters, in particular developing countries and countries with economies in transition.

74. Given that further work is needed throughout the world, there is a special need to provide developing countries with further assistance in:

(a) Strengthening mechanisms and policies designed to reduce the effects of natural disasters, improve preparedness and integrate natural disaster considerations in development planning, through, *inter alia,* access to resources for disaster mitigation and preparedness, response and recovery;

(b) Improving access to relevant technology and training in hazard and risk assessment and early warning systems, and in protection from environmental disasters, consistent with national, subregional and regional strategies;

(c) Providing and facilitating technical, scientific and financial support for disaster preparedness and response in the context of the International Decade for Natural Disaster Reduction.

Major Technological and Other Disasters with an Adverse Impact on the Environment

75. Major technological and other disasters with an adverse impact on the environment can be a substantial obstacle in the way of achieving the goals of sustainable development in many countries. The international community should intensify cooperation in the prevention and reduction of such disasters and in disaster relief and post-disaster rehabilitation in order to enhance the capabilities of affected countries to cope with such situations.

C. Means of Implementation Financial Resources and Mechanisms

76. Financial resources and mechanisms play a key role in the implementation of Agenda 21. In general, the financing for the implementation of Agenda 21 will come from a country's own public and private sectors. For developing countries, official development assistance is a main source of external funding, and substantial new and additional funding for sustainable development and the implementation of Agenda 21 will be required. Hence, all financial commitments of Agenda 21, particularly and the provisions with regard to new and additional resources that are both adequate and predictable need to be urgently fulfilled. Renewed efforts are essential to ensure that all sources of funding contribute to economic growth, social development and environmental protection in the context of sustainable development and the implementation of Agenda 21.

77. For developing countries, particularly those in Africa and the least developed countries, official development assistance remains a main source of external funding; it is essential for the prompt and effective implementation of Agenda 21 and cannot generally be replaced by private capital flows. Developed countries should therefore fulfil the commitments undertaken to reach the accepted United Nations target of 0.7 per cent of gross national product as soon as possible. In this context the present downward trend in the ratio of official development assistance to gross national product causes concern. Intensified efforts should be made to reverse this trend, taking into account the need for improving the quality and effectiveness of official development assistance. In the spirit of global partnership, the underlying factors that have led to this decrease should be addressed by all countries. Strategies should be worked out for increasing donor support for aid programmes and revitalising the commitments that donors made at the United Nations Conference on Environment and Development. Some countries already meet or exceed the 0.7 per cent agreed target. Official financial flows to developing countries, particularly the least developed countries, remain an essential element of the partnership embodied in Agenda 21. Official development assistance plays a significant role, *inter alia,* in capacity-building, infrastructure, combating poverty and environmental protection in developing countries, and a crucial role in the least developed countries. Official development assistance can play an important complementary and catalytic role in promoting economic growth and may, in some cases, play a catalytic role in encouraging private investment and, where appropriate, all aspects of country-driven capacity-building and strengthening.

78. Funding by multilateral financial institutions through their concessional mechanisms is also essential to developing countries in their efforts to fully implement the sustainable development objectives contained in Agenda 21. Such institutions should continue to respond to the development needs and priorities of developing

countries. Developed countries should urgently meet their commitments under the eleventh replenishment of the International Development Association.

79. Continued and full donor commitment to adequate, sustained and predictable funding for Global Environment Facility operations is important for developing countries so that global environmental benefits can be further achieved. Donor countries are urged to engage in providing new and additional resources, with a view to equitable burden-sharing, through the satisfactory replenishment of the Facility, which makes available grant and concessional funding designed to achieve global environmental benefits, thereby promoting sustainable development. Consideration should be given to further exploring the flexibility of the existing mandate of the Facility in supporting activities to achieve global environmental benefits. With regard to the project cycle, further efforts should be made to continue streamlining the decision-making process in order to maintain an effective and efficient, as well as transparent, participatory and democratic framework. The Global Environment Facility, when acting as the operating entity of the financial mechanism of the United Nations Framework Convention on Climate Change and the Convention on Biological Diversity, should continue to operate in conformity with those Conventions and promote their implementation. The Facility implementing agencies, the United Nations Development Programme, the United Nations Environment Programme and the World Bank, should strengthen, as appropriate and in accordance with their respective mandates, their cooperation at all levels, including the field level.

80. The efficiency, effectiveness and impact of the operational activities of the United Nations system must be enhanced by, *inter alia,* a substantial increase in their funding on a predictable, continuous and assured basis, commensurate with the increasing needs of developing countries, as well as through the full implementation of General Assembly resolutions 47/199 of 22 December 1992 and 48/162 of 20 December 1993. There is a need for a substantial increase in resources for operational activities for development on a predictable, continuous and assured basis, commensurate with the increasing needs of developing countries.

81. Private capital is a major tool for achieving economic growth in a growing number of developing countries. Higher levels of foreign private investment should be mobilised given its mounting importance. To stimulate higher levels of private investment, Governments should aim at ensuring macroeconomic stability, open trade and investment policies, and well-functioning legal and financial systems. Further studies should be undertaken, including studies on the design of an appropriate environment, at both the national and international levels, for facilitating foreign private investment, in particular foreign direct investment flows to developing countries, and enhancing its contribution to sustainable development. To ensure

that such investments are supportive of sustainable development objectives, it is essential that the national Governments of both investor and recipient countries provide appropriate regulatory frameworks and incentives for private investment. Therefore further work should be undertaken on the design of appropriate policies and measures aimed at promoting long-term investment flows to developing countries for activities that increase their productive capability, and at reducing the volatility of these flows. Official development assistance donors and multilateral development banks are encouraged to strengthen their commitment to supporting investment in developing countries in a manner that jointly promotes economic growth, social development and environmental protection.

82. The external debt problem continues to hamper the efforts of developing countries to achieve sustainable development. To resolve the remaining debt problems of the heavily indebted poor countries, creditor and debtor countries and international financial institutions should continue their efforts to find effective, equitable, development-oriented and durable solutions to the debt problem, including debt relief in the form of debt rescheduling, debt reduction, debt swaps and, as appropriate, debt cancellation, as well as grants and concessional flows that will help restore creditworthiness. The joint World Bank/International Monetary Fund Heavily Indebted Poor Countries Debt Initiative supported by the Paris Club creditor countries is an important development to reduce the multilateral debt problem. Implementation of the Initiative requires additional financial resources from both bilateral and multilateral creditors without affecting the support required for the development activities of developing countries.

83. A fuller understanding of the impact of indebtedness on the pursuit of sustainable development by developing countries is needed. To this end, the United Nations Secretariat, the World Bank and the International Monetary Fund are invited to collaborate with the United Nations Conference on Trade and Development in further considering the interrelationship between indebtedness and sustainable development for developing countries.

84. While international cooperation is very important in assisting developing countries in their development efforts, in general financing for the implementation of Agenda 21 will come from countries' own public and private sectors. Policies for promoting domestic resource mobilisation, including credit, could encompass sound macroeconomic reforms, including fiscal and monetary policy reforms, review and reform of existing subsidies, and the promotion of personal savings and access to credit, especially micro-credit, in particular for women. Such policies should be decided by each country, taking into account its own characteristics and capabilities and different levels of development, especially as reflected in national sustainable development strategies, where they exist.

85. There is a need for making existing subsidies more transparent in order to increase public awareness of their-actual economic, social and environmental impact, and for reforming or, where appropriate, removing them. Further national and international research in that area should be promoted in order to assist Governments in identifying and considering phasing-out subsidies that have market distorting, and socially and environmentally damaging impacts. Subsidy reductions should take full account of the specific conditions and the different levels of development of individual countries and should consider potentially regressive impacts, particularly on developing countries. In addition, it would be desirable to use international cooperation and coordination to promote the reduction of subsidies where these have important implications for competitiveness.

86. In order to reduce the barriers to the expanded use of economic instruments, Governments and international organisations should collect and share information on their use and introduce pilot schemes that would, *inter alia,* demonstrate how to make the best use of them while avoiding adverse effects on competitiveness and the terms of trade of all countries, particularly developing countries, and on marginalised and vulnerable sectors of society. When introducing economic instruments that raise the cost of economic activities for households and small and medium-sized enterprises, Governments should consider gradual phase-ins, public education programmes and targeted technical assistance as strategies for reducing distributional impacts. Various studies and practical experience in a number of countries, in particular developed countries, indicate that the appropriate use of relevant economic instruments may help generate positive possibilities for shifting consumer and producer behaviour to more sustainable directions in those countries. There is, however, a need to conduct further studies and test practical experience in more countries, taking into account country-specific conditions and the acceptability, legitimacy, equity, efficiency and effectiveness of such economic instruments.

87. Innovative financial mechanisms are currently under discussion in international and national forums but have not yet fully evolved conceptually. The Secretary-General is to submit a report concerning innovative financing mechanisms to the Economic and Social Council at its substantive session of 1997. In view of the widespread interest in those mechanisms, appropriate organisations, including the United Nations Conference on Trade and Development, the World Bank and the International Monetary Fund, are invited to consider conducting forward-looking studies of concerted action on such mechanisms and to share them with the Commission on Sustainable Development, other relevant intergovernmental organisations and non-governmental organisations. In this regard, innovative funding should complement official development assistance, not replace it. New initiatives for cooperative implementation of environment and development objectives under mutually beneficial incentive structures should be further explored.

Transfer of Environmentally Sound Technologies

88. The availability of scientific and technological information and access to and transfer of environmentally sound technologies are essential requirements for sustainable development. There is an urgent need for developing countries to acquire greater access to environmentally sound technologies if they are to meet the obligations agreed at the United Nations Conference on Environment and Development and in the relevant international conventions. The ability of developing countries to participate in, benefit from and contribute to rapid advances in science and technology can significantly influence their development. This calls for the urgent fulfilment of all the Conference commitments concerning concrete measures for the transfer of environmentally sound technologies to developing countries. The international community should promote, facilitate and finance, as appropriate, access to and transfer of environmentally sound technologies and the corresponding know-how, in particular to developing countries, on favourable terms, including concessional and preferential terms, as mutually agreed, taking into account the need to protect intellectual property rights as well as the special needs of developing countries for the implementation of Agenda 21. Current forms of cooperation involving the public and private sectors of developing and developed countries should be built upon and expanded. In this context, it is important to identify barriers and restrictions to the transfer of publicly and privately owned environmentally sound technologies, with a view to reducing such constraints while creating specific incentives, fiscal and otherwise, for the transfer of such technologies. Progress in the fulfilment of all the provisions contained in chapter 34 of Agenda 21 should be reviewed regularly as part of the multi-year work programme of the Commission on Sustainable Development.

89. Technology transfer and the development of the human and institutional capacity to adapt, absorb and disseminate technologies, as well as to generate technical knowledge and innovations, are part of the same process and must be given equal importance. Governments have an important role to play in providing, *inter alia,* research and development institutions with incentives to promote and contribute to the development of institutional and human capacities.

90. Much of the most advanced environmentally sound technology is developed and held by the private sector. The creation of an enabling environment, on the part of both developed and developing countries, including supportive economic and fiscal measures, as well as a practical system of environmental regulations and compliance mechanisms, can help to stimulate private sector investment in and transfer of environmentally sound technology to developing countries. New ways of financial intermediation for the financing of environmentally sound technologies, such as "green credit lines, should be examined. Further efforts should be made by Governments and international development institutions to facilitate the transfer

of privately owned technology on concessional terms, as mutually agreed, to developing countries, especially the least developed countries.

91. A proportion of technology is held or owned by Governments and public institutions or results from publicly funded research and development activities. The Government's control and influence over the technological knowledge produced in publicly funded research and development institutions open up the potential for the generation of publicly owned technologies that could be made accessible to developing countries, and could be an important means for Governments to catalyse private sector technology transfer. Proposals for the further study of the options with respect to those technologies and publicly funded research and development activities are to be welcomed.

92. Governments should create a legal and policy framework that is conducive to technology-related private sector investments and long-term sustainable development objectives. Governments and international development institutions should continue to play a key role in establishing public-private partnerships, within and between developed and developing countries and countries with economies in transition. Such partnerships are essential for linking the advantages of the private-sector-access to finance and technology, managerial efficiency, entrepreneurial experience and engineering expertise—with the capacity of Governments to create a policy environment that is conducive to technology-related private sector investments and long-term sustainable development objectives.

93. The creation of centres for the transfer of technology at various levels, including the regional level, could greatly contribute to achieving the objective of transfer of environmentally sound technologies to developing countries. For this purpose, existing United Nations bodies, including, as appropriate, the Commission on Science and Technology for Development, the United Nations Conference on Trade and Development, the United Nations Industrial Development Organisation, the United Nations Environment Programme and the regional commissions, should cooperate and mechanisms be used, such as technical cooperation among developing countries and economic cooperation among developing countries.

94. Governments and international development institutions can also play an important role in bringing together companies from developed and developing countries and countries with economies in transition so that they can create sustainable and mutually beneficial business linkages. Incentives should be provided to stimulate the building of joint ventures between small and medium-sized enterprises of developed and developing countries and countries with economies in transition, and cleaner production programmes in public and private companies should be supported.

95. Governments of developing countries should take appropriate measures to

strengthen South-South cooperation for technology transfer and capacity-building. Such measures could include the networking of existing national information systems and sources on environmentally sound technologies, and the networking of national cleaner production centres, as well as the establishment of sector-specific regional centres for technology transfer and capacity-building. Interested donor countries and international organisations should further assist developing countries in those efforts through, *inter alia,* supporting trilateral arrangements and contributing to the United Nations Voluntary Trust Fund for South-South Cooperation.

96. Attention must also be given to technology needs assessment as a tool for Governments in identifying a portfolio for technology transfer projects and capacity-building activities to be undertaken to facilitate and accelerate the development, adoption and dissemination of environmentally sound technologies in particular sectors of the national economy. It is also important for Governments to promote the integration of environmental technology assessment with technology needs assessment as an important tool for evaluating environmentally sound technologies and the organisational, managerial and human resource-systems related to the proper use of those technologies.

97. There is a need to further explore and enhance the potential of global electronic information and telecommunication networks. This would enable countries to choose among the available technological options that are most appropriate to their needs. In this respect, the international community should assist developing countries in enhancing their capacities.

Capacity-building

98. Renewed commitment and support from the international community is essential to support national efforts for capacity-building in developing countries and countries with economies in transition.

99. The United Nations Development Programme, *inter alia,* through its Capacity 21 programme, should give priority attention to building capacity for the elaboration of sustainable development strategies based on participatory approaches. In this context, developing countries should be assisted, particularly in the areas of the design, implementation and evaluation of programmes and projects.

100. Capacity-building efforts should pay particular attention to the needs of women in order to ensure that their skills and experience are fully used in decision-making at all levels. The special needs, culture, traditions and expertise of indigenous people must be recognized. International financial institutions should continue to give high priority to funding capacity-building for sustainable development in developing countries and countries with economies in transition. Special attention

should also be given to strengthening the ability of developing countries to absorb and generate technologies. International cooperation needs to be strengthened to promote the endogenous capacity of developing countries to utilize scientific and technological developments from abroad and to adapt them to local conditions. The role of the private sector in capacity-building should be further promoted and enhanced. South-South cooperation in capacity-building should be further supported through "triangular" cooperative arrangements. Both developed and developing countries, in cooperation with relevant international Institutions, need to strengthen their efforts to develop and implement strategies for more effective sharing of environmental expertise, and data.

Science

101. Public and private investment in science, education and training, and research and development should be increased significantly, with emphasis on the need to ensure equal access to opportunities for girls and women.

102. International consensus-building is facilitated by the availability of authoritative scientific evidence. There is a need for further scientific cooperation, especially across disciplines, in order to verify and strengthen scientific evidence and make it accessible to developing countries. This evidence is important for assessing environmental conditions and changes. Steps should also be taken by Governments, academia, and scientific institutions to improve access to scientific information related to the environment and sustainable development. The promotion of existing regional and global networks may be useful for this purpose.

103. Increasing efforts to build and strengthen scientific and technological capacity in developing countries is an extremely important objective. Multilateral and bilateral donor agencies and Governments, as well as specific funding mechanisms, should continue to enhance their support for developing countries. Attention should also be given to countries with economies in transition.

104. The international community should also actively collaborate in promoting innovations in information and communication technologies for the purpose of reducing environmental impacts, *inter alia,* by taking approaches to technology transfer and cooperation that are based on user needs.

Education and Awareness

105. Education increases human welfare, and is a decisive factor in enabling people to become productive and responsible members of society. A fundamental prerequisite for sustainable development is an adequately financed and effective educational system at all levels, particularly the primary and secondary levels, that is accessible to all and that augments both human capacity and well-being. The core

themes of education for sustainability include lifelong learning, interdisciplinary education, partnerships, multicultural education and empowerment. Priority should be given to ensuring women's and girls' full and equal access to all levels of education and training. Special attention should also be paid to the training of teachers, youth leaders and other educators. Education should also be seen as a means of empowering youth and vulnerable and marginalised groups, including those in rural areas, through intergenerational partnerships and peer education. Even in countries with strong education systems, there is a need to reorient education, awareness and training so as to promote widespread public understanding, critical analysis and support for sustainable development. Education for a sustainable future should engage a wide spectrum of institutions and sectors, including but not limited to business/industry, international organisations, youth, professional organisations, non-governmental organisations, higher education, government, educators and foundations, to address the concepts and issues of sustainable development, as embodied throughout Agenda 21, and should include the preparation of sustainable development education plans and programmes, as emphasised in the work programme of the Commission on Sustainable Development on the subject adopted in 1996. The concept of education for a sustainable future will be further developed by the United Nations Educational, Scientific and Cultural Organisation, in cooperation with others.

106. It is necessary to support and strengthen universities and other academic centres in promoting cooperation among them, particularly cooperation between those of developing countries and those of developed countries.

International Legal Instruments and the Rio Declaration on Environment and Development

107. The implementation and application of the principles contained in the Rio Declaration on Environment and Development should be the subject of regular assessment and reporting to the Commission on Sustainable Development by the Secretariat in collaboration with the United Nations Environment, Programme, in particular.

108. Access to information and broad public participation in decision-making are fundamental to sustainable development. Further efforts are required to promote, in the light of country-specific conditions, the integration of environment and development policies, through appropriate legal and regulatory policies, instruments and enforcement mechanisms at the national, state, provincial and local levels. At the national level, each individual should have appropriate access to information concerning the environment that is held by public authorities, including information on hazardous materials and activities in the communities, and the opportunity to

participate in decision-making processes. Governments and legislators, with the support, where appropriate, of competent international organisations, should establish judicial and administrative procedures for legal redress and remedy of actions affecting environment and development that may be unlawful or infringe on rights under the law, and should provide access to individuals, groups and organisations with a recognised legal interest. Access should be provided to effective judicial and administrative channels for affected individuals and groups to ensure that all authorities, both national and local, and other civil organisations remain accountable for their actions in accordance with their obligations, at the appropriate levels for the country concerned, taking into account the judicial and administrative systems of the country concerned.

109. Taking into account the provisions illustration of Agenda 21, it is necessary to continue the progressive development and, as and when appropriate, codification of international law related to sustainable development. Relevant bodies in which such tasks are being undertaken should cooperate and coordinate in this regard.

110. Implementation of and compliance with commitments made under international treaties and other instruments in the field of the environment remain a priority. Implementation can be promoted by secure, sustained and predictable financial support, sufficient institutional capacity, human resources and adequate access to technology. Cooperation on implementation between states on mutually agreed terms may help reduce potential sources of conflict between States. In this context, States should further study and consider methods to broaden and make more effective the range of techniques available at present, taking into account relevant experience under existing agreements and, where appropriate, modalities for dispute avoidance and settlement, in accordance with the Charter of the United Nations. It is also important to further improve reporting and data-collection systems and to further develop appropriate compliance mechanisms and procedures, on a mutually agreed basis, to help and encourage states to fulfil all their obligations, including means of implementation, under multilateral environmental agreements. Developing countries should be assisted to develop these tools according to country-specific conditions.

Information and Tools for Measuring Progress

111. The further development of cost-effective tools for collecting and disseminating information for decision makers at all levels through strengthened data collection, including, as appropriate, gender-disaggregated data and information that makes visible the unremunerated work of women for use in programme planning and implementation, compilation and analysis is urgently needed. In this context, emphasis will be placed on support for national and international scientific and technological data centres with appropriate electronic communication links between them.

112. A supportive environment needs to be established to enhance national capacities and capabilities for information collection, processing and dissemination, especially in developing countries, to facilitate public access to information on global environmental issues through appropriate means, including high-tech information and communication infrastructurc rclated to the global environment, in the light of country-specific conditions, using, where available, such tools as geographic information systems and video transmission technology, including global mapping. In this regard, international cooperation is essential.

113. Environmental impact assessments are an important national tool for sustainable development. In accordance with Principle 17 of the Rio Declaration on Environment and Development, environmental impact assessments should be undertaken for proposed activities that are likely to have a significant adverse impact on the environment and are subject to a decision of a competent national authority; where appropriate, they should be made available early in the project cycle.

114. The work programme of the Commission on Sustainable Development on indicators of sustainable development should result in a practicable and agreed set of indicators, suited to country-specific conditions, including a limited number of aggregated indicators, to be used at the national level, on a voluntary basis, by the year 2000. Such indicators of sustainable development, including, where appropriate, and subject to nationally specific conditions, sector-specific ones, should play an important role in monitoring progress towards sustainable development at the national level and in facilitating national reporting, as appropriate.

115. National reports on the implementation of Agenda 21 have proved to be a valuable means of sharing information at the international and regional levels and, even more important, of providing a focus for the coordination of issues related to sustainable development at the national level within individual countries. National reporting should continue (see also para. 133 (b) and (c) below).

IV. Iinternational Institutional Arrangements

116. The achievement of sustainable development requires continued support from international institutions. The institutional framework and determined by the General Assembly in its resolution 47/191 of 22 December 1992 and other relevant resolutions, including the specific functions and roles of various organs, organisations and programmes within and outside the United Nations system, will continue to be fully relevant in the period after the nineteenth special session of the General Assembly. In the light of the ongoing discussions on reform within the United Nations, international institutional arrangements in the area of sustainable development are intended to contribute to the goal of strengthening the entire

United Nations system. In this context, the strengthening of the institutions for sustainable development, as well as the achievement of the goals and objectives set out below are particularly important.

A. Greater Coherence in Various Intergovernmental Organisations and Processes

117. Given the increasing number of decision-making bodies concerned with various aspects of sustainable development, including international conventions, there is an ever greater need for better policy coordination at the intergovernmental level, as well as for continued and more concerted efforts to enhance collaboration among the secretariats of those decision-making bodies. Under the guidance of the General Assembly, the Economic and Social Council should play a strengthened role in coordinating the activities of the United Nations system in the economic, social and related fields.

118. The conferences of the parties to conventions signed at the United Nations Conference on Environment and Development or as a result of it, as well as other conventions related to sustainable development, should cooperate in exploring ways and means of collaborating in their work to advance the effective implementation of the conventions. There is also a need for environmental conventions to continue to pursue sustainable development objectives consistent with their provisions and be fully responsive to Agenda 21. To this end, *inter alia,* the conferences of the parties to or governing bodies of the conventions signed at the United Nations Conference on Environment and Development, or as a result of it, and of other relevant conventions and agreements should, if appropriate, give consideration to the co-location of secretariats, to improving the scheduling of meetings, to integrating national reporting requirements, to improving the balance between sessions of the conferences of the parties and sessions of their subsidiary bodies, and to encouraging and facilitating the participation of Governments in those sessions, at an appropriate level.

119. Institutional arrangements for the convention secretariats should provide effective support and efficient services, while ensuring the appropriate autonomy necessary for them to be efficient at their respective locations. At the international and national levels there is a need for, *inter alia,* better scientific assessment of ecological linkages between the conventions; identification of programmes that have multiple benefits; and enhanced public awareness-raising with respect to the conventions. Such tasks should be undertaken by the United Nations Environment Programme in accordance with the relevant decisions of its Governing Council and in full cooperation with the conferences of the parties to and governing bodies of relevant conventions. Efforts of convention secretariats, in response to requests from the respective conferences of the parties, to explore, where appropriate, modalities for suitable liaison arrangements in Geneva and/or New York for the purpose of

enhancing linkages with delegations and organisations at those United Nations centres are welcomed and fully supported.

120. It is necessary to strengthen the Inter-Agency Committee on Sustainable Development of the Administrative Committee on Coordination and its system of task managers, with a view to further enhancing system-wide intersectoral cooperation and coordination for the implementation of Agenda 21 and for the promotion of coordinated follow-up to the major United Nations conferences in the area of sustainable development.

121. The Commission on Sustainable Development should promote increased regional implementation of Agenda 21 in cooperation with relevant regional and subregional organisations and the United Nations regional commissions, in accordance with the results of their priority-setting efforts, with a view to enhancing the role such bodies play in the achievement of sustainable development objectives agreed at the international level. The regional commissions could provide appropriate support, consistent with their work programmes, to regional meetings of experts related to the implementation of Agenda 21.

B. Role of Relevant Organisations and Institutions of the United Nations System

122. In order to facilitate the national implementation of Agenda 21, all organisations and programmes of the United Nations system, within their respective areas of expertise and mandates, should strengthen, individually and jointly, the support for national efforts to implement Agenda 21 and make their efforts and actions consistent with national plans, policies and priorities of member States. Coordination of United Nations activities at the field level should be further enhanced through the resident coordinator system in full consultation with national Governments.

123. The role of United Nations Environment Programme, as the principal United Nations body in the field of the environment, should be further enhanced. Taking into account its catalytic role, and in conformity with Agenda 21 and the Nairobi Declaration on the Role and Mandate of the United Nations Environment Programme, adopted on 7 February 1997, the Programme is to be the leading global environmental authority that sets the global environmental agenda, promotes the coherent implementation of the environmental dimension of sustainable development within the United Nations system, and serves as an authoritative advocate for the global environment. In this context, decision 19/32 of 4 April 1997 of the United Nations Environment Programme Governing Council on governance of the Programme and other, related Governing Council decisions are relevant. The role of the United Nations Environment Programme in the further development of international environmental law should be strengthened, including the development of coherent

interlinkages among relevant environmental conventions in cooperation with their respective conferences of the parties or governing bodies. In performing its functions related to the conventions signed at the United Nations Conference on Environment and Development or as a result of it, and other relevant conventions, the United Nations Environment Programme should strive to promote the effective implementation of those conventions in a manner consistent with the provisions of the conventions and the decisions of the conferences of the parties.

124. The United Nations Environment Programme, in the performance of its role, should focus on environmental issues, taking into account the development perspective. A revitalised Programme should be supported by adequate, stable and predictable funding. The Programme should continue providing effective support to the Commission on Sustainable Development, *inter alia,* in the form of scientific, technical and policy information and analysis of and advice on global environmental issues.

125. The United Nations Development Programme should continue to strengthen its contribution to and programmes in sustainable development and the implementation of Agenda 21 at all levels, particularly in the area of promoting capacity-building (including through its Capacity 21 programme) in cooperation with other organisations, as well as in the field of poverty eradication.

126. The United Nations Conference on Trade and Development, in accordance with General Assembly resolution 51/167 of 16 December 1996 and relevant decisions of the Trade and Development Board on the work programme, should continue to play a key role in the implementation of Agenda 21 through the integrated examination of linkages among trade, investment, technology, finance and sustainable development.

127. The Committee on Trade and Environment of the World Trade Organisation, the United Nations Conference on Trade and Development and the United Nations Environment Programme should advance their coordinated work on trade and environment, involving other appropriate international and regional organisations in their cooperation and coordination. In coordination with the World Trade Organisation, the United Nations Conference on Trade and Development and the United Nations Environment Programme should continue to support efforts to promote the integration of trade, environment and development. The Commission on Sustainable Development should continue to play its important role in the deliberations on trade and environment so as to facilitate the integrated consideration of all factors relevant for achieving sustainable development.

128. Implementation of the commitment of the international financial institutions to sustainable development should continue to be strengthened. The World Bank

has a significant role to play, bearing in mind its expertise and the overall volume of resources that it commands.

129. Operationalisation of the global mechanism of the United Nations Convention to Combat Desertification in those Countries Experiencing Serious Drought and/or Desertification, particularly in Africa is also essential.

C. Future Role and Programme of Work of the Commission on Sustainable Development

130. The Commission on Sustainable Development, within its mandate as specified in General Assembly resolution 47/191, will continue to provide a central forum for reviewing progress and for urging further implementation of Agenda 21 and other commitments made at the United Nations Conference on Environment and Development or as a result of it; for conducting a high-level policy debate aimed at consensus-building on sustainable development; and for catalysing action and long-term commitment to sustainable development at all levels. It should continue to undertake these tasks in complementing and providing interlinkages to the work of other United Nations organs, organisations and bodies acting in the field of sustainable development. The Commission has a role to play in assessing the challenges of globalisation as they relate to sustainable development. The Commission should perform its functions in coordination with other subsidiary bodies of the Economic and Social Council and with related organisations and institutions, including making recommendations, within its mandate, to the Economic and Social council, bearing in mind the interrelated outcomes of recent United Nations conferences.

131. The Commission should focus on issues that are crucial to achieving the goals of sustainable development. It should promote policies that integrate economic, social and environmental dimensions of sustainability and should provide for integrated consideration of linkages, both among sectors and between sectoral and cross-sectoral aspects of Agenda 21. In this connection, the Commission should carry out its work in such a manner as to avoid unnecessary duplication and repetition of work undertaken by other relevant forums.

132. In the light of the above, it is recommended that the Commission on Sustainable Development adopt the multi-year programme of work for the period 1998-2002 contained in the appendix below.

D. Methods of Work of the Commission on Sustainable Development

133. Based on the experience gained during the period 1993-1997, the Commission, under the guidance of the Economic and Social Council, should:

(a) Make concerted efforts to attract the greater involvement in its work of ministers and high-level national policy makers responsible for specific economic and social sectors, who, in particular, are encouraged to participate in the annual high-level segment of the Commission, together with the ministers and policy makers responsible for environment and development. The high-level segments of the Commission should become more interactive, and should focus on the priority issues being considered at a particular session. The Bureau of the Commission should conduct timely and open-ended consultations with the view to improving the organisation of the work of the high-level segment;

(b) Continue to provide a forum for the exchange of national experience and best practices in the area of sustainable development, including through voluntary national communications or reports. Consideration should be given to the results of ongoing work aimed at streamlining requests for national information and reporting and to the results of the "pilot phase" relating to indicators of sustainable development. In this context, the Commission should consider more effective modalities for the further implementation of the commitments made in Agenda 21, with appropriate emphasis on the means of implementation. Countries may wish to submit to the Commission, on a voluntary basis, information regarding their efforts to incorporate the relevant recommendations of other United Nations conferences in national sustainable development strategies;

(c) Take into account regional developments related to the implementation of the outcomes of the United Nations Conference on Environment and Development, It should provide a forum for the exchange of experience on regional and subregional initiatives and regional collaboration for sustainable development. This could include the promotion of the voluntary regional exchange of national experience in the implementation of Agenda 21 and, *inter alia,* the possible development of modalities for reviews within regions by and among those countries that voluntarily agree to do so. In this context, the Commission should encourage the availability of funding for the implementation of initiatives related to such reviews;

(d) Establish closer interaction with international financial, development and trade institutions, ai well as with other relevant bodies within and outside the United Nations system, including the World Bank, the Global Environment Facility, the United Nations Development Programme, the World Trade Organisation, the United Nations Conference on Trade and Development and the United Nations Environment Programme, which, in turn, are invited to take full account of the results of policy deliberations in the Commission and to integrate them in their own work programmes and activities;

(e) Strengthen its interaction with representatives of major groups, including through greater and better use of focused dialogue sessions and round tables. These groups are important resources in operationalising, managing and promoting sustainable development and contribute to the implementation of Agenda 21. The major groups are encouraged to adopt arrangements for coordination and interaction in providing inputs to the Commission. Taking into account the Commission's programme of work, this could include inputs from:

(i) The scientific community and research institutions, relating to the greater understanding of the interactions between human activity and natural ecosystems and on how to manage global systems sustainably;

(ii) Women, children and youth, indigenous people and their communities, non-governmental organisations, local authorities, workers and their trade unions and farmers on the elaboration, promotion and sharing of effective strategies, policies, practices and processes to promote sustainable development;

(iii) Business and industry groups on the elaboration, promotion and sharing of sustainable development practices and the promotion of corporate responsibility and accountability;

(f) Organise the implementation of its next multi-year programme of work in the most effective and productive way, including through shortening its annual meeting to two weeks. The inter-sessional ad hoc working groups should help to focus the Commission's sessions by identifying key elements to be discussed and important problems to be addressed within specific items of the Commission's programme of work. Government hosted and funded expert meetings will continue to provide inputs to the work of the Commission.

134. The Secretary-General is invited to review the functioning of the High-level Advisory Board on Sustainable Development and present proposals on ways to promote more direct interaction between the Board and the Commission, with a view to ensuring that the Board contributes to the deliberations on specific themes considered by the Commission in accordance with its programme of work.

135. The work of the Committee on New and Renewable Sources of Energy and on Energy for Development and the Committee on Natural Resources should be more compatible with and supportive of the programme of work of the Commission. The Economic and Social Council, in carrying out its functions related to the implementation of General Assembly resolution 50/227 of 24 Hay 1996, should consider, at its substantive session of 1997, the most effective means of bringing this about.

136. The arrangements for the election of the Bureau should be changed in order to allow the same Bureau to provide guidance in the preparations for and to lead the work during the annual sessions of the Commission. The Commission would benefit from such a change. The Economic and Social Council should take the necessary action at its substantive session of 1997 to ensure that these new arrangements take effect.

137. The next comprehensive review of progress achieved in the implementation of Agenda 21 by the General Assembly will take place in the year 2002. The modalities of this review should be determined at a later stage.

Biodiversity

The Convention on Biological Diversity signed at the Rio Earth Summit in 1992 has been ratified by 169 countries. Signatories to the convention are obliged to conserve and ensure the sustainable use of their own biological diversity. Those countries with the greatest biodiversity are concentrated in the developing world Only one of the eight countries that are home to the largest number of native mammal species is industrial. Of the countries with more than 10,000 species of higher plants, 18 out of 20 are developing countries, and 12 of the 17 countries with more than 500 threatened species of higher plants are developing countries. Developing countries are thus key to meeting the goals set at Rio, and the Convention on Biological Diversity was passed with widespread support from these countries.

Like the Framework Convention on Climate Change, the Biodiversity Convention recognised economic and social development as the top priorities for developing countries. It also stated that the extent to which developing country parties would effectively implement their commitments to preserve biodiversity would depend on industrial country commitments related to financial resources and transfer of technology.

The Benefits of Biodiversity and the Costs of Preservation

As we have seen, the ecosystems (and the species) in which genetic material resides provide valuable services at the national level. For this reason, the primary role of international agencies and bilateral support in the area of biodiversity should be to transfer knowledge and provide technical assistance to help overcome national market failures and create national markets for ecological benefits.

The GEF was chosen as the formal interim financing mechanism for the Convention on Biological Diversity. Total GEF financing for biodiversity projects comes to over $800 million and has already been used to support a range of technical and institutional projects. In Poland, for example, the Forest Biodiversity Protection Project has provided institutional support to the country's environment ministry, funded pilot investments in air-and soil-monitoring equipment and a forest gene bank, and supported farmers in the Bialowieza Primeval Forest who are making the transition to "ecological agriculture." In Algeria the El Kala National Park and Wetlands Management Project introduced actions to stop degradation within the complex and supported assessment activities that included surveys, studies, and public education programmes aimed at bolstering long-term preservation efforts.

While such support may form the backbone of international efforts to preserve biodiversity, the global commons issues connected with existence value and exploitation of genetic resources remain. The economics of these issues is complicated by disagreement on what exactly is being valued—whether it is the tight of plants or animals to exist, the material benefits that diversity offers, or the just the pleasure that the existence of many living organisms brings to people. Even basic facts such as the total number of species on earth and the rate of species extinction worldwide are not fully clear. The UNEP's Global Biodiversity Assessment estimates the number of species on the planet at 7 million-20 million and the expected loss of species over the next 25 years at between 140,000 and 5 million. Combining the lower-bound estimates suggests that 2 per cent of all species are at risk; combining the upper-bound estimates produces an estimate of 25 per cent (although it should be noted that even the lower rate of extinction is approximately 1,000 times the natural rate).

Many of the benefits of preserving genetic material are also difficult to quantify in monetary terms. How is a dollar value to be placed on the rights of organisms to exist or on the pleasure people derive from their existence? Among the more quantifiable benefits is the medicinal use of genetic resources. The United Nations has estimated that medicines originally developed from plant material are worth about $43 billion a year. The rosy periwinkle from Madagascar's rain forest, for example, provided a rare genetic trait that was used in developing pharmaceuticals to treat childhood leukemia. Two of the drugs one company has developed from this plant have sales worth $100 million a year. (None of these proceeds, it should be noted, goes to Madagascar.) But even calculating the marginal benefit of the genetic material in a species is no easy task. The drugs developed from plants must be collected, refined, tested, and developed for the market, and sharing the profits along this value chain is clearly a complex issue. Genetic materials are also likely to be present in more than one species. This fact helps explain why estimates of the marginal value of species existence (put another way, the marginal value of preventing species extinction) are so uncertain. They have been put at anywhere from $44 to $23.7 million for an untested species.

Expanding Participation

Even if the value of genetic material is hard to estimate, it is certainly true that the international community continues to exploit it without paying—a scenario that constitutes a market failure. As a result, biodiversity may be undervalued in developing countries. An additional mechanism for promoting the preservation of generic resources would be the extension of property rights to a country's genetic material. This subject was raised at the Convention on Biological Diversity, but no agreement was reached on what should be done about the situation. One model for resource transfer might be that of Costa Rica's private, nonprofit National Biodiversity Institute (INBio), which struck a deal with U.S.-based pharmaceutical firm Merck and Company to help underwrite INBio's biodiversity prospecting plans. The Merck deal will pay INBio $1.1 million plus royalties for any product Merck develops from Costa Rican resources. In return INBio provides Merck with samples from all over Costa Rica. Ten per cent of the up-front money and 50 per cent of any royalties are allocated to inventory, bioprospecting, and conservation.

A number of doubts have been raised about such mechanisms. The Costa Rica-Merck agreement, for example, does not involve enough resources to pay for significant increases in protected reserves. The scheme might also not be widely replicable. One recent estimate suggests that even in western Ecuador, one of the areas richest in endemic species, the per-hectare value of genetic material to drug companies is only about $20. Furthermore, by claiming royalties on products developed from plants and animals that may be found in more than one country, INBio is effectively reducing the incentive of neighbouring countries to take similar measures to protect their genetic diversity. This problem is likely to be widespread: the rosy periwinkle was not endemic to Madagascar, for example. Moreover, it is unclear how such a scheme could work to protect areas that have already been explored for genetic material. Thus, while establishing limited property rights to genetic material may encourage developing countries to participate in preservation efforts, they represent only a partial solution. If industrial countries feel that additional incentives to preserve genetic material are required (to cover, as it might be, the existence value of species, regardless of their economic uses) the simplest method would be to expand direct international support for this purpose.

Biodiversity covers many different activities, including farming, forestry, coral reef protection, and others. This diversity calls for great flexibility in approach toward agreements on different biodiversity issues, at both the regional and global levels. Technical and institutional support and flexible transfer payments are two such approaches. Sanctions have also been used. When biodiverse habitats are exploited in order to produce a tradable good—including tropical fish, tropical timber, and many of the animals covered by the Convention on International Trade in Endangered Species (CITES)— formal trade limits or certification schemes with strong penalties for noncompliance can play an important role. One way to protect coral, for example, might be to ban trade in fish caught by using cyanide, a significant source of coral degradation. Similar incentives have been widely used in a range of environmental treaties. Although questions remain about the risk of overusing trade measures to counter environmental threats (and thus using the environment as an excuse to strangle trade as a wealth-creating force), trade measures can be a very effective method of pursuing environmental goals under some circumstances.

Customers' Role and Responsibility in Environmental Management

The green consumer is essentially a retail market. But some companies claim that their industrial customers are beginning to put a value on greenery. In a survey in the late 1980s of business attitudes toward suppliers, 3M found that nearly 20% respondents rated "concern for the environment" as the most important quality, followed (17%) by value for money. In some industries the pressure to care about the environment is changing the relationship between companies and their customers. ICI, for instance, now offers to take back contaminated sulphuric acid from its customers in the oil industry and clean it up. Hampson sees this as a chance to strengthen ties with customers: "Customers will increasingly want to deal with suppliers who can solve their environmental problems. Disposing of our customers' waste products is a way for us to link ourselves more closely with them."

Dow Chemical has also turned responsibility into opportunity. It is developing a concept called " product stewardship," designed to ensure that its products are safely distributed, stored, and used by its customers. In 1989, Dow launched an even bolder scheme. Most of the chlorinated solvents the company sells are replacements for ones that have leaked into the air or water. Dow's Chemaware solvent-recovery project is designed to help customers reduce this leakage and to collect and reprocess used solvent, getting rid of the residue in an environmentally acceptable way. The company reckons that the Chemaware programme will eventually mean a big drop in its solvent sales. Even so, it went ahead with the project partly because it saw what had happened with CFCs. A failure to prevent leakage has led to a ban and to the development of other products that will do the same jobs at much higher cost and less efficiency.

Some companies have seen a chance to offer customers a new, greener service for CFCs. Toyota has pioneered a scheme to drain and reprocess the CFCS in their car air-conditioning systems for free, rather than let them leak out to damage the ozone layer. ICI and Du Pont both have schemes to recycle CFCs. Du Pont will ship containers anywhere in America to collect CFC-11 and CFC-12 (the two main kinds used in refrigeration and air conditioning) as long as there is a minimum load of 500 pounds. The customer simply has to meet the cost of filling the containers. Neither Du Pont nor its customers make money from recycling—yet; through Du Pont expects to do so as the supply of CFCs dwindles and recycling becomes more profitable. Suppliers realise that they may be able to offer customers a service that will tie them in more closely, and thus they will eventually increase profits even as the volume of sales diminishes.

Environmental concerns are also encouraging companies to build links with one another. Ed Woolard of Du Pont points out that "One alternative to recycling waste ourselves is to form relations with other companies: we won't lose responsibility for the waste, but they will help us handle it." The search for an alternative to ozone-gobbling CFCs has led to new levels of corporate cooperation: product companies have come together to test toxicity and have run joint recycling ventures. Plastics manufacturers have established some research cooperatives to look for better ways to recycle their materials. The problem of disposing of heavy metals in batteries has led Philips, and electronics group, and other companies in the Netherlands involved in battery manufacture and distribution to consider ways to recycle 80% of nickel cadmium batteries.

Such cooperation may also make it easier for small companies to make room for the environment. Big profitable companies with spare room and management capacity understandably find it easier to be environmentally conscious than small ones, although the most ingenious technology for cleaning the environment often comes from small firms. A survey by Baum a German club for companies that aims to spread environmental ideas, found that may small companies were not even obeying environmental rules, let alone pursuing green policies. One of Baum's aims is to spread environmental awareness by getting bigger firms to persuade their smaller suppliers.

The Role of Corporation in Developing Sustainable Trade Order

However lofty and correct the principles, the problem of how to operationalize them in day-to-day trade activities remains daunting. In part, this is because of profound changes that are occurring in the corporate management of world trade. Alliance formation among multinational corporations is quietly diminishing the power of governments to determine how trade shall be conducted. Thus, at the same time that global environmental threats are calling into question the legal and political viability of national sovereignty, cross-border investment activities of large multinational corporations are calling into question its economic viability.

In 1970, the United Nations identified seven thousand multinational corporations (MNCs) in operation, a majority of them based in the United States and Britain. In a little more than two decades time, that number has increased to over thirty-five thousand—many of them controlled from gleaming new headquarters in Japan, Germany, and Switzerland. Another 170,000 firms serve as foreign affiliates for these MNCs. The top one hundred MNCs are believed to control nearly half of the total foreign assets represented by cross-border trade and direct corporate investment (*Economist,* 1993: 6). In the United States, the top one thousand corporations, most of them MNCs, account for nearly two-thirds of GNP, with the remainder of GNP drawn from eleven million smaller firms. The average size of these MNCs is 16,500 timer larger than the average small—to medium-scale business in America (Hawken, 1993: 8).

From an environmental perspective, the development of the multinational firm has been a very mixed blessing (Choucri, 1991). On the one hand, MNCs tend to be more advanced or sophisticated than many domestic, small-to medium-sized firms in their environmental management. They can transfer environmental technology quickly, draw on large R & D programmes and international information networks for developing waste-management and pollution-control programmes, and they can compare notes on the most promising environmental innovations developed by the countries in which they do business. In addition, they are often in a better financial position than local firms to absorb new environmental mandates, and their environmental impacts are more likely to be scrutinised by host countries (not to mention the countries in which they are headquartered), making them more conscious of their environmental reputations (Choucri, 1993: 205-253; Leonard and Duerksen 1979:23).

On the other hand, MNCs have created an enormous integrated market for goods and services that has had the effect of greatly accelerating the extraction and consumption of natural resources and the pollution output of industrial manufacturing and agriculture. MNCs make possible the massive, coordinated development of heretofore wild or remote parts of the world. They are involved in patterns of exploitation and trade that, in the eyes of many Third World observers (e.g., Raghavan, 1990), amount to nothing less than recolonisation. Moreover, their mobility, corporate secrecy, financial leverage, and ability to exploit ambiguity in national laws often make it possible for them to evade regulation and avoid accountability for their actions. Although there is little evidence today that large numbers of MNCs are actively seeking pollution havens in developing countries (Harrison, 1994), many environmentalists fear that such exploitation will grow in the future as regulation within developed countries is strengthened. This table summarizes the most common regulatory tools and strategies employed to manage the environmental practices and impacts of MNCs.

It is usually assumed that strict environmental standards place firms at a competitive disadvantage and eliminate jobs in regulated industries. While environmental regulation is clearly costly for some firms and industries, empirical studies reveal that the correlation between environmental stringency and economic performance for most industrial sectors is extremely weak and, if anything, positive (Repetto, 1995). Some studies suggest that strict environmental regulation may be a major factor in promoting economic competitiveness (Hall, 1994). It may also reflect the repaid growth of "green technology" industries, now estimated to comprise a world market valued at nearly $200 billion.

Despite the growing evidence that environmental measures contribute to strong economies, many MNCs seem to view environmental goals as major impediments to economic progress. Some have devoted major portion of their public relations and advertising budgets to challenging what they perceive as unreasonably national environmental standards. In these contests of business and government influence, it is not the power or scale of particular MNCs that matters so much as it is their collective ability to form strategic alliances that effectively place them beyond reach of many national governments. In fact, with increasing trade liberalisation and technological advances in worldwide communication and computing, today's strategic alliances of MNCs typically represent more power and influence over particular commodities or services than is wielded by the vast majority of nations in which they do business. The development of these alliances, aided by trade policy and technology, has fostered a subtle transformation of MNCs into TNCs—transactional corporations—and, in a few cases, into SNCs - supranational corporations.

A supranational world is borderless world, and consequently peoples' sense of loyalty is drawn to something larger and separate from the nation-state. To be supranational today, in practical social and economic terms, really means to be glocal. A supranational corporation is one in which economic development is tied to local labour pools, global capital flows, and regional investment opportunities. Consequently, SNCs may perceive existing political institutions and boundaries largely as potential barriers to growth. They will be prone to view international law and diplomacy as costly and inefficient means for protecting cross-border investment and contract enforcement. In the eyes of many SNCs, governance—to the extent it is needed to protect market economies—may have to be entrusted to the World Trade Organisation. It is strategic alliances based on trade, rather than the military alliances of nations, that drives SNC expectations about world affairs in the next country.

TABLE 1
Regulating the Environmental Practices of Transnational Corporations

National Governments	*TNC self-regulations*	*International authority*
Pollution standards	Corporate codes of conduct	International codes and guidelines (e.g., OECD guidelines for MNCs)

Land-use controls	Site selection	Cross border consultation and notification
Disclosure rules and reporting requirements	Corporate health and safety rules	Conventions and protocols (e.g., Montreal Protocol)
Import restrictions	Project-and product-planning goals	Information exchanges and registries (e.g., IRTPC; International Toxics Registry)
Customs inspections	Management contracts and licensing agreements	Trade Policy
Liability for damages	Industry performance objectives and monitoring	Environmental review by international panels/ commissions
Impact assessment and audit requirements	Voluntary audits	Global monitoring and compliance audits
Enforcement Programmes	Inspection and maintenance reviews	Citizen action campaign (e.g., Pesticide Action Network
Penalties/Fines	Promotion and hiring practices	International sanctions

Sources: Charles S Pearson (1987: 3-31), Nazil Choucri (1991, 1993:205-353), Stephan Schmidheiny et al. (1992), Frances Cairncross (1992), and Paul Hawken (1939).

National governments, for their part, are confronted with a rising tide of cross-border investment, managed by so-called "stateless" corporations that have developed "chameleon-like abilities to resemble insiders no matter where they operate," along with abilities to customize their global product line to conform with local tastes (Holstein, 1990:98, 102). Although few corporations are truly stateless in either design or operating philosophy, an increasing number are making decision without regard to national boundaries and interests, excepts as they affect the business climate. This trend has prompted some environmental groups to push for stronger MNC codes of conduct, such as the CERES principles, established in the aftermath of the 1989 *Exxon Valdez* oil spill, and the Global Environmental Management Initiative (GEMI), which promotes greater environmental openness and self-monitoring by its corporate members. These initiatives go beyond past attempts by governments to regulate the international behaviour of corporations, such as the U.N. Code of Conduct on Transnational Corporations, the OECD Guidelines for Multinational Enterprises, and the International Labour Officer's Tripartite Declaration of Principles Concerning Multinational Enterprises and Social Policy.

Fortunately, some corporations are beginning to see environmental protection as a value-added factor in production and service. Concepts such as Total Environmental Quality Management (TQEM) and Industrial Ecology are slowly making their mark on the management philosophy of major corporations. As a consequence, the people who run TNCs are becoming much more aware of the environmental opportunities arising from global interdependence. They are also beginning to realise that the ability to provide stable climates for business and trade depend increasingly on the stability of the earth's physical climate and of its ecological and geochemical systems. In order to understand how the business environment and the natural environment affect each other, a number of private sector leaders have called for changes in the way that nations and corporations define and measures their political economies (e.g., Schmidheiny, 1992; Hawken, 1993). They favour the use of market based environmental indicators and investment indices that will help business, government, and consumers make better informed decisions about economic activities that harm the environment.

Incentives to Participate in Developing Codes for Environmental Management

Firms sign on to a private code of environmental management practice for a variety of reasons. Sometimes they have little choice: Every CMA member, for instance, is required to take part in the Responsible Care programme, and observers predict that ISO 14001 registration may soon be required for doing international business. But there are other motivations as well, including a history of environmental values.

While signing on to Responsible Care or ISO 14001 may not represent a real choice for firms, the degree of their commitment to fulfilling code requirements can and does vary. As a rule, large firms have shown the most commitment to these codes and also to GEMI. Such firms played key roles in developing these codes and, because of their size, have tended to attract the most attention from regulators, environmental groups, and the news media. As a result, they have been the most interested in distinguishing themselves a environmental leaders and in garnering positive press. In addition, large firms often operate a large number of plants in many different locations. Codes offer them a way to ensure consistency across local, state, and even national jurisdictions. Importantly, large firms also have the resources required to implement code practices. For example, many chemical companies employ a full-time Responsible Care coordinator, an expense that is usually not possible for small companies. Large firms may even see codes as a way to undermine their smaller competitors.

Firms with obvious environmental problems may also feel a particular compulsion to become active in private codes. For example, Hampshire Chemical Corporation, a small speciality chemical manufacturers, dramatically changed its environmental programmes after a series of accidents and lawsuits in the late 1980s brought environmental issues to the attention of senior management. In addition to initiating its own corporate environmental programme, Hamsphire's manufacturing manager took on the chairmanship of CMA's community awareness and emergency response task group. Unlike Responsible Care and ISO 14001, participation in CERES and GEMI are purely voluntary. Participating companies tend either to maintain direct sale relationship with the public or hold corporate values that are strongly consistent with code requirements. Domino's pizza offers an example of a firm that may have been motivated to join CERES because of its strong public presence. Even though environmental protection has historically not been a top concern for Domin's, its high public profile made endorsing CERES a good move. Similarly, Procter & Gamble's leading role in GEMI was probably prompted by its large size and the public's familiarity with its many products.

Polaroid, a recent CERES signatory, is an example of a firm that has adopted a code because of its values. Polaroid has had a ling-standing commitment to public reporting of environmental data and was one of the first major companies to publish an annual environmental report—a practice it established in 1988, six years before it endorsed CERES. Joining CERES affirmed the value the company has placed upon disclosure as well as dialogue with environmental groups. These valuer are widely shared among CERES corporate endorsers. For many of these firms, public disclosure and community outreach are explicit corporate goals.

CASE STUDY OF MANAGER—ENVIRONMENTAL FOR CARTER HOLT HARVEY TISSUE, KAWERAU, NZ

Carter Holt Harvey Tissue is a pulp and paper manufacturing company in New Zealand which certified under 14001. For the last 10 years it has been working at continual (environmental) improvement. It established back in 1984 a programme of water reduction and progressively reduced water usage a bit at a time. The consequence is that now it has reduced usage by 55% since that time and plan to be down by 60% by 2000. This has resulted in considerable cost savings from:

1. reduced fibre loss at market value of pulp (50% less)
2. energy savings (15% less)
3. water costs (55% less)
4. less waste fibre disposal costs (50% less)
5. increased productivity on the paper machines from more consistent back water characteristics
6. less chemical usage due to less loss in the waste water.

It has reduced solid waste that goes to land fill by over 30%. Its plan is for a zero waste to land fill. Even if this is not fully realised, the effect of this will be to ensure that it probably will never need to obtain another land fill site. The cost saving of not having to do that is huge and runs into several millions, not to mention the PR benefits. It is also going to land treatment, irrigation, which is expected to have benefits in afforestation by reducing stress from drought conditions and by replacing nutrients in the soil. It is expected that the trees will grow more uniformly through having a steady water supply. This is peculiar to our volcanic ash soil. We treat the sewage from the local town through our effluent treatment plant and that benefits the community and gives us nutrients thereby reducing impact on the environment. It also saves us having to add nutrients to our treatment plant with a consequential cost saving. ISO 14001 just documented the procedures we already had in place, so on its own it is not responsible for any savings. In the future however, it may result in considerable savings by reducing the cost of obtaining discharge permits an enabling the company to obtain longer term permits. It may also reduce the costs of obtaining permits because there could be less opposition to any permit sought and hence less legal procedures to be pursued. These cost savings are seen as being quite large long term.

Green Products and Environment

In view of the constantly increasing amount of goods in circulation, in society, with the accompanying consumption of resources and waste problems, there is reason to question whether current practices provide a basis for an enduring society that can bee passed on to future generations. The conviction that the direction of development that the industrialised age so are has taken is into the right one, from an environmental point of view, is spreading. It is becoming increasingly more apparent that far-reaching changes are needed in society to offset this course. However, changes of this magnitude do not take place spontaneously. It is more likely that a completely new set of strategies for society is required, providing a control of goods aimed at reduced consumption of resources and diminished waste problems. Therefore, there is an urgent need to look into product (goods) related environmental issues with special focus on the policies and instruments that are part of preventive environmental strategies, and which leads to environmentally conscious product development. The concepts that are going to be discussed are *Life Cycle Assessment, Design for Environment, Eco-Labelling, Extended Producer Responsibility and Product Policy.*

1. Life Cycle Assessment (LCA)

The concept of life cycle assessment was born in the late 60s. It was the time when the Report of the Club of Rome raised public awareness of some global problems, such as the depletion of nature's resources, population growth, waste problems, etc.

In 1969 the Coca-Cola company commissioned a study by Midwest Research

Institute to compare different beverage containers from the perspective of emissions and resource consumption. The methodology used, know as Resource an Environmental Profile Analysis (REPA). Later these studies formed the basis for the beginning of the development of life cycle assessment.

According to the International Standard Organisation (ISO), the LCA can be defined as

Life Cycle: "Consecutive and inter-linked stages of a product or service system, from the extraction of natural resources to the final disposal."

Life cycle Assessment: "A systematic set of procedures for compiling and examining the inputs and outputs of materials and the associated environmental impacts directly attributable to the functioning of a product or service system throughout its life cycle."

According to SETAC, the definition of LCA is:

Life Cycle Assessment— a process to evaluate the environmental burdens associated with a product, process, or activity by identifying and quantifying energy and materials used and emissions released to the environment; to assess the impact of those energy and material uses and releases to the environment; and to identifying and evaluate opportunities to effect environmental improvements. The assessment includes the entire life cycle of product, process, or activity, encompassing extracting and processing raw materials; manufacturing, transportation and distribution; use, re-use, maintenance; recycling, and final disposal.

"Life-cycle" assessment (LCA) has evolved over the past 20 years to a point where designers and engineers can effectively use the approach to identify the consequences of their choices of materials, manufacturing processes, and product usage."

2. Eco-Labelling Schemes

An eco-labelling scheme or programme is voluntary participation system, which issues a seal-of-approval, i.e. a third party certificate. It awards a legally protected symbol or logo to products judged to be less environmentally harmful than comparable products, based on a specified set of environmental criteria.

Some of the prominent eco-labels from Asia are:

- *Japan:* The Eco Mark symbol itself represents our desire to protect the earth with our own hands. The eco Mark design was chosen through a public competition sponsored by the JEA in 1988, with the winning entry receiving the Award of Chief of the Environment Agency. A phrase expressing the goal of the Eco Mark Programme 'Friendly to the Earth' appears on the upper half of the symbol; the effect in terms of environmental conservation for a particular product category appears in the lower half.

- *Singapore:* the green leaf stands for the living environment. It symbolises nature and beauty. The outstretching veins portray the interdependence of all living things. The circle symbolises cooperation and commitment to protect and nurture the environment.
- *Taiwan:* The green leaf stands for green consumerism. The green globe represents a clean and unpolluted earth. The integrated figure resembles to the landscape of Taiwan, China, which symbolises the cooperation with international 'green' activities and Taiwan's commitment to conserve natural resources and to protect the environment.
- *China:* The logo of China environmental labelling is chosen from a public design contest. It consists of a mountain, water, the sun, and ten interconnecting rings. The sun, the mountain and water are in the center, which mimics the logo of NEPA, representing the living environment. The ten rings represent 'joint efforts of the nation to protect the environment on which human beings rely. The ring (huan) is the same Chinese work as environment (Huan Jing). The logo is registered as trademark and is protected by law.
- *India:* An earthen pot has been chosen as the logo for the Ecomark scheme in India. The familiar earthen pot uses a renewable resource like earth, does not produce hazardous waste and consumes little energy in making. It solid and graceful form represents both strength and fragility, which also characterises our eco-system.
- *Thailand:* The symbol signifies environmental conservation. The flora (the leaves) and fauna (the bird) depicted are the living wonders of the world. In the myriad of links with our Earth, our hopes for future generations (the smiling face) grows from life respecting nature.

3. *Extended Producer Responsibility*

The principle of Extended Producer Responsibility (EPR) was introduced and developed by Thomas Lindhqvist (*One of the main speakers at the workshop*) for the Swedish Ministry of the Environment in two reports presented in 1990 and 1992.

The definition of EPR (Lindhqvist '92) is:

"Extended producer responsibility is an environmental protection principle to reach an environmental objective of a decreased total environmental impact from

a product, by making the manufacturer of the product responsible for the entire life-cycle of the product and especially for the take-back, recycling and final disposal of the product. The extended producer responsibility is implemented through administrative, economic and informative instruments. The composition of these instruments determines the precise form of the extended producer responsibility."

A Model of EPR

According to Lindhqvist's definition, Extended Producer Responsibility may be divided in five main categories:

1. Ownership
2. Physical responsibility
3. Economic responsibility
4. Liability
5. Informative responsibility

The different forms of responsibility in the model are defined as follows:

Ownership means that the legal ownership of the product remains in the name of the producer or the distributor.

Physical responsibility means that the producer, in a physical sense, takes care of the worn-out product or arranges for someone else to carry out this service upon direct commission.

Economic responsibility means that the producer completely or partially cover the cost of collection and /or cost for the recovery of the product or its final management.

Liability means that the producer is responsible for the damage a product might cause during its lifetime.

Informative responsibility entails a superior form of responsibility, dependent on the reputation of the product on the market. This includes the opinions of the consumers about the environmental qualities of the product and the moral responsibility of the company to show consideration for the environmental concerns of its employees. Education and information are typical examples of informative responsibility.

4. Design for Environment (DfE)

Design for Environment (DfE) is an important part of environmental product research. The environmental impact caused by a product, as well as the environmental awareness of the markets, can be improved through considering environmental issues in product development. By integrating environmental requirements into the earliest stages of product development, adverse environmental impacts associated with the manufacture, use, and end-of-life management of a product may be reduced or eliminated. DFE can also provide significant benefits such as enhanced resource efficiency, reduced liabilities, and enhanced competitiveness. Thus, DFE or life cycle design offers significant opportunities for achieving sustainable development. *Design for Environment* is the term chosen to describe the systematic practice of minimising adverse environmental impact and maximising conservation of valuable resources throughout the product's life cycle, resulting in cost savings and competitive advantage for the producer. This approach is also frequently described as 'Life Cycle Design', 'Eco design', 'Sustainable Design' or Design for Eco-efficiency.

The introduction of market-based instruments of control, such as environmental producer responsibility and environmental labelling, and the growing importance of green consumerism have strengthened incentives for Design for Environment (DFE) as a business strategy at the corporate level.

ISO 14000 Series

As the ISO 14000 series was still in development at the time the ASEAN and UNCTD country studies were being prepared, these studies do not offer much concrete evidence concerning the effects on ASEAN exports. Nevertheless, these "international" standards have potentially significant consequences for international trade and for SMEs. This requires closer examination.

(a) Background

Voluntary standards for environmental management date to 1992 and the British standards BS 7750. Those were followed in 1993 by the European Union eco-management and audit scheme. EMAS, in 1993. During the same period, national and European Union-level environmental labelling schemes were being established. Additionally, industry associations have been developing performance codes such as the Responsible Care Programme of the United States Chemical Manufacturers Association and the Environmental Self-Assessment Programme of the Global Environmental Management Initiative. These developments prompted the ISO to form a technical committee (TC 207) in 1993, to establish uniform international environmental management systems (EMS) standards. This followed the ISO 9000 series which are generic standards for quality management. The ISO 9000 and 14000 series are departures from the traditional ISO function of establishing product technical standards. It is important to note that the ISO 14000 series are not performance standards. They do not mandate environmental performance, nor do they set effluent and emission standards or set design technology. They are process standards, and more specifically standards for the processes of environmental management.

The ISO 14001 draft, which deals with standards for environmental management, requires a firm to commit itself to comply with applicable domestic legislation and regulations, and to continual improvement in its performance, but does not specify performance levels *per se*.

The terminology and structure of ISO 14000 series may be confusing. "ISO 14000" actually comprises some two dozen standards (Table 3). Of these, ISO 14001—Environmental Management Standards—Specification contains the main standards that will be used by firms and will be available for registration (i.e., certification) by third parties. The other standards and guidelines cover such areas as corporate environmental auditing, environmental labelling, life-cycle analysis, and environment performance evaluation.

One purpose of the ISO initiative is to bring some harmonisation to the proliferation of environmental management standards. International uniformity of management process standards, while perhaps less compelling then uniformity in product standards to ensure compatibility of products (e.g., pipe threading), can nevertheless contribute to lower information and transactions costs and greater gains from international trade. This is especially true if environmental management standards are to become preconditions for commercial transactions. In such a situation, a singly uniform international standard has obvious advantages.

The development of ISO 14000 series has generated some controversy. One issue is its relation to WTO and its TBT Agreement. The TBT Agreement encourages members to use relevant international standards (presumably ISO 14000 standards where relevant), and compliance with such standards plays a role in the dispute settlement process. Moreover, the TBT Agreement encourages countries to adhere to a Code of Good Practice defined in the TBT Agreement in relation to voluntary standards. Practice defined in the TBT Agreement in relation to voluntary standards. What is not clear is how the mandatory versus voluntary status of ISO 14000 standards will evolve, and this will have to be clarified within WTO. Although ISO standards originate as voluntary standards, Governments may adopt them and make them mandatory. WTO has different rules with regard to voluntary and mandatory standards. A second issue important for the environmental community is whether the ISO 14000 initiative represents a step back from the more stringent EMAS system (the European Committee for Standardisation is considering whether ISO 14000 is to be recognised under EMAS). EMAS requires disclosure of information that ISO 14000 does not, and represent performance, not conformance, standards.

TABLE 2

International Standards on Environmental Management

ISO number	*Document*
14001	EMS—Specification
14004	EMS—General guidelines
14002	EMA-SME guidelines
14010	Guidelines for EA—General Principles
14011.1	Guidelines for EA—Audit Procedures Part I: Audition of EMS
14011.2	Part II: Compliance audits
14011.3	Part III: Audit of an environmental statement
14012	Guidelines for EA—Qualification Criteria for Environmental Auditors
14013	Guidelines for EA—Management of environmental system audit programmes
14014	Guidelines for initial environmental review
14015	Guidelines for environmental site assessment
14020	Basic principles for all EL
14021	EL—Self declaration and environmental claims—Terms & Definitions
14022	Environmental labelling symbols (Type II)
14023	EL—Testing and Verification Methodologies
14024	EL—Guiding principles, practices and criteria for certification programmes—guide for certification procedures
14031	Generic environmental performance evaluation
14032	Industry—specific environmental performance indicators
14040	LCA—General Principles and Practices
14041	LCA—Life Cycle Inventory analysis
14042	LCA—Life Cycle impact assessment
14043	LCA—Life Cycle improvement assessment
14050	Environmental Management—Terms and Definitions
14060	Guide for the inclusion of environmental aspects in products standards.

Source: ISO/CASCO (1895): Proceedings; Henry J. (1995): *International Standards on Environmental Management,* unpublished), published as Annex 2, *ISO 14000,* UNDP Private Sector Development Programme, New York, 1996.

Note: EMS = Environmental Management System, EA = Environmental Auditing, EL = Environmental Labelling, LCA = Life Cycle Assessment.

Developing Country Concerns

The overriding concerns for developing countries are that ISO 14000 standards may become de facto or even *de jure* requirements for international commercial transactions, and that the costs and difficulties of conforming to ISO 14000 will prove to be a non-tariff trade barrier. Specifically, if firms in OECD marks require ISO 14000 certification from their suppliers in developing countries, or if compliance is mandated by importing Governments, for example, in government procurement, ISO 14000 could become a serious trade obstacle, especially for SMEs.

More specific concerns relate to participation and costs. Thus far participation in writing ISO 14000 standards by developing countries, although limited, has been growing. The UNCTAD study lists only 15 developing countries as participating in TC 207. The Benchmark Environmental Consulting study states large consulting houses and TNCs dominate decision-making within TC 207. The cost of participation in international meetings is a problem. Also, the prospective cost of complying with ISO 14001, the management systems standard, is said to be high for individual companies. The costs include setting up and maintaining and environmental management system, hiring consultants and third party registration. A sample calculation in the UNCTAD study for a small American company (less than 100 workers), without an initial environmental or quality management system in place, but the consultancy and registration costs at US$ 1,22,000 plus internal implementation costs. Other difficulties for developing countries and for SMEs in developing countries may arise from the absence of in-country accreditation or certification bodies for third party certification, or the absence of explicit national environmental legislation. Recall that the management systems standard require a firm to commit itself to comply with national legislation; if such legislation is incomplete or lacking, the obligation itself is called into question.

Status of the ISO 14000 Series

ISO 14001 and 14004, the Environmental Management System specification and guidance standards respectively, were finalised as draft international standards and overwhelmingly approved on a balloting held earlier this year. A second ballot incorporating editorial changes is scheduled, and 14001 and 14004 should become final standards this year. Three standards relating to environmental auditing (ISO 14010, 14011 and 14012) have also passed an initial balloting, and should become final before end-1996. The standards for life-cycle assessment, environmental labelling and environmental performance evaluation are not as far along in the ISO process.

Impact of Product Certification on Consumer Behaviour

Anecdotol evidence suggests that some seals of approval may have a very significant impact on consumer purchasing habits. For example:

- In the two months following the award of a seal of approval by the American Dental Association (ADA) in August 1960, the market share of Procter and Gamble's 'Crest' toothpaste rose from ten per cent to twelve per cent, and within two years it had become the top selling brand, with a 30 per cent share of the market (Bennet and McCrohan, 1993).

- A 1988 survey carried by Good Housekeeping found that, of 18 brands carrying the GH seal for the first time, 16 showed an increase in sales (Kerwin, 1994). In addition, a 1990 survey carried out by the Roper Organisation found that 84 per cent of those surveyed thought that the GH seal signified that the product was of 'superior' or 'fairly good' quality (Schiffman and Kanuk, 1994, p. 191).

Moreover, in what has been described as 'the most comprehensive investigation to date' (Beltrami and Stafford, 1993), Parkinson (1975) found consumers ranked 'seals of approval' highest, above 'friends', 'salespersons', and 'advertisements' in terms of their 'expertise' and 'impartiality' and second highest, behind friends, on trustworthiness'.

However, Parkinson also found that consumers tended to misinterpret seals of approval, a finding confirmed by several subsequent surveys (Beltrami and Stafford, 1993; Laric and Sarel, 1981). One of the main reasons for this is that , with the possible exceptions of Good Housekeeping (which publishes an explanation of its seal in each issue of Good Housekeeping magazine) and NSF International (an independent testing organisation which has advertised the meaning of its seal in such publications as *Business Week*), the sponsors of seals of approval have not typically taken sufficient action to inform the public about the meaning of their seals (Beltrami and Stafford, 1993).

In addition, some researchers have suggested that consumers do not differentiate between certification marks—assuming them all to have equal value. Empirical evidence is not clear cut on this point. A survey commissioned by Better Homes and Gardens (BH&G) found that Whilst 87 per cent of consumers were aware of the Good Housekeeping seal, 68 per cent claimed to be aware of a non-existent 'BH&G seal' (Kerwin, 1994). However, a survey conducted by Beltrami and Stafford (1993) found that, of twelve certification marks examined (including the Good Housekeeping Seal), only the UL seal improved the believability of advertising claims—suggesting that consumers do, in fact, discriminate between certification marks. It is difficult to make any firm inferences from these data—aside from any other consideration many consumers may simply not be interested in certification of information to guide their purchasing decisions. Moreover, it seems likely that some people do use the Good Housekeeping seal as a guide, just as it is likely that some people's decisions to purchase Crest were influenced by the ADA seal. However, given the multitude of different factors affecting purchasing decisions, it is not possible to say with any degree of precision how much impact certification marks have had, even for these relatively well-documented cases.

Thus, consumers rely heavily on the information provided on labels to make their products selection decisions. Product certification marks may help both consumers and producers by reducing information asymmetries. However, there is a considerable possibility that certification marks may be misinterpreted. In conclusion, this brief overview of the impact of product certification on consumer decision-making cautions against giving an overenthusiastic welcome to any new certification scheme. The theoretical merits of each scheme must be weighed in proportion to the ability of the scheme to achieve its ostensive goals, both in theory and in practice, and against any negative consequences, intended or otherwise, which are scheme may have. In addition, the merits of alternative (perhaps mutually exclusive) information-sharing schemes should also be considered. With this in mind, the discussion now turns to the problem of ecolabels.

Challenges for Eco-labels

Arguments against eco-labelling arise from disputes over misleading claims, uninformative labels, erroneous seals of approval, unfair competition, concerns about green consumerism, the feasibility of labelling all products, and methodological difficulties.

Misleading or Fraudulent environmental claims: An eco-label has no value to the environmentally-conscious customer if the label is misleading or fraudulent. Since trust is a major component of a labelling programmes credibility, the label must be above suspicion. Terms such as, "recyclable, "biodegradable, and "ozone friendly", must be used accurately. When claims are used arbitrarily in advertising and labelling, customers will become confused, discouraged, and skeptical even of legitimate claims.

Uninformative claims: Labels that provide trivial or irrelevant environmental information promote environmental benefits which are insignificant and do nothing to improve the status quo.

Unfair Competition: Some companies are concerned about unfair competition. They are reluctant to rely on the word of a foreign company or a foreign eco-labelling programme that specific environmental criteria are being met. Indeed, some companies may choose to internationally misrepresent their products as "environmentally friendly" in order to bolster profits. This amounts to unfair competition for those companies which must spend the time and money to adhere to regulations. Companies seek assurance that a double standard will not emerge.

Green consumerism: Many environmentalists are critical of consumerism and argue that "green consumerism" is a self-contradicting term. The goal is to reduce consumption, not re-define it according to an environmental set of criteria. Individual actions when limited to "green shopping" will do little to bring about the more fundamental economic and social changes that are required to protect the planet. Eco-labels, although complementing government action, are no substitute for it. Consumer preference and market forces cannot, by themselves, guarantee environmental protection.

Feasibility: Another concern is that only a small number of products can realistically be labelled. Since the vast majority of goods will not be considered in eco-labelling programmes, many environmentalists point to regulation as being a more effective tool of public policy than the development of a special case voluntary standard.

Methodologies: Differences in testing and certification methods have created a number of difficulties in the applications of an eco-label to a product category. Should the label represent an overall assessment of a product's environmental burden over its entire life cycle or some subset of it? What techniques can be used to measure environmental impact? Who determines what specific environmental impacts are most important? What criteria are appropriate in rating impacts?

Resource-Based Labels

As observers of international environmental law know well, the U.S. ban on Mexican tuna created an open conflict between international trade law and domestic environmental law. While the bulk of commentary has addressed the implications of the two GATT dispute panels on the issue of PPMs and unilateral trade sanctions (although neither of the decisions were adopted by the GATT parties), environmental labels have played a significant role. The issue has centered on the high incidental dolphin mortality from the tuna fishery in the Eastern Tropical Pacific (ETP). In the late 1980s, recognising a marketing opportunity and faced with threatened consumer boycotts on tuna, major U.S. cannera voluntarily started labelling their cans. Starkist was the first to announce that it would no longer purchase any tuna, caught in association with dolphins and that it would begin labeling cans of tuna with "dolphin safe" symbols, bearing the message 'no harm to dolphins." Other major U.S. canners quickly followed suit. Congress responded as well, passing the Dolphin Protection Consumer Information Act of 1990 (DPCIA), which established a labelling requirement Esty 1994, 268-269). The effect of this restriction has been to close the U.S. market to Latin American tuna imports.

The DPCIA required that all tuna caught in the ETP and labelled dolphin-safe must be verified as not having been caught by internationally deploying purse seine nets around dolphin. Absent this certification, tuna imported into the United States could not be marketed as dolphin-safe on U.S. supermarket shelves. In its Tuna-Dolphin decision, a GATT dispute panel held that labelling provisions were consistent with GATT because they were applied evenhandedly and did not have discriminatory effect (GATT 1991). Indeed, director general of the GATT, Arthur Dunkel, praised ecolabels as the at the time as a means to avoid trade/environment conflicts Esty 1994, 134).

In 1995, recognising that dolphin mortality had been reduced by 95% in the tuna fishery, following a series of negotiations and an informal agreement with Latin American countries, the U.S. Congress proposed legislation to repeal the tuna ban and change the labelling provisions. In their place, the new law would permit the label, "dolphin-safe," so long as no dolphins were observed to have died while catching tuna. In effect, this change would permit the use of dolphin-safe labels on tuna caught by encirclement with purse seine nets, a practice that the DPCIA forbade. Despite strong support and passage in the House of Representatives, the bill died in the Senate without a floor vote as the session expired. Mexico has not ruled out a new challenge before the WTO on both the tuna ban and the labeling provisions. To head this off, the Clinto administration has promised to reintroduce the legislation before the new Congress.

The other resource—based label generating attention concerns sustainably harvested wood. In 1992, Austria passed a law requiring labels on all tropical timber and tropical timber products stating, "made of tropical timber." The law also created a voluntary label identifying "sustainably harvested" timber. Fearful that their timber products would not qualify for the "sustainably harvested" label and that they would suffer lost sales as a result, Malaysia and Indonesia (followed by the Association of Southeast Asian Nations [ASEAN] denounced the plan as protectionist, threatened to challenge Austria before the GATT, and announced to boycott of all Austrian goods and companies. The law was denounced as protectionist because only tropical timber was subject to the labelling requirement while temperate forest timber and products were not covered at all. In the face of this onslaught, the Austrian parliament rescinded the law within a year (Sucharipa-Behrmann 1994, 55).

Since this debacle, resource labels for forestry products have focused more on practical definitions of sustainable forestry and relied on a model of "certifying the certifiers." The Forest Stewardship Council (FSC), based in Oaxaca, Mexico, has developed principles for certification and accredits organisations throughout the world that identify and label wood products from sustainably harvested forests. The council has approved ten Principles of Forest Management, including respect for international environmental agreements, indigenous people's rights, conservation of the forest's biological integrity, and management plans. Once accredited, a producer pays a fee and may label its products with the council's or accreditor's mark. The advantage of the council's approach is its umbrella role, allowing local accreditation organisations to assess issues specific to, for example, boreal tropical, or temperate forests. A similar initiative, "Smart Wood," was founded by the Rainforest Alliance in 1989 and has certified over 7 million acres of environmentally sensitive forest operations in Southeast Asia, Latin America, and the United States. Over 30 companies use the smart 'wood label, including both timber producers and product manufacturers. A similar initiative, called the Marine Stewardship Council, is an alliance between the World Wildlife Fund (WWF) and Unileyer, a large international consumer and food products company. One of the world's largest buyers of frozen fish, Unilever will work with WWF of adapt the Forest Stewardship Council model to fishing standards and fish products Cessford 1996, 3).

The last type of environmental label, the report card label, has had little acceptance in the marketplace. Similar in concept to nutrition labelling on food packaging, the limited use of report card labels by manufacturers is most likely a result of their limited marketing appeal. The labels provide a measure of environmental impact on air, water, solid waste, and energy using a bar graph to compare the labelled product's impact with the impact of standard products in the same category. This graphic is more informative than ecolabels but likely less accessible to the average consumer. The best-known report card label is issued by Scientific Certification Systems, located in the San Francisco Bay area.

Current Ecolabels

The first national ecolabel was used in West Germany in 1978. Known as the "Blue Angel," the seal of consists of a small circle enclosing the symbol for the United Nations Environment Programme with an explanation of why the product is considered less harmful to the environment printed beneath it. This seal is now on several thousand consumer products and is widely recognised by Germans. Although there have been criticisms of the seal no formal studies have been done on its effects, the Blue Angel programme is credited with "stimulating consumer awareness" of environmental issues, inspiring "industrial innovation, and reducing the environmental burdens created by consumer products." For instance, in 1990, the West German government reported that the programme had reduced the amount of household paint solvents entering the waste stream by 40,000 tons and credited Blue Angel with spurring innovations in oil burner standards that reduced sulfur dioxide, carbon monoxide, and nitrogen oxide emissions by 30 per cent, on average. Members of industry also seem to feel they benefit by the programme because manufacturers have been applying for the Blue Angel seal and have often been making changes to meet the environmental criteria to capture a larger market share.

Members of industry, consumer groups, and the public suggest categories of products that may compete for the seal of the Federal Environment Association, the German equivalent of the U.S. Environmental Protection Agency (EPA). These proposals are considered by experts from industry and scientific and consumer organisations, who constitute a special Environmental Label Jury. After a set of public hearings, the criteria a product must meet to merit the seal are finalised, and companies may apply for the right to use the seal. The Institute for Product Safety and Labelling, an independent testing organisation, then evaluates the applications. Generally, about 15 per cent of the products in a given category are able to meet the requirements for a Blue Angel seal. Once awarded, the seal may be used on a product for three years, provided the product continues to meet the Blue Angel standards. Manufacturers who receive the Blue Angel seal pay small fees that help defray the cost of the programme and may reapply for the seal when their initial certification expires.

Since the establishment of the Blue Angel programme, several other countries, including Japan, Canada Sweden and Norway, have established environmental seals of their own, and French, British, Scandinavian, and European Community seals have been proposed. Although Norway's programme is administered by a government-appointed, independent, nonprofit organisation, the Japanese, Canadian, and Swedish programmes, like the German programme, are each administered by a government environmental organisation and a panel composed of industrialists, scientists, environmentalists, and consumer group representatives.

The programmes have become embroiled in debates, however, over a suitable methodology for establishing product standards. A consultants' report prepared for EPA observed: "No country has developed a formal methodology for developing product criteria; the process is typically somewhat *ad hoc*. The Germans have been criticised for their approach of choosing several key attributes of a product and judging its fitness for a seal based on these few factors along. One often-cited case is that of a lawnmower that received the Blue Angel seal because it created less noise than other mowers, though its energy consumption and exhaust levels were not significantly lower than those of other models. Partly in response to the criticism leveled at the German programme, other countries have followed different approaches in evaluating products for their environmental seals. Japan's "Ecomark" label has as yet been given only to products with specific resource-saving functions, such as lowflow shower heads and compoting bins. Canada's Environmental Choice programme, established in 1988, plans to incorporate a broad range of products and will use extensive analyses of each product's impacts, from the inputs used and emissions released for its manufacture to its durability, recyclability, and ultimate disposal. The aim is to make a "life-cycle" analysis of each product's environmental impacts. Many of the difficulties faced by ecolabelling programmes in establishing standards may ultimately be traced to the lack of a scientifically accepted methodology for determining the full life-cycle environmental effects of a product. This scientific problem has had a major influence on the political destiny of environmental labelling in the United States.

The Application of Life-Cycle Assessment (LCA) to Ecolabelling

The Life-Cycle Assessment 'Code of Practice' as defined by the Society of Environmental Toxicology and Chemistry (SETAC) is:

> 'Life-Cycle Assessment is a process to evaluate the environmental burdens associated with a product, process or activity by identifying and quantifying energy and materials used and wastes released to the environment; to assess the impact of those energy and material uses and released to the environment; and to identify and evaluate opportunities to effect environmental improvements. The assessment includes the entire life-cycle of the product, process or activity, encompassing extracting and processing raw materials; manufacturing, transportation and distribution; use, reuse, maintenance, recycling and final disposal.
>
> Quoted from: Groupe des Sages (1994)

In December 1993, the European Commission set up the *Group des Sages* to advise on the role of Life-Cycle Assessment (LCA) in the EU Ecolabelling programme. The objectives of the guidelines are to:

1. Support EU Member States in establishing ecolabelling criteria which are based on a methodology that is both scientifically sound and workable in practice;
2. improve uniformity in the methods applied in different Member State.

The main themes of *Groupe's* final report is reflected in its concluding recommendations:

1. Life-Cycle Assessment can make a significant contribution in providing a scientific, unifying, and transparent basis for the EU Ecolabelling programme.
2. It is central to this programme because it compares different products on the basis of their common function.
3. It relates environmental impacts, at all stages from cradle to grave, to both market changes and technology improvements.
4. It is a methodology still in the process of development, requiring additional research and systematic data collection.
5. Therefore, policy markers, competent bodies and practitioners must remain aware of the current capabilities and limitations of LCA and should support its continuous development.

6. It should be clear, however, that LCA is only a decision support tool; it cannot replace decisions making.

Groupe des Sages (1994)

During 1995 the European Commission published the first phase of LCA guidelines they are to be applied by Competent Bodies to develop criteria for ecolabelling product groups. The guidelines comprise a scientifically sound and workable methodology. The LCA is central to Ecolabel schemes as it compares different products on the basis of their common function; relates environmental impacts, at all stages from cradle to grave, to both market changes and technology improvements; and helps, by identifying well-founded ecological criteria, to minimise data required for applications.

There are two major families of tools for environmental analysis:

1. Environmental Impact Assessment (EIA) tools which concentrate on localised activities. Generally, environmental impacts are studies as site-specific effects. This precludes much attention to life-cycle aspects of the activity in question.
2. Life-Cycle Assessment (LCA) is a cradle-to-grave analysis usually applied to products and services, although it may also be applied to analyse policies and strategies.

Generally, the environmental impacts are studied as non-site specific effects. The main reasons are that many processes cannot be bound to specific locations and that the inclusion of too much local detail would render the analysis impracticable. Impacts at the regional and the local scale can be included, but as generalised non-site specific effects.

LCA as defined by SETAC, consists of four main methodological phases, these are:

1. goal definition and scoping;
2. inventory analysis;
3. impact assessment;
4. setting out proposed criteria

The above steps do not themselves constitute the LCA procedures as it is iterative process, running through the different components (and their steps), increasing the level of detail and thus the reliability of the process at each iteration. It is not the purpose of this paper to review the principles and methodology of LCA as these are well documented in text books on the subject and in more general environmental assessment texts.

Life-Cycle Assessment can make a significant contribution to providing a scientific and transparent base for criteria setting in the EU Ecolabelling programmes for the following reasons:

1. it is central in the programme as it compares products on the basis of their function;
2. it relates environmental impacts, at all stages from cradle to grave, to both market changes and technology improvements;
3. it helps to minimise data required for submission of an application, by identifying well-founded ecological criteria.

Commission Information on Ecolabelling, Issue No 9, May 1995

LCA primarily focuses on quantifiable information; quantitative aspects can be flagged, however, and consequently still play a role in the final decision making. Comprehensive LCA is time-consuming, expensive and very data-intensive. An alternative is a simpler and more 'scoped' approach to LCA where the focus of attention is restricted to the main environmental impact of the product. LCA does not provide ecolabelling criteria but simply provides a coherent basis from which policy makers take decisions. LCA is helpful in identifying the stages in the life-cycle of a product which have the most impact on the environment and in measuring and assessing those impacts.

LCA is a decision support system, which cannot replace actual decision making. Thus a clear distinction should be made between the ecolabelling procedure, under the responsibility of the competent body, and the LCA study under the responsibility of the LCA practitioners. Both are included in a six-phase process, based on the *'Procedural Guidelines for the Establishment of Product Groups and Ecological Criteria'* (Commission Information on Ecolabelling newsletter, issue no 6, June 1994). In phases one and two there is no role for LCA. In phases one and two there is no role for LCA. The results of these phases feed into the LCA which is carried out to formulate proposals for criteria setting. Phases three and four concern the LCA study; in these phases there can be consultations with competent bodies, i.e. with the *Ad Hoc* Working Group which is composed of the different interest groups, including the industrial sectors involved NGOs. Phase five again concerns the policy procedure. In this phase there can also be consultation by the *Ad Hoc* Working Group with the LCA practitioners, because their knowledge about the uncertainties of the LCA study and of the risks connected with the choice of different types of criteria. Phase six also concerns the policy procedure.

TABLE 3

The Different Phases of the Ecolabelling Procedure

	Ecolabel decision phases	*LCA step's in eco-labelling*
Phase 1	Preliminary phase (including product group selection)	
Phase 2	Market study	
Phase 3	Inventory	Goal definition and scoping Inventory analysis
Phase 4	Environmental impact assessment	Impact assessment
Phase 5	Setting Criteria	
Phase 6	Presentation of draft proposal for a commission decision	

Other considerations in the LCA/Ecolabelling procedure include:

1. goal definition and scoping;
2. inventory analysis;
3. impact assessment.

Finally, comes the formulation of proposals for setting criteria by the *Ad Hoc* Working Group under the responsibility of the competent body concerned. This generally implies evaluation, improvement assessment, and validation. Criteria can be derived from different parts of the LCA process. Thus, criteria can be related to:

1. product technology characteristics;
2. individual released to or extracting from the environment (based on the outcome of inventory analysis);
3. impact categories (based on outcome of classification and characterisation); and
4. Scoring systems (based on outcome of a structured valuation step).

The second and third categories together are often referred to as 'hurdle criteria'.

Criteria based on product and technology characteristics include: recyclability absence of certain substances, energy consumption during the use phase, paper bleached without chlorine, etc. These criteria have the advantage that they are relatively easily dealt with during application. A risk, however, is that they may only constitute a part of it. This simple type of criterion therefore risks transferring pollution to the other locations or other impact types. Another problem is that they may fail to give credit to novel products or technologies. Product or technology characteristics can only be used as criteria if they are validated carefully to confirm that they do correspond to well-substantiated environmental impacts, and if they have no side effects of shifting environmental impacts elsewhere or inhibiting development towards future clean technology. The exclusion of a substance is generally acceptable as a criterion only if there is board scientific and political agreement about the future phasing out of the given substance (e.g. halons). Scoring systems may imply a formalised valuation as part of the LCA study, which is as yet not authoritatively available. They can be used if other types of criteria appear to be technologically impossible or undesirable.

Providing Environmental Information

It is possible that some relevant information concerning the use of resources can be provided on product labels. For example, if consumers are charged per unit for the water and electricity that they consume, then they may well look for products which consume less of these inputs for any given level of performance. In addition, if consumers were charged per unit for the solid waste they produce, then they might look for products which produces less waste—for example, those which have less packaging or have packaging which can be reused, recycled or composted at a lower cost than alternative disposal routes. So, labels might also contain information about such things as the type of material used and the weight of the packaging. In many cases, however, the reasons that less material has been used is that this reduces the manufacturer's costs and enables him to charge less for his product. Given that price is usually of greater concern to the consumer than environmental impact, information about the weight of packaging will have little or no impact in such circumstances. Furthermore, even if consumers are charged the marginal cost of disposal, they may have little incentive to make purchasing decisions based on the type of packaging material used, since the personal cost of sorting each item into appropriate piles may be great relative to the reward in terms of reduced cost of disposal. However, experience in the US with 'pay-as-you-throw' trash fees indicates that these programmes do appear to have some effect on purchasing decisions (Skumatz, 1993).

Third Party Verification of Claims

As earlier chapters discussed, one of the primary (if not the primary) justification for ecolabelling programmes has been the assertion that consumers are confused by the plethora of different environmental labelling schemes. But if that were the case, then the role of the ecolabel need be no more than a third-party verifier of claims. However, this is not the role which ecolabelling authorities have chosen; they have preferred to opt for the grander (and more bureaucratically demanding) seal of approval.

In fact, third party verification would be a far superior means of eradicating spurious environmental claims and entrusting that, to the greatest extent possible, consumers were not misinformed. As the experience of Good Housekeeping and Underwriters Laboratories demonstrates, such an accreditation scheme can provide valuable information to the consumer. In the context of environmental labels, a scheme might verity claims regarding such things as the percentage of recycled material used in the product, or the origin of the product—such as 'dolphin friendly' tuna fishing, or sustainably' product timber. Indeed, certification marks for both these product attributes already exist: the Flipper Seal of Approval, an international certification programme run by Earthtrust and endorsed by 23 environmental organisation around the world, is awarded to companies which actively promote dolphin conservation (EPA, 1993, p. 114); under its Forest Conservation Programme, Scientific Certification Systems (SCS) developed a certification mark for timber producers that showed that their timber resources were managed sustainably, that they adequately maintained the surrounding forest ecosystem and that socio-economic benefits were passed on to the surrounding community (EPA, 1993, p. 118).

Of course, it could be argued that even these schemes lack objectivity: what, after all, does it mean to 'actively promote' dolphin conservation; and how does one define 'sustainability' or equitable distribution of benefits? In addition, such single-issue certification scheme might encourage consumers to overlook trade-offs. For example, a label stating that a garbage bag contains 50 per cent recycled material might mislead the consumer into believing that the bag is more environment-friendly. But, having said, that, this is surely not as bad as claiming, as the Canadian Ecology does, that such bags are actually more environment-friendly. Moreover, the manufacturers of light weight garbage bags can advertise the fact that their bags contain less virgin material than the 'recycled' bags, thereby providing the consumer with relevant information upon which to make their decision.

However, attempt to provide multi-attribute certification, such as the SCS Environmental Report Card, typically suffer from the data poverty which affects other product life-cycle inventories (see e.g. Wynne, 1993). For such schemes to be at all useful, they would have to shed the patina of inclusiveness and simply provide information which the consumer can readily comprehend and act upon—in other words, they would simply be aggregations of single issue certificates, telling the consumer, for example, that the outer product wrap, weighing 5g, is made of low density polythene, 50 per cent of which is recycled, whilst the inner product wrap, weighing 40g, is made from aluminium, 70 per cent of which is recycled, and the contents are derived from tuna fish caught by a rod and line.

Voluntary Guidelines

Another alternative to product certification is the development of purchasing guidelines by independent organisations. One model of such guidelines is the Environmental Defence Fund's Paper Task Force recommendations regarding paper purchasing. The paper Task Force recommendations have the virtue of identifying specific product or manufacturing process characteristics that warrant attention in the buying decision while, at the same time, situating all recommendations in a context of the many constraints and trade-offs—both economic and technical—that are important in any purchasing decision. Thus, for example, the recommendation related to recycled content of paper states that 'paper purchases' should maximise their overall use of paper with post-consumer recycled content, *consistent with functional and economic considerations*' (emphasis added). This kind of statement provides guidance to the buyer, but does not present an 'absolute' criterion whereby certain levels of recycled content (or other product attributes) are universally anointed as environmentally preferable. This represents a significant departure from the threshold sort of criteria used to determine whether a product receives a seal of approval in most of the existing ecolabel programmes.

Statutory Definitions of Words

An alternative means of reducing the incidence of misleading claims is to ensure that the wording of manufacturers' claims conforms to strict and generally understood guidelines. This is the approach taken in both the United States and the United Kingdom with regard to most product claims, with the state providing the general guidelines.

In the United States, green advertising is controlled by the deceptive advertising clause of the Federal Trade Commission Act (FTC Act) of 1937, combined with FTC guidelines on the definition of certain words and phrases. These guidelines, adopted in 1992, apply to 'environmental claims included in labelling, advertising, promotional materials and all other forms of marketing, whether asserted directly or by implication, through words, symbols, emblems, logos, depictions, product brand names, or through any other means.' The guidelines provide general principles regarding what statements might be deemed 'deceptive'; establish principles regarding the 'clarity' of statements; prohibit overstatements of environmental attributes; require substantiation of any claim making product comparisons; and establish limits on the kinds of general environmental benefit claims that can be made.

The FTC guidelines are considered interpretative statements, not legislative rules, though Commissioner Mary Azcuenaga dissented from this view, arguing that the guidelines have, indeed, 'defined with specificity a deceptive act or practice' and therefore constitute 'legislative rules' (*Federal Register*, vol. 57, no. 157, 57 FR, 36363, Thursday, August 13, 1992).

As required by the FTC, its guidelines for Environmental Marketing Claims were reviewed in public hearings in July 1995. The purpose of the review was to assess how well the claims were working 'in preventing the false or misleading use of advertising terms such as recyclable, degradable and environmentally friendly', to assess whether 'consumers see "green" claims differently today than they did three years ago;' and to explore whether 'there are changes in environmental technology that should be taken into account in reviewing he guides' (Federal Trade Commisison, FTC news, 'FTC Opens 'green Guides' to Public Comment: Seeks Information about Changes in Industry, Consumer Perception,' July 27, 1995).

A review of the July 1995 FTC hearings shows that the guidelines generally have been viewed as effective in deterring misleading environmental claims. However, two key issues remained unresolved by late—1996. The first issue is whether additional guideline topics ought to be added: the second issue is whether symbolic labels (such as an ecolabel) communication a general environmental benefit claim that might violate the FTC guidelines that strictly limit the use of such claims unless they can be substantiated and verified.

This second issue lies at the heart of current ecolabel debates. Many advocates of these seals, though supportive of the role that FTC guidelines have played in preventing misleading or false environmental claims, still claim the need for ecolabels as a means of providing consumers with information relevant to making product comparisons along environmental dimensions.

However, as argued throughout this study, ecolabels tend to oversimplify, thereby misleading consumers rather than enlightening them as to the actual environmental impacts of different products. How, then, might the concerns of those who argue that consumers need better environmental information best be met?

Mandatory Provision of Information

Many states have mandated that certain kinds of information be supplied with products. For example, under the EU Regulation on Energy Efficiency Labelling, all new washing machines and refrigerators sold in the EU after April 1996 must carry energy efficiency information. Similarly, under the Energy Policy Act 1992, all lights bulbs sold in the US after May 1995 have had to show not only the energy consumption (in Watts) but its brightness (in Lumens) and its (expected) life (Epatko, 1995).

However, it is not clear why such mandates were considered to be necessary. Perhaps it was thought that entrepreneurs, left to their own devices, would not have chosen to label products in this way. This is not a convincing explanation. Entrepreneurs typically respond to the perceived demands of consumers, so if they are not providing energy efficiency information, it must be because they do not perceive a consumer demand for it. Of course, it is possible that the entrepreneurs are wrong. But this seems unlikely, since the profits that accrue to the first manufacturer to add a desired new attribute to a product provide sufficient incentive in most areas, so why not energy efficiency labelling? Moreover, it would only require one enterprising individual to sell products bearing energy efficiency information, and thereby to steal a lead on his competitors, in order fro all manufacturers to be stimulated into action. This leads to the conclusion that, in mandating the provision of energy efficiency information, the state is acting in an arrogant fashion, attempting to dictate what consumers should think, in spite of their reluctance to do so. Indeed, as Whirlpool, the refrigerator manufacturer, found to its cost, consumers do not generally perceive the value of added energy efficiency (Willis, 1994).

Voluntary Information-Sharing

Some manufacturers do provide energy-efficiency information voluntarily. Indeed, there are already several private schemes which certify energy-efficiency information on products. For example, the Waterheater Manufacturers Association began such a scheme in 1995 and the Energy Saving Trust in the United Kingdom plans a scheme in the near future.

This model of information-sharing has some advantages. Rather than purporting to identify 'best' products in a comprehensive sense, with all the intractable problems that such an exercise entails, an information-sharing approach offers consumers specific, verifiable information regarding particular product attributes. Information-sharing approaches can take multiple forms that are not mutually exclusive. Generally information-sharing approaches include:

1. Identification of product attributes about which consumers have expressed particular concern or interest (such as energy efficiency, recycled content, toxic content), and which are sufficiently specific to be verifiable by third parties.
2. Corporate environmental reports about production and distribution processes and regulatory compliance records.

Such programmes do not purport to determine for consumers what vales they 'ought' to have; rather they provide information with which consumers can evaluate products according to their personal values, which may change over time as new environmental issues surface or as consumers' priorities evolve. The 'fluid' nature of this sort of information-sharing avoids the 'lock-in' problems associated with ecolabels. It also avoids the substantial problems associated with attempting, as ecolabels do, to provide claims of general environmental benefit.

Experience in Marketing Eco-labels in India

As the eco-mark scheme has just recently become operational, an assessment of the marketing experience is not yet possible. Nevertheless, from information that is available it would appear that the marketing effects are limited due to the number of independent agencies concerned with policy, promotion, implementation and certification. Moreover, the focus of the Indian scheme is the domestic market. Exports (with the exception of textiles and some food items) are of minimal interest as exporters have already been trying to incorporate eco-labelling requirements of their major markets into their production processes.

A minimal awareness of eco-friendly products and sustainable development is apparent. At present fiscal and financial support is not available to companies or producing to achieve certification. It is noted that companies producing detergents used for textiles are interested in seeking certification due to consumer demands for eco-friendly detergents product textiles. Limited response to certification is attributed to high initial adjustment costs, small domestic market access, non-export product selection and the combination of both quality and environmental requirements. Incompatibility between the existing standards of BIS and the standards the Eco-Marks is trying to achieve have also influenced the response from industry.

Moreover, consultations with industry to establish eco-criteria have been minimal. Initial draft pesticide notification was withdrawn following representations from industry associations who indicated the toxic nature of such substances. Draft notifications for household pesticides are currently under consideration.

The export sector, especially textiles and leather, in their efforts to conform to eco-labelling requirements of importing countries have been facilitated by visiting experts, dissemination seminars; workshops and availability of bilateral governmental support from Germany, Denmark and other countries. Within the textile and leather sectors, exporters have rapidly adjusted with direct investment to meet the required criteria. The need for fulfilling the requirements of a domestic Eco-Mark was therefore not felt.

Problems Encountered

The Eco-Mark scheme has achieved limited success due to several factors.

(a) Costs:

- Increased costs in production and possible loss in export in export competitiveness
- Several resource constraints in meeting requirements, especially in the case of SMEs

- Fiscal and duty incentives not available to assist industry
- Concerns of industry associations relating to cost of implementation, impact on domestic consumer prices and potential loss in profits

(b) Industry:

- Limited consultations in criteria selection especially with industry
- Insufficient research and testing of alternative eco-friendly raw materials and ingredients
- Exporters prefer to adhere to eco-labelling schemes of importing countries to assist with increased market access

(c) Operational impediments:

- Procedural problems due to CPCB being the agency overseeing product and criteria selection, and BIS being the implementing authority
- Confusion related to eco-friendliness as defined by both industry and implementing agencies
- Existing standards as defined by the BIS and standards that the Eco-Mark is attempting to achieve.
- Direct linking of the Eco-Mark to the ISI mark.

Identification of Suitable Practices and Recommendations

Experiences suggest that environmental labelling is required to be flexible in view of the advances in technology. Frequent criteria revision adversely affects manufacturing processes. For example, a minimum period of certification (3 years) with annual renewal could be required. Schemes encouraging design and production which reduce environmental impact, but maintain fitness for use, are also suggested. Labelling conveying the correct environmental message is also to be carefully considered as the term environmentally friendly is subject to several interpretations and may suggest absolute quality which is unattainable. Transparency would increase if the label included a brief description of the rationale for the product label award. For example, "made mostly from recycled materials" or "CFC" free".

Shared experiences from other countries may be useful for Asian countries during eco-labelling scheme development. Examination of methodologies used in other operational schemes, for example, those of Australia, Canada, Denmark, Germany, Norway and Sweden may also significantly contribute towards scheme development.

Experiences of developing countries indicate that eco-labelling schemes implicitly impose domestic environmental conditions which are either irrelevant to the environment or preferences of developing countries, or are contrary of their comparative advantage. For example, textile eco-labelling schemes of some developed countries result in dye specifications being imposed which ignore locally available eco-friendly materials. Similarly, eco-packaging criteria have evolved as a result of developed countries' environmental problems caused by waste volumes, limited disposal facilities and increasing disposal costs. This criteria has overlooked the availability of environmentally friendly packaging material in developing countries. Such issues have prompted the debate in respect of the technical barriers to trade resulting from eco-labelling.

A regional strategy for integrating trade and environmental concerns may focus on a coordinated approach to environmental policies and principles. These may include the concept of proportionality, least cost measures, ease of implementation and environmental effectiveness, avoidance of adverse trade effects and coordination of Foreign Direct Investment (FDI) approaches to ensure use of the best technologies, processes and practices. Dynamism in Asia is largely based on export drives focused on a new sectors and products with a large proportion of export share originating from SMEs. These are vulnerable to domestic and international environmental policies and have limited access to information, raw material and technology sourcing, certification requirements and testing procedures.

To promote cooperation, further consideration should be given to establishing an Asian Environmental Network (AEN) to foster exchange and dissemination of information, as well as to study possible harmonisation of schemes. The AEN may initially include member countries of the region with established eco-labelling schemes with later inclusion of countries establishing their own schemes. An additional consideration is in the creation of favourable market access for eco-products among regional members with a preferential lowering of duties. Member countries may also consider participating in the Asian Environment Fund with initial contributions based on adherence to environmental programmes. Additional supplements may be requested from the Global Environment Fund (GEF) to assist with adjustment costs and regional promotional activities and programmes.

Experience from attempts such as the one between Canada and Taiwan Province of China, to reach mutual recognition may be studied further and considered for application in the ESCAP region with the support of a technical assistance programme. Equivalence of criteria could be considered an instrument of mutual recognition. Publication of a newsletter with updated information on mutual recognition and equivalence, agreements and protocols and details of new technologies and products is also considered to be a valid instrument for mitigating adverse effects arising from various environmental measures.

An Asian action plan for the management of waste for common export products may be considered for implementation by Asian countries which could reduce waste production at source, stimulate development of clean manufacturing technologies and environmentally friendly products.

The encouragement of national green procurement programme could promote intra-Asian trade by governmental and private purchase programmes. Concessional tariffs may be applied to this programme. Success of regional and national programmes will largely depend on efforts to achieve consumer driven demands. Intensive campaigns and open debate on the need to protect the environment and adopt environmentally friendly policies are crucial to achieve successful regional cooperation.

Strengths of LCA

LCA has a number of major strengths. It informs consumer choice, tackles environmental effectiveness, and promotes continual improvement.

Informed Consumer Choice: LCA is a scientific way of advising customers about the environmental burdens of a given product.

Environmental effectiveness: Using LCA techniques products are designed to have a lower impact on the environment than would be the case without careful product scrutiny. It is hoped that LCA will eventually become more available for individual users to analyse emission and energy impacts of various energy processes with timely, region-specific data.

Continual improvement: Companies will have to become more committed to LCA if customers continue to expect on-going product improvements.

Challenges of LCA

Although LCA has many positive aspects, it also presents many challenges, including cost, insufficient or inappropriate data, inconclusive results and the lack of a standard methodology.

Cost: The research required to do an LCA is complex and may be too costly for many products. LCAs must be tailored to particular circumstances and repeated for each set of different circumstances since the impact of certain products or processes vary according to the character of various areas.

Insufficient data: Given the large data requirements in a life cycle assessment, there will inevitably be a number of gaps in the information available. Researchers often have to rely upon typical manufacturing processes, inputs based on national averages, engrineering estimates of processes and professional judgements. As a result, data may be inaccurate, misleading or biased, producing erroneous results.

Inconclusive results: LCA does not provide a customer with conclusive results about the superiority of a given product. This is because many parameters cannot be simplified to the degree necessary to reach a conclusion, data are dated relatively quickly, and assumptions used in weighting may be questioned.

Inadequate methodologies: Translating numbers from life-cycle inventory into human health or ecological impacts is not well-understood. The lack of any single methodology for comparing environmental impact compounds the problem. It is also difficult to compare different types of pollution since no scientifically accepted methods exists for weighing the effects of various pollutants. They are frequently not even measured in the same units.

CONVENTION CONCERNING THE PROTECTION OF WORKERS AGAINST IONISING RADIATIONS: ILO CONVENTION NO. 115

The General Conference of the International Labour Organisation,

Having been convened at Geneva by the Governing Body of the International Labour Office, and having met in its Forty-fourth Session on 1st June 1960, and

Having decided upon the adoption of certain proposals with regard to the protection of workers against ionising radiations, which is the fourth item on the agenda of the session, and

Having determined that these proposals shall take the form of an international Convention,

adopts the twenty-second day of June of the year one thousand nine hundred and sixty, the following Convention, which may be cited as the Radiation Protection Convention, 1960:

Part I. General Provisions

Article 1

Each Member of the International Labour Organisation which ratifies this Convention undertakes to give effect thereto by means of laws or regulations, codes of practice or other appropriate means. In applying the provisions of the Convention the competent authority shall consult with representatives of employers and workers.

Article 2

1. This Convention applies to all activities involving exposure of workers to ionising radiations in the course of their work.

2. This Convention does not apply to radioactive substances, whether sealed or

unsealed, nor to apparatus generating ionising radiation which substances or apparatus, owing to the limited doses of ionising which can be received from them, are exempted from its provisions by one of the methods of giving effect to the Convention mentioned in Article 1.

Article 3

1. In the light of knowledge available at the time, all appropriate steps shall be taken to ensure effective protection of workers, as regards their health and safety, against ionising radiations.

2. Rules and measures necessary for this purpose shall be adopted, and data essential for effective protection shall be made available.

3. With a view to ensuring such effective protection—

(a) measures for the protection of workers against ionising radiations adopted after ratification of the Convention by the Member concerned shall comply with the provision thereof;

(b) the Member concerned shall notify, as soon as practicable, measures adopted by it prior to the ratification of the Convention, so as to comply with the provisions thereof, and shall promote such modification of other measures existing at the time of ratification;

(c) the Member concerned shall communicate to the Director-General of the International Labour Office, when ratifying the Convention, a statement indicating the manner in which and the categories of workers to which the provisions of the Convention are applied, and shall indicate in its reports on the application of the Convention any further progress made in the matter;

(d) at the expiration of three years from the date on which this Convention first enters into force the Governing Body of the International Labour Office shall submit to the Conference a special report concerning the application of subparagraph (b) of this paragraph and containing such proposals as it may think appropriate for further action in regard to the matter.

Part II. Protective Measures

Article 4

The activities referred to in Article 2 shall be so arranged and conducted as to afford the protection envisaged in this Part of the Convention.

Article 5

Every effort shall be made to restrict the exposure of workers to ionising radiations to the lowest practicable level, and any unnecessary exposure shall be avoided by all parties concerned.

Article 6

1. Maximum permissible doses of ionising radiations which may be received from sources external to or internal to or internal to the body and maximum permissible amounts of radioactive substances which can be taken into the body shall be fixed in accordance with Part I of this Convention for various categories of workers.

2. Such maximum permissible doses and amounts shall be kept under constant review in the light of current knowledge.

Article 7

1. Appropriate levels shall be fixed in accordance with Article 6 for workers who are directly engaged in radiation work and are—

(a) aged 18 and over;

(b) under the age of 18;

2. No worker under the age of 16 shall be engaged in work involving ionising radiations.

Article 8

Appropriate levels shall be fixed in accordance with Article 6 for workers who are not directly engaged in radiation work, but who remain or pass where they may be exposed to ionising radiation or radioactive substances.

Article 9

1. Appropriate warnings shall be used to indicate the presence of hazards from ionising radiations. Any information necessary in this connection shall be supplied to the workers.

2. All workers directly engaged in radiation work shall be adequately instructed, before and during such employment, in the precautions to be taken for their protection, as regards their health and safety, and the reaosns therefore.

Article 10

Laws or regulations shall require the notification in a manner prescribed thereby of work involving exposure of workers to ionising radiations in the course of their work.

Article 11

Appropriate monitoring of workers and places of work shall be carried out in order to measure the exposure of workers to ionising radiations and radioactive substances, with a view to ascertaining that the applicable levels are respected.

Article 12

All workers directly engaged in radiation work shall undergo an appropriate medical examination prior to or shortly after taking up such work and subsequently undergo further medical examinations at appropriate intervals.

Article 13

Circumstances shall be specified, by one of the methods of giving effects to the Convention mentioned in Article 1, in which, because of the nature or degree of the exposure or a combination of both, the following action shall be taken promptly:

(a) the workers shall undergo an appropriate medical examination;
(b) the employer shall notify the competent authority in accordance with its requirements;
(c) persons competent in radiation protection shall examine the conditions in which the worker's duties are performed;
(d) the employer shall take any necessary remedial action on the basis of the technical findings and the medical advice.

Article 14

No workers shall be employed or shall continue to be employed in work by reason of which the worker could be subject to exposure to ionising radiations contrary to qualified medical advice.

Article 15

Each Member which ratifies this Convention undertakes to provide appropriate inspection services for the purpose of supervising the application of its provisions, or to satisfy itself that appropriate inspection is carried out.

Part III. Final Provisions

Article 16-23

Standard final provision, except:

Article 18

Denunciation: This provision differs from the standard final provision by substituting "five years" for "ten years" in both paragraph so that after the Convention has been in force for five years, it is open to denunciation at five yearly intervals.

CONVENTION CONCERNING SAFETY AND HEALTH IN CONSTRUCTION: ILO CONVENTION NO. 167

The General Conference of the International Labour Organisation,

Having been convened at Geneva by the Governing Body of the International Labour Office, and having met in its Seventy-fifth Session on 1 June 1988, and

Noting the relevant international labour Conventions and Recommendations and, in particular, the Safety Provisions (Building) Convention and Recommendation, 1937, the Co-operation in Accident Prevention (Building) Recommendation, 1937, the Radiation Protection Convention and Recommendation, 1960, the Guarding of Machinery Convention and Recommendation, 1963, the Maximum Weight Convention and Recommendation, 1967, the Occupational Cancer Convention and Recommendation, 1974, the Working Environment (Air Pollution, Noise and Vibration) Convention and Recommendation, 1977, the Occupational Safety and Health Convention and Recommendation, 1981, the Occupational Health Services Convention and Recommendation, 1985, the Asbestos Convention and Recommendation, 1986, and the list of occupational diseases as revised in 1980 appended to the Employment Injury Benefits Convention, 1964, and

Having decided upon the adoption of certain proposals with regard to safety and health in construction, which is the fourth item on the agenda of the session, and

Having determined that these proposals shall take the form of an international Convention revising the Safety Provisions (Building) Convention, 1937,

adopts the twentieth day of June of the year one thousand nine hundred and eighty-eight, the following Convention, which may be cited as the Safety and Health in Construction Convention, 1988:

I. Scope and Definitions

Article 1

1. This Convention applies to all construction activities, namely building, civil engineering, and erection and dismantling work, including any process, operation or transport on a construction site, from the preparation of the site to the completion of the project.

2. A Member ratifying this Convention may, after consultation with the most representatives organisations of employers and workers concerned, where they

exist, exclude from the application of the Convention, or certain provisions thereof, particular branches of economic activity or particular undertakings in respect of which special problems of a substantial nature arise, on condition that a safe and healthy working environment is maintained.

3. This Convention also applies to such self-employed persons as may be specified by national laws or regulations.

Article 2

For the purpose of this Convention:

(a) The term "construction" covers:

(i) building, including excavation and the construction, structural alteration, renovation, repair, maintenance (including cleaning and painting) and demolition of all types of buildings or structures;

(ii) civil engineering, including excavation and the construction, structural alteration, repair, maintenance and demolition of, for example, airports, docks, harbours, inland waterways, dams, river and avalanche and sea defence works, roads and highways, railways, bridges, tunnels, viaducts and works related to the provision of services such as communications, drainage, sewage, water and energy supplies;

(iii) the erection and dismantling of prefabricated buildings and structures, as well as the manufacturing of prefabricated elements on the construction site;

(b) the term "construction site means any site at which any of the processes or operations described in subparagraph (a) above are carried on;

(c) the term "workplace" means all places where workers need to be or to go by reason of their work and which are under the control of an employer as defined in subparagraph (e) below;

(d) the term "worker" means any person engaged in construction;

(e) the term "employer" means:

(i) any physical or legal person who employs one or more workers on a construction site; and

(ii) as the context requires, the principal contractor, the contractor or the subcontractor;

(f) the term "competent person" means possessing adequate qualifications, such as suitable training and sufficient knowledge, experience and skill for the safe performance of the specific work. The competent authorities may define appropriate criteria for the designation of such persons and may determine the duties to be assigned to them;

(g) the term "scanfold" means any temporary structure, fixed, suspended or mobile, and its supporting components which is used for supporting workers and materials or to gain access to any such structure, and which is not a lifting appliance as defined in subparagraph (h) below;

(h) the term "lifting appliance" means any stationary or mobile appliance used for raising or lowering persons or loads;

(i) the term "lifting gear" means any gear or tackle by means of which a load can be attached to a lifting appliance but which does not form an integral part of the appliance or load.

II. General Provisions

Article 3

The most representative organisations of employers and workers concerned shall be consulted on the measures to be taken to give effect to the provisions of this Convention.

Article 4

Each Member which ratifies this Convention undertakes that it will, on the basis of an assessment of the safety and health hazards involved, adopt and maintain in force laws or regulations which ensure the application of the provisions of the Convention.

Article 5

1. The laws and regulations adopted in pursuance of Article 4 above may provide for their practical application through technical standards or codes of practice, or by other appropriate methods consistent with national conditions and practice.

2. In giving effect to Article 4 above and to paragraph 1 of this Article, each Member shall have due regard to the relevant standards adopted by recognised international organisations in the field of standardisation.

Article 6

Measures shall be taken to ensure that there is co-operation between employers and workers, in accordance with arrangements to be defined by national laws or regulations, in order to promote safety and health at construction sites.

Article 7

National laws or regulations shall require that employers and self-employed persons have a duty to comply with the prescribed safety and health measures at the workplace.

Article 8

1. Whenever two or more employers undertake activities simultaneously at one construction site—

(a) the principal contractor, or other person or body with actual control over or primary responsibility for overall construction site activities, shall be responsible for co-ordinating the prescribed safety and health measures and, in so far as is compatible with national laws and regulations, for ensuring compliance with such measures.

(b) in so far as is compatible with national laws and regulations, where the principal contractor, or other person or body with actual control over or primary responsibility for overall construction site activities, is not present at the site, he shall nominate a competent person or body at the site with the authority and means necessary to ensure on his behalf co-ordination and compliance with the measures, as foreseen in subparagraph (a) above;

(c) each employer shall remain responsible for the application of the prescribed measures in respect of the workers placed under his authority.

2. Whenever employers or self-employed persons undertake activities simultaneously at one construction site they shall have the duty to co-operate in the application of the prescribed safety and health measures, as may be specified by national laws or regulations.

Article 9

Those concerned with the design and planning of a construction project shall take into account the safety and health of the construction workers in accordance with national laws, regulations and practice.

Article 10

National laws or regulations shall provide that workers shall have the right and the duty at any workplace to participate in ensuring safe working conditions to the extent of their control over the equipment and methods of work and to express views on the working procedures adopted as they may affect safety and health.

Article 11

National laws or regulations shall provide that workers shall have the duty to—

(a) co-operate as closely as possible with their employer in the application of the prescribed safety and health measures;

(b) take reasonable care for their own safety and health and that of other persons who may be affected by their acts or omissions at work;

(c) use facilities placed at their disposal and not misuse anything provided for their own protection or the protection of others;

(d) report forthwith to their immediate supervisor, and to the workers' safety representative where one exists, any situation which they believe could present a risk, and which they cannot properly deal with themselves;

(e) comply with the prescribed safety and health measures.

Article 12

1. National laws or regulations shall provide that a worker shall have the right to remove himself from danger when he has good reason to believe that there is an imminent and serious danger to his safety or health, and the duty so to inform his supervisor immediately.

2. Where there is an imminent danger to the safety of workers the employer shall take immediate steps to stop the operation and evacuate workers as appropriate.

III. Preventive and Protective Measures

Article 13

Safety of Workplaces

1. All appropriate precautions shall be taken to ensure that all workplaces are safe and without risk of injury to the safety and health of workers.

2. Safe means of access to and egress from all workplaces shall be provided and maintained, and indicated where appropriate.

3. All appropriate precautions shall be taken to protect persons present at or in the vicinity of a construction site from all risks which may arise from such site.

Article 14

Scaffolds and Ladders

1. Where work cannot safely be done on or from the ground or from part of a building or other permanent structure, a safe and suitable scaffold shall be provided and maintained, or other equally safe and suitable provision shall be made.

2. In the absence of alternative safe means of access to elevated working places, suitable and sound ladders shall be provided. They shall be property secured against inadvertent movement.

3. All scaffolds and ladders shall be constructed and used in accordance with national laws and regulations.

4. Scaffolds shall be inspected by a competent person in such cases and at such times as shall be prescribed by national laws or regulations.

Article 15

Lifting Appliances and Gear

1. Every lifting appliance and item of lifting gear, including their constituent elements, attachments, anchorages and supports, shall—

(a) be of good design and construction, sound material and adequate strength for the purpose for which they are used;

(b) be properly installed and used;

(c) be maintained in good working order;

(d) be examined and tested by a competent person at such times and in such cases as shall be prescribed by national laws or regulations; the results of these examinations and tests shall be recorded;

(e) be operated by workers who have received appropriate training in accordance with national laws and regulations.

2. No person shall be raised, lowered or carried by a lifting appliance unless it is constructed, installed and used for that purpose in accordance with national laws and regulations, except in an emergency situation in which serious personal injury or fatality my occur, and for which the lifting appliance can be safely used.

Article 16

Transport, Earth-Moving and Materials-Handling Equipment

1. All vehicles and earth-moving or materials-handling equipment shall—

(a) be of good design and construction taking into account as far as possible ergonomic principles;

(b) be maintained in good working order;

(c) be properly used;

(d) be operated by workers who have received appropriate training in accordance with national laws and regulations.

2. On all construction sites on which vehicles, earth-moving or materials-handling equipment are used—

(a) safe and suitable access ways shall be provided for them; and

(b) traffic shall be so organised and controlled as to secure their safe operation.

Article 17

Plant, Machinery, Equipment and Hand Tools

1. Plant, machinery and equipment, including hand tools, both manual and power driven, shall—

(a) be of good design and construction, taking into account as far as possible ergonomic principles;

(b) be maintained in good working order.

(c) be used only for work for which they have been designed unless a use outside the initial design purposes has been assessed by a competent person who has concluded that such use is safe;

(d) be operated by workers who have received appropriate training.

2. Adequate instructions for safe use shall be provided where appropriate by the manufacturer or the employer, in a form understood by the users.

3. Pressure plant and equipment shall be examined and tested by a competent person in cases and at times prescribed by national laws or regulations.

Article 18

Work at Heights Including Roofwork

1. Where necessary to guard against danger, or where the height of a structure of its slops exceeds that prescribed by national laws or regulations, preventive measures shall be taken against the fall of workers and tools or other objects or materials.

2. Where workers are required to work on or near roofs or other places covered with fragile material, through which they are liable to fall, preventive measures shall be taken against their inadvertently stepping on or falling through the fragile material.

Article 19

Excavations, Shafts, Earthworks, Underground Works and Tunnels

Adequate precautions shall be taken in any excavation, shaft, earthworks, underground works or tunnel—

(a) by suitable shoring or otherwise to guard against danger to workers from a fall or dislodgement of earth, rock or other material;

(b) to guard against dangers arising from the fall of persons, materials or objects or the inrush of water into the excavation, shaft, earthworks, underground works or tunnel;

(c) to secure adequate ventilation at every workplace so as to maintain an atmosphere fit for respiration and to limit any fumes, gases, vapours, dust or other impurities to levels which are not dangerous or injurious to health and are within limits laid down by national laws or regulations;

(d) to enable the workers to reach safety in the event of fire, or an inrush of water or materials.

(e) to avoid risk to workers arising from possible underground dangers such as the circulation of fluids or the presence of pockets of gas, by undertaking appropriate investigations to locate them.

Article 20

Cofferdams and Caissons

1. Every Cofferdam and caisson shall be—

(a) of good construction and suitable and sound material and of adequate strength;

(b) Provide with adequate means for workers to reach safety in the event of an inrush of water or material.

2. The construction, positioning, modification or dismantling of a cofferdam or caisson shall take place only under the immediate supervision of a competent person.

3. Every cofferdam and caisson shall be inspected by a competent person at prescribed intervals.

Article 21

Work in Compressed Air

1. Work in compressed air shall be carried out only in accordance with measures prescribed by national laws or regulations.

2. Work in compressed air shall be carried out only by workers whose physical aptitude for such work has been established by a medical examination and when a competent person is present to supervise the conduct of the operations.

Article 22

Structural Frames and Framework

1. The erection of structural frames and components, framework, falsework and shoring shall be carried out only under the supervision of a competent person.

2. Adequate precautions shall be taken to guard against danger to workers arising from any temporary state of weakness or instability of a structure.

3. Framework, falsework and shoring shall be so designed, constructed and maintained that it will safely support all loads that may be imposed on it.

Article 23

Work Over Water

Where work is done over or in close proximity to water there shall be adequate provision for—

(a) preventing workers from falling into water;
(b) the rescue of workers in danger of drowning;
(c) safe and sufficient transport.

Article 24

Demolition

When the demolition of any building or structure might present danger to workers or to the public—

(a) appropriate precautions, methods and procedures shall be adopted, including those for the disposal of waste or residues, in accordance with national laws or regulations;
(b) the work shall be planned and undertaken only under the supervision of a competent person.

Article 25

Lighting

Adequate and suitable lighting, including portable lighting where appropriate, shall be provided at every workplace and any other place on the construction site where a worker may have to pass.

Article 26

Electricity

1. All electrical equipment and installation shall be constructed, installed and maintained by a competent person, and so used at to guard against danger.

2. Before construction is commenced and during the progress thereof adequate steps shall be taken to ascertain the presence of and to guard against danger to workers from any live electrical cable or apparatus which is under, over or on the site.

3. The laying and maintenance of electrical cables and apparatus on construction sites shall be governed by the technical rules and standards applied at the national level.

Article 27

Explosives

Explosives shall not be stored, transported, handled or used except—

(a) under conditions prescribed by national laws or regulations; and

(b) by a competent person, who shall take such steps as are necessary to ensure that workers and other persons are not exposed to risk of injury.

Article 28

Health Hazards

1. Where a worker is liable to be exposed to any chemical, physical or biological hazard to such an extent as is liable to be dangerous to health, appropriate preventive measures shall be taken against such exposure.

2. The preventive measures referred to in paragraph 1 above shall comprise—

(a) the replacement of hazardous substances by harmless or less hazardous substances wherever possible; or

(b) technical measures applied to the plant, machinery, equipment or process; or

(c) Where it is not possible to comply with subparagraphs (a) or (b) above, other effective measures, including the use of personal protective equipment and protective clothing.

3. Where workers are required to enter any area in which a toxic or harmful substance may be present, or in which there may an oxygen deficiency, or a flammable atmosphere, adequate measures shall be taken to guard against danger.

4. Waste shall not be destroyed or otherwise disposed of on a construction site in a manner which is liable to be injurious to health.

Article 29

Fire Precautions

1. The employer shall take all appropriate measures to

(a) avoid the risk of fire;

(b) combat quickly and efficiently any outbreak of fire;

(c) bring about a quick and safe evacuation of persons.

2. Sufficient and suitable storage shall be provided for flammable liquids, solids and gases.

Article 30

Personal Protective Equipment and Protective Clothing

1. Where adequate protection against risk of accident or injury to health, including exposure to adverse conditions, cannot be ensured by other means, suitable personal protective equipment and protective clothing, having regard to the type of work and risks, shall be provided and maintained by the employer, without cost of the workers, as may be prescribed by national laws or regulations.

2. The employer shall provide the workers with the appropriate means to enable them to use the individual protective equipment, and shall ensure its proper use.

3. Protective equipment and protective clothing shall comply with standards set by the competent authority taking into account as far as possible ergonomic principles.

4. Workers shall be required to make proper use of and to take good care of the personal protective equipment and protective clothing provided for their use.

Article 31

First Aid

The employer shall be responsible for ensuring that first aid, including trained personnel, is available at all times. Arrangements shall be made for ensuring the removal for medical attention of workers who have suffered an accident or sudden illness.

Article 32

Welfare

1. At or within reasonable access of every construction site an adequate supply of wholesome drinking water shall be provided.

2. At or within reasonable access of every construction site, the following facilities shall, depending on the number of workers and the duration of the work, be provided and maintained—

(a) sanitary and washing facilities;

(b) facilities for changing and for the storage and drying of clothing;

(c) accommodation for taking meals and for taking shelter during interruption of work due to adverse weather conditions.

3. Men and women workers should be provided with separate sanitary and washing facilities.

Article 33

Information and Training

Workers shall be adequately and suitably—

(a) informed of potential safety and health hazards to which they may be exposed at their workplace;

(b) instructed and trained in the measures available for the prevention and control of, and protection against, those hazards.

Article 34

Reporting of Accidents and Diseases

National laws or regulations shall provide for the reporting to the competent authority within a prescribed time of occupational accidents and diseases.

CIV. Implementation

Article 35

Each Member shall—

(a) take all necessary measures, including the provision of appropriate penalties and corrective measures, to ensure the effective enforcement of the provisions of the Convention;

(b) provide appropriate inspection services to supervise the application of the measures to be taken in pursuance of the Convention and provide these services with the resources necessary for the accomplishment of their task, or satisfy itself that appropriate inspection is carried out.

V. Final Provisions

Article 36

The Convention revises the Safety Provisions (Building) Convention, 1937.

Article 37-44

Standard final provisions.

REFERENCES

Almond, Gabriel A. and Powell, G. Bingham, *Comparative Politics: A Developmental Approach* (New Delhi: Oxford & IBH Publishing Co., 1976), 3rd Indian rep.

Basham, A.L., *The Wonder That was India* (London: Sidgwick & Jackson, 1954).

Basu, Durga Das, *Human Rights in Constitutional Law* (New Delhi: Prentice Hall of India, 1994).

Bates, R.E., 'Meaning' of 'Disabled' and 'Handicapped': Their Relationship to Each Other and Specific Defects' (Ph.D. Thesis: University of Houston, 1965), microfilmed copy.

Bennett, A. Le Goy, *International Organisations: Principles and Issues* (Englewood Cliffs, New Jersey, Prentice Hall, 1988).

Berkowitz, Edward D., *Disability Programmes and Government Policies* (New York: Praeger, 1979).

Berkowitz, Monroe et. al., *Public Policy Towards Disability* (New York: Praeger, 1976).

Bhagwati, Justice P.N., *Law as .in Instrument of Change* (Madurai: Society for Community Organisation Trust, 1985).

Bhalla, S.L., *Human Rights: An Institutional Framework for Implementation* (Delhi: Docta-Shelf Publications, 1987).

Bhatt, Usha, *The Physically Handicapped in India: A Growing National Problem* (Bombay: Popular Book Depot, 1963).

Bhattarcharya, Vivek Ranjan, *Some Aspects of Social Security Measures in India* (New Delhi: Metropolitan, 1978).

Black, Nickel et. al., eds., *Health Disease: A Reader* (Open University Press, 1984).

Blackburn, Robert, ed., *Rights of Citizenship* (London: Mansell Publishing Ltd., 1993).

Blaxter, Mildred, *The Meaning of Disability: A Sociological Study of Impairment* (London: Heinemann, 1976).

Borsay, A., *Disabled People in the Community* (London: Bedford, Square Press, 1986).

Bowe, Frank, *Rehabilitating America: Toward Independence for Disabled and Elderly People* (New York: Harper & Row, 1980).

Brechin, A. et.al., *eds., Handicap in a Social World* (London: Hodder A Stoughton, 1981).

Burgdof, Robert L., Jr., ed., *The Legal Rights of Handicapped Persons: Cases, Materials, and Text* (Baltimore: Paul H. Brookes Publishers, 1980).

Campbell, Tom et. al., eds., *Human Rights: From Rhetoric to Reality* (Oxford: Basic Blackwell Ltd., 1986).

Carr, E.H., *The Twenty Years Crisis 1919-1939: An Introduction to the Study of International Relations* (London and Basingstoke: The Macmillan Press Ltd., 1978). rep.

Chowdhry, D. Paul, *A Handbook of Social Welfare* (Delhi: Atma Ram and Sons, 1981).

Clark, June and Henderson, Jill, eds., *Community Health* (Edinburgh: Churchill Livingstone, 1983).

Claude, Inis L., *Swords into Plow Shares: The Problems and Progress of International Organisation* (Delhi: Anupama Publications, 1987). rep.

Collier's Encyclopaedia (New York: 1983), vols. 4, 7 and 21.

Collins Co-build English Language Dictionary (London and Glasgow: Collins, 1989).

Cranston, Maurice, *What are Human Rights?* (London: Bodley Head, 1973).

Cowell, E. B., eds., *The Jataka, or the Stories of the Buddha's Former Births* (Delhi: Low Price, 1805, 1907/1993), rep., vol. 1.

Coxen, M., Overview, *Disability and Employment,* Programme of Research and Actions for the Social Integration of Disabled People (Brussels: EEC, 1982).

Dahi, Robert A., *Modern Political Analysis* (New Delhi: Prentice Hall of India, 1989).

Dak, T.M., ed., *Sociology of Health in India* (Jaipur: Rawat Publication, 1991).

Dartington, T., et. al., *A Life Together: The Distribution of Altitudes; around the Disabled* (London: 1981).

Davidson, Scott, *Human Rights* (Cambridge: 1988), rep.

Davis, K. and Mullender, A., *Ten Turbulent Years: A Review of the Work of the Derbyshire Coalition of Disabled People* (University of Nottingham: 1993).

Davis, Peter, ed., *Human Rights* (London: RKP, 1989).

Deegan, M. and Brooks, N., ed., *Women and Disability: The Double Handicap* (New Brunswick: Transaction Books, 1985).

Dickinson, G.L., *The Greek View of Life* (London: Methuen & Co., 1957), 23rd edn.

Du Bois, C., *The People of Alor* (Cambridge, M.A.: Harvard University Press, 1960).

Dubos, R., *Torch of Life* (New York: Simon & Schuster, 1962),

Dutt, R.C., *A History of Civilisation in Ancient India* (Delhi: Vistar Publishers, 1972), 1st Indian rep.

Durant, Will, *The Life of Greece* (New York: Simon and Schuster 1966), rep.

Edwards, Michael, *Everyday Life in Early India* (London: Batsford 1969).

Eisenberg, Myron G., et. al., eds., *Disabled People as Second Class Citizens* (New York: Springer, 1982).

Evans-Pritchard, E., *Witchcraft, Oracles and Magic among the Azande* (Oxford: Clarendon Press, 1937).

Everson, Stephen, ed., *Aristotle: The Politics* (Cambridge: 1988).

Fiedler, L., *Pity and Fear: Images of the Disabled in Literature and the Popular Arts* (New York: International Centre for the Disabled, 1981).

Finkelstein, V., *Attitude and Disabled People: Issues for Discussion* (New York; World Rehabilitation Fund, 1980).

Fillozat, J., *The Classical Doctrine of Indian Medicine* (Allahabad: Munshi Ram Manohar Lal, 1964).

Fotta, Jeannette R. and Deck, Edith S., eds., *A Sociological Framework for Patient Care* (New York: John Wiley & Sons, 1979), 2nd edn.

Freeden, Michael, *Rights* (Milton Keynes: Open University Press, 1991).

French, Sally, ed., *On Equal Terms: Working with Disabled People* (Oxford: Butterworth Heinemann, 1994).

Friedlander, Walter A., *International Social Welfare* (Englewood-Cliffs, NJ: Prentice Hall, 1975).

Friedson, E., *The Profession of Medicine* (New York ; Dodd-Mead, 1970).

Gajendragadkar, S.N. ed., *Disabled in India* (Bombay: Somaiya Publications Pvt. Ltd., 1983).

Garrison, F. H., *An Introduction to the History of Medicine* (Philadelphia, Pa.: Saunders, 1929).

Gaye, Lancaster, ed., *Personal Relationships, the Handicapped and the Community: Some European Thoughts and Solutions* (London: RKP, 1972).

Glendinning, Caroline, *Unshared Care: Parents and their Disabled Children* (London: RKP, 1983), 1st edn.

Gliedman, John and Roth, William, *The Unexpected Minority Handicapped Children in America* (New York: Carnegie Council on Children, 1980).

Goble, R.E.A. and Nichols, P.J.R., *Rehabilitation of the Severely Disabled* (London: Butterworth, 1971).

Goftman, E., *Stigma: Notes on the Management of Spoiled Identity* (Harmondsworth: Penguin, 1963).

Goldenson, Robert M., et. al., eds., *Disability and Rehabilitation Handbook* (New York: McGraw Hill, 1978).

Gokhale, B.G., *Ancient Indian History and Culture* (Bombay: Asia Publishing House, 1970).

Gore, Madhov S., "Social Work and its Human Rights Aspects" in *Social Welfare and Human Rights,* Proceedings of the XIVth International Conference on Social Welfare held at Helsinki (Finland) August 18-24, 1968 (New York and London: Columbia University Press).

Gray, Stephen E., *Community Health Today* (New York: Macmillan Publishing Co., 1978).

Gramsci, A., *Selections from the Prison Notebooks* (London: Lawrence & Wishart, 1971).

Green, Andrew, *An Introduction to Health Planning in Developing Countries* (Oxford: 1992).

Green, Lawrence W. and Anderson, *Community Health* (St. Louis: Times Mirror Mosby College Publishing, 1986), 5th edn.

Greengross, W., *Entitled to Love: The Sexual and Emotional Needs of the Handicapped* (Sussex: National Fund for Research into Crippling Diseases, 1976).

Greer, John G., el. al., eds., *Strategies for Helping Severely and Multiply Handicapped Citizens* (Baltimore: University Park Press, 1992).

Groce, N., *Everyone here spoke Sign Language*: *Hereditary Deafness on Martha's Vineyard* (Cambridge: MA Harvard University Press, 1985).

Gupta, D.C., *The League of Nations* (New Delhi: Vikas, 1974).

Hale, Glorya, ed., *The Source Book for the Disabled* (New York: Paddington, 1979).

Hamilton, K.W., *Counseling the Handicapped in the Rehabilitation Process* (New York: Ronald, 1950).

Harris, A., *Handicapped and Impaired in Great Britain* (London: HMSO, 1979).

Harrison, Paul, *The Third World Tomorrow: A Report from the Battlefront in War and Poverty* (London: Penguin, 1991).

Hasan, K. A., *The Cultural Frontier of Health in Village India* (Bombay: Manaktalas, 1967).

Haskins, James, *A New Kind of Joy* (New York: Doubleday & Company, 1976).

Henderson, P., *Disability in Childhood and Youth* (London: OUP, 1979).

Henkin, Louis, "The Internationalisation of Human Rights" in *Human Rights: A Symposium,* Proceeding of the General Education Seminar (Columbia: 1977), vol. 6, no. 1, pp. 5-16.

Heslings, K., *Not Made of Stone: The Sexual Problem of Handicapped People* (Cambridge: Woodhead Faulkner, 1974).

Hetzel, B., *Health and Australian Society* (Melbourne: Pelican, 1974).

Hickson, Lynda, *Menial Retardation, Foundations of Educational Programming* (Boston: Allyn and Bacon, 1995).

Hill, Michael, *Understanding Social Security* (Oxford: Basil Blackwell, 1988).

Howards, Irving et. al., *Disability From Social Problem to Federal Program*me (New York: Praeger, 1980).

Humphries, S. and Gordon, P., *Out of Sight: The Experience of Disability 1900-1950* (Plymouth: Northcote, 1992).

Illich, I., *Medical Nemesis: The Expropriation of Health* (New York: Baniom Books, 1975).

International Encyclopedia of the Social Sciences (USA: 1968), Vol. 5.

Irwin, Terence, *Plato: The Republic,* translated by A.D. Lindsay (London: J.M. Dent & Sons Ltd., 1992).

Iyer, Krishna, V.R., *Low, Justice and the Disabled* (New Delhi: Deep & Deep Publications, 1982).

Jaco, E.G., ed., *Patients, Physicians and Illness: A Sourcebook in Behavioral Science and Health* (New York: The Free Press, 1979).

Janowitz, M., ed., *Organisation and Social Personality: Selected Papers* (Chicago: 1996).

Johnson, E., *Love and Sex in Plain Language* (Philadelphia: 1974).

Jolley, J., *Arthashastra of Kautilya* (Delhi: Molilal Banarasi Das, 1923), vol. 1.

Jordon, J.E., *Altitudes Toward Education and Physically Disabled Persons in Eleven Nations* (Michigan: 1968).

Joyce, James Avery, *The New Politics of Human Rights* (London: Macmillan, 1978).

Kangle, R.P., *The Kautilya Arthasastra,* Part III (Bombay: 1965), vol. 3.

Kant, Immanuel, *Perpetual Peace and Other Essays,* trans., by Ted Humphrey (Indianapolis: 1985).

Karna, Gajendra Narayan, 'The United Nation and The Rights of the Disabled Persons'

(Unpublished M.Phil. Dissertation submitt to the Jawaharlal Nehru University, New Delhi: 1990).

Post-Doctoral research work carried out at Jawaharlal Nehru University, (New Delhi: 1998).

Kaushik, B.N., *Viklang Shiksha Sindhu,* Hindi edn. (Jaipur: Rajasthan Hindi Granth Akademi, 1977).

Keith, A.B., *The Sanskrit Drama* (Oxford: Clarendon Press, 1924).

Kent, Howard, *Yoga for the Disabled: A Practical Self-Help Guide to a Happier Healthier Life* (Wellingborough: Thorsons Publishing Group, 1986).

Kessler, Henry H., *Crippled and the Disabled* (New York: Columbia University Press, 1935).

Kuhn, Thomas S., *The Structure of Scientific Revolution* (Chicago: 1970).

Kuper, Adam and Kuper, Jessica, eds., *The Social Science Encyclopedia* (London and New York: RKP, 1989), rev. and rep.

Lanza, Robert, *Medical Science and the Advancement of World Health* (New York: Praeger, 1985).

Laura, R.S. ed., *Problems of Handicap* (London: The Macmillan Company, 1980).

Lecky, William, E., *The History of European Morals* (London: Simon & Schuster, 1911).

Lee, Kenneth and Mills, Anne, *The Economics of Health in Developing Countries* (Oxford: 1983).

Lees, Dennis and Shaw, Stella, *Impairment, Disability and Handicap: A Multi-disciplinary View* (London: Heinemann Educational Books, 1974).

Levitan, Sar A. and Taggart, Robert, *Jobs for the Disabled* (Baltimore: The John Hopkins University Press, 1977).

Locker, David, *Disability and Disadvantage: The Consequences of Chronic Illness* (London and New York: RKP; 1975).

Loring, James and Burns, Graham, eds., *Integration of Handicapped Children in Society* (London: RKP, 1975).

Mac Pherson, Stewart, *Social Policy in the Third World: The Social Dilemmas of Under-development* (Sussex: Wheat-sheaf Books, 1985), rep.

Mani, D. Rama, *The Physically Handicapped in India: Policy and Programme* (New Delhi: Ashish Publishing House, 1988),

Marshall, L., *The King of Nyae Nyae* (Cambridge, MA: Harvard University Press, 1976).

Marshall, T.H., *Social Policy in the Twentieth Century* (London: Hutchinson, 1977), rep.

Mattingly, S., *Rehabilitation Today* (London: Update Books, 1977).

McConkey, Roy and Me Cormack, Bob, *Breaking Barriers: Educating People about Disability* (London: Souvenir, 1983).

McNeill, William H., *A History of the Human Community: Prehistoric to the Present* (Englewood Cliffs, New Jersey: Prentice Hall, 1987).

Mead, G.H., *Mind, Self and Society: From the Standpoint of a Social Behaviourist* (Chicago: 1934).

Mechanic, David, *Medical Sociology* (New York: The Free Press, 1962).

Merriman, A., *An African World: The Besongye* (Bloomington: Indiana University Press, 1979).

Merton, Robert K., *Social Theory and Social Structure* (New York: The Free Press, 1968), rev. and enl. edn.

Miller, E.J. and Gwynne, G.V., *A Life Apart: A Pilot Study of Residential Institutions for the Physically Handicapped and the Young Chronic Sick* (London: Tavistock Publications. 1972), 1st edn.

Morris, J., *Independent Lives: Community Care and Disabled People* (London: The Macmillan Press Ltd., 1993).

Mukhopadhyay, Alok, ed., *State of India's Health* (New Delhi: Voluntary Health Association of India, 1992).

Myers Patrick R., *Succession between International Organisations* (London and New York: Kegan Paul International, 1993).

Myrdal, Gunnar, *Asian Drama: An Enquiry into the Poverty of Nations,* abridged by Seth S. King (London: Allen Lane The Penguin Press, 1972).

Naik, J.P. ed., "An Alternative System of Health Care Services in India: Some General Considerations", in *An Alternative System of Health Care Services in India: Some Proposals* (Bombay: 1977), Sixteenth Annual Conference of the All India Association for the Advancement of Medical Education, Chandigarh, 12 March 1977.

Narasimhan, M.C. and Mukherjee, A.K., *Disability in India: A Continuing Challenge* (New Delhi: Wiley, 1986).

Nathwani, A., *Disability in the Asian Communities* (London: Greater London Association for Disabled People, 1987).

Navarro, V., *Crisis, Health and Medicine: A Social Critique* (New York: Tavistock Publications, 1986).

Nicholas, H.G., *The United Nations as a Political Institution* (London: OUP, 1975).

Nicholas, Philip et. al., *Disabled: An Illustrated Manual of Help and Self-Help* (London: David & Charles, 1981).

Northedge, F.S., *The League of Nations: Its Life and Times 1920-1946* (New York: Leicester University Press, 1988), 2nd imp.

Nussbaum, Martha and Sen, Amartya, eds., *The Quality of Life* (Oxford: 1993).

Oliver, Michael, *Social Work with Disabled People* (London: The Macmillan Press Ltd., 1983).

Pandey, R.S. and Advani, Lal, *Perspectives in Disability and Rehabilitation* (New Delhi: Vikas, 1995).

Parker, Randall M. and Hansen, *Carl, Rehabilitation Counseling: Foundations, Consumers and Service Delivery* (Boston, Massachusetts: Allyn and Bacon, Inc., 1981).

Parsons, T., *The Social System* (New York: The Free Press, 1951).

Phillips, David R. and Verhasselt. Yola, eds., *Health and Development* (London: Routledge, 1994).

Perman, Leonard G. and Austin, Gray P., eds., *Social Influences in Rehabilitation Planning: Blueprint for the 21st Century* (Arlington, Va.: 1985).

Quicke, J., *Disability in Modern Children's Fiction* (London: Croom Helm, 1985).

Quimby, Freeman *H., The Politics of Global Health,* Committee Print in Science, Technology and American Diplomacy Series (USA: 1971).

Rathbone, Frank S. Jr. and Rathbone, Estelle T., *Health and the Nature of Man* (New York: Mc Graw Hill Book, 1971).

Richman, Joel, *Medicine and Health* (London and New York: Longman, 1987).

Rieser. R. and Masun, M., eds., *Disability Equality in the Classroom: A Human Rights Issue* (London: Disability Equality in Education, 1992).

Robertson, A.H., *Human Rights in the World* (Manchester: OUP, 1972).

Robertson, I., *Sociology* (New York: Worth Publishers, 1981).

Rogers, B., *The Demostication of Women* (London: Tavistock Publications, 1980).

Rubin, Jaffrey and Porte, Valerie La, eds., *Alternatives in Rehabilitating the Handicapped: A Policy Analysis* (New York: 1982).

Russo, X. and Willis, George, *Human Services in America* (Englewood-cliffs, N.J.: Prentice Hall, Inc., 1986).

Roessler, Richard and Bolton, Brian, *Psychosocial Adjustment to Disability* (Baltimore: University Park Press, 1978).

Russell, Bertrand, *History of Western Philosophy and, Us Connection with Political and Social Circumstances from the Earliest Times to the Present Day* (London: George Allen and Unwin Ltd., 1971).

Sabine, George, *A History of Political Theory* (New York: Henry Holt, 1937),

Safilios Rothschild, C., *The Sociology and Social Psychology of Disability and Rehabilitaion* (New York: Random House, 1970).

Schwab, Peter and Pollis, Adamantia, eds., *Towards a Human Rights Framework* (New York: Preager, 1982).

Saksena, K.P., ed., *Human Rights: Perspective and Challenges* (New Delhi: Lancer Books, 1994).

Selby, David. *Human Rights* (Cambridge: 1988), rep.

Self, Phyllis S., ed., *Physical Disability: An Annotated Literature Guide* (New York & Basel: Marcel Dennor Inc., 1984).

Sen, Anima, *Psycho-Social Integration of the Handicapped: A Challenge to the Society* (Delhi: Mittal, 1988).

Shamasastry, R., trans., *Kautilya's Arthasastra* (Mysore: Westeyan Mission Press, 1923), 2nd edn.

Sharma, K.L., *Social Assistance System in Ancient India* (Delhi: Macmillan, 1976).

Shearer, Ann, *Disability: Whose Handicap?* (Oxford: Basil Blackwell, 1981).

Shontz, Franklin C., *The Psychological Aspects of Physical Illness and Disability* (New York: Macmillan, 1975).

Shulman, D.D., *The King and the Clown in South Indian Myth and Poetry* (Princeton, NJ: 1985).

Singeton, W.T. and Debney, L.M., eds., *Occupational Disability: The Approaches of Government, Industry and the Universities* (Lancaster: MTP Press Ltd., 1982).

Smelser, Neil J., *Sociology* (New Delhi: Prentice Hall, 1993). Smith, S.M., *The Battered Child Syndrome* (Butterworth: 1975).

Smith, V., Asoka, *The Buddhist Emperor of India* (Oxford: Clarendon Press, 1920), 3rd edn.

Soden, William H., eds., *Rehabilitation of the Handicapped: A Survey of Means and Methods* (New York: The Renold Press Company, 1949).

Solecki, R., *Shanidar; The First Flower People* (New York: 1971).

Stewart, William, *Counseling in Rehabilitation* (London: Croom Helm, 1985).

Stewart, W.F.R., *The Sexual Side of Handicap:* A *Guide for Caring Professions* (Cambridge: Woodhead-Faulkner, 1979).

Stone, Deborah A., *The Disabled Stale (London:* Macmillan, 1985).

Stubbins, Joseph, ed., *Social and Psychological Aspects of Disability: A Handbook for Practitioners* (Battimore: University Park Press, 1977).

Sussman, Marvin B., ed., *Sociology and Rehabilitation* (Washington. D.C.: American Sociological Association, 1965).

Sutherland, Allan T., *Disabled We Stand* (London: Souvenir Press, 1981).

Taylor, Paul and Groom, A.J.R., eds., *International Institutions al Work* (London: Pinter Publishers, 1988).

Taylor, Wallace W. and Taylor, Isabelle Wagner, *Services for the Handicapped in India* (New York: International Society for the Rehabilitation of the Disabled, 1970).

Terence, Irving, *Plato: The Republic,* trans, by A.D. Lindsay (London: J.M. Dent & Sons, 1992).

Thane, Pal, *The Foundations of the Welfare State* (London & New York: Longman, 1982).

Townsend, Peter, *The Social Minority* (London: Allan Lane, 1973).

Trust, Doreen, *Overcoming Disfigurement: Defeating the Problems-Physical, Social and Emotional* (Willingborough: Thorsons Publishing Group, 1987).

Varma, V.P., *Ancient and Medieval Indian Political Thought* (Agra: L.N. Aggrawal Ed. Publishers, 1993), 2nd edn.

Vincent. R.J., *Human Rights and International Relations* (London: Cambridge University Press, 1986).

Wadia, A.R., ed., *History and Philosophy of Social Work in India,* a Souvenir volume of the Tata Institute of Social Sciences, *Bombay* (Bombay: 1961).

Walker, W., "The Remedies of Law of the International Covenant on Civil and Political Rights Current Trends and a Conceptual Framework for the Future," New York University Journal of International Law and Politics. 20(2), 1988, pp. 525-55.

Walters, F.P., *A History of the League of Nations* (London: OUP, 1952), vol. 1.

Weber, Max, *The Theory of Social and Economic Organisation,* edited by Talcot Parsons (New York: The Free Press, 1964).

Webster, C.K., and Sydney, Herbert, *The League of Nations in Theory and Practice* (London: George Allen & Unwin, 1933).

Weissbrodt. D., "The Three 'Theme' Special Rapporteurs of the UN Commission on Human Rights," American Journal of International Law, 80(3), 1986, pp. 685-99.

Wilson, Sir John, ed., *Disability Prevention: The Global Challenge* (Oxford: 1983).

Wolin, Sheldon S., *Politics and Vision: Continuity and Innovation in Western Thought* (Boston: Little, Brown & Co., 1960).

Wright, Beatrice A., *Physical Disability: A Psychological Approach* (New York: Harper & Row, 1960).

Wright, George Nelson, *Total Rehabilitation* (Boston: Little, Brown. & Co., 1980).

Zimmern, Alfred., *The League of Nations and the Rule of Law 1918-35* (London: Macmillan & Co., 1936).

Zola, I.K., *Missing Pieces: A Chronicle of Living with a Disability* (Philadelphia: Temple University Press, 1982).

Barnett, Harold and Chandler Morse. *Scarcity and Growth.* Baltimore: Johns Hopkins University Press. 1963.

Costanza, R. "Three General Policies to Achieve Sustainability" in A. M. Jansson, M. Hammer, C. Folke, and R. Costanza (eds). *Investing in Natural Capital: The Ecological Economics Approach to Sustainability.* Washington DC: Island Press. 1994. pp. 392-407.

Costanza, R. and H. E. Daly. "Natural Capital and Sustainable Development," *Conservation Biology.* No. 6, pp. 37-46. 1992.

Daly, Herman E. *Steady-State Economics.* Washington DC: Island Press. 2nd ed., 1991.

Daly, H. and J. Cobb. *For the Common Good,* Boston: Beacon Press. 2nd ed., 1994.

Georgescu-Roegen, Nicholas. "Comments ..." in V. Kerry Smith, ed., *Scarcity and Growth Reconsidered,* Baltimore: Resources for the Future and Johns Hopkins University Press. 1979. p. 98.

Hawken, Paul. *The Ecology of Commerce: A Declaration of Sustainability.* New York: Harper Business. 1993.

Marshall, Alfred. *Principles of Economics.* New York: Macmillan. 9th ed., 1961 (originally 1920).

Nordhaus, William. Quoted in "Academy Panel Split on Greenhouse Adaptation." *Science.* September 13, 1991. p. 1206.

Oates, Wallace E. "Pollution Charges as a Source of Public Revenues." Working Paper No. 91-22, University of Maryland, Department of Economics. 1991.

Page, Talbot. *Conservation and Economic Efficiency.* Baltimore: Johns Hopkins University Press. 1977.

Passell, P. "Cheapest Protection of Nature May Lie in Taxes, Not Laws." *New York Times,* Nov. 24,1992.

Repetto, R., R. C. Dower, R. Jenkins, and J. Geoghegan. *Green Fees: How A Tax Shift Can Work for the Government and Economy*. Washington, DC: World Resources Institute. 1992.

Thurow, Lester. *The Zero-Sum Society*. New York: Penguin Books. 1980.

von Weizsacker, E. U. and J. Jesinghaus. *Ecological Tax Reform: A Policy Proposal for Sustainable Development*. London: Zed Books. 1992.

World Bank, "Development and the Environment," *World Development Report*. Figure 3.2, p. 69. Washington DC and New York: World Bank and Oxford University Press. 1992.

11

Corporate Management and Environmental Responsibility: Developing a Perspective

GREEN BUSINESS POLICIES, STRATEGIES AND CORPORATE ENVIRONMENTAL MANAGEMENT

Green Business and Environmental Policy Implications

The previous sections have indicated that there is no empirical evidence to suggest that existing environmental policies have widespread effects on market access. However, effects could be more significant for some sectors and for small and medium-sized enterprises. Environmental policies—may have differential competitiveness effects on developed and developing countries. In most cases, however, competitiveness effects of environmental policies can be addressed by appropriate policies at the national, regional and international levels.

Addressing Competitiveness Concerns at the National Level

What policies and measures could the business community and the government in developing countries adopt, at the national level, to mitigate possible adverse trade and competitiveness effects of environmental policies, standards and regulations as well as to allow businesses to take advantage of new trading opportunities? A related question is how programmes such as UNDP's Capacity 21 could assist in implementing such policies and measures.

An important conclusion of a number of the country case studies is that increasing the ability of firms to respond to environmental challenges in the first place requires sustained economic growth and sound macroeconomic and environmental policies. The strong commitment of many developing countries to sustained economic growth and openness through the promotion of trade and investment, together with improved environmental management, are key elements of their sustainable development strategies. The Malaysia study underlines the role of government policies in supporting private sector led growth, for example through support for research and development.

In addition, a number of areas of policy reform that will help address the competitiveness concerns include the following:

Infrastructure Development

Several studies stress the important contribution of improvements in environmental infrastructure in reducing the costs of compliance with environmental regulations, for example with effluent standards. The Colombia and India studies highlight the role of common effluent treatment (GET) plants. The Malaysia study stresses the need to establish a facility for the treatment of hazardous waste. The Philippines study observes that a dollar spent in infrastructural development can make a greater contribution to environmental improvements than a dollar spent in pollution abatement.

Testing and Certification

As noted above, the certification of environment-friendly products may be costly and confusing; in particular when producers in developing countries depend on testing and certification bodies in the developed countries. The creation of standardisation bodies or the expansion of existing bodies in developing countries and steps contributing to their international recognition are of key importance.

Capacity for Innovation

Innovative and cost-saving responses to environmental regulations determine whether compliance costs can be offset or even competitiveness improvements gained. To promote innovation, governments can either provide broad enabling conditions for innovation, or take a more active part in fostering innovation. The study on Malaysia stresses that the central role of technology within Malaysia's development strategy is vital for environmentally sustainable economic growth.

Cooperation between Government and Industry

Cooperation of industry with the authorities in the standard-setting process can ensure the design of feasible standards and facilitate their implementation. For example, the Malaysia study points out that cooperation between government and the private sector has helped to identify cost-effective approaches for the phase-out of ozone depleting substances based on the incorporation of sector-specific elements into a national strategy.

Special Measures for SMEs: Dissemination of Information

As has been indicated above, SMEs tend to have more difficulties than large firms in obtaining information and in adjusting to environmental regulations. The study on India proposes the following steps for improving the channelling of information and availability of environment-friendly input materials to SMEs:

- improve the registration of SMEs so that the government and other agents know where to direct information on environmental regulations;
- improve the provision of timely and accurate information to SMEs;
- provide relevant information to suppliers of input materials, such as dyes and chemicals;
- improve the availability of environment-friendly input materials for SMEs, by providing information, facilitating imports if substitutes are not supplied locally, supporting research and development or the acquisition of appropriate technology for the domestic production of substitutes;
- develop domestic regulatory measures or other policy instruments to support the development of environment-friendly substitutes;
- support the establishment of adequate testing and certification facilities.

Experience indicates that larger firms tend to cooperate with each other, for example in the exchange of information on existing and emerging environmental regulations in external markets. Capacity-building efforts may help promote a process whereby the know-how and experience acquired by larger firms is transmitted to smaller firms.

Operationalising and Implementing Environmental Policies

Placing environmental concerns on the agenda is only the starting point in the sustainable development of business. In parallel with the adoption of a corporate environmental policy it is important that companies should identify and develop methods and procedures that allow for the achievement of their aspirations. Although this book deals in detail with the subject of environmental assessment techniques and procedures, it is important here to outline the need for business to prepare at an early stage for the adoption of a more environmentally responsible stance.

The Leeds Environmental Business Forum

The Mission of the Forum is to promote, encourage and support business practices which are aimed at the improvement of the Leeds and wider environment, and to publicise and disseminate the environmental achievements of the Leeds business community.

In order to achieve this Mission the Leeds Environmental Business Forum has the following broad objectives:

- To raise the awareness of the business community with regard to environmental issues.
- To act as a focal point for the collection, dissemination and exchange of information and actions relating to best practice in the environment, and to act as an interface with organisations and agencies responsible for environmental regulation.
- To share and distribute expertise and resources and to channel advice which will enable positive environmental action to occur.
- To develop a full programme of environmental projects based on partnerships between business, government and other appropriate organisations.
- To publicise the achievement s of the Leeds business community in the environmental field, both within and outside the city.

Techniques such as environmental impact assessment and environmental auditing are beginning to emerge as standard items in the business toolkit. More recent developments have included the introduction of environmental management systems and the use of life-cycle analysis. The importance of these methods of analysis and assessment is that they provide a technical capability to measure the impact of a business upon the environment and to monitor the progress of a company in the achievement of the targets specified in a corporate environmental mission statement.

Views vary as to the desirability of regarding environmental matters as a separate issue in the operation of a business. Welford (1992a) has suggested that total quality management and environmental management should be seen as part of a single system of management and he argues that 'total quality and as such zero defects also means zero negative impact on the environment'. This view is gaining ground and can be seen as a model of best practice to which companies should aspire.

Conclusions

This chapter has provided an introduction to the complex nature of the relationship between business and the environment. It has demonstrated that positive action by business is an essential prerequisite for an improvement in environmental conditions, and it has also indicated the importance of adopting enhanced environmental standards for the future success of business. The relationship between business and the environment can be seen to be symbiotic; both parties can gain advantages through positive action, whilst both parties will suffer if environmental matters fail to become a central plank of business concern.

The scale of the environmental problems confronting business is potentially immense, but there are signs of an improvement. More importantly, an increasing number of companies have now adopted corporate mission statements which place priority on the enhancement of their environmental performance.

Policy Options

Against the background of the linkages detailed in the previous section and the thrust of the chapter, environmental policies can be classified into two broad groups: those which aim at reducing the adverse environmental impacts of economic growth, leading to an improvement in the environment in general and a consequent beneficial impact on the poor, and those which are targeted specifically at the poor and have a positive impact on the environment. These are discussed below.

Enhancing Environment-Friendly Economic Growth and Development

At the United Nations Conference on Environment and Development, held at Rio de Janeiro in 1992, Governments committed themselves to adopting strategies for sustainable development (needs). The challenge has been to achieve convergence between the growth and environment objectives of development through the integration of economic, social and environmental dimensions by the use of appropriate policies. These policies can be broadly classified as economic policies and institutional policies.

Economic Policies

National long-term planning mechanisms, strategic plans and sustainable development plans: the need for coordination

Achieving economic, social and environmental objectives for a country requires coordination between many agents, namely, government, producers, consumers and domestic and foreign investors: policies, namely, sectoral, fiscal and monetary, and trade policies; and institutions, including regulatory agencies and the judiciary. There are various subdivisions within each group, for example, the Government consists of many ministries which deal with many areas, from finance to construction and the environment; and producers may be big corporate entities, SMEs or self-employed. It is only natural that there should be conflict between the objectives of so many entities. Bringing harmony into the functioning of these groups to enhance a country's welfare (of which poverty reduction and improving the environment are major components) calls for the recognition and identification of various trade-offs and the realignment of individual goals and policy instruments under a "win-lose as little as possible" framework.

National planning mechanisms which constitute the articulation of such a framework spell out the objectives, concerns, goals, policy options and strategies of various agents, groups and organisations. Whereas elaborate plans for socio-economic development have a long history, those which deal explicitly with environmental concerns and sustainable development issues are relatively recent and came into being to a considerable extent after the adoption of Agenda 21 at the Rio Conference.

A number of countries have formulated elaborate sustainable development plans. For example, in 1995, the Republic of Korea prepared Green Vision 21, which integrated long-term environmental policies with development needs. Under this broad vision, the relevant ministries, especially the Ministry of Environment, the Ministry of Commerce, Industry and Energy and the Ministry of Construction and Transportation, developed their own annual plans and strategies for action. Viet Nam's Environmental Vision 2020, National Strategy for Environmental Protection, 2001-2010 and National Environmental Action Plan, 2001-2005 address long-medium- and short-term issues of environment and economic growth. Official environment protection visions, integrating economic development and the environment, exist in at least 37 countries of the Asian and Pacific region. Some examples of these are China's Agenda 21, Vision 2020 (Malaysia), the Singapore Green Plan, 2012 National Conservation Strategies (India), the National Policy on the Environment (Uzbekistan), the National Strategy for Ecologically Sustainable Development (Australia) and the National Environment and Development Strategies, 1993 (Samoa). As examples, the major features of three selected sustainable development plans, the Singapore Green Plan, Australia's National Strategy for Ecologically Sustainable Development and China's Agenda 21, are given in table 1. The objectives of the plans are to ensure that economic growth and development take place with minimum damage to the environment. They emphasise the conservation of natural resources and the reduction of environmental pollution. The adoption of clean technologies has often been explicitly considered as a major modality for balancing growth and environment objectives. Strategies for achieving these objectives include raising environmental awareness and education and promotion of the 3P (people, public sector and private sector) partnership. Implementation methods incorporate a broad range of initiatives from the inclusion of environmental issues in school curricula to the formulation of legislation and laws to induce environmental protection. Financial resources for implementing environmental plans come primarily from Governments, and are augmented by bilateral and multilateral donors.

Sectoral Policies

Visions and strategies of sustainable development must be supported by sectoral action plans that fulfil national obligations under global environmental conventions such as the need to contain greenhouse gas emissions, the conservation of biodiversity and the phasing-out of ozone-depleting substances, while ensuring adequate growth of the sector.

Sectoral plans and policies have also been developed to address concerns about agricultural land and water. Against the background of the importance of these sectors in the environment-rural poverty linkage, the associated policies require special mention. For example, forest land is often used to augment agricultural land and trees are cut indiscriminately to satisfy the growing demand for wood. In order to counter deforestation, which has a negative impact on the livelihood and vulnerability of the poor, especially in rural areas, a shift towards plantation forestry is a policy option which can be pursued. This type of shift will relieve the pressure on national forests and at the same time enable the development of reliable sources of industrial raw material and contribute to strengthening the income-earning potential of the persons associated with this sector.

TABLE 1

Selected Sustainable Development Plans

	Objectives	*Strategies*	*Implementation*	*Resource requirements*
Singapore Green Plan 2012	• Aiming for an environmentally conscious nation, through promoting resource conservation and clean technology to protect both the local and global environments • Aiming at strengthening environmental consciousness to promote personal responsibility for and civic commitment to the environment	• Key modalities for achieving the objectives include promoting waste management, conserving nature, ensuring clean air, increasing water supply, improving public health, forging and strengthening strategic partnerships with the private sector, enhancing international collaboration and encouraging, through incentives, the innovation of clean technology for environmental sustainability	• Government level: • Strengthening the message of environmental sustainability among communities (employers, employees, civic groups, labour unions, educators and the media), through training in environmental skills and management, providing incentives and fostering greater collaboration among the communities • Implementing nationwide initiatives such as public transport using compressed natural gas • Including environmental issues in the school curricula to increase awareness • Formulating and enforcing legislative measures to encourage environmental sustainability • Complying with international agreements on the matter • Partnerships:	• Government funding for public utilities development and training

			• Between government agencies, non-governmental organisations, educational institutions, private and international organisations • Close partnership among the 3 Ps (people, public sector and private sector) to create environmental awareness and responsibility	
National Strategy for Ecologically Sustainable Development (Australia)	• Enhancing individual and community well-being and welfare by following a path of economic development that safeguards the welfare of future generations • Providing for equity within and between generations • Protecting biological diversity and maintaining essential ecological processes and life-support systems	• Ensuring that decision-making processes integrate both long-and short-term economic, environmental, social and equity considerations • Incorporating the global dimensions of the environmental impacts of actions and policies in national policy formulation • Developing a strong, growing and diversified economy which can enhance the capacity of environmental protection • Maintaining international competitiveness in an environmentally sound manner	• Government level: • Development of detailed environmental policies to be adopted at all levels of government • National-level strategies on forests, waste management, biodiversity and rangeland management • Legislation and government programmes stress environment and sustainable development objectives and principles, such as waste minimisation and cleaner production Land-use and transport planning and natural resource management to be entrusted to State and local-level authorities Adopting cost-effective and flexible policy instruments, including valuation, pricing and incentive mechanisms • Decisions and actions should provide for community involvement	• State and local government resources
China's Agenda 21	• Rapid economic growth and gradual improvements in the quality of development through relying on scientific and technological advances and improvements in	• Focusing on economic development and the deepening of reforms and openness • Strengthening the foundation for building capacity for sustainable development by: - Establishing a policy framework for developing social and economic	• Government level: Environmental protection has been included in annual and plans for national economic and social development such as the five-year pins - The Government has formulated and implemented a series	• China's own resources to cover roughly 60 per cent of total funding white fee remaining 40 per cent is to be covered from international funding;

Governments willingness to support this associated priority	the quality of labour • Promoting the overall development and progress of society and establishing the social basis for sustainable development • Controlling environmental pollution, improving the environment and protecting the resource base for sustainable development	norms for sustainable development - Developing planning, statistics and information support systems for social and economic development - Improving education and raising awareness of sustainable development issues throughout the country • Gradually popularising sustainable agricultural techniques, while giving consideration to local situations • Vigorously promoting cleaner production technologies, working hard to minimize the production of wastes, encouraging recycling, resource and energy saving and increasing efficiency of production • Arranging for the development and popularisation of important environmental pollution control techniques and equipment through technological skills training, establishing demonstration projects, raising fees for the discharge of pollutants and providing incentives such as preferential taxes for socially beneficial undertakings to control environmental pollution • Emphasising protection of water resources and treatment of waste water by preventing the spread of desertification and developing water-saving technologies that produce little or no waste	of laws and policies concerning environmental protection Support from ministries, departments and local governments - Bilateral and multilateral cooperation with countries and regions through consultation and dialogue, transfer of environmentally sound technologies and environmental protection agreements - Establishment of a comprehensive natural resources and environment monitoring and management system	foreign and international organisations have shown Agenda and its programmes

- Emphasising protection and rational utilisation of natural resources to protect species and improve environmental quality by disseminating scientific findings on environmental problems to raise awareness and promoting the production of "green products" and the use of renewable resources
- Working hard to improve land productivity and reduce the impact of natural disasters

---

Several countries of the region are pursuing the development of plantation forests and simultaneously withdrawing national forests from the production of wood, which is used as industrial raw material. For example, China has one of the roost extensive plantation forestry programmes in the region. It is planned to plant some 26 million hectares of forest in the Yangtze and Yellow river basins by 2030. In India, farmers in three states established a total of 26,000 ha of poplar plantations in 1990 which are now used to provide raw material for industries such as match boxes and plywood.

There have been changes in the policies and strategies for protecting and improving the quality of available freshwater resources. Instead of expanding the supply of freshwater, emphasis has been put on demand management and water-use efficiency, conservation and protection. An integrated approach to water resources management in line with Agenda 21 has been accepted as an effective policy option.

Multisectoral and multidisciplinary approaches are needed to rehabilitate degraded water quality and ensure the provision of a safe drinking water supply. In many countries of the region, this need has been translated into action plans for cleaning up rivers, canals, lakes and other water bodies: the Murray Darling Basin Agreement (Australia), the pollution control plans for three rivers (China), the Ganga Action Plan (India) and the Love Our Rivers Campaign (Malaysia) are some examples. The reduction of pollution loads through proper waste-water treatment, reuse and recycling of domestic sewage and industrial waste water, the introduction of appropriate low-waste technologies and strict control on industrial and municipal effluents are essential elements of these action plans.

Instruments for Implementing Policies: Command and Control and Market-Based Instruments

Command and control-based mechanisms utilize the power of the State in forcing the agents, that is, producers and consumers, to adhere to environmental standards. For example, the management of forest resources in many countries uses a system of awarding licences for logging with the threat of cancellation as a penalty mechanism. In another sector, fishing, in which many of the poor earn their livelihood, a quota system is applied to keep the catch within sustainable limits. A number of countries have used command and control mechanisms to regulate effluents, emissions and disposable wastes.

Command and control measures have often focused on particular sectors which are of primary importance to the poor. For example, Indonesia's PROKASIH Programme, started in 1989, aimed at ensuring that industries install waste-water treatment systems. Initially, attention was focused on the worst industrial pollution occurring near 24 highly polluted rivers with the goal of reducing their pollution load by 50 per cent. The objective was subsequently expanded; more rivers were brought within the purview of the control measures and the pollution standards were also increased. Command and control measures have also been applied to control air pollution. For example, a unique Vehicle Quota System to control the number of vehicles is in operation in Singapore, thereby reducing traffic congestion and emissions from fossil fuel. Under this system the number of new vehicles allowed for registration is predetermined annually. The quota for new vehicles takes into account the prevailing traffic conditions and the number of vehicles taken off the road permanently, while the supply-demand condition of the market determines the price of new vehicles.

The requirement for the success of the command and control system is the ability to use the power of the State to identify violations by continuous policing and then to enforce the associated penalties. These actions require substantial financial and human resources and a well-functioning judicial system. In most countries, resources are scarce and the judicial system is slow in handling non-criminal cases. A market-based fiscal incentive, a disincentive system which seeks to influence the behaviour of producers and consumers by enabling them to choose the most appropriate measures based on their own assessed cost and benefit, is considered a viable and often preferable alternative to the direct command and control system and is implemented in various countries of the region. In contrast to command and control measures, these are self-regulatory in nature and require less policing and fewer court actions. As these instruments are often administered at the local level, like other fiscal instruments, they result in increased revenue/savings for local governments and municipalities. Some of the popular market-based instruments are user charges, an increase/decrease in targeted subsidies and a deposit refund system. These are described in table 2 below.

TABLE 2

Market-based instruments for environmental management[a]

Instrument	*Modus operandi*	*Examples*
Environmental tax	Charged to polluters as the cost of neutralising pollution and environmental damage.	Industries in Singapore are allowed to discharge effluent (containing biodegradable pollutants) directly in public sewage on payment of a tariff which allows the recovery of the additional cost of sewage treatment. In the Philippines, the polluters are required to pay a fee for every unit of pollution they discharge. The emission tax structure consists of 4,500 pesos/ton of particulate matter emitted in excess of the allowance limit and 14,200 pesos/ ton of sulphur dioxide emitted for manufacturing industries. Taxes for power-generating plants are 1,750.86 pesos/ton of particulate matter and 40,725 pesos/ ton of sulphur dioxide emitted. The nominal rates are adjusted periodically for inflation to maintain the deterrent impact in "real terms."
User charges	Charges are levied on the use of public good and natural resources for reducing environmental damage.	Road pricing is practised hi Singapore to reduce air pollution and congestion; peak hour road use in the city centre is taxed. To implement the road tax system efficiently, an electronic road pricing arrangement has been introduced; the system is designed to automate the pay-when-you-use principle. Charges are levied on a per-pass basis and vary according to the time and congestion levels. Indonesia plans to charge ecotourists for the use (enjoyment) of protected areas for the sustainable management of ecosystems.
Increase/decrease in targeted subsidies	Eco-friendly actions are encouraged by providing subsidies/tax exemptions and these are reduced to curb environmentally damaging activities.	Examples of incentives include a preferential low-interest loan for desertification control (China); incentives for wood substitution and subsidies to encourage the use of fuel-saving devices and alternative sources of energy supply; such as biogas and solar energy (India). Examples of disincentives include reduction of the domestic subsidy on coal from 61 per cent in 1984 to 11 per cent in the late 1990s to deter its use (China).
Deposit (fee) refund system	Used to encourage recycling activities; a deposit on the product is levied initially to be refunded to the consumer on the return of the recyclable waste.	Examples include a deposit refund system for beer bottle recycling in Japan. The beer makers levy a fee on the beer bottles and containers, which is ultimately passed on to the consumer. A refund is provided when the bottles and containers are brought back for recycling. A similar system is used for soft drink bottles in India.

[a] ESCAP and ADB, State of the Environment in Asia and the Pacific 200 (ST/ESCAP/2087), p. 269.

Institutional Policies and Institutions

Policies and programmes, whether economic, social or environmental, cannot be implemented effectively without the support of an interconnected network of institutions at different levels national, subnational and local. In addition to the apex institutions for example, parliament and the judiciary, institutions include administrative bodies, such as ministries, departments, councils and committees, which, in general, deal with policies and programmes belonging to specific areas. Institutional policies refer to the legal, normative and regulatory framework which enables the policies to generate the intended impact. For example, the environment ministry, the institution which is primarily responsible for implementing environmental policies, cannot carry out pollution control programmes (either through command and control or by using a market mechanism) without clearly defined pollution norms and legal penalties for not adhering to the norms.

Agenda 21 emphasises the important role of institutions and institutional policies in underscoring the critical importance of reshaping the decision-making process in order to integrate environmental and developmental concerns into decision-making. To achieve this aim, the institutions and policies (which have a bearing on the environment) of various sectors/areas need to be well coordinated not only horizontally but also vertically. Horizontal coordination between institutions will avoid conflict and exploit complimentarities, thus enhancing the impact of policies. Vertical coordination requires the existence of institutions at different levels from the national to the community level, and interaction between them. It ensures people's participation and ownership, without which all policies, especially environmental policies, will not be effective (see diagram).

Although there is room for improvement in the coordination mechanisms, the modalities for horizontal coordination on environmental matters already exist, in many Asian and Pacific countries. For example, in Tonga, the Development Coordination Committee is the paramount economic planning and coordinating body. Its standing members are the Deputy Prime Minister and senior representatives of selected ministries. In Fiji, the National Planning Office is required to have discussions with all ministries during the preparation of policy. If policies have economic or financial implications, the Economic Strategy Committee, chaired by the Prime Minister and comprising such ministers as the Minister of Finance, reviews the policies. In Malaysia, the National Planning Council is the highest planning body in the country and is in charge of the coordination and implementation of development programmes.

The success of vertical coordination between institutions dealing with environmental matters depends on the emphasis given to decentralised decision-making and the creation of decentalised institutions as well as the linkages between them. Effective vertical coordination ensures that environmental issues and concerns at the local level, especially in small towns and rural areas, are given due consideration in designing policies and programmes. The involvement of local institutions in formulating and implementing policies is essential as they are directly in touch with problems in their areas of operation and may have useful insights as to appropriate solutions. Local institutional structures can comprise many forms, from municipal councils or commissions in urban areas to village councils in rural areas.

The role of subnational institutions and local governments in the formulation, implementation and monitoring of policies and programmes on matters related to the environment has expanded in the post-Rio era. In many countries, provincial-level environmental agencies have been set up and local governments have been given the responsibility for formulating and implementing local environmental protection plans (local Agenda 21). For example, in Japan, local authorities are involved in the implementation of environmental laws, regulations and guidelines and the measurement and control of pollution. Out of a total of 248 local governments in the Republic of Korea, 159 are formulating local environmental plans. Provinces, autonomous regions and municipalities in China have also formulated their own local Agenda 21 plans regarding management of the environment. Similar efforts have been made in many other countries, including Indonesia, Malaysia, the Philippines, Thailand and Viet Nam.

The local municipal governments are in principle closest to urban environmental problems, such as air and water pollution and the lack of industrial waste disposal facilities. They are capable of participating effectively in the administration of control measures. In China, there are municipal environmental protection commissions consisting of local leaders, other commissions and bureaucrats. These commissions are responsible for implementing policies and coordinating activities at the local level; these, in turn, are assisted by the environmental protection departments of municipal governments. In Fiji, the Suva City Council is the largest local government entity and deals with environmental issues; in terms of implementation, it relies on different departments to carry out its plans. The Health Services Department in Fiji plays an important role as it is in charge of garbage collection, waste disposal and drainage control. The problem of squatters, with its serious environmental implications, is also addressed by the Health Services Department.

Village councils and committees are common in the rural areas. They can play a significant role in monitoring rural environmental problems, such as soil degradation, deforestation and depletion of fish stocks.

In Samoa, the traditional Polynesian decision-making system is reflected in the structure of most political and social organisations. Village councils decide on all matters pertaining to the village and its land and sea resources. One of the important features of this decision-making system is the emphasis placed on consensus. People avoid conflict by consensus agreement among peers. Meetings involve as many stakeholders as possible. In the villages, the village council discusses all matters concerning village life, especially when there are conflicts over resource use.

In Nepal, there are various district and village development committees. As a result of the decentralisation policy, these committees are empowered to undertake village and district development activities. These activities have had a tremendous positive impact with regard to new road construction, irrigation, drinking water, forest management, etc. For instance, areas of degraded and semi-degraded forest land are allocated to these local bodies for management, controlled exploitation and reforestation. This has improved the quality of land and prevented the massive destruction of forests. The Annapurna Conservation Area Project in Nepal is an illustration of the successful decentralisation of planning and decision-making.

Although there have been several attempts to enhance the importance of local institutions, their role has been mostly confined to participating in environment-related activities in some specific sectors. There are very few mechanisms enabling local institutions to participate and provide inputs, directly or indirectly (through regional institutions), into the formulation of overall development policies and to voice their concerns on environmental matters. In order to overcome the limitations mentioned earlier, the creation of an apex local (urban or rural) council with representation from various local government departments would be beneficial and it could present local views in a regional council. In this way, local issues and concerns, including those related to the environment, would be directly presented at the regional level and indirectly in national policy-making bodies (through the regional entity).

Legislation, laws, rules and guidelines (norm-setting) aimed at improving the environment and conserving natural resources fall under the general rubric of institutional policies. A very good description of the subject can be found in an ESCAP publication and is not reported here.

The Case of Small and Medium-Sized Enterprises (SMEs)

Several studies highlight the point that efforts to reconcile trade and environment policies may have to pay special attention to the situation of small and medium-sized enterprises (SMEs). The studies on China and Colombia point out that SMEs make a relatively large contribution to industrial pollution. At the same time, the Thailand study stresses that there is great potential for improving environmental management in SMEs provided that the proper supporting infrastructure is set up. Many practices which are cleaner than present methods of production are feasible for SMEs, but one important obstacle is lack of information on available options.

Many of the difficulties that SMEs may encounter in responding to either domestic or external environmental requirements are no different from other difficulties arising from the characteristics of their operation. Environment-related requirements, however, may create some additional difficulties. For example, certain installations (such as waste treatment facilities) require certain economies of scale. Several studies show that operating costs (even for common effluent treatment plants) may be relatively high for SMEs. In addition, environment-friendly input materials, which may represent a considerable portion of total variable costs, may be expensive. For example, in the leather tanning sector in India, the costs of chemicals required to meet international standards were approximately three times higher than the costs of conventional chemicals. While large firms may use their bargaining power to obtain inputs at competitive prices, this may not be possible for SMEs. Similar results were found in the study on Brazil.

SMEs may find it difficult to obtain the required input materials (for example dyes or chemicals) or to verify their suppliers. Once a new regulation emerges from an external market, it is often a long time before substitutes become available on the domestic market. While large firms may engage themselves in import activities or can influence their domestic suppliers to switch to environment-friendly materials, such opportunities may not be readily available for SMEs.

The India study expresses concern that rather than incurring cost increases in response to environmental regulations, SMEs often prefer to divert sales to the domestic market or to those external markets where environmental regulations are less stringent. It thus follows that environmental regulations could have an important bearing on export strategies. As an example, take the case of azo dyes which were recently prohibited in Germany, an important market for small-scale textiles exporters from India. Around 70 per cent of dyes used in India are azo dyes and 25 per cent of them are now banned in Germany.

Tenures in Natural Resources

To this point, we have presented a general picture of types of property rights in natural resources in India and regimes for managing them. We now review briefly some specific types of Indian land tenure systems in natural resources, and examine their efficiency, sustainability, and equity consequences.

Private ownership of arable land by its tiller was an established institution in India among the Indo-Aryans in ancient times. In the fifth century B.C., Mann said, "A field is his who cleans it of jungle, game is his who first pierced it" (K.C. Misra 1990). But all natural resources in India from earliest times have been the sovereign's, whether local monarch, British Crown, or Indian government. First, vast areas of land were held by the state as *khas* lands. Later, various lands were granted to big feudal land-lords in return for certain financial or military obligations towards the state. Third, there were lands granted to revenue farmers or rent collectors, commonly called *zamindars,* who had to pay a fixed amount annually as land revenue to the state. Fourth, some lands were held and cultivated by peasants who paid their land revenue directly to the government. These types of property rights and tenures formed a complex system in which the state occupied the position of super landlord. Over time, the *jagirdars* and *zamindars* became the unquestioned owners of the land and land-related resources granted to them. By the middle of the nineteenth century, three major systems of land tenure had become stable: *zamindari, ryotwari,* and *mahalwari.* Concentration of land ownership, high rents, insecurity of tenure, *begari* (forcing tenants to work without wages or compensation), bonded labor, low and; stagnant yields were the hallmarks of the land tenure system when India gained Independence. A national land reform policy was stated for the first time in India in the First Five Year Plan. It accorded high priority to programmes that increased agricultural production and diversification, reduced disparities in distribution of wealth and income, eliminated exploitation, and provided security to tenants and workers (GOI 1976). The first plan also advocated cooperative farming for owners of small and uneconomic plots and cooperative village management for optimal utilisation of land and other village resources (GOI 1976). This policy has been followed in all subsequent five-year plans, with marginal reshuffling of the priorities of its components. It is periodically reviewed and evaluated. One of the common findings of these reviews is that implementation of the policy has been lax. Many glaring gaps occurred between the objectives of the policy and the legislation enacted to achieve them, and between the laws and their enforcement. Property rights and tenures in India for land resources have long been established more firmly and defined more clearly than for other natural resources, but on the recent advice of the central government many state governments have launched tree *patta* (lease) schemes along with land *pattas.* Under these schemes, degraded revenue lands and the lands declared surplus under the land ceiling acts or donated under the *bhoodan* (gift of land) movement are leased to the rural poor for planting trees. The *pattas* range from 10 to 30 years, and the lessees are granted usufruct of trees for subsistence. But the many "catches" in these schemes make them unattractive to target people (Chambers et al. 1989).

Effects of Tenancy

There has been a long-standing controversy about the effects of tenancy, especially sharecropping tenancy, on farm output and income distribution. The traditional (Marshallian) view postulates that sharecropping tenancy is less efficient than owner cultivation because tenants lack incentives to buy and use inputs at their economically optimal levels. The modern view, typified by Cheung (1969), is that under certain conditions tenant cultivation is as efficient as owner cultivation. One example is when the landlord shares the cost of variable inputs in the same proportion in which the output is shared. By sharing the costs, the landlord can influence and monitor the use of inputs by the tenant. If the landlord pays all the costs, the tenant has no vested interest in using the inputs efficiently. However, if the landlord pays for none of the inputs, as often is the case, then the entire risk of crop failure or fall in crop price is borne by the tenant. This latter situation often leads to inefficiency in resource allocation, and indebtedness and impoverishment of the tenant peasants.

To our knowledge, no published empirical research study on land tenancy conducted in India has attempted to determine the long-term effect of tenancy on productivity, efficiency, and sustainability. Raj (1990), however, argues that under certain assumptions, such as perfectly competitive land-lease and labor markets, constant returns to scale, and lack of risk and uncertainty, all households would gain by leasing in or leasing out land to fit their ability to work the land. The total output of land then would be at its maximum. As a result of such land transfers, the marginal productivity of land would be equal across all households. He asserts that in a traditional agrarian society, the problems of managing hired labour, output, and price risks favor leasing out land by big land owners and leasing in land on a sharecropping basis by tenants. On the basis of data collected from eight Indian villages, Shaban (1987) found that the output and input intensities per hectare were both significantly higher on the owned plots than on the share-cropped plots. In other words, sharecropping was inefficient in the sense of "Marshallian productive inefficiency."

Using data collected from a sample of 224 tenant farmers in 12 villages of Midnapur District of West Bengal, Bhaumik (1991) compared patterns of resource allocation under owner cultivation and two alternative tenancy arrangements—sharecropping and fixed rent contracts. He found that both the average quantity of inputs used and the average out-put obtained per hectare were higher on the owner-operated fields than on the fields operated by sharecroppers, but no significant difference was observed in economic performance between the owner-operated fields and the fields operated by fixed-rent tenants. Krishna and Datta (1991), in a small sample survey of 40 farmers conducted in three randomly selected villages in the Kheda District of Gujarat, found that nonformal or oral leases of privately owned farm lands were quite common. There were elements of semifeudalism present in the system, but there seemed to be no social stigma attached to the tenant farmers. Both cash leases and sharecrop leases were common. Among other things, they found that the average yield per hectare was 40 per cent lower on the leased-in land than on lands operated by their owners. This difference in yields is attributed by the authors both to the poor quality of the leased lands and the use of fewer modern inputs. They conclude that the prevailing tenure system was mutually beneficial to the land owner and the tenant.

Pandey (1989) made an in-depth study of a property law, the Oudh Rent Act of 1886, to determine whether the agricultural backwardness of the area (now eastern Uttar Pradesh) could be attributed to the tenure system introduced under the act.

On the basis of his analysis of 72 court cases from 1893 to 1920, he concludes that the property law strengthened essentially precapitalist arrangements that stifled tenants' incentives to invest in land to increase its productivity. The law permitted the existence of various intermediaries (rent collectors) between the landlord and the tenant cultivator. The intermediaries usurped most of the cultivators' surplus in the form of high rents. This led to poverty and indebtedness of the tenant and hence their inability to invest in land and inputs to increase productivity. Pandey argues that the abolition of the *zamindari* system following Independence has changed the tenure system little, explaining the agricultural backwardness of the region.

In almost all Indian states, land tenures are insecure, and seasonal or annual. This is mainly because existing tenancy laws do not provide for long-term leases of land. Under these conditions, as we pointed out earlier, leased lands are most likely to degrade due to lack of needed soil and water conservation measures and application of balanced fertilizers. Consequently, the goal of sustainability will not be achieved; rather, their productivity is likely to decline over time. To remedy this situation, we suggest legalising long-term leases. Also, the lease contract should require the landlord and the tenant to share the cost of purchase inputs in the same proportion, preferably in a 50:50 ratio, in which they share the crop output. Such a policy would be mutually beneficial to both landlords and tenants.

Rural peasants have very limited access to organisations and recourse to effective legal protection. As B. Misra (1990, 72) points out, "probably the only solution to effectively implementing the programme of land reform is to organise the poor peasants." One of the most progressive efforts to organise tenants to exercise their legal rights to the land is Operation Barga in West Bengal. Since 1977 Operation Barga has undertaken to record the names of tenants so they are protected against the illegal practices rampant in rural India. Chadha and Bhaumik (1992) recently completed a thorough study of the success of the programme. They looked at the profiles of both recorded and unrecorded tenants in agriculturally "backward" and more progressive regions. Recorded tenants in both regions were found to be able to exercise their cropsharing rights more effectively than unrecorded tenants. The primary reasons given by tenants in both regions for not being recorded were the seasonal nature of some tenancy arrangements and the desire to maintain good relations with landlords. The recorded tenants typically found that relations with the landlord had deteriorated after being recorded. Most of the unrecorded tenants were families that had worked the land as sharecroppers-cum-laborers for many generations. Recorded tenants were usually relatively new entrants into the land-leasing market Although West Bengal is relatively advanced in providing rights to peasant farmers and improving land use practices, the unwillingness of a large portion of the peasantry to become organised illustrates the extent of the problem.

Corporate Compliance with Law

This Part of the Article considers whether a corporation can have an important voice in arranging—or re-arranging—the rules of the business game. If it can, would it have a responsibility to support environmental laws that might limit profits in the short-term? In brief, this Article contends that MNCs do have a significant voice and that they not only have a responsibility to exercise their freedom of speech to oppose ill-founded laws and regimes, but also to abide by well-considered laws and to support structurally sound proposals for modifying the legal environment of business. If abiding by such laws means reduced profits, corporate civil disobedience would not be warranted just because of such revenue losses.

The idea that corporations may have a duty to support positive changes in the law is somewhat at odds with the prevailing paradigm of a relatively powerless publicly-held company, caught between "the market"—consumer demands, supplier problems, attention to shareholders, competitors, and creditors—and "the government"— local, state, federal, and, for MNCs, nation-states beyond the principal place of business. This paradigm is captured in the comments of auto and oil company CEOs, the outlook of Milton Friedman, the governance principles of the American Law Institute, and the thoughts of business ethicist Norman Bowie. Because this Article pays particular attention to oil companies and industrial democracies' dependence on fossil fuels, Bowie's claims about the ethical obligations of auto companies deserve particular attention.

In *Markets, Morality, and Motor Cars,* Bowie states that business "does not have an obligation to protect the environment over and above what is required by law; however it does have a moral obligation to avoid intervening in the political arena in order to defeat or weaken environmental legislation." The implication of this is that (1) business is relatively powerless relative to the government, and (2) environmental legislation that might be defeated or weakened by business intervention is, on the whole, socially and environmentally desirable. Neither (1) nor (2), however, is invariably true. Much environmental legislation may be misguided or in need of amendment. Also, many MNCs do seem to have significant influence or power relative to governments of developing nations. Even in developed economies, companies often exercise free speech to inform consumers in the market and to make governmental rules and regulations more to their liking. One may reasonably ask, if corporations are powerless to influence public opinion and public policy, why so many spend money on advertising, public relations, and lobbying? Assuming that these expenditures of corporate time and money are meant to enhance shareholder earnings, it is difficult to accept the notion that large corporations possess no power to influence governments and consumers.

Moreover, the implication that all environmental laws are worth abiding by does not stand up to reason or experience, nor does the notion of business non-intervention with poorly-considered proposals "for the environment." For the sake of argument, suppose that Bowie means not only existing legislation, but also environmental bills in the legislative hopper, or even proposed regulatory standards. If he is correct, then all corporate efforts from oil and auto companies to defeat ratification of the Kyoto Protocol, however flawed, would be morally impermissible, as would all efforts from oil companies to defeat additional fossil fuel energy taxes, as would all efforts from auto companies to defeat more stringent fuel efficiency standards. But this cannot be. Oil and auto companies routinely engage in free speech activities, lobby hard to weaken" certain environmental legislation and regulation, show no signs of slacking off, and generally have shareholder—and sometimes union—support. Moreover, some lobbying may result in positive changes. In short, there is nothing about corporate lobbying per se that is morally suspect. Bowie, however, leads us to something important, and this Article hopes to present a slightly more nuanced set of guidelines for corporate obedience and opposition to environmental law.

A. Government Rules, Market Forces, and Corporate Responses

Before revisiting Bowie's thoughts, it may be helpful to look at existing government policies and market forces that affect oil and auto companies. Neither the public nor U.S. politicians give explicit recognition to the interplay between government policies and market forces. The rhetoric of both-left and right contributes to this confusion. From, the left, the notion is that government can capably control the decisions of many market participants. From the right, the notion is that if left to its own devices, the market can allocate resources perfectly well. The words "market failure" seldom issue from the lips of the most ardent advocates of this view, but economists are well aware that government must sometimes act to create and maintain properly functioning, competitive markets. Cass Sunstein—and others—have by now made it clear that by enacting laws, government does set the parameters for the market, and that markets require law; stable laws regarding contract, property, and tort are minimal conditions for assuring that promises are kept, that people are able to utilise land and other property in ways that enrich human life, yet do so without causing uncompensated harm to others.

Governments—and related institutions—have, however, often done more than provide such minimal services to establish markets and continue to do so. For the oil industry, the U.S. government provides subsidies of at least five billion dollars annually, largely in the form of depreciation allowances, accelerated depreciation, and other tax incentives for fossil fuel industries. The OECD has found that U.S. consumption patterns are based on "low density housing, extensive use of motor vehicles, and a high generation of waste," all largely supported by subsidies such as road construction and free parking provided by employers—whose expense is tax deductible— and failure to include the full social and environmental costs in the price of gas. While the United States has imposed some energy taxes, their salutary effect has been "nullified" by energy production tax incentives. The oil industry pays an effective income tax rate of eleven per cent compared to the non-oil industry average of eighteen per cent. Without such a differential rate, the oil industry would have paid an additional two billion dollars in 1991. Moreover, state and local governments tax gasoline at about half the rate they tax other goods—roughly three per cent versus six per cent.

The subsidising of non-renewable energy, however, is just part of the global picture. A 1997 study by the Earth Council entitled "Subsidising Unsustainable Development found at least $700 billion of taxpayer money annually to

> encourage the use of water, the burning of fossil fuels, the use of pesticides, fishing and driving. The report documented countless examples of taxpayers subsidising the use of water in countries where water tables are falling. . . . Additional billions are being spent to expand the world fishing fleet when its capacity is already nearly double the-sustainable catch.

Subsidies are anathema to economists, who understand that subsidies, like negative externalities, represent forms of market failure by undermining efficient resource allocation. For the auto industry, state and federal governments have combined to provide highway construction, bridges, tunnels, traffic signs and signals, policing, traffic courts, civil courts to hear auto accident claims, and a variety of other services that are only partially funded by license taxes, tolls, and gasoline taxes. Oil companies benefit from these government services as well because crude oil is refined into gasoline, which is the primary fuel for U.S. motor vehicles. Both oil and auto companies benefit from substantial and costly U.S. military commitments in the Middle East. The cost of gasoline does not reflect its full social or environmental costs, either; Norman Myers has estimated that internalising all direct environmental and social costs of gasoline would lead to a two dollars per gallon tax and that if Americans internalised "all costs of their car culture" they could be paying at least ten dollars per gallon.

Sensible laws, following sound economic theory, would seek to make markets work better by eliminating the distortions of subsidies that encourage unsustainable behaviours and would seek to make markets work better by letting prices reflect their full social and environmental costs. The strategy is unexceptional, in economic terms, but in political terms, subsidies are sometimes sacrosanct, and the hidden tax of negative externalities is more popular with voters—by default—than any direct tax.

It would be unfair to leave the impression that auto companies get only benefits from the current infrastructure and laws. The auto companies are also subject to a variety of U.S. federal and state laws that dictate engineering and design decisions and raise the cost of any vehicle sold. Yet, some of the most significant legislative and regulatory mandates have been strongly resisted by auto companies; moreover, the overall tendency seems to be resistance to any proposal or initiative that would decrease profits. In the following sections, this Article will posit U.S. policy needs for addressing issues of global warming, energy dependence, and overconsumption, and will then focus on three general policy approaches: energy taxes, fuel efficiency standards, and mandated zero emission vehicles. The conclusion to this Part is that auto companies are clearly justified in opposing ZEV mandates in the law, somewhat justified in opposing fuel efficiency standards, and not justified in opposing higher taxes on fossil fuels. This follows from a broad mandate for corporations to support structural legal reforms that fit sound economic theory and promote structural efficiencies: that the price of a product or service reflects its full social and environmental costs and that barriers to free competition—including subsidies that perpetuate the use of non-renewable resources—are gradually eliminated.

B. Auto Industry Environmental Issues: A Sampling

There are many environmental issues that involve the auto industry, both in the United States and other industrialised nations: global warming, energy dependence, air pollution, and overconsumption must take their place along with toxic byproducts and discharges—such as mercury—recycling, and urban sprawl. For simplicity, the Article will posit that problems of global warming, energy dependence, and overconsumption are problems that the United States needs to address, and that the solution to these problems must come not just from government or consumers—from laws and "the market"—but from corporations as well.

1. Global Warming

The increase in observed temperatures worldwide began to attract serious scientific and political attention in the 1980s and 1990s. Human-based greenhouse gas (GHG) emissions had increased markedly, especially during the latter part of the twentieth century, and many scientists believed there was a causal connection between increased GHG emissions and global warming or global climate change. The cause of this warming is still debated, but most scientists and governments believe it to be the result of human activities, particularly those involving the use of fossil fuels. Fossil fuels provide roughly eighty-five per cent of the world's commercial energy, and the automobile industry—the world's leading industry—is highly dependent on the oil industry—a close global second to the auto industry. Seven of the ten largest industrial corporations in the United States are either oil or auto companies. A substantial percentage of carbon dioxide emissions come from motor vehicle exhaust.

In 1992, the United Nations Framework Convention for Climate Change (UNFCCC) was adopted at Earth Summit in Rio de Janeiro. The UNFCCC set voluntary guidelines for reduction of GHG emissions. In subsequent discussions on the UNFCCC, it became clear that voluntary guidelines were not working, and in 1997, the Kyoto Protocol was adopted, committing industrialised nations to reduce six kinds of greenhouse gas, the most prominent being carbon dioxide. Since the Earth Summit in 1992, U.S. gasoline use has increased markedly. In essence, the U.S. public is consuming more and more, generating more and more GHG emissions, and the transportation sector leads the way. An Energy Department report in 1997 indicated that the United States, with less than one twentieth of the world's population, accounted for almost one fourth of all greenhouse gases; U.S. emissions of carbon dioxide from transportation had grown 8.5 per cent since 1990. The number of vehicles worldwide multiplied by a factor of eight from 1950 to 1989, and by 1999, the transportation sector globally was responsible for twenty-one per cent of worldwide carbon emissions. If current trends continue, the global automobile population could double in as little as twenty-five to fifty years. Moreover, regulatory controls on motor vehicle emissions have focused on carbon monoxide, hydrocarbons, nitrogen oxide, and ozone, and highway emissions of carbon dioxide and other gases "remain largely unaddressed." In both the European Community and the United States, the transportation sector was responsible for the largest increases in carbon dioxide emissions after the Rio de Janeiro Earth Summit. Thus, if the United States and European nations are to reduce GHG emissions, doing so in the transportation sector seems mandatory.

The Kyoto protocol was the result of years of preparation and eleven days of grueling final negotiations. The accord would require industrialised nations to cut greenhouse gas emissions to between six per cent and eight per cent below 1990 levels. The reductions would be achieved between 2008 and 2012. The differing percentage levels reflect different targets for the European Union, the United States, and Japan. The European Union would reduce its emissions by eight per cent below 1990 levels, the United States by seven per cent and Japan by six per cent. Twenty-one other industrialised countries would meet similar binding targets within the same 2008 to 2012 timeframe, and all industrialised nations committed to cutting even deeper after that. Together, these nations will be reducing their greenhouse-gas emissions by slightly more than five per cent from 1990 levels.

The Kyoto Protocol of 1997 was intended to set mandatory reduction targets for GHG emissions from industrialised nations. Considerable opposition to this Protocol emerged in the United States, with "business interests" leading the charge. The Global Climate Coalition (GCC) was created by major MNCs to battle any reduction commitments. Its members initially included the American Petroleum Institute,

Ford, GM, Chrysler, Dow Chemical, DuPont, Exxon, Mobil, the American Automobile Manufacturers Association, Chevron, Shell, Texaco, and Union Carbide. According to CorpWatch, a non-profit organisation promoting corporate environmental accountability, the GCC used various strategies to defeat U.S. ratification of the Protocol. These strategies included: (1) raising public concerns about "unemployment resulting from emissions regulations, (2) releasing reports with-"dubious scientific legitimacy" that questioned whether global warming was taking place, (3) attending climate negotiation meetings "en masse," (4) sending a letter "signed by 119 of the U.S.'s most prominent business leaders" to President Clinton, asking-that all current climate proposals be rejected, and (5) insisting that developing countries commit to the same stringent reductions as industrialised nations. At the same time, however, developing countries are lobbied "to reject any environmental obligations that might hinder their development."

Corporations have been joined by some unions in campaigning against the Kyoto Protocol. The Global Climate Information Project includes trade associations like the American Petroleum Institute and trade unions such as the United Mine Workers of America and the AFL-CIO; their advertising campaign warns of a general increase in prices that would result from a climate agreement and the unfairness of exempting the developing world from binding commitments. Predictions of massive job losses and economic slowdowns are the stock in trade of the American Petroleum Institute, and the Western Fuels Association sponsors "expert witnesses" who contest prevailing scientific opinion and argue that climate change would be a good thing.

Assuming CorpWatch is a credible source, it seems clear that some of these corporate-sponsored organisations will use any arguments, experts, or statistics that will defeat or weaken environmental laws that would impair jobs and profits. Many shareholders—and union members—would even applaud this strategy, particularly if it helped to maintain share prices or preserve jobs. Nor would all agree that deliberate distortions are morally suspect; the experience of advocacy within the rules of courtroom litigation provides a model in which zealous and sometimes extreme claims are made. Even so, deliberate lying is penalised as perjury, and groundless or frivolous claims can be sanctioned. Moreover, in the context of public debate, deliberate distortions may be legal, but are neither honest nor responsible. They may, in the short term, protect profits, but in a society in which people as citizens as much as consumers need accurate information to make sound decisions, strident and distorted advocacy is irresponsible. For auto companies, the primary means of achieving U.S. reductions in carbon dioxide emissions would be alternate fueled vehicles with greatly reduced or no carbon dioxide emissions, much higher taxes on gasoline—to discourage consumption and emissions—increased fuel efficiency mandates for internal combustion vehicles, and mandated use of technologies such as "zero emissions vehicles." Yet, for most of the 1990s, the auto industry has endorsed none of these, while visibly advocating no policy alternatives. At best, this is consistent with the idea that a corporation can be socially responsible in opposing any environmental proposal or law that might impair profits, regardless of its benefits to society.

There could be many good faith reasons for opposing the Kyoto Protocol; it is well beyond the scope of this Article to discuss all these reasons. The involvement of MNCs in the political process, however, is undeniable. If the Protocol has serious flaws, companies have a right—even a duty—to oppose it with arguments based on sound science and economic projections. Preferably, alternative proposals could be offered that would support meaningful re-direction of a fossil-fuel based economy.

Currently, the confluence of corporate political contributions and official U.S. energy policy suggests that the public interest is being ill-served. An international drive to "phase out fossil fuel subsidies and increase financing for nonpolluting energy sources worldwide" is opposed by the Bush administration, which bases its opposition on a "desire to let the marketplace, rather than [the] government, decide how quickly renewable energy resources are adopted worldwide." Given the strong subsidies provided by the U.S. government for promotion of fossil fuel resources, the administration's position seems more driven by ideology—or perhaps campaign contributions—than by practicality. The Bush-Cheney energy plan, released in May 2001, does not mention the Kyoto Protocol, and primarily focuses on increasing energy supply using coal, oil, gas, and nuclear energy. Recommended actions for renewable energy sources, efficiency, equity, and the environment are allegedly "minor, and place all of these issues at the margins of energy policy." Meanwhile, the United States sat on the sidelines in 2001 as delegates from 164 countries decided on the final details of the Protocol's implementation.

2. *Energy Dependence*

Energy dependence will be discussed at length in the next Part. It is enough to note here that U.S. energy independence has been a concern since the "oil shocks" of the 1970s. Complete U.S. energy independence from OPEC is most unlikely, even with stringent conservation and increased domestic production. Nonetheless, purposeful public strategies to wean the United States from Middle East oil supplies are needed, both to defuse Islamic extremists and to protect the economy from another series of oil shocks. Increased reliance on conservation and renewable energy sources could also help to reduce trade deficits. U.S. trade for Saudi or Iraqi oil may also provide revenues used contrary to U.S. foreign policy goals. In a prescient piece published in 1996, Joseph Romm and Charles Curtis noted that

> Persian Gulf nations' oil revenues are likely to almost triple, from $90 billion a year today to $250 billion a year in 2010—a huge geopolitical power shift of great concern, especially since some analysts predict increasing internal and regional pressure on Saudi Arabia to alter is pro-Western·stance. That represents a $1.5 trillion increase in wealth for Persian Gulf producers over the next decade and a half. That money could buy a tremendous amount of weaponry, influence, and mischief in a chronically unstable region.

Moreover, U.S. dependence on oil and current U.S. consumption of oil provides a model of economic development that is clearly unsustainable and has enormous economic repercussions on the U.S. economy as well.

3. Overconsumption

Overconsumption is riot generally recognised as an environmental problem per se, or as a national security problem. Overconsumption does not lead directly to extinction of a particular species, destruction of a specific forest ecosystem, degradation of tillable soil, or loss of fisheries, aquifers, coral reefs, or stable climate patterns. Yet, Overconsumption results in these losses and will continue to do so because it is built into our notions of material progress, standard economic theory, arid law. Moreover, the mentality that promotes Overconsumption can create a society that is restless, edgy, and aggressive, which is contrary to a societal and global goal of a world more at peace. Lester Brown, Paul Hawken, Amory Lovins, and others have made clear that current consumption patterns are depleting natural capital and undermining future well-being. William Rees has pioneered the concept of an "ecological footprint," approximating the varying average impact of humans on the environment, depending on which country they live in. He finds that Japan has a 2.5 hectare/ capita footprint, and that the Netherlands has a 3.3 hectare/capita ecological footprint, giving them "national ecological footprints about ten and twenty-one times larger than their respective domestic productive land bases. Each is running a large eco-deficit and accumulating a non-repayable debt with the rest of the world." If you accept the viewpoint that economic progress is colliding with finite resource limits—and there are those who do not—it is difficult to deny Rees' conclusion that "there is simply not enough natural capital on the planet to sustain present international development trends using prevailing technology." Rising material expectations and an increase over Earth's present population of six billion would require "absolute reduction in material standards of living" or "massive increase in material and energy efficiency." Far greater efficiencies could be found, but only if we minimise perverse subsidies and begin to tax material and energy consumption rather than labour. Instead, major industrialised nations continue to tax labour through income taxes and treat consumption of natural capital as an economic "freebie," retaining perverse subsidies and avoiding consumption taxes that would help reduce negative externalities. Moreover, the current version of corporate global free trade favours the use of machinery over human labour, favours consolidation, and requires long lines of transportation that encourage fossil fuel use. Proponents of the current system must also acknowledge alleged inequities in current consumption patterns, such as the claim that the prosperity of the "new world order" after the fall of the Soviet Union was "limited to a tiny per cent of the world's people, and to an ever smaller number of people even in the United States," even while sacrificing small farmers, farmlands, forests, wetlands, prairies, ecosystems, and watersheds. The U.S. motorists' appetite for oil poses particular problems, yet is still not clearly seen by citizens, corporations, and government as worthy of concerted action.

Thus, overconsumption poses a fundamental problem for global governance and corporate governance: if all nations followed the consumption patterns of the United States, Europe, and Japan, there would be ecological disaster, yet our national and international legal regimes—including the WTO—do not reflect this fundamental premise, while consumers, corporations, and governments appear needlessly locked into overconsumption of fossil fuels in a path-dependent pattern that is resistant to fundamental changes, even changes that are market-oriented and economically sound. These changes may be more mandatory than not. Wes Jackson, of the Land Institute in Salina, Kansas, has commented that "as long as we regard the resources of the world as ours at a minimal cost, there will be children born who, when adults, will willingly lay down their lives to bring us down in the name of social and economic justice." Ideally, science, public interest, and legal thought will converge on solutions that move the U.S. and other economies toward energy independence and consumption at sustainable levels.

C. Environmental Legislation and Auto Companies' Responses

Ever since the OPEC-induced oil price spikes of the 1970s, U.S. auto companies have faced numerous legislative initiatives that would boost fuel economy, eliminate or reduce carbon dioxide and other emissions, and move consumers toward more energy-conscious choices. This Article focuses on three different kinds of legislation: corporate wide fuel-efficiency standards, mandates for a certain percentage of "zero emission vehicles" (or ZEVs), and higher gasoline taxes.

1. Corporate Average Fuel Economy (CAFE) Standards

Even before concerns over global warming, the U.S. Congress was greatly concerned that U.S. energy dependence on oil left the economy vulnerable to "price shocks" based on OPEC price controls or embargoes. Because the U.S. was importing a third of its oil, the Arab oil ministers' 1973 decision to cut supplies to the United States soon led to long lines at gasoline pumps, missed or delayed fuel deliveries, public school closures for lack of heat, and violence on the highways. As a response, the federal government enacted the Energy Policy and Conservation Act (EPCA), which included three measures to promote fuel conservation. In the EPCA, Congress set a national speed limit of fifty-five miles per hour, imposed a "Gas Guzzler Tax," and authorised fuel efficiency standards in Title V. The fuel efficiency standards set in Title V required that all companies selling autos in the United States must reach a sales-weighted corporate average fuel economy (CAFE)-of eighteen miles per gallon by 1978, twenty-two miles per gallon by 1981, and 27.5 miles per gallon by 1985. Miles per gallon averages would include a weighted average of city and highway EPA mileage ratings. The Department of Transportation (DOT) was responsible for setting the precise levels for the 1981-85 model years and for 1986 and beyond. The current standard of 27.5 miles per gallon has not been exceeded since 1990.

When the CAFE standards were set, Congress allowed an exception for "light trucks" and entirely exempted very large trucks. The basic idea was to make some allowances for farmers and ranchers to have pickup trucks with more horsepower and, necessarily, lower fuel efficiency. In the 1980s, Detroit automakers were able to capture a significant market share by producing minivans and light trucks for urban and suburban motorists. For various reasons, throughout the 1980s and 1990s— including lower gasoline prices (adjusted for inflation), an aging "baby boomer" market that increased demand for larger cars, and the increasing use of light trucks for passenger rather than commercial use,—fuel efficiency began to decline. Under pressure from U.S. automakers, and using the discretionary authority provided it by the EPCA, the Department of Transportation rolled back the 27.5 standard to twenty-six for the 1986 model year and held it there for two years. The standard rose to 26.5 miles per gallon for the 1989 model year. Early during the first Bush administration, however, the standards were returned to 27.5 miles per gallon. There were some proposals in the early 1990s to increase CAFE standards to an average of forty miles per gallon, partly in response to the threat of global warming. These proposals, however, were defeated in Congress; during the Clinton presidency, Republicans attached riders to a number of bills that insured a freezing of CAFE standards.

The attacks of September 11, 2001 may provide new impetus for motor vehicle fuel efficiency. Initially, there was not much interest in Congress in exploring conservation and alternative energy sources. There was a continuing push from House Republicans to encourage oil exploration in the Arctic National Wildlife Refuge (ANWR) matched by an alternate energy bill sponsored by Senate Democrats that avoided ANWR exploration. By early 2002, however, legislation was introduced in the Senate that would raise average fuel efficiency standards to 36 mpg by 2015, a standard that would classify minivans and SUVs as passenger vehicles rather than light trucks. Senate debate on a comprehensive energy package began in

March of 2002. The prospect of tightened fuel efficiency standards has resulted in renewed lobbying of legislation and the public by the U.S. auto industry.

There are some strong arguments against mandated fuel efficiency standards. There is also some strident rhetoric that stretches the notion of a "right." Auto company lobbying on this issue has been remarkably effective, at least partly because many of the arguments make practical and economic sense. The principal arguments against fuel economy standards are as follows: (1) lighter vehicles are not as safe; (2) a doubling of fuel-economy in the United States will not appreciably alter GHG emission levels, because moving from twenty-seven miles per gallon to forty miles per gallon would not appreciably reduce GHG emissions if motorists simply drove more; (3) even if vehicle miles traveled did not increase, forty miles per gallon is only half as fuel efficient, making emission reductions almost insignificant; (4) global warming is not a problem, and even if it were, U.S. carbon dioxide emission reductions would be pointless against increased vehicle emissions from developing nations, where large increases in vehicle fleets are predicted; (5) consumers want choice, and do not want to be forced by government regulations into smaller, less safe vehicles; (6) existing CAFE regulations are an inefficient means of enforcing conservation because a bureaucracy must measure miles per gallon for each vehicle the corporation sells and then levy fines; (7) an increase in gasoline taxes would be far more efficient, and tax incentives for purchase of fuel efficient vehicles or alternate fueled vehicles would preserve consumer choice; and (8) market forces will regulate far better, anyway, because when consumers want fuel-efficient cars, they will buy them.

The vigorous efforts to defeat more stringent CAFE standards are partly motivated by reason, and partly motivated by money. U.S. automakers' profitability in the late 1980s and 1990s was due in no small part to the market success of larger vehicles, which produce a far better profit margin than smaller ones. This partly explains why the United Auto Workers would in with automakers to oppose stricter CAFE standards.

Some of the arguments automakers make against CAFE standards are better than others; some of the arguments seemingly lack good faith. For example, the argument that smaller vehicles are not as safe is certainly true in one respect: SUVs as manufactured in the 1990s were unsafe, primarily for small car owners that collided with them. The argument that consumers had a right to choose larger, safer, more comfortable vehicles—SUVs—is a potent one— why should government dictate how comfortable and safe you can be?—but it undercut by the automakers' failure to adequately inform SUV owners of rollover tendencies that drivers would not and did not expect. Moreover, if consumers' right to choose safer vehicles is vitally important, then full disclosure by automakers of safety and crash data would also be important. The industry record in this regard, however, has not been impressive.

There is also some merit to the "market forces" argument: U.S. drivers willingly buy larger, less fuel-efficient vehicles. Seemingly, many drivers also buy for image rather than safety. In polls, the public seems concerned about global warming and favours fuel efficiency standards; in showrooms, they choose size and power. Auto analyst Maryann Keller points out that consumers' car purchases are "dictated by their desires for performance, comfort, and safety, not by any ecological standard." The consumer's choice, however, does not take place on a blank slate. As we have seen, the context of consumer choice for personal transportation takes place in a context of subsidies to oil companies and the market's failure to fully price gasoline.

In addition, insurance companies tend to provide a flat rate insurance that charges drivers the same whether they drive a lot or a little. Bradsher has also shown how current laws favour SUVs and light trucks, and the phenomenon of path-dependence effectively limits consumer choice to different brands of vehicles with internal combustion engines. Generic appeals to "the market," in short, are a lot less straightforward than they seem. Appeais to "the market"— as a self-regulating mechanism always and automatically preferable to government regulation—lack the good faith articulation of structurally efficient principles that this Article recommends as a foundation for responsible corporate advocacy.

The denial of global warming or global climate change as a serious policy problem was part of U.S. automakers' approach to corporate advocacy well into the 1990s. Detroit automakers reacted to the Kyoto Protocol in 1997 by denying that global climate change was a problem that needed to be addressed. Other arguments against CAFE have merit, but are also open to debate. It is true that doubling fuel efficiency for all vehicles—including light trucks—would not necessarily conserve gasoline if people drove twice as much. The majority of driving, however, is done by commuters; presumably, commuters would not drive to work twice as often, but they might live further from work. Still, commuting times may already be approaching their maximums in a number of U.S. metropolitan areas. The safety of smaller cars can be enhanced by decreasing numbers of very large vehicles, and technologies exist to make smaller cars even safer than they are today. In claiming that developing nations' emissions will far outweigh any savings that may be generated in the United States, there is both a denial for past European and North American contributions to carbon emissions and an assumption that the automobile future in nations like China must be similar to the U.S. past. Major automakers will play a part in the kinds of vehicles sold in China, however, and those need not be based on the internal combustion of gasoline unless there are truly no good economic alternatives. That, in turn, depends on what is meant be "economic" if alternate fuels have a viable infrastructure, if the full costs of gasoline are factored into its price, and if subsidies to oil could be phased out, other kinds of vehicles could very well be economical.

The best arguments against CAFE address the relative inefficiency of these mandates compared to more economic-minded alternatives: "feebates" for' those who choose smaller vehicles or other incentives to help automakers sell alternate fueled vehicles. Occasionally, an auto company executive will say that higher fuel taxes are the most sensible economic policy solution, but it is not said too often, or too emphatically. In sum, opposition to CAFE standards has not been linked to advocacy for structurally efficient reforms. The CAFE legislation does have flaws, and even the basic idea is somewhat flawed. It assumes that doubling gas mileage will cut fuel use in half, even though such doubling may just encourage more driving. Moreover, it allows drivers to put the onus on auto companies: people polled about global warming greatly prefer solutions that do not cost them anything and require the auto companies to come up with some new technology that solves the problem. With such flaws, we should certainly expect corporate advocacy directed toward consumers and government; corporate compliance, without strenuous argument, would be disappointing not only to shareholders, but to the driving public. Yet, the real hope is neither silence nor specious or self-serving argument, but instead, the kind of advocacy that provokes searching and honest conversations among corporations, consumers, and the government.

2. Gasoline Tax Increases

If questioned as to how Detroit could help reduce CO_2 emissions, a Big Three CEO or other high-ranking officer might say that only far higher fuel prices would shift consumer demand away from larger vehicles. This can be said, however, without risking the status quo: while several studies have demonstrated that only a steep increase in gasoline taxes will shift gasoline consumption downward, U.S. politicians are not about to risk the public's well-known antipathy toward energy taxes of any kind. They have learned to avoid suggesting even modest increases in gasoline taxes, much less proposing them or voting for them, because U.S. drivers believe-erroneously—that they already pay enough for gasoline. U.S. drivers' preferences for cheaper gas and larger vehicles, however, reflect a number of unusual market, distortions. Norman Myers has pointed out that the total social costs of gasoline use are barely reckoned into the market price, leading to what economists call "negative externalities." In addition, many employers offer free parking as a non-taxable benefit to employees, the oil industry benefits from subsidies under U.S. tax laws, and larger vehicles are now favoured by the law in numerous ways. Finally, both the driving public and the auto industry benefit enormously from having a system of roads and highways built and paid for with public funds, and the costs of maintaining that system—including traffic enforcement—are only partly covered by licensing or registration fees and gasoline taxes. Restructuring the market so that external costs are taken into account is more easily said than done. Eric W. Orts points out at least four "conventional economic approaches" to market-oriented environmental laws: energy taxes or pollution charges, extending rights of ownership in the natural environment, tradeable pollution rights, and providing truthful information about products and their environmental effects. Energy taxes are problematic because government must set a fee based on estimates of potential harm, and reasonable people are bound to disagree; there are also ongoing tasks of oversight to adjust taxes, raising taxes when there are too many emissions and lowering them when there are too few emissions. In the United States, proposals to significantly increase federal gasoline taxes would have to be accompanied by reductions in federal income tax in a process of tax shifting that might prove politically confounding. If enacted, the tax shift could result in underfunding of government in some years, and a surplus in other years.

3. Zero Emission Vehicles

Another strategy for reducing fossil fuel emissions from vehicles is to shift to alternate fueled vehicles. Various choices include electric, natural gas, methane, and fuel cell vehicles. The federal government has provided some limited support for the development of alternate fueled vehicles, while California has mandated that companies selling vehicles there market a certain percentage of zero emission vehicles, or ZEVs. Auto companies have worked hard to defeat and delay this requirement. Indeed, there are good reasons for such opposition; the existing technological means of achieving zero emissions are electric vehicles (EVs). Hybrid cars such as Honda's Insight or the Toyota Prius, while more fuel efficient, still have carbon dioxide and other emissions. While there have been advances in fuel cell research, the marketing of significant numbers of fuel cell powered vehicles is still several years away. As for EVs, the sources of electricity for EVs are mostly coal-fired or nuclear powered electric utilities; coal-fired plants add carbon dioxide emissions to the atmosphere, and nuclear power raises other environmental concerns. Consumers reject vehicles that cannot be conveniently refueled on the road, and drivers cannot find facilities away from home to quickly recharge an electric vehicle. Vehicles powered by natural gas burn clean with no harmful emissions, but drivers do not want the inconvenience of seeking out the few natural gas pumps that could refuel their vehicles; such pumps are hard enough to find in metropolitan areas, and in rural areas are rare indeed. And this is not likely to change: service stations will not put in natural gas pumps until there is enough demand—natural gas-powered cars—and drivers will not buy cars powered by natural gas until there are plenty of pumps to make refueling convenient. The phenomenon of "path dependence" continues to provide a seeming mandate for gasoline powered internal combustion engines that steers both manufacturers and consumers along the path of least resistance.

D. Motor Cars and Morality Revisited

Norm Bowie has written that automakers have no special responsibility toward the environment beyond what is required by law. Yet, many MNCs operate in countries where "law"—as understood in the United States—is so poorly developed that obedience to the law may be ethically empty. For example, Shell Nigeria's actions in Ogoniland did not violate Nigerian law or international law, but were socially and environmentally irresponsible nonetheless. Shell has even acknowledged as much. An oil company or an auto company operating in a nation where environmental laws and regulations are relatively weak or unenforced cannot claim the moral high ground. Bowie also cautions that a business should avoid "intervening in the political arena in order to defeat or weaken environmental legislation." This cannot be entirely correct, either; under Bowie's injunction, auto company campaigns against higher CAFE standards, ratification of the Kyoto Protocol, and zero emission vehicles are all equally suspect. Yet, the previous sections in this Part were meant to show how corporate opposition to poorly conceived lawmaking could be consistent with good corporate governance. This opposition, however, should combine the mandate of profit maximisation with advocacy for sound, market-oriented reforms, and avoiding advocacy for the sole sake of short term profitability.

If California proposes that each automaker must insure that at least fifteen per cent of the vehicles it offers for sale in California have zero emissions by 2003, a strong case for opposition emerges, and not only in terms of maximising shareholder wealth. If a company knows that the law will mandate the production of vehicles that (a) are not likely to be purchased by consumers, (b) will not advance environmental integrity, (c) will significantly reduce profits to the company, and (d) will shift resources away from better, long-term solutions, it surely has a duty to oppose the law. The argument here is that corporate responsibility to oppose unwise or ill-conceived legislation or regulation can be founded on self-interest—short term profit maximisation—but is also grounded in a civic responsibility to ensure that the market is governed by structurally efficient rules. In either case, opposition should be tendered to politicians and the public in good faith, meaning that specious, hypocritical, or solely self-serving arguments are morally suspect.

The argument can be tested in light of the industry's consistent opposition to the CAFE standards. As noted earlier, there are some good faith reasons to oppose government mandates on fuel efficiency. Several of the automakers' arguments against CAFE, however, lack consistency with other aspects of their corporate conduct. Advocacy for increases in energy taxes, or for tax incentives for consumers to buy alternate fueled vehicles, are more appropriate and responsible arguments. Those arguments, in fact, seem to be made with greater frequency.

One of Bowie's arguments fits very well with the paradigm of corporations being squeezed between government regulators and consumer demands. This is the paradigm in which corporations are relatively powerless, warranting Bowie's prescription to obey environmental laws, oppose no environmental laws, and wait on the demands of individuals as citizens and consumers to make needed changes. Business, says Bowie, will "respond to the market." It is the consuming public that has the obligation to make the trade-off between cost and environmental integrity." It is true that if the consuming public demanded zero emission cars, then auto companies would have to make them or go out of business; but, again, the consumer's choice does not take place on a blank slate. The context of consumer choice for motor vehicles is one of subsidised oil, subsidised roads and highways, cheap gas, insurance rates that are the same no matter how much people drive, laws that favour large vehicles, and an understandable consumer preference to drive a vehicle that will not be overpowered by a larger one. It is, therefore, doubtful that "the market" as presently configured could do much of anything to reduce U.S. carbon dioxide emissions from motor vehicles or end U.S. over-dependency on Persian Gulf oil. Markets are not perfectly competitive; they are path dependent, imperfect-externalities and subsidies—and highly politicised—rewarding campaign contributions with "access" or a seat at the table.

A more nuanced version of Bowie's recommendations for ethical corporate governance in auto companies is selective environmental compliance and defiance. Auto companies—and, arguably, all companies—should obey democratically-created laws and regulations, but oppose—in court, in administrative hearings, and in public pronouncements—any law that is contrary to sound economics, and advocate laws that are consistent with making the market more structurally sound. Attention to what Eric Orts has called "reflexive environmental law" will be useful here, as some market-oriented policies will serve particular problems better than others; an emissions-trading system for U.S. motor vehicle owners, for example, is probably unworkable, whereas some form of energy taxes or "feobates" seems more sensible. Thus, there will be room for argument and advocacy in any law or proposal. This Article's recommendation for responsible corporate governance is a good faith articulation of the benefits and pitfalls of any particular proposal and the advancement of alternative proposals, rather than the more "traditional" opposition of a company to any environmental law or proposal that may adversely affect profits.

Telling the truth consistently requires that automakers should be willing to not only criticise government laws and regulations that have misguided aims and ineffective means, but to support laws that foster efficient markets and sound policy goals. If auto and energy companies deny that fossil fuel dependence is a national security problem, or deny that carbon dioxide emissions may be a global climate problem, or plan for a world in which all nations consume gasoline, land, and other resources in order to drive personal vehicles, both efficiency and truth are poorly served.

Auto companies have taken some positive steps. In 2000, the Ford Motor Company acknowledged that its SUVs created more safety and environmental problems than its cars. In December 1999, Ford withdrew its support for the Global Climate Change coalition, an industry-sponsored public relations and lobbying effort that routinely dismissed global warming as a serious problem. In January 2000, Daimler Chrysler also quit the coalition, noting that it had become an impediment to pursuing environmental initiatives in a credible way; while continuing to oppose the Kyoto protocol, GM left the coalition two months later, acknowledging that carbon dioxide buildup could be changing the world's climate.

Detroit automakers have also been working on new, more environmentally friendly technologies. Since 1993, the U.S. government and Detroit automakers have spent $1.6 billion to develop an eighty mile per gallon family car through the Partnership for a New Generation of Vehicles (PNGV). Alliances with companies working on fuel cell technologies may have begun the process of making the internal combustion engine obsolete. Daimler Chrysler expects to market a fuel-cell Mercedes in 2005. These are responses not only to the market and to anticipated rises in gasoline price as supplies diminish, but also to anticipated regulation.

Responsible corporate advocacy requires not only honesty, but also a commitment to addressing legal and regulatory concerns proactively. For example, if there are infrastructure barriers to fuel cell technologies, automakers have the responsibility to address those publicly. If gas prices relative to other markets are encouraging over-consumption and negative externalities, auto companies should say so. If reasonable market-oriented changes in law can encourage reduced fossil fuel consumption, auto companies should say so. If government needs to provide greater incentives to the private sector for exploring new forms of energy, auto companies should say so.

Educating the public and espousing the creation of public goods is a legitimate, even necessary function for large public corporations. Bowie's inclination is to accept a fairly clear separation between public and private spheres, but such a separation seems less and less realistic. Regulators make career moves to private industry-through the "revolving door," and formerly public functions are privatised or de-regulated; companies and trade associations contribute heavily to politicians and parties, lobby extensively for statutory and regulatory changes, and work assiduously at cultivating public opinion. In the context of U.S. law and policy, automakers must acknowledge their influence and use it positively. Because of the U.S. position as a major power, U.S. law and policy will also impact the development of international law and may contribute to—or detract from—the hope of a more peaceful world.

Corporate Environmental Liability

Liability in General

Strict Liability

By virtue of the Interpretation Act 1978 the expression "person" includes a body of persons corporate or unincorporate unless the contrary intention appears in the statute. Whilst the general rule is that criminal-liability is not imposed unless a person intends to, or foresees that he may bring about the constituents of a crime, there is an increasing tendency for legislation relating to the environment to create offences which are absolute and impose strict liability so that the prosecution does not have to prove that the accused had a guilty mind and the accused has no defence that he made a genuine mistake. Offences relating to the licensing system established under Part II of the Food and Environment Protection Act 1985 are thought to be offences creating strict liability.

Vicarious Liability

The wording used is often of a kind which by implication imposes vicarious liability on a master or principal for the acts of his servants or agents acting within the scope of their authority. Frequently the wording refers to a person causing or committing another person to do something. It is well established that to cause something involves some degree of dominance or control or some express or positive mandate from the person doing the causing. A person cannot cause another person to do something unless he actually knows what it is the other person is doing or failing to do and one can have deemed knowledge if a person deliberately chooses to turn a blind eye to what is going on. Likewise, to permit an offence to be committed involves a knowledge of the facts but turning a blind eye, not caring whether something takes place or not, is sufficient. If a company is to be liable for causing or permitting something it must be shown that some person for whose criminal acts the company would be liable caused or permitted the commission of the offence. It has, however, to be shown that a director or someone in authority with power to influence the conduct of a company's affairs had knowledge of the facts. Knowledge on the part of an ordinary employee is not sufficient.

Corporations may be subject to potential liabilities arising from civil actions in trespass, private nuisance and public nuisance, under *Rylands* v. *Fletcher* in negligence and/for statutory nuisance. With regard to trespass a defendant corporation will be liable irrespective of fault if it has in any way caused an encroachment onto the land of the plaintiff. In the case of private nuisance a corporation may be sued by a landowner if his use or enjoyment or some right over or in connection with is interfered with by some activity of the corporation undertaken on its neighbouring land. If a corporation carries on an activity which causes a nuisance to a significant number of people in the surrounding community, a public nuisance might be created which is actionable also as a crime. If a person can show that he suffered special damage, that is to say he has suffered in some specific way over and above the general disturbance to the other members of the community affected, then such person can bring a civil suit against a corporation. Under the rule in *Rylands v Fletcher* if a corporation for its own purposes brings onto its lands and collects and keeps there anything likely to do mischief if it escapes, it must keep it in at its peril and if it does not do so, it is prima facie answerable for all the damage which is the natural consequence of its escape. Such a person is strictly liable for the consequences. The rule does, however, require the defendant corporation to have made some use other than a natural use of its land. The question of non-natural use of land may be a difficult one to determine. Under the tort of negligence where a duty of care is owed by a corporation to another person, any failure to take the requisite care in the circumstances will be actionable if it causes damage.

PALAMPUR DECLARATION ON SUSTAINABLE DEVELOPMENT

(Announced at the 4th World Congress on Environment Management, Palampur (India) on 9th June 2002)

The participants to the 4th World Congress on Environment Management, held during June 7-9, 2002 at Palampur, recognised the importance of action in making the Sustainable development happen. They felt there was too much talk about Sustainable development and in most cases, it was a case of preaching those who are already converted. There is little evidence of action in reaching those beyond the cocoons of converts. The Congress questioned why 10 years after RIO, even the principal product of RIO—'Agenda 21' remains unimplemented.

The Congress noted that according to Mahbub-ul-Haq of Pakistan based Human Development Centre "during the globalisation phase about half a-billion people in South Asia have experienced a decline in their incomes". Also according to UN Report on Human Development (1998), the income ratio between 20% of the world's poorest and 20% richest was 1:30 in 1960. It increased to 1:61 in 1990. In 1999 it was 1:74 and today it stands at 1:87. Similarly even in US the differential between the shop floor worker and CEO which was 1:39 in 1960 has now gone up to 1:254. As for the environment damage there could not be a more convincing evidence of global warming than the shattering of a huge ice-shelf, 'Larsen B', weighing, 500 million billion tonnes in Antarctica on 5 March 2002 and breaking off of another ice-shelf-Thwaites-covering an area of 5500sq. km. within 13 days (on 18 March 2002).

It was realised that Sustainable development cannot take place unless we involve the masses proactively and create a buy-in for all stakeholders viz. government, business and civil society to make the sustainability agenda successful.

The civil society needs to change it's life style and the business needs to change production processes and innovate products, which are eco-efficient and environment friendly. All this will fail, if the government does not support by taxing pollution and stopping subsidies that are endangering the environment.

The conference recognised that environment improvement needed a comprehensive, cohesive and transformational approach involving all activities and that little is achieved by tinkering here and there through isolated and compartmentalised actions. The conference 'blueprint for action' given below is therefore a comprehensive and transformational 11 point agenda, which will henceforth be known as the "Palampur Declaration on Sustainable Development".

1. Partnership In Action for Sustainability (PIAS)

The conference found that despite lack of significant action on Agenda 21 some of the businesses have done excellent work relating to environment management and brought down its impact in many ways. There were several good examples in India. The Congress had the benefit of 18 case studies of excellence and it was felt those stories needed to be shared with others. It was therefore decided to develop a partnership programme to involve all those interested in upgrading the environment. For this an initiative called PIAS (Partners in Action for Sustainability) was formed. A large number of participants joined the initiative by giving a signed undertaking that they will do everything within their power to protect the environment. The PIAS task will be to disseminate good work done by companies and thus provide role models that we badly need and at the same time encourage companies to seek 3rd party certification to 'ISO 14001, Environment Management System' as a first step to demonstrate their commitment to environment. Those already certified are encouraged to go beyond EMS, towards Total Eco-Effective Management.

2. Pricing Natural Capital

The conference felt that an overriding reason why sustainable development has not captured the imagination of people and business is that there is little action on pricing the natural capital. Claims of saving environment sound hollow, if no economic value is accorded to natural capital. It was decided to synergise with like minded organisations worldwide, to develop realistic costs for the use of biological flows and natural capital.

3. Radical increase in Resource Productivity

The industry needs to change its paradigm from increasing labour productivity to radically increasing resource productivity. All effort needs to be directed in this area and special incentives provided to innovators.

4. Fostering Innovation and Creativity

The Congress came to the conclusion that the only way environment can be saved is by nurturing creativity and fostering innovation. This means people need to be given ownership of jobs they are doing and encouraged to make mistakes. The companies should not only recognise success but develop reward system that recognise "good tries".

5. Governance Structure

The Congress participants concluded that poor record of 'action on Sustainability' was the outcome of misgovernance and lack of transparency. We, therefore need to create more transparency, accountability, integrity and equity in our governance structures. Tough legislation and enforcement measures are required to combat corruption.

6. From Cradle to Cradle

All production processes need to be reorganised to pattern themselves on biological model, which has no waste. We need to change our paradigm from minimising waste to eliminating waste by closed loop systems.

7. Practice TEEM (Total Eco-Effective Management)

Everyone in the organisation needs to practice total Eco-Effective Management, which calls for 9-fold action:

- (i) Reduce material intensity,
- (ii) Reduce energy,
- (iii) Eliminate toxicity,
- (iv) Recycle,
- (v) Reuse,
- (vi) Revalourise,
- (vii) Minimise use of scarce natural resources,
- (viii) Use renewables.
- (ix) Provide extension of service wherever possible.

8. Shift Economy from Acquisition to Flow Mode

We must change to life styles that promote flow of goods and services instead of owning them. Car hire, Laundromats and leasing of services are some of the examples. The producer is to be responsible for the entire life cycle of the product.

9. Fiscal Policies

We must change to fiscal and financial policies and adopt the ones that reward those who save environment and punish those who damage. The government must withdraw forthwith all subsides on processes and activities that damage environment. Taxation policy be changed to tax environmental impact instead of labour. Special incentives such as credits be given to those installing renewables.

10. Poor Inclusive Market Frameworks

Poverty is the single largest barrier to Sustainability. We need to develop and implement market frameworks that subsidise business to seek untapped markets in poor areas and develop viable income-generating activities for the poor.

11. Extensive use of Computers and Internet

Extensive use of computers can have a significant effect on the environment. Emission problem can be solved to a significant extent by allowing more and more people to work from home.

Strategy for Energy Sustainability in Germany

Global energy consumption has more than quadrupled since 1950. According to a 1998 estimate by the World Energy Council, economic and population growths will cause global energy consumption to rise from its present level of 12 billion metric tons of hard coal units to between 19 and 25 billion metric tons of hard coal units in 2020.

Germany's energy consumption consists of 500 million metric tons of hard coal units annually. This is equivalent to six million wagons full of coal-a train which would stretch for 75,000 kilometres.

Some 70% of primary energy consumption is accounted for by the industrialised countries alone. Germany's share of global primary energy consumption comes to almost 4%.

Hitherto 90% of global energy needs have been covered by non-renewable energy sources. In Germany, primary energy needs are covered by oil (40%), natural gas (21%), hard coal (14%), brown coal (11%), and nuclear energy (12%). Renewable energy sources account for only 2% of primary energy consumption. On the other hand, in the electricity sector the use of renewable energy sources has steadily increased in recent years. Some 20% of the world's electricity is now generated from renewable energy sources. At 96%, water power easily leads the field among renewable energy sources for generating electricity, way ahead of geothermal and wind power. The total output of all of the world's wind energy plants is currently 14,000 megawatts, chiefly accounted for by Germany (5,000 MW), the USA (2,500 MW) and Denmark (1,750 MW).

In 1999 renewable energy sources accounted for 6% of Germany's electricity production. Water power alone accounted for 4%. The largest share (63%) was accounted for by fossil fuels (coal, oil, and natural gas). 31% of Germany's electricity came from atomic power plants. But the use of wind power in particular is rising rapidly: output from wind energy plants increased by 12% in the first half of 2000 compared with 1999.

The Federal Government has set itself the goal of increasing the share of renewable energy sources used to cover primary energy needs and electricity production to 50 % by the year 2050. Many consider this goal to be unrealistic. Yet reputable studies, such as that conducted by the Shell concern, predict that by 2050 half of the world's energy needs will be supplied from renewable sources.

Environment and Development Debate—A Pradigm Shift

Exclusionist or the Neo-classical Paradigm

The environment-development debate has emerged from the notion that has its roots in the growth oriented developmental pattern being practiced and promoted by the Northern industrialised states. After the Second World War the newly independent states of the South took to ambitious developmental schemes for catching up with the North and this erupted into a major debate about this form of development posing a threat to environment. The dominant social paradigm of the industrialised North believed that free market and unlimited resources are not constrained by the laws of nature because human beings are excluded from them. Technology according to them was the stimulant to sustain this unlimited growth and achieve man's victory over nature. Over the subsequent period of three decades when this paradigm was exposed not only as a big failure but also a major danger to the lives of people inhabiting planet earth the debate on environment and development formalised into a full fledged environmental meet of not only the governments of various states but also the non-official groupings of people, the aborigines and women since they were the most affected groups due to the carnage of this exclusionist developmental paradigm.

In the early sixties with the publication of the book *Silent Springs* by Rachael Carson exposing the hidden threats to the use of pesticides and insecticides the threats to Californian Redwood forests and Grand Canyon from development and the discovery of 'dying' lake Eire led to the publication of the Report *'Limits to Growth'* in 1972 by the Club Of Rome. The Report brought the mad rush to industrialise and capitalize on environment to a sudden halt. This led to the first international initiative at the Stockholm Conference in 1972 to devise an action plan for cooperation on Human Environment. On the recommendation of the 114 signatories to this Conference the United Nations General Assembly created United Nations Environment Programme (UNEP) as a nodal agency for environmental cooperation and coordination. This was a beginning of the paradigm shift in global environmental politics.

Hence the new social paradigm of global environmental politics focuses upon the scientific findings that the resources are limited and the natural environment would be seriously affected if the conventional pattern of growth continues. The concepts of scarcity of natural resources and carrying capacity were floated in the seventies to limit the growth pattern. This alternative paradigm of growth merged well with the theme of sustainability which emerged in the eighties and the protection of environment and natural systems were dependent upon changing the inequitable consumption pattern of the world.

The alternative paradigm extended political support to the redistributive pattern of policies which increases the entitlements of the poor in the South against the institutionalised power centers of the states. Therefore the Environment-Development debate is not one for protecting the environment from the cudgels of the North but to alter the rules of global economy.

The Bretton Woods Institutions

The Bretton Woods economic institutions mainly the World Bank (WB), International Monetary Fund (IMF), International Finance Corporation derive their name from the New Hampshire resort where they were negotiated. WB and IMF were created to maintain monetary stability and help nations to recover from an adverse balance of payment by recycling the global surplus to help deficit nations. Their track record shows just the opposite. In the eighties developing countries faced a severe balance of payments crisis for which they had to borrow from IMF. In later part of the decade just to repay their original debt of $9.5 billion to IMF they were paying back an amount as high as $6.3 billion a year. Between 1983 and 1987, net IMF transfers to developing countries turned from $7.6 billion to $7.9 billion. WB transfers were $1.7 billion of which $500 million were to current borrowers [Human Development Report 1992:53-54]. The WB lending arms IBRD and IDA promote these policies of the Bretton Woods institutions which forces the debtor nations to compromise ecological considerations for higher growth just to repay the loans taken from these institutions. Therefore these institutions had little or no consideration for environment thereby leading to the mass plundering of natural resources possessed by the southern states.

The General Agreement on Trade and Tariff (GATT) as the main regulating framework of international trade emerged in 1947 with the objective of pursuing 'reciprocal and mutually advantageous arrangements dedicated to the substantial reduction of tariffs and other trade barriers to trade and to the elimination of discriminatory treatment in international commerce' [Human Development Report 1992:76]. It appeared for the time being that the change would benefit the poor nations thereby making them more responsible towards their environment. But it was only when environment boiled down on international institutions with the beginning of the decade of seventies that a Group on Environmental Measures was formed within GATT in 1971. The membership of GATT increased to more than 100 and it accounted for more than four-fifths of world trade. Facts ascertain that the percentage of trade fully complying with the prescribed principles of GATT is less than 7 per cent but is trying to expand in the WTO.

G-7 as an elite group of seven most industrialised states Canada, France, Germany, Italy, Japan, UK, and USA have controlled the Bretton Woods institutions and therefore benefitted immensely from them. They provide an organised support to the, policies framed by these multilateral institutions and exercise control over the resources of the southern states. Their support has helped these financial Institutions to politicize the use of surplus also. In 1990-91 the surplus of seven rich countries was over $150 billion which should have been transferred to the developing countries of the South but most of it went to the rich nations and interestingly $100 billion went to just one rich country USA. If that was not enough the WB instead of sending the remaining surplus to the needy in the South it withdrew $500 million from them. The barriers to just and free trade erected by this system continued to harm the interests of the South even in the mid-nineties although the southern states were liberalising their terms to encourage more imports from the North.

GATT as the forum for reducing tariff and other barriers in world trade has only complemented the prevailing system in its 51 year of history. The WTO which has taken over GATT and is just five years old does not intend to change the rules of the game. It brought down labour and environmental standards heavily upon the states of the South so that the undemocratic and unjust system could flourish under the enlightened terms of WTO. 'In fact the threat to sovereignty inherent in instruments such as GATT, would itself be a significant factor to compel rethinking on GATT' [Kothari 1993:317]. Since WTO is erected over GATT the injustice to the Third World continues.

These policies in international trade have helped to maintain the North richer and the South poorer. This has also made the North extravagant about the environment of the South. It is easier to blame the South for its explosive population growth but it would be great disservice to scientific analysis if this is emphasised because it only helps to cover the injustices being done in the name of international trade.

The New Paradigm of the Alternative Growth Strategy

The paradigm shift started surfacing in the early seventies when the UN index for export prices of all primary commodities fell by more than 150 per cent. It was also the period when the concerted efforts of Organisation of Petroleum Exporting Countries (OPEC) led to severe price increases in other areas also. It was a 'Red Signal' [Meadows and Meadows 1972:10]. In April 1968 a group of citizens met in Rome and set up a study entitled 'A project on the predicament of mankind'. The project concerned with five basic factors which limit growth on earth. They were:

1. Population
2. Agricultural production
3. Natural resources
4. Industrial production
5. Pollution

The publication of *'The Limits to Growth'* by the *Club of Rome* warned the American enthusiasts of free economic growth that these economic symptoms were the early manifestations of a period of general scarcity of natural resources. The provocative conclusion of the Report said, 'We have seen the positive feedback loops operating without any constraints generate exponential growth. In a world system two positive feedback loops are dominant now, producing exponential growth of population and of industrial capital... negative feedback loops become stronger and stronger as growth approaches the ultimate limit, or the carrying capacity of the systems environment. . . finally the negative feedback loops balance or dominate the positive ones, and growth comes to an end. In the world system the negative feedback loops involves such processes as pollution of the environment, depletion of the non-renewable resources and famine. The delays inherent in the action of these negative loops tend to allow population and capital to overshoot their ultimately sustainable levels' [Meadows and Meadows 1972:156-157].

The report warned that there is firstly, a natural limit to natural resources, Second, that human population, agricultural production, natural resources and pollution are inter-linked globally, and third, that the economic growth which depends upon the exhaustible natural resources is insured with declining standards of living of people. The stress on natural resources and environment was reaching a point which threatened the Planet Earth. Writings like those of Schumacher [1976:25] had started consolidating the belief that, 'new problems are not the consequences of incidental failure but of technological successes'. Thus the anthropocentric approach to growth and development became the first target of attack in the new paradigm.

Serious negotiations on helping states to observe the basic minimum of environmental regulations began in 1972 with the United Nations Conference on Human Environment (UNCHE) in the Stockholm meet. This evoked a new interest in environmental accounting and sustainable use of natural resources. However, the industry reacted with a sudden impulsive outburst with Lewis Powell, one of the lawyers declared that private companies will have to fight this out in the courts by setting up their own law firms. Board members of the National Legal Centres such as Joe Coors funded key activists against the environmentalists. Andrew Rowell [1996:7-25] has described how the consolidated efforts on a war footing started bringing together all the right wing think-tank against the environmentalists. James Watt, the Secretary of State in the cabinet of Ronald Reagan advocated a policy of 'free will for all' for mining, ranching, drilling companies on public lands, denotification of protected areas, offshore continental shelf exploration and everything else, and was therefore described as the 'apostle of pillage' [Cockburn and Ridgeway 1981 January-February]. He tried to counter the new emerging paradigm with his 'dominion theology' taken from the Bible that 'God made man to subdue the Earth'. Soon Ron Arnold an unknown ex-environmentalist from the Sierra Club and probably a publicity freak was planted to declare the new paradigm of the environmental consciousness as 'anti-humanity, anti civilisation, anti-technology, pro-alarmism and terrorism' and gave open support to Watt in his book *Eye of the Storm* as a hero who would save America from these growth retarders or ecofreaks. Such rabid reaction to the new paradigm did not prevent the early demise of Watts from the cabinet as well as the maturing of the alternative paradigm of *'Sustainable Development'*.

The publication of the Brundtland Commission Report in 1987 named after the Norwegian woman prime minister Gro Harlem Brundtland presented reports of the world Commission on environment and development under the title' *Our Common Future* This report was finally able to define the new paradigm as not an antithesis of growth but as a more lasting and just framework for growth which 'is consistent with the future and the present needs'. In 1989 the Conference of the Heads of States of the Nonaligned Countries for the first time were aroused to protest the external impositions by the North in dealing with environmental issues.

In the Kuala Lumpur Conference of developing countries in April 1992 the participants asked for an additional funding over and above the Official Development Assistance (ODA) to move progressively towards the goal of sustainable development. In the major show of force for the paradigm shift was the United Nations Conference on Trade and Development in 1992 at Rio de Janeiro in Brazil. There were more than 172 participants and thousands of NGOs who focused on the inequitable terms of trade between the North and the South, on traditional rights of tribes and aborigines, on climate change and biodiversity. A separate fund for the implementation of Agenda-21 was also asked for, and these funds have to be necessarily extended by the rich nations of the North since it was laid that their pattern of consumption has been the main cause for the environmental deterioration in the world. Besides this South has also asked the North to democratize institutions for environmental decision making.

The Agenda-21 became the Bible of the new paradigm.

Environment-poverty

Four observations highlight the strength and importance of the interlinkages between the environment and poverty:

- *Observation 1.* The poor live in places which are ecologically more vulnerable and are forced to earn their living from low-productivity natural resources. The rural poor often live in low-lying, flood-prone areas, on steep mountain slopes or on dry land and possess low-productivity marginal land devoid of any irrigation facilities. The number of the rural poor in developing countries living on "marginal" land could be twice the number found on better-developed land. The urban poor are found in the shanty towns of big cities, which are often built on flood-prone, low-lying areas or around city drains; many of the poor earn their livelihood from environmentally hazardous scavenging. Environmental deterioration in the form of land degradation, frequent flooding, increased pollution and other hazards reduces the income of both the rural and urban poor and worsens their health disproportionately by comparison with the rich.
- *Observation 2.* It is commonly observed that poor households, especially in rural areas, derive their livelihood income form natural resources, for example, land resources for agriculture and water resources for fishing. It is also funded that the poorer the household, the greater is the share of its income from environmental resources. In addition to providing a livelihood, the environment lays a very significant part in influencing the health of the poor; while the incidence of disease in poor countries is about twice that of rich countries, the disease burden form environmental risks is 10 times greater in poor countries. Environmental degradation has a disproportionate negative impact on both the livelihood and the health of the poor.
- *Observation 3.* It is apparent that the intensity of suffering of the poor

from the advise impacts of environmental shocks is much higher than that of the rich. However, because of the lack of proper assets, the poor are less capable of coping with those impacts. The vulnerability of the poor to environmental shocks is much higher than that of the rich in both rural and urban areas.

Observation 4. Against the background of the observation that the poor, especially in rural areas, derive a large part of their livelihood income from environmental resources, especially land resources used for agriculture, some of the practices they follow can be damaging to the environment. Clearing forest areas to create land for agricultural use, including slash-and-burn practices, is an example showing that the poor are responsible for environmental degradation. Certain consumption practices of the poor such as damaging the forest to acquire firewood to be used for cooking and heating could also be detrimental to the environment. The urban poor, most of whom live in shanty towns and ghettos, often create unhygienic sanitary conditions because of their lack of access to formal toilet facilities. However, there is overwhelming evidence to show that the impact of poverty on the environment is weak compared with the damage to livelihood and health which the poor suffer owing to environmental degradation not caused by them. Against this background, the impact of poverty on the environment will not be taken up explicitly in the present discussion and analysis.

It can be seen from the above arguments that the environment has strong linkages with the livelihood, health and vulnerability of the poor. These linkages need to be identified in some detail before options for appropriate policy interventions to benefit the poor can be studied.

"Environmental goods and services" which are crucial to all, particularly the poor, can be classified into three broad categories: natural resources, environmental conditions, including environmental stresses, and the ecosystem. The nature and extent of their linkages with poverty, encompassing the livelihood income and health of the poor, are somewhat different. These are discussed below.

Natural Resources

Natural environmental resources can be atmospheric, land-based or sea-based. Of these, the resources which have a major bearing on the poor's livelihood and health are the following:

- Land used for agricultural operations, including grazing land for animal husbandry, provides an important (often the only) source of rural livelihood. Land degradation, either natural or due to the overuse of chemical fertilizers, and the mechanisation or depletion of groundwater, which increases soil salinity, could erode the most important modality of livelihood of the rural population, especially the rural poor, who do not possess the means to counter such adverse impacts.

- The widespread use of chemical fertilizer, pesticides and other chemicals for farming poses a formidable health hazard to the rural poor. Illiterate farm labourers who lack appropriate training in the use of poisonous chemicals are unable to read the instructions written on them and cannot afford protective devices can easily fall prey to a number of associated diseases.
- The poor, in both rural and urban areas, often do not have the luxury of access to safe drinking water. They have to rely on water sources which are frequently contaminated for various reasons. Water-related diseases such as diarrhoea and cholera kill a largo number of people every year in the developing countries. Any improvement in water quality is likely to yield rich dividends in terms of improving, the health standards and productivity of the poor.
- Fishing provides income and protein for the poor living near the sea, rivers, marshy lands and swamps. However, in many developing countries fishing sources are commercially over harvested in an unsustainable manner, which has a negative effect on the livelihood of the poor. Many commercial fishing ventures result in a number of adverse environmental impacts which could constitute health hazards for the poor who are associated with them. Coordinating policies and programmes at the regional and subregional levels aimed at the conservation and sustainable development of fisheries forms an important element of the Plan of Implementation of the World Summit on Sustainable Development, of which poverty reduction to achieve the relevant target is a major objective. Other areas highlighted in the Plan of Implementation that have implications for poverty are given in box 2.
- Forest products often provide livelihood income to the rural poor. Twigs and wood collected from forests provide a major part of the energy used by the poor for cooking and heating. Forests prevent soil erosion, flooding and mud slides in hilly areas during heavy rains. The unsustainable destruction of forests causes much misery to the poor, both directly and indirectly.

Environmental Conditions

Worsening of the quality of atmospheric resources could be extremely harmful to the poorer sections of the population. There are two major environmental conditions which affect both the livelihood and health of the poor:

Environment-poverty linkages and the Plan of Implementation of the World Summit on Sustainable Development

The Plan of Implementation of the World Summit on Sustainable Development specifics that the three components of sustainable development, economic development, social development and environmental protection, are interdependent and mutually reinforcing pillars. It also explicitly recognises that poverty eradication and protecting and managing the natural resource base are overarching objectives of sustainable development. Against the background of the environment-poverty linkages discussed in this section and the large extent of the exposure of the poor to the degradation of environmental resources, which affects their livelihood income and health, the following areas and actions highlighted in the Plan of Implementation assume importance.

Water

One of the major aims in this area is to launch a programme of action, with financial and technical assistance, to achieve the millennium development goal concerning safe drinking water, that is, to halve, by the year 2015, the proportion of people who are unable to reach or to afford safe drinking water.

Other action includes intensifying water pollution prevention to reduce health hazards and promoting ecosystems by introducing technologies for affordable sanitation and industrial and domestic waste-water treatment, by mitigating the effects of groundwater contamination and by establishing monitoring systems and effective legal frameworks at the national level. The need for prevention and protection measures to promote sustainable water use and to address water shortages is also emphasised.

Energy

Governments, as well as relevant regional and international organisations and other relevant stakeholders, are required to implement, taking into account national and regional specificities and circumstances, the recommendations and conclusions of the Commission on Sustainable Development concerning energy for sustainable development adopted at its ninth session. These include integrating energy considerations, including energy efficiency, affordability and accessibility, into socio-economic programmes, especially into the policies of the major energy-consuming sectors, and into the planning, operation and maintenance of long-lived energy-consuming infrastructures, such as the public sector, transport, industry, agriculture, urban land use, tourism and construction. The development and dissemination of alternative energy technologies with the aim of giving a greater share of the energy mix to renewable energies, improving energy efficiency and placing greater reliance on advanced energy technologies, including cleaner fossil fuel technologies, were also recommended.

Health

The objectives in this area include strengthening the capacity of health-care systems to deliver task health services to all, in an efficient, accessible and affordable manner, aimed at preventing, controlling and treating diseases, and to reduce environmental health threats, in conformity with human rights and fundamental freedoms and consistent with national laws and cultural and religious values. Another objective is the implementation, within the agreed time frames, of all commitments made in the Declaration of Commitment on HIV/AIDS adopted by the General Assembly at its twenty-sixth special session, held in June 2001, emphasising in particular the reduction by 2005 of HIV prevalence among young men and women aged 15 to 24 by 25 per cent in the most affected countries and by 25 per cent globally by 2010. Recognising that many communicable diseases spread because of the existence of unhygienic and crowded living conditions, the Plan of Implementation aims at the achievement by 2020 of a significant improvement in the lives of at least 100 million slum dwellers, as proposed in the "Cities without Slums" initiative.

Agriculture

Recognising that sustainable agriculture and rural development are essential to poverty reduction. the implementation of an integrated approach to increase food production and enhance food security and food safety in an environmentally sustainable manner assumes importance. The Plan of Implementation emphasises the need to achieve the Millennium Declaration target of halving by the year 2015 the proportion of the world's people who suffer from hunger. The relevant action includes the development and implementation of integrated land management and water-use plans based on the sustainable use of renewable resources and on integrated assessments of the socio-economic and environmental potentials. Strengthening the capacity of Governments, local authorities and communities to monitor and manage the quantity and quality of land and water resources and the adoption of policies and the implementation of laws that guarantee well-defined and enforceable land and water-use rights and promote legal security of tenure are important steps to be undertaken by Governments. Technical and financial assistance should be provided to developing countries as well as countries with economies in transition that are undertaking land tenure reform in order to enhance sustainable livelihoods.

Biodiversity

Biodiversity, which plays a critical role in overall sustainable development and poverty eradication, is currently being lost at an unprecedented rate owing to human activities. Several measures have been suggested to reverse this trend.

These include the promotion of concrete international support and partnership for the conservation and sustainable use of biodiversity, including in ecosystems, at World Heritage sites and for the protection of endangered species, in particular through the appropriate channelling of financial resources and technology to developing countries and countries with economies in transition, and the provision of financial and technical support to developing countries, including capacity-building, in order to enhance indigenous and community-based biodiversity conservation efforts.

- Indoor air pollution due to the use of biomass fuels (e.g., wood, crop residue) for cooking and heating in poor households affects the health of a large number of people, causing various respiratory diseases. The incidence of this type of health hazard is higher in women and children as they face primary exposure. Nearly 2 million women and children die every year from indoor pollution.
- As many of the poor live in ecologically vulnerable places (e.g., lowlands, mountain slopes, dry areas), atmospheric changes, both gradual (climate change) and sudden (disasters), can cause severe damage to the livelihood and health of the poor. Disasters, which include hurricanes, cyclones, floods and earthquakes, have been known to have a devastating impact on the poverty situation, giving rise a large number of "new poor" almost overnight. These concerns and the extent of such damage have been highlighted in a recent publication.

Ecosystem

Forests, grasslands and the coastal ecosystem, including coral reefs, provide a wide variety of services which contribute to the continuation of economic activities in both rural and urban areas. One important activity which exploits the existence of the natural ecosystem is ecotourism. Ecotourism is often labour-intensive and employs persons from the most vulnerable groups in rural areas, including those in remote and isolated areas and islands. Some examples of other ecosystem services include the provision of natural habitat for wild pollinators that are essential to food crops; watershed protection and the maintenance of hydrological regimes (recharging of water tables) by natural processes, including rainfall; and the natural breakdown of waste products and pollutants. It is apparent that the livelihood and health of a large number of the poor are intimately related to the activities facilitated by the ecosystem and its services, and any deterioration in their availability on quality could be detrimental to the reduction of poverty.

Conservationists have shown concerns on this aspect. For example, it was indicated that the tourism development plan under the Government's economic stimulus scheme for the archipelago of Koh Chang in Thailand could threaten its rich biological diversity and genetic resources. Most of the projects under the tourism master plan are likely to cause damage to the island's ecosystem. A study has found that the island is home to 1,513 species, including ferns, mosses, algae, lichens, fungi, flowering plants and various insects. It is alleged that efforts to develop the area for tourism without addressing environmental concerns adequately, could destroy these rare species.

It has been found in another study that the construction of tourism facilities in Koh Chang would definitely drive away rare insects, particularly fireflies, which are a huge source of income for the local people. Firefly watching is one of the island's most popular ecotourism activities. This perpetuates the dilemma of whether to develop a resort to generate income opportunities or to put-greater emphasis on the protection-of the natural ecology. Such trade-offs could occur in a range of other similar concerns.

Exploiting the Links Between Global Environmental Problems

Climate change and biodiversity are not only serious issues in their own right but are also linked with each other and with a wide range of other environmental concerns. Depending on the rate of climate change, forest species may be unable to adapt fast enough to avoid severe population declines. Aquatic ecosystems such as mangroves and coral reefs adapt even more slowly. The loss of species and genetic material can increase the vulnerability of ecosystems to other environmental stresses, such as pollution. To complete the circle, the destruction of forests has a dramatic impact on climate change because forests release significant quantities of carbon dioxide as they burn.

Exploiting such links can greatly reduce the cost of environmental protection. For example, whether land use activities should be eligible under the Clean Development Mechanism is an issue being decided by the parties to the Kyoto Protocol. But counting the preservation of developing country carbon sinks against emissions commitments could provide powerful synergies with local environmental and biodiversity protection needs.

Costa Rica has already begun trading Certified Tradable of Gets (CTOs)-carbon credits priced at $10 a ton—m ways that exploit such links. The profits from these credits are designed to support sustainable forestry practices on private land or to finance the conservation of land as national parks and bioreserves. So far, sustainable practices have been introduced on 3,000 farms covering 150,000 hectares. The bioreserve project has conserved another 530,000 hectares. Although there have been few early takers for the credits, Costa Rica's experiences with the system, combined with continued international research, could lead ID greater exploitation of this synergy between biodiversity preservation and the prevention of climate change.

These links across global environmental issues suggest that the international community needs to move beyond simply negotiating separate agreements for each environmental issue. As agreements such as CITES demonstrate, treaties are often agreed to only because complex problems are broken up into smaller units. But in some cases, agreements due cover many areas are easier to negotiate because of the potential for tradeoffs or synergies between related issues. This pattern has prevailed in multilateral trade negotiations, for example, when countries that feel strongly about certain provisions have offered concessions in areas that concern them less. Global environmental protection can also be hastened by improving coordination between treaty and convention secretariats, including integration of meetings, scientific assessments, reporting requirements, publicity, training, and capacity-building efforts and improved coordination under the UNEP.

Trade Measures in International Environmental Agreements

The earliest environmental treaty to use trade measures was the International Convention Respecting Measures to be Taken against the Phylloxera Vastatrix, which in 1881 banned trade in torn vines and dried shoots to prevent relocating the plant louse to other vineyards. Other environmental agreements with trade measures include:

- CITES, which allows trade in listed species or products made from them with nonparties to the agreement only when competent authorities in the nonparty country issue documentation comparable to that required of treaty members. The convention also allows members to impose trade bans on other members that do not comply with the restrictions. In 1991 the CITES standing committee recommended that all trade with Thailand in flora and fauna species covered by the convention be stopped because of noncompliance.
- The United Nations agreement on the Conservation and Management of Straddling Fish Stocks and Highly Migratory Fish Stocks, which allows members to prohibit the landing or transshipment of fish caught using methods that undermine the effectiveness of conservation and management measures.
- The Montreal Protocol, which requires parties to ban the import of controlled ozone-depleting substances from non-signatories unless the nonsignatories are found to be in full compliance with the protocol regime.

Trade measures can be an appropriate tool for addressing global environmental problems because they, like the problem, are global in nature. They can also be justified on the grounds that free trade is considered a global good because it maximises welfare. If trade is instead causing serious environmental damage, then, A can be argued, it must be limited. Trade may be especially intertwined with certain environmental risks: damage from relocation; insect infestation of previously unexposed crops from imported infected fruit; negative disposal effects like those posed by imports of toxic waste; negative transport effects such as oil spoils; and negative profit effects— as when trade ends up financing a decline in biodiversity.

Three recent international trade rulings suggest that international trade agreements allow little flexibility on the unilateral introduction of environmental trade bans: the WTO's ruling against U.S. laws banning the import of shrimp caught in nets that also trap turtles, and two rulings by the General Agreement on Tariffs and Trade (GATT) dispute panel on tuna caught in nets that also trap dolphins. But the WTO ruling does support the right of states to impose sanctions as part of an international treaty. Distinguishing between unilateral and multilateral trade bans keeps environmental sanctions from being used as a cover for protectionist interest.

Trade bans should also be limited to areas in which they can be effective. The international ivory trade (as well as mismanagement of elephant stocks) must be seen as an important reason for the drastic decline in elephant populations between 1979 and 1989. In situations in which trade in the dominant outlet for production, as it is for ivory, sanctions can have a major effect. Often, however, trade sanctions are too far from the source of the problem to be effective—which might limit the effectiveness of trade bans to counter greenhouse gas emission.

The world's countries have come far in cooperating to address global environmental issues, and the ozone accords provide a model for future agreements. Although drawing up international agreements on biodiversity and climate change that are as effective as the ozone agreements has been difficult, the basic mechanics of successful international environmental agreements are becoming clearer. Moreover, even taking preliminary steps toward a partial agreement encourages private actors to prepare for stricter agreements and thus lowers the cost of future actions to resolve environmental concerns. Consensus on biodiversity, climate change, and other global environmental issues will only expand over time. Furthermore, the growing understanding of linkages among environmental concerns will create more opportunities to exploit both synergies and trade-offs, helping to foster coalitions that support concerted global action.

Environmental Counselling for Employees' Households as Checklist for Environmentalist Business Management

Environmentalists have calculated that private households account for some 30 per cent of the total environmental pollution in the Federal Republic of Germany. Even if a similar figure may not apply to all the EC countries, it can safely be assumed that private households are to a substantial extent responsible for environmental problems—in the broadest sense, e.g. by producing refuse (approximately 250 kg per person per year), using water (140 to 150 litres per person per day), consuming energy, producing effluent pollution in the form of washing powders, cleansers and various other products (more than 2 million tonnes per annum get into the sewage), polluting the air with the solvents and propellent gases in paints and varnishes, thinners, adhesives and sprays, not to mention having wood/coal/oil heating and producing ever-increasing amounts of exhaust gases from growing numbers of motor vehicles, and, finally, polluting the soil with pesticides, fertilizers and thawing salt.

Faced with this situation, the individual must wonder what he or she can do to protect our environment, which is after all the very basis of our existence. It is very simple. Protection of the environment starts with each person in that (almost) everyone can:

- save many litres of water every day by being careful with it;
- refuse to use pesticides;
- make do with less packaging and without plastic bags;
- use returnable bottles instead of disposable bottles or tins;
- use far fewer chemicals in the household;
- be more environmentally aware when driving;
- purchase articles which help to protect, or at least do not harm, the environment and avoid those which are positively harmful.

If a few million of the people of Europe were to take heed of these words, there would very soon be a considerable improvement in the environment and thus in the quality of life of each and every person. It is vital for the population to be informed and enlightened. At present, there is an all too frequent lack of understanding of what is going on and of advice which can easily be put into practice. It is here that every company can provide vital assistance, by giving its staff the opportunity to benefit from environmental counselling in their own homes.

Provide Opportunities for Environmental Counselling in Employees' Households

1. Appoint environmental counsellors to advise staff, or request them from another body, providing financial support where appropriate.
2. Consider discussing the matter with another company and employing a joint counsellor for the two firms.

3. Contact environmental organisations and urge the appointment of environmental counsellors.
4. Make sure that the counsellor has first-class qualifications and is politically neutral and objective.
5. Introduce the counsellor officially to staff and explain in detail the meaning and purpose of environmental counselling.
6. Collect information on financial assistance from the State. In the FRG, for example, financial support is available through official "job creation programmes".
7. Organise meetings and discussions with environmental counsellors and workers, on such topics as:
   - the use of chemicals in the household;
   - how to avoid creating waste;
   - how to save water and energy;
   - environmentally benign building and do-it-yourself principles.
8. Draw up rules of thumb for objective advice on the environment, taking care not to violate the competition rules.
   - Make advice unbiased and avoid negative criticism.
   - Recommend only those products or services which have been tested by neutral, reliable experts and proved to be environmentally safe (e.g. those with the "Blue Angel" symbol authorised by the Federal Environmental Agency).
   - Make no claims which are untrue or whose veracity cannot be proven.
   - Praise ecologically beneficial products or services rather than criticise those which harm the environment (see PC. Mohr, Vier Faustregeln fur den Umweltberater, Hamburg 1985).
9. Ensure that the information made available to the counsellor as part of his work is handled confidentially.
10. Bring pressure to bear on the city or local authorities to appoint environmental counsellors.
11. Make staff more aware of environmental problems by providing information geared to their specific needs. As far back as 1973, an opinion survey conducted under the auspices of the Commission of the European Communities showed an extraordinary degree of interest in the problem of the environment. In subsequent years, protection of the environment and the fight against pollution remained one of the three most important problems for Europeans—

after unemployment and poverty (see: The Europeans and their Environment, published by the Commission of the European Communities, Brussels 1982, p. 2).

Give Eorkers Tips/Advice on Practical Environmental Protection

12. Save energy
    - Maintain the right room temperature. 1° C less saves 6 per cent energy.
    - Do not use tumble dryers, or, if you must, use machines with low power consumption.
    - Install thermostat radiator valves (up to 15 per cent energy can be saved).
    - Seal/insulate windows.
    - Keep radiators free.
    - Install boilers with low power consumption.
13. Reduce water consumption.
    - Install flow restrictors on taps.
    - Install water-saving devices in toilet cisterns.
    - Install single lever mixer taps.
    - Repair leaking taps and toilet cisterns.
    - Shower instead of having a bath.

    These measures can easily yield a saving of 10,000 litres of water per person per year in the home.
14. Be more pollution-conscious when washing clothes.
    - Do not use the pre-wash.
    - Wash at 60° C instead of boiling.
    - Use one-third less washing powder per cycle.
    - Wash only full loads.
    - Do not use softener or detergent-impregnated clothes.
    - Use washing powder for delicate fabrics rather than heavy duty detergents.
    - Buy washing machines with lower water and power consumption.
    - Treat stains separately.
15. Use cleaning products, etc. more sparingly.

- Use green soap or a mild all-purpose cleanser rather than WC or bathroom cleansers.
- Treat scale deposits with acetic acid.
- Do not use deodorant blocks in toilet pans or cisterns.
- Unblock pipes and U-bends with plungers or specially designed rods rather than with chemical agents.
- Wash non-greasy dishes in hot water, without washing-up liquid.
- Use more green or soft soap as an all-purpose cleanser.
- Avoid cleansers with bleach or disinfectant.
- Clean window panes with water and a dash of vinegar or alcohol.
- Clean floors which are not specially dirty with clear water only; if they are more heavily soiled, use a little green soap or mild all-purpose cleanser.
- Protect untreated furniture with beeswax, linseed oil or shellac.
- Do not use disinfectants.

16. Avoid waste.
    - Do not buy products with unnecessary packaging.
    - Wherever possible, use reusable packaging, e.g. returnable bottles.
    - Avoid plastic packaging.
    - Do not put chemicals into the dustbin.
    - Reuse or collect paper.
    - Collect glass for recycling.
    - Put kitchen waste on a compost heap.
    - Take used oil to special collecting points.
    - Make sure that collecting points are set up for paper, glass, special category refuse and used oil.
17. Use biological paints, dyes and wood preservatives.
    - Use natural resin paints.
    - In the home, use natural resin latex paints or synthetic resin latex paints which are low in pollutants.
    - Inside, avoid using chemical wood preservatives as far as possible.
18. Do not use poisons in the garden.

- Do not use chemical pesticides.
- Treat pesticide residues as special waste and do not put them in the dustbin.
- Use natural manure whenever possible, e.g. compost, green manure or vegetable fertilizer.
- Compost organic waste.
- Do not use peat—bark humus and compost are alternatives.

All these and many more practical tips are taken from the "Oko-Sparbuch fur Haushalt und Familie", Munich 1986. Written by M. Gege, H. Jung, H.J. Pick and G. Winter, it shows how members of the household can, by being more aware of their environment,

- save a great deal of money;
- protect their health, and
- make their own contribution to improving the environment.

An Effective Corporate Policy Regarding Environmental Management

In practice, how does management differ in the most environmentally serious companies from the rest? The answer is complicated by the fact that there is no single set of rules that tells managers how to be truly environmental. (This may be the case because it has taken business schools, at least outside Germany, so long to grasp the importance of environmental management as a subject.) Certainly the number of packed conferences on the subject suggests that managers are eager to learn.

The First and essential step is a clear statement of corporate principles and objectives spelled out with the full backing of the board. These then need to be broken down into detailed rules to cover all activities. Compliance with the rules must be regularly monitored and the results presented to a senior executive who has responsibility for environmental performance. The gathering and dissemination of information are central. A key element in this strategy is the environmental audit and its integration into corporate policy.

But would-be-green companies have other characteristics that set them apart from the dirtier kind. One interesting example is Johnson & Johnson, which demonstrates the close links between a company's environmental policy and its attitude to its employees. Johnson & Johnson's policy grew out of its "credo," a statement of corporate goals laid down in the 1940s by General Johnson, a remarkable man who founded the company in the 1930s. He believed that "factories can be beautiful" and that "we ought to pay our employees what they are worth." His credo said that the company's first responsibility was to its customers, its second to its employees. The community came third, and shareholders last of all.

This fine vision became clouded over the years. In 1978 at a series of meetings, its managers around the world said, "The credo is all very well, but all the chairman want to know about is the profit for the past quarter." In 1982, the company had a traumatic experience when some of its Tylenol was found to have been tampered with and contaminated.

Part of the company's recovery was driven by the development of a new environmental programme based on the "credo." It starts with the proposition that, in the words of Jack Mullen, vice President in charge of corporate affairs, "If you are in the healthcare business, you can't play games with your employees' health and safety." A programme has been set up that involves making sure that all part of the company are obeying government and corporate environmental rules. It is also trying to cut packaging and make as much use of recycled and recyclable materials as possible. New products are studies for their impact on the environment from every point of view, including their use of raw materials, manufacturing process, and packaging.

Johnson & Johnson's experience illustrates a number of features common to companies anxious to improve environmental performance. It has built its environmental programme on the foundations of a corporate disaster. Near-catastrophe influences boards to accept large budgets for environmental protection programmes. It shows, too, how environmentally conscious companies attend to employees' welfare, since these are the people most likely to be hurt by a sloppy environmental policy unless everyone sees that the board backs it up to the hilt.

Environmental Policies

Some companies have achieved a great deal by devising and implementing environmental policies. The pioneers deserve credit for the work they have done to reduce their environmental impact inways that make business sense. The petrochemical industry led this initiative because of their considerable impact on the environment and the unfavourable public image of the industry. Their actions have resulted in a heartening reduction in environmental damage. However, the main reasons for taking action have been to comply with the law, improve safety and health, save money and enhance their reputation. Increased efficiency has been the main achievement but the easy pickings are no longer there for many large companies. The next stage will require a major change in approach and more far reaching improvements. This alone will bring about sustainability. It means integrating environment policy with business strategy in a systematic way.

Some topics which have been included in environmental policy statements of major organisations for some years are listed in Table 3. Few organisations cover all these topics but the list is based on published statements by large organisations. Companies which are still confining their activities to this list are likely to be reacting to the environmental challenges. They may even be defensive in their approach.

TABLE 3

Topics included in Environmental policies

- Commitment to environmental goals.
- Compliance with and/or anticipation of the law.
- Periodic reviews and audits of operations.
- Publishing details of environmental impacts.
- Taking environmental factors into account in decision making.
- Applying the precautionary principle.
- Managing energy, resources and wastes more efficiently.
- Re-using and recycling materials.
- Purchasing policies adjusted for environmental factors.
- Raising standards for health and safety of employees and the community
- Participating in community initiatives.
- Contributing to charitable causes.
- Training employees in environmental matters.
- Requiring suppliers to meet stated environmental standards.
- Reporting progress on achieving environmental goals.
- Using an environmental management system.

Organisations doing most of these things can have confidence that environmental management is an integral part of their philosophy, their strategic thinking and their operating practices. They will be demonstrating how ethical and environmental responsibility contributes to solutions which take us towards sustainable societies.

TABLE 4

Evidence of Strategic Environmental Responsibility

- A Board statement explaining how the strategic thrust of the business contributes to the creation of sustainable societies
- A Board statement showing how the organisation fits into a sustainable society and how progress towards explicit goals will be measured.
- Acknowledgement that the health and safety of employees, customers and the community are the organisation's responsibility and take procedure over profits.
- Social, ecological and economic considerations have equal weight when devising product

strategy, operating processes and buying in services.

- Purchasing policy avoids scarce resources, endangered species, habitat destruction, oppressive regimes and cheating suppliers and growers.
- Operating practices are geared to cut emissions, which are continually monitored, make efficient use of all resources and introduce more severe targets in each successive year. Toxic substances are given special care.
- Waste reduction is an integral part of the way the business is run and employees contribute to implementing 'more or less' and 'reduce, re-use and recycle.'
- Continuous improvement is an accepted goal across all functions.
- Packaging and transportation targets are set to reduce their adverse impact.
- Accounting practices increasingly include the external costs which are typically omitted or ignored.
- All employees receive appropriate environmental training and environmental performance is included in the criteria used for selecting, rewarding and promoting people at all levels.
- Social, environmental and financial auditing are given equal weight and cross impacts are made explicit.

This has looked at the difference between having an environmental policy which is managed alongside business strategy and one that is integrated with it. Both can improve business performance, environmental health and quality of life. However, when social responsibility and environmental health are an integral part of business strategy a more healthy, responsible business emerges. Employees at all levels appreciate and opportunity to contribute to solutions which stop the damage and to offer ideas for better business. They are directly involved in environmental management as part of their work and demonstrate a collective commitment to sustainability. In this way organisations are capable of achieving the dramatic shift that sustainable development requires.

Businesses which adopt an ethical approach and enlightened self-interest can still innovate to bring about significant changes which the community, the environment and the business all benefit. This way not be the only way to achieve sustainability but several organisations are demonstrating that is works and it could be more widely applied.

Designing Green Markets

One of the most promising ways to make decisions about political economy more in keeping with ecological thinking is to incorporate ecological services on the costs of their loss into the design of markets. Although it is customary to speak of market forces in terms of spontaneously arising natural laws and "invisible hands," the fact is that all markets are the design products of governments, business entrepreneurs, and financiers. There are no truly free markets; only designed markets. Unfortunately, most market designs to date have encouraged the systematic destruction of nature's ecological services in return for fleeting forms of profit and capital accumulation. In the future, however, it will be much harder for market designers to hide or defer ecological costs merely by discounting them into oblivion or promising artificial replacements as compensation. Too many ecosystems have been irreparably damages, and too much of the world's remaining capacity for wealth creation is dependent on a healthy biosphere to risk a continuation of past practices.

Green Accounting

If sustainable development is to become more than just a fashionable term, ecological and economic "fund" balances will have to be tightly integrated in the accounting practices of individuals, firms and nations. A major push in this direction has been provided by the new field of "natural income" accounting. Protecting ecological wealth in practical terms requires the preservation of natural capital, as distinct from man-made capital. Natural capital includes a variety of ecological services and vital resources, such as forests, topsoil, breathable air, and drinkable water. Because no distinction was made in the past between natural and man-made capital, much of what counted as growth in GNP and other forms of national income accounting actually represented the destruction of natural example, increased GNP for the moment, but a staggering loss in ecological wealth. Likewise, mining topsoil as a result of intensive agricultural practices led to vast improvements in short-term economic performance, while impovishing a nation's future ability to grown food. The result has been a economic accounting system that highlights short-term benefits and conceals deferred costs. For instance, the severe contamination of hundreds of major groundwater basing represents an enormous deferred cost of modern agriculture and manufacturing. Because natural capital, unlike man-made capital, is seldom reproductive or recoverable by human technology, its loss is irreplaceable and represents a potentially massive constraint on future human development.

Today, as much as one-third of GNP in some countries represents the squandering of natural capital (Heauting, 1990). Studies of Costa Rica (Repetto et al., 1991) and Indonesia (Repettor et al., 1989) suggest that national income and economic growth indicators will have to be substantially revised in order to reflect the natural resource depletion associated with timber, energy, agricultural, and fisheries development. Costa Rica, for example, which is widely regarded as relatively progressive in its environmental thinking, loses 5 per cent of its GNP each year due to overexploitation of natural resources (Repetto et al. 1991:1). The opportunity costs of unsustainable development are increasingly apparent in the degradation of streams and forest lands, which in turn threaten Costa Rica's lucrative ecotourism industry. In many other regions of the world (e.g. East Africa, the Amazon Basin, South Asia) the sustainable development opportunities foregone because of excessive resource exploitation appear to be much greater. Ultimately, most of the world risks severe ecological impoverishment unless the indicators that nations employ to measure economic health are revised to account for natural capital depletion and then acted on. As economist David Pearce observes, sustainability requires the natural capital resource stocks be held constant over time (Pearce and Turner, 1990: 23-58). The implication for future generations is clear; Irreversible processes and irreplaceable uses of natural capital are taking place in ways that entail an unsustainable future and a loss of freedom for those who will inherit the planet.

Market-based Incentives

Modern environmental protection programmes have relied heavily in some countries on government regulations that specify both the standards to be met and the acceptable technologies or other means to meet them. This is especially true of many U.S. pollution-abatement programmes. With the rise of so-called "free-market environmentalism" (Anderson and Leal, 1991), a variety of market -based, "flex-tech" approaches to environmental protection have begun to be employed on a pilot basis, often with favourable results (see, for example, Tietenberg, 1985 and 1991). Using the conventional dichotomy of command-and-control regulation versus market-based incentives, economists and other analysts have argued strongly that "carrots" (incentives are more effective than "sticks" (regulatory enforcement) for protecting the environment. But governments frequently prefer sticks to carrots, and even when carrots are selected, they often end up looking more like orange-colored sticks. According to most economists, environmental control measures that demand extensive governmental coordination and enforcement are likely to fail, while those that rely on the invisible hand of the market have a good chance of succeeding. A preponderance of economic evidence appears to support this conclusions, but it is by no means clear that market based solutions alone can resolve major environmental problems in a timely fashion. Some combination of regulation, taxes, and market mechanisms may indeed be optimal.

The controversy over emission trading as a pollution-abatement strategy illustrates the kind of arguments that arise over the use of green market instruments. Emissions trading operates on the principle that wide differences in pollution-abatement is very expensive to pay those for whom it is relatively cheap to make additional abatement efforts above and beyond what buy credits from firms that can reduce emissions cheaply until the net reduction in total pollution is equal or greater than what would be required of each firm acting alone. The conventional economic wisdom suggests that markets for emissions trading will usually work better than across-the-border regulation in achieving pollution reductions at the lowest cost. A study of seventeen emissions control programmes, in which compliance costs were compared with simulated, market-induced abatement costs, found that emissions trading alternatives in almost all cases produced large savings over command-and-control measures (Tietenberg, 1985). According to some economists, emissions trading could reduce the costs of pollution abatement by at least one-third, without adding any net emissions to the environment (Blinder, 1988).

Mandatory standards, on the other hand, have the advantage of concentrating the attention of polluters on quantifiable objectives an timetables. If consistently and evenly applied, they make it easier to forecast and to plan. Moreover, they are often faster than market approaches in getting results. Emissions trading, by contrast, subjects participants to the changing risks and opportunities of a dynamic market. Some power plant operators, for example, may prefer known regulatory constraints to uncertain opportunities for emissions trading. Another advantage of mandatory standards, theoretically, is that they define the duties and responsibilities of citizens in relatively unambiguous and virtuous ways, while emissions trading is often perceived as a "license to pollute" and motivated by greed.

While there are serious limitations on the use of market-based tools in some environmental problem areas (trade in endangered species, for example), there are many other applications in which such tools appear to offer a promising means of improving environmental conditions at relatively low cost.

Company Vehicles as a Checklist for Environmentalist Business Management

The vehicles operated by a company are there to provide safe and reliable means of transport for the various departments. Cars and lorries play a major role in air pollution and are the cause of what may often be an unbearable noise nuisance for millions of people.

More pollutants enter the air from vehicle exhaust pipes than from all smokestacks in the Federal Republic of Germany put together. These include 6.2 million tonnes of carbon monoxide, 1.3 million tonnes of nitrogen oxides and 650000 tonnes of hydrocarbons, including 50000 tonnes of carcinogenic benzole. The effects on the health of people and the environment are well known: lung diseases, damage to the respiratory organs, blood count disorders, circulatory disorders and forest damage, to name only the most important.

This means that every litre of fuel not consumed by a vehicle and every reduction in the exhaust fumes released by combustion and in the noise caused by vehicles will be to the benefit of man and his environment. The following measures will be of assistance.

Buy Low-Pollution Vehicles

1. Buy vehicles which are equipped with an efficient three-way catalytic

converter. Numerous tests have shown that cars with catalytic converters generate some 90 per cent less noxious substances like carbon monoxide, hydrocarbons and nitrogen oxides. So long as soot filters are not available, diesel vehicles cannot be recommended as an alternative to cars equipped with catalytic converters. While it is true that diesel vehicles consume less fuel, they are also responsible for emitting carcinogenic hydrocarbon compounds. The extent of the cancer hazard is still a matter for discussion, but preventive environmental protection should err on the safe side. The fact remains, though, that a diesel car is still preferable to a petrol car without a catalytic converter.

2. Make exhaust and noise levels one of the company's purchasing criteria.
   - Compare exhaust levels, as there are considerable differences.
   - Obtain information from the Federal Environmental Agency.
3. Do not specify metallic paintwork.

   Aromatic hydrocarbons make metallic paints a health hazard during spraying; pollution is caused by solvent vapours.
4. For long journeys try to use 5-gear vehicles.
5. Fit vehicles with low-noise and long-lifetime exhaust systems. Systems are available which keep noise below specific limits and are guaranteed service-free for two years or 75 000 km.
6. Use approved-quality remould tyres with the "Blue Angel" symbol

   In the Federal Republic of Germany more than 300 000 tonnes of old tyres pile up every year, and burning them pollutes the environment. Remould tyres use 80 per cent of the old tyre, with a significant saving in raw materials.

Re-equip Existing Vehicles and Service them with Environmental Considerations in Mind

7. Fit a catalytic converter or an exhaust recirculation system.
8. Fit petrol vehicles with transistorised ignition systems

   Conventional contact-breakers wear quickly, reducing engine performance and increasing petrol consumption. With a transistorised ignition system, you can be sure that the optimum values as originally set will be maintained for 40 000 km. Conversions sets are also available.
9. Use brake and clutch linings which do not contain asbestos.
10. Fit a device which cuts off fuel supply when the driver takes his foot from the accelerator pedal, saving up to 20 per cent fuel consumption.

11. Fit a device to switch off the engine whilst waiting at traffic lights, etc.
12. Have vehicles serviced regularly.
13. Check tyre pressure at regular intervals.

    A 0.5 bar difference from the prescribed figure can increase fuel consumption by 5 per cent.
14. Change petrol vehicle sparking plugs regularly and if possible use broad-band plugs.

    Sparking plugs play a major role in achieving optimum combustion of the fuel-air mixture.
15. Regular replacement of the air filter reduces fuel consumption by up to 10 per cent.
16. Adjust carburettor properly.

    A badly adjusted carburettor can increase fuel consumption by up to 20 per cent, with a corresponding increase in exhaust fumes.
17. Wash engine only at garages or filling stations.
18. Dispose of used oil properly.
19. Use unleaded petrol.

    Manufacturers, automobile clubs and filling stations can provide information on whether unleaded petrol is suitable for a particular vehicle.

Adopt an environment-minded driving style

20. Do not drive aggressively or at high revs.

    Driving with a high engine speed consumes 25 per cent more fuel than a restrained driving style and produces twice as much carbon monoxide. Engine noise is also greater.
21. Drive smoothly and with foresight.

    Tests have shown a driver in a hurry to consume 40 per cent more fuel (and money) over a 28 km stretch of urban road with 32 sets of traffic lights—saving a mere 7 minutes.
22. Drive at a constant speed, even on motorways.

    On a test journey, a fast driver in a hurry needed 175 litres of fuel for 1550 km, whilst a slower driver consumed only 125 litres, a saving of 50 litres of fuel and 12.5 kg carbon monoxide emission. The difference in time for the journey was 1½ hours.
23. Change gear as soon as possible and use the most economical gear.

Use first gear for setting off only and change rapidly into second, third and fourth gear.

24. Switch off the engine at traffic lights, level crossings, etc. where waiting time is longer than 30 seconds.

25. Drive with foresight to make optimum use of synchronised traffic lights.

26. Do not warm up a cold engine by letting it tick over; set off immediately, as the engine warms up more quickly when the vehicle is being driven and the automatic choke is switched off earlier.

27. Do not fill the petrol tank up to the brim.

28. Use the choke carefully.

    When the outside temperature is 20° C do not use the choke. At 0° C push the choke half way in after 30 seconds and completely in after 2-3 minutes. When the choke is out twice as many hydrocarbons and nine times as much carbon dioxide is produced. (See E.R. Koch, Umweltschutz zu Hause, Munich 1984, p. 233).

29. Switch off fans in traffic jams.

    When the fan is switched on up to 25 per cent more carbon monoxide and up to 36 per cent more hydrocarbons get into vehicles.

Plan Journeys with Environmental Considerations in Mind

30. Avoid short journeys.

    Short journeys consume a very large amount of fuel. When the engine is cold fuel consumption straight after setting off is around 40 litres per 100 km, after 1 km it is still around 20 litres per 100 km, and normal consumption is reached only after 4 km.

31. Use public transport for shopping expeditions into cities; use "park and ride" facilities.

    The German motoring association (ADAC) has calculated that on Saturdays when shops are open 75 per cent of vehicles driving through city centres are looking for a parking space.

32. Offer training in environmentally aware and economical driving, particularly in companies running lorries or a large number of cars and vans.

33. Organise a competition to find the most economical drivers.

34. Remove roofracks when not required.

    The additional fuel consumption of a family saloon travelling at a constant 130 km/h is over 12 per cent with an empty roof-rack, up to 20 per cent with a loaded roofrack and covering sheet and up to 25 per cent with a loaded roofrack without a covering sheet.

35. Encourage car-pooling.

If 4 people share a car for a daily journey up to 70 per cent of travelling costs can be saved, in addition to the substantial reduction in pollution.

Origins of the Major Codes for Environmental Management

The emergence of private codes of environmental management practice is a recent phenomenon. Responsible Care, CERES, and the ICC charter were developed in the late 1980s and early 1990s. ISO 14000 is under development and will likely be finalised later in 1996. These codes have grown out of public concern about corporate environmental performance and corporate fears about a new round of regulation. Large multinational corporations came together over initiative that they felt would improve their performance and demonstrate their commitment to the environment.

The private codes discussed here, while often viewed as unrelated, really represent a progression, with each building upon the foundation laid by its predecessor. Thus, Responsible Care laid the groundwork for each of the subsequent codes with its emphasis upon management practices and continuous improvement. CERES, the only code developed by nonindustry groups, added to this foundation a strong focus upon environmental protection and restoration and a requirement for public disclosure of environmental performance. The ICC charter was developed in large part through the work of the Global Environmental Management Initiative (GEMI). GEMI was founded by large firms from a variety of industries who wanted a forum to stimulate critical thinking about the environment and to generate new strategies. ISO 14000 reflects the globalisation of commerce, the need to develop a "baseline" of minimal environmental performance, and recognition of the importance of third-party verification of environmental practices. Even though the codes were developed during the same period and in response to a similar set of societal concerns, each of them also grew out of unique sets of circumstances.

Responsible Care

The Responsible Care initiative was originally recommended to CMA by its Public Perception Committee, whose mandate was to recommend some industry initiatives that could "improve the legislative, regulatory, market and public interest climate for the industry." The objectives of the initiative are to

promote continuous improvement in member company environmental, health, and safety performance in response to public concern, and to assist members' in performance to critical public audiences.

The programme was created in the wake of the Union Carbide accidents at Bhopal, India, in 1984 and Institute, West Verginia, in 1985, when public distrupt of the chemical industry was strong. Many executives in the chemical industry feared that if they did not respond to public concerns, they would some day be regulated "like the nuclear industry."

In mid-1987, CMA Chairman Bob Clark surveyed CMA Chairman Bob Clark surveyed CMA members and found that "everyone's number-one or number-two problem (was) the negative public perception of the industry." Further, industry polls showed that the public generally did not distinguish among different chemical manufacturers, but rather saw the whole industry in the same light. For this reason, the major companies in the industry felt vulnerable to public perceptions that were based on the poor performances of other (often smaller) companies. They decided that a single plant or company could do only so much to allay public concerns by its own exemplary behaviour, and so CMA devised a mandatory, industrywide programme.

CMA's approximately 175 members companies account for more than 90 per cent of basic chemical production in the United States and Canada. Two companies that have chosen not to participate are Formosa Plastics, which recently completed the newest large petrochemical complex in the united States, and Agrico, which was the second-largest Toxic Release Inventory emitter in 1992.

In creating Responsible Care, U.S. chemical executives relied heavily on the Canadian Responsible Care programme instituted several years earlier. (U.S. chemical manufacturers with Canadian operations first brought Responsible Care to CMA's attention.)

The U.S. programme has several components: a set of "guiding principles," six "codes," a public advisory panel, and executive leadership groups. The codes address community awareness and emergency response, chemical distribution, pollution prevention, process safety, employee health and safety, and product stewardship.

CMA expects member companies to sign the guiding principles, communicate a commitment to Responsible Care to their employees, and make "good faith efforts" to implement the codes. CMA may revoke the membership of a company that has consistently conducted its operations in a manner not in accordance with Responsible Care. To date, no company has had its CMA membership revoked, however.

Responsible Care-type initiatives exist in approximately 30 countries in addition to the United States and Canada. In the United Kingdom, the national associations of chemical distributors have adopted programmes similar to Responsible Care. In France, 360 companies representing 90 per cent of basic chemical sales and 70 per cent of speciality chemical sales have signed up for Responsible Care. The German chemical industry association held its first Responsible Care workshop in November 1993. And although Danish industry first rejected the programme, it is now adopting it.

The CERES Principles

The driving force behind the Coalition for Environmentally Responsible Economics is Joan Bovaria, President of Franklin Research and Development Corporation, a firm dedicated to socially responsible investment. Bavaria and Denis Hayes, an environmental advocate, began to organise CERES in the late 1980s. Bavaria's primary goal was to institutionalise the capability to generate data on corporate environmental management that investors could use in decision-making. She sought "consistent, comparable, and widely disseminated" data that would allow investors to analyse environmental performance in the same way they analyse corporate financial performance:

> *In 1988, when CERES was put together, there was not enough information to make intelligent investment decisions. What we were hearing from environmental advocates like Greenpeace and what we were hearing from companies was often diametrically opposed. There was no way to know about environmental performance without investigating these companies ourselves.*

From the beginning, however, Bavaria's goal was larger than providing information for intelligent investment. She ultimately sought to change the relationship between firms and the public. According to Bavaria, she and the coalition felt that with environmental data publicly available, "trust barriers would be lowered." "The reporting process is a trust-building process," she said.

Bavaria and Hayes began to organise a coalition of investors, environmental advocacy groups, and labour unions. The first order of business for the new group was to develop a common set of environmental principles. This proved to a challenging task, because each environmental group represented different constituencies.

> *For example, the World Wildlife Federation membership overlaps about 75 per cent with the National Rifle Association. This kept them from agreeing on certain restrictions on resource use that other groups such as Sierra were pushing.*

The group named their principles the "Valdez principles" after the *Exxon Valdez* oil spill. The tenth and final principle, considered most important, stated that companies must annually complete and make public a "CERES report" containing detailed information on corporate environmental practices.

In connection with Earth Day 1990, CERES' coordinated a media blitz to encourage a number of Fortune 500 corporations to become founding signatories of the principles. Despite the prominent media coverage, however, fewer than 20 corporations, and none from tthe Fortune 500, signed the principles. Only small companies and those with strong environmental reputations were willing to participate.

The second part of CERES' strategy was less publicized but arguably more effective than the first. U.S. securities law allows any stockholder who owns 1,000 or more shares of a company for a year to introduce a referendum at that company's annual meeting. Beginning in 1990, CERES' investment groups began placing resolutions before the stockholders of the nation's largest companies, with a direct appeal that they adopt the principles. This strategy was not conceived by CERES itself, but by one of its members, the Interfaith Center on Corporate Responsibility. On average, less than 8 per cent of these companies' shareholders supported the Valdez principles. Most resolutions were withdrawn before a vote was reached. Even so, these resolutions offered an opportunity for CERES to air environmental concerns. Many companies promised to review an report on existing corporate environmental programmes and, if appropriate, to implement new ones.

In an effort to attract more members, the coalition became "less adamant and finger-wagging with corporations." Specifically, the principles were amended to eliminate the requirement that each member company have a qualified director responsible for environmental interests. Instead, companies were required to consider demonstrated environmental commitment as a factor in choosing directors. The Valdez principles were also renamed the CERES principles. While CERES had been asking companies to "sign" its principles, it now asked companies to "endorse" them. (Some companies' legal departments had problems with the word "sign" but not with "endorse.") Finally, CERES began to allow companies to make changes to the principles as long as the spirit of those principles was not violated. As Judy Kuszewski, the corporate programmes director fro CERES, explains: "No one in CERES would say the language of the principles is not important. But the language shouldn't be an obstruction to companies (who are) otherwise interested."

These changes have helped secure the participation of several major corporations. Two notable CERES endorsed—Sun Company and General Motors—modified the principles significantly to make them consistent with their own environmental perspectives. Although Sun endorsed the principles as a "generic environmental code of conduct applicable to business behaviour throughout the world, the corporation did not accept them as an explicit code for its own behaviour. Sun's primary concern was that the principles, if taken literally, could eventually force the company out of the fossil fuel business.

Three other large firms have recently joined CERES: H.B. fuller, Polarroid Corporation and Arizona public Service Company. CERES staff members report that they are currently overwhelmed with requests form companies interested in endorsing the principles.

CERES expects endorsing companies to continuously improve their environmental performance, maintain an open relationship with it and other outside stakesholders, and be publicly accountable through publication of an annual CERES report. Endorsers pay annual membership dues of $100 to $15,000, depending upon their volume of sales, and are invited to participate in committees that address issues such as the content of the annual report and the recruitment of few members.

CERES is in the process of developing a protocol by which it can revoke the "endorser" status of companies that seriously violate the letter of spirit of its principles. While CERES hopes never to have to revoke an endorser's role in the organisation, it felt the need to establish the protocol to protect the integrity of the organisation.

GEMI and the ICC Charter

The Global Environmental Management Initiative (GEMI) was formally announced in April 1990, though it had its start months earlier when a group of corporate environmental management issues. Approximately ten people from large firms in the chemical, electronics, consumer products, and pharmaceutical industries were involved initially. Participants recognised that compliance alone was insufficient to convince investors, employees, and other stakeholders that their firms were secure form environmental liability. They wanted a forum to share strategies and "war stories," to stimulate critical thinking, and to strengthen dialogue between themselves and interested publics.

From the start there was interest in making the organisation's membership diverse because more ideas would be generated through a "cross-fertilisation" of thinking. Dorothy Bowers, vice president for environmental an safety policy at Merck & Company, recalls sitting down with a list of industries and picking out a few sectors to target. GEMI then grew slowly through a process of individual contacts; today, it has some 27 members representing large companies from a diverse group of industries.

GEMI members worked closely with the International Chamber of Commerce to draft the Business Charter for Sustainable Development, which contains 16 principles tailored to large multinational corporations. GEMI's participation ensured that the charter received support from U.S. industry. One of GEMI's initial goals was to bring the charter to life by developing a database to track implementation efforts and an environmental and self-assessment programmes to guide companies in this process. Unlike the other code organisations discussed here, however, GEMI has not required its members to adopt or implement the ICC charter. In fact, one of GEMI's founders is the environmental manager of a firm that has decided not to adopt the charter.

Bowers credits the cross-industry composition of GEMI for some of its most important contributions. During its first year, GEMI members brought together the concepts of total quality management and environmental management, coining the term total quality environmental management (TQEM). GEMI soon became the "preeminent facilitator of this discipline," sponsoring a major conference on the topic in January 1991. During its second year, GEMI published a TQEM primer.

A third focus of the group's initial activities was stakeholder communications. GEMI has sponsored conferences and focus groups "to enhance the dialogue between business and its interested publics."

While GEMI does not require firms to adopt the ICC charter, it does require them to be active in GEMI projects. Members must participate in at least three of the four annual membership meetings and take part in at least one working group.

ISO 14000

The International Organisation for Standardisation (ISO) was formed in 1946 and is headquartered in Geneva, Switzerland. Its purpose is to facilitate standardisation as a means of promoting international trade. While its membership consists of the standards organisations of its 100 member nations, ISO has been an "industrydriven" organisation.

ISO standards are documented agreements of technical specifications that companies use as guidelines to ensure that materials and products fit their purpose. For example, the format of credit cards, automatic teller machine (ATM) cards, and phone cards are derived from an ISO standard. Cards that adhere to the standard, which defines such properties as thickness, can be used worldwide.

ISO's first standard, published in 1951, was called "Standard Reference Temperature for Industrial Length Measurement." Later, the organisation developed standards for paper sizes, symbols for automobile controls, and metric screw threads. Work focused upon narrow technical and safety issues. This narrow focus shifted in 1987 when ISO issued the 9000 series of quality standards. The ISO 9000 standard provides guidelines for designing and documenting a firm's quality procedures and practices. Development of ISO 9000 was dominated by European manufacturing firms. U.S. firms largely declined to participate, believing that ISO's foray into management "would fail." But by 1990, ISO 9000 was rapidly gaining acceptance an becoming a defacto requirement for doing business in many areas.

In the early 1990s, ISO came under pressure to develop an environmental management standard. Joe Cascio, chair of the group that is coordinating U.S. participation, in the ISO 14000 negotiations, explains that in 1991, as the United Nations was preparing for its 1992 Conference on Environment and Development in Rio de Janeiro, ISO's director general was approached by the conference leaders. They wanted to know whether anyone from ISO was planning to attend the conference and what ISO "was doing for the environment." This conversation spurred the formation later that year of ISO's Strategic Advisory Group on Environment (SAGE) to consider the appropriateness of an international environmental management standard. In the fall of 1992, SAGE recommended that ISO proceed with a standard, and a technical committee was created to begin negotiations and drafting. SAGE's findings indicated that an environmental management standard would promote a common approach to environmental management, much as the ISO 9000 series had for quality management; enhance firm's ability to attain and measure improvements in environmental performance; and facilitate trade and remove trade barriers.

As ISO set up its environmental effort, many of the world's major manufacturing countries were in the process of developing environmental management standards of their own. Germany, Japan, and Canada were finalising product eco-labelling programmes. By 1992, approximately 15 countries had developed national environmental management standards. Perhaps most important, the European Union was moving rapidly to develop an international environmental management standard that would include strong provisions for pubic disclosure of corporate environmental activities. U.S. multinational corporations with European operations were unhappy about this requirement; they felt that because U.S. law already required disclosure of releases (through the Toxics Release Inventory), the European requirements would be unnecessary and burdensome. SAGE members saw these national efforts as potentially disruptive of international trade. In fact, the primary motivation for ISO 14000 was to rein in the standard-setting activities of individual countries through a programme of international consensus building.

Some 400 representatives of U.S. industry, including ones from the chemical, petroleum, electronics, and consulting industries, have participated actively in the development of ISO 14000. Having seen the impact of ISO 9000, these firms are anxious to influence what they feel will be an important undertaking. In addition, about 20 representatives of government and public interest groups have been involved in developing the standard.

ISO's technical work is highly decentralised, carried out in a hierarchy of technical committees, subcommittees, and working groups. Subcommittees have been established to develop environmental management standards in the following six areas: management systems, auditing, labelling, performance evaluation, lifecycle assessment, and terms and definitions. The entire series of standards is known as ISO 14000. The management systems subcommittee is furthest along and has nearly finalised its work. Its standard, known as ISO 14001, will form the overaching framework within which many of the other standards in the ISO 14000 series will fit.

ISO 14000, like ISO 9000 and other ISO standards, is intended to apply to a very wide range of organisations and to accommodate divers geographical, cultural, and social conditions. The breadth of the standards thus limits how specific it can be. When completed, the ISO 14000 series will provide a very general schematic to firms of the management systems they should design and document. Like the other private codes discussed in the paper, ISO 14000 series will not prescribe specific operational practices or set numerical or other kinds of performance standards that firms must meet. It purpose will be provide firms with the elements of an effective environmental management system, thereby helping them to achieve environmental and economic goals:

> *A system of this kind enables an organisation to establish, and assess the effectiveness of, procedures to set environmental policy and objectives, achieve conformance with them, and demonstrate such conformance to others. The overall aim of the standard is to support environmental protection in balance with socio-economic needs.*

Like that of 9000, ISO 14001's framework, encourage firms to hire third-party contractors to certify that their management systems accord with the standards. Only firms that go through the process of third-party review will be ISO "registered." Firms that opt simply to declare their compliance with the standard will not be so "registered." Because registration is intended to confer a certain legitimacy upon firms, those in the second group may miss out on one of this code's primary benefits.

Environmental Risk Indicators (IRTA) in Environmental Management

An analysis of the current state of the Brazilian chemical industry reveals that the actually used technologies were transferred in a period where industrial waste generation, treatment and disposal represented a significant part of total production costs. A not too strict legislation at that time allowed disconsideration of externalities and poor environmental concern. Environmental management was restricted to the monitoring of the cost and volume of effluents transferred to waste disposal plants and landfills in the observance of sanitary inspection rules, without any regard to inadequate process performance. On the micro-environment scale, action was limited to programmes of accident prevention and use of individual protection equipment, without adequate monitoring of occupational health and programmes for minimising inherent process risk. Rising concern with the environmental impact of processes have led to stricter rules and monitoring parameters with regard to working conditions and waster disposal. In response, industries started to rethink their production processes with an eye to eliminate or minimise industrial process failures. As such, new and easily established parameters were needed to subsidise the decision maker's choice of the best way to allocate resources for improving the environmental performance of the process.

Environmental Management

Environment management is an integrated activity that aims to conform performance with established norms in environmental and labour law.

Accordingly, environmental managers should perform several actions to elucidate the environmental performance of the process, such as: to the decision maker

- Financial aspects of the activity to the technicians
- state of the process as a whole to the community
- process performance with respect to externalities.

Five stages can be identified in environmental management:

Stage (0): without management

Stage (1): observance of legislation

Stage (2): system development and implementation

Stage (3): integration with several management areas

Stage (4): integration with total quality

Environmental Performance Assessment

The following main purposes of the quantitative monitoring of a process by means of an environmental performance indicator can be specified:

- monitoring over time of the condition and evolution of the environment
- assessment of the performance of projects, programmes and plans
- management control of the process for process improvement
- demonstration to all interested parties of the continuous improvement of products and processes
- helping in the development of future plans.

Operational Systems Indicators

Operational system is defined as the set of activities related with the design and operation or physical installations, equipment and mass and energy flow, for the deliverance of products and service. The following categories can be specified for the study of these indicators:

- programme oriented measures
- activity based measures
- mass based measures
- standardised efficiency measures
- concentration based measures.

Management Indicators

Management systems involve all proceedings and activities related to organisation, planning, resource allocation, operational control of process, feedback and checking of performance results, identification of the cause of deficiencies and illustration of the progress made toward the established goals of environmental goals of environmental improvement. The working up of this indicator includes: indicators of the satisfaction and behaviour of interested parties/observers, financial indicators etc.

This indicator refers to the interaction environment-operational system-environment. The organisations should select in function of its potential in environmental improvement. It is not necessary to access directly its effects on the environment. The state of the environment can be divided in three categories: impact on natural resources, ecological resources and on human health. The first category bears respect to the nature and use of land, the consumption of nonrenewable resources, decertification and erosion, impact on productive land, on arqueology, a.s.o.; biological aspects (local, regional and global) include irreversible and reversible impacts, impact on biological diversity, a.s.o. Ecological impacts relate to climate changes, ozone layer depletion, acidising (regional), euthrofication (local an regional), ecotoxicity, etc. Human health refers to impacts that contribute to toxicology, as well as physical and psychological impacts.

Environmental and Technological Risk Indicators

The main purpose of identifying the environmental quality costs is to subsidy the decision make in monitoring the performance of the environmental management system. The aim is to identify possible process failures with direct or indirect impact on the environment and sensible investment areas. The Environmental and technological risk indicator (IRTA) is defined as the rate between total cost of failure (internal and external) (CF) and total expenditure (TE).

Thus, as the insurance of industrial installations seeks to minimise the financial risk from damage due to accidents, the premium to be paid is correlated with the risk level of the installation. The main purpose of an environmental management system being risk of failure minimisation during manufacturing, all related expenses should be considered prevention cost in the sense that an improvement reduces the premium paid and vice-versa. The main purpose of the studied indicators is to represent an effective programme of environmental management. This programme can then be used to check the performance and to help the decision making process in choosing new investment or action areas.

R&D activity involves the following development steps: survey of the state-of-art of the proposed process, preliminary laboratory tests, systematic tests on laboratory scale, economic assessment, pilot studies to establish the project base, process implementation on a semi-commercial or commercial basis acting in several technological segment. In view of this, establishing an environmental management system in a R&D area starts with the indication of a specialist with a knowledge of the various activities and the creation of a management structure embracing segments such as: technical area, maintenance, safety and occupational health. This approach will allow the detailed identification of all costs enabling the formulation of an Environmental an Technological Risk Indicator as efficiency parameter for the referred system.

Eco-Efficiency and Cleaner Production: Charting the Course to Sustainability

Defining the Concepts

Eco-Efficiency starts from issues of economic efficiency which have positive environmental benefits while cleaner production starts from issues of environmental efficiency which have positive economic benefits.

Cleaner Production

The concept of Cleaner Production was introduced by UNEP Industry and Environment in 1989. Cleaner production is the continuous application of an integrated preventive environmental strategy applied to processes, products and services to increase eco-efficiency and reduce risks for humans and the environment. It applies to:

- production processes: conserving raw materials and energy eliminating toxic raw materials and reducing the quantity and toxicity of all emissions and wastes.
- products: reducing negative impacts along the life cycle of a product from raw materials extraction to its ultimate disposal.
- services: incorporating environmental concerns into designing and delivering services.

Cleaner production requires changing attitudes responsible environmental management creating conducive national policy environment and evaluating technology options.

Eco-Efficiency

The concept of eco-efficiency was first coined in 1992 by the Business Council for Sustainable Development (BCSD) in its landmark report Changing Course. Eco-Efficiency was further defined at the first Antwerp Workshop on Eco-efficiency held in November 1993 as being 'reached by the delivery of competitively priced goods and services that satisfy human needs and bring quality of life while progressively reducing ecological impacts and resource intensity throughout the life cycle to a level at least in line with the earth's estimated carrying capacity. The WBCSD has identified seven success factors for eco-efficiency.

- reduce the material intensity of goods and services
- reduce the energy intensity of goods and services
- reduce toxic dispersion
- enhance material recyclability
- maximise sustainable use of renewable resources recluce material durability
- increase the service intensity of goods and services.

These elements clearly build upon and help further the successful United Nations cleaner production initiatives towards sustainable production and consumption patterns:

UNEP and WBCSD: A Joint Commitment

The United Nations Environment Programme (UNEP) and the World Business Council for Sustainable Development (WBCSD) have each been developing and promoting similar concepts, i.e. cleaner production and eco-efficiency. Both organisations have been actively involved in the policy development of these concepts and have decided to work together to disseminate them. The new initiative combines UNEP's public sector interests and WBCSD's industry representation, which complement each other.

This document is the first output of this joint action. It has been prepared for the United Nations Commission on Sustainable Development (UNCSD) annual meeting held in April-May 1996 in New York. The purpose is threefold:

- to catalyse industry/government partnerships to promote eco-efficiency and cleaner production, two related concepts for sustainable development.
- to demonstrate the common efforts of UNEP and the WBCSD to promote these concepts at the international level.
- to exemplify the voluntary actions undertaken by industry to implement these concepts 'from the laboratory into the market', and demonstrate the economic value they bring to corporations and shareholders.

The Broader Sustainable Production and Consumption Vision

Both concepts are integral parts of the macro-vision of Sustainable Production and Consumption (SP&C) which encompasses the entire commercial system and its interrelations.

SP&C was defined at the Oslo Ministerial Roundtable in February 1995 as 'the production and use of goods and services that respond to basic human needs and bring a better quality of life, while minimising the use of natural resources, toxic materials, and emissions of waste and pollutants over the life cycle, so as not to jeopardise the need of future generations.'

The WBCSD identified SP&C more specifically as a concept which:

- forms an integral part of the broader sustainable development agenda, where the focus is on the sustainable production and use of goods and services
- encourage continuous improvement in efficiency of the use of energy and resources (i.e. in the patterns of consumption)
- involves changes in the supply and demand for goods and services
- promote the idea that both consumers durable and non-durable can be compatible with sustainable production and consumption
- emphasises a life cycle perspective in the manufacture, use, recycling, and disposal of goods and services, rather than the traditional focus on discrete activities

The Links between Cleaner Production and Eco-Efficiency

The interlinkages between eco-efficiency and cleaner production and numerous. Like cleaner production, eco-efficiency links the goals of business excellence and environmental excellence, by creating the bridge through which corporate behaviour can support sustainable development, the integration of economic growth and environmental improvement.

Eco-efficiency goes beyond resource use and pollution reduction by emphasising value creation for business and society at large, while providing for competitive needs. By increasing value for the goods and services it creates, business will maximise resource productivity, gain bottom-line benefits, and reward shareholders, rather than simply minimise wastes or pollution.

Eco-efficiency embraces cleaner production concepts such as efficient use of raw materials, pollution prevention, source reduction, waste minimisation, and internal recycling and reuse. It captures the idea of pollution reduction through process change as opposed to the earlier end-of-pipe approaches. It shares characteristics with many environmental management tools such as Environmental Assessment or Design for Environment by including them among the technological options for reducing material and energy intensiveness in production, as well as facilitating reuse through remanufacturing and recycling. Eco-efficiency also features a life cycle perspective which follows products from the raw material through to final disposal stages. It is thus an extension of the Total Quality Management process.

Eco-efficiency is an evolving concept which allows companies to adapt to the changing dynamics of the marketplace. Companies are now faced with new demands from more stakeholders, and those who implement ecoefficient practices will be able to respond more aggressively to competitive pressures and anticipate customer needs, while at the same time protecting the environment and employee health and safety.

Put in simple terms, the vision of eco-efficiency is to 'produce more from less'. Reducing waste and pollution, and using less energy and fewer raw materials is obviously good for the environment. It is also self evidently good for business because it cuts companies' costs, and eventually avoids potential environmental liabilities. It is, therefore, a prerequisite to the long-term sustainability of business.

TABLE 5

Eco-Labelling Programmes Worldwide

Location	*Programme*	*Began*	*Status*
Europe			
Austria	Hundertwasser	1991	Criteria in 28 categories with 40 awards.
European Union	Eco-Labels	1992	Criteria developed for 9 categories. Intended to 'harmonise' national programme across Europe.
France	NF-Environment	1992	Deferring to EU.
Germany	Blue Angel	1978	Historic model of all other programmes. Over 81 categories labelled, and over 3,600 labels awarded.
Netherlands	Stichting Milieukeur	1992	Also deferring to EU, has 17 categories, 25 awards.
Norway	Good Environ-mental choice	1991	Criteria developed in numerous categories. operated by Norwegian Nature Federation (NNF).
Scandinavia	Nordic Swan	1991	Intended to harmonise national programme in the region. 34 categories with 800 awards.
Sweden	Good Environ-mental choice	1988	Criteria developed in numerous categories. Private schemes operated by Swedish Nature Federation (SNF).
U.K	Deferring to EU		UK Eco-Labelling Board running a campaign to persuade industry and consumers to favour ecolabels.

Other countries			All other countries plan to follow EU lead. National programmes in Eastern Europe under consideration.
North America			
Canada	Environmental choice	1988	2nd oldest programme. Has criteria on about 30 categories with over 1,400 awards. Programmes being privatised.
United States	Green Seal	1989	Criteria for about 20 categories with 230 awards. Involved in ISO, and chairs the 'Global Ecolabelling Network'.
	SCS 'Claim Certification'	1989	Validates environmental claims. Not a seal of approval type ecolabel. About 500 awards to date.
South America			
Latin America	None		Mexican government discussing possibility of national ecolabels.
Mercosur	None		Governments of Mercosur countries discussing possibility of regional ecolabel.
Other countries	None		
Far East			
Australia	Environmental choice	1991	Abandoned in 1993 due to industry opposition.
India	Ecomark	1991	Government is reevaluating programme due to lack of market impact.
Japan	Ecomark	1989	Criteria in 49 categories with 2,300 awards. Awards made on the basis of 'inherently improving the environment', and often go to whole categories.
New Zealand	Environmental choice		Abandoned in 1994 due to industry opposition.
South Korea	Ecomark	1992	Criteria in 12 categories with 96 awards.
Singapore	Green Label	1992	Criteria in 5 categories.
China	Green Label	Recent	
Taiwan	Environmental seal	Recent	

Market Responses to Ecolabelling

Ecolabelling reflects a movement towards the use of market-based instruments to protect the environment where possible, in place of regulation or as a supplement to it. Consumer choice through ecolabelling, rather than government intervention, creates pressure for environmental protection. For ecolabels to be effective, consumers need to have confidence in the label and producers need to believe that better environmental performance will increase their sales.

Ecolabelling schemes may be slow to effect market changes, but it is generally accepted that their impact on market could be significant. With markets becoming increasingly global, it is inevitable that successful ecolabelling schemes will have an impact on trade. Even with voluntary ecolabel schemes, competition conditions in the market place and trade can be influenced by decisions about which product categories are to be labelled and the criteria by which they are to be judged.

It is possible that products carrying ecolabels may be more expensive to purchase than products without such a label. Consumer groups in Australia have found that through many people are aware of environmental issues, few are prepared to pay significantly more for environmentally sound products or were prepared to sacrifice convenience.

Norwegian consumer groups have pointed out that the terminology 'environmentally good' should be avoided as every product puts a strain on the environment through using energy and raw materials in the process of production or by causing noise, water and air pollution, finally to be considered as waste, "not even a 'labelled product' is totally without negative side effects on Mother Nature".

Ralph Nader, the US consumer advocate, has commented, "schemes such as Green Seal and Green Cross in the US have not been able to push towards environmental standards which are necessary for better resource management, but have pushed instead towards standards which are easily attainable for otherwise socially and environmentally responsible industry".

However, it is too early to assess what effect the EU Ecolabelling Scheme may have on sales of labelled products. From previous experience with the German 'Blue Angel' Scheme, it was shown that for some product groups it is almost essential to have a label to compete successfully in the marketplace.

Problems with the EU Ecolabelling Scheme

In its present form, the EU Ecolabelling scheme is based on a pass/fail approach. The suggestion of a star rating system, ranging from good (one star) to outstanding (five stars), was considered by the EU but was rejected as too complex. The scheme as now introduced could lead to considerable anomalies. For example, a product which has failed to get into the top 10% in a group in which environmental standards are generally high and for which there is a real need (for example, students note pads) could fail to win an award, while a product at the top of another group with more dubious environmental credentials as a whole (such as hair spray) might be awarded the standards. The situation would not only be confusing to the purchasing public, but could result in companies producing products that fail giving up trying, since there is no advantage in being in the second decile (between 20% to 11% from the top), in comparison with being at the bottom. While the authorities that developed the scheme are aware of this problem, they have not found a solution to it, the scheme having been designed as a carrot-and-stick method of improving environmental, performance. In theory, it should be possible for ever brand in a group to improve its environmental performance sufficiently to win a label. If this should happen, then the criteria would be revised upwards so that in the following year only top few would qualify.

A further problem with this approach is that, as the scheme matures, the criteria which distinguish the winners from the losers will begin to be based on very fine distinctions. There could be a group of brands with high environmental standards-a good outcome from an environmental point of view, but not so good for those brands that marginally fail to win the label.

These problems have led to reservations about the principle of addressing environmental problems through an award scheme. At best, all an award winning product can do is to reduce, slightly, the rate at which the global environment deteriorates. To tackle the real environmental issues associated with product manufacture, use and disposal requires radical steps, which may prove unacceptable to the consuming public in developed countries, but as Fleming points out, "the Ecolabel scheme focuses on a minor problem as a relief from having to tackle major ones".

Ecolabels, Experts and Individuals

Finally, let us return to the question posed are experts really better than consumers at making decisions about which products are most environment friendly? The answer to this question is, clearly, no. Neither the expert nor the consumer is in a position to decide which product has the least overall impact on the environment. However, most of the relevant information concerning the impact of a product during the pre-consumer phases of its life-cycle is embedded in the price of that product: the cost of natural resources is included simply by dint of the fact that manufacturers must purchase these from the owners of the resource, whilst the cost of pollution caused during production and distribution is included to the extent that the producer and distributor must conform to regulations and other legal restrictions on emissions. So consumers need not focus on the use of resources or the emission of chemical into the atmosphere and watercourses at the production and distribution stage of a product life-cycle; it is merely necessary that they find the product which best fits their purpose(s) at the lowest price. In addition, consumers make decisions concerning how much electricity or water they will consume and how much waste they will produce and, as has been argued above, if they pay the appropriate until price for these goods and are provided with information about the amount of them that the product is likely to consume/produce, then they will be able to factor these attributes into their product choice and so make a better choice than the expert about which product is best for them and for the environment.

In the introduction, evidence was presented which suggested that there is a disparity between what people say they are willing to pay for environment-friendlier products and what they actually do pay. Some commentators have argued that this disparity is the result of 'consumer scepticism,' generated by unverified and/or misleading environmental claims. In response, numerous organisations, both private and public, have developed ecolabelling schemes, the ostensive aim of which is to mitigate such scepticism by providing consumers with better information about the environmental impact of products.

Consumers rely heavily on the information provided on labels to make their product selection decisions and that information asymmetries between buyer and seller can drive a wedge between the amount that a consumer is willing to pay for a product and the amount that the owner of the product is willing to accept in payment. It has shown that product certification marks may help consumers and producers reduce these information asymmetries but that is also a considerable possibility that certification marks may be misinterpreted, especially if these act as a seal of approval.

It was noted that experts who are charged with the task of developing an ecolabelling scheme (ideal or otherwise) are likely to encounter a number of more or less intractable problems: they will not be able rationally to select product categories; they will not be able rationally to set product category boundaries; they will not be able to take into consideration all the physical effects which a product has on the environment during its life-cycle, nor will they be able accurately to estimate the impacts of these effects, so they will not be able rationally to set product selection criteria; nor will they be continuously to update these criteria.

At every stage in the ecolabelling process, secondary stakeholders (manufacturers, environmentalists and others) are able to affect the outcome in a distorting manner.

It was argued that ecolabels tend to observe relevant product information and may mislead consumers into purchasing products which, under the circumstances, result in the consumption of more resources and the emission of more chemicals into the atmosphere and into water courses, than other non-ecolabelled products.

Ecolabel can lock firms and industries into particular technologies, distorting the prices and other information which consumers utilise to make product selection decisions, and thereby ensuring that resources are not allocate to their most efficient uses.

Information relating to the use of resources and production of post-consumer waste during the use and disposal phases could be presented to consumers in a manner which might enable them to make better purchasing decisions. These decisions would be further improved if consumers paid the market price for all the inputs to and outputs from the product whilst it was in their possession. In addition, it was argued that certification schemes provide a superior mechanism for reducing consumer scepticism concerning the reliability of environmental information.

In conclusion, ecolabelling schemes result in the distortion of information, the distortion of prices and the distortion of trade flows. They inhabit product development and result in the consumption of more scarce natural resources and more harmful emissions to the environment than would be the case if they were absent. Indeed, as Shimp (1995) notes, '[ecolabels] provide no information that helps the consumer understand what improves the environment or what the consumer can do to help'. Moreover, other, more efficient, means exist by which information concerning the use of resources and emissions to the environment may be conveyed to the consumer in a meaningful way. The policy implications of this seem clear: governments should not operate or support seal-of-approval type environmental labelling schemes.

Though the consumers in India too, like their counterparts in the developed world, would rather prefer better performing as well as cleaner products, while living in the cleanest of environments possible, they have to go for some painful, yet unavoidable trade-offs between the basic necessities of our times and a healthy environment. Most often settling for the former, they are more driven by the price, money saving concerns, brand loyalties and habits owing to their low per capita incomes and affordabilities. There are of course instance when their economic priorities like energy savings or cost savings get involuntarily translated into environmental benefits. For instance, most Indians buy high mileage, fuel-efficient motorbikes and cars because they get a better value for their hard earned, limited cash. Now using less fuel is also a tremendous environmental benefit in terms of both the non-renewable energy savings and pollution reduction. Therefore, it seems that more than Ecolabel comparison of energy use and other product summaries may appear more helpful to consumers.

The fact that a cleaner environment is a market good and commands a particular price like air-conditioning in a hotel, or power supply or purified distilled water supply in a building, or services like telephone, cable TV or the internet, is often missed out in a centralised, expert-centred, socialist view of the environment. For most policy makers, bureaucrats and "experts," understanding that it costs money to provide power supply or telephone connection to everyone in a country like India is relatively easier than understanding and accepting that the same case applies equally for the environment as well.

Thus opting for a cleaner environment is not merely a matter of awareness and choice (as the promulgators and "experts" of the Ecomark Scheme will believe), but more a matter of affordability and willingness to pay for it. The fact that in India, the Ecomarked goods would generally cost 10% higher than the normal ones is largely admitted by the government. Now in the given ground reality discussed earlier in this section, and highly price-sensitive markets in the country, (supposing that the Indian Ecomark does finally take-off) expecting most of the Indian consumers to willingly buy Ecomarked products at a 10% higher price than now would be expecting a miracle. Even in developed countries like USA, where both the Eco-awareness and the general incomes of people are many times higher than in India, only less than 10% are willingly to actually pay higher prices for products due to their environmental benefits.

As both Aristotle and later Adam Smith noted that it is a fundamental and timeless nature of man to think firstly and chiefly of himself, even the modern-day environment conscious man is no different. Just like buying any other product from the market, people pay for a cleaner environment: first, when they can afford to, secondly, if it becomes an urgent necessity (like household waste disposal or putting a stop to the choking smoke from eh cooking stoves or a taxable offence like not fixing a catalytic converter to the car etc.), and lastly, when they choose for a better quality of life due to their growing incomes and affordabilities (for instance when people choose to live or move their offices into spacious, cleaner and greener residential suburbs, or farmhouses away from the dirt and smog of the cities, and yet pay higher prices for commuting to the city etc.). Today's cleaner and greener cities in most developed countries are examples of all the three situations working in unison. The most shocking truth about the whole environmental upsurge in the west is that before 1960s, when the environment was at its worst in most American and West European countries, there was very little talk of environmental issues, either in the media, government circles or among general public. On the contrary, it was precisely when with the rapid technological growth, the environmental quality in most western countries started getting drastically better that the worldwide environmental campaigns started getting launched from the west.

What does this clearly show? It was not the cleaner environment that brought the industrial and technological wealth into US, Canada, Japan or Western Europe, but on the contrary, it was this wealth that made the average person there rich enough, to be able to afford a cleaner environment. This is the stark economic truth, which both the Indian environmentalists and the policy "experts" have to face up to if they seriously want the environmental policy instruments like the current Ecomark to work and yield the desired results.

Even in Indian situation, it is not as hopeless as some officials in the government circles think, that consumers or producers would never volunteer for the environment and hence more government involvement would be needed to activate the scheme. In fact they would readily volunteer if the scheme is designed in such a way that the environmental objectives get built into the individual interests of both consumers and producers.

Since it is in consumers' interest to pay less for a can of Coke, Pepsi or Thumbs Up, the manufacturer would look for a technology to make lighter cans in order to bring down the can-costs and hence the product-price, to sell more cans of the drink. Now lighter cans mean less waste to handle and dispose-off, which is a tremendous environmental benefit. Here the producer did not consciously benefit the environment, but was led by the invisible hand of his self-interest of cost reduction and as a result benefited the environment.

Similarly, consider why most Indians go for fuel-efficient motorbikes or cars. To burn less oil and hence emit must lesser pollutants into the air? Obviously not. Most of them buy fuel-efficient vehicles because they can save money on petrol or diesel. As simple as that. But how come they benefited the environment without much green awareness and not buying ecolabeled bikes or cars approved by the qualified "experts"?

To understand why people don't consciously took for green labelled products, it would be helpful to see why people currently end up making choices that are not as Eco-friendly, as many of them might desire.

People always exchange value of what one owns. One can't exchange something that one doesn't own. Thus the ownership of a resource or a valuable or in short, the property right is the basis of all economic exchanges. The problem with the current pattern of the environmental ownership is that its property rights are not as clearly defined as in the case of a private owned resource like a tea estate, apple orchard, hotel, etc.

Consumers are not charged the real price per unit for the goods such as water, effluent treatment, or solid waste disposal. Often state agencies like municipalities provide these services by tax revenues. As a result, consumers have no incentive to optimise their use. Environment, thus today, is no one's property, neither of an individual or a corporation, thus no one pays for its use or abuse.

In addition to the above problem of environment being treated by most of the modern welfare-based societies as a "free good", another major reason why "experts" can't decide the "right" environmental choices for the consumers is because individuals value environment in a subjective fashion. This subjectivity, as discussed earlier in this section, is most often the product of their real life situation, aspirations and hence priorities at a given point of time. Thus in real world, people always trade off between different choices, while arriving at a final decisions.

For instance, people generally make trade-offs between the atmospheric pollution and other factors that affect their lives: many people choose to live in smog-filled cities, despite the damage that might be done to their lungs. Now why they do this is known to the individual in question rather than the "expert" carrying out a "scientific" LCA. This example especially applies to all developing country situations, including India, where people, inspite of the dirt, grime and pollution of the slum-filled cities, still prefer to move in and make a living there.

Thus, from all previous discussions, it is quite clear that an "expert"-led third-party certification ecolabel, like the now inert Indian Ecomark, cannot achieve the noble environmental objectives that it seeks in principle. On the other hand, such an ecolabel would potentially lead to a worse environment by shifting both product innovation and consumer choice at the marketplace.

A consumer trusts a brand name or the brand reputation, earned through years of trial and error in the marketplace, more than the wisdom of some distant "omniscient" expert, while making a purchase decision. This is because, the consumer knows what is best for him. Secondly, the brand names are earned by the manufacturers or service providers through tough market competition and therefore at their own risk, while the criteria setting experts in a product standard labelling scheme like Ecomark do not face such a risk of losing their goodwill and business share in the marketplace. No wander, people generally look for brand name of the product first, rather than a label like ISI Mark or Agmark on its wrapper, before deciding to buy it. A brand name of a product is like the goodwill of a person, which carries everything desirable about it. And eco-friendliness of a product is also one of those qualities which a brand name could be delighted to carry alongwith it, provided consumers at large see eco-friendliness as something desirable in the goods they seek to buy.

Thus, green concerns cannot be artificially built into the normal consumer choice, through a one capsule panacea like the third-party, product standard ecolabel such as Ecomark. On the contrary, for green choice to get translated into normal consumer choice, it must become an active element in all modes of market-led consumer vouchsafing discussed before. Green must become a quality element which consumers are actively seeking in everything they look out and buy. On one hand government must make the consumers pay proportionally to the electricity and water they consume, and the amount of solid waste, emissions or effluents they let out everyday. This will automatically throw-up a huge demand for products that consumer less energy and water, and those products that waste least, give out the least emissions or effluents. Additionally, under the newly emerged scenario (brought about by the radical shift in the government policy), advertising, image building, personality and institutional endorsements which today promote products, merely on their performance value, would also attach equal importance to the environmental benefits provided by these products, in their promotional. Thus here, green concerns would enter the private domain of each consumer, as being unfriendly to the environment would now cost him or her dearly.

To resolve the authenticity issue of manufacturers' self-declarations about the environmental goodness of their products. There are tow possible options. The green claims may get checked through the same investigative procedures and institutional set-ups as product quality claims of all advertisements are cross-checked toady through the special commissions set up with the joint of industry and the participation government. The other way is by allowing an open regime of multiple green endorsement players in the marketplace who after running some procedural checks on the products, endorse and green claims of companies for their wares.

Product Requirements for Soaps and Detergents Under the Eco-Mark in India

- All products shall meet relevant standards of the Bureau of Indian Standards pertaining to quality, safety and performance
- The manufacturers shall produce documentary evidence with respect to the compliances under the Water, Air and Environmental Protection Acts and other regulations such as the Prevention of Food Adulteration Act and the Drugs and Cosmetic Act and rules made thereunder, as applicable
- The product packaging may display, in brief, the criteria used to label the product as environment friendly
- The material used for product packaging shall be recyclable, reusable or biodegradable
- The product container shall be suitably marked as the Eco-Mark label is applicable to the contents only, if the container is not separately covered under the Eco-Mark scheme
- The product may be accompanied by detailed instructions for proper use of maximize product performance and minimize wastage.

Product Specific Requirements

A. Toilet soaps

(*The Gazette of India,* Extraordinary, Part II-3(I), No. 188, April 28, 1992).

- Free from phosphates and synthetic detergents
- Dermatologically safe as per IS 11601:1990

BIS Amendments Issued: IS 2888:1963, IS 4199:1990;
IS 5784:1970, IS 10523:1983;
IS 11303:1985; IS 11479:1985.

B. Detergents

(*The Gazette of India,* Extraordinary, Part II-3(I), No.188, April 1992).

- Phosphate free
- Dermatologically safe as per IS 11601:1990
- Surfactants shall be readily biodegradable when tested by modified sturm test (IS 13933:1995 under print)

BIS Amendments Issued: IS 4955:1993; IS 4956:1977;
IS 8180:1992; IS 9458:1980.

C. Laundry Soaps

(*The Gazette of India,* Extraordinary, Part II-3(I), No. 4, January 1994).

- Phosphate free
- Dermatologically safe as per IS 11601:1990
- Non-soapy detergent, if used, shall be readily biodegradable when tested by modified sturm test (IS 13933:1995 under print)

BIS Amendments issued IS 285:1992; IS 2887:1974.

Product Requirements for Paper Under the Eco-Mark Scheme in India

Product Group: Paper

(*The Gazette of India,* Extraordinary, Part II-3(I), No. 455, November 13, 1992)

General Requirements

- All products shall meet all relevant standards of Bureau of Indian Standards pertaining to quality, safety and performance
- The manufacturers shall produce documentary evidence with respect to the compliances under the Water, Air and Environmental Protection Acts and other regulations such as Prevention of Food Adulteration Act and Drugs and Cosmetic Act and rules made thereunder, as applicable

- The material used for product packaging shall be recyclable, reusable or biodegradable
- The product container shall be suitably marked that the Eco-Mark label is applicable to contents only, if the container is not separately covered under the Eco-Mark scheme
- The product may be accompanied by detailed instructions for proper use so as to maximize product performance and minimise wastage.

Product Specific Requirements

- 60 per cent of pulp made from material other than bamboo hard wood, soft wood and reed
- Recycled paper made from 100 per cent waste paper.

BIS Amendments Issued

IS 1396:1960;	IS 1774:1986;
IS 1775:1981;	IS 1848:1991;
IS 2483:1986;	IS 2991:1988;
IS 3064:1986;	IS 3302:1986;
IS 3303:1986	IS 3413:1977;
IS 3673:1986;	IS 4658:1988;
IS 4664:1986;	IS 6956;1973;
IS 8431:1986;	IS 8460:1977;
IS 9032:1978;	IS 9033:1978;
IS 10405:1982;	IS 11087:1986;
IS 11687:1986;	IS 11688:1986;
IS 12765:1989;	IS 12766:1989;
IS 10808:1989;	

CONVENTION CONCERNING PROTECTION AGAINST HAZARDS OF POISONING ARISING FROM BENZENE: ILO CONVENTION NO. 136

The General Conference of the International Organisation,

Having been convened a Geneva by the Governing Body of the International Labour Office, and having met in its Fifty-sixth Session on 2 June, 1971, and

Having decided upon the adoption of certain proposals with regard to protection against hazard arising from benzene, which is the sixth item on the agenda of the session, and

Having determined that these proposals shall take the form of an international Convention,

adopts the twenty-third day of June of the year one thousand nine hundred and seventy-one, the following Convention, which may be cited as the Benzene Convention, 1971:

Article 1

This Convention applies to all activities involving exposure of workers to—

(a) the aromatic hydrocarbon benzene C^6H^6, hereinafter referred to as "benzene"

(b) products the benzene content of which exceeds 1 per cent by volume, hereinafter referred to as "products containing benzene."

Article 2

1. Wherever harmless or less harmful substitute products are available, they shall be used instead of benzene or products containing benzene.

2. Paragraph 1 of this Article does not apply to—

(a) the production of benzene;

(b) the use of benzene for chemical synthesis;

(c) the use of benzene in motor fuel;

(d) analytical or research work carried out in laboratories.

Article 3

1. The competent authority in a country may permit temporary derogations from the percentage laid down in Article 1, subparagraph (b), and from the provisions of Article 2, paragraph 1, of this Convention under conditions and within limits of time to be determined after consultation with the most representative organisations of employers and workers concerned, where such exist.

2. In such case the Member in question shall indicate in its reports on the application of this Convention submitted under Article 22 of the Constitution of the International Labour Organisation the position of its law and practice as regards the derogations and any progress made towards complete application of the terms of the Convention.

3. At the expiration of three years from the data on which this Convention first entered into force, the Governing Body of the International Labour Office shall submit to the Conference a special report concerning the application of paragraph 1 and 2 of this Article and containing such proposals as it may think appropriate for further action in regard to the matter.

Article 4

1. The use of benzene and of products containing benzene shall be prohibited in certain work process to be specified by national laws or regulations.

2. This prohibition shall at least include the use of benzene and of products containing benzene as a solvent or diluent, except where the process is carried out in an enclosed system or where there are other equally safe methods of work.

Article 5

Occupational hygiene and technical measures shall be taken to ensure effective protection of workers exposed to benzene or to products containing benzene.

Article 6

1. In premises where benzene or products containing benzene are manufactured, handled or used, all necessary measurers shall be taken to prevent the escape of benzene vapour into the air of places of employment.

2. Where workers are exposed to benzene or to products containing benzene, the employer shall ensure that the concentration of benzene in the air of the places of employment does not exceed a maximum which shall be fixed by the competent authority at a level not exceeding a ceiling values of 25 parts per million (80 mg/m3).

3. The competent authority shall issue direction on carrying out the measurement of the concentration of benzene in the air of places of employment.

Article 7

1. Work processes involving the use of benzene or of products containing benzene shall as far as practicable be carried out in an enclosed system.

2. Where it is not practicable for the work processes to be carried out in an enclosed system, places of work in which benzene or products containing benzene are used shall be equipped with effective means to ensure the removal of benzene vapour necessary for the protection of the health of the workers.

Article 8

1. Workers who may have skin contact with liquid products containing benzene shall be provided with adequate means of personal protection against the risk of absorbing benzene through the skin.

2. Workers who for special reasons may be exposed to concentrations of benzene in the air of places of employment which exceed the maximum referred to in Article 6, paragragh 2, of this Convention shall be provided with adequate means of

personal protection against the risk of inhaling benzene vapour. The duration of exposure shall be limited as far as possible.

Article 9

1. Workers who are to be employed in work processes involving exposure to benzene or to products containing benzene shall undergo—

(a) a thorough pre-employment medical examination for fitness for employment which shall include a blood-test;

(b) periodic re-examinations, which shall include biological tests including a blood-test, at intervals fixed by national laws or regulations.

2. The competent authority in a country may, after consultation with the most representative organisations of employers and workers concerned, where such exist, permit exceptions from the obligations of paragraph 1 of this Article in respect of specified categories of workers.

Article 10

1. The medical examinations provided for in Article 9, paragraph 1, of this Convention shall be.

(a) carried out under the responsibility of a qualified physician, approved by the competent authority, and with the assistance, as appropriate, of a competent laboratory;

(b) certified in an appropriate manner.

2. These medical examinations shall not involve the workers in any expense.

Article 11

1. Women medically certified as pregnant, and nursing mothers, shall not be employed in work processes involving exposure to benzene or products containing benzene.

2. Young persons under 18 years of age shall not be employed in work processes involving exposure to benzene or products containing benzene: Provide that this prohibition need not apply to young persons undergoing education or training who are under adequate technical and medical supervision.

Article 12

The word "Benzene" and the necessary danger symbols shall be clearly visible on any container holding benzene or products containing benzene.

Article 13

Each Member shall take appropriate steps provide that any worker exposed to benzene or products containing benzene receives appropriate instructions on measures to safeguard health and prevent accidents, as well as on the appropriate action if there is any evidence of poisoning.

Article 14

Each Member which ratifies this Convention—

(a) shall, by laws or regulations or any other method consistent with national practice and conditions, take such steps as may be necessary to give effect to the provisions of this Convention;

(b) shall, in accordance with national practice, specify the person or persons on whom the obligation of compliance with the provisions of this Convention rests;

(c) undertaken to provide appropriate inspection services for the purpose of supervising the application of the provisions of this Convention, or to satisfy itself that appropriate inspection is carried out.

Article 15-22

Standard final provisions.

Industry's International Environmental Support Activities

Two new international conventions came out of the UNCED process, one on climate change and one on biological diversity. Both conventions have implications for business and industry especially in terms of patents and energy conservation. *Agenda 21* called for governments to undertake action toward developing new treaties for desertification and land-based sources of marine pollution, again with implications for international business.

As most of the industry guidelines predate *Agenda 21* and UNCED, it is not surprising that the guidelines do not discuss specific post-UNCED follow-up activities. The Banks *Statement* is one of the few that mentions the connection between associations, TNCs and international organisations such as the UN, reflecting the fact that the code was drafted in cooperation with the UNEP.

International governing bodies such as the UN are gradually evolving to allow groups, including industry associations, to participate in the decision-making process. There are, however, few instances where a direct linkage is made between TNCs and the development of environmental conventions, standards and guidelines: activity regarding the control of toxic chemicals. Because of the environmental consequences of accidents and spills, governments and citizens' groups have kept toxic and hazardous chemicals on the international agenda. There are concerns regarding potential and actual exposure to chemicals known to be toxic or hazardous. Many other chemicals in commerce are untested and have unknown health effects. Although heavy metals, radioactive elements and toxic chemicals are found in nature, transnational business and industry is perceived as being responsible for extraction and concentration, manufacture, transport, and use of many such substances. International conventions and guidelines regulate transport and trade, and as *Agenda 21* was being negotiated, UNEP was working toward an expansion of the voluntary as amended 1987 London Guidelines for the Exchange of Information on Chemicals in International Trade into a legally binding mechanism, by which prior informed consent (PIC) of the importing country's government would be necessary before chemicals could be brought in. As a result, some governments and environmental groups pushed for the inclusion of this provision.

The recommendation in *Agenda 21* referring to the development of a code of principles on management of trade in chemicals (item 31) reflects the concern at the time and is idiosyncratic among the more generic recommendations in *Agenda 21* for the sustainable conduct of international business. In 1996, the issue is moot, as the international community has, in some ways, exceeded the narrow request for an internationally agreed upon code of principles for the management of trade in chemicals through action on prior informed consent and pollutant emission and transfer registers as well as ongoing implementation of existing treaties.

Key Features of Industry Guidelines

1. Environmental Conventions: Standards and Guidelines

Many codes pledge note that members will work with governments to identify solutions to environmental problems. In some cases this call for action is linked, by the Keidanren for example, to the need to 'cooperate as much as possible with the various activities conducive to environmental policies, which are scientific as well as rational.' Many TNCs are concerned that international and national regulation can be overly burdensome and costly. Thus, the voluntary codes adopted by associations, such as the ICME, often include items such as 'work with governments and other relevant parties in developing scientifically sound, economic and equitable environmental standards and procedures ...'. There is a delicate balance, because overt attempts to influence the outcome of the international decision-making process will be seen in a negative light by those unable to participate in the process.

As noted above, trade in chemicals was the only area in Agenda 21 which required cooperative action by TNCs in working toward a new international agreement or standard. On the surface, trade in chemicals is only relevant to guidelines for major industrial categories. Member companies of all nearly all associations, however, partake in trade and use of chemicals in their offices and facilities, including pesticides or cleaning solvents, or indirectly through the purchase of products whose manufacturing required international trade in chemicals.

Despite the current absence of certain regulations, such as ones relating the trade in chemicals, companies should have a commitment, as does the WTTC, to 'implement sound environmental principles through self-regulations, recognising that national and international regulation may be inevitable and that preparation is vital.' This work must be done in a cooperative atmosphere; therefore, the International Banks *Statement* suggests 'we ask the United Nations Environment Programme to assist the industry by providing, within its capacity, relevant information relating to sustainable development.'

2. UNCED Follow-up

Several codes suggest that industry can and should play a role in broader environmental protection of the type advocated in *Agenda 21.* One recommendation, *Japan's Role in Post UNCED Global Environmental Issues-On Environmental Problems in the Asian Region,* is 'based on the discussions at the Rio Earth Summit [and] reveals our view on how Japan can and should contribute to the environmental problems of the Asian region'. The IPRA states that its' Green World Communication Programme 1994-1996 will work through IPRA's international membership in a coordinated effort to assist in the analysis, interpretation and response to growing environmental communication needs in the private and public sectors, emphasised by UNCED General Secretary Maurice Strong ... in June 1992.' The *ISO 14000* preparatory documentation cites the UNCED as a major catalyst in its development.

As noted earlier, some associations have passed from a stage in which they react to environmental regulations at the national or international level to a proactive stage. In such cases, associations may even conduct programmes that exceed base level of action identified in *Agenda 21* or the base level in their own codes. The International Fertilizer Industry Association (IFA), in cooperation with the Fertilizer Advisory, Development, and Information Network for Asia and the Pacific (FADINAP), has established *Guidelines on Best Agricultural Practices to Optimize Fertilizer Use in Asia and the Pacific,* the objectives of which 'are particularly appropriate in the aftermath of the UNCED meeting held in Rio de Janeiro in June 1992, and the recommendations put forward under *Agenda 21.*

The international community was active prior to UNCED: UN treaties, trade agreements, negotiated international standards, soft law declarations and other policy statements. None of the industry codes mention action toward implementing other international agreements. In cases where such agreements are ratified and codified into national law, the industry associations may be fulfilling the international norms by acting in accordance with national law.

Many of the industry codes note that their members should work with government. In doing so, they would be indirectly participating in the implementation of *Agenda 21.* Several codes suggest that companies should work with governments to develop law and policy (see Table-6).

TABLE 6

Industry suggestions for working with government

ICC:	'Contribute to the development of public policy and to business, governmental and intergovernmental programmes and educational initiatives that enhance environmental awareness and protection.'
API:	'Participate with government and others in creating responsible laws, regulations and standards to safeguard the community, workplace and environment.'
CMA:	'Participate with governments and others in creating responsible laws, regulations and standards to safeguard the community, workplace and environment,'
CCPE:	'Work with government agencies, downstream users and others in the development of sound, scientifically-based legislation, regulations and product standards that protect and benefit employees, the community and the environment.'

Of course, by working with government, industry can help shape environmental policy in ways that reflect the interest of business. The ATMI encourages its members '[t]o ensure that fair and reasonable environmental policies are enacted to protect your employees, your workplace, your community and the environment, companies must be aware and involved in federal, state and local policy-making decisions.

As discussed above in relation to industry responses to item 1 that recommends that TNCs adopt home country standards as the basis for business activities abroad, leading associations have formulated principles higher than the standards of national law, with the active support of industry associations. The Keidanren counsel that: 'If the standards of the host country are less restrictive that those of our country, or if there are no standards in the host country, a corporation should make efforts towards appropriate environmental protection, with consideration given to the natural environment of the host country, as well as the laws and the status of (environmental) policies in our country and in consultation with those involved in the host country.

In terms of practical activities to follow-up *on Agenda 21,* several industry associations, or groups of leading companies within specific sectors, have worked with multilateral agencies or have indicated their intention to do so. The group of Banks worked from the start with UNEP on the formulation of its *Statement* and has continued to work with them.

Impacting Business Practices for Better Environmental Performance

Environmental information disseminated by social and environmental monitors impacts corporate environmental practices in a number of ways. Some of the impacts are direct while others operate indirectly through a diffuse network of individuals and organisations. Often individuals in companies, even high officials, are not aware of their own environmental record. Often they are also unaware of what their competitors are doing, and how they compare to their competitors. Social monitors help companies directly to improve their environmental performance by providing subscribers with environmental digests and profiles that enable them to make decisions about their own environmental policies and practices. For these consumers, the information serves both as a report card of their own standing on environmental issues, as well as a benchmark against other companies which they want to emulate or surpass.

When a social monitor lists a company among the worst environmental performers in its industry, the group generally provides the company with recommendations for remedial action. At CEP, for instance, each company is invited to meet with the ratings staff and work out solutions to its environmental problems. The judges communicate to the company the various things it needs to do in order to demonstrate an improved environmental record. CEP therefore directly influences corporate decision makers, and thereby impacts corporate environmental practices.

Much of the advice provided to a firm comes from the expertise of experienced advisors, but it is enhanced by knowledge that other companies in the company's industry have developed and are using in their own environmental practices. In this way, social monitors also act as an industry-level disseminator of information about environmental best-practices.

Companies prefer to avoid bad publicity and are generally quick to react to news coverage that contains environmental information on their companies. News reports will therefore typically prompt firms to initiate public relations efforts as well as to take a closer look at their environmental practices—if only to avoid bad press in the future. The news media therefore play a key role in indirectly impacting business practices. Media reports also influence consumers, who react either by mail and phone campaigns to offending firms or by changing their buying patterns.

Social investors also rely on information from social monitors. They use environmental data to channel their investment dollars and vote their proxies. By reacting to environmental information, they send a message to offending companies about their environmental practices. As a result, environmental information compiled and disseminated by social monitors can also indirectly influence the firm's environmental practices.

In this way, then, social monitors play a key role in collecting and disseminating environmental information, and so stand to affect business practices. Despite these developments, however, much work remains to be done if we are to create a truly responsive and responsible business sector. We suggest some avenues to pursue.

Corporate Aims and Strategies as a Checklist for Environmentalist Business Management

If companies are to behave in an ecologically acceptable manner, they must make a critical analysis of their philosophy, formulate a set of objectives which take account of environmental questions and lay down appropriate strategies. An objective analysis of the situation (where do we stand?) must be taken as a starting-point to determine the company's goals (what do we want to achieve?) and decide on medium-term policy (what steps should be taken?).

Identify Key Ecological Issues in the Company's Areas of Activity (Key Issue Analysis)

1. Analyse opportunities and risks in the company's operating environment and on the market (external factors), such as:
   - legislation (e.g. the laws on emission limits and waste disposal)
   - public awareness (e.g. press reports, demonstrations)
   - consumers' awareness (e.g. changes in attitude and behaviour)
   - competitors' activities (e.g. introduction of products which minimise ecological damage)
   - scientific and research developments (e.g. catalyst technology)
2. Analyse the company's strengths and weaknesses (internal factors), such as:
   - finance available for investment geared to environmental questions (e.g. development)
   - open-minded and flexible management attitudes
   - the company's range of products and its relevance to markets in the field of environmental protection (e.g. use of synergetic effects, access to markets arising from ecological concerns)
3. Draw up a consistent ecological strategy for the company, e.g. by:
   - systematic comparison of opportunities/strengths, risks/ strengths, opportunities/weaknesses and risks/weaknesses
   - identification and assessment of the "critical" factors in the strategic

balance (e.g. applicability of existing know-how to environmental markets; lack of adequate finance for the necessary investments)

Determine the Company's Objectives with Due Regard for Ecological Factors

4. Formulate models and principles for the company's strategy in the light of ecological considerations, e.g.:
   - Safeguard the market economy and free enterprise system by demonstrating their compatibility with ecological constraints.
   - Safeguard the company's existence by recognising general trends at an early stage and planning accordingly.
   - Seek competitive advantages by minimising environmental damage.
   - Adopt a dynamic approach to ecological challenges in all areas of the company's activity and help in the search for solutions.
   - Cooperate with business and academic partners to speed up the progress of knowledge in the environmental sector.
5. Lay down long-term objectives for the company as a whole in environmental matters, e.g.:
   - Capitalise on the goodwill of staff, the public and government which will result from an environmentalist commitment.
   - Take advantage of the lower costs resulting from energy savings, economical use of resources and other progress in environment-related technology.
   - Minimise risks arising from management or product liability, sudden changes in statutory limit values or surges in environmentally motivated demand.
   - Cooperate with trade associations, the media and government to generate further knowledge.
   - Influence legislation by lobbying activities at the pre-legislative stage.
6. Assess the relationships between the company's environmental and economic objectives.
   - Assess the implications of environmental objectives for turnover, costs and profits.
   - Identify major areas of conflict between these objectives (e.g. short-term higher costs, lower turnover, cash-flow problems).
   - Lay down long-term priorities within a system of objectives (e.g. environmental aims as main or secondary objectives and possible changes with time).

7. Clarify the ecological dimension in all departmental objectives.
   - Labour market: increase the company's attractiveness to ecologically concerned workers and reinforce staff loyalty.
   - Procurement: adopt an environmentalist posture in the purchase of raw materials, ancillary inputs and consumables.
   - Production: cut costs by saving energy, raw materials and water.
   - Marketing: take environmental considerations into account in logistics (e.g. transport by rail, not road).

Formulate Ecologically Sound Strategies

8. Determine basic strategies.
   - Defensive or offensive stance ? Keep the initiative: better to act than to react!
   - Action on a narrow or a wide front? Seek to concentrate effort on the specific areas in which the company is competent (e.g. as a specialist in the solution of particular environmental problems).
9. Plan environmentalist strategies for the company's various areas of activity.
   - Define areas of activity in product, market and technological terms (describing the product value, market segmentation and technologies with reference to environmental considerations) and lay down priorities for work on specific market segments.
   - Determine the company's basic attitude to other participants in the marketplace (competitors, distributive trades, customers), considering the scope for demarcation of functions or cooperation.
   - Lay down product and product range strategies (improve the environmental acceptability of the product range by introducing new products and changing or replacing existing ones).
   - Establish ecology portfolios and ensure that they are suitably balanced.
10. Formulate environmentalist department strategies. (See "marketing" checklist etc.)

Rules for Green Business

All that has been discussed earlier, other than has been about direct action within companies by managers at all levels. But every company operates within a particular social, cultural and political environment. What managers are able to achieve as leaders depends to a large extent on the attitudes and morale of the employees and the communities in which they operate. Achievement also depends on the rules of the business game that have evolved over decades through the influence of the underlying beliefs and assumptions of the past. It is quite possible to progress some way towards sustainable development under the old rules. But the remainder of the journey cannot be travelled without considerable changes in those rules and the habits of thought and action of people in business.

Of course, it is true that change will be less disruptive if we move in advance, changing direction deliberately in the way that appears right. But where radical change is required, *we* have to be changed before effective action can follow.

This book is about the transformation of management: its main obstacle is habit. The habits of management that need to be overcome are those that have developed over many years as a by-product of the old visions and values of industry and commerce. They are deeply ingrained, and they will not be quickly and easily overcome.

Changing Habits of Thought and Action

To start the transformation in a company there has to be a total commitment by the board and senior managers to a new vision and new values that will create enthusiasm and capture the imagination of the whole work-force; they must also be in tune with sustainable economic development. Often when attempts are made to introduce new schemes, for example quality circles, this necessary preliminary is forgotten. A gardening metaphor is appropriate: new seed cannot be expected to germinate and grow unless the ground has been thoroughly prepared. Before any attempt is made to do anything to break old habits, it is essential for the gardening work to be completed; otherwise the entire project will be counter-productive. When I was responsible for the provision of training programmes for some of our marketing people, the exercise for some trainees was doomed to failure from the start because their senior managers had not shown why the training was needed and how it fitted in with the company's policies. Only when widespread understanding and enthusiasm for the new vision and values have been achieved should the next step be taken.

The second stage of transformation is an integrated and comprehensive programme of management training and development. It must cover all levels of management; its purpose will be to inculcate styles and practices that are appropriate to the new ways in which the company will operate as a result of changes in vision and values. The entire company will need to become a learning organisation. Although there will need to be some class-room work, most will be done through 'action learning'. Such a transformation project cannot hope to succeed if it is just superimposed on an organisation that is already stretched to the limit with its day to day work. Steps must be taken in advance to make room for it, either by increasing capacity or by lightening the load. Furthermore, the project will fail unless senior managers remain fully involved from its beginning to its end. The process of changing the habits of a lifetime is a difficult and painful one for many people. Any sign that senior managers may have withdrawn to do other, 'more important' things is likely to cause the project to fail. Regular reviews of progress are essential.

The demands of such a transformation programme will probably seem fantastic to many readers. They should not be deceived into believing that they can, in the long run, be avoided. The only choice is whether to do it calmly on one's own initiative, or to be forced into it by external circumstances in some future crisis.

Retaining Control When Power is Devolved

One of the main problems facing managements of traditional hierarchically organised, large companies is the retention of proper control as they move to much more decentralised operations. Schumacher drew attention to the 'principle of subsidiarity', whose origin is Catholic social philosophy. A papal encyclical states

> It is an injustice, a grave evil and a disturbance of right order to assign to greater and higher association what lesser and subordinate organisations can do. For every social activity ought of its very nature to furnish help to the members of the body social and never destroy or absorb them.

So much that is amiss with large, bureaucratic and unwieldy businesses is so because they do not conform to that principle. But it is important that lower levels—whether subsidiary companies, divisions, departments or sections—are given the authority and resources that are commensurate with their capacity and competence. Having been given the greatest authority possible, managers need to be told what must be authorized at a higher level after consideration in a wider context. When managers at all levels are convinced that this is the principle upon which the whole organisation operates, nobody need fear that control has been lost as powers of decision are devolved. So it is important to make the commitment to this principle explicit throughout the whole business organisation.

A major and fundamental change took place in the Shell Group in 1959. Previously much of its operation had been centralized in the London and Hague head offices. When the change came, the title of manager was abolished in these offices and the major groupings were called coordinations. As an illustration of the degree of devolution that took place, the marketing coordinator retained only one overriding power: that concerned the integrity of the Shell emblem and colour scheme. At the time the change occurred, the principle governing the allocation of authority in the new organisation was not explained clearly enough at all levels. I believe that the amount of unease that persisted amongst many members of the staff for a long time would have been less, and of shorter duration, if the principle had been widely proclaimed and understood.

Flexible and Responsive Management Information

Many existing information systems are too systematic and rigid to be suitable for very innovative and rapidly changing operation. They frequently do little more than assist in monitoring and control. In so far as they help with the planning process they are seldom capable of contributing anything more than is required for a continuation of the existing business. They may be good administratively, but they lack something that is needed for innovative management. The system that is needed should be capable of alerting managers to new or unused opportunities, and to any danger that lies ahead. By linking information on dangers or opportunities with other data, creative ideas will be initiated that will enable the dangers to be avoided or the opportunities seized.

The information about monitoring, control and conventional business planning is mostly made available in numerical form-as totals, percentages, ratios etc. These are useful mathematical abstractions which, if used carefully and with understanding, are very valuable for certain purposes. But they are not the 'real world'. The kind of information required for creative thinking may sometimes be quantitative, but mostly it will be an association of practical observations. With the greatly increased capacity of computers to store, analyse and relate information, they should be able to help facilitate the generation of creative ideas and initiatives.

Making Team Decisions Effective

As we move from the age of hierarchical, top-down decision-making into an age of team consent, there is an obstacle that must be overcome if the proposed transformation is to be effective. Any rearrangement of work will reveal that there are some people who can sense personal advantage and others who feel that they are threatened. When the decision-making process involves all employees, there is likely to be considerable resistance to change unless the future interests of those who are likely to feel threatened receive at least as much attention and sympathy as their colleagues. Obviously the measures that can be taken will vary from company to company. However, if the future general pattern of working arrangements is seen to be more pluralistic, a carefully planned programme to develop independent enterprise, designed to use the talents of redundant employees, should form an integral part of reorganisation.

Whilst I was involved with the development of local enterprise trusts several of the large companies I was working with, that were undergoing major changes, successfully explored the matching of surplus resources with the skills and abilities of some of their employees for whom they could see no continuing employment in their companies. Another approach was to see what services were being imported from outside the local community that might with advantage be provided locally if a new enterprise was set up. (There are many examples, such as the one mentioned earlier of Xerox employees being helped to work from home with the freedom to take work from other companies as well as from their previous employers.) The shift towards more durable, high-quality products, combined with an expansion of maintenance, repair and reconditioning, will provide another very attractive range of opportunities for people who were previously employed in production. Such people will carry with them invaluable knowledge and skills to apply to their new employment.

Ethical and Environmental Investments

Ethical and environmental investments have shown steady growth in the USA and the UK. Ethical investments tend to avoid specific companies which are deemed to have unethical products or services, while environmental investments tend to select those companies which are taking positive action to help the environment. Holden Meehan have published reports on progress since October 1991 and their most recent guide was published in 1996.

The growth in these investments in the UK has been impressive. In 1990, £280 million was invested. This had risen to over £1000 million by 1996. Figure 2 shows the growth from January 1989 to July 1996.

There are 35 Ethical Unit Trust schemes in the UK and a Green Independent Financial Advisers organisation has been formed. The Ethical Investment Research Service (EIRIS) provides a detailed guide to these investments which cover pensions, Personal Equity Plans (PEPs), endowment mortgages and other ethical investment plans. Their publication *Money & Ethics'* describes 28 of the leading ethical funds, the product range covered by each and a comprehensive analysis of what each fund avoids as well as the types of business activity each supports. Holden Meehan in their guide assess the performance of each fund in terms of:

- ethical criteria;
- environmental criteria;
- the pedigree of the fund manager;
- the launch date and total sum invested to date;
- geographical spread;
- resources applied to screening; and
- products available from each fund.

Although some investors are satisfied with a reasonable return others wish to know that these investments compare favourably with those which are screened only in terms of their financial performance. Holden Meehan believe that there should be 'no special pleading' for ethical investments and the figures in Table 7 show comparisons for 1995.

The general prediction in 1996 is that smaller companies will perform better than larger companies and this will encourage more ethical investment. This seems to contradict the general finding that smaller companies do less to reduce their environmental impact than larger companies. However, when it comes to ethical investment it is possible to select those companies whose business is financially sound and ethically or environmentally responsible in a growing sector of the market. These companies should perform well. Holden Meehan conclude that 'ethical funds can produce above average returns over meaningful time periods.'

TABLE 7
Comparative Performance of Unit Trusts 1994/95

Description	*Growth %*
UK All-Share Index growth	20.14%
FTSE 100 Share Index growth	21.75%
Hoare Govett Smaller Companies Index growth	12.3%
Sector average growth rate for comparable funds	14.2%
Ethical funds growth rate	14.94%

Unit trusts on the international scene had an average growth rate of 6.5 per cent compared with 11.48 per cent for ethical unit trusts. New opportunities for ethical investments are being created. For example the Environmental Value Fund has been launched in Norway. It includes the most environmentally responsible companies in nearly every sector, rather than excluding, whole industries because they cause environmental damage. Kleinwort Benson Investment Management have launched a fund which selects 'Tomorrow's Companies' using the criteria developed in the RSA's Tomorrow's Company Report.

Local Agenda 21 and Business Development

Over two-thirds of local government authorities in the UK are working on their plans for implementing Local Agenda 21 (LA21). An integral part of this planning is the involvement of businesses, voluntary groups and citizens in identifying the priorities, building visions and implementing solutions. There is scope for much more involvement from business. The knowledge and experience of business methods, quite apart from the practical contribution of businesses in different sectors is widely sought.

A typical goal of LA21 is to reduce waste by 25 per cent by the year 2000. If this is not achieved the quantity of waste for disposal will increase. This could mean more lorries taking waste to landfill or the need to build another incinerator. In both cases there is likely to be an increase in road traffic, even more congestion, higher levels of emissions, adverse effects on health and possibly more interference with local trade. If wastes can be reduced this can mean less traffic, less need for landfill sites or an incinerator, and less pollution. The way in which solutions are devised and implemented can help local business. It might be argued that the larger organisations may be the losers and new construction projects could be blocked or become redundant. Against that there is considerable scope to devise ways in which businesses can contribute to the development of new solutions. Local Agenda 21 groups usually welcome more cooperation from businesses.

Applying ESI

Adopting an ethical approach, linked to enlightened self-interest as the basis of innovation has already been emphasised. Using this framework for taking initiatives can be very productive. For example Electrolux is a good example of a company that has done this. Statements from their Chief Executive and the Vice President of Environmental Affairs are used to illustrate their approach and are summarised in Table 8.

TABLE 8

The Electrolux Approach

The Electrolux approach is based on the non-negotiable systems conditions established by The Natural Step described. These provide strategic guidance for every aspect of their business, Johansson says: We wont to integrate environmental *imperatives* into *management* strategy *and into* the *thousands of decisions all our employees make.*

Per Grunewald, Senior Vice President Environmental Affairs, adds:

> Our *environmental strategies have* been *determined. We are now working intensively to integrate them into our operations.*

As a worldwide company Electrolux takes account of different cultures, different stages of maturity and adjusts the pace of change to suit each country.

The Co-operative Bank is another example of successful application of ethical principles for the development of their business.

Beyond Leadership

Warren Bennis, Jagdish Parikh and Ronnie Lessem have written a stimulating book entitled *Beyond Leadership.* All three authors are well known exponents of progressive practices in management, organisation development and leadership. In their book they put forward ideas that integrate concepts from personal development, organisational learning and sustainable development into a leadership approach. They combine the following ideas:

- self mastery—the competence of the *individual manager,*
- group synergy—*good teamwork* to achieve the best results;
- organisational learning—the ability to *learn as an organisation;*
- societal sustainability—emphasises *sustainable development.*

These four elements are seen as the key to achieving a transition from the old paradigm of conventional business to the new paradigm that tomorrow's society will need. Self-mastery combines intuition and vision to provide the basis for effective action. Good teamwork emphasises the importance of managing situations rather than telling people what to do. It includes sound management of conflict as the key to group synergy. The ability of people in organisations to learn together includes self-mastery and group synergy but adds systems thinking and the ability to concentrate on mastering complexity rather than subordinating people. Sustainable development includes the ability to manage in the age of information and emphasises the importance of ecological, social and economic sustainability. Those who achieve this integration leave the world of traditional leadership and move towards the ideas of beyond leadership.

Transitions

The old ways are changing as more people reach out for a sustainable future. This means that in many areas new thinking and new behaviour is needed. The juxtaposition of traditional thinking and innovative ideas can be a helpful way to appreciate what is involved. For example a summary of the main transitions which have been explored in earlier chapters are set out in Table 9.

TABLE 9

Society in Transition

Traditional values and thinking		*Emerging values, innovative Ideas*
Materialism is the key to progress	——>	Materialism alone is not enough
The Earth is resilient	——>	The Earth is threatened
Man can change nature	——>	Natural Laws are immutable
Social trends continue as usual	——>	New social trends are emerging
Profit drives business	——>	Stakeholders want more than profit
Self-centredness prevails	——>	Partnerships are gaining strength
Short-term perspectives	——>	Long-term perspectives
People feel powerless	——>	People fed empowered

Learning occurs mainly at school	——>	Learning is for life
Adversarial approach to conflict		Creative resolution of conflict

Building to Last

Companies seeking lasting success have known for years that they need to be sound financially. Those wanting a competitive edge know that the competence of their people is equally crucial. Market awareness is a third area which is accepted as vital. However, the inclusive approach, adds a new dimension. It is now important to meet the needs of all stakeholders—shareholders, employees, customers, suppliers and the community. This approach shifts the emphasis of the key customer interface from making transactions to developing relationships. The quality of the relationship with all stakeholders is an essential goal for companies which are building to last.

Alongside the inclusive approach these organisations need to explore the implications of sustainable development. Recognising the ecological imperative and the non-negotiable systems conditions established by The Natural Step is another imperative. The first environmental challenge for companies is to establish environmental policies which lead to cutting waste and emissions. The second challenge is to understand the implications of sustainability. This can only be done effectively when ecological considerations permeate every aspect of the business.

Ecological factors will then be considered alongside financial considerations in every decision. Examples of this have been described earlier in this book and there is little doubt that more will emerge in the years ahead.

The inclusive approach and sustainable development are the twin challenges for future leaders. Both the inclusive approach and sustainable development depend on extensive involvement and understanding of people at every level. This means that organisations will need to create participative and adaptable organisation cultures. The future contains many uncertainties and organisational adaptability will be a crucial factor in building an organisation which lasts.

An example of how a large organisation came to terms with the need for radical change is described by Karl-Henrik Robert telling the story about IKEA being introduced to the Natural Step (See Table 10).

TABLE 10

IKEA and The Natural Step

IKEA, the Swedish furniture manufacturer and retailer, is an interesting example of how a large, successful business first doubted their ability to adopt The Natural Step's non-negotiable systems conditions, then changed their mind. Karl-Henrik Robert, founder of The Natural Step, tells the story.

The management team at IKEA agreed that the systems conditions could be applied to farmers and municipalities but not to IKEA. The reason was that they prided themselves on supplying the cheapest possible furniture to the customer. They were convinced that using recycled metals, decreasing the use of chemicals and plastics, and buying only timber grown in sustainable forests would add greatly to their costs. This would mean that others would undercut IKEA's price and ruin their business. As always Karl-Henrik refused to recommend what they should do because he maintains that those who know the business know more than he does about how to accommodate the conditions for their own unique circumstances.

The breakthrough came when one of the IKEA managers realised that The Natural Step was not being prescriptive but suggesting they go right up to the obstacles and push. He suggested that IKEA could manufacture two ranges of the same product. The original range would not change but they could add a range of products which took account of all four conditions. Both ranges could be the cheapest of their kind. The founder of IKEA wants to deal with this issue in front of the customer by saying:

> This is our new *sofa. The quality of this new sofa is not only inherent in its construction but extends right across the entire ecosystem. Sure, it's more expensive, but this quality of sofa cannot be found at a better price anywhere else!*

IKEA have now resolved never again to ignore the systems conditions and have embarked on an extensive training programme for all their employees. Once this commitment is accepted the practical day-to-day matters start being handled differently. When a garbage contractor failed to clear the rubbish the warehouse manager took the opportunity to empty the container and show his team the contents. Each item could be classified in terms of the systems conditions. Wasteful packaging, metals being thrown away, and hazardous chemicals being taken to landfill were all identified. From that moment the operational strategy was being brought in line with the newly agreed business strategy which is to honour the systems conditions.

Leadership

Throughout this book examples of outstanding leadership have been given and quotations have been used to illustrate the forward thinking of several chief executives. In particular the inspirational leadership of businessmen with a successful track record in their own spheres is worth noting. They come from many different countries and have made an impressive contribution towards sustainable development. They include Aurelio Peccei (Italy) cofounder of the Club of Rome, Maurice Strong (Canada) Secretary General of the UN Environmental Conferences in Stockholm (1972) and UNCED in Rio (1992), Stephan Schmidheiny (Switzerland) founder of the BCSD and author of *Changing Course,* Bjorn Stigson (Sweden) Executive Director of the WBCSD and Derek Wanless (UK), former Chairman of ACBE.

It is appropriate to end this chapter and the book by drawing attention to the type of leader that is likely to be successful in the years ahead. Leadership is required in every sector of the community and can come from any level of an organisation.

Three types of situation have been identified in which leadership is required:

- Type 1: The leader can decide using judgement and experience; typically this requires adaptation of existing knowledge and often relates to technical solutions.

- Type 2: The situation can bo defined with some clarity but solutions are not clear cut or require strong commitment from other people for successful implementation.
- Type- 3: The situation is unclear, often complex and hard to define; sound answers or solutions are even harder to establish and often impact on many people in different ways.

Business examples which relate to these three types of situation are:

- Type 1: Making a decision about a complex, negotiated business deal involving one other party. The central issue is a judgement about the terms of the deal and whether they are acceptable.
- Type 2: Setting up a contract with a supplier who obtains raw materials from various sources. The quality of the solution depends on the supplier understanding what is required and being committed to honouring the terms of the agreement.
- Type 3: In a small town there is one major business which provides much of the local employment but is known to cause severe pollution, with some evidence of health risks to children. The situation is complex, there are uncertainties about the severity of the problems and changes to products and processes are expensive. The evidence about the problems has been accumulating steadily and local people are showing increasing concern. Something needs to be done but neither the definition of the problem nor the solution is clear.

In today's world there are an increasing number of situations, similar to Type 3, which are unclear and hard to define. Appropriate solutions are even more difficult to identify and it appears that the different needs of all parties cannot be satisfied. This unexamined assumption may not always be valid. With appropriate leadership it might be possible to find creative solutions which go a long way to meeting the needs of most of the people with an interest in the outcomes. However, this requires a different kind of leadership initiative from the two previous types of situation. In particular it requires the ability to involve people, listen to them and work with them to agree solutions. This kind of leadership can emerge at any level.

In some Type 3 situations the opportunity to take action rests with the chief executive of the business. There are clear ethical dilemmas but it is also in the self-interest of the business to take action sooner rather than later. A combination of ethics and enlightened self-interest creates the basis for innovative solutions. However, those solutions should meet the different needs of the interested parties who need to be committed to implementing the solutions. The best way to achieve this outcome is to involve people at an early stage in defining the situation and developing possible solutions, then evaluating them and agreeing action steps. This is a classic case for use of the ESI approach.

This stresses the importance of taking important initiatives. Businesses have much to contribute to solutions and their initiatives are crucial. Several practical methods are already being used by companies including task forces, awaydays and reviewing their vision and values,

As more organisations identify the need for bringing large groups of people together to develop creative solutions with productive outcomes the methods available for these events are being refined and improved.

In many cases there will be a need to examine the relevance of mindsets for a future that will be different. Changing mindsets is never easy, especially when they seem to have proved successful. However if they are not appropriate for business in the future it is better to revise them than to continue in the belief that all is well. For the smaller business there are added difficulties of finding the time to do this kind of work but the incentive to get it right is often very strong.

Some companies like the idea of being challenged by people who think differently. This has led to creative, open sessions with campaigners from voluntary organisations of various kinds. Knowing how these people think and work helps the business to choose the right people for effective meetings.

There is great scope for applying the ESI approach and to examine the right leadership approach. Building a business that will last well into the next century requires imaginative management of the transition to the new paradigm. This requires leaders who have the courage to bring people together in uncertain situations with no clear outcomes and jointly work towards agreed solutions.

Greenwash of Busines & WBCSD

Global outreach is an area on which WBCSD has been trying to focus, if only because it remains open to criticism for being an association of predominantly northern (European, North American, and Japanese) corporations. For all practical purposes, and for all the obvious reasons, it remains exactly that. Of the 1 23 member companies listed in its 1997 *Annual Review,* only one was from Africa, only three from Central and Eastern Europe, and only eight each from Latin America and Asia (not counting Japan). By way of making up for this imbalance, WBCSD works through a 'global network' of partner organisations with which it has formal and informal links. These range from intergovernmental and non-governmental organisations such as UNDP, the International Institute for Sustainable Development (IISD), and the Stockholm Environment Institute (SEI), to national business groups such as the Confederation of Indian Industry (CII), to a host of national and regional

Business Councils for Sustainable Development. Due to its very nature, however, the network remains a loose confederation with only sporadic and opportunistic links to WBCSD's agendas and activities. Although the WBCSD network has a long way to go to realise its potential, the Council has an obvious interest in strengthening it. Doing so will not only make the WBCSD a more representative global entity but could eventually move it in unintended directions.

For all its success, and partly because of it, WBCSD has also been the target of some scathing criticism, most notably from two NGOs, Greenpeace and Corporate Watch.

Through its various publications and campaigns, Greenpeace has popularised the term 'greenwash', which it defines as 'cynical, superficial, public relations marketing' aimed at projecting a falsely benign environmental corporate image. In fact, it coined the term to describe and expose one of WBCSD's parents, BCSD, and its corporate members. According to *Green or Greenwash? A Greenpeace Detection Kit,* a corporation that fails on any of the CARE criteria—core business, advertising record, research and development funding, and environmental lobbying—is 'probably in the greenwash business'. Like Greenpeace, Corporate Watch, which defines 'greenwashing' as 'the phenomenon of socially and environmentally destructive corporations attempting to preserve and expand their markets by posing as friends of the environment', considers WBCSD to be a 'front' group whose purpose is to greenwash the image of its 'dirty' members. In marking the fifth anniversary of UNCED in 1997, it focused unwanted attention on WBCSD by giving it the so-called Greenwash Award 'for its continuing... efforts to portray itself as the savior of the world's environment and the force that will eliminate poverty'.

Environmental fury directed at large corporations is not a particularly new or novel phenomenon. However, the greenwashing charge now being levelled against WBCSD and its cohorts is not just directed at the environmental harm being caused by certain businesses but suggests a conscious cover-up conspiracy to distract public attention and subvert the environmental agenda. To those who have long considered corporate greed as a leading cause of the environmental problematique, the ultimate insult is that big business is now being allowed to 'co-opt' and even 'define' the meaning of sustainable development. What Maria Elena Hurtado of Consumer International describes as 'a new policy by corporations to engage potential critics' is greeted with grave concern by activists. In noting that 'the WBCSD has been tremendously successful in promoting global market liberalisation and self-regulation by business instead of government intervention as the recipe for sustainable development'. they are outraged that 'the world's governments have allowed corporate greenwash to thwart progress in environmental protection.' Behind what he calls a 'masterful co-optation of ecology', the author Joshua Karliner (1997) sees a 'highly evolved rendition of corporate positioning on issues of environment and development'. His book *The Corporate Planet* elaborates:

> The Earth Summit marked the coming of age of corporate environmentalism—the melding of ecological and economic globalisation into a coherent ideology that has paved the way for the transnational to reconcile, in theory and rhetoric, their ubiquitous hunger for profits and growth with the stark realities of poverty and environmental destruction. In the aftermath of Rio, global corporate environmentalism has helped institutionalise ecological concerns as agenda items in the executive suites and boardrooms of some of the world's largest businesses. It has helped build a public image of transnational corporations as the world's responsible global 'citizens'. It has also, to a certain degree, begun to set the terms of the debate along lines favorable to the transnationals... Indeed, by focusing a relatively small portion of their vast resources on environmental issues, the global corporations have, in many respects, reframed much of the environmental discussion.

To its critics, WBCSD is guilty of wilful duplicity. According to Greenpeace:

> The public message of WBCSD and other groups has been that business now understands and supports the goals of sustainable development and environmental protection, and business will be the leaders achieving both. At the same time, they have been working to avoid regulations of their activities, and working against agreements in the very regimes that UNCED spawned, such as the Climate Convention.

Corporate Watch chimes in with a similar verdict:

> The overall tone of recent WBCSD pronouncements is one of reassurance; to governments and NGOs, reassurance that business understands and is voluntarily taking action; and to their members, reassurance that things are changing but not too fast; that some action is needed but not too much. This carefully Grafted tone of heartening ambiguity masks the reality that WBCSD members, along with many other large corporations, have pushed hard over the last five years for increased corporate power on the global stage... There is simply no evidence that increased corporate rights has led or will lead toward sustainable development or environmental protection, yet this assumption underlies the WBCSD philosophy.

To its credit, the WBCSD has engaged its critics in a dialogue, albeit at a limited level. It has essentially argued that the case laid out by its detractors is oversimplistic, conspiratorial, exaggerated, and based on an unrealistic understanding of how business works. In addressing the critics, the WBCSD Director for External Co-operation, Ted Button, makes the following arguments:

It is inevitable that our proposals on sustainable development will irritate and sometimes annoy pressure groups. Maybe business is not going far enough and fast enough for some participants in the debate, but the Council is genuinely dedicated to making a difference. Also, on some issues we are taking the lead in finding solutions to the sustainable development challenges... This is rather more difficult than our detractors believe, not least because the international business community is as varied in its make-up, ideals, approaches and needs as the various environmental and social groups who urge us to move further and faster.

Green or Mean?

So, is the story of the WBCSD a story about the greening of big business or one about greenwash? The correct answer, as in all such questions, is a little of both.

Indeed, WBCSD has played an important role in making major corporations embrace the concept of sustainable development. While not all are equally enthusiastic about how the concept might change as a result of this bear-hug, the fact remains that there have been lively discussions, if not any deep soul-searching, in at least a few boardrooms. That should be a cause for some satisfaction. At the same time, it is abundantly clear that at least some of what WBCSD does can legitimately be called greenwash. After all, it is part of WBCSD's mandate to put big business's 'greenest' foot forward. In singing the virtues of 'best practice' by member companies it has certainly been guilty of overlooking some of the 'worst practice' by the same corporations. Even if WBCSD itself is not consciously in the business of greenwash, some of its member companies with less than perfect environmental records have probably used their association with it as a means to greenwash their soiled public image.

In the ultimate analysis, the most important thing in this 'Green or Mean?' debate is not to determine which side is right and which is wrong. What is more important is to realise that each has been doing what its institutional form mandates it to do. WBCSD, as the environmental advocate for business that it is, has been highlighting the positive linkages that industry can, and sometimes does, have with sustainable development. Groups such as Greenpeace and Corporate Watch, as the environmental watchdogs that they are, have been focusing public attention on the negative linkages that industry can, and so often does, have with sustainable development. Both are doing what they are supposed, and expected, to do. International efforts for global sustainable development would probably be that much weaker if either were to act significantly differently.

Assessing Impact

In assessing the impact that WBCSD has had on the international environment and development, the most important question relates not to perception, but to impact. In the case of WBCSD, it is important to evaluate its impact separately at two distinct levels: business practice in the corporations whose interests it claims to represent, and international policy at the forums that it seeks to influence.

On the first of these, *business practice related to the environment,* it is not clear what the exact impact has been or is likely to be in the future. This is partly because of self-selection in membership. Companies have found WBCSD useful for improving their public image by highlighting the environmental innovations that they might have undertaken, or simply by associating themselves with other innovative companies. However, the innovations themselves seem to have happened separately from, and not necessarily because of, WBCSD. To be fair, not enough time has passed for the organisation to trigger significant changes in business practices. Yet there are some areas where its work has the clear potential of creating positive change. One of the most significant of these relates to metrics for measuring eco-efficiency. This is one area where the Council is beginning to move from simply highlighting best practice to actually inducing good practice. An initial survey of environmental reports of member companies has been completed. However, at this point their efforts remain focused on the relatively less threatening question of how' companies measure eco-efficiency and shy away from evaluating just how eco-efficient they really are. The link between sustainable development and the financial sector is another area where WBCSD has initiated some interesting discussions. Again, however, the thrust is on documenting rationale rather than holding companies accountable to some framework of evaluation. While the potential for meaningful impact in both areas is immense, the real test will be in the earnestness with which corporations (including WBCSD members) apply these principles.

At a secondary level, there has been some diffusion of information about 'best practice' among WBCSD member companies, and beyond that through its publications programme. More important than that, however, may be the discussions that WBCSD has facilitated and initiated within business, and between business and other stakeholders, on various issues related to operationalising sustainable development. While it is obviously difficult to gauge the direct impact of such activities, it is clear that this has the potential for eventually triggering change in corporate practice. In short, although demonstrable evidence of clear change in corporate practice because of WBCSD actions is hard to come by, the organisation deserves at least a passing grade for initiating activities and programmes that have the potential for bringing about positive environmental innovation.

On the second issue, influencing *international policy and institutions for sustainable development,* the performance has been clearly impressive. What some in 1992 had considered as being no more than Stephan Schmidheiny's personal enthusiasm for the subject has, in fact, been institutionalised. In 1999 WBCSD is a bigger and more influential player than the founders of either BCSD or WICE (or their critics) might have predicted. As an advocacy organisation, WBCSD has been an undeniable success. Owing to the critical constituency that it legitimately represents and the human and financial resources at its command, it has been able to attain an impressive presence and prestige in international forums in a very short period of time.

A clear indicator of this influence is its now undisputed position as the most representative and authentic voice for big business on environmental issues and its ability to use this position to reframe discussions on sustainable development around its chosen parameters. By similar token, the potential for WBCSD's most important influence in the long term lies in the planned WBCSD Virtual University being set up under the aegis of FBSD and the inroads that the organisation is already making into universities and their business curricula. By its own accounts, as well as those of its critics, WBCSD has managed to gain deep penetration into the corridors of global environmental policy and will be a major player in international environmental policy for the foreseeable future. While some will question the quality of this influence, the influence itself is not in doubt.

What does all of this mean for the larger enterprise for global sustainable development? Can WBCSD ever reconcile the practical and conceptual differences between the goals of sustainable development and corporate profit? Can one wear Birkenstocks with a pin stripe suit? Maybe, but not just yet.

Friends of the Earth

Friends of the Earth International (FOEI) is an international network of environmental organisations in 69 countries.

In contrast to many other NGOs operating internationally, Friends of the Earth International is structured from the bottom up as a as a confederation of groups. Each country has its own separate different organisation, which in many cases existed and campaigned in its own right before choosing to affiliate to the global network. The groups conduct their own campaigns and coordinate their activities through the umbrella body Friends of the Earth International. The national groups are often composed of grassroots local groups working in their own areas.

FOEI is assisted by a small secretariat (based in Amsterdam) who provide support for the network and its agreed major campaigns. The executive committee of elected representatives from national groups set policy and oversee the work of the secretariat.

Campaign Issues

Friends of the Earth considers environmental issues in their social, political and human rights contexts. Their campaigns stretch beyond the traditional arena of the conservation movement and seek to address the economic and development aspects of sustainability. Originally based largely in North America and Europe, its membership is now heavily weighted toward groups in the developing world.

The current campaign priorities of Friends of the Earth internationally are:

- climate change
- corporates and Corporate Accountability
- GMOs
- forests
- Public Finance which addresses international financial institutions, such as the IMF, the World Bank, and export credit agencies.
- trade and its impacts on environment and sustainability

The campaign priorities are set at the bi-annual general meeting of Friends of the Earth International.

In addition to the priority campaign areas Friends of the Earth International has a number of other campaign areas which are active internationally, they include:

- desertification
- Antarctica
- water
- maritime
- mining and extractive industries
- no to nuclear power

All FoE International campaigns incorporate elements of three core themes which are:

- protecting human & environmental rights
- protecting the planet's disappearing biodiversity
- the repayment of ecological debt owed by rich countries to those they have exploited.

Friends of the Earth Groups

The Friends of the Earth organisations in each country are themselves many-tiered networks reaching from grassroots activists up to the national pressure group which campaigns for environmentally progressive and sustainable policies. The groups and activists at all levels also carry out educational and research activities.

Friends of the Earth groups are required to act independently of party political, religious or other influences; be open, democratic and non-discriminatory in their internal structures; and be willing to cooperate with other organisations who are working for the same goals. These are conditions of remaining a member of FOEI.

The national groups work on the main issues affecting their own country and choose to participate in the international campaigns of FOEI which are relevant to them. In turn, the local (grassroots) campaigners can work on local, national and/or international campaigns.

Structure of the Network

The member organisation in a particular country may name itself Friends of the Earth or an equivalent translated phrase in the national language, e.g., Friends of the Earth (US), Friends of the Earth (EWNI) (England Wales and Northern Ireland), Amigos de la Tierra (Spain and Argentina). However, roughly half of the member groups work under their own names, sometimes reflecting an independent origin and subsequent accession to the network, such as ERA (FOE Nigeria) and WALHI (FOE Indonesia).

Friends of the Earth International (FoEI) is supported by a secretariat based in Amsterdam, and an executive committee known as ExCom. The ExCom is elected by all member groups at a general meeting held every two years, and it is the ExCom which employs the secretariat. At the same general meeting, overall policies and priority activities are agreed.

In addition to work which is coordinated at the FoEI level, national member groups are free to carry out their own campaigns and to work bi- or multi-laterally as they see fit, as long as this does not go against agreed policy at the international level.

History

Friends of the Earth was founded in 1969 in the United States by David Brower after his split with the Sierra Club based in part on their reluctance to challenge the construction of nuclear power plants. It became an international network in 1971 with a meeting of representatives from the U.S., Sweden, the UK and France. For further historical details see articles on the national FOE organisations.

Notable Supporters of Friends of the Earth

- Radiohead frontman Thom Yorke, well known for his political engagement, is a supporter and member of Friends of the Earth. He has played a benefit concert in support of the organisation and was preparing to meet Prime Minister Tony Blair on the organisation's behalf, though this meeting fell through.

- Dominique Voynet, who ran in the 1995 and 2007 French presidential elections in the "les verts" parti, or Green Party is a member.
- Jay Kay, frontman of the funk/acid jazz group Jamiroquai, is known for donating profit earned from his album sales to Friends of the Earth and Oxfam, among other things.

Support for The Big Ask

Among those present at the launch of Friends of the Earth (EWNI)'s climate change campaign The Big Ask were: Jude Law, Jemima Khan, Edith Bowman, [Sian Lloyd]], Ross Burden, Thom Yorke, David Cameron, David Miliband, Stephen Merchant, Michael Eavis, and Emily Eavis.

Bright Green Environmentalism

Dark Greens, Light Greens and Bright Greens

Contemporary environmentalists are often described as being split into three groups, "dark", "light", and "bright" greens.

"Light greens" see protecting the environment first and foremost as a personal responsibility. They fall in on the transformational activist end of the spectrum, but light greens do not emphasise environmentalism as a distinct political ideology, or even seek fundamental political reform. Instead they often focus on environmentalism as a lifestyle choice. The motto "Green is the new black" sums up this way of thinking, for many. Though many environmentalists of all stripes use "lite green" to describe products or practices they believe are greenwashing.

In contrast, "dark greens" believe that environmental problems are an inherent part of industrialised capitalism, and seek radical political change. Dark greens tend to believe that dominant political ideologies (sometimes referred to as industrialism) are corrupt and inevitably lead to consumerism, alienation from nature and resource depletion. Dark greens claim that this is caused by the emphasis on economic growth that exists within all existing ideologies, a tendency referred to as "growth mania". The dark green brand of environmentalism is associated with ideas of deep ecology, post-materialism, holism, the Gaia hypothesis of James Lovelock and the work of Fritjof Capra as well as support for a reduction in human numbers and/or a relinquishment of technology to reduce humanity's impact on the biosphere. More recently, "bright greens" emerged as a group of environmentalists who believe that radical changes are needed in the economic and political operation of society in order to make it sustainable, but that better designs, new technologies and more widely distributed social innovations are the means to make those changes — and that we can neither shop nor protest our way to sustainability. As Ross Robertson writes, "[B]right green environmentalism is less about the problems and limitations we need to overcome than the "tools, models, and ideas" that already exist for overcoming them. It forgoes the bleakness of protest and dissent for the energising confidence of constructive solutions."

Origin and Evolution of Bright Green Thinking

The term "bright green", first coined in 2003 by writer Alex Steffen, refers to the fast-growing new wing of environmentalism, distinct from traditional forms. Bright green environmentalism aims for a society that relies on new technology and improved design to achieve gains in ecological sustainability thereby increasing the potential for economic growth and attending to human needs.

Its proponents tend to be particularly enthusiastic about green energy, hybrid automobiles, efficient manufacturing systems, bio and nanotechnologies, ubiquitous computing, dense urban settlements, closed loop materials cycles and sustainable product designs. "One-planet living" is a frequently heard buzz-phrase. They tend to focus extensively on the idea that through a combination of well-built communities, new technologies and sustainable living practices, quality of life can actually be improved even while ecological footprints shrink.

The term "bright green" has been used with increased frequency due to the promulgation of its ideas through the Internet and recent coverage in the traditional media.

Currents

The Viridian design movement is an aesthetic movement focused on bright green environmentalist concepts. The name was chosen to refer to a shade of green that does not quite look natural, indicating that the movement is about innovative design and technology, in contrast with the "leaf green" of traditional environmentalism. The movement ties together environmental design, techno-progressivism, and global citizenship. It was founded in 1998 by Bruce Sterling, a postcyberpunk science fiction author. Sterling remains the central figure in the movement to this day, with Alex Steffen perhaps the next best-known. Steffen and Jamais Cascio, along with some other frequent contributors to Sterling's Viridian notes, formed the Worldchanging blog.

Technogaianism is the bright green environmentalist stance that emerging and future technologies can help restore Earth's environment, and that developing safe, clean, alternative technology should therefore be an important goal of environmentalists. Sociologist James Hughes mentions Walter Truett Anderson, author of *To Govern Evolution: Further Adventures of the Political Animal*, as an example of a technogaian political philosopherargues that technogaianism applied to environmental management is found in the reconciliation ecology writings such as Michael Rosenzweig's *Win-Win Ecology: How The Earth's Species Can Survive In The Midst of Human Enterprise*; and considers Bruce Sterling's Viridian design movement to be an exemplary technogaian initiative.

REFERENCES

Hughes, James (2002). *"Democratic Transhumanism 2.0"*. Retrieved on 2007-01-26.

Hughes, James (2004). Citizen Cyborg: Why Democratic Societies Must Respond to the Redesigned Human of the Future. Westview Press.

Rosenzweig, Michael *(2005). Win-Win Ecology: How The Earth's Species Can Survive In The Midst of Human Enterprise. Clarendon Press.*

Steffen, Alex *(2006). Worldchanging: A User's Guide for the 21st Century. Harry N. Abrams.*

Sterling, Bruce (2001). *"Viridian: The Manifesto of January 3, 2000"*. Retrieved on 2007-01-28.

WorldChanging: Tools, Models and Ideas for Building a Bright Green Future: Don't Just Be the Change, Mass-Produce It

WorldChanging: Tools, Models and Ideas for Building a Bright Green Future: Reports from the Team

12

Green Conventions: Towards Regulating Global Science and Technology Towards Environmentally Sustainable Development

Green Conventions

Green conventions or green meetings are conventions which are conducted in ways which minimize the environmental burdens imposed by such activities. Green event planners apply environmentally preferred practices to waste management, resource and energy use, travel and local transportation, facilities selection, siting and construction, food provision and disposal, hotels and accommodations, and management and purchasing decisions.

Green event and convention planning is now an established trend within the global tourism and convention industry. Several cities in the United States and Europe now sport green convention centers designed using green building principles and practices.

Several high visibility events including the Olympic Games in Italy, Sydney, Utah, and Greece; the World Summit on Sustainable Development in Johannesburg, and the 2004 Democratic and Republican National Conventions have implemented green practices with varied success. A more recent example is that of Live Earth, a series of worldwide concerts held on July 7, 2007, that initiated a three-year campaign to combat climate change. Such efforts aim to conserve resources, protect air and water quality, habitat and human health, and showcase sustainability practices and concerns.

Parts of the tourism and convention industry now promote green meetings, conferences, and convention planning as demand for sustainability measures increases. Industry associations have produced standards and guides for green meetings. Government agencies and non-profit organisations also promote these practices with research, recommendations, grants and technical support. Some private consultants in the meeting planning industry specialize in mounting green events, and industry groups and governments now sponsor awards to recognise achievements.

Green conventions, meetings, conferencing and events are part of an international movement to achieve a sustainable world economy and livable planet.

THEORIES OF INTERNATIONAL ENVIRONMENTAL LAW

The revolution in science and technology, the intensive utilisation of natural resources and rapid industrialisation has increased a tremendous pollution that has brought the human race at the brink of serious ecological crisis. It has awakened States, United Nations and other international organisations to take stock of the situation and adopt measures so as to prevent disastrous effect of pollution by evolving pre-emptory norms of international law. But evolving such a law at international level to prevent all types of pollution is a difficult task as it involves the various factors like geographical distribution of States, climatic, economic, and social conditions of States etc. However, evolving an international legal system for prevention of some of serious global pollutants cannot be ruled out. In spite of difficulties in getting a nexus for a uniform international legal system, efforts have been made in this direction. Here it would be pertinent to provide in brief some of the theories relating to adoption of international environmental legal system.

Theory of Cooperation Among States

This theory stresses that States should cooperate in conservation and harmonious utilisation of natural environment and its resources. As to the extent and the application of the theory in environmental matters, Dr. Nagendra has observed as follows:

1. The general duty to cooperate has to be elaborated in several specific obligations concerning the exchange of information like prior notice of planned activities, cooperative activities environmental impact assessment, protection and emergency situations etc.; and
2. Apart from obligation put on States, the legal experts also would like to strengthen the legal position of individuals including private person, who have been or may be detrimentally affected by a transboundary interference with their use of transboundary natural resources or by transboundary environmental interference.

The above view regarding this theory seems to be relevant as it found support in international declaration on conventions and protocols etc. For instance UN Conference on Protection of Human Environment, 1971 (hereafter referred to as UNCHE, 1972/ Stockholm Conference), Vienna Convention for the Protection of the Ozone Layer (Final Act) 1985; Montreal Protocol Substances that Deplete the Ozone Layer, *1987;* Base Convention on the Control of Transboundary Movement of Hazardous Wastes and their Disposal (Final Act) 1987 Convention on the Prevention of Marine Pollution of Dumping of Wastes and other Matter, 1972; Convention on Long-Range Transboundary Air Pollution 1979 have sought the help of the theory of cooperation. The United Nations Environment Programme and Principles of Environmental Law has also adopted this theory.

Though, the theory of cooperation as observed received support in declarations and conventions etc., nevertheless it cannot be said to have been applied completely. The reasons are:

(a) there is no nexus for getting cooperation from States for this purpose;

(b) States cannot be compelled to cooperate and it is at their discretion to do it or not; and

(c) in the way of cooperation comes economic social and political interests of States.

However, the theory of cooperation may prove better at least in case of environmental protection if it is supplemented with some basis for the cooperation of States interse.

Principle of Sovereign Equality

This theory enumerates that all States enjoy the sovereign equality and sovereign right to dispose freely of their resources in conformity with the policy of environment and also implies that each State will respect that right in relation to their States. They have equal rights and duties and are equal members of international community, notwithstanding differences in economic, social/ political or other nature. In particular, Sovereign equality includes:

(a) States are juridically equal;

(b) Each State enjoys the rights inherent in full sovereignty;

(c) Each State has the duty to respect the personality of other states;

(b) The territorial integrity and political independence of States are inviolable;

(e) Each State has the duty to comply fully and in good faith with its international obligation and to live in peace with other States.

According to this theory, therefore, no State may intervene in matters within the domestic jurisdiction of any State as envisaged in UN Charter in its Article 2, Paragraphs 1-4. Further the Charter of Economic Rights and Duties, 1974, provided in its Article 1, that 'Every State has the Sovereign and inalienable right to choose its economic system as well as its political, social, and cultural systems inaccordance with the will of its people without outside interference, coercion, or threat in any form whatsoever'. This is generally believed that the origin of the theory lies in the early modem history of international law which gave importance to the relationship between the law of Nations and the Law of Nature. Christien Wolff's major work emphasised that 'By nature' all nations are equal the one to other. For, nations are considered as individual free persons living in a state of nature. Therefore, since by nature, all men are equal and thus it is inferred that all nations too are equal At present this theory with a minor changes in its concept, subsists, and imports not merely equality of law but also specify for equal rights and equal duties for the utilisation and protection of the environment.

The International Law Commission in its 39th session held in 1987 applied this theory in matters of sharing of water of inter national rivers and their safe use by riparian States. Through the utility of the concept in the adoption of different declarations conventions, and treaties cannot be ruled out, its applicability has been questioned because of the recent trends in change of the rule of unanimity to the rule of majority. Thus, side by side with the principle of equality there were, however *de facto* inequalities which were perforce recognised. It may be observed that the line between the equality of States and their independence tends to become blurred. It may be said that the right of a State freely to choose and develop its political, social, economic and cultural systems appertains to equality but *stricto sensu* this right is merely an expression of States independence.

Thus, theoretically the principle of independence of States, gives each State the sovereign right to dispose freely of its own resources in accordance with its policy of the environmental, and obligates other States to respect that right, but practically it failed due to the evolution of the theory of decision by majority. For instance commenting on the Earth Summit one observer has aptly noted that "the term 'global' as it is emerged in the discussions at the UNCED is not associated with universals humanism. The life of all people including the poor of the third world or the life of the planet are not at the centre of concern in international negotiations on global environmental issues". Instead she notes "The global', in the dominant discourse is the political space in which the dominant 'local' seeks global control. The 'global' does not represent the universal human interest. It represents a particular local reach and control.

Theory of Restrictive Sovereignty

Normally, a State is deemed to possess independence and sovereignty over its subjects and territory, but for the purpose of international law the States have restricted their sovereignty. At present, almost all States have accepted restriction on their liberty of action for the interests of the international community. Not only States but many other international organisations and institutions associated with the United Nations have undertaken obligations limiting their unfettered discretion in matters of international policy. It is therefore, probably correct to say that the sovereignty of States means the residium of power which it possess within the confines laid by international law.

This theory has been applied in the case of enforcement of international environmental law also. In this matter the 'duties' imposed upon States import a wide interpretation and connotes the transboundary pollution, law of States and corporation also. The trial Smelter Tribunal which applies the *maxim sic utero tuo lit alienum laedus* (hereafter referred to as *Sic utero tuo)* expressed the view that under the principles of international law as well as the law of the United States, no State has the right to use or permit the use of territory in such a manner as to cause inquiry by fumes in or to the territory of another or the properties of persons, therein, when the case of serious consequences and the injury was established by clear and convincing evidence. It may be observed that the Tribunal while applying the principle of sic *utero tue* and strict liability had affirmed the scope of restrictive sovereignty. The theory also received recognition in many cases far example, the Corfu Channel Case (1949); The Namibia Case (1971); The Lake Lanoux Arbitration (1955); and the case of India and Pakistan concerning the diversion of the waters of Indus.

In addition to above, theory of restrictions on States sovereignty has found support in different promulgations, declarations, treaties, and conventions etc. The Conference of International Law Association, 1956, promulgated regarding the prevention and abatement of pollution of international water. It lays emphasis on the reasonable use of the water and recognise State's responsibility arising as a result of harm or injury done to another State in the sense that a detrimental change in the environment has been brought about through some resolutions. Further, the well known Helsinki Rules, 1966, have made detailed provisions regarding the question of State responsibility for extra territorial damage. Art. 10 of the Rules, stipulated about the prevention of water pollution and stands to face legal consequences arising out of growing pollution.

One of the issues at Stockholm Conference, 1972 was how to reconcile the sovereign rights of States to develop their own resources in their own way with the necessities of protection of the environment. Principles 21 and 22 of the conference, proclaims, *inter-alia,* that States have sovereign rights to exploit their own resources pursuant to their own environmental policies, while remaining responsible for ensuring that activities within their jurisdiction or control do not cause damage to the environment of other States.

The theory of restrictive sovereignty, have been considered by States to evolve environmental laws to protect environment but the interest of developing countries came in the way. However major gap would have to be filled up as a part of the transition to sustainable development. The World Commission on Environment Development (hereafter referred to as WCED) in its report, calls into question the desirability or even the feasibility of maintaining an international system that cannot prevent one or more States from damaging the ecological basis for development and the very prospects for survival of other or possibly all States. WCED observes the States shall, in principle, prevent or abate any transboundary environmental interference or significant risk thereof which may give rise to substantial harm and hence claim to reparation and damages. It is most interesting to see to what extent this basic susbstantive principle which may be deemed to be generally accepted has been qualified by the legal experts of WCED. The experts propose that even when this basic norm is not breached (e.g. when there is no question of substantial harm or a significant risk)-States should be obligated not to discriminate between the external and internal detrimental effects of activities planned and carried out in their territory. Thus, Article 11, formulated by experts is of fundamental importance in view of the steady increase in ultra-hazardous activities and the incidents to which these activities may give rise, e.g. Bhopal Tragedy (1984), Chernobyl's accident (1986), and the chemical pollution of the Rhine caused by the Swiss Sandoz Company etc.

In addition to above mentioned principle, recently a few conventions and protocols were also adopted to reduce the effect of the 'green house effect, and 'ozone depletion'. For instance, Protocol Concerning Cooperation in Combating Pollution of the Mediterranean Sea by Oil and other Harmful Substances in case of Emergency; Convention for the Protection of the Mediterranean Sea Against Pollution 1976, The World Charter of Nature, 1982 (proclaims many principles for conversation of nautre), `The Vienna Convention for the Protection of Ozone Layer, 1985. The Montreal Protocol on Substances that Deplete the Ozone Layer (Final Act), 1987 and `The Basel Convention on the Control of Transboundary Movement of Hazardous Wastes and Their Disposal, 1989, have adopted the above theory.

The above discussions though show that the theory of restrictive sovereignty has got place in treaties and conventions, relating to environmental protection or the control of transboundary pollution but it could not be recognised as a general theory because of States' unwillingness to restrict their sovereignty for one or the other reasons. Thus, quest for getting a general theory for the environmental law at global regional or state level is still on.

Theory of Common Concern of Mankind

According to this theory the matter of protection of environment from pollution is growing as the 'common concern of the mankind', and as such it is a common global problem and be prevented by all nations collectively, not by each of them going its own way. Presently, "common concern of mankind', about environmental deterioration has been taken as relevant concern of the world community as a whole. The UNEP organised a meeting on 13-15 December 1990 at University of Malta, Malta, to examine the several aspects of the theory, more particularly the debate seems to have concentrated on the following points:

(a) Nature of the concept.
(b) The component elements.
(c) Possible applications.
(d) Extension of the concept.
(e) 'Common concern', 'common heritage' and 'common interest'.
(f) 'Common concern' and 'equitable sharing of burdens' and "environmental protection'.
(g) The right to a healthy environment as a human right; and
(h) Alternative to a convention.

(a) Nature of the Concept

There seems to have been a consensus on the notion that "common concern of mankind" is not a rule of general international law, and has therefore no binding force upon States at present. A predominant opinion considered the concept as evolving one, the importance of which was enhanced by its inclusion in General Assembly Resolutions and other Declarations. It could be useful however for working out principle in treaty law. The concern might become customary taw and thus accepted as a principle of international law.

(b) The Component Elements

The opinion was expressed that it might be undesirable to define or to describe, at present, the concept of "common concern of mankind". On the other hand, the analysis of the elements made by the secretariat paper was found to be acceptable in its general lines. However, and opinion was expressed in the sense that the "concern" element (in common concern of mankind) was elusive since it describes caused as well as responses to problems. Regarding "mankind" and its implications of "intergenerational rights", it was generally agreed that in a strict legal sense the concept of right should only apply to objects of international law. It was difficult to perceive "generations" as such subject. Reference was made to the *"continuum"* of generations and the impossibility of separating specific collective persons *"continuum"*. However, the concept as such of "inter-generational rights" was seen as a forceful and useful one.

(c) Possible Applications

It was generally agreed that the concept was applicable to environmental problems and specifically to those relating to climate protection. At least, as was expressed there has been no opposition to that concept when it was proposed.

(d) Extension of the Concept

The opinion was stared by many members that common concern was not confined in its applications to the protection of climate change. It may be well used in other fields of international law. For example the notion of the "concern" was referred to in a completely different context in the Barcelonia Traction dictum.

(e) "Common Concern", "Common Heritage" and "Common Interest"

It was generally agreed that "common heritage" has some proprietary connotations which are not to be found in the "common concern" notion. There were no question of one concept replacing the other. They were different and applied to different fields. It was correct to apply "common heritage" to the Sea-bed and ocean floors beyond national jurisdiction, but not to climate which was not susceptible of appropriation. The word "interest" has some connotations in common law system makes its use inconvenient in the present context. However in civil law "interest" has no such connotations and in fact the Spanish and French versions refer to "common concern" as *"Interes comun"* and *"interet commun"* respectively.

(f) "Common Concern"and "Equitable Sharing of Burdens"

The predominant opinion in the group was that equitable or fair sharing of burdens meant in the context of climate change, a burden on developed countries which would be proportional either to their historical responsibility in deteriorating the atmosphere or to their present share in the polluting process. The "equitable sharing" concept was found to be necessary instrument in the application of the "common concern" notion, perhaps a subsidiary principle to it in the field of climate change protection. As to the legal basis of the theory, the notion of fault (historical) of developed countries was rejected by some, and considered unnecessary by others, either because, no attention was directed to the past or the present participation of developed countries in polluting the atmosphere is enormous, or because, the modern trend in law, (chiefly domestic, but also noticeable in international law) is to focus attention more on the damage than on the fault. The capacity of States to contribute should also be a guiding principle in this matter, according to some opinion.

(g) "Common Concern" and the Balance of Sovereignty and Environmental Protection

The opinion was voiced that the balance between sovereignty and environmental protection was to be achieved, precisely, by the application of the "common concern" idea. Another ground for that balance was found in Principle 21 of the Stockholm Declaration.

(h) The Right to Healthy Environment as a Human Right

Although there was general agreement that environmental protection should be linked to the philosophy of human rights, on its general frame, the opinion was expressed that it would be convenient to avoid the doctrine of "generations" of human rights. A bridge between human rights protection and environment protection should be established by fundamental rights to life and health in their wide dimension. An opinion was also expressed in the sense that we cannot isolate environmental protection from the general issues of development, poverty makes it impossible for developing countries to pursue enlightened approaches to development. Dr. J.N. Saxena, has rightly observed that the right to an adequate environment as a human right is the synthesis of all human rights. The simple truth that all human rights are interdependent, each conditioning the remaining, offers a new and definitive support to the case of human right to an adequate environment.

(i) Alternative to the Convention

It was suggested that a Declaration of Principles by the General Assembly, or a code of conduct or a body of guidelines and principles might be useful in case a Convention was difficult to achieve in the short time between now and the Brazil Conference on Environment and Development. It was also suggested that if the goal of achieving a Convention by 1992, could not be attained, the process of negotiating a Convention should continue after that date.

In this way the Group of Legal Experts has examined various aspects of the theory of common concern to adjudge its applicability in evolving a general principle of international law for control of environmental pollution. However before it for recognition at Earth Summit 1992, in Agenda 21, it finds its presence in conferences, declarations and conventions also.

Application of the Theory

The theory of common concern has been supported in a number of recent

Declarations adopted by different political groups of countries. The 1989, Caracas Declaration, without explicitly referring to "common concern" expressed in clear terms the philosophy of global interdependence, which underlies the common concern concept. This should be in keeping with contemporary reality, particularly the full recognition of the compulsions of global interdependence. Today's problems can be solved only by nations acting together not by each of them going its own way. No country or group of countries, howsoever powerful it may be, can isolate itself from the consequences of global, social, economic and environmental problems or succeed without detriment to its own interest, in shifting the burden of solving these problems to the weaker members of the international community. Similarly the Ninth Conference of Non-aligned Countries, 1989. The Langkawi Summit Declaration 1989 and the Hague Declaration, 1989 had done accordingly by supporting the theory of common concern of mankind. Another action oriented. "Declaration" has emerged at the "Arch Summit" meeting of seven major industrial nations in July, 1989. The basis for this Declaration was the great concern about the growing pollution of air, lakes, rivers, oceans and seas; acid rain, dangerous substances and the rapid desertification and deforestation, which endanger species and undermines the well-being of individuals and societies. The above Declaration called for joint, actions to achieve the common goals of preserving healthy and balance global environment in order to meet shared economic and social objectives and to carry out obligations to future generations. In addition to above facts the ecological interdependence, which transpierces world community, has obtained not only spatial, but also temporal (Intergenerational) parameters. Possible actions stemming from the common concern, require innovative solutions, new approaches and even introduction of new principles of international law.

The Theory has been manifested at the regional level also in different environmental conferences. For example in Latin America and Carribean Summit on the Environment 1989. It has been recognised by the countries that need to strike a balance between socio-economic development and environmental protection and conservation through the proper management of natural resources and control of environmental impacts as a source of common concern to the countries having the utmost priority. Another case is that of Amazon Declaration 1989, in which State parties have expressed their common interest for the sustainable development of the region and consciousness that the ecological heritage must be preserved through the rational use of the resources of the region, so that the present and future generations may benefit from the legacy of nature. Further the twentieth South Pacific Forum expressed regional concern about the possible effects on Island countries of rising sea levels resulting from global warming and emphasised the importance of a regional approach to the environmental matters. It agrees that forum members take decisive action to draw world attention to the way the environmental problem affected to south-pacific and to represent regional views at appropriate international gatherings, possibly including, by way of a resolution in the U.N. General Assembly.

The concept has been applied in many positive environmental law. The Stockholm Declaration 1972 while adopting the theory, expressly referred in its principle 18, that science and technology as part of their contribution to social development, must be applied to the identification, evidence and control of environmental risks and the solutions of environmental problems and for the common good of mankind. The U.N. Charter on Economic Rights and Duties, 1974 stressed in its Art. 30 that the protection and preservation of the environment for present and future generation were the responsibility of all the states. Similarly, the 1972' World Heritage Convention, has an objective to establish an effective system of collective prevention of world heritage of universal value. 'The 1973 Convention on International Trade in Endangered species of Wild Flora and Fauna', recognises, Wild fauna and flora in their many beautiful and various forms are an replaceable part of the natural system of the earth which just be protected for generations to come.

One more aspect of the "common concern" is a social dimension. Common concern presumes involvement of all structures and sectors of the society into the process of combating global environmental threats-legislative judicial and governmental organisations. This, relatively new phenomenon, has been manifested through comprehensive environmental policies introduced by governments which needs, to be supported with legal guarantees. Some of the policies adopted by States are laid down in various Resolutions, Declarations and Conferences. U.N. General Assembly in 1988, 1989, and 1990 Caracas Declaration, 1989. Report of Ottawa Meeting of Legal and Policy Experts 1989. Noordwijich Declaration 1989 have laid down several policies and directives to protect the environment taking the climate as common concern of mankind. In this regard, the U.N. General Assembly in its resolution 42/11 B of December 1987 has rightly resolved that all those States who have yet not to become the parties to the Vienna Convention for the Protection of Ozone Layer, 1985, and The Montreal, Protocol of Substances that Deplete the Ozone Layer 1987, should do it soon.

From the above description it may be concluded the theory of 'common concern of mankind' is of particular value on account of its general application. It had found support into the jurisprudence of the International Court Justice, particularly in Barcelona Tractions Case (2nd phrase in 1970, wherein the judgement defined it *ergo, amnes.* Therefore it may be said that the theory has been accepted in practice. However the main problem in it is about the economic disparity in between developed and developing countries. The economic growth and other development aspects of developing countries are essentially considered before they are pressed to protect the environment. For this, effective modalities for additional financial resources and transfer of environmentally sound technologies developing countries at concessional and preferential term should be examined.

To conclude the above discussion with the message of Indian President on the eve of Environment Day 5th June, 1992 would rather clarify the concept and its application. His message runs as follows:

"The World Environment Day is being celebrated on
5th of June 1992. This year's (1992) theme is-
"Only one Earth-Care and Share".

Few words emphasise the fact that the environment knows no man made boundaries and Mother Earth is one for all its inhabitant. It is the responsibility of each one of us to take every care to protect our environment. Success in this endeavour depends on our commitment which should find expression through cur thought words, and deeds. Together we can and must save this planet. This is our common concern.

THEORY OF ENVIRONMENTAL DEGRADATION

It follows from this brief exposition of the theory of Himalayan Environmental Degradation, that a series of vicious circles, each linked to the others, is envisaged as operating inexorably to drive a downward spiral. The apparent impossibility to break any of these circles leads to the prediction of widespread environmental and socio-economic ruin in the near future. There is perceived to be a progressive and accelerating shift from *potential instability* to massive *actual instability.* This includes: mountain slopes, from a physical point of view; hill village subsistence agriculture; and the regional or national economy. All these gathering tragedies will put increasing pressure on the already fragile political balances of the wider Himalayan Region.

In the face of the irreversible destructive processes in the Middle Mountains, for instance, outmigration increases. This in ruin not only deprives the source areas - the villages and hamlets of the Middle Mountains - of a proportion of their youngest, fittest, and most creative members, but it adds to the already existing heavy population pressure on the resource base of the Nepal Terai. Population growth in the Terai, for example, including both natural increase and in-migration, is calculated at over 4.0 per cent per annum (1971-81 census; of Goldstein et al. 1983; Hrabovszky and Miyan, this issue, pp. 264-270). Ten-to-fifteen years ago, it was believed that the opening up to settlement of the previously malaria-infested jungle of Terai would provide a breathing space by enabling the absorption of excess population of the Middle Mountains, Nepal's most densily-populated area. Today it can be seen that the "new land" in the Terai is being rapidly used up, a process accelerated by the surge of illegal immigration across the open border with India. And yet the population increase in the Middle Mountains continues at an unacceptable, or unmanageable rate, in excess of 2.0 per cent per annum. Goldstein et al. (1983), for instance, characterise Nepal as being in transition, demographically, from a highland rural country to a lowland, urban one.

The net results of the various destabilising process in the Middle Mountains are perceived as absolute deforestation, lowered crop productivity, both in terms of total national production and as yield per unit area, increase in absolute numbers and percentage of the subsistence farming population whose nutrient intake is below a minimum acceptable level, and progressive mountain desertification. Since the mountain desertification is occurring on steep slopes, the associated process of gullying, soil erosion, and landsliding will have calamitous downstream effects. Thus are envisaged the rapid siltation of reservoirs, excessive shortening of the useful life of major hydroelectric and irrigation projects, increasing flooding on the plains (already an annual disaster for India and Bangladesh), increase in the levels of river beds, and destruction of rich lowland farmland by the spread of sand and gravel as rivers break their banks and change their courses. In short, the worst-case scenario foresees that the terrain of Nepal and that of adjacent areas of the Himalaya, and certainly the very basis of life, the top soil, will virtually flow down the Ganges by the year AD 2000. It has even been suggested that, in preparation for such an event, His Majesty's Government of Nepal should transfer its patronage of the Swiss technical aid system to that of the Dutch. In this manner Nepal can begin the struggle to reclaim (and legally claim) land below sea level and establish polders in the Bay of Bengal, the product of its own top soil. More than a decade ago, however, Eckholm wrote eloquently of the process whereby Nepal was exporting the commodity that it could least afford to part with, namely top soil, to India, in the form that India could least afford to receive it (Eckholm 1976).

The broad theory, nevertheless, is an intellectually satisfying concept which seems so reasonable that it is hardly surprising that it is widely accepted as fact. And, of course, there are further ramifications, such as claims by environmentalists that the deforestation, in turn, is affecting the climate in such a way as to reduce normal annual rainfall amounts. This, of course, would set up yet another vicious circle to accentuate the effects of the others.

The eight-point scenario and brief discussion, presented above, lead to a number of critical implications which further enlarge what can be described loosely as the *perceived* Himalaya-Ganges Problem. It infers that a few million Himalayan hill farmers are responsible for the massive landscape (and climate) changes that are affecting the lives and property of several hundred million people in Gangetic India and Bangladesh. This raises two related points: (1) that the downstream countries, as victims of the unwarranted and irresponsible environmental disruption, could justify reprisals in economic, political, or military terms; and (2) that Nepalese interests are served well (so long as no reprisals are actually taken) by this perceived image of helpless drift into environmental and socio-economic chaos, since it may account for its disproportionate amount of international and bilateral development aid in relation to its total size and population.

A further point is applicable within Nepal itself, and within several of the Indian states that have a plains and a Himalayan component, such as Uttar Pradesh. This popular image of the hill farmer as the cause of the growing environmental disaster makes him a convenient scapegoat; it has been claimed that the relatively few mountain farmers, for instance, are holding hostage the very many more of the plains (World Resources Inst., 1985). Once more, effect is taken for cause, and corrective measures run the risk of being misdirected.

Whether or not the eight-point scenario of disaster for Nepal can be extended along the entire Himalayan system will not be discussed further except for emphasis of several related points. The Kumaun and Garhwal Himalaya appear to fall within this framework with two additional components. One is the excessive commercial cutting of mountain forest stands to meet timber demands of the lowland population centres (until recently checked somewhat by the Chipko Movement). The other, associated with it, is the extensive development of mountain roads, especially as a military response on the part of India resulting from the border war of 1962 with China. Much of the road construction is substandard and has caused a great increase in landslide incidence; the roads also opened up extensive mountain forests to commercial clear-felling.

As we proceed westward into Himachal Pradesh, Jammu and Kashmir, and the Karakoram and Hindu Kush, we enter a series of mountain and highland landscapes with very different climataic regimes. Increasing aridity with distance from the influence of the summer monsoon greatly reduces the value of comparison with the Nepal Himalaya.

Conversely, eastward from Nepal, into Sikkim and the Darjeeling (West Bengal) Himalaya and Bhutan, we enter mountain areas with increasing amounts of summer monsoon precipitation. Bhutan, and probably the even less accessible and less well known Arunachal Pradesh, should be viewed as exceptional cases where assumed deforestation and environmental disturbance are modest or insignificant. However, even these regions may be poised to follow the same road to disaster along which Sikkim, Nepal, and the Central Himalaya of India are travelling.

There are several additional contribution problems that need to be introduced. The first is the political strife that besets the region: the border dispute between India and China; the three Indo-Pakistan wars of 1962, 1966 and 1971, and the continued border tensions, especially along the Kashmir cease-fire line; the frictions between India and Bangladesh, and especially the very slow progress in negotiations over management of the flow of the Ganges and Brahmaputra. There are also the political tensions within India-Punjab, Assam, West Bengal-generated, in part, by the competing demands for access to natural resources by different ethnic groups.

The rapid growth in tourism is also an important contributing process. Once again most data are available for Nepal. However, population access, to the Garhwal Himalaya, especially the valley of Flowers and the Nanda Devi Sanctuary, is seen to be causing a large increase in environmental pressure of an entirely different kind, the result of demands for recreation and adventure from people outside the mountain area. Fifteen years ago the growth of tourism in Nepal was perceived as a panacea for that country's balance of payment deficits. Today that perception has changed. Certainly in specific areas, such as Khumbu Himal and the Kali Gandaki and Annapurna circuit, the numbers of tourist visitors each year now exceed the total indigenous populations. And their demands are seen as increasing the threat to local forest as well as being disruptive of the way of life of the local people, with both positive and negative consequences.

The 1959/60 exodus of approximately 120,000 refugees from Tibet and their impacts on the natural resource base of Khumbu Himal and Mustang, for instance, and the consequences of China's closing of the frontier with Nepal, must be seen as unprecedented disruptive events. Finally, the much more extensive turmoil in Afghanistan, and the impacts of over three million refugees on the Pakistan. Hindu Kush and adjacent northern areas, is a socio-economic problem of world-magnitude in itself, notwithstanding the widespread political implications. Without a doubt, therefore, the broader Himalayan Region must be seen as dynamic in the extreme-geophysically, climatically, and in the socio-economic and political senses.

The presumed linkage between mountain deforestation and downstream effects also has been applied to the Hengduan Mountain-Chengdu Basin system, far to the east of the Himalaya proper. Here extensive deforestation in the "River Gorge Country" of western Sichuan and northwestern Yunnan, and its impact on the hydrology and sediment load of the Jinsha Jiang (Yangtze), is credited with increased flooting, siltation and damage to heavily populated and rich agricultural land downstream in the Chengdu Basin. As with the Himalaya, Ganges system, population growth, deforestation and soil erosion have been perceived as post-1950 phenomena, in this case with the added overtones of mismanagement during the Mao Tse-tung years with the Great Leap Forward and Cultural Revolution.

It must be concluded that the "save-the-world's-forests" lobby has had a powerful influence on large sections of public and scientific opinion. It has been Widely agreed that serious problems will arise from loss of the Tropical Rain Forests, and also from mountain deforestation in the Himalaya and elsewhere. The intent is not to challenge the importance of the world's forest resources but to examine the claims that are made in terms of the assumed effects of mountain forest clearance. It is postulated that this is a vital undertaking, a prerequisite to rational corrective policy development.

Linkages of the Theory and Their Implications

To sum up this overview of the perceived Himalaya-Ganges Problem, it is necessary to point out a number of basic assumptions and salient linkages between the component parts of the eight-point scenario. There is also a basic philosophical issue-what should be our attitude to the ignorant subsistence former who is seen, unthinkingly, to produce swarms of children, and irresponsibly to devastate the mountain forest cover, and so accelerate landslide occurrence on his poorly constructed and badly maintained agricultural terraces or by his catastrophic slash-and-burn (swidden) agriculture? To take the salient linkages and basic assumptions first:

1. That a population explosion was initiated shortly after World War II due to the introduction of modern health care and medicine and the suppression of malaria's and other diseases;
2. That increased population in subsistence mountain societies has led to:
   (a) reduced amount of land per family
   (b) deepening poverty
   (c) massive deforestation;
3. That mountain deforestation, on such a scale, will result in total loss of all accessible forest cover by AD 2000 (World Bank, 1984) and is the cause of accelerating soil erosion and incidence of landsliding;
4. That destabilised mountain slopes resulting from points (1), (2) and (3) above cause;
   (a) increased flooding on the Ganges and Brahmaputra plains,
   (b) extension of the delta and formation of islands in the Bay of Bengal,
   (c) massive siltation and drastic reduction in the useful life of highly expensive water resource projects,
   (d) drying up of wells and springs in the hills and lower dry-season river levels downstream;
5. That deforestation also leads to climate change in general and reduced rainfall amounts in particular.

It is not the intension to dispute the facts, wherever reliable information exists, but the assumptions that so frequently are not based upon facts. Nevertheless, throughout this attempt to analyse the theory of Himalayan Environmental Degradation the causal relationships between the timing and degree of population growth, deforestation, loss of agricultural land, and downstream effects are paramount. An attempt will be made to demonstrate that most of these linkages and assumptions are founded on latter-day myths, or falsely-based intuition, or are not supported by rigorous, replicable, and reliable data. They are the "sacred cows" of the perceived Himalayan Problem, and an effort will be made to strike them down, in part or in whole. However, the claim of lack of reliable data cuts both ways-it cannot be demonstrated unequivocally that *all* the linkages are inoperable in all, or even in most cases. Nevertheless, it is contended that enough of the "sacred cows" can be disposed of and others sufficiently damages to support the claim that the overall theory of Himalayan Environmental Degradation is untenable and that the Himalayan Problem needs to be rigorously defined.

The perceived problem, in this view, is in the minds of the vested interests-whether the World Bank, the Chipko Movement, different national governments, or the scientists. It is likened to a kaleidoscope which will change its pattern depending upon the way in which it is titled, or upon the angle of view. This is the essence of Thompson and Warburton's (1985) *Uncertainty on a Himalayan Scale.* The Uncertainty is a large element of the problem. Thus the present claim to demolish most of the underpinnings of the Theory itself must be qualified by the very nature of the Uncertainty. There must be a qualification: that *in certain instances,* and in *specific areas,* it can be shown that many of the widely preferred assumptions are untenable. The present temptation to extrapolate must be resisted, because to do so would be to commit the same error-unwarranted generalisation. The single and obvious generalisation that can be made, however, is that the Himalayan region is so varied and so complex that generalisation is counterproductive. Hence, the application of broad panaceas by aid and development agencies in most, if not all, instances will not succeed; in some instances they may well exacerbate the Problem.

But what of the ignorant and fecund subsistence farmer whose well-being lies at the crux of the Problem. He had indeed become a convenient scapegoat. This can be exemplified by a single observation, illustrative of many others.

In an otherwise impressive review of the Nepal Agricultural Sector, the following quotation from a report by the Asian Development Bank (1982: II, p-34) indicates the extent of the misunderstanding of the subsistence farmer's role.

> Terraces, especially on rainfed land, are often poorly constructed; they are outward rather than inward sloping and do not have a grassed bund on the edge.

The fact is that *bari* on rain-fed terraces in Nepal usually support maize, millet, buckwheat and other crops. They are constructed usually on the upper, steeper slopes in the Middle Mountains which are inaccessible to irrigation system. They slope outward from the hillside so that these crops are not damaged by waterlogging. In the UNU Kakani field area (Johnson et al. 1982; Kienholz et al., 1983), the local farmers are well aware that an increased accumulation of water on the terraces (such as would result from inward-slopping forms) would greatly exacerbate the problem of landsliding by increasing the degree of soil saturation and adding the weight of the ponded water itself. Furthermore, annual repair of the terraces would require a much larger labour input if they sloped inward. The summer monsoon rain is intended to run-off the outward-sloping terraces.

Although there are undoubtedly poorly maintained terraces in Nepal, and in other areas, many are very well maintained and have vegetated steps; absence of a bund on the *bari* terraces (in contrast to the *khet,* or irrigated terraces) is deliberate and ensures run-off of rainwater. It can be argued that both *bari* and *Khet* terraces are superbly engineered in Nepal. Admittedly, during heavy monsoon downpours available human energy is concentrated on repairing damage to the *Khet* and irrigation systems, and the *bari* terraces may have to be left to collapse; this is because the much higher-yielding *Khet* terraces, usually under paddy rice, are more vital to the survival of the subsistence family. Any apparent neglect of the terraces may be due to shortage of available labour at the particular moment that they were observed by the visiting "expert" rather than a reflection of the ignorance of the farmer. Frequently the subsistence farmer can be shown to be a highly knowledgeable and intelligent land manager with a wealth of accumulated, traditional wisdom of great potential value to the "educated" elites, if only they would listen (Whiteman, 1985). This leads to the claim that there is a need for *gift exchange* in contradistinction to *charity-a* synonym for international and bilateral aid (Hatley and Thompson, 1985). Nevertheless, it is not implied that all subsistence farmers are intelligent indigenous scientists, nor that even the most gifted amongst them can necessarily control the torrent of change which is sweeping them along; and there are ignorant and foolish farmers, just as there are ignorant and foolish factory workers, tradesmen, scientists, and decision makers.

Two further points must be introduced. First, there are no claims being made here for any individual's academic or scientific precedence, despite the intellectual satisfaction of having the opportunity to fault a widely accepted paradigm. Members of the United Nations University/Nepal MAB-Mountain Hazards Mapping Project began to suspect the reliability of some of the pre-existing claims of the theory of Himalayan Environmental Degradation. They began to doubt that deforestation and increased landsliding were linked in a simple cause-and-effect relationship. They also began to understand, as fieldwork progressed over several years and during different parts of the annual agricultural cycle, that the human impacts, principally those of the subsistence fanner, were not all negative. Part of the farmer's coping strategy was to re-terrace landslide scars and stabilise slopes. They also responded to prospects of immediate landslide initiation by such acts as agricultural deintensification (Johnson *et al.*, 1983). Similarly, reconnaissance of the Qinghai-Xizang (Tibetan) Plateau, and in the Hengduan Mountains of western Sichuan and north-western Yunnan (Ives, 1981 1985; Messed, and Ives, 1984) led to the suspicion that the assumptions of post-1950 massive deforestation were also over-simplifications, and that the actual history of deforestation was a very much longer and more complex process. This gradual growth in understanding of the complex nature of the region and the processes operating therein led to this questioning of conventional wisdom. These doubts about recent mountain deforestation were enhanced by contact with the work of forest historians Richard Tucker and John Richards, with ecologists and Chipko activists Jayanta Bandyopadhyaya and Vandana Shiva, with Tej Mahat, David Griffin and his colleagues, with Michael Thompson, Michael Warburton and Tom Hatley, with David Pitt, Lawrence Hamilton and Deepak Bajracharya and many others, together with the spiritual leadership of 'Chipko Messenger, Sundarlal Bahuguna. It was discovered from these contacts that doubts and challenges had been developing simultaneously along similar lines. It is the coming together of this group, facilitated by the United Nations University's support of the highland-lowland Interactive Systems project, that has led to this concerned effort to challenge the theory of Himalaya Environmental Degradation and in turn, to the convening of the Mohonk Mountain Conference.

The second point is equally important. All members of this now considerably enlarged working group do not necessarily agree on all points, or even on any one particular point; nor do they, nor can they, all have the same perspective. But here is agreement that a major arena of enquiry has been opened up that is fraught with an unusual level of uncertainty. It is also emphasised that the enquiry has been encouraged by many individuals within several major agencies of the United Nations Organisation, despite the occasional criticism that appear to be levelled against them. It is contended, however, that the enquiry has a potentially important bearing on the well-being of several hundred million people and on the socio-economic and political stability of a pivotal region of the world. Further exhaustive pursuit of the enquiry should become a major endeavour, not only for the United Nations University, but also for other relevant UN agencies, bilateral aid and development agencies, and the government of the region.

Discussion

Having presented in some detail the theory of Himalayan Degradation it is now necessary to review several of the major challenges brought against it, both during preparation for the Conference agency and as part of the Conference deliberations. many of the points raised for discussion here are analysed in greater detail individually in other parts of these proceedings and in the series of independent papers published in *Mountain Research and Development.* Much additional fieldwork, data analysis, and debate will be necessary before some of the more intractable problems can be resolved. The Mohonk Mountain Conference will have fulfilled one of its major objectives, however, if it builds an atmosphere of healthy sceptism and critical response to the facile argumentation upon which so many perceptions are based.

Deforestation and Uncertainty

Thompson and Warburton (1985a, 1986b), Hatley and Thompson (1985), and Thompson et al. (1986) have highlighted the "uncertainty issue" that has pervaded many of the Conference deliberations. Only one aspect is introduced here: the 67 factor - the measure of difference between the lowest and the highest estimates (supposed measurements) of fuelwood consumption for Nepal. Thompson *et al.* (1986) have demonstrated most effectively the gross unreliability of available data by use of the fuelwood consumption-biomass productivity equation as an illustration. By showing that consumption estimates vary by a *factor* of 67 and that biomass productivity data are even less reliable, or else totally unavailable, they challenge the wisdom of the recurrent tendency of conservationist, the World Bank, and agencies of His Majesty's Government of Nepal alike (this tendency encompasses the entire length of the Himalaya) to use such "data" to demonstrate the proximity of catastrophe.

Mahat *et al.* (1987) explore the fuelwood consumption estimation process in great detail and, in general support the views of Thompson et al. (1986). Nevertheless, Mahat et al. (1987a, 1987b) do appear confident that they have derived a reasonably acceptable average figure for their field area of Sindhu Palchok and Kabhre Palanchok in the Nepalese Middle Mountains east of Kathmandu. This raises the question: have Thompson et al. caused the pendulum to swing back too far in the opposite direction leading to spurious over-dramatisation of the "uncertainty issue"? Any response to this question requires re-examination of Donovan's (1981) original material since this is the source of the "data" that provided Thomspon et al. (1986) with material for their axe.

Certainly, if a possible printing or recording error is eliminated, 67 is replaced with 26 (Donovan, 1981; Thompson et al, 1986). This step does not change the situation, however, since a difference of this magnitude (factor of 26) between the highest and lowest figures for annual per capita fuel consumption is still preposterous. It should deny the "data" of any usefulness except for demonstration of the total unreliability of this and all (?) data deriving from the Himalayan region. This is a prime purpose of Thompson *et al.* (1986).

Prior to re-analyse of the Donovan material (Table 1) it had been assumed that it would proved a grab-bag of fuelwood consumption measurements which grouped separate studies in different areas of Nepal, and for different periods of time, destined to produce widely disparate results. Thus, estimates from Terai villages, where fuel use for household hearing would be relatively unimportant, would provide low consumptions figures. Those from the Middle Mountains would be relatively high as heating requirements increase with greater altitudes. Finally, those from high-altitude villages, such as Khumbu Himal, where household heating would be a continuous requirement throughout the year, would show the highest per capita consumptions. In exceptional cases these high figures would be further inflated where demands for hot showers, cooking, and heating for large numbers of trekkers and mountaineers, who in places greatly exceed the total indigenous population. A brief glance at Table 1 indicates that the three hypothetical clusters-Terai, Middle Mountains, High Himal-do not appear. There is no rational explanation; for instance, some Terai figures are among the highest, some among the lowest. Many are aggregate for the entire country and even this subset nearly covers the entire range of estimates. While there are significant numbers of individual estimates that cluster around 1.0 m³/capita/year, it is suspected that this could be influenced by individual investigators adjusting their computations to more closely match the conclusions of pre-existing publications and reports.

TABLE 1

Various Estimates, 1954-1980, of Average Annual Per Capita Fuelwood Consumption in Nepal (from Donovan, 1981)

	Quantity[2]	
Reference area[1]	*Cubic metres*	*Kilograms*
1	2	3
Nationwide	0.10	60
Bhaktapur	0.13	84
Western hills	0.18-036	108-216
Kathmandu valley	0.18	111
Nationwide	0.20	120
Lalitpur	0.22	136
Kathmandu	0.25	155
Central hills	0.26	156
Nationwide	0.33	198
Eastern hills	0.38	228
Far west Terai	0.44	264
Central hills	0.46	276
Hills	0.52	390
Terai	0.54	324

Nationwide	0.57	342
Nationwide	0.67-0.75	400-450
Nationwide	0.71	442
Nationwide	0.75	450
Central hills	0.75	450
Far west Terai	0.85-0.95	638-713
Central Terai	0.87	521
Far west hills	0.88	527
Far west hills	0.89	535
Far west Terai	0.90-1.80	540-1080
Nationwide	0.91	546
Eastern hills	0.91	547
Central hills	0.91-6.06	545-3636
Western hills	0.93-1.05	656-740
Western hills	0.94	563
Nationwide	1.00	600
Nationwide	1.00	600
Nationwide	1.00	650
Central hills	1.00	600
Central hills	1.08	540
Central hills	1.08	540
Far west terai	1.09-1.49	818-1118
Far west	1.18	708
Central hills	1.23	738
Eastern hills	1.25	780
Western hills	133	687
Central hills	135	810
Eastern hills	1.40	840
Central hills	153	979
Western hills	1.55	928
Far west Terai	1.79-2.42	1074-1452
Far west hills	2.00	1200
Eastern hills	2.08	1300
Western Terai	2.57	1542
Nationwide	3.33-6.67	2000-4000

1. East, central, west, and far west areas correspond to government designated development regions as of 1979.

2. The author assumed an average household size of 5.5 persons and the weight of cubic metre of firewood to be approximately 600 kg. although this varied from 500 to 750 kg. according to individual author's estimates.

It is necessary, therefore, to fall back on the point made by Thompson *et al.* (1986) - the figures are simply unreliable, unrealistic, and unusable. They can only be used to raise the spectre of general unreliability of *all* data from the Himalaya, or at least serve to indicate the need for a very strong scepticism. This constitutes the essence of the Conference's challenge to conventional wisdom.

This seemingly blunt statement applies as much to precipitation data from a angle rain gauge (Byers, unpub., 1987), to hydrograph charts (Vuichard and Zimmermann, 1987) to stream sediment yield (Ramsay, 1986) as it does to infant mortality rates, life expectancy calculations, and GNP per capita (Pitt, 1986). It is an indictment of the methods that have been used as well as the observers who have used them, and especially of the way in which many investigations are conducted in isolation. It could be argued that a new start is needed, or else a new method of policy formulation.

Two independents studies (Bajracharya, 1983; Mahat *et al.*, 1986a, 1986b, 1987a, 1987b) focus on the actual pressures that are, or are perceived to be, inducing loss of forest cover in the Nepalese Middle Mountains. They serve to reduce appreciably the circle of uncertainly by providing a fuller understanding of both the present-day processes and their historical perspective. Bajracharya demonstrates that in Pangma Panchayat in eastern Nepal demand for fuelwood is not the driving force for deforestation. He shows that loss of forest cover is resulting from the rapidly growing population's need to replace forests with agricultural terraces to produce more of the subsistence food crops necessary for their very survival. This has important implications for those who would treat the deforestation problem by introducing efficient wood-burning stoves, solar heaters, and biogas systems.

Mahat *et al.* (1986a, 1986b, 1987a, 1987b) have made a major contribution with their assessment of the subsistence farming system in Sindhu Palchok and Kabhre Palanchok districts and the provision of a historical overview of deforestation. This leads to a substantially different set of conclusions. They demonstrate that for their field areas, adjacent to the Kathmandu Valley, serious deforestation was already well under way more than 200 years ago, that it peaked between about 1890 and 1930 and that the recent trend has been abandonment of marginal agricultural terraces and their *natural* reforestation wherever grazing pressures have permitted.

In Sindhu Palchok and Kabhe Palanchok traditional Nepalese patterns of land ownership and taxation, with government encouragement of deforestation, have been the major driving forces behind widespread landscape modification. Mahat *et al.* demonstrate that the area of forested land has not diminished significantly since 1950, although the *quality* of the remaining forests has; hence their viability as an essential component of successful subsistence agriculture is at risk. Again, the spectre of uncontrolled population growth is raised, and it is argued that the greatest single threat to the remaining forests derives from demands for fodder and bedding for domestic livestock resulting from this population growth. They do argue, nevertheless, that continued acceleration of forest product used for fodder and fuelwood will rapidly destroy the remaining forests, at least to the point of collapse of the local subsistence economy. They also demonstrate, however, that this crisis situation can be turned around as local collective community effort is initiated. This then becomes one of the "virtuous circles" that Thompson et al. (1986) insist do exist throughout the Himalayan region.

While the findings of Bajracharya and of Mahat *et al.* are mutually supportive in some respects, there are also some apparent contradictions. These are most likely due to the relative isolation and distance from Kathmandu of Pangma Panchayat compared with Sindhu Palchoka and Kabhre Palanchok. The latter two districts are also astride the major traditional trade route between Kathmandu and Tibet. It would be interesting to determine how far eastward the Mahat *et al.* findings can be extrapolated and how far both eastward and westward the Pangma Panchayat conditions prevail. This rises another concern of the Conference participants-that of the degree of representativeness of even those studies that have been carried out with a reasonable approach to scientific rigour.

In an entirely different setting. Byers' (this issue: 209-216) has demonstrated that the widespread claims of extensive deforestation and consequent soil erosion in Khumbu Bimal are gross exaggerations and even mis-statements (cf. also Byers, 1987; Houston, 1987). Ives (1985) and Messerli and Ives (1984) have commented on the complex pattern of deforestation (and reforestation) in the Hengduan Mountains of western Sichuan and north-western Yunnan. Many extensive areas from which forests and soil cover have been stripped down to be bedrock appear tɔ have been degraded hundreds, if not more than a thousand, years ago. And of equal significance, some areas have a more complete forest cover today than in the 1920s and 1930s, while still other areas are suffering from current over-cutting.

Continuing with this single theme of deforestation in the context of the viability of the major component of the theory of Himalayan Environmental Degradation, Tucker (this issue: 328-331) and Richards (this issue: 299-304) indicate a prolonged period of deforestation in the Central Indian Himalaya. This supports the many recent publications emanating from Kumaun University, Nainital, U.P. (e.g. Joshi, 1986; Valdiya, 1986). Nevertheless, Moench and Bandopadhyaya (1986), in a most effective study of relatively remote villages in the Garhwal Himalaya north of Mussoore, describe the "nibble effect" whereby forest area is being reduced, not because consumption of forest products exceeds local biomass productivity, but because the forests perimeters nearest the village are receiving the brunt of fodder and fuelwood collection. Karan (this issue: 275-278), in contrast, demonstrate recent and alarming accentuation in the rates of forests clearing in Sikkim. Allan, in an entirely different context, describes the current impact of 3,500,000 Afghan refugees on the forests of northern Pakistan (this issue: 200-204).

In her description of a very different environment, Chatterji (this issue: 217-218 has drawn an effective illustration of Ladakh as an example of a high-altitude arid and semi-arid section of the Himalaya. Here traditional society has evolved with fuelwood scarcity because of climatic constraints, only in turn to feel the pressures of recent population growth. Thus the need has arisen for viable livelihood alternative.

From the foregoing discussion it is concluded that one of the major linkages of the theory of Himalayan Environmental Degradation post-1950 deforestation resulting from rapid population growth and its accompanying demands for fuelwood-is indeed a latter day myth. This conclusion carries the corollary that the lists of figures that have been offered as "data" are more harmful than helpful in that they serve only to extend the area of uncertainty. It remains clear, nevertheless, that the *condition* and *content* of present day forest cover is a critical factor. To maintain a viable subsistence agriculture some major changes in development policy formulation and foreign aid must be introduced in the near future, and changes in forest management are surely needed. However, unless changes are made as a conscious adaptation to the extreme capability, which is the main characteristic of the Himalayan region, and unless the infusion of local, indigenous initiatives is facilitated, success will not come easily. This leads to recognition of the need for plural problem definitions and plural solution definitions.

Is Deforestation Necessarily Bad?

Hamilton (this issue: 256-263) approaches the forest cover component of the theory of Himalayan Environmental Degradation from the point of view of its assumed role as protector of steep slopes from rain-drop impact, rillwash and gullying, overland flow and soil erosion and downstream flooding. He advocates the elimination from the literature of the totally emotional term "deforestation" unless its use is strictly qualified. Of particular importance are the modes of tree felling and evaluation and the use to which the cleared land is subsequently subjected. He maintains that if hand cutting and carrying is the primary mode of clearance, and if the forested slopes are replaced with well-maintained agricultural terraces (a widespread pattern in the Himalaya) then soil losses may even be less than those prevailing under a forest cover.

This line of reasoning is supported by Gilmour et al. (this issue: 239-249). They demonstrate that estimated peak rainfall intensities, when balanced against infiltration capacities, lead to the conclusion that reforestation will not likely have significant beneficial impact on soil loss and local flooding, and certainly not on the hydrological character of the Ganges and Brahmaputra mainstreams. A cautionary note is required here, however Gilmour et al.'s work is preliminary, site specific, one of a kind; it depends, in para, on extrapolated rainfall intensities from Kathmandu. This work does emphasise, however, as do many of the individual studies introduced in these proceedings and in *Mountain Research and Development* in general, that the new generation of Himalayan research is beginning to attain standards of method and rigour of analysis that will withstand critical evaluation. Their results do demonstrate, once again, that it is unwise to depend upon popular notions, such that deforestation is inherently *bad* and that it inevitably leads to large-scale damage for even a thousand kilometres downstream. Hamilton argues (this issue: 256-263) that the reason for serious flooding on the Ganges Plain is heavy precipitation reaching the ground in a short period of time and need not be related to events taking place in the mountains.

The Mountain-Lowland Linkage?

The major linkage-misuse of land and forests in the mountains (whatever its cause and timing) leading to serious disruptions on the plains-is such a large issue that it can only be introduced here briefly. This assumption is being heavily challenged on many fronts (Carson, 1985; Ramsay, 1986; Lauterberg and Messerli, unpub, 1986). According to the best available estimates, the continuing tectonic uplift of the Himalayan ranges may equal or exceed rates of regional denudation which includes soil erosion and sediment transfer. The flood plains of the Brahmaputra, Ganges, and Indus are underlain in places by at least 5,000 metres of sediments derived from the Himalaya over the last several million years. These dramatic geophysical processes, responsible for the very existence of the Himalaya and the plains, are sufficiently impressive that there should be no need to blame the ignorant mountain farmer for the problems facing the lowland land manager *unless there are very reliable data upon which to fix his guilt, and these do not exist.* The point sources of sediments, whether the Tibetan Plateau, the High Himalaya, Middle Mountains, or Lesser Himalaya, are not known. Many, even most, of the undersirable effects of flooding and sedimentation on the plains may result from natural processes aided by human interventions *on the floodplain itself.* In developing this line of reasoning there is no intention to contest the claims that badly engineered roads in the mountain have rapidly accelerated landsliding, or that large increase of sediment transfer can be demonstrated as resulting from land-use changes in small watersheds (Tejwani, this issue: 323-327); Dhruva Narayana, this issue: 287-298). The question is one of scale and historical perspective. Thus, on the Himalayan scale, and in the context of the last several hundred to more than a million years, it is contended that human influences, so far, are insignificant. It could be countered that this is an intellectual irrelevancy if the livelihood of a group of villages in Sindhu Palchok district is endangered in the next decade or so due to continued pressure on existing forests. It is nevertheless maintained that this is yet another confusion of scales and relates directly to the responses that have been adopted by the development and aid agencies.

The Human Component

So far the discussion has been confined to the physical elements of the landscape, whether their modification is caused by long-term geophysical processes or by very recent human interventions. Pitt's (1986) treatment of poverty, women and young people in the Himalaya introduces a parallel set of challenges to standard assessments of socio-economic conditions, such that serious doubt is also cast on both the relevancy and reliability of much of the associated data sets. The derivation of GNP per capita in subsistence economies, the relevance of female literacy rates in oral societies, the reliability of life expectancy at birth estimates, must all be questioned. Similarly, the tendency for aggregation of data often renders them meaningless. The fact remains, however, that the population of the entire region (with certain site specific exceptions, such as Khumbu Himal) is growing and this is leading to increased pressure on diminishing resources. It is necessary to raise the question: is population growth the cause of increased poverty or its result? Certainly the old adage that increased wealth with cause population growth to level off (based upon the model of the industrialisation of western countries) is generally regarded as a myth, or serious misunderstanding. A number of recent village-level studies in the Himalaya (Fisher, 1986, Fricke, 1986) demonstrate, at the village level, that all new-born children are wanted. This, in part, is a reflection that children are capital in predominantly non-monetary society, and that a large family increases the chances of survival. Certainly, the role of women is central to many Himalayan development issues and this has been neglected for too long. It appears that land tenure and taxation play a critical role in the poverty issue, just as they have in the related deforestation issue. Unequal access to resources, therefore, becomes a focal point.

At this point the reader might well ask the question: is there a problem? The attempts here to destroy the "sacred cows" of the theory of Himalayan Environmental Degradation could conceivably lead to this question being answered in the negative. This would contradict the unanimous conclusion of the Conference participants and the very purpose of the Conference would be overtimed. Examination of the arguments of Hrabovszky and Miyan (this issue: 264-270), which illuminate the rapidly narrowing gap between subsistence crop yields and available land, on the one hand, and the growing number of mouths to feed, on the other, should serve to guard against this. Similarly, the thought-provoking discussion by Griffin (this issue: 250-253) on project implementation failure caused by institutional problems further emphasises the difficulties that lie in the path of workable solutions. The general conclusion reached is that a crisis of very large dimensions is developing and that it may be advisable, for political and diplomatic reasons, to regard it as a potential supercrisis. It is postulated that the very real magnitude of the problem has been disguised by its over-simplication and exacerbated by the generalised and bureaucratic approach to its solution. Thus, the uncertainty is indeed part of the problem, and it was the intent of the Conference to expose this. Of equal importance of the demonstration that "virtuous circles" do exist; that, given a community-level approach to resource management and sensitised "outside" co-operation, a village, or district, can move from a downward spiral to an upward mode. This conviction lays the foundation for the claim that there is a very urgent need for a much fuller understanding of Himalaya-Ganges dynamics and for a radical change in "development" policy.

REFERENCES

Asian Development Bank, 1983. *Nepal Agriculture Sector Strategy Study,* Vol. 11. p. 34.

Bandopadhyaya, J., N.D. Jayal, J. Schoettli; and Chhtrapati Singh, 1985. *India's Environment: Crisis and Responses.* Natraj Publishers, Dehra Dun U.P- India, 309 pp.

Bajracharya. D. 1983. Deforestation in the food/fuel context Historical and Political Perspective from Nepal. *Mountain Research and Development,* 3(3): 227-240.

Byers A. 1987a. An assessment of landscape change in the Khumbu region of Mepal using repeat photography. *Mountain Research and Development,* 7(1): 77-81.

Byers, A. 1987b. A geoecological study of landscape change and man-accelerated soil loss: The case of the Sagarmatha (Mt. Everest) National Park, Khumbu, Nepal Unpublished Ph.D. dissertation. University of Colorado, Boulder, May 1987, 299 pp.

Carson B., 1985. Erosion and Sedimentation processes in the Nepalese Himalaya, International Centre for Integrated Mountain Development (ICIMOD). Occasional Paper no. 1, Kathmandu, Nepal. 39pp.

Donovan, D.G., 1981. Fuelwood: How much do we need? Newsletter (DGD14), Inst. of Current World Affairs, Hanover, N H, USA. Unpub. mimeo 23 pp.

Eckholm, E., 1975. The deterioration of mountain environments. *Science,* 189: 764-770.

Eckholm, E., 1976. Cusmg *Ground.* Worldwatch Institute. W.W. Norton & Co., Inc., New York., 223pp.

Fisher, James F., 1986. *Trans-Himalayan Traders: Economy, Society and Culture in Nortlnvest Nepal.* University of California Press, Berkeley, USA. 232 pp.

Fricke, T.E., 1936. Himalayan Households: Tamang Demography and Domestic Process. *Studies in Cultural Anthropology.* No. 11, UMI Research Press, Ann Arbor, Michigan, USA, 227 pp.

Goldstein, M.C., Ross, J.L., and Schuler, S., 1983. From a mountain-rural to a plains-urban society: Implications of the 1981 Nepal Census. *Mountain Research and Development,* 3(1): 61-64.

Houston, C.W. 1987: Deforestation in Solu Khumbu. *Mountain Research and Development,* 7(1): 6.

Ives, J.D., 1981: High Mountains Jnd Plateaus: An Excursion to the rood of the world. *Mountain Research and Development,* 1(1): 79-83.

Ives. J.D., 1985; Yulongxue Shan, Northwest Yunnan, Peoples Republic of China: A geoecological expedition. *Mountain Research and Development,* 5(4): 382-385.

Johnsori, K. Olson, E.A. and Manandhar, S., 1982. Environmental knowledge and response to natural hazards in mountainous Nepal. *Mountain Research ami Development,* 2(2): 175-188.

Joshi, S.C. (ed), 1986. *Nepal Himalaya Geo-ecological Perspectives.* Himalayan Research Group, Nainital, U.P., India, 506 pp.

Mahat, T.B.S., Griffin, D.M. and Shepherd, K.R., 1986-87. Human impact on some forests of the Middle Hills of Nepal 1. Forestry in the context of the traditional resources of the state. 2. Some major human impacts before 1950 on the forests of Sindhu Palchok and Kabhre Palanchok. 3. Forests in the subsistence economy of Sindhu Palchok and Kabhrre Palanchok. 4. A detailed survey of southeast Sindhu Palchok and northeast Kabhre Palanchok, *Mountain Research and Development.* 6(3): 223-232,6(4): 325-334, 7(1): 53-70,7(2): 111-134.

Messerli, B. and Ives, J.D., 1984; Gongga Shan (7556 m) and Yulongxue Shah (5596 m). Geoecological observations in the Hengduan Mountains of Southwestern China. *Erdiuissenschafiliche Forschung,* Bd. XVIII: 55-77, Steiner, Stuttgart, West Germany.

Moench, M. and Bandyopadhyaya, J., 1986. People forest interactions: A neglected parameter in Himalayan forest management. *Mountain Research and Development,* 6(1): 3-16.

Myers. N., 1986. Environmental repercussions of deforestation in the Himalayas. *Journal of World Forest Resource Management,* 2: 63-72.

Pitt. D.C. 1986. Crisis, pseudocrisis, or supercrisis: Poverty, women, and young people in the Himalaya. A survey of recent developments. *Mountain Research and Development,* 6(2): 119-131.

Ramsay, W.J.H.. 1986. Erosion problems in the Nepal Himalaya—an overview. In Joshi, S.C. (ed). *Nepal Himalaya: Cececological Perspective.* Himalayan Research Group, Nainital, U.P, India, pp. 359-395.

Singh, J.S. (ed), 1985. *Environmental Regeneration in Himalaya:* Concepts and Strategies. Central Himalayan Environment Association and Gyanodaya Prakashan, Nainital, U.P., India, p. 468.

Singh T.V. Kauer, J. and Singh, D.P. (eds), 1982. *Studies in Tourism, Wildlife Parks, and Conservation.* Metropolitan, New Delhi, 300 pp.

Thompson, M., and Warburton, M., 1985a. Uncertainty on a Himalayan scale. *Mountain Research and Development,* 5(2): 115-135.

Thompson, M. and Warburton M., 1985b. Knowing where to hit it: A conceptual framework for the sustainable development of the Himalaya. *Mountain Research and Development,* 5(3): 203-220.

Thompson. M., Warburton, M., and Hatley, T., 1986. *Uncertainty on a Himalayan Scale.* Ethnographica, London, 162 pp.

Valdiya, K.S., 1985. Accelerated erosion and landslide-prone zones in the Central Himalayan Region. In Singh, J.S. (ed). *Environmental Regeneration in the Himalaya: Concepts and Strategies.* Central Himalayan Environment Association and Gyanodaya Prakashan, Nainital, U.P., India, pp. 12-38.

Vurchard, D. and Zimmermann, M., 1987. The 1985 catastrophic drainage of a moraine-dammed lake, Khumbu Himal, Nepal: Cause and Consequences. *Mountain Research and Development,* 7(2): 91-110.

Whiteman, P.T.S., 1985. The mountain environment: An agronomist's perspective with a case study from Jumla, Nepal, *Mountain Research and Development,* 5(2): 151-162.

World Bank, 1984. *World Development Report.* World Bank, Washington, DC. 286 pp.

Encyclopaedia of

ENVIRONMENTALLY SUSTAINABLE GREEN SCIENCE AND TECHNOLOGY

Encyclopaedia of

ENVIRONMENTALLY SUSTAINABLE GREEN SCIENCE AND TECHNOLOGY

Volume 4

R.B. Patnaik

ANMOL PUBLICATIONS PVT. LTD.

NEW DELHI - 110 002 (INDIA)

ANMOL PUBLICATIONS PVT. LTD.
H.O.: 4374/4B, Ansari Road, Daryaganj,
New Delhi-110 002 (India)
Ph.: 23278000, 23261597
B.O.: No. 1015, Ist Main Road, BSK IIIrd Stage
IIIrd Phase, IIIrd Block,
Bangalore - 560 085 (India)
Visit us at: www.anmolpublications.com

Encyclopaedia of Environmentally Sustainable Green Science and Technology

First Edition, 2009
ISBN 978-81-261-3842-5 (Set)

PRINTED IN INDIA

Printed at Mehra Offset Press, Delhi.

Contents

VOL. 4

VOL. 5

Preface

This publication entitled "Encyclopaedia of Environmentally Sustainable Green Science & Technology" attempts towards creating a holistic understanding of environment and sustainable development in its totality. It defines environment and sustainable development in all its dimensions. The concepts and methodologies of ecology, environment and sustainable development have been discussed in detail. The ecosystem perspective and environmental concept of sustainable development have been highlighted, so has been the environmental methodology. An effort has been made towards defining green science and its manifestations for understanding pollution. The relevance and importance of environment impact assessment (EIA) for environmental sustainability has been analyzed. Also, the elements and dimensions of green technology have been covered in detail. The relationship among ecomovements, ecodynamics and sustainable development has been discussed in detail. Similarly, the environmental and social aspects of sustainable development, in particular, aspects related to training of people to meet the global environmental needs have been covered. The issues related to green energy, its sources, tools and technologies have been reflected upon briefly. The environmentally sustainable science and technology has been presented in brief with special reflection on climate change and global warming. Current issues and options regarding sustainability aspects of environmental management, corporate management and environmental responsibility have been covered. The aim is to develop a policy perspective. The contemporary global industrial technology has been discussed with focus on evaluating environmental performance and compliance to green laws. The applied aspects of environmentally sustainable science and technology have been outlined. Also, the green conventions enforced towards regulating global science and technology towards environmentally sustainable development has been presented, highlighting the theories of international environmental law and theory of environmental degradation.

The applications of green science and technology have been discussed along with a case of biotechnology. The green technology and environmental monitoring issues have been given considerable coverage. Select case studies have been made in this regard. Also, case studies of environmental engineering and environmental biotechnology have been incorporated with focus on applications and utilities. Future prospect of environmentally sustainable or green technology in today's world has

been analyzed briefly. Similarly, the principles, tools and techniques of environmental management, with particular reference to environmental disasters, have been covered in detail. Glossary of terms, bibliography and links has been provided to make this publication a user friendly one and to aid readers in further research and analysis.

—Editors

13

Contemporary Global Industrial Technology: Evaluating Environmental Performance & Compliance to Green Laws

Industrial Ecology

Industrial Ecology (IE) is an interdisciplinary field that focuses on the sustainable combination of environment, economy and technology. The central idea is the analogy between natural and socio-technical systems. The word 'industrial' does not only refer to industrial complexes but more generally to how humans use natural resources in the production of goods and services. Ecology refers to the concept that our industrial systems should incorporate principles exhibited within natural ecosystems.

Industrial ecology is the shifting of industrial process from linear (open loop) systems, in which resource and capital investments move through the system to become waste, to a closed loop system where wastes become inputs for new processes.

Much of the research focuses on the following areas:

- material and energy flow studies ("industrial metabolism")
- dematerialisation and decarbonisation
- technological change and the environment
- life-cycle planning, design and assessment
- design for the environment ("eco-design")
- extended producer responsibility ("product stewardship")

- eco-industrial parks ("industrial symbiosis")
- product-oriented environmental policy
- eco-efficiency

Industrial ecology proposes not to see industrial systems (for example a factory, an ecoregion, or national or global economy) as being separate from the biosphere, but to consider it as a particular case of an ecosystem - but based on infrastructural capital rather than on natural capital. It is the idea that as natural systems do not have waste in them, we should model our systems after natural ones if we want them to be sustainable.

Along with more general energy conservation and material conservation goals, and redefining commodity markets and product stewardship relations strictly as a service economy, industrial ecology is one of the four objectives of Natural Capitalism. This strategy discourages forms of amoral purchasing arising from ignorance of what goes on at a distance and implies a political economy that values natural capital highly and relies on more instructional capital to design and maintain each unique industrial ecology.

Industrial ecology was popularised in 1989 in a *Scientific American* article by Robert Frosch and Nicholas E. Gallopoulos. Frosch and Gallopoulos' vision was "why would not our industrial system behave like an ecosystem, where the wastes of a species may be resource to another species? Why would not the outputs of an industry be the inputs of another, thus reducing use of raw materials, pollution, and saving on waste treatment?" A notable example resides in a Danish industrial park in the city of Kalundborg. Here several linkages of byproducts and waste heat can be found between numerous entities such as a large power plant, an oil refinery, a pharmaceutical plant, a plasterboard factory, an enzyme manufacturer, a waste company and the city itself. The scientific field Industrial Ecology has grown fast in recent years. The Journal of Industrial Ecology (since 1997), the International Society for Industrial Ecology (since 2001), and the journal Progress in Industrial Ecology (since 2004) give Industrial Ecology a strong and dynamic position in the international scientific community. Industrial Ecology principles are also emerging in various policy realms such as the concept of the Circular Economy that is being promoted in China. Although the definition of the Circular Economy has yet to be formalised, generally the focus is on strategies such as creating a circular flow of materials, and cascading energy flows. An example of this would be using waste heat from one process to run another process that requires a lower temperature. This maximizes the efficiency of exergy use. The hope is that strategy such as this will create a more efficient economy with fewer pollutants and other unwanted by products.

Principles

One of the central principles of Industrial Ecology is the view that societal and technological systems are bounded within the biosphere, and do not exist outside of it. Ecology is used as a *metaphor* due to the observation that natural systems reuse materials and have a largely closed loop cycling of nutrients. Industrial Ecology approaches problems with the hypothesis that by using similar principles as *natural systems, industrial systems* can be improved to reduce their impact on the natural environment as well. The table shows the general metaphor.

Biosphere Technosphere

- Environment
- Organism
- Natural Product
- Natural Selection
- Ecosystem
- Ecological Niche
- Anabolism / Catabolism
- Mutation and Selection
- Succession
- Adaptation
- Food Web
- Market
- Company
- Industrial Product
- Competition
- Eco-Industrial Park
- Market Niche
- Manufacturing / Waste Management
- Design for Environment
- Economic Growth
- Innovation
- Product Life Cycle

The Kalundborg industrial park is located in Denmark. This industrial park is special because companies reuse each others' waste (which then becomes by-products).

For example, the Energy E2 Asnæs Power Station produces gypsum as a by product of the electricity generation process; this gypsum becomes a resource for the BPB Gyproc A/S which produces plasterboards. This is one example of a system inspired by the biosphere-technosphere metaphor: in ecosystems, the waste from one organism is used as inputs to other organisms; in industrial systems, waste from a company is used as a resource by others.

IE examines societal issues and their relationship with both technical systems and the environment. Through this *holistic view* , IE recognises that solving problems must involve understanding the connections that exist between these systems, various aspects cannot be viewed in isolation. Often changes in one part of the overall system can propagate and cause changes in another part. Thus, you can only understand a problem if you look at its parts in relation to the whole. Based on this framework, IE looks at environmental issues with a *systems thinking* approach.

Take a city for instance. A city can be divided into commercial areas, residential areas, offices, services, infrastructures, etc. These are all sub-systems of the 'big city' system. Problems can emerge in one sub-system, but the solution has to be global. Let's say the price of housing is rising dramatically because there is too high a demand for housing. One solution would be to build new houses, but this will lead to more people living in the city, leading to the need of more infrastructure like roads, schools, more supermarkets, etc. This system is a simplified interpretation of reality whose behaviors can be 'predicted'.

In many cases, the systems IE deals with are complex systems. Complexity makes it difficult to understand the behavior of the system and may lead to rebound effects. Due to unforeseen behavioral change of users or consumers, a measure taken to improve environmental performance does not lead to any improvement or may even worsen the situation. For instance, in big cities, traffic can become problematic. Let's imagine the government wants to reduce air pollution and makes a policy stating that only cars with an even license plate number can drive on Tuesdays and Thursdays. Odd license plate numbers can drive on Wednesdays and Fridays. Finally, the other days, both cars are allowed on the roads. The first effect could be that people buy a second car, with a specific demand for license plate numbers, so they can drive every day. The rebound effect is that, the days when all cars are allowed to drive, some inhabitants now use both cars (whereas they only had one car to use before the policy). The policy did obviously not lead to environmental improvement but even made air pollution worse.

Moreover, *life cycle thinking* is also a very important principle in industrial ecology. It implies that all environmental impacts caused by a product, system, or project during its life cycle are taken into account. In this context life cycle includes

- Raw material extraction
- Material processing
- Manufacture
- Use
- Maintenance
- Disposal

The transport necessary between these stages is also taken into account as well as, if relevant, extra stages such as reuse, remanufacture, and recycle. Adopting a life cycle approach is essential to avoid shifting environmental impacts from one life cycle stage to another. This is commonly referred to as problem shifting. For instance, during the re-design of a product, one can choose to reduce its weight, thereby decreasing use of resources. However, it is possible that the lighter materials used in the new product will be more difficult to dispose of. The environmental impacts of the product gained during the extraction phase are shifted to the disposal phase. Overall environmental improvements are thus null.

A final and important principle of IE is its *integrated approach* or *multidisciplinarity*. IE takes into account three different disciplines: social sciences (including economics), technical sciences and environmental sciences. The challenge is to merge them into a single approach.

Future Directions

The ecosystem metaphor popularised by Frosch and Gallopoulos has been a valuable creative tool for helping researchers look for novel solutions to difficult problems. Recently, it has been pointed out that this metaphor is based largely on a model of classical ecology, and that advancements in understanding ecology based on complexity science have been made by researchers such as C. S. Holling and James J. Kay. For industrial ecology, this may mean a shift from a more mechanistic view of systems, to one where sustainability is viewed as an emergent property of a complex system. To explore this further, several researchers are working with agent based modeling techniques.

Triple Bottom Line

The triple bottom line (or "TBL", "3BL", or "People, Planet, Profit") captures an expanded spectrum of values and criteria for measuring organisational (and societal) success; economic, environmental and social. With the ratification of the UN ICLEI TBL standard for urban and community accounting in early 2007, this became the dominant approach to public sector full cost accounting. Similar UN standards apply to natural capital and human capital measurement to assist in measurements required by TBL, e.g. the ecoBudget standard for reporting ecological footprint.

In the private sector, a commitment to corporate social responsibility implies a commitment to some form of TBL reporting. This is distinct from the more limited changes required to deal only with ecological issues.

Definition

In practical terms, kaffers accounting means expanding the traditional reporting framework to take into account environmental and social performance in addition to financial performance. The phrase was coined by John Elkington in 1994. It was later expanded and articulated in his 1998 book *Cannibals with Forks: the Triple Bottom Line of 21st Century Business.* Sustainability, itself, was first defined by the Brundtland Commission of the United Nations in 1987. The concept of TBL demands that a company's responsibility be to 'stakeholders' rather than shareholders. In this case, 'stakeholders' refers to anyone who is influenced, either directly or indirectly, by the actions of the firm. According to the stakeholder theory, the business entity should be used as a vehicle for coordinating stakeholder interests, instead of maximising shareholder(owner) profit.

The Bottom Lines

The Triple Bottom Line is made up of "Social, Economic and Environmental" the "People, Profit, Planet" phrase was coined for Shell by Sustainability

"People, Planet and Profit" is used to succinctly describe the triple bottom lines and the goal of sustainability.

"People" (Human Capital) pertains to fair and beneficial business practices toward labor and the community and region in which a corporation conducts its business. A TBL company conceives a reciprocal social structure in which the well being of corporate, labor and other stakeholder interests are interdependent. A triple bottom line enterprise seeks to benefit many constituencies, not exploit or endanger any group of them. The "upstreaming" of a portion of profit from the marketing of finished goods back to the original producer of raw materials, i.e., a farmer in fair trade agricultural practice, is a not unusual feature. In concrete terms, a TBL business would not knowingly use child labor, would pay fair salaries to its workers, would maintain a safe work environment and tolerable working hours, and would not otherwise exploit a community or its labor force. A TBL business also typically seeks to "give back" by contributing to the strength and growth of its community with such things as health care and education. Quantifying this bottom line is relatively new, problematic and often subjective. The Global Reporting Initiative (GRI) has developed guidelines to enable corporations and NGO's alike to comparably report on the social impact of a business.

"Planet" (Natural Capital) refers to sustainable environmental practices. A TBL company endeavors to benefit the natural order as much as possible or at the least do no harm and curtail environmental impact. A TBL endeavor reduces its ecological footprint by, among other things, carefully managing its consumption of energy and non-renewables and reducing manufacturing waste as well as rendering waste less toxic before disposing of it in a safe and legal manner. "Cradle to grave" is uppermost in the thoughts of TBL manufacturing businesses which typically conduct a life cycle assessment of products to determine what the true environmental cost is from the growth and harvesting of raw materials to manufacture to distribution to eventual disposal by the end user. A triple bottom line company does not produce harmful or destructive products such as weapons, toxic chemicals or batteries containing dangerous heavy metals for example. Currently, the cost of disposing of non-degradable or toxic products is borne financially by governments and environmentally by the residents near the disposal site and elsewhere. In TBL thinking, an enterprise which produces and markets a product which will create a waste problem should not be given a free ride by society. It would be more equitable for the business which manufactures and sells a problematic product to bear part of the cost of its ultimate disposal. Ecologically destructive practices, such as overfishing or other endangering depletions of resources are avoided by TBL companies. Often environmental sustainability is the more profitable course for a business in the long run. Arguments that it costs more to be environmentally sound are often specious when the course of the business is analysed over a period of time. Generally, sustainability reporting metrics are better quantified and standardised for environmental issues than for social ones. A number of respected reporting institutes and registries exist including the Global Reporting Initiative, CERES, Institute 4 Sustainability and others.

"Profit" is the bottom line shared by all commerce, conscientious or not. In the original concept, within a sustainability framework, the "profit" aspect needs to be seen as the economic benefit enjoyed by the host society. It is the lasting economic impact the organisation has on its economic environment. This is often confused to be limited to the internal profit made by a company or organisation. Therefore, a TBL approach cannot be interpreted as traditional corporate accounting *plus* social and environmental impact.

Several books are available on the topic. Among them, *Harvard Business Review on Corporate Responsibility* by Harvard Business School Press; *The Soul of a Business: Managing for Profit and the Common Good* by Tom Chappell; *Capitalism at the Crossroads: The Unlimited Business Opportunities in Solving the World's Most Difficult Problems* by Professor Stuart L. Hart; *The Triple Bottom Line: How Today's*

Best-Run Companies Are Achieving Economic, Social and Environmental Success — and How You Can Too by Andrew W. Savitz and Karl Weber; *The Sustainability Advantage: Seven Business Case Benefits of a Triple Bottom Line (Conscientious Commerce)* by Bob Willard.

Arguments in Favor of the Concept

The following business-based arguments support the concept of TBL companies:

- Reaching untapped market potential: TBL companies can find financially profitable niches which were missed when money alone was the driving factor. Examples include:
  (i) Adding ecotourism or geotourism to an already rich tourism market such as the Dominican Republic
  (ii) Developing profitable methods to assist existing NGOs with their missions such as fundraising, reaching clients, or creating networking opportunities with multiple NGOs
  (iii) Providing products or services which benefit underserved populations and/or the environment which are also financially profitable, i.e. building Sun Ovens for donation and retail sale and creating Fair Trade jobs in the process.
- Adapting to new business sectors: Since many business opportunities are developing in the realm of social entrepreneurialism, businesses hoping to reach this expanding market, must design themselves to be financially profitable, socially beneficial and ecologically sustainable or fail to compete with those companies who do design themselves as such. For example, Fair Trade and Ethical Trade companies require ethical and sustainable practices from all of their suppliers and service providers. A business which is planning to work with Fair Trade or Ethical Trade companies, must design their business model to be TBL.

Fiscal policy of governments usually claims to be concerned with identifying social and natural deficits on a less formal basis. However, in a democracy at least, such choices may be guided more by ideology than by economics. The primary benefit of embedding one approach to measurement of these deficits would be first to direct monetary policy to reduce them, and eventually achieve a global monetary reform by which they could be systematically and globally reduced in some uniform way.

The argument is that the Earth's carrying capacity is itself at risk, and that in order to avoid catastrophic breakdown of climate or nature's services, there is a need for a comprehensive reform in global financial institutions similar in scale to that undertaken at Bretton Woods in 1944, and not since. A major proponent of this has been Marilyn Waring.

With the emergence of an externally-consistent green economics and agreement on definitions of potentially contentious terms such as full-cost accounting, natural capital and social capital, the prospect of formal metrics for ecological and social loss or risk has grown less remote through the 1990s.

In the United Kingdom in particular, the London Health Observatory has undertaken a formal programme to address social deficits via a fuller understanding of what "social capital" is, how it functions in a real community (that being the City of London), and how losses of it tend to require both financial capital and significant political and social attention from volunteers and professionals to help resolve. The data they rely on is extensive, building on decades of statistics of the Greater London Council since World War II. There are some similar studies in North America. Studies of nature's services and assessments of the value of Earth have tried to determine what might constitute an ecological or natural life deficit. The Kyoto Protocol relies on some measures of this sort, and actually relies on some value of life calculations that, among other things, are explicit about the ratio of the price of a human life between developed and developing nations (about 15 to 1). While the motive of this number was to simply assign responsibility for a cleanup, such stark honesty opens not just an economic but political door to some kind of negotiation - presumably to reduce that ratio in time to something seen as more equitable. As it is, people in developed nations can be said to benefit 15 times more from ecological devastation than in developing nations, in pure financial terms. According to the IPCC, they are thus obliged to pay 15 times more per life to avoid a loss of each such life to climate change—Kyoto seeks to implement exactly this formula, and is therefore sometimes cited as a first step towards getting nations to accept formal liability for damage inflicted on ecosystems shared globally.

Advocacy for triple bottom line reforms is common in Green Parties. Some of the measures undertaken in the European Union towards the Euro currency integration standardise the reporting of ecological and social losses in such a way as to seem to endorse in principle the notion of unified accounts, or unit of account, for these deficits.

Arguments Against the Concept

While many people agree with the importance of good social conditions and preservation of the environment, there are also many who disagree with the 'Triple Bottom Line' as the way to enhance these conditions. The main arguments against it may be summarised as;

- Division of labour, which is characteristic of rich societies and a major contributor to their wealth, leading to the view that organisations contribute most to the welfare of society in all respects when they focus on what they do best... the baker exchanges his loaves with the shoemaker rather than making his own shoes to the benefit of both and by extension the whole of society. In the case of business the expertise is in satisfying the needs of society and generating a Value added surplus. Thus the 'triple bottom line' is thought to be harmful by diverting business attention away from its core competency. Just as charitable organisations like the Salvation Army would not be expected to attend to environmental issues or pay a cash dividend, and Greenpeace would not be expected to make a profit or succor the homeless, business should not be expected to take on concerns outside its core expertise.
- Effectiveness: It is observed that concern for social and environmental matters is rare in poor societies (a hungry person would rather eat the whale than photograph it). As a society becomes richer its citizens develop an increasing desire for a clean environment and protected wildlife, and both the willingness and financial ability to contribute to this and to a compassionate society. Indeed support for the concept of the 'Triple Bottom Line' itself is said to be an example of the choices available to the citizens of a society made wealthy by businesses attending to business. Thus by unencumbered attention to business alone, Adam Smith's Invisible Hand will ensure that business contributes most effectively to the improvement of all areas of society, social and environmental as well as economic.
- Nationalism: Some countries adopt a nationalistic approach with the view that they must look after their own citizens first. This view is not confined to one sector of society, having support from elements of business, labor unions, and politicians.
- Libertarian: As it is possible for a socially responsible person to sincerely believe that the 'Triple Bottom Line' is harmful to society, the libertarian view is that it would be arrogant to force them to support a mechanism for the improvement of society that may, or may not, be the best available. That is, those who would not force Greenpeace and the Salvation Army to generate a profit should not force businesses to take responsibilities outside their area of expertise.
- Inertia: The difficulty of achieving global agreement on simultaneous policy may render such measures at best advisory - and thus not enforceable. For example, people may be unwilling to undergo a depression or even sustained recession to replenish lost ecosystems.

- Application: According to Fred Robin's 'The Challenge of TBL: A Responsibility to Whom?' one of the major weaknesses of the TBL framework is its ability to be applied in a monetary-based economic system. Because there is no single way to monetarially measure the benefits on the society and environment as there is with profit, it does not allow for businesses to sum across all three bottom lines. In this regard, it makes it difficult for businesses to recognise the benefits of using TBL for the company, itself.

Legislation

Legislation permitting corporations to adopt a 'Triple Bottom Line' is under consideration in some jurisdictions including Minnesota and Oregon. Some businesses have voluntarily adopted a 'Triple Bottom Line' as part of their articles of incorporation or bylaws, and some have advocated for state laws creating a "Sustainable Corporation" that would grant triple bottom line businesses benefits such as tax breaks. The triple bottom line has been adopted as a part of the State Sustainability Strategy, accepted by the Government of Western Australia.

GREEN BUSINESS MANAGEMENT AND INDUSTRY ENVIRONMENTAL PERFORMANCE & COMPLIANCE

Green Business, Sustainable Development and Companies

Sometimes it seems that it is not merely the environment that is at stake. If business does not change its strategy, business itself may very well be at stake. Many—if not most-consider industry as part of the problem. But many others realise that business is the solution. We believe that if the corporate world does not play an active role in redefining its own operations, moving toward sustainability, the world as a whole will never succeed in that task.

But, on the other hand, business alone cannot overcome the present challenges. The time has come to establish solid cooperation among policy-makers, scientists, opinion leaders, the community, and business. And it is up to business executives to take a pro-active role in the definition of business's new agenda and the priorities it must share with the community. They know by now all too well that if the setting of the agenda is left to other players in society, business cannot get down to business.

Executives know too that doing more business the same way as in the past is a guarantee of failure. By the same token, the time is past when scholars, governments, or environmental organisations could hand down a doctrine from some high pulpit of academic certainty or from emotional distress and fears. Rather, cooperation among all those who have a stake in the future of society is critical.

The objective of defining strategies to steer business toward sustainability is *not* to invent new technologies. Much technology is really not new knowledge; it is, as Peter Drucker has said, "putting together things that no one had thought of putting together before." The reserve of untapped technological breakthroughs, either to be reinvented because we have neglected them or to be discovered because we have not bothered to study the great genius of Nature, must be brought into play, and the faster the better.

The days of the notion of "competitive edge" are numbered.

Business will have to convert its strategic approach to the development of a "sustainable advantage." Traditional economic theory prescribes that the competitive edge of a company depends on the efficient combination of capital, raw materials, and labor. Over the past decades, economists have expanded this to the importance of technology and information.

Companies in crisis wonder how to improve their competitive edge. This was relevant in the sixties, it is relevant today, and it will be relevant in the 21st century. But while the questions remain the same, the answers change. In the sixties, when a corporation felt the pinch from its competitors and looked for a solution, the president was told to stop trying to just *sell* his products-he needed a marketing strategy. The need to listen to the clients' wishes and preferences was the first sign that we were operating in a market characterized by over-supply. If someone were not to adapt to the customer, he would be eliminated. It was the era of Philip Kotler and the "four P's." There is no company today that questions the importance of marketing.

In the early seventies, when the first oil crisis devastated industry, the key question was how to regain a competitive position at a time when increased energy costs affected pricing and caused high inflation, undermining consumer confidence. The answer at the time was *productivity*. There was no doubt that those who could produce more with the same level of input in terms of raw materials, capital, or labor, were to win. It was the first time that Japanese industry was called in to serve as an example. Indeed, the Japanese overcame the first oil crisis faster than anyone else and integrated more robots and information technology, catapulting their production industries to the forefront of worldwide competition.

In the late seventies, the second oil crisis hit the West. Again, industries that lost market share searched for a renaissance of their businesses, and *quality* was identified as the cutting-edge factor that could enable a company to out-compete anyone in the market. It was the time of quality control circles and total quality management. Indeed, as the American management gurus Juran and Deming professed from Japan, bad quality is a cost and good quality leads to a leaner cost structure and higher customer loyalty.

In the early eighties, Western industry again faced decreasing market shares. After having invested massively in quality and productivity programmes, it became obvious that more was needed to succeed long-term in the world competitive game. The concept of *just-in-time* was advanced as the next avenue to success. The Japanese were supposed to have implemented the *kan-ban* system first and foremost, and the West was advised to learn from the Land of the Rising Sun once again. It generated a third wave of industrial tourism.

In the late eighties *service* was presented as the next panacea for success. American corporations embarked on drastic outsourcing, giving rise to a most successful service industry, which is basically the result of shifting people from in-house service providers to outside, cutting employment on industry's payroll and boosting the "success" of services. It became clear that up to 75 per cent of the value added created by industry were services to production. The Swatch was a prime example. It didn't matter if the successful watches of the seventies were made in Taiwan or Macao. Labor costs represent only three per cent of the total value added involved in a sale and as a result, it is possible to pay considerably more, as it does not significantly affect the end price to the customer. Low labor costs did not matter anymore; they were replaced in the North by "no labor costs." The design factor and communication skills represented the real service to the customer.

In the early nineties, when industry was desperately searching for new ideas on how to compete, again a new theory was advanced: re-engineering was the message promoted by the management guru Hammer. The message is simple: start all over again, and start from scratch! And, if someone is not willing to follow, "break a few legs."

Today, companies are expected to excel simultaneously in all six elements: marketing, productivity, quality, just-in-time, services, and re-engineering. Performance in each of these areas is considered a precondition for successful entry into the market. If a business integrates all these elements, then it will have the right to be in the market. If a company wishes to outperform the average businesses, if it is out to gain market share, more will be needed.

What more can be offered? In order to address this effectively, we have to leave supply-side management and study trends in demand. Actually, we are returning to the stage of marketing, to listening to what the customer really wants.

Over the past five years, consumers have enjoyed low levels of inflation. The market wants "more for less." This means that companies have to cut costs even further and offer more. The continuous drive toward better perceived value for less money is difficult to achieve today because we are at a time when business cannot hide increased value through price hikes, or hide increasing costs behind inflation. Today, something more is needed to perform well, to do better in the game. But what is it?

At present, economists cannot agree on any single issue that is to dominate. However, on the basis of thorough market analyses in Europe, Japan, and the U.S., it has been concluded that during the next few years, ethical standards, a moral commitment, and high environmental performance will not only become an integral part of the corporate strategy; these will become the way to outperform the industry and reestablish that unique marketing position so badly desired. It will be the only way to develop a sustainable advantage.

Today, more than 40 per cent of consumers in the U.S. indicate that when the same price and quality are being offered, moral, ethical, and environmental issues considered important by the consumers will determine their choices. Of course, there is a very broad variety of themes that companies can address: biodiversity, AIDS, animal testing, minority rights, Third World development, fair trade, healthy and organic farming, and so on.

Several companies have succeeded in setting an ethical or moral agenda and appealing to a large share of the market. These are concerns that attract the community and have to form an integral part of corporate behavior if business is to deserve long-term respect in society. What is most interesting is that consumers are increasingly well-informed and capable of seeing through marketing slogans that are not based on genuine commitment.

This does not imply that all companies today are unethical, immoral, or unconcerned with the environment. But companies will have to be thoroughly committed to doing more for society than merely playing a role in generating value added and creating profits for their shareholders. The company of the 21st century will have different responsibilities to assume in society from those generally expected today.

It is dramatic to note, for example, that on environmental issues not even ten per cent of the public believes statements made by companies. In a recent survey in Europe, environmental volunteer organisations were identified as the most credible sources of information. This implies that if business does not take a proactive and really credible leap towards converting its operations on the basis of the concept of ecological sustainability, it will lose its legitimacy.

At the time when companies have to re-engineer their operations, it is timely to take the next step forward and retool the company, integrating the most rigid moral, ethical, and ecological standards. A company is a most efficiently organised structure, the only type in society that creates value added based on the principle of minimum input and maximum output. At a time when governments and NGOs are searching for more efficient modes of operation, business practice is welcome indeed. And, if ethical, moral, and environmental considerations are combined with the other cutting-edge factors, this will be a formidable force indeed.

The biggest weakness of industrial conglomerates today is that they have embarked during the last 20 years on programmes to cut costs and remain competitive without taking considerations of the community and the limits of nature into account. Companies only looked at (1) cutting employees and (2) generating more money with less cash.

Productivity increases were first equated with a reduction in personnel. As the president of the Taco Bell restaurant chain remarked, "The ideal company is one where I work on my own with my 10,000 franchisees." All other people are to be considered potentially redundant. This is not exactly the bright picture of the future that communities project for themselves.

The logging communities in the Pacific Northwest have not heard of job creation for decades. The only story they have heard is that it is necessary to cut employment to save the other few jobs that are left. This strategy of business has generated a major weakness. Proctor & Gamble, IBM, and Chrysler have embarked on major cost reduction programmes, cutting tens of thousands of job and closing dozens of factories. The support these companies have built up in their communities is fundamentally based on the provision of local jobs. The systematic loss of employment erodes the basis of trust on which business became established in the community. Whereas no one debates the need for an efficient organisation, the narrow focus on people as the only one of the three critical input factors (people, capital, and raw materials) to bear the brunt of higher productivity is an anomaly.

The second drive toward cutting costs and improving returns is to seek more return for less cash. Corporations desire to maintain their financial performance and to continuously improve the return on their investment. The just-in-time programme frees up a lot of cash previously locked up in raw materials, supplies, half-fabricated goods, and finished products.

"Tax technologies" were introduced, making use of tax havens through complicated holding structures spanning the world, with the objective that companies pay the least amount of taxes to the communities in which they operate. New investment commitments were made subject to bargaining over tax rebates and even straightforward subsidies.

How can a company that does everything to evade taxes—making use of the loopholes in the international laws, soliciting major subsidies, and cutting jobs—ever expect a relationship of trust with society? No loyalty will be generated through this type of behavior, certainly not loyalty sustainable for business over time.

While industry has rightfully been searching for productivity in labor and in capital, it has never sought with the same vigor a dramatically improved level of productivity in the use of virgin raw materials. If industry decreased warehousing of input factors from three months to fifteen minutes—as is the case in the automobile industry—why can't we imagine industry reducing the need for virgin materials by 90 per cent? The redesign of the warehousing systems and delivery systems took less than five years. So, is it unreasonable to expect that industry will eliminate several of its waste streams in a few years' time and reduce the intake of virgin materials by 90 per cent by the end of the century? It is urgent!

Industry will endanger its own future if it does *not* take this route. Industry can reinforce its critical role in society by firmly engaging in massive materials reduction with the same zest as productivity of labor was increased, quality was improved, and just-in-time was established. If in addition business succeeds in integrating moral, ethical, and environmental issues in its strategic approach, then it will become a formidable, respected, and sustainable presence in harmony with the community and the Earth. The companies which achieve this will be the winners, while those which neglect it will be the dinosaurs of tomorrow.

Social Monitors and Companies' Ranking: The Case of CEP

Although there are a growing number of social monitors who issue ratings of companies, we focus here on those of the Council on Economic Priorities (CEP) because of the breadth of their offerings as well as their availability in a variety of formats, including books, disks, reports, and lists. CEP's best-selling books aim to make consumers more socially and environmentally conscious. They include the organisation's annual *Shopping for a Better World* (Ballantine Books 1990-1993; Sierra Club Books, 1994) and *The Better World Investment Guide* (Prentice-Hall). Both books provide user-friendly information on how corporations rate on various social and environmental performance criteria.

Besides publishing occasional books, CEP also regularly circulates 20-90 page reports on the environmental policies and practices of specific corporations. The reports are relatively unique in that they compare each company to others in the industry on a variety of criteria. Another service provides interested customers with more concise quarterly summaries of the social and environmental records of approximately 700 U.S. and international corporations.

Every year, CEP also conducts some more focused campaigns. The annual "America's Corporate Conscience Awards" recognises firms whose policies and practices indicate abiding concern for social and environmental responsibility. The "Campaign for Cleaner Corporations" identifies some of the U.S's *worst* environmental performers relative to their industry peers. The campaign invokes participation from a wide range of consumer, environmental, civic, and investor organisations in a collective effort to improve corporate environmental policies and practices.

Critical to preparing these digests of environmental information are systematic procedures for collecting, analysing, and presenting information. CEP relies on information gathered, not only from the EPA, OSHA, the SEC, and the regional and national media, but from other government publications, legal databases, business and environmental organisations, and corporations themselves. Most of the data are summarized in absolute terms as well as in comparative form to industry rivals.

Environmental ratings of companies are based on industry questionnaires that are prepared by CEP staff with help from industry and environmental experts. Companies provide self-reported information about their environmental policies and practices. Often the data reveal valuable insights that would not otherwise be available from secondary sources. They also provide judges with an insider's view, and so a more complete profile of a company's environmental posture.

Corporate information is combined with information gathered from secondary sources, including government databases, computerized searches of newspapers, periodicals, and other publications, internal CEP files, news clippings, company literature, annual reports, 10-K filings, proxy statements, and other SEC documents. All data are later reviewed and entered into an environmental ratings template. Each subcategory of environmental ratings is given a relative weight based on expert input from CEP's advisors. The subcategories are rated individually based on the ratings template. Finally, through a systematic computation process, an overall rating of the company is developed and communicated to both the company and to advisors, who review the rating and identify areas for clarification, more information or discrepancies. Any new information acquired in the process is input into the rating template and a new rating computed for each subcategory. The subcategories are then recombined to obtain a revised overall rating.

In most of its work, CEP relies on panels of judges to determine which firms to nominate for awards or to highlight on lists. Judges are invariably comprised of prominent experts from the relevant disciplines. Nominated firms are contacted for additional information, based on which the panels determine which companies to include and decide the lists and the winners to announce publicly.

Policy Implications of Ecological Tax Reform

A fundamental economic principle is to maximize the productivity of the limiting factor in the short run and to invest in its increase in the long run. If factors are good substitutes, then the absence of one does not limit the usefulness of the other—i.e. neither can be limiting. But if they are complements, then the one in short supply is limiting. For example, plant growth requires sunlight, soil, and water. These factors are complements, not substitutes. Extra sunlight does not compensate for a lack of water. Plant growth is effectively limited by whichever of the three factors is most scarce. Unless that limiting factor is increased, it does no good to increase the others. The idea is familiar to ecologists and chemists as Liebig's Law of the Minimum.

From the foregoing it is clear that the relationship between value-adding agents (man-made capital, including labor power) and that to which value is added, natural resources (the natural income flow produced by the stock of natural capital), is one of complementarity. Even if man-made and natural capital were good, but imperfect, substitutes, the process of transforming the latter into the former would still reach an optimum extent. But if man-made and natural capital are complements, that optimum extent will be reached much sooner and more dramatically, since the scarcity of the factor in shortest supply will limit the usefulness of the other, more abundant factor. In the empty-world economy of the past, man-made capital was the limiting factor. In today's full-world economy it is remaining natural capital that is scarce and therefore limiting. The fish catch is limited not by the number of fishing boats, but by the remaining populations of fish in the sea. The timber harvest is limited not by the number of sawmills or lumberjacks, but by the remaining standing forests and our desire to preserve the non-timber services of forests (wildlife habitat, flood control, recreation, etc.).

As we move into an era in which natural capital is limiting, we need to economize on it more. That means its price must go up relative to man-made capital. But since much natural capital is common property, outside the market, and since the market itself is very shortsighted temporally, it is necessary to raise the price of natural capital by public policy. As we have seen earlier, this can be done in a general way by shifting the tax base from value added (labor and capital are no longer limiting) on to that to which value is added (the natural resource flow yielded by natural capital). In short, tax throughput, not income.

Consumption limits will be set by differing *national* policies—not by a single global policy. Nations could limit their total consumption (scale) by a strategy of low population and high per capita consumption or by a strategy of high population and low per capita consumption. Different nations will make different choices. Some will not even limit aggregate consumption and those that do will make different choices regarding per capita consumption vs. population. These differences cannot be maintained in a world of free trade, free capital mobility, and increasingly free (or uncontrolled) migration. Compensating tariffs will be necessary to keep these national differences from being homogenized. National policies of controlling consumption and population, and of counting external costs, are more important to sound allocation, distribution, and scale than are the tenuous gains from comparative advantage and free trade, which are currently celebrated beyond reason. This argument does not imply autarchy, but it does throw cold water on the faddish advocacy of global economic integration and consequent homogenisation of policies as an unquestioned good. "Globalisation" ranks with "the demographic transition" and "dematerialisation" as a false panacea-another germ of truth that has been allowed to grow into a whale of a fantasy in order to protect the dominant myth of our culture: that of an ever-growing economy.

Two views of production have been discussed: (a) only value added to indestructible building blocks is consumed, vs. (b) value added by nature as well as human agents is also consumed. There are important policy differences implied by each.

(a) The value-added view of wealth would lead one to reject the very notion of a "global pie" of wealth to be divided justly or unjustly among nations and people. In this view, there is no pie-there are only a lot of separate tarts which some statistician has stupidly aggregated into an abstract pie. The separate tarts are the product solely of value added by the labor and capital of the nations that produced them. If nation A is asked to share some of its large tart with nation B who baked a small tart, the appeal should be made to nation A's generosity and not to any notion of distributive justice, much less exploitation. On the assumption that all value comes from labor and capital, and that nature contributes only a material substratum which is indestructible and superabundant, and hence valueless, this is a quite reasonable view. Are you poor? Well, just add more value by your own labor and capital. There are no limits from nature. Stop whining about this imaginary pie, and get busy adding value. This view is common among neoclassical economists. And if we accept its presuppositions it is not unreasonable.

(b) The natural capital view can reject the imaginary pie also, and look at the tarts different people have baked. Are they really only the product of labor and capital and random atoms? Certainly not. You need flour and sugar and butter and apples. Before that you needed wheat, sugar cane, milk, and apple trees. And before that you needed a gene pool for wheat, sugar cane, cows, and apples, with some minimal degree of genetic diversity, and grass growing on soil whose fertility is maintained by all sorts of worms and

> microbes, and sunlight without too much ultraviolet, and rainfall that is not too acidic, and catchment areas to keep that rain from eroding topsoil, and predictable seasonal temperatures regulated by the mix of gasses in an atmosphere without too much CO_2, etc., etc. In other words, we need natural capital and the flow of resources and services that it renders—a whole lot more than indestructible building blocks! Our dowry of natural capital is more or less given, and is not the product of human labor and ingenuity. It is in many ways systemic and indivisible-more like a vast pie than separate tarts. How it is stewarded and distributed is not an idle question based on some gratuitous aggregation. The demands of justice impinge strongly on the stewardship and distribution of the common life-supporting exosomatic organs we refer to as natural capital, but very little on value added, since the latter rather naturally and reasonably belongs to whoever added it.

A North/South bargain will have to be struck in which the South gets very serious about limiting fertility (and the over-consumption of its elites), while the North gets very serious about limiting over-consumption (and the poverty-reinforcing high fertility of its marginalized underclass). But the North will not get serious about limiting consumption as long as our leaders remain convinced by our economists' view that all wealth comes from value added, that we can only consume what we have produced. We have to recognise that we can and do consume a lot more than we produce, and that what we consume over and above conventional value added is value added by nature. But in addition to consuming value added by nature we are now consuming the very capacity of nature to add value in the future, i.e., natural capital.

Also, in the absence of a North/South bargain not much is likely to happen on either population or consumption. Why should the South control its overpopulation if the resources saved thereby are merely gobbled up by Northern over-consumption? Why should the North control its over-consumption if the saved resources will merely allow a larger number of poor people to subsist at the same level of misery in the South?

The policy implications for the North are that we must economize on and invest in natural capital, because it has become the limiting factor (replacing man-made capital in that role) as we have moved from an empty world to a full world. To force ourselves to economize on natural capital we must raise its price above the market level. A concrete method of accomplishing this is ecological tax reform. Viewed in a North/South context, ecological tax reform takes on even greater relevance as the North's best contribution to a global bargain.

Arguments for Environmentalist Business Management

The following list of arguments is an entirely new departure. For the first time, it sets out in tabular form arguments for and against the introduction of the Integrated System of Environmentalist Business Management (the Winter Model). The Model has been well received in the press, on radio and on television. Judging by the discussions at events organised by chambers of industry and commerce, business associations and economic science faculties, the attitude of specialists (theoretical and practical alike) is generally favourable. It is important, though, to take criticism seriously, especially as it can often be a starting point for improvements. Taking a conscientious look at the pros and cons, it is evident that not all reservations can be dismissed just like that; the fact remains, though, that the positive aspects are in the majority.

The list of arguments sets out to achieve three things:

1. It is intended to help the businessman to decide whether to introduce the Integrated System in whole or in part in his company or in his particular sector.
2. It is intended to give additional help to those responsible for introducing the Model into a company on how to avoid the kind of procedural or psychological mistakes which could jeopardise the chances of success.
3. It is intended as an additional argument for anyone who is determined to work for the more widespread application of the Model, be it on an economic, cultural, religious or political level.

The list of arguments provides answers to the question of what the

Model means in terms of:

- The businessman himself and his view of his work (I.)
- The management team's chances of success (II.)
- Cooperation with the workforce (III.)
- The various management sectors (IV.)
- Public relations work including activities in professional organisations (V.)

The list is set out in the form of a pro/contra balance sheet comprising forty pairs of contradictory arguments. Beginning in each case with the objection, it sets out to help protagonists of the Model to avoid the mistake of overwhelming people with arguments from the outset, and instead to take note of the objections, think them through, defuse them and go onto the offensive with the arguments in favour.

Over and above the ability to listen to the other side's arguments, advocates of the Model must always be receptive to other arguments and be prepared to modify their own opinions in the light of what the other side has to say. After all, the Model is not permanent and unchanging for all time, but has to be constantly tested against new practical and scientific developments and be flexible enough to adapt to the situations in a wide variety of companies.

The word "businessman" is taken to mean both independent and salaried entrepreneurs and managers.

Sectoral Checklists for Environmentalist Business Management

These checklists give concrete form to what the environmentalist management approach means for twenty major corporate departments. They set out clear recommendations as to how each department should be organised in an environmentalist manner.

If what the businessman is most concerned about is short-term success, he should first of all work through the "economising on energy and water" checklist and take a look at all the points in other checklists marked with an asterisk. These recommend measures which are helpful to the environment and to the company as well in that they reduce costs and boost profits. Many of the measures can be introduced with immediate effect and with no significant capital outlay. In other words, even a relatively superficial reading of the checklists can bring the company a cash advantage.

If, however, a businessman is interested in the medium- and long-term viability and profitability of his company, he will realise that—with the escalating environmental problems and their effect on public opinion, markets and legal requirements, and their undoubted importance as the most significant of all future trends—environmental issues should be seriously considered at the planning stage, in which case he should study all the checklists himself or at least have them studied by his departmental heads. Appendix I contains a selection of the relevant databases, publications and organisations. The material has been systematised according to the following 20 sectoral checklists. The databases, publications and organisations are in turn subdivided into International, British, American and German (the latter including Austrian and Swiss sources). The "International" section contains sources associated with international organisations or dealing with the subject from an international viewpoint. The French edition of the book, as yet at the preparatory stage, will also contain French sources. Each subsequent version (beginning with Spanish and Italian) will give the specific sources for the country concerned and additionally include all sources listed in all previous editions. The source lists provide only samples of relevant databases, publications and organisations. They are not comprehensive and do not imply endorsement by the authors, the participants, the editor or the publisher. The differing length of the source lists does not imply that a larger volume of material is available overall on the subject of one list than on that of another. Nor should the number of sources stated under each country be taken as an indication of the total amount of information available in that country. As the manner in which sources are quoted differs slightly from one country to another, no attempt has been made to achieve 100 per cent uniformity. Moreover, differences in the depth of information given on individual sources have been accepted for the sake of practical benefits. Taken together the checklists constitute a large fund of valuable, practical and up-to-date information which will help companies in numerous countries to combine successful business with the responsibility for a healthy environment.

Working Conditions as a Checklist for Environmentalist Business Management

A "humane" working environment is an essential prerequisite for ecologically sound production. Workers can hardly be expected to be creative in the interests of the environment if they themselves are obliged to work under unsatisfactory conditions. The following checklist covers all the items needed to achieve suitable working conditions.

Avoid Stress Caused by Environmental Factors

Any attempt to avoid environmental stresses, whether arising from noise, light and colours, room climate or hazardous working materials, must always be based on the following principles:

- Firstly, implement measures to avoid stress (when technically possible, by altering production conditions and work processes).
- Secondly, isolate stress factors by containment (to reduce noise), de-dusting and detoxification systems (including waste water treatment and incineration plants), etc.
- Only where such measures are not feasible (or are unreasonably expensive), take personal protection measures at the workplace (ear protectors, dust masks, dosimeters, etc.).

1. Reduce noise (noise emissions and vibrations).
   - Reduce noise from the work process itself by selecting suitable procedures and optimising process conditions (e.g. choice of suitable equipment).
   - Reduce noise from subsidiary activities (which are given less attention than the main activities but are often a major component in total noise).
   - Shield off otherwise unattenuable noise sources.

- Demarcate noise areas.
- Ensure that personal protection devices are actually used.
- Make sure that persons exposed to noise undergo health/ medical check-ups.

2. Optimise lighting and colours.
   - Match lighting intensity to the type of work. The relevant regulations (in the Federal Republic of Germany, the Guidelines on Workplaces) specify the minimum lighting requirements for different activities.
   - Make sure that other parameters are complied with, e.g. lighting level, light intensity distribution, dazzle suppression, light direction and shadow, light colour and colour reflection, as laid down in the appropriate provisions (in the Federal Republic of Germany, the Workplace Regulation).
   - Employ indirect lighting at VDU workplaces.

     Lights (if possible white, but certainly as light as possible) should be directed at the ceiling; only indirect light will then fall on the screens, thus obviating reflections (and resulting in less eye strain).
   - Use lamps whose light spectrum is close to that of normal daylight.

     Light has a great physiological and psychological effect on man. With fluorescent light tubes with a very limited light spectrum, combine different lamps to approximate to the spectrum of natural daylight.
   - Use electronic devices or three-phase switching to avoid flickering of fluorescent lamps.
   - Select colour schemes on the basis of the features of the room (size, machines installed, floor structure, ceiling design, pillars and beams, windows, access routes, etc.), existing colours (machines, materials, products, vehicles, etc.), the division into work areas and rest areas, the number and kind of people working there, as well as the jobs to be done and the working times.

     Colour can have a very marked effect: for example, a light colour can make a low ceiling appear higher and a dark colour can make a high ceiling appear lower; green gives the impression of reducing noise and gives maximum light in dim conditions since its wavelength is in the range to which the human eye is most sensitive.

3. Regulate the room climate.
   - Adjust the air temperature, relative humidity, air speed and heat radiation to comfortable levels. Air pressure, being difficult to adjust, is not considered here.

- Pay particular attention to and avoid:
    - draughts (differences in local air speeds)
    - asymmetric distribution of heat radiation
    - floors that are either too cold or too hot
    - excessive temperature gradients between head and foot level

4. Avoid hazardous substances at the workplace.
    - Minimise hazardous emissions by:
        - selecting suitable work processes, materials, additives, process materials, etc.
        - selecting suitable inert gases (if necessary and/or possible)
        - optimising operating parameters
    - Measure whatever hazardous substances are produced and comply with the appropriate legal provisions. Further details are given in "Klima und Luft am Arbeits-platz" (Workplace climate and air), published by the Institute fur angewandte Arbeitswissenschaft (Institute for Applied Labour Studies) in 1986; among other things, steps must be taken to ensure that the M AK values (maximum workplace concentration) and TRK values (technical guide concentrations) are not exceeded, the latter values being laid down for substances for which it is not at present possible to set MAK values (carcinogenic and mutagenic substances).
    - Bear in mind hazardous substances which are not produced by the production process or the resulting products: emissions from the building fabric, wastes, alcohol, cigarettes, preservatives, etc.
    - Demarcate those areas in which hazardous substances are emitted.
    - Ensure that personal protection equipment is used.
    - Give persons exposed to hazardous substances medical check-ups.
    - Ensure that there is adequate ventilation. Ventilation is absolutely essential, even with very low emissions of hazardous substances, in order to avoid any long-term effects.

Ensure that Equipment (Including EDP Hardware) and Workplaces are Economically Designed

5. Ensure that controls operate reliably.
6. Make components easy to use.
7. Ensure that emergency shut-off switches are clearly marked and easily accessible.

8. Design controls and displays and workplace dimensions in accordance with anthropometric principles.

   Pay special attention to space and sequence of operations, height, room for knees and feet, supports and field of vision in order to avoid forced postures and other uncomfortable (static work) or unnatural postures. As a general rule, static work and one-sided and excessive muscular exertion are to be avoided.

9. Ensure that VDUs have adequate luminance, do not flicker excessively, have a stable image, are free of reflection, provide adequate adjustment and have enough contrast.

Improve Safety at Work

10. Ensure that the safety regulations are observed (including the wearing of personal protection equipment).
11. Make sure that the emergency shut-off equipment can be operated quickly, easily and safely and that it operates automatically to preset limits.
12. Install safety fences, barriers, etc. to ensure that workers do not enter the danger area around automatic machines (including digital control equipment and industrial robots) and that any stray objects thrown out by the machine do not endanger personal safety.
13. Take additional precautions if normal safety equipment has to be removed for technical reasons such as installation or maintenance. In such cases, eliminate any danger to personnel (e.g. as a result of equipment operating at crawling speed or being switched to manual operation).
14. Wherever possible, install automatic anti-collision equipment.
15. Demarcate and clearly signpost access routes, etc.
16. Ensure that safety markings (colours, symbols) are always clearly visible.
17. Provide anti-dazzle equipment and spatter protection.

Ensure Ergonomic Design of EDP Programmes

18. Ensure that the software is suited to the job.

    - Avoid unnecessary strain on the operator.
      The software must help the user carry out his task without the extra burden of coping with the idiosyncracies of the system (such as additional management and routine operations - e.g. with text-processing systems, having to remember where or under what code certain passages of text have been stored).

- Match the mask sequence to the normal work step sequence.
- Match the type and form of input to the job in question (e.g. graphic input).
- Ensure that data organisation does not require the active involvement of the user.
- Use programmable function keys for frequently recurring strings of commands.
- Ensure that response times are of appropriate duration (provide a programmable response delay system).

19. Ensure that the system is self-explanatory.
    - Make sure that the system is readily comprehensible. At the request of the user, the purpose, method of operation, performance and requirements of the system must be explained.
    - Match the form of input to the level of knowledge of the user.
    - Provide input aids geared to the situation/process in the form of texts, graphics or images (or combinations thereof).
    - Whenever possible, use physical objects (symbols) for input purposes ("rubber eraser" or "paper basket", depending on the type of deletion, a "brush" for infilling, virtual machines for task reallocation, etc.).
    - Provide facilities for adjusting the information density in a display.
    - Ensure that background processes can be made visible.
20. Provide as wide a range of control facilities as possible.
    - Allow the user to initiate/interrupt/continue the dialogue; the system should take over only in exceptional cases.
    - Provide a choice of dialogue status (beginner, expert, etc.).
    - Conduct test runs to ensure that other data input can be corrected.
21. Pay attention to reliability.
    - Aim to achieve as close a match as possible between system dialogue and user expectations.
    - Ensure that response times are uniform and can be calculated.
    - Ensure confirmation that an input command is being processed.
    - Make sure that information can be obtained at all times on the system status.
    - Ensure that given actions always achieve the same results.

22. Take steps for dealing with errors.
    - Make error display readily comprehensible for correction purposes.
    - Ensure that error messages show the cause of the error and, as far as possible, the manner of correction.

Provide Training in Good Time

23. Provide training to familiarise staff with new systems.

    Group learning has already been successfully employed in numerous companies, and is a good solution.

24. Organise work to take account of what has been learned in training.

    The training process is facilitated if work is organised in a way which solicits the skills and knowledge already learned and allows new tasks to be allocated to staff on an individual basis.

25. Extend training to cover not only professional expertise but also social aspects and encourage staff to develop their own faculties.

26. Clarify the objective of training by explaining hierarchical and parallel systems in addition to the organisational system in which each individual worker finds himself (to enable workers to see relationships within the system).

Observe Six Basic Rules When Designing a Work System

27. Match workers to the demands of the job.

    The aim is to avoid creating stress by making excessive or inadequate demands on staff.

28. Avoid monotony and a rigid work pace.

29. Give staff sufficient room and organisational freedom to carry out their tasks.

30. Make each activity include planning, decision-making and monitoring aspects in addition to executive aspects.

    Each worker must have time to organise his work and monitor his own activities and must be included in decision-making and planning in accordance with his particular abilities, interests and expertise.

31. Avoid social isolation.

32. Organise operations so as to include team work (quality circles, workshop circles, training workshop, etc.).

Industry Environmental Compliance, Training Manual, UNEP, 1996

Why have a Compliance and Enforcement Programme? What are its Objectives? What are its Functions?

Setting the Context

Agenda 21, adopted at the 1992 Earth Summit as the programme of action for the 21st century, emphasised that nations must build their institutional capacity to implement environmental goals to achieve sustainable development. Most governments now have environmental laws and regulations but their implementation often lacks the organised and concerted efforts needed to ensure that they are complied with. Although industry self-regulation, economic incentives, peer and public pressure are important in developing continual environmental improvement by companies, a government compliance and enforcement programme is necessary to:

- Verify compliance and ensure that government environmental objectives, laws and regulations are met fairly and equally by all companies
- Deter non-compliance by creating an atmosphere that encourages and enhances industry, compliance with environmental standards.

Why are Environmental Compliance and Enforcement Programmes Important?

To achieve the Goals of Protecting Public health and Environmental quality envisioned by Environmental Laws. Public health and the environment will be protected only if environmental requirements achieve results. Enforcement programmes are essential to attaining that goal.

To Build and Strengthen the Credibility of Environmental Requirements. To get results, environmental requirements and the government agencies that implement them must be taken seriously. Enforcement is essential to build credibility for environmental requirements and institutions. Credibility means that society perceives its environmental requirements and the institutions that implement them as strong and effective. Credibility encourages compliance by facilities that will be unlikely to comply if environmental requirements and institutions are perceived as weak. The more credible the law, the greater the likelihood of compliance and the likelihood that other government efforts to protect the environment will be taken seriously. To Ensure Fairness. Without enforcement, facilities that violate environmental requirements may benefit, compared with facilities that voluntarily choose to comply. A consistent and effective enforcement programme helps ensure that companies affected by environmental requirements are treated fairly. Facilities will be more likely to comply if they perceive that they will not be economically disadvantaged by doing so.

To Reduce Costs and Liability. Although compliance is often costly in the short term, it can have significant long-term economic benefits for both society and the facility in compliance. The healthier environment created by compliance reduces public health and medical costs, as well as the long-term cost to society of cleaning up the environment. Compliance benefits industry by reducing its liability and long-term cleanup costs. Industry also may realise immediate economic benefits if compliance encourages increasing the efficiency of its processes. No institutional programme can function effectively without a clear mandate and a set of realistic objectives. For a compliance and enforcement programme, there are three broad types of objectives: education and promotional objectives, creating an atmosphere of deterrence, and enforcement objectives. Governments need to determine the right balance and the right kind of emphasis of these objectives for their own country, but some combination of all three is usually for a compliance and enforcement programme to be effective.

Educational and Promotional Objectives

A compliance and enforcement programme should include measures and policies that foster voluntary compliance. Promotion of compliance includes all the actions taken to encourage compliance with environmental requirements. Promotional measures can include:

- Providing technical assistance and education through formal educational processes and informal approaches for companies, industry associations, public interest groups and others.
- Increasing public awareness about the health, environmental and economic costs of non-compliance.
- Disseminating information about technologies to educate industries about the long-term economic and environmental advantages of applying cleaner production measures.
- Demonstrating cleaner production technologies and publicising cleaner production success stories to encourage industries to adopt preventive approaches.
- Providing economic incentives to encourage behaviour change (e.g. tax benefits, ecological tax reform, low interest loans, etc.).
- Encouraging role of industry associations in developing best environmental management practices and industry self-regulation programmes.
- Experimenting with voluntary agreements with industry to achieve jointly agreed environmental targets. Experience shows that these often work best in industries who's members are small numbers of large, publicly

visible companies (e.g. chemical industry), and when backed up by the likelihood that regulations will be applied in the event of failure.

- Providing public with access to information and an opportunity to review facilities' plans and activities, also can play an important role in persuading facilities in their community to reduce the harmful effects they have on the environment and to comply with environmental requirements established in laws and regulations.

Creating an Atmosphere of Deterrence

All measures taken to encourage regulated communities to avoid enforcement penalties are deterrence. Deterrence takes aim at potential violators by sending a clear message that those violators will become subject to legal actions for non-compliance with environmental requirements. An effective deterrence atmosphere usually requires:

- ensuring that potential violators know about the environmental requirements, particularly the liabilities for the damages that they may cause, and that the regulated community has information about penalties, fines, liabilities, and deterrence measures.
- demonstrating that institutions responsible for enforcement can detect violations and violators and can respond to violations swiftly and effectively.
- demonstrating clearly that response includes an appropriate sanction.
- carrying out consistent and fair enforcement responses.
- strategically publicising enforcement measures taken against particular violators.

Enforcement Measures

Enforcement measures are needed to respond to violators who may otherwise benefit over facilities that comply. Enforcement sanctions can remove the economic advantages that facilities may gain from non-compliance. Effective and consistent enforcement gives credibility to a government's enforcement programme and enhances the integrity of the legal system. The more credible the law and the enforcement programme, the greater the likelihood of compliance and the likelihood that other government efforts to protect the environment will be taken seriously. The visibility of enforcement mechanisms is an important factor in building the credibility of a compliance and enforcement programme. If an enforcement programme is to be effective, it must be visible, demonstrating to the public and to regulated communities that environmental requirements are enforced. What should a compliance and enforcement programme actually do? The key functions of an effective compliance and enforcement programme are:

Developing Laws and Regulations that Can be Enforced

Once an environmental need is identified, the legislation and regulation must be developed in a way that can be enforced when required. This may require interpreting broad environmental laws with specific regulations, or providing feed-back to legislatures to revise laws that are unenforceable.

Identifying the Regulated Community

It is necessary to have a clear understanding of who is required to meet the requirements of environmental laws or regulation, and establish priorities based upon the degree of environmental consequences of possible violations, past good or bad company environmental management practices, and any incentives or barriers to compliance. This will likely require an inventory and information management system to identify and keep track of those who make up the regulated community.

Promoting Compliance

No compliance and enforcement programme can expect to be able to monitor compliance everywhere or detect and respond to all violations. Compliance promotion includes:

- disseminating information about environmental requirements to the regulated community, relevant government bodies and public
- providing cleaner production information, education and technical assistance to the regulated community
- building public awareness and support
- publicising success stories
- providing economic incentives and facilitating access to financial resources

Permitting or Licensing Facilities

Once a regulated community has been identified, a permitting system enables environmental requirements to be tailored to the circumstances of specific facilities. This will require developing permit application procedures, processing applications, and issuing permits in co-ordination with other relevant government bodies.

Monitoring Compliance

There are four primary ways of monitoring a regulated community's compliance with environmental laws and regulations:

- formal inspections by government or certified inspectors
- self-monitoring, record-keeping and reporting by the regulated community

- community monitoring and citizen complaints
- sampling of the environmental conditions (air, water, soil) in the vicinity of a facility.

Responding to Violations

When a violation is detected, every compliance and enforcement programme must develop a hierarchy of enforcement responses, consistent with its socio-economic and cultural situation. They may involve taking administrative, civil, and criminal actions, meant to do one or more of the following:

- return violators to compliance
- impose a sanction
- remove the economic benefit of non-compliance
- correct environmental damages
- correct internal facility management problems.

Enforcement responses depend on the severity of the violation, and may involve, for example:

- issuing administrative and legal notices of violation;
- permanently or temporarily closing a facility or a cycle of operation within a facility;
- revoking a permit or refusing to renew a permit;
- conducting emergency response plans and procedures;
- seeking compensation for damage caused by the violation or for the cost of the cleanup;
- imposing a monetary penalty; seeking imprisonment and punitive damages or fines; seizing property;
- requiring community service to improve the environment.

Role of Negotiation

Negotiation is part of a process to achieve compliance and enables both the facility and the concerned party or parties to consider the correctness of the facts, the circumstances of the case, and the variety of alternative responses. Negotiation provides an opportunity to obtain additional information and correct misinterpretations before legal action is pursued. In addition, negotiation provides an opportunity to reach a solution that satisfies all parties. Negotiation can result in a co-operative effort to develop a satisfactory solution that is not intended to alter basic requirements for compliance.

Different Approaches

What position in the government hierarchy? What degree of organisational independence? What degree of centralisation? What degree of integration?

In almost all countries, environmental organisations are relatively young and are still developing. Inevitably, many countries face difficulties related to the position of their environmental management organisations within the government. In general, such difficulties include:

- Overlapping of organisational responsibilities at national, regional, and local levels.
- Costs of staffing and operations.
- Unclear and confusing relationships among organisations on the national level and those on regional and local levels.
- Overly centralised or overly dispersed functions.
- Imbalances in the separation of legislative, executive, and judicial institutions.
- Issues related to establishment of priorities and the phasing in of programme functions.

This section describes the types of organisation that should be considered in a compliance and enforcement programme. It also discusses institutional considerations related to organisational design. The hierarchical position of the environmental agency is determined by the level of emphasis that a government places on environmental issues. In some countries, the environmental agencies began to evolve as a unit within the more traditional ministries of agriculture or natural resources. During the past two decades, governments have begun to pay more attention to the need for a responsible and functional environmental agency. Governments need to consider the placement of environmental institutions within the government apparatus and the level of responsibility and authority assigned to them. In a few countries the environmental management agencies function at the ministerial level or as independent agencies directly responsible to the head of state. In many developing countries, the environmental agencies function within the ministry of natural resources and the heads of the environmental agencies function as deputy ministers or directors. In many countries, the regional, local, and municipal authorities traditionally had the responsibility for licensing and monitoring industrial operations. The following guidelines reflect successful experiences of countries in which effective authorities have been assigned to environmental agencies.

Authority at the Ministerial Level

Many countries have recognised the importance of environmental management to the extent that environmental organisations have the status of a ministry. Environmental issues and concerns are thus can be raised at the highest levels of government. The establishment of the agency at the cabinet level places the head of the agency in direct contact with policy makers, legislative authorities, and the head of state. The advantage of establishing an independent agency or an agency at the ministerial level of authority include:

- Most of the programme, planning, and support functions are housed in one agency.
- The programme functions are performed under a unified strategy, mandate, and common objectives, facilitating long-term planning for the functions of the programme.
- A high level of authority makes it possible for the agency to acquire the resources it needs for various functions or to gain access to the resources of other agencies. It certainly facilitates the delegation of some functions of the programme to regions and municipalities.
- A high level of authority gives the environmental agency the opportunity to establish the regulations concerning environmental requirements and compliance and enforcement programmes.

Permanent Interministrial Commissions

If the environmental agency cannot be placed at cabinet level or function independently, an alternative is the establishment of a commission to bring environmental agendas to the attention of higher levels of government. In several countries, this approach has been used successfully during the early stage of development of environmental institutions. Its purpose is to form an interagency commission mandated by and under the direction of the head of state. The interagency commission comprises key ministers and a chairperson who reports directly to the head of state. The commission can have the following responsibilities:

- Establish a clear mandate, objectives, and a long-term strategy for the environmental functions.
- Define the roles and responsibilities of various ministries in support of environmental functions and the environmental agency.
- Define the role of regional authorities and institutions in carrying out some functions of the environmental programmes.
- Assess the needs and requirements of the environmental agency, identify appropriate ministries to share their resources in the implementation of environmental functions, and secure the support of those ministries.
- Establish a mechanism to enable the environmental agency to expedite

formulation and adoption of regulations by legislative authorities.

- Establish a mechanism to enable the environmental agency to effectively enforce the environmental requirements. Formation of adjudication boards is one vehicle that can facilitate the enforcement process.

For young environmental agencies that do not function at a high level in the hierarchy, the formation of interministrial commissions has been effective. Such commissions have been established in Argentina and earlier in the Philippines, Thailand, Indonesia, Brazil, Zambia, and several other countries. Each country, depending on the goals and priorities of the government, should maximise the use of existing institutions and, when necessary, make modifications to effectively address its environmental concerns. For example, one innovative approach has been taken by Denmark, Luxembourg, Korea, and the Netherlands: to develop environmental expertise within the police force or public prosecution services. Such approaches are examples of methods that maximise resources and provide effective interaction between executive and judicial institutions.

Degree of Organisational Independence

In general, there are three options for the organisation of the various functions of a compliance and enforcement program: a unitary organisation, separated environmental organisations, and an independent environmental organisation. In a unitary organisation, a department of the environment or of environment and natural resources, serves as the executive institution and is responsible for such functions as development of requirements and quality standards, permitting, monitoring, inspection, and planning and support functions. In separated environmental organisations, the authorities of this agency or department is shared with those of legislative and judicial institutions, and the agency assumes some of the functions of those institutions. In an independent environmental organisation such functions are completely the responsibility of agencies and institutions other than the agency responsible for environmental management.

Unitary Organisations

This organisational option embodies most of the three institutions in one organisation and under the supervision of one authority. An analogy is traditional forestry organisations. In some countries, the forestry organisations have responsibility for all the functions of a forestry management cycle. Even forestry inspectors are empowered to arrest violators. The advantages of this type of institution are:

- All resources are directed according to common institutional objectives and in pursuit of a coherent strategy. The links among the permitting, compliance, and inspection functions are strong.

- A high level of co-ordination and centralisation exists.
- Decision making is improved greatly.

The disadvantages of this organisational option for an environmental compliance and enforcement programme are:

- The option is costly and draws heavily on resources.
- It does not use the resources available in existing government institutions.
- Co-ordination with other departments of the government that are involved directly or indirectly in environmental issues typically is performed at higher levels in the hierarchy. This practice can interfere with effective co-ordination.
- The independence of the organisation from judicial institutions may place constraints on the enforcement process.

Separated Environmental Organisations

In this model, the institutionalisation of programme functions is divided among several agencies. For example, the permitting functions may be performed by sectoral or local organisations, and monitoring and inspections by an environmental agency. The advantages of this type of organisation are:

- This option makes the greatest use of the resources of existing organisations in such functions as establishment of requirements, registration and licensing, permitting, inspection, and enforcement.
- If co-ordinated well, this model provides more effective legal enforcement because it facilitates access to law enforcement authorities and resources.

The disadvantages of this organisational option are:

- Opportunities for integrated approaches are limited.
- The existence of several dispersed organisational units within the programme hampers coordination.
- Establishment of common objectives and mandates for the entire compliance and enforcement programme is difficult.

Independent Environmental Organisation

In this model, the environmental management functions, including establishment of requirements and quality standards; permitting; inspection; monitoring of compliance; and certain enforcement responses, such as issuance of administrative notices, are carried out by one agency (environmental), but legal actions and sanctions are carried out by other agencies. This is the organisational model typical of many countries. The most appropriate analogy is the institutional relationship between the Environmental Protection Agency and the Department of Justice in the United States. The advantages of this model include:

- Placing the functions related to establishment of general requirements, permitting, and inspection in one organisation creates more harmony in management and co-ordination among those functions and facilitates implementation of the objectives of those programmes.
- If proper co-ordination and understanding are established between the executive and judicial institutions, the effectiveness of the compliance and enforcement programme will be enhanced significantly.

The disadvantages of this model are:

- The option is relatively costly and draws heavily on resources.
- To be effective, such an organisational option requires the existence of a strong judicial system capable of effective law enforcement.
- The traditional organisation is more appropriate for countries that have well defined federal government systems and strong regional or state legislative executive and judicial institutions.

In all countries, organisational structures for compliance and enforcement programmes fall into one of the three organisational patterns described above. In several countries, in addition to formal organisations, committees and commissions have been formed to co-ordinate the environmental activities of several government organisations. When they are empowered to do so, these committees and commissions can be effective in establishing understanding and co-ordination among legislative, executive, and judicial institutions.

Degree of Centralisation

The institutional arrangements for implementation of the functions of compliance and enforcement programmes are either centralised or decentralised, or they exhibit a combination of the two approaches.

Centralised

In this type of organisation the main functions of the programme are centralised, and branch offices in the provinces carry out the directives of the central agency.

Decentralised

In the decentralised organisation local governments and municipalities have the power to establish the requirements, to permit and monitor the compliance, and to enforce the laws. In such a situation, a great deal of decentralisation and division of responsibilities can take place.

Combined

In many countries, a two-tier or three-tier system divides environmental responsibilities among national and regional governments and local governments or municipalities, combining the national nature and risk of many environmental problems together with the flexibility of regional or local authorities.

Most pollution issues are by nature national issues—for example, pollution of rivers, handling of nuclear waste, and transport of hazardous wastes. It is in the interest of every country to establish a minimum requirement and a set of minimum quality standards that are enforced at the national level. It is important that federal governments establish the objectives and mandate of the compliance and enforcement programme and ensure the integrity and fairness of the programme at the national level. Regional and local authorities may have the flexibility to establish the executive procedures and specific requirements within the framework designed by the federal government. Regional and municipal governments may have the authority to develop more stringent requirements and standards tailored to regional or local conditions, as well as to develop methods and mechanisms for the implementation of enforcement functions.

CO-ORDINATION WITHIN A FEDERAL SYSTEM: ARGENTINA

Argentina's constitution (1853) provides that the federal congress can legislate only on matters explicitly delegated by the provinces to the federal government. No environmental authority has been assigned explicitly to the federal government. Therefore, the provinces play a stronger role in environmental issues. To date, the federal government and the provinces all have enacted different laws concerning water and air pollution and waste disposal in their jurisdictions. They also have their own regulatory institutions to oversee and enforce compliance. In addition, the municipalities in larger cities have their own municipal environmental regulatory institutions. Recently, efforts have been undertaken to provide a national framework for environmental management in the country.

In accordance with government decrees of 1991 and 1992, the Department of Natural Resources and Human Environment (SRNAH) was established, under the office of the president, to "assist the President of Argentina in a nation-wide undertakings for protection, restoration, and management of natural resources and environment." SRNAH's authority is limited by the structure of the federal government. SRNAH may establish federal laws and requirements; however, enforcement rests with the authority of the provincial and municipal governments. To co-ordinate the various functions at national and provincial levels, a Federal Council on the Environment (COFEMA) made up of provincial environmental agencies and SRNAH, has been formed. The council functions within an environmental policy framework agreed upon by the provincial governments.

The council is charged to "promote environmentally-sound development policies and enhance environmental protection at the national level."

Degree of Integration

Environmental laws have often been designed to tackle a single medium problem such as water pollution, air pollution, solid waste. But this division of the environment into separate media fails to recognise that pollutants move from one medium to another. A non integrated approach also tends to encourage traditional, end-of-pipe controls (e.g. filters, scrubbers, cooling towers, electrostatic precipitators) which not only tend to transfer pollutants from one medium to another, but which, despite considerable investment costs, bring no economic payback. An integrated approach, on the other hand, encourages at-source, cleaner production measures, reducing the amount of wastes to be disposed of, minimising energy and raw material consumption, and preventing pollutants from appearing in any medium.

A single medium approach usually means that different agencies are inspecting the same plant, requiring facilities to fill out forms and provide much of the same information. This can cause confusion for a company not to mention added paperwork, duplication of effort and disregard for public authorities' administrative complexity, and inconsistency. To maximise integrated environmental protection, there are generally two approaches:

Fully Integrated Approach

An integrated approach to permitting and inspection is at the centre of the compliance and enforcement efforts of several countries. This approach, with some variations, is being developed or applied in Sweden, the Netherlands, France, Germany, the United Kingdom, and Cote-d'Ivoire. Several other countries are institutionalising this approach. Under this integrated approach, one integrated permit is issued for each facility and is monitored by an integrated or co-ordinated inspection. The integrated approach to permitting and inspection intends to ensure that:

- Pollution problems are not simply transferred from one medium (air, water, or soil) to another.
- Resources are used more efficiently to save costs for both the government and industry—for example, one permit can be issued to the facility to cover all media, or one inspector can monitor both emissions and discharges of effluent.
- The permitting and compliance processes are simplified.
- Monitoring and inspections are conducted to assess the total environmental effect of pollution from a facility.

An integrated approach does not necessarily reduce constraints on enforcement. Under the integrated system, specific discharge and emission limits and standards for each medium still must be met. The concept of integrated permitting and inspection is new and has not been established fully. More discussion on the integrated approach to permitting and inspection is provided in modules 3 and 4.

Interagency Co-ordinated Approaches

The purpose of this approach is to make the best use of resources available within the government system in performing the functions of a compliance and enforcement programme. A number of innovative approaches can be taken to reduce the cost of permitting, inspection, and enforcement. For example, in many countries, inspectors from the ministry of labour, ministry of health, environmental agency, and municipalities inspect facilities. Under the concept of interagency inspection, those inspection functions are integrated, or, at least, the interagency inspections are coordinated. Other examples already examined in several European countries include: use of the local police force to detect non-compliance cases; empowerment of inspectors to participate in judicial action against violators; and empowerment of food, health, and workplace inspectors to perform environmental monitoring and inspections.

Interagency integrated approaches might be considered under the following conditions:

- Established and effective communication and understanding among the agencies involved in integration of permitting, inspection, and enforcement.
- Small and medium-size facilities that have less complicated production processes and single-medium pollution problems.
- Permit and inspection schedules co-ordinated harmoniously among different government agencies.
- Availability of inspectors and permit writers in agencies other than the environmental agency who have been fully trained in monitoring and inspecting pollution in a small facility.

DENMARK AND GERMANY: MAKING USE OF LOCAL AUTHORITIES

In Denmark, the national parliament is responsible for political aspects of environmental policy and the development and enactment of national legislation. The development of environmental law and policy is the responsibility of the Ministry of the Environment and its National Environmental Protection Agency, which employs 350 staff. The country is divided into 14 provinces and 275 municipalities. Permitting and enforcement are undertaken by those provinces and municipalities. The legislation prescribes the industries to be permitted by the provinces and those to be permitted by municipalities.

In Germany, the Federal Agency for Environmental Protection has 292 professional staff who advise the Federal Ministry of Environment. The second tier of government is the Parliament of the Lander in which 16 Lander governments (federal republics) are represented. The structure of the Landers varies. The following structure exists in Niedersachsen:

- 4 districts
- 38 counties
- 16 towns
- Municipalities

In addition, there are 10 industrial control offices (2 or 3 for each district), with approximately 250 inspectors.

Critical Elements of Success

Balance among programme functions Interaction among government, industry and the public Evaluating success

Balance among Programme Functions

In many countries, there are imbalances among programme functions. Typical examples of imbalances in compliance and enforcement programmes are:

- Too much emphasis on development of policies and regulations and not enough emphasis on implementation and enforcement of them. In many countries, statutes concerning air and water quality have been promulgated, in some cases more than two decades ago. Those laws have never been enforced because the institutional functions and resources needed to enforce them were weak or non-existent.
- Inconsistencies in environmental regulations among regions within a country. In the absence of a national policy that establishes a minimum requirement for compliance, violators take advantage of regulations in some states that are less stringent than those of other states by locating their facilities in or transferring the wastes to those states that have less stringent requirements.
- Lack of incentives to promote voluntary compliance. In many countries, the compliance and enforcement programme rests more heavily on enforcement responses than on promotional policies. Because application of enforcement sanctions requires heavy commitment of government resources and is costly, the chance that such sanctions will be implemented is reduced significantly over time.
- Incompatibility between the requirements and the permitting system. In many countries, the laws establish detailed environmental requirements and quality standards. However, permitting systems in some countries

are often general and vague. This imbalance results in poor monitoring of compliance and ineffective enforcement.

- Lack of tools for monitoring" and inspections. An effective compliance monitoring and inspection programme requires sampling and laboratory analysis of wastes. Because laboratory services require skilled personnel and are costly, many governments have great difficulty in providing them to the compliance and enforcement programme.

In designing a compliance and enforcement programme, efforts should be made to create a proper balance among various functions and to avoid significant imbalances and gaps. Several alternative approaches to increasing the effectiveness and balance of the institutional functions are discussed in detail in the following sections of this module and in the subsequent modules of this training manual.

A summary is presented below:

- *Balance between incentives and penalties.* The long-term strategy for any compliance and enforcement programme should rest on both incentives for compliance and sanctions for violations. A balance must be established between the two.
- *Techno-economic compatibility.* The quality standards and requirements for a compliance and enforcement programme should be compatible with both the stage of technological development in the country and the resources available to the programme. It may be useful to establish minimum requirements first and develop more stringent requirements over time.
- *A few strong rather than many weak regulations.* At the early stage of development of a compliance and enforcement programme, it may be useful to establish few regulations and to make those regulations clear and enforceable.
- *National standards, regional applicability. Minimum environmental requirements* and quality standards must be applicable at the national level. However, the programme should provide flexibility so that regional agencies have the authority to design more stringent requirements.
- *Permitting compatible with monitoring capability.* The permitting programme should be harmonised with monitoring and inspection functions in such a way that the resources allocated to the programme could be used to monitor and inspect requirements set forth in the permit.
- *Alternative inspection approaches* can be used in inspections if adequate inspection tools, laboratory services, are not available to the programme. Those approaches should be emphasised at the early stage of design of the programme.

- *Adequate emphasis on education and public outreach.* The compliance and enforcement functions will be enhanced through educating and informing the regulated community and the public about the laws, provisions of the regulations, and rights of the community.

Interaction Among Government, Industry, and the Public

In many countries, the mechanisms for policy makers to interact fully with industries, more specifically the regulated communities, and with the public are poorly defined, and the channels of communication and interaction are not always placed properly. Because limited resources are available to many developing countries, those governments rely more heavily on promotional and voluntary compliance than on sanctions in their compliance and enforcement programmes. Therefore, it is imperative to develop or strengthen effective mechanisms for participation by the public and industry in policy and regulatory functions and decision-making processes. A strategy that combines all three participants-government, industry, and the public—can create a sustainable and self-supporting system for environmental management and compliance. The involvement of the government, industry, and the public will effectively transmit the government's goals, industry's imperatives, and public concerns to the other parties, and will foster understanding among all parties. Such a programme, with contributions from all three sectors, is sufficiently flexible to allow programme managers to adapt policies, regulations, and goals to changing environmental, economic, political, and social conditions. Mechanisms can be developed to ensure that parties outside the government become involved in review of draft policies and requirements. Permanent or temporary special task forces or committees, or private and public forums, can be established to review and comment on draft policies and regulations and requirements. Along with the policy makers and the technical staff of the environmental agency, the representatives of judicial agencies, industries, and the public should comment on those draft requirements and policies and procedures.

STRENGTHENING INSTITUTIONAL FUNCTIONS: THAILAND COMPREHENSIVE REGULATORY AND POLICY SUPPORT TO THE ENVIRONMENTAL SECTOR

In 1992, the Improvement and Conservation of National Environmental Quality Act replaced the 1975-78 National Environmental Act and established a stronger foundation for a national environmental management programme. The act for the first time covers such areas as: (1) assignment of authority for licensing to local administrators; (2) recognition of the rights and role of the public in finding solutions to environmental concerns, including public access to state information and ability to file suits and claim compensation; (3) introduction of heavy monetary penalties and imprisonment for violators; and (4) requirements that the owners of major industries install waste

disposal equipment, according to environmental requirements in place.
The former Ministry of Science, Technology, and Energy was renamed the Ministry of Science, Technology, and Environment (MOSSTE). The National Environment Board (NEB) was promoted to subcabinet level and received more power in decision making. The prime minister becamc thc chairman of the NEB. The NEB is organised into three departments: policy and planning; pollution control; and information, awareness, and promotion. An environmental fund also has been established.
The complementary acts promulgated in 1992 include the Hazardous Substances Act, the Factories Act, and the Energy Conservation Promotion Act.

Role of Industries

Involving the regulated community in developing general requirements helps to create support for the environmental programme and reduce resistance and potential conflict. Involvement of industry communicates the requirements at an early stage and facilitates compliance. Participation also enhances compliance merely by informing the regulated community of its liabilities to the government and the public.

Participation of industries in the development of environmental requirements may be fostered through:

- Informal consultations between policy-makers and representatives of the regulated community and the NGOs.
- A formal process of public and private dialogue.
- Pilot testing of regulations and requirements in selected hot spots or industrial sectors.

Role of NGOs

Non-governmental organisations and other public action groups increasingly are becoming involved in enforcement by detecting and reporting non-compliance, commenting on government enforcement actions, and, when the law allows, taking legal action against a violator for non-compliance or against the government for failure to enforce requirements. A proactive environmental institution should secure the support of NGOs by involving them in the formulation of policy and the development of environmental requirements. It also is important to educate NGOs to enhance their strengths in developing appropriate and informed agendas.

Role of Financial Institutions

Certain industries, such as banking and insurance, can become involved directly in a compliance and enforcement programme by requiring assurance of compliance with environmental requirements before they issue loans or insurance policies to facilities.

Public Support

Strategies to build credibility for and enhance the acceptability of compliance and enforcement programmes vary across political and cultural boundaries. In some cultures, aggressive enforcement will provide credibility. In others, it may be important to establish an initial period of promotion and encouragement to create a spirit of co-operation, followed by a well-publicised and gradual shift to more aggressive enforcement to signal that there will be consequences for non-compliance. In other cultures, a mixed approach at the outset may be more successful. The role of the public can be central in promotion of compliance, deterrence, and enforcement. It should be noted that most environmental programmes in industrial nations were initiated as a result of public pressure. Thus, the relative success of those programmes rests to a large extent on public support and acceptance. However, in those countries in which public institutions are not yet strong, the government should play an active role and enhance public support through education and promotion. Public acceptance ensures the stability and continuity of an environmental programme.

OUTREACH TO THE REGULATED COMMUNITY TO IDENTIFY RESOURCES: U.S.

In the United States, the California Air Quality Management Districts (AQMD) have taken a unique approach to the overall goal of compliance. The AQMDs implement regulations under the Federal Clean Air Act (CAA), as well as their own state laws regarding air emissions. The AQMDs have learned over the years that it is to their benefit to have as much interaction as possible with the regulated community regarding compliance—even before new regulations or rules are passed. The AQMDs believe that notes of compliance are higher if the regulated community actually participated in the rule-making. In such cases, the regulated community is informed of the new rule from the beginning, understands the reasons for the rule, feels a degree of ownership of the rule, and understands what steps must be taken to comply.

Before developing a new regulation or requirement, the AQMDs hold meetings to explain the need for the new requirements and gather the comments of the regulated community on how best to implement the new requirements. For rules that will affect a newly regulated community, the AQMDs manage an aggressive outreach programme to identify those in that community. The AQMDs also manage a similar effort for regulations that affect small businesses.

The AQMDs often rely on "partnering" relationships to help spread the news to the regulated community. Trade associations are a common partner in such an endeavour. The South Coast AQMD has contacted firms that manufacture and distribute equipment and supplies for newly regulated industries and asked for their help in inviting the regulated community to the meetings. The AQMDs explain that it is in the firm's best interest to help its customers stay in compliance because, by avoiding fines, its customers will have more capital to purchase goods from suppliers.

The net effect of such an approach is that the AQMDs can achieve more compliance with less enforcement. Even with minimum enforcement, the AQMDs are confident of a high rate of compliance.

Evaluating Success

Continual evaluation of a compliance and enforcement programme is critical to measure progress the effectiveness of environmental laws and the measures taken to ensure compliance. A number of indicators can be used in combination to evaluate success. These include:

- environmental results-e.g. reduction in air pollution
- compliance rates among the regulated community
- progress in returning significant violators to compliance
- measures of compliance monitoring e.g. number of inspections
- number and timelines of enforcement responses
- total of monetary penalties
- measures of technical assistance-e.g. adoption of cleaner production techniques

Options When Resources are Limited

Establishing Priorities Phasing in Functions

Establishing Priorities

The purpose of setting priorities in environmental management is to:

- Assist the national environmental and natural resource protection agency in systematic delineation of the issues and concerns voiced by the national legislature and the public.
- Develop policies that address priority issues and concerns.
- Enlist public support and participation in implementing a national programme of environmental priorities that effectively assigns scarce resources, while equitably applying consistent environmental standards to both industry and government.

Setting priorities is a process. As the environmental programme unfolds and the compliance and enforcement functions are strengthened, priorities may change. Therefore, it is imperative that priorities be re-examined through appropriate mechanisms at certain intervals. The criteria for setting priorities should be based on:

- The most pressing environmental issues that directly threaten human health and environmental resources.
- Availability of resources.

Once priorities have been set, they should be communicated clearly within the government agency, to the public, and, most important, to the communities that are targeted under the priorities established. Priority status can be assigned to the worst polluters (regulated communities), or it can be based on geographic areas, such as industrial hot spots. In some cases, the decision can be based on both considerations.

Identification of the Regulated Community

A major task in establishing priorities is to identify the polluters and determine the make-up of the regulated community. This group is the focal point for implementation of the environmental compliance and enforcement programme and the priority group for allocation of the resources of the programme. It may include large, privately owned, government-owned, or parastatal facilities and privately owned small facilities (manufacturing, repair, and service industries). A number of risk assessment and hazard ranking models exist that, with modification, can be used for industrial and site ranking and establishing priorities. Identification of regulated communities can be done through:

- Inventory of existing registration and license applications through appropriate channels (census bureaus, tax bureaus, municipalities, government offices, trade unions, and trade associations).
- Survey of facilities in priority areas through personal interviews and questionnaires.
- Remote sensing (interpretation of aerial photographs).

Identification of Priority Geographic Areas

In many developing countries, certain industries or groups of industries are concentrated in particular locations. In almost all large metropolitan areas in developing counties, the main polluting industries are located at or adjacent to the boundary of the municipality. In some countries, clusters of industries can be found in strategically located river basins that cross or run along the outlying areas of municipalities. Such areas, conventionally called hot spots, are excellent priority areas on which to focus compliance and enforcement programmes.

Identification of the Most Affected Polluted Medium

In some cases, an overly polluted environmental medium, e.g. city air, lake or river etc., may require priority attention and co-ordination with other countries.

Identification of the Wastes and Substances that Cause the Greatest Health Risks

Waste disposal problems or the presence of toxic chemicals in the environment may also be a legitimate priority focus.

Phasing-in Functions

In addition to the establishment of priorities, the implementation of planned programme functions should follow a phased approach to ensure that those programme functions that have been determined to have the highest priorities are implemented first. Simply stated, the purpose of a phased approach is to balance priorities with the resources available to the programme. Any country should begin the implementation of its planned compliance and enforcement programme with the resources available, even if those resources appear to be limited. In almost all countries there are resources scattered among various government departments that could be assessed and tapped to initiate the programme.

In the Early Stages: Providing a Solid Foundation

During this phase, efforts should be concentrated on establishing a solid foundation for the compliance and enforcement programme. Actions that typically are required in this phase include:

- Performing an environmental inventory to determine the key environmental issues in the country, establishing priorities among the issues, and developing a framework to implement priority programmes.
- Planning and strategising the compliance and enforcement programme. The strategy should clearly establish the goals, objectives, and functions of the programme and should recognise the political and socio-economic attributes and financial resources of the country.
- Establishing priorities (delineated in several phases, if appropriate) based on the inventory performed earlier, once the strategy has been established.
- Establishing an implementing agency. The agency should establish minimum limits on discharges and emissions and make those limits immediately applicable at the national level. In addition, an interministrial commission should be established to co-ordinate the collaborative efforts of the new agency with the activities of other ministries and industries.
- Establishing a proper balance with regional and municipal authorities.
- Establishing requirements for the highly polluting industries, to be implemented immediately in hot spots.
- Begin registration of the priority industries and establish facility files, in collaboration with the appropriate ministries and municipalities. In collaboration with representatives of industry, the agency should outline the permitting procedures and requirements and phase in the permit application and review process. The agency should use integrated permits to save time and resources and allow a reasonable amount of time for priority industries to comply with permit requirements.

- Develop schedules for compliance self-monitoring and reporting for the priority industries. During the first phase, the agency should concentrate on those monitoring parameters that are easily sampled, observed, and measured, such as acidity of the effluent, total waste, colour, odour, oil sheen, and fish kill. In the initial phase, the agency should conduct few inspections but make them effective and thorough. In addition, an integrated inspection approach should be considered during this phase.
- Define the role of inspectors and prosecuting authorities in implementation of sanctions against violators, in negotiation with the ministry of justice and local police authorities. In the priority geographic areas and hot spots, local police and the citizens' groups should be trained on general compliance requirements and methods for detecting non-compliance. The agency should negotiate with local municipalities to begin registration of non-priority industries in their jurisdictions.
- Develop a few but effective enforcement responses and implement those responses effectively during the first phase to establish credibility and visibility for the programme.
- Plan large-scale public and educational campaign on the priority environmental issues. The resources available at educational institutions and the information and training officers of the resources available at educational institutions and the information and training officers of the relevant government departments should be tapped for this campaign. The campaign should focus initially on priority environmental issues in the priority geographic areas.

A PHASED APPROACH TO ESTABLISHING QUALITY STANDARDS: INDONESIA

In Indonesia, in 1991, a national decree stipulated effluent standards for such industries as metal plating, leather tanning, oil refining, pulp and paper, rubber, fertiliser, and similar industries. Regional governments have been authorised to introduce more strict standards depending on the severity of pollution in their respective regions. The decree applies only to existing facilities. For new facilities stricter standards are being conceived. A decree for less hazardous industries will be stipulated later.

In Later Stages: Gradual Development and Evaluation

During the second phase, the programme managers should re-examine the activities performed during the first phase.

- Fully develop initial phase functions
- Evaluate the programme to revise priorities , assessment of resource

needs for expansion of the programme, regulations, and standards, as appropriate.

- Introduce stronger decentralisation by providing resources to regional offices and establishing guidance and procedures for the regional offices to increase their role in performing permitting, inspection, and enforcement functions.
- Develop more stringent requirements for priority sectors and subsectors of industry.
- Develop requirements tailored to regional conditions.

Industry Environmental Compliance, Training Manual, UNEP, 1996

Environmental Permit?

Contract between the Authorities and Company, Establishing Conditions to Operate

Many countries issue permits (or licenses) to potentially polluting facilities before those facilities are allowed to begin construction and/or to operate. The issuance of such permits provides an opportunity to significantly improve the rate of compliance with environmental laws, regulations, and policies in those countries. In addition, permit issuance improves a government's ability to identify and successfully implement appropriate enforcement responses to incidents of non-compliance among regulated facilities.

Translates National Environmental Standards into Design and Pperating Conditions of a Facility

An environmental permit (or license) is a document that defines the design and operating conditions that a particular facility must meet to be in compliance with one or more environmental laws. Environmental permits are issued by government authorities at the municipal, regional, provincial, or national levels, depending on the legal authorities in a particular country. All environmental permits must contain certain elements designed to ensure that the government will be able to enforce the applicable environmental law or laws. The purposes of an environmental permit are to provide the government control and oversight over the environmental impact of a facility. Such an understanding saves time and resources (that otherwise would be spent on interpreting regulations during each inspection) for both the permittee and the regulatory authority. The regulatory authority benefits from a permit in the time saved by inspectors in identifying the items to be inspected each time they visit a given facility. The inspector can read the permit before each inspection, rather than the more lengthy laws and regulations, many of which do not apply to a particular facility.

Role of Permit

A permit programme is the strategy, guidance, procedures, staff, and resources that are developed and used by the government to implement and sustain the issuance and renewal of environmental permits, pursuant to one or more environmental laws. The role of permitting is to:

- Enable consistent application of laws and regulations in all geographic regions and all sectors of the regulated community;
- Focus government resources on priority facilities that present potentially high environmental risk;
- Stipulate conditions, limits and means that will result in an acceptable level of environmental releases;
- Provide a basis for phasing in the application of regulations for certain types of facilities (if the law allows such phasing). For example, the first permits under a new law may require specific operating procedures only, allowing more time for the government to develop regulations for design standards at each facility. Such design standards could be implemented at a later date through permit modifications.

Contents of a Permit

Environmental permits typically establish a number of provisions that a facility must fulfil to be in compliance with a law and its associated regulations. Some provisions should be standard and included in every permit that is issued under a certain law and regulations. Others are specific to the industry or facility being permitted. Standard provisions usually address four basic elements:

Standard Contents

1. Facility Description

- A description of the corporate environmental policy.
- A general description of the facility and a detailed map of its site and of the surroundings.
- A detailed description of the manufacturing process that will be conducted at the facility (in some cases, some of this information will be considered confidential).
- A detailed description of the pollution prevention measures the facility intends to take for all air, water, and waste emissions, as well as noise and accident prevention measures (for each process or process unit and for the facility as a whole).

- A schedule for the implementation of the measures to be taken (for existing facilities).
- A detailed description of the emergency response plan to be implemented for each type of accident that may occur.

2. *Environmental Impact Assessment and Risk Assessment*

- Information that often constitutes the core of a permit application is set forth in the environmental impact assessment (ELA). The measures to control emissions are defined, and, in many countries, the emission levels are established according to the content of the environmental impact statement (EIS). The risk assessment study, which is included in the EIS, is particularly useful in determining permit conditions that are consistent with the risks posed by the facility.

3. *Information on Discharges*

- The proposed emission levels.
- Hazardous waste treatment and disposal measures (including the name of the firm that will handle the waste, if the waste is to be treated by a company other than the operator of the facility).

4. *Information Related to the Facility's Monitoring and Reporting Measures*

- Proposed monitoring procedures to be used by the facility, parameters to be monitored, frequency of analysis, methodology for monitoring, identity of the laboratory where the analyses are to be performed, and information on record-keeping procedures consistent with or justifying changes to establish protocols and methods.
- Reporting procedures (how and to what particular authority).
- The signature of the responsible officer.

Specific Requirements

Specific requirements in environmental permits are developed according to the activities conducted or constituents of wastes present at a facility that are subject to control under a given law and its regulations. Specific requirements in environmental permits always should establish quantifiable parameters and methods for measuring and verifying compliance with the control standards stated in the applicable law and regulations whenever possible. The methods of measuring those parameters must consider all the factors that may affect the results of the measurement. For example, a law or regulation may specify that combined waste-water discharges at any facility must not exceed 5 milligrams per litre (mg/1) of total chromium. A facility that has numerous individual outfalls may exceed the 5 mg/1 limit at some outfalls, but the total discharge of chromium may be below the limit. In such a case, the permit must include specific methods of measuring the chromium levels at each outfall and combining the results to determine compliance with the standard. If procedures are not specified in the permit, the facility may challenge any procedures used by the government to prove that discharges exceed the limit.

PORTUGAL PERMITTING PROCEDURE

Legislation in Portugal prescribes a "coordinating service" to deal with applications for permits for category A, B, or C installations. Major facility (A) permits are granted by the central departments, and permits for smaller facilities (B and C) are granted by the regional departments. The coordinating body is either the Ministry of Industry and Energy or the Ministry of Agriculture, depending on the type of facility that is applying. The first step for category A and B installations is an application to the municipality for a building permit. The project then is passed to the coordinating service. Advice then is requested of various services—for example, the Directorate General for Environmental Quality in relation to water, environmental assessment, and risk, The project is returned to the coordinating service for a final decision.

Procedures for Issuing Permits

The law or regulations either must require specific procedures that government permitting authorities and permit applicants must follow during the permitting process or must delegate the authority to prepare such requirements to the regulatory stage of the environmental quality management cycle. The key stages of a permit process that need to be addressed are:

- Pre-application-identification and contact with the regulated community information exchange, and identification of any problem areas.
- Facility Application-standard forms and procedures need to be decided upon and disseminated
- Preliminary review by authorities to determine the completeness of the application and to request any additional information required. Ideally, the laws or regulations place time limits on facilities for responding to a permit writer's requests for information and revisions of the application. Otherwise, the permit process may be delayed unnecessarily. Such a provision is especially important when the law allows existing facilities to operate while permit decisions are pending.
- Official /public enquiry by the authorities. The permitting authority must assess whether the applicant's proposed methods of and equipment for attaining compliance are consistent with the laws and regulations and whether they are likely to achieve the required levels of control. This

assessment often requires the advice of specialised technical personnel or explicit guidance document that addresses the technical aspects. Public involvement into the permitting process is often extremely valuable. The public may be aware of information about a facility that is not readily available to the permitting authorities when a permit is issued. In addition, the public can assist the government in tracking a facility's compliance with the terms and conditions of existing permits. The willingness of the public to provide such assistance increases greatly when the public has been properly educated about the terms of a permit through involvement in the permitting process. In providing for public involvement, the law should set forth procedures, timelines, and authorities for resolving any conflicts that may arise.

- Draft permit (application accepted). Permit writers should be provided guidance that describes various industries and how they are likely to be affected by a new law and its regulations.
- Official appeal procedures (application not accepted)
- Stipulate Deadlines. The minimum and maximum time that should be allowed for a facility to submit an application, and for the government's review and response also needs to be determined at all stases.

Different Approaches

Industrial Activities Required to have a Permit

Most laws include licensing or permitting provisions for facilities having the potential to pollute. However, all human activities have the potential to pollute, from a restaurant or garage to a metal finishing plant or a chemical plant. It would be difficult and unnecessary for governments to screen all activities for permitting purposes. Government resources therefore should be focused on those industrial activities that are the most important sources of threats to public health or to the environment.

In setting priorities, some countries have used the following elements to define different levels of permitting:

Important Sources of Threats to Public Health or the Environment

- *Industry sector or type of activity* (for example, agro-industry or leather tanning).
- *Processes used* (for example, acid and halogen processes or combustion processes).
- *Chemicals used* (for example, PCBs, chlorine or ammonia).

- *Size of the facility* (expressed in terms of tons of output or tons of raw materials used).

Location of the Facility

Differentiated by the degree of potential threat to public health or the environment, two levels of licensing procedures often have been established:

Facilities at which the risk or actual environmental effect of pollution is high — for example, large and medium-sized companies or those handling hazardous or toxic materials. Such facilities may be given highest priority. They may be required to obtain a permit to operate and to present another request or approval if there are changes in production or manufacturing processes or capacity. The permit might be issued at the national or local level. Facilities that discharge minor pollutants or pose less risk than high-priority facilities. Such facilities may not be required to obtain permits because their levels of pollution are considered low, but they generally are required to notify the authorities of their activities. In such cases, general codes of practice might be issued for the activity of concern. Such codes should be followed by all those facilities that have having given notice of their operations. Many times, the law will not specifically identify who should receive permits. However, it should contain clauses that indicate the factors the regulatory authority is to use in making such determinations. The regulatory phase must identify these factors, interpret them, and formulate them into specific criteria. For example, the law may state that "permits must be issued to all facilities that generate significant quantities of cyanide-bearing wastes from metal plating operations." In such a case, the regulations must interpret the term "significant" as an exact quantity. As an alternative, the regulations might specify a number of exact quantities, based on concentrations of cyanide. Using this approach, large quantities of wastes having extremely low concentrations of cyanide could be eliminated from the permitting process.

Permits Issuing Authority

Ensuring that levels of pollution and risk are maintained at socially acceptable levels is in the interest of each level of government authority — municipal, regional or provincial, and national. To which, then, does a facility make application, and which grants the permit? The answer is often determined by the structure of the government. Countries respond in different ways, with different systems of delegation of environmental responsibility to different jurisdictions, levels, and departments of government. In general, permits are issued at the national level in the case of selected facilities that are heavy polluters and at the local level for less polluting facilities.

In different countries, the licensing authority may consist of:

- The central government (for example, Ministry of Environment).
- Regional, provincial or local authorities.
- Specialised, independent board, independent of the government.
- A combination of the above.

The law should indicate a lead national permitting authority. In addition, it may set forth conditions under which such authorities may be delegated to state or local governments.

Descriptions of permitting authorities in Sweden, the Netherlands, France, Germany, the United Kingdom, Cote-d'Ivoire and USA

In Sweden, an independent Licensing Board for Environment Protection was established, with the mandate to issue permits for the most polluting activities after a hearing process similar to court procedures. The board consists of one chairperson, one senior engineer, and one representative from the industry. Normally, a facility's technical experts present the case to the public and the board. After the facility's presentation, members of the franchise board, along with scientists from the environmental protection board, ask a number of specific questions about the environmental safeguards that will be built into the application. The public also has an opportunity to question the company's representative. Before a permit application is sent to the licensing board, informal consultations take place between the company and the local authorities, and between the company and the national environmental protection board. During this preliminary stage, the facility receives advice on preparing an application. The country's administrative board issues permits to medium-sized or smaller facilities, with the advice and consultation of the central government.

In the Netherlands, the provinces have the authority to issue permits to the bigger polluters such as chemical waste processing plants with a few exceptions, such as nuclear power stations and facilities that emit pollutants to the nation's main waterways. Other permits are issued by the lower authorities (communities and water boards). In earlier days, the national inspectorate was an influential advisor to the provinces. Today, the provinces have been equipped properly with staff and can handle the technical and jurisdictional consequences of permits themselves, under the aegis of the central authorities. The role of the national inspectorate now is changing, from that of primary advisor to that of primary inspector (inspecting industries as well as the authorities in their role). The role of the inspector with regard to the lower authorities is changing in the same way. Because expertise is limited, the Environmental Health Inspectorate still plays a key role in advising and guiding the provinces toward preparation of proper, enforceable permits. The inspectorate reports to the central government, independently of local authorities.

In France, the representative of the government at the local level *(prefet)* receives the application and issues the permit (except for nuclear power stations and underground gas storage facilities). The industrial inspectorate, which is established at the regional level and under the authority of the Ministry of Environment, acts as the secretary for the procedure, conducting a technical investigation and providing advice on each case.

In Germany, central legislation governs all environmental media. Permits are issued by the local authorities on behalf of the federal minister for environment. There are both formal and simplified licensing procedures. Licensing procedures that include public participation are handled by the chairs of regional councils; others are the responsibility of the factory inspectorates. The local authorities receive the application, make a formal assessment of it, and solicit the opinions of other authorities. Authorities may attach conditions to permits when issuing them.

In the United Kingdom, permits for the most polluting facilities are issued at the national level by the Chief Inspector of Her Majesty's Inspectorate of Pollution in the Department of Environment. Permits are issued at the local level for less polluting facilities.

In Cote-d'Ivoire, permits are issued by the Ministry of Industry and Planning, Division of Environment, Standardisation, and Technology.

In the United States, all water dischargers are required to have a national permit issued in accordance with USEPA's federal environmental guidelines, standards, and criteria. In most instances States may issue permits if their programmes are approved based upon procedures, competency and resources, or the federal EPA issues the permit. USEPA retains veto authority and enforcement authority over state-issued permits and reviews permits issued by the States for adequacy. States also receive funds and training to assist in programme implementation. A similar system is currently being established for air permits, and also exists for waste permits.

Criteria for Issuing Permits

It is the substance of what is required to obtain a permit that will determine much of its environmental impact, and related to this is the ability of the facility and the government to lake the necessary steps to ensure compliance with and enforce any permit conditions. An environmental impact assessment can be used as a tool to determine the potential environmental impacts but it is not in itself the condition of the permit. There are many different types of criteria that can be established as a basis for permitting, each of which poses its own particular challenges in terms of government and industry resources, costs and ability to enforce. These include:

- technology or performance-based criteria/standards. These usually include some cost and feasibility considerations. There is a strong preference internationally for using performance-based criteria instead of specifying a technology where the problem is amenable to measurement of performance.
- ambient quality-based criteria/standards
- cleaner production. Increasingly, to achieve sustainable development goals, governments are recognising the importance of cleaner production rather than controlling pollution at the end of a point of discharge or as a waste to be disposed of. Many permit programmes are beginning to incorporate requirements for cleaner production plans and their implementation as a condition of a permit.
- risk-based permitting has been recently promoted as one means of rationalising private and public expenditures for pollution control in resource-limited settings.
- other: e.g. work practice standards

Single or Integrated Permits

The legislative process for any environmental law may include the consideration of integrated, rather than single-medium (e.g. air, water or waste) permitting.

Integrated permitting takes into consideration the total environmental impact of a facility or activity (water, air pollution, waste...) to ensure that the overall damage to the environment is minimized.

The benefits of an integrated permit include:

- Assessment of total environmental effect of each facility
- Ensures that problems are not simply moved from one medium to another
- Avoids administrative duplication of effort and reduces costs
- Provides more accurate assessment of priority pollution problems & solutions

If a decision is made to create a integrated permitting programme, the new law also must change appropriately wording in existing laws that do not support the integrated approach. In addition, the law must require the reorganisation all existing permitting authorities. It must delineate the ways in which permit writers for various media will interact to accomplish the goal of integrated permitting. The law must anticipate questions that will arise about when and how each permit will be combined into a single integrated permit for each facility. In addition, the regulations should address any changes that are needed in the permit application format and the permit decision-making process.

Other Issues

What Should the Duration of the Permit Be?

Duration of a permit is an important design issue both to programme management and environmental results. The longer the life of a permit, the greater the certainty that industry has about its costs and the lower the permit workload for the issuing authority. But a long term permit may also prevent governments from imposing stricter requirements where they are needed. The shorter the duration of a permit, the greater the opportunity for the permitting authority to respond to new environmental concerns, and problems in permits that have been issued. Permit renewal conditions can also be constructed to provide additional leverage for compliance by making compliance a condition of renewal. All of these factors must be in balance.

EXAMPLE-BELGIUM: MULTI-MEDIA AND SINGLE MEDIA PERMITTING PROGRAMMES

Belgium's two distinct regions have different permitting procedures. In Flanders, the 1991 permitting law requires one single environmental permit for air. noise, water (surface and groundwater), and waste. The plan also includes notification of major risk activities and environmental impact studies. Under the new permitting law, installations are categorized as 1,2, or 3. Permits are required for categories 1 and 2, while category 3 installations are required to register with the authority. Provinces issue category 1 permits, and municipalities deal with category 2 permits and category 3 registration. Seven thousand permits have been granted for category 1 installations. A permit charge of 470 ECU is imposed, and charges are imposed for discharges of wastewater and disposal of waste.

In Wallonia. sectoral legislation applies to air. water, and disposal of waste. All installations categorized as 1 or 2 must have environmental permits. The following permits are required in Wallonia: (1) environmental (air or noise): province (category 1) or municipality (category 2): (2) water: region; and (3) waste: region.

Permits have been granted for 150 category 1 facilities. No permitting charge is imposed in Wallonia, but charges are made for disposal of wastewater and disposal of solid waste.

Should Permitting be Prior to Construction or Operation?

In regard to whether a permit should be a pre-construction, pre-operation review and whether the permit is a construction and/or operating permit, there are various advantages and disadvantages involved in the responses to this question. A permit required prior to construction, applicable to new facilities and major modifications of existing facilities, provides the government and the facility's owners and operators an opportunity to explore siting and land use considerations, and to design options for pollution prevention and risk reduction that may not be practical and available at the time prior to commencement of operations, when the facility has already been built. If properly designed, pre-construction review should enable the parties to fully explore and consider these opportunities within a reasonable period of time and make use of the results of any applicable environmental assessment process. If improperly designed, pre-construction review can lead to substantial delays in construction increasing costs of construction. Legislators and regulators must carefully assess the costs and benefits to be achieved by the permit and how it can be done in the least costly manner.

Critical Elements of Success

Making the Permit Enforceable

The process of writing clear, complete, and enforceable permits begins with understandable and complete laws and regulations. Such laws and regulations should allow the lead agency to incorporate a number of minimum standard provisions into all permits. Those minimum standard provisions include:

1. *Identification of Responsible Parties.* All environmental permits must clearly identify the person or organisation at the permitted facility who is legally responsible for compliance with the terms and conditions of the permit. The laws and regulations governing a given permit usually will specifically identify the person or organisation eligible for permit responsibility.
2. *Provisions Regarding Changes in Operations and Equipment.* Changes in facility operations and equipment at a facility are common during the life of an environmental permit. Therefore, language in the permit must identify for the facility the types of changes that require formal review and approval by the permitting authority and subsequent modifications of the permit.
3. *Requirements for Transfer of Permit.* Requirements for transferring permits must be stated clearly in the permit, so that the government can keep track of who is responsible for carrying out the terms and conditions of the permit. The language of this requirement must clearly identify when a transfer of the permit becomes valid, so that there is never any period of time during which it is unclear who is responsible for compliance at the facility.
4. *Requirements for record-keeping and reporting.* Self-monitoring and records kept by a facility are crucial to government's efforts to assess its

environmental compliance. Self-monitoring requirements should specify: normal or other operating conditions during sampling, and other specifications such as sample locations and frequency, type of sample (e.g. grab, continuous or composite), and text methods. The permit must specify where the facility will store the records and how long each type of record must be kept. The permit also must describe the reporting responsibilities of the facility. In addition, the permit should set forth or make reference to the methods of reporting and the exact types of information to be included in each report and provide deadlines for submittal of each report.

5. *Statement of Duty to Comply.* The permit must state clearly that the permittee must comply with all the terms and conditions set forth in the permit.
6. *Statement of the Duration of the Permit and Duty to Reapply.* The time at which a permit expires must be stated in the permit, with references to the laws and regulations that address the duration of a permit. In addition, the permit must specify the date by which the permittee must reapply for a permit and the consequences of failure to reapply. The permit also must state clearly how the facility must operate if the permit expires while the regulatory authority is determining whether to renew the permit. Most laws and regulations allow such a facility to operate under an existing permit until the permit is renewed or revoked, providing that re-application has been made by a specified time.
7. *Statement of Duly to Allow Entry and Access to the Facility.* To avoid any disagreement about the rights and responsibilities of the government to inspect the facility, the duty to allow regulators entry and access to the facility should be stated in the permit.
8. *Statement of Duty to Provide Information to Regulators.* All permits should contain provisions that describe the types of information that the permittee must make available to the government.

Interactions among the Different Authorities

Separate Permitting, Inspection and Enforcement Functions

What Interactions and Communications Should Occur Among Permitting, Inspection, and Enforcement Authorities? This issue centres on whether permitting authorities should be separate or combined with inspection and enforcement authorities. There are both advantages and disadvantages of having the person who writes the permit also inspect the facility and enforce the permit conditions. If these responsibilities are kept separate it avoids potential conflict of interest. However, it may be necessary to combine permitting, inspection, and enforcement authorities when there is limited funding for these programmes.

Necessary Communications between Permitting, Inspection and Enforcement Authorities

The guidance and internal directives that address this issue are generally included in the regulations that apply to the delegation of permitting authorities. These regulations must specify clearly the communications and interactions that must occur among permitting, inspection, and enforcement authorities at national and local levels to make permits enforceable and to address any on going compliance problems. Examples are:

- what communication procedures inspectors should follow when an inspection reveals a dangerous condition that should be but is not covered by the facility's permit.
- what dialogue should take place between permitting and enforcement officials when compliance schedules must be incorporated into a permit.

Necessary Communications between National / Regional / Local Authorities

These interactions must also incorporate any national/regional/local divisions of responsibility.

Planning and Implementation Considerations

Linking Planning with Resources

The agency responsible for developing permitting procedures and responsibilities must plan on the basis of:

- the numbers and types of personnel required to issue permits.
- the size and experience of permitting staff in existing government agencies.
- the levels of funding available to hire and train new permitting staff.

The planning processes should be linked to the availability of resources. The establishment of deadlines for permit applications, for example, is an area in which the regulations affect the planning process. If the administrative planners find shortages of resources for issuing permits, they may work with the regulatory staff to devise a system for limiting the number of permit applications that must be reviewed at any given time. For example, the permitting regulations may be written to require that, at first, applications be submitted only by facilities within a given industry. Applications from facilities in other industries would be submitted at later dates, as specified in the permit regulations. This phased approach prevents the inundation of the permit authority with permit applications that the authority cannot review in a timely manner.

Policy, Guidance, and Training

Policies, guidance, and training are developed to facilitate actions that are not included in the regulations. Policies often address the same issues that arise during regulatory development and administrative planning, but policies are more flexible and less formal than regulations. Because they are less formal, policies tend to be less publicised and more subject to challenge than regulations. Therefore, the lead agency for development of the permit programme must be carefully selective of the issues chosen for resolution through policy rather than regulation.

Guidance and training are essential to the smooth operation of a permitting programme. The subjects of guidance and training may include:

- Identification of facilities to be permitted.
- Forms and procedures for:
  - communication among state, local, and national permitting authorities.
  - communicating with inspection and enforcement officials (especially when inspections reveal weaknesses in a permit).
  - use of funds that are dedicated to permitting activities,
  - permit applications
- Checklists and procedures for reviewing permit applications from various types of facilities.
- Identification of qualifications of staff for reviewing specific types of permits.
- Public information brochures

These and other subjects for guidance and training should be reviewed regularly by the national permitting authorities. The selection of one or more subjects for future guidance or training should be based on the strengths and weaknesses perceived in the permitting process and the level of funding that will be needed to prepare guidance or to conduct training. In some cases, it may be appropriate that government reduce the annual rate of issuance of permits to temporarily divert resources to guidance or training.

Feedback from Permit Issuance, Inspections, and Enforcement Activities

The quality of environmental permits is crucial to the enforceability of the laws and regulations that they represent. In many countries, a permit replaces the laws and regulations under which it was developed. Thus, the permit often shields the permit holder from penalties stated in the laws and regulations that were not incorporated into the permit. Therefore, it is important that permit writers work closely with enforcement staff to identify problems in existing permits that allow unintended releases of hazardous constituents and other pollutants to the environment. The processes of planning policy and guidance development therefore should address procedures for identifying and resolving problems in individual permits and problems in various types of permits. There are various mechanisms that can be used to improve permit quality and consistency in application of permit criteria within a country. Some options include establishing internal management review of permits prior to issuance by a higher level authority, establishing peer review panels for specific industry sectors, etc. It is important to build into the design of a permit system these approaches to avoid problems. A checklist might provide the kinds of questions policy makers and permit writers might ask themselves to ensure enforceable permits.

Preparations Required When Laws or Regulations are Imprecise

A permit programme based on imprecise or incomplete laws or regulations cannot be effective. When such laws or regulations are identified, it is preferable to spend the time and effort to revise them. Failure to do so could lead to lengthy court challenges that erode public confidence and the ability of the government to control threats to human health and the environment. Steps to take when laws or regulations are imprecise are:

Analyse the Problems with Laws and Regulations

Revision of imprecise or incomplete laws and regulations must begin with an analysis of the problems that the laws and regulations present and the options for resolving each problem. The government should identify members of industry, government, and the public who can form a working group to resolve the issues.

Propose Interim Measures Until They Can be Revised

While the issues are being resolved, the imprecise or incomplete laws and regulations should be excluded from all permitting and enforcement activities. In the interim, in pressing cases, an interim policy could be developed.

Options When Resources are Limited

The amount of funding required will depend on the complexity of the requirements set forth in the law and regulations and the size of the regulated community. For example, one may consider the case of a relatively simple environmental law under which a permit writer may be able to review a permit application and issue a permit in one week. At that rate, the permit writer will be able to issue approximately 50 permits per year. Therefore, a staff of 30 permit writers can issue 1,500 permits a year under the relatively simple environmental law. However, most environmental laws are not simple. Such laws often address a significant number of design considerations and operating practices at regulated facilities. Therefore, permit writers may spend many weeks or months reviewing design drawings, specifications, operating manuals, monitoring plans, and reporting protocols.

Preparations Required When Resources are Limited

- Place a high priority on integrated permitting which will have the advantage of reducing costs by avoiding administrative duplication in permitting and inspection of single media permits.
- Develop simplified permit conditions and consider when general or standardised permit can be used
- Set priorities and phase-in implementation in the regulated communities. Setting priorities in the regulated community can greatly improve the effectiveness of permitting programmes. A system of priorities allows the government to use limited funds for permitting to permit facilities that pose the greatest potential threats to human health and the environment. Facilities posing less of a threat can be gradually phased in as resources become available. Phasing in permit requirements so that facilities are given time to meet new standards also provide the authorities with time to phase in their responsibilities.
- Consider how permitting responsibilities can be shared with other government bodies (central or local). The use of local government entities to issue permits under a new law is often cost-effective, allowing faster issuance of permits than would be possible at the national level of government. Local governments often are already familiar with the facilities in their jurisdictions and thus require less time to become familiar with the waste types and operating practices of those facilities.

Initiating an Integrated Permitting Process

Few governments have the luxury of incorporating an integrated approach into their compliance and enforcement programmes right from the beginning. In most cases, permitting authorities are in a process of development based on single media permits in response to single media laws (e.g. to reduce air or water pollution or waste). Steps that can be taken to initiate an integrated permitting process in the event of an already existing single media approach are:

- Initiate a discussion with policy makers on the importance of a more integrated approach

- Harmonize existing permit schedules of different government agencies
- Begin with trials with one or two mediums (e.g. water & air emission) to establish better coordination
- Stimulate greater awareness of the benefits for government & industry.

INDUSTRY ENVIRONMENTAL COMPLIANCE, TRAINING MANUAL UNEP, 1996

Self-Monitoring, Inspections and Enforcement Measures

Setting the Context

The Monitoring and Response Triangle Issues to Address

The Monitoring and Response Triangle: Compliance self-monitoring, inspections and enforcement responses provide the most visible portion of a government's efforts to implement environmental laws and regulations. They are central to any effort by the government to achieve widespread compliance with environmental requirements and need to be considered together to achieve the optimal balance of voluntary compliance, deterrence and, in the case of violations, penalty. The perception of their efficiency will influence the decisions made by regulated facilities on the measures they will take in response to environmental laws and regulations. The definitions of compliance self-monitoring, inspections and enforcement are:

- Compliance sell-monitoring: the regulated facility regularly checks its own compliance with permit requirements and records and/or reports them to authorities as required.
- Inspections: government-assigned body makes official on-site inspection or file review to assess a regulated facility's compliance status, and identify any equipment, operations, or activities that are not in compliance with one or more laws and regulations
- Enforcement responses: the legal measures (fines, closure orders, law suits) taken in response to violations.

A.2 Key Issues to Address

To achieve the optimal balance of facility self-monitoring, inspections and enforcement measures, each country needs to determine:

- Self-monitoring: Who should be required to self-monitor what, how and when? Integrated monitoring.
- Inspections: Who should inspect what and how? Converting to integrated inspections.

- Enforcement responses: What enforcement actions to take and who should take them?

Compliance Self-Monitoring

Compliance self-monitoring programmes differ from inspection and enforcement programmes in that, under such programmes, the regulated community analyses its own compliance and reports the results to the government. Inspections at regulated facilities that have compliance self-monitoring programmes are then more informative and less costly to the government.

The key issues which authorities need to address are which facilities should be required to self-monitor, what parameters they should be monitoring, the procedures for recording and reporting them, and their frequency. Each of these are briefly discussed below.

- *Specified facilities (size, risk, processes, chemicals, etc.).* Ideally, all facilities would have their own environmental management system that includes regular monitoring of their environmental impact, and would keep the authorities informed as need be. In most cases, however, the authorities would have difficulty in reviewing all self-monitoring information from all facilities in a way that would enable them to take any required action. Consequently, the authorities may to need to prioritise and specify which facilities are required to self-monitor and report, consistent with the priorities set for permitting, inspection, and enforcement activities. Any facility which is likely to be inspected should be required to self-monitor and maintain up-to-date compliance records.
- *Measurable parameters and indicators.* The regulations should specify monitoring parameters that are relevant, and technically and economically feasible to measure. However, the measurement of those parameters must yield results that indicate whether or not the regulated facilities are in compliance with the law. Establishment of parameters that are difficult or expensive to measure may give uncertain results or may exceed the financial resources of some facilities. Indicator parameters could be used instead in a cost effective way to indicate whether the facility or installation is in compliance. For example, at some facilities, measurement of chemical oxygen demand (COD) may be an indicator of the levels of biological oxygen demand (BOD) in an aqueous discharge. The facility could be required or could suggest on its own initiative to self-monitor COD to check compliance with its BOD permit requirement. It should be noted that the use of indicator parameters does not replace a facility's compliance with the required levels for a parameter of concern; however, such

parameters can ensure that gross non-compliance with the given level is not occurring. The regulations can be written in a manner that requires relatively frequent monitoring of indicator parameters, in combination with less frequent analyses of other parameters.

- *Requirements for record-keeping and reporting.* The procedures for the reporting of data generated by compliance self-monitoring systems at individual facilities is an important consideration. The government must ensure that neither the amount of information nor the frequency of reporting exceed the government's ability to process and use the information. The regulations or permit conditions for compliance self-monitoring should specify procedures for storage of information at a facility and the time periods for which information must be retained. One possible approach to reporting requirements is to require more frequent submittal of information by high-risk facilities.
- *Frequency of monitoring and reporting.* The government must attempt to provide regulations for compliance self-monitoring that produce adequate types and amounts of data at minimum cost to the regulated community. The government should require monitoring of parameters or indicators at frequencies that will ensure early identification of significant non-compliance, in a technically and economically feasible way.

Integrated Approach to Self-monitoring

Consolidated self-monitoring and reporting. Applying an integrated approach to a facility's self-monitoring activities involves consolidating the different media (air, water, waste) requirements into one, integrated self-monitoring system so that a single report to the authorities would cover all emissions and wastes of the facility. The administrative benefits of integrated self-monitoring include:

- fewer reports for the facility to prepare
- only one government agency to report to
- identification of the effects of any single environmental medium regulations on other environmental media

Inspections

Developing the Inspection Strategy

Environmental inspection programmes have significant effects on the costs and effectiveness of a government's efforts to implement a new law and regulations. Therefore, planning for inspection programmes should be conducted as early as possible after the enactment of new environmental laws, consistent with the governments priorities.

The inspection strategy must:

- *estimate the numbers and types of facilities* that will be inspected and the items that will be inspected at those facilities. The estimates of the number of facilities that can be inspected annually must consider current and expected annual funding for inspection activities.
- *prioritise inspections at facilities* that pose the greatest threats to human health and the environment. However, the government should consider conducting some inspections at facilities in lower-risk groups, because the combined effect of non-compliance by facilities in such groups may have significant adverse effects on human health and the environment.
- *estimate the number of violations* likely to be found at those facilities and the government's capacity to follow through with an adequate response to each violation. Otherwise, many violations that are identified by inspectors will not be addressed, and both the public and the regulated community will consider the enforcement response inefficient.

Guidance to Inspectors

Once the government has decided on the number of facilities to inspect and what to inspect at each type of facility, it must communicate that strategy to its inspectors. The inspectors must be given specific guidance or training that addresses:

- *Selection of facilities.* The guidance for selecting facilities should allow inspectors to identify groups of facilities within their jurisdictions according to specific criteria and to select for inspection a prescribed number of facilities within each group. By selecting a number of facilities from each type of industry, size category, and geographic location, the government can send a clear message to the regulated community that each facility eventually will be visited by inspectors.
- *Items to be inspected and information required for inspections.* For each group of facilities to be inspected, the items to be inspected should be identified clearly, possibly through the use of simple checklists that make reference so specific laws and regulations. In addition, inspectors should be provided with guidance on the enforcement priorities associated with each item on the checklists. Inspectors should receive guidance on the types of information that should be reviewed before conducting an inspection and on preparation for the inspection (e.g. facility's permit application, permit, and compliance self-monitoring reports). Using that information, the inspector will prepare a simple plan for the sequence in which items are to be inspected at the facility. The types and number of items to be inspected will depend on the law and regulations that apply to the facility.

However, inspection and enforcement policies should clearly separate an inspector's advisory role from his or her enforcement role. In some cases, the inspector may contact the facility to schedule the inspection at a time when key personnel of the facility will be available to answer questions, and at a time when the operations of interest to the inspector will be fully operational. In other cases, particularly when a facility is suspected of engaging in criminal activity, the inspector may need to conduct a surprise inspection (and may require the assistance of a police officer or other appropriate law enforcement official).

- *Rules of conduct.* To avoid conflicts of interest and inappropriate behaviour on the part of inspectors, the rules of conduct must be delineated clearly. For example, most countries prohibit their inspectors from accepting gifts from facilities and many countries enforce strict standards of courtesy and respect. Other examples of specific issues addressed in codes of conduct include requirements that inspectors never inspect dangerous areas alone and always inform the facility owner of the language in the law that allows the government to conduct inspections. In addition, inspectors should be advised of the amount of information they are allowed to disclose to personnel of the facility and the amount of technical or regulatory advice that they may provide during an inspection.
- *Inspection equipment and sampling procedures.* Inspectors must be informed fully about the equipment that they will need to conduct each type of inspection and about any limitations that may apply to the use of such equipment. For example, sampling during inspections often is restricted by the government's capacity to analyse and report the results. Therefore, the government should require the regulated community to implement compliance self-monitoring programmes to the maximum extent possible (and thus applying the principle that "the polluter pays.")
- *Health and safety of inspectors.* The health and safety of its inspectors should be a primary concern of the government. Common health and safety problems that arise during inspections include inhalation of toxic vapours, head injuries from falling debris or overhead equipment, and foot injuries caused by sharp or falling objects. Inspection guidelines and policies should consider the designation of high-risk facilities that may be inspected only by the national authorities, using appropriate safety equipment. All inspectors at all levels of government should be informed of the health and safety equipment and procedures that are necessary at each type of facility to be inspected. If appropriate equipment is not available through the government or at the facility being inspected, the inspector should not inspect the facility. In addition, inspectors should be

provided with lists of questions to ask of representatives of the facility at the beginning of an inspection. The answers to the questions should allow the inspector to determine which operations at the facility can be inspected safely.

- *Preparation of the inspection report.* The inspection report is crucial to the government's ability to respond to violations discovered during an inspection. The report need not be lengthy to be effective, but it must be clear, objective, thorough, accurate, and legally defensible. Inspectors should be provided with guidelines for preparing consistent inspection reports and submitting them to enforcement personnel. The reports should include descriptions of each item that was inspected, what was observed, and what statements facility representatives and the inspector made about each item.

Converting to Integrated Inspections

The government can save considerable resources by implementing an integrated approach to inspections, although the enactment of a new environmental law may be required in consolidating existing, single media regulations.

- First steps government could take in converting to integrated inspections are:
- identify and compare the responsibilities of existing government inspection staff
- consolidate responsibilities or develop interagency teams to inspect more than one media at a time
- incorporate integrated approach into inspector training programmes
- Benefits
- regulated facilities become subject to fewer, more comprehensive, inspections.
- less government resources are needed for inspections (more resources available to respond to compliance promotion or violations).

In addition, the government may benefit by consolidation of the negotiation processes involved when penalties are assessed and compliance schedules are set in response to violations. For example, a facility that has violated several laws would conduct one negotiation with the government that would set priorities for a return to compliance, according to the severity of the violation and the facility's financial capabilities. Otherwise, settlements reached during negotiations over violations under one law may preclude some options for a future settlement of violations of another law.

Enforcement Responses

Enforcement involves a range of actions following detection of a violation either through an inspection, self monitoring, verification of citizen complaints, or other means of observation establishing the offence. Enforcement is designed to correct non compliance. To provide an effective deterrent to violations of the law, enforcement should be timely and appropriate, and carried out in a manner which enhances the awareness within the regulated community. It must be fair and credible to the regulated community as well as to the public.

Authorities can implement and impose sanctions based on the possibilities and competencies laid down in civil, criminal, and administrative laws. They offer a wide range of potential sanctions to the authorities ranging from adverse publicity, to monetary penalties and/or jail terms, permit revocation and shut down, to ordering correction of damages. All actions are specifically authorised and involve legal procedures. Success in pursuing legal actions against non compliance, requires careful, and scientifically supported inspection and evidence from the competent authorities.

Steps in Developing an Enforcement Strategy

- *Review laws to determine the boundaries of government authorities.* A typical first step in developing an enforcement policy is to review the corresponding laws to determine the boundaries of authority of the government for enforcing the law. In most countries, the law provides for only one or two maximum penalties that are applicable to all the provisions of the law and their associated regulations. In those countries, a lead government agency could be authorised to develop various penalties, up to the maximum, to differentiate for different types of non-compliance, depending on the risks posed by each different type of non-compliance.
- *Develop a list of all possible violations of laws and regulations.* The second step in developing a system of government responses to non-compliance is to list all the requirements that are provided in the law and regulations and their possible violations.
- *Group violations according to their actual or potential threats to human health and environment.* In the third step, the government should group the violations according to the risks they pose to human health and the environment (discussed further in D2).
- *Set priorities and determine appropriate response for each level of violation.* Next, the government must determine the general level of response that it may wish to pursue for each group of violations.
- *Balance enforcement strategy with available resources by:*

- Considering the facilities to be inspected in the coming year
- estimating the average government cost of pursuing each type of enforcement response for each type of violation
- estimating the total cost of government response likely to result from planned inspections

Enforcement Responses for Specific Violations

The government can use a number of alternatives in establishing appropriate enforcement actions for various classes of violations. These should:

- correspond to the severity of violation (e.g. failure to apply for permit, failure to self-monitor, failure to treat emissions, wilful dumping, etc.)
- range from informal discussions and official warnings (indicating enforcement actions)
- to fines, closures, imprisonment ...

To develop a system like the one shown below, the government must determine how it will respond to each class of violations. Potential responses include informal discussions with violators as part of the compliance programme, warning letters, notices of violation, administrative orders, fines, closure of the facility, and criminal charges as part of the enforcement action. Government policies on the types of response appropriate for each type of violation may change over time. For example, the initial policy may prescribe informal discussions between inspectors and representatives of the facility for all violations as part of the compliance programme except those that involve wilful and dangerous. The policy may evolve over time through a steady increase in the enforcement action to each type of violation, until a fully evolved response has been identified for each type of violation.

Any alternative selected, however, must include provisions for consistent, timely, and effective response actions for each type of violation. To conserve resources and as part of compliance promotion, the government may opt to stipulate compliance time schedules, with clear penalties in the event of failure.

Like other aspects of a compliance and enforcement programme, the important message is to start doing it, and to evolve and get more sophisticated over time. Most countries, even those with more sophisticated systems of enforcement and a range of authorities did not start at full scale.

One place to begin is to select from several violators severe cases, with the best information and support for success on the part of the governmental judicial support. The case will be highly visible and a costly undertaking, and one that must be supported by time and resources to achieve success.

EXAMPLE OF GOVERNMENT RANKING OF VIOLATIONS AND RESPONSES

	Descripton	*Violations*	*Enforcement Actions*
I.	Potential or actual releases of pollutants that, because of their charges; nature or volume, pose imminent and substantial threats to human health or the environment	Direct discharges of solvents into a river. Wilful dumping of bags of unused pesticides. Leaking drums or tanks. violations.	Temporary or permanent closure of the facility; Criminal Severe fines Repeated failure to address Class II
II.	Potential releases that, because compliance; of their nature or volume, pose substantial threats to human health or the environment violator	Lack of runoff controls to prevent piggery wastes from reaching a river. Lack of surge tanks upstream from treatment systems. Failure to conduct compliance self-monitoring. Repeated failure to address Class III violations	Fines for each day of non- Notice of violation, with fine and issuance of a compliance schedule; Temporary closure of the facility; Informal discussions with the
III.	Improper management practices	Failure to properly report compliance self-monitoring results. Failure to inspect self-monitoring equipment regularly. Failure to properly clean up minor spills. Repeated failure to address one or more Class IV violations	Small fine, with issuance of a compliance schedule; Warning letter, with issuance of a compliance schedule: Verbal warning by inspector; Informal discussions with the violator
IV.	Improper paperwork	Failure to submit a permit application Failure to respond to agency requests for information. Failure to properly label drums and tanks	Warning letter, with issuance of compliance schedule; Informal discussions with the violator

Another possibility for selecting is to run a targeted programme of administrative non compliance checks and imposing time schedules for compliance. Stepping up inspection frequencies and controls will put pressure on the facility to comply. Reporting violations might create public awareness and help the compliance process. At the end sufficient pressure might have been built up to either achieve full compliance by the marked violator or create sufficient political pressure on the legal system to activate and to actually prosecute the offender. In many countries the judicial system is quite slow in acting on new phenomena and this type of action might activate the judicial system.

Enforcement Agency

Clearly Identify Responsible National, Regional or Local Agency

Who should take the enforcement action will depend upon who has the legal authority and how the execution is structured in terms of the degree of decentralisation and integration of functions. The enforcement programme will have to establish who has the lead responsibility for bringing and following through an enforcement action, how to obtain the commitment of supporting organisations that carry additional skills and legal possibilities. Clear responsibility should be established not only for which organisation is responsible for designing and overseeing policies and strategy of enforcement actions but also who should initiate the enforcement action. If the enforcement authority resides within the administering agency responsible for permitting and inspection than the responsibility for the execution of enforcement action is clear. This is the case in the "traditional" enforcement organisation" as described in module 1.

Identify and Develop Co-ordination with the National Judicial System

A more typical situation is that the environmental ministry or department is trying to start-up a programme and gain support of a legal arm of the government which has little or no experience or interest in environmental matters. Isolated and well prepared engagements may open doors to more systematic programmes and relationships across the various organisational boundaries, in this way ensuring effective follow up in future cases. It is recommended in general that as most of the environmental know-how resides in the environmental organisations to involve legal organisations in order to develop a common and consistent set of approaches to enforcement actions.

Options When Resources are Limited

- Set priorities based on greatest risk to environment
- Budget for worst case violation scenarios
- Balance with ability to inspect (and self-monitoring)
- Reduce number of inspections and shift some resources from inspections to enforcement actions.
- Inspect fewer portions of each facility's operations until the government has the resources to respond to violations it expects to find through such inspections
- Phase-in enforcement responses over an extended period of time while maximising the number of annual inspections to educate regulated community
- Widely publicise enforcement action taken to deter other facilities from noncomplying

SELF-MONITORING, INSPECTIONS AND ENFORCEMENT

Setting the Context

The Monitoring and Response Triangle

- Compliance self-monitoring: the regulated facility regularly checks its own compliance with permit requirements and records and/or reports them to authorities as required.

- Inspections: government-assigned body makes official on-site inspection or file review
- Enforcement responses: the legal measures (fines, closure orders, law suits) taken in response to violations.

Key Issues to Address

- Self-monitoring: Who should be required to self-monitor what, how and when? Integrated self-monitoring.
- Inspections: Who should inspect what and how? Converting to integrated inspections.
- Enforcement responses: What enforcement actions to take and who should take them?

Compliance Self-Monitoring

Who, What, How and When?

- All facilities? Specified facilities (size, risk, processes, chemicals, etc.)
- Measurable parameters and indicators.
- Requirements for record-keeping and reporting.
- Frequency of monitoring and reporting.

Integrated Approach

- Consolidated self-monitoring and reporting
- Administrative benefits of integrated self-monitoring include:
- fewer reports for the facility to prepare
- only one government agency to report to
- identification of the effects of any single environmental medium regulations on other environmental media.

GNP and the Index of Sustainable Economic Welfare (ISEW)

GNP measures wealth and economic progress. It is not a good indicator of national well-being because it does not take account of social factors, environmental damage and depletion of resources. A successful economy has increased trade, which usually means more motorised transport, more roads and higher emissions of pollution especially in urban areas. However, when this results in one in seven children suffering from asthma (for example, in London, UK) the medical treatment these children receive increases GNP. If there are more road accidents resulting from increased traffic, all the services required to deal with the accidents are added to GNP. Development, which is widely associated with progress often means clearing away the natural environment, building or widening roads and increasing urbanisation. Many people are fearful when they learn about 'development' close to their own homes.

GNP rises with increased trade. This means that as 'do-it-yourself (DIY) grows in popularity the materials used are included in GNP figures but labour is not. Similarly if people stop growing their own vegetables and cleaning their own homes but pay others to do this for them, GNP rises. Likewise those who start growing their own vegetables, repairing or decorating their home and washing their own car are reducing GNP. When solar panels are fitted to heat water only the purchase of the panels is added to GNP; the energy savings are deducted. These examples highlight the weakness in relying so strongly on GNP as an indicator of social welfare and progress. The New Economics Foundation (NEF) has undertaken a detailed study of the Index of Sustainable Economic Welfare (ISEW) and applied their findings to the UK for the period 1950 to 1990. The ISEW aims to assess the quality of life over time. This is done by deducting expenditure which is incurred in repairing social and environmental damage from GNP. The long-term cost of environmental damage and depreciation of natural capital is taken into account. Changes in the distribution of income are included because £1 means more to the poor than to the rich. Housework, which is excluded from GNP, is included to allow for the non-monetary benefits to people and the economy. A full table of the adjustments to GNP are available from NEF.'

The result of these calculations produces the ISEW for the UK and it is compared with GNP over the period 1950 to 1990. From the 1970s onwards ISEW has been declining even though GNP has continued to rise. Since 1950, long-term environmental damage has increased steadily, as has the value of household labour. Factors like these have produced the drop in ISEW. This analysis provides powerful evidence of the need for change but does not show that it is already happening. It seems plausible that the 'feel good factor' remains illusive because ISEW is declining.

Low Profile Trends

Many people are dissatisfied with society as it is and a growing minority in countries like the UK are unwilling to accept traditional lifestyles without question. Some have started to experiment with new ways of living or changing important aspects of their lives. In several areas these are becoming sufficiently significant to be measurable. They indicate the emergence of new trends—as yet not well known. It is interesting to examine the following:

- traditional beliefs contrasted with emerging ideas;

- ethical and environmental investments;
- energy efficiency and renewable energy;
- householder initiatives;
- diet and health;
- organic farming and food;
- alliances and The Real World Coalition;
- communications; and
- Charter 88 and the campaign for a new UK constitution.

Corporate Environmental Reports (CERs)

Various guidelines have been proposed for the presentation of environmental reports. One example comes from the Public Environmental Reporting Initiative (PERI). Their guidelines spell out what is recommended under the following headings, but leave the selection to individual firms:

1. Organisational Profile
2. Environmental Policy
3. Environmental Management
4. Environmental Releases
5. Resource Conservation
6. Environmental Risk Management
7. Environmental Compliance
8. Product Stewardship
9. Employee Recognition
10. Stakeholder Involvement

Several assessments of environmental reports have been published. *Tomorrow* magazine pays particular attention to their quality, what they cover, how they are verified and how the reporting process is likely to develop. This will depend on the degree to which companies succeed in integrating their environmental policies with their business strategy and demonstrate that they are moving forward in a coherent way. It will also depend on their ability to show how their combined business strategy and environmental management influence their financial performance. The latter is seldom apparent in current environmental reports with the result that financial institutions pay little attention to Corporate Environmental Reports. For the future CERs are likely to become more valuable, more open, include the views of all stakeholders and be independently verified. A summary of developments might include the following:

- most companies will publish a CER;
- the format will be simpler and they will be easier to read and understand;
- progress towards sustainability will be reported with quantitative measures, qualitative assessments and stakeholders' views;
- a standard format, possibly based on or including The Natural Step systems conditions, will emerge;
- the interaction of environmental policy, business strategy and financial performance will be more explicitly stated; and
- social and environmental auditing will complement financial auditing with independent verification.

Conclusion

The search for indicators of sustainable development is taking place worldwide involving governments, public bodies and academics. However, if mindsets remain largely unchanged the focus will confirm present values and orientation. Where proposed indicators omit key aspects of sustainable development, such as the social dimension, they are unlikely to be adequate. However, by bringing together the ideas about indicators from different sources a more comprehensive approach can be developed. Organisations like the Centre for Corporate Environmental Management, at the University of Huddersfield and Ashridge Management College have put forward their views, both of which provide helpful Ideas. Fritjof Capra has added to his already substantial contribution to environmental management with a fresh examination of the web of life. The Natural Stop, a Swedish initiative, has developed a useful set of non-negotiable systems conditions, backed by a consensus of scientists. Capra and The Natural Step emphasise the importance of the living cell—the only truly productive resource on Earth.

These ideas are combined to propose five guiding principles for sustainable development. They are easy to understand and this makes them a valuable basis for a standard approach, which can be used by any organisation. The fact that Electrolux, Ikea and the Co-operative Bank are developing their future business strategy along the lines proposed by The Natural Step demonstrates that this is both feasible and makes business sense.

IBM and Traidcraft have developed their own unique ways of verifying the appropriateness of their policies by doing stakeholder surveys. IBM call this 'consulting the stakeholder', Traidcraft call it 'social accounting', but there is a considerable overlap in the reasons for doing these surveys. In both cases independent organisations are used in addition to internal resources. The question of Corporate Environmental Reports (CERs) is considered and a framework suggested based on the work of the

Public Environmental Reporting Initiative (PERI). How these reports might develop in the future is summarised using guidelines suggested by John Elkington. The choice of measuring and reporting methods rests with each individual firm, but their reports are scrutinised and compared with others, thus driving for higher standards. As the measures and the reporting processes become more coherent there is likely to be some convergence towards a standard approach. Until that happens the most crucial factor will be the commitment of top executives to the indicators, the reporting style adopted and the use made of the findings by each individual organisation.

Social Attitudes and Aspirations

Political attitudes and stances are often seen as a mirror of the views and aspirations of society at large. The increased prominence given to environmental matters by consumers and citizens has been translated into political preferences and this, more than any other factor, can be seen to provide the basic reason for the greening of political attitudes. The surge in the vote for the Green Party in the 1989 elections for the European Parliament reflected this upsurge in popular concern for the environment and it helped to galvanise the traditional political parties into action.

Popular concern for the environment has its origins in a broader-based social concern, which in turn has its origins in the growing fear that the costs associated with the established 'growth at any price' philosophy of the postwar world are too high a price to pay for improved prosperity, and a realisation that the price that has to be paid for this increased prosperity is a deterioration in the quality of life. The clearest manifestation of this change in consumer attitudes and preferences can be seen in the growth in the number of public protests against major infrastructure proposals. The long-running dispute on the line of the route for the fast rail link from the Channel Tunnel to London is a case in point, but other protests have occurred in relation to proposals for motorway building, chemical plants, incinerators, airports, coalmines, new housing estates, industrial and retail parks and many other forms of physical development. The NIMBY (not in my back yard) syndrome is well known, but little understood.

The complexities inherent in many decisions related to the environment mean that, whilst the NIMBY syndrome is a convenient catchphase, it is a poor description of the underlying tensions. In many cases where public protest is raised against a proposed development there are at least two groups involved: those who oppose a project on grounds that it damages their environment or the environment that they seek to protect, and those who either without reservation, or with some notion of a trade-off in mind, support the proposal. This is not a new problem; in the past most planning applications for major projects have attracted both supporters and opponents.

This could be seen as a peacetime equivalent of the guns-or-butter debate. Even within an individual community there may be different views as to the merits, or otherwise, of a development that has a significant effect upon the environment. This argument, was encapsulated in colourful fashion by one of the witnesses at the public inquiry into the proposal for exploratory drilling at Capel Hermon; the witness was of the view that

> There are far too many people who regard the County of Merioneth as some form of Indian Reservation inhabited by the Welsh equivalent of the Blackfoot, Sioux and Apache Indians. We believe that if the Government wishes to conserve certain areas to the detriment of the development of those areas then there should be compensation to the area to be conserved.
>
> (Searle, 1975, p. 93)

It is apparent that there is no single or simple way to define or describe the views of the public, and it is even more difficult to classify those who support or oppose a project into convenient categories.

However, despite the many underlying tensions, a general trend of growing social awareness about, and concern for, the environment can be detected. This has developed from protests that are related to development schemes into a variety of forms of direct consumer action. Increasingly sophisticated and knowledgeable consumers register their protests in the supermarket, the garage forecourt and in their choice of investment trust or personal equity plan. One of the best-known examples relates to the purchase of aerosols that use CFCs as a propellent. In Britain, Friends of the Earth mounted a campaign to demonstrate the harm done by CFCs and published a pamphlet indicating aerosol brands that were CFC free. As Elkington and Burke (1989, p. 137) observe, 'just days before Friends of the Earth followed up with a list of the brands that did contain CFCs, the eight largest aerosol manufacturers announced that they would phase out CFCs by the end of 1989. Those companies alone accounted for 65 per cent of the UK toiletries market'. The power wielded by the consumer had made its mark.

These and other manifestations of changed attitudes indicate the importance for business of understanding and responding to the evolving views of society. At one level this can be accomplished by conducting regular surveys of consumer attitudes (some of the evidence in this field will be reviewed later), but at another level such surveys can miss important underlying trends. It is often difficult to distinguish cause from effect, or to provide a realistic assessment of the strength of feeling expressed by an individual consumer. This problem is at its most severe when you confront an individual with a simple yes-no choice, or when you attempt to measure the strength of environmental opinion among a group of residents who also depend for their livelihood upon the manufacture of an environmentally damaging product. The apparent dichotomy that exists between the role that is played by a citizen as a guardian of the local environment and as an employee of a local company can be resolved if the needs of the community are satisfied through the adoption of an approach to the resolution of environmental problems that also meets the needs of business.

Social attitudes vary between places and what is considered to be an acceptable practice in one locality, region or nation may be unacceptable in another place. This spatial variation in the acceptability of the existence of an environmental hazard or a pollution threat may result from the economic circumstances prevailing in a particular place. The social acceptability, or otherwise, of environmental problems also reflects the long-established images and perceptions of industry that are held by the residents. Some years ago a proposal to remove a large colliery spoil heap in the Stoke on Trent area was opposed by local residents; they perceived the spoil heap as a local asset that provided informal open space and a vantage point from which they could view the adjacent countryside. Whilst the outsider visiting a traditional industrial area may be appalled by some of the effects of dereliction and pollution, the local landscape may be regarded as normal and acceptable by local residents. Hazardous activities are, however, a matter of specific concern to local communities, especially following a major incident at a local industrial installation, or at a plant with a similar configuration that is located elsewhere.

Community attitudes towards industry change over time. For example, the incident that took place at the Associated Octel plant at Ellesmere Port in February 1994, when a leak of ethyl chloride occurred, led to calls from local residents for this plant, which was established on its current site in 1938, to be relocated. The Associated Octel plant forms part of a large petrochemical complex located on the southern bank of the River Mersey; this complex has provided the southern Merseyside area with much-needed jobs and income. The raised environmental awareness that is evident in many local communities in the UK, coupled to the specific effects resulting from the leak of ethyl chloride, caused the role of the plant to be re-examined by local residents and representatives. A local authority councillor concluded that 'the danger is just too great and it is hampering development because we cannot grant planning permission for any new building on that side of town' (Sharratt, 1994, p. 3). Even in traditional industrial areas, where jobs are highly valued and keenly sought, a major incident can result in a change in social attitudes and values. In the past, local communities often gave precedence to the economy in the trade-off between jobs and the environment, but the balance has shifted in recent years and hazardous or environmentally undesirable businesses are no longer tolerated. Such businesses once offered the prospect of jobs at the cost of the environment, but raised aspirations and current legislation have brought about a situation in which environmentally undesirable companies are now viewed as damaging both the economic and environmental health of an area.

Considerable spatial variations exist in social and cultural values, and in the attitude of business towards the environment. These variations are at their extreme when the situation obtaining in less developed nations is contrasted with that in developed countries. Many of the world's most rapidly growing economies are to be found in the developing nations. Economic growth in such nations has often started from a very low base and is relatively unrestricted by social or environmental regulations. The existence of widespread poverty and high levels of unemployment in the developing world, together with the absence of an indigenous industrial base, has left these nations vulnerable to exploitation and they have become the recipients of industries regarded as environmentally unacceptable by developed nations. The operation of such economic activities in the developing world is not only environmentally damaging but it can also lead to a reduction in their potential for economic growth. Pearce (1991) estimates that the developed nations may be losing up to 5 per cent of potential GNP due to environmental degradation, but in the developed world the loss could be as much as 15 per cent. This longer-term loss of environmental and economic value in the less developed world is often ignored by national governments who concentrate their attention on the achievement of short-term gains. Understandably, consumers and citizens in such countries pay relatively little attention to environmental matters and, in any case, there is often little that they can do to influence either the environmental standards set by national governments or the environmental performance of companies.

Spatial variations in social attitudes are reflected in variations in the value that is placed upon the environment in different areas. Such variations may give rise to situations in which the cost to industry of complying with best environmental practice is far in excess of the cost of compensating anyone who is injured or affected by any environmental damage that occurs. However, in some cases, the costs of failing to ensure that the environment is protected can be extremely high and has led to situations in which the overall integrity and financial solvency of a company is in question. Shrivastava (1992a, p. 127) has noted, for example, that 'the Bhopal disaster slowly but steadily sapped the financial strength of Union Carbide. It drained financial resources and adversely affected morale and productivity of the company'. This case may not be typical, but it does demonstrate the consequences for a company of an environmental disaster, even if such a disaster occurs in a developing country. However, while a growing body of evidence suggests that on both environmental and economic grounds sustainable development is the only form of development which should be permitted in both the developed and developing worlds, the developing world is not always in the position to make a choice. Cairncross

(1991) has observed that the costs of environmental damage are high in developing countries because exploiting the environment provides a source of income for many people. Primary production is often the mainstay of the economy of a developing nation and the continued production of timber, ores, agricultural produce and other goods either directly erodes the stock of environmental resources, or makes use of other finite resources in order to ensure that output is maintained. The problem in such a situation, which is recognised by the notion of sustainable development, is that developed countries cannot expect the citizens of developing countries to abandon their existing economic activities without either compensation or, more preferably, alternative activities. The true meaning of intragenerational equity is to be found in this recognition.

Changes in the attitudes of citizens towards the environment in developed nations reflects both the immediate problems associated with the operation of advanced economies and the environmental degradation evident in the developing world. Public concern for the environment is a complex amalgam of many different experiences, perceptions and messages. People are influenced by their personal experiences of pollution and hazard-by the visual portrayal of deforestation, environmental degradation, and the extermination of threatened species on television and film-and by a variety of other messages transmitted via the radio and press. Social perceptions as to what constitutes an environmental problem are also conditioned by the level of education and the availability of information, sometimes supplied by companies, on the operation of a particular plant or process.

Surveys of public attitude reveal that there are many differences between countries in the priority given to the environment. Table 1 illustrates some of these differences. As can be seen from this evidence, whilst the level of priority given to the environment reflects the extent to which economic growth is perceived to be important, the general trend is towards a desire either to prioritise environmental protection or to attempt to balance the needs of the environment with those of the economy.

TABLE 1

Public Opinion on the Environment Versus the Economy (%)

	Priority to environment	*Priority to economy*	*Balance between the two*
USA	71	19	—
Japan	36	3	43
Finland	63	6	26
Norway	48	1	49

Source: Adapted from OECD (1991 a).

The public attitudes revealed by the OECD survey are mirrored in other studies. A MORI study of attitudes towards the environment (Worcester and Corrado, 1991) showed that, irrespective of culture and political system, both the public and opinion leaders were of the view that the condition of the environment had deteriorated over a ten-year period. In most countries opinion leaders were more concerned than the public about the deterioration in the condition of the environment (Table 2). This would seem to reflect the priority given to the environment in global political discussions.

TABLE 2

Rating of environment in the last ten years (%) (Members of the public and opinion leaders were asked: Do you feel that the environment has become better or worse in the last ten years? The percentage who thought it had become worse is shown in the table)

	Public	*Opinion leaders*
All countries	55	69
Hungary	68	69
China	49	58
India	63	98
Japan	39	52
Saudi Arabia	41	38
Argentina	76	92
Brazil	52	82
Kenya	44	82
Senegal	60	70
Norway	62	64
West Germany	59	50

Source: Based upon Worcester and Corrado (1991).

At national level in the UK a similar trend of growing concern for the environment can be seen. In Department of the Environment surveys the number of respondents who have indicated that the environment is a matter of specific concern has risen in recent years. The results of the 1986 survey showed that 8 per cent of respondents thought that the environment was an important issue (sixth of eight issues), whilst in 1989, 30 per cent indicated that the environment was a matter of concern; in this year it was second only to concern about health and social service provision (Department of the Environment, 1992a). There are regional and local differences in the public's perception of the nature and causes of environmental problems. Surveys conducted in England, Wales and Scotland in 1989 and 1990 showed, for example, that whilst pollution of rivers and seas was seen by all respondents as a problem of great significance, the environmental threats associated with the development of the countryside were not as important to Scottish respondents as they were to those surveyed in England and Wales *(ibid.)*. This growth in concern for the environment is also reflected in the growth of membership of voluntary environmental organisations, for example, the membership of Greenpeace UK grew from 8,000 in 1981 to 397,000 in 1991 (Cairncross, 1991).

The increased importance attached to environmental issues by the public and opinion leaders is reflected in the growing priority given to environmental matters by business leaders and individual companies. A survey of 108 large UK companies, conducted by Coopers and Lybrand in 1990, showed that 87 per cent of companies interviewed thought that environmental issues were important, 41 per cent of companies had an environmental policy and 29 per cent had included environmental information in their annual reports (Coopers and Lybrand, 1990b). Although a later survey, by KPMG Management Consulting in 1992, showed that two-thirds of companies considered that the environment was not a trading issue in the past year, many businesses (44 per cent of manufacturing companies) had already been affected by environmental legislation (KPMG, 1992).

A survey conducted by the Advisory Committee on Business and the Environment (ACBE) provides additional evidence on the importance of the environment. This study examined the benefits to companies from the adoption of environmental policies:

- 65 per cent reported an improvement to their perceived image;
- 61 per cent of respondents had achieved direct financial benefits;
- 40 per cent believed that they improved relations with stakeholders, regulators and the local community;
- 28 per cent had gained a competitive advantage;
- 23 per cent had improved their relations with customers and this had enhanced their chances of survival; and
- 12 per cent had observed an improvement in working conditions.

This study (ACBE, 1993b) illustrates the benefits that may accrue to companies if they take action to understand the perceptions of the public, opinion leaders and other business organisations. The companies in this survey demonstrated a relatively high level of concern for the environment; 68 per cent of companies surveyed had developed environmental policies, compared with the average of 25 per cent reported in other recent surveys conducted by the British Institute of Management and the Institute of Directors. The proactive companies in the ACBE survey also reported that they had gained other advantages from the adoption of environmental policies including improvements in the quality of products and services.

Although the changing views of society have clearly had an impact upon larger companies, small and medium enterprises have been slower to respond. A report from the Cranfield School of Management (1990) has shown that most small companies have no policies or procedures on environmental issues and only 10 per cent have carried out an environmental audit. It is this group of companies that is likely to be most vulnerable to the changing attitudes and preferences of consumers, and to increasing supply chain pressures. Despite the virtual absence of action to date, the majority of small firms surveyed by Cranfield accept that environmental concerns are not just 'a passing fad' *(ibid.* p. 4)—this offers a foundation for the development of environmental policies in such companies.

The growing concern of society in general with regard to the environment is reflected in the attitudes of business. At global, national and local levels this concern has been translated into action, including the growth of voluntary organisations and coalitions of interest and, increasingly, in the enactment of stricter environmental legislation. The legal framework of environmental regulation reflects both the views and aspirations of society, and the strength of political conviction that the environment is an important and legitimate field of intervention. The final section of this chapter reviews some of the more important areas of environmental policy and legislation.

Prospects for Cooperative Intergovernmental Policies

Future directions for environmental policy revolve around enduring issues about the role of governments in preventing environmental harms and averting environmental catastrophes. A rethinking of approaches to environmental regulation has been stimulated by acknowledgment of the limitations of current policies for these purposes. The intergovernmental dimensions have taken on cogency with the recognition of the importance for environmental sustainability of the decisions that local governments make about land use and development. Many argue that future directions for environmental policy include less emphasis on regulatory prescription and greater reliance on local governments as partners in pursuing paths to sustainable futures. Bringing this about requires a rethinking of the ways in which higher-level governments influence the decisions of local governments. At issue is how intergovernmental policies can be better designed and implemented to increase shared commitment among different layers of government to environmental goals and to foster practices that promote, rather than undermine, sustainability.

This book has addressed what we label as "cooperative intergovernmental policies" as one promising direction for structuring the intergovernmental approach to environmental management. These policies assume that lower-level governments are willing to serve as regulatory trustees and, with appropriate guidance and assistance from above, are capable of figuring out ways of reaching desired state or national environmental goals. The emphasis is on prescribing goals and planning processes, while supporting efforts of lower-level governments to figure out best practices for reaching those goals. Current ideological preferences for fewer dictates from above, recognition of the conflicts engendered by reliance on intergovernmental coercion, and demands for flexibility by local government, officials all point to emphasis in the future on cooperative forms of governance.

Although this approach appears to have much promise, basic questions remain about the ability to bring about the desired intergovernmental cooperation had the efficacy of the approach. In this book, we have addressed these questions by examining leading examples of this approach to environmental management in New Zealand and to hazard management in New South Wales, Australia. In both of these cases, major policy changes were undertaken with aims of fostering cooperative governmental partnerships for environmental or hazards management. A central aspect of each is empowering local governments to devise paths for managing the environment in a sustainable way. Where appropriate, we have drawn a contrast with the more coercive and prescriptive approaches used in corresponding programmes in the United States as exemplified by Florida's widely heralded growth management programme.

Both cooperative and coercive forms of intergovernmental policies present dilemmas that are discussed in this chapter. Given these dilemmas, the principal challenge in designing intergovernmental policies is to get the policy mix right. This takes on special importance in the environmental arena where the choices by local governments about land use and development management have noteworthy implications for sustainability. These choices can profoundly affect the economic base of communities, the quality of the environment, and the vulnerability of communities to natural hazards.

This chapter summarizes key findings and draws implications from them for the design and implementation of intergovernmental policies. The chapter begins with an assessment of the policy innovations examined in the preceding chapters and summarizes the lessons we draw about cooperative intergovernmental policies. This provides a foundation for discussion of their long-term potential and the general applicability of key features of these policies. Recognition of the limitations of cooperative intergovernmental policies leads to consideration of potential improvements in policy design and implementation.

Organisational Profile of WBCSD

The World Business Council for Sustainable Development (WBCSD) was formed in January 1995 with the merger of the Business Council for Sustainable Development (BCSD) and the World Industry Council for the Environment (WICE).

The BCSD was a direct outgrowth of the 1992 Earth Summit process. It was formed when Maurice Strong— then Secretary General of the United Nations Conference on Environment and Development (UNCED)—asked the Swiss industrialist Stephan Schmidheiny to act as his special advisor on business and environment. Schmidheiny brought together an impressive international group of top business leaders interested in the environment, and produced the book *Changing Course*. Showcasing an array of case studies of best practice, the book coined the term 'eco-efficiency' and argued that sustainable development was not only good for business, it was 'good business'. The ideas discussed in the book became the basis of a formalised BCSD, and soon afterwards UNCED regional and national chapters began to spring up. To its credit, the BCSD was a major breakthrough because it brought together business leaders at the highest level to express a concern for environmental protection, portray it as a common concern, and break the prevailing mould of expressing the issue only in simplistic and antagonistic 'business *versus* environment' terms.

BOX 1

The WBCSD Story

1990 UNCED Secretary-General Maurice Strong asks the industrialist Stephan Schmidheiny to be his advisor on business and environment. This leads to the creation of the Business Council for Sustainable Development (BCSD).

1991 The International Chamber of Commerce (ICC) develops a Business Charter for Sustainable Development, which is launched in April 1991 at the Second World Industry Conference on Environmental Management in Rotterdam.

1992 BCSD represents business at the Rio Earth Summit (UNCED) and releases its report *Changing Course: A Global Business Perspective on Development and the Environment.*

ICC releases its report *From Ideas to Action.*

1993 The World Industry Council for Environment (WICE) is founded as an initiative of the International Chamber of Commerce (ICC).

1994 A number of regional and national BCSDs are formed. BCSD sets up a Sustainable Project Management in partnership with UNDP.

1995 The World Business Council for Sustainable Development (WBCSD) is created through a merger of BCSD and WICE.

1996 WBCSD, UNEP, and the Ceres Initiative launch programme on financial indicators of Sustainable development. *Financing Change* is published.

WBCSD launches its International Business Action Plan on Climate Change (IBAPCC).

The Foundation for Business and Sustainable Development (FBSD) is established.

1997 WBCSD co-hosts a high-level round table of government, NGO, and business leaders on business and environment as part of the Rio+5 celebrations. WBCSD releases its report on the progress since Rio, *Signals of Change*. It also releases a new book, *Eco-Efficiency*, and a report on *Environmental Performance and Shareholder Value*.

WBCSD launches the Sustainable Business Challenge, an Internet-based environmental exam for students preparing for careers in business, finance, and government.

WBCSD signs a memorandum of understanding with UNDP to improve environmental performance of companies in developing countries.

1998 The WBCSD Virtual University is founded in collaboration with the University of Cambridge and the Norwegian School of Management. *The Sustainable Business Challenge* is published.

WBCSD holds a series of Stakeholder Dialogues on Sustainable Consumption, Corporate Social Responsibility, and Sustainable Business.

WBCSD releases its Survey on Corporate Environmental Reports and the results of its Global Scenarios project.

The genesis of WICE, on the other hand, lies in the International Chamber of Commerce (ICC), which boasts more than 7000 member companies in over 130 countries. In 1990 an ICC Working Party for Sustainable Development, led by Peter Scupholme of British Petroleum and W. Ross Stevens III of Du Pont, produced a Business Charter for Sustainable Development. This Charter was launched in 1991 at the Second World Industry Conference on Environmental Management at Rotterdam, and was followed soon afterwards by the report *From Ideas to Action,* which also outlined cases of positive industry response to environmental issues. In 1993 the ICC reconfigured its Environmental Bureau into an expanded WICE to serve as an industry voice in the follow-up to UNCED. The Charter's 16 principles—ranging from recognising environmental management as a corporate priority to affirming the precautionary principle—became the guiding foundation for WICE.

Each of the two parents had already established itself as a credible representative of international business on issues of the environment. The BCSD had been a forceful presence at UNCED and had begun to develop links with international NGOs and governments. The ICC, which had long and deep relationships with international business federations and IGOs, ensured an equally prominent position for WICE. The high degree of similarity in goals, overlap in membership, and complementarity in conceptual principles led to their merger into a single entity, the WBCSD, which immediately became the most authoritative and pre-eminent business voice on sustainable development. Box 1 lists the key events and milestones in the organisation's development.

WBCSD is a pure membership organisation in that its 126 member corporations are the principal source of financing and directly participate in agenda setting and governance. However, with the annual membership subscription being US$30,000 and membership offered only by invitation, it is a fairly exclusive club. For all practical purposes it remains a coalition of big businesses, predominantly from the advanced industrialised countries. Although technically there are no barriers—except hefty annual dues—to the inclusion of smaller companies, especially from developing countries, the priorities and focus of the organisation is decidedly on issues of greater concern to large corporations.

According to the WBCSD leadership, 'a large membership roster is not our overriding concern.' In fact, the number of total members has not grown substantially since 1995, with most new additions made principally to improve geographic or sectoral coverage. For the most part WBCSD tries to address the geographic imbalance through its global network of 'partner' (but non-member) associations. Efforts have been made to strengthen and activate this network, which now comprises 18 national BCSDs and an assortment of national and international business organisations.

Even in taking over many features of WICE, WBCSD closely mirrors the BCSD design. An important component is that, unlike in WICE, membership is by invitation only 'to companies committed to the concepts of sustainable development and responsible environmental management'. Apart from annual dues, members are also expected to provide in-kind and personnel support, including financial backing for individual working groups, and the active participation, including secondment, of their staff in the WBCSD work programme. In return WBCSD offers them the ability to 'exert greater influence on the framework conditions under which [the member companies] operate by being represented by a credible advocate.' Living up to its claim of speaking 'the language of business'. WBCSD makes an appealing case of why major multinational corporations should join its ranks: The collective voice of a cross-sectoral and global body like the WBCSD offers companies greater weight in the sustainable development debate than they would otherwise have singly.'

The promise being offered is that 'WBCSD [will provide] companies with an edge on the competition by being aware, ahead of others, and thus able to anticipate the emerging environmental and social trends which might impact their business.' The appeal of the argument is reflected in the WBCSD membership roster, which, although short, includes some very significant corporate players, including such giants as Assurances Generates de France, Rhone-Poulenc (France), Bayer, Hoechst (Germany), Fiat Auto (Italy), Heineken, Philips Electronics, Unilever (Netherlands), Norsk Hydro (Norway), ABB Asea Brown Boveri, Nestle, Societe Generate (Switzerland), British Petroleum, Glaxo Wellcome, ICI. Shell International (United Kingdom), Ontario Hydro (Canada), AT&T, Cargill, CH2M Hill, Dow Chemicals.

DuPont, Eastman Kodak, General Motors, International Paper, Johnson & Johnson, Monsanto, Proctor & Gamble, 3M Corporation. Xerox (United States), Hitachi, Mitsubishi, Nissan Motors, Seiko, Sony, Tokyo Electric, Toyota (Japan), and Samsung and LG Group (Korea). A few large corporations from developing countries and transition economies, such as Chemical Works Sokolov (Czech Republic). Aracruz Celulose (Brazil), Grupo Vitro (Mexico). China Petro-Chemical (China), and Inti Karya Persada Tehnik (Indonesia), are also members.

Participation is sought at the very highest level, normally from the chief executive officer (CEO) or equivalent. This ensures an earnestness of commitment from the company and allows WBCSD greater credibility in its advocacy activities. WBCSD is governed by a Council, composed of the CEOs of member companies, which meets once a year to determine organisational priorities and direction. It is also responsible for appointing the Executive Committee, determining membership dues, and approving the budget. Council members co-chair WBCSD working groups and are called upon to speak on its behalf at relevant forums. Each Council Member appoints a 'Liaison Delegate', normally a senior member of their staff, who supports them in their WBCSD responsibilities, participates in working groups on their behalf, and interacts with the Secretariat and other member companies. Liaison delegates meet twice a year in plenary sessions and throughout the year in working group meetings. An Executive Committee of up to 14 CEOs from member companies oversees the management of WBCSD, and normally meets three times a year. It is responsible for appointing the President, approving working groups, WBCSD reports and books, and finalising the budget. The day-to-day management is handled by a Secretariat in Geneva, headed by a full-time President (Executive Director until 1997). In 1998 the Secretariat had a staff strength of about 30.

Surprisingly, no financial and budgetary data on the organisation is publicly available. Not only are the WBCSD annual reports silent on this subject (unlike those that its members are more used to), but repeated requests to obtain this information proved unsuccessful. However, some estimates can be made. The bulk of WBCSD's direct funding comes from the annual membership fee of US$ 30,000 (for 1998) that each corporate member pays. For 1998, therefore, membership dues would have amounted to US$3.78 million. The total resources made available by the members would, however, be significantly higher, since they would include in-kind and personnel contributions to various working groups as well as experts seconded to the WBCSD Secretariat. For example, in 1998 as many as six staff members working on various projects at the WBCSD Secretariat were on secondment from member companies. In short, it would be fair to say that, unlike even the biggest environmental NGOs, fund-raising and financial concerns have not been major worries for the WBCSD.

Links between Cleaner Production and Eco-Efficiency

The interlinkages between eco-efficiency and clearer production are numerous. Like cleaner production, eco-efficiency links the goals of business excellence and environmental excellence, by creating the bridge through which corporate behaviour can support sustainable development, the integration of economic growth and environmental improvement. Eco-efficiency goes beyond resource use and pollution reduction by emphasising value creation for business and society at large, while providing for competitive needs. By increasing value for the goods and services it creates, business will maximise resource productivity, gain bottom-line benefits, and reward shareholders, rather than simply minimise wastes or pollution.

Eco-efficiency embraces cleaner production concepts such as efficient use of raw materials, pollution prevention, source reduction, waste minimisation, and internal recycling and reuse. It captures the idea of pollution reduction through process change as opposed to the earlier end-of-pipe approaches. It shares characteristics with many environmental management tools such as Environmental Assessment or Design for Environment by including them among the technological options for reducing material and energy intensiveness in production, as well as facilitating reuse through remanufacturing and recycling Eco-efficiency also features a life cycle perspective which follows products from the raw material through to final disposal stages. It is thus an extension of the Total Quality Management process.

Eco-efficiency an evolving concept which allows companies to adapt to the changing dynamics of the marketplace. Companies are now faced with new demands from more stakeholders, and those who implement ecoefficient practices will be able to respond more aggressively to competitive pressures and anticipate customer needs, while at the same time protecting the environment and employee health and safety.

Put in simple terms, the vision of ecoefficiency is to 'produce more from less'. Reducing waste and pollution, and using less energy and fewer raw materials is obviously good for the environment. It is also selfevidently good for business because it cuts companies' costs, and eventually avoids potential environmental liabilities. It is, therefore, a prerequisite to the long-term sustainability of business.

Assessing environmental impacts for cleaner products An eco-efficient company understands and seeks to minimise the environmental impact of its products throughout their life cycles. Such companies will provide added value from their activities by monitoring and assessing their impact at every stage. Assessing environmental impacts for cleaner products can assist companies in their quest for continuous improvement by identifying ways of maximising profits through reducing waste and liabilities, raising productivity and demonstrating the company's sense e of responsibility towards its customers and the environment. Companies are starting to use environmental assessment in a systematic manner to define goals, collect data, assess impacts, control effects and ultimately communicate to its stakeholders. Indeed, some companies are now requiring environmental impact assessments as part of their product development cycle with action to mitigate findings wherever possible. The Volvo Environmental Priority Strategies (EPS) System, for instance, assesses the environmental impacts of products and processes in terms of ecological and human health consequences. The results are built into a computer system that derives a composite score of the environmental impacts of any given product design. On this basis, alternative materials and product configurations can be evaluated in an iterative process to establish an optimum design.

The EPS system assists design engineers in the selection of environmentally preferable materials for product construction. In one such case, two technically equivalent constructions for the front end of a car, one using a plastic composite and the other galvanised steel, were compared. Environmental Load Units were calculated for production, product use, and product disposal at end of life for each material. The plastic construction proved to have a lower overall environmental impact. The galvanised steel received a less favourable score because its heavier weight increased fuel consumption during product use.

Dow Chemical has developed a 6 point eco-efficiency compass to guide its eco-innovation efforts. It challenges Dow's product managers to:

- Dematerialise in order to reduce the 'mass burdens' resulting from the raw materials, fuels and utilities used in the system to deliver the desired product function.
- Increase energy efficiency to identify where the highest energy is used, and subsequently redesign the product or its use to make significant energy reductions.
- Eliminate negative environmental impacts as well as reduce and control the dispersal of toxic elements introduced at one end of the system.
- Close the loop by recycling effectively and efficiently. Designing for recyclability is essential but recycling becomes unproductive when the energy, materials and pollution used in collecting, preparing and processing the recyclate exceeds those used to produce the goods in the first place.
- Borrow from natural cycles to design the system as part of a larger natural cycle-where materials are borrowed from and returned to nature-without negatively affecting its overall balance.
- Extend service and enhance function, especially at the usage phase, to improve eco-efficiency.

Designing and Producing Cleaner Products

Designers, mostly concerned with product performance and aesthetics, must take into account the effect of design details on energy material requirements for manufacturing, use and secondary use (repairability, remanufacturability and recyclability). Companies should also pay closer attention to energy use and emissions. Major improvements in energy efficiency can often be achieved at little or no cost, even with net savings, through the use of targeted programmes.

Design for Environment has been part of the commitment of Xerox for years. The 'Document Company' has set its environmental goal to be 'waste free products manufactured in waste free factories'. Using 1990 as a base year, Xerox embarked on a five-year effort to create waste-free factories. Operational criteria include 90% minimum reduction in solid waste to landfills, air emissions, hazardous waste, and process wastewater discharges.

Product stewardship, whereby a company considers the upstream and downstream implications of its activities, is a key aspect of Xerox's programme. A number of factors have enabled the company to pursue this goal, among them making environmental considerations a product requirement and developing recycled material specifications for remanufacturing. This shift highlights the contrast between new and old systems. For instance, copy cartridges were disposable when first introduced but, under various pressure s, Xerox has introduced a take back system in which cartridges are taken apart and components recycled. This form of product stewardship illustrates the importance of life cycle thinking, assuming that manufacturers will share responsibility for products from cradle to grave along with suppliers, consumers and others in the life cycle chain.

Sony has responded to environmental expectations and needs in the design of its televisions by introducing, in 1995, 'Green TVs' in its European production lines. With consideration given to disassembly characteristics, weight and cost, the new TV sets we re designed with significant environmental improvements: recyclable materials were promoted, plastics reduced in both type and variety, halogen-free flame retardant materials used and hazardous substances during production eliminated.

Towards Extended Product Responsibility

Extended product responsibility is an emerging concept that uses the life-cycle approach to identify strategic opportunities for cleaner production and eco-efficiency. It also highlights the crucial impact of consumer needs and preferences, government pro curement, and the role played by those on the chain of production and distribution.

In its report released in February 1996, the United States President's Council for

Sustainable Development' identifies extended product responsibility as a means to improve 'the current fragmented approach to waste reduction, resource conservation, and pollution prevention.' This means that manufacturers, suppliers, users, and disposers of products all have a role to play in minimising the negative environmental effects of products and waste streams. Ultimately, the Council believes that sharing responsibility for environmental effects will yield a more efficient use of resources, cleaner products and technologies, improved relations between companies or communities and responsible consumer choices.

Cleaner Production and Eco-Efficiency in Tourism

Tourism is the largest and fastest growing industry in the world, and its rapid growth is fuelling the economy in many developing countries, especially in Asia. The tourism industry's impact on the environment and the potential for change cannot be underestimated in 1995 alone, tourism has produced close to 11% of world GDP and generated an estimated US$ 3.4 trillion in gross output. Because of its 'intangible' nature, the service sector such as tourism has traditionally been behind in embracing the concepts of ecoefficiency and cleaner production.

UNEP, in partnership with the International Hotels Association (IHA) and the International Hotels Environment Ioitiative (IHEI), has published an Environmental Action Pack for Hotels. This publication is intended to serve as a first-step guide for small and medium-sized hotels, which form a large part of the hotel industry, to set up and improve environmental management initiatives.

The Action Pack provides practical information, checklists and examples concerning key issues such as environmental self-audits, priority action areas (energy, solid waste, water, effluents and emissions, choosing contracts and suppliers), the integration of environmental management in daily operations, and progress monitoring.

As a follow-up action of this document, the three organisations have begun to convene regional seminars to provide a showcase for good environmental practices. The first of these meetings, focusing on Asia, was held in Thailand in October 1995. As the resolution of the meeting, hotel business leaders agreed to push forward environmental action through their own hotel companies and to support the efforts of national associations. At the international level, IHA has been entrusted with facilitating the exchange of experiences among national hotel associations. UNEP has pledged to support these efforts by providing technical and educational infomlation through various UN units.

UNEP is now planning to take the Action Pack a step further by assessing the possibility of adapting its contents for the hotel school curriculum, and also by exploring the possibility of a project of identifying existing sources of 'sustainable technologies' for tourism.

Funding and Financing Cleaner Production and Eco-Efficiency

As environmental concerns have become more central to the thinking of industrialists, banks and other financial institutions are starting to realise they can no longer ignore the environmental performance of the businesses they rate or invest in. Bankers and insurers are making attempts at integrating ecoefficiency in their lending decisions as it impacts a company's creditworthiness. Their assumption is that corporations with bad environmental performance pose a high financial risk and have a poor share holder value.

Translating eco-efficiency into numbers within companies is a difficult task. Some indices like waste minimisation and energy intensity are easy to measure but others, such as the expenses for environmental protection, are more critical. Indeed, a company with high expenses does not necessarily have a good environmental performance as its efforts can be focused on end-of-pipe technology.

By integrating environmental criteria in their credit-approval process, banks can avoid lending to customers with unknown or uncovered risks like contaminated sites. This is the first step at looking at the ecological efficiency of a borrower. The second is to know how to calculate the risk premium of the loans. Those who are able to rate the eco-integrated economic risks of a customer will reduce costs by owning a credit-portfolio with higher quality customers, hence with lower bad-debt provisions, and offer very competitive conditions by knowing the customers' risks and managing the 'risk adjusted pricing'.

Swiss Bank Corporation is in the process of integrating environmental criteria into its customer rating system. This will allow it to use 'risk adjusted pricing' by focusing on relevant and reasonable environmental facts, and to assess risks and opportunities in the industrial, financial and management auditing of its borrowers. As a routine, the bank investigates the risk exposure of every project in order to make the risk more transparent to the risk manager. Swiss Bank Corporation has implemented a three-step approach to detect possible risks:

- information provided by customers
- screening carried out by internal bank specialists
- independent risk evaluations by consultants

The European industrial insurer, Gerling, examines very closely the environmental performance of its customers before agreeing to carry their environmental risks- either in third party liability insurance or in fire insurance. The better the environmental safety-performance of its customers, the less chance environmental claims will occur, and this consequently leads to lower insurance premiums. It is a kind of safety-supervision the company offers to its customers by advising them on how to improve environmental safety standards.

Reducing environmental risks enables the achievement of adequate safety and emissions control, especially by ensuring safe disposal of toxic or hazardous wastes-this refers to the traditional end-of-pipe mode. Moreover, it involves cleaner production. Both of these are prerequisites for eco-efficiency as environmental risk declines in inverse proportion to eco-efficiency. In other words, the more eco-efficient the firm, the lower its risk factor and subsequent premium are likely to be.

Production Technology as a Checklist for Environmentalist Business Management

Production and manufacturing technology covers all plant and machinery used for the transformation and processing of raw materials and semi-finished goods and generally plays a decisive role as far as emissions from a factory are concerned. Ecology-oriented innovations help to reduce pollution caused by a factory and have economic advantages in that they reduce the input of materials, water and energy and the disposal costs for waste and sewage. They may also improve product quality and reduce production waste. In the medium term, effective ecological measures can also bring about an improvement in the company's image in the eyes of neighbours, customers and authorities.

Use ecological materials and processes

1. Get rid of particularly hazardous raw materials, processes and parts of the production programme or replace them with others which cause less pollution (see also checklists for materials management and product development). For example:

- instead of the solvent trichlorethylene, which is a notorious health hazard, use trichlorethane or perchlorethylene, which are less polluting and save energy,
- replace cadmium or chrome coatings by zinc,
- in electroplating processes, replace zinc electrolyte containing cyanide by cyanide-free or acidic electrolyte,
- replace conventional paints by low-solvent paints (water-based), high-solid systems or powder coatings,
- use low-emission energy sources, e.g. low-sulphur fuels or light heating oil or gas instead of heavy heating oil.

2. For wet processes reduce the temperature or switch to low-temperature processes, e.g. use cold cleaning solvents and cold bleaching and dying processes

3. Avoid dry processes and the cooling of workpieces. Avoid intermediate storage by switching from non-continuous manufacture to a continuous process.

4. Prepare fuels for optimum combustion, e.g. homogenise heavy heating oil (mechanically breaking down the oil into very fine particles).

5. Use additives which improve the yield from raw materials and/ or reduce emissions, e.g.

- combustion-promoting additives, which act as catalysts to optimise and accelerate combustion processes,
- use steel converters which are oxygen-blown rather than air-blown.

6. Interrupt or change production as little as possible if this means that the plant has to be cleaned or heated.

Use "Clean Technologies"

7. Prevent emissions during delivery, storage, crushing and mixing of dust-producing substances by means of hermetic storage, transport and production facilities.

8. Use technologies which achieve a better input-output ratio.

- Convert combustion equipment, e.g. use fluidised-bed combustion in energy production or back-pressure furnaces in iron production.
- Use modern painting systems, which reduce splashing and boost efficiency from some 50 per cent for normal air-spraying to more than 95 per cent for dipping, rolling or flow coating.
- In wet processes reduce the input of chemicals by using a weaker concentration and/or a smaller bath volume
- The volume of treatment baths can be reduced by avoiding unnecessary voids in treatment vessels and by using a suitable method of getting the liquid to the product (e.g. via jets) or the product through the liquid. Generally speaking, the average chemical concentration depends on the work-piece with the maximum requirements. The average can therefore be reduced by limiting the range of activities, by suitable production planning and/or by a measurement and control technology which adapts the concentration to specific requirements.

9. Use modern measuring and control technology.

- Automatic monitoring and measuring of the concentration of treatment or process chemicals reduces material, personnel and environmental protection costs, ensures consistent product quality and reduces waste and plant downtimes

- Automatic clean water injection for rinsing and washing processes, adapted to the conductivity of the water, not only reduces the amount of clean water needed, but also increases the level of useful substances remaining in the waste water and thus facilitates recycling (cf. 14). With hot processing stages there is also a saving of energy
- In drying processes, automatically controlled production speed and exhaust air humidity, etc. saves energy and improves machinery utilisation and quality
- Optimising combustion processes reduces emissions and improves combustion efficiency in energy production

10. Increase the lifetimes of treatment baths by

- using (ultra-)filtration, flotation, ion exchange, electrodialysis, adsorption, electrolysis or precipitation of impurities
- pre-rinsing workpieces (e.g. with hot water).

11. Reduce leakages and faults causing emissions, by

- regular safety checks,
- using emission-reducing sealing materials and methods,
- more frequent maintenance of equipment, pipes, pipe connections, pumps and shut-off devices,
- information and training for workers.

12. Remove water mechanically from workpieces, e.g. by squeezing.

13. Increase energy efficiency by insulating, sealing or covering equipment with a high working temperature, e.g. reduce the surface of hot treatment baths by means of styrofoam balls (see also checklist "economising on energy and water").

Recover Useful Materials and Heat (See Also "Disposal and Recycling" Checklist)

14. Do not mix effluents containing different substances, as this makes it more difficult to recycle or recover useful substances.

15. Increase the concentration of noxious effluents by sealing the equipment (in the case of air emissions) or cutting down on rinsing and cleaning water (in the case of wet processes), in order to facilitate recycling.

16. Recycle the useful substances contained in effluents for production or convert them into other useful products.

- Re-use effluents in the production process

  In the case of surface treatment, for instance, vaporisation losses in hot treatment baths should be replaced by high-concentration rinsing water. In paper manufacture the water obtained through water removal is used for the preparation of raw materials (to swell the fibres).

- Recover useful substances and use them for specific processing stages, e.g. recovery of

  - o heavy metals, acids, etc. from effluents by means of decentralised ion exchangers, reverse osmosis, ultra-filtration and vaporising plates,
  - o acids and organic solvents from the exhaust air by using adsorbers or absorbers.

- Use slightly soiled water for pre-rinsing
- Re-circulate water several times over

17. Recover heat.

- Use heat exchangers to exploit the exhaust heat in cooling and production water and exhaust gases, and the exhaust heat which accumulates under factory roofs and in certain rooms.
- Circulate hot exhaust air back into the production or combustion process.

Avoid Emissions by Post-Production Environmental Protection Measures

18. Enclose equipment to avoid the release of pollutants.

19. Use more effective environmental protection equipment. In metal surface treatment, the rigorous application of water-saving measures gives scope for replacing the extremely unreliable flow-through plants by safe and efficient charge plants.

20. Use environmental protection equipment which produces little waste. For example, desulphurisation equipment produces reusable substances such as sulphur, sulphur dioxide, sulphuric acid or gypsum instead of sludge, and creates a smaller amount of waste for subsequent dumping.

21. Ensure that pollutant treatment equipment is working properly.

Concentrate on regular maintenance and checks and on monitoring and alarm systems for checking levels in chemical containers, cleaning efficiency and the functioning of measuring erectrodes, injection elements and mixers.

22. Reduce effects of leaks and faults in waste water treatment plant, by

- automatic emergency production shutdown in the event of a fault; emergency catchment tanks, etc.,
- connecting all drains to the waste water treatment system.

23. Recycle or reprocess wastes (see also "disposal and recycling" checklist); e.g. send

- used acid to metal processing plants for ore extraction,
- concentrates containing heavy metals to central processing plants,
- metal or metal-bearing wastes (aluminium, steel, lead, cadmium, chromium, copper, mercury, zinc) to secondary processing plants,
- plastic wastes to the plastics industry,
- waste paper and glass to collection points or users,
- sewage sludge to be used as farm manure' or, where heavy metal concentration is too high, for the recovery of useful substances.

THE BUSINESS CHARTER FOR SUSTAINABLE DEVELOPMENT

Sustainable development involves meeting the needs of the present without compromising the ability of future generations to meet their own needs. Economic growth provides the conditions in which protection of the environment can best be achieved, and environmental protection, in balance with other human goals, is necessary to achieve growth that is sustainable. In turn, versatile, dynamic, responsive and profitable businesses are required as the driving force for sustainable economic development and for providing the managerial, technical and financial resources to contribute to the resolution of environmental challenges. Market economies, characterised by entrepreneurial initiatives, are essential to achieve this.

Business thus shares the view that there should be a common goal, not a conflict, between economic development and environmental protection, both now and for future generations.

Making market forces work in this way to protect and improve the quality of the environment-with the help of standards such as ISO 14000, and judicious use of economic instruments in a harmonious regulatory framework is an on-going challenge that the world faces in entering the 21st century.

This challenge was recognised by the nations of the world at the 1992 United Nations Conference on Environment and Development, which called on the co-operation of business in tackling it. To this end, business leaders have launched initiatives in their individual enterprises as well as through sectoral and cross-sectoral associations.

In order that more businesses join this effort and that their environmental performance continues to improve, the International Chamber of Commerce continues to call upon enterprises and their associations to use the following Principles as a basis for pursuing such improvement and to express publicly their support for them.

Individual programmes to implement these Principles will reflect the wide diversity among enterprises in size and function.

The objective remains that the widest range of enterprises commit themselves to improving their environmental performance in accordance with these Principles, to having in place management practices to effect such improvement, to measuring their progress, and to reporting this progress as appropriate internally and externally.

Note : The term environment as used in this document also refers to environmentally related aspects of health, safety and product stewardship.

Principles

1. Corporate Priority

To recognise environmental management as among the highest corporate priorities and as a key determinant to sustainable development; to establish policies, programmes and practices for conducting operations in an environmentally sound manner.

2. Integrated Management

To integrate these policies, programmes and practices fully into each business as an essential element of management in all its functions.

3. Process of Improvement

To continue to improve corporate policies, programmes and environmental performance, taking into account technical developments, scientific understanding, consumer needs and community expectations, with legal regulations as a starting point; and to apply the same environmental criteria internationally.

4. Employee Education

To educate, train and motivate employees to conduct their activities in an environmentally responsible manner.

5. Prior Assessment

To assess environmental impacts before starting a new activity or project and before decommissioning a facility or leaving a site.

6. Products and Services

To develop and provide products or services that have no undue environmental impact and are safe in their intended use, that are efficient in their consumption of energy and natural resources, and that can be recycled, reused, or disposed of safely.

7. Customer Advice

To advise, and where relevant educate, customers, distributors and the public in the safe use, transportation, storage and disposal of products provided; and to apply similar considerations to the provision of services.

8. Facilities and Operations

To develop, design and operate facilities and conduct activities taking into consideration the efficient use of energy and materials, the sustainable use of renewable resources, the minimisation of adverse environmental impact and waste generation, and the safe and responsible disposal of residual wastes.

9. Research

To conduct or support research on the environmental impacts of raw materials, products, processes, emissions and wastes associated with the enterprise and on the means of minimising such adverse impacts.

10. Precautionary Approach

To modify the manufacture, marketing or use of products or services or the conduct of activities, consistent with scientific and technical understanding, to prevent serious or irreversible environmental degradation.

11. Contractors and Suppliers

To promote the adoption of these principles by contractors acting on behalf of the enterprise, encouraging and, where appropriate, requiring improvements in their practices to make them consistent with those of the enterprise; and to encourage the wider adoption of these principles by suppliers.

12. Emergency Preparedness

To develop and maintain, where significant hazards exist, emergency preparedness plans in conjunction with the emergency services, relevant authorities and the local community, recognising potential transboundary impacts.

13. Transfer of Technology

To contribute to the transfer of environmentally sound technology and management methods throughout the industrial and public sectors.

14. Contributing to the Common Effort

To contribute to the development of public policy and to business, governmental and intergovernmental programmes and educational initiatives that will enhance environmental awareness and protection.

15. Openness to Concerns

To foster openness and dialogue with employees and the public, anticipating and responding to their concerns about the potential hazards and impacts of operations, products, wastes or services, including those of transboundary or global significance.

16. Compliance and Reporting

To measure environmental performance; to conduct regular environmental audits and assessments of compliance with company requirements, legal requirements and these principles; and periodically to provide appropriate information to the Board of Directors, shareholders, employees, the authorities and the public.

Support for the Charter

The ICC undertakes to encourage member companies and others to express their support and implement the Charter and its Principles.

A list of these companies can be obtained from ICC Headquarters. The ICC also publishes regularly a Charter bulletin which provides more specific information on the Charter's Principles and different interpretations possible-an attribute of the Charter that has been widely commended.

The first edition of Business Charter for Sustainable Development was adopted by the ICC Executive Board on 27 November 1990, and first published in April 1991.

It was prepared and revised by the ICC Working Party for Sustainable Development. Chair Peter Scupholme (British Petroleum) Vice-Chair W. Ross Stevens III (Du Pont) The ICC is indebted to numerous companies and business organisations for their input in preparing and revising the Charter. The Business Charter for Sustainable Development provides a basic framework of reference for action by individual corporations and business organisations throughout the world. It has been recognised as a complement to environmental management systems. To this end, the ICC, the United Nations Environment Programme(UNEP) and the International Federation of Consulting Engineers (FIDIC) have developed a kit to help enterprises integrate environmental management systems in the daily management practices, a step consistent with the objectives set out in this Charter.

The Business Charter has been published in over 20 languages, including all the official languages of the United Nations.

It Makes Good Business Sense!

A Green Business...

- Improves employee morale & the health of the workplace
- Holds a marketing edge over the competition
- Strengthens its bottom line through operating efficiencies
- Is recognising an environmental leader

Get recognised for being Green-sign up to become a Green Business today!

The Green Business Programme Recognises Businesses That

- Comply with all environmental regulations
- Take steps to conserve natural resources and prevent pollution, such as using more efficient lighting, purchasing in bulk, watering landscapes efficiently, recycling cardboard, using less toxic products, etc.

Green Businesses Receive Recognition Through

- The Green Business Program
- City and agency newsletters
- Press coverage, promotional events and special recognition
- Window decals, certificates and promotional materials for your business
- Green Business logo to use in your advertising

What Do One has to Do?

- Contact the Green Business Programme *Coordinator* in your county

  Your County Coordinator can send you a packet of materials, including a simple one-page application form, a checklist and a directory of Green Businesses in your county. We'll visit you if you'd like. If you choose to become certified, we will work with you and help you go through the necessary steps to meet the *Green Business Standards*. It's completely voluntary. And, there are no fees involved!

- Pledge to stay Green

  "We believe a successful business is dependent on a healthy environment. We are actively working to show our environmental responsibility to our community by committing to the following objectives:

- *To comply with all applicable regulations and strive to exceed compliance*

- *To conserve energy, water, materials and other resources*
- *To develop and implement practices that prevent pollution and waste*
- *To be an environmentally responsible business within our community*
- *To strive for continuous improvement."*

Staff will work with your business and local agencies to verify that you are complying with any applicable environmental regulations. These could include visits from wastewater, storm water, CUPA (hazardous materials/ waste) and Air, depending on your business type. We pride ourselves on providing technical assistance to ensure your business is in compliance.

- Identify ways to conserve resources & prevent pollution
- Use Green Business checklist to select and record measures to:
- Conserve Water
- Conserve Energy
- Reduce Waste
- Prevention Pollution

You may already be doing enough to qualify. Staff from PG&E, your water company and others will verify what you're doing to meet the Green Business certification criteria.

Householder Initiatives

In many UK communities the strongest shopping trend appears to be the desire to use out-of-town supermarkets and ignore the decline of local producers and smaller shops in towns and villages. The extra traffic this generates and the demise of small traders has led to protests and the government has stated its intention to discourage any further development of large out-of-town supermarkets. In addition to government action there are several ways in which people are finding their own solutions to seemingly intractable problems. Householder initiatives include:

- Cooperatives, housing associations and credit unions;
- Local Exchange Trading Systems (LETS); and
- Global Action Plan for the Earth (GAP).

Cooperatives, Housing Associations and Credit Unions

This is often referred to as the social economy where the primary purpose is to meet a need rather than provide profit to shareholders. Cooperatives and housing associations are organisations which are professionally run by volunteer boards of management. Community enterprise of this kind has been growing steadily since the mid 1970s and has proved invaluable in poorer and more remote areas. Many of the people who benefit were unable to raise even small loans or get support for their initiatives until the advent of community enterprise.

Credit unions make money available to people in local communities at low interest rates to enable them to undertake initiatives which require some capital investment but which high street banks are unwilling to consider. In essence a credit union is a 'save and borrow' cooperative. The members save on a regular basis, then borrow from the common fund when the need arises. The cynical belief that poor people would not save and would fail to repay their debts has not been borne out.

There are now 500 credit unions in the UK and 30 are in Birmingham where the Birmingham Credit Union Development Agency has been particularly successful. The New Economics Foundation has published a paper on community banking which provides a summary of worldwide initiatives. These schemes help people, often in deprived areas, by making money available at low interest rates. The only alternative for many people would be to borrow from loan sharks, who charge exorbitant rates of interest, because conventional banks regard people in deprived areas with few if any assets as unacceptable risks.

Local Exchange Trading Systems (LETS)

LETS were started in Vancouver, Canada by Michael Linton. He was searching for some way to help the unemployed, which included himself. After trying without success to barter his own skills, Linton decided that a new kind of money was needed and this gave birth to LETS. He invented a 'currency' which facilitated local trade but could not be invested, which was community-wide and used for barter trade on a nonprofit basis. Originally the currency was called the 'Green Dollar' but as the LETS idea has spread local communities have invented their own names so in the UK there are now 'Strouds' in Stroud, 'Brights' in Brighton, and 'Olivers' in Bath.

The essence of the scheme is that people who want to trade together make their own arrangements with the help of a starter pack, which in the UK is provided by the UK Letslink office. In each locality where a scheme is initiated someone needs to keep a list of participants and an account of transactions in the local LETS currency. Typically each unit of currency is the equivalent of £1 and people wishing to trade determine the price for their goods and services at their discretion. If they get swamped by demand their price could be too low, or no demand might suggest an unwanted product or service or that the price is too high. The guiding principles for LETS are:

- someone needs to represent the 'agency' and keep a system of accounts;
- people who join the scheme pay a small annual membership fee;
- no money is deposited or issued, all accounts start at zero;
- the agency acts on authority from the account holder in making credit transactions from one account to another;
- there is never any obligation to trade;
- account holders may know the balance and turnover of others;
- no interest is charged or paid on balances;
- administration costs are recovered, using the local currency, from accounts on a cost service basis;
- a catalogue of members of each scheme is drawn up as a trading sheet, showing 'wants' and 'offers', contact names and telephone numbers;
- individuals and businesses can participate; and
- 'payments' for goods and services may be partly in the local LETS currency and partly in sterling.

LETS benefits include the following:

- creates some employment;
- enables unemployed people to recognise that their skills and experience are of
- value to others and thereby regain some self respect;
- helps build community spirit;
- improves the quality of life especially for people with low incomes;
- liberates goods and services within the local community;
- creates a local currency which complements money transactions; and
- reduces the flow of money leaving a community or locality.

It is not easy to find out how successful LETS have been worldwide. There is some indication that they have been particularly successful in the UK and Australia but have not spread rapidly in Canada and the USA. In the UK the Letslink office is swamped by enquiries to start new schemes. The growth of schemes has been striking as is shown by the figures in Table 3.

TABLE 3

Local Exchange Trading Schemes in the UK

Year	*1990*	*1991*	*1992*	*1993*	*1994*	*1995*
Number of schemes	5	10	15	35	200	400

It is estimated that in 1995 there were 30,000 people in the UK taking part in LETS. This figure is not easy to estimate because after a while two people who provide a service to each other on a regular basis may not record every transaction through LETS. For example one family may want a baby sitting service in exchange for working in a neighbour's garden. There may also be some turnover of participation in LETS as individual circumstances change. Some schemes struggle to get started and a few fail. Despite this there is little doubt that LETS show a progressive trend and some local councils are appointing LETS officers to develop schemes. These include Cardiff, Greenwich, Hounslow, Liverpool and Manchester in the UK.

The question of taxation on LETS transactions has been raised and the LEI'S response is that this is a reasonable question. Their answer is that if it is really necessary then it should be paid in the local currency. For example, by cleaning the windows of the council offices or providing some other service to a government department! Any attempt to make tax a formal requirement could result in more transactions reverting to barter deals which are not recorded. Furthermore, the benefits of LETS, as listed above, suggest that these schemes have desirable characteristics for sustainable communities.

Global Action Plan for the Earth (GAP)

GAP started in the USA after Earth Day 1990. GAP is an international charity now being developed in 10 countries in North America and Europe with exploratory discussions taking place in Japan and Korea. GAP'S mission statement is:

> to encourage *individuals to take effective environmental action in their homes, workplace and communities.*

The household programme offers six action packs covering waste, energy, water, transport, shopping and next steps. Each month participants note what they are doing currently, read the ideas in the Action Pack, take action as appropriate, then let the GAP office have details of their savings. A summary of collective achievements is sent to all participants so they know how, with others, they are contributing to an improved environment. News about GAP and the national results are published in a quarterly newsletter.

GAP UK began in 1992 with support from the Department of the Environment, companies, local authorities, voluntary organisations, charities and individuals. GAP UK has recruited more households than the schemes in other countries as the figures in Table 4 show. In the UK individual households join GAP but in other countries the emphasis is on forming 'EcoTeams' so that a group of households work together, using an 'EcoTeam Workbook' rather than Action Packs. In the UK individual households join the scheme because a strong resistance to 'EcoTeams' was encountered.

The best ways to get people involved for lasting results and to assess results is the subject of continuing research. Results to date suggest the six month scheme is sufficient to change habits which tend to continue after the scheme has finished. A significant proportion of those who complete the scheme want to remain involved and encourage others to join. The household scheme in the UK is complemented by other initiatives:

TABLE 4

Global Action Plan—Performance Figures 1996

Schemes were started at different times. The cumulative figures are shown below. Savings are reported by households based on their own measurements, then aggregated. Verification is by sampling.

Country	*Number of Households*	*Garbage savings %*	*Water savings %*	*Home energy savings %*	*Transport fuel savings %*
Belgium	610	N/A	N/A	21	N/A
Denmark	186	49	18	37	N/A
Finland	218	54	22	9	23
Ireland	65	N/A	N/A	N/A	N/A
Netherlands	4907	24	12	12	2
Poland	632	N/A	N/A	N/A	N/A
Sweden	1621	42	18	81	3
Switzerland	741	19	14	71	2
UK	10,043	28	19	9	4
USA	3471	35	41	15	27

Note: N/A means data not available.

- working with the Worldwide Fund for Nature (WWF) through their established groups, using the title 'Action at Home';
- working with several local authorities to help them achieve that part of their waste reduction and recycling targets attributable to households;
- establishing 'Action at Work' based on the same principles as the household programme; and
- starting work on a schools programme to involve children in positive action to improve their environment at school and at home.

The GAP UK programme was tested out by The Consumers' Association who got five very different households in Devon, London (2), Scotland and Gloucestershire, to take part in the full programme but spread over only two months. The findings are reported in *Which?* Magazine. They found that motivated people given detailed information about how to change their lifestyles are able to make some changes while other changes prove difficult if not impossible. It is hard to reduce dependency on the car where public transport is inadequate and cycling on narrow roads is dangerous, but one couple achieved considerablc savings with planned journeys and car sharing.

After doing this survey *Which?* proposed a helpful definition of sustainable development for consumers:

> To achieve a level of consumption that meeets the present and future needs and aspirations of people the world over, from all sectors of society wihtout compromising the sustainability of the environemnt and its life sswupportssystem

GreenBiz.com—A Web Gateway

GreenBiz.com harnesses the Internet through the development and organisation of Web-based information, tools, and resources. It aggregates, filters, and synthesizes the rich lode of available resources available on online and offline and presents them in a highly accessible and user-friendly way. It points businesses to the vast array of organisations and agencies that can help them save money and improve their environmental performance. In just its first year of operation, it has become recognised as the leading Web site for business environmental information. With little formal marketing, the site reaches nearly 30,000 visitors per month. More significantly, visitors spend an average of 13 minutes on the site per visit — an eternity, in Web Time — indicating they find the information accessible and useful and are taking the time to search the site's rich resources.

GreenBiz.com aggregates, catalogs, and disseminates the products and services of many of the leading organisations in the area of business and sustainability-by linking to their web sites, by serving as a distribution center for their reports and tool kits, or by cooperating with them on new products and services. In this role, GreenBiz.com expands the reach and leverage of hundreds of government, professional, and academic organisations.

GreenBiz.com Features

(a) Reference Desk

A significant portion of the site is devoted to compiling resources from a diverse range of source. The site serves as a virtual reference desk. Examples include:

- Government Gateway is a single access point for all federal (and some state and local) programmes related to business and the environment, organised by topic area (pollution prevention, R&D grants, technology transfer, technical assistance, etc.), instead of by government agency. Eventually, the information will be expanded to include a comprehensive and up-to-date directory of state and local government environmental programmes and services.

- Web Guide, featuring descriptions and links to more than 700 handpicked web sites of interest and value to business environmental professionals. The directory will be browsed by subject (á la Yahoo) or searched as a database (seeking, for example, all the web sites dealing with solvent alternatives for printers).
- Awards Directory, with descriptions of and links to more than 100 award and recognition programmes that honor business environmental leaders.

(b) Business Toolbox

There is a vast lode of reports, checklists, handbooks, how-to guides, and other helpful resources on business environmental topics produced by both governmental and nongovernmental organisations. Some of the best such resources remain well-kept secrets. GreenBiz.com plays a vital role by aggregating, filtering, cataloging, and disseminating these tools and resources and making them available from a single source. Currently, GreenBiz.com offers more than 250 such resources, many of which can be downloaded directly from the site, with new ones added weekly.

- *Daily News*

  GreenBiz.com augments the above with a dynamic news and information service, including news and feature stories on business environmental topics culled from both mainstream and niche media, updated daily.
- *GreenBiz JobLink*

  JobLink is a free job- and resumé-posting service for companies searching for qualified job candidates or interns, and individuals seeking job opportunities. It also contains profiles of real-life individuals with environmental positions in mainstream organisations.
- *Future Programmes*
- *"Green Gazelle" Model Companies*—Profiles of successful small and mid-sized businesses that arrived where they are largely as the result of their strong environmental performance. The notion of comparing these companies to a Gazelle stems from the speed and agility they show in achieving their success. Green Gazelles is a programme of GBN partner the Center for Small Business and the Environment.
- *Leadership Practices Database*—a database of leadership examples compiled from the award and voluntary compliance programmes of federal, state,

and local governments, as well as trade associations and the private sector.

- *State and Local Resource Centers*—GBN will develop localised versions of GreenBiz.com specific to cities, counties, states, regions, or bioregions. For example: the Montgomery County Green Business Resource Center, the Ohio Green Business Resource Center, the Region III Green Business Resource Center, the Chesapeake Bay Green Business Resource Center, etc. Each localised center will feature tools, resources, news, and other relevant material specific to the region, but also offer national resources

Technology Partner

GreenBiz.com's technology partner is the Global Environment & Technology Foundation (GETF), a nonprofit organisation that harnesses technology and knowledge to address environmental challenges. Among its many projects, the foundation hosts GNET-the Global Network of Environment and Technology-which provides environmental technology information and encourages community and business interaction online. GETF provides the technical infrastructure for the GreenBiz.com. This allows the site's resources to focus on its core mission of researching, gathering, packaging, and disseminating high-quality information.

Insuranceas a Checklist for Environmentalist Business Management

The aim of company insurance is to insure the company against unquantifiable risks. The most serious risks to a company's financial situation and existence include damage to the environment and liability for such damage. Insurance against pollution and other such damage, however, is still at the experimental stage in all countries. Not enough statistical experience has been obtained yet, mainly because of the constant extension of statutory rules and the rapid development of measuring techniques. Various types of insurance are still being experimented with, aimed at attracting insurers to this market. The following recommendations reflect the position at the end of 1987.

Check What Intrinsic Damage Could Ensue from Environmental Risks

1. Check sickness, accident and life insurance schemes.

   Physical injury, for example, may be caused to the respiratory tract, eyes, skin or internal organs. Such damage may, like a disease, occur slowly and worsen, e.g. asbestosis, pneumoconiosis (covered by sickness insurance)—and frequently result in occupational incapacity and premature death (covered by statutory or mutual insurance schemes and life insurance); however, they may also occur as a result of accidents—as in Bhopal or Seveso—and thus result in permanent or "only" temporary injury (accident insurance, life insurance). Recommendation:

A certain degree of basic protection is already provided on a statutory basis by the advanced social legislation in force in virtually all EEC countries; plants in special-risk sectors (asbestos, mining, chemical industries, etc.) should show social responsibility by supplementary company insurance schemes providing accident and life insurance for their staff.

2. Check fire and machinery insurance policies.

   Material damage, for example, may occur to buildings and installations as a result of acids, gases, vapours or soot following a fire (fire insurance, even if the machines themselves are not destroyed by the fire) and to machines, especially water pumps or filters, as a result of polluted water, contaminated oil or coolants (machinery insurance, technical insurance). Contamination by toxic substances generated by a fire can greatly increase the costs of clearing up.

   In Anglo-american countries, an "all risk policy" will cover these risks. Recommendation:

   Property insurance is optional in most cases, with some exceptions for buildings and, in some areas, for machines. The cost of such policies should be reasonably commensurate with the (subjective) value of the property for the company.

3. Check stoppage insurance policies.

   Site examination resulting from a spillage incident from a hazardous material container or from a fire can result in the firm being closed down by official order and prolong repair time.

   Recommendation:

   Key installations should be insured not just against loss of the equipment itself but above all against the loss-of-production damage arising from stoppages/destruction. On the other hand, for items such as cars, which can be replaced at any time, it is worth contemplating a policy with a retention clause or even no insurance at all. Loss-of-production policies are also optional and are intended to cover the risk of additional demands on the part of official authorities for environmental reasons.

Check Possible Risks to Third Parties Caused by Damage to the Environment

4. Check third-party insurance.

   Third-party insurance protects the businessman from third-party claims for damages, including harm to life and limb and, of course, to third-party property. It may also cover damage to common property such as air, water and soil.

5. Reconsider liability under private law.

   The businessman may become liable under private law. Such liability may or may not be based on negligence and may even take the form of absolute liability. In all EEC countries, however, in order to press claims under private law, there must be a demonstrable causal link between the businessman's behaviour (omission or commission) and the damage which has occurred.

6. Consider the obligation to deal with damage under public law. "Liability" may be linked with an undesirable state of affairs.

   Because of the difficulty in proving either negligence or illegal action, or at least a link between cause and effect, this obligation to eliminate damage under public law is becoming increasingly important. For example, a landowner may be required to decontaminate his land even if it can be proven that he is not responsible for the contamination; the owner is liable for the "condition" of the property. Recommendation:

   Third-party insurance is in most cases optional; insurance is mandatory only in specific professions (doctors), certain sectors (nuclear industry) or on the basis of official regulations for individual cases (operators of waste dumps). In view of the incalculable cost of compensation for damage or its elimination, however, third-party insurance is indispensable. It must cover all conceivable areas of liability for the country in question and be constantly amended to keep pace with the latest legal developments; normal company third-party insurance is not entirely adequate. The need to amend policies is especially relevant in the universally controversial question of when damage starts to occur and, in this case, the inclusion or exclusion of "inherited" pollution.

7. Use special coverage.

   In Germany, most attempts to solve the specific problems of environmental damage and third-party insurance coverage apply the following concepts:

- The risk of liability under the Water Resources Act can be insured against. This act makes the businessman absolutely liable for any changes in the condition of water and soil. This liability can be transferred to an insurer by means of a special policy with different clauses for the different types of liability (plant liability, residual-risk liability, etc.).
- The risk of liability for pollution can be insured against by adding supplementary clauses to normal employers' liability insurance. According

to § 14 of the Federal Law on Atmospheric Pollution and § 906 of the Civil Code (BOB), the businessman is liable for damage caused by emissions from his plant.

Such emissions are understood to include not just dust but also odorous and noise emissions. This liability can be insured against as a supplementary clause to a normal employer's liability insurance policy.

- Radiation damage cannot be insured against. In most insurance sectors such damage is expressly excluded and is dealt with exclusively by the Atomic Energy Act. Other Western countries have developed different liability laws which necessitate different insurance schemes. As there is not yet a uniform codeofliability, utmost attention should be paid to the claims geared to the individual concepts.

Observe Market Trends

8. Find out about new insurance schemes.

   The normal type of employer's liability insurance for an industrial company does not provide adequate coverage for the two main areas of environmental liability: soil and water on the one hand and emissions on the other. It is even more important for specific environment insurance than for general liability that policies should be adapted to the changing scope of liability and keep pace with the latest trends in case law and administrative practice. However, earnings from this branch of insurance have declined so much as a result of a few major cases of environmental damage (USA: Love Channel; Italy: Seveso; India: Bhopal), and above all as a result of the unforeseen number of cases of water pollution from major or minor leaks that obtaining insurance coverage commensurate with the level of risk requires major efforts on the part of the policy holder and his brokers.

9. Make use of insurers' experience.

   With their experience and research facilities, insurers have substantial resources which should be made use of in the interests of ecological risk management.

Environmental Performance Indicators Created for Bay Area Business Best Practices

Bay Area Council Launches Environmental Assessment and Measurement Tool for Local Business

(CSRwire) The Bay Area Council launched a unique, dynamic Environmental Best Practices programme today to assist businesses and employers in enhancing environmental responsibility. The programme allows businesses to confidentially assess their environmental procedures, receive immediate feedback and find out how they compare to others in the region. Participants in the programme complete a Top 40 checklist questionnaire online survey that is then anonymously compiled and displayed in a variety of confidential charts, and points participants to a set of more than 300 Environmental Best Practice resources to help improve their operations and performance. The Bay Area Council will use the composite statistics derived from the surveys to better inform public policy decisions and recommendations and to assist companies in adopting environmental best practices.

"We know that local companies are employing environmental best practices in the Bay Area to improve business productivity, cut waste and reduce costs," said Sunne Wright McPeak, the president and CEO of the Bay Area Council. "What we don't know is what specific areas need work and where Bay Area companies are doing well. This programme will help us, and companies in the region, identify the most pressing environmental issues on which to focus."

There is no current standard for environmental performance indicators, so the programme will help shape and evolve as global consensus on such standards emerges. Business organisations are uniquely positioned to use their resources and the Bay Area culture of innovation to grow their organisations in an environmentally sound and responsible manner. For example, companies in the Bay Area are implementing energy efficiency measures, creating solutions to transportation problems with private enterprise shuttles and using sustainable building materials in their offices.

"Many local business leaders are at the forefront of improving economic performance while enhancing the environment," continued McPeak. "Environmentally-responsible business practices can be a potent source of competitive advantage."

The programme begins with the Top 40 Environmental Best Practices Checklist — an online survey — whose purpose is to provide employers with a user-friendly set of priority actions to become more environmentally responsible. It is organised as a self-assessment tool to guide their implementation of environmental best practices and to let them know confidentially how they compare to other businesses that have completed the checklist.

The survey seeks to help employers discover:

- How will your company's environmental performance stack up in comparison to community peers?
- How will your company answer when consumers or investors ask about your environmental record?
- Does your company have the vision and tools for effective corporate

environmental management that meet the broader challenges of sustainable development?

"Environmental problems pose one of the top five challenges to the region's long-term economic prosperity. Businesses have to play their part, and they are. Indeed, wise business leaders know that environmental quality is a key ingredient for long-term profitability," concluded McPeak.

About the Bay Area Council

The Bay Area Council (www.bayareacouncil.org) is a CEO-led, business-sponsored public policy organisation founded in 1945 to promote regional economic prosperity and quality of life. The Council presents a strong, united voice for more than 275 major Bay Area employers and coalition partners to help shape public policy affecting transportation, housing and land use, environmental quality, economic development, sustainable development, water policy, energy, education and workforce preparation, and telecommunications infrastructure.

The Bay Area Green Business Program

is a cooperative effort that assists businesses and public agencies to come into compliance with all environmental regulations, and take steps to prevent pollution and conserve resources. Certified green businesses and public agencies may display the Green Business logo on their premises and in their advertising. The Green Business Programme markets the logo so that consumers can identify environmentally responsible businesses. The public's growing environmental awareness represents purchasing power that can motivate businesses to become "green." The Bay Area Green Business Programme was developed by the Hazardous Waste Management Facility Allocation Committee of the Association of Bay Area Governments. The Committee championed the Programme as a way to promote pollution prevention and resource conservation. The Programmes businesses with clear consolidated information and technical assistance on:

- complying with environmental regulations (including those pertaining to air quality, wastewater discharge storm water management, chemical storage and handling, and hazardous waste management; and
- going beyond compliance to adopt environmentally-sound practices in four areas energy conservation, solid and hazardous waste reduction, and pollution prevention.

Motivated to be Green

Until recently, most environmental initiatives targeted big businesses. The Bay Area Green Business Programme chose to work with small and medium-sized businessess—auto repair, printing/graphic arts, and wineries. Small operators do not have dedicated environmental specialists on staff and are often not in a position to hire consultants. They lend to avoid seeking help from government and therefore miss opportunities to run environmentally—responsible operations. As the Programme matures, many counties are offering certification to a wider variety of businesses.

The Programme motivates participation by:

- Helping businesses to develop a positive, proactive relationship with environmental agency staff regulatory, pollution prevention, and resource conservation expertise.
- Coordinating outreach so that the businesses receive consistent rather than conflicting information programme materials cover all applicable environmental issues.
- Recognising businesses at a public ceremony where they receive their Green Business certificate and window decal.
- Promoting the Programme to encourage the public to choose Certified green businesses.

Green Seven Years and Counting

The Bay Area Green Business Programme was launched in 1996. The Programme combines the successful compliance oriented Sonoma Green model with the Pollution Prevention and Resource Conservation standards developed by Santa Clara County. Initial outreach was to auto repair shops in Alameda and Napa Counties. Alameda County then extended the programme to printers, hotels and restaurants, and now works with ail businesses Contra Costa began offering the programme to auto repair shops in 1998 extended it to printers in 2000, and now works with businesses. Sonoma County added wineries and printers to their programme in 1999. Santa Clara County programme to all interested businesses in 1999. Marin County joined in 2002 and is open to all businesses San Francisco began certifying businesses early in 2004. Since programme inception, more than 475 businesses including a number of government agencies, have been certified and the pace is accelerating. The Programme is also expanding beyond the Bay Area both within and outside California. Programmes have been established in Arizona and Hawai Santa Cruz, Monterey, Sacramento, and San Diego Counties launched Green Business, Programmes in 2004. Green Business partners are working to achieve the following by the year 2005:

- Recognise 600 Businesses and Public Agencies;
- Offer the programme to more industries;

- Develop and implement a regional marketing campaign to encourage the public to shop "green;"
- Distribute comprehensive environmental information to businesses: and
- Assist in expanding the Programme so that it is available throughout California.

Electrolux

Electrolux is a Swedish company with worldwide operations. They have some impressive achievements as a successful business as well as with their environmental programme and now work with the Natural Step. Leif Johansson, Chief Executive and President of Electrolux, made a powerful statement in their 1994 environmental report:

> *Today the industrialised world a faced with a number of challenges related to the environment and survival For our part at Electrolux it means changes which will have an impact upon the entire company at all levels. In our efforts to satisfy market demands, we assume responsibility to contribute to sustainable development from an environmental perspective... This is a strategic decision given high priority ...I am convinced that we are seeing the birth of a new perspective of the world, where ecology and economics are two sides of the same coin. This means we must strive towards greener solutions for environmental reasons, and also because it's economically profitable and good for us as a Croup.*

Electrolux have a vision which recognises that long-term survival for the individual, for corporations and society means that there must be regard to the limits that nature can accept in the form of resource consumption and pollution. They recognise that growth in consumption of non-renewable raw materials cannot continue indefinitely. They state that their operations and products must be integrated in a cycle, so that customer needs can be satisfied without jeopardising the prospects for future generations. Their environmental policy is summarised in Table 5.

TABLE 5

Electrolux Environmental Policy

Responsibility to fulfil the demands for their products while contributing to sustainable development.
Precaution guides all development and production in order to avoid irrecoverable environmental impact.
The Total Approach from raw materials and production to utilisation and recycling involves life cycle analysis and helps focus attention on important issues.
Preparedness means acquiring new knowledge and readiness to respond in ways which meet future environmental needs.

Priorities are set to address defined goals in cost-effective ways.

Market leader credentials were earned when Electrolux was the first to introduce CFC-free refrigerators and freezers to the European market.

Profitability effective use of resources is a decisive criterion for profitability and a prerequisite for environmental activities.

Electrolux endorse the consensus of scientists achieved by The Natural Step and state in their report:

> *We don't need to guess what nature's limits are. Scientific consensus has [established] the conditions of the ecological system. This frame determines the absolute boundaries for civilisation and every living thing. And it is not negotiable. (ibid p. 8)*

The systems conditions written into the Electrolux Environmental Report are the same as those included in previous chapter and they provide central guiding principles for both business strategy and environmental protection in Electrolux. The Natural Step has devised a diagram to illustrate how an industry adopting a defensive strategy and striving to break the system conditions, could have severe fluctuations in its profitability. By contrast another company, such as Electrolux, which accepts the system conditions could work towards a more consistent approach with more predictable profitability.

In their 1994 report Electrolux describe their approach and quote examples to show how they have made progress in meeting the non-negotiable conditions stipulated by

The Natural Step consensus of scientists. These include:

To Reduce Economic Dependence on Substances from the Earth's Crust

- traditional oil for chain-saws have been replaced by vegetable oils;
- metals such as copper and steel are recycled;
- railways are used for 75 per cent of product transportation within Europe instead of road transport; and
- the Husqvarna lawnmower uses solar cells for power rather than petroleum based fuel.

To Reduce Economic Dependence on Unnatural Substances

- brush-cutters and trimmers are fitted with catalytic converters and use 30-35 per cent less fuel, while emitting less hydrocarbon and nitrous-oxides (supporting systems condition 1);
- CFCs have been phased out from all European white goods and in 1993 a

series of CFC-free refrigerators and freezers were launched; and

- a new generation of ploughs introduces a new system of cultivation which drastically reduces fertiliser usage and waste compared with conventional methods.

To Minimise Exhausting Nature's Biological Diversity

- the Frigidaire Company in Iowa, USA has reduced its water consumption and their water bill by 30 per cent by using a closed loop water cooling system; and
- Electrolux subsidiaries have contributed considerable sums of money in the USA, Canada and Italy to partnership projects with National Parks and the Worldwide Fund for Nature.

Resource Wastage has been Cut to Get More from Less by

- more sensitive electronic controls systems regulate temperature in cocking appliances, use energy more efficiently and utilise leftover heat;
- the amount of material used in packaging is being reduced and higher proportions of packaging are being re-used through cooperation between Electrolux and its suppliers; and
- washing machines are now more efficient in use of both energy and water.

In 1995 Electrolux modified their reporting methods. The Natural Step's system conditions are central to the development of business strategy, but not necessarily the best way to report achievements. In the 1995 report one of the most impressive achievements has been the reduction in the use of substances which have an Ozone Depleting Potential (ODP) and Greenhouse Warming Potential (GWP). Electrolux reduction in use of ODP and GWP substances in Europe are shown in Table 6. A new eco-labelling regulation in Europe requires refrigerators, refrigerator/freezers and freezers to be labelled ranking their energy efficiency in seven categories. Electrolux has about 500 models in the top two classifications and sell some 4.8 million of these products in Europe annually.

TABLE 6

Electrolux: Phasing Out Substances which have Ozone Depleting or Global Warming Potential (European figures).

	1992	*1993*	*1994*	*1995*	*1996*
ODP-thousand tons of CFC11 equivalent	3360	2520	240	0	0
GWP-Carbon dioxide in million kilograms	13584	10766	608	453	390

For the future Electrolux are ensuring that environmental thinking permeat all levels of their organisation. One of the priorities is achieving integration of this thinking across all business functions and with their suppliers. In the coming years they will:

- check performance against stated goals;
- implement a comprehensive change programme which can envisage and identify business opportunities compatible with environmental management;
- strengthen the links between group and business unit strategies; and
- find improved ways of reporting outcomes so that ecological and economic performance is more closely linked.

GHI Best Practice Survey: A Case Study

Building Consumer Demand Through Businesses

The Green Hotel Initiative (GHI) understands that many corporate and organisational purchasers are inclined to use green hotel services, but that they need assistance in accessing the necessary information to include environmental considerations in their purchasing decisions. GHI created the *Best Practice Survey (BPS)* to serve the needs of these institutional purchasers, and to harness their purchasing power to move the hotel industry toward environmentally responsible practices.

Sending the Message

In creating and promoting the *BPS*, the GHI intends to send a wake up call to the hotel industry to reinforce the direct connection between the desire for more environmentally responsible services and the purchasing power of businesses. As an easy-to-use list of criteria that assesses a hotel's environmental performance, the *BPS* allows meeting planners and travel buyers to have access to information on a property's environmental practices during the organisation's siting or contract-setting process. Decision-makers can then choose hotels that meet all of their business needs, including their own organisation's environmental preferences.

Real Impacts

Through this process, companies take action and directly influence the way hotels do business. Organisations that have taken a leadership position and adopted the *BPS* into their procurement rocess include General Motors Corporation, American Airlines, Bethlehem Steel, and Northeast Utilities, and several other *BPS* Adopters. The real impact of the *BPS* can be found in the combined purchasing power represented by companies that are now directly communicating their desire for environmentally responsible hospitality products. Companies and organisations can use the *BPS* as the foundation of a partnership with hotel properties that facilitates environmental improvements at the property, ultimately meeting the organisation's business needs.

CERES has set a leading example of this relationship using the CERES Annual Conference to engage the host hotel into a mutually beneficial agreement that lead to a more environmentally sound property. Hoteliers can also use the *BPS* to benchmark the current environmental impacts of their properties.

Take Leadership

Become involved in the GHI's efforts to aggregate demand for green hotel services. Learn more and commit to using the BPS-the more users, the greater its effectiveness.

Green Events

Green Events is a Green Star programme designed to assist special events in reducing waste and recycling. The program's most visible aspect is the recycling programme but it also encourages waste prevention activities, which help to enhance the efficiency of the recycling programme, as well as reduce the overall waste stream.

Do you have a special event coming up? Do you wish you could recycle at the event but you just don't know where to start? Let Green Star help you make your event a certified Green Event. Green Star will provide all the necessary bins, signs, and materials needed to make your event a Green Event.

As a certified Green Event, you will demonstrate to the community that you care about the environment and our great city. At events, you often end up throwing away a lot more than you would like and wasting time and money.

Green Events makes recycling of a wide range of materials possible, including aluminum cans, glass bottles, cardboard, and paper. We've even arranged with the Anchorage Recycling Center to recycle plastic water and soda bottles.

Best of all, Green Events is free. Green Star loans bins to you for the length of your event. Each bin is marked with attractive recycling labels that let others know you have made that extra effort to recycle at your event.

We even have a list of potential volunteers to help you with your recycling programme. Volunteers can place bins, monitor recycling activities, educate the public about recycling, move materials to storage areas, and transport materials to the recycling center. We also provide information about companies that can transport recyclables if your staff or volunteers are not able.

- Benefits of Green Events
- Past Green Events
- Outreach Materials

- Volunteer Opportunities
- Calendar of Green Events
- Sponsorship Opportunities
- Contact Green Events

SPONSORS: This programme is only possible through the generous financial support of the BP Exploration (Alaska), Inc.

Additional Green Star Programmes

- Green Events
- Green Star Schools
- Site Assessments
- Electronics Recycling

Green Star Makes House Calls!

Green Star's Site Assessment Programme can help you diagnose your workplace's ailments.

Learn where you can make improvements in your business practices. Green Star has recently expanded its assistance capabilities to provide on-site assessments for enrolled businesses and organisations. Green Star staff and consultants will come to your facility and work directly with you to assess your waste prevention progress. The team will:

- Walk through your facility and identify current waste prevention activities. This helps to bring you closer to earning your Green Star Award. Green Star uses a helpful checklist during each site assessment.
- Ask questions to learn more about your processes and activities.
- Make suggestions about what activities could be undertaken to improve waste prevention
- Identify resources that might be helpful in implementing suggestions.
- Assist you in determining the costs and benefits of implementing certain suggestions.
- Provide a written report identifying current activities, suggestions, and resources.

  Use this worksheet to prioritize opportunities suggested in Green Star's assessment report.
- Discuss benefits of Green Star Awardee status so that you can provide information to peers or supervisors.

- Answer any additional questions you might have about the suggestions made.

The first step is to fill out a Pre-assessment survey. Once this is complete, we can arrange a time to conduct the assessment. An assessment can take from one hour to an afternoon, depending on the size of your facility and the number of people you would like to involve in the process. Once the site assessment is complete and Green Star has provided a follow-up report, we request that your organisation submit its Green Star or Air Quality Award application within 60 days.

Examples of suggestions made to Anchorage businesses during site visits include:

- Upgrade fluorescent lighting to include more efficient T-8 lamps. If you replace T-12 lamps with T-8 lamps in a 50,000 sq. ft. facility, you can save $54,000 over the life of the lamps.
- Enhance recycling programmes by adding materials, such as mixed paper.
- Cover solvent sinks/parts washers when not in use to reduce evaporation of solvents. This reduces health problems, air quality impacts, and the cost of purchasing solvents.
- Retrofit EXIT signs with LED lighting. The electricity cost to operate one sign for one year is approximately $2.50 for an LED sign, as compared to an incandescent sign, which costs approximately $35 to operate for a year.
- Purchase supplies with recycled-content, such as copy paper, file folders, trash bags, and more.
- Hold periodic meetings to inform employees about recycling and waste prevention activities
- Develop a feedback mechanism so that employees can share ideas and assess progress.

The Green Star Award

The Green Star Award is the original award on which the Green Star programme was based. It is given to businesses, organisations, schools, and agencies that demonstrate a strong environmental and business ethic by implementing at 12 of the 18 Green Star Standards. These Standards encourage waste reduction within several media, as well as education and outreach. Examples of Standard topics include recycling, improved purchasing, outreach to other organisations education of employees, litter control, toxics reduction, water conservation, and utility monitoring.

Once a business or organisation achieves the Green Star Award, it can display

Green Star Award logo and decals in the workplace, include on its letterhead literature, and publicize its achievements using the Green Star name. Green Star's step-by-step guide walks you through the process involved in earning a Green Star Award.

Air Quality Award

The Air Quality Award is Green Star's newest award, unveiled in 1999. The Award focuses on outdoor air quality improvement activities. It is given to businesses, organisations, schools, and agencies that demonstrate a strong environmental and business ethic by implementing at least 12 of the 18 Air Quality Standards. The Standards encourage the reduction of air emissions through technical improvements, behavior changes, and outreach and education activities. Examples of Standards topics include improving equipment and vehicle maintenance, encouraging the use of alternative forms of transportation, and making vehicle plug-ins available.

Once a business or organisation achieves the Air Quality Award, it can display Air Quality Award decal in the workplace, include the Air Quality logo on its letterhead or literature, and publicize its achievements using the Green Star name Green Star's step-by-step guide walks you through the process involved in earning an Air Quality Award.

IfPeople Admitted to Progressive Business Leaders Network

ifPeople, a leader in providing integrated services to responsible enterprises, has been accepted into the Co-op America Business Network (CABN) in recognition of the company's outstanding business practices and contributions.

As a member of CABN, ifPeople joins a selected group of "green businesses" and progressive enterprises that are contributing to a just, more sustainable world. ifPeople combines a high standard for responsible business practices and policies with quality services to provide:

- Technology Tools, Internet consulting, software development
- Training in technology, sustainability, and leadership development
- Process management consulting for collaboration, business sustainability, and network management.

As a member of CABN, ifPeople will be listed in Co-op America's Green Pages Online, America's premier directory of qualified green companies with demonstrated commitments to social and environmental responsibility. All companies in the Green Pages Online have been screened and approved by Co-op America. See ifPeople's listing at:

http://www.greenpages.org/detail.cfm?List_Key=18321

ifPeople is proud to join the leading socially responsible businesses as part of the CABN. "We are passionate about our work," says Tirza Hollenhorst, ifPeople co-founder and President, "and the CABN is an opportunity to continue to improve our practices and the value of our services. We are dedicated to supporting values-driven organisations and businesses, such as those in the CABN."

ifPeople's achievements in promoting and practicing responsibility to its employees and community, improving environmental impacts, and contributing to the success of its clients were reviewed in the application for the CABN. Among ifPeople's achievements are:

- Carbon neutrality: all of the company's carbon generation is offset by planting native trees in communities of operation.
- ResponsibleIT: ifPeople's innovative standard for social, environmental and client responsibility in IT services.
- FairSource: a programme that extends responsible practices in the supply chain by applying the principles of Fair Trade to trade in services.
- Sustainability education: by providing volunteer education and training in universities and classrooms in Argentina and the USA, ifPeople is helping tomorrow's leaders embrace the principles of sustainability in their work.
- Leadership: People's team continues to receive recognition for their talents and dedication through leadership in community organisations and conference speaking internationally.

Rank Xerox

For many years Rank Xerox (RX) leased all their copiers to their clients. This created a disproportionately large asset on their balance sheet relative to their revenue. It became important to consider how to make more use of these assets. They explored ways in which to re-use and recycle copiers and their components and in due course started to manufacture them to make re-use and recycling easier. Today they sell copiers as well as leasing them and have broadened the base of their business to become 'The Document Company'. Paul Allaire, President and Chief Executive of Xerox Corporation says: 'Our environment policy applies to every aspect of our business operations—*including most emphatically the way we design our products.'* [author's emphasis).

Their Environment Recent States

Protecting Planet Earth begins with a commitment which extends from products and services to where our customers work and live. Environmental *health and* safety concerns take priority over economic considerations.

TABLE 7

Rank Xerox Product Stewardship Notes

Design for the Environment embraces the whole life cycle of products from the design and procurement of parts to end-of-life reprocessing and ensures that an increasing number of equipment components are re-used by being interchangeable and compatible across the different ranges of products.

The Procurement Vision is to work with suppliers to achieve full reprocessing capability of equipment parts and components.

Manufacturing Operations incorporate programmes for energy efficiency and reducing waste to a minimum. Where they have not been eliminated the use of harmful materials is being minimised.

The Integrated Supply Chain (ISC) provides a reliable return process closing the loop for the whole product life cycle. The 'Just in Time' approach reduces the need for storage and avoids waste. End-of-life equipment is returned to one of the RX Asset Recovery Operations (ARO) for further reprocessing. Machine assembly and configuration takes place at the European Logistics Centre to enable 'plug and play' on customer premises.

Customer Operations provide a key focus for RX product stewardship and customer requirements are monitored continually to ensure that operational, environmental and ergonomic needs are responded to. A wide choice of environmental papers are available and take-back schemes recover end-of-life equipment, consumables and used packaging.

Remanufacturing Operations and newly manufactured machines use the same people, assembly lines and testing equipment and carry the same performance guarantees.

Recycling backs up the re-use of parts and components so that as much residue as possible is recycled.

Waste Minimisation and Energy Savings are achieved through the '3-R policy'-Reduce, Re-use, Recycle. RX strives to minimise waste to landfill which helps prevent pollution and resource conservation and is highly cost-effective.

RX goals include product design, minimisation of waste, use of recycled materials and returning components to productive use after their initial lifecycle. The basic principles of Rank Xerox environment product stewardship and Table 7 provides explanatory notes to support the diagram.

As a result of applying these policies Rank Xerox reduced its general waste (mixed packaging from suppliers accounts for 30 per cent) from 2300 tonnes at the end of 1992 to about 800 tonnes a year later and since then the figure has fluctuated between 800 and 1050 tonnes. The analysis of the electronics manufacturing waste stream shows several items are re-used (for example electrical components and recycled assemblies) or recycled (for example cardboard, polystyrene, paper, plastic cups and aluminium cans) but some items are sent to landfill and the quantities have not yet been measured (for example plastic reels, canteen waste, plastic trays and fibre glass dust). The percentage of the waste stream that is re-used..'

Other information and charts in the full report confirm several additional achievements:

- re-use and recycling has made the Product Life Cycle a closed loop;
- components reprocessed through Asset Recovery Operations were fewer than 100,000 in 1988 and had risen to 3,500,000 by the end of 1995;
- the quantity of machines manufactured through Asset Recovery Operations rose from 14,000 in 1989 to 60,000 in 1995;
- waste sent to landfill in 1991 weighed over 2500 tonnes per quarter but by 1995 this had reduced to just over 1000 tonnes per quarter;
- recycled materials represented 55.9 per cent in 1993 but this had risen to 87.6 per cent by the first quarter of 1996;
- environment is fully incorporated into the RX communications strategy; and
- RX have won several awards for their environmental work.

Pierre van Coppernolle, Director, Rank Xerox 2000 Strategy and Environment states the following in his concluding remarks at the end of their Environmental Performance Report, November 1995.

TABLE 8

Rank Xerox Limited

Figures in £ million	*1991*	*1992*	*1993*	*1994*	*1995*
Turnover	2766	2923	3161	2209	4139
Profit after tax	117	81	110	100	332
Fixed assets	726	914	1020	1028	1090
Total capital employed	1371	1567	1640	1730	2039

Our management teams are totally committed to the concept of sustainable development and we believe that Rank Xerox is amongst those leading companies which take the trouble to investigate continuously their environmental performance and then start to make improvements through pragmatic action... Our goal for the Waste Free Office campaign is to develop a showcase to customers by demonstrating that the responsible use of equipment and proper management of waste is not just the domain of The Document Company but an initiative that can be adopted by all companies that use The Document. (pp 38-9)

Rank Xerox Limited business results are given in Table 8.

Importance of Green Business

Everybody is green today. According to the advertisements, you can help the environment by driving your car, stuffing away pizza's, flying to destinations at the other side of the world and buying more of the same products. In the meantime, the world's he alth is guaranteed by a growing number of initiatives taken by private enterprises themselves, so-called self regulation. These initiatives have resulted in an ever increasing jungle of acronyms, which all guarantee a true commitment to a sustainable deve lopment. Green principles, guidelines, audit schemes, labels or cooperation networks vary greatly in scope, language and representation but they all aim to establish themselves as partners in the road to a sustainable development.

Some of these networks, such as the European Chemical Industry Council (CEFIC), were established immediately after the 1972 UN Conference on the Human Environment in Stockholm, but most of them are more recent. Especially during the preparations for the E arth Summit in Rio (1992), fear of tough environmental regulations boosted the number of environmental statements made by individual corporations, business sector or other networks. Stefan Schmidheiny's Business Council for Sustainable Development which r ecently merged with the World Industry Council for the Environment, became the most famous. A few years after the UNCED, the so-called "green business networks" (GBN) are still highly active. It could well be argued that the initiative in the discussion o n which-environmental-obligations should be imposed on the corporate sector, lies with business itself.

NGOs have found two ways to tackle this issue. A major stream within NGOs has taken the position that green policies or guidelines are, at best, marginal. These initiatives are generally too vague to be monitored and they cannot be enforced due to their n on-binding character. Environmental guidelines have been criticized for being nothing more than window dressing. NGOs also argued that GBN statements are a cunning effort to convince governments to refrain from command and control measures. Only measures of this last type can keep business activities within the limits of a sustainable development. Moreover, "green principles" do not question the levels of investment and activities as such. In the words of an employee of the Billiton owned Trombetas mine in the Brazilian part of the Amazon, during a visit from the Netherlands Committee for IUCN: "We are willing to discuss every part of the operation process, but we are not going to discuss the existence of the operation itself." This attitude is typical for the corporate sector. Their approach does not recognise the fact that the current consumption level is the main environmental problem. We cannot solve the excessive use of natural resources by creating more efficient production methods.

On the other hand, a number of NGOs has in different ways worked with corporate guidelines or agreements to improve, on a voluntary basis, the environmental performance of industry. In the Netherlands, the Panda-symbol of the World Wide Fund for Nature ha s appeared on boxes with washing powder. Some environmental organisations have joined so-called covenants between governments and representatives of the production sector. The joint project in the United States between the quite radical Environmental Defense Fund and McDonalds also should be mentioned. Basically, this approach is guided by the 'best practices' philosophy. If every company would adhere to the best available environmental practice, environmental problems would be reduced substantially. Furt hermore, promoting best practices stimulates the awareness of environmental issues at the workplace. An ideal way to tackle this new face of the corporate sector has not yet been found by NGOs. One important factor is that NGOs are usually less knowledgeable about the environmental performance of the company than the employees of the company themselves. Most larger companies have hired environmental specialists with a level of knowledge on the effects of the firm's activities on soil, water and land which is hard to match by NGOs. For a balanced and democratic society, an independent party with a high level of knowledge of the impact of industry would be indispensable.

One of the new challenges for NGOs is to claim their position as stakeholders in the societal debate about the regulation of business. Regrettably, this is not a dialogue between equals. Genuine equity demands expert knowledge and financial resources for all parties-exactly what the NGOs often lack. Take for instance the NGO participation in decision making procedures by the International Standardisation Organisation (ISO) which amongst other tasks defines quality and environmental management standards for the environmental performance of the companies. In principle the NGOs can claim access to the meetings of the technical committees where the most crucial decisions are taken. In reality, the transnational corporations have resources to send their expe rts to these meetings-NGOs do not. Following the agenda as it is currently set by industry itself, therefore, puts NGOs almost inevitably in a disadvantaged position.

So, how should NGOs approach the "green business networks"? Of course, diversity in NGO strategies should be maintained but there is one approach which could have a considerable added value. First of all, we need to bring some clarity in the various green business networks, guidelines or principles which currently exist. They have to be reviewed, not only in order to identify differences, similarities, but also gaps. Rather than to ignore these eco-initiatives categorically, or to support them wholeheartedly, it is better to define exactly what to expect from the private initiatives. Apart from environmental principles developed by corporations, the role of similar initiatives from international organisations such as the UN or the OECD must be defined. So me issues are not included in these guidelines which obviously have to be, such as a commitment to continuous improvement and measurement of progress. On the other hand, some principles which are included are nevertheless questionable. It is, for instance , inappropriate if the corporate sector establishes itself as a policy-making actor in order to guarantee a sustainable development.

Careful reading of the statements and principles which are the basis of green business networks does not only give us opportunity to learn something about business initiatives, it also serves as useful guidance in a better understanding of the translation of the principles of sustainable development to concrete activities on all levels of society.

The following guidelines are being reviewed:

- The ICC Business Charter for Sustainable Development
- The Business Council on Sustainable Development (BCSD)
- The OECD Guidelines for Multinational Enterprises
- ECCR et al, Principles for Global Corporate Responsibility: Bench Marks for Measuring Business Performance
- The Coalition for Environmentally Responsible Economies (CERES)
- Keidanren global environmental charter
- The Declaration of cleaner production in Poland
- The International Council on Metals and the Environment (ICME)
- The IISI Environmental Principles
- The Forest Stewardship Council (FSC)
- European Chemical Industry Council (CEFIC)

The Best Practice Basic Foundation is designed to provide broad, base-level coverage of best practices. The section begins by defining the term *best practice*. The discussion differentiates between best practices and benchmarking. The composition of a best practice is also discussed, as well as best practice enablers. This portion of the knowdule closes with some benefits of implementing a best practice.

Environmental Awakening of Business

Thirty years ago this book would have been regarded by most business leaders, at best, as a contribution to science fiction or, at worst, as a futile demonstration of an unnecessary concern with irrelevant issues and arguments that would hinder economic efficiency. Today, many of the views that were advanced during the 1960s and 1970s regarding the possible finite nature of the capability of the environment to absorb the damage imposed by pollution, excessive resource consumption and the other consequences of industrialisation (see, for example, Goldsmith *et al.*, 1972; Meadows *et al.*, 1972) are now accepted as proven facts.

The scientific demonstration of the existence of very real limits to growth has been accompanied by a marked change in the attitudes of governments and businesses towards the environment. This change in attitude is typified by the changes that have occurred in the boardrooms of many major companies. In the most advanced companies environmental directors now share a boardroom table that was once the exclusive domain of production engineers, marketing directors, personnel managers and accountants. In these companies the environmental dimension is no longer viewed as an expensive luxury; rather, it is seen as a vital component in responsible corporate behaviour. This implies a transformation in attitudes, away from the old style approach of how little can we get away with in order to comply with minimalist environmental regulations, and towards a view that places environment at the centre of corporate planning and business activity.

This awakening of business to the needs and values of the environment has taken place in a number of stages. During the 1960s the doomwatchers challenged business with the prospect of environmental catastrophe, but they failed to achieve anything but a marginal shift in attitudes. The first significant moves towards the adoption by business of a more responsible attitude towards the environment occurred in the early 1970s. A first wave of environmental pioneers, chiefly major enterprises in the USA and West Germany, began increasingly to identify and adopt operational practices which placed greater value upon the environment. By the late 1970s some of the largest multinational companies were applying methods of environmental impact assessment and environmental audit (Royston, 1979), whilst small and medium enterprises in Germany were reporting their positive achievements with regard to a wide range of environmental concerns. It is difficult to judge whether state regulation (or the threat of it) or changing consumer attitudes and preferences have provided the major stimulus for this change in attitudes. Some business leaders claim that commonsense business economics and clear moral dictates mattered more. Whatever the precise reason, by the late 1970s the greening of business was well under way.

During the late 1970s and throughout the 1980s the response of businesses was, in part, coordinated and further stimulated by the actions of various business interest groups. The International Chamber of Commerce, working with the United Nations Environment Programme, organised the first World Industry Conference on Environmental Management in 1984. Such actions helped to ensure that the messages of early best practice elsewhere were transmitted to the UK and to other less environmentally aware nations. By the mid-1980s the Confederation of British Industry and other organisations had accepted the logic of moving towards accepting a greater sense of responsibility for the environment and were actively advocating the benefits of adopting enhanced environment attitudes to their membership. Other initiatives followed, including the establishment of Business in the Community and the gradual evolution of a network of local environmental business clubs and forums (Marshall and Roberts, 1992).

A second strand in the development of enhanced business attitudes towards the environment can be identified in the shift which occurred within the environmental movement during the late 1970s. For most of the preceding period the environmental movement had maintained a strong anti-business stance but, during the 1970s in the USA and elsewhere, environmental campaigners realised the power which they possessed. This realisation led to a series of campaigns to influence business behaviour and, in turn, to the forging of alliances among business, government' and the environmental movement. The argument behind this move was that business could and would adapt to more environmentally acceptable forms if consumer and campaigner power was exerted in a targeted manner. Following the success of this movement in North America, green consumerism emerged in Europe during the 1980s, becoming a powerful force behind the changed images, if not products, developed and marketed by companies in order to gain a share of this sizeable and growing market. Although some of the initial enthusiasm and support for green processes and products diminished, especially under recession conditions, a number of companies changed their production methods and product profiles in response to changed consumer attitudes and the increased power and influence of the environmental movement.

From this analysis of the rise of enhanced environmental attitudes in business it is possible to identify a number of important trends. These trends reflect the change in priority and emphasis that is placed by some businesses upon a wide range of environmental matters. Attitudinal and other changes can be seen to have occurred in relation to a number of important issues:

- There has been a shift in the attitude of some companies away from a grudging minimum level of compliance with a prevailing regime of environmental regulation and towards a more proactive role in setting and maintaining new standards of behaviour.
- Some businesses have gone further and have moved beyond the requirements for environmental performance as specified in current legislation and towards a willing participation in both enhancing their own level of performance and stimulating a higher degree of environmental responsibility in other companies.
- The arena within which the above changes have occurred has widened with companies becoming more concerned about the overall performance of their business, rather than restricting environmental concerns to, for example, the minimisation of any pollution associated with production processes.

- The above changes have been linked to a shift in attitudes, from a concern solely with the environmental problems and potentials of an individual company, to a more general concern with the role of the company in the local environment within which it is located.
- There has also been a move from a position whereby the operation of a company, especially with regard to environmental matters, is cloaked with secrecy, to a greater degree of openness and willingness to disclose information.

Rating of Industry Guidelines Regarding Environmental Management

All the industry guidelines address elements of the 9 global corporate environmental management recommendations of *Agenda 21.* Some are more comprehensive that others. In general, there is a stronger commitment in the principles to a vision than to measurable targets or a system to monitor change. They tend to be especially active in the need to establish a vision and to incorporate a concept of sustainable development; in commitments to responsible and ethical management of processes; in the adoption of global standards; and in the establishment of corporate environmental policies and environmental management systems. The recommendation in item 1, that TNCs voluntarily introduce policies and commitments to adopt equivalent or not less stringent standards of operation as in the country of origin is accepted by several associations.

Overall, the ratings show good recognition of global corporate environmental management. Among the 9 industry codes in the rating chart, there is most activity around the broad areas of environmental management, i.e. items 2-5. The requirement that members operate internationally at equivalent standards as they operate domestically (item 1) does have a good number of adherents-full agreement or distinguished contributions from the Banks, the CCPA, the ICC, the Keidanren and the Keizai Doyukai. There is scant interest in the items on cooperation and sharing of information and technology (items 6-8); and there are very few commitments to the requirement for annual environmental reporting (item 9).

The CERES Principles exceed the guidelines of *Agenda 21* on items 2-5; the Keidanren and Keizai Doyukai exceed in three out of four and are in full agreement on the fourth; the ICC and the International Banks are in full agreement or exceed the requirements of the four points; the ICME is fully committed to three of the four and, considering its strong commitment on item 3, is excluded from item 2 on semantic grounds.

On items 6-8, it is conspicuous that the CERES Principles and the WTTC code show no commitments. It is particularly interesting that the CERES Principles, otherwise so ahead of *Agenda 21* in the previous items in this chapter, are silent on these subjects. In sharp contrast are the strong statements coming from the Japanese associations on the same items. Apart from a partial commitment from the Keidanren to 'strengthen partnerships to implement the principles and criteria for sustainable development,' (item 6), both associations meet or exceed *Agenda 21* commitments on all three points. This reflects their strong commitment to technology transfer, both of 'hardware' elements such as technology and of 'software' elements such as skills training and human resource development.

The Response: Environmentalist Business Management

The European economy can only accomplish its new task of making a major contribution to the reconstruction of the ravaged environment if individual companies are prepared to go along with the environmentalist concept.

1. Reasons for Environmentalist Business Management

There are six major reasons why any businessman or business manager with a sense of responsibility (and with his own interests at heart) should apply the concept of environmentalist business management in his firm:

- Without environmentalist-minded companies there can be no environmentalist--minded economy and without an environmentalist--minded economy the human race cannot expect to survive for a life that is worth living.
- Without environmentalist-minded companies there can be no public consensus with the business community--and without a public consensus with the business community there can be no free market economy.
- Without environmentalist business management there will be a loss of fast-growth market openings and a risk of the company being held liable for environmental damage for enormous sums of money, thus jeopardising the future of the company and of all the jobs dependent on it.
- Without environmentalist business management, boards of directors, management executives, heads of department and other members of staff are more likely to risk liability for environmental damage, with the consequent jeopardisation of jobs and careers.
- Without environmentalist business management, substantial potential for reducing costs will go by the board.
- Without environmentalist business management, businessmen will be in conflict with their own consciences and without self--respect, there can be no real sense of identification with one's job.

2. The Philosophy of Environmentalist Business Management

The concept of environmentalist business management canenrich and strengthen the corporate philosophy, and help to attain the classic corporate aims (provided they are not in conflict with the environment).

The essential principles on which the long-term success of a properly and responsibly run company rests are quality, creativity, humanity, profitability, continuity and loyalty. All six are easier to comply with if the management takes an environmentalist view of things:

- *Quality and the environment*

  A product can only be said to be high-quality if it is produced in a way which is not inimical to the environment and can be used and disposed of without causing damage to the environment.

- *Creativity and the environment*

  The creativity of the company's workforce can be encouraged by working conditions which take account of human biological needs, e.g. low noise levels at the workplace, healthy air temperature and humidity levels, ergonomically designed office furniture and vitamin-rich canteen food.

- *Humanity and the environment*

  The general working atmosphere can be made more "humane" by gearing the corporate aims, strategies and activities not only to the economic criteria, but also to a sense of responsibility vis-a-vis all forms of life.

- *Profitability and the environment*

  Company profitability can be improved by adopting cost-reducing environmental protection measures covering such things as raw materials and energy--and water--saving programmes, and by exploiting market openings for environmentally sound products.

- *Continuity and the environment*

  In the interests of continuity, it is becoming more and more important to make sure that the company is not adversely affected by liability risks under the increasingly stringent environmental legislation and by the market risks resulting from the reduced level of demand for environmentally damaging products.

- *Loyalty and the environment*

  In the final analysis, the managers and staff of a company can only be loyal to the laws and aims of the state if the sentiment is a genuinely felt one, and that in turn is only possible as long as the country has not sacrificed its intrinsic value to the forces of environmental destruction.

3. The Starting Point for Environmentalist Business Management

If a businessman has taken the plunge and decided to take environmentalism seriously, there is no reason why the company should hinder him in any way. Indeed, the company's connections and potential can act as a genuinc tool for the achievement of environmentalist aims.

Without abandoning the aim of making a reasonable profit, a company can still act as a spearhead in the drive for environmental protection. By committing itself as an "environmentalist organisation", it can complement the work of other like-minded organisations; after all, it can boast unusually favourable starting conditions:

- The company has access to particularly potent—often international—information channels.
- It can exert a direct influence on chambers of commerce and industry and professional organisations.
- A company which has espoused the environmentalist approach cannot be branded as an other-wordly set-up if it is patently successful.
- The company, as a non-party organisation, is not subject to the factional losses of political conflict.
- The effectiveness of a management team trained and experienced in setting itself targets and seeing that they are achieved can be extended to the environmental context.
- The workforce is in effect a "representative in-house cross-section of the population", which is ideally suited to environmentalist pilot projects (e.g. environmental counselling for households).
- There is such a thing as a "company bonus", meaning that an otherwise sceptical member of the staff is likely to be more open to environmentalist matters if the initiative comes from the company.
- Staff creativity can be mobilised for the environmentalist cause (e.g. through the staff suggestions scheme).
- Planning practical pilot projects (e.g. an environmental information, suggestion and invention exchange) can be facilitated by an in-house ecology/economy dialogue.
- An industrial company often has access to a wide range of specialists who can bring their knowledge to bear on environmentalist pilot projects.

Overview Checklist for Environmentalist Business Management

This checklist answers the question as to which areas of management are covered by the Integrated System of Environmentalist Business Management, and introduces a step-by-step approach to the checklists drawn up for individual sectors.

The numbering sequence is not binding in terms of the practical introduction of the system, nor does it set out the various points in order of importance. Likewise, no priority judgement is involved in the various categories to which the points are allocated (e.g. person-related or technology-related points). The idea is simply to make the system easier to follow.

The Integrated System can be introduced into a company regardless of its organisational set-up. For instance, it is immaterial whether research and development and materials management are separate departments or are subordinated to a technical department; whether waste disposal comes under technical services or materials management; whether the company's transport section is a sub-unit of the technical, materials management or administration department; and whether foreign subsidiaries are managed under the individual departments of the company or as a separate department for foreign subsidiaries.

1. If necessary, acquaint the management team with environment-oriented perception, thought and action processes. (See "management motivation" checklist.)
2. Make protection of the natural environment one of the corporate aims and strategies. (See "corporate aims and strategies" checklist.)
3. Exploit the opportunities and eliminate the risks from environment-induced market trends. (See "marketing" checklist.)
4. Create in-house environmental protection institutions and appoint appropriate members of the staff to manage them. (See "in-house environmental protection arrangements" checklist.)
5. Foster a balanced approach to environment and quality-consciousness by staff motivation and training. (See "staff motivation and training" checklist.)
6. Create humane working conditions in factories, offices and at other workplaces. (See "working conditions" checklist.)
7. Provide good (including organically produced) food in company canteens. (See "canteen food" checklist.)
8. Promote environment-conscious behaviour and health of workers by environmental counselling in their private households. (See "environmental counselling in employees' households".)
9. Achieve a spectacular reduction in costs and help the environment by energy- and water-saving. (See "economising on energy and water" checklist.)

10. Develop ecological products at the behest of, or to mobilise, the marketing department. (See "product development" checklist.)
11. Add the "environmental impact" element to the materials management department's conventional purchasing criteria. (See "materials management" checklist.)
12. Reduce pollution and waste disposal costs by modern production technology. (See "production technology" checklist.)
13. Reduce pollution and material costs by recycling and ensure that waste is properly disposed of. (See "disposal and recycling" checklist.)
14. Select and equip company vehicles with a view to environmental considerations. (See "company vehicles" checklist.)
15. Construct and maintain office and factory premises by reference to environmentally benign principles. (See "building" checklist.)
16. Ensure a healthy financing situation, not least as a precondition for non-profit environmental protection measures. (See "finance" checklist.)
17. Ensure that environmental protection legislation and official instructions are complied with. (See "law" checklist)
18. Systematically detect and insure against unavoidable and incalculable risks arising from environmental liability on the part of the company or its management. (See "insurance" checklist.)
19. Extend environmentalist management principles to exports, imports and foreign subsidiaries. (See "international business relations" checklist.)
20. Ensure undistorted publicity for company environmental efforts and pilot projects, with a view to attracting imitators. (See "PR work" checklist.)

Successful Environmental Mentoring Tips

- *Define your roles:* From the beginning, mentors and mentees should know their roles in the mentoring relationship. For a formal one-to-one mentoring relationship, the mentor should prepare a letter of commitment outlining the roles of both parties. The letter should include specific project objectives — so that the outcome can later be measured. Any security, legal, and liability issues that might affect the relationship between the mentor and mentee must be addressed, understood, and agreed upon by all parties. In most cases, the mentee must agree not to hold the mentor liable if a problem occurs. Nevertheless, mentors must be careful about the advice that they give their mentees. Additionally, a confidentiality agreement may be required if the mentee is concerned about trade secrets.

- *Evaluate before you act:* Conducting a "baseline assessment" allows the mentor to develop realistic exceptions of the first visit. Before the mentoring process starts, the mentor should contact the mentee to assess his or her knowledge of the environmental issues that are the focus of the mentoring project. This should not be seen as a test or inspection. Rather, an assessment provides the mentor with some background regarding the company, facility, staff, etc. This enables the mentor to match "skill sets" with the mentee's staff; plan for a self-directed project; and develop options for the mentee. Some mentoring programmes ask potential mentees to complete a questionnaire; other programmes may conduct a casual phone interview.
- *Conduct a needs assessment:* Mentees should conduct a needs assessment of their companies to identify the areas where they need the most help.
- *Define your commitment:* Both companies in a mentoring relationship — particularly the mentor company — must have a strong commitment to seeing the process through. Uncertainty on the part of either party may yield disappointing results.
- *Identify what you hope to achieve:* Clear, realistic goals — that include timelines and meeting schedules – are essential to a successful mentoring initiative. The mentor and mentee should prepare a written document listing their project goals at the beginning of the project.
- *Identify the obstacles to success:* As in any business endeavor, success often relies on the anticipation of problems that will arise. Identify in advance any obstacles that would prevent the implementation of any proposed recommendations – such as limited financial resources and contractual relationships – and consider possible solutions.
- *Maximize success by matching business skills and interests:* Businesses from the same industry sector, or those that use similar processes, are more likely to have a successful mentoring relationship. Matching "skill sets" between individuals from the mentor company to those in the mentee's staff is also an effective strategy.
- *Test your new knowledge with self-directed projects:* Self-directed projects provide opportunities for the mentee company to put what it has learned in place.
- *Be willing to try several approaches:* The strength of mentoring relationships is that they allow one company to provide individualized assistance to another. This means flexibility is the key. Mentors should be willing to adapt their mentoring approaches to the needs of the mentee.

- *Keep it simple:* Mentors cannot assume that the mentee has the same level of experience, education, and understanding of environmental issues. Therefore, mentors should be able to explain complex concepts and avoid using technical jargon.
- *Focus on practical lessons:* Practical information about how to set a system in place or think about doing a specific task is more useful than abstract discussions about materials flows and industrial ecology.
- *Understand time constraints:* Both mentor and mentee must have realistic expectations regarding the time spent in the mentoring process. Talk about time commitments before you start.
- *Eliminate preconceived notions of what the mentee needs:* Mentors will be more successful if they listen to the needs of the mentee. Allow the mentee to tell you about the company's expectations and needs.
- *Use business language:* Frame advantages in terms of cost savings, efficiency, and profits. This makes it easier for the mentee to justify potential changes – and demonstrates that environmental success can align with business success.

Best Practice Benchmarking

A best practice is a method which has been judged to be superior to other methods. Many times it is the most efficient way to perform a task ["Arthur Andersen-Contract Services: Control Self-Assessment Technique"]. In other words, it is the straightest line between raw material, through processing, to finished product.

Best Practices v. Benchmarking

It is important to form a clear distinction between best practices (BP) and benchmarking. Robert Camp defines benchmarking as "the search for industry best practices that lead to superior performance" [Camp, 1989]. It is through this comparison that executives, technicians, specialists, et cetera determine the best practice [Watson, 1992].

What makes a Best Practice?

The company attempting to identify a best practice is the entity that actually determines if the discovered process is indeed a best practice [Young, 1997]. The discovered process must first be assessed within the company's environment, judged in light of its financial and human resources, corporate growth rate, technological prowess, et cetera ["Arthur Andersen-Global Best Practices: A Definition"]. It is not uncommon for the discovered practice to become a catalyst for developing a suitable, sometimes similar process. The end result of examining a found process within the context of an organisation becomes the best practice.

Best Practice Enablers

Best practices are determined through the benchmarking phase of a business process reengineering project (BPR). BPR, or process transformation, is more than just changing the way people and corporations conduct business. It also entails changing the paradigm in the workplace, as well as the behind the scenes mechanics of the industry, to bring about dramatic improvements in output and performance [Hammer & Stanton, 1995]. Although there are many components of a reengineering effort, integral parts of this process transformation are best practices identified through the benchmarking process. Naturally, the BPR success factors are similar, if not the same, as the factors leading to successfully implementing a BP. Before further discussing BP, it is important to reiterate the fact that, although a BP may have been borrowed from another entity, a process is not a best practice until it has been adapted to fit the total environment of the adopting organisation.

Financial Resources

The BP/BPR project will not get the green light without proper financial resources. Totally changing core business processes is not an inexpensive proposal. Key executives and other personnel will be required to shift their focuses from revenue-producing endeavors to participating in a project, which will not have an immediate return on investment. Further, existing equipment will need to be refitted or new equipment will need to be purchased. In addition, consultants may be brought in to assist with the project. With all of these high cost areas involved, it is important to properly plan the project so that the proper amount of resources may be allocated.

Full Management Support

The upper management must fully support the BP/BPR effort, or it will never succeed ["Reengineering Success Factors: BPR Principles"]. A project of the size and scope of a reengineering effort, no matter how small, requires strong leadership. The project management must sell the concept to the other members of the organisation, from the CEO to the Shop Foreman. Upper level support is the key to insuring, or at least greatly enhancing, the availability of appropriate resources throughout the reengineering effort. However, the promise of additional resources is not enough. The staff members involved in the actual legwork of the effort must believe that they are working toward a mutually beneficial end. In an age where downsizing has become a dirty word, everyone will be uneasy about his or her future. Project managers must create a positive environment to boost morale and quell rumors. Proper communication aids in this task also.

Open Communication Channels

Open communication channels are paramount to conducting business. It is crucial that accurate information be shared across all lines during the project so that effective decisions can be made. A process is only as good as the input-garbage in, garbage out. Further, communication between all levels will boost morale by promoting project ownership. The employees are more likely to support the project if they are given the opportunity to play a part in its development. This communication and increased interaction between the employee groups will aid in managing the change created by implementing a BP.

Change Management

Strong change management personnel must be in place from the onset to counteract unnecessary anxiety. Many employees begin looking for new jobs when the BP/BPR project is undertaken. Personnel may also be lost to industry rivals that are perceived to be more secure. BPR will have limited success if the staff does not sustain high levels of morale throughout the organisation. Depression or despair, especially in the lower strata of employees, may present problems as staff reductions and sweeping changes become more apparent. It is important to fully consider the needs of those ultimately affected by the changes [Hammer & Stanton, 1995].

End Determined Process

The BP/BPR effort should be targeted in an end-justifies-the-means structure. Clear goals should be defined before embarking on a mission to find a best practice. Obviously if a process does not approach the goal in mind, it can not be considered a best practice. By developing process criteria in this manner, the effects of traditional business methods fall by the wayside, allowing the development team to explore more innovative methods.

Diverse and Knowledgeable Personnel

In most cases, a best practice will involve all areas within the target process ["Reengineering Success Factors: BPR Principles"]. It is vital to include personnel from all facets of the process to ensure complete coverage during the revision planning and solution implementation stages. Not only will the base of available knowledge be comprehensive, but by including all levels of personnel the team can better assess the morale of the organisation as the effort progresses. Should the organisation determine that it is lacking in an area of expertise, a consulting firm may be a viable solution. Consulting firms have specialized knowledge that many organisations do not. In addition, consultants are arguably the most effective enabler of process transformation. In fact, many projects depend on utilising the skilled personnel of firms like Andersen Consulting, Ernst & Young, and McKinsey & Co. Unlike most corporate personnel, consultants bring previous BPR experience to the project. Working in specialized functional areas, consultants consistently experience similar situations . Many of the firms have established centralized knowledge bases from which proven solutions may be gleaned. Further, consultants operate on a rotating team basis. Members rotate on and off of the team, as the skills are needed, thereby providing necessary skill sets all of the time. This enables the consulting and project teams to stay up to date on the latest technological advancements.

Technology

Perhaps the most important BP enabler is technology. Information sharing technology has become a prominent factor in business today. It must be considered essential when engaging in a process transformation undertaking. Today's technology allows corporations to share information from locations around the world in a matter of seconds. In fact, new advances are being made daily. An entity that has a well-developed technical infrastructure, particularly technical staff, is in a much better position to squeeze every ounce of benefit from its BP/BPR project. One area of technology that is evolving at a rapid pace is Enterprise Resource Planning (ERP) software. ERP facilitates information sharing from most, if not all, business processes no matter the geographical dispersion of the organisation ["TechWeb-ERP"]. Although SAP was the first widely used ERP package, several competitors are gaining ground, including PeopleSoft, Baan, and Oracle. The ERP market has exploded in recent years following the development of SAP. As more and more industry leaders turn to ERP systems to improve business, the packages have become a necessary tool to remain competitive in the market. Corporations have come to rely on the one-stop best practices provided by an ERP [Stein & McGee, 1989]. One of the most beneficial aspects of an ERP system is consolidation of various best practices into one location. SAP incorporates over 800 proven best practices into its R/3 package, which offers R/3 customers instant access to a deep knowledge base of BP.

Rise of Green Business

A growing number of companies around the world are making efforts to become greener, or more environmentally friendly. They include Aveda, Baxter International, BP, Ericsson, Ford, Green Mountain Energy, Herman Miller, Interface, IKEA, and many others that represent a variety of industries. Once nearly unheard of in the business world, most U.S. companies now consider some level of greening to be a priority. This trend is occurring for several reasons and is likely to continue. Though the idea of green business is fairly new, the language and concepts behind it have taken hold across multiple industries around the world. For instance, BP changed its name from British Petroleum to Beyond Petroleum as it looks to a future that includes renewable sources of power. The company advertises itself as "the first oil company to voluntarily introduce gasoline with reduced sulfur...the equivalent of taking 100,000 cars off U.S. roads every day". Automotive giants Honda, Toyota, Volkswagen, BMW, and Ford are not only working on cars that get better gas mileage but are working to develop a more environmentally friendly transportation industry. Visteon, the world's second largest automotive supplier and a spin-off from Ford Motor Company, follows 11 principles to reduce its environmental impacts. The auto industry, which has been under scrutiny for decades for the environmental impacts of its manufacturing processes and products, is not the only industry to invest in greener technologies and activities. With an interest in environmental transparency, or being candid about its environmental impacts, multi-billion-dollar medical products company Baxter International was chosen as an industry leader by the Dow Jones Sustainability Group Index (DJSGI). The company also is a member of the Coalition for Environmentally Responsible Economies (CERES) and a pilot member of CERES' Global Reporting Initiative (GRI), both of which are discussed later in this study. The textiles industry also has joined the ranks with companies working to go green. Carpet manufacturer Collins & Aikman takes back used carpets as part of its recycling programme, in which it reuses carpet fibers in the production of new carpet products. Commercial flooring behemoth Interface, Inc. sells replaceable carpet tiles so that only the worn ones can be replaced rather than the entire carpet. Clothing designer Giorgio Armani fashions garments made of hemp, a fiber with several environmental benefits. Armani explains,

> "We began using ecologically sound fabrics and finishes in 1995, and, every year, new innovations are brought in. Hemp doesn't require pesticides or weed killers and is one of the most effective photosynthetic converts of carbon dioxide into oxygen. It also produces a highly resilient yarn that requires very little processing. When I discovered how efficient hemp is, it was an easy decision to use it, since it's my belief that clothing should not hurt the environment...."

Furniture manufacturers IKEA and Herman Miller, representing the residential and commercial markets, respectively, also have worked to become more environmentally friendly. IKEA requires all of its stores to reuse, recycle, or produce energy from 75% of their waste. Herman Miller, a manufacturer of office furniture, has won several environmental awards over the past decade. In 1999, *Fortune* magazine named it one of America's most admired companies and ranked it third in social responsibility. Industrialist 3M (Minnesota Mining & Manufacturing Co.) also has made efforts to minimize its harmful environmental impacts. In 1975, 3M began running its Pollution Prevention Pays (3P) programme. In the first year, alone, the programme resulted in savings of over $750 million. Since that time, the company has prevented a considerable amount of air, water, and other pollution. In the building and architecture industries, green design is growing in popularity and importance as resource inefficiency, ecological degradation, and persistent exposure to toxins are viewed as symptoms of poor design that is ultimately expensive to maintain and redress. The number of green architecture firms continues to grow as does the number of mainstream firms that are struggling to catch up to learn green design and offer related services. Companies of all types and sizes are demanding more environmentally friendly buildings that save money through both efficiency and increased productivity as a result of better design. These have included the Chesapeake Bay Foundation, the Gap, Herman Miller, and Oberlin College. Green landscape design is a sub-discipline of landscape design. Like green building, it is setting a new, more environmentally friendly standard for its parent field. The field generally emphasises the use of native plants to restore or maintain natural habitats and the healthy functioning of ecosystems. Because it is interested in maintaining or enhancing the health of soils and plant and animal communities, it also avoids using harmful chemicals, such as pesticides that will be detrimental to the flora and fauna locally, downwind, and downstream.

With the development of these green industries and a growing number of green business tools, the role that the environment is playing in several companies' business practices and products has become so central that some refer to this period as "the latest industrial revolution" as companies search for environmentally friendly, even nature-based, models that can create cutting-edge business opportunities while limiting inefficiency. In this era of corporate competition and of increasing productivity and regulation, a growing number of companies are realising that they no longer can afford to conduct business as they used to. Edgar Woolard, past-CEO of DuPont, predicted in 1990,The green economies and lifestyles of the twenty-first century may be conceptualized by environmental thinkers, but they can only be actualized by industrial corporations. Industry has a next-century vision of integrated environmental performance. Not every company is there yet, but most are trying. Those that aren't trying won't be a problem long-term, simply because they won't be around long-term. That is the new competitive reality. Less than a decade ago, however, it seemed as though businesses paid little attention to the environment. One study conducted by Arthur D. Little in 1995 indicated that just four percent of 187 companies that responded took environmental issues seriously in their business decisions.

Historically, the environment has not played a large part in business thinking beyond simple compliance with environmental regulations. Many companies were created in order for their founders and shareholders to earn a profit with little consideration for environmental or social issues. Believing that the company's primary purpose is to make money for its shareholders, Milton Friedman, Nobel laureate in economics, insisted, Few trends could so thoroughly undermine the very foundations of our free society as the acceptance by corporate officials of a social responsibility other than to make as much money for their stockholders as possible. This is a fundamentally subversive doctrine. However, that view of business is changing. Tachi Kuichi (1997), CEO of Mitsubishi Corp., explained that businesses should provide a service greater than merely making their shareholders wealthy:

> "I am often told the needs of business and environment conflict-that the highest mission of a corporation is to maximize profits. But in the long term, there is no incompatibility. For example, a large market will arise for photovoltaic solar cells, especially in the developing world. Mitsubishi's work on such devices serves the company but also the global environment. Ultimately, profit is just money-a medium of exchange. You always trade it for something else. We don't run our business to earn profits. We earn profits to run our business.

Kiuchi is not alone in understanding that the environment and business are not mutually exclusive. Some companies that once had ignored or resisted environmental issues took on a more proactive stance in the mid-1980s, evidenced by an increase in corporate environmental policies, environmental managers and other environmental staff positions, new relationships with environmental organisations, and environmentally friendlier innovations. By the end of that decade, academic studies were being conducted – first conceptual and, then, empirical – on organisational greening, moving the environmental field into the realm of management theory and practice. And by 1998, the number of businesses considering the environment in their activities had grown to 90% of 287 businesses polled by *Industry Week* magazine. Over 80% of Fortune 500 companies have created environmental charters, and most multinational firms, responding to stakeholder pressures, have designed environmental strategies. Additionally, industrial codes of conduct, such as the CERES Principles have been developed to improve industrial environmental behaviors and are increasingly common. Some companies also have adopted the idea of a triple bottom line, which embraces not only economic but also social and environmental issues as the three facets of sustainability.

Overview

Since the late 1980s, sustainability, or going green (i.e., becoming environmentally friendly), has become a goal for many businesses. This has occurred for several reasons. Some businesses have realised that their operations result in negative environmental impacts while some view increased efficiency in energy and resource use as a way to increase profits and improve public image. Still, others see going green as a means of staying ahead of regulatory pressures and avoiding legal hassles. Companies use a variety of principles, tools, and certification schemes designed to increase workplace or manufacturing efficiency, improve corporate social responsibility, decrease waste, and save money. Although intended to help companies become greener than they were previously or than their competitors, many of them are not based on ecological principles or on a clear understanding of how natural systems work or what they require in order to maintain ecological integrity. As a result, some of these methods used to green companies could result in exacerbating ecological problems over time. Concurrently, there is a tendency in many disciplines to learn from and model nature in human design, organisation, and activity. Systems theorist Margaret Wheatley discusses the importance of learning from the principles of chaos theory, self-organising systems, and quantum mechanics in how we humans organise ourselves and work together. Science writer Janine Benyus talks of the extraordinary models that nature provides for human design, from the leaf that generates its own energy using free, solar income to the tiny cell, which somehow manages to store data and function using spirals of efficiently organised information. In the business arena, corporate systems consultant and founder/director of MIT's Sloan School of Management Peter Senge proposes that businesses are really living systems that have the capacity to learn and evolve just like natural organisms do; this understanding, he proposes, should and can create the basis for a new and profound change in the way that businesses organise and learn. Ray Anderson, CEO of a global, multi-billion-dollar company, writes of the need for humans to consider their impacts upon nature in their business operations and to model their business systems after natural ones. Entrepreneur Paul Hawken and efficiency experts Amory Lovins and L. Hunter Lovins discuss the wealth inherent in nature, which they refer to as "natural capital," and offer examples as to how businesses can increase their own wealth and productivity through new technologies and practices that work with and support natural systems. This essay sets out to explain the reasons why companies seek to go green and examine the principles, tools, and certification schemes that they use toward that end. It also offers a critique of these principles, tools, and schemes, stating that they fail to offer a holistic, systems-based view that would help companies to achieve global ecological sustainability. Finally, it makes the argument for the development of a holistic set of principles, tools, and measures that would help companies to design more ecologically benign or regenerative goals, operations, and activities throughout their entire life cycles. Such a holistic methodology, which I refer to as business ecology, would be based upon an understanding of a company's global context, accounting for the needs and boundaries of both global and local systems. Thus, it would further integrate the work being done by those involved in understanding and protecting the integrity of natural systems as well as those seeking to ensure that their businesses follow the Hippocratic corollary by doing no harm.

Corporate Environmental Awards

Environmental awards have become increasingly popular tools in the U.S. for recognising companies that try to promote energy efficiency and minimize pollution. Sponsors hope that awards will encourage other companies to imitate them, and will thereby foster greater ecological awareness and interest in questions of global sustainability. There are now many environmental awards. Some are backed by government agencies, some are given by watchdog groups, others are sponsored by companies themselves. Two of the more prestigious awards are the *Gold Medal for*

International Corporate Achievement, and the *Global 500 Roll of Honour for Environmental Achievement.*

- The *Gold Medal* Award honors industrial companies that have shown outstanding, sustained, and well-implemented environmental management policies in their international operations. In the late 1980s, the Award was repeatedly singled out by U.S. President George Bush, adding to its visibility.
- The *Global 500* is sponsored by the United Nations Environmental Programme. Nominations are made by third parties and awards are given in 27 areas of environmental accomplishment. The awards identify individuals and NGOs, and a few companies, that have demonstrated sustainable development practices; that have mobilized public attention and support, or taken action toward solving an environmental protection issue; or that have contributed significantly to intellectual, scientific, or theoretical approaches to environmental concerns. Other environmental awards given in the U.S. include:
- *The DuPont/Conoco Environmental Leadership Award:* Based on nominations from customers of DuPont or Conoco Mining Services, companies with mining operations in North America are assessed on their success in reclaiming mines, protecting land-use and water quality, and in local environmental leadership.
- *The Edison Award for Environmental Achievement:* Sponsored by the American Marketing Association, the award champions American companies whose commercial products contribute significantly to source reduction.
- *The Environmental Achievement Award:* Given to companies whose environmental projects exceed regulatory requirements and who develop creative and innovative solutions with proven economic and environmental benefits. The award is sponsored by the National Wildlife Federation's Corporate Conservation Council.
- *The Safety Award for Excellence (SAFE):* Targets oil companies that achieve the highest levels of safety and environmental compliance in U.S. off-shore drilling operations. Nominees are picked from inspection reports, and the award is sponsored by the Minerals Management Service of the U.S. Department of the Interior.
- *Searching for Success National Environmental Achievement Award:* Sponsored by the not-for-profit group RENEW America, awards are made in twenty environmental categories to individuals, community groups, schools, companies, or government agencies with outstanding environmental programmes.

- *America's Corporate Conscience Awards:* Sponsored by the Council on Economic Priorities, a non-profit public policy research organisation based in New York, a group of both small and large companies are recognised annually for their achievements in five categories of social and environmental responsiveness. Despite their apparent diversity, all of these awards, in fact, rely on common sources of information about firms and their environmental performance. In this chapter, we review:
- the principal *sources* of environmental information in the U.S.,
- the *process through which environmental information gets conveyed to outside observers, and*
- the *effects* that these environmental data have on the business practices of U.S. firms.

Ecolabeling and Certification

Products

How does a company convey a product's environmental superiority over that of competing products? One way is through advertising. However, consumers have become wary of product environmental claims and, without knowledge of product life-cycle information and corporate accountability, have few easy ways to compare products against one another. In an effort to help consumers identify greener products from their less green counterparts, some organisations have developed systems to certify product environmental claims. Using their own sets of criteria, companies pay these organisations to certify their products as green. Certified products generally carry a label by the certifying organisation, identifying the product's environmental preferability to consumers through the use of a familiar logo. This logo is an environmental label, or ecolabel, and it is intended to reward companies for being environmental leaders in their product categories.

Criteria

There are several ecolabeling organisations worldwide, and those that belong to the Global Ecolabeling Network (GEN) a non-profit membership organisation, must follow the criteria listed below as well as their own:

- Be based solely on voluntary participation for potential licensees
- Run by not-for-profit organisations without commercial interests
- Exhibit independence from undue commercial interest
- The source of funding shall not create a conflict of interest
- Seek advice from, and consult with, stakeholder interests
- Legally protected logo

- Determination of criteria based on an assessment of the overall life of a product category
- Open access to potential licensees from all countries
- Criteria levels established to encourage the production and use of products and services that are significantly less damaging to the environment that other products
- Periodic review, and if necessary, update of both environmental criteria and categories, taking into account technological and market place development.

This section examines eight product ecolabeling programmes. They are:

- Blue Angel
- Ecolabel
- Eco Mark
- Energy Star
- Environmental Choice
- Green Seal
- NF Environnement
- Scientific Certification Systems

Germany's Blue Angel was the world's first ecolabeling programme, introduced in 1977. It relies on the voluntary cooperation of manufacturers to provide information about their products. It also relies on individuals and retailers to prchase more environmentally friendly products. Unlike other ecolabel programmes, all Blue Angel products must be reclaimed by their manufacturers at the end of their useful lives because of a federal mandate. In 1991, Germany mandated a take-back programme (of products and their packaging), which is credited with cutting 1 million tons of waste between 1991 and 1993. The Blue Angel organisation's Eco-Label Jury, composed of industry, environmental associations, trade unions, churches, and public authorities, scrutinizes product groups twice yearly in light of developments in product design and environmental technology. Products are allowed to carry the ecolabel when they are demonstrated to surpass most others in their peer groups. Approved products are reviewed every two to three years to determine whether or not they should continue to be approved as part of the programme. Current criteria for the Blue Angel award include resource conservation, low greenhouse gas emissions, and fossil fuel efficiency, resulting in products with minimal climate impact. As applied to personal computers, the criteria used focus on power consumption, durability, and recyclability. To date, over 4000 products have been approved in 71 product groups.

The European Union's Eco-label programme was developed in 1993 to foster the production and purchase of products that are more environmentally friendly than their peers. All countries with membership in the European Union can take part in this programme. One benefit of membership is that once one EU member country certifies a product with the ecolabel, it can sell its ecolabeled product throughout the EU, reaching a potential market of 340 million consumers. The logo is comprised of a flower with the EU's symbol – a circle of stars with the letter "E" in the center. Each awarded product must undergo a life-cycle analysis. At present, 10-20% of EU products on the market have been manufactured in accordance with ecolabel criteria. Products that have been awarded Eco-label certification are reviewed after about three years, when acceptance standards may be raised. This is done to encourage improvements in products' environmental performances. Manufacturers are charged an annual license fee. This fee is based on 0.15% of product sales.

Japan's Environmental Association runs the country's Eco Mark programme. This programme was developed to encourage more environmentally friendly manufacturing and help consumers and retailers to identify products that do less environmental harm. The logo shows someone protecting the earth with her/his hands. Across the top are the words "Friendly to the Earth" while the product category is shown across the bottom. Products awarded the Eco Mark label must meet the following criteria: be less environmentally unfriendly in their manufacture, use, and disposal than their product peers AND be more environmentally friendly in other ways (not specified) that result in improved conservation.

Energy Star is a voluntary government programme that was developed in 1992 by the U.S. Environmental Protection Agency to encourage energy efficiency in office equipment. Because it is a government programme, Energy Star is unlike the other ecolabeling programmes already mentioned. Those involved in the programme are the U.S. Department of Energy, U.S. Environmental Protection Agency, local utilities, retailers, and manufacturers. The year after the program's inception, an Executive Order was issued requiring all federal government agencies in the United States to purchase only Energy Star labeled computers, monitors, and printers which helped to create an immediate market for Energy Star-labeled equipment. The programme has grown to cover other products, including buildings (see section on Energy Star Buildings). It also has inspired similar programmes around the world, including New Zealand Energy Star, Japan Energy Star, and Sweden Energy Star. The Energy Star programme focuses on energy efficiency to the exclusion of other environmental criteria. In addition to labeling energy-efficient products, the programme has partnerships that encourage energy conservation, saving more than $1 billion in energy costs while reducing air pollution.

Like the other ecolabels, Canada's Environmental Choice programme was designed in 1988 to assist consumers in identifying products and services that are less environmentally harmful than similar products. The logo is comprised of three doves (consumers, industry, and government); forming a maple leaf when combined, the doves are working together to improve Canada's environment. Before a product can be awarded a logo, there must be guidelines in place for its product category. The guidelines for certification are created with the help of industry, environmental groups, independent scientific/technical advisors, and universities. Guidelines are updated as new information and technology become available. Criteria for the Environmental Choice ecolabel include: energy efficiency, low production of hazardous byproducts and waste, recycled material use, product recyclability. Environmental Choice awards last for only one year. During that time, certified products are monitored for compliance with the guidelines relevant to their product categories. Following the one-year award, products are reviewed for renewal. To date, over 1400 products have been approved in the Environmental Choice programme. Additionally, 119 companies have been licensed to certify products, and there are 29 guidelines for certification. Product manufacturers/importers apply for a license to use the Environmental Choice logo, and Environment Canada's independent Technical Agency works with companies in the application process.

Green Seal was developed in 1992 to award a "Green Seal of Approval" to U.S. products that are environmentally superior to their peers. The logo, a green check mark over the globe, helps consumers and retailers to identify environmentally preferable products. Like the other ecolabels, Green Seal establishes environmentally preferable product standards and offers its logo for use on those products that meet its criteria. The organisation takes suggestions regarding the categories for which to develop standards. Categories are chosen depending upon both the environmental impacts associated with the products that fall within them and the range of products within the suggested category. When a category is chosen, it is reviewed for the environmental impacts of its products with a focus upon product characteristics and the places in product manufacturing, use, and disposal that have the most significant environmental impacts. For example, for the category "Paper Towels and Paper Napkins," Green Seal has established specific performance, environmental, packaging, and labeling requirements. Then, Green Seal proposes a standard, which is circulated for review and comment by the public and organisations, including manufacturers, consumer groups, government agencies, and environmental groups. Following the review, Green Seal publishes the standard for the category in question. Green Seal's standards are not intended to modify or supercede government labeling requirements. Most of the organisation's product testing and evaluation is undertaken by Underwriters Laboratories Inc. (UL). In addition to establishing product standards, Green Seal also has an Environmental Partners programme. Environmental Partners are organisations (over 450) that agree to purchase environmentally preferable products within their performance and price requirements. Environmental Partners pay an annual fee in exchange for an Office Green Buying Guide, Choose Green Reports, Environmental Partners logo, and certificate of membership. Intended to help businesses, universities, and other organisations from having to wade through product environmental claims, Green Seal recommends products based on factors that include packaging, product performance, and environmental impacts.

Like the other ecolabels, France's NF Environnement label is found on products that are less environmentally detrimental than (while performing on par with) their product peers. This programme, founded in 1992, is voluntary and, like the others, was developed for businesses that want to make their products stand out for their environmental superiority. The logo shows a single leaf draped over a globe. Several criteria are used to determine environmental preferability within each of the product categories (e.g., trash bags, paints, automobile coolants). These criteria are developed largely by NF Environnement's Eco-Logic Committee, which is comprised of partners from consumer and environmental organisations, industry, distributors, and government. This committee also is involved with determining new categories for criteria development. An additional steering group is made of Committee members and of industry individuals involved with the categories under review. After the steering group uses consensus to develop criteria for each product category, NF Environnement publishes the new environmental guidelines.

Scientific Certification Systems (SCS) began in 1984 as the United States' first third-party certifier for pesticide residues on food. Today, SCS is a third-party testing and certification organisation that tests and certifies food safety and environmental claims. This is intended to help consumers make environmentally informed choices and give companies with environmentally preferable products an edge in the marketplace. Although still involved in the food industry, SCS also certifies environmental claims regarding forestry and marine management as well as business practices related to natural resource extraction and product manufacturing. It certifies individual product attributes, such as recycled content and biodegradability, as well as larger environmental preferability claims by looking at product life cycles. Well-managed forests are audited and certified by SCS under the Forest Stewardship Council (FSC) label. In the marine arena, SCS works with the Marine Stewardship Council to develop standards for well-managed fisheries around the world.

What Product Ecolabels Do & Do Not Do

The ecolabel/certification programmes discussed above are similar in that their stated goal is to alert consumers (and, sometimes, retailers) to environmentally superior products. Their methods vary, ranging from conducting product life cycle analyses to establishing limited criteria (energy efficiency in the case of Energy

Star) against which to measure product performance. Thus, all ecolabels are not equal. Germany's Blue Angel is the only ecolabel that requires manufacturers to take back labeled products, assuming environmental responsibility for their products not only during manufacture but throughout the lives of the products. However, there are no facilities checks, and the ecolabels, including the Blue Angel, depend on the honesty and voluntary disclosure of businesses' environmental behaviors. Although the voluntary nature of product disclosure can lead to more companies seeking ecolabels for their products, it also can result in inaccurate environmental reports by companies that use nefarious accounting methods. Additionally, some companies that manufacture more environmentally friendly products may find it prohibitively expensive to pay for the testing, certification, and logo use associated with some ecolabeling programmes. As a result, those products that carry ecolabels do not necessarily represent the greenest of their peers – they only represent the greenest of their peers that have applied for and been certified to use an ecolabel.

Buildings

If one product's manufacture, use, and disposal/rebirth can cause environmental damage, imagine the potential impact of thousands of various products. Such is the case with a building, a designed product in its own right that is comprised of numerous component products ranging from windows and insulation to wallboard and paint. How does one rate a building? This section will look at four programmes that have attempted to do just that. They are:

- BEPAC (Canada)
- BREEAM (UK)
- Energy Star Buildings Programmes (US)
- LEED (US)

Like individual product certifiers, organisations involved in certifying buildings are developing standards to help consumers choose buildings that do less environmental damage than their peers. They also want to encourage those companies that are creating better buildings by awarding them a logo boasting their environmental superiority. Building Environmental Performance Assessment Criteria (BEPAC), founded in 1987, was modeled after the United Kingdom's BREEAM programme. Created at the University of British Columbia, BEPAC is a voluntary programme that measures the environmental performance of commercial (non-residential) buildings. Its goal is to improve the impacts of building on indoor, local, and global environments. BEPAC looks at five primary criteria in determining a building's environmental performance. These are:

- Environmental impact of energy use

- Indoor environmental quality
- Ozone layer protection
- Resource conservation
- Site and transportation (www.bepac.org)

The BEPAC Foundation also developed a set of environmental building performance measures that enables companies to do their own building evaluations rather than requiring the expertise of a third party. This tool contains software spreadsheets as well as BEPAC documentation and a training programme for users. BEPAC has about 200 members (either individual or group).

Another rating programme is Great Britain's Building Research Establishment Environmental Assessment Method (BREEAM) programme. BREEAM was established in 1990 to assess indoor environmental quality and the environmental impacts of commercial buildings. BREEAM helps to determine a building's contributions to climate change, ozone depletion, and the degradation of rainforests and other ecosystems. It also factors in air and noise pollution, lighting quality, hazardous materials, and other issues specific to building design, construction, and occupation. Although BREEAM was created by a third-party organisation, it has been developed with input from the construction and property industries as well as government and building regulators. BREEAM is a tool that allows building owners, users, and designers to review and improve environmental performance throughout the life of a building. It establishes a benchmark for environmental performance and provides a wide range of benefits. Environmental assessments can be carried out in three parts: building fabric/services, design quality and procurement quality (optional), and building management and operating methods (optional and done only in occupied buildings). BREEAM is updated regularly to stay current with new design methodologies and building and materials research. Although Building Research Establishment does all BREEAM certification, the actual assessment is carried out by a licensed assessment organisation. Such organisations can do official assessment reviews, write assessment reports, and offer consultation under BREEAM's label. Procedures have been implemented to ensure that both approach and service are consistent across the assessment organisations.

Like its Energy Star programme for products, the U.S. Environmental Agency developed Energy Star programmes for buildings. The Energy Star Homes Programme was developed for home builders who want to build energy-efficient new homes. From the consumer standpoint, an Energy Star Home is one with such efficiency that it will save on utility bills, increase resale value, and decrease air pollution. Energy Star Small Business was designed to help small businesses save money and reduce energy usage through increased efficiency. Finally, Greenlights and Energy

Star Buildings were developed as volunteer programmes for companies interested in using energy-efficient lighting and creating efficient buildings, respectively. Both of these programmes encourage U.S. businesses, where financially compelling, to invest in more energy-efficient lighting, heating, cooling, and ventilation systems, and other technologies, including computer monitoring and automated systems. This volunteer programme between the U.S. EPA and U.S. organisations stresses that an Energy Star building may be more attractive to potential customers, tenants, investors, and others. It seeks to reduce energy use and help organisations to monitor their efficiency-related practices. These organisations vary in the types of buildings that they own, including:

- Commercial Real Estate
- Energy Service And Product Providers
- Federal Government
- Higher Education And K-12 Schools
- Hospitals
- Hotels And Motels
- Industry
- Property Managers
- Restaurants
- Retail Stores
- State And Local Government
- Supermarkets

For companies, the emphasis is on saving money through increased energy efficiency. In exchange for using the Energy Star Buildings logo, organisations simply sign a Memorandum of Agreement that they will follow energy-efficient steps whenever it makes financial sense to do so. The EPA provides technical information, support, help with public relations, and access to a variety of tools and information. The building efficiency plan is based on three steps – planning and benchmarking success, developing an integrated approach to reducing energy consumption by focusing on the building as a system of interrelated functions and tools, and communicating success to the media and public. If done correctly, the EPA claims that organisations can reduce their energy use by one-third while gaining a 20% or higher internal rate of return by investing in efficiency. In the late 1990s, the U.S. Green Building Council (USGBC), a non-profit coalition of individuals and companies dedicated to supporting the creation of efficient and healthy buildings, began to develop the Leadership in Energy Environmental Design (LEED) Green Building Rating System™. Under the LEED system, interested persons or organisations can self-assess institutional, commercial, and high-rise residential buildings using LEED software and documentation. LEED is a building rating system designed to assess asbestos, daylighting, energy conservation, indoor air quality, and temperature in buildings. By assessing these factors, LEED can be used to determine a building's environmental performance throughout its life cycle. LEED-certified buildings may have a higher resale value and lower operating costs. Green building certification is awarded on different levels (gold, silver, or bronze) depending on the total points gained in the LEED checklist.

The costs for U.S. Green Building Council services vary, as members receive a discount. To register projects under development (providing minimal assistance), the prices are $350 (member) and $500 (non-member). To certify completed buildings, the costs are $1200 (member) and $1500 (non-member). To offer technical assistance, the cost ranges from being free to $220, depending on the service and membership.

What Building Ecolabels Do & Do Not Do

BEPAC is a voluntary programme that looks at the environmental performance of commercial buildings against five factors – energy use, indoor environmental quality, ozone protection, resource conservation, and on site and transportation issues. It allows either third-party building assessment or internal building assessment with the assistance of BEPAC tools, including software and documentation. The fact that assessment can be carried out internally by those not experienced with it raises some concern with the consistency of buildings carrying the BEPAC label; however, the provision of BEPAC tools to businesses helps to increase the likelihood for consistent measures and methodologies. BREEAM's set of environmental performance criteria differ from BEPAC's. Climate change, ozone depletion, rainforest/ecosystem degradation, air and noise pollution, lighting quality, and hazardous materials were cited as being some of the main outdoor and indoor environmental concerns. BREEAM works with companies to establish a benchmark against which to measure environmental performance. Buildings are assessed by licensed organisations with a focus on consistency of style and method. However, like BEPAC, BREEAM is not all-inclusive in its building and site assessment, leaving room for improvement. The Energy Star building programmes are useful because they are voluntary (companies sign an MOA) and free of charge. Because businesses agree to make efficiency-related changes only when it is financially compelling, it increases a company's likelihood of engaging with the programme. Moreover, businesses with Energy Star buildings receive technical support as well as assistance with public relations and the media from the Energy Star programme. The Energy Star logo is limited in scope to energy efficiency. In addition, companies are not required to maintain their buildings as efficiently as possible, so the Energy Star logo does not indicate that a building is as efficient as it could be. Moreover, the Energy Star three-stage programme is not as rigorous as it could be. For example, internal benchmarking (rather than a certified third-party audit) may be less expensive, making it more likely that a company would undertake the effort, but it may produce inconsistent or inaccurate results. The LEED building rating system has four strong points. It is fairly comprehensive in scope (from landscaping to renewable energy to employee transportation to recycling building materials), it uses a point system, it offers technical support, and it is updated regularly. Although the point system may not be perfect, it offers a consistent metric against which a building's environmental performance can be measured. Compared against some of the product ecolabels, the cost for carrying the LEED certification and logo is inexpensive, particularly with respect to commercial buildings. However, measuring some things may be challenging – like how much of a building shell is reused or the percentage of stormwater that is retained. And, while LEED covers broad topics, it does miss a few, like offering points for planting native species.

Companies

Few certifications have been established for entire companies. One exception is the ISO 14000 family of standards, which was developed by the International Organisation for Standardisation's Technical Committee 207 in 1996 to provide a consistent way to monitor and carry out environmental management, providing both environmental and economic gains to companies wishing to participate. Through ISO certification, companies and organisations that use ISO standards also gain public, government, and industry recognition for their efforts to ameliorate their effects on the environment. ISO/TC 207 subcommittees (whose members represent ISO national committees from around the world) prepared the environmental management standards in working groups. ISO 14000 was created in an attempt to create a set of environmental management standards that reflected current environmental and design practices and technologies. This set of standards is known as a family because there are 23 standards (each with accompanying guidelines and technical reports) that cover a broad range of environmental management concerns. The standards can be divided into seven main groups, or types, of environmental standard. See the following Table 9 for a listing of those seven groups.

TABLE 9

ISO 14000 Family of Standards – Seven Main Groups

Environmental Management Systems
ISO 14001
ISO 14004
ISO/TR 14061

Environmental Auditing

ISO 14010

ISO 14011

ISO 14012

ISO 14015

ISO 19011

Environmental Labeling

ISO 14020

ISO 14021

ISO 14024

ISO/TR 14025

Environmental Performance Evaluation

ISO 14031

ISO/TR 14032

Life Cycle Assessment

ISO 14040

ISO 14041

ISO 14042

ISO 14043

ISO/TR 14047

ISO/TR 14048

ISO/TR 14049

Environmental Management Vocabulary

ISO 14050

Environmental Aspects in Product Standards

ISO 14062

ISO Guide 64

The ISO 14000 family of standards is used around the world. Although this section is intended to provide a brief overview of this tool, it will be helpful to illustrate one of the standards below. I will outline ISO 14001 because of its increasing regularity in environmental management systems literature and its growing application. In March 1998, there were 4,045 organisations that were certified as ISO 14001-compliant. In July 2000, there were 18,052 ISO 14001 certifications with many countries represented.

ISO 14001 is the first step in an organisation's development of an environmental management system. It helps an organisation to

1. establish quantifiable environmental management goals;
2. identify individual roles in meeting, measuring, and documenting progress toward achieving the goals;
3. use this information to inform the system of the activities that contributed to the success of achieving the goals;
4. review activities and goals regularly to ensure that they are in sync with the organisation's vision; and
5. enable strong group communication and good management. According to ISO, ISO 14001 is useful to any organisation that wants to do the following:
   - implement, maintain and improve an environmental management system;
   - assure itself of its conformance with its stated environmental policy;
   - demonstrate such conformance to others;
   - seek certification/registration of its environmental management system by an external organisation;
   - make a self-determination and self-declaration of conformance with this International Standard (ANSI/ISO 14001-1996)

Although ISO states up front that the extent of to which an organisation applies the ISO 14001 requirements depends upon its activities, conditions, and environmental policy, it also states that the ISO 14001 requirements are intended to be embedded in any environmental management system.

What ISO 14000 Do & Do Not Do

The ISO 14000 family of standards is helpful to companies because it provides specific guidelines and steps for those companies interested in improving their overall environmental performance. Additionally, there are individuals and companies trained in ISO who, for a fee, will help a company to attain an ISO rating. As helpful as ISO is, however, it fails to reconnect a company with the place that it inhabits. It does not appear to be grounded in fairly well-accepted natural laws (e.g., thermodynamics) like The Natural Step is, nor does it take into account the entire living, breathing life of a company. For that, we need to turn to nature to find an appropriate model.

Procter and Gamble

Procter & Gamble (P&G) celebrated their 150th anniversary in 1986 which gave John G Smale, former Chief Executive of P&G the opportunity to look forward to the next 150 years and say: 'Our commitment must be to continue the vitality of this company ... so that this company, this institution, will last through another 150 years'. A claim of that kind in a changing world establishes a remarkable challenge for all employees. However, P&G pride themselves on a long history of striving for excellence in their products, their honest and fair trading practices, continuous self-improvement and respect and concern for the individual. They positively relish challenging goals and say: 'we like to try the impractical and impossible and prove it to be practical and possible'. Environmental issues first began to emerge in P&G in the 1950s and were at the forefront of business strategy long before they became relevant for many other companies. P&G is a good example of a company using the virtuous triangle described. P&G Environmental Policy affirms their:

- commitment to operating in an environmentally responsible manner;
- continuously improving their products, packages and processes; and
- establishing innovative technology, processes and communications.

They have established their own way of integrating environmental factors into their business decision making through three specialist departments, a corporate-wide coordinating group and a unifying philosophy:

Professional and Regulatory Services makes sure that products are safe for people and the environment they disseminate research findings to scientists, medical profession, government and universities.

Environmental Science Department conducts basic research to establish how products and packages affect the environment.

Product Supply Environmental Department ensures P&G is consistent worldwide in measuring environmental performance and meeting regulatory requirements.

Environmental Quality Control Group coordinates the P&G worldwide corporate environmental programme and helps identify improvement opportunities across business sectors.

Environmental Stewardship provides the corner stone for thinking and action throughout the business on environmental issues.

The P&G Environmental Management Framework is described by Peter White in a paper presented to the Royal Academy of Engineers. P&G's Environmental Stewardship approach. The P&G philosophy empowers people to be innovative in the development of ideas and action including cross-fertilisation of best practice between all sites worldwide. Long-term thinking is encouraged and even when progress is accomplished in small steps the cumulative benefits are impressive. P&G carry these principles forward when working with others as they often do when tackling environmental problems in a societal context. Environmental Management Structure depicted above ensures that environmental considerations are built into every aspect of the business. Such integreation is vial to the overall environmental performance. The areas in which improvements have been accomplished through the application of P&G's long-standing slogan more from *less,* include:

- Product design —> re-thinking formulation
- Raw materials —> using less
- Manufacturing —> using less water and energy
- Packaging —> smaller packaging
- Transportation —> less energy for smaller, compact products
- Consumer use —> smaller dosage

The crucial aspect of this work is that consumers are given greater performance and value ('more') while the environmental effects are reduced. 'More from less' can be applied by designers, manufacturers, waste managers and policy makers to help create sustainable product life cycles. In 1994 the USA Superfund Act (SARA) required companies to report releases annually of more than 300 chemicals (rising to 600 from 1995). Since 1991 these releases have been reduced from 18.9 million to 6.1 million pounds (8.6 to 2.8 kilotonnes) and P&G is on track to achieve their targets. Equally impressive results are being achieved with the use of post-consumer recycled content (PCR) of major packaging materials. Among the achievements which P&G claim are the following:

- integrating environmental policies with business strategy so that they impact directly on product design, manufacturing processes and purchasing decisions;
- being the first company to bring recycled plastic bottles and refill packs to the laundry detergent market which helped to eliminate 80 million cartons per year;
- getting consumers, manufacturers, retailers, environmentalists and government officials at all levels to work together to seek environmental solutions;
- recognising that consumer decisions are based primarily on performance, but environmental attributes of products are also important and customers need accurate, science-based information;
- publicly declaring the environmental targets, measuring performance against these targets and publishing the results; and
- over the period 1984 to 1994 P&G reduced the energy usage per production unit by 60 per cent by improving the efficiency of processes.

Supply Chain Management for Environmental Improvement

Outreach and Assistance to Suppliers

Many suppliers and contractors are willing to work with customers, and meet their environmental standards or requirements. After all, good environmental practices increase efficiency and competitiveness, and save money. Furthermore, meeting customer requirements usually means continued business. On the other hand, rigorous environmental requirements, or a different set of requirements from each major customer, may prove unattractive or overwhelming to suppliers and contractors. Some smaller suppliers may not be able to devote the necessary resources on their own, to be able to implement necessary changes. As a customer interested in greening your supply chain, be ready to actively participate in the supplier's efforts by investing time, resources, and other assistance. Examples of assistance to suppliers include providing access to tools, information, or programmes that your company has had success with, offering training and mentorship from your staff or through third-party consultants, and being an available resource to key suppliers when needed.

Suggestions on Outreach and Assistance to Suppliers

Make Suppliers or Contractors Aware of Your Requirements, and Well in Advance of the Effective Date

Examples:

- Develop and publish your environmental criteria
- Put yourself in your supplier's shoes to help understand how they will accept and meet the new requirements
- Send advanced notification explaining expectations for their level of participation and commitment, timelines, benefits, and compensation (if applicable)

Offer Education, Assistance, and/or Training to Suppliers on How to Meet Your Environmental Standards

Examples to consider:

- Interact directly with the pertinent supplier contacts
- Offer training seminars or workshops
- Share tools you have successfully employed
- Offer assistance in the startup or transition phases
- Establish a mentoring relationship
- Offer to help bear the transitional costs

Conduct Assessments and Provide Recommendations

(Also see Supplier Evaluation and Certification)

Examples:

- Review supplier and contractor facilities, operations, and policies; provide recommendations on improvement opportunities and ways to help meet your environmental requirements.
- Assess the products or services you purchase from the supplier or contractor. In the case of product assessments, additional information from second or third-tier suppliers may be necessary.
- Provide recommendations for greening suppliers' products, or processes used to manufacture the product.

Offer Technical Support Services and Effective Means of Communication

Examples: Set up one (or all) of the following to provide needed support and answer supplier questions on meeting your environmental requirements

- supplier environmental advisory council
- web help screens
- e-mail, hotline and/or other direct methods of communication
- supplier listserv

Educate Your Customers and Encourage them to Support Your Efforts

Examples: Volvo provides customers with environmental information on their products.

Maintenance

- Use latex instead of oil-based paints.
- Use refillable containers for cleaning supplies.
- Keep water in swimming pools year round.
- Sweep down drive ways instead of hosing them.

Purchase and Inventory

- Work with your supplier to reduce and reuse their packaging.
- Purchase in bulk when appropriate.
- Purchase recycled content products when possible.

Basic Operations

Reuse or eliminate plastic trash can liners in guest rooms.

Instead of automatically providing giveaway toiletry items such as soap, shampoo, shower cap, toothbrush, and toothpaste, post a sign in the room that states they are available upon request. Make an effort to get off junk mailing lists. *Junkbusters* web site can give you some tips. Use xeriscape landscaping. Xeriscape landscaping is based on planning and design, soil analysis, appropriate plant selection, practical turf areas, efficient irrigation, use of mulches, and appropriate maintenance. To learn more, download this free software, *Water Efficient Landscape Planner* or contact your county extension service or local garden/nursery centers. conserves water decreases energy use because less pumping and treatment of water is required reduces building heating and cooling costs because of placement of trees and plants minimizes stormwater and irrigation runoff

Equipment Changes

Install energy-efficient lighting. To learn more, check out *EPA's Green Lights* web site or call (888)STAR YES. save money on electric bills and reduce energy usage improve comfort levels and lighting quality reduce creation of air pollutants Install efficient-flow shower heads, low-flush toilets, and faucet aerators. To learn more about reducing water use, check out *EPA's Water Alliances for Voluntary Efficiency* (WAVE) web site or call (202)260-7288. reduce volume of water used and purchase costs avoid wastewater treatment costs and energy for heating water Purchase high quality, durable equipment with good service contracts. Install refillable soap, shampoo, and lotion dispensers in guest bathrooms

Material Substitution

Consider carpet made of recycled plastic. Replace single-use food service supplies (cups, plates, bowls, forks, spoons, knives, napkins, etc.) with reusable ones. Replace plastic cups with durable water glasses in guest rooms. Purchase reusable cloth coffee filters. Replace single-serving breakfast cereal boxes with larger bulk dispensers.Use bulk dispensers for condiments and serve in small dishes for room service. Use cream pitchers, sugar pourers, and jelly servers rather than individual packets. Use chips or coins instead of disposable paper tickets for coat checking and auto parking. Use less hazardous cleaning supplies and laundry detergents. Use rechargeable batteries.

Process Change

Offer optional linen and towel reuse in guest rooms. Post signs in rooms asking guests to consider using their towels and linens more than once. Let your guests know that this will help the environment by reducing water and energy usage and reducing soap residues in our waterways. *The"Green" Hotels Association®* sells these cards and has devoted itself to researching other water and energy saving ideas. Provide a video segment on the in-room hotel/motel channel that highlights your pollution prevention efforts, improve company image

Recycling & Waste Exchange

Develop an in-house recycling programme. Place recycling containers where they are conveniently located for guest use (in guest rooms, next to soda machines, etc.) Recycle cardboard packaging and pallets or arrange with your supplier to reuse them. Recycle newspapers, office paper, and telephone directories. Recycle plastic, glass, and aluminum from restaurant, bar areas, and guest rooms. Donate banquet leftovers to homeless shelters or other charitable organisations. Grind up food scraps from restaurant in food mulchers and give or sell to farmers for composting or compost on-site. For more information on composting, contact the *Composting Council,* 114 S. Pitt St., Alexandria Virginia 22314, (703)739-2401. Reduce landscaping wastes by mulching/chipping grass clippings, branches, and downed trees use mulch in flower beds or spread along trails Donate used soap, bottled shampoo, linens, blankets, mattresses, furnishings, carpet, and equipment to charitable organisations such as homeless shelters Recycle stained tablecloths and napkins-they can be reused as napkins, aprons, cleaning rags and even neckties!

Reuse old sheets as cloth laundry bags.

Auto Body
Auto Repair
Dry Cleaning
Fiberglass Operations
Furniture Manufacturing and Refinishing
Health Care
Hotels & Motels
Metal Finishing
Painting
Printing

CII-Sohrabji Godrej Green Business Centre

The Centre is a joint initiative of the Government of Andhra Pradesh, the house of Godrej and Confederation of Indian Industry (CII), with the technical support from USAID-an unique model of public-private partnership. The CII-Sohrabji Godrej GBC is the centre of Excellence for Energy, Environment and climate change activities in India.

What is "GREEN"?

It is the voluntary pursuit of any activity that encompasses concern for energy efficiency, environment, water, conservation and the use of recycled/recycled products & renewable energy. In addition to making the business profitable, these activities lead to sustainable development and a healthy economy.

Objective

The core objective of CII-Sohrabji Godrej GBC is to make the world a better place to live in, by providing world-class 'green' services. The centre will promote 'green' concepts leading to higher efficiency, equitable growth and sustainable development.

Services

World-class energy efficiency: Facilitates industry to achieve the lowest specific Energy Consumption levels in the world, by implementing world-class practices and technologies. Facilitation tools include benchmarking with international norms and identification of ways and means to achieve the same.

Green Audits: Green audits would focus on the root cause of inefficiency instead of focussing on mere treatment and compliance. This, in turn presents an opportunity for better utilisation of resources (raw materials, consumables, water, energy etc) at every stage of the manufacturing process.

Green Building rating: A 'green' building is one that is environmentally responsible, profitable and a healthy place to live and work in. The Centre facilitates construction of 'green buildings' and recognition through the LEED (Leadership in Energy and Environmental Design) 'green building' rating system of USGBC.

Technology Centre: To provide a platform to showcase and demonstrate innovative 'green' products and technologies. The centre is open to business visitors from India and abroad, which will facilitate greater visibility and market access to the participants.

Green business incubation: Handholding of entrepreneurs to develop green products/ technologies till they reach the stage of commercialisation.

Information Centre: To provide access to the latest information and trends on 'green concepts and technologies'.

Coverage

The service of CII-Sohrabji Godrej GBC will focus on various sectors that include:

- Manufacturing industries like cement, ceramic, chemicals, engineering, fertilizers, food, foundry, glass, metallurgy, paper, pharmaceuticals, power, refineries, steel, sugar and textiles
- Commercial buildings and establishments.

Benefits

Your future business opportunities will depend on how 'Green" you are!

- A 'Green' corporate image can leverage access to local and international markets
- 'Green' activities result in reduction of manufacturing costs and improve profitability.
- Being ' Green' can attract environmentally conscious customers

Energy Efficiency

Energy conservation practices have acquired top priority, in the present context of increasing energy prices, acute energy shortage and the ever-widening demand supply gap. A multi-pronged approach is advocated for achieving energy conservation. This approach comprises of pursuing/ adopting the following three-pronged approach

1. Capacity utilisation
2. Fine tuning
3. Technology upgradation

Capacity utilisation involves optimally utilising the available resources. This is a least cost option. Fine-tuning is incorporating small changes in the existing system to get major gains in energy efficiency. These fine tuning methods normally do not involve high investments. With the advent of new technologies, Technology upgradation can result in mega savings in energy. Though this involves higher investments to be made, they can prove to be highly beneficial in the long run. Capacity utilisation and fine tuning can result in energy savings to the extent of 5-15 % while technology upgradation could yield up to 30-40 %. Several successful energy efficient companies have adopted the three pronged approach and reaped rich benefits. The Indian industry and individual beneficiary units can utilise the energy efficiency aspects covered in this website, by fine-tuning to their individual needs and achieve energy efficiency.

Green Leadership

The 'first mover' in any product category can often select the environmental attributes it wants to offer, and mould consumer expectations. Future competitors must then play by the same rules. Companies can assume the mantle of green leaders without greening themselves completely. They need only be sincere and green in a few substantial ways. They can retain market leadership by introducing incremental improvements over time a strategy termed 'greening as we go'. Companies can often create an effective presence in green markets by incorporating a number of 'green' improvements into a new or refurbished product. As these improvements are perfected and their consumer appeal tested, the new attributes can be cost-effectively incorporated into the company's other product lines. This leadership position can be protected for months or years. Market barriers such as patents, distribution rights and strict product standards can be used to keep competitors at buy, allowing the market-leader to reap the rewards. Many companies have found that an environmentally-improved product, repositioned for a green-conscious market, can successfully renew an ageing brand. Green improvements made to a 'youth' line can also be readily applied to 'adult' product lines. Similarly, manufacturers of products traditionally sold to men can increase their sales by marketing 'green' improvements to women. Female consumers, who are traditionally more sensitive to environmental considerations, are increasingly purchasing items like automobiles and paints, as well as men's clothes and toiletries. Success in the marketplace can also be enhanced by tying a product's environmental attributes to the lifestyle of target It is only in recent years that more and more people in the wealthiest nations have come around to the belief, now firmly lodged in public consciousness, that high environmental standards are not just desirable but their birthright. To set the stage, Reinhardt reminded the group that pollution is not a new phenomenon in human existence, nor is fear of overwhelming population growth and depletion of natural resources. He thinks one of the main reasons people in the United States are more concerned about the environment is that incomes are higher. "Although people who think of themselves as environmentalists tend not to like it hear it characterized in this way, environmental quality does behave like a luxury good," he said Rich people tend to want more of it than poor, according to Reinhardt, and rich countries demand it whereas poor countries do not. In addition, thanks to progress in the treatment of infectious diseases over the last fifty years so, more people are living longer—but eventually dying of diseases that could plausibly have environmental component. Just as there has been progress in scientifically detecting foreign contaminants in the air and water, the public clamors for protection from involuntarily risks. "Finally, there are emerging global problems with very long timelines and possibly irreversible effects," Reinhardt told the group.

More Governance Best Practices

Effective governance promotes resourcefulness. Here are four more best practices of this year's CIO-100 winners.

By Alice Dragoon

Identify relative strategic value (even if you're comparing apples to oranges). When faced with a stack of potential projects, it's important to find a way to see beyond their differences and compare them, accounting for their business value as well as cost and risk. Many IT shops develop their own evaluation and ranking tools, which help them focus on the factors that matter most to their organisation.

At Northeastern University, CIO Bob Weir wanted the senior VPs to prioritize the backlog of IT requests. "The trouble with asking them is that there isn't necessarily a common language, or set of priorities, or common much of anything," he says. "You get into advocacy mode. Whoever waves their arms the most and screams the loudest [gets their project approved]. But that isn't any way to make a business decision." So he and one of his direct reports sat down with more than 40 university leaders and asked them to identify critical success factors for IT projects. They then developed a model with six weighted value categories (including "competitive advantage" and "service to the NEU community") and compiled 60 key questions to ask when "interviewing" a project under consideration. They ran several old projects through the new tool to calibrate it. The answers to those 60 questions yield a score that enables comparisons between dissimilar projects like applications and network investments. This year, NEU's tool helped university leaders allocate the 15 per cent of the IT budget earmarked for new projects relatively quickly, whittling a list of 36 potential projects down to just 12. In the past, university leaders requested IT projects as the need arose, and Weir would say yes or no to each one. "The 15 per cent got decided as we went," says Weir. "It was haphazard." Now, he says, "the whole senior executive team looks at the whole pile" of potential projects, so the new project budget gets allocated according to overall institutional priorities, not functional priorities. Thanks to the new value-based model, "everybody was on the same page talking about the same criteria list," says Weir. "Everybody got serious more quickly." The comparison tool is also effective for remediation. Champions of low-scoring projects can tweak them so they'll score higher in important value categories.

Yellow Technologies takes the concept of weighted questions a step further by plotting out the results on a grid showing cost and financial return. Projects with large risks are depicted by large bubbles; small-risk projects get small bubbles. The color of the bubble indicates how strategic the project is (green is highly strategic, red is not very strategic, yellow is in between). Roadway simply borrows the forced weighting technique long favored by eye doctors ("Which is clearer, A or B?"), forcing executives to choose between pairs of projects until a ranked list of priorities emerges.

GE Industrial Systems relies on a cause and effect matrix to help sort out which IT projects it will pursue. First, business leaders identify strategic focuses for the year, such as customer centricity, business simplification, increasing productivity or reducing inventory. Then would-be project owners must identify whether their project has low, medium or high impact for each strategic initiative on the matrix. An algorithm factors in payback on financials, strategic value and the proposing team's success of delivering promised value on time and on budget to score each project on a scale of 1 to 100. "It helps take bad projects easily off the table and get obvious good projects into the programme deck," says CIO Stuart Scott. Then the

Investment Council spends most of its time discussing the projects that are on the fence.

Top business execs should set IT priorities. When it comes to the final decision about how to spend the IT budget, most companies rely on a committee of business and IT leaders. Of this year's CIO-100 winners, 51 per cent use an executive council to vet ideas for new projects and business transformation and 38 per cent use a cross-functional IT steering committee. Giving the business control over the IT purse strings (or at least sharing that control) not only helps IT align with business objectives, it also encourages business executives to share accountability for realising the benefits. So it's not surprising that 73 per cent of CIO-100 honorees say the CIO and business unit sponsors share responsibility for achieving value from IT projects. Project-specific steering committees with relatively high-level executives are also fairly common.

At Farmers Insurance, for example, the top business executives who compose the IT Governance Board meet quarterly and provide overall strategic direction for IT. Funding decisions, however, are made by the Capital Appropriations Committee (CAC), which includes the IT Governance Board members plus some of their key direct reports. The CAC meets twice a month to review all capital appropriation requests with a total cost of ownership of $50,000 or more. Requesters must submit what CIO Jan Franklin calls "a very, very detailed standard document" that outlines the project's purpose and an in-depth cost/benefit analysis. Once a project is approved, a project-specific business and technology integration (BTI) group of line vice presidents from IT and the business meets weekly to review project progress and touch base with the project management office (PMO). Farmers has been convening BTI groups for nine years. "We completely revamped our applications starting in the early '90s, and when we started the effort, we built in control points," explains Franklin. Such governance practices help Farmers achieve its goal of cutting expenses or improving capacity 5 per cent annually.

When AARP's CIO John Sullivan established AARP's overall governance committee three years ago, he initially wanted the executive team on the committee. But they insisted he'd be better served with a handpicked group of their direct reports, who are one step closer to the organisation's day-to-day operations. In retrospect, Sullivan agrees they were right. He chairs the committee, but doesn't vote on funding decisions unless there's a tie. And so far, he hasn't had to, since the committee tends to have unanimous up or down votes. All major IT investments get steering committees, and a group of IT staff serve as dedicated business liaisons to promote business/IT alignment and help craft IT project proposals.

Communicate priorities and progress clearly. Once IT and business unit leaders have established priorities, they must communicate them clearly to the rank and file. Executives at The MONY Group, an insurance and financial services company, agree on companywide priorities monthly, then post them, along with the name of the person responsible for each one, on every floor. Employees' incentive compensation is tied to meeting 90 per cent of the monthly priorities, and achieving a customer satisfaction score or 3 out of 5 or better on those projects. Since MONY shifted to a new business model with separate lines of business in 2001, IT has operated on a shared services model. The lines of business "have a choice," says CIO E. P. Rogers. "They can go outside and do business with someone else. So we have to earn our stripes and prove our value every day." To keep the IT staff focused on value and thinking resourcefully, Rogers posts a monthly IT scorecard that tracks innovative ideas, flawless execution (measured by whether key projects are implemented on time, within budget and with 90 per cent of customer satisfaction ratings at 3 or better), operational excellence (the goal is to meet at least 90 per cent of companywide monthly priorities) and IT transformation (how much progress IT has made in shifting from a corporate IT mindset to that of a full-service IT service bureau). "If you don't keep score, then people don't really know how they're doing," he says. But he's careful to avoid beating people up, yelling or being punitive. "We're doing it more in terms of, here's the game we're playing, we're trying to get to shoot under a hundred. After 9 holes, are we at 45 or 55? I think it's all about how you communicate things."

Good communication that sets the proper tone and ensures that people understand how your governance processes work is also critical. "Governance by definition is somewhat constraining," says GE's Scott. "It's not a free for all, it's meant to control. I try to communicate a lot and effectively up front about what is the governance process, how does it work, and how people can be successful as project leaders." His goal is to avoid ambiguity and surprises.

Monitor projects regularly. Once IT projects are chosen, funded and launched, CIOs need a way to stay on top of their progress to protect the value of these investments. Monthly "traffic reports" provide quick summaries and highlight potential problems early on, enabling early intervention to prevent those snags from escalating into major snafus. At MONY, for example, Rogers reviews monthly dashboard documents for each project—about 20 in all. Each one-page dashboard summarizes progress against milestones, resource usage, user involvement and issues/barriers, using green, yellow or red traffic lights to indicate the status of each at a glance. Senior vice president and chief service and information officer of insurance company Humana Bruce Goodman relies on a similar monthly IT traffic report (which also classifies projects as green, yellow or red.) The PMO at Farmers gives Franklin a weekly report identifying IT risks in every project; she also has access to the PMO weekly oversight tracking database, which lets her drill down to get more detail on schedule, effort and budget. "I'm a metric maniac," says Franklin. "We measure everything around here. Just by measuring, you improve."

The Forum for Best Management Practices is sponsored by the Northeast Business Environmental Network (NBEN). By becoming an NBEN member, a company can take full advantage of the Forum. NBEN encourages companies of all sizes to participate in the Forum and the Network.

To learn more about the benefits of the Forum, follow these steps:

1. Quick Guide

A process, a framework, a network, a set of proven real-world examples: the Forum for Best Management Practices combines these features into an effective tool for improving your company's employee health, safety and environmental management system (EMS). If you are not already familiar with the Forum, then the Quick Guide is the place to start. Find out what the Forum is—and what it isn't.

2. EMS Matrix

NBEN's EMS Matrix is based on a comparison of ISO 14001, CMA's Responsible Care Codes©, EPA's Self-Policing & Federal Sentencing Guidelines and Massachusett's TURA Planning Guidelines. Based on these regulations and guidelines, NBEN has developed a model EMS that is a framework for a comprehensive environmental health and safety management system. The model consists of four common themes and ten basic features. To make the matrix a practical tool, we have collected best management practices that implement each feature of the model EMS. Use the EMS matrix and its associated BMPs to benchmark your company's EMS.

3. Best Management Practices Table of Contents

Best Management Practices are real-world examples of the policies, procedures, and forms used by a broad range of companies to implement their employee health, safety and environmental management systems. The Table of Contents is a complete list of best management practices, organised according to the ten basic features of the model EMS. Whenever possible, we have included abstracts of BMPs so you can learn more about the purpose and content of the BMP.

4. Learn More

To help you appreciate the value of a comprehensive EMS, we have included a set of background materials. If you've read the Quick Guide and still have questions, you'll find the answers here.

5. Index to the Best Management Practices

The index is a simple, practical cross-reference of the Best Management Practices (BMPs). Use it to find help for a specific problem, or to scan a list of topics covered by the BMPs.

6. Peer Assistance

One of the greatest benefits of membership in the Forum is access to EHS professionals with whom you can discuss problems and share solutions. Learn how NBEN members can take advantage of peer assistance through networking and a structured, pilot peer assistance programme. Networking occurs at events sponsored by NBEN and the Forum, as well as informally among NBEN members. NBEN's peer assistance programme is a pilot project to demonstrate that small and medium size companies can improve their EMS through the Forum. Find out how your company can benefit by participating in this innovative programme, either as a peer advisor or as a recipient of BMP examples that can help fill gaps in your company's EMS.

Development of Green Policy 21 Philosophy by FUJITSU

Since its establishment in 1935, Fujitsu has been placing environmental issues at the top of its management priorities, having the idea of "manufacturing in harmony with nature" in mind.

And we meet 'The FUJITSU Way' and the 'FUJITSU group environmental policy', and are performing environmental activity in all enterprise domains.

Towards the 21st century, Although we had hung up the "Green Life 21" as corporate philosophy, from the 2004 fiscal year, to compensate for decision and execution of the 4th stage of "Environmental Protection Programme of Fujitsu Group", we developed the previous concept and changed, saying "The green policy 21. We make every activity green"

Under the slogan of "We make every activity green" and the concept of "Green Policy21", The FUJITSU group serves as a body and tackles strengthening of the further environmental activity.

Green Policy Products	We are promoting production of a product which considered environment in the whole life cycle through creation of new technologies for environmentally conscious products, and providing "Super Green Products" which boast at least one of various best-of-breed environmentally conscious qualities, and also reusing.

Green Policy Factories	We strive toward actualisation of environmentally conscious manufacturing, through "Green Process" activities that limit resource consumption that leads to environmental burden, and through "Green Facility" activities to reduce emissions.
Green Policy Solutions	We are contributing to reducing the environmental burden imposed by society through "Environmentally Conscious Solutions" which are highly effective in reducing environmental burden. We help reduce the environmental burden by improving our client companies' operating efficiency.
Green Policy Earth	We are promoting enhancement of environmental education and the voluntary environmental activities by employees. Every employee is playing a leading role and contributing to the global environment.
Green Policy Management	We established an environmental management system in accordance with the international standard (ISO14001), as the basis of all activities. We are strengthening Group-Wide "Sustainable manage-ment" and aim to improvement in the environmental performance.

The GBCP can grow in a number of directions. It can attract More Businesses in the targeted sectors to the programme, it can add New Sectors, it can Expand Geographically, it can add New Partners and it can engage in Continual Improvement.

More Businesses

To attract more businesses, the programme must get beyond the "low-hanging fruit". According to the Bay Area GBPs, about 10 per cent of businesses in a selected sector tend to participate in their programmes. Three forms of pressure exist to induce additional business participation—peer pressure, customer pressure, and employee pressure.

In order leverage peer pressure, to get participation to be the norm in an industry sector, at least 14% of a group needs to make the change, according to behavior change research. To reach this "tipping point" requires the assistance of "mavens"", individuals who are well respected within the industry and support the programme, and "persuaders/salesmen" to help spread the word about the programme. *The Tipping Point* by Malcolm Gladwell describes this phenomenon.

Targeted outreach programmes can help achieve increase the number of participating businesses. For example, through Greening South Asian Restaurants (GSAR) individuals from the same cultural and language background as the restaurant owners, managers and employees work one-on-one to help them become certified green businesses.

Customer pressure can encourage those businesses that directly serve the public to become certified as in the case of auto repair shops and landscapers. Publicising health and environmental concerns that businesses can minimize by undertaking measures suggested in GBCP checklists may induce customers to use GBCP certification as a way to choose one business over another.

Industries that serve business clients (b-2-b), such as printers, are motivated by different incentives. According to EFC9 research, these businesses, especially small ones, are reluctant to make environmentally beneficial changes if they see no immediate need or payoff and perceive the changes as troublesome. EFC9 surveyed printers and found that they need incentives to be motivated to do more than just comply with environmental regulations. For example, requiring ISO14001 certification may be an incentive for original equipment manufacturers (OEMs) to request their suppliers to meet ISO 14001 requirements. GBCP certification may meet the ISO requirement. According to EFC9 research, the preferred incentive by printers was for government agencies and businesses to offer a selection preference for certified green businesses. More specifically, they indicated that they would be willing to get certified if it would give them a leg up when bidding for contract work. Alameda County General Services Agency Purchasing Department adopted just such a policy in December 2002. The language follows:

> "The County is committed to ensuring that companies providing printing services to the County are in compliance with environmental protection laws and regulations and are taking additional steps to conserve resources and prevent waste. We encourage contractors wishing to provide printing services to the County become Bay Area Green Business certified as an assurance to the County that environmental laws and regulations are being complied with however, certification, although highly recommended is not a requirement of this RFQ.

Employee concern about health and environmental issues may cause a business to look into GBCP certification, especially in sectors where trained labor is in short supply. Certification then becomes a way for the business to attract and retain good employees.

More Sectors

In order to decide which new sectors to add, the programme administration should develop selection methodology and assemble a decision-making team. EFC9 suggests starting this process by identifying sectors for which checklists are available.

Alameda County Green Business Programme uses a worksheet to identify the next target industry group for the development of industry-specific Green Business Standards. They use six criteria assigning each weighting factors based on their relative importance for programme success. These criteria include:

- Interest/commitment from programme partners
- Interest/commitment from industry group
- Direct relationship with public
- Significant number of businesses in county
- Fixed place of business
- Environmental impact
- Stable compliance issues

Various industries are proposed and the Steering Committee then discusses the pros and cons of each industry group coming to consensus on the numerical scores.

Greater Geographical Coverage

Depending upon the programme, it may be able to grow by expanding its geographical coverage. Programmes that are developed at the regional level or state level, but implemented locally may expand into new locales. This requires a programme champion or two in the "new" community who is/are willing to move the process along.

Adding New Programme Partners

As the programme credibility grows, agencies that had first rejected the concept may become interested. To encourage new partner participation, the programme can demonstrate how it will help the agency achieve its goals and therefore increase its effectiveness. In addition, as more business sectors are added, the programme may wish to add partners that address specific issues in those industries, including industry trade associations.

Continual Improvement

To remain relevant and effective, the programme needs to be reviewed on a regular basis to ensure that it continues to contribute to the achievement of the desired environmental protection outcomes and that the goals of the programme remain appropriate in light of experience, knowledge and changing public values. Towards this end EFC9 has developed this website/information clearinghouse where you can learn from others—duplicating their successes.

Industrial Revolution

The Industrial Revolution was a period in the late 18th and early 19th centuries when major changes in agriculture, manufacturing, and transportation had a profound effect on the socioeconomic and cultural conditions in Britain. The changes subsequently spread throughout Europe and North America and eventually the world, a process that continues as industrialisation. The onset of the Industrial Revolution marked a major turning point in human society; almost every aspect of daily life was eventually influenced in some way. In the later part of the 1700s the manual labour-based economy of some parts of Great Britain began to be replaced by one dominated by the manufacture by machinery. It started with the mechanisation of the textile industries, the development of iron-making techniques and the increased use of refined coal. Trade expansion was enabled by the introduction of canals, improved roads and railways. The introduction of steam power (fuelled primarily by coal) and powered machinery (mainly in textile manufacturing) underpinned the dramatic increases in production capacity. The development of all-metal machine tools in the first two decades of the 19th century facilitated the manufacture of more production machines for manufacturing in other industries. The effects spread throughout Western Europe and North America during the 19th century, eventually affecting most of the world. The impact of this change on society was enormous.

The First Industrial Revolution, which began in the eighteenth century, merged into the Second Industrial Revolution around 1850, when technological and economic progress gained momentum with the development of steam-powered ships, railways, and later in the nineteenth century with the internal combustion engine and electrical power generation. The period of time covered by the Industrial Revolution varies with different historians. Eric Hobsbawm held that it 'broke out' in the 1780s and was not fully felt until the 1830s or 1840s, while T.S. Ashton held that it occurred roughly between 1760 and 1830. Some twentieth century historians such as John Clapham and Nicholas Crafts have argued that the process of economic and social change took place gradually and the term *revolution* is not a true description of what took place. This is still a subject of debate amongst historians. GDP per capita was broadly stable before the Industrial Revolution and the emergence of the modern capitalist economy. The Industrial Revolution began an era of per-capita economic growth in capitalist economies.

Causes

The causes of the Industrial Revolution were complicated and remain a topic for debate, with some historians feeling the Revolution as an outgrowth of social and institutional changes brought by the end of feudalism in Britain after the English Civil War in the 17th century. As national border controls became more effective, the spread of disease was lessened, therefore preventing the epidemics common in previous times. The percentage of children who lived past infancy rose significantly, leading to a larger workforce. The Enclosure movement and the British Agricultural Revolution made food production more efficient and less labour-intensive, forcing the surplus population who could no longer find employment in agriculture into cottage industry, for example weaving, and in the longer term into the cities and the newly developed factories. The colonial expansion of the 17th century with the accompanying development of international trade, creation of financial markets and accumulation of capital are also cited as factors, as is the scientific revolution of the 17th century.

Technological innovation was the heart of the Industrial Revolution and the key enabling technology was the invention and improvement of the steam engine. Historian Lewis Mumford has proposed that the Industrial Revolution had its origins in the early Middle Ages, much earlier than most estimates. He explains that the model for standardised mass production was the printing press and that "the archetypal model for the industrial era was the clock". He also cites the monastic emphasis on order and time-keeping, as well as the fact that mediaeval cities had at their centre a church with bell ringing at regular intervals as being necessary precursors to a greater synchronisation necessary for later, more physical, manifestations such as the steam engine.

The presence of a large domestic market should also be considered an important driver of the Industrial Revolution, particularly explaining why it occurred in Britain. In other nations, such as France, markets were split up by local regions, which often imposed tolls and tariffs on goods traded amongst them. Governments' grant of limited monopolies to inventors under a developing patent system (the Statute of Monopolies 1623) is considered an influential factor. The effects of patents, both good and ill, on the development of industrialisation are clearly illustrated in the history of the steam engine, the key enabling technology. In return for publicly revealing the workings of an invention the patent system rewards inventors by allowing, e.g, James Watt to monopolise the production of the first steam engines, thereby enabling inventors and increasing the pace of technological development. However, monopolies bring with them their own inefficiencies which may counterbalance, or even overbalance, the beneficial effects of publicising ingenuity and rewarding inventors. Watt's monopoly may have prevented other inventors, such as Richard Trevithick, William Murdoch or Jonathan Hornblower, from introducing improved steam engines thereby retarding the industrial revolution by up to 20 years.

Causes for Occurrence in Europe

One question of active interest to historians is why the industrial revolution occurred in Europe and not in other parts of the world in the 18th century, particularly China, India, and the Middle East, or at other times like in Classical Antiquity or the Middle Ages. Numerous factors have been suggested, including ecology, government, and culture. Benjamin Elman argues that China was in a high level equilibrium trap in which the non-industrial methods were efficient enough to prevent use of industrial methods with high costs of capital. Kenneth Pomeranz, in the *Great Divergence*, argues that Europe and China were remarkably similar in 1700, and that the crucial differences which created the Industrial Revolution in Europe were sources of coal near manufacturing centres, and raw materials such as food and wood from the New World, which allowed Europe to expand economically in a way that China could not. However, most historians contest the assertion that Europe and China were roughly equal because modern estimates of per capita income on Western Europe in the late 18th century are of roughly 1,500 dollars in purchasing power parity (and Britain had a per capita income of nearly 2,000 dollars) whereas China, by comparison, had only 450 dollars. Also, the average interest rate was about 5% in Britain and over 30% in China, which illustrates how capital was much more abundant in Britain; capital that was available for investment.

Some historians such as David Landes and Max Weber credit the different belief systems in China and Europe with dictating where the revolution occurred. The religion and beliefs of Europe were largely products of Judaeo-Christianity, and Greek thought. Conversely, Chinese society was founded on men like Confucius, Mencius, Han Feizi (Legalism), Lao Tzu (Taoism), and Buddha (Buddhism). The key difference between these belief systems was that those from Europe focused on the individual, while Chinese beliefs centred around relationships between people. The family unit was more important than the individual for the large majority of Chinese history, and this may have played a role in why the Industrial Revolution took much longer to occur in China. There was the additional difference as to whether people looked backwards to a reputedly glorious past for answers to their questions or looked hopefully to the future. Furthermore, Western European peoples had experienced the Renaissance, Reformation and Enlightenment; other parts of the world had not had a similar intellectual breakout, a condition that holds true even into the 21st century. Regarding India, the Marxist historian Rajani Palme Dutt said: "The capital to finance the Industrial Revolution in India instead went into financing the Industrial Revolution in England." In contrast to China, India was split up into many competing kingdoms, with the three major ones being the Marathas, Sikhs and the Mughals. In addition, the economy was highly dependent on two sectors—agriculture of subsistence and cotton, and technical innovation was non-existent. The vast amounts of wealth were stored away in palace treasuries, and as such, were easily moved to Britain.

Causes for Occurrence in Britain

The debate about the start of the Industrial Revolution also concerns the massive lead that Great Britain had over other countries. Some have stressed the importance of natural or financial resources that Britain received from its many overseas colonies or that profits from the British slave trade between Africa and the Caribbean helped fuel industrial investment. It has been pointed out, however, that slavery provided only 5% of the British national income during the years of the Industrial Revolution.

Alternatively, the greater liberalisation of trade from a large merchant base may have allowed Britain to produce and use emerging scientific and technological developments more effectively than countries with stronger monarchies, particularly China and Russia. Britain emerged from the Napoleonic Wars as the only European nation not ravaged by financial plunder and economic collapse, and possessing the only merchant fleet of any useful size (European merchant fleets having been destroyed during the war by the Royal Navy). Britain's extensive exporting cottage industries also ensured markets were already available for many early forms of manufactured goods. The conflict resulted in most British warfare being conducted overseas, reducing the devastating effects of territorial conquest that affected much of Europe. This was further aided by Britain's geographical position — an island separated from the rest of mainland Europe.

Another theory is that Britain was able to succeed in the Industrial Revolution due to the availability of key resources it possessed. It had a dense population for its small geographical size. Enclosure of common land and the related Agricultural Revolution made a supply of this labour readily available. There was also a local coincidence of natural resources in the North of England, the English Midlands, South Wales and the Scottish Lowlands. Local supplies of coal, iron, lead, copper, tin, limestone and water power, resulted in excellent conditions for the development and expansion of industry. Also, the damp, mild weather conditions of the North West of England provided ideal conditions for the spinning of cotton, providing a natural starting point for the birth of the textiles industry. The stable political situation in Britain from around 1688, and British society's greater receptiveness to change (compared with other European countries) can also be said to be factors favouring the Industrial Revolution. In large part due to the Enclosure movement, the peasantry was destroyed as significant source of resistance to industrialisation, and the landed upper classes developed commercial interests that made them pioneers in removing obstacles to the growth of capitalism. (This point is also made in Hilaire Belloc's The Servile State.)

Protestant work Ethic

Another theory is that the British advance was due to the presence of an entrepreneurial class which believed in progress, technology and hard work. The existence of this class is often linked to the Protestant work ethic (see Max Weber) and the particular status of the Baptists and the dissenting Protestant sects, such as the Quakers and Presbyterians that had flourished with the English Civil War. Reinforcement of confidence in the rule of law, which followed establishment of the prototype of constitutional monarchy in Britain in the Glorious Revolution of 1688, and the emergence of a stable financial market there based on the management of the national debt by the Bank of England, contributed to the capacity for, and interest in, private financial investment in industrial ventures.

Dissenters found themselves barred or discouraged from almost all public offices, as well as education at England's only two universities at the time (although dissenters were still free to study at Scotland's four universities). When the restoration of the monarchy took place and membership in the official Anglican Church became mandatory due to the Test Act, they thereupon became active in banking, manufacturing and education. The Unitarians, in particular, were very involved in education, by running Dissenting Academies, where, in contrast to the universities of Oxford and Cambridge and schools such as Eton and Harrow, much attention was given to mathematics and the sciences —areas of scholarship vital to the development of manufacturing technologies.

Historians sometimes consider this social factor to be extremely important, along with the nature of the national economies involved. While members of these sects were excluded from certain circles of the government, they were considered fellow Protestants, to a limited extent, by many in the middle class, such as traditional financiers or other businessmen. Given this relative tolerance and the supply of capital, the natural outlet for the more enterprising members of these sects would be to seek new opportunities in the technologies created in the wake of the scientific revolution of the 17th century.

Innovations

The commencement of the Industrial Revolution is closely linked to a small number of innovations, made in the second half of the 18th century:

- Textiles - Cotton spinning using Richard Arkwright's water frame, James Hargreaves's Spinning Jenny, and Samuel Crompton's Spinning Mule (a combination of the Spinning Jenny and the Water Frame). This was patented in 1769 and so came out of patent in 1783. The end of the patent was rapidly followed by the erection of many cotton mills. Similar technology was subsequently applied to spinning worsted yarn for various textiles and flax for linen.

- Steam power - The improved steam engine invented by James Watt was initially mainly used for pumping out mines, but from the 1780s was applied to power machines. This enabled rapid development of efficient semi-automated factories on a previously unimaginable scale in places where waterpower was not available.
- Iron founding - In the Iron industry, coke was finally applied to all stages of iron smelting, replacing charcoal. This had been achieved much earlier for lead and copper as well as for producing pig iron in a blast furnace, but the second stage in the production of bar iron depended on the use of potting and stamping (for which a patent expired in 1786) or puddling (patented by Henry Cort in 1783 and 1784).

These represent three 'leading sectors', in which there were key innovations, which allowed the economic take off by which the Industrial Revolution is usually defined. This is not to belittle many other inventions, particularly in the textile industry. Without some earlier ones, such as spinning jenny and flying shuttle in the textile industry and the smelting of pig iron with coke, these achievements might have been impossible. Later inventions such as the power loom and Richard Trevithick's high pressure steam engine were also important in the growing industrialisation of Britain. The application of steam engines to powering cotton mills and ironworks enabled these to be built in places that were most convenient because other resources were available, rather than where there was water to power a mill.

In the textile sector, such mills became the model for the organisation of human labour in factories, epitomised by Cottonopolis, the name given to the vast collection of cotton mills, factories and administration offices based in Manchester. The assembly line system greatly improved efficiency, both in this and other industries. With a series of men trained to do a single task on a product, then having it moved along to the next worker, the number of finished goods also rose significantly.

Also important was the 1756 rediscovery of concrete (based on hydraulic lime mortar) by the British engineer John Smeaton, which had been lost for 13 centuries.

Transfer of Knowledge

Knowledge of new innovation was spread by several means. Workers who were trained in the technique might move to another employer or might be poached. A common method was for someone to make a study tour, gathering information where he could. During the whole of the Industrial Revolution and for the century before, all European countries and America engaged in study-touring; some nations, like Sweden and France, even trained civil servants or technicians to undertake it as a matter of state policy. In other countries, notably Britain and America, this practice was carried out by individual manufacturers anxious to improve their own methods. Study tours were common then, as now, as was the keeping of travel diaries. Records made by industrialists and technicians of the period are an incomparable source of information about their methods.

Another means for the spread of innovation was by the network of informal philosophical societies, like the Lunar Society of Birmingham, in which members met to discuss 'natural philosophy' (*i.e.* science) and often its application to manufacturing. The Lunar Society flourished from 1765 to 1809, and it has been said of them, "They were, if you like, the revolutionary committee of that most far reaching of all the eighteenth century revolutions, the Industrial Revolution". Other such societies published volumes of proceedings and transactions. For example, the London-based Royal Society of Arts published an illustrated volume of new inventions, as well as papers about them in its annual *Transactions*.

There were publications describing technology. Encyclopaedias such as Harris's *Lexicon Technicum* (1704) and Dr Abraham Rees's *Cyclopaedia* (1802-1819) contain much of value. *Cyclopaedia* contains an enormous amount of information about the science and technology of the first half of the Industrial Revolution, very well illustrated by fine engravings. Foreign printed sources such as the *Descriptions des Arts et Métiers* and Diderot's *Encyclopédie* explained foreign methods with fine engraved plates.

Periodical publications about manufacturing and technology began to appear in the last decade of the 18th century, and many regularly included notice of the latest patents. Foreign periodicals, such as the Annales des Mines, published accounts of travels made by French engineers who observed British methods on study tours.

Technological Developments in Britain

Textile Manufacture

In the early 18th century, British textile manufacture was based on wool which was processed by individual artisans, doing the spinning and weaving on their own premises. This system is called a cottage industry. Flax and cotton were also used for fine materials, but the processing was difficult because of the pre-processing needed, and thus goods in these materials made only a small proportion of the output.

Use of the spinning wheel and hand loom restricted the production capacity of the industry, but incremental advances increased productivity to the extent that manufactured cotton goods became the dominant British export by the early decades of the 19th century. India was displaced as the premier supplier of cotton goods.

Lewis Paul patented the Roller Spinning machine and the flyer-and-bobbin system for drawing wool to a more even thickness, developed with the help of John Wyatt in Birmingham. Paul and Wyatt opened a mill in Birmingham which used their new rolling machine powered by a donkey. In 1743, a factory was opened in Northampton with fifty spindles on each of five of Paul and Wyatt's machines. This operated until about 1764. A similar mill was built by Daniel Bourn in Leominster, but this burnt down. Both Lewis Paul and Daniel Bourn patented carding machines in 1748. Using two sets of rollers that travelled at different speeds, it was later used in the first cotton spinning mill. Lewis's invention was later developed and improved by Richard Arkwright in his water frame and Samuel Crompton in his spinning mule.

Other inventors increased the efficiency of the individual steps of spinning (carding, twisting and spinning, and rolling) so that the supply of yarn increased greatly, which fed a weaving industry that was advancing with improvements to shuttles and the loom or 'frame'. The output of an individual labourer increased dramatically, with the effect that the new machines were seen as a threat to employment, and early innovators were attacked and their inventions destroyed.

To capitalise upon these advances, it took a class of entrepreneurs, of which the most famous is Richard Arkwright. He is credited with a list of inventions, but these were actually developed by people such as Thomas Highs and John Kay; Arkwright nurtured the inventors, patented the ideas, financed the initiatives, and protected the machines. He created the cotton mill which brought the production processes together in a factory, and he developed the use of power — first horse power and then water power — which made cotton manufacture a mechanised industry. Before long steam power was applied to drive textile machinery.

Metallurgy

The major change in the metal industries during the era of the Industrial Revolution was the replacement of organic fuels based on wood with fossil fuel based on coal. Much of this happened somewhat before the Industrial Revolution, based on innovations by Sir Clement Clerke and others from 1678, using coal reverberatory furnaces known as cupolas. These were operated by the flames, which contained carbon monoxide, playing on the ore and reducing the oxide to metal. This has the advantage that impurities (such as sulphur) in the coal do not migrate into the metal. This technology was applied to lead from 1678 and to copper from 1687. It was also applied to iron foundry work in the 1690s, but in this case the reverberatory furnace was known as an air furnace. The foundry cupola is a different (and later) innovation.

This was followed by Abraham Darby, who made great strides using coke to fuel his blast furnaces at Coalbrookdale in 1709. However, the coke pig iron he made was used mostly for the production of cast iron goods such as pots and kettles. He had the advantage over his rivals in that his pots, cast by his patented process, were thinner and cheaper than theirs. Coke pig iron was hardly used to produce bar iron in forges until the mid 1750s, when his son Abraham Darby II built Horsehay and Ketley furnaces (not far from Coalbrookdale). By then, coke pig iron was cheaper than charcoal pig iron.

Bar iron for smiths to forge into consumer goods was still made in finery forges, as it long had been. However, new processes were adopted in the ensuing years. The first is referred to today as potting and stamping, but this was superseded by Henry Cort's puddling process. From 1785, perhaps because the improved version of potting and stamping was about to come out of patent, a great expansion in the output of the British iron industry began. The new processes did not depend on the use of charcoal at all and were therefore not limited by charcoal sources.

Up to that time, British iron manufacturers had used considerable amounts of imported iron to supplement native supplies. This came principally from Sweden from the mid 17th century and later also from Russia from the end of the 1720s. However, from 1785, imports decreased because of the new iron making technology, and Britain became an exporter of bar iron as well as manufactured wrought iron consumer goods.

Since iron was becoming cheaper and more plentiful, it also became a major structural material following the building of the innovative The Iron Bridge in 1778 by Abraham Darby III.

An improvement was made in the production of steel, which was an expensive commodity and used only where iron would not do, such as for the cutting edge of tools and for springs. Benjamin Huntsman developed his crucible steel technique in the 1740s. The raw material for this was blister steel, made by the cementation process.

The supply of cheaper iron and steel aided the development of boilers and steam engines, and eventually railways. Improvements in machine tools allowed better working of iron and steel and further boosted the industrial growth of Britain.

Mining

Coal mining in Britain, particularly in South Wales started early. Before the steam engine, pits were often shallow bell pits following a seam of coal along the surface, which were abandoned as the coal was extracted. In other cases, if the geology was favourable, the coal was mined by means of an adit or drift mine driven into the side of a hill. Shaft mining was done in some areas, but the limiting factor was the problem of removing water. It could be done by hauling buckets of water up the shaft or to a sough (a tunnel driven into a hill to drain a mine). In either case, the water had to be discharged into a stream or ditch at a level where it could flow away by gravity. The introduction of the steam engine greatly facilitated the removal of water and enabled shafts to be made deeper, enabling more coal to be extracted. These were developments that had begun before the Industrial Revolution, but the adoption of James Watt's more efficient steam engine from the 1770s reduced the fuel costs of engines, making mines more profitable. Coal mining was very dangerous owing to the presence of firedamp in many coal seams. Some degree of safety was provided by the safety lamp which was invented in 1816 by Sir Humphrey Davy and independently by George Stephenson. However, the lamps proved a false dawn because they became unsafe very quickly and provided a weak light. Firedamp explosions continued , often setting off coal dust explosions, so casualties grew during the entire nineteenth century. Conditions of work were very poor, with a high casualty rate from rock falls.

Steam Power

The development of the stationary steam engine was an essential early element of the Industrial Revolution; however, for most of the period of the Industrial Revolution, the majority of industries still relied on wind and water power as well as horse and man-power for driving small machines.

The first real attempt at industrial use of steam power was due to Thomas Savery in 1698. He constructed and patented in London a low-lift combined vacuum and pressure water pump, that generated about one horsepower (hp) and was used as in numerous water works and tried in a few mines (hence its "brand name", *The miner's Friend*), but it was not a success since it was limited in pumping height and prone to boiler explosions.

The first safe and successful steam power plant was introduced by Thomas Newcomen from 1719. Newcomen apparently conceived his machine quite independently of Savery, but as the latter had taken out a very wide-ranging patent, Newcomen and his associates were obliged to come to an arrangement with him, marketing the engine until 1733 under a joint patent. Newcomen's engine appears to have been based on Papin's experiments carried out 30 years earlier, and employed a piston and cylinder, one end of which was open to the atmosphere above the piston. Steam just above atmospheric pressure (all that the boiler could stand) was introduced into the lower half of the cylinder beneath the piston during the gravity-induced upstroke; the steam was then condensed by a jet of cold water injected into the steam space to produce a partial vacuum; the pressure differential between the atmosphere and the vacuum on either side of the piston displaced it downwards into the cylinder, raising the opposite end of a rocking beam to which was attached a gang of gravity-actuated reciprocating force pumps housed in the mineshaft. The engine's downward power stroke raised the pump, priming it and preparing the pumping stroke. At first the phases were controlled by hand, but within ten years an escapement mechanism had been devised worked by of a vertical *plug tree* suspended from the rocking beam which rendered the engine self-acting.

A number of Newcomen engines were successfully put to use in Britain for draining hitherto unworkable deep mines, with the engine on the surface; these were large machines, requiring a lot of capital to build, and produced about 5 hp (3.7 kW). They were extremely inefficient by modern standards, but when located where coal was cheap at pit heads, opened up a great expansion in coal mining by allowing mines to go deeper. Despite their disadvantages, Newcomen engines were reliable and easy to maintain and continued to be used in the coalfields until the early decades of the nineteenth century. By 1729, when Newcomen died, his engines had spread (first) to Hungary in 1722 ,Germany, Austria, and Sweden. A total of 110 are known to have been built by 1733 when the joint patent expired, of which 14 were abroad. In the 1770s, the engineer John Smeaton built some very large examples and introduced a number of improvements. A total of 1,454 engines had been built by 1800.

A fundamental change in working principles was brought about by James Watt. With the close collaboration Matthew Boulton, he had succeeded by 1778 in perfecting his steam engine, which incorporated a series of radical improvements, notably the closing off of the upper part of the cylinder thereby making the low pressure steam drive the top of the piston instead of the atmosphere, use of a steam jacket and the celebrated separate steam condenser chamber. All this meant that a more constant temperature could be maintained in the cylinder and that engine efficiency no longer varied according to atmospheric conditions. These improvements increased engine efficiency by a factor of about five, saving 75% on coal costs.

Nor could the atmospheric engine be easily adapted to drive a rotating wheel, although Wasborough and Pickard did succeed in doing so towards 1780. However by 1783 the more economical Watt steam engine had been fully developed into a double-acting rotative type, which meant that it could be used to directly drive the rotary machinery of a factory or mill. Both of Watt's basic engine types were commercially very successful, and by 1800, the firm Boulton & Watt had constructed 496 engines, with 164 driving reciprocating pumps, 24 serving blast furnaces, and 308 powering mill machinery; most of the engines generated from 5 to 10 hp (7.5 kW).

The development of machine tools, such as the lathe, planing and shaping machines powered by these engines, enabled all the metal parts of the engines to be easily and accurately cut and in turn made it possible to build larger and more powerful engines.

Until about 1800, the most common pattern of steam engine was the beam engine, built as an integral part of a stone or brick engine-house, but soon various patterns of self-contained portative engines (readily removable, but not on wheels) were developed, such as the table engine. Towards the turn of the 19th century, the Cornish engineer Richard Trevithick, and the American, Oliver Evans began to construct higher pressure non-condensing steam engines, exhausting against the atmosphere. This allowed an engine and boiler to be combined into a single unit compact enough to be used on mobile road and rail locomotives and steam boats.

In the early 19th century after the expiration of Watt's patent, the steam engine underwent many improvements by a host of inventors and engineers.

Chemicals

The large scale production of chemicals was an important development during the Industrial Revolution. The first of these was the production of sulphuric acid by the lead chamber process invented by the Englishman John Roebuck (James Watt's first partner) in 1746. He was able to greatly increase the scale of the manufacture by replacing the relatively expensive glass vessels formerly used with larger, less expensive chambers made of riveted sheets of lead. Instead of a few pounds at a time, he was able to make a hundred pounds (45 kg) or so at a time in each of the chambers.

The production of an alkali on a large scale became an important goal as well, and Nicolas Leblanc succeeded in 1791 in introducing a method for the production of sodium carbonate. The Leblanc process was a reaction of sulphuric acid with sodium chloride to give sodium sulphate and hydrochloric acid. The sodium sulphate was heated with limestone (calcium carbonate) and coal to give a mixture of sodium carbonate and calcium sulphide. Adding water separated the soluble sodium carbonate from the calcium sulphide. The process produced a large amount of pollution (the hydrochloric acid was initially vented to the air, and calcium sulphide was a useless waste product). Nonetheless, this synthetic soda ash proved economical compared to that from burning certain plants (barilla) or from kelp, which were the previously dominant sources of soda ash, and also to potash (potassium carbonate) derived from hardwood ashes.

These two chemicals were very important because they enabled the introduction of a host of other inventions, replacing many small-scale operations with more cost-effective and controllable processes. Sodium carbonate had many uses in the glass, textile, soap, and paper industries. Early uses for sulphuric acid included pickling (removing rust) iron and steel, and for bleaching cloth.

The development of bleaching powder (calcium hypochlorite) by Scottish chemist

Charles Tennant in about 1800, based on the discoveries of French chemist Claude Louis Berthollet, revolutionised the bleaching processes in the textile industry by dramatically reducing the time required (from months to days) for the traditional process then in use, which required repeated exposure to the sun in bleach fields after soaking the textiles with alkali or sour milk. Tennant's factory at St Rollox, North Glasgow, became the largest chemical plant in the world.

In 1824 Joseph Aspdin, a British brick layer turned builder, patented a chemical process for making portland cement which was an important advance in the building trades. This process involves sintering a mixture of clay and limestone to about 1400 °C, then grinding it into a fine powder which is then mixed with water, sand and gravel to produce concrete. Portland cement was used by the famous English engineer Marc Isambard Brunel several years later when constructing the Thames Tunnel. Cement was used on a large scale in the construction of the London sewerage system a generation later.

Machine Tools

The Industrial Revolution could not have developed without machine tools, for they enabled manufacturing machines to be made. They have their origins in the tools developed in the 18th century by makers of clocks and watches and scientific instrument makers to enable them to batch-produce small mechanisms. The mechanical parts of early textile machines were sometimes called 'clock work' because of the metal spindles and gears they incorporated. The manufacture of textile machines drew craftsmen from these trades and is the origin of the modern engineering industry.

Machines were built by various craftsmen—carpenters made wooden framings, and smiths and turners made metal parts. A good example of how machine tools changed manufacturing took place in Birmingham, England, in 1830. The invention of a new machine by William Joseph Gillott, William Mitchell and James Stephen Perry allowed mass manufacture of robust, cheap steel pen nibs; the process had been laborious and expensive. Because of the difficulty of manipulating metal and the lack of machine tools, the use of metal was kept to a minimum. Wood framing had the disadvantage of changing dimensions with temperature and humidity, and the various joints tended to rack (work loose) over time. As the Industrial Revolution progressed, machines with metal frames became more common, but they required machine tools to make them economically. Before the advent of machine tools, metal was worked manually using the basic hand tools of hammers, files, scrapers, saws and chisels. Small metal parts were readily made by this means, but for large machine parts, production was very laborious and costly.

Apart from workshop lathes used by craftsmen, the first large machine tool was the cylinder boring machine used for boring the large-diameter cylinders on early steam engines. The planing machine, the slotting machine and the shaping machine were developed in the first decades of the 19th century. Although the milling machine was invented at this time, it was not developed as a serious workshop tool until during the Second Industrial Revolution.

Military production had a hand in the development of machine tools. Henry Maudslay, who trained a school of machine tool makers early in the 19th century, was employed at the Royal Arsenal, Woolwich, as a young man where he would have seen the large horse-driven wooden machines for cannon boring made and worked by the Verbruggans. He later worked for Joseph Bramah on the production of metal locks, and soon after he began working on his own. He was engaged to build the machinery for making ships' pulley blocks for the Royal Navy in the Portsmouth Block Mills. These were all metal and were the first machines for mass production and making components with a degree of interchangeability. The lessons Maudslay learned about the need for stability and precision he adapted to the development of machine tools, and in his workshops he trained a generation of men to build on his work, such as Richard Roberts, Joseph Clement and Joseph Whitworth.

James Fox of Derby had a healthy export trade in machine tools for the first third of the century, as did Matthew Murray of Leeds. Roberts was a maker of high-quality machine tools and a pioneer of the use of jigs and gauges for precision workshop measurement.

Gas Lighting

Another major industry of the later Industrial Revolution was gas lighting. Though others made a similar innovation elsewhere, the large scale introduction of this was the work of William Murdoch, an employee of Boulton and Watt, the Birmingham steam engine pioneers. The process consisted of the large scale gasification of coal in furnaces, the purification of the gas (removal of sulphur, ammonium, and heavy hydrocarbons), and its storage and distribution. The first gaslighting utilities were established in London between 1812-20. They soon became one of the major consumers of coal in the UK. Gaslighting had in impact on social and industrial organisation because it allowed factories and stores to remain open longer than with tallow candles or oil. Its introduction allowed night life to flourish in cities and towns as interiors and street could be lighted on a larger scale than before.

Glass Making

A new method of producing glass, known as the cylinder process, was developed in Europe during the early 19th century. In 1832, this process was used by the Chance Brothers to create sheet glass. They became the leading producers of window and plate glass. This advancement allowed for larger panes of glass to be created without interruption, thus freeing up the space planning in interiors as well as the fenestration of buildings. The crystal palace is the supreme example of the use of sheet glass in a new and innovative structure.

Agriculture

In the mid nineteenth century John Fowler, an engineer and inventor, began to look at the possibility of using steam engines for ploughing and digging drainage channels. The system that he invented involved either a single stationary engine at the corner of a field drawing a plough via sets of winches and pulleys, or two engines placed at either end of a field drawing the plough backwards and forwards between them by means of a cable attached to winches. Fowler's ploughing system vastly reduced the cost of ploughing farmland compared with horse-drawn ploughs. Also his ploughing system, when used for digging drainage channels, made possible the cultivation of previously unusable swampy land. The traction engine later became a common sight in working threshing machines during haymaking time and ploughing fields.

Transport in Britain

At the beginning of the Industrial Revolution, inland transport was by navigable rivers and roads, with coastal vessels employed to move heavy goods by sea. Railways or wagon ways were used for conveying coal to rivers for further shipment, but canals had not yet been constructed. Animals supplied all of the motive power on land, with sails providing the motive power on the sea.

The Industrial Revolution improved Britain's transport infrastructure with a turnpike road network, a canal, and waterway network, and a railway network. Raw materials and finished products could be moved more quickly and cheaply than before. Improved transportation also allowed new ideas to spread quickly.

Coastal Sail

Sailing vessels had long been used for moving goods round the British coast. The trade transporting coal to London from Newcastle had begun in mediaeval times. The major international seaports such as London, Bristol, and Liverpool, were the means by which raw materials such as cotton might be imported and finished goods exported. Transporting goods onwards within Britain by sea was common during the whole of the Industrial Revolution and only fell away with the growth of the railways at the end of the period.

Navigable Rivers

All the major rivers of the United Kingdom were navigable during the Industrial Revolution. Some were anciently navigable, notably the Severn, Thames, and Trent.

Some were improved, or had navigation extended upstream, but usually in the period before the Industrial Revolution, rather than during it.

The Severn, in particular, was used for the movement of goods to the Midlands which had been imported into Bristol from abroad, and for the export of goods from centres of production in Shropshire (such as iron goods from Coalbrookdale) and the Black Country. Transport was by way of trows—small sailing vessels which could pass the various shallows and bridges in the river. The trows could navigate the Bristol Channel to the South Wales ports and Somerset ports, such as Bridgwater and even as far as France.

Canals

Canals began to be built in the late eighteenth century to link the major manufacturing centres in the Midlands and north with seaports and with London, at that time itself the largest manufacturing centre in the country. Canals were the first technology to allow bulk materials to be easily transported across country. A single canal horse could pull a load dozens of times larger than a cart at a faster pace. By the 1820s, a national network was in existence. Canal construction served as a model for the organisation and methods later used to construct the railways. They were eventually largely superseded as profitable commercial enterprises by the spread of the railways from the 1840s on.

Britain's canal network, together with its surviving mill buildings, is one of the most enduring features of the early Industrial Revolution to be seen in Britain.

Roads

Much of the original British road system was poorly maintained by thousands of local parishes, but from the 1720s (and occasionally earlier) turnpike trusts were set up to charge tolls and maintain some roads. Increasing numbers of main roads were turnpiked from the 1750s to the extent that almost every main road in England and Wales was the responsibility of some turnpike trust. New engineered roads were built by John Metcalf, Thomas Telford and John Macadam. The major turnpikes radiated from London and were the means by which the Royal Mail was able to reach the rest of the country. Heavy goods transport on these roads was by means of slow, broad wheeled, carts hauled by teams of horses. Lighter goods were conveyed by smaller carts or by teams of pack horse. Stage coaches carried the rich, and the less wealthy could pay to ride on carriers carts.

Railways

Wagonways for moving coal in the mining areas had started in the 17th century and were often associated with canal or river systems for the further movement of coal. These were all horse drawn or relied on gravity, with a stationary steam engine to haul the wagons back to the top of the incline. The first applications of the steam locomotive were on wagon or plate ways (as they were then often called from the cast iron plates used). Horse-drawn public railways did not begin until the early years of the 19th century. Steam-hauled public railways began with the Stockton and Darlington Railway in 1825 and the Liverpool and Manchester Railway in 1830. Construction of major railways connecting the larger cities and towns began in the 1830s but only gained momentum at the very end of the first Industrial Revolution.

After many of the workers had completed the railways, they did not return to their rural lifestyles but instead remained in the cities, providing additional workers for the factories.

Railways helped Britain's trade enormously, providing a quick and easy way of transport.

Social Effects

In terms of social structure, the Industrial Revolution witnessed the triumph of a middle class of industrialists and businessmen over a landed class of nobility and gentry. Ordinary working people found increased opportunities for employment in the new mills and factories, but these were often under strict working conditions with long hours of labour dominated by a pace set by machines. However, harsh working conditions were prevalent long before the Industrial Revolution took place. Pre-industrial society was very static and often cruel—child labour, dirty living conditions and long working hours were just as prevalent before the Industrial Revolution.

Factories and Urbanisation

Industrialisation led to the creation of the factory. Arguably the first was John Lombe's water-powered silk mill at Derby, operational by 1721. However, the rise of the factory came somewhat later when cotton spinning was mechanised.

The factory system was largely responsible for the rise of the modern city, as large numbers of workers migrated into the cities in search of employment in the factories. Nowhere was this better illustrated than the mills and associated industries of Manchester, nicknamed "Cottonopolis", and arguably the world's first industrial city. For much of the 19th century, production was done in small mills, which were typically water-powered and built to serve local needs. Later each factory would have its own steam engine and a chimney to give an efficient draft through its boiler.

The transition to industrialisation was not without difficulty. For example, a group of English workers known as Luddites formed to protest against industrialisation and sometimes sabotaged factories. In other industries the transition to factory production was not so divisive. Some industrialists themselves tried to improve factory and living conditions for their workers. One of the earliest such reformers was Robert Owen, known for his pioneering efforts in improving conditions for workers at the New Lanark mills, and often regarded as one of the key thinkers of the early socialist movement.

By 1746, an integrated brass mill was working at Warmley near Bristol. Raw material went in at one end, was smelted into brass and was turned into pans, pins, wire, and other goods. Housing was provided for workers on site. Josiah Wedgwood and Matthew Boulton were other prominent early industrialists, who employed the factory system.

Child Labour

The Industrial Revolution led to a population increase, but the chance of surviving childhood did not improve throughout the industrial revolution (although *infant* mortality rates were improved markedly). There was still limited opportunity for education, and children were expected to work. Employers could pay a child less than an adult even though their productivity was comparable; there was no need for strength to operate an industrial machine, and since the industrial system was completely new there were no experienced adult labourers. This made child labour the labour of choice for manufacturing in the early phases of the Industrial Revolution between the 18th and 19th centuries.

Child labour had existed before the Industrial Revolution, but with the increase in population and education it became more visible. Before the passing of laws protecting children, many were forced to work in terrible conditions for much lower pay than their elders.

Reports were written detailing some of the abuses, particularly in the coal mines and textile factories and these helped to popularise the children's plight. The public outcry, especially among the upper and middle classes, helped stir change in the young workers' welfare.

Politicians and the government tried to limit child labour by law, but factory owners resisted; some felt that they were aiding the poor by giving their children money to buy food to avoid starvation, and others simply welcomed the cheap labour. In 1833 and 1844, the first general laws against child labour, the Factory Acts, were passed in England: Children younger than nine were not allowed to work, children were not permitted to work at night, and the work day of youth under the age of 18 was limited to twelve hours. Factory inspectors supervised the execution of the law. About ten years later, the employment of children and women in mining was forbidden. These laws decreased the number of child labourers; however, child labour remained in Europe up to the 20th century.

Housing

Living conditions during the Industrial Revolution varied from the splendour of the homes of the owners to the squalor of the lives of the workers. Cliffe Castle, Keighley, is a good example of how the newly rich chose to live. This is a large home modelled loosely on a castle with towers and garden walls. The home is very large and was surrounded by a massive garden, the Cliffe Castle is now open to the public as a museum.

Poor people lived in very small houses in cramped streets. These homes would share toilet facilities, have open sewers and would be at risk of damp. Disease was spread through a contaminated water supply. Conditions did improve during the 19th century as public health acts were introduced covering things such as sewage, hygiene and making some boundaries upon the construction of homes. Not everybody lived in homes like these. The Industrial Revolution created a larger middle class of professionals such as lawyers and doctors. The conditions for the poor improved over the course of the 19th century because of government and local plans which led to cities becoming cleaner places, but life had not been easy for the poor before industrialisation. However, as a result of the Revolution, huge numbers of the working class died due to diseases spreading through the cramped living conditions. Chest diseases from the mines, cholera from polluted water and typhoid were also extremely common, as was smallpox. Accidents in factories with child and female workers were regular. Dickens' novels perhaps best illustrate this; even some government officials were horrified by what they saw. Strikes and riots by workers were also relatively common.

Luddites

The rapid industrialisation of the English economy cost many craft workers their jobs. The movement started first with lace and hosiery workers near Nottingham and spread to other areas of the textile industry owing to early industrialisation. Many weavers also found themselves suddenly unemployed since they could no longer compete with machines which only required relatively limited (and unskilled) labour to produce more cloth than a single weaver. Many such unemployed workers, weavers and others, turned their animosity towards the machines that had taken their jobs and began destroying factories and machinery. These attackers became known as Luddites, supposedly followers of Ned Ludd, a folklore figure. The first attacks of the Luddite movement began in 1811. The Luddites rapidly gained popularity, and the British government took drastic measures using the militia or army to protect industry. Those rioters who were caught were tried and hanged, or transported for life.

Unrest continued in other sectors as they industrialised, such as agricultural labourers in the 1830s, when large parts of southern Britain were affected by the Captain Swing disturbances. Threshing machines were a particular target, and rick burning was a popular activity. The riots led however, to the first formation of trade unions, and further pressure for reform.

Organisation of Labour

The Industrial Revolution concentrated labour into mills, factories and mines, thus facilitating the organisation of *combinations* or trade unions to help advance the interests of working people. The power of a union could demand better terms by withdrawing all labour and causing a consequent cessation of production. Employers had to decide between giving in to the union demands at a cost to themselves or suffer the cost of the lost production. Skilled workers were hard to replace, and these were the first groups to successfully advance their conditions through this kind of bargaining.

The main method the unions used to effect change was strike action. Many strikes were painful events for both sides, the unions and the management. In England, the Combination Act forbade workers to form any kind of trade union from 1799 until its repeal in 1824. Even after this, unions were still severely restricted.

In 1832, the year of the Reform Act which extended the vote in England but did not grant universal suffrage, six men from Tolpuddle in Dorset founded the Friendly Society of Agricultural Labourers to protest against the gradual lowering of wages in the 1830s. They refused to work for less than 10 shillings a week, although by this time wages had been reduced to seven shillings a week and were due to be further reduced to six shillings. In 1834 James Frampton, a local landowner, wrote to the Prime Minister, Lord Melbourne, to complain about the union, invoking an obscure law from 1797 prohibiting people from swearing oaths to each other, which the members of the Friendly Society had done. James Brine, James Hammett, George Loveless, George's brother James Loveless, George's brother in-law Thomas Standfield, and Thomas's son John Standfield were arrested, found guilty, and transported to Australia. They became known as the Tolpuddle martyrs. In the 1830s and 1840s the Chartist movement was the first large scale organised working class political movement which campaigned for political equality and social justice. Its *Charter* of reforms received over three million signatures but was rejected by Parliament without consideration.

Working people also formed friendly societies and co-operative societies as mutual support groups against times of economic hardship. Enlightened industrialists, such as Robert Owen also supported these organisations to improve the conditions of the working class.

Unions slowly overcame the legal restrictions on the right to strike. In 1842, a General Strike involving cotton workers and colliers was organised through the Chartist movement which stopped production across Great Britain.

Eventually effective political organisation for working people was achieved through the trades unions who, after the extensions of the franchise in 1867 and 1885, began to support socialist political parties that later merged to became the British Labour Party.

Other Effects

The application of steam power to the industrial processes of printing supported a massive expansion of newspaper and popular book publishing, which reinforced rising literacy and demands for mass political participation. During the Industrial Revolution, the life expectancy of children increased dramatically. The percentage of the children born in London who died before the age of five decreased from 74.5% in 1730 - 1749 to 31.8% in 1810 - 1829. Also, there was a significant increase in worker wages during the period 1813-1913.

Industrial Revolution Elsewhere

United States

As in Britain, the United States originally used water power to run its factories, with the consequence that industrialisation was essentially limited to New England and the rest of the Northeastern United States, where fast-moving rivers were located. However, the raw materials (cotton) came from the Southern United States. It was not until after the American Civil War in the 1860s that steam-powered manufacturing overtook water-powered manufacturing, allowing the industry to fully spread across the nation.

Samuel Slater (1768–1835) is popularly known as the founder of the American cotton industry. As a boy apprentice in Derbyshire, England he learnt of the new techniques in the textile industry and defied laws against the emigration of skilled workers by leaving for New York in 1789, hoping to make money with his knowledge. Slater started Slater's mill at Pawtucket, Rhode Island, in 1793 and went on to own thirteen textile mills. Daniel Day established a wool carding mill in the Blackstone Valley at Uxbridge, Massachusetts in 1810, the third woollen mill established in the U.S. (The first was in Hartford, CT, and the second at Watertown, MA). The John H. Chafee Blackstone River Valley National Heritage Corridor retraces the history of

"America's Hardest working River', the Blackstone. The Blackstone River and its tributaries, which cover more than 45 miles (72 km) from Worcester to Providence, was the birthplace of America's Industrial Revolution. At its peak over 1100 mills operated in this valley, including Slater's mill, and with it the earliest beginnings of America's Industrial and Technological Development. (see also Blackstone River Valley National Heritage Corridor)

While on a trip to England in 1810, Newburyport, Massachusetts merchant Francis Cabot Lowell was allowed to tour the British textile factories, but not take notes. Realising the War of 1812 had ruined his import business but that a market for domestic finished cloth was emerging in America, he memorised the design of textile machines, and on his return to the United States, he set up the Boston Manufacturing Company. Lowell and his partners built America's first cotton-to-cloth textile mill at Waltham, Massachusetts. After his death in 1817, his Associates built America's first planned factory town, which they named after him. This enterprise was capitalised in a public stock offering, one of the first uses of it in the United States. Lowell, Massachusetts, utilising 5.6 miles (9.0 km) of canals and ten thousand horsepower delivered by the Merrimack River, is considered the 'Cradle of the American Industrial Revolution'. The short-lived utopia-like Lowell System was formed, as a direct response to the poor working conditions in Britain. However, by 1850, especially following the Irish Potato Famine, the system had been replaced by poor immigrant labour.

Continental Europe

The Industrial Revolution on Continental Europe came later than in Great Britain. In many industries, this involved the application of technology developed in Britain in new places. Often the technology was purchased from Britain or British engineers and entrepreneurs moved abroad in search of new opportunities. By 1809 part of the Ruhr Valley in Westphalia were being called *Miniature England* because of its similarities to the industrial areas of England. The German, Russian and Belgian governments all provided state funding to the new industries. In some cases (such as iron), the different availability of resources locally meant that only some aspects of the British technology were adopted.

Japan

In 1871 a group of Japanese politicians known as the Iwakura Mission toured Europe and the USA to learn western ways. The result was a deliberate state led industrialisation policy to prevent Japan from falling behind. The Bank of Japan, founded in 1877, used taxes to fund model steel and textile factories. Education was expanded and Japanese students were sent to study in the west.

Second Industrial Revolutions and Later Evolution

The insatiable demand of the railways for more durable rail led to the development of the means to cheaply mass-produce steel. Steel is often cited as the first of several new areas for industrial mass-production, which are said to characterise a "Second Industrial Revolution", beginning around 1850, although a method for mass manufacture of steel was not invented until the 1860s, when Sir Henry Bessemer invented a new furnace which could make wrought iron and steel in large quantities. However, it only became widely available in the 1870s. This second Industrial Revolution gradually grew to include the chemical industries, petroleum refining and distribution, electrical industries, and, in the twentieth century, the automotive industries, and was marked by a transition of technological leadership from Britain to the United States and Germany.

The introduction of hydroelectric power generation in the Alps enabled the rapid industrialisation of coal-deprived northern Italy, beginning in the 1890s. The increasing availability of economical petroleum products also reduced the importance of coal and further widened the potential for industrialisation.

Marshall McLuhan analysed the social and cultural impact of the electric age. While the previous age of mechanisation had spread the idea of splitting every process into a sequence, this was ended by the introduction of the instant speed of electricity that brought simultaneity. This imposed the cultural shift from the approach of focusing on "specialized segments of attention" (adopting one particular perspective), to the idea of "instant sensory awareness of the whole", an attention to the "total field", a "sense of the whole pattern". It made evident and prevalent the sense of "form and function as a unity", an "integral idea of structure and configuration". This had major impact in the disciplines of painting (with cubism), physics, poetry, communication and educational theory.

By the 1890s, industrialisation in these areas had created the first giant industrial corporations with burgeoning global interests, as companies like U.S. Steel, General Electric, and Bayer AG joined the railroad companies on the world's stock markets.

Intellectual Paradigms and Criticism

Capitalism

The advent of The Enlightenment provided an intellectual framework which welcomed the practical application of the growing body of scientific knowledge — a factor evidenced in the systematic development of the steam engine, guided by scientific analysis, and the development of the political and sociological analyses, culminating in Adam Smith's The Wealth of Nations. One of the main arguments for capitalism, presented for example in the book *The Improving State of the World*, is that industrialisation increases wealth for all, as evidenced by raised life expectancy, reduced working hours, and no work for children and the elderly.

Marxism

Marxism is essentially a reaction to the Industrial Revolution. According to Karl Marx, industrialisation polarised society into the bourgeoisie (those who own the means of production, the factories and the land) and the much larger proletariat (the working class who actually perform the labour necessary to extract something valuable from the means of production). He saw the industrialisation process as the logical dialectical progression of feudal economic modes, necessary for the full development of capitalism, which he saw as in itself a necessary precursor to the development of socialism and eventually communism. ;

Romanticism

During the Industrial Revolution an intellectual and artistic hostility towards the new industrialisation developed. This was known as the Romantic movement. Its major exponents in English included the artist and poet William Blake and poets William Wordsworth, Samuel Taylor Coleridge, John Keats, Byron and Percy Bysshe Shelley. The movement stressed the importance of "nature" in art and language, in contrast to 'monstrous' machines and factories; the "*Dark satanic mills*" of Blake's poem *And did those feet in ancient time*. Mary Shelley's short story *Frankenstein* reflected concerns that scientific progress might be two-edged.

History of the Name

The term *Industrial Revolution* applied to technological change was common in the 1830s. Louis-Auguste Blanqui in 1837 spoke of *la révolution industrielle*. Friedrich Engels in *The Condition of the Working Class in England in 1844* spoke of "an industrial revolution, a revolution which at the same time changed the whole of civil society."

In his book *Keywords: A Vocabulary of Culture and Society*, Raymond Williams states in the entry for Industry: *The idea of a new social order based on major industrial change was clear in Southey and Owen, between 1811 and 1818, and was implicit as early as Blake in the early 1790s and Wordsworth at the turn of the century.*

Credit for popularising the term may be given to Arnold Toynbee, whose lectures given in 1881 gave a detailed account of the process.

REFERENCES

Acharya, H.P. 1984. Management of forest resources in Nepal: A case study of Madan Pokhara Village Panchayat. Ms. thesis, Cornell University.

Arnold, J.E.M., and W.C Stewart. 1991. *Common property resource management in India.* Tropical Forestry Paper no. 24. Oxford: Oxford Forestry Institute, University of Oxford.

Ashton, Thomas S., *The Industrial Revolution (1760-1830)*, Oxford University Press, 1948,.

B. Trinder, *The Industrial Revolution in Shropshire* (3rd edn, 2000).

Barrington Moore, Jr., *Social Origins of Dictatorship and Democracy: Lord and Peasant in the Making of the Modern World*, pp. 29-30, Boston, Beacon Press, 1966.

Berkes, Fikret, ed. 1989. *Common property resources: Ecology and community-based sustainable development.* London: Belhaven Press.

Berkes, Fikret. 1985. Fishermen and the 'tragedy of the commons.' *Environmental Conservation* 12:199-206.

Berlanstein; Lenard R. *The Industrial Revolution and work in nineteenth-century Europe* Routledge, 1992 online edition

Bernal, John Desmond, *Science and Industry in the Nineteenth Century*, Bloomington: Indiana University Press, 1970.

Bhaumik, Sankar Kumar. 1991. Tenancy and resource allocation: A review with fresh data. *Indian Journal of Agricultural Economics* 46(4):568-75.

Bromley, Daniel W. 1986. Closing comments at the conference on common property resource management. In *Proceedings of the conference on common* property *resource management.* The National Research Council. Washington, D.C.: National Academy Press.

Bromley, Daniel W., and M.M. Cernea. 1989. *The management of common property resources: Some conceptual and operational fallacies.* Discussion Paper no. 57. Washington, D.C: World Bank.

Bromley,, Daniel W. 1989. *Economic interests and institutions: The conceptual foundations of public policy.* New York: Basil Blackwell.

Bromley,, Daniel W. 1991. *Environment and economy. Property rights and public policy.* Cambridge Massachusetts: Basil Blackwell.

C. K. Hyde, *Technological change and the British iron industry 1700-1870* (Princeton University Press, Princeton NJ 1977).

Chadha, G.K., and S.K. Bhaumik. 1992. Changing tenancy relations in West Bengal: Popular notions and grassroots realities. *Economic and Political Weekly* 27(19):1009-17 and 27(20-21):1089-98.

Chambers, Robert, N.C. Saxena, and Tushaar Shah. 1989. To *the hands of the poor: Water and trees.* New Delhi, Oxford & IBH Publishing Co.

Cheung, Steven N.S. 1969. *The theory of share tenancy.* Chicago: University of Chicago Press.

Cheung, Steven N.S. 1970. The structure of a contract and the theory of a non-exclusive resource. *Journal of Loss and Economics* 13 (April): 49-70.

Chopra, Kanchan, G.K. Kadekodi, and M.N. Murty. 1990. *Participatory development: People and common property resources.* New Delhi: Sage Publications.

Ciriacy-Wantrup, S.V. 1968. *Resource conservation, economics and policies.* Berkeley: University of California Press.

Ciriacy-Wantrup, S.V., and Richard C. Bishop. 1975. 'Common property'—a concept in natural resource policy, *'Natural Resources Journal* 15 (October): 715-27.

Clow, Archibald and Clow, Nan L. (1952). *Chemical Revolution,* (Ayer Co Pub, June 1952), pp. 65-90.

Cobb-Douglas in pre-modern Europe1 - Simulating early modern growthPDF (254 KiB) Jan Luiten van Zanden, International Institute of Social History/University of Utrecht. May 2005. Accessed January 2007

Commons, John R. 1934. *Myself.* New York: Macmillan.

Constance McLaughlin Green, *Holyoke, Massachusetts: A Case History of the Industrial Revolution in America* Yale University Press, 1939 online edition

Dahlman, Carl J. 1980. *The open field system and beyond.* Cambridge, Massachusetts: Cambridge University Press.

Dahlman, Carl J. 1991. The tragedy of the commons that wasn't: On technical solutions to the institutions gama, *Population and Environment* 12:285-96.

Dasgupta, Partha, 1982. *The control of resources.* Delhi: Oxford University Press.

David S. Landes. *The Wealth and Poverty of Nations: Why Some Are So Rich and Some So Poor* (1999)

Deane, Phyllis. *The First Industrial Revolution,* Cambridge University Press.

Demsetz, Harold. 1967. Towards a theory of property rights. *American Economic Review* (May): 347-59.

Derry, Thomas Kingston and Trevor I. Williams, *A Short History of Technology: From the Earliest Times to A.D. 1900,* New York: Dover Publications, 1993.

Dunham; Arthur Louis. *The Industrial Revolution in France, 1815-1848* Exposition Press, 1955 online edition

Easter, K.W., and K. Palanisami. 1986. Tank irrigation in India: An example of common property management. In *Proceedings of the conference on common property resource management.* The National Research Council, Washington, D.C.: National Academy Press.

Eggertsson, T. 1990. *Economic behavior and institutions.* New York: Cambridge University Press.

Encyclopedia Britannica. 1980. *The new encyclopedia Britannica macropaedia.* 15th ed. Vol. 15. London.

Eric Hobsbawm, *The Age of Revolution: Europe 1789–1848,* Weidenfeld & Nicolson Ltd.

Eric Schiff, *Industrialisation without national patents: the Netherlands, 1869-1912; Switzerland, 1850-1907,* Princeton University Press, 1971.

Frosch, R.A.; Gallopoulos, N.E. (1989) "Strategies for Manufacturing" *Scientific American* 261:3, pp 144-152.

Frosch, R.A.; Gallopoulos, N.E. (1989) "Strategies for Manufacturing" *Scientific American* 261:3, pp 144-152.

GOI (Government of India). 1976. Agrarian reforms. Part 15 of the *Report of thc national commission on* agriculture, New Delhi: Ministry of Agriculture and Irrigation.

Gordon, H. Scott. 1954. The economic theory of a common property resource: The fishery. *Journal of Political Economy* 62(2):124-42.

Graeme Gill; "Farm to Factory: A Reinterpretation of the Soviet Industrial Revolution, " *Economic Record,* Vol. 80, 2004 online edition

Hardin, G. 1968. The tragedy of the commons. *Science* 162:1243-8.

Hardin, G. 1978. Political requirements for preserving our common heritage. In *Wildlife and America* (H.P. Bokaw, ed.). Washington, D.C.: Council on Environmental Quality.

Hayek, Friedrich A., *Capitalism and the Historians*, The University of Chicago Press, 1963,.

Herbert Kisch, *From Domestic Manufacture to Industrial Revolution The Case of the Rhineland Textile Districts.* Oxford US, 1989 online edition

Hirshleifer, Jack. 1987. Evolutionary models in economics and law: Co-operation versus conflict strategies. *In Economic behavior in adversity (Jack* Hirshleifer, ed.). Chicago: University of Chicago Press.

Hobsbawm, Eric J., *Industry and Empire: From 1750 to the Present Day*, W. W. Norton, 1999.

Howe, Charles W. 1979. *Natural resource economics.* New York: John Wiley & Sons.

Hudson, Pat. *The Industrial Revolution*, Oxford University Press US.

Hughes, Thomas Parke. *Development of Western Technology Since 1500* (1980)

Hulse, David H: The Early Development of the Steam Engine; TEE Publishing, Leamington Spa, U.K., 1999.

Immanuel Chung-Yueh Hsu. *The Rise of Modern China*, Oxford University Press US.

Industrial Revolution and the Standard of Living From www.econlib.org, downloaded 17 July 2006.

Ivor Blashka Hart. *James Watt and the History of Steam Power* (1949)

J. Ehrenfeld (2004), "Can Industrial Ecology be the Science of Sustainability?",*Journal of Industrial Ecology*, 8:1-2, pp 1-3

J. Ehrenfeld (2007), "Would Industrial Ecology Exist without Sustainability in the Background?", *Journal of Industrial Ecology*,11:1

J. H. Clapham; *An Economic History of Modern Britain: The Early Railway Age, 1820-1850.* Cambridge University Press, 1926 online edition

Jodha, N.S. 1986. Common property resources and the rural poor in dry regions of India. *Economic and Political Weekly* 21(27):169-81.

Jodha, N.S. 1989. Management of common property resources in selected areas of India. In *Local Institutions and resource management* (A. Dani and J.G. Campbell, eds.). Kathmandu, Nepal: ICIMOD.

Kay, J.J. "On Complexity Theory, Exergy and Industrial Ecology: Some Implications for Construction Ecology.",*Construction Ecology: Nature as the Basis for Green Buildings*,edited by Kibert, C., Sendzimir, J., Guy, B., pp 72-107, Spon Press, 2002

Kranzberg, Melvin and Carroll W. Pursell, Jr. eds. *Technology in Western civilisation*, Oxford University Press, 1967.

Krishna, K.VS.M., and Samar K. Datta. 1991. Some dynamics of tenancy cultivation: A short study from Gujarat. Institute of Rural Management, Anand. Mimeo.

L. T. C. Rolt and J. S. Allen, *The Steam Engine of Thomas Newcomen*, Landmark Publishing Ltd, (1997),

Landes, David S., *The Unbound Prometheus: Technical Change and Industrial Development in Western Europe from 1750 to the Present* 2nd ed. New York: Cambridge University Press, 2003

Levine, B.L. 1986. The tragedy of the commons and the comedy of community: The commons in history. *Journal of Community Psychology* 14:81-99.

Lines, Clifford, *Companion to the Industrial Revolution*, London, New York etc., Facts on File, 1990, ISBN 0-8160-2157-0

Lucas, Robert (2003). The Industrial Revolution *Past and Future*. "[consider] annual growth rates of 2.4 percent for the first 60 years of the 20th century, of 1 percent for the entire 19th century, of one-third of 1 percent for the 18th century"

M. J. Daunton; *Progress and Poverty: An Economic and Social History of Britain, 1700-1850,* Oxford University Press, 1995 online edition

Mabel C. Buer, *Health, Wealth and Population in the Early Days of the Industrial Revolution*, London: George Routledge & Sons, 1926,.

Magrath, William B. 1986. *The challenge of the commons: Nonexclusive resources and economic development: Theoretical issues.* Working Paper. Washington, D.C.: World Resources Institute.

Michele Boldrin and David K. Levine, Economic and Game Theory Against Intellectual Monopoly, Chapter 1, November 11, 2005 revisionPDF (143 KiB), page 3.

Misra, B. 1990. *Agricultural development: Problems and prospects.* New Delhi: Ashish Publishing.

Misra, K.C. 1990. *Land systems and land reforms.* Delhi: Himalaya Publishing.

Mokyr; Joel. *The British Industrial Revolution: An Economic Perspective* (1999) online edition

NIRD (National Institute of Rural Development). 1990. *Rural development statistics, 1989.* Hyderabad: NIRD.

P. W. King, 'The production and consumption of iron in early modern England and Wales' *Economic History Review* LVIII (2005), 1-33.

Pandey, Vikash N. 1989. *Historical roots of agricultural backwardness: An essay on property law in colonial Oudh.* Research Paper no. 7. Anand: Institute of Rural Management.

Paul Bairoch, *Economics and World History: Myths and Paradoxes.* University of Chicago Press, 1993

Paul Mantoux, *The Industrial Revolution in the Eighteenth Century*, First English translation 1928, revised edition 1961 online edition

Pawson, E., *Transport and Economy: the turnpike roads of 18th century England*, 1977.

Properties of Concrete Published lecture notes from University of Memphis Department of Civil Engineering, accessed 2007-10-17

R.L. Axtell, C.J. Andrews, M.J. Small (2002), "Agent-Based Modeling and Industrial Ecology",*Journal of Industrial Ecology*, 5:4, pp 10-13

R.M. Hartwell, *The Industrial Revolution and Economic Growth*, Methuen and Co., 1971, page 339-341.

R.M. Hartwell, *The Rising Standard of Living in England, 1800-1850*, Economic History Review, 1963, page 398.

Raj, K.N-. 1990. *Organisational issues in Indian agriculture.* Delhi: Oxford-University Press.

Randall, Allan. 1975. Property rights and social microeconomics. *Natural Resources Journal* 15:729-40.

Rehabilitating the Industrial Revolution Maxine Berg, Pat Hudson, Economic History Review, New Series, Vol. 45, No. 1 (Feb., 1992), pp. 24-50 doi:10.2307/2598327

Rev. Dr. Richard L. Hills, *Life and Inventions of Richard Roberts, 1789-1864*, Landmark Publishing Ltd, 2002,.

Robert Lucas, Jr. (2003). The Industrial Revolution. Federal Reserve Bank of Minneapolis. Retrieved on 2007-11-14. "it is fairly clear that up to 1800 or maybe 1750, no society had experienced sustained growth in per capita income. (Eighteenth century population growth also averaged one-third of 1 percent, the same as production growth.) That is, up to about two centuries ago, per capita incomes in all societies were stagnated at around $400 to $800 per year."

Roe, Joseph Wickham (1916), English and American Tool Builders, New Haven, Connecticut, USA: Yale University Press, LCCN 16-011753 . Reprinted by McGraw-Hill, New York and London, 1926 (LCCN 27-024075); and by Lindsay Publications, Inc., Bradley, IL, USA (ISBN 978-0-917914-73-7). Also available online via Google Book Search.

Roemer, J.E. 1989. A public ownership resolution of the tragedy of the commons. *Social Philosophy and Policy* 6:74-92.

Runge, C.F. 1986. Common property and collective action in economic development. In *Proceedings of the conference on common property resource management.* The National Research Council. Washington, D.C.: National Academy Press.

S. Kraines and D. Wallace (2006), "Applying Agent-based Simulation in Industrial Ecology", *Journal of Industrial Ecology*, 10:1-2, pp 15-18

ScienceDirect - Explorations in Economic History: Trends in Real Wages in Britain, 1750-1913 from www.sciencedirect.com, downloaded 17 July 2006.

Scully, Gerald W. 1988. The institutional framework and economic development. *Journal of Political Economy,* 96(3):652-62.

Shaban, Radwan Ali. 1987. Testing between competing models of sharecropping. *Journal of Political Economy* 95(5): 893-920.

Singh Chhatrapathi. 1991. *Water rights and principles of water resources management.* Water Project Series, Indian Law Institute. Bombay: N.M. Tripathi.

Singh Chhatrapathi. *Common property and common poverty: India's forests, forest dwellers and the law.* Delhi: Oxford University Press.

Singh, Katar. 1992. *Managing common pool irrigation tanks: A case study in Andhra Pradesh and West Bengal.* Case Study no. 9. Anand: Institute of Rural Management.

Singh, Katar. 1993. *Managing common pool resources: Principles and case studies.* Delhi: Oxford University Press. Sann, Hans-Werner, 1984. Common property resources, storage facilities and ownership structures: A Cournot model of the oil market. Economics 51 (August): 235-52.

Smith, R.J. 1981, Resolving the tragedy of the commons by creating private property rights in wildlife. CATO *journal* 1:439-68.

Stanfield J.. David. 1985. *Projects that title land in Central and South America and the Caribbean: Expectations and problems.* Land Tenure Center Paper no. 126. Madison: University of Wisconsin.

Stearns; Peter N. *The Industrial Revolution in World History* Westview Press, 1998 online version

Szostak; Rick. *The Role of Transportation in the Industrial Revolution: A Comparison of England and France* McGill-Queens University Press, 1991 online edition

Thompson, E. P., *The Making of the English Working Class*, London: Penguin Books, 1980. (First published 1963.)

Toynbee, Arnold *(1884). Lectures on the Industrial Revolution of the Eighteenth Century in England. Whitefish, Montana: Kessinger Publishing (paperback edition 2004).*

Usher; Abbott Payson. *An Introduction to the Industrial History of England* (1920) online edition

Vaclav Smil; *Energy in World History.* Westview Press, 1994 online edition

Veeman, Terrence. 1978. Water policy and water institutions in northern India: The case of groundwater rights. *Natural Resources Journal* 18(3)569-87.

Wade, R 1986. Common property resource management in South Indian villages. In *Proceedings of the conference on common property resource management.* The National Research Council. Washington, D.C.: National Academy Press.

Watt steam engine image: located in the lobby of the Superior Technical School of Industrial Engineers of the UPM (Madrid)

Welch, W.P. 1983. The political feasibility of full ownership property rights: The cases of pollution and fisheries. *Policy Sciences* 16:165-80.

Yuan, Z; Bi, J; Moriguichi, Y "The Circular Economy: A New Development Strategy in China" *Journal of Industrial Ecology* Vol 10:1-2, pp 4-8

14

Environmentally Sustainable Science and Technology: Applied Aspects

Sustainable Living

Sustainable living refers to an individual or society's lifestyle that can be sustained with limited exhaustion of natural resources. Its adherents most often hold true sustainability as a goal or guide, and make lifestyle tradeoffs, such as transport, housing, energy, and diet favor sustainability.

Most often these tradeoffs involve making more environmentally-friendly lifestyle choices. Lester R. Brown concisely summarizes the situation as "sustaining progress depends on shifting from a fossil fuel-based, automobile-centered, economy to a renewable energy-based, diversified transport, reuse/recycle economy".

Overview

Sustainable living is a sub-division of sustainability where the prerequisites of a modern, industrialised society are left unexercised by choice for a variety of reasons. The practices and motives overlap somewhat between the movements. Sustainable living in urban areas requires a sustainable urban infrastructure.

Self-sufficiency is the principle of consuming only those things produced by oneself or one's family. It is generally a stricter lifestyle than a sustainable lifestyle in that an effort is made to limit trade with others regardless of the sustainability of such trade.

Permaculture is a design philosophy that emphasises sustainability in land use and landscaping, as well as fields such as architecture and economics (for example, encouraging the spread of Local Exchange Trading Systems (LETS)). In terms of agriculture, food production and building materials, permaculture emphasises use of well-adapted plant materials that require few inputs, especially trees, hemp, and other edible and useful perennials.

Some people are opposed to furthering mechanisation and technology in society for any reason. Adherents of sustainable living, in contrast, are willing to accept appropriate technology.

Carbon Footprint

An individual's carbon footprint represents the amount of carbon required to maintain that person's lifestyle, and what impact that lifestyle has on the earth. A higher footprint represents a lifestyle that is less sustainable, and a lower footprint represents a more sustainable lifestyle. Most carbon footprint websites have suggestions for decreasing a person's carbon footprint, and therefore increasing their sustainability.

History

- 1854 Henry David Thoreau's work *Walden* represents the earliest literature that specifically addresses the sustainable lifestyle in simple living.
- 1954 The publication of *Living the Good Life* by Helen and Scott Nearing is the modern-day beginning of the sustainability movement. The book fostered the back-to-the-land movement in the late 1960s and early 1970s. As the back-to-the-landers realised the difficulty of copying the Nearings' lifestyle, they returned to more conventional lifestyles yet incorporated self-sufficiency where they could.
- 1962 The publication of *Silent Spring* by Rachel Carson was another major milestone in the sustainability movement.
- 1972 Donella Meadows wrote the international bestselling book *The Limits to Growth*, which reported on a study of long-term global trends in population, economics, and the environment. It sold millions of copies and was translated into 28 languages.
- 1973E. F. Schumacher published a collection of essays on shifting towards sustainable living through the appropriate use of technology in his book *Small Is Beautiful*.
- The United Nations had an a series of conferences called Earth Summit in 1992, 1997, and 2002. Each of these conferences focused on increasing sustainability within societies, with the goal of conserving the world's natural resources.
- 2007 the U.N. published *Sustainable Consumption and Production, Promoting Climate-Friendly Household Consumption Patterns*, a paper that promotes sustainable lifestyles in communities and homes.

Sustainable Forest Management

Sustainable forest management (SFM) is the management of forests according to the principles of sustainable development. Sustainable forest management uses very broad social, economic and environmental goals. A range of forestry institutions now practice various forms of sustainable forest management and a broad range of methods and tools are available that have been tested over time.

The Forest Principles adopted at The United Nations Conference on Environment and Development (UNCED) in Rio de Janeiro in 1992 captured the general international understanding of sustainable forest management at that time. A number of sets of criteria and indicators have since been developed to evaluate the achievement of SFM at both the country and management unit level. These were all attempts to codify and provide for independent assessment of the degree to which the broader objectives of sustainable forest management are being achieved in practice. In 2007, the United Nations General Assembly adopted the Non-Legally Binding Instrument on All Types of Forests. The instrument is the first of its kind, and reflects the strong international commitment to promote implementation of sustainable forest management through a new approach that brings all stakeholders together.

Definition

A definition of the present day understanding of the term sustainable forest management was developed by the Ministerial Conference on the Protection of Forests in Europe (MCPFE), and has since been adopted by the Food and Agriculture Organisation (FAO). It defines sustainable forest management as:the stewardship and use of forests and forest lands in a way, and at a rate, that maintains their biodiversity, productivity, regeneration capacity, vitality and their potential to fulfill, now and in the future, relevant ecological, economic and social functions, at local, national, and global levels, and that does not cause damage to other ecosystems.

In simpler terms, the concept can be described as the attainment of balance - balance between society's increasing demands for forest products and benefits, and the preservation of forest health and diversity. This balance is critical to the survival of forests, and to the prosperity of forest-dependent communities.

For forest managers, sustainably managing a particular forest tract means determining, in a tangible way, how to use it today to ensure similar benefits, health and productivity in the future. Forest managers must assess and integrate a wide array of sometimes conflicting factors - commercial and non-commercial values, environmental considerations, community needs, even global impact - to produce sound forest plans. In most cases, forest managers develop their forest plans in consultation with citizens, businesses, organisations and other interested parties in and around the forest tract being managed.

Because forests and societies are in constant flux, the desired outcome of sustainable forest management is not a fixed one. What constitutes a sustainably managed forest will change over time as values held by the public change.

Criteria and Indicators

Criteria and indicators are policy instruments by which progress towards implementing sustainable forest management may be evaluated and reported on. Criteria define and characterize the essential elements, as well as a set of conditions or processes, by which sustainable forest management may be assessed. Periodically measured indicators reveal the direction of change with respect to each criterion.

Criteria and indicators of sustainable forest management are widely used and many countries produce national reports that assess their progress toward sustainable forest management. There are nine international and regional criteria and indicators initiatives, which collectively involve more than 150 countries. Three of the more advanced initiatives are those of the Working Group on Criteria and Indicators for the Conservation and Sustainable Management of Temperate and Boreal Forests (also called the Montreal Process), the Ministerial Conference for the Protection of Forests in Europe, and the International Tropical Timber Organisation. Countries who are members of the same initiative usually agree to produce reports at the same time and using the same indicators. Within countries, at the management unit level, efforts have also been directed at developing local level criteria and indicators of sustainable forest management. The Center for International Forestry Research, the International Model Forest Network and researchers at the University of British Columbia have developed a number of tools and techniques to help forest-dependent communities develop their own local level criteria and indicators. Criteria and Indicators also form the basis of the Canadian Standards Association certification standard for sustainable forest management.

There appears to be growing international consensus on the key elements of sustainable forest management. Seven common thematic areas of sustainable forest management have emerged based on the criteria of the nine ongoing regional and international criteria and indicators initiatives. The seven thematic areas are:

- Extent of forest resources
- Biological diversity
- Forest health and vitality
- Productive functions and forest resources
- Protective functions of forest resources
- Socio-economic functions
- Legal, policy and institutional framework.

This consensus on common thematic areas (or criteria) effectively provides a common, implicit definition of sustainable forest management. The seven thematic areas were acknowledged by the international forest community at the fourth session of the United Nations Forum on Forests and the 16th session of the Committee on Forestry.. These thematic areas have since been enshrined in the Non-Legally Binding Instrument on All Types of Forests as a reference framework for sustainable forest management to help achieve the purpose of the instrument.

Ecosystem Approach

The Ecosystem Approach has been prominent on the agenda of the Convention on Biological Diversity (CBD) since 1995 . The CBD definition of the Ecosystem Approach and a set of principles for its application were developed at an expert meeting in Malawi in 1995, known as the Malawi Principles. The definition, 12 principles and 5 points of "operational guidance" were adopted by the fifth Conference of Parties (COP5) in 2000. The CBD definition is as follows

The ecosystem approach is a strategy for the integrated management of land, water and living resources that promotes conservation and sustainable use in an equitable way. Application of the ecosystem approach will help to reach a balance of the three objectives of the Convention. An ecosystem approach is based on the application of appropriate scientific methodologies focused on levels of biological organisation, which encompasses the essential structures, processes, functions and interactions among organisms and their environment. It recognises that humans, with their cultural diversity, are an integral component of many ecosystems.

Sustainable forest management was recognised by parties to the Convention on Biological Diversity in 2004 (Decision VII/11 of COP7) to be a concrete means of applying the Ecosystem Approach to forest ecosystems. The two concepts, sustainable forest management and the ecosystem approach, aim at promoting conservation and management practices which are environmentally, socially and economically sustainable, and which generate and maintain benefits for both present and future generations. In Europe, the MCPFE and the Council for the Pan-European Biological and Landscape Diversity Strategy (PEBLDS) jointly recognised sustainable forest management to be consistent with the Ecosystem Approach in 2006.

Independent Certification

Growing environmental awareness and consumer demand for more socially responsible businesses helped third-party forest certification emerge in the 1990s as a credible tool for communicating the environmental and social performance of forest operations.

There are many potential users of certification, including: forest managers, investors, environmental advocates, business consumers of wood and paper, and individuals. With forest certification, an independent organisation develops standards of good forest management, and independent auditors issue certificates to forest operations that comply with those standards. This certification verifies that forests are well-managed—as defined by a particular standard—and ensures that certain wood and paper products come from responsibly managed forests. This rise of certification led to the emergence of several different systems throughout the world. As a result, there is no single accepted forest management standard worldwide, and each system takes a somewhat different approach in defining standards for sustainable forest management. Third-party forest certification is an important tool for those seeking to ensure that the paper and wood products they purchase and use come from forests that are well-managed and legally harvested. Incorporating third-party certification into forest product procurement practices can be a centerpiece for comprehensive wood and paper policies that include factors such as the protection of sensitive forest values, thoughtful material selection and efficient use of products. There are more than 50 certification standards worldwide. Some common certification standards are:

- Canada's National Sustainable Forest Management Standard (CSA)
- Forest Stewardship Council (FSC)
- Sustainable Forestry Initiative (SFI)
- Programme for the Endorsement of Forest Certification schemes (PEFC)

The area of forest certified worldwide is growing rapidly. As of December 2006, there were over 2,440,000 square kilometres of forest certified under the CSA, FSC or SFI standards, with over 1,237,000 square kilometres certified in Canada alone. While certification is intended as a tool to enhance forest management practices throughout the world, to date most certified forestry operations are located in Europe and North America. A significant barrier for many forest managers in developing countries is that they lack the capacity to undergo a certification audit and maintain operations to a certification standard.

Sustainable Agriculture

Sustainable agriculture integrates three main goals: environmental stewardship, farm profitability, and prosperous farming communities. These goals have been defined by a variety of disciplines and may be looked at from the vantage point of the farmer or the consumer. "It's easy to understand why key individuals and organisations in agriculture have flocked to this term. After all, who would advocate a 'non-sustainable agriculture?'" - Charles A. Francis.

Description

Sustainable agriculture refers to the ability of a farm to produce food indefinitely, without causing irreversible damage to ecosystem health. Two key issues are biophysical (the long-term effects of various practices on soil properties and processes essential for crop productivity) and socio-economic (the long-term ability of farmers to obtain inputs and manage resources such as labor).

The physical aspects of sustainability are partly understood (Altieri 1995). Practices that can cause long-term damage to soil include excessive tillage (leading to erosion) and irrigation without adequate drainage (leading to accumulation of salt in the soil). Long-term experiments provide some of the best data on how various practices affect soil properties essential to sustainability.

Although air and sunlight are available everywhere on Earth, crops also depend on soil nutrients and the availability of water. When farmers grow and harvest crops, they remove some of these nutrients from the soil. Without replenishment, the land would suffer from nutrient depletion and be unusable for further farming. Sustainable agriculture depends on replenishing the soil while minimising the use of non-renewable resources, such as natural gas (used in converting atmospheric nitrogen into synthetic fertilizer), or mineral ores (e.g., phosphate). Possible sources of nitrogen that would, in principle, be available indefinitely, include:

1. recycling crop waste and livestock or human manure
2. growing legume crops and forages such as, peanuts, or alfalfa that form symbioses with nitrogen-fixing bacteria called rhizobia
3. industrial production of nitrogen by the Haber Process uses hydrogen, which is currently derived from natural gas, (but this hydrogen could instead be made by electrolysis of water using electricity (perhaps from solar cells or windmills)) or
4. genetically engineering (non-legume) crops to form nitrogen-fixing symbioses or fix nitrogen without microbial symbionts.

The last option was proposed in the 1970s, but would be well beyond the capability of early 21st century technology, even if various concerns about biotechnology were addressed. Sustainable options for replacing other nutrient inputs (phosphorus, potassium, etc.) are more limited.

In some areas, sufficient rainfall is available for crop growth, but many other areas require irrigation. For irrigation systems to be sustainable they must be managed properly (to avoid salt accumulation) and not use more water from their source than is naturally replenished, otherwise the water source becomes, in effect, a non-renewable resource. Improvements in water well drilling technology and the development of submersible pumps have made it possible for large crops to be regularly grown where reliance on rainfall alone previously made this level of success unpredictable. However, this progress has come at a price, in that in many areas where this has occurred, such as the Ogallala Aquifer, the water is being used at a greater rate than its rate of recharge. Socioeconomic aspects of sustainability are also partly understood. Regarding less concentrated farming, the best known analysis is Netting's (1993) study on smallholder systems through history. Sustainable agriculture was also addressed by the 1990 farm bill [Food, Agriculture, Conservation, and Trade Act of 1990 (FACTA), Public Law 101-624, Title XVI, Subtitle A, Section 1603].

It was defined as follows:

Stated by: "the term sustainable agriculture means an integrated system of plant and animal production practices having a site-specific application that will, over the long term:

- satisfy human food and fiber needs
- enhance environmental quality and the natural resource base upon which the agricultural economy depends
- make the most efficient use of nonrenewable resources and on-farm resources and integrate, where appropriate, natural biological cycles and controls
- sustain the economic viability of farm operations
- enhance the quality of life for farmers and society as a whole."

Economics

Given the finite supply of natural resources at any specific cost and location, agriculture that is inefficient or damaging to needed resources may eventually exhaust the available resources or the ability to afford and acquire them. It may also generate negative externality, such as pollution as well as financial and production costs. The way that crops are sold must be accounted for in the sustainability equation. Food sold locally requires little additional energy, aside from that necessary for cultivation, harvest, and transportation (including consumers). Food sold at a remote location, whether at a farmers' market or the supermarket, incurs a different set of energy cost for materials, labour, and transport. The most important factors for an individual site are sun, air, soil and water. Of the four, water and soil quality and quantity are most amenable to human intervention through time and labour. What grows and how and where it is grown are a matter of choice. Two of the many possible practices of sustainable agriculture are crop rotation and soil amendment, both designed to ensure that crops being cultivated can obtain the necessary nutrients for healthy growth.

Methods

Monoculture, a method of growing only one crop at a time in a given field, is a very widespread practice, but there are questions about its sustainability, especially if the same crop is grown every year. Growing a mixture of crops (polyculture) sometimes reduces disease or pest problems Nature 406, 718-722 Genetic diversity and disease control in rice, Environ. Entomol. 12:625) but polyculture has rarely, if ever, been compared to the more widespread practice of growing different crops in successive years crop rotation with the same overall crop diversity. For example, how does growing a corn-bean mixture every year compare with growing corn and bean in alternate years? Cropping systems that include a variety of crops (polyculture and/or rotation) may also replenish nitrogen (if legumes are included) and may also use resources such as sunlight, water, or nutrients more efficiently (Field Crops Res. 34:239).

Replacing a natural ecosystem with a few specifically chosen plant varieties as is done in farming results in an artificial ecosystem that lacks the genetic diversity found in wildlife and is thus more susceptible to widespread disease. The Great Irish Famine (1845-1849) is a well-known example of the dangers of monoculture. Many scientists, farmers, and businesses have debated how to make agriculture farming sustainable. One of the many practices includes growing a diverse number of perennial crops in a single field, each of which would grow in separate season so as not to compete with each other for natural resources. This system would result in increased resistance to diseases and decreased effects of erosion and loss of nutrients in soil. Nitrogen fixation from legumes, for example, used in conjunction with plants that rely on nitrate from soil for growth, helps to allow the land to be reused annually. Legumes will grow for a season and replenish the soil with ammonium and nitrate, and the next season other plants can be seeded and grown in the field in preparation for harvest.

In practice, there is no single approach to sustainable agriculture, as the precise goals and methods must be adapted to each individual case. There may be some techniques of farming that are inherently in conflict with the concept of sustainability, but there is widespread misunderstanding on impacts of some practices. For example, the slash-and-burn techniques that are the characteristic feature of shifting cultivators are often cited as inherently destructive, yet slash-and-burn cultivation has been practiced in the Amazon for at least 6000 years (Sponsel 1986); serious deforestation did not begin until the 1970s, largely as the result of Brazilian government programmes and policies (Hecht and Cockburn 1989).

There are also many ways to practice sustainable animal husbandry. Some of the key tools to grazing management include fencing off the grazing area into smaller areas called paddocks, lowering stock density, and moving the stock between paddocks frequently. Several attempts have been made to produce an artificial meat, using isolated tissues to produce it in vitro; Jason Matheny's work on this topic, whichin the New Harvest project, is one of the most commented.

Off-Farm Impacts

What if a farm is able to "produce perpetually", yet has negative effects on environmental quality elsewhere? Most people concerned with sustainability take a global view, so they try to avoid negative off-farm impacts. For example, over-application of synthetic fertilizer or animal manures can pollute nearby rivers and coastal waters. On the other hand, if crop yields are too low, because of soil exhaustion of nutrients or reduced ability to retain water, farmers would need to access new lands for agriculture, leading to the decimation of the rainforest, draining wetlands, etc. We must also consider the impact of sustainability on overall production. If the human population is to continue its growth, it will need food and fiber to do so. The United Nations estimates the world will inhabit 9.3 billion people by 2050, which will necessitate a dramatic increase in production capabilities. The increased production will likely come from one of two different ways. First, you can either break out virgin land to grow crops, though concerns over global warming make this option unsavory. Second, you can increase yields on existing land, which will likely require adopting technologies many sustainable advocates are opposed to, principally Genetically modified organism crops. Many advocates of sustainable agriculture think that organic agriculture is the only system which can be sustained over the long-term. However, organic production methods, especially in transition, yield less than their conventional counterparts. While there is evidence which gives organic an advantage during periods of drought one must be careful not to place too much emphasis on these figures. While drought is a real threat to agriculture production, it is hardly the norm. Even when drought strikes, stockpiles help mitigate food security concerns. Regardless of reduced yields under optimal conditions, it should be noted that during periods of prolonged drought as predicted by global warming scientists, organic production methods should be considered as a way to adapt to a changing climate.

Urban Planning

There has been considerable debate about which form of human residential habitat may be a better social form for sustainable agriculture. Generally, it is thought that village communities can improve sustainability in that such communities tend to provide a cooperative environment that supports farming.

Many environmentalists pushing for increased population density to preserve agricultural land point out that urban sprawl is less sustainable and more damaging to the environment than living in the cities where cars are not needed because food and other necessities are within walking distance. However, others have theorized that sustainable ecocities, or ecovillages which combine habitation and farming with close proximity between producers and consumers, may provide greater sustainability. The use of available city space (e.g., rooftop gardens and community gardens) for cooperative food production is another way to achieve greater sustainability. One of the latest ideas in achieving sustainable agricultural involves shifting the production of food plants from major factory farming operations to large, urban, technical facilities called vertical farms. The advantages of vertical farming include year-round production, isolation from pests and diseases, controllable resource recycling, and on-site production that eliminates the need for transportation costs. While a vertical farm has yet to become a reality, the idea is gaining momentum among those who believe that current sustainable farming methods will be insufficient to provide for a growing global population. For vertical farming to become a reality, billions of dollars in tax credits and subsidies will need to be made available to the operation. It may be difficult to justify spending billions of dollars on a vertical farm that will only feed 50,000 people when agriculture land remains abundant.

Environmental Effects of Fishing

The environmental effects of fishing can be divided into issues that involve the availability of fish to be caught, such as overfishing, sustainable fisheries, and fisheries management; and issues that involve the impact of fishing on the environment, such as by-catch. These conservation issues are part of marine conservation, and are addressed in fisheries science programmes. There is a growing gap between how many fish are available to be caught and humanity's desire to catch them, a problem that gets worse as the world population grows. Similar to other environmental issues, there can be conflict between the fishermen who depend on fishing for their livelihoods and fishery scientists who realise that if future fish populations are to be sustainable then some fisheries must reduce or even close. The journal *Science* published a four-year study in November 2006, which predicted that, at prevailing trends, the world would run out of wild-caught seafood in 2048. The scientists stated that the decline was a result of overfishing, pollution and other environmental factors that were reducing the population of fisheries at the same time as their ecosystems were being degraded. Yet again the analysis has met criticism as being fundamentally flawed, and many fishery management officials, industry representatives and scientists challenge the findings, although the debate continues. Many countries, such as Tonga, the United States, Australia and New Zealand, and international management bodies have taken steps to appropriately manage marine resources.

Effects on Habitat

Some fishing techniques also may cause habitat destruction. Dynamite fishing and cyanide fishing, which are illegal in many places, harm surrounding habitat. Bottom trawling, the practice of pulling a fishing net along the sea bottom behind trawlers, removes around 5 to 25% of an area's seabed life on a single run. A 2005 report of the UN Millennium Project, commissioned by UN Secretary-General Kofi Annan, recommended the elimination of bottom trawling on the high seas by 2006 to protect seamounts and other ecologically sensitive habitats. In mid October 2006, U.S. President Bush joined other world leaders calling for a moratorium on deep-sea trawling, a practice shown to often have harmful effects on sea habitat and, hence, on fish populations.

Overfishing

Overfishing has also been widely reported due to increases in the volume of fishing hauls to feed a quickly growing number of consumers. This has led to the breakdown of some sea ecosystems and several fishing industries whose catch has been greatly diminished. The extinction of many species has also been reported. According to an FAO estimate, over 70% of the world's fish species are either fully exploited or depleted. According to Nitin Desai, Secretary General of the 2002 World Summit on Sustainable Development, "Overfishing cannot continue, the depletion of fisheries poses a major threat to the food supply of millions of people." The cover story of the May 15, 2003 issue of the science journal *Nature* – with Dr. Ransom A. Myers, an internationally prominent fisheries biologist (Dalhousie University, Halifax, Canada) as the lead author – was devoted to a summary of the scientific information. The story asserted that, as compared with 1950 levels, only a remnant (in some instances, as little as 10%) of all large ocean-fish stocks are left in the seas. These large ocean fish are the species at the top of the food chains (e.g., tuna, cod, among others). However, this article was subsequently criticized as being fundamentally flawed, although much debate still exists (Walters 2003; Hampton et al. 2005; Maunder et al. 2006; Polacheck 2006;Sibert et al. 2006) and the majority of fisheries scientists now consider the results irrelevant with respect to large pelagics (the open seas). The environmental impact of recreational fishing may be alleviated to some extent by catch and release fishing.

Ecological Disruption

Fishing may disrupt food webs by targeting specific, in-demand species. There might be too much fishing of prey species such as sardines and anchovies, thus reducing the food supply for the predators. It may also cause the increase of prey species when the target fishes are predator species such as salmon and tuna.

By-Catch

By-catch is the portion of the catch that is not the target species. These are either kept to be sold or discarded. In some instances the discarded portion is known as discards.

Possible Remedies

Many governments have implemented fisheries management policies designed to curb the environmental impact of fishing. Fishing conservation aims to control the human activities that may completely decrease a fish stock or washout an entire aquatic environment. These laws include the quotas on the total catch of particular species in a fishery, limits on the number of vessels allowed in specific areas, and the imposition of seasonal restrictions on fishing. Fish farming has also been proposed as a more sustainable alternative to traditional capture of wild fish. However, fish farming has been found to have negative impacts on nearby wild fish.

What are Biopesticides?

Biopesticides are certain types of pesticides derived from such natural materials as animals, plants, bacteria, and certain minerals. For example, canola oil and baking soda have pesticidal applications and are considered biopesticides. At the end of 2001, there were approximately 195 registered biopesticide active ingredients and 780 products. Biopesticides fall into three major classes:

1. Microbial pesticides consist of a microorganism (e.g., a bacterium, fungus, virus or protozoan) as the active ingredient. Microbial pesticides can control many different kinds of pests, although each separate active ingredient is relatively specific for its target pest[s]. For example, there are fungi that control certain weeds, and other fungi that kill specific insects. The most widely used microbial pesticides are subspecies and strains of Bacillus thuringiensis, or Bt. Each strain of this bacterium produces a different mix of proteins, and specifically kills one or a few related species of insect larvae. While some Bt's control moth larvae found on plants, other Bt's are specific for larvae of flies and mosquitoes. The target insect species are determined by whether the particular Bt produces a protein that can bind to a larval gut receptor, thereby causing the insect larvae to starve.

2. Plant-Incorporated-Protectants (PIPs) are pesticidal substances that plants produce from genetic material that has been added to the plant. For example, scientists can take the gene for the Bt pesticidal protein, and introduce the gene into the plant's own genetic material. Then the plant, instead of the Bt bacterium, manufactures the substance that destroys the pest. The protein and its genetic material, but not the plant itself, are regulated by EPA.

3. Biochemical pesticides are naturally occurring substances that control pests by non-toxic mechanisms. Conventional pesticides, by contrast, are generally synthetic materials that directly kill or inactivate the pest. Biochemical pesticides include substances, such as insect sex pheromones, that interfere with mating, as well as various scented plant extracts that attract insect pests to traps. Because it is sometimes difficult to determine whether a substance meets the criteria for classification as a biochemical pesticide, EPA has established a special committee to make such decisions.

What are the Advantages of Using Biopesticides?

Biopesticides are usually inherently less toxic than conventional pesticides. Biopesticides generally affect only the target pest and closely related organisms, in contrast to broad spectrum, conventional pesticides that may affect organisms as different as birds, insects, and mammals. Biopesticides often are effective in very small quantities and often decompose quickly, thereby resulting in lower exposures and largely avoiding the pollution problems caused by conventional pesticides. When used as a component of Integrated Pest Management (IPM) programmes, biopesticides can greatly decrease the use of conventional pesticides, while crop yields remain high. To use biopesticides effectively, however, users need to know a great deal about managing pests.

How Does EPA Encourage the Development and use of Biopesticides?

In 1994, the Biopesticides and Pollution Prevention Division was established in the Office of Pesticide Programmes to facilitate the registration of biopesticides. This Division promotes the use of safer pesticides, including biopesticides, as components of IPM programmes. The Division also coordinates the Pesticide Environmental Stewardship Programme (PESP). Since biopesticides tend to pose fewer risks than conventional pesticides, EPA generally requires much less data to register a biopesticide than to register a conventional pesticide. In fact, new biopesticides are often registered in less than a year, compared with an average of more than 3 years for conventional pesticides. While biopesticides require less data and are registered in less time than conventional pesticides, EPA always conducts rigorous reviews to ensure that pesticides will not have adverse effects on human health or the environment. For EPA to be sure that a pesticide is safe, the Agency requires that registrants submit a variety of data about the composition, toxicity, degradation, and other characteristics of the pesticide.

Regulating Biopesticides

Before a pesticide can be marketed and used in the United States, the Federal Insecticide, Fungicide, and Rodenticide Act (FIFRA) requires that EPA evaluate the proposed pesticide to assure that its use will not pose unreasonable risks of harm to human health and the environment. This regulation involves an extensive review of health and safety information. Biopesticides include naturally occurring substances that control pests (biochemical pesticides), microorganisms that control pests (microbial pesticides), and pesticidal substances produced by plants containing added genetic material (plant-incorporated protectants) or PIPs. The Biopesticide Registration Tools provide links to information tools to assist applicants. The Fact Sheets sections provides links to information about each of the biopesticide active ingredients. The Product Lists section provides various lists of individual biopesticide products to assist the public in identifying appropriate biopesticide product for pest problems. Finally, the PIPs section provides extensive information regarding the regulation of genetically engineered plants.

Biopesticide Registration Tools

The federal pre-marketing approval of pesticides - termed registration — is a complex process. The documents linked from this page augment the general registration process as they relate specifically to the registration of biopesticides. The e-mail address bppdconsistency@epa.gov has been created to respond to issues concerning biopesticide registration inconsistency that affect processing of submissions. Biopesticide Regulatory Action Leaders will meet to discuss these issues and provide an answer usually within two to three weeks. Resolution of these issues will be posted to this site. Since this e-mail address is only for the submission of generic issues regarding consistency in the regulation of biopesticides. Other questions regarding biopesticides should be directed to the appropriate biopesticide Regulatory Action Leader or the Biopesticide Ombudsman, Brian Steinwand (steinwand.brian@epa.gov). If you have questions about conventional pesticides or antimicrobials, please contact the Ombudsman for the Registration Division or the Antimicrobial Division.

Biopesticide Active Ingredient Fact Sheets

This collection of fact sheets contains chemical specific information about biopesticide active ingredients. Additionally, some of the fact sheets include a detailed technical document, bibliographies, regulatory history, Federal Register notices, and/or registrant and product lists.

Biopesticide Ingredient & Product Lists

Lists of EPA registered (approved) biopesticide ingredients & products are provided, including a listing by active ingredient PDF (43 pp, 378 K, about PDF), and lists of biopesticide active ingredients by year first approved.

Plant Incorporated Protectants (PIPs)

Plant-Incorporated Protectants are pesticidal substances produced by plants and the genetic material necessary for the plant to produce the substance. For example, scientists can take the gene for a specific Bt pesticidal protein, and introduce the gene into the plant's genetic material. Then the plant manufactures the pesticidal protein that controls the pest when it feeds on the plant. Both the protein and its genetic material are regulated by EPA; the plant itself is not regulated.

Biopesticides

Biopesticides use naturally occurring organisms, such as fungi, bacteria, viruses and nematodes to control plant diseases and arthropod pests. Many of our projects deal with the necessary fundamental research to identify useful isolates and their practical use, while others have followed right through to mass production and commercial field use. We have a multi-disciplinary team of scientists with experience in all aspects of the development and commercialisation of biopesticide products.

Biopesticides

What are Biopesticides?

Biopesticides (also known as biological pesticides) are certain types of pesticides derived from such natural materials as animals, plants, bacteria, and certain minerals. For example, garlic, mint, and baking soda all have pesticidal applications and are considered biopesticides. At the end of 1998, there were approximately 175 registered biopesticide active ingredients and 700 products. Biopesticides fall into three major categories:

(1) Microbial pesticides contain a microorganism (bacterium, fungus, virus, protozoan or alga) as the active ingredient. The most widely known microbial pesticides are varieties of the bacterium *Bacillus thuringiensis*, or Bt, which can control certain insects in cabbage, potatoes, and other crops. Bt produces a protein that is harmful to specific insect pests. Certain other microbial pesticides act by out-competing pest organisms. Microbial pesticides need to be continuously monitored to ensure they do not become capable of harming non-target organisms, including humans.

(2) Plant-pesticides are pesticidal substances that plants produce from genetic material that has been added to the plant. For example, scientists can take the gene for the Bt pesticidal protein, and introduce the gene into the plants= own genetic material. Then the plantBinstead of the Bt bacterium—manufactures the substance that destroys the pest. Both the

protein and its genetic material are regulated by EPA; the plant itself is not regulated.

(3) Biochemical pesticides are naturally occurring substances that control pests by non-toxic mechanisms. Conventional pesticides, by contrast, are synthetic materials that usually kill or inactivate the pest. Biochemical pesticides include substances that interfere with growth or mating, such as plant growth regulators, or substances that repel or attract pests, such as pheromones. Because it is sometimes difficult to determine whether a natural pesticide controls the pest by a non-toxic mode of action, EPA has established a committee to determine whether a pesticide meets the criteria for a biochemical pesticide.

What are the advantages of using biopesticides?

- Biopesticides are inherently less harmful than conventional pesticides.
- Biopesticides are designed to affect only one specific pest or, in some cases, a few target organisms, in contrast to broad spectrum, conventional pesticides that may affect organisms as different as birds, insects, and mammals.
- Biopesticides often are effective in very small quantities and often decompose quickly, thereby resulting in lower exposures and largely avoiding the pollution problems caused by conventional pesticides.
- When used as a component of Integrated Pest Management (IPM) programmes, biopesticides can greatly decrease the use of conventional pesticides, while crop yields remain high.

To use biopesticides effectively, however, users need to know a great deal about managing pests. How does EPA encourage the development and use of biopesticides? In 1994, the Biopesticides and Pollution Prevention Division was established in the Office of Pesticide Programmes to facilitate the registration of biopesticides. This Division promotes the use of safer pesticides, including biopesticides, as components of IPM programmes. The Division also coordinates the Pesticide Environmental Stewardship Programme (PESP). (See related fact sheets on IPM and PESP.) Since biopesticides tend to pose fewer risks than conventional pesticides, EPA generally requires much less data to register a biopesticide than to register a conventional pesticide. In fact, new biopesticides are often registered in less than a year, compared with an average of more than 3 years for conventional pesticides.

While biopesticides require less data and are registered in less time than conventional pesticides, EPA must always conduct rigorous reviews to ensure that pesticides will not have adverse effects on human health or the environment. For EPA to be sure that a pesticide is safe, the Agency requires that registrants submit a variety of data about the composition, toxicity, degradation, and other characteristics of the pesticide.

Biofertilizers

Abstract

Biofertilizers include microorganisms and their metabolites that are capable of enhancing soil fertility, crop growth, and/or yield. These include both indigenous microbes and microbial inoculants, that is, microorganisms that replace fertilizers or increase a crop's fertilizer use efficiency. Soil microorganisms such as bacteria, ectomycorhiza, arbuscular mycorrhizal fungi, and soil algae, especially the N_2-fixing cyanobacteria have potential as biofertilizers. Nitrogen-fixing inoculants based on *Rhizobium* species were among the first biofertilizers introduced into agroecosystems back in the nineteenth century. In the twenty-first century, biofertilizers will become an increasingly important area of research and development (1). The use of fertilizers and pesticides has increased steadily since the 1970s; consequently, concerns about the impacts of these chemicals on land, air, and water have become significant environmental issues. Biofertilizers provide an alternative to agricultural chemicals as more sustainable and ecologically sound practices to increase crop productivity. Biofertilizer sales forecasts in the United States for the years 2001 and 2006 represent $690 million and $1.6 billion, respectively. Examples of some biofertilizers currently in use worldwide.

Biofertilizers

1. What is biofertilizer?

Biofertilizers are ready to use live formulates of such beneficial microorganisms which on application to seed, root or soil mobilize the availability of nutrients by their biological activity in particular, and help build up the micro-flora and in turn the soil health in general.

2. Why should we use biofertilizers?

With the introduction of green revolution technologies the modern agriculture is getting more and more dependent upon the steady supply of synthetic inputs (mainly fertilizers), which are products of fossil fuel (coal+ petroleum). Adverse effects are being noticed due to the excessive and imbalanced use of these synthetic inputs. This situation has lead to identifying harmless inputs like biofertilizers. Use of such natural products like biofertilizers in crop cultivation will help in safeguarding the soil health and also the quality of crop products.

3. What are the benefits from using biofertilizers?

- Increase crop yield by 20-30%.
- Replace chemical nitrogen and phosphorus by 25%.
- Stimulate plant growth.
- Activate the soil biologically.
- Restore natural soil fertility.
- Provide protection against drought and some soil borne diseases.

4. What are the advantages of bio-fertilizers?

1. Cost effective.
2. Suppliment to fertilizers.
3. Eco-friendly (Friendly with nature).
4. Reduces the costs towards fertilizers use, especially regarding nitrogen and phosphorus.

5. What types of biofertilizers are available?

1. For Nitrogen
   - o Rhizobium for legume crops.
   - o Azotobacter/Azospirillum for non legume crops.
   - o Acetobacter for sugarcane only.
   - o Blue –Green Algae (BGA) and Azolla for low land paddy.
2. For Phosphorous
   - o Phosphatika for all crops to be applied with Rhizobium, Azotobacter, Azospirillum and Acetobacter
3. For enriched compost
   - o Cellulolytic fungal culture
   - o Phosphotika and Azotobacter culture

6. What biofertilizers are recommended for crops?

- o Rhizobium + Phosphotika at 200 gm each per 10 kg of seed as seed treatment are recommended for pulses such as pigeonpea, green gram, black gram, cowpea etc, groundnut and soybean.
- o Azotobacter + Phosphotika at 200 gm each per 10 kg of seed as seed treatment are useful for wheat, sorghum, maize, cotton, mustard etc.
- o For transplanted rice, the recommendation is to dip the roots of seedlings

for 8 to 10 hours in a solution of Azospirillum + Phosphotika at 5 kg each per ha.

7. How biofertilizers are applied to crops?
   1. Seed treatment: 200 g of nitrogenous biofertilizer and 200 g of Phosphotika are suspended in 300-400 ml of water and mixed thoroughly. Ten kg seeds are treated with this paste and dried in shade. The treated seeds have to be sown as soon as possible.
   2. Seedling root dip: For rice crop, a bed is made in the field and filled with water. Recommended biofertilizers are mixed in this water and the roots of seedlings are dipped for 8-10 hrs.
   3. Soil treatment: 4 kg each of the recommended biofertilizers are mixed in 200 kg of compost and kept overnight. This mixture is incorporated in the soil at the time of sowing or planting.

8. How could one get good response to biofertilizer application?

- o Biofertilizer product must contain good effective strain in appropriate population and should be free from contaminating microorganisms.
- o Select right combination of biofertilizers and use before expiry date.
- o Use suggested method of application and apply at appropriate time as per the information provided on the label.
- o For seed treatment adequate adhesive should be used for better results.
- o For problematic soils use corrective methods like lime or gypsum pelleting of seeds or correction of soil pH by use of lime.
- o Ensure the supply of phosphorus and other nutrients.

9. What would be probable reasons for not getting response from the application of biofertilizers?
   1. On account of quality of product
      - o Use of ineffective strain.
      - o Insufficient population of microorganisms.
      - o High level of contaminants.
   2. On account of inadequate storage facilities
      - o May have been exposed to high temperature.
      - o May have been stored in hostile conditions.
   3. On account of usage
      - o Not used by recommended method in appropriate doses.

- o Poor quality adhesive.
- o Used with strong doses of plant protection chemicals.

4. On account of soil and environment
   - o High soil temperature or low soil moisture.
   - o Acidity or alkalinity in soil.
   - o Poor availability of phosphorous and molybdenum.
   - o Presence of high native population or presence of bacteriophages.

10. What precautions one should take for using biofertilizers?
    - o Biofertilizer packets need to be stored in cool and dry place away from direct sunlight and heat.
    - o Right combinations of biofertilizers have to be used.
    - o As Rhizobium is crop specific, one should use for the specified crop only.
    - o Other chemicals should not be mixed with the biofertilizers.
    - o While purchasing one should ensure that each packet is provided with necessary information like name of the product, name of the crop for which intended, name and address of the manufacturer, date of manufacture, date of expiry, batch number and instructions for use.
    - o The packet has to be used before its expiry, only for the specified crop and by the recommended method of application.
    - o Biofertilizers are live product and require care in the storage
    - o Both nitrogenous and phosphatic biofertilizers are to be used to get the best results.
    - o It is important to use biofertilizers along with chemical fertilizers and organic manures.
    - o Biofertilizers are not replacement of fertilizers but can supplement plant nutrient requirements.

Biofertilizers

One of the major concerns in today's world is the pollution and contamination of soil. The use of chemical fertilizers and pesticides has caused tremendous harm to the environment. An answer to this is the biofertilizer, an environmentally friendly fertilizer now used in most countries. Biofertilizers are organisms that enrich the nutrient quality of soil. The main sources of biofertilizers are bacteria, fungi, and cynobacteria (blue-green algae). The most striking relationship that these have with plants is symbiosis, in which the partners derive benefits from each other.

Plants have a number of relationships with fungi, bacteria, and algae, the most common of which are with mycorrhiza, rhizobium, and cyanophyceae. These are known to deliver a number of benefits including plant nutrition, disease resistance, and tolerance to adverse soil and climatic conditions. These techniques have proved to be successful biofertilizers that form a health relationship with the roots. Biofertilizers will help solve such problems as increased salinity of the soil and chemical run-offs from the agricultural fields. Thus, biofertilizers are important if we are to ensure a healthy future for the generations to come.

Mycorrhiza

Mycorrhizae are a group of fungi that include a number of types based on the different structures formed inside or outside the root. These are specific fungi that match with a number of favourable parameters of the the host plant on which it grows. This includes soil type, the presence of particular chemicals in the soil types, and other conditions. These fungi grow on the roots of these plants. In fact, seedlings that have mycorrhizal fungi growing on their roots survive better after transplantation and grow faster. The fungal symbiont gets shelter and food from the plant which, in turn, acquires an array of benefits such as better uptake of phosphorus, salinity and drought tolerance, maintenance of water balance, and overall increase in plant growth and development. While selecting fungi, the right fungi have to be matched with the plant. There are specific fungi for vegetables, fodder crops, flowers, trees, etc. Mycorrhizal fungi can increase the yield of a plot of land by 30%-40%. It can absorb phosphorus from the soil and pass it on to the plant. Mycorrhizal plants show higher tolerance to high soil temperatures, various soil- and root-borne pathogens, and heavy metal toxicity.

Legume-Rhizobium Relationship

Leguminous plants require high quantities of nitrogen compared to other plants. Nitrogen is an inert gas and its uptake is possible only in fixed form, which is facilitated by the rhizobium bacteria present in the nodules of the root system. The bacterium lives in the soil to form root nodules (i.e. outgrowth on roots) in plants such as beans, gram, groundnut, and soybean.

Blue-Green Algae

Blue-green algae are considered the simplest, living autotrophic plants, i.e. organisms capable of building up food materials from inorganic matter. They are microscopic. Blue-green algae are widely distributed in the aquatic environment. Some of them are responsible for water blooms in stagnant water. They adapt to extreme weather conditions and are found in snow and in hot springs, where the water is 85 °C. Certain blue-green algae live intimately with other organisms in a symbiotic relationship. Some are associated with the fungi in form of lichens. The ability of blue-green algae tophotosynthesize food and fix atmospheric nitrogen accounts for their symbiotic associations and also for their presence in paddy fields. Blue-green algae are of immense economic value as they add organic matter to the soil and increasc soil fcrtility. Barren alkaline lands in India have been reclaimed and made productive by inducing the proper growth of certain blue-green algae.

REFERENCES

Altieri, Miguel A. (1995) Agroecology: The science of sustainable agriculture. Westview Press, Boulder, CO.

Dedicated double issue of *Philosophical Transactions B* on Sustainable Agriculture. Some articles are freely available.

Dore, J. 1997. *Sustainability Indicators for Agriculture: Introductory Guide to Regional/ National and On-farm Indicators*, Rural Industries Research and Development Corporation, Australia.

Gold, Mary. 1999. *Sustainable Agriculture: Definitions and Terms*. Special Reference Briefs Series no. SRB 99-02 Updates SRB 94-05 September 1999. National Agricultural Library, Agricultural Research Service, U.S. Department of Agriculture.

Hecht, Susanna and Alexander Cockburn (1989) The Fate of the Forest: developers, destroyers and defenders of the Amazon. New York: Verso.

Jahn, GC, B. Khiev, C. Pol, N. Chhorn, S. Pheng, and V. Preap. 2001. Developing sustainable pest management for rice in Cambodia. pp. 243-258, In S. Suthipradit, C. Kuntha, S. Lorlowhakarn, and J. Rakngan [eds.] "Sustainable Agriculture: Possibility and Direction" Proceedings of the 2nd Asia-Pacific Conference on Sustainable Agriculture 18-20 October 1999, Phitsanulok, Thailand. Bangkok (Thailand): National Science and Technology Development Agency. 386 p.

Lindsay Falvey (2004) Sustainability - Elusive or Illusion: Wise Environmental Management. Institute for International Development, Adelaide pp259.

Netting, Robert McC. (1993) Smallholders, Householders: Farm Families and the Ecology of Intensive, Sustainable Agriculture. Stanford Univ. Press, Palo Alto.

Sponsel, Leslie E. (1986) Amazon ecology and adaptation. Annual Review of Anthropology 15: 67-97.

15

Green Technology and Environment and Environmentally Monitoring: Select Case Studies

CHICAGO GREEN TECH

The Chicago Center for Green Technology is located at 445 North Sacramento Boulevard in Chicago. It is open to the public Monday, Wednesday, and Friday from 9:00 a.m. to 5:00 p.m., Tuesday and Thursday from 9:00 a.m. to 8:00 p.m., and Saturday from 9:00 a.m. to 4:00 p.m. The Center is closed on Sunday. Visitors are welcome to explore the public areas of the building and campus. Brochures for self-guided tours are available at the building. The Chicago Department of Environment offers a variety of free educational programmes at Chicago Green Tech for professionals and the public. See our calendar of events for more information. Guided and self-guided tours of the building and grounds are also available. Please call (312) 746-9642 for more information. Free guided tours of Chicago Green Tech are available for groups of 10 or more persons and are subject to staff availability.

Chicago Green Tech Tour Request Form

By Bike

The Chicago Center for Green Technology is located at 445 N. Sacramento Blvd (between Chicago Ave. and Lake St.), 60612. Bike racks are located in front and behind the building. Find your route at http://www.ci.chi.il.us/Transportation/bikemap/keymap.html .

By Car

From the Eisenhower Expressway (I-290) eastbound: Exit at Sacramento Blvd.

Turn left (north) onto Sacramento and travel about one mile to the Chicago Center for Green Technology. It is located on the right (east) side of the street, at 445 N. Sacramento, between Lake St. and Chicago Ave. Visitor parking lots are located directly in front and behind the building.

From the Eisenhower Expressway (I-290) westbound: Exit at California Ave. Turn right (north) onto California and travel 8 blocks to Lake St. Turn left (west) onto Lake St. and travel three blocks to Sacramento Blvd. Turn right (north) onto Sacramento and travel four blocks to the Chicago Center for Green Technology. It is on the right (east) side of the street, at 445 N. Sacramento, between Lake St. and Chicago Ave. Visitor parking lots are located directly in front and behind the building.

From the Kennedy Expressway (1-90) eastbound: Exit at California Ave. Turn right (south) onto California and travel four blocks to Fullerton Ave. Turn right (west) onto Fullerton and travel four blocks to Sacramento Blvd. Turn left (south) onto Sacramento and travel about 2.5 miles, through Humboldt Park, to the Chicago Center for Green Technology. The Center is on the left (east) side of the street, at 445 N. Sacramento, about .5 mile past Chicago Ave. Visitor parking lots are located directly in front and behind the building.

From the Kennedy Expressway (1-90) westbound: Exit at North Ave. Turn left (west) onto North Ave. and travel approximately 2 miles to North Humboldt Blvd. Turn left (south) onto North Humboldt Blvd. After driving south through Humboldt Park, the name of the street will change to Sacramento Ave. Continue south on Sacramento for about one mile to the Chicago Center for Green Technology, which is on the left (east) side of the street, at 445 N. Sacramento, about .5 mile past Chicago Ave. Visitor parking lots are located directly in front and behind the building.

ORBIT: A CASE STUDY

Products and Services

The Association targets the following areas:

- Organisation, assistance and conduct of conferences, workshops and exhibitions;
- Production of publications;
- International co-operation;
- Organisation and conduct of training courses.

The activities currently undertaken are:

- The ORBIT Biennial Internation Conferences;
- ORBIT Special Events
- Meeting Proceedings
- Establishment of the European Compost Network ECN
- Co-operation of the Biowaste Networks worldwide;
- ECN Workshops for practitioners
- ORBIT Biennial International Conferences

The ORBIT biennial conferences has established itself as one of the most significant conferences of the world by becoming a one-stop destination in communicating information relative-'to bioprocessing of waste and recovery of organic residues. This conference series is the major regular meeting organised by the ORBIT Association. Regular events have found place in Europe, i.e. ORBIT-97' in UK, the ORBIT-99 in Germany, QRBIT-2001 in Spain and in Australia with the ORBIT-2003 in Perth, in Germany with the ORBIT - 2006 (www.orbit2006.de) in Weimar and in the Netherlands with the ORBIT - 2008 (www.orbit2008.de) in Wageningen. If you are interested in (co-)organising an ORBIT biennial venue, please read the.relevant terms and conditions. Also the staff of the ORBIT Association is always at your disposal, should you have any inquiries. The next ORBIT conference is scheduled for September 2008.

- European Compost Network ECN

  The fast development in biological waste treatment in Europe and the EU Commission's work has to be assisted. Therefore the ORBIT Association has established as a branch the "European Compost Network ECN" in 2002 with more than 20 European and Candidate Countries as members. ECN is a collaboration of partners promoting sustainable practices in composting, anaerobic digestion and other treatment procedures of organic waste. It represents researschers, experts and more than 1000 plants with nearly 20 million tons annual capacity. It serves first of all for the need of the practical operation and (he decision makers by exchange of knowledge and experiences. Visit the ECN website www.compostne work.info for more information.

- ORBIT Special Events / ECN Workshops

  Meetings can be organised to discuss specific issues; concerning the biological treatment of waste and/or the recovery of organic waste. Such meetings my appeal to a local, regional, or international audience depending on the topics covered.

  The ORBIT Association has a substanial capacity in timely detecting demand for information exchange by a global network of contacts.

International conferences were organised covering issues like biodegradable polymers, innovative biotechnologies and sustainable management of organic waste in Eastern European Countries. The focus of the workshops organised in co-operation with the European Compost Network IECN is dedicated to issues like odour management or the EU Biowaste Directive which are related to the practice of waste treatment. Training courses may be organised to transfer professional knowledge. The ORBIT Association has access to an excellent pool of experienced professional and commercial entities.

- Do You Wish to Organise an ORBIT Meeting?

  Please load down the description "Organisation and Conduct of ORBIT Conferences - Guidelines, Terms and Conditions" from its webpage www.orbit-onlire.net/cr.iwnloads/conference.....guides,pdf

CENTER FOR ENVIRONMENTAL BIOTECHNOLOGY

The Center for Environmental Biotechnology (CEB) scientists and engineers provide expertise to ESD in microbial ecology and environmental engineering.

The personnel within CEB are supported by the Environmental Remediation Technology Programme, the Energy Resources Programme, and the Climate Change and Carbon Management Programme. The Center for Environmental Biotechnology, houses all of the Ecology offices and laboratory facilities. The Department focuses on research in real-time direct environmental assessment and biological treatment, bioremediation, and natural attenuation.

The Center for Environmental Biotechnology maintains the highest quality and highest visibility for its research and development in these areas:

- HydroEcological Engineering Advanced Decision Support (HEADS)
- Molecular Microbial Ecology
- Real-time Direct Environmental Assessment
- Biological Treatment, Bioremediation, and Natural Attenuation

For additional information about bioremediation, check out the PBS programme "Intimate Strangers: Unseen Life on Earth."

Want biodiversity? Look no further than the air around you. It could be teeming with more than 1,800 types of bacteria, according to a first-of-its-kind census of airborne microbes recently conducted by Berkeley Lab. Earth scientist Gary Andersen and colleagues used an innovative DNA test to catalog the bacteria in air samples taken from the Texas cities of San Antonio and Austin. Surprisingly, they found a widely varied bacterial population that rivals the diversity found in soil. They also found naturally occurring relatives of microbes that could be used in bioterrorist attacks — although many of these relatives are harmless.

ENVIRONMENTAL BIOTECHNOLOGY!

Environmental Biotechnology Cooperative Research Centre (EBCRC) develops advanced technologies based on biological systems to improve efficiency and reduce or utilise waste to benefit a wide range of industries and the environment. EBCRC concentrates its research and commercial efforts in three core areas: biofilm prevention and dispersal, rapid in-field microbial detection and control and bioprocesses such as bioremediation and industrial wastewater treatment.

EBCRC brings together the multidisciplinary skills of researchers, engineers and industry participants to bring out novel technologies in environmental, industrial, agricultural, veterinary, mining and medical applications.

EBCRC discovers new biofilm dispersal technology.

Current technologies used to deal with biofilms are largely ineffective, costly and based on toxic biocides. EBCRC has discovered a natural cell signalling system that initiates biofilm dispersal in established biofilms and prevents the formation of new biofilms.

This technology will provide significant improvements to many industries in performance and cost. For example it reduces corrosion, operational costs and capital required for equipment and improves efficiency as well as human, animal and environmental health. » Download video [WMV 8.7MB]

What does EBCRC do?

Many industry processes are based around biological systems. The Environmental Biotechnology Cooperative Research Centre (EBCRC) uses advanced biotechnology to make natural systems work harder to benefit industry, the environment as well as human and animal health.

The technologies can be applied to many areas of industrial processes such as management of waste materials, recycling waste into useful products and cleaner production methods.

EBCRC is helping industry work smarter to reduce operational costs, improve efficiency and reduce the ecological footprint.

Improvements in environmental management can be leveraged from advances in biotechnology that have benefited medical science and the pharmaceutical industry

Only 4.5% of Australia's biotech research is in environmental work. EBCRC will more than double this by spending over $60 million. Potential outcomes could be worth billions of dollars.

Australia has a strong research base in biotechnology, but is yet to realise its potential for the development of new industries producing new products in the environment sector.

Through its strong industry focus, EBCRC will harness these skills for the benefit of the economy and the environment. R&D projects are selected to meet the criteria of needs-based research, and having significant potential for investment return.

Outputs include valuable chemicals from waste, improvements in the efficiency of waste treatment, control of biofilms in water supply, wastewater and industrial processes, diagnostic devices, odour control, bioremediation and transforming low value by-products into high valued products.

What is Environmental Biotechnology?

Environmental Biotechnology is the use of living organisms or biological processes in a wide variety of applications across many industries. Technologies can be applied to industrial processes to improve them and to sustain and manage the environment as well as solve environmental problems.

Industries that Benefit

- Agriculture and intensive livestock production
- Aquaculture and other marine industries
- Water supply and wastewater treatment
- Oil and Chemical waste management
- Food processing industries

COOPERATIVE RESEARCH CENTRES (CRCS)

The CRC programme is an initiative of the Australian Commonwealth Government. The programme was launched in 1990 to strengthen collaborative research between industry, research organisations, education and relevant government agencies.

There are currently 56 cooperative research centres around Australia. They cover various industries including manufacturing technology, information and communication technology, mining and energy, agriculture and rural based manufacturing, environment and medical science and technology. Click here to learn more about the programme.

INTELLECTUAL PROPERTY

EBCRC is actively pursuing commercialisation via two primary routes:

1) Development

2) Active technology transfer of process improvements or practices to EBCRC members and Australian businesses.

- Patent applications

EBCRC currently has a total of ten patent applications in different stages of development.

Some of the protected intellectual property covers:

- The control of biofilms using cell signals.
- An assay for the electrochemical detection of a micro-organism.
- A process for treating high nutrient wastes.
- A method for remediating chlorinated hydrocarbons
- An efficient method of producing multisubunit enzymes.
- A diagnostic platform

TECHNOLOGY TRANSFER

Rapid Detection Pty Ltd

Rapid Detection Pty Ltd has been established to commercialise EBCRC's advanced microbial detection technologies. EBCRC has filed four patents in the area of microbial detection, two of which will be exclusively commrcialised by Rapid Detection.

Media Release

Rapid Detection Pty Ltd has been awarded a COMET grant for the development of its business in field based rapid detection technology of micro-organisms and chemicals. The technology uses a combination of sophisticated biotechnologies to create a novel reactor and sensor system. These can detect the presence of pathogenic organisms or chemical contamination in water, food, industrial or point-of-care situations in less than six hours. The technology platform can be also applied to many existing test systems as well as new test systems. More

Demonstration Projects

The purpose of the demonstration projects is to provide practical user and commercial evaluation of the technologies developed by EBCRC. Demonstration projects are funded primarily through in-kind support from industry partners.

COMMERCIALISATION

Environmental biotechnology is the latest technological revolution following on the heels of medical, industrial and agricultural biotechnology. EBCRC is one of very few organisations in Australia and globally who are not only researching but also commercialising environmental biotechnology innovations. Environmental biotechnology brings together the multi-disciplinary skills of scientists and engineers in the application of natural and modified biological mechanisms to solve industrial problems. EBCRC has set for itself the challenge of generating at least seven novel technologies or innovations to be adopted by industry over the seven year life of EBCRC. EBCRC has aligned its research and commercial activities across three broad areas: Biofilms, Microbial detection and control and Bioprocesses. The commercial objectives fall into three areas:

- short term tangible products or service outcomes
- long term tangible products or service outcomes
- exploratory or 'blue sky' products or service outcomes

Each activity builds upon existing expertise or skills within the organisation. EBCRC is actively pursuing commercial partners in fields outside of environmental biotechnology, eg the industrial, agricultural, medical and veterinary areas. It is also targeting global markets and opportunities with the potential for significant returns to EBCRC and its members. EBCRC aims to commercialise discoveries through licensing or spinout companies. EBCRC welcomes enquiries from any industry with interests in biofilm or biofouling control, micro-organism detection needs or with bioprocess problems. Enquiries may be directed to Dr. Brenton Hamdorf (IP and Commercialisation Manager)

MOLECULAR MICROBIAL ECOLOGY (MME)

Our research group uses molecular approaches for the study of a variety of natural microbial communities. This includes the development of new techniques to dissect the microbial diversity of complex ecosystems. The long-term goal of this research is to integrate different fields of biology (i.e., genomics, ecology, molecular biology, proteomics, bioinformatics and metagenomics) to provide insight into the survival of environmental microorganisms under stressful conditions. Current research in the lab focuses on the microbial ecology of a number of environments including the air, water, soil, and deep subsurface with an emphasis on the natural distribution of pathogens in the environment. By understanding processes that allow specific bacteria to exploit unique environments we hope to better understand critical mechanisms for survival.

Our laboratory is a center for microarray research at LBNL. We are working on advanced technologies to measure microbial species composition, whole-genome transcriptional analysis, and SNP discovery for epidemiology studies. We use Affymetrix high-density microarrays for much of our work. The combinatorial power of at least 500,000 oligonucleotide probes has greatly increased precision and reproducibility in our measurements. We have designed arrays for the broad characterisation of species in a microbial community using the 16S rRNA gene, pathogen specific diagnostics using unique signatures, bacterial gene expression using whole genome sequence, and SNP discovery for genotyping of bacterial strains.

Current projects in our lab include understanding mechanisms of bacterial diversity using 16S rRNA gene sequence to measure relative abundance of individual members of microbial communities. We have developed novel microarray systems to measure dynamic changes and we are working with the Joint Genome Institute in Walnut Creek, CA to develop a rapid system for classifying the thousands of individual sequences from clone libraries that we a constructing. We are also measuring whole genome transcriptional expression of bacteria subjected to different environmental stresses. With collaborators at Stanford University and the University of California at Berkeley, we are developing systems for Caulobacter crescentus, Dehalococcoides ethenogenes, and Thiobacillus ferrooxidans.

We are also interested in the survival and spread of pathogens in the environment. We are studying two related pathogens with very distinctive disease lifecycles, Salmonella enterica serovar Enteritidis and Yersinia pestis. We have completed a 10X draft genome sequence of a phage type 4 strain of Salmonella enteritidis and have recently finished two strains of Yersinia pestis, Nepal516, and Antiqua. To better understand the epidemiology of these pathogens, we are identifying novel SNPs for high accuracy genotyping in a partnership with Perlegen Sciences from Mountain View, CA. We are also assisting Callida Genomics in a project to develop a diagnostic system for the identification of plague (Yersinia pestis) and anthrax (Bacillus anthracis) using a revolutionary SBH system. Details of individual projects may be seen by selecting the Research Projects link on the left.

How an EVAPORATIVE COOLING System Works

An evaporative cooling produces effective cooling by combining a natural process — water evaporation — with a simple, reliable air-moving system. Fresh outside air is filtered through the saturated evaporative media, cooled by evaporation, and circulated by a blower.

Cooling Temperatures

An AIR WASHER will nearly always deliver air cooler than 80 degrees F.

Energy Savings

An evaporative cooling System consumes only one-fourth of the electrical energy required to operate a refrigerated air conditioning unit.

Health Benefits

With evaporative cooling, a complete air change occurs every one-to-three minute. This offers a greathealth advantage over traditional refrigerated air conditioning, which employs a complicated "closed"system that recirculates the same stale dry air over and over.Constant cool air movement pushes heat out — along with stale air, smoke, odors and pollution.The high volume of fresh, cool air produced by the evaporative cooler helps your body ventilate naturally .Evaporative cooling helps maintain natural humidity level.

INDUSTRIAL WASTEWATER TREATMENT SYSTEMS

Physical chemical are techniques to remove the coarse fraction.

Oil, fatty acids and suspended solids could be removed by the use of the following techniques:

- Screens
- Coaqulation
- Flocculation
- Centrifugation
- Fluidisation
- Settling
- Precipittion

Aircon manufactures high efficiency clarifiers for precipitates and other non-biological solids settling applications. PVC structured fill media create a high settling area compared to the floor space occupied by the clarifier by 5.5 mtr2/m2 for 500 mm deep module. Aircon clarifiers have no moving parts and virtually no wear. Few solids will adhere to this material which is virtually impervious to chemical attack.

General features include:-

- Flows from 100 to over 25,000 LPH
- Standard units only 10' overall height
- Enhances The Particle agglomeration growth
- Exclusive surface skimming feature

ACOUSTIC HOOD FOR DG SETS, FLASH DRYER

Aircon Systems is a manufacturer and exporter in India of ACOUSTIC HOOD FOR DG SETS, FLASH DRYER.

We offer Acoustic Hoods designed for quick assembly and dismantling for meeting critical requirements of noise suppression of DG Sets.

The salient features of our Acoustic Canopy are as under:

- Enough space to house panel and Fuel-Tank inside canopy (optional)
- Heavy-duty sturdy steel base frame for D.G. Set.
- Provision of Air-Intake and Air-Exhaust silencer(s) for best results
- Open able doors for easy access to virtually every part of D.G. Set for comfortable maintenance.

We provide

a) 1st Layer of Tissue Paper.

b) Rock-Wool 64 Kg/m3,50 mm Thk.

c) Total thickness of panel-100mm High Performance most comfortable Sound Reduction level to 70-75 db at a distance of 3 meters.

- Axial Fan(s) to supply air.
- Separate Duct for ventilating hot air coming out of engine (In case of Air Cooled Engines)/Radiator duct for water cooled engines.
- Weather proof structure that can also be erected in open space, totally water proof.
- Anti-Corrosion treatment to the full structure and body for long-life under adverse weather conditions.
- 2 coats of Epoxy primer followed by 2 coats of Enamel Paint.

Flash Dryer is a pneumatic system primarily used to dry products requiring the removal of free moisture. Drying takes place in a matter of seconds. Wet material is dispersed into a stream of heated air (or gas) which conveys it through a drying duct.

Using the heat from the air stream, the material dries as it is conveyed. Product is separated using cyclones, and/or bag filters. Typically, cyclones are followed by scrubbers or bag filters for final cleaning of the exhaust gases to meet current emission requirements. Elevated drying temperatures can be used with many products since the flashing off of surface moisture instantly cools the drying gas without appreciably increasing the product temperature. For even greater thermal efficiency and where inertisation is required, recycling of exhaust gases can be used. This can be implemented on all our air stream drying systems and retrofitted on customer's existing drying operations.

DIVISION OF AIR POLLUTION CONTROL

The Division of Air Pollution Control is directed to maintain the purity of the air resources of the State of Tennessee consistent with the protection of normal health, general welfare, and physical property of the people while preserving maximum employment and enhancing the industrial development of the State.

The Division directly serves 91 counties within the state and oversees and assists in the actions of Davidson, Hamilton, Knox, and Shelby counties, which have their own local air pollution control programmes. If you have a question, concern or complaint about air pollution in Davidson, Hamilton, Knox or Shelby counties, please contact the appropriate local air programme.

The Division establishes emission standards and procedure requirements to monitor industries in the State through the issuance of construction and operating permits. Established to carry out control and abatement of air pollution, the Tennessee Air Pollution Control Board adopts regulations, holds hearings, and initiates court actions to enforce regulations. Division staff function as the administrative agency of the Board. Other duties include conducting source visits and compliance inspections, developing enforcement cases on violations of the regulations, maintaining surveillance of the state's ambient air sampling stations, performing and observing stack tests, certifying persons as Visible Emissions Readers, and collecting and disseminating information relative to the control of air pollution. For more information regarding the permits issued by the Division of Air pollution, check out the Environmental Permitting Handbook.

PHYTOREMEDIATION—USING PLANTS TO CLEAN UP POLLUTED SOIL

Polluted soil poses a severe problem for both ecosystem health and land development. Because soil lies at the confluence of many natural systems, soil pollution can be spread to other parts of the natural environment. Groundwater, for instance, percolates through the soil and can carry the soil pollutants into streams, rivers, wells and drinking water. Erosion can create the same problem. Plants growing on polluted soil may contain harmful levels of pollutants themselves, and this can be passed on to the animals and people that eat them. Dust blown from polluted soil can be inhaled directly by passersby. Additionally, in an urban setting such as Fairfax County, polluted soil makes valuable open land unusable for parks, recreation or commercial development.

Despite the benefits of cleaning polluted soil, remediation often never takes place because of the cost and effort of the work. Both soil minerals and soil pollutants carry small electric charges that can cause each to bond with each other, thus making polluted soil very hard to clean. Additionally, soil is a dense medium. This causes excavation of polluted soil for off site treatment or disposal to be very expensive because of the time, labor and heavy machinery necessary to do the job. Therefore, cheaper on-site, or in-situ, remediation techniques have been the focus of much attention and research lately. One of the most interesting and promising of these in-situ techniques is phytoremediation.

Phytoremediation is the use of specialized plants to clean up polluted soil. While most plants exposed to high levels of soil toxins will be injured or die, scientists have discovered that certain plants are resistant, and an even smaller group actually thrive. Both groups of plants are of interest to researchers, but the thriving plants show a particular potential for remediation because it has been shown that some of them actually transport and accumulate extremely high levels of soil pollutants within their bodies. They are therefore aptly named hyper-accumulators.

Hyper-accumulators already are being used throughout the country to help clean up heavy metal polluted soil. Heavy metals are some of the most stubborn soil pollutants. They can bond very tightly to soil particles, and they cannot be broken down by

microbial processes. Most heavy metals are also essential plant nutrients, so plants have the ability to take up the metals and transport them throughout their bodies. However on polluted soil, the levels of heavy metals are often hundreds of times greater than normal, and this overexposure is toxic to the vast majority of plants. Hyper accumulators, on the other hand, actually prefer these high concentrations. Essentially, hyper-accumulators are acting as natural vacuum cleaners, sucking pollutants out of the soil and depositing them in their above ground leaves and shoots. Removing the metals is as simple as pruning or cutting the hyper-accumulators above ground mass, not excavating tons of soil.

Resistant, but not hyper-accumulating, plants also have a role in phytoremediation. Organic toxins, those that contain carbon such as the hydrocarbons found in gasoline and other fuels, can be broken down by microbial processes. Plants play a key role in determining the size and health of soil microbial populations. All plant roots secrete organic materials that can be used as food for microbes, and this creates a healthier, larger, more diverse and active microbial population, which in turn causes a faster breakdown of pollutants. Resistant plants can thrive on sites that are often too toxic for other plants to grow. They in turn give the microbial processes the boost they need to remove organic pollution more quickly from the soil. Both forms of phytoremediation have the added benefit of not disturbing the soil. While excavation is an effective way to get rid of pollution, it removes the organic matter rich topsoil and. because of the use of heavy machinery, compacts the soil that is left behind. Phytoremediation does not degrade the physical or chemical health of the soil. Actually, it creates a more fertile soil. Soil organic matter is increased as a result of root secretions and falling stems and leaves, and the roots create pores through which water and oxygen can flow. Additionally few would argue that a dusty excavation site is more aesthetically pleasing than a nicely planted field.

There are. however, many limitations to phytoremediation. It is a slow process that may take many growing seasons before an adequate reduction of pollution is seen, whereas soil excavation and treatment cleans up the site quickly. Also, hyper-accumulators can be a pollution hazard themselves. For instance, animals can cat the metal rich hyper accumulators and cause the toxins to enter the food chain. II the concentration of metals in the plants is thought to be high enough to cause toxicity, there must be a way to segregate the plants from humans and wildlife, which may not be an easy task. Additionally, phytoremediation is in its infancy, and its effectiveness in cleaning up v arums toxins compared to conventional means of treatment is not always known. However, with more research and practice, the practicality of using phytoremedialion should increase.

The growing popularity of phytorcmcdiation can be seen in the number of businesses beginning to appear in urban areas of the United States that arc involved in the field. In New Jersey, a company named Phytotech is using mustard greens to remove lead from polluted yards in Boston and hydroponically grown sunflowers to take radioactive metals out of the water surrounding the Chernobyl nuclear power plant. Right here in Fairfax Country. Edenspace Systems Corp. of Chantilly has more than two dozen current and future contracts to utilize phytoremediation in the field, and they even sell an arsenic accumulating Edenfern that can be planted in your backyard today.

ECOLOGY

Scientific Focus Areas

The Ecology Department (ED) intends to maintain the highest quality and highest" visibility for its research and development in five areas:

•

- Bioenergy
- Hydroecologcal engineering advanced decision support
- Molecular microbial ecology
- Real-time assessment of bioavailability and biokinetics
- Bioremediation and natural attenuation

These five R&D areas are largely integrated, but contain some domains that are not inclusive. These five areas are considered ED's core competencies.

Bioenergy

Concerns about global warming, dependence or foreign oil, and rising gasoline prices have spurred a strong interest in bioenergy in the United States. ED scientists will be contributing their expertise in microbiology, molecular microbial ecology, and bacterial metabolism to a scientific: challenge of great national importance: the efficient production of biofuels. LBNL has been a key strategic partner in large research initiatives for devebping bioenergy, including the Joint BioEnergy Institute; (JBEI), a Department of Energy-funded research center that will receive $125 million over five years, and the Energy Biosciences Institute (EBI), which is being funded by the energy company BP at a level of $500 million over ten years; both JBEI and EBI began operations in late 2007. The Ecology Department will play an important role in both JBEI and EBI. For example, two ED personnel will serve as Directors in JBEI: the Director of Microbial Communities (responsible for studying the structural and functional dynamics of microbial communities that efficiently degrade lignocellulosic biomass) and the Director of Biofuels Pathways (responsible for discovering genes that can be engineered into metabolic pathways resulting in efficient production of various transportation fuels, from cellulose degradation products). In EBI, ED scientists will be involved in the study of Microbially Enhanced Hydrocarbon Recovery and will co-leed a Systems Biology programme, which will entail the intensive, genome-enabled study of certain bioenergy-related microorganisms ("omics" studies, including transcriptomics, proleomics, and metabolomics).

Hydroecological Engineering Advanced Decision Support (HEADS)

The HEiADS research focus area has established a strong track record in the rapidly growing, new subject area of Ecological Engineering. The term "hydroecological engineering" signifies the group's concentration on water resources and wastewater engineering. Recognising the growth in the field of

environmental informatics and the application of computer-based models in the development of decision support systems, the group is active in this niche area. Interest in decision support interfaces well with the group's expertise in the deployment of real time flow and water quality sensors, rapid laboratory assessment techniques, and mathematical models to develop an early warning system for contaminant management and containment.

Molecular Microbial Ecology (MME)

Understanding microbial interactions is key to the study of global warming, biodegradation of harmful cormounds, and the exploration of complex microbial communities in their natural environment. The DOE has placed an increased emphasis on the role microbes play in modifying their environment and their impact on energy security. The MME group has responded to these needs by agcressively seeking out new projects and expanding its staff to develop new core capabilities. One of the key challenges has-been to harness the explosion of microbial DNA sequence information to accurately measu'e the microbial dynamics in extreme environments. Since less than one percent of the microbial species can be cultured from these environments, our knowledge of what these organisms may be doing is limited to where they are observed and the similarity of their genomes to studied organisms. By understanding the ecological structure: of microbial communities and the fine-scale dynamics resulting from subtle perturbations, it may be possible to identify novel functional pathways and use the diverse microbial capabilities to assist in key DOE missions. The molecular tools being developed in the ESD's Center for Environmental Biot2chnology will position us to be leaders in this area.

Real-Time Assessment of Bioavailability and Biokinetics (RABB)

Interactions between environmental pollutants and ecological receptors begin when the pollutants become available to the target sites of a live ecological receptor. The ab lity to characterize the dynamics of the bioavailability of pollutants, their transformation kinetics, and the subsequent ecological response is a keystone to advancing the science in relevant DOE areas, including biogeochemistry, bioremediation, and exposure and risk iissessrnent. Since 1999, the RABEJ research group at ED has pioneered the development and application of several cutting-edge technologies, such as synchrotron radiation-based spectromicroscopy, in vitro human gastrointestinal mimetic reactors, and in vivo mouse protocols that allow for the real-time assessment of bioavailability and biokinetics of environmental pollutants. The RABB group intends to establish this capability further by seeking oit new projects and new collaborators, as well as expanding our staff, to position us as leaders in the areas of biological -and environmental sciences.

Bioremediation and Natural Attenuation

Bioremediation and natural attenuation have been rapidly growing areas of so identific study over the past cecade. The acceptance of natu -al attenuation as a solution For cleaning up contaminated sites and DOE's recognition that they will have long-term steward; hip concerns to address at the most contaminated sites has greatly increased the urgency for research related to microbial ecology, biogeochemistry, biochemistry, and molecular detection techniques. This type of research is truly enabling for natural attenuation, since characterisation, transport-and-fa'te prediction, and verification monitoring require a strong scientific basis. Natjral attenuation is viewed as the best solution for cleaning up many waste sites and will save billior s of dollars in cleanup costs. ED scientists and engineers are recognised leaders in the field of bioremediation and natural attenuation. The Center for Environmental Biotschnology provides the primary facilities used by ED, including state-of-the-art equipment for microbiology and environmental engineering. ED investigators have extensive experience in both water treatment and aquifer bioremediation. In addition to basic research, ED investigators have been involved in various aspects of more than 100 field demonstrations and deployments, and have five patents in this area that are licensed to more than 30 companies. The types of contaminants in which ED investigators have expertise include chlorinated solvent;, petroleum hydrocarbons (including monoaromatic and polynuclear aromatic hydrocarbons), methyl tert-butyl ether (MTBE), high explosives (especially TNT and RDX), nitrate, plutonium, neptunium, chromium, and uranium. The Bioremediation and Natural Attenuation area has both basic research and field application foci for the ED. The basic research foci are metabolism and physiology (including the genetics and biochemistry underlying pollutant metabolism), biotransformation kinetics, and modeling of bioyeochemical processes. Field-application foci are in situ monitoring techniques (including "molecular, mass spectrometric, and biogeochemical approaches), co-metabolic techniques, and modeling of attenuation and environmental fate.

Funding

ED personnel are funded by DOE Programmes in (1.) the Office of Science, Office of Biological and Environmental Research (OBER) (Natural and Accelerated Bioremediation Research Programme, Genomics:GTL, and Medical Sciences); (2) the Office of Environmental Management, Environmental Restoration Programmes; (3) the Office of Fossil Research, the Petroleum Environmental Research Forum; and (4) the National Nuclear Security Administration, Office of Nonproliferation Research and Engineering (NN20). In addition, support is provided by the U.S. Department of Homeland Security, the Department of Agriculture; the Department of the Interior, Bureau of Land Management, and Bureau of Reclamation under the CALFED programme, NASA Astrobiology Instititutes, as well as several projects with remediation companies using DOE-patented technologies for in situ bioremediation. ED personnel are also funded by Berkeley Lab's Laboratory Directed-Research and Development (LORD) Programme in the area of microbial fuel cells, fungal rDNA arrays, and FTIR biokinetic analaysis.

PHYTOREMEDIATION

Overview: Phytoremediation is the use of certain plants to clean up soil, sediment, and water contaminated with metals and/or organic contaminants such as crude oil, solvents, and polyaromatic hydrocarbons (PAHs). It is a name for the expansion of an old process that occurs naturally in ecosystems as both inorganic and organic constituents cycle through plants. Plant physiology, agronomy, microbiology, hydrogeology, and engineering are combined to select the proper plant and conditions for a specific site. Phytoremediation is an aesthetically pleasing mechanism that can reduce remedial costs, restore habitat, and clean up contamination in place rather than entombing it in place or transporting the problem to another site.

Phytoremediation can be used to clean up contamination in several ways:

- Phytovolatilisation: Plants take up water and organic contaminants through the roots, transport them to the leaves, and release the contaminants as a reduced of detoxified vapor into the atmosphere.
- Microorganism stimulation: Plants excrete and provide enzymes and organic substances from their roots that stimulate growth of microorganisms such as fungi and bacteria. The microorganisms in the root zone then metabolize the organic contaminants.
- Phytostabilisation: Plants prevent contaminants from migrating by reducing runoff, surface erosion, and ground-water flow rates. "Hydraulic pumping" can occur when tree roots reach ground water, take up large amounts of water, control the hydraulic gradient, and prevent lateral migration of contaminants within a ground water zone.
- Phytoaccumulation/extraction: Plant roots can remove metals from contaminated sites and transport them to leaves and stems for harvesting and disposal or metal recovery through smelting processes.
- Phytodegradation by plants: Organic contaminants are absorbed inside the plant and metabolized (broken down) to non-toxic molecules by natural chemical processes within the plant.

Advantages and Disadvantages of Phytoremediation

When using phytoremediation there are many positive and negative aspects to consider. The advantages and disadvantages are listed below.

Advantages

- Works on a variety on organic and inorganic compounds
- Can be either *In Situ* / *Ex Situ*

- Easy to implement and maintain
- Low-cost compared to other treatment methods
- Environmentally Friendly and aesthetically pleasing to the public
- Reduces the amount wastes to be landfilled

Disadvantages

- May take several years to remediate
- May depend on climatic conditions
- Restricted to sites with shallow contamination within rooting zone
- Harvested biomass from phytoextraction may be classified as a RCRA hazardous waste
- Consumption of contaminated plant tissue is also a concern.
- Possible effect on the food chain

A major advantage that is listed above is the low cost. For example, the cost of cleaning up one acre of sandy loam soil at a depth of 50cm with plants is estimated at $60,000-$100,000 compared to $400,000 for the conventional excavation and disposal method. One reason for this low cost is phytoremediation may not require expensive equipment or highly specialized personnel, and can be relatively easy to implement.

One major concern with phytoremediation is the possible affects on the food chain. For example vegetation is used that absorbs toxic or heavy metals and moles or voles eat the metal contaminated plants. The predators of the moles or voles then become victims of intoxication. All though the possibilities of such scenarios are being looked at, more fieldwork and analysis is necessary to understand the possible effects phytoremediation can have.

Regulatory Issues

As of now phytoremediation is too new to be approved by regulatory agencies such as the EPA. Eventually the main question that regulators will focus on is will phytoremediation remediate the site to the standards and reduce the risk to human health and the environment. In developing regulations for phytoremediation the following questions will need answering.

- Can it cleanup the site below action levels? On what scale?
- Does it create any toxic intermediate or products?
- Is it cost effective as alternative methods?
- Does the public accept the technology?

PHYTOREMEDIATION

Phytoremediation is a set of processes that uses plants to remove, transfer, stabilize and destroy organic/inorganic contamination in ground water, surface water, and leachate.

Description

Typical In Situ Phytoremediation System Phytoremediation is a set of processes that uses plants to clean contamination in ground water and surface water. The treatment of metals or other inorganic contamination has been discussed in Section 4.6 (Phytoremediation for Soil). There are several ways plants can be used for the phytoremediation. These mechanisms include enhanced rhizosphere biodegradation, hydraulic control, phyto-degradation and phyto-volatilisation.

Enhanced Rhizosphere Biodegradation

Enhanced rhizosphere biodegradation takes place in the soil surrounding plant roots. Natural substances released by plant roots supply nutrients to microorganisms, which enhances their ability to biodegrade organic contaminants. Plant roots also loosen the soil and then die, leaving paths for transport of water and aeration. This process tends to pull water to the surface zone and dry the lower saturated zones.

Hydraulic Control

Depending on the type of trees, climate, and season, trees can act as organic pumps when their roots reach down towards the water table and establish a dense root mass that takes up large quantities of water.

Phyto-degradation

Phyto-degradation is the metabolism of contaminants within plant tissues. Plants produce enzymes, such as dehalogenase and oxygenase, that help catalyze degradation. Investigations are proceeding to determine if both aromatic and chlorinated aliphatic compounds are amenable to phyto-degradation.

Phyto-volatilisation

Phyto-volatilisation occurs as plants take up water containing organic contaminants and release the contaminants into the air through their leaves. Plants can also break down organic contaminants and release breakdown products into air through leaves.

Synonyms

Vegetation-enhanced bioremediation.

Applicability

Phytoremediation can be used to clean up organic contaminants from surface water, ground water, leachate, and municipal and industrial wastewater.

Plants also produce enzymes, such as dehalogenase and oxygenase, which help catalyze degradation.

Limitations

There are a number of limitations to phytoremediation

- It is limited to shallow soils, streams, and ground water.
- High concentrations of hazardous materials can be toxic to plants.
- It involves the same mass transfer limitations as other biotreatments.
- Climatic or seasonal conditions may interfere or inhibit plant growth, slow remediation efforts, or increase the length of the treatment period.
- It can transfer contamination across media, e.g., from soil to air.
- It is not effective for strongly sorbed (e.g., PCBs) and weakly sorbed contaminants.
- Phytoremediation will likely require a large surface area of land for remediation.
- The toxicity and bioavailability of biodegradation products is not always known. Products may be mobilized into ground water or bioaccumulated in animals. More research is needed to determine the fate of various compounds in the plant metabolic cycle to ensure that plant droppings and products manufactured by plants do not contribute toxic or harmful chemicals into the food chain or increase risk exposure to the general public.

Data Needs

A detailed discussion of these data elements is provided in Subsection 2.2.2. (Data Requirements for Ground Water, Surface Water, and Leachate). In addition, detailed information is needed to determine the kinds of soil used for phytoremediation projects. Water movement, reductive oxygen concentrations, root growth, and root structure all affect the growth of plants and should be considered when implementing phytoremediation.

Performance Data

The U.S. Air Force used poplar trees to contain a ground water TCE plume. TCE was shown to be degraded in the tissues of the poplar trees. The trees pumped a sufficient amount of water to produce a cone of depression limiting the spread of the TCE plume. In Iowa, EPA demonstrated the usage of phytoremediation by planting poplar trees along a stream bank between a corn field and the stream. These trees acted as natural pumps to keep toxic herbicides, pesticides, and fertilizers out of the streams and ground water. After three years, while the nitrate concentration in ground water at the edge of the corn field was measured at 150 mg/L, the ground water among the poplar trees along the stream bank had nitrate concentration of only 3 mg/L. USAEC is also leading the team of experts from EPA, Tennessee Valley Authority (TVA) and the Waterways Experimental Station (WES) to successfully demonstrate phytoremediation of explosive contaminated sites in Milan Army Ammunition Plant in Milan, TN.

Cost

The key cost driver information and cost analysis was developed in 2006 using the Remedial Action Cost Engineering and Requirements (RACER) software.

Key Cost Drivers

- Scale of effort
  - o Area of contamination is the primary cost driver
- Tree size (maturity) is the secondary cost driver.

BIOINDICATORS

What are Bioindicators?

Bioindicators are organisms, such as lichens, birds, and bacteria, that are used to monitor the health of the environment. The organisms are monitored for changes that may indicate a problem within their ecosystem. The changes can be chemical, physiological, or behavioural.

Uses and Types of Bioindicators

Each organism within an ecosystem has the ability to report on the health of its environment. Bioindicators are used to:

- detect changes in the natural environment;
- monitor for the presence of pollution and its effect on the ecosystem in which the organism lives;
- monitor the progress of environmental cleanup; and,
- test substances, like drinking water, for the presence of contaminants.

Types of bioindicators and their uses include the following:

- *Plant Indicators* — The presence or absence of certain plant or other vegetative life in an ecosystem can provide important clues about the health of the environment.
- Lichens, often found on rocks and tree trunks, are organisms consisting of both fungi and algae. They respond to environmental changes in forests, including changes in forest structure, air quality, and climate. The disappearance of lichens in a forest may indicate environmental stresses, such as high levels of sulfur dioxide, sulfur-based pollutants, and nitrogen.
- *Animal Indicators* — An increase or decrease in an animal population may indicate damage to the ecosystem caused by pollution. For example, if pollution causes the depletion of important food sources, animal species dependent upon these food sources will also be reduced in number. In addition to monitoring the size and number of certain species, other mechanisms of animal indication include monitoring the concentration of toxins in animal tissues, or monitoring the rate at which deformities arise in animal populations.
- *Microbial Indicators* — Microorganisms can be used as indicators of aquatic or terrestrial ecosystem health. Found in large quantities, microorganisms are easier to sample than other organisms. Some microorganisms will produce new proteins, called stress proteins, when exposed to contaminants like cadmium and benzene. These stress proteins can be used as an early warning system to detect low levels of pollution.

The Science - How do Bioindicators Work?

Specific physiological and behavioural changes in bioindicators are used to detect changes in environmental health. The specific changes differ from organism to organism. The use of organisms as bioindicators encompasses many areas of science. Wildlife conservation genetics is an example of how traditional approaches can be combined with emerging biotechnologies to improve accuracy, and to collect information not available through conventional methods. Wildlife conservation genetics combines traditional monitoring of wildlife populations, like racoons, with the scientific discipline of genetics, to gain information about the health of ecosystems.

Behavioural and population changes in a species can be observed by scientists, but physiological changes must be detected using special tests. Bioassays require samples from organisms to detect changes in the environment. These tests may be used to ensure drinking water safety or to measure river health. In the future, as research identifies new ways to use microbes, these uses will expand to include testing of soil and air.

Bioassays can be carried out in traditional ways and with new biotechnology-derived methods. These methods are outlined in more detail below.

CASE STUDY: TESTING WATER

Bioluminescent bacteria are being used to test water for environmental toxins. If there are toxins present in the water, the cellular metabolism of the bacteria is inhibited or disrupted. This affects the quality or amount of light emitted by the bacteria. Unlike traditional tests, this one is very quick – taking from five to 30 minutes to complete. However, it only indicates the presence of a toxin and cannot identify the specific toxin causing the change in the organism.

Traditional Bioassays

In traditional bioassays, a bioindicator organism is introduced to environmental samples, such as soil or water, and researchers observe any changes that occur as a result of exposure. These methods are based primarily on observation to detect changes. Examples of traditional bioassay methods include the following:

- measurement of plant root growth in suspected polluted environments and comparison of the measured growth rate against normal root growth rates.
- exposure of microorganisms to an environment and observation of any changes in the organism related to toxin exposure, such as the presence of stress proteins produced when cells are exposed to harmful environmental conditions.

Biotechnology-Based Bioassays

Several biotechnology-based methods use microorganisms to test environmental health. Unlike traditional methods, biotechnology-based bioassays do not rely on observation alone but set out to create specific reactions that indicate the presence of a specific pollutant or an unwanted microorganism. In this way they are similar to traditional chemical analysis of environmental samples.

DNA Microarray Technology — Using DNA microarray technology, environmental samples, such as water, are tested for the actual genetic material of an organism. This form of testing is used to detect dangerous microorganisms in the environment, such as *E. Coli* bacteria in water.

DNA microarrays are stamp-sized glass or silicon microchips that are embedded with thousands of single-stranded DNA or RNA. In this case, the DNA is that of the microorganisms being tested for. The microarrays are manufactured using samples of microorganisms. If the same type of microorganisms are present in the water sample, the DNA or RNA on the array will react with the complementary DNA or RNA of the microorganism in the sample. This identifies its presence in the sample. When these tests are fully developed, it will take as little as four hours to test for microbial presence in environmental samples, such as drinking water and soil. Traditional chemical-based tests take an average of 48 hours.

Fluorescence In-Situ Hybridisation (FISH) — This is a method for detecting the presence of particular genes in a sample. As a bioindicator, FISH can determine whether or not specific microorganisms are polluting certain areas. It does this by testing environmental samples for the presence of microbial genes. A fluorescent marker is attached to the DNA of the type of microorganism being testing. This marked DNA is now called a 'probe.' Environmental samples are fixed onto a slide, and the slide is exposed to the flourescent DNA probe. If the polluting microorganism is present on the slide, its DNA will bind to the fluorescent probe, causing the slide to glow with ultraviolet light. Detection of this ultraviolet light by a special fluorescent microscope demonstrates the presence of polluting microorganisms in the sample.

Biotechnology and Bioindicators

Currently, biotechnology-based tests are being used to identify changes in indicator species to gauge the presence of pollutants in the environment. Many of the tests being developed are designed to detect pollutants in rivers and drinking water sources. These tests will be faster and more accurate than conventional tests in detecting metabolic changes within microorganisms. Microbial bioindicators are also being researched to detect pollution in other substances, such as soil.

Current Research Areas in Bioindicators

Bioindicator research is currently focussed on developing more rapid and reliable tests for the presence of microorganisms in water and soil. Tests for drinking water are a special area of concern for both developed and developing countries. Although biotechnology-based tests currently exist for drinking water, there are still many pollutants that are not detectable. Scientists are busy trying to replace time consuming, traditional methods with newer, faster, and more reliable tests based on biotechnology.

Sustainable Development and Bioindicators

Bioindicators are a method of monitoring or detecting the negative impacts that industrial activity has on the environment. This information helps develop strategies that will prevent or lower such effects and make industry more sustainable. The role of bioindicators in sustainable development will help ensure that industry leaves the smallest footprint possible on the environment.

BIOLOGICAL INDICATORS

Welcome to the Biological Indicators home page. This site contains information about the Biological Indicators Programme in the Environmental Sciences Division (ESD) at Oak Ridge National Laboratory (ORNL). Biological indicators have been developed and applied at a variety of field sites in the eastern United States to assess and evaluate the effects that environmental stressors such as contaminants have on the health of aquatic organisms, populations, and communities.

Concepts and Use

- What are bioindicators?
- Biological responses to stress
- Types and characteristics of bioindicators
- Design & use of bioindicators

Research Projects

- Streams Impacted by Mixed Contaminants
- River Contamination by Paper Mill Effluents
- PCBs in Reservoirs
- Super Fund Site Studies

Staff Directory

Personnel

- S. Marshall Adams
- M. S. Greeley
- Kai-Lin Lee
- M. S. Bevelhimer
- R. A. Efroymson

Relevant Staff Publications

Capabilities

- Laboratory
- Field
- Staff Expertise

Bioindicator

Bioindicators are species or chemicals used to monitor the health of an environment or ecosystem. They are any biological species or group of species whose function, population, or status can be used to determine ecosystem or environmental integrity. An example of such a group are the copepods and other small water crustaceans present in many water bodies. Such organisms are monitored for changes (chemical, physiological, or behavioural) that may indicate a problem within their ecosystem. Depending on the organism selected and their use, there are three types of bioindicators:

1. Plant indicators
2. Animal indicators
3. Microbial and chemical indicators

Plant Indicators

Plant Indicators—The presence or absence of certain plant or other vegetative life in an ecosystem can provide important clues about the health of the environment-(environmental preservation). Lichens-(not a *plant*), found on rocks and tree trunks, are organisms comprising both fungi and algae. They respond to environmental changes in forests, including changes in forest structure-(conservation biology), air quality, and climate. The disappearance of lichens in a forest may indicate environmental stresses, such as high levels of sulfur dioxide, sulfur-based pollutants, and nitrogen-oxides. The composition and total biomass of algal species in aquatic systems serves as an important metric for organic pollution and nutrient loading such as nitrogen and phosphorus.

Animal Indicators, and Toxins

Animal Indicators—An increase or decrease in an animal population may indicate damage to the ecosystem caused by pollution. For example, if pollution causes the depletion of important food sources, animal species dependent upon these food sources will also be reduced in number-population decline. Overpopulation, can be the result of opportunistic species growth. In addition to monitoring the size and number of certain species, other mechanisms of animal indication include monitoring the concentration of toxins in animal tissues, or monitoring the rate at which deformities arise in animal populations.

Microbial Indicators and Chemicals

Microbial Indicators—Microorganisms can be used as indicators of aquatic or terrestrial ecosystem health. Found in large quantities, microorganisms are easier to sample than other organisms. Some microorganisms will produce new proteins, called stress proteins, when exposed to contaminants like cadmium and benzene. These stress proteins can be used as an early warning system to detect low levels of pollution.

Bioindicators

The term bioindicator is used for organisms or organism associations which respond to pollutant load with changes in vital functions, or which accumulate pollutants (Arndt et al. 1987). Information about specific biological effects supplements data on air pollutions generated by technical analysis methods. The most important reasons for using bioindicators are:

- the direct determination of biological effects,
- the determination of synergetic and antagonistic effects of multiple pollutants on an organism,
- the early recognition of pollutant damage to plants as well as toxic dangers to humans and
- relatively low cost compared to technical measuring methods.

The great potential of bioindicators for environmental monitoring is often confronted with difficult questions of methodology resulting from the use of "living measuring instruments". The effects of environmental load cannot always be clearly differentiated from natural stress factors. Lack of practical experience with certain bioindicators sometimes makes clear interpretion of findings more difficult, especially if no comparable pollutant measurements are available.

Intensive research over the last decades has resulted in the availability of numerous bioindicators which satisfy the requirements of convenience, standardisation, cost, and evaluative capability (cf. Arndt 1987, Zimmermann u. Umlauff-Zimmermann 1994).

Bioindicators are commonly grouped into accumulation indicators and response indicators. Accumulation indicators store pollutants without any evident changes in their metabolisms. Response indicators react with cell changes or visible symptoms of damage when taking up even small amounts of harmful substances.

Biomonitoring is divided into passive and active. Passive biomonitoring is the use of organisms, organism associations, and parts of organisms which are a natural component of the ecosystem and appear there spontaneously. Active biomonitoring includes all methods which insert organisms under controlled conditions into the site to be monitored.

Cadastre of Ecological Pollution Effects

The Berlin Department of Urban Development, Environmental Protection, and Technology has conducted comprehensive studies for the determination of air pollution effects since 1991. These studies emphasised the complex of pollutants typical for metropolitan areas (SenStadtUm 1993a). All bioindicator methods employed have advantages and disadvantages. This means in practice that not only individual bioindicators, but a whole series of bioindicators was employed, according to the particular goal. This series is as finely adjusted to each other as possible. The systematic use of bioindicators was used for the Cadastre of Ecological Pollution Effects, conducted for the Ecological Monitoring Berlin and Hinterland. The findings are important supplements to and extensions of air quality monitoring data, especially for the year-long random sample measurement programme of the Berlin Air Quality Monitoring System - BLUME.

A great amount of data and knowledge has been generated by the Cadastre of Ecological Pollution Effects. The size of this data requires presentation in the Environmental Atlas to be limited to findings which have been sufficiently secured. These findings allow a differentiation of pollutant load ranges and could serve as a particularly significant basis for planning.

The Environmental Atlas took active and passive bioindicator methods, as well as accumulation and reaction indicators almost equally into consideration. The selected results allow an overview of general air pollution in Berlin, manifested by effects upon naturally occuring lichen and exposed lichen transplants, as well as an overview of regional pollutant patterns, indexed by accumulation in pine needles, rye grass, and green kale.

Findings of the Cadastre of Ecological Pollution Effects not presented here can be found in various studies and publications.

The evaluation of study findings can, in some cases, be supplemented by earlier research. This enables a chronological comparison of general pollutant effects and certain pollutant materials (cf. Cornelius et al. 1984). These kinds of reference data are valuable yardsticks for evaluation of the current pollutant situation in Berlin, for changes in air quality over recent decades can thus be documented.

Investigated Pollutants

The selection of analysed pollutants took into consideration human toxicity, phytotoxicity, and ecosystem aspects. The information can thus be used for an overview of environmental loads in the Berlin area. The Cadastre of Ecological Pollution Effects focused mainly on the pollutants described below.

Until the end of the 80's, Berlin air had relatively high SO_2 and dust pollutions, especially during low-exchange winter weather periods. This was mainly due to the large amounts of sulfur-rich lignite brown coal used for heating in older residential areas (cf. Map 03.01, SenStadtUm 1994).

SO_2 in the air enters plants as a gas or in water solution and can cause damage. Chemical reactions of SO_2 create acids. Precipitation can transport these acids into the soil and trigger indirect damages, such as lack of nutrients and acid stress. Value limits for direct damaging effects on plants are clearly lower than for animals and humans. High sulfur dioxide concentrations lead to air pollution damages on coniferous trees, including non-specific chloroses, necroses, growth inhibition, and impairment of reproduction.

Lead is a heavy metal and has special environmental relevance because it is toxic. Leaded gasoline is still the main source. Most lead particles remain on the surface of leaves. Only a small amount becomes physiologically active and damages plants. This process can lead to accumulation in the food chain. Toxic effects for grazing animals or humans cannot be ruled out.

Fluorine is released from various industrial processes, waste burning, and by fossil fuel energy production. Hydrogen fluoride damages in plants appear primarily as necroses on leaf edges (marginal) and tips (terminal). Fluoride is accumulated in plants, damages leaves and inhibits growth.

Polycyclic aromatic hydrocarbons (PAH) are emitted primarily as incomplete combustion products, such as from internal combustion motors and heating plants. They also enter the environment as wash-outs from surfaces covered by tar, asphalt and bitumen. These substances are taken up by the human organism mainly by breathing, but also in food and by skin contact. The considerable carcinogenic and mutagenic potential of this group makes it more important for humans than for plants. Benzo(a)pyrene (BaP) is a leading component of polycyclic aromatic hydrocarbons.

Polychlorinated biphenyls (PCB) are ubiquitous in the environment today. Production of these chemicals was stopped in Germany in 1983, but considerable diffuse emissions continue from waste disposal sites, burning of waste-oil and waste, and leakage. PCBs enter the human organism primarily in the form of animal products, through accumulation along the food chain. Accumulation in plants seems to be of little phytotoxic significance. Great attention must be given to the exposure of the population to these substances, in view of their damaging effects on embryos and a well-founded suspicion of carcinogenic effects.

Polychlorinated dibenzodioxin (PCDD) and polychlorinated dibenzofuran (PCDF) are undesired side products of chlorine created in chemical and thermic processes. The main sources of emissions in Berlin are industrial (chemical cleaners, metallurgical processes) waste combustion plants, power plants, metal recycling plants, traffic, and domestic burning. Secondary sources are contaminated areas. The group known as "dioxins" accumulate in the food chain. They primarily enter the human organism in foods containing animal fats. The behavior of this material in the environment as well as its toxicity for human is not yet sufficiently known. Studies on these materials conducted for the Cadastre of Ecological Pollution Effects thus represent an important contribution to the environmental relevance of these materials in Berlin.

Bioindicators were also used to study the effects of highly volatile chlorinated hydrocarbons, photooxidants (ozone), and nitrogen oxides. These results are not presented, but can be found in the studies.

Bioindicator Methods

Maps 03.07.1 - 4 contain findings on pollution effects obtained by bioindicator use in Berlin since 1991. They are differentiated between 1) the determination of biological effects caused by a general complex of air pollutant factors working as a sum and 2) the analytical detection of specific pollutants.

The response of lichen to pollutants in the course of a "screening" programme was the first indication for the presence of a pollutant load (Map 03.07.1). The *Hypogymnia physodes* lichen was employed for differentiation of small-area pollutant patterns in areas where a rich variety of natural lichen species can no longer exist (Map 03.07.2). Additionally, relevant pollutants were isolated with suitable accumulation indicators and evaluated for human toxicity, phytotoxicity, and ecosystem effects (Map 03.07.3/4).

BIOMARKER (MEDICINE)

In medicine, a biomarker is an indicator of a particular disease state or a particular state of an organism.

An NIH study group committed to the following definition in 1998: "a characteristic that is objectively measured and evaluated as an indicator of normal biologic processes, pathogenic processes, or pharmacologic responses to a therapeutic intervention."

In the past, biomarkers were primarily physiological indicators such as blood pressure or heart rate. More recently, biomarker is becoming a synonym for molecular biomarker, such as elevated prostate specific antigen as a molecular biomarker for prostate cancer, or using enzyme assays as liver function tests. Biomarkers also cover the use of molecular indicators of environmental exposure in epidemiologic studies such as human papilloma virus or certain markers of tobacco exposure such as 4-(methylnitrosamino)-1-(3-pyridyl)-1-butanone (NNK).

BIOLOGICAL MARKERS: THEIR USE IN QUANTITATIVE ASSESSMENTS

Biological markers can be conceptualized in terms of categories of markers that form a continuum representing a sequence of events from exposure to disease. These categories include markers of internal dose, biologically effective dose, early response, and disease. Outside of this sequence are susceptibility factors that can act at any point along the way to modify the effects of external exposures on disease outcomes. Examples of the use of these different types of markers in epidemiologic research are provided. There are many factors that one must consider when selecting a biological marker for use in an epidemiologic study. These factors include: the objectives of the study, the availability and specificity of potential markers, the feasibility of measuring the markers in various biological media, the invasiveness of the techniques necessary to measure the markers, the amount of biological specimen needed for analysis, the time to appearance of the markers in the biological media, the persistence of the markers in biological media, the variability of marker levels within and between individuals, the stability of markers in storage, as well as the cost, sensitivity, specificity, and reliability of the assays used to measure the markers. Each of these characteristics is discussed. The usefulness of biological markers in an epidemiologic study depends on the objectives of the study and whether the properties of the markers fulfill the objectives of the study in a feasible and cost-effective manner.

BIOLOGICAL TREATMENT, BIOREMEDIATION, AND NATURAL ATTENUATION

Biological treatment, bioremediation and natural attenuation has been a rapidly growing area of science over the past decade. The acceptance of natural attenuation as a solution for cleaning up contaminated sites and DOE's recognition that they will have long-term stewardship issues that they must address at the most contaminated sites has greatly increased the urgency for basic and applied research related to microbial ecology and biogeochemistry. This type of research is truly enabling for natural attenuation since characterisation, predictions, and verification monitoring require a strong scientific basis. Natural attenuation is viewed as the best solution for cleaning up many waste sites and will save billions of dollars in cleanup costs. Bioremediation, both *in situ* and *ex situ* have also enjoyed strong scientific growth, in part due to the increased use of natural attenuation, since most natural attenuation is due to biodegradation. Bioremediation and Natural Attenuation are also seen as a solution for emerging contaminant problems, e.g. MTBE, endocrine disrupters, landfill stabilisation, mixed waste biotreatment, and biological carbon sequestration.The Center for Environmental Biotechnology is a recognised leader in the field of biological treatment, bioremediation, and natural attenuation. The Center includes a world-class bioengineering laboratory and CEB investigators have extensive experience in both water treatment and bioremediation, especially co-metabolic biodegradation and the treatment of inhibitory compounds. In addition to basic research, CEB investigators have been involved in various aspects of more than 60 field demonstrations and deployments, and have 5 patents in this area that are licensed to more than 30 companies. The types of contaminants that CEB investigators have expertise with include chlorinated solvents, petroleum hydrocarbons, polynuclear aromatic hydrocarbons, ketones, MTBE, TNT, inorganic nitrogen (NO_3, NH_4), tritium, Pu, Np, Cr, and U. The Biotreatment, Bioremediation and Natural Attenuation area has both basic research and field application foci for the CEB. The basic research foci are co-metabolism, biotreatability, biotransformation kinetics, and modeling of biogeochemical processes. The field application foci are co-metabolic techniques, biogeochemical assessment techniques, and modeling of attenuation and environmental fate.

Basic Research Areas

Co-Metabolism. CEB scientists are recognised leaders in the field of co-metabolic pollutant transformation. Research conducted by scientists at CEB has demonstrated that co-metabolism is a dominant process for the degradation of PAHs, chlorinated solvents, and fuel oxygenates. Co-metabolic processes are difficult to study and require novel experimental approaches. One approach under development is the use of genetic probes to identify the presence of metabolic pathways implicated in co-metabolic processes. In this research, physiological responses are being linked to gene probe signatures to develop methods for identifying co-metabolic potential in environmental samples. The second approach uses a kinetic evaluation of partial transformation reactions to compare and characterize bacterial enzyme systems implicated in co-metabolism. Co-metabolic processes are competitive enzyme reactions and can be tested and modeled as such. Both these approaches are yielding unique and valuable advances in our understanding and application of co-metabolic and partial transformations.*Biotreatability*. CEB does a number of types of treatability tests for contaminants in soil and water, using soil columns, respirometers, bioreactors, and field respiration tests. Real-time direct biogeochemical techniques are being developed to provide the most direct possible methods to measure the rates of biodegradation and the effect stimulants and environmental conditions have on both the functional microbial components and the biogeochemistry of the environment being studied. For example, mixed waste biotreatment includes both the engineering of bioreactors for the treatment of high strength waste streams and understanding how natural attenuation or active bioremediation can be applied to mitigate the impact of the DOE weapons legacy. Past projects have examined how bacterial blooms on solvents may effect the environmental fate of actinides in the subsurface

Current studies are examining the biotreatment of tritiated solvent wastes and the transformation of uranium on bacteria surfaces. Biological processes are increasingly being recognised by DOE as important factors effecting the fate of radionuclides in the environment and are more often being considered as useful tools in the clean-up of mixed wastes and waste sites. *Biotransformation Kinetics.* CEB is linking engineering and microbiology in an integrated programme to examine biotransformation kinetic for pollutant clean-up and microbial product formation. The kinetic programme examines microbial response and activity as a function of substrate concentrations and time. The results of kinetic studies are being used in reactor design and operations analysis. Future research will focus on the application of kinetic approaches to predicting microbial population response to pollutant inputs.

Modeling. CEB currently does not have a modeling programme, but is developing collaboration with a modeling groups within ESD. Modeling is targeted as an area for increased collaboration in the next three years. In particular, there is an need to link biodegradation kinetic information with subsurface fate and transport of pollutants. Modeling is also needed to further understand the control and operation of co-metabolising biological reactors.

Field Applications

Co-Metabolic Techniques. Increasingly we are finding that contaminants can be deleterious at extremely low concentrations, eg. endocrine disrupters at parts per trillion concentrations. These concentrations of contaminants are too low to allow the contaminant to be a carbon or energy source for the microflora. Thus, a secondary carbon and energy source is needed to stimulate the microbes that are capable of degrading the contaminants of concern, since the excess enzymes produced will degrade the contaminants co-metabolically or fortuitously. CEB scientists have demonstrated the ability of methane to be injected with air into soil and groundwater contaminated with chlorinated solvents and petroleum compounds and shown how this stimulation of methanotrophs can cleanup sites to non-detect levels of these contaminants. They have also shown how various other gases can be added to supplement nutrient deficiencies in both nitrogen and phosphorus, thus allowing the microbes to achieve higher biomass levels and higher rates of biodegradation of the contaminants. Current studies are focusing on chlorinated solvent contaminants at DOD and DOE sites, MTBE at commercial sites, PAHs at DOD sites and municipal solid water landfills. CEB is also focusing other nutrients and surfactants that might stimulate bioremediation of more recalcitrant and strongly sorbed contaminants.Electrokinetic treatment of soil, ground water and ex-situ treatment systems. This technology being studied in collaboration with EETD is particularly applicable to metal ion contamination, which may be removed without damage to the indigenous microorganisms. The electrokinetic treatment also provides gentle heat to soil, which is also biocompatible, and the current can also supply nutrients and even transport organisms themselves.*Biogeochemical Assessment Techniques.* Our ability to determine the rates of bioremediation or biodegradation in the field is quite often hampered by the need to collect samples and transport them back to the laboratory for analysis. This obviously introduces bottle effect, collection contamination, and changes the in situ conditions so that the final measurements may not reflect accurately what is going on in that environment. In order to get a truer understanding of conditions as they occur we have been studying techniques that allow direct assessment, eg. helium tracer tests, soil gas measurements, isotopic analysis of contaminants and daughter products, and immediate lipid and nucleic acid extraction from sediment and groundwater. These techniques are being applied at a number of field sites to determine their efficiency over conventional baseline laboratory analyses.

Modeling of Attenuation & Environmental Fate. Using the powerful codes developed for fate and transport by the Yucca Mt. Programme (TOUGH & ITOUGH) CEB is applying kinetic information relevant to biodegradation but especially aerobic processes that have received little attention. Using the computing facilities at LBNL (NERSC) and collaborations with the Nuclear Waste Programme, we are currently developing contaminant fate and transport codes that will be used to model aerobic injection into municipal solid waste landfills. These models may allow us to better control the composting type processes that are occurring and discover better parameters to monitor and characterize the process.

Collaborations

LBNL (EETD, ALS, CSD, LSD), UC Berkeley, Washington State University, Geokinetics, Earth Tech, North Carolina State, Utah State University, USGS, NSF, Yolo County, INEEL, ORNL, PNNL, SRTC, IconGenetics, Maxygen, Physical Optics Systems, American Home Products, DuPont Agrobusiness, UMBI, VECTOR (Russia)) International Institute for Applied Microbiology (Ukraine); Utah Water Research Laboratory at Utah State University, Sandia National Lab, Department of Genetics at Cambridge University in the Great Britain, Institute of Physics at Georgian Academy of Sciences, Brain Tumor Research Center UCSF, Department of Molecular and Cell Biology UC Davis, Department of Chemical and Biochemical Engineering UC Irvine,; Department of Pathology at Stanford University Medical Center, Department of Physics and School of Medicine at Boston University. EETD/CEB collaborates closely with Geokinetic/EDA in Berkeley for electrochemical/electrokinetic technology. EETD also has several close collaborations with UCB Chemistry, Materials Science and Chemical Engineering Departments with CEB associated investigators. EETD investigators collaborating with CEB have several labs in the 70 area that can be utilized for CEB purposes plus labs in 62, which contain polymer synthesis and characterisation activities for fuel cells, batteries, polymer LEDs and electrochromic windows. The laboratories are well equipped with chemical, thermal, rheological and electrochemical analytical equipment. CEB also possesses up to six electrokinetic lab cell systems and all appropriate power supplies and monitoring equipment. These facilities may also be applied to bioprocessing.

About CCGT

The Chicago Center for Green Technology is:

1 A model for the nation.Chicago Green Tech is only the third building in the United States to be designed according to the LEED rating system using the highest standards of green technology available. It is the only one of the three that is a renovation of an existing building and the only one accessible by public transportation.

2 Home to organisations and businesses committed to the environment. Tenants who occupy Chicago Green Tech provide environmental products and services. Greencorps Chicago, the city's community gardening and job training programme and WRD Environmental, an urban landscape company, all have offices at Chicago Green Tech.

3 A place to learn. Chicago Green Tech's building and campus are open for visitors to explore and to learn how green buildings are good for people and good for the environment. Visitors leave knowing how to incorporate environmentally friendly, cost saving features into their home or business.

CCGT also offers free seminars for the public 2-3 times per week. CCGT's Green Building Resource Center also has a small library that includes green product samples. The library open to visitors whenever CCGT is open.

PHYTOREMEDIATION—USING PLANTS TO CLEAN UP POLLUTED SOIL

Polluted soil poses a severe problem for both ecosystem health and land development. Because soil lies at the confluence of many natural systems, soil pollution can be spread to other parts of the natural environment. Groundwater, for instance, percolates through the soil and can carry the soil pollutants into streams, rivers, wells and drinking water. Erosion can create the same problem. Plants growing on polluted soil may contain harmful levels of pollutants themselves, and this can be passed on to the animals and people that eat them. Dust blown from polluted soil can be inhaled directly by passersby. Additionally, in an urban setting such as Fairfax County, polluted soil makes valuable open land unusable for parks, recreation or commercial development.

Despite the benefits of cleaning polluted soil, remediation often never takes place because of the cost and effort of the work. Both soil minerals and soil pollutants carry small electric charges that can cause each to bond with each other, thus making polluted soil very hard to clean. Additionally, soil is a dense medium. This causes excavation of polluted soil for off site treatment or disposal to be very expensive because of the time, labor and heavy machinery necessary to do the job. Therefore, cheaper on-site, or in-situ, remediation techniques have been the focus of much attention and research lately. One of the most interesting and promising of these in-situ techniques is phytoremediation.

Phytoremediation is the use of specialized plants to clean up polluted soil. While most plants exposed to high levels of soil toxins will be injured or die, scientists have discovered that certain plants are resistant, and an even smaller group actually thrive. Both groups of plants are of interest to researchers, but the thriving plants show a particular potential for remediation because it has been shown that some of them actually transport and accumulate extremely high levels of soil pollutants within their bodies. They are therefore aptly named hyper-accumulators.

Hyper-accumulators already are being used throughout the country to help clean up heavy metal polluted soil. Heavy metals are some of the most stubborn soil pollutants. They can bond very tightly to soil particles, and they cannot be broken down by microbial processes. Most heavy metals are also essential plant nutrients, so plants have the ability to take up the metals and transport them throughout their bodies. However, on polluted soil, the levels of heavy metals are often hundreds of times greater than normal, and this overexposure is toxic to the vast majority of plants. Hyper-accumulators, on the other hand, actually prefer these high concentrations. Essentially, hyper-accumulators are acting as natural vacuum cleaners, sucking pollutants out of the soil and depositing them in their above ground leaves and shoots. Removing the metals is as simple as pruning or cutting the hyper-accumulators above ground mass, not excavating tons of soil.

Resistant, but not hyper-accumulating, plants also have a role in phytoremediation. Organic toxins, those that contain carbon such as the hydrocarbons found in gasoline and other fuels, can be broken down by microbial processes. Plants play a key role in determining the size and health of soil microbial populations. All plant roots secrete organic materials that can be used as food for microbes, and this creates a healthier, larger, more diverse and active microbial population, which in turn causes a faster breakdown of pollutants. Resistant plants can thrive on sites that are often too toxic for other plants to grow. They in turn give the microbial processes the boost they need to remove organic pollution more quickly from the soil.

Both forms of phytoremediation have the added benefit of not disturbing the soil.

While excavation is an effective way to get rid of pollution, it removes the organic matter rich topsoil and, because of the use of heavy machinery, compacts the soil that is left behind. Phytoremediation does not degrade the physical or chemical health of the soil. Actually, it creates a more fertile soil. Soil organic matter is increased as a result of root secretions and falling stems and leaves, and the roots create pores through which water and oxygen can flow. Additionally, few would argue that a dusty excavation site is more aesthetically pleasing than a nicely planted field.

There are, however, many limitations to phytoremediation. It is a slow process that may take many growing seasons before an adequate reduction of pollution is seen, whereas soil excavation and treatment cleans up the site quickly. Also, hyper-accumulators can be a pollution hazard themselves. For instance, animals can eat the metal rich hyper-accumulators and cause the toxins to enter the food chain. If the concentration of metals in the plants is thought to be high enough to cause toxicity, there must be a way to segregate the plants from humans and wildlife, which may not be an easy task. Additionally, phytoremediation is in its infancy, and its effectiveness in cleaning up various toxins compared to conventional means of treatment is not always known. However, with more research and practice, the practicality of using phytoremediation should increase.

The growing popularity of phytoremediation can be seen in the number of businesses beginning to appear in urban areas of the United States that are involved in the field. In New Jersey, a company named Phytotech is using mustard greens to remove lead from polluted yards in Boston and hydroponically grown sunflowers to take radioactive metals out of the water surrounding the Chernobyl nuclear power plant.

PHYTOREMEDIATION

Description

Phytoremediation is a bioremediation process that uses various types of plants to remove, transfer, stabilize, and/or destroy contaminants in the soil and groundwater. There are several different types of phytoremediation mechanisms. These are:

1. *Rhizosphere biodegradation.* In this process, the plant releases natural substances through its roots, supplying nutrients to microorganisms in the soil. The microorganisms enhance biological degradation.
2. *Phyto-stabilisation.* In this process, chemical compounds produced by the plant immobilize contaminants, rather than degrade them.
3. *Phyto-accumulation (also called phyto-extraction).* In this process, plant

roots sorb the contaminants along with other nutrients and water. The contaminant mass is not destroyed but ends up in the plant shoots and leaves. This method is used primarily for wastes containing metals. At one demonstration site, water-soluble metals are taken up by plant species selected for their ability to take up large quantities of lead (Pb). The metals are stored in the plant's aerial shoots, which are harvested and either smelted for potential metal recycling/recovery or are disposed of as a hazardous waste. As a general rule, readily bioavailable metals for plant uptake include cadmium, nickel, zinc, arsenic, selenium, and copper. Moderately bioavailable metals are cobalt, manganese, and iron. Lead, chromium, and uranium are not very bioavailable. Lead can be made much more bioavailable by the addition of chelating agents to soils. Similarly, the availability of uranium and radio-cesium 137 can be enhanced using citric acid and ammonium nitrate, respectively.

4. *Hydroponic Systems for Treating Water Streams (Rhizofiltration).* Rhizofiltration is similar to phyto-accumulation, but the plants used for cleanup are raised in greenhouses with their roots in water. This system can be used for *ex-situ* groundwater treatment. That is, groundwater is pumped to the surface to irrigate these plants. Typically hydroponic systems utilize an artificial soil medium, such as sand mixed with perlite or vermiculite. As the roots become saturated with contaminants, they are harvested and disposed of.
5. *Phyto-volatilisation.* In this process, plants take up water containing organic contaminants and release the contaminants into the air through their leaves.
6. *Phyto-degradation.* In this process, plants actually metabolize and destroy contaminants within plant tissues.
7. *Hydraulic Control.* In this process, trees indirectly remediate by controlling groundwater movement. Trees act as natural pumps when their roots reach down towards the water table and establish a dense root mass that takes up large quantities of water. A poplar tree, for example, pulls out of the ground 30 gallons of water per day, and a cottonwood can absorb up to 350 gallons per day.

The plants most used and studied are poplar trees. The U.S. Air Force has used poplar trees to contain trichloroethylene (TCE) in groundwater. In Iowa, EPA demonstrated that poplar trees acted as natural pumps to keep toxic herbicides, pesticides, and fertilizers out of the streams and groundwater. The US Army Corps of Engineers has experimented with wetland plants to destroy explosive compounds in the soil and groundwater. Submersed and floating-leafed species (coontail and pondweed, and arrowhead, respectively) decreased trinitrotoluene (TNT) to 5% of original concentration. Submersed plants were able to decrease Royal Demolition Explosive (RDX) levels by 40%, and when microbial degradation was added, RDX decreased by 80%. Sunflowers, using rhizofiltration, were used successfully to remove radioactive contaminants from pond water in a test at Chernobyl, Ukraine.

Limitations and Concerns

The toxicity and bioavailability of biodegradation products is not always known. Degradation by-products may be mobilized in groundwater or bio-accumulated in animals. Additional research is needed to determine the fate of various compounds in the plant metabolic cycle to ensure that plant droppings and products do not contribute toxic or harmful chemicals into the food chain. Scientists need to establish whether contaminants that collect in the leaves and wood of trees are released when the leaves fall in the autumn or when firewood or mulch from the trees is used. Disposal of harvested plants can be a problem if they contain high levels of heavy metals. The depth of the contaminants limits treatment. The treatment zone is determined by plant root depth. In most cases, it is limited to shallow soils, streams, and groundwater. Pumping the water out of the ground and using it to irrigate plantations of trees may treat contaminated groundwater that is too deep to be reached by plant roots. Where practical, deep tilling, to bring heavy metals that may have moved downward in the soil closer to the roots, may be necessary. Generally, the use of phytoremediation is limited to sites with lower contaminant concentrations and contamination in shallow soils, streams, and groundwater. However, researchers are finding that the use of trees (rather than smaller plants) allows them to treat deeper contamination because tree roots penetrate more deeply into the ground. The success of phytoremediation may be seasonal, depending on location. Other climatic factors will also influence its effectiveness. The success of remediation depends in establishing a selected plant community. Introducing new plant species can have widespread ecological ramifications. It should be studied beforehand and monitored. Additionally, the establishment of the plants may require several seasons of irrigation. It is important to consider extra mobilisation of contaminants in the soil and groundwater during this start-up period.

If contaminant concentrations are too high, plants may die. Some phytoremediation transfers contamination across media, (e.g., from soil to air). Phytoremediation is not effective for strongly sorbed contaminants such as polychlorinated biphenyls (PCBs). Phytoremediation requires a large surface area of land for remediation.

Applicability

Phytoremediation is used for the remediation of metals, radionuclides, pesticides, explosives, fuels, volatile organic compounds (VOCs) and semi-volatile organic compounds (SVOCs). Research is underway to understand the role of phytoremediation to remediate perchlorate, a contaminant that has been shown to be persistent in surface and groundwater systems. It may be used to cleanup contaminants found in soil and groundwater. For radioactive substances, chelating agents are sometimes used to make the contaminants amenable to plant uptake.

Technology Development Status

Phytoremediation is a broad technology type that has been successfully demonstrated for some contaminants and is experimental for others.

Web Sites

- http://www.itrcweb.org/PHYTO2.pdf
- http://www.frtr.gov/matrix2/section4/4_36.html (groundwater)
- http://www.frtr.gov/matrix2/section4/4_5.html (soil)
- http://www.envirotools.org/factsheets/phytoremediation.shtml
- http://www.rtdf.org/public/phyto/protocol/protocol99.htm
- http://www.cluin.org/products/citguide/phyto2.htm
- http://www.wes.army.mil/el/phyto/index.html
- http://www.environmental.usace.army.mil/library/success/itsucc/rhizosph/rhizosph.html (cold weather sites)
- http://www.estcp.org/projects/cleanup/199809v.cfm
- http://www.estcp.org/projects/cleanup/199520v.cfm
- http://www.estcp.org/projects/cleanup/199519o.cfm
- http://www.estcp.org/projects/cleanup/1011o.cfm

PHYTOREMEDIATION

The Air Force has developed the Environmental Restoration Programme to identify, investigate, and cleanup environmental contamination associated with past activities. There are various technologies available to address remediation of contaminated sites, one such method is phytoremediation.

Phytoremediation is a biological process in which living plants are used to remove, accumulate, degrade, or contain environmental contaminants. This passive remediation technique is based on the natural ability of vegetation to utilize nutrients, which are transported by capillary action from the soil and groundwater through a plant's root system. The use of a plant's own biological mechanisms to contain and reduce concentrations of inorganic and organic contaminants in soils, sediments, and groundwater is a slow process relying on a plants growth rate. However, with advances in biological, chemical, and engineering technologies, phytoremediation has the potential to serve as a sustained, ecologically sound method to remediate contaminated soil and groundwater.

Biological Mechanisms

In addition to transporting nutrients, certain plants are capable of transporting environmental pollutants such as metals, radionuclides, chlorinated solvents, petroleum hydrocarbons, and ammunition wastes through biological means.

- *Hydraulic Control or phytohydraulics* - the use of a plant's root system to contain contaminant migration and infiltration by creating a hydraulic barrier, which also provides a vegetative cover that reduces contaminant exposure.
- *Phytoextraction or phytoaccumulation* - the use of a plant's root system to extract and transport inorganic contaminants such as metals or radionuclides from soils, which accumulate in aboveground plant tissue.
- *Rhizofiltration* - the use of a plant's root system to absorb inorganic contaminants dissolved in groundwater, which accumulate in a plant's roots.
- *Phytostabilisation* - the use of a plant's root system to precipitate and absorb inorganic contaminants from soils and groundwater, which accumulate within the plant's roots.
- *Phytodegradation or phytotransformation* - the use of a plant's root system to extract organic contaminants such as solvents, pesticide and herbicide residues, and hydrocarbons from soils and groundwater, which degrade within the plant.
- *Rhizodegradation, also known as rhizosphere biodegradation, phytostimulation, or plant-assisted bioremediation* - is a result of a plant's root system releasing chemicals that enhance organic contaminant biodegradation by soil microorganisms.
- *Phytovolatilisation* - the use of a plant's root system to extract and transport organic contaminants from groundwater to the leaves that transpire, evaporate, or volatilize the contaminants into the atmosphere.

Site & Plant Selection

Phytoremediation is not applicable at all contaminated sites. This remediation technique requires a large land area that has low to moderate contaminant concentrations within shallow soils and water table. Generally, phytoremediation is limited to within 3 feet of the surface for contaminated soils and within 10 feet of the surface for contaminated groundwater.

Aside from site conditions, phytoremediation is dependent on selecting the right plant for contaminants. Plants are complex systems whose biological mechanisms are dependent on various environmental factors; however, a fundamental trait in selecting a plant for phytoremediation is an expansive root system. Other plant selection parameters include the rate of plant and root growth, whether a plant freely transpires, and if a plant is suitable for the local environment and the contaminant of concern.

Fate of Contaminants

Once plants extract environmental contaminants, these contaminants are either broken down by the plant into non-hazardous substances or concentrated within plant tissues. Concentration of contaminants in plant material gives rise to the possibility of animals and insects consuming the contaminated material thereby introducing contaminants into the food chain. To date, there is limited information regarding this issue due to the complex biological mechanisms involved. However, remedial sites implementing phytoremediation are taking measures to restrict grazing animals and birds by erecting fences and overhead nets. In addition, biodegradable pesticides are used to eliminate rodents and insects, and plants are harvested prior to seeding or flowering, which limits the availability of food.

Harvesting/Disposal

To sustain viable plant populations and operation of the system, plant material must occasionally be harvested. Harvested plant tissue, which has metabolized and accumulated environmental contaminants requires testing to determine if the tissue is hazardous. If testing confirms harvested material has metabolized the contaminant into non-hazardous substances, the plant material can be mulched or composted and reused on site. However, if the plant has accumulated contaminants and testing indicates plant tissues are hazardous, then those tissues which are hazardous (roots, stems, leaves, etc.) must be disposed of as such. Although harvested plant material may be considered hazardous, phytoremediation is a cost effective remediation method, which causes little environmental disturbance while successfully reducing soil and groundwater contaminants.

Phytoremediation:

Using Plants yo Clean Up Soils

When it comes to helping clean up soils contaminated with heavy and toxic metals, nature has ARS plant physiologist Leon V. Kochian to thank.

During 13 years of research at the U.S. Plant, Soil, and Nutrition Laboratory at Ithaca, New York, Kochian has become an authority on mechanisms used by certain plants to take up essential mineral nutrients and toxic heavy metals from soils. He has also characterized strategies some plants use to tolerate toxic soil environments.

Kochian is an international expert on plant responses to environmental stress, plant mineral nutrition, and use of plants to clean up or remediate soils contaminated with heavy metals and radioisotopes.

Besides providing important new information on how to use plants in this practical way, Kochian's research may also shed light on an important nutritional concern: how to prevent toxic metals from entering the food chain.

"One of the primary ways toxic heavy metals, such as cadmium, get in food is through plant uptake the metal is taken up by the roots and deposited in edible portions," he says.

The lack of vegetation in the barren area above is a result of the soil's high zinc content and low pH. This site in Palmerton, Pennyslvania, was contaminated by a zinc smeltry operated from 1890 to 1980.

(K6057-11) "Contaminated soils and waters pose major environmental, agricultural, and human health problems worldwide," says Kochian. "These problems may be partially solved by an emerging new technology phytoremediation."

"Green" Technology: Simple Concept and Cost-Effective

Phytoremediation is the use of green plants to remove pollutants from the environment or render them harmless.

"Current engineering-based technologies used to clean up soils like the removal of contaminated topsoil for storage in landfills are very costly," Kochian says, "and dramatically disturb the landscape."

Kochian's cost-effective "green" technology uses plants to "vacuum" heavy metals from the soil through their roots. He says, "Certain plant species known as metal hyperaccumulators have the ability to extract elements from the soil and concentrate them in the easily harvested plant stems, shoots, and leaves. These plant tissues can be collected, reduced in volume, and stored for later use."

While acting as vacuum cleaners, the unique plants must be able to tolerate and survive high levels of heavy metals in soils like zinc, cadmium, and nickel. Plant physiologist Leon Kochian(right) and Cornell Universitysupport scientist Jon Shaff analyse compounds released from sorghum roots. "Phytoremediation has been hampered historically by our inadequate understanding of transport and tolerance mechanisms," says Kochian. To address this deficit, Kochian working with ARS research associate Deborah L. Lethman, Cornell University postdoctora postdoctoral associates Mitch Lasat and Paul B. Larsen, and graduate students Nicole S. Pence and Stephen D. Ebbs has been studying a unique and promising metal hyperaccumulator. The plant is Thlaspi caerulescens, commonly known as alpine pennycress. "Thlaspi is a small, weedy member of the broccoli and cabbage family," Kochian says. "It thrives on soils having high levels of zinc and cadmium." His lab has been trying to discover the underlying mechanism that enables T. caerulescens to accumulate excessive amounts of heavy metals.

How Plants Clean Up?

"Hyperaccumulators like Thlaspi are a marvelous model system for elucidating the fundamental mechanisms of and ultimately the genes that control metal hyperaccumulation," says Kochian. "These plants possess genes that regulate the amount of metals taken up from the soil by roots and deposited at other locations within the plant.

"There are a number of sites in the plant that could be controlled by different genes contributing to the hyperaccumulation trait," says Kochian. "These genes govern processes that can increase the solubility of metals in the soil surrounding the roots as well as the transport proteins that move metals into root cells. From there, the metals enter the plant's vascular system for further transport to other parts of the plant and are ultimately deposited in leaf cells."

Alpine pennycress doesn't just thrive on soils contaminated with zinc and cadmium it cleans them up by removing the excess metals. Kochian's team has gained insights into how, at the molecular level, Thlaspi accumulates these metals in its shoots at astoundingly high levels. "A typical plant may accumulate about 100 parts per million (ppm) zinc and 1 ppm cadmium. Thlaspi can accumulate up to 30,000 ppm zinc and 1,500 ppm cadmium in its shoots, while exhibiting few or no toxicity symptoms," he says. "A normal plant can be poisoned with as little as 1,000 ppm of zinc or 20 to 50 ppm of cadmium in its shoots."

The research also suggests an approach for economically recovering these metals. "Zinc and cadmium are metals that can be removed from contaminated soil by harvesting the plant's shoots and extracting the metals from them," he says. Above, Cornell University research associate Miguel Pineros (left) and plant physiologist Leon Kochian study some of these mechanisms in corn. After investigating the molecular physiology of zinc hyperaccumulation in Thlaspi, Kochian's group found that several key sites for zinc transport were greatly stimulated in this plant. To get at the mechanism underlying the stimulation, they cloned a zinc transport gene one of the first such accomplishments achieved with any plant. This breakthrough enabled the researchers to discover that zinc transport is regulated differently in normal and hyperaccumulator plants. "In normal plants, the activity of zinc transporter genes is regulated by the zinc levels in the plant," he says. "In Thlaspi, however, these genes are maximally active at all times independent of plant zinc levels until you raise the tissue zinc levelsto very high concentrations. This results in very high rates of zinc transport from the soil and movement of this metal to the leaves." For soil contaminated with uranium, Kochian found that adding the organic acid citrate to soils greatly increases both the solubility of uranium and its bioavailability for plant uptake and translocation. Citrate does this by binding to insoluble uranium in the soil. Thlaspi caerulescens genes responsible for heavy-metal transport. (K8785-1) "With the citrate treatment, shoots of test plants increased their uranium concentration to over 2,000 ppm"1 00 times higher than the control plants," he says. This demonstrates the possibility of using citrate" an inexpensive soil amendment to help plants reduce uranium contamination. Recently, Kochian, with colleagues Lasat and Ebbs, identified specific agronomic practices and plant species to remediate soils contaminated with radioactive cesium or cesium-137.

"Although the cause of cesium-137 contamination aboveground nuclear testing has been reduced, large land areas are still polluted with radiocesium," Kochian says. "Cesium is a long-lived radioisotope with a half-life of 32.2 years. It contaminates soils at several U.S. Department of Energy (DOE) sites in the United States. Projected costs of cleaning up these soils is very high over $300 billion." Phytoremediation is an attractive alternative to current cleanup methods that are energy intensive and very expensive. Leon Kochian (left) and molecular biologist David Garvin check wheat plants for aluminum tolerance. Some wheat and corn plants can tolerate aluminum by excluding the metal from the root tip.

In initial lab and greenhouse studies, Kochian's team showed that the primary limitation to removing cesium from soils with plants was its bioavailability. The form of the element made it unavailable to the plants for uptake. In a series of soil extraction studies, Kochian's team found the ammonium ion was most effective in dissolving cesium-137 in soils. This treatment increased the availability of cesium-137 for root uptake and significantly stimulated radioactive cesium accumulation in plant shoots. Later, Kochian did field studies with six different plant species in collaboration with Mark Fuhrmann, a DOE scientist at Brookhaven National Laboratory in Upton, New York. They found significant variation in the effectiveness of plant species for cleaning up contaminated sites. "One species, a pigweed called Amaranthus retroflexus, was up to 40 times more effective than others tested in removing radiocesium from soil. We were able to remove 3 percent of the total amount in just one 3-month growing season," says Kochian. "With two or three yearly crops, the plant could clean up the contaminated site in less than 15 years."

As a result of Kochian's findings, DOE is performing pilot studies at Brookhaven using this technology. Hyperaccumulators like Thlaspi possess genes that regulate the amount of metals taken up from the soil by roots and deposited at other locations within the plant. Kochian's lab is also working on finding ways to grow crops on marginal lands such as acid soils, where toxic levels of aluminum limit crop production. Aluminum is the third most abundant element in the Earth's crust; it is a major component of clays in soil.

At neutral or alkaline pH values, aluminum is not a problem for plants. However, in acid soils a form of aluminum Al+3 is solubilized into a soil solution that is quite toxic to plant roots. For years, scientists have been baffled by the causes of aluminum toxicity in plants. "Aluminum toxicity limits crop production on acid soils, which cover well over half of the world's 8 billion acres of otherwise arable land, including about 86 million acres in the United States," Kochian says. "When soils become acid, the toxic aluminum damages plant root systems, which greatly reduces yields."

Kochian's research in collaboration with ARS plant molecular biologist David F. Garvin uses an interdisciplinary approach integrating molecular, genetic, and physiological research to provide insights into how particular genetic types of some plant species including wheat, corn, and sorghum can tolerate high levels of the metal in acid soils. "We found that the root tip is the key site of injury, leading to inhibited root growth, a stunted root system, and reduced yields or crop failures from decreased uptake of water and nutrients," Kochian says. "Aluminum triggers the release of protective organic acids, specifically from the root tip into adjacent soil. When released, these acids form a complex with the toxic aluminum, preventing the metal s entry into the root. Wheat and corn tolerate aluminum by excluding the metal from the root tip," Kochian says.

Kochian is also conducting research on an aluminum tolerance mechanism in collaboration with plant molecular biologist Steve H. Howell of Boyce Thompson Institute at Cornell, using thale cress, Arabidopsis thaliana, a diminutive, weedy member of the mustard family.

He and colleagues have successfully identified Arabidopsis mutants that are aluminum tolerant. Kochian is studying differences between these mutants and a wild type of Arabidopsis to identify the molecular basis of tolerance. The ultimate goal of this research is to isolate the genes conferring aluminum tolerance. It should then be possible to improve the tolerance of relatively aluminum-sensitive crop species, such as barley, or to further enhance the tolerance of existing al aluminum-tolerant germplasm. "One of the major goals for agricultural scientists for the immediate future is to increase food production to keep up with an ever-growing world population," Kochian says. "As much of the world's best agricultural land is already under cultivation or is being lost to industrialisation, there is increasing pressure for farmers to cultivate marginal lands such as the huge expanses of acid soils that are not currently used for production."

He continues, "Research aimed at producing crop genotypes that tolerate the suboptimal conditions of these marginal lands is one way global food production can be increased significantly. Being able to produce a wider range of crop species with increased aluminum tolerance will make a major contribution to these efforts to cultivate marginal, stressed soil environments." Besides helping farmers who grow crops on acid soils, Kochian's phytoremediation research findings are used by other scientists in government and academia and by environmental consultants, government, and industry groups complying with cleanup of contaminated sites. For his landmark phytoremediation research, Kochian has received two awards: in 1999, the U.S. Department of Agriculture Secretary's Honor Award for Environmental Protection and an award as ARS 1999 Outstanding Senior Scientist of the Year. By Hank Becker, Agricultural Research Service Information Staff.

Gold rush miners might have been better off using plants to find gold rather than panning streams for the precious metal. Early prospectors in Europe used certain weeds as indicator plants that signaled the presence of metal ore. These weeds are the only plants that can thrive on soils with a high content of heavy metals. One such plant is alpine pennycress, Thlaspi caerulescens, a wild perennial herb found on zinc- and nickel-rich soils in many countries. This plant occurs in alpine areas of Central Europe as well as in our Rocky Mountains. Most varieties grow only 8 to 12 inches high and have small, white flowers.

In 1998, ARS agronomist Rufus L. Chaney and colleagues in ARS, at the University of Maryland, and in England patented a method to use such plants to "phyto-mine" nickel, cobalt, and other metals. Chaney says biomining is the use of plants to mine valuable heavy-metal minerals from contaminated or mineralized soils, as opposed to decontaminating soils. "The crops would be grown as hay. The plants would be cut and baled after they'd taken in enough minerals," Chaney says. "Then they'd be burned and the ash sold as ore. Ashes of alpine pennycress grown on a high-zinc soil in Pennsylvania yielded 30 to 40 percent zinc which is as high as high-grade ore. Electricity generated by the burning could partially offset biomining costs."

USDA has signed a cooperative research and development agreement with Viridian Environmental, a technology company based in Houston, Texas. The CRADA involves

Scott Angle at the University of Maryland; Alan J.M. Baker at the University of Sheffield, United Kingdom; plant breeder Yin-Ming Li with Viridian; and a cooperator at Oregon State University. Viridian is funding the CRADA's phyto-mining research and development to the tune of $1 million over 5 years. Chaney says that to make phyto-mining as well as phytoremediation worthwhile requires, at a minimum, a plant with very high annual intake of minerals, such as the high-cadmium-accumulating pennycress variety for which they have filed a patent application. "Better still, the traits of plants like pennycress could be incorporated into a high-yielding commercial crop like canola grown for hay," Chaney says. His idea of the best hyperaccumulators? "They'd have all the characteristics of a hay crop: They should be tall, high yielding, fast growing, easy to harvest, and deep rooted. And they should hold onto their mineral-rich leaves so they can be harvested along with the plant stems." By Don Comis, Agricultural Research Service Information Staff.

What is Green Technology?

The term "technology" refers to the application of knowledge for practical purposes.

The field of "green technology" encompasses a continuously evolving group of methods and materials, from techniques for generating energy to non-toxic cleaning products.

The present expectation is that this field will bring innovation and changes in daily life of similar magnitude to the "information technology" explosion over the last two decades. In these early stages, it is impossible to predict what "green technology" may eventually encompass.

The goals that inform developments in this rapidly growing field include:

Sustainability - meeting the needs of society in ways that can continue indefinitely into the future without damaging or depleting natural resources. In short, meeting present needs without compromising the ability of future generations to meet their own needs.

"Cradle to cradle" design - ending the "cradle to grave" cycle of manufactured products, by creating products that can be fully reclaimed or re-used.

Source reduction - reducing waste and pollution by changing patterns of production and consumption.

Innovation - developing alternatives to technologies - whether fossil fuel or chemical intensive agriculture - that have been demonstrated to damage health and the environment. Viability - creating a center of economic activity around technologies and products that benefit the environment, speeding their implementation and creating new careers that truly protect the planet. Examples of green technology subject areas:

Energy

Perhaps the most urgent issue for green technology, this includes the development of alternative fuels, new means of generating energy and energy efficiency.

Green Building

Green building encompasses everything from the choice of building materials to where a building is located.

Environmentally Preferred Purchasing

This government innovation involves the search for products whose contents and methods of production have the smallest possible impact on the environment, and mandates that these be the preferred products for government purchasing.

Green Chemistry

The invention, design and application of chemical products and processes to reduce or to eliminate the use and generation of hazardous substances.

Green Nanotechnology

Nanotechnology involves the manipulation of materials at the scale of the nanometer, one billionth of a meter. Some scientists believe that mastery of this subject is forthcoming that will transform the way that everything in the world is manufactured. "Green nanotechnology" is the application of green chemistry and green engineering principles to this field.

Environmental Technology

Environmental technology (abbreviated as EnviroTech) or green technology (abbreviated as GreenTech) or clean technology (abbreviated as CleanTech) is the application of the environmental sciences to conserve the natural environment and resources, and to curb the negative impacts of human involvement. Sustainable development is the core of *environmental technologies*. When applying *sustainable development* as a solution for *environmental issues*, the solutions need to be socially equitable, economically viable, and environmentally sound.

Related Technologies

Some *environmental technologies* that retain *sustainable development* are; recycling, water purification, sewage treatment, remediation, flue gas treatment, solid waste management, and renewable energy. Some technologies assist directly with energy conservation, while other technologies are emerging that help the environment by reducing the amount of waste produced by human activities. Energy sources such as solar power create less problems for the environment than traditional sources of energy like coal and petroleum.

Scientists continue to search for clean energy alternatives to our current power production methods. Some technologies such as anaerobic digestion produce renewable energy from waste materials. The global reduction of greenhouse gases is dependent on the adoption of energy conservation technologies at industrial level as well as this clean energy generation. That includes using unleaded gasoline, solar energy and alternative fuel vehicles, including plug-in hybrid and hybrid electric vehicles.

Since electric motors consume 60% of all electricity generated, advanced energy efficient electric motor (and electric generator) technology that are cost effective to encourage their application, such as the brushless wound-rotor doubly-fed electric machine and energy saving module, can reduce the amount of carbon dioxide (CO_2) and sulfur dioxide (SO_2) that would otherwise be introduced to the atmosphere, if electricity is generated using fossil fuels. Greasestock is an event held yearly in Yorktown Heights, New York which is one of the largest showcases of environmental technology in the United States.

Criticism

Some groups, including green anarchists, have criticised the concept of environmental technology. From their viewpoint, technology is seen as a system rather than a specific physical tool. Technology, it is argued, *requires* the exploitation of the environment through the creation and extraction of resources, and the exploitation of people through labour, specialisation and the division of labor. There is no "neutral" form of technology, as things are always created in a certain context with certain aims and functions. Thus, green technology is rejected as an attempt to reform this exploitative system, merely changing it on the surface to make it seem environmentally friendly, despite continued unsustainable levels of human and natural exploitation.

REFERENCES

Biomarkers and surrogate endpoints: preferred definitions and conceptual framework. Clinical Pharmacology & Therapeutics, Volume 69, Issue 3, March 2001, Pages 89-95 (2001). Retrieved on 2006-04-01.

Boyajian, G. E. and Devedjian, D. L., 1997. "Phytoremediation: It Grows on You", *Soil & Groundwater Cleanup*, February/March, pp. 22-26.

Brownfields Technology Primer: Selecting and Using Phytoremediation for Site Cleanup,

July 2001, U.S. Environmental Protection Agency Office of Solid Waste and Emergency Response Technology Innovation Office

California Base Closure Environmental Committee (CBCEC), 1994. *Treatment Technologies Applications Matrix for Base Closure Activities, Revision 1*, Technology Matching Process Action Team, November, 1994.

Environment Canada. "Testing the World's Drinking Water." Science and Technology Bulletin.www.ec.gc.ca/science/sandemar99/article6_e.html December 2001.

EPA, 1996. *A Citizen's Guide to Bioremediation*, Technology Fact Sheet, EPA NCEPI, EPA/542/F-96/007.

EPA, 1996. *A Citizen's Guide to Phytoremediation*, Technology Fact Sheet, EPA NCEPI, EPA/542/F-96/014.

EPA, 1996. *Recent Developments for In Situ Treatment of Metal Contaminated Soils*, EPA/542/R-96/008.

EPA, 1998. *A Citizen's Guide to Phytoremediation*, Technology Fact Sheet, EPA NCEPI, EPA/542/F-98/011.

European Centre for Nature Conservation. "ELISA: Environmental Indicators for Sustainable Agriculture Final Report." European Centre for Nature Conservation. www.ecnc.nl/doc/projects/elisa.html December 2001.

European Initiative for Biotechnology Education (EIBE). "Biotechnology and the Environment." EIBE. www.rdg.ac.uk/EIBE/ENGLISH/U16.HTM December 2001.

Groundwater Cleanup: Overview of Operating Experience at 28 Sites, September 1999, EPA 542-R-99-006,

Guide to Documenting and Managing Cost and Performance Information for Remediation Projects - Revised Version, October, 1998, EPA 542-B-98-007

Innovative Remediation Technologies: Field Scale Demonstration Project in North America, 2nd Edition

Jacobson, K. Bruce. "Biosensors and Other Medical and Environmental Probes." Oak Ridge National Laboratory. www.ornl.gov/ORNLReview/rev29_3/text/biosens.htm December 2001.

Phytoremediation Decision Tree, November 1999, The Interstate Technology and Regulatory Cooperation Work Group Phytoremediation Work Team

Phytoremediation of TCE in Groundwater using Populus, February 1998, U.S. EPA Technology Innovation

Phytoremediation Resource Guide, June 1999, U.S. Environmental Protection Agency Office of Solid Waste and Emergency Response Technology Innovation Office

Potential Applicability of Assembled Chemical Weapons Assessment Technologies to RCRA Waste Streams and Contaminated Media, August 2000, EPA 542-R-00-004

Remediation Technology Cost Compendium - Year 2000

Stevens, Dr. Melita, Dr. Nicholas Ashbolt, and DR. David Cunliffe. “Microbial Indicators of Water Quality – An NHMRC Discussion Paper.” National Health and Medical Research Council. < P>

Treatment Experiences at RCRA Corrective Actions, December 2000, EPA 542-F-00-020

Treatment Technologies for Site Cleanup: Annual Status Report (ASR), Tenth Edition, EPA 542-R-01-004

Veal, Duncan ed. “Microbial Indicators of River Health – 1997 Workshop.” Australian Land and Water Resources Research and Development Corporation. January 1998.

Encyclopaedia of

ENVIRONMENTALLY SUSTAINABLE GREEN SCIENCE AND TECHNOLOGY

Encyclopaedia of

ENVIRONMENTALLY SUSTAINABLE GREEN SCIENCE AND TECHNOLOGY

Volume 5

R.B. Patnaik

ANMOL PUBLICATIONS PVT. LTD.

NEW DELHI - 110 002 (INDIA)

ANMOL PUBLICATIONS PVT. LTD.

H.O.: 4374/4B, Ansari Road, Daryaganj,
New Delhi-110 002 (India)
Ph.: 23278000, 23261597

B.O.: No. 1015, Ist Main Road, BSK IIIrd Stage
IIIrd Phase, IIIrd Block,
Bangalore - 560 085 (India)
Visit us at: www.anmolpublications.com

Encyclopaedia of Environmentally Sustainable Green Science and Technology

First Edition, 2009

ISBN 978-81-261-3842-5 (Set)

PRINTED IN INDIA

Printed at Mehra Offset Press, Delhi.

Contents

VOL. 4

VOL. 5

Preface

This publication entitled "Encyclopaedia of Environmentally Sustainable Green Science & Technology" attempts towards creating a holistic understanding of environment and sustainable development in its totality. It defines environment and sustainable development in all its dimensions. The concepts and methodologies of ecology, environment and sustainable development have been discussed in detail. The ecosystem perspective and environmental concept of sustainable development have been highlighted, so has been the environmental methodology. An effort has been made towards defining green science and its manifestations for understanding pollution. The relevance and importance of environment impact assessment (EIA) for environmental sustainability has been analyzed. Also, the elements and dimensions of green technology have been covered in detail. The relationship among ecomovements, ecodynamics and sustainable development has been discussed in detail. Similarly, the environmental and social aspects of sustainable development, in particular, aspects related to training of people to meet the global environmental needs have been covered. The issues related to green energy, its sources, tools and technologies have been reflected upon briefly. The environmentally sustainable science and technology has been presented in brief with special reflection on climate change and global warming. Current issues and options regarding sustainability aspects of environmental management, corporate management and environmental responsibility have been covered. The aim is to develop a policy perspective. The contemporary global industrial technology has been discussed with focus on evaluating environmental performance and compliance to green laws. The applied aspects of environmentally sustainable science and technology have been outlined. Also, the green conventions enforced towards regulating global science and technology towards environmentally sustainable development has been presented, highlighting the theories of international environmental law and theory of environmental degradation.

The applications of green science and technology have been discussed along with a case of biotechnology. The green technology and environmental monitoring issues have been given considerable coverage. Select case studies have been made in this regard. Also, case studies of environmental engineering and environmental biotechnology have been incorporated with focus on applications and utilities. Future prospect of environmentally sustainable or green technology in today's world has

been analyzed briefly. Similarly, the principles, tools and techniques of environmental management, with particular reference to environmental disasters, have been covered in detail. Glossary of terms, bibliography and links has been provided to make this publication a user friendly one and to aid readers in further research and analysis.

—Editors

16

Applications of Green Science & Technology: A Case of Biotechnology

Introduction

The principle goals of our "green science" project are:

- Educate the community about "green science", i.e. conservation, alternative/ renewable energy, and sustainable practices.
- Reduce the environmental footprint of Earlham's campus.

In order to accomplish those goals we plan to implement the following initiatives in the near term:

- Prototype wind and solar energy systems on-campus.
- A system to record, display, visualize, and analyze all energy consumption data for the campus.
- The Green Zone, an area on the first floor of Dennis Hall where students, faculty, visitors, etc. can see alternative/renewable energy gear in-use and learn about all the Green Science projects on campus.

Each initiative in some detail

Wind

- Small tower on the roof. The difference in performance and ease of integration will make the extra cost and the fact that it won't be easily visible from the ground worth it.
- Instrumented for real-time performance data collection.

Solar

- Roof mount or anwing on South parapet. Should be visible from the ground if possible.
- Instrumented for real-time performance data collection.

Energy Consumption Monitoring and Recording

- This would be integrated with our existing weather station/database and the wind/solar systems. This combination would providing a comprehensive look at energy consumption and

The Green Zone

- Signage and wall paint to distinguish an area in the Dennis entrance lobby lounge as the "Green Zone". One of the display cases on the North wall and the area where the glass cluster case is now.
- Power Control Unit
- Inverter
- Batteries
- Visual indicators for battery level, solar rate, wind rate
- Emergency wall charge capability and indicator?
- Arranged and labeled in such a way as to explain the function of each piece and the whole. Very neat wiring.
- Data Visualization Tool
- energy consumption, total, by fuel type (with cost?)
- energy production, total, by technology
- weather
- network traffic, total, by protocol, by subnet
- camera view of the wind and solar view on the roof
- control box
- choose data set, choose display panel or full
- time scale adjustment: 1d, 1w, 1m, 3m, 6m, 12m, 24m, 48m,
- mounted 6' or so back from the display, corded | | wireless?
- Displays - One poster panel for each of the Green Science projects, all using common framework/background, giving the highlights, some background, and where to learn more.
- The Green Zone

- Wind electric production
- Solar electric production
- Energy consumption monitoring and recording
- Data visualization (with HIP URL)
- Weather (with HIP URL)
- Green Kiosk
- Email and web browsing, built out of recycled computer components and a recycled plastic enclosure. Powered by the renewable energy plant.

Long Term Ideas

- Green Kiosks around campus (Runyon, cafeteria, Carpenter, the heart, Wellness Center, etc.) Solar panels on top for outdoor installs (like MuniMeters), on the roof of the building for indoor units.
- Wind turbine farm using the land Mike Nocton leases South of the horsebarn and West of Miller Farm. Some back of the envelope calculations indicate it might be financially viable over the long term.
- Environmental House or Miller Farm "off the grid".
- New construction at Earlham done to the best current thinking about green construction and energy sustainability.
- Series of course modules on "green science" that can be dropped into existing science division classes; or just a separate course?

What is Biotechnology?

Biotechnology can be summarized as the manipulation of living organisms to produce goods and services. Although the technology has received widespread media coverage in the past few years, it is a technology with a long history, dating as far back as 6000 B.C. Advances in science and technology have transformed traditional biotechnology techniques, such as selective breeding, hybridisation and mutagenesis, into modern ones, such as recombinant DNA techniques and tissue culture. This transformation has opened the door to more varied applications in areas such as health care, the environment, forestry, industrial processes and others. Some developments to watch for include research into nutritionally enhanced GM foods and transgenic animals, bio chips and protein drugs.

The Canadian Environmental Protection Act defines biotechnology as "the application of science and engineering in the direct and indirect use of living organisms or parts or products of living organisms in their natural or modified forms."

When did biotechnology begin?

Biotechnology began in early farming societies with practices such as selective breeding to improve plants and animal breeds, and with fermentation to produce such commodities as beer, bread, and wine. However, it wasn't until 1919 that the term biotechnology was given to these and other practices that manipulated living organisms to produce goods. The science of biotechnology was made popular with the discovery of genes and the study of genetics in the 20th century. To read more, go to A Brief History of Biotechnology.

What are some techniques of biotechnology?

Biotechnology techniques include both traditional and modern techniques (for the difference between the two, go to Traditional versus Modern Biotechnology). Some examples are selective breeding, fermentation, mutagenesis and recombinant DNA technology. To learn more, go to Biotech Tools. Has biotechnology only been used to produce genetically modified (GM) foods and new drugs for human health care?

No. In fact, biotechnology has been used in more areas than food and agriculture and health care. It has been used in such areas as the environment, forestry, and industrial processes. Even in health care, biotechnology has been used to achieve other benefits beyond the production of drugs. It has been used to develop new vaccines, diagnostic tools, and gene therapy.

What are some areas in which biotechnology could be used in the future? Transgenic animals, nutritionally enhanced food crops, novel biocatalysts, biosensors, stem cell medicine, tissue engineering, and DNA probes are some areas that are currently under research and development.

Industrial Biotechnology

This section looks at how biotechnology contributes to new and enhanced industrial processes. It examines how biotechnology enables the use of renewable resources in industry, thereby increasing efficiency, decreasing pollution, and reducing energy usage and waste production during various types of industrial processing.

Related Technologies

- Biochemicals
- Bioplastics and Biopolymers
- Biotechnology in the Textiles Industry
- Enzymes in Industrial Applications

- Microbial Enhanced Oil Recovery
- Starch Bioprocessing

Microorganism

A microorganism (also can be spelled as micro organism) or microbe is an organism that is microscopic (too small to be seen by the naked human eye). The study of microorganisms is called microbiology, a subject that began with Anton van Leeuwenhoek's discovery of microorganisms in 1675, using a microscope of his own design. Microorganisms are incredibly diverse and include bacteria, fungi, archaea, and protists, as well as some microscopic plants and animals such as plankton, and popularly-known animals such as the planarian and the amoeba. They do not include viruses and prions, which are generally classified as non-living. Most microorganisms are single-celled, or unicellular, but some multicellular organisms are microscopic, while some unicellular protists, and a bacteria called *Thiomargarita namibiensis* are visible to the naked eye. Microorganisms live in all parts of the biosphere where there is liquid water, including hot springs, on the ocean floor, high in the atmosphere and deep inside rocks within the Earth's crust. Microorganisms are critical to nutrient recycling in ecosystems as they act as decomposers. As some microorganisms can fix nitrogen, they are a vital part of the nitrogen cycle, and recent studies indicate that airborne microbes may play a role in precipitation and weather. Microbes are also exploited by people in biotechnology, both in traditional food and beverage preparation, as well as modern technologies based on genetic engineering. However, pathogenic microbes are harmful, since they invade and grow within other organisms, causing diseases that kill millions of people, other animals, and plants.

Evolution

Single-celled microorganisms were the first forms of life to develop on earth, approximately 3–4 billion years ago. Further evolution was slow, and for about 3 billion years in the Precambrian eon, all organisms were microscopic. So, for most of the history of life on Earth the only form of life were microorganisms. Bacteria, algae and fungi have been identified in amber that is 220 million years old, which shows that the morphology of microorganisms has changed little since the triassic period. Most microorganisms can reproduce rapidly and microbes such as bacteria can also freely exchange genes by conjugation, transformation and transduction between widely-divergent species. This horizontal gene transfer, coupled with a high mutation rate and many other means of genetic variation, allows microorganisms to swiftly evolve (via natural selection) to survive in new environments and respond to environmental stresses. This rapid evolution is important in medicine, as it has led to the recent development of 'super-bugs' — pathogenic bacteria that are resistant to modern antibiotics

Pre-Microbiology

The possibility that microorganisms might exist was discussed for many centuries before their actual discovery in the 17th century. The first ideas about microorganisms were those of the Roman scholar Marcus Terentius Varro in a book titled *On Agriculture* in which he warns against locating a homestead near swamps:

> "...and because there are bred certain minute creatures which cannot be seen by the eyes, which float in the air and enter the body through the mouth and nose and there cause serious diseases"

This passage seems to indicate that the ancients were aware of the possibility that diseases could be spread by yet unseen organisms. When the Black Death bubonic plague reached al-Andalus in the 14th century, Ibn Khatima wrote that infectious diseases were caused by "contagious entities" that enter the human body. Later, in 1546, Girolamo Fracastoro proposed that epidemic diseases were caused by transferable seedlike entities that could transmit infection by direct or indirect contact, or even without contact over long distances. All these early claims about the existence of microorganisms were speculative in nature and not based on any data or science. Microorganisms were neither proven, observed, nor correctly and accurately described until the 17th century. The reason for this was that all these early inquiries lacked the most fundamental tool in order for microbiology and bacteriology to exist as a science, and that was the microscope.

Discovery

Anton van Leeuwenhoek was the first person to observe microorganisms, using a microscope of his own design, thereby making him the first microbiologist. In doing so Leeuwenhoek would make one of the most important contributions to biology and open up the fields of microbiology and bacteriology. Prior to Leeuwenhoek's discovery of microorganisms in 1675, it had been a mystery as to why grapes could be turned into wine, milk into cheese, or why food would spoil. Leeuwenhoek did not make the connection between these processes and microorganisms, but using a microscope, he did establish that there were forms of life that were not visible to the naked eye. Leeuwenhoek's discovery, along with subsequent observations by Lazzaro Spallanzani and Louis Pasteur, ended the long-held belief that life spontaneously appeared from non-living substances during the process of spoilage.

Lazzarro Spallanzani found that microorganisms could only settle in a broth if the broth was exposed to the air. He also found that boiling the broth would sterilise it and kill the microorganisms. Louis Pasteur expanded upon Spallanzani's findings by exposing boiled broths to the air, in vessels that contained a filter to prevent all particles from passing through to the growth medium, and also in vessels with no filter at all, with air being admitted via a curved tube that would not allow dust particles to come in contact with the broth. By boiling the broth beforehand, Pasteur ensured that no microorganisms survived within the broths at the beginning of his experiment. Nothing grew in the broths in the course of Pasteur's experiment. This meant that the living organisms that grew in such broths came from outside, as spores on dust, rather than spontaneously generated within the broth. Thus, Pasteur dealt the death blow to the theory of spontaneous generation and supported germ theory.

In 1876, Robert Koch established that microbes can cause disease. He did this by finding that the blood of cattle who were infected with anthrax always had large numbers of *Bacillus anthracis*. Koch also found that he could transmit anthrax from one animal to another by taking a small sample of blood from the infected animal and injecting it into a healthy one, causing the healthy animal to become sick. He also found that he could grow the bacteria in a nutrient broth, inject it into a healthy animal, and cause illness. Based upon these experiments, he devised criteria for establishing a causal link between a microbe and a disease in what are now known as Koch's postulates. Though these postulates cannot be applied in all cases, they do retain historical importance in the development of scientific thought and can still be used today.

Classification and Structure

Microorganisms can be found almost anywhere in the taxonomic organisation of life on the planet. Bacteria and archaea are almost always microscopic, while a number of eukaryotes are also microscopic, including most protists, some fungi, as well as some animals and plants. Viruses are generally regarded as not living and therefore are not microbes, although the field of microbiology also encompasses the study of viruses.

Prokaryotes

Prokaryotes are organisms that lack a cell nucleus and the other organelles found in eukaryotes. Prokaryotes are almost always unicellular, although some species such as myxobacteria can aggregate into complex structures as part of their life cycle. These organisms are divided into two groups, the archaea and the bacteria.

Bacteria

Bacteria are the most diverse and abundant group of organisms on Earth. Bacteria inhabit practically all environments where some liquid water is available and the temperature is below +140 °C. They are found in sea water, soil, air, animals' gastrointestinal tracts, hot springs and even deep beneath the Earth's crust in rocks. Practically all surfaces which have not been specially sterilized are covered in bacteria. The number of bacteria in the world is estimated to be around five million trillion trillion, or 5×10^{30}.

Bacteria are practically all invisible to the naked eye, with a few extremely rare exceptions, such as *Thiomargarita namibiensis*. They are unicellular organisms and lack membrane-bound organelles. Their genome is usually a single loop of DNA, although they can also harbor small pieces of DNA called plasmids. These plasmids can be transferred between cells through bacterial conjugation. Bacteria are surrounded by a cell wall, which provides strength and rigidity to their cells. They reproduce by binary fission or sometimes by budding, but do not undergo sexual reproduction. Some species form extraordinarily resilient spores, but for bacteria this is a mechanism for survival, not reproduction. Under optimal conditions bacteria can grow extremely rapidly and can double as quickly as every 10 minutes.

Archaea

Archaea are also single-celled organisms that lack nuclei. In the past, the differences between bacteria and archaea were not recognised and archaea were classified with bacteria as part of the kingdom Monera. However, in 1990 the microbiologist Carl Woese proposed the three-domain system that divided living things into bacteria, archaea and eukaryotes. Archaea differ from bacteria in both their genetics and biochemistry. For example, while bacterial cell membranes are made from phosphoglycerides with ester bonds, archaean membranes are made of ether lipids. Archaea were originally described in extreme environments, such as hot springs, but have since been found in all types of habitats. Only now are scientists beginning to appreciate how common archaea are in the environment, with crenarchaeota being the most common form of life in the ocean, dominating ecosystems below 150 m in depth. These organisms are also common in soil and play a vital role in ammonia oxidation.

Eukaryotes

All living things which are *individually* visible to the naked eye are eukaryotes (with few exceptions, such as *Thiomargarita namibiensis*), including humans. However, a large number of eukaryotes are also microorganisms. Unlike bacteria and archaea, eukaryotes contain organelles such as the cell nucleus, the Golgi apparatus and mitochondria in their cells. The nucleus is an organelle which houses the DNA that makes up a cell's genome. DNA itself is arranged in complex chromosomes.

Mitochondria are organelles vital in metabolism as they are the site of the citric acid cycle and oxidative phosphorylation. They evolved from symbiotic bacteria and retain a remnant genome. Like bacteria, plant cells have cell walls, and contain organelles such as chloroplasts in addition to the organelles in other eukaryotes. Chloroplasts produce energy from light by photosynthesis, and were also originally symbiotic bacteria.

Unicellular eukaryotes are those eukaryotic organisms that consist of a single cell throughout their life cycle. This qualification is significant since most multicellular eukaryotes consist of a single cell called a zygote at the beginning of their life cycles. Microbial eukaryotes can be either haploid or diploid, and some organisms have multiple cell nuclei (see coenocyte). However, not all microorganisms are unicellular as some microscopic eukaryotes are made from multiple cells.

Protists

Of eukaryotic groups, the protists are most commonly unicellular and microscopic. This is a highly diverse group of organisms that are not easy to classify. Several algae species are multicellular protists, and slime molds have unique life cycles that involve switching between unicellular, colonial, and multicellular forms. The number of species of protozoa is uncertain, since we may have identified only a small proportion of the diversity in this group of organisms.

Animals

All animals are multicellular, but some are too small to be seen by the naked eye. Microscopic arthropods include dust mites and spider mites. Microscopic crustaceans include copepods and the cladocera, while many nematodes are too small to be seen with the naked eye. Another particularly common group of microscopic animals are the rotifers, which are filter feeders that are usually found in fresh water. Micro-animals reproduce both sexually and asexually and may reach new habitats as eggs that survive harsh environments that would kill the adult animal. However, some simple animals, such as rotifers and nematodes, can dry out completely and remain dormant for long periods of time.

Fungi

The fungi have several unicellular species, such as baker's yeast (*Saccharomyces cerevisiae*) and fission yeast (*Schizosaccharomyces pombe*). Some fungi, such as the pathogenic yeast *Candida albicans*, can undergo phenotypic switching and grow as single cells in some environments, and filamentous hyphae in others. Fungi reproduce both asexually, by budding or binary fission, as well by producing spores, which are called conidia when produced asexually, or basidiospores when produced sexually.

Plants

The green algae are a large group of photosynthetic eukaryotes that include many microscopic organisms. Although some green algae are classified as protists, others such as charophyta are classified with embryophyte plants, which are the most familiar group of land plants. Algae can grow as single cells, or in long chains of cells. The green algae include unicellular and colonial flagellates, usually but not always with two flagella per cell, as well as various colonial, coccoid, and filamentous forms. In the Charales, which are the algae most closely related to higher plants, cells differentiate into several distinct tissues within the organism. There are about 6000 species of green algae.

Habitats and Ecology

Microorganisms are found in almost every habitat present in nature. Even in hostile environments such as the poles, deserts, geysers, rocks, and the deep sea, some types of microorganisms have adapted to the extreme conditions and sustained colonies; these organisms are known as extremophiles. Extremophiles have been isolated from rocks as much as 7 kilometres below the earth's surface, and it has been suggested that the amount of living organisms below the earth's surface may be comparable with the amount of life on or above the surface. Extremophiles have been known to survive for a prolonged time in a vacuum, and can be highly resistant to radiation, which may even allow them to survive in space. Many types of microorganisms have intimate symbiotic relationships with other larger organisms; some of which are mutually beneficial (mutualism), while others can be damaging to the host organism (parasitism). If microorganisms can cause disease in a host they are known as pathogens.

Extremophiles

Extremophiles are microorganisms which have adapted so that they can survive and even thrive in conditions that are normally fatal to most lifeforms. For example, some species have been found in the following extreme environments:

- Temperature: as high as 130 °C (266 °F), as low as "17 °C (1.4 °F)
- Acidity/alkalinity: less than pH 0,up to pH 11.5
- Salinity: up to saturation
- Pressure: up to 1,000-2,000 atm, down to 0 atm (e.g. vacuum of space)
- Radiation: up to 5kGy

Extremophiles are significant in different ways. They extend terrestrial life into much of the Earth's hydrosphere, crust and atmosphere, their specific evolutionary adaptation mechanisms to their extreme environment can be exploited in bio-technology, and their very existence under such extreme conditions increases the potential for extraterrestrial life.

Soil Microbes

The nitrogen cycle in soils depends on the fixation of atmospheric nitrogen. One way this can occur is in the nodules in the roots of legumes that contain symbiotic bacteria of the genera *Rhizobium*, *Mesorhizobium*, *Sinorhizobium*, *Bradyrhizobium*, and *Azorhizobium*.

Symbiotic Microbes

Importance

Microorganisms are vital to humans and the environment, as they participate in the Earth's element cycles such as the carbon cycle and nitrogen cycle, as well as fulfilling other vital roles in virtually all ecosystems, such as recycling other organisms' dead remains and waste products through decomposition. Microbes also have an important place in most higher-order multicellular organisms as symbionts. Many blame the failure of Biosphere 2 on an improper balance of microbes.

Use in Food

Microorganisms are used in brewing, winemaking, baking, pickling and other food-making processes. They are also used to control the fermentation process in the production of cultured dairy products such as yogurt and cheese. The cultures also provide flavour and aroma, and inhibit undesirable organisms.

Use in Water Treatment

Microbes are used in the biological treatment of sewage and industrial waste effluents..

Use in Energy

Microbes are used in fermentation to produce ethanol.

Use in Science

Microbes are also essential tools in biotechnology, biochemistry, genetics, and molecular biology. The yeasts (*Saccharomyces cerevisiae*) and fission yeast (*Schizosaccharomyces pombe*) are important model organisms in science, since they are simple eukaryotes that can be grown rapidly in large numbers and are easily manipulated. They are particularly valuable in genetics, genomics and proteomics. Microbes can be harnessed for uses such as creating steroids and treating skin diseases. Scientists are also considering using microbes for living fuel cells, and as a solution for pollution.

Use in warfare

In the Middle Ages, dead corpses were thrown over walls during sieges, this meant that any bacteria carrying the disease that killed the person/creature would multiply in the vicinity of the opposing side.

Importance in Human Health

Human Digestion

Microorganisms can form an endosymbiotic relationship with other, larger organisms. For example, the bacteria that live within the human digestive system contribute to gut immunity, synthesise vitamins such as folic acid and biotin, and ferment complex indigestible carbohydrates.

Diseases and Immunology

Microorganisms are the cause of many infectious diseases. The organisms involved include pathogenic bacteria, causing diseases such as plague, tuberculosis and anthrax; protozoa, causing diseases such as malaria, sleeping sickness and toxoplasmosis; and also fungi causing diseases such as ringworm, candidiasis or histoplasmosis. However, other diseases such as influenza, yellow fever or AIDS are caused by pathogenic viruses, which are not living organisms and are not therefore microorganisms. As of 2007, no clear examples of archaean pathogens are known, although a relationship has been proposed between the presence of some methanogens and human periodontal disease.

Hygiene

Hygiene is the avoidance of infection or food spoiling by eliminating microorganisms from the surroundings. As microorganisms, particularly bacteria, are found practically everywhere, this means in most cases the reduction of harmful microorganisms to acceptable levels. However, in some cases it is required that an object or substance be completely sterile, i.e. devoid of all living entities and viruses. A good example of this is a hypodermic needle.

In food preparation microorganisms are reduced by preservation methods (such as the addition of vinegar), clean utensils used in preparation, short storage periods or by cool temperatures. If complete sterility is needed, the two most common methods are irradiation and the use of an autoclave, which resembles a pressure cooker. There are several methods for investigating the level of hygiene in a sample of food, drinking water, equipment etc. Water samples can be filtrated through an extremely fine filter. This filter is then placed in a nutrient medium. Microorganisms on the filter then grow to form a visible colony. Harmful microorganisms can be detected in food by placing a sample in a nutrient broth designed to enrich the organisms in question. Various methods, such as selective media or PCR, can then be used for detection. The hygiene of hard surfaces, such as cooking pots, can be tested by touching them with a solid piece of nutrient medium and then allowing the microorganisms to grow on it. There are no conditions where all microorganisms would grow, and therefore often several different methods are needed. For example, a food sample might be analysed on three different nutrient mediums designed to indicate the presence of "total" bacteria (conditions where many, but not all, bacteria grow), molds (conditions where the growth of bacteria is prevented by e.g. antibiotics) and coliform bacteria (these indicate a sewage contamination).

In Fiction

Microorganisms have frequently played an important part in science fiction, both as agents of disease, and as entities in their own right.

Some notable uses of microorganisms in fiction include:

- *The War of the Worlds*, where microorganisms play important thematic and plot-related roles.
- *Fantastic Voyage*, in which some scientists are miniaturised to microscopic size and observe micro-organisms from a new perspective
- *Blood Music*, in which a colony of microorganisms is given intelligence
- *The Andromeda Strain*, in which extraterrestrial microorganisms kill several people
- *The White Plague*, is created and released in vengeance by John Roe O'Neill for the death of his wife and children, it is designed to kill only women.
- *Twelve Monkeys*, James Cole (Bruce Willis) searches for a pure germ in the past, which creates a deadly plague in the future. Also, Brad Pitt (as Jeffery Goines) discusses his germaphobia.

What is a Microbe?

Microbes are single-cell organisms so tiny that millions can fit into the eye of a needle. They are the oldest form of life on earth. Microbe fossils date back more than 3.5 billion years to a time when the Earth was covered with oceans that regularly reached the boiling point, hundreds of millions of years before dinosaurs roamed the earth. Without microbes, we couldn't eat or breathe. Without us, they'd probably be just fine. Understanding microbes is vital to understanding the past and the future of ourselves and our planet.

Microbes <*my-crobes*> are everywhere. There are more of them on a person's hand than there are people on the entire planet!

Microbes are in the air we breathe, the ground we walk on, the food we eat—they're even inside us!

We couldn't digest food without them—animals couldn't, either. Without microbes, plants couldn't grow, garbage wouldn't decay and there would be a lot less oxygen to breathe.

In fact, without these invisible companions, our planet wouldn't survive as we know it!

Our Challenge: How are we going to solve the case of what a microbe is when they're so small they're invisible to us?

Microbe is a term for tiny creatures that individually are too small to be seen with the unaided eye. Microbes include bacteria (*back-tear-ee-uh*), archaea (*are-key-uh*), fungi (*fun-jeye*) and protists (*pro-tists*). You've probably heard of bacteria and fungi before. Archaea are bacteria-like creatures that have some traits not found in any true bacteria. Protists include primitive algae (*al-gee*), amoebas (*ah-me-buhs*), slime molds and protozoa (*pro-toe-zoh-uh*). We can also include viruses (*vye-rus-is*) as a major type of microbe, though there is a debate as to whether viruses can be considered living creatures or not.

To solve the case of what a microbe is, we have to use tools such as high-power microscopes. Let's zoom in on some microbes and see what a few of these strange creatures look like.

As you can see, microbes come in many varieties. They may live as individuals or cluster together in communities.

Every Day Roles

Some people think it would be great if scientists could wipe out all the microbial bugs! Should we do it, and why or why not?

We get a lot of requests to track down the nasty bugs that are making people sick. It's true that some microbes cause health problems such as strep throat, chickenpox and the common cold. Unfortunately, because a small number of microbes cause disease, all microbes have got a bad rap.

It seems only a few bad microbes get almost all the press. You just don't hear as much about all the many GOOD things microbes do every day.

So we've created our own bug gallery of some microbe heroes who make our lives better.

Microbial intelligence

Microbial intelligence (popularly known as bacterial intelligence) is the intelligence shown by microorganisms. The concept encompasses complex adaptive behaviour shown by single cells, and altruistic and/or cooperative behavior in populations of like or unlike cells mediated by chemical signalling that induces physiological or behavioral changes in cells and influences colony structures.

Complex cells, like protozoa or algae, show remarkable abilities to organise themselves in changing circumstances. Shell-building by amoebae, reveals complex discrimination and manipulative skills that are ordinarily thought to occur only in multicellular organisms.

Even bacteria, which show primitive behavior as isolated cells, can display more sophisticated behavior as a population. These behaviors occur in single species populations, or mixed species populations. Examples are colonies of Myxobacteria, quorum sensing, and biofilms.

It has been suggested that a bacterial colony loosely mimics a biological neural network. The bacteria can take inputs in form of chemical signals, process them and then produce output chemicals to signal other bacteria in the colony.

The mechanisms that enable single celled organisms to coordinate in populations presumably carried over in those lines that evolved multicellularity, and were co-opted as mechanisms to coordinate multicellular organisms.

Examples of Microbial Intelligence

- The formation of biofilms requires joint decision by the whole colony.
- Under nutritional stress bacterial colonies can organise themselves in such a way so as to maximise nutrient availability.
- Bacteria reorganise themselves under antibiotic stress.
- Bacteria can swap genes(such as genes coding antibiotic ressistance) between members of mixed species colonies.
- Individual cells of myxobacteria and cellular slime moulds coordinate to produce complex structures or move as multicellular entities.
- Populations of bacteria use Quorum sensing to judge their own densities and change their behaviors accordingly. This occurs in the formation of biofilms, infectious disease processes, and the light organs of bobtail squid.
- It's known that for any bacterium to enter a host's cell, the cell must display receptors to which bacteria can adhere and be able to enter the

cell. Some strains of *E. coli* are able to internalize themselves into a host's cell even without the presence of specific receptors as they bring their own receptor to which they then attach and enter the cell.

- Under rough circumstances, some bacteria transform into endospores to resist heat and dehydration.
- A huge array of microorganisms have the ability to overcome being recognised by the immune system as they change their surface antigens so that any defense mechanisms directed against previously present antigens are now useless with the newly expressed ones.

How Microbes Do Some of the Amazing Things They Do?

As you read the pages of this Web site, you might have come across things that made you pause and say, "now how do they do that?" Well, here's where you can find out at least some of the answers. Click on the links below to find out how microbes manage to do some of the amazing things they do.

- Hot! Hot! Hot! But How?
- Beat the Cold
- Survive Lethal Radiation Blasts
- Repair DNA
- Microbial Spore Formation

Special Enzymes

Thermophiles have certain proteins, or enzymes, that are specially geared to working in high temperatures even as hot at 284° F (140° C). Keep in mind that water boils at 212° F (100° C). The normal-temperature proteins and enzymes in your body would start unfolding and breaking apart long before it got as hot as 284° F.

You see, proteins are strings of molecules called amino *<ah-mee-no>* acids (imagine a string of pearls with each pearl representing an amino acid). But for proteins to work right, their amino acid strings can't just stretch out in a line.

They have to be twisted and folded in just the right three-dimensional patterns (imagine bunching that pearl necklace up in your hand). High heat can cause these 3-D structures to unfold and the links holding the pattern together to break. And when they unfold, enzymes no longer work right.

Enzymes in thermophiles are called extremozymes *<ex-treem-oh-zimes>*. One thing that these extremozymes do to hold their 3-D shapes together in high heat is folding themselves tighter than normal temperature enzymes do. As you know, a tight knot in your shoelace is harder to get out than a loose one. Also, lots of chemical bonds formed between amino acids on the outer surfaces of extremozymes help keep these proteins intact and folded just right. These bonds act as fasteners holding together the shape of the enzymes.

Another trick is having a smaller amount of the amino acid glycine *<gly-seen>* than normal-temperature proteins have. Of the 20 different amino acids that make proteins, glycine is the second most commonly found in regular proteins. It helps them be flexible and bend a bit. Having less glycine makes extremozymes more rigid and sturdier.

One really cool tool hyperthermophiles use to keep their enzymes working right is a special type of fix-it protein called "chaperonin" *<shap-er-oh-nin>*. These chaperonin proteins fix enzymes beginning to fall apart by latching onto them and refolding them back into their active patterns—sort of like a microscopic back doctor except that chaperonins put the kinks back in instead of taking them out.

How Some Microbes Beat the Cold

In recent years, scientists have turned up microbes living in the frigid waters of Antarctic lakes that are permanently covered with snow and a thick layer of ice. Scientists have found microbes thriving in the wind-blasted rocks and soil atop high mountain summits and in polar snow. Now maybe all this doesn't seem quite so amazing at first: after all, penguins waddle across the Antarctic ice floes and polar bears and walruses meander the Arctic tundra and even we humans can trek to the poles given enough layers of down, fleece and wool. But how in the world do microbes do it without the aid of blubber, fur or other obvious insulating protection? After all, cold kills or at least slows down lots of microbes—that's why we put our food in refrigerators to keep it fresh longer.

Scientists studying microbes that thrive in the cold (which go by the fancy name of psychrophiles *<sigh-crow-files>*) have not yet discovered all the secrets to these microbes' amazing cold-loving abilities. However, they have learned a couple of neat things.

Cold-loving microbes may not have fur or down-filled coats, but they do have special proteins or enzymes that help them shrug off the frigid chill. What are enzymes, you say? These are proteins that help make lots of chemical reactions take place inside cells and basically help cells do all the things they need to do to stay alive. (In case your next question was, what's a protein, go here for a quick, easy primer on proteins). So, psychrophilic *<sigh-crow-fill-ick>* enzymes are similar to the enzymes in microbes that live at normal temperatures, but the psychrophilic enzymes continue to function at temperatures below which normal enzymes get sluggish or stop working. How? It appears that enzymes in cold-adapted microbes have a little more flexibility in them because they've lost a few of the fastener molecules that keep these protein folded rigidly into a specific three-dimensional shape. (Proteins bunch up into 3-D shapes that are very important to how well they work.) Since these enzymes can flex a bit, they don't need as much energy to do their thing, which is good in an environment where you can't count on heat for energy.

Another way cold-loving microbes thrive in their chilly environments is keeping the membranes of their cells loose with molecules called fatty acids. Membranes are like plastic sacks with holes that surround the goopy fluid and all the parts inside of cells. A fatty acid is a long chain of carbon atoms strung together. Having a lot of these fatty acids keeps membranes fluid in low temperatures so they can continue to easily take in and release molecules—"food" and "wastes" basically—through their cells.

There's a lot still to learn about how microbes live comfortably in super-cold temperatures. These are just a couple of the microbial tricks for beating the cold. Scientists are studying ways that we humans could put these psychrophilic enzymes to work for us, say in preserving foods without freezer burn, improving microbial clean up of pollution, and making new medicines.

How "Conan the Bacterium" Survives Lethal Radiation Blasts

When you go to the dentist to have your jaws X-rayed, you'll notice that you have to wear a heavy lead apron—and the person who takes the X-ray leaves the room to do it. X-rays are safe, but these precautions give you some sense of how careful we have to be around even relatively harmless levels of radiation. The problem is that unsafe levels of radiation can mess with the cells and molecules in our body, warping them and even breaking them apart. It does the same thing to microbial cells.

But there's one bacterium called *Deinococcus radiodurans <die-no-cah-cuss radio-dur-anz>* that can live through blasts of radiation thousands of times greater than the level that would kill a human being. Radiation is measured in units called rads. A dose of 500 to 1,000 rads is enough to kill a person. *D. radiodurans* thrives even after being hit by up to 1,500,000 rads—yikes!

The bacterium's Latin name means "strange berry that withstands radiation." Scientists who study *D. radiodurans* have nicknamed it "Conan the Bacterium." The Guinness Book of World Records dubbed it the world's toughest bacterium.

So how does *D. radiodurans* manage to live through such intense blasts of radiation?

If you've read other pages on this site, you may have learned that some bacteria form protective spores to survive through drought, heat and radiation (see How Microbes Form Spores). So could it be that *D. radiodurans* forms extra tough spores?

Actually no, it's not a spore-former. Anyway, while bacterial spores can endure radiation, none can take nearly as much as *D. radiodurans* can. So, without any special protective coating, what then does *D. radiodurans* do to withstand all that radiation?

Let's first take a quick look at what radiation does to a living thing. When a creature gets hit by a high dose of radiation, the intense energy causes the large DNA molecule in each cell—the collection of all the genes that make a living thing what it is—to fall apart. No creature can survive without its genes in working order.

Most microbes have tools they use to repair occasional damage to their DNA. For example, when a break happens in the DNA of an *E. coli* bacterium (one of the common bacteria living in your gut), the bug can usually repair it and get on with daily life. However, *E. coli* usually can't survive more than two or three major breaks in its DNA. *D. radiodurans*, on the other hand, can stitch back together DNA shattered to bits by radiation within a matter of a few hours.

One reason is that it has lots of extra copies of its genes. *D. radiodurans* cells have four to ten copies of their DNA molecule. Most bacteria have only one copy. These copies serve as back-ups, kind of like the back-up copies you make of your important computer files in case one day your computer crashes. So when radiation hits and *D. radioduran's* DNA gets busted up, the microbe has lots more chances of finding an intact copy of each gene to use as it stitches its DNA back together. A special protein called RecA does the stitching. Also, it appears that *D. radiodurans* may have more of the cell repair tools that most bugs have.

Still, scientists aren't entirely clear exactly how and why *D. radiodurans* is so much more radiation-resistant than other microbes. After all, other microbes have many of the same tools, although not in as high numbers and variety as *D. radiodurans* does. And some other bacteria have more than one copy of their genes, although none have quite as many copies as *D. radiodurans*. Scientists are sifting through the genes of *D. radiodurans*, trying to out if the bacterium has any unique tools or genes that give it extra radiation protection.

They're also trying to out why *D. radiodurans* evolved this super radiation-resistance in the first place. After all, there aren't any places on the planet where the microbe would have been exposed naturally to such incredibly high radiation blasts. It turns out that *D. radiodurans* is also amazingly able to live through long periods with absolutely no water without turning into a tiny dried-out husk. Some researchers think that the bacterium's radiation resistance is a lucky side effect of the ability the bug evolved to withstand long periods without water—a common natural occurrence. This is because dehydration causes the same kinds of breaks in DNA as radiation does and requires the same patching process to fix these breaks.

However it does what it does, *D. radiodurans* is indeed the toughest of the tough in the microbial world.

DNA Repair

Developing tricks and tools to keep their enzymes in order is one way thermophiles survive. They also use techniques to keep their DNA from falling apart under intense heat. Like proteins, the parts of the long, spiral ladder-shaped DNA molecule start to unravel and break apart under high heat. One way thermophiles keep that from happening is with a helper enzyme called "reverse DNA gyrase" *<jeye-race>*. This enzyme makes DNA coil up and twist upon itself in a certain way that makes the DNA more stable in high heat. (If you've ever seen a telephone handset cord that gets twisted and bunched up on itself, then you know what DNA coiling is like, only coiled up DNA is thousands of times more tightly twisted and bunched.) Microbes that live at normal temperatures have regular "DNA gyrase" instead, which also makes their DNA coil up, but in a different, looser way.

Thermophiles also have lots of DNA binding proteins. These binding proteins do just what their name suggests, running around gluing the pieces and parts of the DNA molecule back together when they start to come undone, much like the chaperonin proteins discussed above.

Holding it All Together

Of course, tools that keep the enzymes, DNA and other inside parts of the cell from breaking up will do no good if the outside or surface of the cell is falling apart. So these heat-loving creatures have cell membranes—the rubbery lining just inside the cell wall that surrounds the cell fluid—that are formed differently than those of microbes living at normal temperatures. Normal temperature microbes have membranes that are formed by two layers of molecules called lipids which join together to create what's called a bilayer (for a picture of what a lipid bilayer looks like, see this page). In thermophiles, the parts of each lipid layer that point inward are chemically glued together so that instead of a bilayer that could be pulled apart in intense heat, thermophiles have a thick single or monolayer.

There's a lot that scientists still are learning about how these amazing heat-loving microbes live happily at such high temperatures. They may well have other tools and tricks that we don't know about yet. The more we learn, the better for us because some of these techniques and tools could become useful products for us. For example, an extremozyme called Taq <*tack*> from one thermophilic bacterium is what makes DNA testing and DNA fingerprinting possible. Thanks in part to Taq, scientists have been able to sequence the entire human genome—everything on the whole humongous human DNA molecule—and the genomes of lots of microbes and other living things, too.

Microbial Spore Formation

You may have read elsewhere in this site that bacteria sometimes form protective spores to help them survive through tough times. Some other kinds of microbes do, too. Here's how that transformation takes place.

First off, you might think of a bacterial spore roughly as a mummified bacterium. The spore has a hard protective coating that encases the key parts of the bacterium—think of this coating as the sarcophagus that protects a mummy. The spore also has layers of protective membranes, sort of like the wrappings around a mummy. Within these membranes and the hard coating, the dormant bacterium is able to survive for weeks, even years, through drought, heat and even radiation. When conditions become more favorable again—when there's more water or more food available—the bacterium "comes to life" again, transforming from a spore back to a cell. Some bacterial spores have possibly been revived after they lay underground for more than 250 million years!

Ok, so how do bacteria turn themselves into spores? First, the bacterium senses that its home or habitat is turning bad: food is becoming scarce or water is disappearing or the temperature is rising to uncomfortable levels. So it makes a copy of its chromosome, the string of DNA that carries all its genes.

Then, the rubbery cell membrane that surrounds the bacterial cell fluid begins pinching inward around this chromosome copy, until there's a little cell within the larger bacterial cell. This little cell is called the "daughter cell" and the bigger, original one, what starts out as the "vegetative cell" in this illustration, is now called the "mother cell." Next, the membrane of the mother cell surrounds and swallows up the smaller cell, so that now two membrane layers surround the daughter cell. Between these two membranes a thick wall forms made out of stuff called peptidoglycan <*pep-tid-oh-gly-can*>, the same stuff found in bacteria's rigid cell walls. Finally, a tough outer coating made up of a bunch of proteins forms around all this, closing off the entire daughter cell, which is now a spore. As the mother cell withers away or gets blasted by all kinds of environmental damage, the spore lies dormant, enduring it all, just waiting for things to get better.

Not all bacteria can form spores. But several types that live in the soil can.

Bacteria in the *Bacillus* *<buh-sill-us>* and *Clostridium* *<clah-strih-dee-um>* groups are spore-formers. Their spores are called endospores.

Another group of bacteria called *Methylosinus* *<meth-ill-oh-sigh-nus>* produces spores called exospores. The difference between endospores and exospores is mainly in how they form. Endospores form inside the original bacterial cell, as described above. Exospores form outside by growing or budding out from one end of the cell. Exospores also don't have all the same building blocks as endospores, but they're similarly durable.

Microsporidia spores are usually round, oval or rod-shaped, although many species have elaborately shaped spores that may help hide them from their host immune systems. The spores help the protozoa survive while outside of a host's body. Typically, hosts are infected when they swallow *Microsporidia* spores. Once the spores reach the gut, they poke a tube through their spore coats. This tube stabs through the host's gut wall and other tissues. Then the *Microsporidia* cell fluid and nucleus—a cell's central command center—move through the hollow tube from the spore into the host cells. As *Microsporidia* reproduce in the host cells, new spores are formed that are typically passed out of the body with feces.

The spores we're talking about on this page are protective spores. A group of bacteria called *Actinomycetes* *<ack-tin-oh-my-see-tees>* and many kinds of fungi produce seed-like structures during reproduction that are also called spores. If you've ever stomped on a puffball mushroom, the brown cloud that jets out is a cloud of these reproductive spores. Like seeds, reproductive spores have tough outer coatings on them, but they aren't as durable as protective spores or cysts.

Microbial Record Holders

The biggest of the big! The baddest of the bad! The oldest of the ancients! This is where you'll find all the microbial record holders, microbes that beat all others hands down in a variety of categories. This is our microbial Book of World Records.

How Small Can You Get?

As you can tell by their name, microbes are so teensy-weensy you almost always need a microscope to see them (exceptions are described in the Biggest Microbes page). But just how tiny can microbes get? Well, the smallest of the small are the viruses. They're so tiny, we need a special kind of microscope called an electron microscope to see even the largest of them. Viruses come in a range of tiny sizes, but the smallest of all are members of a group called the Parvoviruses *<par-voh-vye-rus-iz>*. Some of these spherical viruses can be as little as 18 nanometers in diameter. How small is that, you ask? Well, a nanometer (abbreviated nm) is one-billionth (1/1,000,000,000th) of a meter. There are about 25,000,000 nanometers in 1 inch.

To give you a comparison, an itty-bitty *E. coli* bacterium is a giant next to a parvovirus. This capsule-shaped bacterium is about 2,000 nm long and 500 nm wide. A single hair from your head is anywhere from about 17,000 to 180,000 nm in diameter, depending on what kind of hair you have. Even though a hair sounds gargantuan compared to a parvovirus, just try staring down at the tip of one of your hairs—you'll go cross-eyed trying to focus on its tiny width. (To get an even better sense of how different microbes compare in size to a hair, try the Let's Get Small activity in the Experiments section.)

So parvoviruses are pretty darn tiny! But are they the tiniest living things? Well, since it's uncertain whether you can call viruses "living" or not—and at the very least viruses definitely aren't free-living things since they can't reproduce on their own—we have to give the crown for smallest known free-living things to a group of bacteria called the *Mycoplasmas <my-coh-plaz-muhs>*. Some of the little bacteria in this family measure only about 150 nm wide. They are actually smaller than some viruses. There are some 90 different *Mycoplasmas* out there, some of which infect animals, others plants and others insects. Fourteen are known that infect people.

Now, notice I said that *Mycoplasmas* are the smallest "known" free-living things. There may be even tinier bacteria out there. In 1998, a team of Australian scientists announced that they had found incredibly tiny bacteria in ancient sandstone dredged up from an oil drilling site about three miles below the ocean floor off of western Australian (see this news article).

Viewed through powerful electron microscopes, the things they found look like fuzzy tangles of threads. Individual "threads" measure only 20 to 150 nm in length. At the largest, that would make them the size of the smallest *Mycoplasmas*. At the tiniest, that makes them no bigger than little viruses. The researchers nicknamed their find "nanobes."

Because these nanobes are so incredibly tiny, they are devilishly hard for scientists to work with. But the Australian scientists reported that they have seen these evidence that these little things reproduce and grow and their studies have shown signs of DNA, the genetic recipe of all living things, in them. Also, when the scientists cut through some of the tangles, they saw different patterns that looked like the outside and inside of a cell.

Still, other scientists are doubtful that these nanobes are living things. After all, as some scientists point out, proteins and molecules—all the tools and stuff needed for a living cell to function—take up space. A single ribosome *<rye-boh-zome>*, a molecule that reads the DNA recipe and builds proteins from it, would fill up a sphere 50 to 60 nm in diameter. So how could you have a free-living cell that uses DNA as its code of life that is only 20 or 30 nm wide?

The scientific jury is still out on whether these nanobes are truly free-living super-tiny bacteria or not. That means that officially, *Mycoplasmas* still get to claim the title of smallest living things on Earth—for now.

Oldest Living Microbes

In the fairy tale, Sleeping Beauty went to sleep for 100 years in a castle protected by giant thorns and then was revived by the kiss of a prince. In 2000, scientists told a microbial version of this fairy tale, announcing that they had revived bacteria that had lain in suspended animation for 250 million years, encased in salt crystals deep in the Earth. (Of course, the scientists did it by putting the bacteria in special nutrients, not with a kiss! See this news report for the details.) If these scientists' claims are proved true, these microbes will take the prize as the oldest living things on Earth by far.

250 million years old—Wow, now that's old! The first dinosaurs were just starting to appear on Earth at that time. It would be another 45 million years before Tyrannosaurus rex roared its first roar. The scientists believe that the bacteria got trapped in the salt crystals, which were buried 1,850 feet down under what's now Carlsbad, New Mexico, in the southwestern United States.

How could the bugs possibly survive all that time, century after century, millennium after millennium, and suddenly wake up all perky when put into a nutrient-rich petri dish? The answer could be spores. The bacteria appear to be similar to modern day soil bacteria that form protective spores when things get tough. Spores are kind of like plant seeds—a tough protective shell insulates and protects the genes and basic cell parts, which are in an inactive state (that means they aren't doing anything but just sitting there). Spores are much, much tougher than plant seeds, however. Spores can even survive blasts of radiation as well as years of going with no water or nutrients. If you want to know more about spores, visit this page on our site: How Bacteria Form Spores.

Scientists know that modern-day soil bacteria spores can revive after many years. But 250 million years??? Some scientists think that these microbial sleeping beauties are a fairy tale themselves. These scientists say that there needs to be more proof that the bacteria found in the salt crystals didn't seep into the salt beds at a more recent time. Or perhaps the bacteria found in the crystals were microbes that contaminated the lab or the instruments used to drill into the salt. (See this news article for more details on why some scientists doubt the ancient age of these microbes.)

The scientists who did the experiments say that they took extra care to make sure there was no chance for contamination. Even so, their work will have to be repeated by other teams of scientists if everyone is to accept the idea that 250 million-year-old bacteria spores can be revived. This is what the scientific process is all about: piling up evidence upon evidence until you have eliminated all possible explanations but one.

However, earlier work on raising ancient microbes from their dormant sleep makes it a little easier to believe these latest claim—and the latest work makes it easier to believe the earlier findings. In 1995, scientist Raúl Cano and his coworkers announced that they'd revived 30-million-year-old bacteria that had lived in the gut of an ancient bee. The bee and its microbial guests got trapped and preserved in a drop of tree sap that became amber. To learn more about this discovery and about Dr. Cano, read his profile in the Careers section of my site. Cano's team also ran into skepticism from others when they announced their findings. It's tough to prove something this amazing beyond any shadow of a doubt. But if more examples of revived ancient microbes occur, it may well become accepted by everyone that prehistoric microbes are now dwelling happily in petri dishes in scientists' labs.

Final note: Is there any danger in reviving ancient bacterial spores? Not likely. There's no more risk that scientists will grow dangerous ancient microbes than they'll grow dangerous modern-day ones from samples they are taking from all over the world. In fact, scientists are finding and growing lots of previously unknown modern microbes all the time from all kinds of environments and so far no one has grown up any deadly microbes (at least not to my or my colleagues' knowledge). You've got lots more to worry about from the microbes floating in the air from your pal's sneeze than any revived ancient microbes. Also, scientists take great care in growing and storing these ancient microbes in secure lab environments so that they don't ever come into contact with the outside world. But that's because the scientists have to protect the ancient microbes from modern microbes or other things outside the lab, not the other way around.

The Deadliest, Scariest of All

Nature sure has created some tiny monsters when it comes to microbes—bugs that make us throw up, ooze pus, bleed out of our eyes and cough up blood. But which take top honors (or dishonors) for being the most lethal of all?

HIV/AIDS, the Global Tragedy

The Human Immunodeficiency Virus or HIV causes the disease we know as AIDS. AIDS is a pandemic—a disease outbreak that is happening all over the world. In 2000 alone, 3,000,000 people died of AIDS. Since the AIDS pandemic began around 1985, almost 22,000,000 people have died from the disease. Right now, HIV infects more than 36,000,000 people across the world, so that death toll will definitely get a lot bigger over the next several years.

HIV doesn't kill people itself. Instead, the virus shuts down a person's immune defenses—the tools used to fight off invading germs—by infecting and destroying important immune cells called T cells. Once a person loses too many T cells, his or her body can no longer deal with other microbes that cause infections. HIV merely opens the floodgates. Eventually HIV-infected people become overrun by germs and die of lung infections, skin infections or other diseases.

We now have several drugs that fight HIV. They cannot cure the infection, but they can keep it in check. Unfortunately, victims of HIV have to take several of these drugs every day for life and the drugs are very expensive. In Africa and other developing parts of the world where HIV is spreading most rapidly, most people don't have enough money to buy these drugs. That's why we can expect the death toll from HIV/AIDS to get a lot bigger over the next several years.

The Flu Virus—Even Deadlier than HIV?

Oh come on now, I'm sure you're saying. The *flu virus* here in a list of the most lethal microbes? Sure, every year during what's called "the flu season" tens of thousands of people get the flu. Despite feeling all achy and lousy for several days, most people eventually beat the virus and recover just fine. But that's not how it went during the 1918 flu season. That year, World War I was raging so the effects of the flu virus sort of got overshadowed by the bigger, more obvious effects of bullets and bombs. However, in 1918 the flu virus killed at least 675,000 American people. Worldwide, it killed at least 21,000,000 people. These are low estimates. Some research now suggests that the flu may have killed 40,000,000 to 50,000,000 people around the world that year. Researchers still aren't sure why the 1918 flu virus was so deadly. Victims of the flu that year died in a gruesome way, the virus causing so much fluid to build up in their lungs so rapidly that it was like drowning.

If higher estimates of how many people died from the 1918 flu are true, then the 1918 flu is the deadliest microbe ever in a single year. Thank goodness we haven't seen a flu bug that deadly again—but will we someday?

The Plague

The 1918 flu virus and HIV are the biggest killers of modern times. But back in the 14th century, the bacterium that causes bubonic plague, or the Black Death as it was also known, was the baddest bug of all.

In just a few years, from 1347 to 1351, the plague killed off about 75,000,000 people worldwide, including one-third of the entire population of Europe at that time. It spread through Asia, Italy, North Africa, Spain, Normandy, Switzerland, and eastward into Hungary. After a brief break, it crossed into England, Scotland, and then to Norway, Sweden, Denmark, Iceland and Greenland.

The plague bacterium is called *Yersinia <yer-sin-ee-uh> pestis*. There are two main forms of the disease. In the bubonic *<boo-bah-nick>* form, the bacteria cause painful swellings as large as an orange to form in the armpits, neck and groin. These swellings, or buboes, often burst open, oozing blood and pus. Blood vessels leak blood that puddles under the skin, giving the skin a blackened look. That's why the disease became known as the Black Death. At least half of its victims die within a week. The pneumonic *<new-mon-ick>* form of plague causes victims to sweat heavily and cough up blood that starts filling their lungs. Almost no one survived it during the plague years.

Yersinia pestis is the deadliest microbe we've ever known, although HIV might catch up to it. *Yersinia pestis* is still around in the world. Fortunately, with bacteria-killing antibiotics and measures to control the pests—rats and mice—that spread the bacteria, we've managed to conquer this killer.

Ebola, the Bloody Virus

Some of you at the beginning of this page might have thought of the Ebola virus as the scariest, deadliest microbe you've ever heard of. Those of you who did so have probably seen the movie Outbreak or read The Hot Zone, which spotlighted this grisly virus.

Ebola is definitely a nasty killer. It is part of a group of viruses that, among other effects on the body, cause the blood to stop clotting. Victims begin oozing blood from their mouths, noses, internal organs, even their eyes. It kills up to almost 90% of those who get infected.

With that kind of death rate, Ebola would be the deadliest microbe of all if it was more common. Fortunately, infections by this virus are pretty rare. There have only been seven outbreaks in humans. It has killed just over 800 people since the first outbreak in the Democratic Republic of the Congo (formerly called Zaire) in 1976.

While outbreaks of this virus are rare and relatively small, Ebola is still one mean microbe that I wouldn't ever want to tangle with. If you want to meet someone who has tackled Ebola, visit the interview with Dr. Ali Kahn in the Careers section.

Why Are Microbes Such an Evolutionary Success Story?

Which traits do microbes have that make them so good at survival?

Microbes have been around longer than ANYthing else on earth—longer than dinosaurs, plants or people. Here's the clue to prove it: fossils of single-celled creatures that date back at least 3.5 BILLION years! We humans only came on the scene about 2 million years ago.

If you imagine all time since the Earth began as a single day...

...and we humans don't in until just seconds before midnight!

Microbes developed when the Earth was still relatively new, a mass of steaming lava, boiling water and a swirl of methane, ammonia and other smelly gases. Out of these basic elements, proteins—the building blocks of life—developed. These proteins came together to form bacteria, the first creatures that could reproduce themselves and contained a genetic blueprint, or DNA. Early cells eventually developed into more complex single-celled creatures, such as the first algae and protozoa. For more details on how all this might have happened and the impact microbes have had on the development of our planet, check out this detailed web site on the Origins of Life made by the folks at NASA.

Microbes became so good at survival that they've continued to live on the planet through today and they are now EVERYwhere. If all the microbes on and in the Earth could be gathered in one place, they would take up more space than all the animals put together!

There are a lot of clues that tell us why microbes have been such an evolutionary success. Here are the top three:

1. DNA: The Blueprint for Life
2. Microbial Reproduction
3. Mutations Make a Hardy Microbe

DNA: The Blueprint for Life

DNA is short for deoxyribonucleic (*dee-ox-ee-rye-bo-new-clay-ick*) acid. (Try saying that 3 times real fast!) It is the genetic blueprint, or recipe, for making all living things. Almost every cell in your body contains DNA and all the information needed to make you what you are, from the way you look to which hand you write with.

DNA is shaped like a long ladder that's twisted into a spiral. If all the DNA in just one of your cells was unpacked and stretched out straight, it would be two yards long. Since you have about five trillion (5,000,000,000,000) cells in your body, the total length of DNA packed into you would stretch from here to the sun and back 30 times!

DNA is wound and coiled into structures called chromosomes (*crow-mo-somes*). In animals, plants, fungi and protists, these chromosomes are enclosed in a compartment within their cells called a nucleus (*new-klee-us*). In bacteria and archaea, the chromosome is not enclosed in a nucleus and floats free in their cytoplasm (*sigh-toe-plazm*), the fluid inside cells.

The sides of the DNA ladder are long chains of sugars and phosphates (*fahs-fates*). The rungs of the ladder are four chemical bases: adenine (*add-n-een*), guanine (*gwa-neen*), cytosine (*sigh-tuh-seen*) and thymine (*thigh-meen*). Adenine (A) always pairs with thymine (T) and guanine (G) always pairs with cytosine (C).

It's the sequence of these bases that makes up the genetic code or blueprint that determines all of the traits of living things. It may be hard to believe at first—how can things as different as grasshoppers and elephants be built from the arrangement of just four chemical bases?

The genetic code is like a musical score. In music, the instructions for all songs and tunes are created from just eight different notes. These eight notes have been arranged and rearranged in a wide variety of combinations to create the scores for every piece of music ever produced, from a Beethoven symphony to a Beastie Boys rap.

Similarly, the four chemical bases of DNA have been arranged and rearranged in numerous combinations to create the scores, or genomes (*gee-nomes*), of almost every creature that has ever lived.

Some viruses don't have DNA. Instead, their genetic blueprint is RNA, or ribonucleic (*rye-boh-new-clay-ick*) acid. RNA looks like one half of the DNA ladder. It has a single long chain of sugars and phosphates and four chemical bases, though not in pairs. Instead of thymine, RNA has uracil (*your-uh-sill*) (U). When RNA viruses infect cells, they first have to use the host cell's machinery to turn their RNA into DNA before they can begin reproducing.

Microbial Reproduction

One of the reasons microbes have survived and flourished for billions of years is the different ways they reproduce and how fast they do it.

Bacteria usually reproduce by simply dividing in two. Each new bacterium is a clone of the original—they each contain a copy of the same DNA. This is called binary fission (*bye-nair-ee fish-un*). If conditions are just right, one bacterium could become a BILLION (1,000,000,000) bacteria in just 10 hours through binary fission!

Sometimes bacteria do have a form of sex called conjugation (con-ju-gay-shun). One bacterium reaches out to another using a thread-like structure called a pilus. The first bacterium transfers part of its DNA to the other bacterium through the pilus.

By this exchange, bacteria can quickly create or pass along new traits that help them adapt to different environments.

Some yeasts, which are a type of fungus, also reproduce through binary fission. Other yeasts reproduce by budding. A parent yeast cell forms a growth, or bud, on its surface. As this bud gets bigger, the parent cell's nucleus divides into two by a process called mitosis (my-toe-sis). One of these nuclei transfers into the growing bud, which eventually pinches off. One yeast cell can produce up to 24 daughter cells.

Other fungi reproduce through spore formation. Fungi can produce spores either without sex or when two fungal cells fuse and share DNA. Spores are kind of like seeds in that they can be carried on the wind or rain, spreading the fungus to new places. Some species can do both sexual and asexual reproduction, depending on environmental conditions.

Dennis KunkelMicrobes in the protist category reproduce in a number of ways. Algae, for example, can reproduce without sex. The nucleus of the algal cell divides by mitosis and moves to opposite ends of the cell. Then the cell divides into two new cells—this is called cytokinesis (sigh-toe-kin-ee-sis). Some species can reproduce sexually. Some do both. Protozoa reproduce by binary fission, budding or a process called schizogony (skiz-og-uh-knee). Schizogony is multiple fission: the cell's nucleus divides several times before the cell itself divides into multiple new cells, each with one of those new nuclei.

Viruses cannot reproduce on their own, which is one reason people debate whether they should be counted as living creatures. But when they're inside a suitable host cell, all they do is reproduce and they do it well. They take over the host cell's machinery and use it to make copies of their genes and proteins. These genes and proteins come together to make new viruses, which eventually escape the host cell and drift off to infect new cells.

So why aren't we up to our elbows in bacteria, fungi and protists if microbes can reproduce so quickly? It's because conditions are almost never perfect in the real world. Once there are too many microbes in one place, food runs out, their wastes pile up, they crowd each other and eventually they start dying off. But there are still billions of them under your feet and all around you!

Mutations Make a Hardy Microbe

Mutations (*mew-tay-shuns*) are misspellings or changes in the genetic code. They're pretty rare; only about one mutation occurs for every 1-10 million DNA bases.

But microbes reproduce very quickly. For example some bacteria can divide every nine minutes. And the entire DNA string or genome (*gee-nome*) of microbes is relatively small. For example, the average bacterium has only about five million DNA bases. So in each new generation of bacteria, there may be one or two mistakes in the genetic code. In other words, these random mutations occur often enough in microbes to play a significant role in their ability to survive and adapt.

Some mutations are random "mistakes." They can occur when a microbe is exposed to radiation or chemicals that cause changes in DNA. Microbes, like us, have DNA repair proteins that, like a mini construction crew, work hard to repair any such mistakes. They don't always catch every mistake, however.

Other mutations result from a change in a creature's circumstances. For example, if the temperature starts getting hotter and hotter, "jumping genes"—pieces of DNA that can move around—may pick up and move to a new place in the microbe's DNA. This switch may result in giving the microbe a new genetic trait—like greater heat tolerance—that helps it adapt and survive better. It's like a homebuilder who, upon realising that he's building in a hurricane-prone area, decides that the heavy oak wood he planned to use for hardwood floors is more urgently needed to build the beams that support the house, so he changes it on the blueprint. Jumping genes cause genetic changes in all kinds of microbes.

Another way bacteria can undergo mutation is to take up bits of free-floating DNA in the environment. This is called transformation. Bits of DNA litter the environment when it's released from dying bacterial cells or bacteria burst open by a virus infection. DNA in the form of plasmids—circular bits of DNA that contain genes but exist outside the main chromosome—are also released by bacteria into the environment where they can be taken up by other bacteria.

Sometimes mutations are harmful and the microbes die off before they can reproduce and pass the mutations on to offspring. But other times a mutation gives rise to a new trait that helps the microbe better survive. Take antibiotic resistance among bacteria.

Antibiotics (*an-tea-bye-ah-ticks*) are life-saving drugs that kill bacteria. We discovered and began using them about 60 years ago. In that relatively short amount of time, bacteria have developed new traits through mutations that help protect them against antibiotics—sort of like a helmet protects you against injury.

Other mutations have helped microbes adapt to all sorts of environments from salty to icy to extremely hot and live off everything from decaying leaves to sunlight to bubbling sulfur.

Bioinvasion and Disease in Fishes at Allahabad

Aristichthys nobilis (Big head) and *Cyprinus carpio cunis* have initiated havoc in the riverine waters of Rivers Yamuna and Ganges. The dreadful "rule of Ten" is to the advantage of species of biota that have resulted into total disappearance of our local stock of *Clarias batrachus* ("magur"). What remains now is, of course, an *hybrid* of *Clarias* produced by new entrants into the aquatic ecosystems at Allahabad. Several other local species of fish *viz. Heteropneustes fossilis ("singhi"), Mastocembelus armatus ("Bam"), M. pancalus, Labeo dero ("rohu"), L. bata* are under duress of biotic stress.

What happens to ecosystem balance & biodiversity?

The ecosystem imbalance is writ large as the marine fish fauna (juveniles of *Trichurus savala*) have commonly co-occurred (Fig. 15) in the food of local catfishes (*Rita rita*) inhabiting Gangetic waters. The loss of biodiversity in terms of zooplankton and phytoplankton, that are the common constituents of our local fishes, is immense. The process of ecosystem alteration has thus been set in.

Is this an Indicator of forthcoming Cultural Revolution?

The local fish farmers have merrily transported these exotic species from rivers to ponds in villages. It must act as an eye opener before the conditions are ripe for a cultural revolution to the detriment of human population.

Have the exotic invaders brought in exotic paratitic disease organisms?

The greatest danger, however, is the explosive spread of marine roundworms, *Rostellascaris* and *Dichelyne* that have frequented in freshwater fishes in Gangetic water system at Allahabad for the past 2 years. It is a cause of concern that *Rostellascaris* never occurred in the last century from fishes of North India.

Bio-pollution has overtaken Water Pollution

The hue and cry was all about water pollution during the past few decades. The control of water pollution perhaps did not yield the desired results.

How has marine fish and parasite fauna survived in the Gangetic ecosystem?- is the moot question.

The answer to this intriguing question shall perhaps raise as many questions as it answered! Is the inherent chemistry of Rivers Ganges and Yamuna fast becoming identical to marine waters of Arabian Sea?

Have the human beings unconsciously or subconsciously contributed to introduction of parasitic "germs" through man-made alterations in natural environment?

The entry of thriving roundworms of catrfishes from Dona Paula *viz. Rostellascaris* and *Dichelyne* clearly rules out the probability of "vagile host" or "migratory fish (anadromous)" transfer routes from marine fishes at Goa to Rivers Ganges and Yamuna at Allahabad. The seriousness of man-made alterations in the natural water systems due to human intervention requires attention. The shipping activity by Waterways Authority of India, between Hubli at Kolkata (West Bengal) and Ganges at Allahabad (Uttar Pradesh) is a case in point. Slowly but surely the inherent chemistry of waters of Rivers

Ganges and Yamuna has been altered to suit flourishing intermediate arthropod hosts to parasitic organisms that have enabled Bio-pollution to overpower Water Pollution now while the statistics of pollutant contamination in hydrological compositions of Rivers Ganges and Yamuna still reign supreme in the minds of planners, fishery managers and politicians of this country.

A Much Broader Vision Required

The nature of aquatic ecosystems of Rivers Ganges and Yamuna has already given way to identical marine ecosystems. This can only verify survival of fish and parasite species from high saline environments of Arabian Sea in the riverine freshwater ecosystems at Allahabad. The need of the hour is to look beyond pollution abatement and surveys for quantification of contaminants from industrial effluents and the like. The application of advanced technology of assessment of diagnostic disease potential should receive foremost attention if fish wealth in our natural wetlands is to survive and retain its lost glory.

Bio-invasion

This term seems to be new, but has recently been popularised through a book by Chris Bright of the Worldwatch Institute in Washington, DC, called *Life Out of Bounds, Bio-invasion in a Borderless World*. He describes how the traditional geographical barriers between species have broken down as a result of human activity. Plants and animals are transported on purpose or by accident all around the world by carriage on ships and aircraft. Ecosystems that have been isolated for millions of years are being invaded by exotic species, with catastrophic effects. The Everglades are being taken over by the Australian melaleuca tree, the Asian tiger mosquito is greatly expanding its range, bringing dengue fever and other diseases, the Amazonian water hyacinth has spread along the margins of Lake Victoria, where it impedes fishing, and Leidy's comb jelly has been introduced to the Black Sea, where it has caused the collapse of the ecosystem. The consequences of *bioinvasions* are sometimes called *biopollution*. The term may be new, but the concept has been well understood among ecologists for many years; some are even predicting that in a century or so all we shall have left are the versatile and aggressive species.

What is Biosorption

Biosorption is a property of certain types of inactive, dead, microbial biomass to bind and concentrate heavy metals from even very dilute aqueous solutions. Biomass exhibits this property, acting just as a chemical substance, as an ion exchanger of biological origin. It is particularly the cell wall structure of certain algae, fungi and bacteria which was found responsible for this phenomenon. Opposite to biosorption is metabolically driven active bioaccumulation by living cells. That is an altogether different phenomenon requiring a different approach for its exploration.

Pioneering research on biosorption of heavy metals at McGill University in Montreal has led to identification of a number of microbial biomass types which are extremely effective in concentrating metals. Some of the biomass types come as a waste by-product of large-scale industrial fermentations (the mold Rhizopus or the bacterium Bacillus subtilis). Other metal-binding biomass types, certain abundant seaweeds (particularly brown algae e.g. Sargassum, Ecklonia), can be readily collected from the oceans. These biomass types, serving as a basis for metal biosorption processes, can accumulate in excess of 25% of their dry weight in deposited heavy metals: Pb, Cd, U, Cu, Zn, even Cr and others. Research on biosorption is revealing that it is sometimes a complex phenomenon where the metallic species could be deposited in the solid biosorbent through different sorption processes of ion exchange, complexation, chelation, microprecipitation, etc.

A whole new family of suitably "formulated" biosorbents can be used in the process of metal removal and detoxification of industrial metal-bearing effluents. The sorption packed-column configuration is the most effective mode of application for the purpose. Recovery of the deposited metals from saturated biosorbent can be accomplished because they can often be easily released from the biosorbent in a concentrated wash solution which also regenerates the biosorbent for subsequent multiple reuse. This and extremely low cost of biosorbents makes the process highly economical and competitive particularly for environmental applications in detoxifying effluents of e.g.

- metal-plating and metal-finishing operations,
- mining and ore processing operations,
- metal processing, battery and accumulator manufacturing operations,
- thermal power generation (coal-fired plants in particular),
- nuclear power generation, (etc.)

Different types of science background, from engineering to biochemistry, can make a significant contribution in elucidating the biosorption phenomenon. Interdisciplinary efforts are mandatory and represent quite a challenge. Preparing biosorption for application as a process requires mainly chemical engineering background. Good understanding of the sorption operation is mandatory. While ongoing research is essential for improving and optimising metal biosorption effectiveness, wastewater purification applications of the biosorption process are readied for pilot testing of this alternative new technology. Optimisation of specific biosorption process applications has to be done in conjunction with industrial users/clients and requires specific process engineering expertise and a serious development capital commitment.

U.S Geological Survey

Bioremediation: Nature's Way to a Cleaner Environment

The USGS provides maps, reports, and information to help others meet their needs to manage, develop, and protect America's water, energy, mineral, and land resources. We help find natural resources needed to build tomorrow, and supply scientific understanding needed to help minimize or mitigate the effects of natural hazards and environmental damage caused by human activities. The results of our efforts touch the daily lives of almost every American.

The Problem

The problem in Hanahan, South Carolina, a quiet suburb of Charleston, was not particularly unusual. In 1975, a massive leak from a military fuel storage facility released about 80,000 gallons of kerosine-based jet fuel. Immediate and extensive recovery measures managed to contain the spill, but could not prevent some fuel from soaking into the permeable sandy soil and reaching the underlying water table. Soon, ground water was leaching such toxic chemicals as benzene from the fuel-saturated soils and carrying them toward a nearby residential area.

By 1985, contamination had reached the residential area, and the facility was faced with a serious environmental problem. Removing the contaminated soils was technically impractical, and removing contaminated ground water did not address the source of the contaminants. How could contaminated ground water be kept from seeping toward the residential area in the future?

One possible solution was a new technology called bioremediation. Studies by the U.S. Geological Survey (USGS) had shown that microorganisms naturally present in the soils were actively consuming fuel-derived toxic compounds and transforming them into harmless carbon dioxide. Furthermore, these studies had shown that the rate of these biotransformations could be greatly increased by the addition of nutrients. By "stimulating" the natural microbial community through nutrient addition, it was theoretically possible to increase rates of biodegradation and thereby shield the residential area from further contamination.

In 1992, this theory was put into practice by USGS scientists. Nutrients were delivered to contaminated soils through infiltration galleries, contaminated ground water was removed by a series of extraction wells, and the arduous task of monitoring contamination levels began. By the end of 1993, contamination in the residential area had been reduced by 75 percent. Nearer to the infiltration galleries (the source of the nutrients), the results were even better. Ground water that once had contained more than 5,000 parts per billion toluene now contained no detectable contamination.

Bioremediation Works

The success of the Hanahan Bioremediation Project was no accident. It was the result of many years of intensive effort by many USGS scientists.

In the early 1980's, little was known about how toxic wastes interact with the hydrosphere. This lack of knowledge was crippling efforts to remediate environmental contamination under the new Superfund legislation—the Comprehensive Environmental Response, Compensation, and Liability Act. Faced with this problem, Congress directed the USGS to conduct a programme to provide this critically needed information. By means of this programme, known as the Toxic Substances Hydrology Programme, the most important categories of wastes were systematically investigated at sites throughout the United States. One of the principal findings of this programme was that microorganisms in shallow aquifers affect the fate and transport of virtually all kinds of toxic substances. For example:

Crude Oil Spill, Bemidji, Minnesota

In 1979, a pipeline carrying crude oil burst and contaminated the underlying aquifer. USGS scientists studying the site found that toxic chemicals leaching from the crude oil were rapidly degraded by natural microbial populations. Significantly, it was shown that the plume of contaminated ground water stopped enlarging after a few years as rates of microbial degradation came into balance with rates of contaminant leaching. This was the first and best-documented example of intrinsic bioremediation in which naturally occurring microbial processes remediates contaminated ground water without human intervention.

Sewage Effluent, Cape Cod, Massachusetts

Disposal of sewage effluent in septic drain fields is a common practice throughout the United States. Systematic studies of a sewage effluent plume at Massachusetts Military Reservation (formerly known as Otis Air Force Base) led to the first accurate field and laboratory measurements of how rapidly natural microbial populations degrade nitrate contamination (denitrification) in a shallow aquifer.

Chlorinated Solvents, New Jersey

Chlorinated solvents are a particularly common contaminant in the heavily industrialised Northeast. Because their metabolic processes are so adaptable, microorganisms can use chlorinated compounds as oxidants when other oxidants are not available. Such transformations, which can naturally remediate solvent contamination of ground water, has been extensively documented by USGS scientists at Picatinny Arsenal, New Jersey.

Pesticides, San Francisco Bay Estuary

Pesticide contamination of rivers and streams is a matter of concern throughout the United States. Field and laboratory studies in the Sacramento River and San Francisco Bay have shown the effects of biological and non-biological processes in degrading commonly used pesticides, such as molinate, thiobencarb, carbofuran, and methyl parathion.

Agricultural Chemicals in the Midcontinent

Agricultural chemicals affect the chemical quality of ground water in many Midwestern States. Studies in the midcontinent have traced the fate of nitrogen fertilizers and pesticides in ground and surface waters. These studies have shown that many common contaminants, such as the herbicide atrazine, are degraded by biological (microbial degradation) and non-biological (photolytic degradation) processes.

Gasoline Contamination, Galloway, New Jersey

Gasoline is probably the most common contaminant of ground water in the United States. Studies at this site have demonstrated rapid microbial degradation of gasoline contaminants and have shown the importance of processes in the unsaturated zone (the zone above the water table) in degrading contaminants.

Creosote Contaminants, Pensacola, Florida

Creosote and chlorinated phenols have been used extensively as wood preservatives throughout the United States. Contaminants leaked to the underlying aquifer through several unlined ponds and were transported toward nearby Pensacola Bay. Studies at this site have demonstrated that microorganisms can adapt to extremely harsh chemical conditions and that microbial degradation was restricting migration of the contaminant plume. Together, these studies laid the technical foundation that enabled bioremediation to be applied at Hanahan.

Technology Transfer

The Hanahan Bioremediation Project is just one of many successful bioremediation experiments that can be traced to basic research carried out by USGS scientists. Methods and technology developed in the Toxic Substances Hydrology Programme are now being used by private contractors, State environmental managers, and other Federal agencies to address contaminant problems throughout the United States.

Stretching Remediation Dollars

Cleaning up existing environmental contamination in the United States could cost as much as $1 *trillion* dollars. Bioremediation can help contain costs as follows:

Treating contamination in place —Most of the cost associated with traditional cleanup technologies is associated with physically removing and disposing of contaminated soils. Because engineered bioremediation can be carried out in place by delivering nutrients to contaminated soils, it does not incur removal-disposal costs.

Harnessing natural processes—At some sites, natural microbial processes can remove or contain contaminants without human intervention. In these cases where intrinsic bioremediation (natural attenuation) is appropriate, substantial cost savings can be realised.

Reducing environmental stress—Because bioremediation methods minimize site disturbance compared with conventional cleanup technologies, post-cleanup costs can be substantially reduced.

Future Challenges

Although bioremediation holds great promise for dealing with intractable environmental problems, it is important to recognise that much of this promise has yet to be realised. Specifically, much needs to be learned about how microorganisms interact with different hydrologic environments. As this under-standing increases, the efficiency and applicability of bioremediation will grow rapidly. Because of its unique interdisciplinary expertise in microbiology, hydrogeology, and geochemistry, the USGS will continue to be at the forefront of this exciting and rapidly evolving technology.

Biofertilizers

One of the major concerns in today's world is the pollution and contamination of soil. The use of chemical fertilizers and pesticides has caused tremendous harm to the environment. An answer to this is the biofertilizer, an environmentally friendly fertilizer now used in most countries. Biofertilizers are organisms that enrich the nutrient quality of soil. The main sources of biofertilizers are bacteria, fungi, and cynobacteria (blue-green algae). The most striking relationship that these have with plants is symbiosis, in which the partners derive benefits from each other. Plants have a number of relationships with fungi, bacteria, and algae, the most common of which are with mycorrhiza, rhizobium, and cyanophyceae. These are known to deliver a number of benefits including plant nutrition, disease resistance, and tolerance to adverse soil and climatic conditions. These techniques have proved to be successful biofertilizers that form a health relationship with the roots. Biofertilizers will help solve such problems as increased salinity of the soil and chemical run-offs from the agricultural fields. Thus, biofertilizers are important if we are to ensure a healthy future for the generations to come.

Mycorrhiza

Mycorrhizae are a group of fungi that include a number of types based on the different structures formed inside or outside the root. These are specific fungi that match with a number of favourable parameters of the the host plant on which it grows. This includes soil type, the presence of particular chemicals in the soil types, and other conditions. These fungi grow on the roots of these plants. In fact, seedlings that have mycorrhizal fungi growing on their roots survive better after transplantation and grow faster. The fungal symbiont gets shelter and food from the plant which, in turn, acquires an array of benefits such as better uptake of phosphorus, salinity and drought tolerance, maintenance of water balance, and overall increase in plant growth and development. While selecting fungi, the right fungi have to be matched with the plant. There are specific fungi for vegetables, fodder crops, flowers, trees, etc. Mycorrhizal fungi can increase the yield of a plot of land by 30%-40%. It can absorb phosphorus from the soil and pass it on to the plant. Mycorrhizal plants show higher tolerance to high soil temperatures, various soil- and root-borne pathogens, and heavy metal toxicity.

Legume-Rhizobium Relationship

Leguminous plants require high quantities of nitrogen compared to other plants. Nitrogen is an inert gas and its uptake is possible only in fixed form, which is facilitated by the rhizobium bacteria present in the nodules of the root system. The bacterium lives in the soil to form rootnodules (i.e. outgrowth on roots) in plants such as beans, gram, groundnut, and soybean.

Blue-Green Algae

Blue-green algae are considered the simplest, living autotrophic plants, i.e. organisms capable of building up food materials from inorganic matter. They are microscopic. Blue-green algae are widely distributed in the aquatic environment. Some of them are responsible for water blooms in stagnant water. They adapt to extreme weather conditions and are found in snow and in hot springs, where the water is 85°C. Certain blue-green algae live intimately with other organisms in a symbiotic relationship. Some are associated with the fungi in form of lichens. The ability of blue-green algae tophotosynthesize food and fix atmospheric nitrogen accounts for their symbiotic associations and also for their presence in paddy fields. Blue-green algae are of immense economic value as they add organic matter to the soil and increase soil fertility. Barren alkaline lands in India have been reclaimed and made productive by inducing the proper growth of certain blue-green algae.

Phosphate Fertilizers & Water Pollution

Phosphates are a major source of pollution in lakes and streams, and high phosphate levels support over-production of algae and water weeds. However, many of us have misconceptions regarding the source of polluting phosphates, and many homeowners unknowingly contribute to the problem. Lawn and Garden fertilizers often are implicated as the major source of phosphate pollution. However, research clearly demonstrates that with proper application, fertilizer does not pollute. When phosphates are applied to soils, they quickly bind to soil particles, much like a magnet picks up paper clips. Soil-bound phosphates contribute to pollution only when soil erosion occurs. Research studies found little or no difference in the phosphate content of storm runoff from lawn fertilizers with and without phosphate.

Problems arise, however, when fertilizer is over-spread or spilled onto hard surfaces like streets, driveways and sidewalks. Here, the phosphate washes with rain through the storm drains into lakes and rivers.

Likewise, grass clippings and leaves that fall on hard surfaces release their phosphorous into water sources. Research studies indicate that 80 percent of the phosphorous from urban settings comes from lawn clippings and leaves that end up in street gutters. While a few grass clippings mowed into the street look rather innocent, collectively they have a major impact on our water quality. Using a mulching lawn mower to keep lawn clippings on your lawn is especially useful.

Another important source of phosphate pollution comes from soil erosion caused by water or wind. When soil moves, it takes the soil-bound phosphates with it. Construction sites and sparsely vegetated ground also cause erosion.

You can take an active part in protecting Colorado's water quality by keeping fertilizers, leaves and grass clippings off streets and driveways. Prevent soil erosion by planting slopes with grass or other plants. We all have a part in protecting our water, one of Colorado's most valuable resources!

For more information, see the following Colorado State Extension fact sheet(s).

- Soil, Water and Plant Testing
- Lawn Care
- Composting Yard Waste
- Organic Fertilizers
- Xeriscaping: Ground Cover Plants
- Ornamental Grasses
- Xeriscaping: Retrofit Your Yard
- Vegetable Garden: Soil Management and Fertilisation

Biofertilizers

Biofertilizers are ready to use live formulates of such beneficial microorganisms which on application to seed, root or soil mobilize the availability of nutrients by their biological activity in particular, and help build up the micro-flora and in turn the soil health in general. With the introduction of green revolution technologies the modern agriculture is getting more and more dependent upon the steady supply of synthetic inputs (mainly fertilizers), which are products of fossil fuel (coal+ petroleum). Adverse effects are being noticed due to the excessive and imbalanced use of these synthetic inputs. This situation has lead to identifying harmless inputs like biofertilizers. Use of such natural products like biofertilizers in crop cultivation will help in safeguarding the soil health and also the quality of crop products.

Benefits From Using Biofertilizers?

- Increase crop yield by 20-30%.
- Replace chemical nitrogen and phosphorus by 25%.
- Stimulate plant growth.
- Activate the soil biologically.
- Restore natural soil fertility.
- Provide protection against drought and some soil borne diseases.

Advantages of Bio-fertilizers

1. Cost effective.
2. Suppliment to fertilizers.
3. Eco-friendly (Friendly with nature).
4. Reduces the costs towards fertilizers use, especially regarding nitrogen and phosphorus.

Types of Biofertilizers

1. For Nitrogen
   - o Rhizobium for legume crops.
   - o Azotobacter/Azospirillum for non legume crops.
   - o Acetobacter for sugarcane only.
   - o Blue –Green Algae (BGA) and Azolla for low land paddy.
2. For Phosphorous
   - o Phosphatika for all crops to be applied with Rhizobium, Azotobacter, Azospirillum and Acetobacter
3. For enriched compost
   - o Cellulolytic fungal culture
   - o Phosphotika and Azotobacter culture

Biofertilizers Recommended for crops:

- Rhizobium + Phosphotika at 200 gm each per 10 kg of seed as seed treatment are recommended for pulses such as pigeonpea, green gram, black gram, cowpea etc, groundnut and soybean.
- Azotobacter + Phosphotika at 200 gm each per 10 kg of seed as seed treatment are useful for wheat, sorghum, maize, cotton, mustard etc.
- For transplanted rice, the recommendation is to dip the roots of seedlings for 8 to 10 hours in a solution of Azospirillum + Phosphotika at 5 kg each per ha.

Ways Biofertilizers are Applied to Crops:

1. Seed treatment:

   200 g of nitrogenous biofertilizer and 200 g of Phosphotika are suspended in 300-400 ml of water and mixed thoroughly. Ten kg seeds are treated with this paste and dried in shade. The treated seeds have to be sown as soon as possible.
2. Seedling root dip:

   For rice crop, a bed is made in the field and filled with water. Recommended biofertilizers are mixed in this water and the roots of seedlings are dipped for 8-10 hrs.
3. Soil treatment:

   4 kg each of the recommended biofertilizers are mixed in 200 kg of

compost and kept overnight. This mixture is incorporated in the soil at the time of sowing or planting.

Geting Good Response to Biofertilizer Application:

- Biofertilizer product must contain good effective strain in appropriate population and should be free from contaminating microorganisms.
- Select right combination of biofertilizers and use before expiry date.
- Use suggested method of application and apply at appropriate time as per the information provided on the label.
- For seed treatment adequate adhesive should be used for better results.
- For problematic soils use corrective methods like lime or gypsum pelleting of seeds or correction of soil pH by use of lime.
- Ensure the supply of phosphorus and other nutrients.

Probable Reasons for not Getting Response From the Application of Biofertilizers:

1. On account of quality of product
   - o Use of ineffective strain.
   - o Insufficient population of microorganisms.
   - o High level of contaminants.
2. On account of inadequate storage facilities
   - o May have been exposed to high temperature.
   - o May have been stored in hostile conditions.
3. On account of usage
   - o Not used by recommended method in appropriate doses.
   - o Poor quality adhesive.
   - o Used with strong doses of plant protection chemicals.
4. On account of soil and environment
   - o High soil temperature or low soil moisture.
   - o Acidity or alkalinity in soil.
   - o Poor availability of phosphorous and molybdenum.
   - o Presence of high native population or presence of bacteriophages.

Precautions For Using Biofertilizers :

- Biofertilizer packets need to be stored in cool and dry place away from direct sunlight and heat.
- Right combinations of biofertilizers have to be used.
- As Rhizobium is crop specific, one should use for the specified crop only.
- Other chemicals should not be mixed with the biofertilizers.
- While purchasing one should ensure that each packet is provided with necessary information like name of the product, name of the crop for which intended, name and address of the manufacturer, date of manufacture, date of expiry, batch number and instructions for use.
- The packet has to be used before its expiry, only for the specified crop and by the recommended method of application.
- Biofertilizers are live product and require care in the storage
- Both nitrogenous and phosphatic biofertilizers are to be used to get the best results.
- It is important to use biofertilizers along with chemical fertilizers and organic manures.
- Biofertilizers are not replacement of fertilizers but can supplement plant nutrient requirements.

Biopesticides

Biopesticides (also known as biological pesticides) are certain types of pesticides derived from such natural materials as animals, plants, bacteria, and certain minerals. For example, garlic, mint, and baking soda all have pesticidal applications and are considered biopesticides. At the end of 1998, there were approximately 175 registered biopesticide active ingredients and 700 products. Biopesticides fall into three major categories:

(1) Microbial pesticides contain a microorganism (bacterium, fungus, virus, protozoan or alga) as the active ingredient. The most widely known microbial pesticides are varieties of the bacterium Bacillus thuringiensis, or Bt, which can control certain insects in cabbage, potatoes, and other crops. Bt produces a protein that is harmful to specific insect pests. Certain other microbial pesticides act by out-competing pest organisms. Microbial pesticides need to be continuously monitored to ensure they do not become capable of harming non-target organisms, including humans.

(2) Plant-pesticides are pesticidal substances that plants produce from genetic

material that has been added to the plant. For example, scientists can take the gene for the Bt pesticidal protein, and introduce the gene into the plants= own genetic material. Then the plantBinstead of the Bt bacterium—manufactures the substance that destroys the pest. Both the protein and its genetic material are regulated by EPA; the plant itself is not regulated.

(3) Biochemical pesticides are naturally occurring substances that control pests by non-toxic mechanisms. Conventional pesticides, by contrast, are synthetic materials that usually kill or inactivate the pest. Biochemical pesticides include substances that interfere with growth or mating, such as plant growth regulators, or substances that repel or attract pests, such as pheromones. Because it is sometimes difficult to determine whether a natural pesticide controls the pest by a non-toxic mode of action, EPA has established a committee to determine whether a pesticide meets the criteria for a biochemical pesticide.

Avantages of using biopesticides :

- Biopesticides are inherently less harmful than conventional pesticides.
- Biopesticides are designed to affect only one specific pest or, in some cases, a few target organisms, in contrast to broad spectrum, conventional pesticides that may affect organisms as different as birds, insects, and mammals.
- Biopesticides often are effective in very small quantities and often decompose quickly, thereby resulting in lower exposures and largely avoiding the pollution problems caused by conventional pesticides.
- When used as a component of Integrated Pest Management (IPM) programmes, biopesticides can greatly decrease the use of conventional pesticides, while crop yields remain high.

To use biopesticides effectively, however, users need to know a great deal about managing pests.

The Ways in Which EPA encourage the development and use of biopesticides:

In 1994, the Biopesticides and Pollution Prevention Division was established in the Office of Pesticide Programmes to facilitate the registration of biopesticides. This Division promotes the use of safer pesticides, including biopesticides, as components of IPM programmes. The Division also coordinates the Pesticide Environmental Stewardship Programme (PESP). (See related fact sheets on IPM and PESP.)

Since biopesticides tend to pose fewer risks than conventional pesticides, EPA generally requires much less data to register a biopesticide than to register a conventional pesticide. In fact, new biopesticides are often registered in less than a year, compared with an average of more than 3 years for conventional pesticides.

While biopesticides require less data and are registered in less time than conventional pesticides, EPA must always conduct rigorous reviews to ensure that pesticides will not have adverse effects on human health or the environment. For EPA to be sure that a pesticide is safe, the Agency requires that registrants submit a variety of data about the composition, toxicity, degradation, and other characteristics of the pesticide.

Biopesticides

Biopesticides are certain types of pesticides derived from such natural materials as animals, plants, bacteria, and certain minerals. For example, canola oil and baking soda have pesticidal applications and are considered biopesticides. At the end of 2001, there were approximately 195 registered biopesticide active ingredients and 780 products. Biopesticides fall into three major classes:

1. Microbial pesticides consist of a microorganism (e.g., a bacterium, fungus, virus or protozoan) as the active ingredient. Microbial pesticides can control many different kinds of pests, although each separate active ingredient is relatively specific for its target pest[s]. For example, there are fungi that control certain weeds, and other fungi that kill specific insects.

The most widely used microbial pesticides are subspecies and strains of Bacillus thuringiensis, or Bt. Each strain of this bacterium produces a different mix of proteins, and specifically kills one or a few related species of insect larvae. While some Bt's control moth larvae found on plants, other Bt's are specific for larvae of flies and mosquitoes. The target insect species are determined by whether the particular Bt produces a protein that can bind to a larval gut receptor, thereby causing the insect larvae to starve.

2. Plant-Incorporated-Protectants (PIPs) are pesticidal substances that plants produce from genetic material that has been added to the plant. For example, scientists can take the gene for the Bt pesticidal protein, and introduce the gene into the plant's own genetic material. Then the plant, instead of the Bt bacterium, manufactures the substance that destroys the pest. The protein and its genetic material, but not the plant itself, are regulated by EPA.

3. Biochemical pesticides are naturally occurring substances that control pests by non-toxic mechanisms. Conventional pesticides, by contrast, are generally synthetic materials that directly kill or inactivate the pest. Biochemical pesticides include

substances, such as insect sex pheromones, that interfere with mating, as well as various scented plant extracts that attract insect pests to traps. Because it is sometimes difficult to determine whether a substance meets the criteria for classification as a biochemical pesticide, EPA has established a special committee to make such decisions.

What are the Advantages of Using Biopesticides?

Biopesticides are usually inherently less toxic than conventional pesticides.

Biopesticides generally affect only the target pest and closely related organisms, in contrast to broad spectrum, conventional pesticides that may affect organisms as different as birds, insects, and mammals.

Biopesticides often are effective in very small quantities and often decompose quickly, thereby resulting in lower exposures and largely avoiding the pollution problems caused by conventional pesticides.

When used as a component of Integrated Pest Management (IPM) programmes, biopesticides can greatly decrease the use of conventional pesticides, while crop yields remain high.

To use biopesticides effectively, however, users need to know a great deal about managing pests.

How does EPA encourage the development and use of biopesticides?

In 1994, the Biopesticides and Pollution Prevention Division was established in the Office of Pesticide Programmes to facilitate the registration of biopesticides. This Division promotes the use of safer pesticides, including biopesticides, as components of IPM programmes. The Division also coordinates the Pesticide Environmental Stewardship Programme (PESP).

Since biopesticides tend to pose fewer risks than conventional pesticides, EPA generally requires much less data to register a biopesticide than to register a conventional pesticide. In fact, new biopesticides are often registered in less than a year, compared with an average of more than 3 years for conventional pesticides.

While biopesticides require less data and are registered in less time than conventional pesticides, EPA always conducts rigorous reviews to ensure that pesticides will not have adverse effects on human health or the environment. For EPA to be sure that a pesticide is safe, the Agency requires that registrants submit a variety of data about the composition, toxicity, degradation, and other characteristics of the pesticide.

Regulating Biopesticides

Before a pesticide can be marketed and used in the United States, the Federal Insecticide, Fungicide, and Rodenticide Act (FIFRA) requires that EPA evaluate the proposed pesticide to assure that its use will not pose unreasonable risks of harm to human health and the environment. This regulation involves an extensive review of health and safety information.

Biopesticides include naturally occurring substances that control pests (biochemical pesticides), microorganisms that control pests (microbial pesticides), and pesticidal substances produced by plants containing added genetic material (plant-incorporated protectants) or PIPs.

The Biopesticide Registration Tools provide links to information tools to assist applicants. The Fact Sheets sections provides links to information about each of the biopesticide active ingredients. The Product Lists section provides various lists of individual biopesticide products to assist the public in identifying appropriate biopesticide product for pest problems. Finally, the PIPs section provides extensive information regarding the regulation of genetically engineered plants.

Biopesticide Registration Tools

The federal pre-marketing approval of pesticides - termed registration — is a complex process. The documents linked from this page augment the general registration process as they relate specifically to the registration of biopesticides. The e-mail address bppdconsistency@epa.gov has been created to respond to issues concerning biopesticide registration inconsistency that affect processing of submissions.

Biopesticide Regulatory Action Leaders will meet to discuss these issues and provide an answer usually within two to three weeks. Resolution of these issues will be posted to this site. Since this e-mail address is only for the submission of generic issues regarding consistency in the regulation of biopesticides. Other questions regarding biopesticides should be directed to the appropriate biopesticide Regulatory Action Leader or the Biopesticide Ombudsman, Brian Steinwand (steinwand.brian@epa.gov). If you have questions about conventional pesticides or antimicrobials, please contact the Ombudsman for the Registration Division or the Antimicrobial Division.

Biopesticide Active Ingredient Fact Sheets

This collection of fact sheets contains chemical specific information about biopesticide active ingredients. Additionally, some of the fact sheets include a detailed technical document, bibliographies, regulatory history, Federal Register notices, and/or registrant and product lists.

Biopesticide Ingredient & Product Lists

Lists of EPA registered (approved) biopesticide ingredients & products are provided, including a listing by active ingredient PDF (43 pp, 378 K, about PDF), and lists of biopesticide active ingredients by year first approved.

Plant Incorporated Protectants (PIPs)

Plant-Incorporated Protectants are pesticidal substances produced by plants and the genetic material necessary for the plant to produce the substance. For example, scientists can take the gene for a specific Bt pesticidal protein, and introduce the gene into the plant's genetic material. Then the plant manufactures the pesticidal protein that controls the pest when it feeds on the plant. Both the protein and its genetic material are regulated by EPA; the plant itself is not regulated.

Applications of Composting

Composting is the aerobic decomposition of biodegradable organic matter, producing compost. (Or in a simpler form: Composting is the decaying of food, mostly vegetables or manure.) The decomposition is performed primarily by facultative and obligate aerobic bacteria, yeasts and fungi, helped in the cooler initial and ending phases by a number of larger organisms, such as springtails, ants, nematodes and oligochaete worms. Composting can be divided into home composting and industrial composting. Essentially the same biological processes are involved in both scales of composting, however techniques and different factors must be taken into account.

Importance

Composting recycles or "downcycles" organic household and yard waste and manures into an extremely useful humus-like, soil end-product called compost. Examples are fruits, vegetables and yard clippings. Ultimately this permits the return of needed organic matter and nutrients into the foodchain and reduces the amount of "green" waste going into landfills. Composting is widely believed to speed up the natural process of decomposition appreciably as a result of the raised temperatures that often accompany it. The elevated heat results from exothermic processes, and the heat in turn reduces the generational time of microorganisms and thereby speeds the energy and nutrient exchanges taking place. It is a popular misconception that composting is a "controlled" process; if the right environmental circumstances are present the process virtually runs itself. Hence a popular expression, "compost happens". It is nonetheless very necessary to provide as optimal circumstances as possible for large amounts of organic waste to break down properly. This is especially so when it is accompanied by heating, since at elevated temperatures oxygen within the piles is consumed more rapidly, and if not controlled, will lead to malodor.

Decomposition similar to composting occurs throughout nature as garbage dissolves in the absence of all the conditions that modern composters talk about; however, the process can be slow. For example, in the forest bark, wood and leaves break down into humus over 3-7 years. In restricted environments, for example, vegetables in a plastic trash container, decomposition with a lack of air encourages growth of anaerobic microbes, which produce disagreeable odors. Another form of degradation practiced deliberately in absence of oxygen is called anaerobic digestion- an increasingly popular companion to composting as it enables capture of residual energy in the form of biogas, whereas composting releases the majority of bound carbon-energy as excess heat (which helps sanitize the material) as well as copious amounts of biogenic CO_2 to the atmosphere.

It is important to distinguish between terms such as "biodegradable", "compostable", and "compost-compatible".

- A biodegradable material is capable of being broken down completely under the action of microorganisms into carbon dioxide, water and biomass. It may take a very long time for a material to biodegrade depending on its environment (e.g. hardwood in an arid area), but it ultimately breaks down completely.
- A compostable material biodegrades substantially under composting conditions, into carbon dioxide, methane, water and compost biomass. Compost biomass refers to the portion of the material that is metabolized by the microorganisms and which is incorporated into the cellular structure of the organisms or converted into humic acids etc. Compost biomass residues from a compostable material are fully biodegradable. "Compostable" is thus a subset of "biodegradable". The size of the material is a factor in determining compostability because it affects the rate of degradation. Large pieces of hardwood may not be compostable under a specific set of composting conditions, whereas sawdust of the same type of wood may be.
- A compost-compatible material does not have to be compostable or even biodegradable. It may biodegrade too slowly to be compostable itself, or it may not biodegrade at all. However, it is not readily distinguishable from the compost on a macroscopic scale and does not have a deleterious effect on the compost (e.g. it is not a biocide). Compost-compatible materials are generally inert and are present in compost at relatively low levels. Examples of compost-compatible materials include sand particles and inert particles of plastic.

Reducing Waste

Although composting has historically focussed on creating garden-ready soil, it is becoming more important as a tool for reducing solid waste. More than 60 percent of household waste in the United States is recyclable or compostable. But Americans only compost 8 percent of their waste. Surveys have shown that the #1 reason Americans don't compost their waste is because they feel the process is complicated, time-consuming or requires special equipment. However, especially in rural areas, much of the solid waste could be removed from the waste stream by promoting "extremely passive composting" where consumers simply discard their yard waste and kitchen scraps on their own land, regardless of whether the material is ever re-used as "compost". (http://www.nrdc.org/cities/recycling/fover.asp)

Materials

Many different materials are suitable for composting organisms. Composters often refer to "C:N" requirements; some materials contain high amounts of carbon in the form of cellulose which the bacteria need for their energy. Other materials contain nitrogen in the form of protein, which provide nutrients for the energy exchanges. It would however be an over-simplification to describe composting as about carbon and nitrogen, as is often portrayed in popular literature. Elemental carbon - such as charcoal - is not compostable nor is a pure form of nitrogen, even in combination with carbon. Not only this, but a great variety of man-made, carbon-containing products, including many textiles and polyethylene, are not compostable - hence the push for biodegradable plastics.

Suitable ingredients with relatively high carbon content include:

- Dry, straw-type material, such as cereal straws
- Autumn leaves
- Sawdust and wood chips
- paper and cardboard (such as corrugated cardboard or newsprint with soy-based inks)

Ingredients with relatively high nitrogen content include:

- Green plant material (fresh or wilted) such as crop residues, hay, grass clippings, weeds
- Manure of poultry and herbivorous animals such as horses, cows and llamas
- Fruit and vegetable trimmings

The most efficient composting occurs by seeking to obtain an initial C:N mix of 25-30 to 1 by dry chemical weight. Grass clippings have an average ratio of 10-19 to

1 and dry autumn leaves from 55-100 to 1. Mixing equal parts by volume approximates the ideal range.

Poultry manure provides much nitrogen but with a ratio to carbon that is imbalanced. If composted alone, this results in excessive N-loss in the form of ammonia - and some odor. Horse manure provides a good mix of both, although in modern stables, so much bedding may be used as to make the mix too carbonaceous.

For home-scale composting, mixing the materials as they are added increases the rate of decomposition, but it can be easier to place the materials in alternating layers, approximately 15 cm (6 in) thick, to help estimate the quantities. Keeping carbon and nitrogen sources separated in the pile can slow down the process, but decomposition will still occur.

Some people put special materials and activators into their compost. A light dusting of agricultural lime (not on animal manure layers) can curb excessive acidity, especially with food waste. Seaweed meal provides a ready source of trace elements. Finely pulverized rock (rock flour or rock dust) can also provide minerals, while clay and leached rock dust are poor in trace minerals.

Composting in the form of bioremediation can break down petroleum hydrocarbons, TNT and a variety of toxic compounds. It is the bacterial and in some cases fungal content of the compost that possess the enzymatic properties to de-polymerize the complex man-made molecules. In other words, there is nothing about the composting process *per se* that adds or detracts from this, unless as noted above, by warming, to increase the metabolic rate of the constituent organisms. Some materials are best left to a high-rate thermophilic composting system, as they decompose slower, attract vermin and require higher temperatures to kill pathogens than backyard composting provides. These materials include meat, dairy products, eggs, restaurant grease, cooking oil, manure and bedding of non-herbivores, and residuals from the treatment of wastewater and drinking water. Meat and dairy products can be recycled using bokashi, a fermentation method. Human waste can be composted by industrial, high-heat methods and also composting toilets, even though most composting toilets do not allow for the thermophilic decomposition believed to be necessary for a rapid kill of pathogens, such as Salmonella. This is not a problem, however, since composting toilets also incorporate the essential element of time required to reduce the available substrate on which pathogens can feed, while increasing the growth of competing microbes. If high temperatures are reached, the resulting compost can be safely used as a fertilizer for food crops and even directly edible crops provided it is not illegal in the regions where the humanure is applied. Careful filtration of the compost also prevents contamination. Humanure fertilizer is, however, used throughout the less developed world and is becoming more accepted as a garden amendment in the developed world (see humanure).

Approaches

There are two major approaches to composting: active and passive. These terms are somewhat of a misnomer since both active and passive composts can attain high heating, which increases the rate of biochemical processes. But the terms active and passive are appropriate descriptions for the nature of human intervention used.

Active

Active (hot) composting is composting at close to ideal conditions, allowing aerobic bacteria to thrive. Aerobic bacteria break down material faster and produce less odor and fewer pathogens and destructive greenhouse gases than anaerobic bacteria. Commercial-grade composting operations actively control the composting conditions such as the carbon-to-nitrogen ratio. For backyard composters, the charts of carbon and nitrogen ratios in various ingredients and the calculations required to get the ideal mixture can be intimidating, so many rules of thumb exist for approximating it.

Pasteurisation is a misnomer in composting, as no compost will become truly sterilized by high temperatures alone. Rather, in a very hot compost where the temperature exceeds 55 °C (130 °F) for several days, the ability of organisms to survive is greatly compromised. Nevertheless, there are many organisms in nature that can survive extreme temperatures, including the group of pathogenic Clostridium, and so no compost is completely safe. To achieve the elevated temperatures, the compost bin must be kept warm, insulated and damp. Aerated Composting is an efficient form of composting from the chemical point of view as it produces ultimately only energy in the form of waste heat and CO_2 and H_2O . With aerated composting, fresh air (i.e. oxygen) is introduced throughout the mix of materials using any appropriate mechanism. The air stimulates the microorganisms that are already in the mix, and their by-product is heat. In a properly operated compost system, pile temperatures are sufficient to stabilize the raw material, and the oxygen-rich conditions within the core of the pile eliminate offensive odors. High temperatures also destroy fly larvae and weed seeds, yielding a safe, high-quality finished product.

Finally, aeration expedites the composting process through the mechanism of heating insofar as the elevated heat will drive biochemical processes faster, so that a finished product can be rendered in 60 to 120 days. Aerated compost is an excellent source of macro- and micro-nutrients as well as stable organic matter, all of which support healthy plant growth. In addition, the micro-organisms in compost aid in the suppression of plant pathogens. Finally, compost retains water extremely well resulting in improved drought resistance, a longer growing season, and reduced soil erosion. In Thailand this system has been used by farmer groups for more than 445 sites (May, 2008). The process needs only 30 days to finish without turning need and 10 metric tons of compost is obtained each time. A blower (15 inch squirrel-caged blower with 3 hp motor) is needed to force the air through 10 static piles of compost twice a day and 15 minutes each. The raw materials consist of agricultural wastes and animal manure in the ratio of 3 : 1 by volume. For more information please visit www.compost.mju.ac.th/eng

Passive

Passive composting is composting in which the level of physical intervention is kept to a minimum, and often as a result the temperatures never reach much above 30°C (86 °F). It is slower but is the more common type of composting in most domestic garden compost bins. Such composting systems may be either enclosed (home container composting, industrial in-vessel composting) or in exposed piles (industrial windrow composting). Kitchen scraps are put in the garden compost bin and left untended. This scrap bin can have a very high water content which reduces aeration, and so becomes odorous. To improve drainage and airflow, a gardener can mix in wood chips, small pieces of bark, leaves or twigs, or make physical holes through the pile.

Natural

An unusual form of natural composting in nature is seen in the case of the mound-builders (megapodes) of eastern Indonesia, New Guinea, and Australia as well as in the case of bowerbirds of New Guinea and Australia. These Megapodes are fowl-sized birds famous for building nests in the form of huge compost heaps containing leaf litter, in which they incubate their eggs. The birds work constantly to maintain the correct, almost exact, incubation temperatures, by adding and removing leaves from the compost pile. In effect, this teaches us that thermophilic high-temperature composting is not man-made.

Home Composting

Home composters use a range of techniques, varying from extremely passive (throw everything in a pile and leave it for a year or two) to extremely active (monitor the temperature, turn the pile regularly, and adjust the ingredients over time). Some composters use mineral powders to absorb smells, although a well-maintained pile seldom has bad odors. It is usually located in the back garden.

Moisture and Heat

An effective compost pile is about as damp as a well wrung-out sponge. This provides the moisture that all life requires. Microorganisms vary by their ideal temperature and the heat they generate as they digest. Mesophilic bacteria survive best at temperatures of 20 to 44 °C (70 to 120 °F). thermophilic (heat-surviving) bacteria grow optimally at around 55°C (130 °F), and can attain the fastest decomposition, since metabolic processes proceed more rapidly under higher temperatures. Elevated temperature is also preferred since it causes the most rapid pathogen reduction, and is more destructive of weed seeds. To minimally achieve it, the heap should be about 1 m (3 ft) wide, 1 m (3 ft) tall, and as long as is practicable. This provides enough insulating mass to build up heat but also allows aeration. The center of the pile heats up the most.

If the pile does not heat up, common reasons include that:

- The heap is too wet, limiting the oxygen which bacteria require
- The heap is too dry for the bacteria to survive and reproduce
- There is insufficient protein (nitrogen-rich material)

The necessary material should be added, or the pile should be turned to aerate it and bring the outer layers inside and vice versa. You should add water at this time to help keep the pile damp. One guideline is to turn the pile when the high temperature has begun to drop, indicating that the food source for the fastest-acting bacteria (in the center of the pile) has been largely consumed. When turning the pile does not cause a temperature rise, it brings no further advantage. When all the material has turned into dark brown crumbly matter, it is ready to use.

Worm Composting

Worm composting or vermicomposting is a method of composting using Red Wiggler worms in a container. Food waste and moistened bedding are added, and the worms and micro-organisms eventually convert them to rich compost. The worms excrete a soil-nutrient material called worm castings. Worm composting can be done indoors, allowing year-round composting, and providing apartment dwellers with a means of composting. Worms are low in the food chain, and so are critical to healthy soil. This is why farmers have historically wanted healthy worm populations to live in their fields. The nutrients and micro-organisms can be concentrated in liquid form called worm tea, made by running distilled water through worm castings. When it is poured into the soil, the microorganisms multiply, creating a healthy growing environment for plants.

Industrial Composting

Industrial composting systems are increasingly being installed as a waste management alternative to landfills, along with other advanced waste processing systems. Industrial composting or anaerobic digestion combined with mechanical sorting of mixed waste streams is called mechanical biological treatment increasingly used in Europe due to stringent new regulations controlling the amount of organic matter allowed in landfills. Treating biodegradable waste before it enters a landfill reduces global warming from fugitive methane; untreated waste breaks down anaerobically in a landfill, producing landfill gas that contains methane, a greenhouse gas even more potent than carbon dioxide. Most commercial and industrial composting operations use active composting techniques. These ensure that the process does not get out of control especially with the high through-put demand imposed by contracted, incoming waste. This means that as short as possible a processing time must be maintained to keep the facility properly functioning (see compost windrow turner). Partly for this reason composters have declined to support compost maturity standards if it would increase the required holding time. The greatest amount of technological control of composting is seen in systems using an enclosed vessel and controlling its temperature, air flow, moisture and other parameters.

Large-scale composting systems are used by many urban centers around the world. Co-composting is a technique which combines solid waste with de-watered biosolids, which originated in the 1960s and has fallen somewhat out of favor due to difficulties controlling inert and plastic contamination from MSW. In Europe, mixed waste composting is virtually illegal. The world's largest MSW co-composter is the Edmonton Composting Facility in Edmonton in Alberta, Canada, which turns 220,000 tonnes of residential solid waste and 22,500 dry tonnes of biosolids per year into 80,000 tonnes of compost. The facility is 38,690 square metres (416,500 ft^2) large (equivalent to 4½ Canadian football fields), and the aeration building alone is the largest stainless steel building in North America, the size of 14 NHL rinks.

Biosorption

Biosorption is the binding and concentration of heavy metals from aqueous solutions (even very dilute ones) by certain types of inactive, dead, microbial biomass. Pioneering research on biosorption of heavy metals has led to the identification of a number of microbial biomass types that are extremely effective in concentrating metals. Some types of biomass are waste by products of large-scale industrial fermentations (*e.g.*, the mold *Rhizopus* or the bacterium *Bacillus subtilis*). Other metal-binding biomass

Bacillus subtilis). Other metal-binding biomass types, such as certain abundant seaweeds (particularly brown algae, *e.g.*, *Sargassum*, *Ecklonia*), can be readily harvested from the oceans. These biomass types can accumulate in excess of 25% of their dry weight in deposited heavy metals: Pb, Cd, U, Cu, Zn, Cr and others. Research on biosorption is revealing that it is sometimes a complex phenomenon where the metallic species could be deposited in the solid biosorbent through various sorption processes, such as ion exchange, complexation, chelation, microprecipitation, etc. Individuals with different backgrounds, from engineering to biochemistry, can make significant contributions to the understanding of biosorption. Interdisciplinary efforts are essential to exploit this technology commercially. A chemical engineering background is particularly useful for expanding the application of this technology in large-scale process industries.

Threat from the Environment

The greatest demand for metal sequestration today comes from the need to immobilize the metals released to the environment (or mobilized) by and partially lost through human technological activities. It has been established that dissolved metals (particularly heavy metals) escaping into the environment pose a serious health hazard. They accumulate in living tissues throughout the food chain, which has humans at its top, multiplying the danger. Thus, it is necessary to control emissions of heavy metals into the environment. Am example of one method for prioritising the recovery of ten metals is presented in Table 1. This may be simplistic, but it provides a useful direction by ranking metals into three general priority categories:

(1) Environmental Risk (ER)

(2) Reserve Depletion Rate (RDR)

(3) Combination of ER and RDR.

Environmental risk assessment could be based on a number of different factors, which could also be weighted. The Need for Novel Technology Conventional techniques to remove toxic metals and radionuclides, such as ion exchange and precipitation, lack specificity and are ineffective at low metal ion concentrations. The need for effective and economically viable technologies is driven by environmental pressures such a:

- Stricter regulations with regard to the metal discharges are being enforced, particularly in industrialised countries.
- Toxicology studies confirm the dangerous impacts of heavy metals.
- Current technologies for the removal of heavy metals from industrial effluents often create secondary problems with metal-bearing sludge.

Biosorption Mechanisms

Various metal-binding mechanisms have been postulated to be active in biosorption, such as:

- Chemisorption by ion exchange, complexation, coordination and/or chelation
- Physical Adsorption

- Microprecipitation
- Oxidation/Reduction.

Due to the complexity of the biomaterials used, it is possible that at least some of these mechanisms are acting simultaneously to varying degrees, depending on the biosorbent and the solution environment.

Ion Exchange

Ion exchange is a reversible chemical reaction wherein an ion in a solution is exchanged for a similarly charged ion attached to an immobile solid particle. These solid ion-exchange particles are either naturally occurring inorganic zeolites or synthetically produced organic resins. Synthetic organic resins are the predominant type used today because their characteristics can be tailored to specific applications.

Ion exchange reactions are stoichiometric and reversible, and as such they are similar to other solution-phase reactions. For example, in the reaction

$$NiSO_4 + Ca(OH)_2 — Ni(OH)_2 + CaSO_4$$

the nickel ions of the nickel sulfate ($NiSO_4$) are exchanged for the calcium ions of the calcium hydroxide $Ca(OH)_2$ molecule.

Chelation

The word *chelation* is derived from the Greek word *chele,* which means *claw,* and is defined as the firm binding of a metal ion with an organic molecule (ligand) to form a ring structure. The resulting ring structure protects the mineral from entering into unwanted chemical reactions. Examples include the carbonate (CO_3^{2-}) and oxalate ($C_2O_4^{2-}$) ions:

Coordination (Complex Formation)

A coordination complex is any combination of cations with molecules or anions containing free pairs of electrons. Bonding may be electrostatic, covalent or a combination of both; the metal ion is coordinately bonded to organic molecules. Example of the formation of a coordination compound are:

$$Cu^{2+} + 4H_2O \rightarrow [Cu(H_2O)]_4^{2+}$$
$$Cu^{2+} + 4Cl^- \rightarrow [CuCl_4]^{2-}$$

where coordinate covalent bonds are formed by donation of a pair of electrons from H_2O and Cl^- (Lewis bases) to Cu^{2+} (Lewis acid).

In general, biosorption of toxic metals and radionuclides is based on non-enzymatic processes such as adsorption. Adsorption is due to the non-specific binding of ionic species to polysaccharides and proteins on the cell surface or outside the cell. Bacterial cell walls and envelopes, and the walls of fungi, yeasts and algae, are efficient metal biosorbents that bind charged groups. The cell walls of gram-positive bacteria bind larger quantities of toxic metals and radionuclides than the envelopes of gram-negative bacteria.

Bacterial sorption of some metals can be described by the linearized Freundlich adsorption equation:

$$\log S = \log K + n \log C$$

where:

S is the amount of metal absorbed in μmol/g
C is the equilibrium solution concentration in μmol/L
K and n are the Freundlich constants.

Biomass deriving from several industrial fermentations may provide an economical source of biosorptive materials. Many species have cell walls with high concentrations of chitin, a polymer of *N*-acetyl-glucosamine that is an effective biosorbent.

Biosorption uses biomass raw materials that are either abundant (*e.g.*, seaweeds) or wastes from other industrial operations (*e.g.*, fermentation wastes). The metal-sorbing performance of certain types of biomass can be more or less selective for heavy metals, depending on the type of biomass, the mixture in the solution, the type of biomass preparation, and the chemical-physical environment.

It is important to note that the concentration of a specific metal in solution can be reduced either during the sorption uptake by manipulating the properties of the biosorbent or upon desorption during the regeneration cycle of the biosorbent.

Sources of Biomass for Biosorption

Sources of biomass include:

- Seaweeds
- Microorganisms (bacteria, fungi, yeast, molds)
- Activated sludge
- Fermentation waste
- Other specially propagated biomasses.

Biosorbents must be hard enough to withstand the application pressures, porous and/or "transparent" to metal ion sorbate species, and have high and fast sorption uptake even after repeated regeneration cycles. Granulation of biomass materials into suitable cost-effective biosorbents is a crucial step for the successful application of biosorption processes. The objectives of granulation are to:

- Establish the behavior of native biomass in a packed-bed reactor
- Establish the effectiveness of biomass granulation and reinforcement
- Determine the effect of size reduction on sorption capacity
- Determine the feasibility of biomass processing.

Conventional granulation technologies are rather advanced, and their adaptation will likely yield desirable biosorbent granules. Because of the wide variety of biomass types, extensive experimentation will undoubtedly be required. The need to transport raw biomass may also present some logistical problems. Microbial biomass has a high water content and is prone to decay, so drying may be required if it cannot be processed and/or granulated directly on location in the wet state.

Equilibrium Modeling

Biosorption has been studied as simplified sorption systems, usually containing one heavy metal. This is an appropriate simplification for effective experimentation. Sumarizes some of the simple sorption isotherm models that are most frequently applied. A particular model may not apply to a particular situation, and in some cases more than one model may explain the biosorption mechanism. There is no critical reason to use a more-complex model if a two-parameter model (such as the Langmuir and Freundlich isotherm models) can fit the data reasonably well.

where:

C_e is the equilibrium solute concentration in the fluid

K,n are the Freundlich isotherm constants

a_i, b_i are the Langmuir isotherm parameters

e is the column bed porosity; Polanyi's adsorption potential

q_m is the Langmuir maximum metal uptake in mg/g

W_o, W_m are the initial and final volumes, respectively, in L

Q is the meta uptake in mg/g

b is the Polanyi scaling factor in the Polanyi models.

The sorption uptake, q, can be expressed in different units depending on the purpose of the exercise:

- For practical and engineering process evaluation purposes eventually concerned with process mass balances, it is customary to use weight per (dry) weight (*e.g.*, mg of metal sorbed per gram of the (dry) sorbent material).
- Ultimately, mainly because of reactor volume considerations (*e.g.*, a packed-bed column), the uptake may also be expressed on a per volume basis (*e.g.*, mg/L). However, the porosity may complicate the quantitative comparison of biosorption performance.
- Only when working on the stoichiometry of the process and when studying the functional groups and metal-binding mechanisms might it be useful to express q on a molar or charge equivalent basis — again, per unit weight or volume of the sorbent (*e.g.*, mmol/g or mequiv/g).

It is relatively easy to convert among these units; the only problem may arise with the sorbent weight-volume conversions. For scientific interpretations, the sorbent material dry-weight basis is thus preferred. The use of "wet biomass weight" should be discouraged, unless the wet-weight-to-dry-weight conversion is well specified. Different biomass types are likely to retain different moisture contents, intracellular as well as that trapped in the interstitial space between the cells or tissue particles (*e.g.*, seaweed particles). Different types of biomass obviously compact in a different ways. When centrifuging biomass, the g-force and time need to be specified, and even then it is difficult to make any comparisons. All this makes the "wet biomass weight" citation very approximate at best and generally undesirable.

Biomass Types

The assessment of the metal-binding capacity of some types of biomass has gained momentum since 1985. Indeed, some biomass types are very effective in accumulating heavy metals. Availability is a major factor to be taken into account to select biomass for clean-up purposes. The economics of environmental remediation dictate that the biomass must come from nature, or even be a waste material. Seaweeds, molds, yeasts, bacteria, and crab shells, among other kinds of biomass (3), have been tested for metal biosorption with very encouraging results.

Biosorbents

Some biosorbents can bind and collect a wide range of heavy metals with no specific priority, whereas others are specific for certain types of metals. When choosing the biomass for metal biosorption experiments, its origin is a major factor to be considered. Biomass can come from:

- industrial wastes which should be obtained free of charge

- organisms that can be obtained easily in large amounts in nature (*e.g.*, bacteria, yeast, algae)
- fast-growing organisms that are specifically cultivated or propagated for biosorption purposes (crab shells, seaweeds).

Organisms for Biosorption

There is a wide variety of microorganisms (Table 4), including bacteria, fungi, yeast, and algae, that can interact with metals and radionuclides and transform them through several mechanisms.

Cost-effectiveness is the main attraction of metal biosorption. This cost-effectiveness can be maintained by using the microbial biomass directly where possible. In addition, biosorbents derived from microbial biomass through a simple process are expected to be the lowest-priced and most-economical for metal removal. It has been suggested that numerous chemical groups contribute to biosorption metal binding, by either whole organisms such as algae and bacteria or by molecules such as biopolymers. These include hydroxyl, carbonyl, carboxyl, sulfhydryl, thioether, sulfonate, amine, imine, amide, imidazole, phosphonate, and phosphodiester groups. The importance of any given group for biosorption of a certain metal by a certain biomass depends on such factors as the number of sites in the biosorbent material, the accessibility of the sites, the chemical state of the sites (*i.e.*, availability), and the affinity between the site and the metal (*i.e.*, binding strength). For covalent metal binding, even an occupied site is theoretically available; the extent to which the site can be used by a given metal depends on its binding strength and concentration compared to the metal already occupying the site. Some types of industrial fermentation waste biomass are excellent metal sorbers. It is necessary to realise that some "waste" biomass is actually a commodity, not a waste. This applies particularly to the ubiquitous brewer's yeasts sold on the open market, usually as animal fodder. Activated sludge from wastewater treatment plants has not demonstrated high enough metal-sorbing capacities. Some types of seaweed biomass offer excellent metal-sorbing properties, and sometimes a local economy can benefit from turning seaweeds into a resource. As a fallback, biomass with a high metal-sorbing capacity can be specifically grown relatively cheaply in fermenters using low-cost or even waste carbohydrate-containing growth media such as molasses or cheese whey.

Experimental Sorption Isotherms

It is relatively easy to obtain equilibrium sorption data for a single sorbate in the laboratory. A small amount of the sorbent is brought into contact with a solution containing the sorbate of interest. The conditions of the sorption system, particularly pH, must be carefully controlled at the required values over the entire period of contact until the sorption equilibrium is reached. This may take a few hours or much longer, depending on the size of the sorbent particles and the time it takes until they attain sorption equilibrium. A simple preliminary sorption kinetics test will establish the exposure time necessary for the given sorbent particles to reach the equilibrium state. The following procedure provides an example for obtaining the experimental sorption equilibrium data points for the isotherm:

1. Prepare the sorbatc in solution at the highest concentration of interest.
2. Prepare dilutions covering the entire concentration range, from 0 (blank) to the maximum.
3. Adjust the conditions, *e.g.*, pH, ionic strength, etc.
4. Determine the sorbate initial concentrations (C_i) in all the liquid samples.
5. Distribute the samples into containers of appropriate volumes (30–150 mL of liquid) such as flasks or test tubes; prepare samples in duplicate, triplicate or as required.
6. Accurately weigh each quantity of the biosorbent solids to be used in the tests and record the weights (*S*, mg). It may help to be able to roughly estimate the anticipated sorption uptake so that there is an easily detectable final sorbate concentration in each sample solution at equilibrium. If too much sorbent is added, there may be virtually no sorbate left in the solution, precluding a reliable analysis. Varying the initial concentration could cause the sorbent weight to fluctuate, which has to be precisely known for each sample. Metal depletion in the solution must be avoided because it renders such samples useless.
7. Add the sorbent solids into each sample solution and provide rather gentle mixing over the contact period.
8. Make sure the conditions (especially pH) are controlled at constant values during the contact period. Use an appropriate acid or base for this; do not dilute the sorption system by adding excessive volume.
9. At the end of the contact period, separate the solids from the liquid by decantation, filtration, centrifugation, etc.
10. Analyse the liquid portion to determine the residual final sorbate concentration (C_f).
11. Calculate the sorbate uptake: $q = V$ [L] ´ $(C_i - C_f)$ [mg/L] / S [g]. Note that q could also be determined directly by analysing the separated solids and thus closing the material balance on the sorbate in the system. However, this usually presents analytical difficulties (digestion-liquefaction of solids, and/or very sophisticated analytical methods may be required).
12. Plot the sorption isotherm q vs. (C_f).

Comparison of Sorption Performance

The performance of sorbing materials needs to be evaluated and often compared. The simplest situation is when there is only one sorbate species in the system, in which case it is best to base the single-sorbate sorption performance on a complete single-sorbate sorption isotherm curve. To fairly compare two or more sorbents, the comparison must be done under uniform conditions. These may be restricted by the environmental factors under which sorption may have to take place (pH, temperature, ionic strength, etc.), which may not necessarily be easily or widely adjustable. In particular, it is important to compare sorption performance under the same pH conditions, since isotherms can vary with pH. The performance of the sorbent is usually gauged by its uptake (q). Sorbents can be compared based on their respective maximum uptake values (q_{max}), which can be calculated by fitting the Langmuir isotherm model to the actual experimental data (if it fits). This approach is feasible if q_{max} reaches a plateau. Some isotherms might not exhibit the asymptotic plateau represented by the Langmuir equation.

In general, one is looking for a "good" sorbent with a high sorption uptake capacity (q_{max}). Surface area in biosorption is not particularly important.

Types of Biosorption

Biosortpion can be carried out as a batch process, a continuous process, or a two-stage process with continuous metal recovery. Biomass should be defrosted and washed with deionized water. To ensure equal quality of the biomass during all experiments, different kinds of biomass should be mixed together to obtain a uniform mixture.

Batch Process

Batch biosorption experiments can be done in a stirred vessel (4) with a working volume of approximately 100 mL. A minimal amount of concentrated solution of $Pb(NO_3)_2$ (metal) can be added into a suspension of fungal pellets in water of various concentrations (25, 50, 100, 150 and 200 g of wet biomass per L of biomass suspension) to produce the desired initial metal concentrations of 10, 20, 50, 100 and 300 mg/L Pb^{2+} (metal). The decreasing metal concentration can be recorded as a function of the initial metal concentration and the biomass loading.

Continuous Process

Continuous process experiments can be carried out in a glass column having an inner diameter of 5–8 cm and filled with a packed bed of biomass pellets of varying heights (20, 40 and 55 cm), set with an adjustable plug. The effluent solution of metal ions can be fed from the top of the column with the help of a pump using varying flowrates. An inert bed of glass spheres can be placed at the bottom of the column below the active biomass bed to ensure homogenous distribution of the feed. The remaining metal concentration can be measured online in the effluent at the top of the column. The breakthrough curves can be recorded as a function of the flowrate and bed height. Measurements of metal ion concentrations in the solution can be made online with metal-detecting electrodes or ion-selective electrodes, and may be verified with an atomic absorption spectrometer.

Two-Stage Process with Continuous Metal Recovery

Two-stage continuous biosorption and metal recovery can be carried out as shown . This process is similar to continuous biosorption, although the metal solution is adsorbed in two stages. After initial adsorption and filtration in stage one, the effluent is fed with fresh biosorbent into stage two, where further biosorption of the metal ion takes place. The effluent from the second stage is filtered to recover the metal ions and biosorbents. The effluent sample can be analysed using an ion meter or by adsorption spectroscopy.

Desorption

Regeneration of loaded biosorbent is critical to keeping costs down and to recovering the metal(s) extracted from the liquid phase. The deposited metals are washed out (desorbed) and the biosorbent is regenerated for another cycle. The desorption process should result in:

- high-concentration metal effluent
- undiminished metal uptake upon re-use
- no physico-chemical damage to the biosorbent.

The desorption and sorbent regeneration studies might require somewhat different methodologies, beginning with screening for the most effective regenerating solution. Because different metal ions have different affinities for the biosorbent, the uptake has some degree of metal selectivity. The selectivity of the elution-desorption operation may be different, which may serve as another means of eventually separating metals from one another if desirable. The concentration ratio (*CR*) is used to evaluate the overall concentration effectiveness of the whole sorption-desorption process: Obviously, the higher the *CR,* the better the overall performance of the sorption process, making the eventual recovery of the metal more feasible with higher eluate concentrations. Recovery of the metal from these concentrated desorption solutions is carried out in a different plant by electrowinning. Following desorption of the metal(s), the column may be further pre-treated (*e.g.,* pre-saturated with protons such as Ca, K, etc.) for optimum operation in the next metal uptake cycle.

The specific type of pre-treatment used to optimize the column performance may vary.

Feasibility of Biosorption

For successful application on a large scale, any operation needs to be economically viable. The feasibility of a biosorption process depends on such factors as:

- biosorbent uptake performance
- the source of the raw biomass
- biomass granulation and treatment
- the desorption and regeneration processes used.

Often, the source of the biosorbent has a major impact on the feasibility of the operation. Biosorbents (biomass) should always be obtained from the least-expensive source, such as from the effluent of a fermenter, seaweeds from nearby bodies of water, algae, etc. The spent biosorbents can be regenerated at very low cost using water, so the material can be reused many times. Hence, considering the overall unit operations involved in biosorption, we can conclude that the process is generally economically viable.

Advantages of Biosorption

Biosorption is highly competitive with the presently available technologies like ion exchange, electrodialysis, reverse osmosis, etc. Some of the key features of biosorption compared to conventional processes include:

- competitive performance
- heavy metal selectivity
- cost-effectiveness
- regenerative
- no sludge generation.

Biosorption is particularly economical and competitive for environmental applications in detoxifying effluents from, for example:

- metal plating and metal finishing operations
- mining and ore processing operations
- metal processing
- battery and accumulator manufacturing operations

- thermal power generation (coal-fired plants in particular)
- nuclear power generation.

There appear to be many modes of non-active metal uptake by microbial biomass. Any one or a combination of them can be functional in immobilising metallic species on biosorbents. A number of anionic ligands participate: phosphoryl, carbonyl, sulfhydryl and hydroxyl groups can all be active to various degrees in binding the metal. Many scientific studies are currently underway to provide a deeper understanding of biosorption and to support its effective application. Some pollution seems inevitable, and one might wonder what should be done to minimize it. Human populations need methods and technologies to clean waters and diminish the environmental dangers related to technological progress. Biosorption can be one such solution to clean up heavy metal contamination.

Biosorption of Heavy Metals

The discharge of heavy metals into aquatic ecosystems has become a matter of concern in India over the last few decades. These pollutants are introduced into the aquatic systems significantly as a result of various industrial operations. Industrialisation in India gained a momentum with initiation of five year developmental plan in the early 50's. The pollutants of concern include lead, chromium, mercury, uranium, selenium, zinc, arsenic, cadmium, gold, silver, copper and nickel. These toxic materials may be derived from mining operations, refining ores, sludge disposal, fly ash from incinerators, the processing of radioactive materials, metal plating, or the manufacture of electrical equipment, paints, alloys, batteries, pesticides or preservatives. Heavy metals such as zinc, lead and chromium have a number of applications in basic engineering works, paper and pulp industries, leather tanning, organochemicals, petrochemicals fertlisers, etc. Major lead pollution is through automobiles and battery manufacturers. For zinc and chromium the major application is in fertliser and leather tanning respectively (Trivedi, 1989). Over the few decades, several methods have been devised for the treatment and removal of heavy metals.

The commonly used procedures for removing metal ions from aqueous streams include chemical precipitation, lime coagulation, ion exchange, reverse osmosis and solvent extraction (Rich and Cherry, 1987). The process description of each method is presented below. Reverse Osmosis: It is a process in which heavy metals are separated by a semi-permeable membrane at a pressure greater than osmotic pressure caused by the dissolved solids in wastewater. The disadvantage of this method is that it is expensive. Electrodialysis: In this process, the ionic components (heavy metals) are separated through the use of semi-permeable ion-selective membranes. Application of an electrical potential between the two electrodes causes a migration of cations and anions towards respective electrodes. Because of the alternate spacing of cation and anion permeable membranes, cells of concentrated and dilute salts are formed. The disadvantage is the formation of metal hydroxides, which clog the membrane.

Ultrafiltration: They are pressure driven membrane operations that use porous membranes for the removal of heavy metals. The main disadvantage of this process is the generation of sludge.

Ion-exchange: In this process, metal ions from dilute solutions are exchanged with ions held by electrostatic forces on the exchange resin. The disadvantages include: high cost and partial removal of certain ions.

Chemical Precipitation: Precipitation of metals is achieved by the addition of coagulants such as alum, lime, iron salts and other organic polymers. The large amount of sludge containing toxic compounds produced during the process is the main disadvantage.

Phytoremediation: Phytoremediation is the use of certain plants to clean up soil, sediment, and water contaminated with metals. The disadvantages include that it takes a long time for removal of metals and the regeneration of the plant for further biosorption is difficult.

Hence the disadvantages like incomplete metal removal, high reagent and energy requirements, generation of toxic sludge or other waste products that require careful disposal has made it imperative for a cost-effective treatment method that is capable of removing heavy metals from aqueous effluents.

Biosorption

The search for new technologies involving the removal of toxic metals from wastewaters has directed attention to biosorption, based on metal binding capacities of various biological materials. Biosorption can be defined as the ability of biological materials to accumulate heavy metals from wastewater through metabolically mediated or physico-chemical pathways of uptake (Fourest and Roux, 1992). Algae, bacteria and fungi and yeasts have proved to be potential metal biosorbents (Volesky, 1986). The major advantages of biosorption over conventional treatment methods include (Kratochvil and Volesky, 1998 a):

- Low cost;
- High efficiency;
- Minimisation of chemical and lor biological sludge;
- No additional nutrient requirement;

- Regeneration of biosorbent; and
- Possibility of metal recovery.

The biosorption process involves a solid phase (sorbent or biosorbent; biological material) and a liquid phase (solvent, normally water) containing a dissolved species to be sorbed (sorbate, metal ions). Due to higher affinity of the sorbent for the sorbate species, the latter is attracted and bound there by different mechanisms. The process continues till equilibrium is established between the amount of solid-bound sorbate species and its portion remaining in the solution. The degree of sorbent affinity for the sorbate determines its distribution between the solid and liquid phases.

Biosorbent material: Strong biosorbent behaviour of certain micro-organisms towards metallic ions is a function of the chemical make-up of the microbial cells. This type of biosorbent consists of dead and metabolically inactive cells.

Some types of biosorbents would be broad range, binding and collecting the majority of heavy metals with no specific activity, while others are specific for certain metals. Some laboratories have used easily available biomass whereas others have isolated specific strains of microorganisms and some have also processed the existing raw biomass to a certain degree to improve their biosorption properties;

Recent biosorption experiments have focused attention on waste materials, which are by-products or the waste materials from large-scale industrial operations. For e.g. the waste mycelia available from fermentation processes, olive mill solid residues (Pagnanelli, et al 2002), activated sludge from sewage treatment plants (Hammaini et aI. 2003), biosolids (Norton et al 2003), aquatic macrophytes (Keskinkan et aI. 2003), etc. Norton et aI. 2003, used dewatered waste activated sludge from a sewage treatment plant for the biosorption of zinc from aqueous solutions. The adsorption capacity was determined to be 0.564 mM/g of biosolids. The use of biosolids for zinc adsorption was favourable compared to the bioadsorption rate of 0.299 mM/g by the seaweed *Durvillea potatorum* (Aderhold et aI. 1996). Keskinkan et al. 2003 studied the adsorption characteristics of copper, zinc and lead on submerged aquatic plant *Myriophyllum spicatum.* The adsorption capacities were 46.69 mg/g for lead, 15.59 mg/g for zinc and 10.37 mg/g for copper. Table 1 gives a comparison of heavy metal uptakes of various macrophytes.

Pagnanelli, et al 2002 have carried out a preliminary study on the 'Use of oli ve mill residues as hea vy metal sorbent material The results revealed that copper was maximally adsorbed in the range of 5.0 to 13.5 mg/g under different operating conditions.

The simultaneous biosorption capacity of copper, cadmium and zinc on dried activated sludge (Hammaini et al. 2003) were 0.32 mmol/g for metal system such as Cu-Cd; 0.29 mmol/g for Cu-Zn and 0.32 mmol/g for Cd-Zn. The results showed that the biomass had a net preference for copper followed by cadmium and zinc. Another inexpensive source of biomass where it is available in copious quantities is in oceans as seaweeds, representing many different types of marine macro-algae. However most of the contributions studying the uptake of toxic metals by live marine and to a lesser extent freshwater algae focused on the toxicological aspects, metal accumulation, and pollution indicators by live, metabolically active biomass. Focus on the technological aspects of metal removal by algal biomass has been rare. Although abundant natural materials of cellulosic nature have been suggested as biosorbents, very less work has been actually done in that respect. The mechanism of biosorption is complex, mainly ion exchange, chelation, adsorption by physical forces, entrapment in inter and intrafibrilliar capillaries and spaces of the structural polysaccharide network as a result of the concentration gradient and diffusion through cell walls and membranes.

There are several chemical groups that would attract and sequester the metals in biomass: acetamido groups of chitin, structural polysaccharides of fungi, amino and phosphate groups in nucleic acids, amido, amino, sulphhydryl and carboxyl groups in proteins, hydroxyls in polysaccharide and mainly carboxyls and sulphates in polysaccharides of marine algae that belong to the divisions Phaeophyta, Rhodophyta and Chlorophyta. However, it does not necessarily mean that the presence of some functional group guarantees biosorption, perhaps due to steric, conformational or other barriers.

Choice of metal for biosorption process: The appropriate selection of metals for biosorption studies is dependent on the angle of interest and the impact of different metals, on the basis of which they would be divided into four major categories: (i) toxic heavy metals (ii) strategic metals (iii) precious metals and (iv) radio nuclides. In terms of environmental threats, it is mainly categories (i) and (iv) that are of interest for removal from the environment and/or from point source effluent discharges.

Apart from toxicological criteria, the interest in specific metals may also be based on how representative their behaviour may be in terms of eventual generalisation of results of studying their biosorbent uptake. The toxicity and interesting solution chemistry of elements such as chromium, arsenic and selenium make them interesting to study. Strategic and precious metals though not environmentally threatening are important from their recovery point of view.

Mechanisms

Biosorption Mechanisms: The complex structure of microorganisms implies that there are many ways for the metal to be taken up by the microbial cell. The biosorption mechanisms are various and are not fully understood. They may be classified according to various criteria.

According to the dependence on the cell's metabolism, biosorption mechanisms can be divided into:

1. Metabolism dependent and
2. Non -metabolism dependent.

According to the location where the metal removed from solution is found, biosorption can be classified as

1. Extra cellular accumulation/ precipitation
2. Cell surface sorption/ precipitation and
3. Intracellular accumulation.

Transport of the metal across the cell membrane yields intracellular accumulation, which is dependent on the cell's metabolism. This means that this kind of biosorption may take place only with viable cells. It is often associated with an active defense system of the microorganism, which reacts in the presence of toxic metal.

During non-metabolism dependent biosorption, metal uptake is by physico-chemical interaction between the metal and the functional groups present on the microbial cell surface. This is based on physical adsorption, ion exchange and chemical sorption, which is not dependent on the cells' metabolism. Cell walls of microbial biomass, mainly composed of polysaccharides, proteins and lipids have abundant metal binding groups such as carboxyl, sulphate, phosphate and amino groups. This type of biosorption, i.e., non-metabolism dependent is relatively rapid and can be reversible (Kuyucak and Volesky, 1988).

In the case of precipitation, the metal uptake may take place both in the solution and on the cell surface (Ercole, et al. 1994). Further, it may be dependent on the cell's' metabolism if, in the presence of toxic metals, the microorganism produces compounds that favour the precipitation process. Precipitation may not be dependent on the cells' metabolism, if it occurs after a chemical interaction between the metal and cell surface.

Transport across cell membrane: Heavy metal transport across microbial cell membranes may be mediated by the same mechanism used to convey metabolically important ions such as potassium, magnesium and sodium. The metal transport systems may become confused by the presence of heavy metal ions of the same charge and ionic radius associated with essential ions. This kind of mechanism is not associated with metabolic activity. Basically biosorption by living organisms comprises of two steps. First, a metabolism independent binding where the metals are bound to the cell walls and second, metabolism dependent intracellular uptake, whereby metal ions are transported across the cell membrane. (Costa, et.al., 1990, Gadd et.al., 1988, Ghourdon et.al., 1990, Huang et.al., 1990., Nourbaksh et.al., 1994)

Physical adsorption: In this category, physical adsorption takes place with the help of van der Waals' forces. Kuyucak and Volesky 1988, hypothesized that uranium, cadmium, zinc, copper and cobalt biosorption by dead biomasses of algae, fungi and yeasts takes place through electrostatic interactions between the metal ions in solutions and cell walls of microbial cells. Electrostatic interactions have been demonstrated to be responsible for copper biosorption by bacterium *Zoogloea ramigera* and alga *Chiarella vulgaris* (Aksu et al. 1992), for chromium biosorption by fungi *Ganoderma lucidum* and *Aspergillus niger* .

Ion Exchange: Cell walls of microorganisms contain polysaccharides and bivalent metal ions exchange with the counter ions of the polysaccharides. For example, the alginates of marine algae occur as salts of K+, Na+, Ca^2+, and Mg^2+. These ions can exchange with counter ions such as CO^2+, Cu^2+, Cd^2+ and Zn^2+ resulting in the biosorptive uptake of heavy metals (Kuyucak and Volesky 1988). The biosorption of copper by fungi *Ganoderma lucidium* (Muraleedharan and Venkobachr, 1990) and *Aspergillus niger* was also up taken by ion exchange mechanism.

Complexation: The metal removal from solution may also take place by complex formation on the cell surface after the interaction between the metal and the active groups. Aksu et al. 1992 hypothesized that biosorption of copper by C. *vulgaris* and Z. *ramigera* takes place through both adsorption and formation of coordination bonds between metals and amino and carboxyl groups of cell wall polysaccharides. Complexation was found to be the only mechanism responsible for calcium, magnesium, cadmium, zinc, copper and mercury accumulation by *Pseudomonas syringae.* Micro-organisms may also produce organic acids (e.g., citric, oxalic, gluonic, fumaric, lactic and malic acids), which may chelate toxic metals resulting in the formation of metallo-organic molecules. These organic acids help in the solubilisation of metal compounds and their leaching from their surfaces. Metals may be biosorbed or complexed by carboxyl groups found in microbial polysaccharides and other polymers.

Precipitation: Precipitation may be either dependent on the cellular metabolism or independent of it. In the former case, the metal removal from solution is often associated with active defense system of the microorganisms. They react in the presence of a toxic metal producing compounds, which favour the precipitation process. In the case of precipitation not dependent on the cellular metabolism, it may be a consequence of the chemical interaction between the metal and the cell surface. The various biosorption mechanisms mentioned above can take place simultaneously.

Use of Recombinant bacteria for metal removal: Metal removal by adsorbents from water and wastewater is strongly influenced by physico-chemical parameters such as ionic strength, pH and the concentration of competing organic and inorganic compounds. Recombinant bacteria are being investigated for removing specific metals from contaminated water. For example a genetically engineered *E.coli,* which expresses Hg^{2+} transport system and metallothionin (a metal binding protein) was able to selectively accumulate 8 mmole Hg^{2+}/g cell dry weight. The presence of chelating agents Na^{+}, Mg^{2+} and Ca^{2+} did not affect bioaccumulation.

Factors affecting Biosorption : The investigation of the efficacy of the metal uptake by the microbial biomass is essential for the industrial application of biosorption, as it gives information about the equilibrium of the process which is necessary for the design of the equipment.

The metal uptake is usually measured by the parameter 'q' which indicates the milligrams of metal accumulated per gram of biosorbent material and 'qH' is reported as a function of metal accumulated, sorbent material used and operating conditions.

The following factors affect the biosorption process:

1. Temperature seems not to influence the biosorption performances in the range of 20-35 ^{0}C (Aksu et al. 1992)
2. pH seems to be the most important parameter in the biosorptive process: it affects the solution chemistry of the metals, the activity of the functional groups in the biomass and the competition of metallic ions (Friis and Myers-Keith, 1986, Galun et al. 1987)
3. Biomass concentration in solution seems to influence the specific uptake: for lower values of biomass concentrations there is an increase in the specific uptake (Fourest and Roux, 1992; Gadd et al. 1988). Gadd et al. 1988 suggested that an increase in biomass concentration leads to interference between the binding sites. Fourest and Roux, 1992 invalidated this hypothesis attributing the responsibility of the specific uptake decrease to metal concentration shortage in solution. Hence this factor needs to be taken into consideration in any application of microbial biomass as biosorbent.
4. Biosorption is mainly used to treat wastewater where more than one type of metal ions would be present; the removal of one metal ion may be influenced by the presence of other metal ions. For example: Uranium

uptake by biomass of bacteria, fungi and yeasts was not affected by the presence of manganese, cobalt, copper, cadmium, mercury and lead in solution (Sakaguchi and Nakajima, 1991). In contrast, the presence of Fe^{2+} and Zn^{2+} was found to influence uranium uptake by *Rhizopus arrhizus* (Tsezos and Volesky, 1982) and cobalt uptake by different microorganisms seemed to be completely inhibited by the presence of uranium, lead, mercury and copper (Sakaguchi and Nakajima, 1991).

Biosorption equilibrium models - Assessment of sorption performance: Examination and preliminary testing of solid-liquid sorption system are based on two types of investigations: (a) equilibrium batch sorption tests and (b) dynamic continuous flow sorption studies. The equilibrium of the biosorption process is often described by fitting the experimental points with models (Gadd, et al. 1988) usually used for the representation of isotherm adsorption equilibrium.

But the above said adsorption isotherms may exhibit an irregular pattern due to the complex nature of both the sorbent material and its varied multiple active sites, as well as the complex solution chemistry of some metallic compounds (Volesky and Holan, 1995). Evaluation of equilibrium sorption performance needs to be supplemented by process-oriented studies of its kinetics and eventually by dynamic continuous flow tests. Biosorption by immobilized cells: Microbial biomass consists of small particles with low density, poor mechanical strength and little rigidity. The immobilisation of the biomass in solid structures Qeates a material with the right size, mechanical strength and rigidity and porosity necessary for metal accumulation. Immobilisation can also yield beads and granules that can be stripped of metals, reactivated and re-used in a manner similar to ion exchange resins and activated carbon. Various applications are available for biomass immobilisation. The principal techniques that are available in literature for the application of biosorption are based on adsorption on inert supports, on entrapment in polymeric matrix, on covalent bonds in vector compounds, or on cell cross-linking. Adsorption on inert supports: Support materials are introduced prior to sterilisation and inoculation with starter culture and are left inside the continuous culture for a period oftime, after which a film of microorganisms is apparent on the support surfaces. This technique has been used by Zhou and Kiff, 1991 for the immobilisation of *Rhizopus arrhizus* fungal biomass in reticulated foam biomass support particles; Macaskie et al.. 1987, immobilised the bacterium *Citrobacter sp.* by this technique. Scott and Karanjakar 1992, used activated carbon as a support for *Enterobacter aerogens* biofilm. Bai and Abraham, 2003 immobilized *Rhizopus nigricans* on polyurethane foam cubes and coconut fibres.

Entrapment in polymeric matrices: The polymers used are calcium alginate (Babu et al. 1993, Costa and Leite, 1991, Peng and Koon, 1993, Gulay Bayramoglu et al. 2002), polyacrylamide (Macaskie et aI., 1987, Michel et al. 1986, Sakaguchi and Nakajima et al. 1991, Wong and Kwok, 1992), polysulfone (Jeffers et al. 1991, Bai and Abraham, 2003) and polyethylenimine (Brierley and Brierley, 1993). The materials obtained from immobilisation in calcium alginate and polyacrylamide are in the form of gel particles. Those obtained from immobilisation in polysulfone and polyethyleneimine are the strongest. Covalent bonds to vector compounds: The most common vector compound (carrier) is silica gel. The material obtained is in the form of gel particles. This technique is mainly used for algal immobilisation (Holan et al. 1993, Mah:!mn and Holocombe, 1992). Cross-linking: The addition of the cross-linker leads to the formation of stable cellular aggregates. This technique was found useful for the immobilisation of algae (Holan et al. 1993). The most common cross linkers are: formaldehyde, glutaric dialdehyde, divinylsulfone and formaldehyde - urea mixtures.

Desorption: If the biosorption process were to be used as an alternative to the wastewater treatment scheme, the regeneration of the biosorbent may be crucially important for keeping the process costs down and in opening the possibility of recovering the metals extracted from the liquid phase. For this purpose it is desirable to desorb the sorbed metals and to regenerate the biosorbent material for another cycle of application.

Bioremediation

Bioremediation can be defined as any process that uses microorganisms, fungi, green plants or their enzymes to return the natural environment altered by contaminants to its original condition. Bioremediation may be employed to attack specific soil contaminants, such as degradation of chlorinated hydrocarbons by bacteria. An example of a more general approach is the cleanup of oil spills by the addition of nitrate and/or sulfate fertilisers to facilitate the decomposition of crude oil by indigenous or exogenous bacteria.

Overview and Applications

Naturally occurring bioremediation and phytoremediation have been used for centuries. For example, desalination of agricultural land by phytoextraction has a long tradition. Bioremediation technology using microorganisms was reportedly invented by George M. Robinson. He was the assistant county petroleum engineer for Santa Maria, California. During the 1960's, he spent his spare time experimenting with dirty jars and various mixes of microbes. Bioremediation technologies can be generally classified as *in situ* or *ex situ*. *In situ* bioremediation involves treating the contaminated material at the site while *ex situ* involves the removal of the contaminated material to be treated elsewhere. Some examples of bioremediation technologies are bioventing, landfarming, bioreactor, composting, bioaugmentation, rhizofiltration, and biostimulation. Not all contaminants, however, are easily treated by bioremediation using microorganisms. For example, heavy metals such as cadmium and lead are not readily absorbed or captured by organisms. The assimilation of metals such as mercury into the food chain may worsen matters. Phytoremediation is useful in these circumstances, because natural plants or transgenic plants are able to bioaccumulate these toxins in their above-ground parts, which are then harvested for removal. The heavy metals in the harvested biomass may be further concentrated by incineration or even recycled for industrial use. The elimination of a wide range of pollutants and wastes from the environment is an absolute require increasing our understanding of the relative importance of different pathways and regulatory networks to carbon flux in particular environments and for particular compounds and they will certainly accelerate the development of bioremediation technologies and biotransformation processes.

Genetic Engineering Approaches

The use of genetic engineering to create organisms specifically designed for bioremediation has great potential. The bacterium Deinococcus radiodurans (the most radioresistant organism known) has been modified to consume and digest toluene and ionic mercury from highly radioactive nuclear waste.

Advantages

There are a number of cost/efficiency advantages to bioremediation, which can be employed in areas that are inaccessible without excavation. For example, hydrocarbon spills (specifically, petrol spills) or certain chlorinated solvents may contaminate groundwater, and introducing the appropriate electron acceptor or electron donor amendment, as appropriate, may significantly reduce contaminant concentrations after a lag time allowing for acclimation. This is typically much less expensive than excavation followed by disposal elsewhere, incineration or other *ex situ* treatment strategies, and reduces or eliminates the need for "pump and treat", a common practice at sites where hydrocarbons have clean groundwater.

Monitoring Bioremediation

The process of bioremediation can be monitored indirectly by measuring the *Oxidation Reduction Potential* or redox in soil and groundwater, together with pH, temperature, oxygen content, electron acceptor/donor concentrations, and concentration of breakdown products (e.g. carbon dioxide). This table shows the (decreasing) biological breakdown rate as function of the redox potential. This, by itself and at a single site, gives little information about the process of remediation.

1. it is necessary to sample enough points on and around the contaminated

site to be able to determine contours of equal redox potential. Contouring is usually done using specialised software, e.g. using Kriging interpolation.

2. if all the measurements of redox potential show is that electron acceptors have been used up, it's in effect an indicator for total microbial activity. Chemical analysis is also required to determine when the levels of contaminants and their breakdown products have been reduced to below regulatory limits.

Bioremediation

Types of Bioremediation

There are two main types of bioremediation

In-Situ Remediation

In-situ Bioremediation treats the contaminated soil or groundwater in the location in which it is found. In this technology oxygen and occasionally nutrients are pumped under pressure into the soil through wells. The nutrients are spread on the surface to infiltrate into the contaminated area of material or the saturated zone.

Ex-Situ Remediation

Ex-situ Bioremediation requires pumping of the groundwater or excavation of contaminated soil prior to remediation treatments. Ex-situ Bioremediation can be further broken down into two main components or processes; Slurry-phase and solid-phase treatment. Slurry-phase- This treatment involves the initial combination of water with the contaminated soil and later degradation in a bioreactor. Solid-phase- This treatment achieves the similar goal of the former treatment yet, in this process, the contaminated soil is placed in a bed and nourished with nutrients, moisture and oxygen in hopes that decomposition will occur.

Bioremediation Services

Bioremediate.com uses a range of microbial probiotic products that are suitable for the bioremediation of terrestrial, wetland, marine, farm land and industrial areas and is specifically suited to the restoration of natural ecological environments. The pace of the Bioremediation Industry has moved quickly. From the discovery that natural indigenous microbes had the ability to remediate pollutants, to the identification, isolation and commercial production of viable products, it continues to progress. Manufacturers are now able to isolate and mass-produce standardised bacteria and fungi in concentrated liquid formats and dry endosporic (dormant) state, a tremendous advancement from the 1st generation liquids.

Recently, traceable scientific studies for these new, standardised, CFU count microbial products have been conducted and demonstrate the efficacy of microbes for the safe remediation of numerous contaminants. Remediation is achieved by using a combination of microbial products that remove hydrocarbon spills, biodegrades organic pollutants, reduce phosphorous levels and then, by adding microbial supplements and biostimulants, promote natural soil conditions. This process of bio-augmentation enhances the viability of all planting, minimizes transplant shock, encourages germination and regrowth to the treated areas.

Our team of scientists will formulate a product line and methodology specific to customer requirements. This assures that you will achieve project closure approvals more quickly and economically than most mechanical or hard chemical technologies.

A Citizen's Guide to Bioremediation

What is Bioremediation?

Bioremediation is a treatment process that uses naturally occurring microorganisms (yeast, fungi, or bacteria) to break down, or *degrade*, hazardous substances into less toxic or nontoxic substances. Microorganisms, just like humans, eat and digest organic substances for nutrients and energy. In chemical terms, "organic" compounds are those that contain carbon and hydrogen atoms. Certain microor- ganisms can digest organic substances such as fuels or solvents that are hazardous to humans. The microorganisms break down the organic contaminants into harmless products—mainly carbon dioxide and water. Once the contaminants are degraded, the microorganism population is reduced because they have used all of their food source. Dead microorganisms or small populations in the absence of food pose no contamination risk.

How Does it Work?

Microorganisms must be active and healthy in order for bioremediation to take place. Bioremediation technologies assist microorganisms' growth and increase microbial populations by creating optimum environmental conditions for them to detoxify the maximum amount of contaminants. The specific bioremediation technology used is determined by several factors, for instance, the type of micro-organisms present, the site conditions, and the quantity and toxicity of contaminant chemicals. Different microorganisms degrade different types of compounds and survive under different conditions.

Indigenous microorganisms are those microorganisms that are found already living at a given site. To stimulate the growth of these indigenous microorganisms, the proper soil temperature, oxygen, and nutrient content may need to be provided.

If the biological activity needed to degrade a particular contaminant is *not* present in the soil at the site, microorganisms from other locations, whose effectiveness has been tested, can be added to the contaminated soil. These are called *exogenous* microorganisms. The soil conditions at the new site may need to be adjusted to ensure that the exogenous microorganisms will thrive.

Bioremediation can take place under *aerobic* and *anaerobic* conditions. In aerobic conditions, microorganisms use available atmospheric oxygen in order to function. With sufficient oxygen, microorganisms will convert many organic contaminants to carbon dioxide and water. *Anaerobic* conditions support biological activity in which no oxygen is present so the microorganisms break down chemical compounds in the soil to release the energy they need. Sometimes, during aerobic and anaerobic processes of breaking down the original contaminants, intermediate products that are less, equally, or more toxic than the original contaminants are created.

Bioremediation can be used as a cleanup method for contaminated soil and water. Bioremediation applications fall into two broad categories: *in situ* or *ex situ*. In situ biore-mediation treats the contaminated soil or groundwater in the location in which it was found. Ex situ bioremediation processes require excavation of contaminated soil or pumping of groundwater before they can be treated.

In Situ Bioremediation of Soil

In situ techniques do not require excavation of the contaminated soils so may be less expensive, create less dust, and cause less release of contaminants than ex situ techniques. Also, it is possible to treat a large volume of soil at once. In situ techniques, however, may be slower than exsitu techniques, may be difficult to manage, and are most effective at sites with *permeable* (sandy or uncompacted) soil.

The goal of aerobic in situ bioremediation is to supply oxygen and nutrients to the microorganisms in the soil. Aerobic in situ techniques can vary in the way they supply oxygen to the organisms that degrade the contaminants.

Two such methods are bioventing and injection of hydrogen peroxide. Oxygen can be provided by pumping air into the soil above the water table (bioventing) or by delivering the oxygen in liquid form as hydrogen peroxide. In situ bioremediation may not work well in clays or in highly layered subsurface environments because oxygen cannot be evenly distributed throughout the treatment area.

In situ remediation often requires years to reach cleanup goals, depending mainly on how biodegradable specific contaminants are. Less time may be required with easily degraded contaminants.

Bioventing. Bioventing systems deliver air from the atmosphere into the soil above the water table through injection wells placed in the ground where the contamination exists.

The number, location, and depth of the wells depend on many geological factors and engineering considerations. An air blower may be used to push or pull air into the soil through the injection wells. Air flows through the soil and the oxygen in it is used by the microorganisms. Nutrients may be pumped into the soil through the injection wells.

Nitrogen and phosphorous may be added to increase the growth rate of the microorganisms.

Injection of Hydrogen Peroxide.

This process delivers oxygen to stimulate the activity of naturally occurring microorganisms by circulating hydrogen peroxide through contaminated soils to speed the bioremediation of organic contaminants. Since it involves putting a chemical (hydrogen peroxide) into the ground (which may eventually seep into the groundwater), this process is used only at sites where the groundwater is already contaminated.

What is an Innovative Treatment Technology?

Treatment technologies are processes applied to the treatment of hazardous waste or contaminated materials to permanently alter their condition through chemical, biological, or physical means. Innovative treatment technologies are those that have been tested, selected or used for treatment of hazardous waste or contaminated materials but lack well-documented cost and performance data under a variety of operating conditions. A system of pipes or a sprinkler system is typically used to deliver hydrogen peroxide to shallow contaminated soils. Injection wells are used for deeper contaminated soils.

In Situ Bioremediation of Groundwater

In situ bioremediation of groundwater speeds the natural biodegradation processes that take place in the water-soaked underground region that lies below the water table. For sites at which both the soil and groundwater are con- taminated, this single technology is effective at treating both. Generally, an in situ groundwater bioremediation system consists of an extraction well to remove groundwater from the ground, an above-ground water treatment system where nutrients and an oxygen source may be added to the contaminated groundwater, and injection wells to return the "conditioned" groundwater to the subsurface where the microorganisms degrade the contaminants. One limitation of this technology is that differences in underground soil layering and density may cause reinjected conditioned groundwater to follow certain preferred flow paths. Consequently, the conditioned water may not reach some areas of contamination. Another frequently used method of in situ groundwater treatment is *air sparging,* which means pumping air into the groundwater to help flush out contaminants. Air sparging is used in conjunction with a technology called soil vapor extraction and is described in detail in the document entitled *A Citizen's Guide to Soil Vapor Extraction and Air Sparging* (see page 4).

Ex Situ Bioremediation of Soil

Ex situ techniques can be faster, easier to control, and used to treat a wider range of contaminants and soil types than in situ techniques. However, they require excavation and treatment of the contaminated soil before and, sometimes, after the actual bioremediation step. Ex situ tech- niques include slurry-phase bioremediation and solid- phase bioremediation.

Slurry-phase bioremediation. Contaminated soil is combined with water and other additives in a large tank called a "bioreactor" and mixed to keep the microorganisms— which are already present in the soil—in contact with the contaminants in the soil. Nutrients and oxygen are added, and conditions in the bioreactor are controlled to create the optimum environment for the microorganisms to degrade the contaminants. Upon completion of the treatment, the water is removed from the solids, which are disposed of or treated further if they still contain pollutants.

Slurry-phase biological treatment can be a relatively rapid process compared to other biological treatment processes, particularly for contaminated clays. The success of the process is highly dependent on the specific soil and chemical properties of the contaminated material. This technology is particularly useful where rapid remediation is a high priority.

Solid-phase bioremediation. Solid-phase bioremediation is a process that treats soils in above-ground treatment areas equipped with collection systems to prevent any con- taminant from escaping the treatment. Moisture, heat, nutrients, or oxygen are controlled to enhance biodegradation for the application of this treatment. Solid-phase systems are relatively simple to operate and maintain, require a large amount of space, and cleanups require more time to complete than with slurry-phase processes. Solid-phase soil treatment processes include *landfarming, soil biopiles,* and *composting.*

Landfarming. In this relatively simple treatment method, contaminated soils are excavated and spread on a pad with a built-in system to collect any "leachate" or contaminated liquids that seep out of contaminant-soaked soil. The soils are periodically turned over to mix air into the waste. Moisture and nutrients are controlled to enhance bio- remediation. The length of time for bioremediation to occur will be longer if nutrients, oxygen or temperature are not properly controlled. In some cases, reduction of contaminant concentrations actually may be attributed more to volatilisation than biodegradation. When the process is conducted in enclosures controlling escaping volatile contaminants, volatilisation losses are minimized.

Soil biopiles. Contaminated soil is piled in heaps several meters high over an air distribution system. Aeration is provided by pulling air through the heap with a vacuum pump. Moisture and nutrient levels are maintained at levels that maximize bioremediation. The soil heaps can be placed in enclosures. Volatile contaminants are easily controlled since they are usually part of the air stream being pulled through the pile.

Composting. Biodegradable waste is mixed with a bulking agent such as straw, hay, or corn cobs to make it easier to deliver the optimum levels of air and water to the microorganisms. Three common designs are *static pile composting* (compost is formed into piles and aerated with blowers or vacuum pumps), *mechanically agitated in-vessel composting* (compost is placed in a treatment vessel where it is mixed and aerated), and *windrow composting* (compost is placed in long piles known as windrows and periodically mixed by tractors or similar equipment).

Will it Work at Every Site?

Biodegradation is useful for many types of organic wastes and is a cost-effective, natural process. Many techniques can be conducted on-site, eliminating the need to transport hazardous materials.

The extent of biodegradation is highly dependent on the toxicity and initial concentrations of the contaminants, their biodegradability, the properties of the contaminated soil, and the particular treatment system selected.

Contaminants targeted for biodegradation treatment are non-halogenated volatile and semi-volatile organics and fuels. The effectiveness of bioremediation is limited at sites with high concentrations of metals, highly chlorinated organics, or inorganic salts because these compounds are toxic to the microorganisms.

Petro-Clear Hydrocarbon and Pesticide Remediation - Microbial Solutions

We have bioremediation solutions for of all weights of petroleum hydrocarbons, sites contaminated with solvents and other volatile organic compounds (such as trichloroethane, trichloroethene, benzene, toluene, ethylbenzene, and xylenes, BTEX) and fuels such as diesel or gasoline. Converts harmful petroleum products into carbon dioxide and water. Many of the most toxic environmental contaminants are now candidates for bioremediation. Bioremediation of hydrocarbons is best accomplished with a process called Bio-augmentation. That is, the addition of a large number of selected, laboratory grown, standardised microorganisms to a contaminated matrix. These contaminant specific microbes capable of degrading these compounds breaking them down into carbon dioxide and water. The microbes will survive and consume their contaminant food source until the unwanted pollutant is remediated.

Successful Bioaugmentation with Microbes

Bioaugmentation with microbes requires a few criteria to be successful.

- The first is the ability for the added microbes to reach the contaminant.
- The second is oxygen, microbes require oxygen.
- In the bio-degradation of petroleum based products, microbial activity is most frequently limited by insufficient oxygen due to slow rates of diffusion into the interior of the soil layers or piles and into the center of soil aggregates. Generally, the greater the mass of oxygen that can be distributed the more rapid and complete the cleanup. Much of their requirement will come from the atmosphere. However, if the ground is compacted, the ground should be tilled or raked to allow as much oxygen as possible to permeate the soil. If this is not possible, please contact us for other suggestions on how this can be achieved. We have had success with Oxy-Gen, an organic oxygen generating compound, that releases oxygen so that indigenous and augmented aerobic microbes may remediate the organic containments.
- The third is temperature. The working range is between 5-45° C (with 28° C being the optimum)
- The fourth is pH. The microbes a have a working range of pH 6.5 to 8.5
- The fifth is the addition of nutrients to the soil and moisture. We will advise you of the necessary nutrients required for each remediation project.

As long as the microbes can reach the contaminants with the necessary oxygen, nutrients, moisture, and the temperature and pH are within the microbes working range, then remediation will be successful.

How long will a typical clean up take?

This depends on the contaminant involved and it's concentration levels. It can range anywhere from a week for simple gasoline spills to several months for complicated crude oils. Contamination will either be dissolved in the groundwater, floating on the groundwater as free product, "sorbed" (bound) to the soil. In each case, we offer a specific product whose microbial types and levels have been optimized to cope with the various scenarios.

Hydrocarbons in Water

A series of all-natural microbial systems has been isolated to bioremediate hydrocarbon contamination in either salt or fresh water environments. These microbial systems contain high concentrations of safe, non-pathogenic, microorganisms. Will reduce and eliminate hydrocarbons from contributing to contamination in:

- stormwater runoff
- at commercial and industrial sites
- oil spill cleanup in lakes, rivers, groundwater, and ocean
- treatment of bilge and ballast water
- marina spills

These products are ideally suited for immediate application in treating the typical contamination from boats on a marina's surface water. Specifically formulated for use in soil applications, landfarming, underground storage tanks, large industrial clean-ups and spills on turf - soil - grass. Also, formulas for concrete, pavement and parking lots clean-ups.

Bioremediation

"Remediate" means to solve a problem, and "bio-remediate" means to use biological organisms to solve an environmental problem such as contaminated soil or groundwater.

In a non-polluted environment, bacteria, fungi, protists, and other microorganisms are constantly at work breaking down organic matter. What would occur if an organic pollutant such as oil contaminated this environment? Some of the microorganisms would die, while others capable of eating the organic pollution would survive. Bioremediation works by providing these pollution-eating organisms with fertilizer, oxygen, and other conditions that encourage their rapid growth. These organisms would then be able to break down the organic pollutant at a correspondingly faster rate. In fact, bioremediation is often used to help clean up oil spills.

Bioremediation of a contaminated site typically works in one of two ways. In the case described above, ways are found to enhance the growth of whatever pollution-eating microbes might already be living at the contaminated site. In the second, less common case, specialized microbes are added to degrade the contaminants.

Bioremediation provides a good cleanup strategy for some types of pollution, but as you might expect, it will not work for all. For example, bioremediation may not provide a feasible strategy at sites with high concentrations of chemicals that are toxic to most microorganisms. These chemicals include metals such as cadmium or lead, and salts such as sodium chloride.

Nonetheless, bioremediation provides a technique for cleaning up pollution by enhancing the same biodegradation processes that occur in nature. Depending on the site and its contaminants, bioremediation may be safer and less expensive than alternative solutions such as incineration or landfilling of the contaminated materials. It also has the advantage of treating the contamination in place so that large quantities of soil, sediment or water do not have to be dug up or pumped out of the ground for treatment.

Bioremediation

On-site remediation of soils and media contaminated with petrochemicals, pesticides, explosives or hazardous organic materials can be an expensive and time-consuming process. There are also serious legal and regulatory issues to consider when decontaminating soils on-site – issues that can jeopardize the viability of a construction or redevelopment project.

Waste Management, North America's largest solid waste services company, has developed a collection of innovative off-site remediation technologies to help companies deal effectively with contaminated soils. These bioremediation services include:

- TOSS^{SM} (Two-Step Static System) –TOSS^{SM} is a two-stage, solid-phase bioremediation technology that involves both anaerobic and aerobic treatment stages. In the first stage, explosives-contaminated soil is combined with a carbon source, an inoculum, vitamins and water to achieve anaerobic conditions. The resulting mixture is formed into a static pile or placed in a bermed construction or box to facilitate the chemical reduction of nitroaromatic and nitramine explosives. In the second stage, the anaerobically treated soil is combined with yard waste compost and built into an aerated biopile. The biopile may be aerated by forced air conveyed through perforated piping buried within the pile or by turning the pile with a compost turner. Previous testing of TOSS^{SM} has demonstrated TNT removal efficiencies of greater than 99 percent.
- The BioSite^{SM} System – The BioSite^{SM} System is Waste Management's proprietary system for the large-scale bioremediation of soils contaminated with:
  - o petrochemicals including, but not limited to:

- acetone, alcohols, benzene, ethylbenzene, methyl ethyl ketone (MEK), methyl isobutyl ketone (MIBK), petroleum hydrocarbons, toluene, two- and three-ring PAHs, xylene

o other contaminants, including:

- aliphatic chlorinated hydrocarbons (e.g., trichloroethylene), spent molecular sieve from packing towers, chemical manufacturing wastes, pesticides

Regulated compounds including underlying hazardous constituents (UHC) are screened prior to acceptance. Soils co-contaminated with metals may be accepted depending on their concentration.

- Bio-In-A-Box^{SM} – Bio-In-A-Box^{SM} is based on the same principles as both BioSite^{SM} and TOSS^{SM}, but is designed to operate indoors on a relatively smaller scale. Instead of being formed into long earthen mounds, the contaminated soil is moistened, mixed with nutrients and custom-grown microorganisms, and then placed in enclosed containers called "solid phase bioreactors" for incubation. These containers may or may not be linked to aeration and vacuum pipes, depending on the contaminants being processed. In just a few weeks, the decontaminated soil will be ready for landfill disposal or reintroduction into the environment.

REFERENCES

Aldor, I.; Fourest, E.; Volesky, B. *Can. J. Chem. Eng.* 1995, *73,* 516-522.

Benedict, B.; Pigford, T.H.; Levi, H.W. *Nuclear Chemical Engineering*, McGraw-Hill: New York, NY 1981.

Byerley, J.J.; Scharer, J.M.; Charles, A.M. *Chem. Eng. Journal* 1987, *36,* B49-B59.

Cavicchioli R., *Extremophiles and the search for extraterrestrial life.* Astrobiology. 2002 Fall;2(3):281-92.

Crist, D.R.; Crist, R.H.; Martin, J.R.; Watson, J. In *Metals-Microorganisms Relationships and Applications, FEMS Symposium Abstracts, Metz, France, May*; Bauda, P., ed.; Societe Francaise de Microbiologie: Paris, France, 1993, p. 13.

Edgington, D.N.; Gorden, S.A.; Thommes, M.M.; Almodovar, L.R. *Limnol. Ocean.* 1970, *15,* 945-955.

Fourest, E.; Volesky, B. *Environ. Sci. Technol.* 1996, *30,* 277-282.

Guibal, E.; Roulph, C.; Le Cloirec, P. *Water Res.* 1992, *26,* 1139-45.

Horikoshi, T.; Nakajima, A.; Sakaguchi, T. *Agric. Biol. Chem.* 1979, *332,* 617.

Hu, M.Z.-C.; Norman, J.M.; Faison, N.B.; Reeves, M. *Biotechnol. Bioeng.* 1996, *51,* 237-47.

Kuyucak, N.; Volesky, B. In *Biosorption of Heavy Metals*; Volesky, B., ed.; CRC Press: Boca Raton, FL, 1990, pp. 173-198.

Laul, J.C. *Radioanal. Nucl. Chem. Articles* 1992, *156,* 235.

Leusch, A.; Holan, Z.R.; Volesky, B. *J. Chem. Tech. Biotechnol.* 1995, *62,* 279-288.

Macaskie, L.E.; Empson, R.M.; Cheetham, A.K.; Grey, C.P.; Skarnulis, A.J. *Science* 1992, *257,* 782-784.

Mullen, M.D.; Wolf, D.C.; Beveridge, T.J.; Bailey, G.W. "Sorption of heavy metals by soil fungi *Aspergillus niger* and *Mucor Rouxii,*" In *Soil Biol. Biochem.* 1992, *24,* 129-135.

Munroe, N.D.H.; Bonner, J.D.; Williams, R.; Pattison, K.F.; Norman, J.M.; Faison, B.D. In *Abstracts, American Society for Microbiology Annual Meeting*, 1993.

Strain 121, a hyperthermophilic archaea, has been shown to reproduce at 121 °C (250 °F), and survive at 130 °C (266 °F).

Thomas, D. 2002. *Seaweeds.* The Natural History Museum, London. ISBN 0 565 09175 1

Volesky, B.; Holan, Z.R. *Biotechnol. Prog.* 1995, *11,* 235-250.

Volesky, B.; Tsezos, M. U.S. Patent 4320093, 1981. Canadian Patent 1143007, 1983.

White, S.K. *J. Am. Water Works Assoc.* 1983, *75,* 374.

17

Case Studies of Environmental Engineering & Environmental Biotechnology: Application & Utilities

Environmental Engineering

Environmental engineering is the application of science and engineering principles to improve the environment (air, water, and/or land resources), to provide healthy water, air, and land for human habitation and for other organisms, and to remediate polluted sites. Environmental engineering involves water and air pollution control, recycling, waste disposal, and public health issues as well as a knowledge of environmental engineering law. It also includes studies on the environmental impact of proposed construction projects. Environmental engineers conduct hazardous-waste management studies to evaluate the significance of such hazards, advise on treatment and containment, and develop regulations to prevent mishaps. Environmental engineers also design municipal water supply and industrial wastewater treatment systems as well as being concerned with local and worldwide environmental issues such as the effects of acid rain, ozone depletion, water pollution and air pollution from automobile exhausts and industrial sources.

Educational Licensing Requirements

To become an environmental engineer, at least a Bachelor's degree in engineering (usually civil or chemical, and more frequently environmental engineering) is required, usually followed by specialized training at the Master's or Doctoral level. Unlike traditional Engineering disciplines like Civil Engineering, Electrical Engineering or Chemical Engineering, Environmental Engineering evolved as a branch of Chemical and Civil Engineering over the past several decades to deal with increasingly complex environmental challenges. At many universities, Environmental Engineering programmes follow either the Department of Civil Engineering or The Department of Chemical Engineering at Engineering faculties. Environmental "civil" engineers focus on hydrology, water resources management and water treatment plant design. Environmental "chemical" engineers, on the other hand, focus on environmental chemistry, advanced air and water treatment technologies and separation processes. Additionally, engineers are more frequently obtaining specialized training in law (J.D.) and are utilising their technical expertise in the practices of Environmental engineering law. Most jurisdictions also impose licensing and registration requirements.

Development of Environmental Engineering

Ever since people first recognised that their health and well-being were related to the quality of their environment, they have applied thoughtful principles to attempt to improve the quality of their environment. The ancient Harappan civilisation utilized early sewers in some cities. The Romans constructed aqueducts to prevent drought and to create a clean, healthful water supply for the metropolis of Rome. In the 15th century, Bavaria created laws restricting the development and degradation of alpine country that constituted the region's water supply. Modern environmental engineering began in London in the mid-19th century when it was realised that proper sewerage could reduce the incidence of waterborne diseases such as cholera. The introduction of drinking water treatment and sewage treatment in industrialised countries reduced waterborne diseases from leading causes of death to rarities.

In many cases, as societies grew, actions that were intended to achieve benefits for those societies had longer-term impacts which reduced other environmental qualities. One example is the widespread application of DDT to control agricultural pests in the years following World War II. While the agricultural benefits were outstanding and crop yields increased dramatically, thus reducing world hunger substantially, and malaria was controlled better than it ever had been, numerous species were brought to the verge of extinction due to the impact of the DDT on their reproductive cycles. The story of DDT as vividly told in Rachel Carson's "Silent Spring" is considered to be the birth of the modern environmental movement and the development of the modern field of "environmental engineering."

Conservation movements and laws restricting public actions that would harm the environment have been developed by various societies for millennia. Notable examples are the laws decreeing the construction of sewers in London and Paris in the 19th century and the creation of the U.S. national park system in the early 20th century. Briefly speaking, the main task of environmental engineering is to protect public health by protecting (from further degradation), preserving (the present condition of), and enhancing the environment.

Scope of Environmental Engineering

Pollutants may be chemical, biological, thermal, radioactive, or even mechanical. Environmental engineering emphasises several areas: process engineering, environmental chemistry, water and sewage treatment (sanitary engineering), waste reduction/management, and pollution prevention/cleanup. Environmental engineering is a synthesis of various disciplines, incorporating elements from the following:

- Agricultural engineering
- Civil engineering
- Chemical engineering
- Public health
- Mechanical engineering
- Chemistry
- Biology
- Geology
- Ecology

Environmental engineering is the application of science and engineering principles to the environment. Some consider environmental engineering to include the development of sustainable processes. There are several divisions of the field of environmental engineering.

Environmental Impact Assessment and Mitigation

In this division, engineers and scientists assess the impacts of a proposed project on environmental conditions. They apply scientific and engineering principles to evaluate if there are likely to be any adverse impacts to water quality, air quality, habitat quality, flora and fauna, agricultural capacity, traffic impacts, social impacts, ecological impacts, noise impacts, visual(landscape) impacts, etc. If impacts are expected, they then develop mitigation measures to limit or prevent such impacts. An example of a mitigation measure would be the creation of wetlands in a nearby location to mitigate the filling in of wetlands necessary for a road development if it is not possible to reroute the road.

Water Supply and Treatment

Engineers and scientists work to secure water supplies for potable and agricultural use. They evaluate the water balance within a watershed and determine the available water supply, the water needed for various needs in that watershed, the seasonal cycles of water movement through the watershed and they develop systems to store, treat, and convey water for various uses. Water is treated to achieve water quality objectives for the end uses. In the case of potable water supply, water is treated to minimize risk of infectious disease transmittal, risk of non-infectious illness, and create a palatable water flavor. Water distribution systems are designed and built to provide adequate water pressure and flow rates to meet various end-user needs such as domestic use, fire suppression, and irrigation.

Wastewater Conveyance and Treatment

Most urban and many rural areas no longer discharge human waste directly to the land through outhouse, septic, and/or honey bucket systems, but rather deposit such waste into water and convey it from households via sewer systems. Engineers and scientists develop collection and treatment systems to carry this waste material away from where people live and produce the waste and discharge it into the environment. In developed countries, substantial resources are applied to the treatment and detoxification of this waste before it is discharged into a river, lake, or ocean system. Developing nations are striving to obtain the resources to develop such systems so that they can improve water quality in their surface waters and reduce the risk of water-borne infectious disease.

There are numerous wastewater treatment technologies. A wastewater treatment train can consist of a primary clarifier system to remove solid and floating materials, a secondary treatment system consisting of an aeration basin followed by flocculation and sedimentation or an activated sludge system and a secondary clarifier, a tertiary biological nitrogen removal system, and a final disinfection process. The aeration basin/activated sludge system removes organic material by growing bacteria (activated sludge). The secondary clarifier removes the activated sludge from the water. The tertiary system, although not always included due to costs, is becoming more prevalent to remove nitrogen and phosphorus and to disinfect the water before discharge to a surface water stream or ocean outfall.

Air Quality Management

Engineers apply scientific and engineering principles to the design of manufacturing and combustion processes to reduce air pollutant emissions to acceptable levels. Scrubbers, electrostatic precipitators, catalytic converters, and various other processes are utilized to remove particulate matter, nitrogen oxides, sulfur oxides, volatile organic compounds (VOC), reactive organic gases (ROG) and other air pollutants from flue gases and other sources prior to allowing their emission to the atmosphere.

Scientists have developed air pollution dispersion models to evaluate the concentration of a pollutant at a receptor or the impact on overall air quality from vehicle exhausts and industrial flue gas stack emissions.

To some extent, this field overlaps the desire to decrease carbon dioxide and other greenhouse gas emissions from combustion processes.

Environmental Engineering Law

Environmental engineering law is a profession that requires an expertise in both environmental engineering and law. This field includes professionals with both a legal and environmental engineering education. This dual educational requirement is typically satisfied through an ABET accredited degree in environmental engineering and an ABA accredited law degree. Likewise, this profession requires both licensure in professional environmental engineering and admittance to one bar.

Environmental engineering law is the professional application of law, science and engineering principles to improve the environment (air, water, and/or land resources), to provide healthy water, air, and land for human habitation and for other organisms, and to remediate polluted sites. Environmental engineering lawyers seek to promote the advancement of technical engineering knowledge in the legal profession and to enhance informed legal analysis of complex environmental matters.

Practice Areas

Environmental engineering law professionals offer a sound knowledge base in the fields of both environmental engineering and law to address complex environmental problems which demand both professional technical practice and legal expertise. Areas of practice are continually expanding, but frequently include complex land transactions, such as:

- Brownfields redevelopment
- Asbestos Baseline Survey and Building Revaluation due to forthcoming Asbestos Abatements
- Soil Contamination Assessment & Remediation, the development of a Remedial Action WorkPlan (RAWP) and Engineering Controls, including an Environmental Land Use Restriction (ELUR)
- TMDL Nutrient Loading Studies (ex. for NPDES Wastewater Discharges) and regulatory negotiation of Nutrients Discharge Limits from waste treatment plants, such as phosphorus and nitrogen.

Society of Environmental Engineers

The Society of Environmental Engineers (SEE) is a British professional engineering institution founded in 1959. It is licenced by the Engineering Council UK to assess candidates for inclusion on ECUK's Register of Professional Engineers and Technicians. The Society publishes the quarterly journal "Environmental Engineering", as well as a regular newsletter. Its address is The Manor House, High Street, Buntingford, Herts, SG9 9PL, United Kingdom.

Environmental Biotechnology

Biotechnology is "the integration of natural sciences and engineering in order to achieve the application of organisms, cells, parts thereof and molecular analogues for products and services" (EFB General Assembly, 1989). Environmental biotechnology as discussed in this briefing paper is the application of these processes for the protection and restoration of the quality of our environment.

Biotechnological processes to protect the environment have been used for almost a century now, even longer than the term'biotechnology' exists. Municipal sewage treatment plants and filters to purify town gas were developed around the turn of the century. They proved very effective although at the time, little was known about the biological principles underlying their function. Since that time our knowledge base has increased enormously. This briefing paper describes the state-of-the-art and possibilities of environmental biotechnology. It also deals with the societal aspects of environmental biotechnology.

Biotechnological techniques to treat waste before or after it has been brought into the environment are described and exemplified in the section on bioremediation. Biotechnology can also be used to develop products and processes that generate less waste and use less non-renewable resources and energy. In this respect biotechnology is well positioned to contribute to the development of a more sustainable society, a principle which was advocated in the Brundtland report in 1987 and in Agenda 21 of the second Earth Summit in Rio de Janeiro in 1992 and which has been widely accepted in the mean time. The section on prevention deals with this subject.

Biotechnological techniques to monitor the quality of the environment are presented in the section on detection and monitoring. Recombinant DNA technology has improved the possibilities for the prevention of pollution and holds a promise for a further development of bioremediation. These topics will be discussed in the section on genetic modification. The development of modern biotechnology has been accompanied by the establishment or adaptation of regulations to deal with genetically modified organisms.

Bioremediation

Bioremediation is the use of biological systems for the reduction of pollution from air or from aquatic or terrestrial systems. Micro-organisms and plants are the biological systems which are generally used. Biodegradation with micro-organisms is the most frequently occurring bioremediation option. Micro-organisms can break down most compounds for their growth and/or energy needs. These biodegradation processes may or may not need air. In some cases, metabolic pathways which organisms normally use for growth and energy supply may also be used to break down pollutant molecules. In these cases, known as co-metabolism, the micro-organism does not benefit directly.

Researchers have taken advantage of this phenomenon and use it for bioremediation purposes. A complete biodegradation results in detoxification by mineralising pollutants to carbon dioxide, water and harmless inorganic salts. Incomplete biodegradation will yield breakdown products which may or may not be less toxic than the original pollutant.

Incomplete biodegradation of tri- or tetrachloroethylene for instance can yield vinylchloride, which is more toxic and carcinogenic than the original compounds. Biodegradation may occur spontaneously, in which case the expressions "intrinsic bioremediation" or "natural attenuation" are often used. In many cases however the natural circumstances are not favourable enough for this to happen due to the lack of enough nutrients, oxygen or suitable bacteria. Such situations may be improved by supplementing one or more of these prerequisites. Extra nutrients were for instance disseminated to speed up the breakdown of the oil spilled on 1000 miles of Alaskan shoreline by the super tanker Exxon Valdez in 1989. The future trend in bioremediation increasingly is to look at the speed of unaided biodegradation first and act only if there is insufficient activity to remove the pollutant quickly enough to prevent any expected risks of the pollutant.

Bioremediation techniques can be used to reduce or to remove hazardous waste which has already polluted the environment. They can also be used to treat waste streams before they leave production facilities: end-of-pipe- processes. Some applications of bioremediation are discussed below. Waste water and industrial effluents: Micro-organisms in sewage treatment plants remove the more common pollutants from waste water before it is discharged into rivers or the sea. Increasing industrial and agricultural pollution has led to a greater need for processes that remove specific pollutants such as nitrogen and phosphorus compounds, heavy metals and chlorinated compounds.

New methods include aerobic, anaerobic and physico-chemical processes in fixed-bed filters and in bioreactors in which the materials and microbes are held in suspension. The costs of waste water treatment can be reduced by the conversion of wastes into useful products. For example, heavy metals and sulphur compounds can be removed from waste streams of the galvanisation industry by the aid of sulphur metabolising bacteria and reused. Another example is the production of animal feed from the fungal biomass which remains after the production of penicillin. Most anaerobic waste water treatment systems produce useful biogas.

Drinking and Process Water

Abundant

supplies of water are vital for modern urban and industrial development. By the turn of this century, it is estimated that two-thirds of the world's nations will be water stressed - using clean water faster than it is replenished in aquifers or rivers. A very important aspect of biotechnology is therefore its potential for the reclamation and purification of waste waters for re-use. Public concern has also increased over the quality of drinking water. Not only does water need to be recycled in the development of sustainable use of resources, overall quality must also be improved to satisfy consumers. In many agricultural regions of the world, animal wastes and excess fertilisers result in high levels of nitrates in drinking water. Biotechnology has provided successful methods by which these compounds can be removed from processed water before it is delivered to customers.

Air and Waste Gases

Originally, industrial waste gas treatment systems were based on cheap compost-filled filters that removed odours. Such systems still exist. However, slow processing rates and the short life of such filters drove research into better methods such as bioscrubbers, in which the pollutants are washed out using a cell suspension and biotrickling filters, in which the pollutant is degraded by micro-organisms immobilised on an inert matrix and provided with an aqueous nutrient film trickling through the device. The selection of micro-organisms that are more efficient at metabolising pollutants has also led to better air and gas purifying biofilters.

Examples are a bioscrubber based system for the simultaneous removal of nitrogen and sulphur oxides from the flue gas of blast- furnaces which has been developed as an alternative to the classical limestone gypsum process, and the elimination of styrene from the waste gas of polystyrene processing industries by a biofilter containing fungi. Soil and land treatment:

Both in situ (in its original place) and ex situ (somewhere else) methods are commercially exploited for the leanup of soil and the associated groundwater. In situ treatments may include the introduction of micro-organisms (bio-augmentation), ventilation and/or adding nutrient solutions (biostimulation). Ex situ treatment involves removing the soil and groundwater and treating it above ground. The soil may be treated as compost, in soil banks, or in specialised slurry bioreactors. Groundwater is treated in bioreactors and either pumped back into the ground or drained into surface water. Bioremediation of land(biorestoration) is often cheaper than physical methods and its products are harmless if complete mineralisation takes place. Its action can however, be time-consuming, tying up capital and land.

The in situ bioremediation of the ground under petrol stations has already become common practice but also for chlorinated solvents like tri- and tetrachloroethylene in situ bioremediation is possible. The applicability of in situ bioremediation is and probably will remain dependent on the physical parameters of the soil, mainly its transport properties.

Bioremediation using plants is called phyto-remediation. This technique is presently already used to remove metals from contaminated soils and groundwater and is being further explored for the remediation of other pollutants. The combined use of plants and bacteria may also be possible. Certain bacteria live closely associated with the roots of plants and depend on substances excreted by the roots. Such rhizobacteria, whose numbers are much higher than those of other soil bacteria, may be genetically modified to break down pollutants. Research is being conducted to test this hypothesis.

Solid Waste

Domestic solid wastes are a major problem in our consumption society.

Their elimination is both costly and warrants constant surveillance in terms of groundwater and air pollution. Yet, for a major part they are composed of readily biodegradable organics. In this respect, source separated bio-wastes can be converted to a valuable resource by composting or anaerobic digestion. In recent years, both processes have seen remarkable developments in terms of process design and control. Particularly, anaerobic digestion of solid wastes in high-rate anaerobic digesters has gained increasing public acceptance because it permits the recovery of substantial amounts of high-value biogas together with a high quality stable organic residue and this without giving rise to environmental nuisance. Moreover, anaerobic digestion of mixed solid wastes is under intensive development because in the near future it may be an important step in recycling of solid wastes and constitute an alternative to incineration.

Prevention

Progressively more industrial companies are developing processes with reduced environmental impact responding to the international call for the development of a sustainable society. There is a pervading trend towards less harmful products and processes; away from "end-of-pipe" treatment of waste streams. Biotechnology is pre-eminently suitable to contribute to this trend and it has already done so in many cases, both by the improvement of existing processes and by the development of new ones.

Many industrial processes have been made more environmentally friendly by the use of enzymes. Enzymes are biological catalysts that are highly efficient and have numerous advantages over non-biological catalysts. They are non-toxic and biodegradable, work best at moderate temperatures and in mild conditions, and have fewer side reactions than traditional methods because they are highly specific. Production methods that employ enzymes are generally not only cleaner and safer compared with other methods, but mostly also more economic in energy and resource consumption (see box). Their specificity does however mean that it is not always easy to find the appropriate enzyme for a given application. Enzymes are already widely employed in industry and have been for many years. New techniques and approaches.to protein design and molecular modelling are enabling researchers to develop novel enzymes active at high temperatures, in non-aqueous solvents and as solids. The leather processing industry has introduced enzymes to replace harsh chemicals traditionally used for cleaning the hide. In textile production, enzymes have superseded chemicals for bleaching, including the "stone washing" of jeans. Chlorine consumption by the pulp and paper industry may soon also be reduced considerably by the use of enzymes.

The grease and protein digesting enzymes in washing powders significantly reduce the quantity of detergents needed for a given washing effect. They also mean that the washing temperature can be reduced. Lowering the temperature 20°C saves more than a third of the energy used by the machine. Since in many Western European countries up to 5% of household energy consumption was used for washing, these molecules have made a significant contribution to energy conservation.

Detection and Monitoring

Detection and monitoring of pollutants:

A wide range of biological methods are already in use to detect pollution incidents and for the continuous monitoring of pollutants. Long established measures include: counting the number of plant, animal and microbial species, counting the numbers of individuals in those species or analysing the levels of oxygen, methane or other compounds in water. More recently, biological detection methods using biosensors and immunoassays have been developed and are now being commercialised.

Most biosensors are a combination of biological and electronic devices - often built onto a microchip. The biological component might be simply an enzyme or antibody, or even a colony of bacteria, a membrane, neural receptor, or an entire organism. Immobilised on a substrate, their properties change in response to some environmental effect in a way that is electronically or optically detectable. It is then possible to make quantitative measurements of pollutants with extreme precision or to very high sensitivities.

The sensors can be designed to be very selective, or sensitive to a broad range of compounds. For example, a wide range of herbicides can be detected in river water using algal-based biosensors; the stresses inflicted on the organisms being measured as changes in the optical properties of the plant's chlorophyll.

Microbial biosensors are micro-organisms which produce a reaction upon contact with the substance to be sensed. Usually they produce light but cease to do so upon contact with substances which are toxic to them. Both naturally occurring light emitting micro-organisms as well as specially developed ones are used. Positively acting bacterial biosensors have been constructed which start emitting light upon contact (and subsequent reaction) with a specific pollutant. In the USA such a light emitting bacterium has been approved for the detection of polyhalogenated aromatic hydrocarbons in field tests.

Biotechnological solutions for pollution Pigs and chickens cannot utilize phosphate from phytate in their feed, which therefore ends up in their manure. By adding the enzyme phytase to their feed the amount of phosphate which is excreted by these animals can be reduced by more than 30 %.

In South Africa bacteria are used for the isolation of gold from gold-ore. This so-called biomining saves an enormous amount of smelting energy and generates much less waste. The chemical production of indigo, the dye which is used for blue jeans, takes eight steps, the use of very toxic chemicals and special protection measures for the process operators and the environment. The biotechnological production of indigo, which uses a genetically modified bacterium containing the right enzymes, takes only three steps, proceeds in water, uses simple raw materials like sugar and salts and generates only indigo, carbon dioxide and biomass which is biodegradable.

Detection and monitoring of micro-organisms used for bioremediation:

When laboratory grown micro-organisms are inoculated into a bioremediation site (bio-augmentation) it often becomes necessary to monitor their presence and/or multiplication to check the progress of the process. This is especially true and even required when genetically modified micro-organisms are involved.

The traditional technique to detect the presence of micro-organisms in soil is direct plating on selective media. This is greatly facilitated if the organism contains a marker which can be selected for. Newer techniques include the abovementioned immunological and light-based bioreporter techniques. The spatial distribution of specific micro-organisms in a sample can be determined microscopically and non-invasively by using fluorescent in situ hybridisation (FISH) of micro-organisms. The most sensitive and specific technique is the direct isolation and amplification of DNA from soil, which is increasingly being used.

Detection and monitoring of ecological effects:

Bioremediation is aimed at improving the quality of the environment by removing pollutants. However, the disappearance of the original pollutant is not the only criterion by which the success of a bioremediation operation is determined. (Even more) toxic metabolites may be produced from the pollutant or the biodegrading bacterium may cause diseases or produce substances that are harmful to useful micro-organisms, plants, animals or humans. All these negative effects, are of course, excluded as much as possible in advance by getting as familiar as possible with the organism through extensive literature searches and microcosm studies in which the bioremediation process is simulated in the laboratory. To avoid unexpected effects, especially after the release of new member of the eco-system like a genetically modified organism, the monitoring of the ecological effects of a bioremediation operation may be required. The problem with monitoring ecological effects is what to monitor.

Numerous ecological effects are possible but not all of them may be relevant or permanent or even the result of the bioremediation operation. The parameters to be monitored are usually determined case-by-case. Monitoring techniques may include all of those mentioned in the two previous subsections on detection and monitoring.

Genetic Engineering

Recombinant DNA technology has had amazing repercussions in the last few years. Molecular biologists have mapped entire genomes, many new medicines have been developed and introduced and agriculturists are producing plants with novel types of disease resistance that could not be achieved through conventional breeding. Several of the previously mentioned examples like the amylose-free potato and the indigo-producing bacterium also involve the use of organisms genetically modified by recombinant DNA technology. Many enzymes are routinely produced by genetically modified organisms too.

Given the overwhelming diversity of species, biomolecules and metabolic pathways on this planet, genetic engineering can in principle be a very powerful tool in creating environmentally friendlier alternatives for products and processes that presently pollute the environment or exhaust its non-renewable resources. Politics, economics and society will ultimately determine which scientific possibilities will become reality.

Nowadays organisms can also be supplemented with additional genetic properties for the biodegradation of specific pollutants if naturally occurring organisms are not able to do that job properly or not quickly enough. By combining different metabolic abilities in the same micro-organism bottlenecks in environmental cleanup may be circumvented. Until now this has not been done on any significant scale. The main reason being the fact that in most cases naturally occurring organisms can be found or selected for, which are able to clean up a polluted site. Examples have been found where soil bacteria have developed new properties in response to the introduction of xenobiotics (that is, man- made chemicals that are normally not found in nature). In some cases they even appear to have acquired properties from other species. In the USA some genetically modified bacteria have been approved for bioremediation purposes but large scale applications have not yet been reported. In Europe only controlled field tests have been authorized.

Because new organisms can be created by genetic engineering that may never be produced by spontaneous or selection driven evolution, concerns exist about the unpredictability of their possible interactions with the eco-system. Genetically modified organisms which are properly kept within the confines of their approved production facilities are much less a concern than genetically modified organisms which are meant to be released into the environment like disease-resistant plants or soil bacteria for bioremediation.

The possible ecological effects of the latter are even more difficult to evaluate due to the fact that it is well known that soil bacteria frequently exchange genetic material (also between species). This together with the fact that we know little about the great majority of soil inhabiting bacterial species, makes it almost impossible to predict the fate of every DNA copy of a newly introduced genetic property in a soil bacterium. If the extra DNA is derived from another soil bacterium, it may on the other hand be reasonable to argue that the genetically modified bacterium might also have evolved spontaneously some day due to the frequent exchange of genetic material in the soil.

Legislation

Regulation to ensure safe application of novel or modified organisms in the environment is important, not least to maintain public confidence. The European Union has two Directives(1,2)on the contained use of genetically modified micro-organisms, and on the deliberate release of genetically modified organisms into the environment. These have been implemented in the national legislation of most EU Member States. They require that a detailed experimental protocol, including assessment of potential risks, is approved by competent authorities before a genetically modified organism is released into the environment. The nature and sometimes even the site of the release has to be published in the local press in some countries. After several years' experience using the legislation, the procedures involved are now being revised.

Ammendments to clarify and revise Directive 90/219/EEC were published in

December 1998. The aim of the European Commission is to maintain the EU's competitiveness globally-both in research and commercial applications- without compromising safety.

Public Opinion, Dialogue and Debate

In spite of the fact that traditional biotcchnology already is of great value to bioremediation and modern biotechnology may enhance this even further, there are no recent data on what Europeans specifically think about environmental biotechnology.

Generally speaking, Europeans tend to take an"optimistic" view of the developments they expect from modern biotechnology, according to the most recent European Commission public opinion survey which was published in 1997. Unfortunately this survey did not investigate the attitude of the public towards environmental biotechnology. The only environmentally related question in this survey was whether people believed that modern biotechnology would substantially reduce environmental pollution, which 47 % did. Whether or not this is only wishful thinking remains to be determined. Ultimately, hard proof expessed in the form of improved environmental parameters will be needed for full acceptance of environmental biotechnology.

Conferences, public debates, seminars and round table meetings have been held to bring people from the public, government, environmental organisations, science and industry together to discuss critical issues.

These lively debates do not always lead to consensus, but they can provide a fuller appreciation of all the aspects in a particular issue, facilitating a better understanding of the problems involved. A recent example is the workshop 'How can biotechnology benefit the environment'.

Public information aimed at advancing dialogue and debate is provided by many organisations. A compilation of these can be found in the handbook of information sources which has been published by and can be ordered from the EFB Taskgroup on Public perceptions of Biotechnology.

Environmental biotechnology has a career extending back into the last century. As the need is better appreciated to move towards less destructive patterns of economic activity, while maintaining improvement of social conditions in spite of increasing population, the role of biotechnology grows as a tool for remediation and environmentally sensitive industry. Already, the technology has been proven in a number of areas and future developments promise to widen its scope. Some of the new techniques now under consideration make use of genetically modified organisms designed to deal efficiently with specific tasks. As with all situations where there is to be a release of new technology into the environment. concerns exist. There is a potential for biotechnology to make a further major contribution to protection and remediation of the environment. Hence biotechnology is well positioned to contribute to the development of a more sustainable society. As we move into the next millennium this will become even more vitally important as populations, urbanisation and industrialisation will continue to climb.

Incineration

Incineration is a waste treatment technology that involves the combustion of organic materials and/or substances. Incineration and other high temperature waste treatment systems are described as "thermal treatment". Incineration of waste materials converts the waste into incinerator bottom ash, flue gases, particulates, and heat, which can in turn be used to generate electric power. The flue gases are cleaned of pollutants before they are dispersed in the atmosphere.

Incineration with energy recovery is one of several waste-to-energy (WtE) technologies such as gasification, pyrolysis and anaerobic digestion. Incineration may also be implemented without energy and materials recovery.

In several countries there are still expert and local community concerns about the environmental impact of incinerators (see The argument against incineration).

In some countries, incinerators built just a few decades ago often did not include a materials separation to remove hazardous, bulky or recyclable materials before combustion. These facilities tended to risk the health of the plant workers and the local environment due to inadequate levels of gas cleaning and combustion process control. Most of these facilities did not generate electricity.

Incinerators reduce the volume of the original waste by 95-96 %, depending upon composition and degree of recovery of materials such as metals from the ash for recycling. This means that while incineration does not completely replace landfilling, it reduces the necessary volume for disposal significantly.

Incineration has particularly strong benefits for the treatment of certain waste types in niche areas such as clinical wastes and certain hazardous wastes where pathogens and toxins can be destroyed by high temperatures. Examples include chemical multi-product plants with diverse toxic or very toxic wastewater streams, which cannot be routed to a conventional wastewater treatment plant.

Waste combustion is particularly popular in countries such as Japan where land is a scarce resource. Denmark and Sweden have been leaders in using the energy generated from incineration for more than a century, in localised combined heat and power facilities supporting district heating schemes. In 2005, waste incineration produced 4.8 % of the electricity consumption and 13.7 % of the total domestic heat consumption in Denmark. A number of other European Countries rely heavily on incineration for handling municipal waste, in particular Luxemburg, The Netherlands, Germany and France.

Technology

Types of Incinerators

An incinerator is a furnace for burning waste. Modern incinerators include pollution mitigation equipment such as flue gas cleaning. There are various types of incinerator plant design: moving grate, fixed grate, rotary-kiln, fluidised bed.

Moving Grate

The typical incineration plant for municipal solid waste is a moving grate incinerator. The moving grate enables the movement of waste through the combustion chamber to be optimised to allow a more efficient and complete combustion. A single moving grate boiler can handle up to 35 tonnes of waste per hour, and can operate 8,000 hours per year with only one scheduled stop for inspection and maintenance of about one months duration. Moving grate incinerators are sometimes referred to as Municipal Solid Waste Incinerators (MSWIs). The waste is introduced by a waste crane through the "throat" at one end of the grate, from where it moves down over the descending grate to the ash pit in the other end. Here the ash is removed through a water lock. Part of the combustion air (primary combustion air) is supplied through the grate from below. This air flow also has the purpose of cooling the grate itself. Cooling is important for the mechanical strength of the grate, and many moving grates are also water cooled internally. Secondary combustion air is supplied into the boiler at high speed through nozzles over the grate. It facilitates complete combustion of the flue gases by introducing turbulence for better mixing and by ensuring a surplus of oxygen. In multiple/stepped hearth incinerators, the secondary combustion air is introduced in a separate chamber downstream the primary combustion chamber.

According to the European Waste Incineration Directive, incineration plants must be designed to ensure that the flue gases reach a temperature of at least 850 °C for 2 seconds in order to ensure proper breakdown of organic toxins. In order to comply with this at all times, it is required to install backup auxiliary burners (often fueled by oil), which are fired into the boiler in case the heating value of the waste becomes too low to reach this temperature alone.

The flue gases are then cooled in the superheaters, where the heat is transferred to steam, heating the steam to typically 400 °C at a pressure of 40 bar for the electricity generation in the turbine. At this point, the flue gas has a temperature of around 200 °C, and is passed to the flue gas cleaning system. At least in Scandinavia scheduled maintenance is always performed during summer, where the demand for district heating is low. Often incineration plants consist of several separate 'boiler lines' (boilers and flue gas treatment plants), so that waste receival can continue at one boiler line while the others are subject to revision.

Fixed Grate

The older and simpler kind of incinerator was a brick-lined cell with a fixed metal grate over a lower ash pit, with one opening in the top or side for loading and another opening in the side for removing incombustible solids called clinkers. Many small incinerators formerly found in apartment houses have now been replaced by waste compactors.

Rotary-kiln

The rotary-kiln incinerator used by municipalities and by large industrial plants. This design of incinerators have 2 chambers a primary chamber and secondary chamber. The primary chamber in a rotary kiln incinerator consist of an inclined refractory lined cylindrical tube. Movement of the cylinder on its axis facilitates movement of waste. In the primary chamber, there is conversion of solid fraction to gases, through volatilisation, destructive distillation and partial combustion reactions. The secondary chamber is necessary to complete gas phase combustion reactions

The clinkers spill out at the end of the cylinder. A tall flue gas stack, fan, or steam jet supplies the needed draft. Ash drops through the grate, but many particles are carried along with the hot gases. The particles and any combustible gases may be combusted in an "afterburner". A diagram of a rotary-kiln incinerator can be found here.

Fluidized Bed

A strong airflow is forced through a sandbed. The air seeps through the sand until a point is reached where the sand particles separate to let the air through and mixing and churning occurs, thus a fluidised bed is created and fuel and waste can now be introduced.

The sand with the pre-treated waste and/or fuel is kept suspended on pumped air currents and takes on a fluid-like character. The bed is thereby violently mixed and agitated keeping small inert particles and air in a fluid-like state. This allows all of the mass of waste, fuel and sand to be fully circulated through the furnace.

Specialized Incineration

Furniture factory sawdust incinerators need much attention as these have to handle resin powder and many flammable substances. Controlled combustion, burn back prevention systems are very essential as dust when suspended resembles the fire catch phenomenon of any liquid petroleum gas.

Use of Heat

The heat produced by an incinerator can be used to generate steam which may then be used to drive a turbine in order to produce electricity. The typical amount of net energy that can be produced per ton municipal waste is about 0.67 MWh of electricity and 2 MJ of district heating. Thus, incinerating about 600 tonnes per day of waste will produce about 17 MW of electrical power and 1200 MJ district heating each day.

Pollution

Incineration has a number of outputs such as the ash and the emission to the atmosphere of flue gas. Before the flue gas cleaning system, the flue gases may contain significant amounts of particulate matter, heavy metals, dioxins, furans, sulfur dioxide, and hydrochloric acid. In a study from 1994, Delaware Solid Waste Authority found that, for same amount of produced energy, incineration plants emitted fewer particles, hydrocarbons and less SO_2, HCl, CO and NO_x than coal-fired power plants, but more than natural gas fired power plants. According to Germany's Ministry of the Environment, waste incinerators reduce the amount of some atmospheric pollutants by substituting power produced by coal-fired plants with power from waste-fired plants.

Gaseous Emissions

Dioxin and Furans

The most publicized concerns from environmentalists about the incineration of municipal solid wastes (MSW) involve the fear that it produces significant amounts of dioxin and furan emissions. Dioxins and furans are considered by many to be serious health hazards. Older generation incinerators that were not equipped with adequate gas cleaning technologies were indeed significant sources of dioxin emissions. Today, however, due to advances in emission control designs and stringent new governmental regulations, incinerators emit virtually no dioxins. In 2005, The Ministry of the Environment of Germany, where there were 66 incinerators at that time, estimated that "...whereas in 1990 one third of all dioxin emissions in Germany came from incineration plants, for the year 2000 the was less than 1 %. Chimneys and tiled stoves in private households alone discharge approximately twenty times more dioxin into the environment than incineration plants.". According to the U.S. EPA, incineration plants are no longer significant sources of dioxins and furans. In 1987, before the governmental regulations required the use of emission controls, there was a total of 10,000 grams of dioxin emissions from U.S. incinerators. Today, the total emissions from the 87 plants are only 10 grams yearly, a reduction of 99.9 %. Backyard barrel burning of household and garden wastes, still allowed in some rural areas, generates 580 grams of dioxins yearly. Studies conducted by EPA demonstrate that the emissions from just one family using a burn barrel produces more emissions than an incineration plant disposing of 200 tonnes of waste per day.

Generally the breakdown of dioxin requires exposure of the molecule to a sufficiently high temperature so as to trigger thermal breakdown of the molecular bonds holding it together. When burning of plastics outdoors in a burn barrel or garbage pit such temperatures are not reached, causing high dioxin emissions as mentioned above. While the plastic does burn in an open-air fire, the dioxins remain after combustion and float off into the atmosphere.

Modern municipal incinerator designs include a high temperature zone, where the flue gas is ensured to sustain a temperature above 850 °C for at least 2 seconds befores it is cooled down. They are equipped with auxiliary heaters to ensure this at all times. These are often fueled by oil, and normally only active for a very small fraction of the time. A side effect controlling dioxin is the potential for generation of reactive oxides (NO_x) in the flue gas, which must be removed with SCR or SNCR (see below). For very small municipal incinerators, the required temperature for thermal breakdown of dioxin may be reached using a high-temperature electrical heating element, plus an SCR stage.

CO_2

As for other complete combustion processes, nearly all of the carbon content in the waste is emitted as CO_2 to the atmosphere. MSW contain approximately the same mass fraction of carbon as CO_2 itself (27%), so incineration of one tonne of MSW produce approximately 1 tonne of CO_2. In the event that the waste was landfilled, one tonne of MSW would produce approximately 62 m^3 methane via the anaerobic decomposition of the biodegradable part of the waste. This amount of methane has more than twice the global warming potential than the one tonne of CO_2, which would have been produced by incineration. In some countries, large amounts of landfill gas are collected, but still the global warming potential of the landfill gas emitted to atmosphere in the US in 1999 was approximately 32 % higher than the amount of CO_2 that would have been emitted by incineration.

In addition, nearly all biodegradable waste has biological origin. This material has been formed by plants using atmospheric CO_2 typically within the last growing season. If these plants are regrown the CO_2 emitted from their combustion will be taken out from the atmosphere once more. Such considerations are the main reason why several countries administrate incineration of the biodegradable part of waste as renewable energy. The rest - mainly plastics and other oil and gas derived products - is generally treated as non-renewables. Different results for the CO_2 footprint of incineration can be reached with different assumptions. Local conditions (such as limited local district heating demand, no fossil fuel generated electricity to replace or high levels of aluminum in the waste stream) can decrease the CO_2 benefits of incineration. The methology and other assumptions may also influence the results significantly. For example the methane emissions from landfills occurring at a later date may be neglected or given less weight, or biodegradable waste may not be considered CO_2 neutral. A recent study by Eunomia Research and Consulting on potential waste treatment technologies in London demonstrated that by applying several of these (according to the authors) unusual assumptions the average existing incineration plants performed poorly for CO_2 balance compared to the theoretical potential of other emerging waste treatment technologies.

Other Emissions

Other gaseous toxins in the flue gas from incinerator furnaces include sulfur dioxide, hydrochloric acid, heavy metals and fine particles. The steam content in the flue may produce visible fume from the stack, which can be perceived as a visual pollution. It may be avoided by decreasing the steam content by flue gas condensation, or by increasing the flue gas exit temperature well above its dew point. Flue gas condensation allows the latent heat of vaporisation of the water to be recovered, subsequently increasing the thermal efficiency of the plant.

Flue Gas Cleaning

The quantity of pollutants in the flue gas from incineration plants is reduced by several processes. Particulate is collected by particle filtration, most often electrostatic precipitators (ESP) and/or baghouse filters. The latter are generally very efficient for collecting fine particles. In an investigation by the Ministry of the Environment of Denmark in 2006, the average particulate emissions per energy content of incinerated waste from 16 Danish incinerators were below 2.02 g/GJ (grams per energy content of the incinerated waste). Detailed measurements of fine particles with sizes below 2.5 micrometres ($PM_{2.5}$) were performed on three of the incinerators: One incinerator equipped with an ESP for particle filtration emitted 5.3 g/GJ fine particles, while two incinerators equipped with baghouse filters emitted 0.002 and 0.013 g/GJ $PM_{2.5}$. Acid gas scrubbers are used to remove hydrochloric acid, nitric acid, hydrofluoric acid, mercury, lead and other heavy metals. Basic scrubbers remove sulfur dioxide, forming gypsum by reaction with lime. Waste water from scrubbers must subsequently pass through a waste water treatment plant. Sulfur dioxide may also be removed by dry desulfurisation by injection limestone slurry into the flue gas before the particle filtration. NO_x is either reduced by catalytic reduction with ammonia in a catalytic converter (selective catalytic reduction, SCR) or by a high temperature reaction with ammonia in the furnace (selective non-catalytic reduction, SNCR). Heavy metals are often adsorbed on injected active carbon powder, which is collected by the particle filtration.

Solid Outputs

Incineration produces fly ash and bottom ash just as is the case when coal is combusted. The total amount of ash produced by municipal solid waste incineration ranges from 4-10 % by volume and 15-20 % by weight of the original quantity of waste, and the fly ash amounts to about 10-20 % of the total ash. The fly ash, by far, constitutes more of a potential health hazard than does the bottom ash because the fly ash often contain high concentrations of heavy metals such as lead, cadmium, copper and zinc as well as small amounts of dioxins and furans. The bottom ash seldom contain significant levels of heavy metals. While fly ash is always regarded as hazardous waste, bottom ash is generally considered safe for regular landfill after a certain level of testing defined by the local legislation. Ash, which is considered hazardous, may generally only be disposed of in landfills which are carefully designed to prevent pollutants in the ash from leaching into underground aquifers - or after chemical treatment to reduce its leaching characteristics. In testing over the past decade, no ash from an incineration plant in the USA has ever been determined to be a hazardous waste. At present although some historic samples tested by the incinerator operators' group would meet the being ecotoxic criteria at present the EA say "we have agreed" to regard incinerator bottom ash as "non-hazardous" until the testing programme is complete.

Other Pollution Issues

Odour pollution can be a problem with old-style incinerators, but odours and dust are extremely well controlled in newer incineration plants. They receive and store the waste in an enclosed area with a negative pressure with the airflow being routed through the boiler which prevents unpleasant odours from escaping into the atmosphere. However, not all plants are implemented this way, resulting in inconveniences in the locality. An issue that affects community relationships is the increased road traffic of waste collection vehicles to transport municipal waste to the incinerator. Due to this reason, most incinerators are located in industrial areas.

The Debate Over Incineration

Use of incinerators for waste management is controversial. The debate over incinerators typically involves business interests (representing both waste generators and incinerator firms), government regulators, environmental activists and local citizens who must weigh the economic appeal of local industrial activity with their concerns over health and environmental risk. People and organisations professionally involved in this issue include the U.S. Environmental Protection Agency and a great many local and national air quality regulatory agencies worldwide.

The Argument for Incineration

- The concerns over the health effects of dioxin and furan emissions have been significantly lessened by advances in emission control designs and very stringent new governmental regulations that have resulted in large reductions in the amount of dioxins and furans emissions.
- Incineration plants generate electricity and heat that can substitute power plants powered by other fuels at the regional electric and district heating grid, and steam supply for industrial customers.
- The bottom ash residue remaining after combustion has been shown to be a non-hazardous solid waste that can be safely landfilled or recycled as construction aggregate.
- In densely populated areas, finding space for additional landfills is becoming increasingly difficult.
- Fine particles can be efficiently removed from the flue gases with baghouse filters. Even though approximately 40 % of the incinerated waste in Denmark was incinerated at plants with no baghouse filters, estimates based on measurements by the Danish Environmental Research Institute showed that incinerators were only responsible for approximately 0.3 % of the total domestic emissions of particulate smaller than 2.5 micrometres ($PM_{2.5}$) to the atmosphere in 2006.
- Incineration of municipal solid waste avoids the release of methane. Every ton of MSW incinerated, prevents about one ton of carbon dioxide equivalents from being released to the atmosphere.
- Incineration of medical waste and sewage sludge produces an end product ash that is sterile and non-hazardous.

The Argument Against Incineration

- The highly toxic fly ash must be safely disposed of. This usually involves additional waste miles and the need for specialist toxic waste landfill elsewhere, sometimes with concerns for local residents. This has been the case in Bishops Cleeve, Gloucestershire, UK. Dr Dick van Steenis MBBS,industrial toxicologist
- There are still concerns by many about the health effects of dioxin and furan emissions into the atmosphere from old incinerators; especially during start up and shut down events, or where filter bypass events are required.

- Incinerators emit varying levels of heavy metals such as vanadium, manganese, chromium, nickel, arsenic, mercury, lead, and cadmium, which can be toxic at very minute levels.
- Incinerator Bottom Ash (IBA) has high levels of heavy metals with ecotoxicity concerns if not reused properly. Some people have the opinion that IBA reuse is still in its infancy and is still not considered to be a mature or desirable product, despite additional engineering treatments.
- Alternative technologies are available or in development such as Mechanical Biological Treatment, Anaerobic Digestion (MBT/AD), Autoclaving or Mechanical Heat Treatment (MHT) using steam or Plasma arc gasification PGP, or combinations.
- Building and operating an incinerator requires long contract periods to recover initial investment costs, causing a long term lock-in. Incinerator lifetimes normally range from 25-30 years.
- Incinerators produce fine particles in the furnace. Even with modern particle filtering of the flue gases, a fraction of these are emitted to the atmosphere. As an example, the baghouse filters in an incineration plant planned for erection in the UK, are only specified to capture 65-70 % (in weight) particulate smaller than 2.5 micrometres ($PM_{2.5}$), if the significant filtration in the filter cake is not accounted for. $PM_{2.5}$ is not separately regulated in the European Waste Incineration Directive, even though they are suspected to be linked to infant mortality in the UK,Dr Dick van Steenis MBBS,industrial toxicologist and $PM_{2.5}$ emissions from local incinerators to be a significant $PM_{2.5}$ source here.
- Local communities are often opposed to the idea of locating incinerators in their vicinity. (The Not In My Back Yard phenomenon). Studies in Andover, Massachusetts linked 10 % property devaluations with close incinerator proximity.
- Prevention, waste minimisation, reuse and recycling of waste should all be preferred to incineration according to the waste hierarchy. Supporters of zero waste consider incinerators and other waste treatment technologies as barriers to recycling and separation beyond particular levels, and that waste resources are sacrificed for energy producion.
- A recent Eunomia report found that incinerators under some circumstances and assumptions can be seen as contributing to climate change, and are less energy efficient than emerging technologies for treating residual mixed waste.
- Some incinerators are architecturally monstrous and ugly. In many countries they require a visually intrusive chimney stack.

Trends in Incinerator Use

The history of municipal solid waste (MSW) incineration is linked intimately to the history of landfills and other waste treatment technology. The merits of incineration are inevitably judged in relation to the alternatives available. Since the 1970s, recycling and other prevention measures have changed the context for such judgements. Since the 1990s alternative waste treatment technologies have been maturing and becoming viable. Incineration is a key process in the treatment of hazardous wastes and clinical wastes. It is often imperative that medical waste be subjected to the high temperatures of incineration to destroy pathogens and toxic contamination it contains.

Incineration in North America

The first full-scale waste-to-energy facility in the U.S. was the Arnold O. Chantland Resource Recovery Plant, built in 1975 located in Ames, Iowa. This plant is still in operation and produces refuse-derived fuel that is sent to local power plants for fuel. The first commercially-successful incineration plant in the U.S. was built in Saugus, Massachusetts in October 1975 by Wheelabrator Technologies, and is still in operation today. Several older generation incinerators have been closed; of the 186 MSW incinerators in 1990, only 89 remained by 2007, and of the 6200 medical waste incinerators in 1988, only 115 remained in 2003. Between 1996 and 2007, no new incinerators were built. The main reasons for lack of activity have been:

- Economics. With the increase in the number of large inexpensive regional landfills and, up until recently, the relatively low price of electricity, incinerators were not able to compete for the 'fuel', i.e., waste. By contrast, a number of Canadian cities are working toward installation of incinerators.
- Tax Policies. Tax credits for plants producing electricity from waste were rescinded in the 1990s. In Europe, some of the electricity generated from waste is deemed to be from a 'Renewable Energy Source (RES)'. A new law granting tax credits for such plants was implemented in the U.S. in 2004.

Despite these problems, there has been renewed interest in waste-to-energy in the U.S., Canada, and the UK. Projects to add capacity to existing plants are underway, and municipalities are once again evaluating the option of building incinerators rather than continue landfilling municipal wastes.

Incineration in Europe

In Europe, with the ban on landfilling untreated waste, scores of incinerators have been built in the last decade, with more under construction. Recently, a number of municipal governments have begun the process of contracting for the construction and operation of incinerators. In Europe, some of the electricity generated from waste is deemed to be from a 'Renewable Energy Source (RES)' and is thus eligible for tax credits if privately operated.

Incineration in the United Kingdom

The technology employed in the UK waste management industry has been greatly lagging behind that of Europe due to the wide availablility of landfills. The Landfill Directive set down by the European Union led to the Government of the United Kingdom imposing waste legislation including the landfill tax and Landfill Allowance Trading Scheme. This legislation is designed to reduce the release of greenhouse gases produced by landfills through the use of alternative methods of waste treatment. It is the UK Government's position that incineration will play an increasingly large role in the treatment of municipal waste and supply of energy in the UK.

Small incinerator units

Small scale incinerators exist for special purposes. For example, the small scale incinerators are aimed for hygienically safe destruction of medical waste in developing countries. Simple, mobile incinerators are becoming more widely used in developing countries where the threat of avian influenza is high. Small incinerators can be quickly deployed to remote areas where an outbreak has occurred to dispose of infected animals quickly and without the risk of cross contamination.

Landfill

A landfill, is a site for the disposal of waste materials by burial and is the oldest form of waste treatment. Historically, landfills have been the most common methods of organised waste disposal and remain so in many places around the world. Landfills may include internal waste disposal sites (where a producer of waste carries out their own waste disposal at the place of production) as well as sites used by many producers. Many landfills are also used for other waste management purposes, such as the temporary storage, consolidation and transfer, or processing of waste material (sorting, treatment, or recycling). A landfill also may refer to ground that has been filled in with soil and rocks instead of waste materials, so that it can be used for a specific purpose, such as for building houses. Unless they are stabilized, these areas may experience severe shaking or liquefaction of the ground in a large earthquake.

Site Construction Requirements

The construction of a landfill requires a staged approach. Landfill designers are primarily concerned with the viability of a site. To be commercially and environmentally viable a landfill must be constructed in accord with specific requirements, which are related to:

- Location
  - o Easy access to transport by road
  - o Transfer stations if rail network is preferred
  - o Land value
  - o Cost of meeting government requirements, such as the Environment Agency in England and Wales
  - o Location of community served
  - o Type of construction (more than one may be used at single site)
    - Pit - filling existing holes in the ground, typically left behind by mining
    - Canyon - filling in naturally occurring valleys or canyons
    - Mound - piling the waste up above the ground
- Stability
  - o Underlying geology
  - o Nearby earthquake faults
  - o Water table
  - o Location of nearby rivers, streams, and flood plains
- Capacity The available voidspace must be calculated by comparison of the landform with a proposed restoration profile.

  This calulation of capacity is based on,
  - o Density of the wastes
  - o Amount of intermediate and daily cover
  - o Amount of settlement that the waste will undergo following tipping
  - o Thickness of capping
  - o Construction of lining and drainage layers.
- Protection of soil and water through:
  - o Installation of liner and collection systems.
  - o Storm water control
  - o Leachate management.
  - o Landfill gas management.
- Nuisances and hazards management.

- Costs
  - o Feasibility studies
  - o Site after care
  - o Site investigations (costs involved may make small sites uneconomical).

Operations

Typically, in non hazardous waste landfills, in order to meet predefined specifications, techniques are applied by which the wastes are:

1. Confined to as small an area as possible.
2. Compacted to reduce their volume.
3. Covered (usually daily) with layers of soil.

During landfill operations the waste collection vehicles are weighed at a weighbridge on arrival and their load is inspected for wastes that do not accord with the landfill's waste acceptance criteria. Afterwards, the waste collection vehicles use the existing road network on their way to the tipping face or working front where they unload their load. After loads are deposited, compactors or dozers are used to spread and compact the waste on the working face. Before leaving the landfill boundaries, the waste collection vehicles pass through the wheel cleaning facility. If necessary, they return to weighbridge in order to be weighed without their load. Through the weighing process, the daily incoming waste tonnage can be calculated and listed in databases. In addition to trucks, some landfills may be equipped to handle railroad containers. The use of 'rail-haul' permits landfills to be located at more remote sites, without the problems associated with many truck trips.

Typically, in the working face, the compacted waste is covered with soil daily. Alternative waste cover materials are several sprayed on foam products and temporary blankets. Blankets can be lifted into place with tracked excavators and then removed the following day prior to waste placement. Chipped wood and chemically 'fixed' biosolids, may also be used as an alternate daily cover. The space that is occupied daily by the compacted waste and the cover material is called daily cell. Waste compaction is critical to extending the landfill life. Factors such as waste compressibility, waste layer thickness and the number of passes of the compactor over the waste affect the waste densities.

Land Reclamation

As human overcrowding of developed areas intensified during the 20th century, it has become important to develop land re-use strategies for completed landfills. Some of the most common usages are for parks, golf courses and other sports fields.

Increasingly, however, office buildings and industrial uses are made on a completed landfill. In these latter uses, methane capture is customarily carried out to minimize explosive hazard within the building. An example of a Class A office building constructed over a landfill is the Dakin Building at Sierra Point, Brisbane, California. The underlying fill was deposited from 1965 to 1985, mostly consisting of construction debris from San Francisco and some municipal wastes. Aerial photographs prior to 1965 show this area to be tidelands of the San Francisco Bay. A clay cap was constructed over the debris prior to building approval. Another strategy for landfill reclamation is the incineration of landfill trash at high temperature via the plasma-arc gasification process, which is currently used at two facilities in Japan, and will be used at a planned facility in St. Lucie County, Florida.

Impacts

A number of adverse impacts occur from landfill operations. These impacts can vary: fatal accidents (e.g., scavengers buried under waste piles); infrastructure damage (e.g., damage to access roads by heavy vehicles); pollution of the local environment (such as contamination of groundwater and/or aquifers by leakage and residual soil contamination during landfill usage, as well as after landfill closure); offgassing of methane generated by decaying organic wastes (methane is a greenhouse gas many times more potent than carbon dioxide, and can itself be a danger to inhabitants of an area;) harbouring of disease vectors such as rats and flies, particularly from improperly operated landfills, which are common in Third-world countries; injuries to wildlife; and simple nuisance problems (e.g., dust, odour, vermin, or noise pollution). Environmental noise and dust are generated from vehicles accessing a landfill as well as from working face operations. These impacts are best to intercept at the planning stage where access routes and landfill geometrics can be used to mitigate such issues. Vector control is also important, but can be managed reasonably well with the daily cover protocols. Most modern landfills in industrialised countries are operated with controls to attempt manage problems such as these. Analysis of common landfill operational problems are available in. Some local authorities have found it difficult to locate new landfills. Communities may charge a fee or levy in order to discourage waste and/or recover the costs of site operations. Some landfills are operated for profit as commercial businesses. Many landfills, however, are publicly operated and funded.

Impacts to People Near Landfills in the U.S.

Communities near landfills are increasingly facing health consequences from air and water contamination, particularly from landfills that are poorly constructed and operated. Environmental contamination from landfills is entering waterways and underground aquifer at alarming rates. Liner breaches are not uncommon. Liners can delay contamination but they do not prevent it. With large amounts of toxic solid waste entering landfills today, ground and air contamination pose a significant threat to public health for those living within three to five miles of a landfill, and will eventually degrade the environment far beyond those limits. Poorly constructed and operated landfills persist with leachate breaks, uncovered trash, and unchecked banned hazardous compounds. Federal laws to protect the public in Sec. 4001, Subtitle D of the Resource Conservation and Recovery Act (RCRA) can be unenforceable to citizens without adequate legal funding. The U.S. Environmental Protection Agency generally relies on the states to enforce their own operating permits and federal laws. If state agencies are not aggressive, violations can worsen, multiplying negative environmental impacts exponentially. There are some notable recorded violations in the U.S., such as for a landfill in Hawaii that was fined $2.8 million in 2006 for operating violations, but this is not common.

Regional practice

United Kingdom

Landfilling practices in the UK have had to change in recent years to meet the challenges of the European Landfill Directive. The UK now imposes landfill tax upon biodegradable waste which is landfilled. In addition to this the Landfill Allowance Trading Scheme has been established for local authorities to trade landfill quotas.

United States

In the U.S., landfills are regulated by the state's environmental agency that establishes minimum guidelines; however, none of these standards may fall below those set by the United States Environmental Protection Agency (EPA); such as was the case with the Fresh Kills Landfill in Staten Island, which is claimed by many to not only be the world's largest landfill, but the world's largest manmade structure. The Fresno Municipal Sanitary Landfill, opened in Fresno, California in 1937, is considered to have been the first modern, sanitary landfill in the United States, innovating the techniques of trenching, compacting, and the daily covering of waste with soil. It has been designated a National Historic Landmark, underlining the significance of waste disposal in urban society. Before the advent of modern landfills in America, most Americans lived in sparsely populated rural farming communities and most burned their garbage. Due to environmental and safety concerns, burning garbage by civilians has been outlawed by most municipalities and can only be performed by landfill managers or people who have obtained permits from the municipality. More information on landfill history in the United States can be found at. The Resource Conservation and Recovery Act is a US federal law that is designed to protect the public from harm caused by waste disposal. The EPA runs a Landfill Methane Outreach Programme (LMOP), a voluntary assistance programme that helps to reduce methane emissions from landfills by encouraging the recovery and use of landfill gas as an energy resource. U.S. landfills consist of 40% to 50% paper waste, 20% to 30% construction debris, and 1.2% disposable diapers.

Alternatives

The obvious alternative to landfills are waste reduction and recycling strategies. Secondary to not creating waste, there are various alternatives to landfills. In the late 20th century, alternative methods to waste disposal to landfill and incineration have begun to gain acceptance. Anaerobic digestion, composting, mechanical biological treatment, pyrolysis and plasma arc gasification have all began to establish themselves in the market. In recent years, some countries, such as Germany, Austria, and Switzerland, have banned the disposal of untreated waste in landfills. In these countries, only the ashes from incineration or the stabilised output of mechanical biological treatment plants may still be deposited.

Methanogenesis

Methanogenesis or biomethanation is the formation of methane by microbes known as methanogens. Organisms capable of producing methane have been identified only from the kingdom Archaea, a group phylogenetically distinct from both eukaryotes and bacteria, although many live in close association with anaerobic bacteria. The production of methane is an important and widespread form of microbial metabolism. In most environments, it is the final step in the decomposition of biomass.

Recently, it has been demonstrated that leaf tissues of living plants emit methane. Although the mechanism by which such methane production occurs is, as yet, unknown, the implications are far-reaching; this is an example of methanogenesis occurring in non-microbes, presumably under aerobic conditions. Most of what is known about methanogenesis comes from microbial studies.

Biochemistry of Methanogenesis

Methanogenesis in microbes is a form of anaerobic respiration. Methanogens do not use oxygen to breathe; in fact, oxygen inhibits the growth of methanogens. The terminal electron acceptor in methanogenesis is not oxygen, but carbon. The carbon can occur in a small number of organic compounds, all with low molecular weights. The two best described pathways involve the use of carbon dioxide and acetic acid as terminal electron acceptors:

$$CO_2 + 4\,H_2 \text{ '! } CH_4 + 2H_2O$$
$$CH_3COOH \text{ '! } CH_4 + CO_2$$

However, methanogenesis has been shown to use carbon from other small organic compounds, such as formic acid (formate), methanol, methylamines, dimethyl sulfide, and methanethiol. The biochemistry of methanogenesis is relatively complex, involving the following coenzymes and cofactors: F430, coenzyme B, coenzyme M, methanofuran, and methanopterin.

Importance in Carbon Cycle

Methanogenesis is the final step in the decay of organic matter. During the decay process, electron acceptors (such as oxygen, ferric iron, sulfate, nitrate, and manganese) become depleted, while hydrogen (H_2) and carbon dioxide accumulate. Light organics produced by fermentation also accumulate. During advanced stages of organic decay, all electron acceptors become depleted except carbon dioxide. Carbon dioxide is a product of most catabolic processes, so it is not depleted like other potential electron acceptors.

Only methanogenesis and fermentation can occur in the absence of electron acceptors other than carbon. Fermentation only allows the breakdown of larger organic compounds, and produces small organic compounds. Methanogenesis effectively removes the semi-final products of decay: hydrogen, small organics, and carbon dioxide. Without methanogenesis, a great deal of carbon (in the form of fermentation products) would accumulate in anaerobic environments.

In Ruminants

Methanogenesis occurs in the guts of humans and other animals, especially ruminants. In the rumen, anaerobic organisms including methanogens digest cellulose into forms usable by the animal, without them, livestock such as cattle would not be able to graze grass. The useful products of methanogenesis are absorbed by the gut, but the methane is released from the animal mainly by belching (eructation). The average cow emits around 600 litres of methane per day.

Role in Global Warming

Methane in the Earth's atmosphere is an important greenhouse gas with a global warming potential 25 times greater than carbon dioxide (averaged over 100 years), and methanogenesis in livestock and the decay of organic material is thus a considerable contributor to global warming. It may not be a net contributor in the sense that it works on organic material which used up atmospheric carbon dioxide when it was created, but its overall effect is to convert the carbon dioxide into methane which is a much more potent greenhouse gas. Methanogenesis can also be beneficially exploited, to treat organic waste, to produce useful compounds, and the methane can be collected and used as biogas, a fuel.

Methanogenesis and Extra-Terrestrial Life

The presence of atmospheric methane has a role in the scientific search for extraterrestrial life. The argument being that methane in the atmosphere will eventually dissipate, unless something is replenishing it. This something could then be the decomposition of organic matter (Methanogenesis). So, if it can be detected (by using a spectrometer for example) then that means there is, or relatively recently was, life present. This was debated when methane was discovered in the Martian atmosphere by (among others) the Mars Express Orbiter (2004) and in Titan's atmosphere by the Huygens probe (2005). It is also argued that atmospheric methane can come from volcanoes or other fissures in the planet's crust and that without an Isotopic signature it is difficult to say what exactly was the origin.

Bioleaching

Bioleaching is the extraction of specific metals from their ores through the use of bacteria. Bioleaching is one of several applications within biohydrometallurgy and several methods are used to recover copper, zinc, lead, arsenic, antimony, nickel, molybdenum, gold, cobalt.

The Process

The extraction of gold from its ore can involve numerous ferrous and sulfur oxidisingbacteria, including *Acidithiobacillus ferrooxidans* and *Acidithiobacillus thiooxidans* (formerly known as *Thiobacillus*). For example, bacteria catalyse the breakdown of the mineral arsenopyrite (FeAsS) by oxidising the sulfur and metal (in this case arsenic ions) to higher oxidation states whilst reducing dioxygen by H_2 and Fe^{3+}. This allows the soluble products to dissolve.

$$FeAsS_{(s)} \text{ '! } Fe^{2+}{}_{(aq)} + As^{3+}{}_{(aq)} + S^{6+}{}_{(aq)}$$

This process actually occurs at the cell membrane of the bacteria. The electrons pass into the cells and are used in biochemical processes to produce energy for the bacteria to reduce oxygen molecules to water.

In stage 2, bacteria oxidise Fe^{2+} to Fe^{3+} (whilst reducing O_2).

$$Fe^{2+} \text{ '! } Fe^{3+}$$

They then oxidise the metal to a higher positive oxidation state. With the electrons gained, they reduce Fe^{3+} to Fe^{2+} to continue the cycle.

$$M^{3+} \text{ '! } M^{5+}$$

The gold is now separated from the ore and in solution.

The process for copper is very similar. The mineral chalcopyrite ($CuFeS_2$) follows the two stages of being dissolved and then further oxidised, with Cu^{2+} ions being left.

Extraction from mixture

Copper (Cu^{2+}) ions are removed from the solution by ligand exchange solvent extraction which leaves other ions in the solution. The copper is removed by bonding to a ligand, which is a large molecule consisting of a number of smaller groups each possessing a lone pair. The ligand is dissolved in an organic solvent such as kerosene and shaken with the solution producing this reaction:

$$Cu^{2+}_{(aq)} + 2LH(organic) \text{ '! } CuL_2(organic) + 2H^+_{(aq)}$$

The ligand donates electrons to the copper, producing a complex - a central metal atom (copper) bonded to 2 molecules of the ligand. Because this complex has no charge, it is no longer attracted to polar water molecules and dissolves in the kerosene, which is then easily separated from the solution. Because the initial reaction is reversible, it is determined by pH. Adding concentrated acid reverses the equation, and the copper ions go back into an aqueous solution.

Then the copper is passed through an electro-winning process to increase its purity: an electric current is passed through the resulting solution of copper ions. Because copper ions have a 2+ charge, they are attracted to the negative cathodes and collect there.

The copper can also be concentrated and separated by displacing the copper with Fe from scrap iron:

$$Cu^{2+}_{(aq)} + Fe_{(s)} \text{ '! } Cu_{(s)} + Fe^{2+}_{(aq)}$$

The electrons lost by the iron are taken up by the copper. Copper is the oxidising agent (it accepts electrons), and iron is the reducing agent (it loses electrons). Traces of precious metals such as gold may be left in the original solution. Treating the mixture with sodium cyanide in the presence of free oxygen dissolves the gold. The gold is removed from the solution by adsorbing (taking it up on the surface) to charcoal.

Bioleaching with Fungi

Several species of fungi can be used for bioleaching. Fungi can be grown on many different strata, as with electronic scrap, catalytic converters, and fly ash from municipal waste incineration. Experiments have shown that two fungal strains (*Aspergillus Niger, Penicillium simplicissimum*) were able to mobilize Cu and Sn by 65%, and Al, Ni, Pb, and Zn by more than 95% *Aspergillus Niger* can produce some organic acids such as citric acid. So it can be used for bioleaching sulfides .

Bioleaching Compared with other Extraction Techniques

Traditional extractions involve many expensive steps such as roasting and smelting, which require sufficient concentrations of elements in ores and are environmentally unfriendly. Low concentrations are not a problem for bacteria because they simply ignore the waste which surrounds the metals, attaining extraction yields of over 90% in some cases. These microorganisms actually gain energy by breaking down minerals into their constituent elements. The company simply collects the ions out of the solution after the bacteria have finished.

Some advantages associated with bioleaching are:

- economical: bioleaching is generally simpler and therefore cheaper to operate and maintain than traditional processes, since fewer specialists are needed to operate complex chemical plants.
- environmental: The process is more environmentally friendly than traditional extraction methods. For the company this can translate into profit, since the necessary limiting of sulfur dioxide emissions during smelting is expensive. Less landscape damage occurs, since the bacteria involved grow naturally, and the mine and surrounding area can be left relatively untouched. As the bacteria breed in the conditions of the mine, they are easily cultivated and recycled.

Some disadvantages associated with bioleaching are:

- economical: the bacterial leaching process is very slow compared to smelting. This brings in less profit as well as introducing a significant delay in cash flow for new plants.
- environmental: Toxic chemicals are sometimes produced in the process. Sulfuric acid and H^+ ions which have been formed can leak into the ground and surface water turning it acidic, causing environmental damage. Heavy ions such as iron, zinc, and arsenic leak during acid mine drainage. When the pH of this solution rises, as a result of dilution by fresh water, these ions precipitate, forming "Yellow Boy" pollution. For these reasons, a setup of bioleaching must be carefully planned, since the process can lead to a biosafety failure.

Currently it is more economical to smelt copper ore rather than to use bioleaching, since the concentration of copper in its ore is generally quite high. The profit obtained from the speed and yield of smelting justifies its cost. However, the concentration of gold in its ore is generally very low. The lower cost of bacterial leaching in this case outweighs the time it takes to extract the metal.

BIOHEAP – A BIOLEACHING PROCESS FOR NICKEL EXTRACTION

Background

Titan Resources, the Australian-based nickel producer, has recently confirmed the economics of its BioHeap process, following a pre-feasibihty cost analysis of the exploration of two small low-grade nickel sulphide deposits in northern Western Australia.

Who Developed the Process?

The process has been developed over the past four years by a 75%-owned subsidiary of Titan, Pacific Ore Technology Ltd, with capital to help the project from the Australian Government's R & D Start Grant programme.

Bioleaching

The BioHeap process is a form of bioleaching. Traditionally, this is carried out by naturally occurring micro organisms, usually thermophiles, which are micro organisms usually found in acidic environments produced by the oxidation of sulphur - for example in and around hot springs, volcanic regions and sulphide-rich areas. While most living organisms derive energy for growth and reproduction from organic carbon, thermophiles grow on inorganic matter and are harmless to living creatures. Their diet consists of pyrite, arsenopyrite and other metal suphides, such as chalcocite and chalcopyrite.

How Does the Bioleaching Work?

The exact mechanisms by which these micro-organisms oxidise sulphidised materials is not precisely known, although both chemical and biological forces work together to oxidise the metal sulphide to form acid-soluble sulphates. Precious metals, which are not soluble, remain with the residue. Iron, arsenic and base metals, such as copper, cobalt and zinc pass into solution, which can then be separated from the residue and be treated by conventional processing methods. The residue created by the process may contain precious metals, which can be recovered by cyanidation.

Types of Bioleaching

There are currently two methods of bioleaching - tank bioleaching and heap bioleaching. The latter involves crushing the ore, stacking it on plastic mats and spraying it with a dilute sulphuric acid solution containing bacteria and nutrients. The solution drains through the heap and is recovered and resprayed over the heap. When the solution is considered rich enough, it is drained off and the metals extracted by conventional processes.

Advantages of the BioHeap Process

Although both bacterial oxidation and non-bacterial heap leaching are established in the gold and copper industries, their combined use in nickel extraction is unique. The BioHeap process treats crushed ore, so avoiding the need for fine grinding and concentration steps usually needed for most bacterial oxidation techniques. This in turn leads to cost savings which allows the processing of lower grade nickel sulphide resources, which may currently be uneconomic to exploit.

How Does the BioHeap Process Perform?

Small samples of the disseminated nickel sulphide ore from the company's Radio Hill mine were subjected to bacterial leach amenability treating during 1998. The results demonstrated that greater than 90% of the contained nickel, copper and cobalt could be extracted. The BioHeap process was then tested using disseminated ore - ore that had a content of 0.73% Ni and 0.87% Cu.

*How Does the Bacteria Work?*T

he bacterial culture works in the temperature range 45-60°C, at pH levels of less than 2 and requires air, from which it fixes carbon from carbon dioxide and oxygen for oxidation reactions. This occurs both directly and indirectly - in the direct method the bacteria attaches to the sulphide mineral and oxidises Fe and S moieties in the mineral to release metal ions into solution. The indirect bacterial attack on the mineral involves the bacteria in solution oxidising ferrous ion to ferric ion. Ferric ion is a strong oxidising agent and this in turn attacks the sulphide mineral.

Evaluation of the BioHeap Process

The actual BioHeap test was constructed on the apron of the Radio Hill tailings dam using ore from the Mount Sholl deposit, and consists of three pads, sand filtration bunds, drainage gutters from each pad and a holding pond. The entire area is lined with HPDE.Two heaps have initially been erected, one a 5,000 tonne ore heap and the second a 5,000 tonne waste heap. The former was inoculated with the bacterial cell culture, while the later was used in conditioning the pregnant liquor. Aeration pipes are located one metre from the base of the heaps and run the length of the heap (1). These are connected to a manifold and low-pressure blowers used to aerate the heaps. Covering the heaps is a network of irrigation pipes and drippers used for acid irrigation, bacterial inoculation and nutrient irrigation of both heaps.

Following bacterial inoculation, the heaps are continually irrigated with the pregnant liquor stream and with a nutrient solution. As the oxidation of the ore progresses, so the base metal tenure in the pregnant liquor stream increases. Once a desired level of metal in solution has been achieved, a bleed stream is removed from the system and the dissolved metals removed using conventional precipitation, electrowinning and solvent extraction methods. By 22 weeks of operation, nickel recovery was 74%, well above the company's estimate of 70% after nine months.

Establishment Costs

Total capital costs for the venture were estimated at Aus$15 million, including preparation of the leach pads, bacterial breeding facilities and ion exchange and precipitation equipment. This capital cost equates to a unit comparison of less than US$2 per pound of nickel produced per annum over the two year life of the project.

Summary

While the actual production costs of nickel using BioHeap will vary for different ore deposits and for different scales of operation, the results of the study are consistent with Titan's projections for BioHeap as a process with low capital and operating expenditures and inherent technical and environmental advantages over conventional flotation and smelting.

Bioleaching Microbes

The microbes that are found in bioleaching environments are of two types: those that produce leaching chemicals ("true" bioleaching microbes) and those that do not produce leaching chemicals but support "true" bioleaching microbes - bioleaching supporters. In a consortium, bioleaching supporters assist those microbes producing leaching reagents by for example feeding on, and thereby removing, organic waste products that otherwise would build up and be detrimental to said "true" bioleaching microbes.

Prokaryotes are Common While Eukaryotes are Rare

Even though all (?) microbes used in commercial plants for bioleaching are prokaryotes (bacteria and archaea), bioleaching eukaryotes exist too. These organisms often excrete organic acids which may act as leaching agents. Research on metal leaching of electronic scrap by bioleaching fungi has been performed with a certain degree of success. {Brandl, 2001 #9}

Reasons why "True" Bioleachers may not Thrive

Sub-optimal growth conditions are attributable to at least the following:

1. incorrect pH conditions;
2. a lack of critical macro- and micro-nutrients;
3. a high ionic strength or total salt content

4. the presence of dissolved or entrained organic compounds with inhibitory effects on microbial growth;
5. carbon- or oxygen-limiting conditions

Biopolymers and Bioplastics

Many everyday products used by Canadians are made of plastic. Pens, computers, cars, food packaging, and clothing are all examples of products which contain plastic. Plastic is a material made up of one or more polymers. The main source of the chemicals needed to manufacture plastics are fossil fuels. These petrochemical plastics are very durable, but take a long time to biodegrade when disposed. Rising concern about the cost of fossil fuels, and their impact on the environment has resulted in a search for alternatives to petrochemical plastics, namely biopolymers and bioplastics.

What are Biopolymers and Bioplastics?

Biopolymers and bioplastics go by many different names. They are often referred to as bio-based plastics and polymers, or as biodegradable plastics or polymers. They are defined below: Biopolymers are polymers which are present in, or created by, living organisms. These include polymers from renewable resources that can be polymerized to create bioplastics.

Bioplastics are plastics manufactured using biopolymers, and are biodegradable.

Biopolymers and bioplastics are not new products. Henry Ford developed a method of manufacturing plastic car parts from soybeans in the mid-1900s. However, World War II side-tracked the production of bioplastic cars. Today, bioplastics are gaining popularity once again as new manufacturing techniques developed through biotechnology are being applied to their production.

Types of Biopolymers

There are two main types of biopolymers: those that come from living organisms; and, those which need to be polymerized but come from renewable resources. Both types are used in the production of bioplastics.

Using Fermentation to Produce Plastics

Fermentation, used for hundreds of years by humans, is even more powerful when coupled with new biotechnology techniques. Fermentation is the use of microorganisms to break down organic substances in the absence of oxygen. Today, fermentation can be carried out with genetically engineered microorganisms, specially designed for the conditions under which fermentation takes place, and for the specific substance that is being broken down by the microorganism. There are two ways fermentation can be used to create biopolymers and bioplastics:

Bacterial Polyester Fermentation – Bacteria are one group of microorganisms that can be used in the fermentation process. Fermentation, in fact, is the process by which bacteria can be used to create polyesters. Bacteria called Ralstonia eutropha are used to do this. The bacteria use the sugar of harvested plants, such as corn, to fuel their cellular processes. The by-product of these cellular processes is the polymer. The polymers are then separated from the bacterial cells. Lactic Acid Fermentation – Lactic acid is fermented from sugar, much like the process used to directly manufacture polymers by bacteria. However, in this fermentation process, the final product of fermentation is lactic acid, rather than a polymer. After the lactic acid is produced, it is converted to polylactic acid using traditional polymerisation processes.

Growing Plastics in Plants

Plants are becoming factories for the production of plastics. Researchers created a Arabidopis thaliana plant through genetic engineering. The plant contains the enzymes used by bacteria to create plastics. Bacteria create the plastic through the conversion of sunlight into energy. The researchers have transferred the gene that codes for this enzyme into the plant, as a result the plant produces plastic through its cellular processes. The plant is harvested and the plastic is extracted from it using a solvent. The liquid resulting from this process is distilled to separate the solvent from the plastic.

Biotechnology and Biopolymers and Bioplastics

Biotechnology is driving the production of new bioplastics. Biotechnology techniques used to produce bioplastics include fermentation, and genetic engineering. For example, fermentation is used to release the cellulose from plants, so the cellulose can be used to create plastics. Also, genetic engineering can be used to create plants, such as soybean, specifically designed to be used as a raw material for the production of bioplastics.

Current Research Areas in Biopolymers and Bioplastics

Improving efficiency is a major concern for the production of plastics and bioplastics. Currently, fossil fuel is still used as an energy source during the production process. This has raised questions by some regarding how much fossil fuel is actually saved by manufacturing bioplastics. Only a few processes have emerged that actually use less energy in the production process. Therefore, researchers are still working on refining the processes used in order to make bioplastics viable alternatives to petrochemical plastics.

Energy use is not the only concern when it comes to biopolymers and bioplastics. There are also concerns about how to balance the need to grow plants for food, and the need to grow plants for use as raw materials. Agricultural space needs to be shared. Researchers are looking into creating a plant that can be used for food, but also as feedstock for plastic production. One group is attempting to genetically engineer corn to contain the bacterial enzyme responsible for plastic production. Eventually, they are hoping to create the plant in a way which would restrict the plastic production to the stem, and leaves of the plant. This would leave the edible part of the corn plastic free. The edible part of the corn would be used as food, or as livestock feed. The plastic would be removed from the remaining part of the corn plant.

Sustainable Development and Biopolymers and Bioplastics

Biopolymers and bioplastics are the main components in creating a sustainable plastics industry. These products reduce the dependence on non-renewable fossil fuels, and are easily biodegradable. Together, this greatly limits the environmental impacts of plastic use and manufacture. Also, characteristics such as being biodegradable make plastics more acceptable for long term use by society. It is likely that in the long term, these products will mean plastics will remain affordable, even as fossil fuel reserves diminish.

Bioplastic

Bioplastics are a form of plastics derived from renewable biomass sources, such as vegetable oil, corn starch, pea starch or microbiota, rather than traditional plastics which are derived from petroleum. They are used either as a direct replacement for traditional plastics or as blends with traditional plastics. There is no international agreement on how much bio-derived content is required to use the term bioplastic.

Bioplastics and Biodegradation

The terminology used in the bioplastics sector is sometimes misleading. Most in the industry use the term bioplastic to mean a plastic produced from a biological source. One of the oldest plastics, cellulose film, is made from wood cellulose. All bio- and petroleum-based plastics are technically biodegradable, meaning they can be degraded by microbes under suitable conditions. However many degrade at such slow rates as to be considered non-biodegradable. Some petrochemical-based plastics are considered biodegradable, and may be used as an additive to improve the performance of many commercial bioplastics. Non-biodegradable bioplastics are referred to as durable. The degree of biodegradation varies with temperature, polymer stability, and available oxygen content. Consequently, most bioplastics will only degrade in the tightly controlled conditions of commercial composting units. An internationally agreed standard, EN13432, defines how quickly and to what extent a plastic must be degraded under commercial composting conditions for it to be called biodegradable. This is published by the International Organisation for Standardisation ISO and is recognised in many countries, including all of Europe, Japan and the US. However, it is designed only for the aggressive conditions of commercial composting units. There is no standard applicable to home composting conditions.

The term "biodegradable plastic" is often also used by producers of specially modified petrochemical-based plastics which appear to biodegrade. Traditional plastics such as polyethylene are degraded by ultra-violet (UV) light and oxygen. To prevent this process manufacturers add stabilising chemicals. However with the addition of a degradation initiator to the plastic, it is possible to achieve a controlled UV/ oxidation disintegration process. This type of plastic may be referred to as *degradable plastic* or *oxy-degradable plastic* or *photodegradable plastic* because the process is not initiated by microbial action. While some degradable plastics manufacturers argue that degraded plastic residue will be attacked by microbes, these degradable materials do not meet the requirements of the EN13432 commercial composting standard.

Environmental Impacts

The production and use of bioplastics is generally regarded as a more sustainable activity when compared with plastic production from petroleum, because it relies less on fossil fuel as a carbon source and also introduces less, net-new greenhouse emissions if it biodegrades. However, manufacturing of bioplastic materials is often still reliant upon petroleum as an energy and materials source. This comes in the form of energy required to power farm machinery and irrigate growing crops, to produce fertilisers and pesticides, to transport crops and crop products to processing plants, to process raw materials, and ultimately to produce the bioplastic. Italian bioplastic manufacturer Novamont states in its own environmental audit that producing one kilogram of its starch-based product uses 500g of petroleum and consumes almost 80% of the energy required to produce a traditional polyethylene polymer. Environmental data from NatureWorks, the only commercial manufacturer of PLA (polylactic acid) bioplastic, says that making its plastic material delivers a fossil fuel saving of between 25 and 68 per cent compared with polyethylene, in part due to its purchasing of renewable energy certificates for its manufacturing plant. A detailed study examining the process of manufacturing a number of common packaging items in several traditional plastics and polylactic acid carried out by US-group and published by the Athena Institute shows the bioplastic to be less environmentally damaging for some products, but more environmentally damaging for others. While production of most bioplastics results in reduced carbon dioxide emissions compared to traditional alternatives, there are some real concerns that the creation of a global bioeconomy could contribute to an accelerated rate of deforestation if not managed effectively. There are associated concerns over the impact on water supply and soil erosion.

Cost

With the exception of cellulose, most bioplastic technology is relatively new and is currently not as cost competitive with petroleum-based plastics. Many bioplastics are reliant on fossil fuel-derived energy for their manufacturing, reducing the cost advantage over petroleum-based plastic.

Performance and Usage

Many bioplastics also lack the performance and ease of processing of traditional materials albeit materials such as Bioplast from Stanelco have closed this performance gap. Polylactic acid plastic is being used by a handful of small companies for water bottles. But shelf life is limited because the plastic is permeable to water - the bottles lose their contents and slowly deform. However, bioplastics are seeing some use in Europe, where they account for 60% of the biodegradable materials market. The most common end use market is for packaging materials. Japan has also been a pioneer in bioplastics, incorporating them into electronics and automobiles.

Recycling

There are also fears that bioplastics will damage existing recycling projects. Packaging such as HDPE milk bottles and PET water and soft drinks bottles is easily identified and hence setting up a recycling infrastructure has been quite successful in many parts of the world. Polylactic acid and PET do not mix - as bottles made from polylactic acid cannot be distinguished from PET bottles by the consumer there is a risk that recycled PET could be rendered unusable. This could be overcome by ensuring distinctive bottle types or by investing in suitable sorting technology. However, the first route is unreliable and the second costly.

Genetically Modified Bioplastics

Genetic modification (GM) is also a challenge for the bioplastics industry. None of the currently available bioplastics - which can be considered first generation products - require the use of GM crops. However, it is not possible to ensure corn used to make bioplastic in North America is GM-free. European consumers are hostile to any products that are linked to the GM industry. As a result, some UK retailers such as Sainsbury's will not use bioplastic manufactured in the US, such as Natureworks polylactic acid. There is currently no commercial European source of polylactic acid bioplastic. There is also concern that the route from corn to bioplastics is not the most efficient. Looking further ahead, some of the second generation bioplastics manufacturing technologies under development employ the "plant factory" model, using genetically modified crops or genetically modified bacteria to optimise efficiency. However, a change in consumer perception of GM technology in Europe will be required for these to be widely accepted.

Market Size

Because of the fragmentation in the market it is difficult to estimate the total market size for bioplastics, but estimates by SRI Consulting put global consumption in 2006 at around 85,000 tonnes. In contrast, global consumption of all flexible packaging is estimated at around 12.3 million tonnes.

COPA (Committee of Agricultural Organisation in the European Union) and COGEGA (General Committee for the Agricultural Cooperation in the European Union) have made an assessment of the potential of bioplastics in different sectors of the European economy:

- Catering products: 450,000 tonnes per year
- Organic waste bags: 100,000 tonnes per year
- Biodegradable mulch foils: 130,000 tonnes per year
- Biodegradable foils for diapers 80,000 tonnes per year
- Diapers, 100% biodegradable: 240,000 tonnes per year
- Foil packaging: 400,000 tonnes per year
- Vegetable packaging: 400,000 tonnes per year
- Tyre components: 200,000 tonnes per year
- Total 2,000,000 tonnes per year

Certification

Biodegradability - EN 13432, ASTM D6400

The EN 13432 industrial standard is arguably the most international in scope and compliance with this standard is required to claim that a product is compostable in the European marketplace. In summary, it requires biodegradation of 90% of the materials in a commercial composting unit within 90 days. The ASTM 6400 standard is the regulatory framework for the United States and sets a less stringent threshold of 60% biodegradation within 180 days, again within commercial composting conditions. The "compostable" marking found on many items of packaging indicates that the package complies with either of the two standards mentined above. However, the marking is not owned by either regulatory body but by third party trade associations representing companies making or selling biodegradable plastics. In Europe, this is European Bioplastics, in the U.S. it is the Biodegradable Products Institute. Many starch based plastics, PLA based plastics and certain aliphatic-aromatic co-polyester compounds such as succinates and adipates, have obtained these certificates. Additivated plastics sold as fotodegradable or oxobiodegradable do not comply with these standards in their current form.

Biobased - ASTM D6866

The ASTM D6866 method has been developed to certify the biologically derived content of bioplastics. There is an important difference between biodegradability and biobased content. A bioplastic such as high density polyethylene (HDPE) can be 100% biobased (i.e. contain 100% renewable carbon), yet be non-biodegradable. These bioplastics such HDPE play nonetheless an important role in greenhouse gas abatement, particularly when they are combusted for energy production. The biobased component of these bioplastics is considered carbon-neutral since their origin is from biomass.

Applications

Because of their biological biodegradability, the use of bioplastics is especially popular for disposable items, such as packaging and catering items (crockery, cutlery, pots, bowls, straws). The use of bioplastics for shopping bags is already very common. After their initial use they can be reused as bags for organic waste and then be composted. Trays and containers for fruit, vegetables, eggs and meat, bottles for soft drinks and dairy products and blister foils for fruit and vegetables are also already widely manufactured from bioplastics. Non-disposable applications include mobile phone casings (NEC), carpet fibres (Dupont Sorona), and car interiors (Mazda). The French company, Arkema, produces a grade of bioplastic called Rilsan, which is being used in fuel line and plastic pipe applications. In these areas, the goal is obviously not biodegradability, but to create items from sustainable resources.

Plastic Types

Starch Based Plastics

Constituting about 50 percent of the bioplastics market, thermoplastic starch, such as Plastarch Material, currently represents the most important and widely used bioplastic. Pure starch possesses the characteristic of being able to absorb humidity and is thus being used for the production of drug capsules in the pharmaceutical sector. Flexibiliser and plasticiser such as sorbitol and glycerine are added so that starch can also be processed thermo-plastically. By varying the amounts of these additives, the characteristic of the material can be tailored to specific needs (also called “thermo-plastical starch”).

Polylactide Acid (PLA) Plastics

Polylactide acid (PLA) is a transparent plastic produced from cane sugar or corn starch. It not only resembles conventional petrochemical mass plastics (like PE or PP) in its characteristics, but it can also be processed easily on standard equipment that already exists for the production of conventional plastics. PLA and PLA-Blends (such as the CompostablesTM by Cereplast,Inc. generally come in the form of granulates with various properties and are used in the plastic processing industry for the production of foil, moulds, tins, cups, bottles and other packaging.

Poly-3-Hydroxybutyrate (PHB)

The biopolymer poly-3-hydroxybutyrate (PHB) is a polyester produced produced by certain bacteria processing glucose or starch. Its characteristics are similar to those of the petrochemical-produced plastic polypropylene. The South American sugar industry, for example, has decided to expand PHB production to an industrial scale. PHB is distinguished primarily by its physical characteristics. It produces transparent film at a melting point higher than 130 degrees Celsius, and is biodegradable without residue.

Polyamide 11 (PA 11)

PA 11 is a biopolymer derived from natural oil. It is also known under the tradename Rilsan B commercialized by Arkema. PA 11 belongs to the technical polymers family and is not biodegradable. Its properties are similar than PA 12 although emissions of greenhouse gases and consumption of non-renewable resources are reduced during its production. Its thermal resistance is also superior than PA 12. It is used in high performance applications as automotive fuel lines, pneumatic airbrake tubing, electrical anti-termite cable sheathing, oil & gas flexible pipes & control fluid umbilicals, sports shoes, electronic device components, catheters, etc.

Bio-Derived Polyethylene

The basic building block (monomer) of polyethylene is ethylene. This is just one small chemical step from ethanol, which can be produced by fermentation of agricultural feedstocks such as sugar cane or corn. Bio-derived polyethylene is chemically and physically identical to traditional polyethylene - it does not biodegrade but can be recycled. It can also considerably reduce greenhouse gas emissions. Brazilian chemicals group Braskem claims that using its route from sugar cane ethanol to produce one tonne of polyethylene captures (removes from the environment) 2.5 tonnes of carbon dioxide while the traditional petrochemical route results in emissions of close to 3.5 tonnes. Braskem plans to introduce commercial quantities of its first bio-derived high density polyethylene, used in a packaging such as bottles and tubs, in 2010 and has developed a technology to produce bio-derived butene, required to make the linear low density polethylene types used in film production.

Developments

- In the early 1950s, Amylomaize (>50% starch content corn) was successfully bred and commercial bioplastics applications started to be explored.
- In 2004, NEC developed a flame retardant plastic, polylactic acid, without using toxic chemicals such as halogens and phosphorus compounds.
- In 2005, Fujitsu became one of the first technology company to make personal computer cases from bioplastics, which are featured in their FMV-BIBLO NB80K line.
- In 2007 Braskem of Brazil announced it had developed a route to manufacture high density polyethylene (HDPE) using ethylene derived from sugar cane.

Biopolymers I

Many of the most important and most interesting naturally occuring molecules are polymeric. In general these biopolymers belong to one of five classes of materials:

Polyisoprenoids

Natural rubber, is a polymeric form of isoprene, 2-methyl-1,3-butadiene. 1 presents the structure of a tetrameric fragment of polyisoprene. The red lines connecting the 4th carbon atom of one isoprene unit with the 1st carbon of the next indicate that latex is an addition polymer that results from the 1,4-addition of one isoprene unit to the next. Note the head-to-tail pattern in which the isoprene units are connected. Note, too, that the stereochemistry is the same at each double bond, namely *cis*. The regularity that is evident in the structure of polyisoprene is characteristic of terpenes as well. Terpenes are oligomers of isoprene, typically containing 2, 4, or 6 isoprene units. Whether oligomeric or polymeric, these compounds all arise from a common biosynthetic path. Exercise 1 Assume that you polymerized a sample of isoprene by treating it with a catalytic amount of sulfuric acid. Use curved arrows to indicate how the polymer fragment shown in 1 would be formed. Draw the structures of the intermediate carbocations and show the resonance interaction between each cation and the double bond that is conjugated to it.

Unnatural Polyisoprenoids

Synthetic Rubber

In 1839 Charles Goodyear discovered, literally by accident, that heating natural rubber with elemental sulfur altered the properties of the polymer, most notably making it tougher and more elastic. Goodyear's discovery led to the development of synthetic rubber, a material that found its most profitable application in the manufacture of automobile tires. Investigation of the structure of synthetic rubber revealed that the sulfur had formed disulfide bonds that linked one polyisoprene chain to the next. As 2 demonstrates, these cross-links serve to restore the polymer to its original shape after it has been deformed by the application of a force.

Neoprene

Not surprisingly, the desireable properties of natural as well as synthetic rubber led to investigations of the polymerisation of structural analogs of isoprene. One notable success came from the polymerisation of 2-chloro-1,3-butadiene, sometimes called chloroprene. Polychloroprene is known commercially as neoprene rubber. It is widely used in the automotive industry for the manufacture of oil-resistant hoses. Neoprene that contains entrapped air has good insulating properties and is used in the production of wet suits. Exercise 2 Draw a tetrameric fragment of polychlorprene similar to that shown for polyisoprene in 1. Identify the repeat unit.

Polyphenols

Some of the structurally most complex natural polymers are the polyphenols that occur in woody plants. These materials are referred to as lignins. Degradation of lignins has revealed that they are composed of three basic building blocks, p-coumaric acid, ferulic acid, and syringyl alcohol. These structures are shown in 3 along with the structure of gallic acid, the principal phenolic component in tannins. 4 suggests a possible fragment of a lignin co-polymer based on these three monomers. The point here is simply that lignins are structurally very complex and, therefore, very difficult to characterize completely. Exercise 3 Cellulose is a polysaccharide. As such it contains many alcohol OH groups. While there are many alcohol OH groups in lignins, they also contain phenolic OH groups. The pKa values of alcohols and phenols are 16 and 10, respectively. The paper industry exploits this difference in pKa values to separate cellulose from lignins during the pulping process. Can you explain how?

Unnatural Polyphenols

Bakelite

In the presence of a catalytic quantity of acid, a mixture of phenol, C_6H_5OH, and formaldehyde, $H_2C{=}O$, condenses to form a polymeric material known as Bakelite. This was one of the first commercial polymers. It was used for many years in the manufacture of cases for radios and televisions as well as electronic circuit boards. The formation of this co-polymer involves several of the types of reactions we have discussed:

- nucleophilic addition
- nucleophilic aliphatic substitution
- electrophilic aromatic substitution

Biopolymer

Biopolymers are a class of polymers produced by living organisms. Starch, proteins and peptides, DNA, and RNA are all examples of biopolymers, in which the monomer units, respectively, are sugars, amino acids, and nucleotides.

Biopolymers Versus Polymers

A major but defining difference between polymers and biopolymers can be found in their structures. Polymers, including biopolymers, are made of repetitive units called monomers. Biopolymers inherently have a well defined structure: The exact chemical composition and the sequence in which these units are arranged is called the primary structure. Many biopolymers spontaneously fold into characteristic compact shapes (see also "protein folding" as well as secondary structure and tertiary structure), which determine their biological functions and depend in a complicated way on their primary structures. Structural biology is the study of the structural properties of the biopolymers. In contrast most synthetic polymers have much simpler and more random (or stochastic) structures. This fact leads to a molecular mass distribution that is missing in biopolymers. In fact, as their synthesis is controlled by a template directed process in most in vivo systems all biopolymers of a type (say one specific protein) are all alike: they all contain the same sequence and number of monomers and thus all have the same mass. This phenomenon is called monodispersity in contrast to the polydispersity encountered in synthetic polymers. As a result biopolymers have a polydispersity index of 1.

Conventions and Nomenclature

Polypeptides

The convention for a polypeptide is to list its constituent amino acid residues as they occur from the amino terminus to the carboxylic acid terminus. The amino acid residues are always joined by peptide bonds. Protein, though used colloquially to refer to any polypeptide, refers to larger or fully functional forms and can consist of several polypeptide chains as well as single chains. Proteins can also be modified to include non-peptide components, such as saccharide chains and lipids.

Nucleic Acids

The convention for a nucleic acid sequence is to list the nucleotides as they occur from the 5' end to the 3' end of the polymer chain, where 5' and 3' refer to the numbering of carbons around the ribose ring which participate in forming the phosphate diester linkages of the chain. Such a sequence is called the primary structure of the biopolymer.

Sugars

Structural Characterisation

There are a number of biophysical techniques for determining sequence information. Protein sequence can be determined by Edman degradation, in which the N-terminal residues are hydrolyzed from the chain one at a time, derivatized, and then identified. Mass spectrometer techniques can also be used. Nucleic acid sequence can be determined using gel electrophoresis and capillary electrophoresis. Lastly, mechanical properties of these biopolymers can often be measured using optical tweezers or atomic force microscopy.

Biopolymers as Materials

Some biopolymers- such as polylactic acid, naturally occurring zein, and poly-3-hydroxybutyrate can be used as plastics, replacing the need for polystyrene or polyethylene based plastics. Some plastics are now referred to as being 'degradable', 'oxy-degradable' or 'UV-degradable'. This means that they break down when exposed to light or air, but these plastics are still primarily (as much as 98 per cent) oil-based and are not currently certified as 'biodegradable' under the European Union directive on Packaging and Packaging Waste (94/62/EC). Biopolymers, however, will break down and some are suitable for domestic composting.

Biopolymers as Packaging

Biopolymers (also called renewable polymers) are produced from biomass for use in the packaging industry. Biomass comes from crops such as sugar beet, potatoes or wheat: when used to produce biopolymers, these are classified as non food crops. These can be converted in the following pathways:

Sugar beet > Glyconic acid > Polyglonic acid
Starch > (fermentation) > Lactic acid > Polylactic acid (PLA)
Biomass > (fermentation) > Bioethanol > Ethene > Polyethylene

Many types of packaging can be made from biopolymers: food trays, blown starch pellets for shipping fragile goods, thin films for wrapping.

Biopolymers are renewable, sustainable, and can be carbon neutral

Biopolymers are renewable, because they are made from plant materials which can be grown year on year indefinitely. These plant materials come from agricultural non food crops. Therefore, the use of biopolymers would create a sustainable industry. In contrast, the feedstocks for polymers derived from petrochemicals will eventually run out. In addition, biopolymers have the potential to cut carbon emissions and reduce CO_2 quantities in the atmosphere: this is because the CO_2 released when they degrade can be reabsorbed by crops grown to replace them: this makes them close to carbon neutral.

Some biopolymers arc biodegradable: they are broken down into CO_2 and water by microorganisms. In addition, some of these biodegradable biopolymers are compostable: they can be put into an industrial composting process and will break down by 90% within 6 months. Biopolymers that do this can be marked with a 'compostable' symbol, under European Standard EN 13432 (2000). Packaging marked with this symbol can be put into industrial composting processes and will break down within 6 months (or less). An example of a compostable polymer is PLA film under 20ìm thick: films which are thicker than that do not qualify as compostable, even though they are biodegradable. A home composting logo may soon be established: this will enable consumers to dispose of packaging directly onto their own compost heap. The standards for such a home composting logo have not yet been developed.

Biosurfactants and their roles in bioremediation

Introduction

The practical applications of biotechnology are vast and diverse; ranging from the cultivation of plant hybrids to the breeding of mice specifically designed as research instruments. It is a new world of possibilities, in which a spark of imagination one day can become the life-saving reality of the next. Many industries whose activities were traditionally remote from any living or biological systems have now recognised the tremendous potential of living cells in many areas, e.g. pre-treatment of raw material, product modifications, waste management, energy recycling and conservation and etc. Microbes and their products are nothing new to us as we have been using them since the ancient Egyptian era. Today, some of these products are used as specific foodstabilising agents, emulsifiers, vitamins, amino acids, proteins, food enzymes, specific fatty acids, water-binding agents, flavours, foaming agents and etc. Their applications also extend to the petroleum industries where various processes, like oil recovery, petroleum de-asphalting, viscosity control, desulphurisation, oil-spill control, waste management and detoxification, are enhanced. Our reliance on microbial systems to produce alternative liquid and gaseous fuels from renewable world biomass will become greater, in view of the rapid deletion of the petroleum reserves at current stage.

Recent advances in industrial biotechnology have generated a lot of interest, from the perspectives of both research scientists and investors. One of the many reasons for the interest in biote chnology is the prediction of sales of by-products. The world market was about U.S. $25 million worth some twenty years back, but it was U.S. $1.7 billion in 1992 and reached U.S. $500 million just last year. That was an increase at an amazing rate of 3 to 5 % annually (Lin, 1996; Desai and Banat, 1997).

Surfactants constitute an important class of industrial chemicals that are widely used in almost every sector of modern industry. During the last decade, the growing demand of surfactants within the U.S. chemical industry was noted. The global production in 1999 was 4*10 tonnes and this figure has been projected to increase many folds as we proceed to the 21st century. The various types of modern surfactants used in industries currently.

At present, majority of the surfactant that one can find in the market is chemically synthesised. The manufacture of surfactants is on such a large scale that their production has always been considered to lie within the realms of organic chemistry and chemical engineering. Nevertheless, much interest and attention have been directed toward biosurfactants in recent years, due to their broad-range functional properties and the diverse synthetic capabilities of microbes. The structural analysis of biosurfactants has also opened possibilities for their chemical synthesis. They are likely to gain wide acceptance since they are readily biodegradable and have lower toxicity as compared to their chemically synthesised counterparts.

Some facts about surfactants and biosurfactants Surfactants are amphiphilic molecules consisting of a hydrophilic and a hydrophobic domain. The former can be non-ionic, positively or negatively charged or amphoteric, but the latter is usually a hydrocarbon (Georgiou *et al.*, 1992; Desai *et al.*, 1994). Common non-ionic surfactants include ethoxylates, ethylene and propylene oxide copolymers and sorbitan esters, while fatty acids, ester sulphonates or sulphates(anionic) and quaternary ammonium salts (cationic) are some examples of commercially available ionic surfactants. Since both the hydrophilic and hydrophobic groups reside within the same molecule, surfactants tend to partition preferentially at the interfaces between fluid phases with a different degree of polarity and hydrogen bonding (like that of oil/water or air/water interfaces). Unique properties of the surfactant molecules were driven by the reduction of interfacial energy (interfacial tension) and surface tension through the formation of an ordered molecular film at the interface. These properties allow them to be used extensively in industrial applications involving emulsification, foaming, detergence, wetting and phase dispersion or solubilisation.

Many biological molecules are amphiphilic and partition preferentially at interphases. Microbial compounds that exhibit particularly high surface activity and emulsifying activity are classified as biosurfactants. At present, many types of biosurfactant are being utilised but they have been unable to compete economically with their chemically synthesised counterparts in the market, due to high production costs involved. However, this problem can be overcome by improving the efficiency of our current bioprocessing methodology and strain productivity, and the use of cost-effective substrates.

A variety of biosurfactants, of a broad spectrum of industrial applications, is synthesised by a wide range of microbes, particularly during their growth on water-immiscible substrates. They have definite structure: the lipophilic portion is usually the hydrocarbon (alkyl) tail of one or more fatty acids, which may be saturated, unsaturated, hydroxylated or branched; the fatty acid is linked to the hydrophilic group by a glycosidic ester or amide bond. Most of them are either neutral or negatively charged, the anionic character being due to carboxylate groups. According to Desai and Desai (1993), biosurfactants can be grouped into five major classes, namely glycolipids, phospholipids and fatty acids, lipopeptides or lipoproteins, polymeric surfactants and particulate surfactants. The types of biosurfactants, which are produced by microbes.

Physiological roles of biosurfactants

The main physiological role of biosurfactants is to permit microorganisms to grow on water-immiscible substrates by reducing the surface tension at the phase boundary, therefore making the substrate more readily available for uptake and metabolism. However, the molecular mechanisms of the uptake process of these substrates (e.g. alkanes) are still unclear. In addition to emulsification of the carbon source, they are also involved in the adhesion of microbial cells to the hydrocarbon. The cellular adsorption of the hydrocarbon-degrading microbes to water-immiscible substrates, and the excretion of surface-active compounds together allow growth on such carbon sources. In general, negatively charged biosurfactants inhibit, whereas positively charged biosurfactants promote microbial adhesion to hydrophobic phases. Furthermore, biosurfactants have been shown to be involved in antagonistic effect towards other microbes in the environment.

Biosynthesis of biosurfactant and its regulation The hydrophilic and hydrophobic moieties of biosurfactants are synthesised by two metabolic pathways: the hydrocarbon and carbohydrate pathways (Desai and Banat, 1997). These pathways are usually diverse in nature and utilise specific sets of enzymes. According to Syldatk and Wagner (1987), who documented various possibilities for the synthesis of different moieties of biosurfactants and their association, any of the following possibilities could occur:

(1) both hydrophilic and hydrophobic moieties are synthesised *de novo* by two independent pathways,

(2) the hydrophilic moiety is synthesised de novo while synthesis of the hydrophobic moieties induced by the substrate,

(3) the hydrophobic moiety is synthesised de novo while synthesis of the hydrophilic moieties induced by the substrate,

(4) both hydrophilic and hydrophobic moieties have substrate-dependent synthesis.

Generally, the regulation of biosurfactant production is governed by three main mechanisms, namely induction, repression and nitrogen and multivalentions. Induction of biosurfactant synthesis can be achieved by addition of long chain fatty acids, hydrocarbons or glycerides to the growth medium; while addition of D-glucose, acetate or tricarboxylic acids represses their production. Nitrogen or metal iondependent regulation also played a prominent role in biosurfactant synthesis. It was observed that rhamnolipid synthesis in *Pseudomonas aeruginosa* occurs upon exhaustion of nitrogen and commencement of the stationary phase of growth (Guerra-Santos *et al.*, 1984; Ramana and Karanth, 1989). Lastly, the limitation of multivalent cations causes overproduction of biosurfactants.

Large Scale Production of Biosurfactants

Most of the biosurfactants are released into the culture medium either at the stationary phase or throughout the exponential phase. The production of biosurfactant has been carried out in batch fermentation or continuous fermentation at low dilution rates. Immobilised catalysis has been used for the biosurfactants that could be produced by resting cells. Alternatively, enhanced production of biosurfactants can be achieved by the application of an airlift fermentor and aqueous two-phase fermentation.

Application of Biosurfactants and Future Potential

A number of industrial of industrial applications of biosurfactants has been envisaged. One of the potential uses is in the oil industry with minimum purity specification so that whole cell-broth could be used. They are very selective, required in small amounts, effective under broad ranges of oil and reservoir conditions and environmentally friendly; especially in the safeguarding the coastal areas from additional damage caused by synthetic chemicals. An estimated 30% increase in total oil recovery from underground sandstone by using trehalolipids from *Norcardia rhodochrous* has been documented. Recently, Multi-biotech (a subsidiary of Geodyne Technology) has commercialised biosurfactants for enhanced oil recovery applications. *Bacillus licheniformis* JF-2, an isolate from oilfield injection water which, in addition to producing the most effective biosurfactants, has other characteristics like being anaerobic, halotolerant, and thermotolerant, produces biosurfactants that are potentially useful for in situ microbially enhanced oil recovery. As such, the viscosity of heavy crude oil can be greatly reduced when it is treated with this biosurfactant. In addition, the use of a biosurfactant for desludging of a crude oil storage tank for Kuwait Oil Co. has been reported (Banat *et al.*, 1991)

Apart from these, biosurfactants are also used in emerging technologies like microbial remediation of hydrocarbon and crude oil–contaminated soils. Hydrocarbon contaminants are removed from the environment, primarily, as a result of their biodegradation, which is performed by native microbial populations. Such biodegradation is known to be time-consuming and new technologies have been developed; for example bioaugmentation, biosurfactants and etc. While, the effectiveness of enhancing hydrocarbon degradation through addition of microbial inocula of nonindigenous populations (bioaugmentation in essence) has been ambiguous, the use of a combination of new approaches has shown much success. For instance, the addition of biosurfactant help to stimulate the indigenous microbial population to degrade hydrocarbons at rates higher than those which could be achieved through addition of nutrients alone. A good example of this is the use of rhamnolipid from *Pseudomonas aeruginosa* to remove large quantities of oil from contaminated

Alaskan gravel a decade back, in the *Exxon Valdez* oil spill (Harvey *et al.*, 1990). Since then, Van Dyke *et al.* (1993) and Bragg *et al.* (1994) have demonstrated increased hydrocarbon recovery from the contaminated sites by rhamnolipid and the effectivenes of *in situ* bioremediation on the *Exxon Valdez* oil spill respectively. The application of biosurfactant on the remediation of metal, phenanthrene and polychlorinated biphenyl (PCB) contaminated soil has recently been showed. For instance, Golyshin *et al.* (1999) conducted a study to determine the potential positive effect of novel biosurfactants on the enhancement of Aroclor 1248 metabolisation in both *in vitro* and *in situ* experiments. Among two lipopeptides tested the highest activity was found in experiments with a hydrolytically opened form of lichenysin A. Lichenysin A itself did not enhance the degradation activity of chosen microorganismdegraders and in most cases inhibited their PCB mineralisation rates. Glucolipid surfactant from marine bacterium *Alcanivorax borkumensis* (*A. borkumiensis*) showed in several tests a strong enhancing effect on microbial metabolisation of Aroclor 1248 congeners. Biosurfactants appeared to act very specifically, i.e. depending on strain and concentration used. According to Zhang and Miller (1995), the three-way interaction between biosurfactant, substrate and cells is very critical to the enhanced rate of biodegradation. Experiments set up with soil samples did not give a clear answer whether bioemulsifiers applied at low concentration could sufficiently increase the rates of biodegradation *in situ*. Only *A. borkumiensis* glucose lipid caused the most marked enhancement of Aroclor 1248 metabolisation in soil microcosm.

Taking into account the specificity of surface- and biological activities of various biosurfactants, it was suggested that they might promote the mineralisation of sorbed PCBs in polluted soils, when the optimised biosurfactant-degrader combination is used. The ability to emulsify hydrocarbon-water mixtures is another established property of microbial biosurfactant. It has been shown to increase hydrocarbon degradation significantly and is therefore potentially applicable to oil spill management. Schulz *et al.* (1991) isolated biosurfactant-producing *Alcaligenes sp.* strain MM-1, *Arthrobacter sp.* strains EK1 and S-II from the North Sea, while screening for oil-degrading marine microorganisms. Dave *et al.* (1994) demonstrated that the use of a mixture of hydrocarbon-degrading microbes for bioaugmentation of soil contaminated with slop oil from a petrochemical industry resulted in the bioreclamation of soil, while Ghosh *et al.* (1995) showed an enhancement in bioremediation of polyaromatic hydrocarbons and soil contaminated with PCBs.

In addition to these, Deziel *et al.* (1996) has shown biosurfactant production by polycyclic aromatic hydrocarbon-metabolising *Pseudomonas aeruginosa* 19SJ. Biosurfactants are also of great value to bioremediation of sites heavily contaminated with toxic heavy metals such as uranium, cadmium, lead and etc. (Miller, 1995). These observations suggest the application of natural microbial consortia and their products is a feasible approach to resolve certain environmentalrelated problems caused by mankind.

Moving away from the petrochemical industry, biosurfactants also found several applications in the food, healthcare and cosmetic industries. In the food industry, they are potential candidates in the search for functionally different products because they meet the requirements of functional food additives. Furthermore, lecithin and its derivatives, fatty acid esters containing glycerol, sorbitan, or ethylene glycol and ethoxylated derivatives of monoglycerides are currently in use as emulsifiers in the food industries globally. In the healthcare industry, biosurfactants are used commercially as a skin moisturiser due to their excellent skin compatibility (Yamane, 1987). Antibiotic effects and inhibition of growth of human immunodeficiency virus (HIV), herpes simplex virus and influenza virus by biosurfactants have been documented (Uchida *et al.*, 1989; Kitamoto *et al.*, 1993). In the cosmetic industry, biosurfactants play a pivotal role in the production of many compounds for cosmetic applications. Kao Co. Ltd., for instance, uses sophorolipid as a humectant for cosmetic makeup brands such as Sofina. This organisation has developed a fermentation process for sophorolipid production, and after a two-step esterification process, the product can be used in lipsticks and as moisturiser for skin and hair products. Other potential areas of application include that of the pulp and paper, coal, textile, ceramic processing and uranium ore-processing industries. For example, a heteropolysaccharide containing D-mannuronic acid and D-guluronate from *Macrocystis pyrifera* (kelp) and *Azobacter vinelandii* and polyglutamate from *Bacillus licheniformis* have been used as effective dispersants in the ceramic processing industry.

Recent interest has been generated in the production of biosurfactants due to their environment-friendly composition and wide range of applications. Their market is extremely competitive and manufacturers will have to focus on product development to capture the market for the year 2000 and beyond. Ultimately, it is the net economic balance between their production cost and functional benefits that controls the future of biosurfactant. To strike a fine balance between the two, one has to develop a costeffective process that utilises low-cost materials and, on the other hand, achieving products of high yield and activity. Since biosurfactants and the microbes that produce them have numerous industrial, medical and environmental applications that often involve exposure to extremes of temperatures, pressure, pH and etc., there will be an urgent need to isolate extremophiles or to generate super-active genetically engineered bacteria that are capable of functioning under extreme conditions.

A Look to the Future

As we review the history of xenobiotic metabolism the rapid acceleration of the pace of discovery becomes quite clear. From the beginning of the 19th century reactions were enumerated at a slow deliberate pace. As the 20th century began the basics of transformation-oxidation, reduction, hydrolysis, and synthesis were understood but the coherence of applications was unclear.

As the new millennium dawns we find xenobiotic disposition playing a major role in understanding individual response to drugs and to the chemical environment. It takes little imagination to see the day when a patient's knowledge of their genetic make-up and degree of expression is as important as the knowledge of the disease in shaping appropriate therapy. An understanding of how minor genetic changes can alter an individual response to the environment will build our database for controlling our response to new agents.

The combination of genomics and proteomics and powerful analytical techniques will allow us to probe human responses at the molecular level. The use of high powered computer programmes should allow us to assemble the molecular data into predictive paradigms relating to the individual organism's response to its surroundings. The dawn of the new millennium holds great promise and challenge for the field of xenobiotic metabolism.

Xenobiotic

A xenobiotic is a chemical which is found in an organism but which is not normally produced or expected to be present in it. It can also cover substances which are present in much higher concentrations than are usual. Specifically, drugs such as antibiotics are xenobiotics in humans because the human body does not produce them itself nor would they be expected to be present as part of a normal diet. However, the term is also used in the context of pollutants such as dioxins and polychlorinated biphenyls and their effect on the biota. Natural compounds can also become xenobiotics if they are taken up by another organism, such as the uptake of natural human hormones by fish found downstream of sewage treatment plant outfalls, or the chemical defenses produced by some organisms as protection against predators. Some xenobiotics can induce apoptosis at lower doses and necrosis at higher doses. Xenobiotics at low concentration can induce precommited reversible apoptotic phase which is followed by irreversible apoptotic phase.

Xenobiotic Metabolism

The body removes xenobiotics by xenobiotic metabolism. This consists of the deactivation and the secretion of xenobiotics, and happens mostly in the liver. Secretion routes are urine, feces, breath, and sweat. Hepatic enzymes are responsible for the metabolism of xenobiotics by first activating them (oxidation, reduction, hydrolysis and/or hydration of the xenobiotic), and then conjugating the active secondary metabolite with glucuronic or sulphuric acid, or glutathione, followed by excretion in bile or urine. An example of a group of enzymes involved in xenobiotic metabolism is hepatic microsomal cytochrome P450. These enzymes that metabolize xenobiotics are very important for the pharmaceutical industry, because they are responsible for the breakdown of medications. Organisms can also evolve to tolerate xenobiotics. An example is the co-evolution of the production of tetrodotoxin in the rough-skinned newt and the evolution of tetrodotoxin resistance in its predator, the common garter snake. In this predator-prey pair, an evolutionary arms race has produced high levels of toxin in the newt and correspondingly high levels of resistance in the snake. This evolutionary response is based on the snake evolving modified forms of the ion channels that the toxin acts upon, so becoming resistant to its effects.

Xenobiotics in the Environment

Xenobiotic substances are becoming an increasingly large problem in Sewage Treatment systems, since they are relatively new substances and are very difficult to categorize. Antibiotics, for example, were derived from plants originally, and so mimic naturally occurring substances. This, along with the natural monopoly nature of municipal Waste Water Treatment Plants makes it nearly impossible to remove this new pollutant load. Some xenobiotics are resistant to degradation. For example, they may be synthetic organochlorides such as plastics and pesticides, or naturally occurring organic chemicals such as polyaromatic hydrocarbons (PAHs) and some fractions of crude oil and coal. However, it is believed that microorganisms are capable of degrading all the different complex and resistant xenobiotics found on the earth.

Inter-Secies Ogan Tansplantation

The term xenobiotic is also used to refer to organs transplanted from one species to another. For example, some researchers hope that hearts and other organs could be transplanted from pigs to humans. Many people die every year whose lives could have been saved if a critical organ had been available for transplant. Kidneys are currently the most commonly transplanted organ. Xenobiotic organs would need to be developed in such a way that they would not be rejected by the immune system. With the development of vitrification transplantable organs could be stored in organ banks for long periods.

The Composting Toilet - an Ecologically Responsible Solution

The composting toilet solves with one blow the problems of "black waste water" and loaded nutrient salts in the recipient and overuse of our precious water resources.

Water Consumption

Our daily consumption of water is 180-200 litres which translates to approximately 70,000 litres of drinking water/person/year. Ordinary flush toilets are responsible for 15-20,000 litres of this amount. Composting toilets do away with this water wastage and at the same time keep feces, paper and even urine out of the sewage system.

The Composting Toilet

When one mentions the word "composting toilet" or "mulch latrine", people immediately protest and come up with such comments as, "you don't mean you expect us to go back to the use of stinky latrines ?!" and "doesn't it smell like slurry ?!" But on this matter I must be firm - today we have an odourless and comfortable solution very suitable for installation inside the home. In contrast to many so-called "ingenious" solutions such as incineration toilets, freezer-bag toilets, dry toilets and the like, which often require the use of large amounts of energy, a genuine composting toilet consists of a large container in which a composting process takes place before the contents are removed, and a toilet stool + seat very similar in appearance to that of an ordinary flush toilet. The end product of the composting toilet is an odourless humus material totally unlike manure or the contents of a latrine. (there is no danger in handling it and it can be used as fertilizer for shrubs and decorative plants as long as it has had soil contact for 19 months)

Denmark's Neighboring Countries

During the past few decades, individuals in Norway and Sweden have been continuously working on the development of composting toilets, so that today we have a new generation of composting toilets. In 1982, according to the Nordic Ministercounsil, there were 300,000 composting toilets installed throughout Scandinavia. I suppose that Denmark accounts for only about 100 of these. Development has continued at a fast rate in the neighbouring countries and the latest encouraging news is that Tannum "municipality", near Goteborg, Sweden, has recently forbidden flush toilets from the year 2000 and currently builds 70% of their new housing without flush toilets.

Development in Denmark

Because of the fact that 7% of Danish housing (in rural areas) aren't connected to central sewage systems, some research on composting toilets has been undertaken at The Danish Technical University and the most current news in this field is that a project called "Kompost-toiletter 1994" by Simon Wris-Berg and Peter Friis Moller has been awarded a prize by Denmark's agricultural school. Some Danish firms have experienced an increase in inquiries about composting toilets during the past few years due both to increasing environmental awareness as well as ever increasing water costs.

The Design of a Composting Toilet

Composting toilets vary in appearance depending upon the manufacturer, but common for them all, is that they have a collection container for feces, paper, in some cases kitchen garbage, while some separate urine into a second container. The contents of the feces container are composted by aerobic process, which takes about one year, into a humus like material. In the company "Renewable Energy", we have developed a year-round model called the "Eco-composter" for family homes. The collection container of our toilets is made of fibre reinforced concrete which makes for a reduction in use of fibreglass and epoxy. The toilet stool, which is made of either fibre-concrete or porcelain, is designed so that urine and feces are kept separate. The feces, paper and kitchen garbage drop vertically down into one tank, while the urine is collected at the front - a design appropriate for both female and male use. In this way the urine can, without coming into contact with the feces, be collected in a tank and because there is no danger of disease, be used as liquid fertilizer when mixed with 4-5 times its amount in water. The urine can also, of course, be mixed with household "grey waste water". But this must be regarded as the second solution. Our porcelain stool is designed in such a way that the urine compartment can be flushed with 0.1 litres of water. The tank has ventilation channels to enable the oxidising process to take place, and all air is drawn up through a vent which reaches up through and over the roof of the house. This means that the composting toilet itself functions as a ventilation system for the entire toiletroom, keeping out unwanted smell. This makes for quite an advantage - there is never any unpleasant smell connected with a "visit" to the toiletroom.

The components of our composting toilet take up the same amount of space as a traditional flush toilet in the toiletroom itself, while the collection-container and ante-chamber are located beneath the floor of the toiletroom or from a connected "lightbox" located outdoors.

Infection and the Authorities

The Danish authorities require that the same minimum criteria are fullfilled as with sludge delivery, due to a risk of disease-carrying bacteria and parasites, although the compost can always be spread on the ground because the amount of heavy metals in human feces + urine is extremely low when not mixed with other wastewater (see table 1) The compost-humus could of course be pasteurized at 70 degrees for a minimum of 1 hour, but this holds no interest for us, because we want to avoid the use of large amounts of energy and retain the good balance with nutrients which are present in the compost-humus. We don't want to use energy for no reasons. We require that the compost is brought in contact with soil bacteria and kept behind a little fence for yet another year before being spread out on gardens or fields. This requirement is easy to live with in cases where here is a complete plan for the composting toilet and the "grey wastewater", for instance first via a microbiological cleaning in a sandfilter and finally by means of gravitation in an "energy-forest draining system" so that we really can put the compost in the "energy-forest". Today we are no longer confronted with the same resistance to this installation design as 10 years ago, due to greater environmental awareness and stricter wastewater treatment requirements.

The Future

We will experience fast and hopefully creative development in the area of composting toilets combined with more research which initially can find support in the results of our neighbouring countries. Because the lack of fresh water is a global problem, these solutions are quite suitable for export to other countries.

Sewage Treatment

The wastes generated by some 60% of the U.S. population are collected in sewer systems and carried along by some 14 billion gallons of water a day. Of this enormous volume, some 10% is allowed to pass untreated into rivers, streams, and the ocean. The rest receives some form of treatment to improve the quality of the water (which makes up 99.9% of sewage) before it is released for reuse.

Biochemical Oxygen Demand (BOD)

The BOD is an important measure of water quality. It is a measure of the amount of oxygen needed (in milligrams per liter or parts per million) by bacteria and other microorganisms to oxidize the organic matter present in a water sample over a period of 5 days. The BOD of drinking water should be less than 1. That of raw sewage may run to several hundred. It is also called the "biological" oxygen demand.

Primary Treatment

The simplest, and least effective, method of treatment is to allow the undissolved solids in raw sewage to settle out of suspension forming sludge. Such primary treatment removes only one-third of the BOD and virtually none of the dissolved minerals. Attempts to use digested sludge as a fertilizer have been hampered by its frequent contamination by toxic chemicals derived from industrial wastes.

Secondary Treatment

However, many treatment plants in North America then pass the effluent from primary treatment to secondary treatment. Here the effluent is brought in contact with oxygen and aerobic microorganisms. They break down much of the organic matter to harmless substances such as carbon dioxide. Primary and secondary treatment together can remove up to 90% of the BOD. After chlorination to remove its content of bacteria, the effluent from secondary treatment is returned to the local surface water.

Advanced Waste Treatment

The combination of primary and secondary treatment removes most of the organic matter in sewage and thus lowers the BOD. However, most of the nitrogen and phosphorus in sewage remains in the effluent from secondary treatment. These inorganic nutrients can cause eutrophication of surface water receiving the effluent causing blooms of algae. To avoid this, a few communities add a third stage of treatment called tertiary or advanced waste treatment. Several techniques are available to remove dissolved salts from sewage effluent, but all are quite expensive.

Sewage Treatment

There are few aspects that are more responsible for moving civilisation out of the Middle Ages and into the Modern Age than is the treatment of our municipal and industrial sewage. This and the treatment of our drinking water (potable water) comprise THE two major advances in public health that have contributed to greatly extended life expectancies due to the control of infectious diseases. (What non-infectious disease kills the most people worldwide?) Anyway, if you do not recognise the significance of this technology to your life, go hit your head on a brick wall until you do appreciate its importance. IT IS VERY IMPORTANT, and, as a citizen, you should know about it, unpleasant as it may be to your nose.

In essence, sewage treatment has two functions:

- to kill pathogens (whatever they are) and
- to eliminate harmful chemicals from the water.
  - o chemicals that are toxic per se, and
  - o the bulk of organic chemicals that cause the depletion of life-supporting oxygen in the receiving waters of a stream, river or lake.

Let's take a look at the process and equipment your locality possesses. While you follow the steps, take note of what "primary treatment" is, and what "secondary treatment" is.

1. The first part of sewage treatment is concerned with the removal of large items that enter the treatment plant. These items can jam the system, can consist of twigs, Volkswagens, gravel, animals, tampons and condoms. The raw influent first goes through a self-cleaning screen (B = Bar screen), and then into one end of a shallow and rather fast moving basin (S = Sand basin) so that sand and gravel can settle out. Often skimmers rotate around the surface of the basin to remove oils that may have been flushed into the system.

2. In order to provide for the further processes - such as the addition of flocculants, the volume of the incoming raw sewage water must be measured (M = Meter). One of the most common types of meters that has no moving parts to become clogged or jammed is the kind shown here, and is known as a Parshal Flume. As the volume of water increases, the difference in water level on each side of the constriction increases. This difference is then translated into flow-through volume. With today's electronics, signals are transferred downstream for the needed adjustments for chlorine, flocculants, etc.

3. Finally the waste water enters the first step of sanitisation and is directed into the Primary Clarifier (PC). The theory behind this device is that it creates an ecosystem that is very alien to cellular pathogens (e.g.: bacteria) of humans. It does this by supporting anaerobic growth of microbes that also digest a small amount of the dissolved organic material (BOD, see later) while they produce lots of sulfides and other chemicals that are noxious to both noses and to the pathogens of humans. The construction is a huge circular vat with a metal curtain that does not quite reach the bottom. The water comes in between the curtain and the outer wall; gently and slowly flows under the curtain and rises to spill into a receiving trough at the

top of the tank. As microbes digest the organic materials they grow and floc together. Large quantities settle to the bottom of this tank. If you are lucky to find the treatment plant, which you are visiting, has spare, empty tanks, you can see scrapers on the bottom that slowly sweep the sludge into a hole. The sludge is then augered into a sludge holding tank. You have now finished what is called "Primary Treatment." The effluent is more or less safe with regard to bacterial pathogens, but it is a disaster to dump this into a stream or river because it contains so much dissolved organic material. In the receiving body of water, which is filled with lots of its own microbes, all that organic material is digested at the expense of the oxygen that is dissolved in that river or lake. This is called biological oxygen demand (BOD) which is directly correlated with the amount of rapidly biodegradable organic material. It is disasterous to dump this "high BOD" water into a lake because soon all the fish and other aerobic creatures will die, and you will get a fetid, stagnant cess pool.

4. Rather than dumping the "primary effluent" into a beautiful natural stream, let us make an artificial lake and vigorously aerate the water to promote the rapid growth of microbes that will "eat up" the BOD before the water gets to the river. Hence we have an "aeration chamber" (AC). In the days before electric power, the water was allowed to sprinkle out over the surface of a basin that was filled with rocks. The water would trickle down through the rocks before exiting at the bottom. The water was thus exposed to a great amount of surface area, and the rocks became coated with a slime of all sorts of aerobic microbes that rapidly dropped the BOD. On the sunlit top rocks, algae vigorously grew, but just below them the slime turned to a vast microbial dog-eat-dog ecosystem of bacteria, fungi, grazing amoebae, snails. An alternative way of doing this is by making huge silos filled with plastic honeycombs. The water is sprayed on top to trickle down through the honeycomb while air is forced upwards through the honeycomb. Yet a third way of promoting aeration is shown in this diagram: giant "egg beaters" churn and splash the water. A fourth way is even more direct: bubble air up through the water from the bottom. All these alternative devices are with the purpose of converting the BOD to either carbon dioxide (which wafts away) or to cell matter (for the next step).

5. We take the outflow from the aeration chamber, and try to recover as much of the microbial growth as possible to feed it back into the beginning of the aeration process. This is done with another clarifier (SC = Secondary Clarifier). Many treatment plants have secondary clarifiers that are almost identical to the primary clarifier's design. But, for change of pace, here is shown one where the incoming water flows down a central shaft and slowly works its way back up to the top, where is spills over at the edges into a circular trough. Often alum or other flocculant is added to the incoming water to assist in the aggregation of the microbial cells into globs that a big enough to sink. The water that spills out of this tank has passed "secondary treatment." It is surprisingly clear water.

6. Before allowing the effluent to reach the river or other receiving natural body of water, a final safety step is taken - the disinfection of the water. Shown here is a maze into which chlorine is added at the beginning, and remains in "contact" until it reaches the end of the tank. (CC = Contact Chamber)

Because excess chlorine is harmful to the receiving body of water, the amount added must be carefully monitored at the outfall. Because chlorine is so extremely toxic, alternative methods are used elsewhere. Ozone is commonly used in Europe, which is generally more advanced in its treatment of sewage than is North America. (The reason is that Europe is much more crowded and the tolerance of receiving bodies of water cannot absorb overages there, while much less crowded North American watersheds can absorb and "mistakes." Other processes are being considered: ultraviolet radiation; high intensity sound (sonic disruption of cells), and very intense 1,000 cps electrical current. It is always better to add "physics" to a system because there is no residue to be monitored. It is easy to get rid of "physics" - just turn it off!

There are two other components of sewage that must be considered that often escape the above "secondary treatment" plants.

- Nitrogen compounds, and
- Viruses

"Tertiary Treatment Plants" do one further step after the foregoing: they allow the secondary effluent to flow into large ponds where algae grows and uses up the ammonia and nitrates in the water. This is a slow process and requires very large holding ponds. The algal growth is harvested and used as a nitrogen-rich organic fertilizer.

The elimination of viruses from sewage is problematic. Addition of chlorine is known to have little effect on many types of viruses. The European usage of ozone is known to damage and thus inactivate most viruses, and so that process is better than the use of chlorine. Ultraviolet irradiation of the effluent should be highly effective, but there is a problem that even the slightest amount of any organic chemical containing double-bonds absorbs and renders UV rather ineffective.

A class project to test one's local sewage treatment plant's efficacy in eliminating viruses is rather simple. Just as public health use usually harmless E.coli as an indicator of fecal pollution, you can use those viruses that infect only E.coli as your indicator viruses. Of course, the treatment plant's raw influent waters contain huge amounts of E.coli. Therefore, these waters ought also contain even greater amounts of those viruses that infect E.coli. Those viruses have a nickname "coliphages" (after the bacteriophages of E.coli, where the term bacteriophage means a virus of bacteria).

1. Obtain 5 ml samples of treatment plant waters taken at various stages along the process.
2. Add 1 ml of chloroform to the samples; shake vigorously for a few minutes. The chloroform emulsion kills all the cells in the samples, but does not harm the coliphages.
3. Allow the chloroform to settle to the bottom of the tubes.
4. Make some dilutions of the overlying aqueous phases of these tubes.
5. Mix 0.05 ml of those dilutions with 0.05 ml of an overnight culture of a lab strain of E.coli (such as K-12, C600, B, etc.)
6. Transfer to the middle of the agar in a nutrient agar filled petri plate, and spread the droplet evenly over the surface of the agar. (Use a presterilized bent glass or metal rod for this purpose.) The droplet will soon be absorbed into the agar leaving the bacteria and coliphages stranded high and dry on the surface.
7. Incubate at body temperature; results often appear within 6 to 9 hours.

As the thousands of stranded bacteria begin to grow, some cells come in contact with a coliphage, which infects the contacting cell. In about 40 minutes, that cell bursts releasing a hundred or so progeny coliphages, which infect surrounding cells. The infection and bursting continues upon the "lawn" of growing E.coli resulting in holes or "plaques" in the lawn. By holding the plate up to the light, the plaques become visible. You should see plaques in the samples derived from raw influent water. Do you see any decrease in numbers of plaques as the water moved through the treatment process? Probably. But is there complete removal of "plaque-forming units" (pfu's)? If not, what does that mean? While coliphages are not a health hazard to you, they do indicate that other viruses - flu, hepatitis, HIV - might also escape the treatment process. If you find that your treatment plant is not removing the coliphages, it is something to think about very seriously. But do not run off to the news reporters with your information. You will not accomplish anything constructive. Instead, try to work with the plant's engineers to discover a way to treat the viruses. You will make history in a constructive way. Society needs this discovery.

Hyperion Sewage Treatment Plant

The Hyperion Wastewater Treatment plant is located in southwest Los Angeles, California next to Dockweiler State Beach on Santa Monica Bay. The largest such facility in the Los Angeles Metropolitan Area, Hyperion is operated by the City of Los Angeles, Department of Public Works, Bureau of Sanitation.

Until 1925, raw sewage from the city of Los Angeles was discharged untreated directly into Santa Monica Bay in the region of today's Hyperion Treatment Plant.

With the population increase, the amount of sewage became a major problem to the beaches, so in 1925 the city of Los Angeles built a simple screening plant in the 200 acres the city had acquired in 1892. Even with the screening plant, the quality of the water in the Santa Monica Bay was unacceptable, and in 1950 the city of Los Angeles opened the Hyperion Treatment Plant with full secondary treatment processes. In addition, the new plant included capture of biogas from anaerobic digesters to produce heat dried fertilizer. In order to keep up with the increase of influent wastewater produced by the ever growing city of Los Angeles, by 1957 the plant engineers had cut back treatment levels and increased the discharge of a blend of primary and secondary effluent through a five-mile pipe into the ocean. They also opted to halt the production of fertilizers and started discharging digested sludge into the Santa Monica Bay through a seven-mile pipe.

Environmental Forces

The discharge of sewage effluent into Santa Monica Bay changed the ecosystem so dramatically that by the 1970s, only worms and hardy clams existed in the ocean floor of the bay. The increasing volume of treated sewage being dumped into Santa Monica Bay captured the attention of a group of concerned users of the bay, who started to observe a decrease in the number and quality of fish in the bay, sick dolphins and people. From this movement, Heal the Bay was founded by Dorothy Green in 1985.

Heal the Bay joined the Environmental Protection Agency in a lawsuit against the city of Los Angeles to force the sewage treatment done by Hyperion to comply with the Clean Water Act. Heal the Bay and EPA succeeded and by November 1987, Hyperion stopped discharging sludge into the Santa Monica Bay and by December 1998, Hyperion had a full secondary treatment plant. For its significance and impact to quality of life, in 2001 the Hyperion Treatment Plan was named one of the Top Ten Public Works project of the 20th century by the American Public Works Association.

Treatment System

The plant treats approximately 350-450 million gallons per day of raw sewage. The sewage undergoes both primary and secondary treatment. The wastewater produced by houses, businesses, and industries, flows in the sewer system, separately from the storm drain system to the Hyperion treatment plant. Once the wastewater reached the plant, the first treatment stage is to remove large objects such as plastic, rags, metals, and wood. The wastewater then flows into sedimentation tanks were about 15 tons of sand and other materials settle to the bottom of the tank and are removed every day. The chemical treatment starts with the addition of coagulants to the wastewater to improve the settling of small particles. At Hyperion, the wastewater flows during one hour in underground tanks after the addition of coagulants. The settled solids are pumped to digesters and oil and grease are skimmed off the top of the wastewater. After the removal of solids, the wastewater is pumped to tanks where aerators deliver 96% oxygen to the wastewater, providing microorganisms with the necessary conditions to decompose organic solids in the wastewater. The wastewater spends about one to two hours in the oxygen reactors. Now that the microorganisms fed on most of the organics, they are able to settle to the bottom in quiescent conditions. The wastewater is pumped into settling, clarifying tanks where it stays for about four hours. After this time, about 90-95% of the solids have been removed from the wastewater and the effluent is clean enough to be discharged or recycled. About 94% of the water is discharged into the Santa Monica Bay through a pipe 5 miles out from the shore and 190 feet deep. Approximately 30 million gallons of secondary effluent per day is pumped to the West Basin Municipal Water District where the water is filtered further and reused for various purposes.

All the solids settled throughout the several stages in the treatment are treated in anaerobic digesters. The idea is to destroy pathogens before the sludge can be dumped in landfills. At Hyperion, the biosolids stay about 15 days in the digesters – enough time for the anaerobic bacteria to consume pathogens. One by-product of this process is the production of methane gas, which is captured and processed by a nearby power plant. After the "sterilisation", the biosolids have their water removed to reduce volume and thus reduce transportation costs. The dewatering is achieved by using centrifuges with variable settings and addition of polymers (Horenstein et al, 1990). Each 1% decrease in volume results in an economy of $1 million in disposal costs (Horenstein et al, 1990). The dewatered biosolids are then transported to landfills or to locations where it can be reused.

Recently, the public has become aware of the presence of pharmaceutical products in the nation's waterways. Several studies, such as (Jones-Lepp and Stevens, 2007) have shown the presence of pharmaceuticals and personal care products in biosolids/sewage sludge from wastewater treatment plant. How to reduce the presence of pharmaceuticals and personal care products may become a major challenge in the future not only for Hyperion, but for all wastewater treatment plants. However, currently there are no requirements for Hyperion to remove such products.

Sewage Treatment Systems

Changes to the Sewage Treatment Systems Law and Rules

Amended Substitute House Bill 119 (Am. Sub. HB 119), passed by the 127th General Assembly, contains substantial amendments to the Ohio Revised Code (ORC) and the Ohio Administrative Code (OAC) regarding the regulation of household and small flow on-site sewage systems in Ohio. The sewage treatment system rules adopted by the Public Health Council that became effective on Jan. 1, 2007, have been rescinded as required by the bill. The bill also enacts several uncodified provisions into state law that take effect July 1, 2007. These uncodified provisions are effective until July 1, 2009, and have substantial impact on the sewage programmes implemented by the Ohio Department of Health (ODH) and local health districts.

Excerpts from Am. Sub. HB 119

Am. Sub. HB 119 suspended operation of most portions of ORC Chapter 3718, the sewage law, until July 1, 2009. This law, which had allowed for continued state oversight and the ability in the future to effectively treat sewage for homes and businesses not served by public sewer systems, is suspended in the interim. During this time, further work and research will be done to assess the available technologies, consider the current types of systems and rates of failure across Ohio and determine the best ways to ensure that future systems achieve treatment of sewage to protect the public health, the environment and homeowners' investment in their property in a cost-effective manner.

In compliance with Am. Sub. HB 119, the director of Health adopted statewide interim sewage rules effective July 2, 2007. The Public Health Council, at its July 25, 2007 meeting, adopted these rules as minimum standards through July 1, 2009. Household sewage disposal systems (HSDS) serving one-, two-, or three-family dwellings and small flow on-site sewage treatment systems (SFOSTS) are regulated under OAC Chapter 3701-29 as minimum standards. Local health districts are responsible for code enforcement, and are permitted to adopt more stringent rules during the same time period.

Reports Prepared by ODH as required by Am. Sub. H.B. 119

Section 737.12. (A) of Am. Sub. H.B. 119 required the director of Health to issue a report to the Household Sewage and Small Flow On-Site Sewage Treatment study Commission created in Section 737.11 of this act that includes recommendations regarding standards for the siting, design, installation, operation, monitoring, maintenance and abandonment of household sewge treatment systems and small flow on-site sewage treatment systems. The recommendations include information concerning the cost and state of technology currently utilized in household sewage treatment systems and small flow on-site sewage treatment systems and the nature and economics of available alternatives to that technology. The Director was required to issue the report to the Commission not later than January 1, 2008. The links below provide access to an executive summary, the full report and appendices of the report.

Am. Sub. H.B. 119 also required the director to prepare and submit a second report under section 737.12 (B) no later than June 1, 2008. For this report, the director was required to conduct a survey of boards of health in this state concerning household sewage treatment system operations and the failure rates of those systems. Boards of health were instruced to provide relevant information pertaining to the household sewage treatment system programme to assist with completion of the survey.

Report 2 - Survey of Household Sewage Treatment Systems Operation and Failure Rates in Ohio

Through July 1, 2009, ODH along with local health districts, the Ohio Environmental Protection Agency, the Ohio Department of Natural Resources, industry and with many other interested parties will work through the Home Sewage and Small Flow Onsite System Sewage Treatment System Study Commission whose directive is to recommend efficient and effective ways to treat sewage to ensure protection of public health with consideration of economic impacts.

ODH will continue to work with local health districts and others to collect data regarding the performance of traditional and new technology systems in Ohio. Past surveys show more than 25 percent of sewage systems in Ohio experience failure related to performance and treatment. Research will be conducted on how other states in the Midwest have addressed these same issues of balancing proper treatment and performance against cost. ODH intends to provide quality data to determine the best course of action regarding sewage treatment for homes and small businesses to effectively protect our streams, lakes and ground water supplies.

ODH will continue to work in partnership with local health districts, industry professionals, academia and sewage treatment system (STS) owners to assure proper STS management for the protection of public health and the prevention of surface and ground water contamination due to failing or poorly operating systems.

The Health Information Technology Adoption Initiative

The dissemination and adoption of health information technology (HIT), particularly Electronic Health Records (EHR), is widely regarded as critical to enhancing the performance of the nation's health care system. President Bush and Secretary Leavitt have recognised the essential role of HIT in improving health care quality and efficiency. The Office of the National Coordinator for Health Information Technology (ONC) has partnered with the George Washington University, Partners / Massachusetts General Hospital Institute for Health Policy and Brigham and Women's Hospital through a contract on the Health IT (HIT) Adoption Initiative. The new initiative is aimed at better characterising and measuring the state of EHR adoption and determining the effectiveness of policies aimed at accelerating adoption of EHRs and interoperability. Despite the promises it offers health care and quality improvement, only a small minority of U.S. physicians have embraced electronic health records (EHR) as a routine part of practice, says a study in the June 19 online edition of the New England Journal of Medicine. The survey of 2,758 physicians – the most up-to-date and comprehensive picture of EHR adoption trends – shows that only 4 percent have a fully functional EHR system and 13 percent have a basic one.

- Electronic Health Records Still Not Routine Part of Medical Practice, Says New Study
- General — Oct 10, 2006
- In the most comprehensive study to date that reliably measures the state of electronic health record (EHR) use by doctors and hospitals, researchers from Massachusetts General Hospital (MGH) and George Washington University (GWU) estimate that one in four doctors (24.9 percent) use EHRs to improve how they deliver care to patients.

The study reflects one year of examination of dozens of studies and surveys by some of the nation's leading experts on health IT and illustrates the EHR adoption environment among physicians and hospitals, what predicts whether or not a provider will adopt an EHR, where the gaps in adoption are, how much adoption depends on location, practice size, specialty, or kinds of patients treated; and how the U.S. can collect more precise and timely data on adoption to better enlighten policymakers. In addition, the report points to four key things that drive adoption: financial incentives and barriers, laws and regulations, the state of the technology and organisational influences such as the size of a practice or hospital or payer mix; and how integrated a health care system is.

Health Information Technology Home

Health Information Technology

Health information technology (Health IT) allows comprehensive management of medical information and its secure exchange between health care consumers and providers. Broad use of health IT will:

- Improve health care quality;
- Prevent medical errors;
- Reduce health care costs;
- Increase administrative efficiencies;
- Decrease paperwork; and
- Expand access to affordable care.

Interoperable health IT will improve individual patient care, but it will also bring many public health benefits including:

- Early detection of infectious disease outbreaks around the country;
- Improved tracking of chronic disease management; and
- Evaluation of health care based on value enabled by the collection of de-identified price and quality information that can be compared.

Making Health Information Technology Personal

Health information technologies can be tools that help individuals maintain their health through better management of their health information. Health IT will help consumers gather all of their health information in one place so they can thoroughly understand it and share it securely with their health care providers so they get the care that best fits their individual needs.

Health IT can help to improve public health one individual at a time by building partnerships between health care consumers and providers across the country.

Health IT

Health IT initiatives harness current and emerging information technologies to improve patient safety and convenience while reducing the cost of providing care. Some of the numerous benefits of health IT initiatives will include: a reduction in medical errors, avoidance of costly duplicate testing, and elimination of unnecessary hospitalisations.

The benefits of health IT initiatives range from consumer convenience, as patients will not have to fill out repetitive paper work, to life-saving early detection of an infectious disease outbreak, as anonymous data from emergency rooms is sent to public health systems instantly. These initiatives aid in fulfilling the President's goal for most Americans to have electronic health records by the year 2014.

American Health Information Community (AHIC) Topics

The American Health Information Community (AHIC) is a federal advisory body, chartered to make recommendations to the Secretary of HHS on how to accelerate the development and adoption of Health IT. The Community will make recommendations to the Secretary of HHS to enable advancement in four areas of focus by the end of 2006.

- Consumer Empowerment - Make available a consumer-directed and secure electronic record of health care registration information and a medication history for patients.

- Chronic Care - Allow the widespread use of secure messaging, as appropriate, as a means of communication between doctors and patients about care delivery.
- Biosurveillance - Enable the transfer of standardised and anonymized health data from the point of health care delivery to authorized public health agencies within 24 hours of its collection.
- Electronic Health Records - Create an electronic health record that includes laboratory results and interpretations, that is standardised, widely available and secure.

Highlighted Federal Efforts

Health & Human Services

Health & Human Services (HHS) - HHS is facilitating the development of standards for Health IT systems that will improve patient care and increase efficiency across the health care system. HHS, through several of its agencies, also provides funding to organisations engaged in building and testing Health IT systems, standards and projects.

Department of Defense

Department of Defense (DoD) - Currently, thousands of military medical providers use the DoD's electronic health record system, AHLTA, and nearly 300,000 outpatient visits are captured digitally every week. DoD's vision is to provide each patient with a continuously updated digital medical record from the point of injury or care on the battlefield to discharge from military clinics and hospitals in the United States. These records would be completely transferable electronically to the Veterans Health Administration as part of the Joint Patient Electronic Health Record (JPEHR).

Department of Veterans Affairs

Veterans Health Administration (VHA) - The VHA is a division of the U.S. Department of Veteran'•s Affairs, and provides care for over five million veterans of the United States Armed Services. The VHA's electronic health record system, My HealtheVet, allows patients to refill prescriptions online, and provides access to health information, links to Federal and VA benefits and resources, and the patient's Personal Health Journal. The VHA continues to add capabilities to My HealtheVet, to empower consumers to take a more active role in managing their health and health care.

Nanotechnology/Environmental

How will nanotechnology affect our lives - this part will not look in terms of the technological impact such as faster and cheaper computers, but at the very important health and environmental effects that necessarily must be considered.

When will it help cure cancer or and when might it cause it? Will the apparent ecological benefit of a nanoparticle that improves catalytic reactions be futile when we consider the ecological footprint of the nanoaparticles' lifecycle? There are many open questions where the search for answers has only just begun!

Nanotechnology has been proclaimed to imply a wealth of potential benefits for society many of which will have direct or indirect positive effects on the environment. Of specific environmental importance is the potential improved efficiency of energy production, reduced energy use of products, reduced use of chemicals due to e.g. functionalisation of surfaces, remediation of soil pollution, improved sensoring and monitoring of the environment and improved functionality of materials.

Additionally, principles of green engineering and chemistry are beginning to be integrated into the development of nanomaterials, meaning that nanotechnology may also lead to more environmental beneficial production methods (Schmidt, 2007).

There are a lot of opportunities but as with many other nanotechnologies there is still a large gap from the research labs to manufacturing and real life use of nanotechnological products and solutions. The focus here will however be at the potential health and environmental impacts. For a more thorough review of ecopotentials see e.g. Malsch , Dionysiou , Masciangioli ,and Schmidt.

Natural Environment

The natural environment, commonly referred to simply as the environment, is a terminology that comprises all living and non-living things that occur naturally on Earth or some region thereof. This term includes a few key components:

1. Complete ecological units that function as natural systems without massive human intervention, including all vegetation, animals, microorganisms, rocks, atmosphere and natural phenomena that occur within their boundaries.
2. Universal natural resources and physical phenomena that lack clear-cut boundaries, such as air, water, and climate, as well as energy, radiation, electric charge, and magnetism, not originating from human activity.

The natural environment is contrasted with the built environment, which comprises the areas and components that are strongly influenced by man. A geographical area is regarded as a natural environment (with an indefinite article), if the human impact on it is kept under a certain limited level (similar to section 1 above). This level depends on the specific context, and changes in different areas and contexts. The term wilderness, on the other hand, refers to areas without human intervention.

Challenges

It is the common understanding of *natural environment* that underlies environmentalism — a broad political, social, and philosophical movement that advocates various actions and policies in the interest of protecting what nature remains in the natural environment, or restoring or expanding the role of nature in this environment. While true wilderness is increasingly rare, *wild* nature (e.g., unmanaged forests, uncultivated grasslands, wildlife, wildflowers) can be found in many locations previously inhabited by humans.

Goals commonly expressed by environmental scientists include:

- reduction and clean up of pollution, with future goals of zero pollution;
- cleanly converting nonrecyclable materials into energy through direct combustion or after conversion into secondary fuels;
- reducing societal consumption of non-renewable fuels;
- development of alternative, green, low-carbon or renewable energy sources;
- conservation and sustainable use of scarce resources such as water, land, and air;
- protection of representative or unique or pristine ecosystems;
- preservation of threatened and endangered species extinction;
- the establishment of nature and biosphere reserves under various types of protection; and, most generally, the protection of biodiversity and ecosystems upon which all human and other life on earth depends.

Very large development projects - also called megaprojects - pose special challenges and risks to the natural environment. Major dams and power plants are cases in point. The challenge to the environment from such projects is growing because more and bigger megaprojects are being built, in developed and developing nations alike.

Recently, there has been a strong concern about climate change such as global warming caused by anthropogenic releases of greenhouse gases, most notably carbon dioxide, and their interactions with humans and the natural environment. Efforts here have focused on the mitigation of greenhouse gases that are causing climatic changes (e.g. through the Climate Change Convention and the Kyoto Protocol), and on developing adaptative strategies to assist species, ecosystems, humans, regions and nations in adjusting to the Effects of global warming.

A more profound challenge, however, is to identify the natural environmental dynamics in contrast to environmental changes not within natural variances. A common solution is to adapt a static view neglecting natural variances to exist. Methodologically this view could be defended when looking at processes which change slowly and short time series, while the problem arrives when fast processes turns essential in the object of the study.

REFERENCES

Banat I.M., Samarah N., Murad M., Horne R. and Benerjee S. (1991). Biosurfactant production and use in oil tank clean-up. *World J. Microbiol. Biotechnol.*, 7: 80-84

Behaviour of metals in MSW fly ash during roasting with chlorinating agents, PhD Thesis by Chan, C.C., Chemical Engineering Department at University of Toronto, 1997.

Braag J.R., Prince R.C., Harner E.J. and Atlas R.M. (1994). Effectiveness of bioremediation for the Exxon Valdez oil spill. *Nature*, 368: 413-418

Chea, Terence. "From Fields to Factories: Plant-Based Materials Replace Oil-Based Plastics, Polyesters." Washington Post. Friday May 03, 2002. Page E01.

Cheong Jit Kong (972748N02) Page 11 04/15/2000

Cheong Jit Kong (972748N02) Page 12 04/15/2000 Miller R.M. (1995). Biosurfactant-facilitated remediation of metal-contaminated soils. *Environ. Health Perspect.*, 103: 59-62

Council Directive on the Contained Use of Genetically Modified Micro-organisms.

Council Directive on the Deliberate Release into the Environment of Genetically Modified Organisms. 1990, 90/220/EEC

D. Hogg, *et al* (2008) Greenhouse Gas Balances of Waste Management Scenarios, Report for the Greater London Authority, Eunomia

Danish Energy Statistics 2005 by the Danish Ministry of Energy.

Dave H., Ramakrishna C., Bhatt B.D. and Desai J.D. (1994). Biodegradation of slop oil from a petrochemical industry and bioremediation of slop oil contaminated soil. *World J. Microbiol.*, 10: 653-656

Desai A.J., Patel R.M. and Desai J.D. (1994). Advances in the production of biosurfactants and their commercial applications. *J. Sci. Ind. Res.*, 53: 619-629

Desai J.D. and Banat I.M. (1997). Microbial production of surfactants and their commercial potential. *Microbiol. Mol. Biol. Rev.*, 61(1): 47-64

Desai J.D. and Desai A.J (1993). Production of surfactants. In: *Biosurfactants production, properties and applications*, pp. 65-97. Kosaric N. (ed.). New York: Dekker

Determination of an Application for a PPC Permit under the Pollution Prevention and Control, IPPC public registers, November 6. 2006

Deziel E., Paquette G., Villemur R., Lepine F. and Bisaillon J. (1996). Biosurfactant production by soil *Pseudomonas* strain growing on polycyclic aromatic hydrocarbons.

Dionysiou, D. D. Environmental Applications and Implications of Nanotechnology and Nanomaterials. Journal of Environmental Engineering, Jul, 2004, pp 723-724.

Economic Analysis and Land Use Policy, Session of Workshop sponsored by EPA and NCERQA, Washington 1999

Emissionsfaktorer og emissionsopgørelse for decentral kraftvarme, Kortlægning af emissioner fra decentrale kraftvarmeværker, Ministry of the Environment of Denmark 2006 (in Danish)

Energy From Waste, from the homepage of the UK Renewable Energy Association

Environmental and Health Risks Associated with the Use of Processed Incinerator Bottom Ash in Road Construction, Report AEAT/ENV/R/0716 by BREWEB October 2003

Eurobarometer 46.1, *The Europeans and modern biotechnology,* CEC DG XII, 1997

European Space Agency, Methane in Martian Atmosphere http://www.esa.int/esaMI/Mars_Express/SEMZ0B57ESD_0.html

Evaluation of Emissions from the Burning of Household Waste in Barrels, EPA November 1997

Fujitsu Global Service. "Fujitsu Pioneers Use of Biodegradable Plastic in Notebook Computers." Fujitsu.com. www.fujitsu.com.en/news/2002/06/05/5-1.html. Accessed June, 2002.

Georgiou G., Chyr S. and Sharma M. (1992). Surface active compounds from microorganisms. *Biotechnology*, 10: 60-65 MC4203: Environmental Biotechnology Supervisor: A/Prof. Tan H.M.

Gerngross, Tillman U., and Steven C. Slater. "How Green are Green Plastics?" Scientific American. www.sciam.com/article.cfm?articleID=0000D61F-E193-1C73-9B81809EC588EF21&pageNumber=5&catID=2 Accessed June 2002.

Ghosh M.M., Yeon I.T., Shi Z., Cox C.D. and Robinson K.G. (1995). Surfactantenhanced bioremediation of PAH- and PCB-contaminated soils. In: *In situ and on-site bioreclamation symposium*, pp. 15-23. Hinchee R.E., Vogel C.M. and Brockman F.J.(ed.). Columbus, Ohio: Battelle Press

Golyshin PM, Fredrickson HL, Giuliano L, Rothmel R, Timmis KN, Yakimov MM. (1999). Effect of novel biosurfactants on biodegradation of polychlorinated biphenyls by pure and mixed bacterial cultures. *New Microbiol.*, 22(3):257-67.

Goswami P. and Singh H.D. (1991). Different modes of hydrocarbon uptake by two different *Pseudomonas sps. Biotechnol. Bioeng.*, 37: 1-11 Guerra-Santos L.H., Kappeli O. and Fiechter (1984). *Pseudomonas aeruginosa* biosurfactant production in continuous culture with glucose as carbon source. *Appl. Environ. Microbiol.*, 48: 301-305

Haferburg D., Hommel R., Kleber H.P., Klug S., Schuster G. and Zchienger H.J. (1987). Antiphytovirale aktivität von rhamnolipid aus *Pseudomonas aeruginosa. Acta Biotechnol.*, 7: 353-356

Harrison Weir, Emily. "Plastic + Time = Water? Sure!" Alternate Views, Mount Holyoke College. www.mtholoyoke.edu/offices/comm/vista/9709/2.html. Accessed June 2002.

Harvey S., Elashvili I., Valdes J., Kamely D. and Chakrabarty A.M. (1990). Enhanced removal of Exxon Valdez spilled oil from Alaskan gravel by a microbial surfactant. *Bio/ Technology*, 8: 228-230

Incinerators - Weapons of Mass Destruction, Country Doctor Magazine January 31. 2005, by Dick van Steenis

Iowa State University, College of Agriculture. "Soy-Based Plastics Finding Their Way to Market." Iowa State University. www.ag.iastate.edu/aginfo/news/1998releases/soyplastic.html. Accessed June 2002.

Kitamoto D., Yanagishita H., Shinbo T., Nakane T., Kamisawa C. and Nakahara T. (1993). Surface active properties and antimicrobial activities of mannosylerythritol lipids as biosurfactants produced by *Candida antarctica*. *J. Biotechnol.*, 29: 91-96

Kleis, Heron and Dalager, Søren (2004) 100 Years of Waste Incineration in Denmark, A historical review of incineration in Denmark

Knox, A. (2005) Overview of incineration, An Overview of Incineration and EFW Technology as Applied to the Management of Municipal Solid Waste (MSW), University of Western Ontario, Canada

Kraftvärmeverket, avfall blir el och värme, brochure on the SYSAV incinerator in Malmö, Sweden (in Swedish)

Lin S.C. (1996). Biosurfactants: recent advances. *Chem. Technol. Biotechnol.*, 66: 109-120 MC4203: Environmental Biotechnology Supervisor: A/Prof. Tan H.M.

Malsch, I. Benefits, Risks, Ethical, Legal and Social Aspects of Nanotechnology 4th Nanoforum report; Nanoforum: Jun, 04.

Masciangioli, T.; Zhang, W. X. Environmental technologies at the nanoscale. Environmental Science & Technology 2003, 37 (5), 102A-108A.

National Geographic Article about methane as sign of life http://news.nationalgeographic.com/news/2004/10/1007_041007_mars_methane.html

Ramana K.V. and Karanth N.G. (1989). Factors affecting biosurfactants production using *Pseudomonas aeruginosa* CFTR-6 under submerged conditions. *J. Chem. Technol. Biotechnol.*, 45: 249-257

Rotary-kiln incinerators An excellent detailed description of rotary-kiln incinerators

Sanger D.B. and Vulfson E.N. (1995). Application of enzymes to the synthesis of surfactants. *Trends Biotechnol.*, 13: 164-172

Schmidt, K.: Green Nanotechnology: It's easier than you think. Woodrow Wilson International Center for Scholars. PEN 8 April 2007

Schulz D., Passeri A., Schmidt M., Lang S., Wagner F., Wray V. and Gunkel W.(1991). Marine biosurfactants. I.

Surfactant Science Series, 25: 89-120. Kosaric N., Cairns W.L., Gray N.C.C (ed.). New York: Dekker

Thauer, R. K., "Biochemistry of Methanogenesis: a Tribute to Marjory Stephenson", Microbiology, 1998, volume 144, pages 2377-2406.

Themelis, Nickolas J. An overview of the global waste-to-energy industry, Waste Management World 2003

Uchida Y., Misava S., Nakahara T. and Tabuchi T. (1989). Factors affecting the production of succinoltrehalose lipids by *Rhodococcus erythropolis* SD-74 grown on nalkanes. *Agric. Biol. Chem.*, 53: 765-769

Van Dyke M.I., Gulley S.L., Lee H. and Trevors J.T. (1993). Evaluation of microbial surfactants for recovery of hydrophobic pollutants from soil. *J. Ind. Microbiol.*, 11: 163-170

Vestforbrænding anlæg 6 - Danmarks største forbrændingsovn, Brochure (in Danish) on the largest incineration line in Europe, 2004

Yamane T. (1987). Enzyme technology for the lipid industry: an engineering overview. *J. Am. Oil. Chem. Soc.*, 64: 1657-1662 Zhang Y. and Miller R.M. (1995). Effect of rhamnolipid (biosurfactant) structure on solubilisation and biodegradation of n-alkanes. *Appl. Environ. Microbiol.*, 61: 2247- 2251

18

Environmental or Green Technology

Environmental Technology

Environmental technology or "green technology" is the application of the environmental sciences to conserve the natural environment and resources, and by curbing the negative impacts of human involvement. Sustainable development is the core of *environmental technologies*. When applying *sustainable development* as a solution for *environmental issues*, the solutions need to be socially equitable, economically viable, and environmentally sound. Some *environmental technologies* that retain *sustainable development* are; recycling, water purification, sewage treatment, remediation, flue gas treatment, solid waste management, renewable energy, and others.

Related Technologies

Some technologies assist directly with energy conservation, while other technologies are emerging that help the environment by reducing the amount of waste produced by human activities. Energy sources such as solar power create less problems for the environment than traditional sources of energy like coal and petroleum. Scientists continue to search for clean energy alternatives to our current power production methods. Some technologies such as anaerobic digestion produce renewable energy from waste materials. The global reduction of greenhouse gases is dependent on the adoption of energy conservation technologies at industrial level as well as this clean energy generation. The introduction of an energy saving module to common industrial refrigeration or air conditioning can dramatically cut the amount of carbon dioxide (CO_2) and sulphur dioxide (SO_2)that would otherwise be introduced to the atmosphere.

Examples

- Alternative energy
- Energy Saving Modules
- Green building
- Green syndicalism
- Hybrid vehicle
- Solar power
- Solar cell
- Solar heating

Energy Efficiency Trends in the United States

The U.S. is currently the biggest consumer of energy in the world, although at current levels of growth, it is possible that in the future China could become the leading energy consumer. The U.S. Department of Energy categorizes national energy use in four broad sectors: transportation, residential, commercial, and industrial. Energy usage in the transportation and residential sectors (about half of U.S. energy consumption) is largely controlled by individual domestic consumers. Commercial and industrial energy expenditures are determined by businesses, government entities and other facility managers. National energy policy has a significant effect on energy usage across all four sectors.

Energy Conservation

Energy conservation is the practice of decreasing the quantity of energy used while achieving a similar outcome of end use. This practice may result in increase of financial capital, environmental value, national security, personal security, and human comfort. Individuals and organisations that are direct consumers of energy may want to conserve energy in order to reduce energy costs and promote economic, political and environmental sustainability. Industrial and commercial users may want to increase efficiency and thus maximize profit. On a larger scale, energy conservation is an important element of energy policy. In general, energy conservation reduces the energy consumption and energy demand per capita, and thus offsets the growth in energy supply needed to keep up with population growth. This reduces the rise in energy costs, and can reduce the need for new power plants, and energy imports. The reduced energy demand can provide more flexibility in choosing the most preferred methods of energy production. By reducing emissions, energy conservation is an important part of lessening climate change. Energy conservation facilitates the replacement of non-renewable resources with renewable energy. Energy conservation is often the most economical solution to energy shortages, and is a more environmentally benign alternative to increased energy production.

Transportation Sector

The transportation sector includes all vehicles used for personal or freight transportation. Of the energy used in this sector, approximately 65% is consumed by gasoline-powered vehicles, primarily personally owned. Diesel-powered transport (trains, merchant ships, heavy trucks, etc.) consumes about 20%, and air traffic consumes most of the remaining 15%. The oil supply crises of the 1970s spurred the creation, in 1975, of the federal Corporate Average Fuel Economy (CAFE) programme, which required auto manufacturers to meet progressively higher fleet fuel economy targets. The next decade saw dramatic improvements in fuel economy, mostly the result of reductions in vehicle size and weight. These gains eroded somewhat after 1990 due to the growing popularity of sport utility vehicles, pickup trucks and minivans, which fall under the more lenient "light truck" CAFE standard.

In addition to the CAFE programme, the U.S. government has tried to encourage better vehicle efficiency through tax policy. Since 2002, taxpayers have been eligible for income tax credits for gas/electric hybrid vehicles. A "gas-guzzler" tax has been assessed on manufacturers since 1978 for cars with exceptionally poor fuel economy. While this tax remains in effect, it currently generates very little revenue as overall fuel economy has improved. Another focus in gasoline conservation is reducing the number of miles driven.

An estimated 40% of American automobile use is associated with daily commuting. Many urban areas offer subsidized public transportation to reduce commuting traffic, and encourage carpooling by providing designated high-occupancy vehicle lanes and lower tolls for cars with multiple riders. In recent years telecommuting has also become a viable alternative to commuting for some jobs, but as of 2003 only 3.5% of workers were telecommuters.

Ironically, hundreds of thousands of American and European workers have been replaced by workers in Asia who telecommute from thousands of miles away. A vehicle's gas mileage normally decreases rapidly at speeds above 55 miles per hour. A car or truck moving at 55 miles an hour can get about 15 percent better fuel economy than the same car going 65 mph. According to the U.S. Department of Energy (DOE), as a rule of thumb, each 5 mph you drive over 60 mph is similar to paying an additional $0.21 per gallon for gas (at $3.00 per gallon).

Residential Sector

The residential sector refers to all private residences, including single-family homes, apartments, manufactured homes and dormitories. Energy use in this sector varies significantly across the country, due to regional climate differences and different regulation. On average, about half of the energy used in the U.S. homes is expended on space conditioning (i.e. heating and cooling). The efficiency of furnaces and air conditioners has increased steadily since the energy crises of the 1970s.

The 1987 National Appliance Energy Conservation Act authorized the Department of Energy to set minimum efficiency standards for space conditioning equipment and other appliances each year, based on what is "technologically feasible and economically justified". Beyond these minimum standards, the Environmental Protection Agency awards the Energy Star designation to appliances that exceed industry efficiency averages by an EPA-specified percentage. Despite technological improvements, many American lifestyle changes have put higher demands on heating and cooling resources.

The average size of homes built in the United States has increased significantly, from 1500 ft² in 1970 to 2300 ft² in 2005. The single-person household has become more common, as has central air conditioning: 23% of households had central air conditioning in 1978, that figure rose to 55% by 2001. As a cheaper alternative to the purchase of a new furnace or air conditioner, most public utilities encourage smaller changes the consumer can make to lessen space conditioning usage. Weatherisation is frequently subsidized by utilities or state/federal tax credits, as are programmable thermostats. Consumers have also been urged to adopt a wider indoor temperature range (e.g. 65 °F in the winter, 80 °F in the summer).

Home Energy Consumption Averages

- space conditioning, 44%
- water heating, 13%
- lighting, 12%
- refrigeration, 8%
- home electronics, 6%
- laundry appliances, 5%
- kitchen appliances, 4%
- other uses, 8%

Energy usage in some homes may vary widely from these averages. For example, milder regions such as the southern U.S. and Pacific coast of the USA need far less energy for space conditioning than New York City or Chicago. In milder climates, lighting energy may easily consume up to 40% of total energy. Certain appliances such as a waterbed, hot tub, or pre-1990 refrigerator use significant amounts of electricity. In most residences no single appliance dominates, and any conservation efforts must be directed to numerous areas in order to achieve substantial energy savings. However, Ground Source Heat Pump systems are the more energy efficient, environmentally clean, and cost-effective space conditioning systems available (Environmental Protection Agency), and can achieve reductions in energy consumptions of up to 70%.

Best Building Practices

Current best practices in building design and construction result in homes that are profoundly more energy conserving than average new homes. See Passive house, Superinsulation, Self-sufficient homes, Earthship, Straw-bale construction, MIT Design Advisor, Energy Conservation Code for Indian Commercial Buildings.

Commercial Sector

The commercial sector consists of retail stores, offices (business and government), restaurants, schools and other workplaces. Energy in this sector has the same basic end uses as the residential sector, in slightly different proportions. Space conditioning is again the single biggest consumption area, but it represents only about 30% of the energy use of commercial buildings. Lighting, at 25%, plays a much larger role than it does in the residential sector. Lighting is also generally the most wasteful component of commercial use. A number of case studies indicate that more efficient lighting and elimination of over-illumination can reduce lighting energy by approximately fifty percent in many commercial buildings.

Commercial buildings can greatly increase energy efficiency by thoughtful design, with today's building stock being very poor examples of the potential of systematic (not expensive) energy efficient design (Steffy, 1997). Commercial buildings often have professional management, allowing centralized control and coordination of energy conservation efforts. As a result, fluorescent lighting (about four times as efficient as incandescent) is the standard for most commercial space, although it may produce certain adverse health effects. Potential health concerns can be mitigated by using newer fixtures with electronic ballasts rather than older magenetic ballasts. As most buildings have consistent hours of operation, programmed thermostats and lighting controls are common. However, too many companies believe that merely having a computer controlled Building automation system guarantees energy efficiency. As an example one large company in Northern California boasted that it was confident its state of the art system had optimized space heating. A more careful analysis by Lumina Technologies showed the system had been given programming instructions to maintain constant 24 hour temperatures in the entire building complex. This instruction caused the injection of nighttime heat into vacant buildings when the daytime summer temperatures would often exceed 90 degrees Fahrenheit. This mis-programming was costing the company over $130,000 per year in wasted energy(Lumina Technologies, 1997). Many corporations and governments also require the Energy Star rating for any new equipment purchased for their buildings.

Solar heat loading through standard window designs usually leads to high demand for air conditioning in summer months. An example of building design overcoming this excessive heat loading is the Dakin Building in Brisbane, California, where fenestration was designed to achieve an angle with respect to sun incidence to allow maximum reflection of solar heat; this design also assisted in reducing interior over-illumination to enhance worker efficiency and comfort.

Industrial Sector

The industrial sector represents all production and processing of goods, including manufacturing, construction, farming, water management and mining. Increasing costs have forced energy-intensive industries to make substantial efficiency improvements in the past 30 years. For example, the energy used to produce steel and paper products has been cut 40% in that time frame, while petroleum/aluminum refining and cement production have reduced their usage by about 25%. These reductions are largely the result of recycling waste material and the use of cogeneration equipment for electricity and heating.

The energy required for delivery and treatment of fresh water often constitutes a significant percentage of a region's electricity and natural gas usage (an estimated 20% of California's total energy use is water-related. In light of this, some local governments have worked toward a more integrated approach to energy and water conservation efforts.

Unlike the other sectors, total energy use in the industrial sector has declined in the last decade. While this is partly due to conservation efforts, it's also a reflection of the growing trend for U.S. companies to move manufacturing operations offshore.

Energy Conservation in the United Kingdom

Energy conservation in the United Kingdom has been receiving increased attention over recent years. Key factors behind this are the Government's commitment to reducing carbon emissions, the projected 'energy gap' in UK electricity generation, and the increasing reliance on imports to meet national energy needs. Domestic housing and road transport are currently the two biggest problem areas.

Issues with Energy Conservation

Critics and advocates of some forms of energy conservation make the following arguments:

- It may be difficult for home owners or small business to justify investment in some energy saving measures. Condensing boilers are very much more efficient than older types. Energy savings are achieved by venting less heat externally and heating water for showers etc, as it is used. Refrigeration is also a major factor of energy consumption, electronic Energy saving modules (ESM) can be added to existing HVAC and refrigeration systems at little cost to conserve electricity.

- For transportation the same financial payback versus energy savings arguement can be made. For example, new hybrid vehicles have a better miles per gallon rating than an equivalent conventional vehicle, but they have a higher initial cost due to the greater costs of manufacturing the vehicle. The higher costs of maintenance, insurance, and depreciation can outweigh the savings on fuel and tax credits (depending on the particular vehicle, and extent of the tax credit). The hybrid car may give 45 mpg, saving the average driver 278 gallons of fuel per year. The hybrid engine cost $3000 extra, so with gasoline at $2.50 per gallon, it pays for itself in 4.3 years, and saves over 1000 gallons of gasoline from being used. The resale value of a fuel-efficient vehicle may be higher than a conventional gas-powered vehicle.
- Some retailers argue that bright lighting stimulates purchasing. Health studies have demonstrated that headache, stress (medicine), blood pressure, fatigue and worker error all generally increase with the common over-illumination present in many workplace and retail settings (Davis, 2001), (Bain, 1997). It has been shown that natural daylighting increases productivity levels of workers, while reducing energy consumption. Consumers are also motivated by a number of factors, and corporate stewardship may provide an incentive for shoppers to visit stores who conserve energy. Lower overhead costs may allow retailers to lower prices, stimulating consumption.

Jevons Paradox

The Jevons paradox is an observation made by William Stanley Jevons who stated that as technological improvements increase the efficiency with which a resource is used, total consumption of that resource may increase, rather than decrease. It may also be that some conservation efforts have the same effect. For example, if 10% of a country's population reduces its use of gasoline, the price of gasoline may drop, and the remaining 90% of the population may use more gasoline as a result. This hypothesis has not been sufficiently studied. Compare Parkinson's law.

Overview of Technology

Hydropower plants harness water's energy and use simple mechanics to convert that energy into electricity. Hydropower plants are actually based on a rather simple concept — water flowing through a dam turns a turbine, which turns a generator.

Here are the basic components of a conventional hydropower plant:

- Dam — Most hydropower plants rely on a dam that holds back water, creating a large reservoir.
- Intake — Gates on the dam open and gravity pulls the water through the penstock, a pipeline that leads to the turbine. Water builds up pressure as it flows through this pipe.
- Turbine — The water strikes and turns the large blades of a turbine, which is attached to a generator above it by way of a shaft.
- Generators — As the turbine blades turn, so do a series of magnets inside the generator. Giant magnets rotate past copper coils, producing alternating current (AC) by moving electrons.
- Transformer — The transformer inside the powerhouse takes the AC and converts it to higher-voltage current.
- Power lines — Out of every power plant come four wires: the three phases of power being produced simultaneously plus a neutral or ground common to all three.
- Outflow — Used water is carried through pipelines, called tailraces, and re-enters the river downstream.

The water in the reservoir is considered stored energy. When the gates open, the water flowing through the penstock becomes kinetic energy because it's in motion.

Techbology: Turbines

The turbine is the heart of the hydro system, where water power is converted into the rotational force that drives the generator. It is arguably the most important component in the system, because its efficiency determines how much electricity is generated.

Hydro Green Energy's kinetic Hydropower Turbine Arrays operate differently than a traditional hydropower plant. Like a traditional hydropower station, the electricity that we produce is clean and renewable, however, there are significant differences. Hydro Green Energy's Krouse Turbines are kinetic turbines. This means that the renewable power that is generated comes from the energy in the "motion" of the moving water, i.e. the velocity of the moving water be it river, tidal or ocean current to generate river, tidal energy or ocean energy, respectively. Since the Hydro Green Energy Hydropower Turbine Arrays are inexpensive to manufacture, they can be engineered and designed for many different types of conditions and they produce the lowest cost hydropower comercially available. Hydropower Turbine Arrays can be used to generate renewable energy in rivers, irrigation canals, tidal energy, and ocean enery. Another significant advantage to using Hydro Green Energy's Hydropower Turbine Arrays is that there is no harm to the environment. The amount of environmental damage due to infrastructure creation and modification is significantly reduced with Hydropower Turbine Arrays compared to other kinetic hydropower designs and especially when compared to traditional hydropower.

Green Syndicalism

Green syndicalism has been used as a name for the philosophy of the green guild or sustainable trades movement. It is related to Anarcho-syndicalism and Eco-anarchism methodologically, but lacks specifically anarchism and Marxism as ideologies. (However, the term has also been used by those embracing both anarchism and syndicalism, such as author Graham Purchase). Its methods are a fusion of the trade union, more robust methods from formal syndicalism, and the direct action and workplace democracy movements (the most prominent proponent of green syndicalism is the IWW).

It may work closely with green movement and fair trade groups. However, it does not usually adopt the larger political goals of these movements. It is more likely to be an outgrowth of craft and heritage worker movements, e.g. to preserve wooden ship building, or traditional organic ranching methods. Some think the eco-syndalicalists are closely related to the back to the land or survivalist or hippie movements, or to ethical traditions that resemble those movements such as Mennonites, e.g. the Amish, especially for their focus on appropriate technology. However, they are not usually unwilling to adopt new technologies, and quite willing to work with other such groups worldwide. Eco-syndicalism is probably more closely related to the eco-gastronomy movement.

Green Building

Green building or bioclimatic building is the practice of increasing the efficiency with which buildings and their sites use and harvest energy, water, and materials, and reducing building impacts on human health and the environment, through better siting, design, construction, operation, maintenance, and removal — the complete building life cycle. Green building is also sometimes known as "sustainable building" or "environmental building", although there are slight differences in the definitions.

The practice of green building can lead to benefits including reduced operating costs by increasing productivity and using less energy and water, improved public and occupant health due to improved indoor air quality, and reduced environmental impacts by, for example, lessening storm water runoff and the heat island effect. Green building is an essential component of the related concepts of sustainable design, sustainable development and general sustainability. Practitioners of green building often seek to achieve not only ecological but aesthetic harmony between a structure and its surrounding natural and built environment. The appearance and style of sustainable homes and buildings can be nearly indistinguishable from their less sustainable counter-parts. Green design often emphasises taking advantage of renewable resources, e.g., using sunlight through passive solar, active solar, and photovoltaic techniques and using plants and trees through green roofs, rain gardens, and for reduction of rainwater run-off. Many other techniques, such as using packed gravel for parking lots instead of concrete or asphalt to enhance replenishment of ground water, are used as well.

Types of Green Buildings

Straw Bale Houses

Straw bale is actually the dry leftover part of the plant after harvest. Wheat, oats, barley, rye, rice and others are mostly used in straw bale houses to construct walls which are then covered by stucco.... Any type of hay or straw can be used for this for construction, although the less roughage that's still on the straw stems, the better (because moisture in the roughage can cause mold, which can spontaneously combust). The preferred straw is actually clean, white straw, like oat straw, barley straw. Three string bales are optimal to build with as they are more solid.

Extruded Straw Panel Houses

Straw Panels are a commercial alternative to erecting labor-intensive Straw Bale Houses and require less deviation from the present home construction techniques that the existing workforce is comfortable with. Several workers can erect a single-family straw panel home in less than a day.

Built from straw that is milled and super-compressed in a fiber extrusion mill, straw panels have the durability of concrete, the finish surface of gypsum, the insulating value of conventional glass insulation product and the sound absorption characteristics of sound insulating panels. As fiber extrusion takes advantage of the plant fiber's own cellulosic 'glue' no MDI or polymeric chemicals are used in the panel and therefore there is no chemical offgassing into the environment where the panel is deployed. Upwards of twenty different agricultural waste fibers can be extruded into straw building panels, including switchgrass, rice and wheat. Because the fiber extrusion process makes high-quality building material so inexpensively it is being reviewed for large housing projects around the world.

Standards and Ratings

Many countries have developed their own standards of energy efficiency for buildings.

- PassivHaus, Germany and Austria
- EnerGuide for Houses, Canada
- Green Star, Green Building Council of Australia
- Haute_qualité_environnementale Haute Qualité Environnementale, France
- House Energy Rating, Australia
- Leadership in Energy and Environmental Design (LEED), USA and Canada
- Green Globes, USA, Canada and United Kingdom
- National Association of Home Builders Green Building Guidelines, USA

Green Building Worldwide

India

The Confederation of Indian Industry plays an active role in promoting sustainability in the Indian construction sector. There are many energy efficient buildings in India, situtated in all of the following climatic zones:

- cold and cloudy
- cold and sunny
- composite
- hot and dry
- moderate
- warm and humid

Canada

Canada has implemented "r2000" guidelines for new buildings built after the year 2000. Incentives are offered to builders to meet the r2000 standard in an effort to increase energy efficiency and promote sustainability. In December 2002, Canada formed the Canada Green Building Council and in July 2003 obtained an exclusive licence from the US Green Building Council to adapt the LEED rating system to Canadian circumstances.

- Beamish-Munro Hall at Queen's University features sustainable construction methods such as high fly-ash concrete, triple-glazed windows, dimmable fluorescent lights and a grid-tied photovoltaic array.

TermoDeck Canadais rated Green Tip #1 in The Ashrae Green Guide. TermoDeck® specialists in thermal mass and energy storage in the structure design and install mechanical systems linked with the building structure. This proven method allows ordinary heating and cooling systems to deliver sustainability and value at no premium cost. TermoDeck Canada is the winner of the 2006 Ontario Concrete

Award. Construct Canada "Canada's largest building & construction show" award TermoDeck for Sheridan College in Brampton. It won for its sustainable construction in the Construct Canada of 2006.

The TermoDeck Principle: The effects of using the heat storage capacity of hollow core slabs varies between summer and winter conditions. Surplus heat, generated from body heat, lighting, computers, sun radiation, etc, can be stored in the slabs increasing their temperature by 2-3°C during the day without affecting the comfort of the occupants. In the summer this excess heat is dissipated from the slabs by cooling the slabs with cool night air in temperate climates, or using conventional chillers, but with up to a 50% reduction in capacity when compared to conventional technologies, in hot climates. During the winter, in cold climates we retain the heat stored in the slabs overnight, to maintain comfortable internal conditions for the occupants the next day. In some hot climates just using the cool night air, without the need for chillers, can cool the structure to provide comfortable conditions by the following morning.

Germany

German developments that employ green building techniques include:

- The Solarsiedlung (Solar Village) in Freiburg, Germany, which features energy-plus houses.
- The Vauban development, also in Freiburg.
- Houses designed by Baufritz, incorporating passive solar design, heavily insulated walls, triple-glaze doors and windows, non-toxic paints and finishes, summer shading, heat recovery ventilation, and greywater treatment systems.
- The new Reichstag building in Berlin, which produces its own energy.

Australia

The Green Building Council of Australia has its own green buildings standard known as Green Star. In 2007, the GBCA will host an Australiasian green building conference and expo called "Green Cities — Where Our Future Lives".

United Kingdom

The Association for Environment Conscious Building (AECB) has promoted sustainable building in the UK since 1989. This organisation was founded by Keith Hall, who now runs The (Green Building Press). The Green Building Press publishes Building for a Future magazine and The Green Building Bible. The UK Building Regulations set requirements for insulation levels and other aspects of sustainability in building construction.

Malaysia

The Standards and Industrial Research Institute of Malaysia (SIRIM) promotes green building techniques. Malaysian architect Ken Yeang is a prominent voice in the area of ecological design.

United States

The U.S. Green Building Council has developed definitions of what constitutes sustainable design of green buildings through the Leadership in Energy and Environmental Design or LEED ((R) USGBC) green building rating system. USGBC has attracted over 7,000 organisations as members to date. The U.S. Green Building Council maintains a list of state and local green building programmes and initiatives in the U.S. The Green Building Initiative is a non-profit network of building industry leaders committed to bringing green to mainstream residential and commercial construction. The GBI believes in building approaches that are environmentally progressive, but also practical and affordable for builders to implement. The GBI has developed an easy to use, inexpensive and ANSI standard web-based rating tool called Green Globes.

The United States Environmental Protection Agency's EnergyStar programme rates commercial buildings for energy efficiency and provides EnergyStar qualifications for new homes that meet their standards for energy efficient building design.

In 2005, Washington became the first state in the U.S. to enact green building legislation. According to the law, all major public agency facilities with a floor area exceeding 5,000 square feet (465 m^2), including state funded school buildings, are required to meet or exceed LEED standards in construction or renovation. The projected benefits from such a law are

- 20% annual savings in energy costs
- 20% reduction in water costs
- 38% reduction in waste water production
- 22% reduction in construction waste

Noted Green Designers & Builders

- Advanced Environmental, Australia
- J. Baldwin
- Steve Baer
- Tom Bender
- Patrick Blanc
- John Boecker

- Peter Calthorpe
- Eric Corey Freed
- Buckminster Fuller
- William McDonough
- Glenn Murcutt
- Rocky Mountain Institute
- Sim Van der Ryn
- Walter Segal
- Michael Sorkin
- Thompson, Ventulett, Stainback & Associates
- Brenda and Robert Vale
- James Wines
- Ken Yeang
- Paganelli Construction

Introduction to Green Technology

The term "technology" refers to the application of knowledge for practical purposes. The field of "green technology" encompasses a continuously evolving group of methods and materials, from techniques for generating energy to non-toxic cleaning products. The present expectation is that this field will bring innovation and changes in daily life of similar magnitude to the "information technology" explosion over the last two decades. In these early stages, it is impossible to predict what "green technology" may eventually encompass.

The goals that inform developments in this rapidly growing field include: Sustainability — meeting the needs of society in ways that can continue indefinitely into the future without damaging or depleting natural resources. In short, meeting present needs without compromising the ability of future generations to meet their own needs.

"Cradle to cradle" design — ending the "cradle to grave" cycle of manufactured products, by creating products that can be fully reclaimed or re-used.

Source reduction — reducing waste and pollution by changing patterns of production and consumption.

Innovation — developing alternatives to technologies — whether fossil fuel or chemical intensive agriculture — that have been demonstrated to damage health and the environment.

Viability — creating a center of economic activity around technologies and products that benefit the environment, speeding their implementation and creating new careers that truly protect the planet.

Examples of Green Technology Subject Areas

Energy

Perhaps the most urgent issue for green technology, this includes the development of alternative fuels, new means of generating energy and energy efficiency.

Green building

Green building encompasses everything from the choice of building materials to where a building is located.

Environmentally Preferred Purchasin

This government innovation involves the search for products whose contents and methods of production have the smallest possible impact on the environment, and mandates that these be the preferred products for government purchasing.

Green Chemistry

The invention, design and application of chemical products and processes to reduce or to eliminate the use and generation of hazardous substances.

Green Nanotechnology

Nanotechnology involves the manipulation of materials at the scale of the nanometer, one billionth of a meter. Some scientists believe that mastery of this subject is forthcoming that will transform the way that everything in the world is manufactured. "Green nanotechnology" is the application of green chemistry and green engineering principles to this field.

This web page and its content are the result of the work of Landscape Architecture students at the University of Washington who completed our LARC 433 construction course in Winter quarter of 2001. Our intent was to begin to develop a base of information about green technologies that could serve as a primer to inform our students and others desirous of using environmentally friendly approaches to site planning and design. We see this effort only as a beginning. The topics, while compelling, are limited in scope for this first step. It is our hope that subsequent classes can build upon this research and expand the range topics covered. The goal is to develop a rich information base about existing and emerging green technologies and materials. Our objective is to format this information in a way that makes it a convenient tool for those wanting basic information on the design concerns, applicability criteria, costs, and other practical concerns for the implementation and use of these technologies and materials. We know there are many of you out there. Ultimately we would like to see this page become a forum for sharing green tech information and know-how. We hope this first step proves useful to you. Enjoy!

Green Technology Research

Below you'll find links to our research on green technology and it's application to landscape architecture

Art & Water Infrastructure

Landscape architects are blending art and infrastructure to create fascinating results. In this report, two examples of artful reinterpretations of infrastructure are explored.

Black Water Wetlands

Blackwater wetlands are one type of constructed wetland used to treat wastewater. Examples of benefits are explored in two case studies.

Cisterns

A cistern is a vessel used to collect and store water for later use. While sometimes used in areas of the world experiencing drought, cisterns are also used as a means of conserving water where sustainability is a goal. Green Roofs This report provides a synopsis of the applications, components, construction and cost benefits of green roofs, as well as a few case studies of successful installation.

Grey Water Systems

Greywater systems take non-fecal water from showers, bathtubs, sinks, washers and dishwashers and reuse that water for irrigation, cooling devices and in some cases toilet water.

Photovoltaic Systems

Photovoltaic systems make solar energy an alternative energy source for some landscape design applications. The expertise of landscape architects provides a number of opportunities for more progressive professionals to take the lead in designing solar-powered landscapes.

Urban BMPs

Urban Best Management Practices are one method to increase the quality of storm water runoff. This report explores the ways in which experts test and apply the most reliable BMPs.

Solar Aquatics

Solar aquatics is a form of wastewater treatment which uses plants and animals to create a microenvironment which works well with bacteria used in more traditional systems.

Urban Soils

In urban areas, soil is an important resource whose quality is essential in maintaining the health of urban ecosystems.

Wind Energy

Wind energy is one of today's most promising alternatives to society's dependence on fossil fuels. Because wind energy is often harnessed by windmills and turbines it is becoming an important design issue in landscape architecture.

Urban Trees

Some of the most critical components of supporting healthy trees in our urban environments include: physical elements, construction requirements, best management practices, and appropriate species selection.

On-line Resources

If you're interested in further exploring green tech, these are some of the more important links we've gathered in our attempt to build an on-line community for green technology in landscape architecture. If you have any websites to add to the list, please let us know about them!

Constructed Wetland Wastewater Treatment and Wildlife Habitat

Cisterns To Collect Non-Potable Water For Domestic Use

Roofscapes, Inc. ZinCo Green Roof Systems

The Relationship between Soil and Salmon (King County Solid Waste Division Organics and Composting)

Putting New Orleans On the Green Line

Laura Bush couldn't have been more effective if she'd waved a magic wand, but it's hard to compete with Brad Pitt when trying to turn the nation's attention to "green" design. At the White House on Monday, the first lady appealed to a gathering of National Design Award honorees, urging them to get involved in rebuilding the Gulf Coast.

Their legacy, she suggested, would be determined by their efforts to improve the devastated region. She also put in a surprise plug for sustainable design, saying that Hurricane Katrina "gives us the chance to build 'green' buildings and to build energy-saving buildings" — which made her sound like an early-warning siren for global warming. By Thursday, Thom Mayne — recipient of the Smithsonian's architecture prize and last year's Pritzker Architecture Prize — had traded a gold chair in the East Room for a seat on a bus roaming the Big Easy.

Mayne's firm, Morphosis, is already a major player, having recently completed the master plan for a $716 million, 20-acre jazz park near the Louisiana Superdome, the first major redevelopment proposal to emerge since Katrina struck in August. On this occasion, Mayne, a rock star among architects, was touring the Lower Ninth Ward and talking small-scale housing and child-care centers with Brad Pitt, aka the world's most recognisable actor-turned-architectural activist.

"We want to rebuild intelligently," Pitt said yesterday at a news conference.

The timing of their New Orleans visit was coincidental, but the purpose was right in line with the first lady's plea, which might be all that matters to the displaced residents. Pitt invited Mayne to judge a contest to create affordable housing and a community center as models of energy-efficient, ecologically friendly design.

The contest sponsor, Global Green USA, has committed to building such a "green" complex in the Holy Cross neighborhood. Pitt, an architecture fan and a green-design advocate, had met the group's president, Matt Petersen, through the Clinton Global Initiative, and signed on as head of the jury.

Winners of the contest ("Sustainable Design Competition for New Orleans — Advancing the Sustainable Rebuilding of New Orleans and the Gulf Coast") were to be notified over the weekend and their designs unveiled Monday on the "Today" show. (Beginning Monday, Global Green will open the voting to the public at Globalgreen.org.)

Mayne called during the judging to put the week's imagery in perspective. A leading practitioner of "green" technology, Mayne says he was eager to respond to Laura Bush personally but didn't get the chance.

"I was next in the receiving line when they pulled her away," he said. "I was going to say to her: 'You should encourage your husband. This administration isn't exactly famous for that.' "

In her remarks, Bush acknowledged that many designers were already engaged in such work, but she encouraged them with textbook-worthy savvy about urban design.

"The planning of buildings, city blocks and public spaces determines how businesses, governments, civic organisations and neighbors come together and interact as a community," she noted. By creating communities that "integrate schools, homes, cultural centers and commercial enterprises," designers also would entice displaced families back home.

Green Government

Government agencies at state and local levels are working to build sustainable communities. The following are examples of noteworthy initiatives.

(Submissions welcome — write to editor@green-technology.org.)

State

- Green California
  http://www.green.ca.gov
- Illinois Green Government Council
  http://www.standingupforillinois.org/green/
- New York State Green Building Initiative
  http://www.dec.state.ny.us/website/ppu/grnbldg/index.html
- North Carolina Project Green
  http://www.sustainablenc.org/
- Pennsylvania Green Government Council
  http://www.gggc.state.pa.us/gggc/site/default.asp

County

- Santa Barbara County Green Award Program
  http://www.countyofsb.org/pwd/swud/GreenAwardProg.htm

City

- City of Los Angeles
  http://www.ladwp.com/ladwp/areaHomeIndex.jsp?contentId=LADWP_GREENLA_SCID
- City of Palo Alto
  http://www.pafd.org/sustainability/
- City of Santa Monica
  http://santa-monica.org/EPD/EnvironmentalProgrammes
- City and County of San Francisco

http://www.sfenvironment.com

- City of Portland
  http://www.sustainableportland.org/
- City of Seattle
  http://www.seattle.gov/environment/
- City of Oakland
  http://www.sustainableoakland.com

Greening Across the Chemistry Curriculum

A Project with major funding from The Camille and Henry Dreyfus Foundation Special Grant Programme in the Chemical Sciences. Additional funding was provided by the ACS/EPA Green Chemistry Educational Materials Development Project and the University of Scranton.

Introduction to Green Chemistry

The late sixties and early seventies were times when the environment received a great deal of attention including the formation of the Environmental Protection Agency (EPA) and the celebration of the first Earth Day, both of which occurred in 1970. In the intervening years in excess of 100 environmental laws have been passed. These include the twelve major laws listed below.

- 1970 Clean Air Act. Regulates air emissions.
- 1972 National Environmental Policy Act. Requires in part that EPA review environmental impact statements of proposed major federal projects (e.g. highways, buildings, airports, parks and military complexes).
- 1972 Clean Water Act. Establishes the sewage treatment construction grants programme and a regulatory and enforcement programme for discharges of pollutants into U.S. waters.
- 1972 Federal Insecticide, Fungicide & Rodenticide Act. Governs distribution, sale and use of pesticide products. All pesticides must be registered (licensed) by EPA.
- 1972 Ocean Dumping Act. Regulates the intentional disposal of materials into ocean waters.
- 1974 Safe Drinking Water Act. Establishes primary drinking water standards.
- 1976 Toxic Substances Control Act. Requires the testing, regulating, and screening of all chemical produced or imported in the U.S.
- 1976 Resource Conservation & Recovery Act. Regulates solid and hazardous waste form “cradle to grave.”

- 1976 Environmental Research & Development Demonstration Act. Authorizes all EPA research programmes.
- 1980 Comprehensive Environmental Response, Compensation & Liability Act, better known as Superfund. Provides for a federal "superfund" to clean up abandoned hazardous waste sites, accidental spills and other emergency releases of pollutants in the environment.
- Emergency Planning & Community Right-to-Know Act. Requires that industries report toxic releases and encourages planning by local communities to respond to chemical emergencies.
- 1990 Pollution Prevention Act. Seeks to prevent pollution by encouraging companies to reduce the generation of pollutants through cost-effective changes in production, operation, and raw material use.

All of these acts, with one exception, deal with pollution after it is formed. These laws are in general focused on the treatment or abatement of pollution and have become know as "command and control" laws. In many instances these laws, which were passed by the US congress, place limits on pollution and timetables for compliance, with little regard to whether the science/technology could attain these goals and with little regard to the economic costs of these laws. Risk associated with a toxic chemical is a function of Hazard and Exposure. The "end of the pipe" laws attempt to control Risk by dealing with the prevention of the Exposure to toxic hazardous chemicals. Of course all to often prevention of Exposure has failed.

Risk = f (Hazard, Exposure)

While these laws have accomplished a great deal in terms of improving our environment by controlling our exposure to hazardous substances, we still have a long way to go. For example under the Toxic Release Inventory (TRI), which is part of the Emergency Planning and Community Right to Know Act (EPCRA), companies are required to report the use and/or release of certain hazardous substances. In 1997 industries reported that 23.85 billion pounds of hazardous substances were treated, recycled, used for energy production, disposed of or released to the environment. This act covers only 650 of the 75,000 chemicals in use in US commerce today and only companies that manufacture or proce ss more that 25,000 pounds or use more than 10,000 pounds of a listed substance are required to report.

Companies traditionally have viewed environmental regulations with disdain and as an economic hardship. In order to comply with environmental regulations it is estimated to cost U.S. industries between $100 to 150 billion per year. Since the EPA is charged with the implementation and enforcement of these laws, the relationship between industry and the EPA has been adversarial and one of mistrust.

In the last decade a new paradigm has emerged at the EPA, ushered in, in part by the Pollution Prevention Act of 1990. This is the first and only act that is focused on pollution prevention rather than the typical treatment and remediation. The EPA is now attempting to partner with industry to find more flexible and cost effective ways of not only meeting exi sting regulations but also preventing pollution at the source. In 1991 green chemistry became a formal focus of EPA (green chemistry at EPA). *Green Chemistry or environmentally benign chemistry is the design of chemical products and processes that reduce of eliminate the use and generation of hazardous substances.* Thus instead of limiting Risk by control ling our Exposure to hazardous chemicals, green chemistry attempts to reduce and preferentially eliminate the Hazard thus negating the necessity to control Exposure. The bottom line is, if we don't use or produce hazardous substances then the Risk is zero, and we don't not have to worry about the treatment of hazardous substances or limiting our exposure to them.

Green chemistry has gained a strong foothold in the areas of research and development in both industry and academia. Several conferences and meetings (e.g. the Green Chemistry and Engineering Conference) are held each year with green chemistry/technology as their focus. The journal Green Chemistry made it debut in 1999, the Green Chemistry Institute was recently created and the Presidential Green Chemistry Challenge Awards were established in 1995.

The Presidential Green Chemistry Challenge Awards

The Presidential Green Chemistry Challenge Awards were announced in 1995 by the Clinton administration and the first awards were presented in 1996. These awards are a means of recognising outsta nding achievements in applied green chemistry/technology and are the only awards in chemistry given out on the presidential level. Nominees for these awards must demonstrate how their work has met one or more of the following criteria:

- Greener reaction conditions for an old synthesis (e.g., replacement of an organic solvent with water or the use of no solvent at all).
- A greener synthesis for an old chemical (e.g., a synthesis which us es biomass rather than petrochemical feedstocks or the use of catalytic rather than stoichiometric reagents).
- The synthesis of a new compound that is less toxic but has the same desirable properties as an existing compound (e.g., a new pesticide that is toxic only to target organisms and biodegrades to environmentally benign substances).

Examples of green chemistry/technology that have been developed encompass most all areas of chemistry including organic, biochemistry, inorganic, polymer, toxicology, environmental, physical, industrial etc. Some cases of green chemistry/ technology that have won the Presidential Green Chemistry Challenge Award consist of:

- Barry Trost's concept of atom economy which looks at utilized and wasted atoms in a reaction.
- A new synthesis of ibuprofen which has a much better record of atom economy and pollution prevention.
- The use of waste carbon dioxide as a blowing agent (which is non ozone depleting unlike the traditional CFC blowing agents) for foam polystyrene.
- Development of surfactan ts for carbon dioxide enabling CO_2 to be used as a solvent (for example in dry cleaning).
- Development of oxidant activators for hydrogen peroxide. This allows, for example, the replacement of chlorine containing (ozone depleting) bleaches with hydrogen peroxide in the manufacture of paper.
- The develop ment of new insecticides that are more specific to target organisms.

The Twelve Principles of Green Chemistry

Anastas and Warner have developed the Twelve Principles of Green Chemistry to aid one in assessing how green a chemical, a reaction or a process is.

1. It is better to prevent waste than to treat or clean up waste after it is formed.
2. Synthetic methods should be designed to maximize the incorporation of all materials used in the process into the final product.
3. Wherever practicable, synthetic methodologies should be designed to use and generate substances that possess little or no toxicity to human health and the environment.
4. Chemical products should be designed to preserve efficacy of function while reducing toxicity.
5. The use of auxiliary substances (e.g. solvents, separation agents, etc.) should be made unnecessary whenever possible and, innocuous when used.
6. Energy requirements should be recognised for their environmental and economic impacts and should be minimized. Synthetic methods should be conducted at ambient temperature and pressure.

7. A raw material feedstock should be renewable rather than depleting whenever technically and economically practical.
8. Unnecessary derivatisation (blocking group, protection/deprotection, temporary modification of physical/chemical processes) should be avoided whenever possible.
9. Catalytic reagents (as selective as possible) are superior to stoichiometric reagents.
10. Chemical products should be designed so that at the end of their function they do not persist in the environment and break down into innocuous degradation products.
11. Analytical methodologies ne ed to be further developed to allow for real-time in-process monitoring and control prior to the formation of hazardous substances.
12. Substances and the form of a substance used in a chemical process should chosen so as to minimize the potential for chemical accidents, including releases, explosions, and fires.

Exposing Students to Green Chemistry

In 2000, Daryle Busch, former president of the American Chemical Society said "Green chemistry represents the pillars that hold up our sustainable future. It is imperative to teach the value of green chemistry to tomorrow's chemists."

It is clear that many industries and the research of many academics recognise the significance of green chemistry. However very little discussion of green chemistry has found its way into the chemistry curriculum. Although we and others have made some isolated attempts to bring green chemistry into the classroom, the EPA and ACS have recognised the need to make a concerted and sustained effort to green the curriculum so that future chemists are taught to "think green." The EPA/ACS Green Chemistry Educational Materials Development Project was begun at a workshop in October 1998. The thrust of this project is to develop materials that will aid in the infusion of green chemistry into the curriculum. The major foci of this project are the development of an Annotated Bibliography of Green Chemistry, Green Chemistry Laboratory Experiments, Real-World Cases in Green Chemistry and short courses on green chemistry. John Warner of U. Mass. Boston is in charge of the first two projects and the third is a project accomplished by Michael Cann and Marc Connelly of the University of Scranton.

Real-World Cases in Green Chemistry was published in March of 2000. This work, which is published by ACS, is an attempt to compile and edit information on green chemistry so that chemistry instructors may use this information to green the chemistry curriculum at their institution. Each case focuses on a Presidential Green Chemistry Challenge award winner or nominee.

In order to take the greening of the chemistry curriculum to its next logical step we are developing green chemistry modules for insertion of green chemistry into specific chemistry courses. These modules are being developed by instructors who teach these courses. The introduction to green chemistry that you are now reading is part of this undertaking.

Our Green Future: Cultivating Change

The 21st century has been called the "century of the environment." Governments —and individual citizens — can no longer assume that social challenges such as pollution, dwindling natural resources and climate change can be set aside for future generations.

Through policy, research, education, incentives and forward-looking relationships with industry, government can play a central role in building a green future, community by community.

The prospects for success have never been greater. A dawning era of creativity and innovation in "green technology" (also known as "clean technology") is bringing the promise of a healthier planet — as well as the prospect of growing businesses that can sustain its health. The excitement building around this sector is reminiscent of the early years of the information technology revolution.

Green Technology is a non-profit initiative designed to inform government efforts toward sustainability, providing a forum in which government officials can communicate with those in the private sector who are developing and distributing green technologies.

With headquarters in California, Green Technology is ideally situated to stay on the leading edge. California, with the world's sixth-largest economy (its annual gross product exceeds that of China), is setting the standard for state support for green technologies.

With mandates for green building, renewable energy and "environmentally preferable" purchasing, California has made an unequalled commitment to a green futur.

REFERENCES

Addison, M., and D. Nall. 2001. "Cooling via underfloor air distribution: Current design issues and analysis options." From: Cooling Frontiers: The Advanced Edge of Cooling Research and Applications in the Built Environment. College of Architecture and Environmental Design, Arizona State University.

Akimoto, T., Nobe, T., Tanabe, S., and Kimura, K. 1996. "Experimental study on indoor thermal environment and air quality of the floor-supply displacement ventilation system." Proceedings, Indoor Air 1996.

Akimoto, T., T. Nobe, and Y. Takebayashi. 1995. Experimental study on the floor-supply displacement ventilation system. ASHRAE Transactions, 101(2).

ArchitectureWeek. 2000. "Building for 'Harmony with nature'." ArchitectureWeek, June 14, http://www.architectureweek.com/2000/0614/building 1-1.html.

Arens, E., M. Fountain, T. Xu, K. Miura, H. Zhang, and F. Bauman. 1995. "A study of occupant cooling by two types of personally controlled air movement."

Arens, E.A., and F.S. Bauman. 1994. "Improving the performance of task conditioning systems." Proceedings, International Symposium: Issues on Task-Ambient Conditioning. Nagoya University, Nagoya, Japan, 11 January, pp. 77-94.

Arens, E.A., F. Bauman, L. Johnston, and H. Zhang. 1991. "Testing of localised ventilation systems in a new controlled environment chamber." Indoor Air, No. 3, pp. 263-281.

Argon Corporation. 2002. Product information. Argon Corporation, Naples, FL, www.argonair.com.

Arnold, D. 1990. "Raised floor air distribution — a case study." ASHRAE Transactions, Vol. 96, Pt. 2. Early case study on a building from design stage through to construction highlights problem areas such as the construction and sealing of underfloor plenums. Other issues include how to conceal extract ducts in a building without ceiling voids, co-ordination of the floor services installation and finishes, and optimum supply air temperatures.

ASHRAE. 1990. ANSI/ASHRAE Standard 113-1990, "Method of testing for room air diffusion." Atlanta: American Society of Heating, Refrigerating, and Air-Conditioning Engineers, Inc.

ASHRAE. 1992. ANSI/ASHRAE Standard 55-1992, "Thermal environmental conditions for human occupancy." Atlanta: American Society of Heating, Refrigerating, and Air-Conditioning Engineers, Inc.

ASHRAE. 1993. Air-Conditioning Systems Design Manual. Atlanta: ASHRAE, Inc.

ASHRAE. 1997. ASHRAE Standard 129-1997, "Measuring air change effectiveness." Atlanta: American Society of Heating, Refrigerating, and Air-Conditioning Engineers, Inc.

ASHRAE. 1999. ANSI/ASHRAE Standard 62-1999, "Ventilation for acceptable indoor air quality." Atlanta: American Society of Heating, Refrigerating, and Air-Conditioning Engineers, Inc.

ASHRAE. 1999. HVAC Applications volume of the ASHRAE Handbook. Atlanta: ASHRAE, Inc.

ASHRAE. 2000. HVAC Systems and Equipment volume of the ASHRAE Handbook. Atlanta: ASHRAE, Inc.

ASHRAE. 2001. ANSI/ASHRAE Standard 62-2001, "Ventilation for acceptable indoor air quality." Atlanta: American Society of Heating, Refrigerating, and Air-Conditioning Engineers, Inc.

ASHRAE. 2001. Fundamentals volume of the ASHRAE Handbook. Atlanta: ASHRAE, Inc.

ASHRAE. 2002. Refrigeration volume of the ASHRAE Handbook. Atlanta: ASHRAE, Inc.

ASTM. 2000. ASTM E779-99: Test Method for Determining Air Leakage Rate by Fan Pressurisation. American Society for Testing and Material, Philadelphia, PA.

Ballinger, R., (1997). Coastal Management in Wales: Looking to the Future. Cardiff, University of Wales, Countryside Council for Wales.

Ballinger, R., H. Smith, et al. (1994). The Management of the Coastal Zone of Europe. Ocean and Coastal Management 22: 45-85

Banks, R. and James, C., (1998). Fisheries Offences - The Rationale, The Perspectives and Control Options, EEZ Technology, ICG Publishing n. 3, London, pp. 193-199

Barnes, R.S.K. (ed), (1977). The coastline. John Wiley & Sons, London.

Barragan-Munoz, J., (n.d.). Planning and Management in the Coastal Zone in Spain. Coastline: 28-30.

Barrangán Muñoz, J. M.(1997). Medio ambiente y desarrollo en las áreas litorales: Guia práctica para la planificación y gestión integradas. Barcelona, OIKOS-TAU, 160 pp.

Bauer, B.O., Davidson-Arnott, G.D., Nordstrom, K.F., Ollerhead, J., Jackson, N.L., (1996). Indeterminacy in Aeolian Sediment Transport Across Beaches. J. of Coastal Research, v. 12, n. 3, pp. 641-653.

Bayliss R., et al, (1998). Participation Techniques in Coastal Zone Management - Case Study of Dorset, Devon and Cornwall.

Beatley, T., Brower, T., et al. (1994). An Introduction to Coastal Zone Management. Washington, D.C, Island Press.

Bennett, R., (1996). Challenges in Norwegian Coastal Zone Planning. Geojournal 39(2): 153-165.

Bijlsma, L., Ehler, C.N., Klein, R.J.T., et al., (1996). Coastal zones and small islands. In: Climate Change 1995—Impacts, Adaptations and Mitigation of Climate Change: Scientific-Technical Analyses [Watson, R.T., M.C. Zinyowera, and R.H. Moss (eds.)]. Contribution of Working Group II to the Second Assessment Report of the Inter-governmental Panel on Climate Change, Cambridge University Press, Cambridge, UK, pp. 289–324.

Bijvoet, L., Strating, K., Roos, H., (1991). European Coastal Conference, proceedings, Scheveningen (The Hague), The Netherlands, 19-21 November 1991, EUCC, Leiden & Dutch Ministry of Agriculture, Nature Management & Fisheries, The Hague.

Bird, E.C.F. and Schwartz, M.L., (eds). (1985). The Worlds coastlines. Van Nortrand Rheinhold, New York.

Bird, E.C.F., (1984). Coasts - an introduction to coastal geomorphology. 3rd Edition, Basil Blackwell.

Boelaert-Suominen, S. and Cullinan, C., (1994). Legal and Institutional Aspects of Integrated Coastal Area Management in National Legislation, FAO.

Boesch, D.F., Josselyn, M.N., et al., (1994). Scientific assessment of coastal wetland loss, restoration and management in Louisiana. Journal of Coastal Research, special issue #20.

Bolsius, E.C.A., Clark, G., Groenendijk, J.G. (eds), (1993). The Retreat. Rural Land-use and European Agriculture. Faculty of Environmental Sciences, University of Amsterdam, pp. 170.

Bondesan, M., Castiglioni, G.B., et al., (1995). Coastal Areas at Risk from Storm Surges and Sea-level Rise in Northeastern Italy. J. of Coastal Research, v. 11, n. 4, pp. 1354-1379.

Bonnot, Y., (1994). Pour une Politique Globale et Coherente du Littoral en France.

Born, S. and Miller, A., (1988). Assessing Networked Coastal Zone Management Programmes. Coastal Management 16: 229-243.

Bosman, J., Flemming, N.C., Holden, N., Taylor, K., (1998). The EuroGOOS Marine Technology Survey, EuroGOOS Pub. No. 4, Southampton Oceanography Centre, Southampton. ISBN 0-904175-29-4.

Brady, Shipman, Martin, (1997). Coastal Zone Management, a draft policy for Ireland - Discussion document. Government of Ireland.

Brimicombe, A.J., (1992). Flood Risk Assessment Using Spatial Decision Support Systems. Simulation, December 1992, pp.379-380.

Bryant, D., E. Rodenburg, et al., (1996). Coastlines at Risk. An Index of Potential Development-Related Threats to Coastal Systems. Washington, D.C., WRI.

BSM, (1997). Coastal Zone Management: A Draft Policy for Ireland. Dublin, Government of Ireland.

Burbridge, P., (1998). A Generic Framework for Measuring Success in Integrated Coastal Management. Ocean and Coastal Management 37(2): 175-190.

Burrough, P.A., (1989). Principles of Geographic Information Systems for Land Resources Assessment, Clarendon Press, Oxford.

Cakiroylu, C., Yurteri, C., (1998). Methodology for Predicting Cooling Water Effects on Fish, Journal of Environmental Engineering, v. 124, n. 7, pp. 612-618.

Capobianco, M, Hde Vriend,.J., Nicholls, R.J., Stive, M.J.F., (1999). Coastal area impact and vulnerability assessment: a morphodynamic modeller's point of view. Journal of Coastal Research, in press.

Capobianco, M. and Furlanetto, L., (1996). Impact of Climatic Change on Po Delta. The Normative Aspects Regulating Land Use and Socio-Economic Changes. MEDDALT Final Workshop, Venezia, Italy, October 2-6.

Capobianco, M. and Otter H.S., (1996). Modelling Land-use and Land-cover Changes in Deltaic Areas. Int. Workshop on the State-of-the-Art of ICZM in the Mediterranean and Black Sea: Immediate Needs for Research, Education & Training, & Implementation, Iberotel Sarigerme Park, Turkey, November 2-5

Capobianco, M. and Stive, M.J.F., (1997). Soft Protection Technologies as a Tool for Integrated Coastal Zone Management. MEDCOAST '97, Malta, November 11-14

Capobianco, M., (1998). Discussion Document on "Fundamentals" for ICZM. Tecnomare Report 636A0753-RAP-A200-002.

Capobianco, M., (1998). Discussion Document on EIA&ICZM. Tecnomare Report 636A0753-RAP-D100-004.

Capobianco, M., (1998). Peculiarities of the Coastal Zone. Tecnomare Report 636A0753-RAP-A200-001.

Capobianco, M., (1998). Predictability for Long-term Coastal Evolution - Handling the Limiting Factors. In Proc. 3rd Marine Science and Technology Conf., 23-27 May, Lisbon, Portugual.

Capobianco, M., (1998). Twenty Questions on ICZM. Tecnomare Report 636A0753-RAP-D100-003.

Capobianco, M., (1999). The Role and Use of GIS in relation to ICZM. Tecnomare Report 636A0753-RAP-D100-006.

Carter, R.W.G. and Woodroffe, C.D., (1994). Coastal Evolution - Late Quaternary shoreline morphodynamics. Cambridge University Press, Cambridge.

Carter, R.W.G., (1988). Coastal environments. An introduction to the physical, ecological and cultural systems of coastlines. Academic Press, London.

Carter, R.W.G., Curtis, T.G.F., Sheehy-Skeffington, M.J., (eds), (1992). Coastal Dunes. Geomorphology, ecology and management for conservation. Proceedings of the third European dune congress Galway, Ireland, 17-21 June 1991. A.A. Balkema, Rotterdam.

Cencini, C. and Varani, L., (1989). Degradation of coastal dune systems through anthropogenic action. In Coastlines of Italy. Ed. Paolo Fabri. American Society of Civil Engineers, New York, 55-69.

Cencini, C., Marchi, M., Torresani, S., Varani, L., (1988). The impact of tourism on Italian deltic coastlands: four case studies. Ocean & Shoreline Management, 11, 353-374.

Cervantes M., (1998). Public/Private Partnerships in Science and Technology: An Overview. STI Review No. 23, OECD, Paris, 12 pp.

Chao, P.T. and Hobbs, B.F., (1997). Decision analysis of shoreline protection under climate change uncertainty. Water Resources Research, 33(4), 817–829.

Chapman, V.J., (1976). Coastal vegetation. 2nd ed. Oxford, Pergamon.

Cheltenham Countryside Commission, (1992). European Workshop on Coastal Zone Management. Held in Poole, Dorset 21-27 April 1991.

Christie, P. and White A.T., (1997). Trends in Development of Coastal Area Management in Tropical Countries: From Central to Community Organisation. Coastal Management 25: 155-181.

Chua, T., (1993). Essential Elements of Integrated Coastal Management. Ocean and Coastal Management 21 (1-3)

Cican-Sain, B., (1993). Sustainable Development and Integrated Coastal Management. Ocean and Coastal Management 21(1-3): 11-44.

Cican-Sain, B., (1998). Integrated Coastal and Ocean Management: Concepts and Practices. Washington, D.C, Island Press.

Clark, J., (1998). Coastal Zone Planning for the New Century. Ocean and Coastal Management 37(2): 191-216.

Coastal Zone Management Act, (1994). Untitled Unites States Code. Title 16: Conservation; Chapter 33: Coastal Zone Management; Section 1453: Definitions.

Coccossis, H., (1997). Integrated Coastal Area and River Basin Management. ICARM Technical Reports Series No. 1. UNEP, Priorities Action programme.

Colecchia A. and Papaconstantinou G., (1996). The Evolution of Skills in OECD Countries and the Role of Technology. STI Working Papers, No. 8, OCDE/GD(96)183, OECD, Paris, 36 pp.

Commission of the European Communities (CEC), (1995). An Energy Policy for the European Union, White Paper. (COM-95-682). 47 pp.

Commission of the European Communities (CEC), (1996). COM (96) 511 final, Brussels, 4.12.1996.

Commission of the European Communities (CEC), (1996). Communication from the Commission - Energy for the Future: Renewable Sources of Energy. Green Paper for a Community Strategy, (COM-96-576). 48 pp.

Commission of the European Communities (CEC), (1998). Amended Proposal for a Council Directive establishing a framework for Community action in the field of Water Policy, (COM-98-76-final). 57 pp.

Commission of the European Communities (CEC), (1998). Communication from the Commission to the European Council. Partnership for Integration - A Strategy for Integrating Environment into European Union Policies. 10 pp.

Commission of the European Communities (CEC), (1998). Communication from the Commission to the European Council. The Future for the Market in Fisheries Products in the European Union: Responsibility, Partnership and Competitiveness. 22 pp.

Commission of the European Communities (CEC), (1999). Lessons from the European Commission's Demonstration Programme on Integrated Coastal Zone Management (ICZM), CEC, Luxembourg.

Commission of the European Communities (CEC), (1999). Towards a European Integrated Coastal Zone Management (ICZM) Strategy - General Principles and Policy Options, CEC, Luxembourg.

Commission of the European Communities (CEC), DG XII, MAST and IOC/IODE, (1993). Manual of Quality Control Procedures for validation of Oceanographic Data. UNESCO Manual and Guides n. 26, 436 pp.

Commission of the European Communities (CEC), DGXIII, (1996). Towards a European Policy Framework for Geographic Information - A Working Document. November, 1996.

Commission of the European Communities (CEC), DGXVI, (1997). European Spatial Development Perspective – First Official Draft. 72 pp.

Commission of the European Communities (CEC), DGXVII, (1996). European Energy to 2020: a Scenario Approach. 120 pp.

Committee on Beach Nourishment and Protection (CBNP), (1995). Beach Nourishment and Protection. National Research Council, National Academy Press, Washington, D.C., 334 pp.

CORINE. (1998) Coastal Erosion Database. CEC, Luxembourg, 79pp plus maps.

Costanza, R. and Ruth, M., (1997). Using Dynamic Modeling to Scope Environmental Problems and Build Consensus. Environmental Management,Vol. 22, pp. 183-196.

Costanza, R., Sklar, F.H., White, M.L., (1990). Modelling Coastal Landscape Dynamics. BioScience, Vol. 40, No. 2, pp. 91-107.

Costanza, R., Wainger, L., Folke, C., Mäler, K.G., (1993). Modelling Complex Ecological Economic Systems. BioScience, Vol. 43, No. 8, pp. 545-555.

Council of Europe, (1991). (Special on the Coast). Naturopa No 67. Council of Europe, Strasbourg

Council of Europe, (1998). Coastal zones - towards sustainable management. Naturopa, No 88. Council of Europe, Strasbourg.

Countryside Council for Wales, (1996). Seas, Shores and Coastal Areas: Maritime Policy. Bangor, Countryside Council for Wales.

Department of the Environment, (1992). Coastal zone protection and planning: the Government's response to the second report from the House of Commons Select Committee on the Environment, Cm 2011, London: HMSO.

Department of the Environment, (1993). Development Below Low Water Mark: A Review of Legislation in England and Wales. UK, Department of the Environment and Welsh Office.

Department of the Environment, (1995). Policy Guidelines for the Coast. UK, Department of the Environment.

Department of the Environment, (1996). Coastal Zone Management - Towards Best Practice. UK, Department of the Environment.

DOD, (1994). Remediation Technologies Screening Matrix and Reference Guide - Second Edition. Environmental Technology Transfer Committee, 597 pp.

Drees, J.M., (1997). Coastal Dunes, Recreation and Planning. Proceedings of European Seminar, Castricum, The Netherlands, EUCC, Leiden

ECOPRO, (1996). Environmentally Friendly Coastal Protection - Code of Practice, Government Publications, Dublin, 320 pp.

Edwards, S.D., Jones, P.J.S. and Nowell, D.E. (1997). Participation in coastal zone management initiatives: a review and analysis of examples from the UK. Ocean and Coastal Management 36(1-3), pp 143-165 See http://www.ucl.ac.uk/~ucfwpej/cpuk97.htm for abstract.

Environment Service, (1995). Delivering Coastal Management in Northern Ireland: A Consultation Paper. Belfast, Northern Ireland, Environment Service.

Environmental Protection Agency (EPA), (1996). Treasured Waters: Protecting Our Coastal And Marine Resources, EPA/842/B-96/001, Office of Water.

Environmental Protection Agency (EPA), (1997). Environmental Technology Verification Programme - Verification Strategy, EPA/600/K-96/003.

Environmental Protection Agency (EPA), (1997). Identification of Pollution Prevention (P2) Technologies for Possible Inclusion in Enforcement Agreements Using Supplemental Environmental Projects (SEPs) and Injunctive Relief. Environmental Protection Agency Publ. No. 300-R-97-001, 61 pp. + ann.

Environmental Protection Agency (EPA), (1997). Reducing Risks from Waste, EPA530-k-97-04. 41 pp.

Environmental Protection Agency (EPA), (1998). EPA's Contaminated Sediment Management Strategy, Office of Water, EPA-823-R-98-001, 100 pp. + xiv.

ESA, (1995). Coastal Zone Earth Workshops, ESTEC 28, 29-03-95, ESRIN 5, 6-4-95, Workshop for the Definition of ENVISAT Coastal Zone Mission, ESA, Paris.

Essaid, H.I., (1990). A multilayered sharp interface model of coupled freshwater and saltwater flow in coastal systems: model development and application. Water Resources Research, v. 26, n. 7, pp. 1431-1454.

European Environment Agency (EEA), (1998). Framework on Integrated Models and Indicators for European Coastal Zone Management. European Environment Agency. European Topic Centre on Marine and Coastal Environment. February 1998.

Eurostat (1999). Towards Environmental Pressure Indicators for the EU. CEC, Luxembourg, 181pp

Fankhauser, S., (1995). Protection versus Retreat: the Economic Cost of Sea Level Rise. Environment and Planning, v. 27, pp. 299-319.

Farinha J.C., Costa L., Zalidis G., Matzavelas A., Fitoka E., Hecher N., Tomàas Vives P., (1996). Mediterranean Wetland Inventory: Habitat Description System. MedWet/Instituo da Conservacao da Naturaleza (ICN)/Wetlands International/Greek Biotope/Wetland Center.

Federal Emergency Management Agency (FEMA), (1994). Mitigation of Flood and Erosion Damage to Residential Buildings in Coastal Areas. FEMA-257, Mitigation Directorate, Federal Emergency Management Agency, Washington, DC.

Ferreira Marques, J., (1991). Portugal. Naturopa 67: 22-23.

Funnell, C., (1998). Europe's ocean industry must be ready for the EC's FP5. International Ocean Systems Design, 2(1), 15–18.

Funtowicz, S.O. and Ravetz, J.R., (1995). Planning and Decision Making in an Uncertain World: the Challenge of Post-Normal Science, in: Natural Risk and Civil Protection, Horlick-Jones T., A. Amendola and R. Casale (eds), Chapman & Hall, London, pp. 554.

Garbuglia, E., Berti, D., Capobianco, M., (1994). Marine Science: Prospective and Technological Needs for the year 2000 and Beyond. Oceanology International '94, Brighton, U.K., March 8-11.

Garcia Nova, F., Crawford, R.M.M., Diaz Barradas, (1997). The ecology and conservation of European dunes ("Sevilla Procedings"). Universidad de Sevilla, Spain.

Georges, T.M. and Harlan, J.A., (1994). Military over-the-horizon radars turn to ocean monitoring. Mar. Tech. Soc. J., 27,4, pp. 31-38.

Georges, T.M., Harlan, J.A., Leben, R. R., Lematta, R. A., (1998). A Test of Ocean Surface-

Current Mapping with Over-the-Horizon Radar. IEEE Trans. Geosci. Remote Sensing, vol. 36(1), pp. 101-110

GESAMP, (1996). The Contributions of Science to Coastal Zone Management. Rep. Stud. GESAMP, IMO/FAO/UNESCO-IOC/WMO/WHO/IAEA/UN/UNEP Joint Group of Experts on the Scientific Aspects of Marine Environmental Protection. 61: 66.

GESAMP, (1996). The Contributions of Science to Coastal Zone Management. Rep. Stud. GESAMP, IMO/FAO/UNESCO-IOC/WMO/WHO/IAEA/UN/UNEP Joint Group of Experts on the Scientific Aspects of Marine Environmental Protection. 61: 66.

Gibson, J., (1993). Coastal Zone Planning Law. Marine Policy 17(2): 118-129.

Goldberg, E.D., (1994). Coastal Zone Space - Prelude to Conflict? IOC Ocean Forum I, UNESCO Publishing, Paris, France, 138 pp.

Gorman, L., A. Morang, Larson, R., (1998). Monitoring the coastal environment; part IV: mapping, shoreline changes, and bathymetric analysis. Journal of Coastal Research, 14(1), 61–92.

Gudelis, V., Povilanskas, R. Roepstorff, A., (1995). Coastal Conservation and Management in the Baltic Region. Proceedings EUCC - WWF Conference, 3-7 May 1994, Riga - Klaipeda - Kaliningrad, EUCC Baltic Office.

Guillot, S. and Rousseau, M., (1998). Recensement des constraintes juridiques et outils de gestion qui interviennent dans le perimetre du programme TERRA-Symphonie: Problematiques en cours, Conseil General de la Gironde.

Haward, M., (1995). Institutional design and policy making 'down under': developments in Australian and New Zealand coastal management. Ocean & Coastal Management 26: 87-117.

Healy M.G. and Doody, J.P. (eds), (1995). Directions in European coastal management. EUCC, Samara Publishing, Cardigan.

Heaton G.R., Banks R.D., Ditz, D.W., (1994). Missing Links: Technology and Environmental Improvement in the Industrialising World, World Resources Institute, 51 pp.

Huggett, D.J. (1998) The Role of Federal Government Intervention in Coastal Zone Planning and Management. Ocean and Coastal Management 39: 33-50.

International Institute for Sustainable, Development, (1996). Global green standards: ISO 14000 and sustainable development, 106 pp.

IOC, (1997). GOOS Coastal Module Planning Workshop Report, Workshop Report No. 131, UNESCO, 57 pp.

IPCC CZMS, (1992). A common methodology for assessing vulnerability to sea-level rise - second revision. In: Global Climate Change and the Rising Challenge of the Sea. Report of the Coastal Zone Management Subgroup, Response Strategies Working Group of the Intergovernmental Panel on Climate Change, Ministry of Transport, Public Works and Water Management, The Hague, The Netherlands, Appendix C, 27 pp.

IWICM, (1996). Enhancing the Success of Integrated Coastal Management in Tropical Developing Countries: Good Practices in the Formulation, Design, and Implementation of Integrated Coastal Management Initiatives. Report of the International Workshop on Integrated

Coastal Management in Tropical Developing Countries: Lessons Learned From Successes and Failures, Xiamen, China, May 1996. Quezon City, Philippines, GEF/UNDP/IMO Regional Programme for the Prevention and Management of Marine Pollution in the East Asian Seas and Coastal management Center.

Jeftic, L., Milliman, J., Sestini, G., (eds), (1992). Climate Change and the Mediterranean—Environmental and Societal Impacts of Climatic Change and Sea-Level Rise in the Mediterranean Region. Edward Arnold, London, UK, xii+673 pp.

Jemmett, A, K. Keneddy, D. Maters, & R. Witt. (1999). Towards Sustainable Estuary Management. English Nature Research Reports No. 329, English Nature, Peterborough.

Jones, A.R., (1995). GIS in coastal management: a progress review. In: CoastGIS '95 [Furness, R. (ed.)]. Proceedings of the International Symposium on GIS and Computer Mapping for Coastal Zone Management, Cork, Ireland, 3–5 February 1995, University College Cork, Cork, Ireland, pp. 165–178.

Jones, P.S., Healy M.G. and Williams, A.T. (eds), (1996). Studies in European Coastal Management. EUCC, Samara Publishing, Cardigan.

Kay, R. and Alder J., (1999). Coastal Planning and Management. E & FN Spon.

Kay, R. and Lester, C., (1997). Benchmarking the Future Direction for Coastal Management in Australia. Coastal Management 25(3): 265-292.

Kenchington, R. and Crawford, D., (1993). On the Meaning of Integration in Coastal Zone Management. Ocean and Coastal Management 21(1-3): 109-128.

Keokebakker, P. and Peet, G., (1987). Coastal Zone Planning and Management in the Netherlands. Coastal Management 15: 121-134.

Kildow, J., (1997). The Roots and Context of the Coastal Zone Movement. Coastal Management 25(3): 231-263.

Klein, R.J.T. and Nicholls, R.J., (1999). Assessment of coastal vulnerability to climate change. Ambio, in press.

Knecht, R., Cican-Sain, B., et al., (1996). Perceptions of the Performance of State Coastal Zone Management Programmes in the United States. Coastal Management 24: 141-163.

Larson, R., Morang, A., Gorman, L., (1997). Monitoring the coastal environment; part II: sediment sampling and geotechnical methods. Journal of Coastal Research, v. 13, n.2, pp. 308–330.

Leafe, R., Pethick, J., Townend, I., (1998). Realising the benefits of shoreline management. The Geographical Journal, 164(3), 282–290.

Lee, E., (1993). Coastal Planning and Management: A Review prepared by Rendel Geotechnics for the Department of the Environment. London, HMSO.

Lefebvre, C., (1998). Coastal Conservation Policy - Guidelines for Political Actors, Programme 1998-2000. IUCN, Commission on Environmental, Economic and Social Policy Coastal Working Group.

Leick, A., (1995). GPS Satellite Surveying. 2nd. ed. John Wiley & Sons, New York.

Lillycrop, W.J. and Estep, L.L., (1995). Generational advancements in coastal surveying, mapping. Sea Technology, v. 36, n. 6, pp. 10–16.

Local Government Management Board (LGMB-UK), (1995). 'Community Participation', and 'Action on the Coast'.

Maglia, S. and Santoloci, M., (1996). Il Codice dell'Ambiente, Casa Editrice la Tribuna, Piacenza, pp. 2431.

Marine Institute, (1996). Towards a Marine Policy for Ireland. Dublin, Marine Institute.

Martuccelli, A.M., (1994). Il Ruolo della Bonifica per la Costruzione, Tutela e Gestione del Territorio Polesano. in "Il Fiume e la sua Terra. Tutela e Gestione del Territorio a Quarant'Anni dall'Alluvione del Polesine. 1951-1991. Atti del Convegno di Studi.

Mayer, T. and d'Ozouville, L., (1998). Europe's Grand Challenge: Achieving Coordination in Marine Research. Ocean Challenge 8(2): 20-21.

Merrifield, D.B., (1990). Industrial competitiveness. In: Technology Commercialisation and Competitiveness. IC2 Institute, University of Texas, Austin, TX, pp. 23–29.

METAP, (1997). Assessment of Integrated Coastal Area Assessment Initiaives in the Mediterranean: Experiences from METAP and MAP (1988-1996). Split, Croatia, PAP.

Ministry of Agriculture, Fisheries and Food (MAFF), Welsh Office, Association of District Councils, English Nature, and National Rivers Authority, (1995). Shoreline Management Plans—A Guide for Coastal Defence Authorities. Her Majesty's Stationery Office, 24 pp.

Ministry of Environment and Energy, (1997). Spatial Planning in the Coastal Zone of Denmark. Copenhagen, Denmark, Ministry of Environment and Energy, Spatial Planning Department.

Mintzberg, H., (1994). The Rise and Fall of Strategic Planning. Prentice Hall.

Mitsch, W.J. and Gosselink, J.G., (1993). Wetlands. 2nd Edition. New York: Van Nostrand Reinhold.

Mitsch, W.J. and Reeder, B.C., (1991). Modelling nutrient retention of a freshwater coastal wetland: estimating the roles of primary productivity, sedimentation, resuspension and hydrology. Ecological Modelling 54, pp. 151-187.

Montoya, F., (1991). An Administrative Regulation Pattern of Coastal Management for Mediterranean Sea: Spanish Shores Act 1988. Marine Pollution Bulletin 23: 769-771.

Morang, A., Larson, R., Gorman, L., (1997). Monitoring the Coastal Environment; Part I: Waves and Currents. Journal of Coastal Research, v. 13, n. 1, pp. 111-133.

Morang, A., Larson, R., Gorman, L., (1997). Monitoring the coastal environment; Part III: geophysical and research methods. Journal of Coastal Research, v. 13, n. 4, pp. 1064–1085.

Mullard, S., (1995). Towards an EU Strategy for Integrated Coastal Zone Management. European Environmental Law Review 4(January): 16-20.

National Research Council (NRC), (1987). Responding to Changes in Sea Level: Engineering Implications. National Acad-emy Press, Washington, DC, 148 pp.

National Research Council (NRC), (1990). Managing Coastal Erosion. National Academy Press, Washington, DC, 522 pp.

National Research Council (NRC), (1995). Science, Policy and the Coast: Improving Decisionmaking. National Academy Press, Washington, DC, 85 pp.

National Science and Technology Council (NSTC), (1995). Bridge to a Sustainable Future. Interagency Environmental Technologies Office, Washington, DC, vi+87 pp.

Nicholls, R.J. and Branson, J., (eds), (1998). Enhancing coastal resilience - planning for an uncertain future. The Geographical Journal, v. 3, n. 164, pp. 255–343.

Nordstrom, K.F. and Roman, C.T., (1996). Estuarine Shores - Evolution, Environments and Human Activities. John Wiley & Sons, New York.

O'Regan, P.R., (1996). The use of contemporary information technologies for coastal research and management—a review. Journal of Coastal Research, v. 12, n. 1, pp. 192–204.

Odum, H.T., (1983, 1993). Systems Ecology (reprinted as Ecological and General Systems). Univ. Press of Colorado, Niwot, CO, 644 pp.

Odum, W., (1982). Environmental Degradation and the Tyranny of Small Decisions. Bioscience 32(9): 728-729.

Office of Technology Assessment (OTA), U.S., (1995). Renewing Our Energy Future, OTA Renewable Energy & Energy Efficiency Reports, 300 pp.

Olsen, S., (1993). Will Integrated Coastal Management Programmes be Sustainable; the Constituency Problem. Ocean and Coastal Management 21(1-3): 201-226.

Olsen, S., Tobey, J., et al., (1998). A Common Framework for Learning from ICM Experience. Ocean and Coastal Management 37(2): 155-174.

Olsen, S.B., (1993). A Learning Approach to the Management of Coastal Ecosystems: Lessons from High and Low Income Nations. Preparing to Meet the Challenges for the 21st Century: Proceedings of the World Coast Conference, Netherlands.

Organisation for Economic Co-operation and Development (OECD), (1995). Guidelines for Aid Agencies on Global and Regional Aspects of the Development and Protection of the Marine And Coastal Environment. OECD Development Assistance Committee, Guidelines on Aid and Environment, No. 1, 51 pp.

Organisation for Economic Co-operation and Development (OECD), (1997). The Evaluation of Scientific Research: Selected Experiences Report, OCDE/GD(97)194. 112 pp.

Organisation for Economic Co-operation and Development (OECD), (1995). Technology and Environment - Policies to Promote Technologies for Cleaner Production and Products: Guide for Government Self-Assessment, Report OCDE/GD(95)21. 48 pp.

Organisation for Economic Co-operation and Development (OECD), (1996). Technology, Productivity and Job Creation. OECD Publication, Paris 35 pp.

Organisation for Economic Co-operation and Development (OECD), (1997). Information Technology Outlook. OECD Publication, Paris, 234 pp.

Organisation for Economic Co-operation and Development (OECD), (1992). Good Practices for Environmental Impact Assessment of Development Projects. OECD Development Assistance Committee, Guidelines on Aid and Environment, No. 1, 15 pp.

Organisation for Economic Cooperation and Development (OECD), (1992). Recommendation of the Council on Integrated Coastal Zone Management, OECD.

Organisation for Economic Cooperation and Development (OECD), (1993). Coastal Zone Management: Integrated policies, OECD, Paris.

Organisation for Economic Cooperation and Development (OECD), (1997). Integrated Coastal Zone Management: Review of Progress in Selected OECD Countries, OECD, Paris.

Ovesen, C.L and Vestergaard, P. (eds), (1992). Danish dunes - monitoring, management and research. Report from a seminar in Northern Jutland, 25th – 27th May 1992. Ministry of the Environment, the National Forest and Nature Agency, Copenhagen.

Ovesen, C.L., (1998). Coastal Dunes - management, protection and research. EUCC Seminar, 19th-22nd August 1997, Skagen, Denmark. National Forest and nature Agency and the Geological Survey of Denmark and Greenland.

Parson, L.E., Lillycrop, W.J., Klein, C.J., Ives, R.C., Orlando, S.P., (1996). Use of Lidar Technology for collecting Shallow Bathymetry of Florida Bay. Journal of Coastal Research, Vol. 13, No. 4.

Penning-Rowsell E., Green, C.H., Thompson, P.M., et al., (1992). The Eco-nomics of Coastal Management - A Manual of Benefit Assessment Techniques. Belhaven Press, London, UK, 140 pp.

Penning-Rowsell, E., (1997). A General Model for Promoting the Integration of National Natural Resources Management. GeoJournal, v. 43, n. 3, pp. 247-262.

Penning-Rowsell, E., Winchester, P., Gardiner, G., (1998). New approaches to sustainable hazard man

Pernetta, J., and Elder, D., (1993). Cross-sectoral, Integrated Coastal Planning: Guidelines and Principles for Coastal Area Development. Gland, Switzerland, IUCN.

Post, J. and Lundin, G. (eds), (1996). Guidelines for Integrated Coastal Zone Management. Environmentally Sustainable Development Studies and Monographs Series. Washington, D.C., The World Bank.

Rautalahti-Miettinen, E., Ripatti, V., (1998). Development of Environmental Awareness. Workshop on Education and Learning for Sustainable Consumption, 14-15 September, OECD Headquarters, Paris.

Rigg, K., Salman, A., et al., (1997). Threats and opportunities in the coastal areas of the European Union: A scoping study. The Netherlands, Ministry of Housing, Spatial Planning and the Environment.

RIKZ. 1998. Report of the Transdisciplinary Euroconference on Coastal Management Research, San Feliu de Guixols, Spain, 6-10 December 1997. RIKZ, 100pp.

Salman, A., Berends H., Bonazountas M. (eds), (1995). Coastal management and habitat conservation. Vol. I: Proceedings of the 4th EUCC Congress, Marathon, Greece, April, 1993. EUCC, Leiden.

Salman, A., Langeveld, M.J., Bonazountas M. (eds), (1996). Coastal management and habitat conservation. Vol. II: Proceedings of the 4th EUCC Congress, Marathon, Greece, April, 1993. EUCC, Leiden.

Science and Technology Awareness Network (STAN), (1998). Planning and Implementing

Science Awareness Activities. Centre for Marine Geology, Dalhousie University, Halifax, Canada, 90 pp.

Scura, L., Chua, T.E., et al., (1992). Lessons for Integrated Coastal Zone Management: The ASEAN Experience. Integrative framework and methods for coastal area management. ICLARM Conf. Proc. 37. T.E. Chua and L. Scura: 169.

Smith, L.G., (1993). Impact Assessment and Sustainable Resource Management, Longman.

Sorensen, J. and McCreary, S., (1990). Institutional Arrangements for Managing Coastal Resources and Environments, National Park Service, US Department of the Interior & USAID.

Sorensen, J., (1993). The International Proliferation of Integrated Coastal Zone Management Efforts. Ocean and Coastal Management 21(1-3): 45-80.

Steers, J.A., (1969). The coastline of England and Wales. 2nd ed. Cambridge. At the University Press.

Steers, J.A., (1969). The sea coast. 4th ed. London, Collins. New Naturalist Series No. 25.

Steers, J.A., (1973). The coastline of Scotland. Cambridge, Cambridge University Press.

Suman, D., (1997). The Florida Keys National Marine Sanctuary: A Case Study of an Innovative Federal-State Partnership in Marine Resource Management. Coastal Management. Vol. 25, No. 3, PP. 293-324.

Symes, D. (1998) The Integration of Fisheries Management and Marine Wildlife Conservation. JNCC Report No. 287. Joint Nature Conservation Committee, Peterborough.

Tekke, R.M.H., (1996). Management of Coastal Lagoons in Albania. Proceedings of an International Seminar, Tirana, 29th May – 4th June 1994. EUCC, Leiden.

Tol, R., Klein, R., Jansen, H., Verbruggen, H., (1996). Some economic considerations on the importance of pro-active integrated coastal zone management. Ocean & Coastal Management, v. 32, n. 1, pp.39-55.

Turner, R.K. and Adger, W.N., (1996). Coastal Zone Resources Assessment Guidelines. LOICZ Reports and Studies No. 4, IGBP/LOICZ, Texel, The Netherlands, iv+101 pp.

Vallega, A., (1992). Sea Management - A Theoretical Approach. Elsevier Applied Science, pp. 259.

Van Alphen, J., (1995). The Voordelta Integrated Policy Plan: Administrative Aspects of Coastal Zone Management in the Netherlands. Ocean and Coastal Management 26(2): 133-150.

Van der Meulen, F., and de Haes, U., (1996). Nature Conservation and Integrated Coastal Zone Management in Europe: Present and Future. Landscape and Urban Planning 34: 401-410.

Van der Meulen, F., Jungerius, P.D., Visser, J. (eds), (1989). Perspectives in Coastal Dune Management. The Hague: SPB Academic Publishing.

Van Dijk, H., (1994). Integrated Management and Conservation of Dutch Coastal Areas: Premises, Practice and Solutions. Marine Pollution Bulletin 29(6-12): 609-616.

Verhagen, H.J., (1989). Sand waves along the Dutch coast. Coastal Engineering, v. 13, n. 2, pp. 129-147.

Verrone, (1996). Sea level in the Mediterranean: a first step towards separating crustal movements and absolute sea-level variations. Global and Planetary Change, v. 14, n. 1–2, pp. 1-48.

Von Alt, C., De Luca, M.P., Glenn, S.M., Grassle, J.F., Haidvogel, D.B., (1997). LEO-15: Monitoring & Managing Coastal Resources. Sea Technology, v. 38, n. 8, pp. 10-16.

Welp, M. (1999) Defining integrated coastal zone management for the Baltic Sea Region. In Hedegaard, L. & Lindström, B. (eds.) The NEBI Yearbook 1999: North European and Baltic Sea Integration. Springer-Verlag, Berlin, Heidelberg, 491 p. 101-119 (ISBN 3-540-66407-6).

Woods H., Dahlin L., Droppert M., Glass S., Vallerga S., Flemming N.C., (1996). The Strategy for EuroGOOS. EuroGOOS Pub. No. 1, Southampton Oceanography Center, Southampton.

World Coast 2000, (1993). Preparing to meet the coastal challenges of the 21st century. Noordwijk, The Netherlands, 5th November 1993.

Worm, K., (1998). Coastal Zone Planning in Denmark. Ocean and Coastal Management 37(2): 253- 268.

Glossary

Adequately Wet: Asbestos containing material that is sufficiently mixed or penetrated with liquid to prevent the release of particulates.

Aeration: A process which promotes biological degradation of organic matter in water. The process may be passive (as when waste is exposed to air), or active (as when a mixing or bubbling device introduces the air).

Aerosol: A finely divided material suspended in air or other gaseous environment.

Age Tank: A tank used to store a chemical solution of known concentration for feed to a chemical feeder. Also called a day tank.

Agricultural Pollution: Farming wastes, including runoff and leaching of pesticides and fertilizers; erosion and dust from plowing; improper disposal of animal manure and carcasses; crop residues, and debris.

Agroecosystem: Land used for crops, pasture, and livestock; the adjacent uncultivated land that supports other vegetation and wildlife; and the associated atmosphere, the underlying soils, groundwater, and drainage networks.

Air Cleaning: Indoor-air quality-control strategy to remove various airborne particulates and/or gases from the air. Most common methods are particulate filtration, electrostatic precipitation, and gas sorption.

Air Contaminant: Any particulate matter, gas, or combination thereof, other than water vapor.

Air Curtain: A method of containing oil spills. Air bubbling through a perforated pipe causes an upward water flow that slows the spread of oil. It can also be used to stop fish from entering polluted water.

Air Exchange Rate: The rate at which outside air replaces indoor air in a given space.

Air Handling Unit: Equipment that includes a fan or blower, heating and/or cooling coils, regulator controls, condensate drain pans, and air filters.

Air/Oil Table: The surface between the vadose zone and ambient oil; the pressure of oil in the porous medium is equal to atmospheric pressure.

Air Padding: Pumping dry air into a container to assist with the withdrawal of liquid or to force a liquefied gas such as chlorine out of the container.

Air Permeability: Permeability of soil with respect to air. Important to the design of soil-gas surveys. Measured in darcys or centimeters-per-second.

Air Plenum: Any space used to convey air in a building, furnace, or structure. The space above a suspended ceiling is often used as an air plenum.

Air Pollutant: Any substance in air that could, in high enough concentration, harm man, other animals, vegetation, or material. Pollutants may include almost any natural or artificial composition of airborne matter capable of being airborne. They may be in the form of solid particles, liquid droplets, gases, or in combination thereof. Generally, they fall into two main groups: (1) those emitted directly from identifiable sources and (2) those produced in the air by interaction between two or more primary pollutants, or by reaction with normal atmospheric constituents, with or without photoactivation. Exclusive of pollen, fog, and dust, which are of natural origin, about 100 contaminants have been identified. Air pollutants are often grouped in categories for ease in classification; some of he categories are: solids, sulfur compounds, volatile organic chemicals, particulate matter, nitrogen compounds, oxygen compounds, halogen compounds, radioactive compound, and odors.

Air Pollution: The presence of contaminants or pollutant substances in the air that interfere with human health or welfare, or produce other harmful environmental effects.

Air Pollution Control Device: Mechanism or equipment that cleans emissions generated by a source (e.g. an incinerator, industrial smokestack, or an automobile exhaust system) by removing pollutants that would otherwise be released to the atmosphere.

Air Pollution Episode: A period of abnormally high concentration of air pollutants, often due to low winds and temperature inversion, that can cause illness and death.

Air Quality Control Region:

Air Quality Criteria: The levels of pollution and lengths of exposure above which adverse health and welfare effects may occur.

Air Quality Standards: The level of pollutants prescribed by regulations that are not be exceeded during a given time in a defined area.

Air Sparging: Injecting air or oxygen into an aquifer to strip or flush volatile contaminants as air bubbles up through The ground water and is captured by a vapor extraction system.

Air Stripping: A treatment system that removes volatile organic compounds (VOCs) from contaminated ground water or surface water by forcing an airstream through the water and causing the compounds to evaporate.

Air Toxics: Any air pollutant for which a national ambient air quality standard (NAAQS) does not exist (i.e. excluding ozone, carbon monoxide, PM-10, sulfur dioxide, nitrogen oxide) that may reasonably be anticipated to cause cancer; respiratory, cardiovascular, or developmental effects; reproductive dysfunctions, neurological disorders, heritable gene mutations, or other serious or irreversible chronic or acute health effects in humans.

Airborne Particulates: Total suspended particulate matter found in the atmosphere as solid particles or liquid droplets. Chemical composition of particulates varies widely, depending on location and time of year. Sources of airborne particulates include: dust, emissions from industrial processes, combustion products from the burning of wood and coal, combustion products associated with motor vehicle or non-road engine exhausts, and reactions to gases in the atmosphere.

Algae: Simple rootless plants that grow in sunlit waters in proportion to the amount of available nutrients. They can affect water quality adversely by lowering the dissolved oxygen in the water. They are food for fish and small aquatic animals.

Algal Blooms: Sudden spurts of algal growth, which can affect water quality adversely and indicate potentially hazardous changes in local water chemistry.

Algicide: Substance or chemical used specifically to kill or control algae.

Alternative Compliance: A policy that allows facilities to choose among methods for achieving emission-reduction or risk-reduction instead of command-and control regulations that specify standards and how to meet them. Use of a theoretical emissions bubble over a facility to cap the amount of pollution emitted while allowing the company to choose where and how (within the facility) it complies.

Alternative Fuels: Substitutes for traditional liquid, oil-derived motor vehicle fuels like gasoline and diesel. Includes mixtures of alcohol-based fuels with gasoline, methanol, ethanol, compressed natural gas, and others.

Alternative Remedial Contract Strategy Contractors: Government contractors who provide project management and technical services to support remedial response activities at National Priorities List sites.

Ambient Air: Any unconfined portion of the atmosphere: open air, surrounding air.

Ambient Measurement: A measurement of the concentration of a substance or pollutant within the immediate environs of an organism; taken to relate it to the amount of possible exposure.

Ambient Medium: Material surrounding or contacting an organism (e.g. outdoor air, indoor air, water, or soil, through which chemicals or pollutants can reach the organism.

Ambient Temperature: Temperature of the surrounding air or other medium.

Amprometric Titration: A way of measuring concentrations of certain substances in water using an electric current that flows during a chemical reaction.

Antarctic "Ozone Hole": Refers to the seasonal depletion of ozone in the upper atmosphere above a large area of Antarctica.

Anti-Degradation Clause: Part of federal air quality and water quality requirements prohibiting deterioration where pollution levels are above the legal limit.

Anti-Microbial: An agent that kills microbes.

Applicable or Relevant and Appropriate Requirements (ARARs): Any state or federal statute that pertains to protection of human life and the environment in addressing specific conditions or use of a particular cleanup technology at a Superfund site,

Applied Dose: In exposure assessment, the amount of a substance in contact with the primary absorption boundaries of an organism (e.g. skin, lung tissue, gastrointestinal track) and available for absorption.

Aqueous: Something made up of water.

Aqueous Solubility: The maximum concentration of a chemical that will dissolve in pure water at a reference temperature.

Aquifer: An underground geological formation, or group of formations, containing water. Are sources of groundwater for wells and springs.

Aquifer Test: A test to determine hydraulic properties of an aquifer.

Aquitard: Geological formation that may contain groundwater but is not capable of transmitting significant quantities of it under normal hydraulic gradients. May function as confining bed.

Architectural Coatings: Coverings such as paint and roof tar that are used on exteriors of buildings.

Area of Review: In the UIC program, the area surrounding an injection well that is reviewed during the permitting process to determine if flow between aquifers will be induced by the injection operation.

Area Source: Any source of air pollution that is released over a relatively small area but which cannot be classified as a point source. Such sources may include vehicles and other small engines, small businesses and household activities, or biogenic sources such as a forest that releases hydrocarbons.

Aromatics: A type of hydrocarbon, such as benzene or toluene, with a specific type of ring structure. Aromatics are sometimes added to gasoline in order to increase octane. Some aromatics are toxic.

Artesian (Aquifer or Well): Water held under pressure in porous rock or soil confined by impermeable geological formations.

Asbestos: A mineral fiber that can pollute air or water and cause cancer or asbestosis when inhaled. EPA has banned or severely restricted its use in manufacturing and construction.

Asbestos Abatement: Procedures to control fiber release from asbestos-containing materials in a building or to remove them entirely, including removal, encapsulation, repair, enclosure, encasement, and operations and maintenance programs.

Asbestos Assessment: In the asbestos-in-schools program, the evaluation of the physical condition and potential for damage of all friable asbestos containing materials and thermal insulation systems.

Asbestos Program Manager: A building owner or designated representative who supervises all aspects of the facility asbestos management and control program.

Asbestos-Containing Waste Materials (ACWM): Mill tailings or any waste that contains commercial asbestos and is generated by a source covered by the Clean Air Act Asbestos NESHAPS.

A-Scale Sound Level: A measurement of sound approximating the sensitivity of the human ear, used to note the intensity or annoyance level of sounds.

Back Pressure: A pressure that can cause water to backflow into the water supply when a user's waste water system is at a higher pressure than the public system.

Backflow/Back Siphonage: A reverse flow condition created by a difference in water pressures that causes water to flow back into the distribution pipes of a drinking water supply from any source other than the intended one.

Background Level: 1. The concentration of a substance in an environmental media (air, water, or soil) that occurs naturally or is not the result of human activities. 2. In exposure assessment the concentration of a substance in a defined control area, during a fixed period of time before, during, or after a data-gathering operation..

Backwashing: Reversing the flow of water back through the filter media to remove entrapped solids.

Backyard Composting: Diversion of organic food waste and yard trimmings from the municipal waste stream by composting hem in one's yard through controlled decomposition of organic matter by bacteria and fungi into a humus-like product. It is considered source reduction, not recycling, because the composted materials never enter the municipal waste stream.

Barrel Sampler: Open-ended steel tube used to collect soil samples.

BACT - Best Available Control Technology: An emission limitation based on the maximum degree of emission reduction (considering energy, environmental, and economic impacts) achievable through application of production processes and available methods, systems, and techniques. BACT does not permit emissions in excess of those allowed under any applicable Clean Air Act provisions. Use of the BACT concept is allowable on a case by case basis for major new or modified emissions sources in attainment areas and applies to each regulated pollutant.

Baling: Compacting solid waste into blocks to reduce volume and simplify handling.

Ballistic Separator: A machine that sorts organic from inorganic matter for composting.

Band Application: The spreading of chemicals over, or next to, each row of plants in a field.

Banking: A system for recording qualified air emission reductions for later use in bubble, offset, or netting transactions.

BEN: EPA's computer model for analyzing a violator's economic gain from not complying with the law.

Benefit-Cost Analysis: An economic method for assessing the benefits and costs of achieving alternative health-based standards at given levels of health protection.

Benthic/Benthos: An organism that feeds on the sediment at the bottom of a water body such as an ocean, lake, or river.

Bentonite: A colloidal clay, expansible when moist, commonly used to provide a tight seal around a well casing.

Beryllium: An metal hazardous to human health when inhaled as an airborne pollutant. It is discharged by machine shops, ceramic and propellant plants, and foundries.

Bioaccumulants: Substances that increase in concentration in living organisms as they take in contaminated air, water, or food because the substances are very slowly metabolized or excreted.

Bioassay: A test to determine te relative strength of a substance by comparing its effect on a test organism with that of a standard preparation.

Bioavailabiliity: Degree of ability to be absorbed and ready to interact in organism metabolism.

Biochemical Oxygen Demand (BOD): A measure of the amount of oxygen consumed in the biological processes that break down organic matter in water. The greater the BOD, the greater the degree of pollution.

Bioconcentration: The accumulation of a chemical in tissues of a fish or other organism to levels greater than in the surrounding medium.

Biological Integrity: The ability to support and maintain balanced, integrated, functionality in the natural habitat of a given region. Concept is applied primarily in drinking water management.

Biological Magnification: Refers to the process whereby certain substances such as pesticides or heavy metals move up the food chain, work their way into rivers or lakes, and are eaten by aquatic organisms such as fish, which in turn are eaten by large birds, animals or humans. The substances become concentrated in tissues or internal organs as they move up the chain.

Biological Measurement: A measurement taken in a biological medium. For exposure assessment, it is related to the measurement is taken to related it to the established internal dose of a compound.

Biological Medium: One of the major component of an organism; e.g. blood, fatty tissue, lymph nodes or breath, in which chemicals can be stored or transformed.

Biological Oxidation: Decomposition of complex organic materials by microorganisms. Occurs in self-purification of water bodies and in activated sludge wastewater treatment.

Biological Oxygen Demand (BOD): An indirect measure of the concentration of biologically degradable material present in organic wastes. It usually reflects the amount of oxygen consumed in five days by biological processes breaking down organic waste.

Biological pesticides: Certain microorganism, including bacteria, fungi, viruses, and protozoa that are effective in controlling pests. These agents usually do not have toxic effects on animals and people and do not leave toxic or persistent chemical residues in the environment.

Biological Stressors: Organisms accidentally or intentionally dropped into habitats in which they do not evolve naturally; e.g. gypsy moths, Dutch elm disease, certain types of algae, and bacteria.

Biological Treatment: A treatment technology that uses bacteria to consume organic waste.

Biologically Effective Dose: The amount of a deposited or absorbed compound reaching the cells or target sites where adverse effect occur, or where the chemical interacts with a membrane.

Biologicals: Vaccines, cultures and other preparations made from living organisms and their products, intended for use in diagnosing, immunizing, or treating humans or animals, or in related research.

Biomass: All of the living material in a given area; often refers to vegetation.

Biome: Entire community of living organisms in a single major ecological area.

Bioremediation: Use of living organisms to clean up oil spills or remove other pollutants from soil, water, or wastewater; use of organisms such as non-harmful insects to remove agricultural pests or counteract diseases of trees, plants, and garden soil.

Biota: The animal and plant life of a given region.

Biotechnology: Techniques that use living organisms or parts of organisms to produce a variety of products (from medicines to industrial enzymes) to improve plants or animals or to develop microorganisms to remove toxics from bodies of water, or act as pesticides.

Biotic Community: A naturally occurring assemblage of plants and animals that live in the same environment and are mutually sustaining and interdependent.

Biotransformation: Conversion of a substance into other compounds by organisms; includes biodegradation.

Blackwater: Water that contains animal, human, or food waste.

Blood Products: Any product derived from human blood, including but not limited to blood plasma, platelets, red or white corpuscles, and derived licensed products such as interferon.

Bloom: A proliferation of algae and/or higher aquatic plants in a body of water; often related to pollution, especially when pollutants accelerate growth.

BOD5: The amount of dissolved oxygen consumed in five days by biological processes breaking down organic matter.

Body Burden: The amount of a chemical stored in the body at a given time, especially a potential toxin in the body as the result of exposure.

Bog: A type of wetland that accumulates appreciable peat deposits. Bogs depend primarily on precipitation for their water source, and are usually acidic and rich in plant residue with a conspicuous mat of living green moss.

Boiler: A vessel designed to transfer heat produced by combustion or electric resistance to water. Boilers may provide hot water or steam.

Boom: 1. A floating device used to contain oil on a body of water. 2. A piece of equipment used to apply pesticides from a tractor or truck.

Borehole: Hole made with drilling equipment.

Botanical Pesticide: A pesticide whose active ingredient is a plant-produced chemical such as nicotine or strychnine. Also called a plant-derived pesticide.

Bottle Bill: Proposed or enacted legislation which requires a returnable deposit on beer or soda containers and provides for retail store or other redemption. Such legislation is designed to discourage use of throw-away containers.

Bottom Ash: The non-airborne combustion residue from burning pulverized coal in a boiler; the material which falls to the bottom of the boiler and is removed mechanically; a concentration of non-combustible materials, which may include toxics.

Bottom Land Hardwoods: Forested freshwater wetlands adjacent to rivers in the southeastern United States, especially valuable for wildlife breeding, nesting and habitat.

Bounding Estimate: An estimate of exposure, dose, or risk that is higher than that incurred by the person in the population with the currently highest exposure, dose, or risk. Bounding estimates are useful in developing statements that exposures, doses, or risks are not greater than an estimated value.

Brackish: Mixed fresh and salt water.

Breakpoint Chlorination: Addition of chlorine to water until the chlorine demand has been satisfied.

Breakthrough: A crack or break in a filter bed that allows the passage of floc or particulate matter through a filter; will cause an increase in filter effluent turbidity.

Breathing Zone: Area of air in which an organism inhales.

Brine Mud: Waste material, often associated with well-drilling or mining, composed of mineral salts or other inorganic compounds.

British Thermal Unit: Unit of heat energy equal to the amount of heat required to raise the temperature of one pound of water by one degree Fahrenheit at sea level.

Broadcast Application: The spreading of pesticides over an entire area.

Brownfields: Abandoned, idled, or under used industrial and commercial facilities/sites where expansion or redevelopment is complicated by real or perceived environmental contamination. They can be in urban, suburban, or rural areas. EPA's Brownfields initiative helps communities mitigate potential health risks and restore the economic viability of such areas or properties.

Bubble: A system under which existing emissions sources can propose alternate means to comply with a set of emissions limitations; under the bubble concept, sources can control more than required at one emission point where control costs are relatively low in return for a comparable relaxation of controls at a second emission point where costs are higher.

Buffer: A solution or liquid whose chemical makeup is such that it minimizes changes in pH when acids or bases are added to it.

Buffer Strips: Strips of grass or other erosion-resisting vegetation between or below cultivated strips or fields.

Building Cooling Load: The hourly amount of heat that must be removed from a building to maintain indoor comfort (measured in British thermal units (Btus).

Building Envelope: The exterior surface of a building's construction—the walls, windows, floors, roof, and floor. Also called building shell.

Building Related Illness: Diagnosable illness whose cause and symptoms can be directly attributed to a specific pollutant source within a building (e.g. Legionnaire's disease, hypersensitivity, pneumonitis.)

Bulk Sample: A small portion (usually thumbnail size) of a suspect asbestos-containing building material collected by an asbestos inspector for laboratory analysis to determine asbestos content.

Bulky Waste: Large items of waste materials, such as appliances, furniture, large auto parts, trees, stumps.

Burial Ground (Graveyard): A disposal site for radioactive waste materials that uses earth or water as a shield.

Buy-Back Center: Facility where individuals or groups bring reyclables in return for payment.

By-product: Material, other than the principal product, generated as a consequence of an industrial process or as a breakdown product in a living system.

Cadmium (Cd): A heavy metal that accumulates in the environment.

Cancellation: Refers to Section 6 (b) of the Federal Insecticide, Fungicide and Rodenticide Act (FIFRA) which authorizes cancellation of a pesticide registration if unreasonable adverse effects to the environment and public health develop when a product is used according to widespread and commonly recognized practice, or if its labeling or other material required to be submitted does not comply with FIFRA provisions.

Cap: A layer of clay, or other impermeable material installed over the top of a closed landfill to prevent entry of rainwater and minimize leachate.

Capacity Assurance Plan: A statewide plan which supports a state's ability to manage the hazardous waste generated within its boundaries over a twenty year period.

Capillary Action: Movement of water through very small spaces due to molecular forces called capillary forces.

Capillary Fringe: The porous material just above the water table which may hold water by capillarity (a property of surface tension that draws water upwards) in the smaller void spaces.

Capillary Fringe: The zone above he water table within which the porous medium is saturated by water under less than atmospheric pressure.

Capture Efficiency: The fraction of organic vapors generated by a process that are directed to an abatement or recovery device.

Carbon Adsorption: A treatment system that removes contaminants from ground water or surface water by forcing it through tanks containing activated carbon treated to attract the contaminants.

Carbon Monoxide (CO): A colorless, odorless, poisonous gas produced by incomplete fossil fuel combustion.

Carboxyhemoglobin: Hemoglobin in which the iron is bound to carbon monoxide(CO) instead of oxygen.

Carcinogen: Any substance that can cause or aggravate cancer.

Carrier: 1.The inert liquid or solid material in a pesticide product that serves as a delivery vehicle for the active ingredient. Carriers do not have toxic properties of their own. 2. Any material or system that can facilitate the movement of a pollutant into the body or cells.

Carrying Capacity: 1. In recreation management, the amount of use a recreation area can sustain without loss of quality. 2. In wildlife management, the maximum number of animals an area can support during a given period.

CAS Registration Number: A number assigned by the Chemical Abstract Service to identify a chemical.

Case Study: A brief fact sheet providing risk, cost, and performance information on alternative methods and other pollution prevention ideas, compliance initiatives, voluntary efforts, etc.

Cask: A thick-walled container (usually lead) used to transport radioactive material. Also called a coffin.

Catalyst: A substance that changes the speed or yield of a chemical reaction without being consumed or chemically changed by the chemical reaction.

Catalytic Converter: An air pollution abatement device that removes pollutants from motor vehicle exhaust, either by oxidizing them into carbon dioxide and water or reducing them to nitrogen.

Catalytic Incinerator: A control device that oxidizes volatile organic compounds (VOCs) by using a catalyst to promote the combustion process. Catalytic incinerators require lower temperatures than conventional thermal incinerators, thus saving fuel and other costs.

Categorical Exclusion: A class of actions which either individually or cumulatively would not have a significant effect on the human environment and therefore would not require preparation of an environmental assessment or environmental impact statement under the National Environmental Policy Act (NEPA).

Categorical Pretreatment Standard: A technology-based effluent limitation for an industrial facility discharging into a municipal sewer system. Analogous in stringency to Best Availability Technology (BAT) for direct dischargers.

Cathodic Protection: A technique to prevent corrosion of a metal surface by making it the cathode of an electrochemical cell.

Cavitation: The formation and collapse of gas pockets or bubbles on the blade of an impeller or the gate of a valve; collapse of these pockets or bubbles drives water with such force that it can cause pitting of the gate or valve surface.

Chemical Case: For purposes of review and regulation, the grouping of chemically similar pesticide active ingredients (e.g. salts and esters of the same chemical) into chemical cases.

Chemical Compound: A distinct and pure substance formed by the union or two or more elements in definite proportion by weight.

Chemical Element: A fundamental substance comprising one kind of atom; the simplest form of matter.

Chemical Oxygen Demand (COD): A measure of the oxygen required to oxidize all compounds, both organic and inorganic, in water.

Chemical Stressors: Chemicals released to the environment through industrial waste, auto emissions, pesticides, and other human activity that can cause illnesses and even death in plants and animals.

Chemical Treatment: Any one of a variety of technologies that use chemicals or a variety of chemical processes to treat waste.

Chemnet: Mutual aid network of chemical shippers and contractors that assigns a contracted emergency response company to provide technical support if a representative of the firm whose chemicals are involved in an incident is not readily available.

Chemosterilant: A chemical that controls pests by preventing reproduction.

Chemtrec: The industry-sponsored Chemical Transportation Emergency Center; provides information and/or emergency assistance to emergency responders.

Child Resistant Packaging (CRP): Packaging that protects children or adults from injury or illness resulting from accidental contact with or ingestion of residential pesticides that meet or exceed specific toxicity levels. Required by FIFRA regulations. Term is also used for protective packaging of medicines.

Chiller: A device that generates a cold liquid that is circulated through an air-handling unit's cooling coil to cool the air supplied to the building.

Chilling Effect: The lowering of the Earth's temperature because of increased particles in the air blocking the sun's rays.

Chisel Plowing: Preparing croplands by using a special implement that avoids complete inversion of the soil as in conventional plowing. Chisel plowing can leave a protective cover or crops residues on the soil surface to help prevent erosion and improve filtration.

Chlorinated Solvent: An organic solvent containing chlorine atoms(e.g. methylene chloride and 1,1,1-trichloromethane). Uses of chlorinated solvents are include aerosol spray containers, in highway paint, and dry cleaning fluids.

Chlorination: The application of chlorine to drinking water, sewage, or industrial waste to disinfect or to oxidize undesirable compounds.

Chlorinator: A device that adds chlorine, in gas or liquid form, to water or sewage to kill infectious bacteria.

Chlorine-Contact Chamber: That part of a water treatment plant where effluent is disinfected by chlorine.

Chlorofluorocarbons (CFCs): A family of inert, nontoxic, and easily liquefied chemicals used in refrigeration, air conditioning, packaging, insulation, or as solvents and aerosol propellants. Because CFCs are not destroyed in the lower atmosphere they drift into the upper atmosphere where their chlorine components destroy ozone.

Chlorophenoxy: A class of herbicides that may be found in domestic water supplies and cause adverse health effects.

Chlorosis: Discoloration of normally green plant parts caused by disease, lack of nutrients, or various air pollutants.

Cholinesterase: An enzyme found in animals that regulates nerve impulses by the inhibition of acetylcholine. Cholinesterase inhibition is associated with a variety of acute symptoms such as nausea, vomiting, blurred vision, stomach cramps, and rapid heart rate.

Chronic Effect: An adverse effect on a human or animal in which symptoms recur frequently or develop slowly over a long period of time.

Chronic Exposure: Multiple exposures occurring over an extended period of time or over a significant fraction of an animal's or human's lifetime (Usually seven years to a lifetime.)

Chronic Toxicity: The capacity of a substance to cause long-term poisonous health effects in humans, animals, fish, and other organisms.

Circle of Influence: The circular outer edge of a depression produced in the water table by the pumping of water from a well.

Cistern: Small tank or storage facility used to store water for a home or farm; often used to store rain water.

Clarification: Clearing action that occurs during wastewater treatment when solids settle out. This is often aided by centrifugal action and chemically induced coagulation in wastewater.

Clay Soil: Soil material containing more than 40 percent clay, less than 45 percent sand, and less than 40 percent silt.

Clean Coal Technology: Any technology not in widespread use prior to the Clean Air Act Amendments of 1990. This Act will achieve significant reductions in pollutants associated with the burning of coal.

Clean Fuels: Blends or substitutes for gasoline fuels, including compressed natural gas, methanol, ethanol, and liquified petroleum gas.

Cleaner Technologies Substitutes Assessment: A document that systematically evaluates the relative risk, performance, and cost trade-offs of technological alternatives; serves as a repository for all the technical data (including methodology and results) developed by a DfE or other pollution prevention or education project.

Cleanup: Actions taken to deal with a release or threat of release of a hazardous substance that could affect humans and/or the environment. The term "cleanup" is sometimes used interchangeably with the terms remedial action, removal action, response action, or corrective action.

Clear Cut: Harvesting all the trees in one area at one time, a practice that can encourage fast rainfall or snowmelt runoff, erosion, sedimentation of streams and lakes, and flooding, and destroys vital habitat.

Clear Well: A reservoir for storing filtered water of sufficient quantity to prevent the need to vary the filtration rate with variations in demand. Also used to provide chlorine contact time for disinfection.

Climate Change (also referred to as 'global climate change'): The term 'climate change' is sometimes used to refer to all forms of climatic inconsistency, but because the Earth's climate is never static, the term is more properly used to imply a significant change from one climatic condition to another. In some cases, 'climate change' has been used synonymously with the term, 'global warming'; scientists however, tend to use the term in the wider sense to also include natural changes in climate.

Cloning: In biotechnology, obtaining a group of genetically identical cells from a single cell; making identical copies of a gene.

Closed-Loop Recycling: Reclaiming or reusing wastewater for non-potable purposes in an enclosed process.

Closure: The procedure a landfill operator must follow when a landfill reaches its legal capacity for solid ceasing acceptance of solid waste and placing a cap on the landfill site.

Co-fire: Burning of two fuels in the same combustion unit; e.g., coal and natural gas, or oil and coal.

Coagulation: Clumping of particles in wastewater to settle out impurities, often induced by chemicals such as lime, alum, and iron salts.

Coal Cleaning Technology: A precombustion process by which coal is physically or chemically treated to remove some of its sulfur so as to reduce sulfur dioxide emissions.

Coal Gasification: Conversion of coal to a gaseous product by one of several available technologies.

Coastal Zone: Lands and waters adjacent to the coast that exert an influence on the uses of the sea and its ecology, or whose uses and ecology are affected by the sea.

Code of Federal Regulations (CFR): Document that codifies all rules of the executive departments and agencies of the federal government. It is divided into fifty volumes, known as titles. Title 40 of the CFR (referenced as 40 CFR) lists all environmental regulations.

Cogeneration: The consecutive generation of useful thermal and electric energy from the same fuel source.

Coke Oven: An industrial process which converts coal into coke, one of the basic materials used in blast furnaces for the conversion of iron ore into iron.

Cold Temperature CO: A standard for automobile emissions of carbon monoxide (CO) emissions to be met at a low temperature (i.e. 20 degrees Fahrenheit). Conventional automobile catalytic converters are not efficient in cold weather until they warm up.

Coliform Index: A rating of the purity of water based on a count of fecal bacteria.

Coliform Organism: Microorganisms found in the intestinal tract of humans and animals. Their presence in water indicates fecal pollution and potentially adverse contamination by pathogens.

Collector: Public or private hauler that collects nonhazardous waste and recyclable materials from residential, commercial, institutional and industrial sources.

Collector Sewers: Pipes used to collect and carry wastewater from individual sources to an interceptor sewer that will carry it to a treatment facility.

Colloids: Very small, finely divided solids (that do not dissolve) that remain dispersed in a liquid for a long time due to their small size and electrical charge.

Combined Sewer Overflows: Discharge of a mixture of storm water and domestic waste when the flow capacity of a sewer system is exceeded during rainstorms.

Combined Sewers: A sewer system that carries both sewage and storm-water runoff. Normally, its entire flow goes to a waste treatment plant, but during a heavy storm, the volume of water may be so great as to cause overflows of untreated mixtures of storm water and sewage into receiving waters. Storm-water runoff may also carry toxic chemicals from industrial areas or streets into the sewer system.

Combustion: 1. Burning, or rapid oxidation, accompanied by release of energy in the form of heat and light. 2. Refers to controlled burning of waste, in which heat chemically alters organic compounds, converting into stable inorganics such as carbon dioxide and water.

Combustion Chamber: The actual compartment where waste is burned in an incinerator.

Combustion Product: Substance produced during the burning or oxidation of a material.

Command Post: Facility located at a safe distance upwind from an accident site, where the on-scene coordinator, responders, and technical representatives make response decisions, deploy manpower and equipment, maintain liaison with news media, and handle communications.

Command-and-Control Regulations: Specific requirements prescribing how to comply with specific standards defining acceptable levels of pollution.

Comment Period: Time provided for the public to review and comment on a proposed EPA action or rulemaking after publication in the Federal Register.

Commercial Waste: All solid waste emanating from business establishments such as stores, markets, office buildings, restaurants, shopping centers, and theaters.

Commercial Waste Management Facility: A treatment, storage, disposal, or transfer facility which accepts waste from a variety of sources, as compared to a private facility which normally manages a limited waste stream generated by its own operations.

Commingled Recyclables: Mixed recyclables that are collected together.

Comminuter: A machine that shreds or pulverizes solids to make waste treatment easier.

Comminution: Mechanical shredding or pulverizing of waste. Used in both solid waste management and wastewater treatment.

Common Sense Initiative: Voluntary program to simplify environmental regulation to achieve cleaner, cheaper, smarter results, starting with six major industry sectors.

Community: In ecology, an assemblage of populations of different species within a specified location in space and time. Sometimes, a particular subgrouping may be specified, such as the fish community in a lake or the soil arthropod community in a forest.

Community Relations: The EPA effort to establish two-way communication with the public to create understanding of EPA programs and related actions, to ensure public input into decision-making processes related to affected communities, and to make certain that the Agency is aware of and responsive to public concerns. Specific community relations activities are required in relation to Superfund remedial actions.

Community Water System: A public water system which serves at least 15 service connections used by year-round residents or regularly serves at least 25 year-round residents.

Compact Fluorescent Lamp (CFL): Small fluorescent lamps used as more efficient alternatives to incandescent lighting. Also called PL, CFL, Twin-Tube, or BIAX lamps.

Compaction: Reduction of the bulk of solid waste by rolling and tamping.

Comparative Risk Assessment: Process that generally uses the judgement of experts to predict effects and set priorities among a wide range of environmental problems.

Complete Treatment: A method of treating water that consists of the addition of coagulant chemicals, flash mixing, coagulation-flocculation, sedimentation, and filtration. Also called conventional filtration.

Compliance Coal: Any coal that emits less than 1.2 pounds of sulfur dioxide per million Btu when burned. Also known as low sulfur coal.

Compliance Coating: A coating whose volatile organic compound content does not exceed that allowed by regulation.

Compliance Cycle: The 9-year calendar year cycle, beginning January 1, 1993, during which public water systems must monitor. Each cycle consists of three 3-year compliance periods.

Compliance Monitoring: Collection and evaluation of data, including self-monitoring reports, and verification to show whether pollutant concentrations and loads contained in permitted discharges are in compliance with the limits and conditions specified in the permit.

Compliance Schedule: A negotiated agreement between a pollution source and a government agency that specifies dates and procedures by which a source will reduce emissions and, thereby, comply with a regulation.

Composite Sample: A series of water samples taken over a given period of time and weighted by flow rate.

Compost: The relatively stable humus material that is produced from a composting process in which bacteria in soil mixed with garbage and degradable trash break down the mixture into organic fertilizer.

Composting: The controlled biological decomposition of organic material in the presence of air to form a humus-like material. Controlled methods of composting include mechanical mixing and aerating, ventilating the materials by dropping them through a vertical series of aerated chambers, or placing the compost in piles out in the open air and mixing it or turning it periodically.

Compressed Natural Gas (CNG): An alternative fuel for motor vehicles; considered one of the cleanest because of low hydrocarbon emissions and its vapors are relatively non-ozone producing. However, vehicles fueled with CNG do emit a significant quantity of nitrogen oxides.

Concentration: The relative amount of a substance mixed with another substance. An example is five ppm of carbon monoxide in air or 1 mg/l of iron in water.

Condensate: 1.Liquid formed when warm landfill gas cools as it travels through a collection system. 2. Water created by cooling steam or water vapor.

Condensate Return System: System that returns the heated water condensing within steam piping to the boiler and thus saves energy.

Conditional Registration: Under special circumstances, the Federal Insecticide, Fungicide, and Rodenticide Act (FIFRA) permits registration of pesticide products that is "conditional" upon the submission of additional data. These special circumstances include a finding by the EPA Administrator that a new product or use of an existing pesticide will not significantly increase the risk of unreasonable adverse effects. A product containing a new (previously unregistered) active ingredient may be conditionally registered only if the Administrator finds that such conditional registration is in the public interest, that a reasonable time for conducting the additional studies has not elapsed, and the use of the pesticide for the period of conditional registration will not present an unreasonable risk.

Conditionally Exempt Generators (CE): Persons or enterprises which produce less than 220 pounds of hazardous waste per month. Exempt from most regulation, they are required merely to determine whether their waste is hazardous, notify appropriate state or local agencies, and ship it by an authorized transporter to a permitted facility for proper disposal.

Conductance: A rapid method of estimating the dissolved solids content of water supply by determining the capacity of a water sample to carry an electrical current. Conductivity is a measure of the ability of a solution to carry and electrical current.

Conductivity: A measure of the ability of a solution to carry an electrical current.

Cone of Depression: A depression in the water table that develops around a pumped well.

Cone of Influence: The depression, roughly conical in shape, produced in a water table by the pumping of water from a well.

Cone Penterometer Testing (CPT): A direct push system used to measure lithology based on soil penetration resistance. Sensors in the tip of the cone of the DP rod measure tip resistance and side-wall friction, transmitting electrical signals to digital processing equipment on the ground surface.

Construction and Demolition Waste: Waste building materials, dredging materials, tree stumps, and rubble resulting from construction, remodeling, repair, and demolition of homes, commercial buildings and other structures and pavements. May contain lead, asbestos, or other hazardous substances.

Construction Ban: If, under the Clean Air Act, EPA disapproves an area's planning requirements for correcting nonattainment, EPA can ban the construction or modification of any major stationary source of the pollutant for which the area is in nonattainment.

Consumptive Water Use: Water removed from available supplies without return to a water resources system, e.g. water used in manufacturing, agriculture, and food preparation.

Contact Pesticide: A chemical that kills pests when it touches them, instead of by ingestion. Also, soil that contains the minute skeletons of certain algae that scratch and dehydrate waxy-coated insects.

Contaminant: Any physical, chemical, biological, or radiological substance or matter that has an adverse effect on air, water, or soil.

Contamination: Introduction into water, air, and soil of microorganisms, chemicals, toxic substances, wastes, or wastewater in a concentration that makes the medium unfit for its next intended use. Also applies to surfaces of objects, buildings, and various household and agricultural use products.

Contamination Source Inventory: An inventory of contaminant sources within delineated State Water-Protection Areas. Targets likely sources for further investigation.

Contingency Plan: A document setting out an organized, planned, and coordinated course of action to be followed in case of a fire, explosion, or other accident that releases toxic chemicals, hazardous waste, or radioactive materials that threaten human health or the environment.

Continuous Discharge: A routine release to the environment that occurs without interruption, except for infrequent shutdowns for maintenance, process changes, etc.

Continuous Sample: A flow of water, waste or other material from a particular place in a plant to the location where samples are collected for testing. May be used to obtain grab or composite samples.

Contour Plowing: Soil tilling method that follows the shape of the land to discourage erosion.

Contour Strip Farming: A kind of contour farming in which row crops are planted in strips, between alternating strips of close-growing, erosion-resistant forage crops.

Contract Labs: Laboratories under contract to EPA, which analyze samples taken from waste, soil, air, and water or carry out research projects.

Control Technique Guidelines (CTG): EPA documents designed to assist state and local pollution authorities to achieve and maintain air quality standards for certain sources (e.g. organic emissions from solvent metal cleaning known as degreasing) through reasonably available control technologies (RACT).

Controlled Reaction: A chemical reaction under temperature and pressure conditions maintained within safe limits to produce a desired product or process.

Conventional Pollutants: Statutorily listed pollutants understood well by scientists. These may be in the form of organic waste, sediment, acid, bacteria, viruses, nutrients, oil and grease, or heat.

Conventional Site Assessment: Assessment in which most of the sample analysis and interpretation of data is completed off-site; process usually requires repeated mobilization of equipment and staff in order to fully determine the extent of contamination.

Conventional Systems: Systems that have been traditionally used to collect municipal wastewater in gravity sewers and convey it to a central primary or secondary treatment plant prior to discharge to surface waters.

Conventional Tilling: Tillage operations considered standard for a specific location and crop and that tend to bury the crop residues; usually considered as a base for determining the cost effectiveness of control practices.

Conveyance Loss: Water loss in pipes, channels, conduits, ditches by leakage or evaporation.

Cooling Electricity Use: Amount of electricity used to meet the building cooling load.

Cooling Tower: A structure that helps remove heat from water used as a coolant; e.g., in electric power generating plants.

Cooling Tower: Device which dissipates the heat from water-cooled systems by spraying the water through streams of rapidly moving air.

Cooperative Agreement: An assistance agreement whereby EPA transfers money, property, services or anything of value to a state, university, non-profit, or not-for-profit organization for the accomplishment of authorized activities or tasks.

Core: The uranium-containing heart of a nuclear reactor, where energy is released.

Core Program Cooperative Agreement: An assistance agreement whereby EPA supports states or tribal governments with funds to help defray the cost of non-item-specific administrative and training activities.

Corrective Action: EPA can require treatment, storage and disposal (TSDF) facilities handling hazardous waste to undertake corrective actions to clean up spills resulting from failure to follow hazardous waste management procedures or other mistakes. The process includes cleanup procedures designed to guide TSDFs toward in spills.

Corrosion: The dissolution and wearing away of metal caused by a chemical reaction such as between water and the pipes, chemicals touching a metal surface, or contact between two metals.

Corrosive: A chemical agent that reacts with the surface of a material causing it to deteriorate or wear away.

Cost/Benefit Analysis: A quantitative evaluation of the costs which would have incurred by implementing an environmental regulation versus the overall benefits to society of the proposed action.

Cost Recovery: A legal process by which potentially responsible parties who contributed to contamination at a Superfund site can be required to reimburse the Trust Fund for money spent during any cleanup actions by the federal government.

Cost Sharing: A publicly financed program through which society, as a beneficiary of environmental protection, shares part of the cost of pollution control with those who must actually install the controls. In Superfund, for example, the government may pay part of the cost of a cleanup action with those responsible for the pollution paying the major share.

Cost-Effective Alternative: An alternative control or corrective method identified after analysis as being the best available in terms of reliability, performance, and cost. Although costs are one important consideration, regulatory and compliance analysis does not require EPA to choose the least expensive alternative. For example, when selecting or approving a method for cleaning up a Superfund site, the Agency balances costs with the long-term effectiveness of the methods proposed and the potential danger posed by the site.

Cover Crop: A crop that provides temporary protection for delicate seedlings and/or provides a cover canopy for seasonal soil protection and improvement between normal crop production periods.

Cover Material: Soil used to cover compacted solid waste in a sanitary landfill.

Cradle-to-Grave or Manifest System: A procedure in which hazardous materials are identified and followed as they are produced, treated, transported, and disposed of by a series of permanent, linkable, descriptive documents (e.g. manifests). Commonly referred to as the cradle-to-grave system.

Criteria: Descriptive factors taken into account by EPA in setting standards for various pollutants. These factors are used to determine limits on allowable concentration levels, and to limit the number of violations per year. When issued by EPA, the criteria provide guidance to the states on how to establish their standards.

Criteria Pollutants: The 1970 amendments to the Clean Air Act required EPA to set National Ambient Air Quality Standards for certain pollutants known to be hazardous to human health. EPA has identified and set standards to protect human health and welfare for six pollutants: ozone, carbon monoxide, total suspended particulates, sulfur dioxide, lead, and nitrogen oxide. The term, "criteria pollutants" derives from the requirement that EPA must describe the characteristics and potential health and welfare effects of these pollutants. It is on the basis of these criteria that standards are set or revised.

Critical Effect: The first adverse effect, or its known precursor, that occurs as a dose rate increases. Designation is based on evaluation of overall database.

Crop Consumptive Use: The amount of water transpired during plant growth plus what evaporated from the soil surface and foliage in the crop area.

Crop Rotation: Planting a succession of different crops on the same land rea as opposed to planting the same crop time after time.

Cross Contamination: The movement of underground contaminants from one level or area to another due to invasive subsurface activities.

Cross-Connection: Any actual or potential connection between a drinking water system and an unapproved water supply or other source of contamination.

Crumb Rubber: Ground rubber fragments the size of sand or silt used in rubber or plastic products, or processed further into reclaimed rubber or asphalt products.

Cryptosporidium: A protozoan microbe associated with the disease cryptosporidiosis in man. The disease can be transmitted through ingestion of drinking water, person-to-person contact, or other pathways, and can cause acute diarrhea, abdominal pain, vomiting, fever, and can be fatal as it was in the Milwaukee episode.

Cubic Feet Per Minute (CFM): A measure of the volume of a substance flowing through air within a fixed period of time. With regard to indoor air, refers to the amount of air, in cubic feet, that is exchanged with outdoor air in a minute's time; i.e. the air exchange rate.

Cullet: Crushed glass.

Cultural Eutrophication: Increasing rate at which water bodies "die" by pollution from human activities.

Cultures and Stocks: Infectious agents and associated biologicals including cultures from medical and pathological laboratories; cultures and stocks of infectious agents from research and industrial laboratories; waste from the production of biologicals; discarded live and attenuated vaccines; and culture dishes and devices used to transfer, inoculate, and mix cultures.

Cumulative Ecological Risk Assessment: Consideration of the total ecological risk from multiple stressors to a given eco-zone.

Cumulative Exposure: The sum of exposures of an organism to a pollutant over a period of time.

Cumulative Working Level Months (CWLM): The sum of lifetime exposure to radon working levels expressed in total working level months.

Curb Stop: A water service shutoff valve located in a water service pipe near the curb and between the water main and the building.

Curbside Collection: Method of collecting recyclable materials at homes, community districts or businesses.

Cutie-Pie: An instrument used to measure radiation levels.

Cuttings: Spoils left by conventional drilling with hollow stem auger or rotary drilling equipment.

Cyclone Collector: A device that uses centrifugal force to remove large particles from polluted air.

Data Call-In: A part of the Office of Pesticide Programs (OPP) process of developing key required test data, especially on the long-term, chronic effects of existing pesticides, in advance of scheduled Registration Standard reviews. Data Call-In from manufacturers is an adjunct of the Registration Standards program intended to expedite re-registration.

Data Quality Objectives (DQOs): Qualitative and quantitative statements of the overall level of uncertainty that a decision-maker will accept in results or decisions based on environmental data. They provide the statistical framework for planning and managing environmental data operations consistent with user's needs.

Day Tank: Another name for deaerating tank.

DDT: The first chlorinated hydrocarbon insecticide chemical name: Dichloro-Diphenyl-Trichloroethane. It has a half-life of 15 years and can collect in fatty tissues of certain animals. EPA banned registration and interstate sale of DDT for virtually all but emergency uses in the United States in 1972 because of its persistence in the environment and accumulation in the food chain.

Dead End: The end of a water main which is not connected to other parts of the distribution system.

Deadmen: Anchors drilled or cemented into the ground to provide additional reactive mass for DP sampling rigs.

Decant: To draw off the upper layer of liquid after the heaviest material (a solid or another liquid) has settled.

Decay Products: Degraded radioactive materials, often referred to as "daughters" or "progeny"; radon decay products of most concern from a public health standpoint are polonium-214 and polonium-218.

Dechlorination: Removal of chlorine from a substance.

Decomposition: The breakdown of matter by bacteria and fungi, changing the chemical makeup and physical appearance of materials.

Decontamination: Removal of harmful substances such as noxious chemicals, harmful bacteria or other organisms, or radioactive material from exposed individuals, rooms and furnishings in buildings, or the exterior environment.

Deep-Well Injection: Deposition of raw or treated, filtered hazardous waste by pumping it into deep wells, where it is contained in the pores of permeable subsurface rock.

Deflocculating Agent: A material added to a suspension to prevent settling.

Defluoridation: The removal of excess flouride in drinking water to prevent the staining of teeth.

Defoliant: An herbicide that removes leaves from trees and growing plants.

Degasification: A water treatment that removes dissolved gases from the water.

Degree-Day: A rough measure used to estimate the amount of heating required in a given area; is defined as the difference between the mean daily temperature and 65 degrees Fahrenheit. Degree-days are also calculated to estimate cooling requirements.

Delegated State: A state (or other governmental entity such as a tribal government) that has received authority to administer an environmental regulatory program in lieu of a federal counterpart. As used in connection with NPDES, UIC, and PWS programs, the term does not connote any transfer of federal authority to a state.

Delist: Use of the petition process to have a facility's toxic designation rescinded.

Demand-side Waste Management: Prices whereby consumers use purchasing decisions to communicate to product manufacturers that they prefer environmentally sound products packaged with the least amount of waste, made from recycled or recyclable materials, and containing no hazardous substances.

Demineralization: A treatment process that removes dissolved minerals from water.

Denitrification: The biological reduction of nitrate to nitrogen gas by denitrifying bacteria in soil.

Dense Non-Aqueous Phase Liquid (DNAPL): Non-aqueous phase liquids such as chlorinated hydrocarbon solvents or petroleum fractions with a specific gravity greater than 1.0 that sink through the water column until they reach a confining layer. Because they are at the bottom of aquifers instead of floating on the water table, typical monitoring wells do not indicate their presence.

Density: A measure of how heavy a specific volume of a solid, liquid, or gas is in comparison to water. depending on the chemical.

Depletion Curve: In hydraulics, a graphical representation of water depletion from storage-stream channels, surface soil, and groundwater. A depletion curve can be drawn for base flow, direct runoff, or total flow.

Depressurization: A condition that occurs when the air pressure inside a structure is lower that the air pressure outdoors. Depressurization can occur when household appliances such as fireplaces or furnaces, that consume or exhaust house air, are not supplied with enough makeup air. Radon may be drawn into a house more rapidly under depressurized conditions.

Dermal Absorption/Penetration: Process by which a chemical penetrates the skin and enters the body as an internal dose.

Dermal Toxicity: The ability of a pesticide or toxic chemical to poison people or animals by contact with the skin.

DES: A synthetic estrogen, diethylstilbestrol is used as a growth stimulant in food animals. Residues in meat are thought to be carcinogenic.

Desalination: [Desalinization] (1) Removing salts from ocean or brackish water by using various technologies. (2) Removal of salts from soil by artificial means, usually leaching.

Desiccant: A chemical agent that absorbs moisture; some desiccants are capable of drying out plants or insects, causing death.

Design Capacity: The average daily flow that a treatment plant or other facility is designed to accommodate.

Design Value: The monitored reading used by EPA to determine an area's air quality status; e.g., for ozone, the fourth highest reading measured over the most recent three years is the design value.

Designated Pollutant: An air pollutant which is neither a criteria nor hazardous pollutant, as described in the Clean Air Act, but for which new source performance standards exist. The Clean Air Act does require states to control these pollutants, which include acid mist, total reduced sulfur (TRS), and fluorides.

Designated Uses: Those water uses identified in state water quality standards that must be achieved and maintained as required under the Clean Water Act. Uses can include cold water fisheries, public water supply, and irrigation.

Designer Bugs: Popular term for microbes developed through biotechnology that can degrade specific toxic chemicals at their source in toxic waste dumps or in ground water.

Destination Facility: The facility to which regulated medical waste is shipped for treatment and destruction, incineration, and/or disposal.

Destratification: Vertical mixing within a lake or reservoir to totally or partially eliminate separate layers of temperature, plant, or animal life.

Destroyed Medical Waste: Regulated medical waste that has been ruined, torn apart, or mutilated through thermal treatment, melting, shredding, grinding, tearing, or breaking, so that it is no longer generally recognized as medical waste, but has not yet been treated (excludes compacted regulated medical waste).

Destruction and Removal Efficiency (DRE): A percentage that represents the number of molecules of a compound removed or destroyed in an incinerator relative to the number of molecules entering the system (e.g. a DRE of 99.99 percent means that 9,999 molecules are destroyed for every 10,000 that enter; 99.99 percent is known as "four nines." For some pollutants, the RCRA removal requirement may be as stringent as "six nines").

Destruction Facility: A facility that destroys regulated medical waste.

Desulfurization: Removal of sulfur from fossil fuels to reduce pollution.

Detectable Leak Rate: The smallest leak (from a storage tank), expressed in terms of gallons- or liters-per-hour, that a test can reliably discern with a certain probability of detection or false alarm.

Detection Criterion: A predetermined rule to ascertain whether a tank is leaking or not. Most volumetric tests use a threshold value as the detection criterion.

Detection Limit: The lowest concentration of a chemical that can reliably be distinguished from a zero concentration.

Detention Time: 1. The theoretical calculated time required for a small amount of water to pass through a tank at a given rate of flow. 2. The actual time that a small amount of water is in a settling basin, flocculating basin, or rapid-mix chamber. 3. In storage reservoirs, the length of time water will be held before being used.

Detergent: Synthetic washing agent that helps to remove dirt and oil. Some contain compounds which kill useful bacteria and encourage algae growth when they are in wastewater that reaches receiving waters.

Development Effects: Adverse effects such as altered growth, structural abnormality, functional deficiency, or death observed in a developing organism.

Dewater: 1. Remove or separate a portion of the water in a sludge or slurry to dry the sludge so it can be handled and disposed of. 2. Remove or drain the water from a tank or trench.

Diatomaceous Earth (Diatomite): A chalk-like material (fossilized diatoms) used to filter out solid waste in wastewater treatment plants; also used as an active ingredient in some powdered pesticides.

Diazinon: An insecticide. In 1986, EPA banned its use on open areas such as sod farms and golf courses because it posed a danger to migratory birds. The ban did not apply to agricultural, home lawn or commercial establishment uses.

Dibenzofurans: A group of organic compounds, some of which are toxic.

Dicofol: A pesticide used on citrus fruits.

Diffused Air: A type of aeration that forces oxygen into sewage by pumping air through perforated pipes inside a holding tank.

Diffusion: The movement of suspended or dissolved particles (or molecules) from a more concentrated to a less concentrated area. The process tends to distribute the particles or molecules more uniformly.

Digester: In wastewater treatment, a closed tank; in solid-waste conversion, a unit in which bacterial action is induced and accelerated in order to break down organic matter and establish the proper carbon to nitrogen ratio.

Digestion: The biochemical decomposition of organic matter, resulting in partial gasification, liquefaction, and mineralization of pollutants.

Dike: A low wall that can act as a barrier to prevent a spill from spreading.

Diluent: Any liquid or solid material used to dilute or carry an active ingredient.

Dilution Ratio: The relationship between the volume of water in a stream and the volume of incoming water. It affects the ability of the stream to assimilate waste.

Dimictic: Lakes and reservoirs that freeze over and normally go through two stratifications and two mixing cycles a year.

Discharge: Flow of surface water in a stream or canal or the outflow of ground water from a flowing artesian well, ditch, or spring. Can also apply tp discharge of liquid effluent from a facility or to chemical emissions into the air through designated venting mechanisms.

Disinfectant: A chemical or physical process that kills pathogenic organisms in water, air, or on surfaces. Chlorine is often used to disinfect sewage treatment effluent, water supplies, wells, and swimming pools.

Disinfectant By-Product: A compound formed by the reaction of a disinfenctant such as chlorine with organic material in the water supply; a chemical byproduct of the disinfection process..

Disinfectant Timc: Thc time it takes water to move from the point of disinfectant application (or the previous point of residual disinfectant measurement) to a point before or at the point where the residual disinfectant is measured. In pipelines, the time is calculated by dividing the internal volume of the pipe by he maximum hourly flow rate; within mixing basins and storage reservoirs it is determined by tracer studies of an equivalent demonstration.

Dispersant: A chemical agent used to break up concentrations of organic material such as spilled oil.

Displacement Savings: Saving realized by displacing purchases of natural gas or electricity from a local utility by using landfill gas for power and heat.

Disposables: Consumer products, other items, and packaging used once or a few times and discarded.

Disposal: Final placement or destruction of toxic, radioactive, or other wastes; surplus or banned pesticides or other chemicals; polluted soils; and drums containing hazardous materials from removal actions or accidental releases. Disposal may be accomplished through use of approved secure landfills, surface impoundments, land farming, deep-well injection, ocean dumping, or incineration.

Disposal Facilities: Repositories for solid waste, including landfills and combustors intended for permanent containment or destruction of waste materials. Excludes transfer stations and composting facilities.

Dissolved Oxygen (DO): The oxygen freely available in water, vital to fish and other aquatic life and for the prevention of odors. DO levels are considered a most important indicator of a water body's ability to support desirable aquatic life. Secondary and advanced waste treatment are generally designed to ensure adequate DO in waste-receiving waters.

Dissolved Solids: Disintegrated organic and inorganic material in water. Excessive amounts make water unfit to drink or use in industrial processes.

Distillation: The act of purifying liquids through boiling, so that the steam or gaseous vapors condense to a pure liquid. Pollutants and contaminants may remain in a concentrated residue.

Disturbance: Any event or series of events that disrupt ecosystem, community, or population structure and alters the physical environment.

Diversion: 1. Use of part of a stream flow as water supply. 2. A channel with a supporting ridge on the lower side constructed across a slope to divert water at a non-erosive velocity to sites where it can be used and disposed of.

Diversion Rate: The percentage of waste materials diverted from traditional disposal such as landfilling or incineration to be recycled, composted, or re-used.

DNA Hybridization: Use of a segment of DNA, called a DNA probe, to identify its complementary DNA; used to detect specific genes.

Dobson Unit (DU): Units of ozone level measurement. measurement of ozone levels. If, for example, 100 DU of ozone were brought to the earth's surface they would form a layer one millimeter thick. Ozone levels vary geographically, even in the absence of ozone depletion.

Domestic Application: Pesticide application in and around houses, office buildings, motels, and other living or working areas.

Dosage/Dose: 1. The actual quantity of a chemical administered to an organism or to which it is exposed. 2. The amount of a substance that reaches a specific tissue (e.g. the liver). 3. The amount of a substance available for interaction with metabolic processes after crossing the outer boundary of an organism.

Downgradient: The direction that groundwater flows; similar to "downstream" for surface water.

Downstream Processors: Industries dependent on crop production (e.g. canneries and food processors).

Drainage Well: A well drilled to carry excess water off agricultural fields. Because they act as a funnel from the surface to the groundwater below. Drainage wells can contribute to groundwater pollution.

Dredging: Removal of mud from the bottom of water bodies. This can disturb the ecosystem and causes silting that kills aquatic life. Dredging of contaminated muds can expose biota to heavy metals and other toxics. Dredging activities may be subject to regulation under Section 404 of the Clean Water Act.

Drilling Fluid: Fluid used to lubricate the bit and convey drill cuttings to the surface with rotary drilling equipment. Usually composed of bentonite slurry or muddy water. Can become contaminated, leading to cross contamination, and may require special disposal. Not used with DP methods

Drinking Water Equivalent Level: Protective level of exposure related to potentially non-carcinogenic effects of chemicals that are also known to cause cancer.

Drinking Water State Revolving Fund: The Fund provides capitalization grants to states to develop drinking water revolving loan funds to help finance system infrastructure improvements, assure source-water protection, enhance operation and management of drinking-water systems, and otherwise promote local water-system compliance and protection of public health.

Drive Casing: Heavy duty steel casing driven along with the sampling tool in cased DP systems. Keeps the hole open between sampling runs and is not removed until last sample has been collected.

Drive Point Profiler: An exposed groundwater DP system used to collect multiple depth-discrete groundwater samples. Ports in the tip of the probe connect to an internal stainless steel or teflon tube that extends to the surface. Samples are collected via suction or airlift methods. Deionized water is pumped down through the ports to prevent plugging while driving the tool to the next sampling depth.

Drop-off: Recyclable materials collection method in which individuals bring them to a designated collection site.

Dual-Phase Extraction: Active withdrawal of both liquid and gas phases from a well usually involving the use of a vacuum pump.

Duplicate: A second aliquot or sample that is treated the same as the original sample in order to determine the precision of the analytical method.

Dustfall Jar: An open container used to collect large particles from the air for measurement and analysis.

Dynamometer. A device used to place a load on an engine and measure its performance.

Ecology: The relationship of living things to one another and their environment, or the study of such relationships.

Economic Poisons: Chemicals used to control pests and to defoliate cash crops such as cotton.

Ecosphere: The "bio-bubble" that contains life on earth, in surface waters, and in the air.

Ecosystem: The interacting system of a biological community and its non-living environmental surroundings.

Ecosystem Structure: Attributes related to the instantaneous physical state of an ecosystem; examples include species population density, species richness or evenness, and standing crop biomass.

Ecotone: A habitat created by the juxtaposition of distinctly different habitats; an edge habitat; or an ecological zone or boundary where two or more ecosystems meet.

Electromagnetic Geophysical Methods: Ways to measure subsurface conductivity via low-frequency electromagnetic induction.

Electrostatic Precipitator (ESP): A device that removes particles from a gas stream (smoke) after combustion occurs. The ESP imparts an electrical charge to the particles, causing them to adhere to metal plates inside the precipitator. Rapping on the plates causes the particles to fall into a hopper for disposal.

Eligible Costs: The construction costs for wastewater treatment works upon which EPA grants are based.

EMAP Data: Environmental monitoring data collected under the auspices of the Environmental Monitoring and Assessment Program. All EMAP data share the common attribute of being of known quality, having been collected in the context of explicit data quality objectives (DQOs) and a consistent quality assurance program.

Emergency and Hazardous Chemical Inventory: An annual report by facilities having one or more extremely hazardous substances or hazardous chemicals above certain weight limits.

Emergency Exemption: Provision in FIFRA under which EPA can grant temporary exemption to a state or another federal agency to allow the use of a pesticide product not registered for that particular use. Such actions involve unanticipated and/or severe pest problems where there is not time or interest by a manufacturer to register the product for that use. (Registrants cannot apply for such exemptions.)

Emergency Removal Action: 1. Steps take to remove contaminated materials that pose imminent threats to local residents (e.g. removal of leaking drums or the excavation of explosive waste.) 2. The state record of such removals.

Emergency Response Values: Concentrations of chemicals, published by various groups, defining acceptable levels for short-term exposures in emergencies.

Emergency Suspension: Suspension of a pesticide product registration due to an imminent hazard. The action immediately halts distribution, sale, and sometimes actual use of the pesticide involved.

Emission: Pollution discharged into the atmosphere from smokestacks, other vents, and surface areas of commercial or industrial facilities; from residential chimneys; and from motor vehicle, locomotive, or aircraft exhausts.

Emission Cap: A limit designed to prevent projected growth in emissions from existing and future stationary sources from eroding any mandated reductions. Generally, such provisions require that any emission growth from facilities under the restrictions be offset by equivalent reductions at other facilities under the same cap.

Emission Factor: The relationship between the amount of pollution produced and the amount of raw material processed. For example, an emission factor for a blast furnace making iron would be the number of pounds of particulates per ton of raw materials.

Emission Inventory: A listing, by source, of the amount of air pollutants discharged into the atmosphere of a community; used to establish emission standards.

Emission Standard: The maximum amount of air polluting discharge legally allowed from a single source, mobile or stationary.

Emissions Trading: The creation of surplus emission reductions at certain stacks, vents or similar emissions sources and the use of this surplus to meet or redefine pollution requirements applicable to other emissions sources. This allows one source to increase emissions when another source reduces them, maintaining an overall constant emission level. Facilities that reduce emissions substantially may "bank" their "credits" or sell them to other facilities or industries.

Emulsifier: A chemical that aids in suspending one liquid in another. Usually an organic chemical in an aqueous solution.

Encapsulation: The treatment of asbestos-containing material with a liquid that covers the surface with a protective coating or embeds fibers in an adhesive matrix to prevent their release into the air.

Enclosure: Putting an airtight, impermeable, permanent barrier around asbestos-containing materials to prevent the release of asbestos fibers into the air.

End User: Consumer of products for the purpose of recycling. Excludes products for re-use or combustion for energy recovery.

End-of-the-pipe: Technologies such as scrubbers on smokestacks and catalytic convertors on automobile tailpipes that reduce emissions of pollutants after they have formed.

End-use Product: A pesticide formulation for field or other end use. The label has instructions for use or application to control pests or regulate plant growth. The term excludes products used to formulate other pesticide products.

Endangered Species: Animals, birds, fish, plants, or other living organisms threatened with extinction by anthropogenic (man-caused) or other natural changes in their environment. Requirements for declaring a species endangered are contained in the Endangered Species Act.

Endangerment Assessment: A study to determine the nature and extent of contamination at a site on the National Priorities List and the risks posed to public health or the environment. EPA or the state conducts the study when a legal action is to be taken to direct potentially responsible parties to clean up a site or pay for it. An endangerment assessment supplements a remedial investigation.

Endrin: A pesticide toxic to freshwater and marine aquatic life that produces adverse health effects in domestic water supplies.

Energy Management System: A control system capable of monitoring environmental and system loads and adjusting HVAC operations accordingly in order to conserve energy while maintaining comfort.

Energy Recovery: Obtaining energy from waste through a variety of processes (e.g. combustion).

Enforceable Requirements: Conditions or limitations in permits issued under the Clean Water Act Section 402 or 404 that, if violated, could result in the issuance of a compliance order or initiation of a civil or criminal action under federal or applicable state laws. If a permit has not been issued, the term includes any requirement which, in the Regional Administrator's judgement, would be included in the permit when issued. Where no permit applies, the term includes any requirement which the RA determines is necessary for the best practical waste treatment technology to meet applicable criteria.

Enforcement: EPA, state, or local legal actions to obtain compliance with environmental laws, rules, regulations, or agreements and/or obtain penalties or criminal sanctions for violations. Enforcement procedures may vary, depending on the requirements of different environmental laws and related implementing regulations. Under CERCLA, for example, EPA will seek to require potentially responsible parties to clean up a Superfund site, or pay for the cleanup, whereas under the Clean Air Act the Agency may invoke sanctions against cities failing to meet ambient air quality standards that could prevent certain types of construction or federal funding. In other situations, if investigations by EPA and state agencies uncover willful violations, criminal trials and penalties are sought.

Enforcement Decision Document (EDD): A document that provides an explanation to the public of EPA's selection of the cleanup alternative at enforcement sites on the National Priorities List. Similar to a Record of Decision.

Engineered Controls: Method of managing environmental and health risks by placing a barrier between the contamination and the rest of the site, thus limiting exposure pathways.

Enhanced Inspection and Maintenance (I&M): An improved automobile inspection and maintenance program—aimed at reducing automobile emissions—that contains, at a minimum, more vehicle types and model years, tighter inspection, and better management practices. It may also include annual computerized or centralized inspections, under-the-hood inspection—for signs of tampering with pollution control equipment—and increased repair waiver cost.

Enrichment: The addition of nutrients (e.g. nitrogen, phosphorus, carbon compounds) from sewage effluent or agricultural runoff to surface water, greatly increases the growth potential for algae and other aquatic plants.

Entrain: To trap bubbles in water either mechanically through turbulence or chemically through a reaction.

Environment: The sum of all external conditions affecting the life, development and survival of an organism.

Environmental Assessment: An environmental analysis prepared pursuant to the National Environmental Policy Act to determine whether a federal action would significantly affect the environment and thus require a more detailed environmental impact statement.

Environmental Audit: An independent assessment of the current status of a party's compliance with applicable environmental requirements or of a party's environmental compliance policies, practices, and controls.

Environmental/Ecological Risk: The potential for adverse effects on living organisms associated with pollution of the environment by effluents, emissions, wastes, or accidental chemical releases; energy use; or the depletion of natural resources.

Environmental Equity/Justice: Equal protection from environmental hazards for individuals, groups, or communities regardless of race, ethnicity, or economic status. This applies to the development, implementation, and enforcement of environmental laws, regulations, and policies, and implies that no population of people should be forced to shoulder a disproportionate share of negative environmental impacts of pollution or environmental hazard due to a lack of political or economic strength levels.

Environmental Exposure: Human exposure to pollutants originating from facility emissions. Threshold levels are not necessarily surpassed, but low-level chronic pollutant exposure is one of the most common forms of environmental exposure.

Environmental Fate: The destiny of a chemical or biological pollutant after release into the environment.

Environmental Fate Data: Data that characterize a pesticide's fate in the ecosystem, considering factors that foster its degradation (light, water, microbes), pathways and resultant products.

Environmental Impact Statement: A document required of federal agencies by the National Environmental Policy Act for major projects or legislative proposals significantly affecting the environment. A tool for decision making, it describes the positive and negative effects of the undertaking and cites alternative actions.

Environmental Indicator: A measurement, statistic or value that provides a proximate gauge or evidence of the effects of environmental management programs or of the state or condition of the environment.

Environmental Justice: The fair treatment of people of all races, cultures, incomes, and educational levels with respect to the development and enforcement of environmental laws, regulations, and policies.

Environmental Lien: A charge, security, or encumbrance on a property's title to secure payment of cost or debt arising from response actions, cleanup, or other remediation of hazardous substances or petroleum products.

Environmental Medium: A major environmental category that surrounds or contacts humans, animals, plants, and other organisms (e.g. surface water, ground water, soil or air) and through which chemicals or pollutants move.

Environmental Monitoring for Public Access and Community Tracking: Joint EPA, NOAA, and USGS program to provide timely and effective communication of environmental data and information through improved and updated technology solutions that support timely environmental monitoring reporting, interpreting, and use of the information for the benefit of the public.

Environmental Response Team: EPA experts located in Edison, N.J., and Cincinnati, OH, who can provide around-the-clock technical assistance to EPA regional offices and states during all types of hazardous waste site emergencies and spills of hazardous substances.

Environmental Site Assessment: The process of determining whether contamination is present on a parcel of real property.

Environmental Sustainability: Long-term maintenance of ecosystem components and functions for future generations.

Environmental Tobacco Smoke: Mixture of smoke from the burning end of a cigarette, pipe, or cigar and smoke exhaled by the smoker.

Epidemiology: Study of the distribution of disease, or other health-related states and events in human populations, as related to age, sex, occupation, ethnicity, and economic status in order to identify and alleviate health problems and promote better health.

Epilimnion: Upper waters of a thermally stratified lake subject to wind action.

Episode (Pollution): An air pollution incident in a given area caused by a concentration of atmospheric pollutants under meteorological conditions that may result in a significant increase in illnesses or deaths. May also describe water pollution events or hazardous material spills.

Equilibrium: In relation to radiation, the state at which the radioactivity of consecutive elements within a radioactive series is neither increasing nor decreasing.

Equivalent Method: Any method of sampling and analyzing for air pollution which has been demonstrated to the EPA Administrator's satisfaction to be, under specific conditions, an acceptable alternative to normally used reference methods.

Erosion: The wearing away of land surface by wind or water, intensified by land-clearing practices related to farming, residential or industrial development, road building, or logging.

Established Treatment Technologies: Technologies for which cost and performance data are readily available.

Estimated Environmental Concentration: The estimated pesticide concentration in an ecosystem.

Estuary: Region of interaction between rivers and near-shore ocean waters, where tidal action and river flow mix fresh and salt water. Such areas include bays, mouths of rivers, salt marshes, and lagoons. These brackish water ecosystems shelter and feed marine life, birds, and wildlife.

Ethanol: An alternative automotive fuel derived from grain and corn; usually blended with gasoline to form gasohol.

Ethylene Dibromide (EDB): A chemical used as an agricultural fumigant and in certain industrial processes. Extremely toxic and found to be a carcinogen in laboratory animals, EDB has been banned for most agricultural uses in the United States.

Eutrophic Lakes: Shallow, murky bodies of water with concentrations of plant nutrients causing excessive production of algae.

Eutrophication: The slow aging process during which a lake, estuary, or bay evolves into a bog or marsh and eventually disappears. During the later stages of eutrophication the water body is choked by abundant plant life due to higher levels of nutritive compounds such as nitrogen and phosphorus. Human activities can accelerate the process.

Exemption: A state (with primacy) may exempt a public water system from a requirement involving a Maximum Contaminant Level (MCL), treatment technique, or both, if the system cannot comply due to compelling economic or other factors, or because the system was in operation before the requirement or MCL was instituted; and the exemption will not create a public health risk.

Exotic Species: A species that is not indigenous to a region.

Experimental Use Permit: Obtained by manufacturers for testing new pesticides or uses thereof whenever they conduct experimental field studies to support registration on 10 acres or more of land or one acre or more of water.

Experimental Use Permit: A permit granted by EPA that allows a producer to conduct tests of a new pesticide, product and/or use outside the laboratory. The testing is usually done on ten or more acres of land or water surface.

Explosive Limits: The amounts of vapor in the air that form explosive mixtures; limits are expressed as lower and upper limits and give the range of vapor concentrations in air that will explode if an ignition source is present.

Exports : In solid waste program, municipal solid waste and recyclables transported outside the state or locality where they originated.

Exposure: The amount of radiation or pollutant present in a given environment that represents a potential health threat to living organisms.

Fabric Filter: A cloth device that catches dust particles from industrial emissions.

Facilities Plans: Plans and studies related to the construction of treatment works necessary to comply with the Clean Water Act or RCRA. A facilities plan investigates needs and provides information on the cost-effectiveness of alternatives, a recommended plan, an environmental assessment of the recommendations, and descriptions of the treatment works, costs, and a completion schedule.

Facility Emergency Coordinator: Representative of a facility covered by environmental law (e.g, a chemical plant) who participates in the emergency reporting process with the Local Emergency Planning Committee (LEPC).

Facultative Bacteria: Bacteria that can live under aerobic or anaerobic conditions.

Feasibility Study: 1. Analysis of the practicability of a proposal; e.g., a description and analysis of potential cleanup alternatives for a site such as one on the National Priorities List. The feasibility study usually recommends selection of a cost-effective alternative. It usually starts as soon as the remedial investigation is underway; together, they are commonly referred to as the "RI/FS". 2. A small-scale investigation of a problem to ascertain whether a proposed research approach is likely to provide useful data.

Fecal Coliform Bacteria: Bacteria found in the intestinal tracts of mammals. Their presence in water or sludge is an indicator of pollution and possible contamination by pathogens.

Federal Implementation Plan: Under current law, a federally implemented plan to achieve attainment of air quality standards, used when a state is unable to develop an adequate plan.

Federal Motor Vehicle Control Program: All federal actions aimed at controlling pollution from motor vehicles by such efforts as establishing and enforcing tailpipe and evaporative emission standards for new vehicles, testing methods development, and guidance to states operating inspection and maintenance programs. Federally designated area that is required to meet and maintain federal ambient air quality standards. May include nearby locations in the same state or nearby states that share common air pollution problems.

Feedlot: A confined area for the controlled feeding of animals. Tends to concentrate large amounts of animal waste that cannot be absorbed by the soil and, hence, may be carried to nearby streams or lakes by rainfall runoff.

Fen: A type of wetland that accumulates peat deposits. Fens are less acidic than bogs, deriving most of their water from groundwater rich in calcium and magnesium.

Ferrous Metals: Magnetic metals derived from iron or steel; products made from ferrous metals include appliances, furniture, containers, and packaging like steel drums and barrels. Recycled products include processing tin/steel cans, strapping, and metals from appliances into new products.

FIFRA Pesticide Ingredient: An ingredient of a pesticide that must be registered with EPA under the Federal Insecticide, Fungicide, and Rodenticide Act. Products making pesticide claims must register under FIFRA and may be subject to labeling and use requirements.

Fill: Man-made deposits of natural soils or rock products and waste materials.

Filling: Depositing dirt, mud or other materials into aquatic areas to create more dry land, usually for agricultural or commercial development purposes, often with ruinous ecological consequences.

Filter Strip: Strip or area of vegetation used for removing sediment, organic matter, and other pollutants from runoff and wastewater.

Filtration: A treatment process, under the control of qualified operators, for removing solid (particulate) matter from water by means of porous media such as sand or a man-made filter; often used to remove particles that contain pathogens.

Financial Assurance for Closure: Documentation or proof that an owner or operator of a facility such as a landfill or other waste repository is capable of paying the projected costs of closing the facility and monitoring it afterwards as provided in RCRA regulations.

Finished Water: Water is "finished" when it has passed through all the processes in a water treatment plant and is ready to be delivered to consumers.

First Draw: The water that comes out when a tap is first opened, likely to have the highest level of lead contamination from plumbing materials.

Fix a Sample: A sample is "fixed" in the field by adding chemicals that prevent water quality indicators of interest in the sample from changing before laboratory measurements are made.

Fixed-Location Monitoring: Sampling of an environmental or ambient medium for pollutant concentration at one location continuously or repeatedly.

Flammable: Any material that ignites easily and will burn rapidly.

Flare: A control device that burns hazardous materials to prevent their release into the environment; may operate continuously or intermittently, usually on top of a stack.

Flash Point: The lowest temperature at which evaporation of a substance produces sufficient vapor to form an ignitable mixture with air.

Floc: A clump of solids formed in sewage by biological or chemical action.

Flocculation: Process by which clumps of solids in water or sewage aggregate through biological or chemical action so they can be separated from water or sewage.

Floodplain: The flat or nearly flat land along a river or stream or in a tidal area that is covered by water during a flood.

Floor Sweep: Capture of heavier-than-air gases that collect at floor level.

Flow Rate: The rate, expressed in gallons -or liters-per-hour, at which a fluid escapes from a hole or fissure in a tank. Such measurements are also made of liquid waste, effluent, and surface water movement.

Flowable: Pesticide and other formulations in which the active ingredients are finely ground insoluble solids suspended in a liquid. They are mixed with water for application.

Flowmeter: A gauge indicating the velocity of wastewater moving through a treatment plant or of any liquid moving through various industrial processes.

Flue Gas: The air coming out of a chimney after combustion in the burner it is venting. It can include nitrogen oxides, carbon oxides, water vapor, sulfur oxides, particles and many chemical pollutants.

Flue Gas Desulfurization: A technology that employs a sorbent, usually lime or limestone, to remove sulfur dioxide from the gases produced by burning fossil fuels. Flue gas desulfurization is current state-of-the art technology for major SO2 emitters, like power plants.

Fluidized: A mass of solid particles that is made to flow like a liquid by injection of water or gas is said to have been fluidized. In water treatment, a bed of filter media is fluidized by backwashing water through the filter.

Fluidized Bed Incinerator: An incinerator that uses a bed of hot sand or other granular material to transfer heat directly to waste. Used mainly for destroying municipal sludge.

Flume: A natural or man-made channel that diverts water.

Fluoridation: The addition of a chemical to increase the concentration of fluoride ions in drinking water to reduce the incidence of tooth decay.

Fluorides: Gaseous, solid, or dissolved compounds containing fluorine that result from industrial processes. Excessive amounts in food can lead to fluorosis.

Fly Ash: Non-combustible residual particles expelled by flue gas.

Fogging: Applying a pesticide by rapidly heating the liquid chemical so that it forms very fine droplets that resemble smoke or fog. Used to destroy mosquitoes, black flies, and similar pests.

Food Chain: A sequence of organisms, each of which uses the next, lower member of the sequence as a food source.

Food Processing Waste: Food residues produced during agricultural and industrial operations.

Food Waste: Uneaten food and food preparation wastes from residences and commercial establishments such as grocery stores, restaurants, and produce stands, institutional cafeterias and kitchens, and industrial sources like employee lunchrooms.

Food Web: The feeding relationships by which energy and nutrients are transferred from one species to another.

Formaldehyde: A colorless, pungent, and irritating gas, CH20, used chiefly as a disinfectant and preservative and in synthesizing other compounds like resins.

Formulation: The substances comprising all active and inert ingredients in a pesticide.

Fossil Fuel: Fuel derived from ancient organic remains; e.g. peat, coal, crude oil, and natural gas.

Fracture: A break in a rock formation due to structural stresses; e.g. faults, shears, joints, and planes of fracture cleavage.

Free Product: A petroleum hydrocarbon in the liquid free or non aqueous phase.

Freeboard: 1. Vertical distance from the normal water surface to the top of a confining wall. 2. Vertical distance from the sand surface to the underside of a trough in a sand filter.

Fresh Water: Water that generally contains less than 1,000 milligrams-per-liter of dissolved solids.

Friable: Capable of being crumbled, pulverized, or reduced to powder by hand pressure.

Friable Asbestos: Any material containing more than one-percent asbestos, and that can be crumbled or reduced to powder by hand pressure. (May include previously non-friable material which becomes broken or damaged by mechanical force.)

Fuel Economy Standard: The Corporate Average Fuel Economy Standard (CAFE) effective in 1978. It enhanced the national fuel conservation effort imposing a miles-per-gallon floor for motor vehicles.

Fuel Efficiency: The proportion of energy released by fuel combustion that is converted into useful energy.

Fugitive Emissions: Emissions not caught by a capture system.

Fume: Tiny particles trapped in vapor in a gas stream.

Fumigant: A pesticide vaporized to kill pests. Used in buildings and greenhouses.

Functional Equivalent: Term used to describe EPA's decision-making process and its relationship to the environmental review conducted under the National Environmental Policy Act (NEPA). A review is considered functionally equivalent when it addresses the substantive components of a NEPA review.

Fungicide: Pesticides which are used to control, deter, or destroy fungi.

Fungistat: A chemical that keeps fungi from growing.

Fungus (Fungi): Molds, mildews, yeasts, mushrooms, and puffballs, a group of organisms lacking in chlorophyll (i.e. are not photosynthetic) and which are usually non-mobile, filamentous, and multicellular. Some grow in soil, others attach themselves to decaying trees and other plants whence they obtain nutrients. Some are pathogens, others stabilize sewage and digest composted waste.

Furrow Irrigation: Irrigation method in which water travels through the field by means of small channels between each groups of rows.

Future Liability: Refers to potentially responsible parties' obligations to pay for additional response activities beyond those specified in the Record of Decision or Consent Decree.

Game Fish: Species like trout, salmon, or bass, caught for sport. Many of them show more sensitivity to environmental change than "rough" fish.

Garbage: Animal and vegetable waste resulting from the handling, storage, sale, preparation, cooking, and serving of foods.

Gas Chromatograph/Mass Spectrometer: Instrument that identifies the molecular composition and concentrations of various chemicals in water and soil samples.

Gasahol: Mixture of gasoline and ethanol derived from fermented agricultural products containing at least nine percent ethanol. Gasohol emissions contain less carbon monoxide than those from gasoline.

Gasification: Conversion of solid material such as coal into a gas for use as a fuel.

Gasoline Volatility: The property of gasoline whereby it evaporates into a vapor. Gasoline vapor is a mixture of volatile organic compounds.

General Permit: A permit applicable to a class or category of dischargers.

General Reporting Facility: A facility having one or more hazardous chemicals above the 10,000 pound threshold for planning quantities. Such facilities must file MSDS and emergency inventory information with the SERC, LEPC, and local fire departments.

Generally Recognized as Safe (GRAS): Designation by the FDA that a chemical or substance (including certain pesticides) added to food is considered safe by experts, and so is exempted from the usual FFDCA food additive tolerance requirements.

Genetic Engineering: A process of inserting new genetic information into existing cells in order to modify a specific organism for the purpose of changing one of its characteristics.

Genotoxic: Damaging to DNA; pertaining to agents known to damage DNA.

Geographic Information System (GIS): A computer system designed for storing, manipulating, analyzing, and displaying data in a geographic context.

Geological Log: A detailed description of all underground features (depth, thickness, type of formation) discovered during the drilling of a well.

Geophysical Log: A record of the structure and composition of the earth encountered when drilling a well or similar type of test hold or boring.

Geothermal/Ground Source Heat Pump: These heat pumps are underground coils to transfer heat from the ground to the inside of a building.

Germicide: Any compound that kills disease-causing microorganisms.

Giardia Lamblia: Protozoan in the feces of humans and animals that can cause severe gastrointestinal ailments. It is a common contaminant of surface waters.

Glass Containers: For recycling purposes, containers like bottles and jars for drinks, food, cosmetics and other products. When being recycled, container glass is generally separated into color categories for conversion into new containers, construction materials or fiberglass insulation.

Global Warming: An increase in the near surface temperature of the Earth. Global warming has occurred in the distant past as the result of natural influences, but the term is most often used to refer to the warming predicted to occur as a result of increased emissions of greenhouse gases. Scientists generally agree that the Earth's surface has warmed by about 1 degree Fahrenheit in the past 140 years. The Intergovernmental Panel on Climate Change (IPCC) recently concluded that increased concentrations of greenhouse gases are causing an increase in the Earth's surface temperature and that increased concentrations of sulfate aerosols have led to relative cooling in some regions, generally over and downwind of heavily industrialized areas.

Global Warming Potential: The ratio of the warming caused by a substance to the warming caused by a similar mass of carbon dioxide. CFC-12, for example, has a GWP of 8,500, while water has a GWP of zero.

Glovebag: A polyethylene or polyvinyl chloride bag-like enclosure affixed around an asbestos-containing source (most often thermal system insulation) permitting the material to be removed while minimizing release of airborne fibers to the surrounding atmosphere.

Gooseneck: A portion of a water service connection between the distribution system water main and a meter. Sometimes called a pigtail.

Grab Sample: A single sample collected at a particular time and place that represents the composition of the water, air, or soil only at that time and place.

Grain Loading: The rate at which particles are emitted from a pollution source. Measurement is made by the number of grains per cubic foot of gas emitted.

Granular Activated Carbon Treatment: A filtering system often used in small water systems and individual homes to remove organics. Also used by municipal water treatment plantsd. GAC can be highly effective in lowering elevated levels of radon in water.

Grasscycling: Source reduction activities in which grass clippings are left on the lawn after mowing.

Grassed Waterway: Natural or constructed watercourse or outlet that is shaped or graded and established in suitable vegetation for the disposal of runoff water without erosion.

Gray Water: Domestic wastewater composed of wash water from kitchen, bathroom, and laundry sinks, tubs, and washers.

Greenhouse Effect: The warming of the Earth's atmosphere attributed to a buildup of carbon dioxide or other gases; some scientists think that this build-up allows the sun's rays to heat the Earth, while making the infra-red radiation atmosphere opaque to infra-red radiation, thereby preventing a counterbalancing loss of heat.

Greenhouse Gas: A gas, such as carbon dioxide or methane, which contributes to potential climate change.

Grinder Pump: A mechanical device that shreds solids and raises sewage to a higher elevation through pressure sewers.

Gross Alpha/Beta Particle Activity: The total radioactivity due to alpha or beta particle emissions as inferred from measurements on a dry sample.

Gross Power-Generation Potential: The installed power generation capacity that landfill gas can support.

Ground Cover: Plants grown to keep soil from eroding.

Ground Water: The supply of fresh water found beneath the Earth's surface, usually in aquifers, which supply wells and springs. Because ground water is a major source of drinking water, there is growing concern over contamination from leaching agricultural or industrial pollutants or leaking underground storage tanks.

Ground Water Under the Direct Influence (UDI) of Surface Water: Any water beneath the surface of the ground with: 1. significant occurence of insects or other microorganisms, algae, or large-diameter pathogens; 2. significant and relatively rapid shifts in water characteristics such as turbidity, temperature, conductivity, or pH which closely correlate to climatological or surface water conditions. Direct influence is determined for individual sources in accordance with criteria established by a state.

Ground-Penetrating Radar: A geophysical method that uses high frequency electromagnetic waves to obtain subsurface information.

Ground-Water Discharge: Ground water entering near coastal waters which has been contaminated by landfill leachate, deep well injection of hazardous wastes, septic tanks, etc.

Ground-Water Disinfection Rule: A 1996 amendment of the Safe Drinking Water Act requiring EPA to promulgate national primary drinking water regulations requiring disinfection as for all public water systems, including surface waters and ground water systems.

Gully Erosion: Severe erosion in which trenches are cut to a depth greater than 30 centimeters (a foot). Generally, ditches deep enough to cross with farm equipment are considered gullies.

Habitat: The place where a population (e.g. human, animal, plant, microorganism) lives and its surroundings, both living and non-living.

Habitat Indicator: A physical attribute of the environment measured to characterize conditions necessary to support an organism, population, or community in the absence of pollutants; e.g. salinity of estuarine waters or substrate type in streams or lakes.

Halogen: A type of incandescent lamp with higher energy-efficiency that standard ones.

Halon: Bromine-containing compounds with long atmospheric lifetimes whose breakdown in the stratosphere causes depletion of ozone. Halons are used in firefighting.

Hammer Mill: A high-speed machine that uses hammers and cutters to crush, grind, chip, or shred solid waste.

Hard Water: Alkaline water containing dissolved salts that interfere with some industrial processes and prevent soap from sudsing.

Hauler: Garbage collection company that offers complete refuse removal service; many will also collect recyclables.

Hazard Assessment: Evaluating the effects of a stressor or determining a margin of safety for an organism by comparing the concentration which causes toxic effects with an estimate of exposure to the organism.

Hazard Communication Standard: An OSHA regulation that requires chemical manufacturers, suppliers, and importers to assess the hazards of the chemicals that they make, supply, or import, and to inform employers, customers, and workers of these hazards through MSDS information.

Hazard Evaluation: A component of risk evaluation that involves gathering and evaluating data on the types of health injuries or diseases that may be produced by a chemical and on the conditions of exposure under which such health effects are produced.

Hazard Identification: Determining if a chemical or a microbe can cause adverse health effects in humans and what those effects might be.

Hazard Quotient: The ratio of estimated site-specific exposure to a single chemical from a site over a specified period to the estimated daily exposure level, at which no adverse health effects are likely to occur.

Hazard Ratio: A term used to compare an animal's daily dietary intake of a pesticide to its LD 50 value. A ratio greater than 1.0 indicates that the animal is likely to consume an a dose amount which would kill 50 percent of animals of the same species.

Hazardous Air Pollutants: Air pollutants which are not covered by ambient air quality standards but which, as defined in the Clean Air Act, may present a threat of adverse human health effects or adverse environmental effects.Such pollutants include asbestos, beryllium, mercury, benzene, coke oven emissions, radionuclides, and vinyl chloride.

Hazardous Chemical: An EPA designation for any hazardous material requiring an MSDS under OSHA's Hazard Communication Standard. Such substances are capable of producing fires and explosions or adverse health effects like cancer and dermatitis. Hazardous chemicals are distinct from hazardous waste.

Hazardous Ranking System: The principal screening tool used by EPA to evaluate risks to public health and the environment associated with abandoned or uncontrolled hazardous waste sites. The HRS calculates a score based on the potential of hazardous substances spreading from the site through the air, surface water, or ground water, and on other factors such as density and proximity of human population. This score is the primary factor in deciding if the site should be on the National Priorities List and, if so, what ranking it should have compared to other sites on the list.

Hazardous Waste: By-products of society that can pose a substantial or potential hazard to human health or the environment when improperly managed. Possesses at least one of four characteristics (ignitability, corrosivity, reactivity, or toxicity), or appears on special EPA lists.

Hazardous Waste Landfill: An excavated or engineered site where hazardous waste is deposited and covered.

Hazardous Waste Minimization: Reducing the amount of toxicity or waste produced by a facility via source reduction or environmentally sound recycling.

Hazards Analysis: Procedures used to (1) identify potential sources of release of hazardous materials from fixed facilities or transportation accidents; (2) determine the vulnerability of a geographical area to a release of hazardous materials; and (3) compare hazards to determine which present greater or lesser risks to a community.

Hazards Identification: Providing information on which facilities have extremely hazardous substances, what those chemicals are, how much there is at each facility, how the chemicals are stored, and whether they are used at high temperatures.

Headspace: The vapor mixture trapped above a solid or liquid in a sealed vessel.

Health Advisory Level: A non-regulatory health-based reference level of chemical traces (usually in ppm) in drinking water at which there are no adverse health risks when ingested over various periods of time. Such levels are established for one day, 10 days, long-term and life-time exposure periods. They contain a wide margin of safety.

Health Assessment: An evaluation of available data on existing or potential risks to human health posed by a Superfund site. The Agency for Toxic Substances and Disease Registry (ATSDR) of the Department of Health and Human Services (DHHS) is required to perform such an assessment at every site on the National Priorities List.

Heat Island Effect: A "dome" of elevated temperatures over an urban area caused by structural and pavement heat fluxes, and pollutant emissions.

Heat Pump: An electric device with both heating and cooling capabilities. It extracts heat from one medium at a lower (the heat source) temperature and transfers it to another at a higher temperature (the heat sink), thereby cooling the first and warming the second.

Heavy Metals: Metallic elements with high atomic weights; (e.g. mercury, chromium, cadmium, arsenic, and lead); can damage living things at low concentrations and tend to accumulate in the food chain.

Heptachlor: An insecticide that was banned on some food products in 1975 and in all of them 1978. It was allowed for use in seed treatment until 1983. More recently it was found in milk and other dairy products in Arkansas and Missouri where dairy cattle were illegally fed treated seed.

Herbicide: A chemical pesticide designed to control or destroy plants, weeds, or grasses.

Herbivore: An animal that feeds on plants.

Heterotrophic Organisms: Species that are dependent on organic matter for food.

High End Exposure (dose) Estimate: An estimate of exposure, or dose level received anyone in a defined population that is greater than the 90th percentile of all individuals in that population, but less than the exposure at the highest percentile in that population. A high end risk descriptor is an estimate of the risk level for such individuals. Note that risk is based on a combination of exposure and susceptibility to the stressor.

High Intensity Discharge: A generic term for mercury vapor, metal halide, and high pressure sodium lamps and fixtures.

High-Density Polyethylene: A material used to make plastic bottles and other products that produces toxic fumes when burned.

High-Level Nuclear Waste Facility: Plant designed to handle disposal of used nuclear fuel, high-level radioactive waste, and plutonium waste.

High-Level Radioactive Waste (HLRW): Waste generated in core fuel of a nuclear reactor, found at nuclear reactors or by nuclear fuel reprocessing; is a serious threat to anyone who comes near the waste without shielding.

High-Line Jumpers: Pipes or hoses connected to fire hydrants and laid on top of the ground to provide emergency water service for an isolated portion of a distribution system.

High-Risk Community: A community located within the vicinity of numerous sites of facilities or other potential sources of envienvironmental exposure/health hazards which may result in high levels of exposure to contaminants or pollutants.

High-to-Low-Dose Extrapolation: The process of prediction of low exposure risk to humans and animals from the measured high-exposure-high-risk data involving laboratory animals.

Highest Dose Tested: The highest dose of a chemical or substance tested in a study.

Holding Pond: A pond or reservoir, usually made of earth, built to store polluted runoff.

Holding Time: The maximum amount of time a sample may be stored before analysis.

Hollow Stem Auger Drilling: Conventional drilling method that uses augurs to penetrate the soil. As the augers are rotated, soil cuttings are conveyed to the ground surface via augur spirals. DP tools can be used inside the hollow augers.

Homeowner Water System: Any water system which supplies piped water to a single residence.

Homogeneous Area: In accordance with Asbestos Hazard and Emergency Response Act (AHERA) definitions, an area of surfacing materials, thermal surface insulation, or miscellaneous material that is uniform in color and texture.

Hood Capture Efficiency: Ratio of the emissions captured by a hood and directed into a control or disposal device, expressed as a percent of all emissions.

Hydraulic Conductivity: The rate at which water can move through a permeable medium. (i.e. the coefficient of permeability.)

Hydraulic Gradient: In general, the direction of groundwater flow due to changes in the depth of the water table.

Hydrocarbons (HC): Chemical compounds that consist entirely of carbon and hydrogen.

Hydrogen Sulfide (H_2S): Gas emitted during organic decomposition. Also a by-product of oil refining and burning. Smells like rotten eggs and, in heavy concentration, can kill or cause illness.

Hydrogeological Cycle: The natural process recycling water from the atmosphere down to (and through) the earth and back to the atmosphere again.

Hydrogeology: The geology of ground water, with particular emphasis on the chemistry and movement of water.

Hydropneumatic: A water system, usually small, in which a water pump is automatically controlled by the pressure in a compressed air tank.

Hypersensitivity Diseases: Diseases characterized by allergic responses to pollutants; diseases most clearly associated with indoor air quality are asthma, rhinitis, and pneumonic hypersensitivity.

Hypolimnion: Bottom waters of a thermally stratified lake. The hypolimnion of a eutrophic lake is usually low or lacking in oxygen.

Hypoxia/Hypoxic Waters: Waters with dissolved oxygen concentrations of less than 2 ppm, the level generally accepted as the minimum required for most marine life to survive and reproduce.

Identification Code or EPA I.D. Number: The unique code assigned to each generator, transporter, and treatment, storage, or disposal facility by regulating agencies to facilitate identification and tracking of chemicals or hazardous waste.

Ignitable: Capable of burning or causing a fire.

IM240: A high-tech, transient dynamometer automobile emissions test that takes up to 240 seconds.

Imhoff Cone: A clear, cone-shaped container used to measure the volume of settleable solids in a specific volume of water.

Immediately Dangerous to Life and Health (IDLH): The maximum level to which a healthy individual can be exposed to a chemical for 30 minutes and escape without suffering irreversible health effects or impairing symptoms. Used as a "level of concern."

Imminent Hazard: One that would likely result in unreasonable adverse effects on humans or the environment or risk unreasonable hazard to an endangered species during the time required for a pesticide registration cancellation proceeding.

Imminent Threat: A high probability that exposure is occurring.

Immiscibility: The inability of two or more substances or liquids to readily dissolve into one another, such as soil and water. Immiscibility The inability of two or more substances or liquids to readily dissolve into one another, such as soil and water.

Impermeable: Not easily penetrated. The property of a material or soil that does not allow, or allows only with great difficulty, the movement or passage of water.

Imports: Municipal solid waste and recyclables that have been transported to a state or locality for processing or final disposition (but that did not originate in that state or locality).

Impoundment: A body of water or sludge confined by a dam, dike, floodgate, or other barrier.

In Situ: In its original place; unmoved unexcavated; remaining at the site or in the subsurface.

In-Line Filtration: Pre-treattment method in which chemicals are mixed by the flowing water; commonly used in pressure filtration installations. Eliminates need for flocculation and sedimentation.

In-Situ Flushing: Introduction of large volumes of water, at times supplemented with cleaning compounds, into soil, waste, or ground water to flush hazardous contaminants from a site.

In-Situ Oxidation: Technology that oxidizes contaminants dissolved in ground water, converting them into insoluble compounds.

In-Situ Stripping: Treatment system that removes or "strips" volatile organic compounds from contaminated ground or surface water by forcing an airstream through the water and causing the compounds to evaporate.

In-Situ Vitrification: Technology that treats contaminated soil in place at extremely high temperatures, at or more than 3000 degrees Fahrenheit.

In Vitro: Testing or action outside an organism (e.g. inside a test tube or culture dish.)

In Vivo: Testing or action inside an organism.

Incident Command Post: A facility located at a safe distance from an emergency site, where the incident commander, key staff, and technical representatives can make decisions and deploy emergency manpower and equipment.

Incident Command System (ICS): The organizational arrangement wherein one person, normally the Fire Chief of the impacted district, is in charge of an integrated, comprehensive emergency response organization and the emergency incident site, backed by an Emergency Operations Center staff with resources, information, and advice.

Incineration: A treatment technology involving destruction of waste by controlled burning at high temperatures; e.g., burning sludge to remove the water and reduce the remaining residues to a safe, non-burnable ash that can be disposed of safely on land, in some waters, or in underground locations.

Incineration at Sea: Disposal of waste by burning at sea on specially-designed incinerator ships.

Incompatible Waste: A waste unsuitable for mixing with another waste or material because it may react to form a hazard.

Indemnification: In the pesticide program, legal requirement that EPA pay certain end-users, dealers, and distributors for the cost of stock on hand at the time a pesticide registration is suspended.

Indicator: In biology, any biological entity or processes, or community whose characteristics show the presence of specific environmental conditions. 2. In chemistry, a substance that shows a visible change, usually of color, at a desired point in a chemical reaction. 3.A device that indicates the result of a measurement; e.g. a pressure gauge or a moveable scale.

Indirect Discharge: Introduction of pollutants from a non-domestic source into a publicly owned waste-treatment system. Indirect dischargers can be commercial or industrial facilities whose wastes enter local sewers.

Indirect Source: Any facility or building, property, road or parking area that attracts motor vehicle traffic and, indirectly, causes pollution.

Indoor Air: The breathable air inside a habitable structure or conveyance.

Indoor Air Pollution: Chemical, physical, or biological contaminants in indoor air.

Indoor Climate: Temperature, humidity, lighting, air flow and noise levels in a habitable structure or conveyance. Indoor climate can affect indoor air pollution.

Industrial Pollution Prevention: Combination of industrial source reduction and toxic chemical use substitution.

Industrial Process Waste: Residues produced during manufacturing operations.

Industrial Sludge: Semi-liquid residue or slurry remaining from treatment of industrial water and wastewater.

Industrial Source Reduction: Practices that reduce the amount of any hazardous substance, pollutant, or contaminant entering any waste stream or otherwise released into the environment. Also reduces the threat to public health and the environment associated with such releases. Term includes equipment or technology modifications, substitution of raw materials, and improvements in housekeeping, maintenance, training or inventory control.

Industrial Waste: Unwanted materials from an industrial operation; may be liquid, sludge, solid, or hazardous waste.

Inert Ingredient: Pesticide components such as solvents, carriers, dispersants, and surfactants that are not active against target pests. Not all inert ingredients are innocuous.

Inertial Separator: A device that uses centrifugal force to separate waste particles.

Infectious Agent: Any organism, such as a pathogenic virus, parasite, or or bacterium, that is capable of invading body tissues, multiplying, and causing disease.

Infectious Waste: Hazardous waste capable of causing infections in humans, including: contaminated animal waste; human blood and blood products; isolation waste, pathological waste; and discarded sharps (needles, scalpels or broken medical instruments).

Infiltration: 1. The penetration of water through the ground surface into sub-surface soil or the penetration of water from the soil into sewer or other pipes through defective joints, connections, or manhole walls. 2. The technique of applying large volumes of waste water to land to penetrate the surface and percolate through the underlying soil.

Infiltration Gallery: A sub-surface groundwater collection system, typically shallow in depth, constructed with open-jointed or perforated pipes that discharge collected water into a watertight chamber from which the water is pumped to treatment facilities and into the distribution system. Usually located close to streams or ponds.

Infiltration Rate: The quantity of water that can enter the soil in a specified time interval.

Inflow: Entry of extraneous rain water into a sewer system from sources other than infiltration, such as basement drains, manholes, storm drains, and street washing.

Influent: Water, wastewater, or other liquid flowing into a reservoir, basin, or treatment plant.

Information Collection Request (ICR): A description of information to be gathered in connection with rules, proposed rules, surveys, and guidance documents that contain information-gathering requirements. The ICR describes what information is needed, why it is needed, how it will be collected, and how much collecting it will cost. The ICR is submitted by the EPA to the Office of Management and Budget (OMB) for approval.

Information File: In the Superfund program, a file that contains accurate, up-to-date documents on a Superfund site. The file is usually located in a public building (school, library, or city hall) convenient for local residents.

Inhalable Particles: All dust capable of entering the human respiratory tract.

Initial Compliance Period (Water): The first full three-year compliance period which begins at least 18 months after promulgation.

Injection Well: A well into which fluids are injected for purposes such as waste disposal, improving the recovery of crude oil, or solution mining.

Injection Zone: A geological formation receiving fluids through a well.

Integrated Pest Management (IPM): A mixture of chemical and other, non-pesticide, methods to control pests.

Integrated Waste Management: Using a variety of practices to handle municipal solid waste; can include source reduction, recycling, incineration, and landfilling.

Interceptor Sewers: Large sewer lines that, in a combined system, control the flow of sewage to the treatment plant. In a storm, they allow some of the sewage to flow directly into a receiving stream, thus keeping it from overflowing onto the streets. Also used in separate systems to collect the flows from main and trunk sewers and carry them to treatment points.

Interface: The common boundary between two substances such as a water and a solid, water and a gas, or two liquids such as water and oil.

Interfacial Tension: The strength of the film separating two immiscible fluids (e.g. oil and water) measured in dynes per, or millidynes per centimeter.

Interim (Permit) Status: Period during which treatment, storage and disposal facilities coming under RCRA in 1980 are temporarily permitted to operate while awaiting a permanent permit. Permits issued under these circumstances are usually called "Part A" or "Part B" permits.

Internal Dose: In exposure assessment, the amount of a substance penetrating the absorption barriers (e.g. skin, lung tissue, gastrointestinal tract) of an organism through either physical or biological processes.

Interstate Carrier Water Supply: A source of water for drinking and sanitary use on planes, buses, trains, and ships operating in more than one state. These sources are federally regulated.

Interstate Commerce Clause: A clause of the U.S. Constitution which reserves to the federal government the right to regulate the conduct of business across state lines. Under this clause, for example, the U.S. Supreme Court has ruled that states may not inequitably restrict the disposal of out-of-state wastes in their jurisdictions.

Interstate Waters: Waters that flow across or form part of state or international boundaries; e.g. the Great Lakes, the Mississippi River, or coastal waters.

Interstitial Monitoring: The continuous surveillance of the space between the walls of an underground storage tank.

Intrastate Product: Pesticide products once registered by states for sale and use only in the state. All intrastate products have been converted to full federal registration or canceled.

Inventory (TSCA): Inventory of chemicals produced pursuant to Section 8 (b) of the Toxic Substances Control Act.

Inversion: A layer of warm air that prevents the rise of cooling air and traps pollutants beneath it; can cause an air pollution episode.

Ion: An electrically charged atom or group of atoms.

Ion Exchange Treatment: A common water-softening method often found on a large scale at water purification plants that remove some organics and radium by adding calcium oxide or calcium hydroxide to increase the pH to a level where the metals will precipitate out.

Ionizing Radiation: Radiation that can strip electrons from atoms; e.g. alpha, beta, and gamma radiation.

IRIS: EPA's Integrated Risk Information System, an electronic data base containing the Agency's latest descriptive and quantitative regulatory information on chemical constituents.

Irradiated Food: Food subject to brief radioactivity, usually gamma rays, to kill insects, bacteria, and mold, and to permit storage without refrigeration.

Irradiation: Exposure to radiation of wavelengths shorter than those of visible light (gamma, x-ray, or ultra- violet), for medical purposes, to sterilize milk or other foodstuffs, or to induce polymerization of monomers or vulcanization of rubber.

Irreversible Effect: Effect characterized by the inability of the body to partially or fully repair injury caused by a toxic agent.

Irrigation: Applying water or wastewater to land areas to supply the water and nutrient needs of plants.

Irrigation Efficiency: The amount of water stored in the crop root zone compared to the amount of irrigation water applied.

Irrigation Return Flow: Surface and subsurface water which leaves the field following application of irrigation water.

Irritant: A substance that can cause irritation of the skin, eyes, or respiratory system. Effects may be acute from a single high level exposure, or chronic from repeated low-level exposures to such compounds as chlorine, nitrogen dioxide, and nitric acid.

Isoconcentration: More than one sample point exhibiting the same isolate concentration.

Isopleth: The line or area represented by an isoconcentration.

Isotope: A variation of an element that has the same atomic number of protons but a different weight because of the number of neutrons. Various isotopes of the same element may have different radioactive behaviors, some are highly unstable..

Isotropy: The condition in which the hydraulic or other properties of an aquifer are the same in all directions.

Jar Test: A laboratory procedure that simulates a water treatment plant's coagulation/ flocculation units with differing chemical doses, mix speeds, and settling times to estimate the minimum or ideal coagulant dose required to achieve certain water quality goals.

Joint and Several Liability: Under CERCLA, this legal concept relates to the liability for Superfund site cleanup and other costs on the part of more than one potentially responsible party (i.e. if there were several owners or users of a site that became contaminated over the years, they could all be considered potentially liable for cleaning up the site.)

Karst: A geologic formation of irregular limestone deposits with sinks, underground streams, and caverns.

Kinetic Energy: Energy possessed by a moving object or water body.

Kinetic Rate Coefficient: A number that describes the rate at which a water constituent such as a biochemical oxygen demand or dissolved oxygen rises or falls, or at which an air pollutant reacts.

Laboratory Animal Studies: Investigations using animals as surrogates for humans.

Lagoon: 1. A shallow pond where sunlight, bacterial action, and oxygen work to purify wastewater; also used for storage of wastewater or spent nuclear fuel rods. 2. Shallow body of water, often separated from the sea by coral reefs or sandbars.

Land Application: Discharge of wastewater onto the ground for treatment or reuse.

Land Ban: Phasing out of land disposal of most untreated hazardous wastes, as mandated by the 1984 RCRA amendments.

Land Disposal Restrictions: Rules that require hazardous wastes to be treated before disposal on land to destroy or immobilize hazardous constituents that might migrate into soil and ground water.

Land Farming (of Waste): A disposal process in which hazardous waste deposited on or in the soil is degraded naturally by microbes.

Landfills: 1. Sanitary landfills are disposal sites for non-hazardous solid wastes spread in layers, compacted to the smallest practical volume, and covered by material applied at the end of each operating day. 2. Secure chemical landfills are disposal sites for hazardous waste, selected and designed to minimize the chance of release of hazardous substances into the environment.

Landscape: The traits, patterns, and structure of a specific geographic area, including its biological composition, its physical environment, and its anthropogenic or social patterns. An area where interacting ecosystems are grouped and repeated in similar form.

Landscape Characterization: Documentation of the traits and patterns of the essential elements of the landscape.

Landscape Ecology: The study of the distribution patterns of communities and ecosystems, the ecological processes that affect those patterns, and changes in pattern and process over time.

Landscape Indicator: A measurement of the landscape, calculated from mapped or remotely sensed data, used to describe spatial patterns of land use and land cover across a geographic area. Landscape indicators may be useful as measures of certain kinds of environmental degradation such as forest fragmentation.

Langelier Index (LI): An index reflecting the equilibrium pH of a water with respect to calcium and alkalinity; used in stabilizing water to control both corrosion and scale deposition.

Large Quantity Generator: Person or facility generating more than 2200 pounds of hazardous waste per month. Such generators produce about 90 percent of the nation's hazardous waste, and are subject to all RCRA requirements.

Laser Induced Fluorescence: A method for measuring the relative amount of soil and/or groundwater with an in-situ sensor.

Latency: Time from the first exposure of a chemical until the appearance of a toxic effect.

Lateral Sewers: Pipes that run under city streets and receive the sewage from homes and businesses, as opposed to domestic feeders and main trunk lines.

Laundering Weir: Sedimention basin overflow weir.

LC 50/Lethal Concentration: Median level concentration, a standard measure of toxicity. It tells how much of a substance is needed to kill half of a group of experimental organisms in a given time.

LD 50/ Lethal Dose: The dose of a toxicant or microbe that will kill 50 percent of the test organisms within a designated period. The lower the LD 50, the more toxic the compound.

Ldlo: Lethal dose low; the lowest dose in an animal study at which lethality occurs.

Leachate: Water that collects contaminants as it trickles through wastes, pesticides or fertilizers. Leaching may occur in farming areas, feedlots, and landfills, and may result in hazardous substances entering surface water, ground water, or soil.

Leachate Collection System: A system that gathers leachate and pumps it to the surface for treatment.

Leaching: The process by which soluble constituents are dissolved and filtered through the soil by a percolating fluid.

Lead (Pb): A heavy metal that is hazardous to health if breathed or swallowed. Its use in gasoline, paints, and plumbing compounds has been sharply restricted or eliminated by federal laws and regulations.

Lead Service Line: A service line made of lead which connects the water to the building inlet and any lead fitting connected to it.

Legionella: A genus of bacteria, some species of which have caused a type of pneumonia called Legionaires Disease.

Lethal Concentration 50: Also referred to as LC50, a concentration of a pollutant or effluent at which 50 percent of the test organisms die; a common measure of acute toxicity.

Lethal Dose 50: Also referred to as LD50, the dose of a toxicant that will kill 50 percent of test organisms within a designated period of time; the lower the LD 50, the more toxic the compound.

Level of Concern (LOC): The concentration in air of an extremely hazardous substance above which there may be serious immediate health effects to anyone exposed to it for short periods

Life Cycle of a Product: All stages of a product's development, from extraction of fuel for power to production, marketing, use, and disposal.

Lifetime Average Daily Dose: Figure for estimating excess lifetime cancer risk.

Lifetime Exposure: Total amount of exposure to a substance that a human would receive in a lifetime (usually assumed to be 70 years).

Limnology: The study of the physical, chemical, hydrological, and biological aspects of fresh water bodies.

Lindane: A pesticide that causes adverse health effects in domestic water supplies and is toxic to freshwater fish and aquatic life.

Liner: 1. A relatively impermeable barrier designed to keep leachate inside a landfill. Liner materials include plastic and dense clay. 2. An insert or sleeve for sewer pipes to prevent leakage or infiltration.

Lipid Solubility: The maximum concentration of a chemical that will dissolve in fatty substances. Lipid soluble substances are insoluble in water. They will very selectively disperse through the environment via uptake in living tissue.

Liquefaction: Changing a solid into a liquid.

Liquid Injection Incinerator: Commonly used system that relies on high pressure to prepare liquid wastes for incineration by breaking them up into tiny droplets to allow easier combustion.

List: Shorthand term for EPA list of violating facilities or firms debarred from obtaining government contracts because they violated certain sections of the Clean Air or Clean Water Acts. The list is maintained by The Office of Enforcement and Compliance Monitoring.

Listed Waste: Wastes listed as hazardous under RCRA but which have not been subjected to the Toxic Characteristics Listing Process because the dangers they present are considered self-evident.

Lithology: Mineralogy, grain size, texture, and other physical properties of granular soil, sediment, or rock.

Litter: 1. The highly visible portion of solid waste carelessly discarded outside the regular garbage and trash collection and disposal system. 2. leaves and twigs fallen from forest trees.

Littoral Zone: 1. That portion of a body of fresh water extending from the shoreline lakeward to the limit of occupancy of rooted plants. 2. A strip of land along the shoreline between the high and low water levels.

Local Education Agency (LEA): In the asbestos program, an educational agency at the local level that exists primarily to operate schools or to contract for educational services, including primary and secondary public and private schools. A single, unaffiliated school can be considered an LEA for AHERA purposes.

Local Emergency Planning Committee (LEPC): A committee appointed by the state emergency response commission, as required by SARA Title III, to formulate a comprehensive emergency plan for its jurisdiction.

Low Density Polyethylene (LOPE): Plastic material used for both rigid containers and plastic film applications.

Low Emissivity (low-E) Windows: New window technology that lowers the amount of energy loss through windows by inhibiting the transmission of radiant heat while still allowing sufficient light to pass through.

Low NO_x Burners: One of several combustion technologies used to reduce emissions of Nitrogen Oxides (NOx.)

Low-Level Radioactive Waste (LLRW): Wastes less hazardous than most of those associated with a nuclear reactor; generated by hospitals, research laboratories, and certain industries. The Department of Energy, Nuclear Regulatory Commission, and EPA share responsibilities for managing them.

Lower Detection Limit: The smallest signal above background noise an instrument can reliably detect.

Lower Explosive Limit (LEL): The concentration of a compound in air below which the mixture will not catch on fire.

Lowest Acceptable Daily Dose: The largest quantity of a chemical that will not cause a toxic effect, as determined by animal studies.

Lowest Achievable Emission Rate: Under the Clean Air Act, the rate of emissions that reflects (1) the most stringent emission limitation in the implementation plan of any state for such source unless the owner or operator demonstrates such limitations are not achievable; or (2) the most stringent emissions limitation achieved in practice, whichever is more stringent. A proposed new or modified source may not emit pollutants in excess of existing new source standards.

Lowest Observed Adverse Effect Level (LOAEL): The lowest level of a stressor that causes statistically and biologically significant differences in test samples as compared to other samples subjected to no stressor.

Macropores: Secondary soil features such as root holes or desiccation cracks that can create significant conduits for movement of NAPL and dissolved contaminants, or vapor-phase contaminants.

Magnetic Separation: Use of magnets to separate ferrous materials from mixed municipal waste stream.

Major Modification: This term is used to define modifications of major stationary sources of emissions with respect to Prevention of Significant Deterioration and New Source Review under the Clean Air Act.

Major Stationary Sources: Term used to determine the applicability of Prevention of Significant Deterioration and new source regulations. In a nonattainment area, any stationary pollutant source with potential to emit more than 100 tons per year is considered a major stationary source. In PSD areas the cutoff level may be either 100 or 250 tons, depending upon the source.

Majors: Larger publicly owned treatment works (POTWs) with flows equal to at least one million gallons per day (mgd) or servicing a population equivalent to 10,000 persons; certain other POTWs having significant water quality impacts.

Man-Made (Anthropogenic) Beta Particle and Photon Emitters: All radionuclides emitting beta particles and/or photons listed in Maximum Permissible Body Burdens and Maximum Permissible Concentrations of Radonuclides in Air and Water for Occupational Exposure.

Managerial Controls: Methods of nonpoint source pollution control based on decisions about managing agricultural wastes or application times or rates for agrochemicals.

Mandatory Recycling: Programs which by law require consumers to separate trash so that some or all recyclable materials are recovered for recycling rather than going to landfills.

Manifest: A one-page form used by haulers transporting waste that lists EPA identification numbers, type and quantity of waste, the generator it originated from, the transporter that shipped it, and the storage or disposal facility to which it is being shipped. It includes copies for all participants in the shipping process.

Manifest System: Tracking of hazardous waste from "cradle-to-grave" (generation through disposal) with accompanying documents known as manifests.

Manual Separation: Hand sorting of recyclable or compostable materials in waste.

Manufacturer's Formulation: A list of substances or component parts as described by the maker of a coating, pesticide, or other product containing chemicals or other substances.

Manufacturing Use Product: Any product intended (labeled) for formulation or repackaging into other pesticide products.

Margin of Safety: Maximum amount of exposure producing no measurable effect in animals (or studied humans) divided by the actual amount of human exposure in a population.

Marine Sanitation Device: Any equipment or process installed on board a vessel to receive, retain, treat, or discharge sewage.

Marsh: A type of wetland that does not accumulate appreciable peat deposits and is dominated by herbaceous vegetation. Marshes may be either fresh or saltwater, tidal or non-tidal.

Material Category: In the asbestos program, broad classification of materials into thermal surfacing insulation, surfacing material, and miscellaneous material.

Material Safety Data Sheet (MSDS): A compilation of information required under the OSHA Communication Standard on the identity of hazardous chemicals, health, and physical hazards, exposure limits, and precautions. Section 311 of SARA requires facilities to submit MSDSs under certain circumstances.

Material Type: Classification of suspect material by its specific use or application; e.g., pipe insulation, fireproofing, and floor tile.

Materials Recovery Facility (MRF): A facility that processes residentially collected mixed recyclables into new products available for market.

Maximally (or Most) Exposed Individual: The person with the highest exposure in a given population.

Maximum Acceptable Toxic Concentration: For a given ecological effects test, the range (or geometric mean) between the No Observable Adverse Effect Level and the Lowest Observable Adverse Effects Level.

Maximum Contaminant Level: The maximum permissible level of a contaminant in water delivered to any user of a public system. MCLs are enforceable standards.

Maximum Contaminant Level Goal (MCLG): Under the Safe Drinking Water Act, a non-enforceable concentration of a drinking water contaminant, set at the level at which no known or anticipated adverse effects on human health occur and which allows an adequate safety margin. The MCLG is usually the starting point for determining the regulated Maximum Contaminant Level.

Maximum Exposure Range: Estimate of exposure or dose level received by an individual in a defined population that is greater than the 98th percentile dose for all individuals in that population, but less than the exposure level received by the person receiving the highest exposure level.

Maximum Residue Level: Comparable to a U.S. tolerance level, the Maximum Residue Level the enforceable limit on food pesticide levels in some countries. Levels are set by the Codex Alimentarius Commission, a United Nations agency managed and funded jointly by the World Health Organization and the Food and Agriculture Organization.

Maximum Tolerated Dose: The maximum dose that an animal species can tolerate for a major portion of its lifetime without significant impairment or toxic effect other than carcinogenicity.

Measure of Effect/ Measurement Endpoint: A measurable characteristic of ecological entity that can be related to an assessment endpoint; e.g. a laboratory test for eight species meeting certain requirements may serve as a measure of effect for an assessment endpoint, such as survival of fish, aquatic, invertebrate or algal species under acute exposure.

Measure of Exposure: A measurable characteristic of a stressor (such as the specific amount of mercury in a body of water) used to help quantify the exposure of an ecological entity or individual organism.

Medium-size Water System: A water system that serves 3,300 to 50,000 customers.

Meniscus: The curved top of a column of liquid in a small tube.

Mercury (Hg): Heavy metal that can accumulate in the environment and is highly toxic if breathed or swallowed.

Mesotrophic: Reservoirs and lakes which contain moderate quantities of nutrients and are moderately productive in terms of aquatic animal and plant life.

Metabolites: Any substances produced by biological processes, such as those from pesticides.

Metalimnion: The middle layer of a thermally stratified lake or reservoir. In this layer there is a rapid decrease in temperature with depth. Also called thermocline.

Methane: A colorless, nonpoisonous, flammable gas created by anaerobic decomposition of organic compounds. A major component of natural gas used in the home.

Methanol: An alcohol that can be used as an alternative fuel or as a gasoline additive. It is less volatile than gasoline; when blended with gasoline it lowers the carbon monoxide emissions but increases hydrocarbon emissions. Used as pure fuel, its emissions are less ozone-forming than those from gasoline. Poisonous to humans and animals if ingested.

Method 18: An EPA test method which uses gas chromatographic techniques to measure the concentration of volatile organic compounds in a gas stream.

Method 24: An EPA reference method to determine density, water content and total volatile content (water and VOC) of coatings.

Method 25: An EPA reference method to determine the VOC concentration in a gas stream.

Methoxychlor: Pesticide that causes adverse health effects in domestic water supplies and is toxic to freshwater and marine aquatic life.

Methyl Orange Alkalinity: A measure of the total alkalinity in a water sample in which the color of methyl orange reflects the change in level.

Microbial Growth: The amplification or multiplication of microorganisms such as bacteria, algae, diatoms, plankton, and fungi.

Microbial Pesticide: A microorganism that is used to kill a pest, but is of minimum toxicity to humans.

Microclimate: 1. Localized climate conditions within an urban area or neighborhood. 2. The climate around a tree or shrub or a stand of trees.

Microenvironmental Method: A method for sequentially assessing exposure for a series of microenvironments that can be approximated by constant concentrations of a stressor.

Microenvironments: Well-defined surroundings such as the home, office, or kitchen that can be treated as uniform in terms of stressor concentration.

Million-Gallons Per Day (MGD): A measure of water flow.

Minimization: A comprehensive program to minimize or eliminate wastes, usually applied to wastes at their point of origin.

Mining of an Aquifer: Withdrawal over a period of time of ground water that exceeds the rate of recharge of the aquifer.

Mining Waste: Residues resulting from the extraction of raw materials from the earth.

Minor Source: New emissions sources or modifications to existing emissions sources that do not exceed NAAQS emission levels.

Minors: Publicly owned treatment works with flows less than 1 million gallons per day.

Miscellaneous ACM: Interior asbestos-containing building material or structural components, members or fixtures, such as floor and ceiling tiles; does not include surfacing materials or thermal system insulation.

Miscellaneous Materials: Interior building materials on structural components, such as floor or ceiling tiles.

Miscible Liquids: Two or more liquids that can be mixed and will remain mixed under normal conditions.

Missed Detection: The situation that occurs when a test indicates that a tank is "tight" when in fact it is leaking.

Mist: Liquid particles measuring 40 to 500 micrometers (pm), are formed by condensation of vapor. By comparison, fog particles are smaller than 40 micrometers (pm).

Mitigation: Measures taken to reduce adverse impacts on the environment.

Mixed Funding: Settlements in which potentially responsible parties and EPA share the cost of a response action.

Mixed Glass: Recovered container glass not sorted into categories (e.g. color, grade).

Mixed Liquor: A mixture of activated sludge and water containing organic matter undergoing activated sludge treatment in an aeration tank.

Mixed Metals: Recovered metals not sorted into categories such as aluminum, tin, or steel cans or ferrous or non-ferrous metals.

Mixed Municipal Waste: Solid waste that has not been sorted into specific categories (such as plastic, glass, yard trimmings, etc.)

Mixed Paper: Recovered paper not sorted into categories such as old magazines, old newspapers, old corrugated boxes, etc.

Mixed Plastic: Recovered plastic unsorted by category.

Mobile Incinerator Systems: Hazardous waste incinerators that can be transported from one site to another.

Mobile Source: Any non-stationary source of air pollution such as cars, trucks, motorcycles, buses, airplanes, and locomotives.

Model Plant: A hypothetical plant design used for developing economic, environmental, and energy impact analyses as support for regulations or regulatory guidelines; first step in exploring the economic impact of a potential NSPS.

Modified Bin Method: Way of calculating the required heating or cooling for a building based on determining how much energy the system would use if outdoor temperatures were within a certain temperature interval and then multiplying the energy use by the time the temperature interval typically occurs.

Modified Source: The enlargement of a major stationary pollutant sources is often referred to as modification, implying that more emissions will occur.

Moisture Content: 1.The amount of water lost from soil upon drying to a constant weight, expressed as the weight per unit of dry soil or as the volume of water per unit bulk volume of the soil. For a fully saturated medium, moisture content indicates the porosity. 2. Water equivalent of snow on the ground; an indicator of snowmelt flood potential.

Molecule: The smallest division of a compound that still retains or exhibits all the properties of the substance.

Molten Salt Reactor: A thermal treatment unit that rapidly heats waste in a heat-conducting fluid bath of carbonate salt.

Monitoring: Periodic or continuous surveillance or testing to determine the level of compliance with statutory requirements and/or pollutant levels in various media or in humans, plants, and animals.

Monitoring Well: 1. A well used to obtain water quality samples or measure groundwater levels. 2. A well drilled at a hazardous waste management facility or Superfund site to collect ground-water samples for the purpose of physical, chemical, or biological analysis to determine the amounts, types, and distribution of contaminants in the groundwater beneath the site.

Monoclonal Antibodies (Also called MABs and MCAs): 1. Man-made (anthropogenic) clones of a molecule, produced in quantity for medical or research purposes. 2. Molecules of living organisms that selectively find and attach to other molecules to which their structure conforms exactly. This could also apply to equivalent activity by chemical molecules.

Monomictic: Lakes and reservoirs which are relatively deep, do not freeze over during winter, and undergo a single stratification and mixing cycle during the year (usually in the fall).

Montreal Protocol: Treaty, signed in 1987, governs stratospheric ozone protection and research, and the production and use of ozone-depleting substances. It provides for the end of production of ozone-depleting substances such as CFCS. Under the Protocol, various research groups continue to assess the ozone layer. The Multilateral Fund provides resources to developing nations to promote the transition to ozone-safe technologies.

Moratorium: During the negotiation process, a period of 60 to 90 days during which EPA and potentially responsible parties may reach settlement but no site response activities can be conducted.

Morbidity: Rate of disease incidence.

Mortality: Death rate.

Most Probable Number: An estimate of microbial density per unit volume of water sample, based on probability theory.

Muck Soils: Earth made from decaying plant materials.

Mudballs: Round material that forms in filters and gradually increases in size when not removed by backwashing.

Mulch: A layer of material (wood chips, straw, leaves, etc.) placed around plants to hold moisture, prevent weed growth, and enrich or sterilize the soil.

Multi-Media Approach: Joint approach to several environmental media, such as air, water, and land.

Multiple Chemical Sensitivity: A diagnostic label for people who suffer multi-system illnesses as a result of contact with, or proximity to, a variety of airborne agents and other substances.

Multiple Use: Use of land for more than one purpose; e.g., grazing of livestock, watershed and wildlife protection, recreation, and timber production. Also applies to use of bodies of water for recreational purposes, fishing, and water supply.

Multistage Remote Sensing: A strategy for landscape characterization that involves gathering and analyzing information at several geographic scales, ranging from generalized levels of detail at the national level through high levels of detail at the local scale.

Municipal Discharge: Discharge of effluent from waste water treatment plants which receive waste water from households, commercial establishments, and industries in the coastal drainage basin. Combined sewer/separate storm overflows are included in this category.

Municipal Sewage: Wastes (mostly liquid) orginating from a community; may be composed of domestic wastewaters and/or industrial discharges.

Municipal Sludge: Semi-liquid residue remaining from the treatment of municipal water and wastewater.

Municipal Solid Waste: Common garbage or trash generated by industries, businesses, institutions, and homes.

Mutagen/Mutagenicity: An agent that causes a permanent genetic change in a cell other than that which occurs during normal growth. Mutagenicity is the capacity of a chemical or physical agent to cause such permanent changes.

National Ambient Air Quality Standards (NAAQS): Standards established by EPA that apply for outdoor air throughout the country.

National Emissions Standards for Hazardous Air Pollutants (NESHAPS): Emissions standards set by EPA for an air pollutant not covered by NAAQS that may cause an increase in fatalities or in serious, irreversible, or incapacitating illness. Primary standards are designed to protect human health, secondary standards to protect public welfare (e.g. building facades, visibility, crops, and domestic animals).

National Environmental Performance Partnership Agreements: System that allows states to assume greater responsibility for environmental programs based on their relative ability to execute them.

National Estuary Program: A program established under the Clean Water Act Amendments of 1987 to develop and implement conservation and management plans for protecting estuaries and restoring and maintaining their chemical, physical, and biological integrity, as well as controlling point and nonpoint pollution sources.

National Municipal Plan: A policy created in 1984 by EPA and the states in 1984 to bring all publicly owned treatment works (POTWs) into compliance with Clean Water Act requirements.

National Oil and Hazardous Substances Contingency Plan (NOHSCP/NCP): The federal regulation that guides determination of the sites to be corrected under both the Superfund program and the program to prevent or control spills into surface waters or elsewhere.

National Pollutant Discharge Elimination System (NPDES): A provision of the Clean Water Act which prohibits discharge of pollutants into waters of the United States unless a special permit is issued by EPA, a state, or, where delegated, a tribal government on an Indian reservation.

National Priorities List (NPL): EPA's list of the most serious uncontrolled or abandoned hazardous waste sites identified for possible long-term remedial action under Superfund.

The list is based primarily on the score a site receives from the Hazard Ranking System. EPA is required to update the NPL at least once a year. A site must be on the NPL to receive money from the Trust Fund for remedial action.

National Response Center: The federal operations center that receives notifications of all releases of oil and hazardous substances into the environment; open 24 hours a day, is operated by the U.S. Coast Guard, which evaluates all reports and notifies the appropriate agency.

National Response Team (NRT): Representatives of 13 federal agencies that, as a team, coordinate federal responses to nationally significant incidents of pollution—an oil spill, a major chemical release, or a - superfund response action—and provide advice and technical assistance to the responding agency(ies) before and during a response action.

National Secondary Drinking Water Regulations: Commonly referred to as NSDWRs.

Navigable Waters: Traditionally, waters sufficiently deep and wide for navigation by all, or specified vessels; such waters in the United States come under federal jurisdiction and are protected by certain provisions of the Clean Water Act.

Necrosis: Death of plant or animal cells or tissues. In plants, necrosis can discolor stems or leaves or kill a plant entirely.

Negotiations (Under Superfund): After potentially responsible parties are identified for a site, EPA coordinates with them to reach a settlement that will result in the PRP paying for or conducting the cleanup under EPA supervision. If negotiations fail, EPA can order the PRP to conduct the cleanup or EPA can pay for the cleanup using Superfund monies and then sue to recover the costs.

Nephelometric: Method of of measuring turbidity in a water sample by passing light through the sample and measuring the amount of the light that is deflected.

Netting: A concept in which all emissions sources in the same area that owned or controlled by a single company are treated as one large source, thereby allowing flexibility in controlling individual sources in order to meet a single emissions standard.

Neutralization: Decreasing the acidity or alkalinity of a substance by adding alkaline or acidic materials, respectively.

New Source: Any stationary source built or modified after publication of final or proposed regulations that prescribe a given standard of performance.

New Source Performance Standards (NSPS): Uniform national EPA air emission and water effluent standards which limit the amount of pollution allowed from new sources or from modified existing sources.

New Source Review (NSR): A Clean Air Act requirement that State Implementation Plans must include a permit review that applies to the construction and operation of new and modified stationary sources in nonattainment areas to ensure attainment of national ambient air quality standards.

Nitrate: A compound containing nitrogen that can exist in the atmosphere or as a dissolved gas in water and which can have harmful effects on humans and animals. Nitrates in water can cause severe illness in infants and domestic animals. A plant nutrient and inorganic fertilizer, nitrate is found in septic systems, animal feed lots, agricultural fertilizers, manure, industrial waste waters, sanitary landfills, and garbage dumps.

Nitric Oxide (NO): A gas formed by combustion under high temperature and high pressure in an internal combustion engine; it is converted by sunlight and photochemical processes in ambient air to nitrogen oxide. NO is a precursor of ground-level ozone pollution, or smog..

Nitrification: The process whereby ammonia in wastewater is oxidized to nitrite and then to nitrate by bacterial or chemical reactions.

Nitrilotriacetic Acid (NTA): A compound now replacing phosphates in detergents.

Nitrite: 1. An intermediate in the process of nitrification. 2. Nitrous oxide salts used in food preservation.

Nitrogen Dioxide (NO_2): The result of nitric oxide combining with oxygen in the atmosphere; major component of photochemical smog.

Nitrogen Oxide (NO_x): The result of photochemical reactions of nitric oxide in ambient air; major component of photochemical smog. Product of combustion from transportation and stationary sources and a major contributor to the formation of ozone in the troposphere and to acid deposition.

Nitrogenous Wastes: Animal or vegetable residues that contain significant amounts of nitrogen.

Nitrophenols: Synthetic organopesticides containing carbon, hydrogen, nitrogen, and oxygen.

No Further Remedial Action Planned: Determination made by EPA following a preliminary assessment that a site does not pose a significant risk and so requires no further activity under CERCLA.

No Observable Adverse Effect Level (NOAEL): An exposure level at which there are no statistically or biologically significant increases in the frequency or severity of adverse effects between the exposed population and its appropriate control; some effects may be produced at this level, but they are not considered as adverse, or as precurors to adverse effects. In an experiment with several NOAELs, the regulatory focus is primarily on the highest one, leading to the common usage of the term NOAEL as the highest exposure without adverse effects.

No Till: Planting crops without prior seedbed preparation, into an existing cover crop, sod, or crop residues, and eliminating subsequent tillage operations.

No-Observed-Effect-Level (NOEL): Exposure level at which there are no statistically or biological significant differences in the frequency or severity of any effect in the exposed or control populations.

Noble Metal: Chemically inactive metal such as gold; does not corrode easily.

Noise: Product-level or product-volume changes occurring during a test that are not related to a leak but may be mistaken for one.

Non-Aqueous Phase Liquid (NAPL): Contaminants that remain undiluted as the original bulk liquid in the subsurface, e.g. spilled oil.

Non-Attainment Area: Area that does not meet one or more of the National Ambient Air Quality Standards for the criteria pollutants designated in the Clean Air Act.

Non-Binding Allocations of Responsibility (NBAR): A process for EPA to propose a way for potentially responsible parties to allocate costs among themselves.

Non-Community Water System: A public water system that is not a community water system; e.g. the water supply at a camp site or national park.

Non-Compliance Coal: Any coal that emits greater than 3.0 pounds of sulfur dioxide per million BTU when burned. Also known as high-sulfur coal.

Non-Contact Cooling Water: Water used for cooling which does not come into direct contact with any raw material, product, byproduct, or waste.

Non-Conventional Pollutant: Any pollutant not statutorily listed or which is poorly understood by the scientific community.

Non-Degradation: An environmental policy which disallows any lowering of naturally occurring quality regardless of preestablished health standards.

Non-Ferrous Metals: Nonmagnetic metals such as aluminum, lead, and copper. Products made all or in part from such metals include containers, packaging, appliances, furniture, electronic equipment and aluminum foil.

Non-ionizing Electromagnetic Radiation: 1. Radiation that does not change the structure of atoms but does heat tissue and may cause harmful biological effects. 2. Microwaves, radio waves, and low-frequency electromagnetic fields from high-voltage transmission lines.

Non-Methane Hydrocarbon (NMHC): The sum of all hydrocarbon air pollutants except methane; significant precursors to ozone formation.

Non-Methane Organic Gases (NMOG): The sum of all organic air pollutants. Excluding methane; they account for aldehydes, ketones, alcohols, and other pollutants that are not hydrocarbons but are precursors of ozone.

Non-Point Sources: Diffuse pollution sources (i.e. without a single point of origin or not introduced into a receiving stream from a specific outlet). The pollutants are generally carried off the land by storm water. Common non-point sources are agriculture, forestry, urban, mining, construction, dams, channels, land disposal, saltwater intrusion, and city streets.

Non-potable: Water that is unsafe or unpalatable to drink because it contains pollutants, contaminants, minerals, or infective agents.

Non-Road Emissions: Pollutants emitted by combustion engines on farm and construction equipment, gasoline-powered lawn and garden equipment, and power boats and outboard motors.

Non-Transient Non-Community Water System: A public water system that regularly serves at least 25 of the same non-resident persons per day for more than six months per year.

Nondischarging Treatment Plant: A treatment plant that does not discharge treated wastewater into any stream or river. Most are pond systems that dispose of the total flow they receive by means of evaporation or percolation to groundwater, or facilities that dispose of their effluent by recycling or reuse (e.g. spray irrigation or groundwater discharge).

Nonfriable Asbestos-Containing Materials: Any material containing more than one percent asbestos (as determined by Polarized Light Microscopy) that, when dry, cannot be crumbled, pulverized, or reduced to powder by hand pressure.

Nonhazardous Industrial Waste: Industrial process waste in wastewater not considered municipal solid waste or hazardous waste under RARA.

Notice of Deficiency: An EPA request to a facility owner or operator requesting additional information before a preliminary decision on a permit application can be made.

Notice of Intent to Cancel: Notification sent to registrants when EPA decides to cancel registration of a product containing a pesticide.

Notice of Intent to Deny: Notification by EPA of its preliminary intent to deny a permit application.

Notice of Intent to Suspend: Notification sent to a pesticide registrant when EPA decides to suspend product sale and distribution because of failure to submit requested data in a timely and/or acceptable manner, or because of imminent hazard.

Nuclear Reactors and Support Facilities: Uranium mills, commercial power reactors, fuel reprocessing plants, and uranium enrichment facilities.

Nuclear Winter: Prediction by some scientists that smoke and debris rising from massive fires of a nuclear war could block sunlight for weeks or months, cooling the earth's surface and producing climate changes that could, for example, negatively affect world agricultural and weather patterns.

Nuclide: An atom characterized by the number of protons, neturons, and energy in the nucleus.

Nutrient: Any substance assimilated by living things that promotes growth. The term is generally applied to nitrogen and phosphorus in wastewater, but is also applied to other essential and trace elements.

Nutrient Pollution: Contamination of water resources by excessive inputs of nutrients. In surface waters, excess algal production is a major concern.

Ocean Discharge Waiver: A variance from Clean Water Act requirements for discharges into marine waters.

Odor Threshold: The minimum odor of a water or air sample that can just be detected after successive dilutions with odorless water. Also called threshold odor.

OECD Guidelines: Testing guidelines prepared by the Organization of Economic and Cooperative Development of the United Nations. They assist in preparation of protocols for studies of toxicology, environmental fate, etc.

Off-Site Facility: A hazardous waste treatment, storage or disposal area that is located away from the generating site.

Office Paper: High grade papers such as copier paper, computer printout, and stationary almost entirely made of uncoated chemical pulp, although some ground wood is used. Such waste is also generated in homes, schools, and elsewhere.

Offsets: A concept whereby emissions from proposed new or modified stationary sources are balanced by reductions from existing sources to stabilize total emissions.

Offstream Use: Water withdrawn from surface or groundwater sources for use at another place.

Oil and Gas Waste: Gas and oil drilling muds, oil production brines, and other waste associated with exploration for, development and production of crude oil or natural gas.

Oil Desulfurization: Widely used precombustion method for reducing sulfur dioxide emissions from oil-burning power plants. The oil is treated with hydrogen, which removes some of the sulfur by forming hydrogen sulfide gas.

Oil Fingerprinting: A method that identifies sources of oil and allows spills to be traced to their source.

Oil Spill: An accidental or intentional discharge of oil which reaches bodies of water. Can be controlled by chemical dispersion, combustion, mechanical containment, and/or adsorption. Spills from tanks and pipelines can also occur away from water bodies, contaminating the soil, getting into sewer systems and threatening underground water sources.

Oligotrophic Lakes: Deep clear lakes with few nutrients, little organic matter and a high dissolved-oxygen level.

On-Scene Coordinator (OSC): The predesignated EPA, Coast Guard, or Department of Defense official who coordinates and directs Superfund removal actions or Clean Water Act oil- or hazardous-spill response actions.

On-Site Facility: A hazardous waste treatment, storage or disposal area that is located on the generating site.

Onboard Controls: Devices placed on vehicles to capture gasoline vapor during refueling and route it to the engines when the vehicle is starting so that it can be efficiently burned.

Onconogenicity: The capacity to induce cancer.

One-hit Model: A mathematical model based on the biological theory that a single "hit" of some minimum critical amount of a carcinogen at a cellular target such as DNA can start an irreversible series events leading to a tumor.

Opacity: The amount of light obscured by particulate pollution in the air; clear window glass has zero opacity, a brick wall is 100 percent opaque. Opacity is an indicator of changes in performance of particulate control systems.

Open Burning: Uncontrolled fires in an open dump.

Open Dump: An uncovered site used for disposal of waste without environmental controls.

Operable Unit: Term for each of a number of separate activities undertaken as part of a Superfund site cleanup. A typical operable unit would be removal of drums and tanks from the surface of a site.

Operating Conditions: Conditions specified in a RCRA permit that dictate how an incinerator must operate as it burns different waste types. A trial burn is used to identify operating conditions needed to meet specified performance standards.

Operation and Maintenance: 1. Activities conducted after a Superfund site action is completed to ensure that the action is effective. 2. Actions taken after construction to ensure that facilities constructed to treat waste water will be properly operated and maintained to achieve normative efficiency levels and prescribed effluent limitations in an optimum manner. 3. On-going asbestos management plan in a school or other public building, including regular inspections, various methods of maintaining asbestos in place, and removal when necessary.

Operator Certification: Certification of operators of community and nontransient noncommunity water systems, asbestos specialists, pesticide applicators, hazardous waste transporter, and other such specialists as required by the EPA or a state agency implementing an EPA-approved environmental regulatory program.

Optimal Corrosion Control Treatment: An erosion control treatment that minimizes the lead and copper concentrations at users' taps while also ensuring that the treatment does not cause the water system to violate any national primary drinking water regulations.

Oral Toxicity: Ability of a pesticide to cause injury when ingested.

Organic: 1. Referring to or derived from living organisms. 2. In chemistry, any compound containing carbon.

Organic Chemicals/Compounds: Naturally occuring (animal or plant-produced or synthetic) substances containing mainly carbon, hydrogen, nitrogen, and oxygen.

Organic Matter: Carbonaceous waste contained in plant or animal matter and originating from domestic or industrial sources.

Organism: Any form of animal or plant life.

Organophosphates: Pesticides that contain phosphorus; short-lived, but some can be toxic when first applied.

Organophyllic: A substance that easily combines with organic compounds.

Organotins: Chemical compounds used in anti-foulant paints to protect the hulls of boats and ships, buoys, and pilings from marine organisms such as barnacles.

Original AHERA Inspection/Original Inspection/Inspection: Examination of school buildings arranged by Local Education Agencies to identify asbestos-containing-materials, evaluate their condition, and take samples of materials suspected to contain asbestos; performed by EPA-accredited inspectors.

Original Generation Point: Where regulated medical or other material first becomes waste.

Other Ferrous Metals: Recyclable metals from strapping, furniture, and metal found in tires and consumer electronics but does not include metals found in construction materials or cars, locomotives, and ships.

Other Glass: Recyclable glass from furniture, appliances, and consumer electronics. Does not include glass from transportation products (cars trucks or shipping containers) and construction or demolition debris.

Other Nonferrous Metals: Recyclable nonferrous metals such as lead, copper, and zinc from appliances, consumer electronics, and nonpackaging aluminum products. Does not include nonferrous metals from industrial applications and construction and demolition debris.

Other Paper: For Recyclable paper from books, third-class mail, commercial printing, paper towels, plates and cups; and other nonpackaging paper such as posters, photographic papers, cards and games, milk cartons, folding boxes, bags, wrapping paper, and paperboard. Does not include wrapping paper or shipping cartons.

Other Plastics: Recyclable plastic from appliances, eating utensils, plates, containers, toys, and various kinds of equipment. Does not include heavy-duty plastics such as yielding materials.

Other Solid Waste: Recyclable nonhazardous solid wastes, other than municipal solid waste, covered under Subtitle D of RARA.

Other Wood: Recyclable wood from furniture, consumer electronics cabinets, and other nonpackaging wood products. Does not include lumber and tree stumps recovered from construction and demolition activities, and industrial process waste such as shavings and sawdust.

Outdoor Air Supply: Air brought into a building from outside.

Outfall: The place where effluent is discharged into receiving waters.

Overburden: Rock and soil cleared away before mining.

Overdraft: The pumping of water from a groundwater basin or aquifer in excess of the supply flowing into the basin; results in a depletion or "mining" of the groundwater in the basin.

Overfire Air: Air forced into the top of an incinerator or boiler to fan the flames.

Overflow Rate: One of the guidelines for design of the settling tanks and clarifers in a treatment plant; used by plant operators to determine if tanks and clarifiers are over or under-used.

Overland Flow: A land application technique that cleanses waste water by allowing it to flow over a sloped surface. As the water flows over the surface, contaminants are absorbed and the water is collected at the bottom of the slope for reuse.

Oversized Regulated Medical Waste: Medical waste that is too large for plastic bags or standard containers.

Overturn: One complete cycle of top to bottom mixing of previously stratified water masses. This phenomenon may occur in spring or fall, or after storms, and results in uniformity of chemical and physical properties of water at all depths.

Oxidant: A collective term for some of the primary constituents of photochemical smog.

Oxidation Pond: A man-made (anthropogenic) body of water in which waste is consumed by bacteria, used most frequently with other waste-treatment processes; a sewage lagoon.

Oxidation: The chemical addition of oxygen to break down pollutants or organizac waste; e.g., destruction of chemicals such as cyanides, phenols, and organic sulfur compounds in sewage by bacterial and chemical means.

Oxidation-Reduction Potential: The electric potential required to transfer electrons from one compound or element (the oxidant) to another compound (the reductant); used as a qualitative measure of the state of oxidation in water treatment systems.

Oxygenated Fuels: Gasoline which has been blended with alcohols or ethers that contain oxygen in order to reduce carbon monoxide and other emissions.

Oxygenated Solvent: An organic solvent containing oxygen as part of the molecular structure. Alcohols and ketones are oxygenated compounds often used as paint solvents.

Ozonation/Ozonator: Application of ozone to water for disinfection or for taste and odor control. The ozonator is the device that does this.

Ozone (O_3): Found in two layers of the atmosphere, the stratosphere and the troposphere. In the stratosphere (the atmospheric layer 7 to 10 miles or more above the earth's surface) ozone is a natural form of oxygen that provides a protective layer shielding the earth from ultraviolet radiation.In the troposphere (the layer extending up 7 to 10 miles from the earth's surface), ozone is a chemical oxidant and major component of photochemical smog. It can seriously impair the respiratory system and is one of the most wide- spread of all the criteria pollutants for which the Clean Air Act required EPA to set standards. Ozone in the troposphere is produced through complex chemical reactions of nitrogen oxides, which are among the primary pollutants emitted by combustion sources; hydrocarbons, released into the atmosphere through the combustion, handling and processing of petroleum products; and sunlight.

Ozone Depletion: Destruction of the stratospheric ozone layer which shields the earth from ultraviolet radiation harmful to life. This destruction of ozone is caused by the breakdown of certain chlorine and/or bromine containing compounds (chlorofluorocarbons or halons), which break down when they reach the stratosphere and then catalytically destroy ozone molecules.

Ozone Hole: A thinning break in the stratospheric ozone layer. Designation of amount of such depletion as an "ozone hole" is made when the detected amount of depletion exceeds fifty percent. Seasonal ozone holes have been observed over both the Antarctic and Arctic regions, part of Canada, and the extreme northeastern United States.

Ozone Layer: The protective layer in the atmosphere, about 15 miles above the ground, that absorbs some of the sun's ultraviolet rays, thereby reducing the amount of potentially harmful radiation that reaches the earth's surface.

Packaging: The assembly of one or more containers and any other components necessary to ensure minimum compliance with a program's storage and shipment packaging requirements. Also, the containers, etc. involved.

Packed Bed Scrubber: An air pollution control device in which emissions pass through alkaline water to neutralize hydrogen chloride gas.

Packed Tower: A pollution control device that forces dirty air through a tower packed with crushed rock or wood chips while liquid is sprayed over the packing material. The pollutants in the air stream either dissolve or chemically react with the liquid.

Packer: An inflatable gland, or balloon, used to create a temporary seal in a borehole, probe hole, well, or drive casing. It is made of rubber or non-reactive materials.

Palatable Water: Water, at a desirable temperature, that is free from objectionable tastes, odors, colors, and turbidity.

Pandemic: A widespread epidemic throughout an area, nation or the world.

Paper: In the recycling business, refers to products and materials, including newspapers, magazines, office papers, corrugated containers, bags and some paperboard packaging that can be recycled into new paper products.

Paper Processor/Plastics Processor: Intermediate facility where recovered paper or plastic products and materials are sorted, decontaminated, and prepared for final recycling.

Parameter: A variable, measurable property whose value is a determinant of the characteristics of a system; e.g. temperature, pressure, and density are parameters of the atmosphere.

Paraquat: A standard herbicide used to kill various types of crops, including marijuana. Causes lung damage if smoke from the crop is inhaled..

Parshall Flume: Device used to measure the flow of water in an open channel.

Participation Rate: Portion of population participating in a recycling program.

Particle Count: Results of a microscopic examination of treated water with a special "particle counter" that classifies suspended particles by number and size.

Particulate Loading: The mass of particulates per unit volume of air or water.

Particulates: 1. Fine liquid or solid particles such as dust, smoke, mist, fumes, or smog, found in air or emissions. 2. Very small solids suspended in water; they can vary in size, shape, density and electrical charge and can be gathered together by coagulation and flocculation.

Partition Coefficient: Measure of the sorption phenomenon, whereby a pesticide is divided between the soil and water phase; also referred to as adsorption partition coefficient.

Passive Smoking/Secondhand Smoke: Inhalation of others' tobacco smoke.

Passive Treatment Walls: Technology in which a chemical reaction takes place when contaminated ground water comes in contact with a barrier such as limestone or a wall containing iron filings.

Pathogens: Microorganisms (e.g., bacteria, viruses, or parasites) that can cause disease in humans, animals and plants.

Pathway: The physical course a chemical or pollutant takes from its source to the exposed organism.

Pay-As-You-Throw/Unit-Based Pricing: Systems under which residents pay for municipal waste management and disposal services by weight or volume collected, not a fixed fee.

Peak Electricity Demand: The maximum electricity used to meet the cooling load of a building or buildings in a given area.

Peak Levels: Levels of airborne pollutant contaminants much higher than average or occurring for short periods of time in response to sudden releases.

Percent Saturatiuon: The amount of a substance that is dissolved in a solution compared to the amount that could be dissolved in it.

Permissible Exposure Limit: Also referred to as PEL, federal limits for workplace exposure to contaminants as established by OSHA.

Permit: An authorization, license, or equivalent control document issued by EPA or an approved state agency to implement the requirements of an environmental regulation; e.g. a permit to operate a wastewater treatment plant or to operate a facility that may generate harmful emissions.

Persistence: Refers to the length of time a compound stays in the environment, once introduced. A compound may persist for less than a second or indefinitely.

Persistent Pesticides: Pesticides that do not break down chemically or break down very slowly and remain in the environment after a growing season.

Personal Air Samples: Air samples taken with a pump that is directly attached to the worker with the collecting filter and cassette placed in the worker's breathing zone (required under OSHA asbestos standards and EPA worker protection rule).

Personal Measurement: A measurement collected from an individual's immediate environment.

Personal Protective Equipment: Clothing and equipment worn by pesticide mixers, loaders and applicators and re-entry workers, hazmat emergency responders, workers cleaning up Superfund sites, et. al., which is worn to reduce their exposure to potentially hazardous chemicals and other pollutants.

Pest: An insect, rodent, nematode, fungus, weed or other form of terrestrial or aquatic plant or animal life that is injurious to health or the environment.

Pest Control Operator: Person or company that applies pesticides as a business (e.g. exterminator); usually describes household services, not agricultural applications.

Pesticide: Substances or mixture there of intended for preventing, destroying, repelling, or mitigating any pest. Also, any substance or mixture intended for use as a plant regulator, defoliant, or desiccant.

Pesticide Regulation Notice: Formal notice to pesticide registrants about important changes in regulatory policy, procedures, regulations.

Pesticide Tolerance: The amount of pesticide residue allowed by law to remain in or on a harvested crop. EPA sets these levels well below the point where the compounds might be harmful to consumers.

PETE (Polyethylene Terepthalate): Thermoplastic material used in plastic soft drink and rigid containers.

Petroleum: Crude oil or any fraction thereof that is liquid under normal conditions of temperature and pressure. The term includes petroleum-based substances comprising a complex blend of hydrocarbons derived from crude oil through the process of separation, conversion, upgrading, and finishing, such as motor fuel, jet oil, lubricants, petroleum solvents, and used oil.

Petroleum Derivatives: Chemicals formed when gasoline breaks down in contact with ground water.

pH: An expression of the intensity of the basic or acid condition of a liquid; may range from 0 to 14, where 0 is the most acid and 7 is neutral. Natural waters usually have a pH between 6.5 and 8.5.

Pharmacokinetics: The study of the way that drugs move through the body after they are swallowed or injected.

Phenolphthalein Alkalinity: The alkalinity in a water sample measured by the amount of standard acid needed to lower the pH to a level of 8.3 as indicated by the change of color of the phenolphthalein from pink to clear.

Phenols: Organic compounds that are byproducts of petroleum refining, tanning, and textile, dye, and resin manufacturing. Low concentrations cause taste and odor problems in water; higher concentrations can kill aquatic life and humans.

Phosphates: Certain chemical compounds containing phosphorus.

Phosphogypsum Piles (Stacks): Principal byproduct generated in production of phosphoric acid from phosphate rock. These piles may generate radioactive radon gas.

Phosphorus: An essential chemical food element that can contribute to the eutrophication of lakes and other water bodies. Increased phosphorus levels result from discharge of phosphorus-containing materials into surface waters.

Phosphorus Plants: Facilities using electric furnaces to produce elemental phosphorous for commercial use, such as high grade phosphoric acid, phosphate-based detergent, and organic chemicals use.

Photochemical Oxidants: Air pollutants formed by the action of sunlight on oxides of nitrogen and hydrocarbons.

Photochemical Smog: Air pollution caused by chemical reactions of various pollutants emitted from different sources.

Photosynthesis: The manufacture by plants of carbohydrates and oxygen from carbon dioxide mediated by chlorophyll in the presence of sunlight.

Physical and Chemical Treatment: Processes generally used in large-scale wastewater treatment facilities. Physical processes may include air-stripping or filtration. Chemical treatment includes coagulation, chlorination, or ozonation. The term can also refer to treatment of toxic materials in surface and ground waters, oil spills, and some methods of dealing with hazardous materials on or in the ground.

Phytoplankton: That portion of the plankton community comprised of tiny plants; e.g algae, diatoms.

Phytoremediation: Low-cost remediation option for sites with widely dispersed contamination at low concentrations.

Phytotoxic: Harmful to plants.

Phytotreatment: The cultivation of specialized plants that absorb specific contaminant from the soil through their roots or foliage. This reduces the concentration of contaminant in the soil, but incorporates them into biomasses that may be released back into th environment when the plant dies or is harvested.

Picocuries Per Liter pCi/L): A unit of measure for levels of radon gas; becquerels per cubi meter is metric equivalent.

Piezometer: A nonpumping well, generally of small diameter, for measuring the elevation of a water table.

Pilot Tests: Testing a cleanup technology under actual site conditions to identify potentia problems prior to full-scale implementation.

Plankton: Tiny plants and animals that live in water.

Plasma-Arc Reactor: An incinerator that operates at extremely high temperatures; trea highly toxic wastes that do not burn easily.

Plasmid: A circular piece of DNA that exists apart from the chromosome and replicat independently of it. Bacterial plasmids carry information that renders the bacteria resistar to antibiotics. Plasmids are often used in genetic engineering to carry desired genes in organisms.

Plastics: Non-metallic chemoreactive compounds molded into rigid or pliable constructi materials, fabrics, etc.

Plate Tower Scrubber: An air pollution control device that neutralizes hydrogen chlori gas by bubbling alkaline water through holes in a series of metal plates.

Plug Flow: Type of flow the occurs in tanks, basins, or reactors when a slug of water mov through without ever dispersing or mixing with the rest of the water flowing through.

Plugging: Act or process of stopping the flow of water, oil, or gas into or out of a formation through a borehole or well penetrating that formation.

Plume: 1. A visible or measurable discharge of a contaminant from a given point of origin. Can be visible or thermal in water, or visible in the air as, for example, a plume of smoke. 2 The area of radiation leaking from a damaged reactor. 3. Area downwind within which a release could be dangerous for those exposed to leaking fumes.

Plutonium: A radioactive metallic element chemically similar to uranium.

PM-10/PM-2.5: PM 10 is measure of particles in the atmosphere with a diameter of less than ten or equal to a nominal 10 micrometers. PM-2.5 is a measure of smaller particles in the air. PM-10 has been the pollutant particulate level standard against which EPA has been measuring Clean Air Act compliance. On the basis of newer scientific findings, the Agency is considering regulations that will make PM-2.5 the new "standard".

Pneumoconiosis: Health conditions characterized by permanent deposition of substantial amounts of particulate matter in the lungs and by the tissue reaction to its presence; can range from relatively harmless forms of sclerosis to the destructive fibrotic effect of silicosis.

Point Source: A stationary location or fixed facility from which pollutants are discharged; any single identifiable source of pollution; e.g. a pipe, ditch, ship, ore pit, factory smokestack.

Point-of-Contact Measurement of Exposure: Estimating exposure by measuring concentrations over time (while the exposure is taking place) at or near the place where it is occurring.

Point-of-Disinfectant Application: The point where disinfectant is applied and water downstream of that point is not subject to recontamination by surface water runoff.

Point-of-Use Treatment Device: Treatment device applied to a single tap to reduce contaminants in the drinking water at the one faucet.

Pollen: The fertilizing element of flowering plants; background air pollutant.

Pollutant: Generally, any substance introduced into the environment that adversely affects the usefulness of a resource or the health of humans, animals, or ecosystems..

Pollutant Pathways: Avenues for distribution of pollutants. In most buildings, for example, HVAC systems are the primary pathways although all building components can interact to affect how air movement distributes pollutants.

Pollutant Standard Index (PSI): Indicator of one or more pollutants that may be used to inform the public about the potential for adverse health effects from air pollution in major cities.

Pollution: Generally, the presence of a substance in the environment that because of its chemical composition or quantity prevents the functioning of natural processes and produces undesirable environmental and health effects.Under the Clean Water Act, for example, the term has been defined as the man-made or man-induced alteration of the physical, biological, chemical, and radiological integrity of water and other media.

Pollution Prevention: 1. Identifying areas, processes, and activities which create excessive waste products or pollutants in order to reduce or prevent them through, alteration, or eliminating a process. Such activities, consistent with the Pollution Prevention Act of 1990, are conducted across all EPA programs and can involve cooperative efforts with such agencies as the Departments of Agriculture and Energy. 2. EPA has initiated a number of voluntary programs in which industrial, or commercial or "partners" join with EPA in promoting activities that conserve energy, conserve and protect water supply, reduce emissions or find ways of utilizing them as energy resources, and reduce the waste stream. Among these are: Agstar, to reduce methane emissions through manure management. Climate Wise, to lower industrial greenhouse-gas emissions and energy costs. Coalbed Methane Outreach, to boost methane recovery at coal mines. Design for the Environment, to foster including environmental considerations in product design and processes. Energy Star programs, to promote energy efficiency in commercial and residential buildings, office equipment, transformers, computers, office equipment, and home appliances. Environmental Accounting, to help businesses identify environmental costs and factor them into management decision making. Green Chemistry, to promote and recognize cost-effective breakthroughs in chemistry that prevent pollution. Green Lights, to spread the use of energy-efficient lighting technologies. Indoor Environments, to reduce risks from indoor-air pollution. Landfill Methane Outreach, to develop landfill gas-to-energy projects. Natural Gas Star, to reduce methane emissions from the natural gas industry. Ruminant Livestock Methane, to reduce methane emissions from ruminant livestock. Transportation Partners, to reduce carbon dioxide emissions from the transportation sector. Voluntary Aluminum Industrial Partnership, to reduce perfluorocarbon emissions from the primary aluminum industry. WAVE, to promote efficient water use in the lodging industry. Wastewi$e, to reduce business-generated solid waste through prevention, reuse, and recycling.

Polychlorinated Biphenyls: A group of toxic, persistent chemicals used in electrical transformers and capacitors for insulating purposes, and in gas pipeline systems as lubricant. The sale and new use of these chemicals, also known as PCBs, were banned by law in 1979.

Portal-of-Entry Effect: A local effect produced in the tissue or organ of first contact between a toxicant and the biological system.

Polonium: A radioactive element that occurs in pitchblende and other uranium-containing ores.

Polymer: A natural or synthetic chemical structure where two or more like molecules are joined to form a more complex molecular structure (e.g. polyethylene in plastic).

Polyvinyl Chloride (PVC): A tough, environmentally indestructible plastic that releases hydrochloric acid when burned.

Population: A group of interbreeding organisms occupying a particular space; the number of humans or other living creatures in a designated area.

Population at Risk: A population subgroup that is more likely to be exposed to a chemical, or is more sensitive to the chemical, than is the general population.

Porosity: Degree to which soil, gravel, sediment, or rock is permeated with pores or cavities through which water or air can move.

Post-Chlorination: Addition of chlorine to plant effluent for disinfectant purposes after the effluent has been treated.

Post-Closure: The time period following the shutdown of a waste management or manufacturing facility; for monitoring purposes, often considered to be 30 years.

Post-Consumer Materials/Waste: Recovered materials that are diverted from municipal solid waste for the purpose of collection, recycling, and disposition.

Post-Consumer Recycling: Use of materials generated from residential and consumer waste for new or similar purposes; e.g. converting wastepaper from offices into corrugated boxes or newsprint.

Potable Water: Water that is safe for drinking and cooking.

Potential Dose: The amount of a compound contained in material swallowed, breathed, or applied to the skin.

Potentially Responsible Party (PRP): Any individual or company—including owners, operators, transporters or generators—potentially responsible for, or contributing to a spill or other contamination at a Superfund site. Whenever possible, through administrative and legal actions, EPA requires PRPs to clean up hazardous sites they have contaminated.

Potentiation: The ability of one chemical to increase the effect of another chemical.

Potentiometric Surface: The surface to which water in an aquifer can rise by hydrostatic pressure.

Precautionary Principle: When information about potential risks is incomplete, basing decisions about the best ways to manage or reduce risks on a preference for avoiding unnecessary health risks instead of on unnecessary economic expenditures.

Pre-Consumer Materials/Waste: Materials generated in manufacturing and converting processes such as manufacturing scrap and trimmings and cuttings. Includes print overruns, overissue publications, and obsolete inventories.

Pre-Harvest Interval: The time between the last pesticide application and harvest of the treated crops.

Prechlorination: The addition of chlorine at the headworks of a treatment plant prior to other treatment processes. Done mainly for disinfection and control of tastes, odors, and aquatic growths, and to aid in coagulation and settling,

Precipitate: A substance separated from a solution or suspension by chemical or physical change.

Precipitation: Removal of hazardous solids from liquid waste to permit safe disposal; removal of particles from airborne emissions as in rain (e.g. acid precipitation).

Precipitator: Pollution control device that collects particles from an air stream.

Precursor: In photochemistry, a compound antecedent to a pollutant. For example, volatile organic compounds (VOCs) and nitric oxides of nitrogen react in sunlight to form ozone or other photochemical oxidants. As such, VOCs and oxides of nitrogen are precursors.

Preliminary Assessment: The process of collecting and reviewing available information about a known or suspected waste site or release.

Prescriptive: Water rights which are acquired by diverting water and putting it to use in accordance with specified procedures; e.g. filing a request with a state agency to use unused water in a stream, river, or lake.

Pressed Wood Products: Materials used in building and furniture construction that are made from wood veneers, particles, or fibers bonded together with an adhesive under heat and pressure.

Pressure Sewers: A system of pipes in which water, wastewater, or other liquid is pumped to a higher elevation.

Pressure, Static: In flowing air, the total pressure minus velocity pressure, pushing equally in all directions.

Pressure, Total: In flowing air, the sum of the static and velocity pressures.

Pressure, Velocity: In flowing air, the pressure due to velocity and density of air.

Pretreatment: Processes used to reduce, eliminate, or alter the nature of wastewater pollutants from non-domestic sources before they are discharged into publicly owned treatment works (POTWs).

Prevalent Level Samples: Air samples taken under normal conditions (also known as ambient background samples).

Prevalent Levels: Levels of airborne contaminant occurring under normal conditions.

Prevention of Significant Deterioration (PSD): EPA program in which state and/or federal permits are required in order to restrict emissions from new or modified sources in places where air quality already meets or exceeds primary and secondary ambient air quality standards.

Primacy: Having the primary responsibility for administering and enforcing regulations.

Primary Drinking Water Regulation: Applies to public water systems and specifies a contaminant level, which, in the judgment of the EPA Administrator, will not adversely affect human health.

Primary Effect: An effect where the stressor acts directly on the ecological component of interest, not on other parts of the ecosystem.

Primary Standards: National ambient air quality standards designed to protect human health with an adequate margin for safety.

Primary Treatment: First stage of wastewater treatment in which solids are removed by screening and settling.

Primary Waste Treatment: First steps in wastewater treatment; screens and sedimentation tanks are used to remove most materials that float or will settle. Primary treatment removes about 30 percent of carbonaceous biochemical oxygen demand from domestic sewage.

Principal Organic Hazardous Constituents (POHCs): Hazardous compounds monitored during an incinerator's trial burn, selected for high concentration in the waste feed and difficulty of combustion.

Prions: Microscopic particles made of protein that can cause disease.

Prior Appropriation: A doctrine of water law that allocates the rights to use water on a first-come, first-served basis.

Probability of Detection : The likelihood, expressed as a percentage, that a test method will correctly identify a leaking tank.

Process Variable: A physical or chemical quantity which is usually measured and controlled in the operation of a water treatment plant or industrial plant.

Process Verification: Verifying that process raw materials, water usage, waste treatment processes, production rate and other facts relative to quantity and quality of pollutants contained in discharges are substantially described in the permit application and the issued permit.

Process Wastewater: Any water that comes into contact with any raw material, product, byproduct, or waste.

Process Weight: Total weight of all materials, including fuel, used in a manufacturing process; used to calculate the allowable particulate emission rate.

Producers: Plants that perform photosynthesis and provide food to consumers.

Product Level: The level of a product in a storage tank.

Product Water: Water that has passed through a water treatment plant and is ready to be delivered to consumers.

Products of Incomplete Combustion (PICs): Organic compounds formed by combustion. Usually generated in small amounts and sometimes toxic, PICs are heat-altered versions of the original material fed into the incinerator (e.g. charcoal is a P.I.C. from burning wood).

Public Health Context: The incidence, prevalence, and severity of diseases in communities or populations and the factors that account for them, including infections, exposure to pollutants, and other exposures or activities.

Public Hearing: A formal meeting wherein EPA officials hear the public's views and concerns about an EPA action or proposal. EPA is required to consider such comments when evaluating its actions. Public hearings must be held upon request during the public comment period.

Public Notice: 1. Notification by EPA informing the public of Agency actions such as the issuance of a draft permit or scheduling of a hearing. EPA is required to ensure proper public notice, including publication in newspapers and broadcast over radio and television stations. 2. In the safe drinking water program, water suppliers are required to publish and broadcast notices when pollution problems are discovered.

Public Water System: A system that provides piped water for human consumption to at least 15 service connections or regularly serves 25 individuals.

Publicly Owned Treatment Works (POTWs): A waste-treatment works owned by a state, unit of local government, or Indian tribe, usually designed to treat domestic wastewaters.

Pumping Station: Mechanical device installed in sewer or water system or other liquid-carrying pipelines to move the liquids to a higher level.

Pumping Test: A test conducted to determine aquifer or well characteristics.

Purging: Removing stagnant air or water from sampling zone or equipment prior to sample collection.

Putrefaction: Biological decomposition of organic matter; associated with anaerobic conditions.

Putrescible: Able to rot quickly enough to cause odors and attract flies.

Pyrolysis: Decomposition of a chemical by extreme heat.

Qualitative Use Assessment: Report summarizing the major uses of a pesticide including percentage of crop treated, and amount of pesticide used on a site.

Quality Assurance/Quality Control: A system of procedures, checks, audits, and corrective actions to ensure that all EPA research design and performance, environmental monitoring and sampling, and other technical and reporting activities are of the highest achievable quality.

Quench Tank: A water-filled tank used to cool incinerator residues or hot materials during industrial processes.

Radiation: Transmission of energy though space or any medium. Also known as radiant energy.

Radiation Standards: Regulations that set maximum exposure limits for protection of the public from radioactive materials.

Radioactive Decay: Spontaneous change in an atom by emission of of charged particles and/or gamma rays; also known as radioactive disintegration and radioactivity.

Radioactive Substances: Substances that emit ionizing radiation.

Radioactive Waste: Any waste that emits energy as rays, waves, streams or energetic particles. Radioactive materials are often mixed with hazardous waste, from nuclear reactors, research institutions, or hospitals.

Radioisotopes: Chemical variants of radioactive elements with potentially oncogenic, teratogenic, and mutagenic effects on the human body.

Radionuclide: Radioactive particle, man-made (anthropogenic) or natural, with a distinct atomic weight number. Can have a long life as soil or water pollutant.

Radius of Vulnerability Zone: The maximum distance from the point of release of a hazardous substance in which the airborne concentration could reach the level of concern under specified weather conditions.

Radius of Influence: 1. The radial distance from the center of a wellbore to the point where there is no lowering of the water table or potentiometric surface (the edge of the cone of depression); 2. the radial distance from an extraction well that has adequate air flow for effective removal of contaminants when a vacuum is applied to the extraction well.

Radon: A colorless naturally occurring, radioactive, inert gas formed by radioactive decay of radium atoms in soil or rocks.

Radon Daughters/Radon Progeny: Short-lived radioactive decay products of radon that decay into longer-lived lead isotopes that can attach themselves to airborne dust and other particles and, if inhaled, damage the linings of the lungs.

Radon Decay Products: A term used to refer collectively to the immediate products of the radon decay chain. These include Po-218, Pb-214, Bi-214, and Po-214, which have an average combined half-life of about 30 minutes.

Rainbow Report: Comprehensive document giving the status of all pesticides now or ever in registration or special reviews. Known as the "rainbow report" because chapters are printed on different colors of paper.

Rasp: A machine that grinds waste into a manageable material and helps prevent odor.

Raw Agricultural Commodity: An unprocessed human food or animal feed crop (e.g., raw carrots, apples, corn, or eggs.)

Raw Sewage: Untreated wastewater and its contents.

Raw Water: Intake water prior to any treatment or use.

Re-entry: (In indoor air program) Refers to air exhausted from a building that is immediately brought back into the system through the air intake and other openings.

Reactivity: Refers to those hazardous wastes that are normally unstable and readily undergo violent chemical change but do not explode.

Reaeration: Introduction of air into the lower layers of a reservoir. As the air bubbles form and rise through the water, the oxygen dissolves into the water and replenishes the dissolved oxygen. The rising bubbles also cause the lower waters to rise to the surface where they take on oxygen from the atmosphere.

Real-Time Monitoring: Monitoring and measuring environmental developments with technology and communications systems that provide time-relevant information to the public in an easily understood format people can use in day-to-day decision-making about their health and the environment.

Reasonable Further Progress: Annual incremental reductions in air pollutant emissions as reflected in a State Implementation Plan that EPA deems sufficient to provide for the attainment of the applicable national ambient air quality standards by the statutory deadline.

Reasonable Maximum Exposure: The maximum exposure reasonably expected to occur in a population.

Reasonable Worst Case: An estimate of the individual dose, exposure, or risk level received by an individual in a defined population that is greater than the 90th percentile but less than that received by anyone in the 98th percentile in the same population.

Reasonably Available Control Measures (RACM): A broadly defined term referring to technological and other measures for pollution control.

Reasonably Available Control Technology (RACT): Control technology that is reasonably available, and both technologically and economically feasible. Usually applied to existing sources in nonattainment areas; in most cases is less stringent than new source performance standards.

Recarbonization: Process in which carbon dioxide is bubbled into water being treated to lower the pH.

Receiving Waters: A river, lake, ocean, stream or other watercourse into which wastewater or treated effluent is discharged.

Receptor: Ecological entity exposed to a stressor.

Recharge: The process by which water is added to a zone of saturation, usually by percolation from the soil surface; e.g., the recharge of an aquifer.

Recharge Area: A land area in which water reaches the zone of saturation from surface infiltration, e.g., where rainwater soaks through the earth to reach an aquifer.

Recharge Rate: The quantity of water per unit of time that replenishes or refills an aquifer.

Reclamation: (In recycling) Restoration of materials found in the waste stream to a beneficial use which may be for purposes other than the original use.

Recombinant Bacteria: A microorganism whose genetic makeup has been altered by deliberate introduction of new genetic elements. The offspring of these altered bacteria also contain these new genetic elements; i.e. they "breed true."

Recombinant DNA: The new DNA that is formed by combining pieces of DNA from different organisms or cells.

Recommended Maximum Contaminant Level (RMCL): The maximum level of a contaminant in drinking water at which no known or anticipated adverse effect on human health would occur, and that includes an adequate margin of safety. Recommended levels are nonenforceable health goals.

Reconstructed Source: Facility in which components are replaced to such an extent that the fixed capital cost of the new components exceeds 50 percent of the capital cost of constructing a comparable brand-new facility. New-source performance standards may be applied to sources reconstructed after the proposal of the standard if it is technologically and economically feasible to meet the standards.

Reconstruction of Dose: Estimating exposure after it has occurred by using evidence within an organism such as chemical levels in tissue or fluids.

Record of Decision (ROD): A public document that explains which cleanup alternative(s) will be used at National Priorities List sites where, under CERCLA, Trust Funds pay for the cleanup.

Recovery Rate: Percentage of usable recycled materials that have been removed from the total amount of municipal solid waste generated in a specific area or by a specific business.

Recycle/Reuse: Minimizing waste generation by recovering and reprocessing usable products that might otherwise become waste (.i.e. recycling of aluminum cans, paper, and bottles, etc.).

Recycling and Reuse Business Assistance Centers: Located in state solid-waste or economic-development agencies, these centers provide recycling businesses with customized and targeted assistance.

Recycling Economic Development Advocates: Individuals hired by state or tribal economic development offices to focus financial, marketing, and permitting resources on creating recycling businesses.

Recycling Mill: Facility where recovered materials are remanufactured into new products.

Recycling Technical Assistance Partnership National Network: A national information-sharing resource designed to help businesses and manufacturers increase their use of recovered materials.

Red Border: An EPA document undergoing review before being submitted for final management decision-making.

Red Tide: A proliferation of a marine plankton toxic and often fatal to fish, perhaps stimulated by the addition of nutrients. A tide can be red, green, or brown, depending on the coloration of the plankton.

Redemption Program: Program in which consumers are monetarily compensated for the collection of recyclable materials, generally through prepaid deposits or taxes on beverage containers. In some states or localities legislation has enacted redemption programs to help prevent roadside litter.

Reduction: The addition of hydrogen, removal of oxygen, or addition of electrons to an element or compound.

Reentry Interval: The period of time immediately following the application of a pesticide during which unprotected workers should not enter a field.

Reference Dose (RfD): The RfD is a numerical estimate of a daily oral exposure to the human population, including sensitive subgroups such as children, that is not likely to cause harmful effects during a lifetime. RfDs are generally used for health effects that are thought to have a threshold or low dose limit for producing effects.

Reformulated Gasoline: Gasoline with a different composition from conventional gasoline (e.g., lower aromatics content) that cuts air pollutants.

Refueling Emissions: Emissions released during vehicle re-fueling.

Refuse Reclamation: Conversion of solid waste into useful products; e.g., composting organic wastes to make soil conditioners or separating aluminum and other metals for recycling.

Regeneration: Manipulation of cells to cause them to develop into whole plants.

Regional Response Team (RRT): Representatives of federal, local, and state agencies who may assist in coordination of activities at the request of the On-Scene Coordinator before and during a significant pollution incident such as an oil spill, major chemical release, or Superfund response.

Registrant: Any manufacturer or formulator who obtains registration for a pesticide active ingredient or product.

Registration: Formal listing with EPA of a new pesticide before it can be sold or distributed. Under the Federal Insecticide, Fungicide, and Rodenticide Act, EPA is responsible for registration (pre-market licensing) of pesticides on the basis of data demonstrating no unreasonable adverse effects on human health or the environment when applied according to approved label directions.

Registration Standards: Published documents which include summary reviews of the data available on a pesticide's active ingredient, data gaps, and the Agency's existing regulatory position on the pesticide.

Regulated Asbestos-Containing Material (RACM): Friable asbestos material or nonfriable ACM that will be or has been subjected to sanding, grinding, cutting, or abrading or has crumbled, or been pulverized or reduced to powder in the course of demolition or renovation operations.

Regulated Medical Waste: Under the Medical Waste Tracking Act of 1988, any solid waste generated in the diagnosis, treatment, or immunization of human beings or animals, in research pertaining thereto, or in the production or testing of biologicals. Included are cultures and stocks of infectious agents; human blood and blood products; human pathological body wastes from surgery and autopsy; contaminated animal carcasses from medical research; waste from patients with communicable diseases; and all used sharp implements, such as needles and scalpels, and certain unused sharps.

Relative Ecological Sustainability: Ability of an ecosystem to maintain relative ecological integrity indefinitely.

Relative Permeability: The permeability of a rock to gas, NAIL, or water, when any two or more are present.

Relative Risk Assessment: Estimating the risks associated with different stressors or management actions.

Release: Any spilling, leaking, pumping, pouring, emitting, emptying, discharging, injecting, escaping, leaching, dumping, or disposing into the environment of a hazardous or toxic chemical or extremely hazardous substance.

Remedial Action (RA): The actual construction or implementation phase of a Superfund site cleanup that follows remedial design.

Remedial Design: A phase of remedial action that follows the remedial investigation/ feasibility study and includes development of engineering drawings and specifications for a site cleanup.

Remedial Investigation: An in-depth study designed to gather data needed to determine the nature and extent of contamination at a Superfund site; establish site cleanup criteria; identify preliminary alternatives for remedial action; and support technical and cost analyses of alternatives. The remedial investigation is usually done with the feasibility study. Together they are usually referred to as the "RI/FS".

Remedial Project Manager (RPM): The EPA or state official responsible for overseeing on-site remedial action.

Remedial Response: Long-term action that stops or substantially reduces a release or threat of a release of hazardous substances that is serious but not an immediate threat to public health.

Remediation: 1. Cleanup or other methods used to remove or contain a toxic spill or hazardous materials from a Superfund site; 2. for the Asbestos Hazard Emergency Response program, abatement methods including evaluation, repair, enclosure, encapsulation, or removal of greater than 3 linear feet or square feet of asbestos-containing materials from a building.

Remote Sensing: The collection and interpretation of information about an object without physical contact with the object; e.g., satellite imaging, aerial photography, and open path measurements.

Removal Action: Short-term immediate actions taken to address releases of hazardous substances that require expedited response.

Renewable Energy Production Incentive (REPI): Incentive established by the Energy Policy Act available to renewable energy power projects owned by a state or local government or nonprofit electric cooperative.

Repeat Compliance Period: Any subsequent compliance period after the initial one.

Reportable Quantity (RQ): Quantity of a hazardous substance that triggers reports under CERCLA. If a substance exceeds its RQ, the release must be reported to the National Response Center, the SERC, and community emergency coordinators for areas likely to be affected.

Repowering: Rebuilding and replacing major components of a power plant instead of building a new one.

Representative Sample: A portion of material or water that is as nearly identical in content and consistency as possible to that in the larger body of material or water being sampled.

Reregistration: The reevaluation and relicensing of existing pesticides originally registered prior to current scientific and regulatory standards. EPA reregisters pesticides through its Registration Standards Program.

Reserve Capacity: Extra treatment capacity built into solid waste and wastewater treatment plants and interceptor sewers to accommodate flow increases due to future population growth.

Reservoir: Any natural or artificial holding area used to store, regulate, or control water.

Residential Use: Pesticide application in and around houses, office buildings, apartment buildings, motels, and other living or working areas.

Residential Waste: Waste generated in single and multi-family homes, including newspapers, clothing, disposable tableware, food packaging, cans, bottles, food scraps, and yard trimmings other than those that are diverted to backyard composting.

Residual: Amount of a pollutant remaining in the environment after a natural or technological process has taken place; e.g., the sludge remaining after initial wastewater treatment, or particulates remaining in air after it passes through a scrubbing or other process.

Residual Risk: The extent of health risk from air pollutants remaining after application of the Maximum Achievable Control Technology (MACT).

Residual Saturation: Saturation level below which fluid drainage will not occur.

Residue: The dry solids remaining after the evaporation of a sample of water or sludge.

Resistance: For plants and animals, the ability to withstand poor environmental conditions or attacks by chemicals or disease. May be inborn or acquired.

Resource Recovery: The process of obtaining matter or energy from materials formerly discarded.

Response Action: 1. Generic term for actions taken in response to actual or potential health-threatening environmental events such as spills, sudden releases, and asbestos abatement/management problems. 2. A CERCLA-authorized action involving either a short-term removal action or a long-term removal response. This may include but is not limited to: removing hazardous materials from a site to an EPA-approved hazardous waste facility for treatment, containment or treating the waste on-site, identifying and removing the sources of ground-water contamination and halting further migration of contaminants. 3. Any of the following actions taken in school buildings in response to AHERA to reduce the risk of exposure to asbestos: removal, encapsulation, enclosure, repair, and operations and maintenance.

Responsiveness Summary: A summary of oral and/or written public comments received by EPA during a comment period on key EPA documents, and EPA's response to those comments.

Restoration: Measures taken to return a site to pre-violation conditions.

Restricted Entry Interval: The time after a pesticide application during which entry into the treated area is restricted.

Restricted Use: A pesticide may be classified (under FIFRA regulations) for restricted use if it requires special handling because of its toxicity, and, if so, it may be applied only by trained, certified applicators or those under their direct supervision.

Restriction Enzymes: Enzymes that recognize specific regions of a long DNA molecule and cut it at those points.

Retrofit: Addition of a pollution control device on an existing facility without making major changes to the generating plant. Also called backfit.

Reuse: Using a product or component of municipal solid waste in its original form more than once; e.g., refilling a glass bottle that has been returned or using a coffee can to hold nuts and bolts.

Reverse Osmosis: A treatment process used in water systems by adding pressure to force water through a semi-permeable membrane. Reverse osmosis removes most drinking water contaminants. Also used in wastewater treatment. Large-scale reverse osmosis plants are being developed.

Reversible Effect: An effect which is not permanent; especially adverse effects which diminish when exposure to a toxic chemical stops.

Ribonucleic Acid (RNA): A molecule that carries the genetic message from DNA to a cellular protein-producing mechanism.

Rill: A small channel eroded into the soil by surface runoff; can be easily smoothed out or obliterated by normal tillage.

Ringlemann Chart: A series of shaded illustrations used to measure the opacity of air pollution emissions, ranging from light grey through black; used to set and enforce emissions standards.

Riparian Habitat: Areas adjacent to rivers and streams with a differing density, diversity, and productivity of plant and animal species relative to nearby uplands.

Riparian Rights: Entitlement of a land owner to certain uses of water on or bordering the property, including the right to prevent diversion or misuse of upstream waters. Generally a matter of state law.

Risk: A measure of the probability that damage to life, health, property, and/or the environment will occur as a result of a given hazard.

Risk (Adverse) for Endangered Species: Risk to aquatic species if anticipated pesticide residue levels equal one-fifth of LD10 or one-tenth of LC50; risk to terrestrial species if anticipated pesticide residue levels equal one-fifth of LC10 or one-tenth of LC50.

Risk Assessment: Qualitative and quantitative evaluation of the risk posed to human health and/or the environment by the actual or potential presence and/or use of specific pollutants.

Risk Characterization: The last phase of the risk assessment process that estimates the potential for adverse health or ecological effects to occur from exposure to a stressor and evaluates the uncertainty involved.

Risk Communication: The exchange of information about health or environmental risks among risk assessors and managers, the general public, news media, interest groups, etc.

Risk Estimate: A description of the probability that organisms exposed to a specific dose of a chemical or other pollutant will develop an adverse response, e.g., cancer.

Risk Factor: Characteristics (e.g., race, sex, age, obesity) or variables (e.g., smoking, occupational exposure level) associated with increased probability of a toxic effect.

Risk for Non-Endangered Species: Risk to species if anticipated pesticide residue levels are equal to or greater than LC50.

Risk Management: The process of evaluating and selecting alternative regulatory and non-regulatory responses to risk. The selection process necessarily requires the consideration of legal, economic, and behavioral factors.

Risk-based Targeting: The direction of resources to those areas that have been identified as having the highest potential or actual adverse effect on human health and/or the environment.

Risk-Specific Dose: The dose associated with a specified risk level.

River Basin: The land area drained by a river and its tributaries.

Rodenticide: A chemical or agent used to destroy rats or other rodent pests, or to prevent them from damaging food, crops, etc.

Rotary Kiln Incinerator: An incinerator with a rotating combustion chamber that keeps waste moving, thereby allowing it to vaporize for easier burning.

Rough Fish: Fish not prized for sport or eating, such as gar and suckers. Most are more tolerant of changing environmental conditions than are game or food species.

Route of Exposure: The avenue by which a chemical comes into contact with an organism, e.g., inhalation, ingestion, dermal contact, injection.

Rubbish: Solid waste, excluding food waste and ashes, from homes, institutions, and workplaces.

Run-Off: That part of precipitation, snow melt, or irrigation water that runs off the land into streams or other surface-water. It can carry pollutants from the air and land into receiving waters.

Running Losses: Evaporation of motor vehicle fuel from the fuel tank while the vehicle is in use.

Sacrifical Anode: An easily corroded material deliberately installed in a pipe or intake to give it up (sacrifice it) to corrosion while the rest of the water supply facility remains relatively corrosion-free.

Safe: Condition of exposure under which there is a practical certainty that no harm will result to exposed individuals.

Safe Water: Water that does not contain harmful bacteria, toxic materials, or chemicals, and is considered safe for drinking even if it may have taste, odor, color, and certain mineral problems.

Safe Yield: The annual amount of water that can be taken from a source of supply over a period of years without depleting that source beyond its ability to be replenished naturally in "wet years."

Safener: A chemical added to a pesticide to keep it from injuring plants.

Salinity: The percentage of salt in water.

Salt Water Intrusion: The invasion of fresh surface or ground water by salt water. If it comes from the ocean it may be called sea water intrusion.

Salts: Minerals that water picks up as it passes through the air, over and under the ground, or from households and industry.

Sampling Frequency: The interval between the collection of successive samples.

Sanctions: Actions taken by the federal government for failure to provide or implement a State Implementation Plan (SIP). Such action may include withholding of highway funds and a ban on construction of new sources of potential pollution.

Sand Filters: Devices that remove some suspended solids from sewage. Air and bacteria decompose additional wastes filtering through the sand so that cleaner water drains from the bed.

Sanitary Sewers: Underground pipes that carry off only domestic or industrial waste, not storm water.

Sanitary Survey: An on-site review of the water sources, facilities, equipment, operation and maintenance of a public water system to evaluate the adequacy of those elements for producing and distributing safe drinking water.

Sanitary Water (Also known as gray water): Water discharged from sinks, showers, kitchens, or other non-industrial operations, but not from commodes.

Sanitation: Control of physical factors in the human environment that could harm development, health, or survival.

Saprolite: A soft, clay-rich, thoroughly decomposed rock formed in place by chemical weathering of igneous or metamorphic rock. Forms in humid, tropical, or subtropical climates.

Saprophytes: Organisms living on dead or decaying organic matter that help natural decomposition of organic matter in water.

Saturated Zone: The area below the water table where all open spaces are filled with water under pressure equal to or greater than that of the atmosphere.

Saturation: The condition of a liquid when it has taken into solution the maximum possible quantity of a given substance at a given temperature and pressure.

Science Advisory Board (SAB): A group of external scientists who advise EPA on science and policy.

Scrap: Materials discarded from manufacturing operations that may be suitable for reprocessing.

Scrap Metal Processor: Intermediate operating facility where recovered metal is sorted, cleaned of contaminants, and prepared for recycling.

Screening: Use of screens to remove coarse floating and suspended solids from sewage.

Screening Risk Assessment: A risk assessment performed with few data and many assumptions to identify exposures that should be evaluated more carefully for potential risk.

Scrubber: An air pollution device that uses a spray of water or reactant or a dry process to trap pollutants in emissions.

Secondary Drinking Water Regulations: Non-enforceable regulations applying to public water systems and specifying the maximum contamination levels that, in the judgment of EPA, are required to protect the public welfare. These regulations apply to any contaminants that may adversely affect the odor or appearance of such water and consequently may cause people served by the system to discontinue its use.

Secondary Effect: Action of a stressor on supporting components of the ecosystem, which in turn impact the ecological component of concern.

Secondary Materials: Materials that have been manufactured and used at least once and are to be used again.

Selective Pesticide: A chemical designed to affect only certain types of pests, leaving other plants and animals unharmed.

Semi-Confined Aquifer: An aquifer partially confined by soil layers of low permeability through which recharge and discharge can still occur.

Semivolatile Organic Compounds: Organic compounds that volatilize slowly at standard temperature (20 degrees C and 1 atm pressure).

Senescence: The aging process. Sometimes used to describe lakes or other bodies of water in advanced stages of eutrophication. Also used to describe plants and animals.

Septic System: An on-site system designed to treat and dispose of domestic sewage. A typical septic system consists of tank that receives waste from a residence or business and a system of tile lines or a pit for disposal of the liquid effluent (sludge) that remains after decomposition of the solids by bacteria in the tank and must be pumped out periodically.

Septic Tank: An underground storage tank for wastes from homes not connected to a sewer line. Waste goes directly from the home to the tank.

Service Connector: The pipe that carries tap water from a public water main to a building.

Service Line Sample: A one-liter sample of water that has been standing for at least 6 hours in a service pipeline and is collected according to federal regulations.

Service Pipe: The pipeline extending from the water main to the building served or to the consumer's system.

Set-Back: Setting a thermometer to a lower temperature when the building is unoccupied to reduce consumption of heating energy. Also refers to setting the thermometer to a higher temperature during unoccupied periods in the cooling season.

Settleable Solids: Material heavy enough to sink to the bottom of a wastewater treatment tank.

Settling Chamber: A series of screens placed in the way of flue gases to slow the stream of air, thus helping gravity to pull particles into a collection device.

Settling Tank: A holding area for wastewater, where heavier particles sink to the bottom for removal and disposal.

7Q10: Seven-day, consecutive low flow with a ten year return frequency; the lowest stream flow for seven consecutive days that would be expected to occur once in ten years.

Sewage: The waste and wastewater produced by residential and commercial sources and discharged into sewers.

Sewage Sludge: Sludge produced at a Publicly Owned Treatment Works, the disposal of which is regulated under the Clean Water Act.

Sewer: A channel or conduit that carries wastewater and storm-water runoff from the source to a treatment plant or receiving stream. "Sanitary" sewers carry household, industrial, and commercial waste. "Storm" sewers carry runoff from rain or snow. "Combined" sewers handle both.

Sewerage: The entire system of sewage collection, treatment, and disposal.

Shading Coefficient: The amount of the sun's heat transmitted through a given window compared with that of a standard 1/8- inch-thick single pane of glass under the same conditions.

Sharps: Hypodermic needles, syringes (with or without the attached needle), Pasteur pipettes, scalpel blades, blood vials, needles with attached tubing, and culture dishes used in animal or human patient care or treatment, or in medical, research or industrial laboratories. Also included are other types of broken or unbroken glassware that were in contact with infectious agents, such as used slides and cover slips, and unused hypodermic and suture needles, syringes, and scalpel blades.

Shock Load: The arrival at a water treatment plant of raw water containing unusual amounts of algae, colloidal matter. color, suspended solids, turbidity, or other pollutants.

Short-Circuiting: When some of the water in tanks or basins flows faster than the rest; may result in shorter contact, reaction, or settling times than calculated or presumed.

Sick Building Syndrome: Building whose occupants experience acute health and/or comfort effects that appear to be linked to time spent therein, but where no specific illness or cause can be identified. Complaints may be localized in a particular room or zone, or may spread throughout the building.

Signal Words: The words used on a pesticide label—Danger, Warning, Caution—to indicate level of toxicity.

Significant Deterioration: Pollution resulting from a new source in previously "clean" areas.

Significant Municipal Facilities: Those publicly owned sewage treatment plants that discharge a million gallons per day or more and are therefore considered by states to have the potential to substantially affect the quality of receiving waters.

Significant Potential Source of Contamination: A facility or activity that stores, uses, or produces compounds with potential for significant contaminating impact if released into the source water of a public water supply.

Significant Violations: Violations by point source dischargers of sufficient magnitude or duration to be a regulatory priority.

Silt: Sedimentary materials composed of fine or intermediate-sized mineral particles.

Silviculture: Management of forest land for timber.

Single-Breath Canister: Small one-liter canister designed to capture a single breath. Used in air pollutant ingestion research.

Sink: Place in the environment where a compound or material collects.

Sinking: Controlling oil spills by using an agent to trap the oil and sink it to the bottom of the body of water where the agent and the oil are biodegraded.

SIP Call: EPA action requiring a state to resubmit all or part of its State Implementation Plan to demonstrate attainment of the require national ambient air quality standards within the statutory deadline. A SIP Revision is a revision of a SIP altered at the request of EPA or on a state's initiative.

Site: An area or place within the jurisdiction of the EPA and/or a state.

Site Assessment Program: A means of evaluating hazardous waste sites through preliminary assessments and site inspections to develop a Hazard Ranking System score.

Site Inspection: The collection of information from a Superfund site to determine the extent and severity of hazards posed by the site. It follows and is more extensive than a preliminary assessment. The purpose is to gather information necessary to score the site, using the Hazard Ranking System, and to determine if it presents an immediate threat requiring prompt removal.

Site Safety Plan: A crucial element in all removal actions, it includes information on equipment being used, precautions to be taken, and steps to take in the event of an on-site emergency.

Siting: The process of choosing a location for a facility.

Skimming: Using a machine to remove oil or scum from the surface of the water.

Slow Sand Filtration: Passage of raw water through a bed of sand at low velocity, resulting in substantial removal of chemical and biological contaminants.

Smoke: Particles suspended in air after incomplete combustion.

Soft Detergents: Cleaning agents that break down in nature.

Soft Water: Any water that does not contain a significant amount of dissolved minerals such as salts of calcium or magnesium.

Soil Adsorption Field: A sub-surface area containing a trench or bed with clean stones and a system of piping through which treated sewage may seep into the surrounding soil for further treatment and disposal.

Soil and Water Conservation Practices: Control measures consisting of managerial, vegetative, and structural practices to reduce the loss of soil and water.

Soil Conditioner: An organic material like humus or compost that helps soil absorb water, build a bacterial community, and take up mineral nutrients.

Soil Erodibility: An indicator of a soil's susceptibility to raindrop impact, runoff, and other erosive processes.

Soil Gas: Gaseous elements and compounds in the small spaces between particles of the earth and soil. Such gases can be moved or driven out under pressure.

Soil Moisture: The water contained in the pore space of the unsaturated zone.

Soil Sterilant: A chemical that temporarily or permanently prevents the growth of all plants and animals,

Solder: Metallic compound used to seal joints between pipes. Until recently, most solder contained 50 percent lead. Use of solder containing more than 0.2 percent lead in pipes carrying drinking water is now prohibited.

Sole-Source Aquifer: An aquifer that supplies 50-percent or more of the drinking water of an area.

Solid Waste: Non-liquid, non-soluble materials ranging from municipal garbage to industrial wastes that contain complex and sometimes hazardous substances. Solid wastes also include sewage sludge, agricultural refuse, demolition wastes, and mining residues. Technically, solid waste also refers to liquids and gases in containers.

Solid Waste Disposal: The final placement of refuse that is not salvaged or recycled.

Solid Waste Management: Supervised handling of waste materials from their source through recovery processes to disposal.

Solidification and Stabilization: Removal of wastewater from a waste or changing it chemically to make it less permeable and susceptible to transport by water.

Solubility: The amount of mass of a compound that will dissolve in a unit volume of solution. Aqueous Solubility is the maximum concentration of a chemical that will dissolve in pure water at a reference temperature.

Soot: Carbon dust formed by incomplete combustion.

Sorption: The action of soaking up or attracting substances; process used in many pollution control systems.

Source Area: The location of liquid hydrocarbons or the zone of highest soil or groundwater concentrations, or both, of the chemical of concern.

Special Local-Needs Registration: Registration of a pesticide product by a state agency for a specific use that is not federally registered. However, the active ingredient must be federally registered for other uses. The special use is specific to that state and is often minor, thus may not warrant the additional cost of a full federal registration process. SLN registration cannot be issued for new active ingredients, food-use active ingredients without tolerances, or for a canceled registration. The products cannot be shipped across state lines.

Special Review: Formerly known as Rebuttable Presumption Against Registration (RPAR), this is the regulatory process through which existing pesticides suspected of posing unreasonable risks to human health, non-target organisms, or the environment are referred for review by EPA. Such review requires an intensive risk/benefit analysis with opportunity for public comment. If risk is found to outweigh social and economic benefits, regulatory actions can be initiated, ranging from label revisions and use-restriction to cancellation or suspended registration.

Special Waste: Items such as household hazardous waste, bulky wastes (refrigerators, pieces of furniture, etc.) tires, and used oil.

Spring: Ground water seeping out of the earth where the water table intersects the ground surface.

Spring Melt/Thaw: The process whereby warm temperatures melt winter snow and ice. Because various forms of acid deposition may have been stored in the frozen water, the melt can result in abnormally large amounts of acidity entering streams and rivers, sometimes causing fish kills.

Stabilization: Conversion of the active organic matter in sludge into inert, harmless material.

Stable Air: A motionless mass of air that holds, instead of dispersing, pollutants.

Stack: A chimney, smokestack, or vertical pipe that discharges used air.

Stack Effect: Air, as in a chimney, that moves upward because it is warmer than the ambient atmosphere.

Stack Effect: Flow of air resulting from warm air rising, creating a positive pressure area at the top of a building and negative pressure area at the bottom. This effect can overpower the mechanical system and disrupt building ventilation and air circulation.

Stage II Controls: Systems placed on service station gasoline pumps to control and capture gasoline vapors during refuelling.

Stagnation: Lack of motion in a mass of air or water that holds pollutants in place.

Stakeholder: Any organization, governmental entity, or individual that has a stake in or may be impacted by a given approach to environmental regulation, pollution prevention, energy conservation, etc.

Standard Industrial Classification Code: Also known as SIC Codes, a method of grouping industries with similar products or services and assigning codes to these groups.

Standard Sample: The part of finished drinking water that is examined for the presence of coliform bacteria.

Standards: Norms that impose limits on the amount of pollutants or emissions produced. EPA establishes minimum standards, but states are allowed to be stricter.

Start of a Response Action: The point in time when there is a guarantee or set-aside of funding by EPA, other federal agencies, states or Principal Responsible Parties in order to begin response actions at a Superfund site.

State Emergency Response Commission (SERC): Commission appointed by each state governor according to the requirements of SARA Title III. The SERCs designate emergency planning districts, appoint local emergency planning committees, and supervise and coordinate their activities.

Storm Sewer: A system of pipes (separate from sanitary sewers) that carries water runoff from buildings and land surfaces.

Stratification: Separating into layers.

Stratigraphy: Study of the formation, composition, and sequence of sediments, whether consolidated or not.

Stratosphere: The portion of the atmosphere 10-to-25 miles above the earth's surface.

Stressors: Physical, chemical, or biological entities that can induce adverse effects on ecosystems or human health.

Strip-Cropping: Growing crops in a systematic arrangement of strips or bands that serve as barriers to wind and water erosion.

Strip-Mining: A process that uses machines to scrape soil or rock away from mineral deposits just under the earth's surface.

Structural Deformation: Distortion in walls of a tank after liquid has been added or removed.

Subchronic: Of intermediate duration, usually used to describe studies or periods of exposure lasting between 5 and 90 days.

Subchronic Exposure: Multiple or continuous exposures lasting for approximately ten percent of an experimental species lifetime, usually over a three-month period.

Submerged Aquatic Vegetation: Vegetation that lives at or below the water surface; an important habitat for young fish and other aquatic organisms.

Subwatershed: Topographic perimeter of the catchment area of a stream tributary.

Sulfur Dioxide (SO2): A pungent, colorless, gasformed primarily by the combustion of fossil fuels; becomes a pollutant when present in large amounts.

Sump: A pit or tank that catches liquid runoff for drainage or disposal.

Superchlorination: Chlorination with doses that are deliberately selected to produce water free of combined residuals so large as to require dechlorination.

Surface Runoff: Precipitation, snow melt, or irrigation water in excess of what can infiltrate the soil surface and be stored in small surface depressions; a major transporter of non-point source pollutants in rivers, streams, and lakes..

Surface Water: All water naturally open to the atmosphere (rivers, lakes, reservoirs, ponds, streams, impoundments, seas, estuaries, etc.)

Surface-Water Treatment Rule: Rule that specifies maximum contaminant level goals for Giardia lamblia, viruses, and Legionella and promulgates filtration and disinfection requirements for public water systems using surface-water or ground-water sources under the direct influence of surface water. The regulations also specify water quality, treatment, and watershed protection criteria under which filtration may be avoided.

Surfacing ACM: Asbestos-containing material that is sprayed or troweled on or otherwise applied to surfaces, such as acoustical plaster on ceilings and fireproofing materials on structural members.

Surfacing Material: Material sprayed or troweled onto structural members (beams, columns, or decking) for fire protection; or on ceilings or walls for fireproofing, acoustical or decorative purposes. Includes textured plaster, and other textured wall and ceiling surfaces.

Surfactant: A detergent compound that promotes lathering.

Surrogate Data: Data from studies of test organisms or a test substance that are used to estimate the characteristics or effects on another organism or substance.

Surveillance System: A series of monitoring devices designed to check on environmental conditions.

Susceptibility Analysis: An analysis to determine whether a Public Water Supply is subject to significant pollution from known potential sources.

Suspect Material: Building material suspected of containing asbestos; e.g., surfacing material, floor tile, ceiling tile, thermal system insulation.

Suspended Loads: Specific sediment particles maintained in the water column by turbulence and carried with the flow of water.

Suspended Solids: Small particles of solid pollutants that float on the surface of, or are suspended in, sewage or other liquids. They resist removal by conventional means.

Suspension: Suspending the use of a pesticide when EPA deems it necessary to prevent an imminent hazard resulting from its continued use. An emergency suspension takes effect immediately; under an ordinary suspension a registrant can request a hearing before the suspension goes into effect. Such a hearing process might take six months.

Suspension Culture: Cells growing in a liquid nutrient medium.

Swamp: A type of wetland dominated by woody vegetation but without appreciable peat deposits. Swamps may be fresh or salt water and tidal or non-tidal..

Synergism: An interaction of two or more chemicals that results in an effect greater than the sum of their separate effects.

Synthetic Organic Chemicals (SOCs): Man-made (anthropogenic) organic chemicals. Some SOCs are volatile; others tend to stay dissolved in water instead of evaporating.

System With a Single Service Connection: A system that supplies drinking water to consumers via a single service line.

Systemic Pesticide: A chemical absorbed by an organism that interacts with the organism and makes the organism toxic to pests.

Tail Water: The runoff of irrigation water from the lower end of an irrigated field.

Tailings: Residue of raw material or waste separated out during the processing of crops or mineral ores.

Tampering: Adjusting, negating, or removing pollution control equipment on a motor vehicle.

Technical Assistance Grant (TAG): As part of the Superfund program, Technical Assistance Grants of up to $50,000 are provided to citizens' groups to obtain assistance in interpreting information related to clean-ups at Superfund sites or those proposed for the National Priorities List. Grants are used by such groups to hire technical advisors to help them understand the site-related technical information for the duration of response activities.

Technical-Grade Active Ingredient (TGA): A pesticide chemical in pure form as it is manufactured prior to being formulated into an end-use product (e.g. wettable powders, granules, emulsifiable concentrates). Registered manufactured products composed of such chemicals are known as Technical Grade Products.

Technology-Based Limitations: Industry-specific effluent limitations based on best available preventive technology applied to a discharge when it will not cause a violation of water quality standards at low stream flows. Usually applied to discharges into large rivers.

Technology-Based Standards: Industry-specific effluent limitations applicable to direct and indirect sources which are developed on a category-by-category basis using statutory factors, not including water-quality effects.

Teratogen: A substance capable of causing birth defects.

Teratogenesis: The introduction of nonhereditary birth defects in a developing fetus by exogenous factors such as physical or chemical agents acting in the womb to interfere with normal embryonic development.

Terracing: Dikes built along the contour of sloping farm land that hold runoff and sediment to reduce erosion.

Tertiary Treatment: Advanced cleaning of wastewater that goes beyond the secondary or biological stage, removing nutrients such as phosphorus, nitrogen, and most BOD and suspended solids.

Theoretical Maximum Residue Contribution: The theoretical maximum amount of a pesticide in the daily diet of an average person. It assumes that the diet is composed of all food items for which there are tolerance-level residues of the pesticide. The TMRC is expressed as milligrams of pesticide/kilograms of body weight/day.

Therapeutic Index: The ratio of the dose required to produce toxic or lethal effects to the dose required to produce nonadverse or therapeutic response.

Thermal Pollution: Discharge of heated water from industrial processes that can kill or injure aquatic organisms.

Thermal Stratification: The formation of layers of different temperatures in a lake or reservoir.

Thermal System Insulation (TSI): Asbestos-containing material applied to pipes, fittings, boilers, breeching, tanks, ducts, or other interior structural components to prevent heat loss or gain or water condensation.

Thermal Treatment: Use of elevated temperatures to treat hazardous wastes.

Thermocline: The middle layer of a thermally stratified lake or reservoir. In this layer, there is a rapid decrease in temperatures in a lake or reservoir.

Threshold: The lowest dose of a chemical at which a specified measurable effect is observed and below which it is not observed.

Threshold: The dose or exposure level below which a significant adverse effect is not expected.

Threshold Level: Time-weighted average pollutant concentration values, exposure beyond which is likely to adversely affect human health.

Threshold Limit Value (TLV): The concentration of an airborne substance to which an average person can be repeatedly exposed without adverse effects. TLVs may be expressed in three ways: (1) TLV-TWA—Time weighted average, based on an allowable exposure averaged over a normal 8-hour workday or 40-hour work- week; (2) TLV-STEL—Short-term exposure limit or maximum concentration for a brief specified period of time, depending on a specific chemical (TWA must still be met); and (3) TLV-C—Ceiling Exposure Limit or maximum exposure concentration not to be exceeded under any circumstances. (TWA must still be met.)

Threshold Planning Quantity: A quantity designated for each chemical on the list of extremely hazardous substances that triggers notification by facilities to the State Emergency Response Commission that such facilities are subject to emergency planning requirements under SARA Title III.

Thropic Levels: A functional classification of species that is based on feeding relationships (e.g. generally aquatic and terrestrial green plants comprise the first thropic level, and herbivores comprise the second.)

Tidal Marsh: Low, flat marshlands traversed by channels and tidal hollows, subject to tidal inundation; normally, the only vegetation present is salt-tolerant bushes and grasses.

Tillage: Plowing, seedbed preparation, and cultivation practices.

Time-weighted Average (TWA): In air sampling, the average air concentration of contaminants during a given period.

Tire Processor: Intermediate operating facility where recovered tires are processed in preparation for recycling.

Tires: As used in recycling, passenger car and truck tires (excludes airplane, bus, motorcycle and special service military, agricultural, off-the-road and-slow speed industrial tires). Car and truck tires are recycled into rubber products such as trash cans, storage containers, rubberized asphalt or used whole for playground and reef construction.

Tolerance Petition: A formal request to establish a new tolerance or modify an existing one.

Tolerances: Permissible residue levels for pesticides in raw agricultural produce and processed foods. Whenever a pesticide is registered for use on a food or a feed crop, a tolerance (or exemption from the tolerance requirement) must be established. EPA establishes the tolerance levels, which are enforced by the Food and Drug Administration and the Department of Agriculture.

Tonnage: The amount of waste that a landfill accepts, usually expressed in tons per month. The rate at which a landfill accepts waste is limited by the landfill's permit.

Topography: The physical features of a surface area including relative elevations and the position of natural and man-made (anthropogenic) features.

Total Dissolved Phosphorous: The total phosphorous content of all material that will pass through a filter, which is determined as orthophosphate without prior digestion or hydrolysis. Also called soluble P. or ortho P.

Total Dissolved Solids (TDS): All material that passes the standard glass river filter; now called total filtrable residue. Term is used to reflect salinity.

Total Petroleum Hydrocarbons (TPH): Measure of the concentration or mass of petroleum hydrocarbon constituents present in a given amount of soil or water. The word "total" is a misnomer—few, if any, of the procedures for quantifying hydrocarbons can measure all of them in a given sample. Volatile ones are usually lost in the process and not quantified and non-petroleum hydrocarbons sometimes appear in the analysis.

Total Recovered Petroleum Hydrocarbon: A method for measuring petroleum hydrocarbons in samples of soil or water.

Total Suspended Particles (TSP): A method of monitoring airborne particulate matter by total weight.

Total Suspended Solids (TSS): A measure of the suspended solids in wastewater, effluent, or water bodies, determined by tests for "total suspended non-filterable solids."

Toxaphene: Chemical that causes adverse health effects in domestic water supplies and is toxic to fresh water and marine aquatic life.

Toxic Chemical: Any chemical listed in EPA rules as "Toxic Chemicals Subject to Section 313 of the Emergency Planning and Community Right-to-Know Act of 1986."

Toxic Chemical Release Form: Information form required of facilities that manufacture, process, or use (in quantities above a specific amount) chemicals listed under SARA Title III.

Toxic Chemical Use Substitution: Replacing toxic chemicals with less harmful chemicals in industrial processes.

Toxic Cloud: Airborne plume of gases, vapors, fumes, or aerosols containing toxic materials.

Toxic Concentration: The concentration at which a substance produces a toxic effect.

Toxicological Profile: An examination, summary, and interpretation of a hazardous substance to determine levels of exposure and associated health effects.

Transboundary Pollutants: Air pollution that travels from one jurisdiction to another, often crossing state or international boundaries. Also applies to water pollution.

Transfer Station: Facility where solid waste is transferred from collection vehicles to larger trucks or rail cars for longer distance transport.

Transient Water System: A non-community water system that does not serve 25 of the same nonresidents per day for more than six months per year.

Transmission Lines: Pipelines that transport raw water from its source to a water treatment plant, then to the distribution grid system.

Transmissivity: The ability of an aquifer to transmit water.

Transpiration: The process by which water vapor is lost to the atmosphere from living plants. The term can also be applied to the quantity of water thus dissipated.

Transportation Control Measures (TCMs): Steps taken by a locality to reduce vehicular emission and improve air quality by reducing or changing the flow of traffic; e.g. bus and HOV lanes, carpooling and other forms of ride-shairing, public transit, bicycle lanes.

Transporter: Hauling firm that picks up properly packaged and labeled hazardous waste from generators and transports it to designated facilities for treatment, storage, or disposal. Transporters are subject to EPA and DOT hazardous waste regulations.

Trash: Material considered worthless or offensive that is thrown away. Generally defined as dry waste material, but in common usage it is a synonym for garbage, rubbish, or refuse.

Trash-to-Energy Plan: Burning trash to produce energy.

Treatability Studies: Tests of potential cleanup technologies conducted in a laboratory.

Treated Regulated Medical Waste: Medical waste treated to substantially reduce or eliminate its pathogenicity, but that has not yet been destroyed.

Treated Wastewater: Wastewater that has been subjected to one or more physical, chemical, and biological processes to reduce its potential of being health hazard.

Treatment: (1) Any method, technique, or process designed to remove solids and/or pollutants from solid waste, waste-streams, effluents, and air emissions. (2) Methods used to change the biological character or composition of any regulated medical waste so as to substantially reduce or eliminate its potential for causing disease.

Treatment Plant: A structure built to treat wastewater before discharging it into the environment.

Treatment, Storage, and Disposal Facility: Site where a hazardous substance is treated, stored, or disposed of. TSD facilities are regulated by EPA and states under RCRA.

Tremie: Device used to place concrete or grout under water.

Trial Burn: An incinerator test in which emissions are monitored for the presence of specific organic compounds, particulates, and hydrogen chloride.

Trichloroethylene (TCE): A stable, low boiling-point colorless liquid, toxic if inhaled. Used as a solvent or metal degreasing agent, and in other industrial applications.

Trickle Irrigation: Method in which water drips to the soil from perforated tubes or emitters.

Trickling Filter: A coarse treatment system in which wastewater is trickled over a bed of stones or other material covered with bacteria that break down the organic waste and produce clean water.

Trihalomethane (THM): One of a family of organic compounds named as derivative of methane. THMs are generally by-products of chlorination of drinking water that contains organic material.

Troposhpere: The layer of the atmosphere closest to the earth's surface.

Trust Fund (CERCLA): A fund set up under the Comprehensive Environmental Response, Compensation and Liability Act (CERCLA) to help pay for cleanup of hazardous waste sites and for legal action to force those responsible for the sites to clean them up.

Tube Settlcr: Device using bundles of tubes to let solids in water settle to the bottom for removal by conventional sludge collection means; sometimes used in sedimentation basins and clarifiers to improve particle removal.

Tuberculation: Development or formation of small mounds of corrosion products on the inside of iron pipe. These tubercules roughen the inside of the pipe, increasing its resistance to water flow.

Tundra: A type of treeless ecosystem dominated by lichens, mosses, grasses, and woody plants. Tundra is found at high latitudes (arctic tundra) and high altitudes (alpine tundra). Arctic tundra is underlain by permafrost and is usually water saturated.

Turbidimeter: A device that measures the cloudiness of suspended solids in a liquid; a measure of the quantity of suspended solids.

Turbidity: 1. Haziness in air caused by the presence of particles and pollutants. 2. A cloudy condition in water due to suspended silt or organic matter.

Ultra Clean Coal (UCC): Coal that is washed, ground into fine particles, then chemically treated to remove sulfur, ash, silicone, and other substances; usually briquetted and coated with a sealant made from coal.

Ultraviolet Rays: Radiation from the sun that can be useful or potentially harmful. UV rays from one part of the spectrum (UV-A) enhance plant life. UV rays from other parts of the spectrum (UV-B) can cause skin cancer or other tissue damage. The ozone layer in the atmosphere partly shields us from ultraviolet rays reaching the earth's surface.

Uncertainty Factor: One of several factors used in calculating the reference dose from experimental data. UFs are intended to account for (1) the variation in sensitivity among humans; (2) the uncertainty in extrapolating animal data to humans; (3) the uncertainty in extrapolating data obtained in a study that covers less than the full life of the exposed animal or human; and (4) the uncertainty in using LOAEL data rather than NOAEL data.

Unconfined Aquifer: An aquifer containing water that is not under pressure; the water level in a well is the same as the water table outside the well.

Underground Injection Control (UIC): The program under the Safe Drinking Water Act that regulates the use of wells to pump fluids into the ground.

Underground Injection Wells: Steel- and concrete-encased shafts into which hazardous waste is deposited by force and under pressure.

Underground Sources of Drinking Water: Aquifers currently being used as a source of drinking water or those capable of supplying a public water system. They have a total dissolved solids content of 10,000 milligrams per liter or less, and are not "exempted aquifers."

Underground Storage Tank (UST): A tank located at least partially underground and designed to hold gasoline or other petroleum products or chemicals.

Unreasonable Risk: Under the Federal Insecticide, Fungicide, and Rodenticide Act (FIFRA), "unreasonable adverse effects" means any unreasonable risk to man or the environment, taking into account the medical, economic, social, and environmental costs and benefits of any pesticide.

Unsaturated Zone: The area above the water table where soil pores are not fully saturated, although some water may be present.

Upper Detection Limit: The largest concentration that an instrument can reliably detect.

Uranium Mill Tailings Piles: Former uranium ore processing sites that contain leftover radioactive materials (wastes), including radium and unrecovered uranium.

Uranium Mill-Tailings Waste Piles: Licensed active mills with tailings piles and evaporation ponds created by acid or alkaline leaching processes.

Urban Runoff: Storm water from city streets and adjacent domestic or commercial properties that carries pollutants of various kinds into the sewer systems and receiving waters.

Urea-Formaldehyde Foam Insulation: A material once used to conserve energy by sealing crawl spaces, attics, etc.; no longer used because emissions were found to be a health hazard.

Use Cluster: A set of competing chemicals, processes, and/or technologies that can substitute for one another in performing a particular function.

Used Oil: Spent motor oil from passenger cars and trucks collected at specified locations for recycling (not included in the category of municipal solid waste).

User Fee: Fee collected from only those persons who use a particular service, as compared to one collected from the public in general.

Vadose Zone: The zone between land surface and the water table within which the moisture content is less than saturation (except in the capillary fringe) and pressure is less than atmospheric. Soil pore space also typically contains air or other gases. The capillary fringe is included in the vadose zone.

Valued Environmental Attributes/Components: Those aspects(components/processes/functions) of ecosystems, human health, and environmental welfare considered to be important and potentially at risk from human activity or natural hazards. Similar to the term "valued environmental components" used in environmental impact assessment.

Vapor: The gas given off by substances that are solids or liquids at ordinary atmospheric pressure and temperatures.

Vapor Capture System: Any combination of hoods and ventilation system that captures or contains organic vapors so they may be directed to an abatement or recovery device.

Vapor Dispersion: The movement of vapor clouds in air due to wind, thermal action, gravity spreading, and mixing.

Vapor Plumes: Flue gases visible because they contain water droplets.

Vapor Pressure: A measure of a substance's propensity to evaporate, vapor pressure is the force per unit area exerted by vapor in an equilibrium state with surroundings at a given pressure. It increases exponentially with an increase in temperature. A relative measure of chemical volatility, vapor pressure is used to calculate water partition coefficients and volatilization rate constants.

Vapor Recovery System: A system by which the volatile gases from gasoline are captured instead of being released into the atmosphere.

Variance: Government permission for a delay or exception in the application of a given law, ordinance, or regulation.

Volatile Organic Compound (VOC): Any organic compound that participates in atmospheric photochemical reactions except those designated by EPA as having negligible photochemical reactivity.

Volatile Solids: Those solids in water or other liquids that are lost on ignition of the dry solids at 550° centigrade.

Waste Exchange: Arrangement in which companies exchange their wastes for the benefit of both parties.

Waste Minimization: Measures or techniques that reduce the amount of wastes generated during industrial production processes; term is also applied to recycling and other efforts to reduce the amount of waste going into the waste stream.

Waste Piles: Non-containerized, lined or unlined accumulations of solid, nonflowing waste.

Waste-Heat Recovery: Recovering heat discharged as a byproduct of one process to provide heat needed by a second process.

Water Quality Criteria: Levels of water quality expected to render a body of water suitable for its designated use. Criteria are based on specific levels of pollutants that would make the water harmful if used for drinking, swimming, farming, fish production, or industrial processes.

Water Quality Standards: State-adopted and EPA-approved ambient standards for water bodies. The standards prescribe the use of the water body and establish the water quality criteria that must be met to protect designated uses.

Water Quality-Based Limitations: Effluent limitations applied to dischargers when mere technology-based limitations would cause violations of water quality standards. Usually applied to discharges into small streams.

Water Quality-Based Permit: A permit with an effluent limit more stringent than one based on technology performance. Such limits may be necessary to protect the designated use of receiving waters (e.g. recreation, irrigation, industry or water supply).

Water Solubility: The maximum possible concentration of a chemical compound dissolved in water. If a substance is water soluble it can very readily disperse through the environment.

Water Storage Pond: An impound for liquid wastes designed to accomplish some degree of biochemical treatment.

Water Supply System: The collection, treatment, storage, and distribution of potable water from source to consumer.

Watershed: The land area that drains into a stream; the watershed for a major river may encompass a number of smaller watersheds that ultimately combine at a common point.

Watershed Area: A topographic area within a line drawn connecting the highest points uphill of a drinking waterintake into which overland flow drains.

Weight of Scientific Evidence: Considerations in assessing the interpretation of published information about toxicity—quality of testing methods, size and power of study design, consistency of results across studies, and biological plausibility of exposure-response relationships and statistical associations.

Well Monitoring: Measurement by on-site instruments or laboratory methods of well water quality.

Well Plug: A watertight, gastight seal installed in a bore hole or well to prevent movement of fluids.

Wetlands: An area that is saturated by surface or ground water with vegetation adapted for life under those soil conditions, as swamps, bogs, fens, marshes, and estuaries.

Whole-Effluent-Toxicity Tests: Tests to determine the toxicity levels of the total effluent from a single source as opposed to a series of tests for individual contaminants.

Wildlife Refuge: An area designated for the protection of wild animals, within which hunting and fishing are either prohibited or strictly controlled.

Zero Air: Atmospheric air purified to contain less than 0.1 ppm total hydrocarbons.

Bibliography

Aber, J.D., and J.M., Melillo (1982). "Nitrogen immobilisation in decaying hardwood leaf litter as a function of initial nitrogen and lignin content". In: *Cannadian Journal of Botany* 60:2263-2269.

Almond, Gabriel A. and Powell, G. Bingham (1976).*Comparative Politics: A Developmental Approach,* New Delhi: Oxford & IBH Publishing Co., 3rd Indian rep.

Anil Agarwal (ed.), *State of India's Environment—A Citizen's Report* (1982 and 1984-85), Centre for Science and Environment, New Delhi.

Asian Development Bank (1983). *Nepal Agriculture Sector Strategy Study,* Vol. 11. p. 34.

Auffenburg, W. (1981). *The Behavorial Ecology of the Komodo Monitor: Gainesville,* Florida, University of Florida Presses.

Bajracharya. D. (1983). Deforestation in the food/fuel context Historical and Political Perspective from Nepal. *Mountain Research and Development,* 3(3): 227-240.

Bakker, R. T. (1971). *Ecology of the brontosaurs: Nature*, v. 229, p. 172-174.

Bandopadhyaya, J., N.D. Jayal, J. Schoettli; and Chhtrapati Singh (1985). *India's Environment: Crisis and Responses.* Natraj Publishers, Dehra Dun U.P- India, 309 pp.

Barford, C.C., (2001). "Factors controlling long and short term sequestration of atmospheric CO2 in a mid-latitude forest". In: *Science* 294: 1688-1691

Barnett, Harold and Chandler Morse. *Scarcity and Growth.* Baltimore: Johns Hopkins University Press. 1963.

Bartlett, A. (1998). "Reflections on Sustainability, Population Growth, and the Environment—Revisited" revised version (January 1998) of paper first published in Population & Environment, Vol. 16, No. 1, September 1994, pp. 5-35.

Bartlett, M. S. (1960). *Stochastic Population Models in Ecology and Epidemiology,* London, Methuen, 90 p.

Basham, A.L. (1954). *The Wonder That was India* (London: Sidgwick & Jackson).

Basu, Durga Das (1994). *Human Rights in Constitutional Law* (New Delhi: Prentice Hall of India).

Belovsky, G.E. and J.B. Slade (2000). "Insect herbivory accelerates nutrient cycling and increases plant production". In: *Proceedings of the national academy of sciences* (USA). 97:14412-14417.

Bennett, A. Le Goy (1988). *International Organisations: Principles and Issues* (Englewood Cliffs, New Jersey, Prentice Hall).

Benyus, J. (1997). *Biomimicry: Innovations Inspired by Nature*, William Morrow, New York

Berry, R. J., and Jakobson, M. E. (1975). Ecological genetics of an island population of the house mouse *(Mus musculus)*: Journal of Zoology, v. 175, p. 532-540.

Bingham, S. (2007), Pesticides in rivers and groundwater. Environment Agency, UK. Retrieved on 2007-10-12.

Birch, L.C. (1948). The intrinsic rate of natural increase of an insect population: Journal of Animal Ecology, v. 16, p. 15-26.

Black, C. A. (1968). Soil-Plant Relationships [2nd ed.]: New York, Wiley, 792 p.

Blankenship, Karl (1997-10). NOx in the Air: Multiple Effects. Chesapeake Bay Journal. Retrieved on 2008-06-04.

Bob Joynt & Stephen Wu (2000). *Nitrogen oxides emissions standards for domestic gas appliances background study* Combustion Engineering Consultant.

Boer, P.J. den, and J. Reddingius (1996). Regulation and stabilisation paradigms in population ecology. Population and Community Biology Series 16. Chapman and Hall, New York, p. 397.

Bookchin, M. (2005). The Ecology of Freedom: the emergence and dissolutioni of hierarchy. Oakland, CA: AK Press.

Brown, J.L. (1970). Seeds of change. The green revolution and development in the 1970's: New York, Praeger.

Brown, M.T. and Ulgiati, S. (1999). *Emergy Evaluation of Natural Capital and Biosphere Services*, AMBIO, Vol.28, No.6, Sept. 1999.

Brundtland, G.H. (ed.), (1987), *Our Common Future: The World Commission on Environment and Development*, Oxford, Oxford University Press.

Buol, S. W.; Hole, F. D. and McCracken, R. J. (1973). *Soil Genesis and Classification*, First, Ames, IA: Iowa State University Press.

Byers A. (1987a). An assessment of landscape change in the Khumbu region of Mepal using repeat photography. *Mountain Research and Development,* 7(1): 77-81.

Byers, A. (1987b). A geoecological study of landscape change and man-accelerated soil loss: The case of the Sagarmatha (Mt. Everest) National Park, Khumbu, Nepal Unpublished Ph.D. dissertation. University of Colorado, Boulder, 299 pp.

Cailliet, G., Setzer, P., and Love, M. (1971). Everyman's Guide to Ecological Living: New York, 119 p.

Carpenter, S.A. (1981). "Decay of heterogeneous detritus: a general model". In: *Journal of theoretical biology,* 89:539-547.

Carson B. (1985). Erosion and Sedimentation processes in the Nepalese Himalaya, International Centre for Integrated Mountain Development (ICIMOD). Occasional Paper no. 1, Kathmandu, Nepal. 39pp.

Caswell, H., Koenig, H. E., Resh, J. A., and Ross, Q. E. (1972). An Introduction to Systems Science for Ecologists, *in* Patten, B., ed., Systems Analysis and Simulation in Ecology: New York, Academic Press, v. 2, p. 4-78; 592 pp.

Caswell, H., Reed, F., Stephenson, S. N., and Werner, P. A. (1973). Photosynthetic pathways and selective herbivory: a hypothesis: American Naturalist, v. 107, p. 465-480.

Caughley, G. (1966). Mortality patterns in mammals: Ecology, v. 47, p. 906-918.

Chapin F.S. III, Matson, P.A., and Mooney, H.A. (2003). *Principles of terrestrial ecosystem ecology*. Springer-Verlag, New York, N.Y.

Chapin, F.S. III, B.H., Walker, R.J., Hobbs, D.U.,Hooper, J.H., Lawton, O.E., Sala, and D., Tilman (1997). "Biotic control over the functioning of ecosystems". in: *Science* 277:500-504.

Chapman, S.K., Hart, S.C., Cobb, N.S., Whitham, T.G., and Koch, G.W. (2003). "Insect herbivory increases litter quality and decomposition: an extension of the acceleration hypothesis". in: *Ecology* 84:2867-2876.

Chrispeels, M.J. and Sadava, D. (1977). *Plants, food, and people*. W. H. Freeman and Company, San Francisco.

Clark, L. R., Geier, P. W., Hughes, R. D., and Morris, R. F. (1967). The Ecology of Insect Populations in Theory and Practice: London, Methuen, 232 p.

Clarke, G. (1954). *Elements of Ecology*, New York, Wiley, 560 p.

Clements, F. E. (1949). *Dynamics of Vegetation*, New York, Hafner, 296 p.

Colinvaux, P. A. (1973). *Introduction to Ecology*, New York, Wiley, 621 p.

Collier, B., Cox, G. W., Johnson, A. W., and Miller, P. C. (1973). *Dynamic Ecology,* Englewood Cliffs, New Jersey, Prentice-Hall, 563 p.

Colwell, R.K. and Futuyma, D.J. (1971). On the measurement of niche breadth and overlap: Ecology, v. 52, p. 567-576.

Cone M. (2000), A wind-borne threat to Sierra frogs: A study finds that pesticides used on farms in the San Joaquin Valley damage the nervous systems of amphibians in Yosemite and elsewhere. *L.A. Times* Retrieved on **September** 17, 2007.

Conz, B.W. (2004). Continuity and Contestation: Conservation Landscapes in Totonicapán , Guatemala. University of Massachussetss Masters of Science thesis.

Costanza, R. (1994). "Three General Policies to Achieve Sustainability" in A. M. Jansson, M. Hammer, C. Folke, and R. Costanza (eds). *Investing in Natural Capital: The Ecological Economics Approach to Sustainability*. Washington DC: Island Press. pp. 392-407.

Costanza, Robert, Lisa J. Graumlich, and Will Steffen (eds.) (2007). *Sustainability or Collapse? An Integrated History and Future of People on Earth*, The MIT Press. ISBN 978-0-262-03366-4.

Cowles, R. P., and Brambel, C. E. (1936). A study of the environmental conditions in a bog pond with special reference to the diurnal vertical distribution of *Gonyostomum semen*: Biological Bulletin, Marine Biological Laboratory, Woods Hole, Mass., v. 71, p. 286-298.

D. Fowler, et al. (1998). *"The atmospheric budget of oxidised nitrogen and its role in ozone formation and deposition". New Phytologist 139: 11-23.* doi:10.1046/j.1469-8137.1998.00167.x.

D.A. Crossley (1970). *Roles of Microflora and fauna in soil systems*, International Symposium on Pesticides in Soils, Feb. 25, University of Michigan

Dahi, Robert A. (1989). *Modern Political Analysis* (New Delhi: Prentice Hall of India).

Dale, M.B. (1970). Systems analysis and ecology: Ecology, v. 51, p. 2-16.

Daly H, Doyen JT, and Purcell AH III (1998). *Introduction to insect biology and diversity*, 2nd edition. Oxford University Press. New York, New York. Chapter 14, Pages 279-300.

Daly H. (1996). *Beyond Growth: The Economics of Sustainable Development*. Boston: Beacon Press. ISBN 0-8070-4709-0

Daly H. and J. Cobb (1989). *For the Common Good: Redirecting the Economy Toward Community, the Environment, and a Sustainable Future*. Boston: Beacon Press.

Dean, J.W. (2006). *Conservatives Without Conscience*. New York: Viking Penguin.

Defries, R.S., J.A. Foley, and G.P. Asner (2004). "Land-use choices: balancing human needs and ecosystem function". in: *Frontiers in ecology and environmental science*. 2:249-257.

E.A. Davidson & W. Kingerlee (1997). *"A global inventory of nitric oxide emissions from soils"*. Nutrient Cycling in Agroecosystems 48: 37-50.

Eckholm, E. (1975). The deterioration of mountain environments. *Science*, 189: 764-770.

Edwards, Michael (1969). *Everyday Life in Early India* (London: Batsford).

Ehrenfeld, J.G. and Toth, L.A. (1997). "Restoration ecology and the ecosystem perspective". in: *Restoration Ecology* 5:307-317.

Ehrlich, Paul; Walker, Brian "Rivets and Redundancy".BioScience.vol.48.no.5. May 1998. pp. 387. American Institute of Biological Sciences.

Ekins, P. (Ed.). (1986). *The Living Economy*. London: Routledge and Kegan Paul.

Elton, C. S. (1927). Animal ecology: London, Sidgwick and Jackson, 209 p.

F. Powledge (1982). *Water, The Nature, Uses and Future of our most precious and Abused Source,* Farra, Straus, Giroux.

Fillozat, J., *The Classical Doctrine of Indian Medicine* (Allahabad: Munshi Ram Manohar Lal, 1964).

Fisher, James F., 1986. *Trans-Himalayan Traders: Economy, Society and Culture in Nortlnvest Nepal.* University of California Press, Berkeley, USA. 232 pp.

Foster, D. R. (1992). "Land-use history (1730-1990) and vegetation dynamics in central New England, USA". In: *Journal of Ecology* 80: 753-772.

Fotta, Jeannette R. and Deck, Edith S., eds., *A Sociological Framework for Patient Care* (New York: John Wiley & Sons, 1979), 2nd edn.

Fox, JE, Gulledge, J, Engelhaupt, E, Burrow, ME, and McLachlan, JA (2007). *"Pesticides reduce symbiotic efficiency of nitrogen-fixing rhizobia and host plants".* Proceedings of the National Academy of Sciences of the USA 104: 10282-10287.

Freeden, Michael, *Rights* (Milton Keynes: Open University Press, 1991).

Fricke, T.E., 1936. Himalayan Households: Tamang Demography and Domestic Process. *Studies in Cultural Anthropology.* No. 11, UMI Research Press, Ann Arbor, Michigan, USA, 227 pp.

Friedson, E., *The Profession of Medicine* (New York ; Dodd-Mead, 1970).

Gannon, RW; Osmond, DL; Humenik, FJ; Gale, JA; Spooner, J Agricultural Water Quality WATER RESOUR BULL, vol. 32, no. 3, pp. 437-450, 1996

Garrison, F. H., *An Introduction to the History of Medicine* (Philadelphia, Pa.: Saunders, 1929).

Georgescu-Roegen, Nicholas. "Comments ..." in V. Kerry Smith, ed., *Scarcity and Growth Reconsidered,* Baltimore: Resources for the Future and Johns Hopkins University Press. 1979. p. 98.

Gerrard, Michael B. (2007-09-25). *Global Climate Change and U.S. Law. American Bar Association,* 38. ISBN 1590318161.

Gilliom, RJ, Barbash, JE, Crawford, GG, Hamilton, PA, Martin, JD, Nakagaki, N, Nowell, LH, Scott, JC, Stackelberg, PE, Thelin, GP, and Wolock, DM (February

15, 2007), The Quality of our nation's waters: Pesticides in the nation's streams and ground water, 1992–2001. Chapter 1, Page 4. US Geological Survey. Retrieved on September 13, 2007.

Goftman, E., *Stigma: Notes on the Management of Spoiled Identity* (Harmondsworth: Penguin, 1963).

Gokhale, B.G., *Ancient Indian History and Culture* (Bombay: Asia Publishing House, 1970).

Goldstein, M.C., Ross, J.L., and Schuler, S., 1983. From a mountain-rural to a plains-urban society: Implications of the 1981 Nepal Census. *Mountain Research and Development,* 3(1): 61-64.

Goulden, M. L., J. W. Munger, S.-M. Fan, B. C. Daube, and S. C. Wofsy, (1996). "Effects of interannual climate variability on the carbon dioxide exchange of a temperate deciduous forest". In: *Science* 271:1576-1578

Graeme Murphy (December 1 2005), Resistance Management - Pesticide Rotation. Ontario Ministry of Agriculture, Food and Rural Affairs. Retrieved on September 15, 2007.

Gramsci, A., *Selections from the Prison Notebooks* (London: Lawrence & Wishart, 1971).

Gray, Stephen E., *Community Health Today* (New York: Macmillan Publishing Co., 1978).

Green, Andrew, *An Introduction to Health Planning in Developing Countries* (Oxford: 1992).

Green, Lawrence W. and Anderson, *Community Health* (St. Louis: Times Mirror Mosby College Publishing, 1986), 5th edn.

Greenwood, N. N.; Earnshaw, A. (1997). *Chemistry of the Elements, 2nd Edition,* Oxford:Butterworth-Heinemann. p. 700

Grime, J.P. "Biodiversity and Ecosystem Function: The Debate Deepens." Science Vol. 277. no. 533029 Aug 1997 pp. 1260 - 1261. 25 May 2007

Groce, N., *Everyone here spoke Sign Language*: *Hereditary Deafness on Martha's Vineyard* (Cambridge: MA Harvard University Press, 1985).

Groom , Martha J., and Gary K. Meffe. Principles of Conservation Biology. 3. Sunderland, MA: Sinauer Associates, Inc, 2006.

Groundwater – the processes and global significance of aquifer degradation

Gupta, D.C., *The League of Nations* (New Delhi: Vikas, 1974).

H.C. Perkins, *Air Pollution,* McGraw-Hill, New York, 1974.

Hackenberg D (2007-03-14). Letter from David Hackenberg to American growers from March 14, 2007 (English). Plattform Imkerinnen — Austria. Retrieved on 2007-03-27.

Haefeker, Walter (2000-08-12). Betrayed and sold out – German bee monitoring (English). Retrieved on 2007-10-10.

Hagen, J.B. (1992). *An Entangled Bank: The origins of ecosystem ecology*. Rutgers University Press, New Brunswick, N.J.

Hargroves, K. and M. Smith (Eds.) 2005. *The Natural Advantage of Nations: Business Opportunities, Innovation and Governance in the 21st Century.* ISBN 1-84407-121-9, 525 pages. Earthscan/James&James. (See the books online companion at www.thenaturaladvantage.info)

Harrison, Paul, *The Third World Tomorrow: A Report from the Battlefront in War and Poverty* (London: Penguin, 1991).

Hasan, K. A., *The Cultural Frontier of Health in Village India* (Bombay: Manaktalas, 1967).

Haskins, James, *A New Kind of Joy* (New York: Doubleday & Company, 1976).

Hättenschwiler S. and P.M. Vitousek (2000). "The role of polyphenols in terrestrial ecosystem nutrient cycling". In: *Trends in Ecology and Evolution* 15: 238-243

Hawken, Paul, Lovins, Amory and Lovins, L. H. 1999. *Natural Capitalism: Creating the Next Industrial Revolution*, Earthscan, London (Downloadable from www.natcap.org)

Hawken, Paul. *The Ecology of Commerce: A Declaration of Sustainability.* New York: Harper Business. 1993.

Health and Environmental Impacts of NOx. United States Environmental Protection Agency. Retrieved on 2007-12-26.

Helfrich, LA, Weigmann, DL, Hipkins, P, and Stinson, ER (June 1996), Pesticides and aquatic animals: A guide to reducing impacts on aquatic systems. Virginia Cooperative Extension. Retrieved on 2007-10-14.

Henkin, Louis, "The Internationalisation of Human Rights" in *Human Rights: A Symposium,* Proceeding of the General Education Seminar (Columbia: 1977), vol. 6, no. 1, pp. 5-16.

Hetzel, B., *Health and Australian Society* (Melbourne: Pelican, 1974).

Hickson, Lynda, *Menial Retardation, Foundations of Educational Programming* (Boston: Allyn and Bacon, 1995).

Hogan,, CM, Patmore L, Latshaw, G, Seidman, H, et al. (1973), *Computer modeling of pesticide transport in soil for five instrumented watersheds*, U.S. Environmental Protection Agency Southeast Water laboratory, Athens, Ga. by ESL Inc., Sunnyvale, California.

Holleman, A. F.; Wiberg, E. "Inorganic Chemistry" Academic Press: San Diego, 2001.

Homer-Dixon, Thomas. "Environment, Scarcity, and Violence." Princeton University Press. (1999).

Houstan, D. C., 1979, The Adaptations of Scavengers, *in* Sinclair, A. R. E., and Norton-Griffiths, M., eds., Serengeti: Dynamics of an Ecosystem: Chicago, University of Chicago Press, p. 263-286.

Houston, C.W. 1987: Deforestation in Solu Khumbu. *Mountain Research and Development,* 7(1): 6.

How is high-level nuclear waste managed in Canada?. *The Canadian Nuclear FAQ*. Retrieved on June 28, 2006.

Hunter, M.D. (2001). "Insect population dynamics meets ecosystem ecology: effects of herbivory on soil nutrient dynamics". In: *Agricultural and Forest Entomology* 3:77-84.

IAEA Waste Management Database: Report 3 - L/ILW-LL. International Atomic Energy Agency (2000-03-28).

International Atomic Energy Agency, *The radiological accident in Goiânia*, 1988, retrieved September 2007.

International Institute for Sustainable Development (1996) *Global Tomorrow Coalition Sustainable Development Tool Kit: A Sample Policy Framework*, Chapter 4.

Irwin, Terence, *Plato: The Republic,* translated by A.D. Lindsay (London: J.M. Dent & Sons Ltd., 1992).

Ives, J.D., 1981: High Mountains Jnd Plateaus: An Excursion to the rood of the world. *Mountain Research and Development,* 1(1): 79-83.

Ives. J.D., 1985; Yulongxue Shan, Northwest Yunnan, Peoples Republic of China: A geoecological expedition. *Mountain Research and Development,* 5(4): 382-385.

J.M. Moran and others, *Introduction to Environmental Science* (2nd ed.), Freeman, New York, 1976.

Jablonski, D., 1986, Larval ecology and macroevolution in marine invertebrates: Bulletin of Marine Science, v. 39, p. 565-587.

Jaco, E.G., ed., *Patients, Physicians and Illness: A Sourcebook in Behavioral Science and Health* (New York: The Free Press, 1979).

James A. Danoff-Burg, Columbia University The Terrestrial Influence: Geology and Soils

Jarzombek, Mark. "Sustainability—Architecture: between Fuzzy Systems and Wicked Problems," *Blueprints* 21/1 (Winter 2003), pp. 6-9; and LOG 8 (Summer 206) 7-13.

Johnsori, K. Olson, E.A. and Manandhar, S., 1982. Environmental knowledge and response to natural hazards in mountainous Nepal. *Mountain Research and Development,* 2(2): 175-188.

Johnston, AE (1986). *"Soil organic-matter, effects on soils and crops"*. Soil Use Management 2: 97-105.

Jolley, J., *Arthashastra of Kautilya* (Delhi: Molilal Banarasi Das, 1923), vol. 1.

Joshi, S.C. (ed), 1986. *Nepal Himalaya Geo-ecological Perspectives.* Himalayan Research Group, Nainital, U.P., India, 506 pp.

Joyce, James Avery, *The New Politics of Human Rights* (London: Macmillan, 1978).

Kangle, R.P., *The Kautilya Arthasastra,* Part III (Bombay: 1965), vol. 3.

Kant, Immanuel, *Perpetual Peace and Other Essays,* trans., by Ted Humphrey (Indianapolis: 1985).

Kaushik, B.N., *Viklang Shiksha Sindhu,* Hindi edn. (Jaipur: Rajasthan Hindi Granth Akademi, 1977).

Keeling, C.D. and T.P. Whorf. (2005). "Atmospheric CO2 records from sites in the SIO air sampling network". In: *Trends: A Compendium of Data on Global Change*. Carbon Dioxide Information Analysis Center, Oak Ridge National Laboratory, U.S. Department of Energy, Oak Ridge, Tenn., U.S.A.

Keith, A.B., *The Sanskrit Drama* (Oxford: Clarendon Press, 1924).

Kellogg RL, Nehring R, Grube A, Goss DW, and Plotkin S (February 2000), Environmental indicators of pesticide leaching and runoff from farm fields. United States Department of Agriculture Natural Resources Conservation Service. Retrieved on 2007-10-03.

Kerbs JR, Wilson JD, Bradbury RB, and Siriwardena GM (August 12, 1999), The second silent spring. Commentary in *Nature*, Volume 400, Pages 611-612.

Kriegman, O. 2006. *Dawn of the Cosmopolitan: The Hope of a Global Citizens Movement*. Boston: Tellus Institute.

Krivtsov, A.I., 2006, Geoenvironmental Problems of Mineral Resources Development, in *Geology and Ecosystems*, Zekster (Ru), Marker (UK), Ridgeway (UK), Rogachevskayarochmaninoff (Ru), & Vartanyan (Ru), 2006 Springer Inc.,

Kuhn, Thomas S., *The Structure of Scientific Revolution* (Chicago: 1970).

Kull, Kalevi; Kukk, Toomas; Lotman, Aleksei 2003. When culture supports biodiversity: The case of the wooded meadow. In: Roepstrorff, Andreas; Bubandt, Nils; Kull, Kalevi (eds.), *Imagining Nature: Practices of Cosmology and Identity*. Aarhus: Aarhus University Press, 76-96.

L. Hodges, *Environmental Pollution* (2nd ed.), Holt, Reinhardt and Winston, New York, 1977.

Ladd, H. S., 1959, Ecology, paleontology and stratigraphy: Science, v. 129, p. 69-78.

Lanza, Robert, *Medical Science and the Advancement of World Health* (New York: Praeger, 1985).

Lawton, John H., *What Do Species Do in Ecosystems?*], Oikos, December, 1994. vol.71,no.3.

Leather, D., 1986, The tale of the Tasaday: The Geographical Magazine, v. 58, no. 10, p. 490-491.

Lecky, William, E., *The History of European Morals* (London: Simon & Schuster, 1911).

Lee, Kenneth and Mills, Anne, *The Economics of Health in Developing Countries* (Oxford: 1983).

Lester R. Brown and others, *State of the World* (continuing series), World Watch Report (1984 to 1991) W.W. Norton, New York.

Likens, G. E., F. H. Bormann, N. M. Johnson, D. W. Fisher and R. S. Pierce. (1970). "Effects of forest cutting and herbicide treatment on nutrient budgets in the Hubbard Brook watershed-ecosystem". in: *Ecological Monographs* 40:23-47.

Lindeman, R.L. 1942. The trophic-dynamic aspect of ecology. *Ecology* 23: 399-418.

Lindsey, T; Neese, S; Thomas, D Pollutin Prevention WATER QUAL. INT., pp. 32-36, 1996

Lutgens, Frederick K. and Edward J. Tarbuck (1995) *The Atmosphere*, Prentice Hall, 6th ed., pp14-17, ISBN 0-13-350612-6

M. I. Ojovan, W.E. Lee. *An Introduction to Nuclear Waste Immobilisation*, Elsevier Science Publishers B.V., Amsterdam, 315pp. (2005)

M.N. Sastri, *History of Planet Earth,* Himalaya Publishing House, Bombay, 1993.

MacDonnell, LJ Water Quality LAND WATER LAW REV., vol. 31, no. 2, pp. 329-348, 1996

MacKenzie, SH The Ecosystem ISLAND PRESS, CA (USA), 1996, 240 pp

Mahat, T.B.S., Griffin, D.M. and Shepherd, K.R., 1986-87. Human impact on some forests of the Middle Hills of Nepal 1. Forestry in the context of the traditional resources of the state. 2. Some major human impacts before 1950 on the forests of Sindhu Palchok and Kabhre Palanchok. 3. Forests in the subsistence economy of Sindhu Palchok and Kabhrre Palanchok. 4. A detailed survey of southeast Sindhu Palchok and northeast Kabhre Palanchok, *Mountain Research and Development*. 6(3): 223-232,6(4): 325-334, 7(1): 53-70,7(2): 111-134.

Marathon Resources Ltd :: Our Business :: Uranium Industry :: Nuclear Waste

Marks, N., Simms, A., Thompson, S., and Abdallah, S. (2006).The (Un)happy Planet Index. London: New Economics Foundation. Downloadable from www.neweconomics.org

Marshall, Alfred. *Principles of Economics*. New York: Macmillan. 9th ed., 1961 (originally 1920).

Marshall, L., *The King of Nyae Nyae* (Cambridge, MA: Harvard University Press, 1976).

Mattingly, S., *Rehabilitation Today* (London: Update Books, 1977).

McDonough, W. & Braungart, M. (2002). Cradle to Cradle. North Point Press

McNeill, William H., *A History of the Human Community: Prehistoric to the Present* (Englewood Cliffs, New Jersey: Prentice Hall, 1987).

Mechanic, David, *Medical Sociology* (New York: The Free Press, 1962).

Meentemeyer, V. 1978 "Macroclimate and lignin control of litter decomposition rates". in: *Ecology* 59:465-472.

Meersman T (October 25, 1999), Studies link frog deformities to pesticides. *Star Tribune* Retrieved on September 18, 2007.

Melillo, J.M., Aber, J.D., and Muratore, J.F. (1982). "Nitrogen and lignin control of hardwood leaf litter decomposition dynamics". In: *Ecology* 63:621-626.

Merriman, A., *An African World: The Besongye* (Bloomington: Indiana University Press, 1979).

Messerli, B. and Ives, J.D., 1984; Gongga Shan (7556 m) and Yulongxue Shah (5596 m). Geoecological observations in the Hengduan Mountains of Southwestern China. *Erdiuissenschafiliche Forschung,* Bd. XVIII: 55-77, Steiner, Stuttgart, West Germany.

Michael Hogan, Leda Patmore, Gary Latshaw and Harry Seidman *Computer modelng of pesticide transport in soil for five instrumented watersheds*, prepared for the U.S. Environmental Protection Agency Southeast Water laboratory, Athens, Ga. by ESL Inc., Sunnyvale, California (1973)

Miller GT (2004), *Sustaining the Earth*, 6th edition. Thompson Learning, Inc. Pacific Grove, California. Chapter 9, Pages 211-216.

Moench, M. and Bandyopadhyaya, J., 1986. People forest interactions: A neglected parameter in Himalayan forest management. *Mountain Research and Development,* 6(1): 3-16.

Morman, R. H., 1979, Distribution and ecology of lampreys in the lower peninsula of Michigan, 1957-1975. Great Lakes Fishery Commission, Technical Report No. 33, 59 pp.

Motzkin, G., D. R. Foster, A. Allen, J. Harrod, and R. D. Boone. (1996). "Controlling site to evaluate history: vegetation patterns of a New England sand plain". In: *Ecological Monographs* 66: 345-365.

Muckenfuss AE, Shepard BM, Ferrer ER, Natural mortality of diamondback moth in coastal South Carolina Clemson University, Coastal Research and Education Center.

Mukhopadhyay, Alok, ed., *State of India's Health* (New Delhi: Voluntary Health Association of India, 1992).

Myers Patrick R., *Succession between International Organisations* (London and New York: Kegan Paul International, 1993).

Myers. N., 1986. Environmental repercussions of deforestation in the Himalayas. *Journal of World Forest Resource Management,* 2: 63-72.

Myrdal, Gunnar, *Asian Drama: An Enquiry into the Poverty of Nations,* abridged by Seth S. King (London: Allen Lane The Penguin Press, 1972).

National Park Service. US Department of the Interior. (August 1, 2006), Sequoia & Kings Canyon National Park: Air quality — Airborne synthetic chemicals. Nps.gov. Retrieved on September 19, 2007.

National Research Council *(1995). Technical Bases for Yucca Mountain Standards. Washington, D.C.: National Academy Press.* cited in in The Status of Nuclear Waste Disposal. The American Physical Society (January 2006).

National Trends in Sulfur Dioxide Levels, United States Environmental Protection Agency.

Nelson, E. H. (1986). *New Values and Attitudes Throughout Europe*. Epsom, England: Taylor-Nelson.

Newswise: National Study Examines Health Risks of Coarse Particle Pollution

Nicholas, H.G., *The United Nations as a Political Institution* (London: OUP, 1975).

Nordhaus, William. Quoted in "Academy Panel Split on Greenhouse Adaptation." *Science.* September 13, 1991. p. 1206.

Northedge, F.S., *The League of Nations: Its Life and Times 1920-1946* (New York: Leicester University Press, 1988), 2nd imp.

Nussbaum, Martha and Sen, Amartya, eds., *The Quality of Life* (Oxford: 1993).

Oates, Wallace E. "Pollution Charges as a Source of Public Revenues." Working Paper No. 91-22, University of Maryland, Department of Economics. 1991.

Odum, E.P 1969. "The strategy of ecosystem development". in: *Science* 164:262-270.

Odum, H.T. (1971). *Environment, Power, and Society*. Wiley-Interscience New York, N.Y.

Olson, J.S. (1963). "Energy storage and the balance of producers and decomposers in ecological systems". In: *Ecology* 44:322-331.

Page, Talbot. *Conservation and Economic Efficiency*. Baltimore: Johns Hopkins University Press. 1977.

Papendick RI, Elliott LF, and Dahlgren RB (1986), Environmental consequences of modern production agriculture: How can alternative agriculture address these issues and concerns? *American Journal of Alternative Agriculture*, Volume 1, Issue 1, Pages 3-10. Retrieved on 2007-10-10.

Parker, Randall M. and Hansen, *Carl, Rehabilitation Counseling: Foundations, Consumers and Service Delivery* (Boston, Massachusetts: Allyn and Bacon, Inc., 1981).

Passell, P. "Cheapest Protection of Nature May Lie in Taxes, Not Laws." *New York Times,* Nov. 24,1992.

Patten, B.C. 1959. An Introduction to the Cybernetics of the Ecosystem: The Trophic-Dynamic Aspect. *Ecology* 40, no. 2.: 221-231.

Pedersen, TL (June 1997), Pesticide residues in drinking water. extoxnet.orst.edu. Retrieved on September 15, 2007.

Pesticide Action Network North America (June 4, 1999), Pesticides threaten birds and fish in California. PANUPS. Retrieved on 2007-09-17.

Peters, R. H., 1976, Tautology in Evolution and Ecology: American Naturalist, v. 110, p. 1-12.

Phillips, David R. and Verhasselt. Yola, eds., *Health and Development* (London: Routledge, 1994).

Pope, C Arden*; et al. (2002).* "Cancer, cardiopulmonary mortality, and long-term exposure to fine particulate air pollution"*. J. Amer. Med. Assoc. 287: 1132-1141.* doi:10.1001/jama.287.9.1132.

Poppe, Wayne; Hurst, Renee Water Pollution WATER QUAL INT, pp. 39-43, 1997

Quinones, M.A., N.E. Borlaug, C.R. Dowswell. (1997). "A fertilizer-based green revolution for Africa".. In: *Replenishing soil fertility in Africa*. Soil Science Society of America special publication number 51. Soil Science Society of America, Madison, WI.

R. B. Brown (September 2003). Soil Texture. *Fact Sheet SL-29*. University of Florida, Institute of Food and Agricultural Sciences. Retrieved on 2007-12-02.

R.E. Long, *The Problem of Waste Disposal,* Wilson, 1989.

Rainer Stegmann, *Treatment of Contaminated Soil: Fundamentals, Analysis, Applications*, Springer Verlag, Berlin 2001

Rajan, C. 2006. *Global Politics and Institutions*. Boston: Tellus Institute.

Raloff, J (September 5, 1998) Common pesticide clobbers amphibians. Science News, Volume 154, Number 10, Page 150. Retrieved on 2007-10-15.

Ramsay, W.J.H.. 1986. Erosion problems in the Nepal Himalaya—an overview. In Joshi, S.C. (ed). *Nepal Himalaya: Cececological Perspective.* Himalayan Research Group, Nainital, U.P, India, pp. 359-395.

Ranganathan, J & Irwin, F. (2007, May 7). Restoring Nature's Capital: An Action Agenda to Sustain Ecosystem Services

Raskin, P., Banuri, T., Gallopin, G., Gutman, P., Hammond, A., Kates, R., and Swart, R. 2002. *Great Transition: The Promise and Lure of the Times Ahead.* Boston: Tellus Institute.

Rasler, Karen A. and W. R. Thompson."Contested Territory, Strategic Rivalries, and Conflict Escalation." International Studies Quarterly. 50. 1.(2006): 145-168.

Rathbone, Frank S. Jr. and Rathbone, Estelle T., *Health and the Nature of Man* (New York: Mc Graw Hill Book, 1971).

Raven, J. (1995). The New Wealth of Nations: A New Enquiry into the Nature and Origins of the Wealth of Nations and the Societal Learning Arrangements Needed for a Sustainable Society. Unionville, New York: Royal Fireworks Press; Sudbury, Suffolk: Bloomfield Books.

Reich, P.B., Grigal, D.F., Aber, J.D., Gower, S.T. (1997). "Nitrogen mineralisation and productivity in 50 hardwood and conifer stands on diverse soils". In: *Ecology* 78:335-347.

Repetto, R., R. C. Dower, R. Jenkins, and J. Geoghegan. *Green Fees: How A Tax Shift Can Work for the Government and Economy.* Washington, DC: World Resources Institute. 1992.

Reynolds, JD (1997), International pesticide trade: Is there any hope for the effective regulation of controlled substances? *Florida State University Journal of Land Use & Environmental Law*, Volume 131. Retrieved on 2007-10-16.

Richardson, B.J. and Wood, S. (eds) (2006). *Environmental Law for Sustainability: A Reader*, Hart Publishing, Oxford.

Richman, Joel, *Medicine and Health* (London and New York: Longman, 1987).

Richman, M IND. Water Pollution WASTEWATER, vol. 5, no. 2, pp. 24-29, 1997

Robert, Karl-Henrik et al. (2002). Strategic Sustainable Development: Selection, Design and Synergies of applied tools. Journal of Cleaner Production 10(3):197-214

Robèrt, Karl-Henrik. (2002). *The Natural Step Story: Seeding a Quiet Revolution.* Gabriola Island, BC: New Society Publishers.

Robertson, I., *Sociology* (New York: Worth Publishers, 1981).

Rockets, Rusty (June 8, 2007), Down On The Farm? Yields, Nutrients And Soil Quality. Scienceagogo.com. Retrieved on September 15, 2007.

Russo, X. and Willis, George, *Human Services in America* (Englewood-cliffs, N.J.: Prentice Hall, Inc., 1986).

S.K. Gupta, C.T. Kincaid, P.R. Mayer, C.A. Newbill and C.R. Cole, "A multidimensional finite element code for the analysis of coupled fluid, energy and solute transport", Battelle Pacific Northwest Laboratory PNL-2939, EPA contract 68-03-3116 (1982)

Sabine, George, *A History of Political Theory* (New York: Henry Holt, 1937),

Science Daily (February 3, 2006), Pesticide combinations imperil frogs, probably contribute to amphibian decline. Sciencedaily.com. Retrieved on 2007-10-16.

Science Daily (May 4, 1998), Pesticides linked to widespread cases of deformed frogs. Sciencedaily.com. Retrieved on 2007-10-12.

Science Daily (November 19, 1999), Evergreens help block spread of pesticide from crop fields. Sciencedaily.com. Retrieved on September 19, 2007.

Shah, H., & Marks, N. (2004). *A Well-being Manifesto for a Flourishing Society*. London: New Economics Foundation.

Shulman, D.D., *The King and the Clown in South Indian Myth and Poetry* (Princeton, NJ: 1985).

Singh T.V. Kauer, J. and Singh, D.P. (eds), 1982. *Studies in Tourism, Wildlife Parks, and Conservation*. Metropolitan, New Delhi, 300 pp.

Singh, J.S. (ed), 1985. *Environmental Regeneration in Himalaya:* Concepts and Strategies. Central Himalayan Environment Association and Gyanodaya Prakashan, Nainital, U.P., India, p. 468.

Smith, V., Asoka, *The Buddhist Emperor of India* (Oxford: Clarendon Press, 1920), 3rd edn.

Soil Survey Division Staff (1993). Soil Structure. *Handbook 18. Soil survey manual*. Retrieved on 2006-04-11.

Solecki, R., *Shanidar; The First Flower People* (New York: 1971).

Stadler, B., Solinger, St., and Michalzik, B. (2001). "Insect herbivores and the nutrient flow from the canopy to the soil in coniferous and deciduous forests". In: *Oecologia* 126:104-113

Stanley, S. M., 1973, An ecological theory for the sudden origin of multicellular life in the Late Precambrian: Proceedings of the National Academy of Sciences, v. 70, p. 1486-1489.

Stewart, William, *Counseling in Rehabilitation* (London: Croom Helm, 1985).

Swank, W.T., Waide, J.B., Crossley, D.A., and Todd R.L. (1981). "Insect defoliation enhances nitrate export from forest ecosystems". In: *Oecologia* 51:297-299.

Taylor, Paul and Groom, A.J.R., eds., *International Institutions al Work* (London: Pinter Publishers, 1988).

Terence, Irving, *Plato: The Republic,* trans, by A.D. Lindsay (London: J.M. Dent & Sons, 1992).

Thane, Pal, *The Foundations of the Welfare State* (London & New York: Longman, 1982).

Thompson. M., Warburton, M., and Hatley, T., 1986. *Uncertainty on a Himalayan Scale.* Ethnographica, London, 162 pp.

Tibbetts, J Anon. Water ENVIRON. SOLUTIONS, vol. 9, no. 8, p. 26, 1996

Trainer, F. E. (Ed.). (1990). A rejection of the Brundtland Report. *International Foundation for the Development of Alternatives Dossier,* 77, May-June, 71-85.

U.S. Department of Energy Environmental Management - "Department of Energy Five Year Plan FY 2007-FY 2011 Volume II." Retrieved on 8 April 2007.

Unruh, G. (2000). *Understanding Carbon Lock-in*, Energy Policy, Volume 28, Issue 12, October, 817–830.

Uranium Information Centre, Synroc, *Nuclear Issues Briefing Paper* 21, retrieved September 2007

Vannote, R.I., Minshall, G.W., Cummings, K.W., Sedell, J.R., and Cushing, C.E. (1980). "The river continuim concept". in: *Cannadian Journal of Fisheries and Aquatic Science* 37:120-137.

Whiteman, P.T.S., 1985. The mountain environment: An agronomist's perspective with a case study from Jumla, Nepal, *Mountain Research and Development,* 5(2): 151-162.

Will Allen. 2007."Learning for Sustainability: Sustainable Development."

Index